Comprehensive Organometallic Chemistry II

A Review of the Literature 1982–1994

Comprehensive Organometallic Chemistry II

A Review of the Literature 1982–1994

Editors-in-Chief

Edward W. Abel
University of Exeter, UK

F. Gordon A. Stone
Baylor University, Waco, TX, USA

Geoffrey Wilkinson
Imperial College of Science, Technology and Medicine, London, UK

Volume 14

CUMULATIVE INDEXES

PERGAMON

UK — Elsevier Science Ltd., The Boulevard, Langford Lane, Kidlington, Oxford OX5 1GB, UK

USA — Elsevier Science Inc., 660 White Plains Road, Tarrytown, New York 10591-5153, USA

JAPAN — Elsevier Science Japan, Tsunashima Building Annex, 3-20-12 Yushima, Bunkyo-ku, Tokyo 113, Japan

---

First edition 1995

Library of Congress Cataloging in Publication Data
Comprehensive organometallic chemistry II : a review of the literature 1982–1994 / editors-in-chief, Edward W. Abel, F. Gordon A. Stone, Geoffrey Wilkinson
p. cm.
Includes indexes.
1. Organometallic chemistry. I. Abel, Edward W. II. Stone, F. Gordon A. III. Wilkinson, Geoffrey.
QD411.C652 1995
547'.05—dc20 95–7030

British Library Cataloguing in Publication Data
A catalogue record for this book is available from the British Library.

ISBN 0–08–040608–4 (set : alk. paper)
ISBN 0–08–042321–3 (Volume 14)

Important note
For safety reasons, readers should always consult the list of abbreviations on p. xi before making use of the experimental details provided.

∞™ The paper used in this publication meets the minimum requirements of the American National Standard for Information Sciences—Permanence of Paper for Printed Library Materials, ANSI Z39.48–1984.

Chemical structures drawn by Synopsys Scientific Systems Ltd., Leeds, UK.

Printed and bound in Great Britain by BPC Wheatons Ltd., Exeter, UK.

Contents

Preface	vii
Preface to 'Comprehensive Organometallic Chemistry'	vii
Contributor to Volume 14	ix
Abbreviations	xi
Contents of All Volumes	xv
Cumulative Formula Index J. NEWTON, *David John (Services), Slough, UK*	1
Cumulative Subject Index J. NEWTON, *David John (Services), Slough, UK*	223

Preface

'Comprehensive Organometallic Chemistry', published in 1982, was well received and remains very highly cited in the primary journal literature. Since its publication, studies on the chemistry of molecules with carbon–metal bonds have continued to expand rapidly. This is due to many factors, ranging from the sheer intellectual challenge and excitement provided by the continuing production of novel results, which demand new ideas, through to the successful application of organometallic species in organic syntheses, the generation of living catalysts for polymerization, and the synthesis of precursors for materials employed in the electronic and ceramic industries. For many reasons, therefore, we judged it timely to update 'Comprehensive Organometallic Chemistry' with a new work.

Due to the scope and depth of this area of chemistry, to have merely revised each of the original nine volumes did not seem the most user-friendly or cost-effective procedure to follow. As a consequence of the sheer bulk of the literature of the subject, a revised edition would necessarily require either the elimination of much chemistry of archival value but which is still important, or the production of a set of volumes significantly larger in number than the original nine. Accordingly, we decided it would be best to use the original work as a basis for new volumes focusing on organometallic chemistry reported since 1982, with reference back to the original work when necessary. For ease of use the new volumes maintain the same general structure as employed previously but reflect the changes in substance and direction the field has undergone in the last ten years. Thus it is not surprising that the largest volume in the new work concerns the role of the transition elements in metal-mediated organic syntheses.

The expansion of organometallic chemistry since the early 1980s also led us to decide that an updating of 'Comprehensive Organometallic Chemistry' would be more effectively accomplished if each volume had one or two editors who would be responsible both for recruiting experts for the Herculean task of writing the many chapters of each volume and for overseeing the content. We are deeply indebted to the volume editors and their authors for the time and effort they have given to the project.

As with the original 'Comprehensive Organometallic Chemistry', published some thirteen years ago, we hope this new version will serve as a pivotal reference point for new work and will function to generate new ideas and perceptions for the continued advance of what will surely continue as a vibrant area of chemistry.

Edward W. Abel
Exeter, UK

F. Gordon A. Stone
Waco, Texas, USA

Geoffrey Wilkinson
London, UK

Preface to 'Comprehensive Organometallic Chemistry'

Although the discovery of the platinum complex that we now know to be the first π-alkene complex, $K[PtCl_3(C_2H_4)]$, by Zeise in 1827 preceded Frankland's discovery (1849) of diethylzinc, it was the latter that initiated the rapidly developing interest during the latter half of the nineteenth century in compounds with organic groups bound to the elements. This era may be considered to have reached its apex in the discovery by Grignard of the magnesium reagents which occupy a special place because of their ease of synthesis and reactivity. With the exception of trimethylplatinum chloride discovered by Pope, Peachy and Gibson in 1907 by use of the Grignard reagent, attempts to make stable transition metal alkyls and aryls corresponding to those of main group elements met with little success, although it is worth recalling that even in 1919 Hein and his co-workers were describing the 'polyphenyl-chromium' compounds now known to be arene complexes.

The other major area of organometallic compounds, namely metal compounds of carbon monoxide, originated in the work starting in 1868 of Schützenberger and later of Mond and his co-workers and was subsequently developed especially by Hieber and his students. During the first half of this century, aided by the use of magnesium and, later, lithium reagents the development of main group organo chemistry was quite rapid, while from about 1920 metal carbonyl chemistry and catalytic reactions of carbon monoxide began to assume importance.

In 1937 Krause and von Grosse published their classic book 'Die Chemie der Metallorganischen Verbindungen'. Almost 1000 pages in length, it listed scores of compounds, mostly involving metals of the main groups of the periodic table. Compounds of the transition elements could be dismissed in 40 pages. Indeed, even in 1956 the stimulating 197-page monograph 'Organometallic Compounds' by Coates adequately reviewed organo transition metal complexes within 27 pages.

Although exceedingly important industrial processes in which transition metals were used for catalysis of organic reactions were developed in the 1930s, mainly in Germany by Reppe, Koch, Roelen, Fischer and Tropsch and others, the most dramatic growth in our knowledge of organometallic chemistry, particularly of transition metals, has stemmed from discoveries made in the middle years of this century. The introduction in the same period of physical methods of structure determination (infrared, nuclear magnetic resonance, and especially single-crystal X-ray diffraction) as routine techniques to be used by preparative chemists allowed increasingly sophisticated exploitation of discoveries. Following the recognition of the structure of ferrocene, other major advances quickly followed, including the isolation of a host of related π-complexes, the synthesis of a plethora of organometallic compounds containing metal–metal bonds, the characterization of low-valent metal species in which hydrocarbons are the only ligands, and the recognition from dynamic NMR spectra that ligand site exchange and tautomerism were common features in organometallic and metal carbonyl chemistry. The discovery of alkene polymerization using aluminium alkyl–titanium chloride systems by Ziegler and Natta and of the Wacker palladium–copper catalysed ethylene oxidation led to enormous developments in these areas.

In the last two decades, organometallic chemistry has grown more rapidly in scope than have the classical divisions of chemistry, leading to publications in journals of all national chemical societies, the appearance of primary journals specifically concerned with the topic, and the growth of annual review volumes designed to assist researchers to keep abreast of accelerating developments.

Organometallic chemistry has become a mature area of science which will obviously continue to grow. We believe that this is an appropriate time to produce a comprehensive review of the subject, treating organo derivatives in the widest sense of both main group and transition elements. Although advances in transition metal chemistry have appeared to dominate progress in recent years, spectacular progress has, nevertheless, also been made in our knowledge of organo compounds of main group elements such as aluminium, boron, lithium and silicon.

In these Volumes we have assembled a compendium of knowledge covering contemporary organometallic and carbon monoxide chemistry. In addition to reviewing the chemistry of the elements individually, two Volumes survey the use of organometallic species in organic synthesis and in catalysis, especially of industrial utility. Within the other Volumes are sections devoted to such diverse topics as the nature of carbon–metal bonds, the dynamic behaviour of organometallic compounds in solution, heteronuclear metal–metal bonded compounds, and the impact of organometallic compounds on the environment. The Volumes provide a unique record, especially of the intensive studies conducted during the past 25 years. The last Volume of indexes of various kinds will assist readers seeking information on the properties and synthesis of compounds and on earlier reviews.

As Editors, we are deeply indebted to all those who have given their time and effort to this project. Our Contributors are among the most active research workers in those areas of the subject that they have reviewed and they have well justified international reputations for their scholarship. We thank them sincerely for their cooperation.

Finally, we believe that 'Comprehensive Organometallic Chemistry', as well as providing a lasting source of information, will provide the stimulus for many new discoveries since we do not believe it possible to read any of the articles without generating ideas for further research.

E. W. ABEL
Exeter

F. G. A. STONE
Bristol

G. WILKINSON
London

Contributor to Volume 14

Dr. J. Newton
David John (Services) Ltd., 221 Wentworth Avenue, Farnham Royal, Slough, SL2 2AP, UK

Abbreviations

The abbreviations used throughout 'Comprehensive Organometallic Chemistry II' are consistent with those used in 'Comprehensive Organometallic Chemistry' and with other standard texts in this area. The abbreviations in some instances may differ from those commonly used in other branches of chemistry.

Ac	acetyl
acac	acetylacetonate
AIBN	2,2'-azobisisobutyronitrile
Ar	aryl
arphos	1-(diphenylphosphino)-2-(diphenylarsino)ethane
Azb	azobenzene
9-BBN	9-borabicyclo[3.3.1]nonyl
9-BBN-H	9-borabicyclo[3.3.1]nonane
BHT	2,6-di-*t*-butyl-4-methylphenol (butylated hydroxytoluene)
bipy	2,2'-bipyridyl
t-BOC	*t*-butoxycarbonyl
bsa	*N,O*-bis(trimethylsilyl)acetamide
bstfa	*N,O*-bis(trimethylsilyl)trifluoroacetamide
btaf	benzyltrimethylammonium fluoride
Bz	benzyl
can	ceric ammonium nitrate
cbd	cyclobutadiene
1,5,9-cdt	cyclododeca-1,5,9-triene
chd	cyclohexadiene
chpt	cycloheptatriene
[Co]	cobalamin
(Co)	cobaloxime $[Co(DMG)_2]$ derivative
cod	1,5-cyclooctadiene
cot	cyclooctatetraene
Cp	η^5-cyclopentadienyl
Cp*	pentamethylcyclopentadienyl
18-crown-6	1,4,7,10,13,16-hexaoxacyclooctadecane
CSA	camphorsulfonic acid
csi	chlorosulfonyl isocyanate
Cy	cyclohexyl
dabco	1,4-diazabicyclo[2.2.2]octane
dba	dibenzylideneacetone
dbn	1,5-diazabicyclo[4.3.0]non-5-ene
dbu	1,8-diazabicyclo[5.4.0]undec-7-ene
dcc	dicyclohexylcarbodiimide
dcpe	1,2-bis(dicyclohexylphosphino)ethane
ddq	2,3-dichloro-5,6-dicyano-1,4-benzoquinone
deac	diethylaluminum chloride
dead	diethyl azodicarboxylate
depe	1,2-bis(diethylphosphino)ethane
depm	1,2-bis(diethylphosphino)methane
det	diethyl tartrate (+ or –)

DHP	dihydropyran
diars	1,2-bis(dimethylarsino)benzene
dibal-H	diisobutylaluminum hydride
dien	diethylenetriamine
DIGLYME	bis(2-methoxyethyl)ether
diop	2,3-*O*-isopropylidene-2,3-dihydroxy-1,4-bis(diphenylphosphino)butane
dipt	diisopropyl tartrate (+ or –)
dma	dimethylacetamide
dmac	dimethylaluminum chloride
DMAD	dimethyl acetylenedicarboxylate
dmap	4-dimethylaminopyridine
DME	dimethoxyethane
DMF	*N,N'*-dimethylformamide
DMG	dimethylglyoximate
DMI	*N,N'*-dimethylimidazalone
dmpe	1,2-bis(dimethylphosphino)ethane
dmpm	bis(dimethylphosphino)methane
DMSO	dimethyl sulfoxide
dmtsf	dimethyl(methylthio)sulfonium fluoroborate
dpam	bis(diphenylarsino)methane
dppb	1,4-bis(diphenylphosphino)butane
dppe	1,2-bis(diphenylphosphino)ethane
dppf	1,1'-bis(diphenylphosphino)ferrocene
dpph	1,6-bis(diphenylphosphino)hexane
dppm	bis(diphenylphosphino)methane
dppp	1,3-bis(diphenylphosphino)propane
eadc	ethylaluminum dichloride
edta	ethylenediaminetetraacetate
eedq	*N*-ethoxycarbonyl-2-ethoxy-1,2-dihydroquinoline
en	ethylene-1,2-diamine
Et_2O	diethyl ether
F_6 acac	hexafluoroacetylacetonate
Fc	ferrocenyl
Fp	$Fe(CO)_2Cp$
HFA	hexafluoroacetone
hfacac	hexafluoroacetylacetonate
hfb	hexafluorobut-2-yne
HMPA	hexamethylphosphoramide
hobt	hydroxybenzotriazole
$IpcBH_2$	isopinocampheylborane
Ipc_2BH	diisopinocampheylborane
kapa	potassium 3-aminopropylamide
K-selectride	potassium tri-*s*-butylborohydride
LAH	lithium aluminum hydride
LDA	lithium diisopropylamide
LICA	lithium isopropylcyclohexylamide
LITMP	lithium tetramethylpiperidide
L-selectride	lithium tri-*s*-butylborohydride
LTA	lead tetraacetate
mcpba	*m*-chloroperbenzoic acid
MeCN	acetonitrile
MEM	methoxyethoxymethyl
MEM-Cl	β-methoxyethoxymethyl chloride

Mes	mesityl
mma	methyl methacrylate
mmc	methylmagnesium carbonate
MOM	methoxymethyl
Ms	methanesulfonyl
MSA	methanesulfonic acid
MsCl	methanesulfonyl chloride
nap	1-naphthyl
nbd	norbornadiene
NBS	*N*-bromosuccinimide
NCS	*N*-chlorosuccinimide
nmo	*N*-methylmorpholine *N*-oxide
NMP	*N*-methyl-2-pyrrolidone
Nu^-	nucleophile
ox	oxalate
pcc	pyridinium chlorochromate
pdc	pyridinium dichromate
phen	1,10-phenanthroline
phth	phthaloyl
ppa	polyphosphoric acid
ppe	polyphosphate ester
$[PPN]^+$	$[(Ph_3P)_2N]^+$
ppts	pyridinium *p*-toluenesulfonate
py	pyridine
pz	pyrazolyl
Red-Al	sodium bis(2-methoxyethoxy)aluminum dihydride
sal	salicylaldehyde
salen	*N,N'*-bis(salicylaldehydo)ethylenediamine
SEM	β-trimethylsilylethoxymethyl
tas	tris(diethylamino)sulfonium
tasf	tris(diethylamino)sulfonium difluorotrimethylsilicate
tbaf	tetra-*n*-butylammonium fluoride
TBDMS	*t*-butyldimethylsilyl
TBDMS-Cl	*t*-butyldimethylsilyl chloride
TBDPS	*t*-butyldiphenylsilyl
tbhp	*t*-butyl hydroperoxide
TCE	2,2,2-trichloroethanol
TCNE	tetracyanoethene
TCNQ	7,7,8,8-tetracyanoquinodimethane
terpy	2,2':6',2''-terpyridyl
tes	triethylsilyl
Tf	triflyl (trifluoromethanesulfonyl)
TFA	trifluoracetic acid
TFAA	trifluoroacetic anhydride
tfacac	trifluoroacetylacetonate
THF	tetrahydrofuran
THP	tetrahydropyranyl
tipbs-Cl	2,4,6-triisopropylbenzenesulfonyl chloride
tips-Cl	1,3-dichloro-1,1,3,3-tetraisopropyldisiloxane
TMEDA	tetramethylethylenediamine [1,2-bis(dimethylamino)ethane]
TMS	trimethylsilyl
TMS-Cl	trimethylsilyl chloride
TMS-CN	trimethylsilyl cyanide
Tol	tolyl
tpp	*meso*-tetraphenylporphyrin

Tr	trityl (triphenylmethyl)
tren	2,2',2''-triaminotriethylamine
trien	triethylenetetraamine
triphos	1,1,1-tris(diphenylphosphinomethyl)ethane
Ts	tosyl
TsMIC	tosylmethyl isocyanide
ttfa	thallium trifluoroacetate

Contents of All Volumes

Volume 1 Lithium, Beryllium, and Boron Groups
1 Alkali Metals
2 Beryllium
3 Magnesium, Calcium, Strontium and Barium
4 Compounds with Three- or Four-coordinate Boron, Emphasizing Cyclic Systems
5 Boron Rings Ligated to Metals
6 Polyhedral Carbaboranes
7 Main-group Heteroboranes
8 Metallaboranes
9 Transition Metal Metallacarbaboranes
10 Aluminum
11 Gallium, Indium and Thallium, Excluding Transition Metal Derivatives
12 Transition Metal Complexes of Aluminum, Gallium, Indium and Thallium
Author Index
Subject Index

Volume 2 Silicon Group, Arsenic, Antimony, and Bismuth
1 Organosilanes
2 Carbacyclic Silanes
3 Organopolysilanes
4 Silicones
5 Germanium
6 Tin
7 Lead
8 Arsenic, Antimony and Bismuth
Author Index
Subject Index

Volume 3 Copper and Zinc Groups
1 Gold
2 Copper and Silver
3 Mercury
4 Cadmium and Zinc
Author Index
Subject Index

Volume 4 Scandium, Yttrium, Lanthanides and Actinides, and Titanium Group
1 Zero Oxidation State Complexes of Scandium, Yttrium and the Lanthanide Elements
2 Scandium, Yttrium, and the Lanthanide and Actinide Elements, Excluding their Zero Oxidation State Complexes
3 Titanium Complexes in Oxidation States Zero and Below
4 Titanium Complexes in Oxidation States +2 and +3
5 Titanium Complexes in Oxidation State +4
6 Zirconium and Hafnium Complexes in Oxidation States Zero and Below
7 Metallocene(II) Complexes of Zirconium and Hafnium
8 Zirconium and Hafnium Compounds in Oxidation State +3
9 Bis(cyclopentadienyl)zirconium and -hafnium Halide Complexes in Oxidation State +4
10 Bis(cyclopentadienyl) Metal(IV) Compounds with Si, Ge, Sn, N, P, As, Sb, O, S, Se, Te or Transition Metal-centred Ligands
11 Zirconium and Hafnium Complexes in Oxidation State +4

12 Cationic Organozirconium and Organohafnium Complexes
13 Cyclooctatetraene Complexes of Zirconium and Hafnium
Author Index
Subject Index

Volume 5 Vanadium and Chromium Groups
1 Vanadium
2 Niobium and Tantalum
3 Hexacarbonyls and Carbonyl Complexes of Carbon σ-Bonded Ligands of Chromium, Molybdenum and Tungsten
4 Carbonyl Complexes of Noncarbon σ-Bonded Ligands of Chromium, Molybdenum and Tungsten
5 Organometallic Complexes of Chromium, Molybdenum and Tungsten without Carbonyl Ligands
6 π-Complexes of Chromium, Molybdenum and Tungsten, Excluding those of Cyclopentadienyls and Arenes
7 Cyclopentadienyl Complexes of Chromium, Molybdenum and Tungsten
8 Arene and Heteroarene Complexes of Chromium, Molybdenum and Tungsten
Author Index
Subject Index

Volume 6 Manganese Group
1 Manganese Carbonyls and Manganese Carbonyl Halides
2 Manganese Alkyls and Hydrides
3 Manganese Complexes Containing Nonmetallic Elements
4 Manganese Hydrocarbon Complexes Excluding Cyclopentadienyl
5 Cyclopentadienyl Manganese Complexes
6 Manganese Nitrosyl and Isonitrile Complexes
7 High-valent Organomanganese Compounds
8 Technetium
9 Low-valent Organorhenium Compounds
10 High-valent Organorhenium Compounds
Author Index
Subject Index

Volume 7 Iron, Ruthenium, and Osmium
1 Iron Compounds without Hydrocarbon Ligands
2 Mononuclear Iron Compounds with η^1–η^6 Hydrocarbon Ligands
3 Dinuclear Iron Compounds with Hydrocarbon Ligands
4 Polynuclear Iron Compounds with Hydrocarbon Ligands
5 Introduction to Organoruthenium and Organoosmium Chemistry
6 Mononuclear Complexes of Ruthenium and Osmium Containing η^1 Carbon Ligands
7 Complexes of Ruthenium and Osmium Containing η^2–η^6 Hydrocarbon Ligands: (i) Complexes not Containing Cyclobutadiene, Cyclopentadienyl or η-Arene Coligands
8 Complexes of Ruthenium and Osmium Containing η^2–η^6 Hydrocarbon Ligands: (ii) Complexes Containing Four- and Five-membered Rings (Including MCp(arene) Complexes)
9 Complexes of Ruthenium and Osmium Containing η^2–η^6 Hydrocarbon Ligands: (iii) Complexes Containing Six-, Seven- and Eight-membered Rings
10 Ruthenocenes and Osmocenes
11 Binuclear Complexes of Ruthenium and Osmium Containing Metal–Metal Bonds
12 Trinuclear Clusters of Ruthenium and Osmium: (i) Introduction and Simple Neutral, Anionic and Hydrido Clusters
13 Trinuclear Clusters of Ruthenium and Osmium: (ii) Hydrocarbon Ligands on Metal Clusters
14 Trinuclear Clusters of Ruthenium and Osmium: (iii) Clusters with Metal–Carbon Bonds to Heteroatom Ligands
15 Tetranuclear Clusters of Ruthenium and Osmium
16 Medium- and High-nuclearity Clusters of Ruthenium and Osmium
Author Index
Subject Index

Volume 8 Cobalt, Rhodium, and Iridium
1 Cobalt
2 Rhodium
3 Iridium
4 Cluster Complexes of Cobalt, Rhodium, and Iridium
Author Index
Subject Index

Volume 9 Nickel, Palladium, and Platinum
1 Nickel Complexes with Carbonyl, Isocyanide, and Carbene Ligands
2 Nickel–Carbon σ-Bonded Complexes
3 Nickel–Carbon π-Bonded Complexes
4 Palladium Complexes with Carbonyl, Isocyanide and Carbene Ligands
5 Palladium–Carbon σ-Bonded Complexes
6 Palladium–Carbon π-Bonded Complexes
7 Platinum Complexes with Carbonyl, Isocyanide and Carbene Ligands
8 Platinum–Carbon σ-Bonded Complexes
9 Platinum–Carbon π-Bonded Complexes
Author Index
Subject Index

Volume 10 Heteronuclear Metal–Metal Bonds
1 Synthesis of Compounds Containing Heteronuclear Metal–Metal Bonds
2 Heterodinuclear Compounds
3 Heteronuclear Clusters Containing C_1, C_2, C_3, ..., C_n Acyclic Hydrocarbyl Ligands
4 Binary Carbonyls, Carbonyls plus Hydrides, Carbonyls plus Phosphines, Cyclic Hydrocarbyls and Main-group Ligands without Acyclic Hydrocarbyls
5 Cluster Complexes with Bonds Between Transition Elements and Copper, Silver and Gold
6 Cluster Complexes with Bonds Between Transition Elements and Zinc, Cadmium, and Mercury
7 Catalysis and Related Reactions with Compounds Containing Heteronuclear Metal–Metal Bonds
Author Index
Subject Index

Volume 11 Main-group Metal Organometallics in Organic Synthesis
1 Lithium
2 Sodium and Potassium
3 Magnesium
4 Zinc and Cadmium
5 Boron
6 Aluminum
7 Silicon
8 Tin
9 Mercury
10 Thallium
11 Lead
12 Antimony and Bismuth
13 Selenium
14 Tellurium
Author Index
Subject Index

Volume 12 Transition Metal Organometallics in Organic Synthesis
1 Introduction and Fundamentals
2 Transition Metal Hydrides: Hydrocarboxylation, Hydroformylation, and Asymmetric Hydrogenation
3.1 Transition Metal Alkyl Complexes from Hydrometallation
3.2 Transition Metal Alkyl Complexes from RLi and CuX
3.3 Transition Metal Alkyl Complexes: Main-group Transmetallation and Insertion Chemistry
3.4 Transition Metal Alkyl Complexes: Oxidative Addition and Transmetallation

3.5 Transition Metal Alkyl Complexes: Oxidative Addition and Insertion
3.6 Transition Metal Alkyl Complexes: Multiple Insertion Cascades
3.7 Transition Metal Alkyl Complexes: Reductive Dimerization of Alkenes and Alkynes
4 Transition Metal Carbonyl Complexes
5.1 Transition Metal Carbene Complexes: Cyclopropanation
5.2 Transition Metal Carbene Complexes: Diazodecomposition, Ylide, and Insertion
5.3 Transition Metal Carbene Complexes: Alkyne and Vinyl Ketene Chemistry
5.4 Transition Metal Carbene Complexes: Photochemical Reactions of Carbene Complexes
5.5 Transition Metal Carbene Complexes: Tebbe's Reagent and Related Nucleophilic Alkylidenes
6.1 Transition Metal Alkene, Diene, and Dienyl Complexes: Nucleophilic Attack on Alkene Complexes
6.2 Transition Metal Alkene, Diene, and Dienyl Complexes: Complexation of Dienes for Protection
6.3 Transition Metal Alkene, Diene, and Dienyl Complexes: Nucleophilic Attack on Diene and Dienyl Complexes
7.1 Transition Metal Alkyne Complexes: Transition Metal-stabilized Propargyl Systems
7.2 Transition Metal Alkyne Complexes: Pauson–Khand Reaction
7.3 Transition Metal Alkyne Complexes: Transition Metal-catalyzed Cyclotrimerization
7.4 Transition Metal Alkyne Complexes: Zirconium–Benzyne Complexes
8.1 Transition Metal Allyl Complexes: Telomerization of Dienes
8.2 Transition Metal Allyl Complexes: Pd, W, Mo-assisted Nucleophilic Attack
8.3 Transition Metal Allyl Complexes: Intramolecular Alkene and Alkyne Insertions
8.4 Transition Metal Allyl Complexes: Trimethylene Methane Complexes
8.5 Transition Metal Allyl Complexes: π-Allylnickel Halides and Other π-Allyl Complexes Excluding Palladium
9.1 Transition Metal Arene Complexes: Nucleophilic Addition
9.2 Transition Metal Arene Complexes: Ring Lithiation
9.3 Transition Metal Arene Complexes: Side-chain Activation and Control of Stereochemistry
10 Synthetically Useful Coupling Reactions Promoted by Ti, V, Nb, W, Mo Reagents
11.1 Transition Metal-catalyzed Oxidations: Asymmetric Epoxidation
11.2 Transition Metal-catalyzed Oxidations: Asymmetric Hydroxylation
11.3 Transition Metal-catalyzed Oxidations: Other Oxidations
12.1 Transition Metals in Polymer Synthesis: Ziegler–Natta Reaction
12.2 Transition Metals in Polymer Synthesis: Ring-opening Metathesis Polymerization and Other Transition Metal Polymerization Techniques
Author Index
Subject Index

Volume 13 Structure Index

Structures of Organometallic Compounds Determined by Diffraction Methods

Volume 14 Cumulative Indexes

Cumulative Formula Index
Cumulative Subject Index

Cumulative Formula Index

JOHN NEWTON
David John (Services), Slough, UK

This Cumulative Formula Index contains entries for compounds that occur in Volumes 1–12. The main criterion for inclusion is that a compound must contain a metal–carbon bond. Compounds that have been indexed may occur in the text, equations, schemes, figures, or tables. However, limitations of space could not permit every organometallic compound to be indexed each time it was mentioned. Compounds have therefore been indexed only when considered important in the context of the topic under discussion.

Where a general compound or reaction type is given with large numbers of specific examples (usually in tables but sometimes in equations or text), a selection has been made on the basis of choosing simple or common representative examples. Threrefore a chloro compound will be indexed, but not the bromo or iodo equivalents; the methyl derivative will be included in preference to higher homologues; the PPh_3 complex will be indexed but not the PPh_2Me equivalent, etc. Readers should therefore bear this in mind when searching the Index for specific compounds.

The element symbols within each formula are arranged according to a modified Hill system, in which the order is metal(s), C, H, and then any remaining symbols arranged alphabetically. Compounds containing two (or more) different metals appear twice (or more), being listed under each metal. Common ligand abbreviations that occur in the text have also been used in the Index. These include cdt (cyclododecatriene), dmpe ($Me_2PCH_2CH_2PMe_2$), Fe (ferrocenyl), Fp ($Fe(CO)_2Cp$), nbd (norbornadiene), etc.

A

$AgBF_4$
  $Ag[BF_4]$, **10**, 2
$AgB_2Ru_8C_{60}H_{34}NO_{24}P_2$
  $[PPN][HRu_4(CO)_{12}BH]_2Ag(Cl)]Cl$, **1**, 304
$AgB_{11}IrC_{38}H_{42}ClOP_2$
  $IrCl(CO)(PPh_3)_2Ag(B_{11}CH_{12})$, **10**, 377
$AgCF_3$
  $AgCF_3$, **3**, 72
AgC_3F_5
  $Ag[C(CF_3)CF_2]$, **3**, 72
AgC_4F_{12}
  $[Ag(CF_3)_4]^-$, **3**, 72
$AgC_4H_9F_3P$
  $AgCF_3(PMe_3)$, **3**, 72
$AgC_5H_3F_7N$
  $Ag[CF(CF_3)_2](NCMe)$, **3**, 110
AgC_8H_7
  $Ag(CHCHPh)$, **3**, 72
$AgC_8H_{12}F_6N_4P$
  $[Ag(NCMe)_4][PF_6]$, **10**, 42
$AgC_9H_{12}N$
  $Ag(C_6H_4CH_2NMe_2)$, **3**, 125
$AgC_{24}H_{17}$
  $Ag(C_6H_2Ph_3)$, **3**, 82
$AgC_{25}H_{17}F_5P$
  $Ag(C_6F_5)(CH_2PPh_3)$, **3**, 111, 115
$AgLiSi_6C_{36}H_{86}O_4$
  $[Ag[C(TMS)_3]_2][Li(THF)_4]$, **3**, 106
$AgOs_2Si_4C_{26}H_{54}N_2$
  $[Ag[NOs(CH_2TMS)_2Cp]_2]^+$, **7**, 375
$AgOs_6C_{22}H_2O_{22}$
  $[Os_3H(CO)_{11}]_2Ag$, **7**, 720
$AgPtC_{15}H_{24}BrN_3$
  $Pt[C_6H_3(CH_2NMe_2)_2](NTolCHMe)AgBr$, **10**, 71
$AgPtC_{53}H_{47}Cl_2P_4$
  $Pt(MeCC)(Cl)(dppm)_2(AgCl)$, **8**, 239
$AgPtC_{84}H_{60}Br_4Cl_{10}P_2$
  $[(Ph_3P)Ag(Br)PtBr(C_6Cl_5)(PPh_3)]_2$, **9**, 474
$AgPt_2C_{56}H_{46}F_{30}NO$
  $[Bu_4N][Pt_2Ag(C_6F_5)_2(C_6F_5)_4]\cdot O\cdot Et_2$, **9**, 468
$AgRhC_{60}H_{62}N_2P_4$
  $Rh(Bu^tNC)_2(dppm)_2Ag$, **10**, 68
$AgRu_2SbC_{19}H_{37}F_6N_3O_{13}P_5$
  $[Ru_2[Ag(MeCN)](CO)_5[(MeO)_2PNEtP(OMe)_2]_2]SbF_6$, **7**, 627
$AgRu_3C_{35}H_{22}F_3O_{10}P_2$
  $Ru_3(AgO_2CCF_3)(CO)_8(dppm)$, **7**, 720
$AgRu_4C_{30}H_{18}O_{12}P$
  $AgRu_4(H)_3(CO)_{12}(PPh_3)$, **10**, 41
$AgSi_3TiC_{18}H_{31}NO_2$
  $Ti[C_5H_4(TMS)][C_2(TMS)]_2AgNO_2$, **10**, 37
$Ag_2Au_2C_{68}H_{50}P_2$
  $Au_2Ag_2(CCPh)_4(PPh_3)_2$, **3**, 75
Ag_2CO_3
  Ag_2CO_3, **12**, 306
$Ag_2C_8H_{20}P_2$
  $Ag_2[(CH_2)_2PMe_2]_2$, **3**, 79
$Ag_2C_{26}H_{24}P_2S_2$
  $Ag_2(CH_2PPh_2S)_2$, **3**, 79
$Ag_2C_{40}H_{44}O_8P_2$
  $[Ag[CH(CO_2Et)]_2PPh_2]_2$, **3**, 115
$Ag_2Cu_2C_{32}H_{20}$
  $Ag_2Cu_2(CCPh)_4$, **3**, 109
$Ag_2Li_2C_{32}H_{40}O_2$
  $Li_2Ag_2(Ph)_4(OEt_2)_2$, **3**, 125
$Ag_2Li_2C_{36}H_{48}N_4$
  $Li_2Ag_2(C_6H_4CH_2NMe_2)_4$, **3**, 58, 124, 125
$Ag_2Mn_2C_{43}H_{39}ClO_{10}P_3$
  $Ag_2Mn_2(H)_2(Ph_2PCH_2PPh_2)(CO)_6(PPh_3)(OClO_3)$, **10**, 43
Ag_2O
  Ag_2O, **12**, 1183
$Ag_2Pt_2C_{38}H_{20}F_5$
  $[Pt_2Ag_2(C_6F_5)(CCPh)_4]^{2-}$, **3**, 75
$Ag_2Pt_2C_{48}H_{72}$
  $Pt_2Ag_2(CCBu^t)_8$, **3**, 75
$Ag_2Pt_4C_{80}H_{72}Cl_{44}$
  $[Bu_4N]_2[Pt(C_6Cl_5)_2]_4(Cl)_4(Ag)_2$, **9**, 475
$Ag_2Ru_4C_{38}H_{26}O_{12}P_2$
  $Ag_2Ru_4(H)_2[Ph_2P(CH_2)_2PPh_2](CO)_{12}$, **10**, 41
$Ag_2Ru_4C_{54}H_{44}O_{12}P_2$
  $Ag_2Ru_4H_2(CO)_{12}[P(Bz)_3]_2$, **10**, 39
$Ag_2Ru_8C_{50}H_{30}O_{24}P_2$
  $[AgRu_4(H)_3(CO)_{12}]_2[Ph_2P(CH_2)_2PPh_2]$, **10**, 41
$Ag_2Si_2C_{12}H_{23}N_2$
  $Ag_2[C(py)(TMS)_2]_2$, **3**, 77
$Ag_3C_{75}H_{63}P_6$
  $Ag_3[CH(PPh_2)_2]_3$, **3**, 80
$Ag_3Co_3C_{108}H_{108}N_{12}$
  $Ag_3Co_3(CNC_6H_3Me_2)_{12}$, **10**, 43
$Ag_3Li_2C_{36}H_{30}$
  $[Li_2Ag_3Ph_6]^-$, **3**, 109
$Ag_3Nb_3C_{30}H_{48}O_{12}P_6$
  $Ag_3Nb_3(CO)_{12}[Me_2P(CH_2)_2PMe_2]_3$, **10**, 43
$Ag_3OsRh_3C_{178}H_{171}F_3P_9S$
  $Ag_3Rh_3(H)_9[MeC(CH_2PPh_3)_3]_3[OsSCF_3]$, **10**, 35
$Ag_4B_2Ru_8C_{60}H_{34}Cl_2NO_{24}P_2$
  $[PPN][[HRu_4(CO)_{12}BH]_2Ag_4(Cl)]Cl$, **1**, 304
$Ag_4C_{26}H_{24}P_2S_2$
  $[Ag_2(CH_2PPh_2S)]_2$, **3**, 115
$Ag_4C_{36}H_{44}$
  Ag_4Mes_4, **3**, 82
$Ag_6C_{32}H_{40}Br_2N_4$
  $Ag_6Br_2(C_6H_4NMe_2)_4$, **3**, 74
$Ag_6Cu_7C_{112}H_{70}$
  $[Ag_6Cu_7(CCPh)_{14}]^-$, **3**, 75, 109
$Ag_{12}Au_{13}SbC_{180}H_{150}Br_8F_6$
  $[Ag_{12}Au_{13}(PPh_3)_{10}Br_8][SbF_6]$, **10**, 37
$Ag_{13}Fe_8C_{32}O_{33}$
  $[Ag_{13}Fe_8(CO)_{32}]^{4-}$, **7**, 80
$Ag_{20}Au_{18}C_{360}H_{576}Cl_{14}P_{12}$
  $Ag_{20}Au_{18}[P(C_6H_4Me_4)_3]_{12}Cl_{14}$, **10**, 36
$AlAs_2B_{11}C_{49}H_{54}$
  $(AsPh_4)_2[B_{11}H_{11}AlMe]$, **1**, 258, 271
$AlBTi_2C_{20}H_{28}$
  $(Cp_2TiH_2)_2AlH_2BH_2$, **1**, 290–2
$AlB_4C_{10}H_{30}$
  $(Et_3N)(AlH)(EtC)_2B_4H_4$, **1**, 258, 259
$AlB_8C_8H_{25}O$
  $AlEt(OEt_2)C_2B_8H_{10}$, **1**, 259
$AlB_9C_4H_{15}Cl_4P$
  $ClP(MeC)_2B_9H_9AlCl_3$, **1**, 265
$AlB_{12}C_4H_{16}$
  $Al(C_2B_6H_8)_2$, **1**, 259
$AlB_{16}C_4H_{20}$
  $Al(C_2B_8H_0)_2$, **1**, 259
$AlBeC_{21}H_{42}P$
  $Cp^*BePBu^t{}_2\cdot AlMe_3$, **1**, 43
$AlCH_2$
  $AlCH_2{}^+$, **1**, 441
$AlCH_3$
  $AlMe^+$, **1**, 441
  MeAl, **1**, 441
$AlCH_3O$
  $[AlMeO]_n$, **12**, 1195
AlC_2H_3
  $HC{\equiv}CAlH_2$, **1**, 444
AlC_2H_5
  $H_2C{=}CHAlH_2$, **1**, 444
AlC_2H_6
  $AlMe_2{}^+$, **1**, 441
AlC_2O_2
  $Al(CO)_2$, **1**, 442
AlC_3H_8

$[Me_2Al{=}CH_2]^-$, **1**, 443
AlC_3H_9
$AlMe_3$, **1**, 67, 434, 441, 447, 486; **11**, 285, 294; **12**, 156, 585
AlC_3H_9O
Me_2AlOMe, **11**, 295
AlC_4H_{11}
Et_2AlH, **11**, 294
AlC_5H_5
CpAl, **1**, 445
AlC_6H_5
PhAl, **1**, 435, 437
AlC_6H_6
$Al(C_6H_6)$, **1**, 442
$AlC_6H_{10}Cl$
$ClAlCH_2CMe{=}CMeCH_2$, **1**, 444, 445
AlC_6H_{13}
$(CH_2{=}CMeCH_2)Me_2Al$, **1**, 496
AlC_6H_{15}
Et_3Al, **1**, 463, 486; **11**, 279, 284, 294; **12**, 592
$AlC_6H_{15}O$
Et_2AlOEt, **11**, 295
$AlC_6H_{15}O_3$
$Al(OEt)_3$, **11**, 295
AlC_7H_{11}
Me_2CpAl, **1**, 449
AlC_8H_{17}
$(MeMeCH{=}CH)Et_2Al$, **1**, 465
$AlC_8H_{18}Cl$
Bu^i_2AlCl, **1**, 446
AlC_8H_{19}
Bu^i_2AlH, **1**, 447, 467; **11**, 294, 299, 302, 307
AlC_9H_{15}
$(C_3H_5)_3Al$, **1**, 448
AlC_9H_{21}
Pr_3Al, **1**, 486
Pr^i_3Al, **11**, 288
$AlC_9H_{21}O_3$
$Al(OPr^i)_3$, **11**, 295
$AlC_{10}H_{15}$
Cp*Al, **1**, 436, 441, 445
$AlC_{12}H_{27}$
Bu_3Al, **1**, 446, 463, 464, 468, 486, 493
Bu_3^iAl, **11**, 284, 294
Bu_3^tAl, **11**, 288
$AlC_{12}H_{27}O_3$
$Al(OBu^s)_3$, **11**, 295
$Al(OBu^t)_3$, **11**, 295
$AlC_{13}H_{29}N$
$Et_2Al(C_9H_{19}N)$, **11**, 299
$AlC_{14}H_{30}N_2O_6$
$[N(CH_2CH_2O)_3]_2Al_2(Me)_3$, **1**, 433, 434
$AlC_{15}H_{33}$
$[Bu^tCH_2]_3Al$, **1**, 434
$AlC_{16}H_{39}$
Bu^i_4Al, **1**, 446
$AlC_{17}H_{27}O$
$Me_2Al(Bu^t_2MeC_6H_2O)$, **1**, 448
$AlC_{18}H_{15}$
Ph_3Al, **1**, 434, 447, 448, 466; **11**, 291
$AlC_{18}H_{39}$
$(C_6H_{13})_3Al$, **1**, 486
$AlC_{20}H_{15}$
$Ph_2(PhC{\equiv}C)Al$, **1**, 448
$AlC_{21}H_{21}$
Bz_3Al, **1**, 434, 448
$(Tol)_3Al$, **1**, 448
$AlC_{24}H_{51}$
$(C_8H_{17})_3Al$, **1**, 486
$AlC_{27}H_{33}$
$(Mes)_3Al$, **1**, 448
$AlC_{29}H_{43}Br_2O_2$
$MeAl(BrBu^t_2C_6H_2O)_2$, **1**, 492
$AlC_{30}H_{63}$
$(C_{10}H_{21})_3Al$, **1**, 486
$AlC_{31}H_{49}O_2$
$MeAl(Bu^t_2MeC_6H_2O)_2$, **1**, 449
$AlC_{37}H_{29}O_2$
$MeAl(Ph_2C_6H_3O)_2$, **1**, 492
$AlC_{54}H_{39}O_3$
$Al(Ph_2C_6H_3O)_3$, **11**, 287, 296
AlH_3
AlH_3, **11**, 279
$AlKC_{12}H_{28}$
$K(AlBu^i_3H)$, **1**, 447
$AlKC_{16}H_{36}$
$[Bu^i_4Al]^-K^+$, **1**, 446
$AlLiC_9H_{22}$
$LiAlBu^i_2MeH$, **11**, 293, 302
$AlLiC_{24}H_{20}$
$LiAlPh_4$, **11**, 293
$AlLiH_4$
$LiAlH_4$, **11**, 279; **12**, 135, 1073
$AlLiNiC_{24}H_{52}N_4$
$[Li(TMEDA)_2][Ni(cdt)(AlH_4)]$, **1**, 551
$AlLiNi_2C_{48}H_{84}N_4$
$[Li(TMEDA)_2][[Ni(cdt)]_2$-$[(H)_2AlBu^i_2]$, **1**, 551
$AlLiSi_4C_{12}H_{36}$
$(TMS)_4AlLi$, **2**, 22
$AlNaSi_4C_{12}H_{36}$
$(TMS)_4AlNa$, **2**, 23
$AlNa_2B_{11}CH_{14}$
$Na_2[B_{11}H_{11}AlMe]$, **1**, 258
$AlNiC_7H_{18}$
$[Ni(HAlMe_3)(C_2H_4)_2]^-$, **9**, 110
$AlNiC_{12}H_{20}$
$Ni(cdt)[AlH_4]$, **1**, 551
$AlNiC_{21}H_{38}N$
$NiHAlMe_2(cdt)(C_7H_{13}N)$, **9**, 111
$AlOs_3RuC_{12}H_3O_{12}$
$[Al][H_3RuOs_3(CO)_{12}]$, **10**, 52
$AlReC_{10}H_{11}Cl_3$
$Re(HAlCl_3)Cp_2$, **6**, 184
$AlReC_{18}H_{34}P_2$
$ReH_4(PMe_2Ph)_2[(H)_2AlMe_2]$, **1**, 550
$AlReC_{26}H_{43}P_3$
$ReH_2(PMe_2Ph)_3[(H)_2AlMe_2]$, **1**, 550
$AlReC_{42}H_{44}Cl_2OP_2$
$Re[C(H)Me](OAlCl_2)Et_2$-$(PPh_3)_2$, **6**, 244
$AlRhC_{14}H_{32}P_2$
$Rh(PMe_3)_2(AlMe_3)Cp$, **1**, 552
$AlRuC_{13}H_{28}P$
$Ru(PMe_3)Cp^*[AlH_4]$, **1**, 551
$AlSiC_{12}H_{27}$
$(Bu^t_3Si)Al$, **1**, 445
$AlSi_3C_{54}H_{45}O_3$
$(Ph_3SiO)_3Al$, **2**, 34
$AlSi_4C_9H_{27}Cl_3$
$[(TMS)_3SiAlCl_3]^-$, **2**, 23
$AlSi_4C_{14}H_{38}Cl$
$[(TMS)_2CH]_2AlCl$, **1**, 435
$AlSi_6C_{21}H_{57}$
$[(TMS)_2CH]_3Al$, **1**, 446
$AlSmC_{12}H_{18}Cl_4$
$Sm(C_6Me_6)(AlCl_4)$, **4**, 4
$AlSmC_{28}H_{50}$
$Cp^*_2Sm(Et)_2AlEt_2$, **4**, 88
$AlTaC_{20}H_{51}ClP_4$
$[Ta(CBu^t)(H)(dmpe)_2]$-$[AlMe_3Cl]$, **5**, 86
$AlTaC_{28}H_{50}$
$Cp^*_2Ta(C_2H_4{\cdot}AlEt_3)(H)$, **5**, 92
$AlTiCH_2Cl_5$
$Cl_2AlMeTiCl_3$, **12**, 579
$AlTiC_{10}H_{10}Cl_4$
$TiCp_2(Cl)_2Al(Cl)_2$, **4**, 249
$AlTiC_{10}H_{14}$
$TiCp_2(AlH_4)$, **4**, 260, 261
$TiCp_2(H)_2AlH_2$, **1**, 546; **4**, 260
$AlTiC_{13}H_{18}Cl$
$Cp_2Ti(Cl)CH_2AlMe_2$, **11**, 296, 308
$TiCp_2(CH_2)(Cl)AlMe_2$, **4**, 374
$AlTiC_{21}H_{27}$
$(TiCp_2)_2(AlH_4Me)$, **4**, 262
$AlTiC_{22}H_{36}Cl_2$
$TiCp_2(Cl)_2Al(hexyl)_2$, **4**, 248
$AlTi_2C_{21}H_{27}$
$(TiCp_2)_2(AlH_4Me)$, **4**, 260
$AlUC_{15}H_{19}$
$Cp_3U(AlH_4)$, **4**, 158
$AlVC_{18}H_{24}Cl_4$
$[(Me_3C_6H_3)_2V]AlCl_4$, **5**, 46
$AlWC_{12}H_{28}Cl_4O_2P_3$
$WCl(CO)(PMe_3)_3$-$(HCCOAlCl_3)$, **5**, 372
$AlWC_{12}H_{41}Cl_2NP_3$
$WH_2(PMe_3)_3[(H)_3AlCl_2$-$(NMe_3)]$, **1**, 549
$AlWC_{13}H_{21}$
$WCp_2[(H)_2AlMe_3]$, **1**, 549
$AlW_2C_{18}H_{62}P_6$
$[WH_3(PMe_3)_3]_2[(H)_2AlH(H)_2]$, **1**, 548
$AlYC_{14}H_{22}O$
$Cp_2Y(AlH_4)(THF)$, **4**, 49
$AlYbC_{12}H_{16}Cl_2$
$Cp_2Yb(Cl)_2AlMe_2$, **4**, 36
$AlYbC_{20}H_{30}Cl_4$
$Cp^*_2Yb(Cl_2)AlCl_2$, **4**, 75
$AlYbC_{30}H_{53}O$
$Cp^*_2Yb(Et)AlEt_2(THF)$, **4**, 88
$AlZrC_{21}H_{31}$

$ZrCp_3(AlHEt_3)$, **1**, 548
$AlZrC_{25}H_{38}$
$[Zr[CH_2CH(AlEt_2)_2]Cp_2]Cp$, **4**, 590
$Al_2BeC_{16}H_{24}$
$Me_4(MeC{\equiv})_4Al_2Be$, **1**, 39
$Al_2CH_2Cl_4$
$(Cl_2Al)_2CH_2$, **1**, 456
$Al_2C_2H_4Cl_2$
$(Cl_2Al)_2C_2H_4$, **1**, 437
$Al_2C_2H_6Br_2O$
$Me_2Br_2Al_2O$, **1**, 452
$Al_2C_4H_4$
$\overline{AlCH{=}CHAlCH{=}CH}$, **1**, 443
$Al_2C_6H_5Cl_4N$
$(Cl_2Al)_2NPh$, **1**, 493
$Al_2C_{12}H_{27}ClO$
$Bu^i_3ClAl_2O$, **1**, 452
$Al_2C_{16}H_{36}$
$Bu^i_4Al_2$, **1**, 435
$Al_2Cr_2C_{48}H_{138}P_{16}$
$[CrH(dmpe)_4]_2[(H)_2AlH\text{-}(H)_2AlH(H)_2]$, **1**, 548
$Al_2MgC_{16}H_{24}$
$(Me_2Al)_2Mg(C{\equiv}CMe)_4$, **1**, 107
$Al_2Mn_2C_{22}H_{36}N_6O_{10}$
$Mn_2(CO)_9[C(NMe_2)OAl_2\text{-}(NMe_2)_5]$, **6**, 8
$Al_2Mn_2C_{24}H_{72}P_8$
$[Mn(dmpe)_2]_2[(H)_2AlH\text{-}(H)_2AlH(H)_2]$, **1**, 550
$Al_2MoC_{10}H_{24}Cl_2NO_4$
$Mo(NO)_2(OPr^iAlCl_2Et)_2$, **1**, 552
$Al_2Re_2C_{48}H_{78}P_6$
$[ReH_2(PMe_2Ph)_3]_2[(H)_2AlH\text{-}(H)_2AlH(H)_2]$, **1**, 550
$Al_2Re_2C_{56}H_{94}O_2P_6$
$[ReH_2(PMe_2Ph)_3]_2[(H)_2AlH\text{-}(OBu)_2AlH(H)_2]$, **1**, 550
$Al_2Ru_2C_6H_{28}P_2$
$[Ru(PMe_3)]_2[(H)_2AlH(H)_2AlH\text{-}(H)_2]$, **1**, 550
$Al_2Si_4C_{12}H_{32}$
$(Me_2SiCH_2)_4Al_2$, **1**, 462
$Al_2SmC_{26}H_{50}O_2$
$Sm(C_5H_4Bu^t)_2[(H)_3AlH\text{-}(THF)]_2$, **1**, 551
$Al_2Sm_2C_{44}H_{76}O_2$
$[(Bu^tC_5H_4)_2Sm]_2[(H)_3AlH\text{-}(THF)]_2$, **4**, 49
$Al_2Sm_2C_{48}H_{84}$
$Cp^*{}_2Sm[Me_2AlMe_2]_2SmCp^*{}_2$, **4**, 87
$Al_2TiC_{10}H_{15}Cl_8$
$TiAl_2Cl_8Cp^*$, **4**, 244
$Al_2TiC_{12}H_{20}Cl_7$
$TiAl_2Cl_7EtCp^*$, **4**, 244
$Al_2W_2C_{18}H_{65}$
$[WH_3(PMe_3)_3]_2[(H)_2AlH\text{-}(H)_2AlH(H)_2]$, **1**, 548
$Al_2W_2C_{22}H_{74}N_2P_6$
$[WH_3(PMe_3)_3]_2[(H)_2AlH\text{-}(N{=}CHMe)_2AlH(H)_2]$, **1**, 549
$Al_2Y_2C_{32}H_{58}N_2$
$[YCp_2[(H)_2AlH_2(NEt_3)]]_2$, **1**, 551
$Al_2ZrC_{15}H_{38}$
$[Zr[CH_2CH(AlEt_2)_2]Cp_2]Cp$, **4**, 599
$Al_3C_{12}H_{27}O_3$
$Bu^t_3(Al_3O_3)$, **1**, 453
$Al_3UC_6H_6Cl_{12}$
$(C_6H_6)U(AlCl_4)_3$, **4**, 178
$Al_3UC_{12}H_{18}Cl_{12}$
$(C_6Me_6)U(AlCl_4)_3$, **4**, 178
$Al_4B_{18}C_8H_{32}$
$Al[(H)_2(AlEt_2)AlC_2B_9H_9]\text{-}[AlC_2B_9H_{11}]$, **1**, 258
$Al_4C_{32}H_{72}O_2$
$[Bu^t_2AlOAlBu^t_2]_2$, **1**, 453
$Al_4Si_4C_{16}H_{36}$
$[Bu^tSiAl]_4$, **2**, 23
$Al_6Nd_{12}C_{62}H_{154}Cl_{12}O_2$
$[Nd_6Al_3(Cl)_6(Et)_{14}(OPr^i)]_2$, **4**, 23
$Al_{12}K_2C_{48}H_{108}$
$[Bu^iAl]_{12}K_2$, **1**, 446
$AmC_{16}H_{16}$
$[Am(C_8H_8)_2]^-$, **4**, 181
$AsB_9C_4H_{15}Cl$
$ClAs(MeC)_2B_9H_9$, **1**, 265
$AsB_{10}CH_{15}$
$B_{10}H_{12}AsMe$, **1**, 564
$AsB_{10}C_4H_{24}N$
$NMe_4[B_{10}H_{12}As]$, **1**, 265
$AsB_{10}C_6H_{16}$
$[PhAsB_{10}H_{11}]^-$, **1**, 266
$AsB_{10}C_6H_{17}$
$PhAsB_{10}H_{12}$, **1**, 266
$AsB_{10}H_{10}P$
$PAsB_{10}H_{10}$, **1**, 264
$AsB_{10}H_{12}$
$[B_{10}H_{12}As]^-$, **1**, 266
$AsB_{11}H_{11}$
$[B_{11}H_{11}As]^-$, **1**, 266
$AsB_{20}C_6H_{26}Cl$
$(C_2B_{10}H_{10}Me)_2AsCl$, **1**, 241
$AsCH_3Cl_4$
$MeAcCl_4$, **2**, 343
$AsC_3H_9Cl_2$
Me_3AsCl_2, **2**, 337
$AsC_3H_9F_2$
Me_3AsF_2, **2**, 337
$AsC_5H_{10}Cl$
$(CH_2)_5AsCl$, **2**, 341
$AsC_5H_{11}O_2$
$(CH_2)_5As(O)(OH)$, **2**, 341
$AsC_6H_5Cl_2$
$PhAsCl_2$, **2**, 332
AsC_6H_{15}
$MeEtPrAs$, **2**, 331
$AsC_9H_{12}N$
$PhAsCH(NMe_2)$, **2**, 323
$AsC_{12}H_{10}Cl$
Ph_2AsCl, **2**, 332
$AsC_{12}H_{20}N$
$BuPhAsCH_2CH_2NH_2$, **2**, 331
$AsC_{18}H_{15}I$
$[Ph_3AsI]^+$, **2**, 337
$AsC_{18}H_{15}I_2$
Ph_3AsI_2, **2**, 337
$AsC_{25}H_{24}$
$[(Me)(Bz)(Tol)(nap)As]^+$, **2**, 336
$AsC_{30}H_{25}$
Ph_5As, **2**, 335
$AsCoFeMoWC_{19}H_{16}O_7S$
$FeCoMoW(S)(AsMe_2)Cp_2\text{-}(CO)_7$, **10**, 37, 244
$AsCo_2MoC_{17}H_{21}O_6P$
$MoCo_2(PBu^t)(AsMe_2)(CO)_6Cp$, **10**, 222
$AsCo_2MoC_{18}H_{15}O_3$
$MoCo_2(As)(CO)_3Cp_3$, **10**, 218
$AsCrC_{29}H_{20}O_5$
$Cr(CO)_5(AsPh)_4$, **5**, 260
$AsCrMnC_{15}O_{15}$
$As[Mn(CO)_5][Cr(CO)_5]_2$, **6**, 86
$AsCr_2MnC_{14}O_{14}$
$As[Cr_2(CO)_9][Mn(CO)_5]$, **6**, 86
$AsCu_2Ru_4C_{38}H_{26}O_{12}P$
$Cu_2Ru_4(H)_2[Ph_2As(CH_2)_2PPh_2]\text{-}(CO)_{12}$, **10**, 299
$AsFeC_7H_8O_5$
$(CO)_4FeAs(Me)(OEt)$, **7**, 55
$AsFeC_{25}H_{21}O_4$
$Fe(CO)_4As(Tol)_3$, **7**, 55
$AsFeMnC_{11}H_6O_9$
$(CO)_5Mn(AsMe_2)Fe(CO)_4$, **10**, 39
$AsFeMoC_{14}H_{11}O_7$
$Cp(CO)_3Mo(AsMe_2)Fe(CO)_4$, **10**, 39
$AsFeMo_3C_{25}H_{15}O_{10}$
$Mo_3Fe(As)(CO)_{10}Cp_3$, **10**, 218
$AsFeW_3C_{25}H_{15}O_{10}$
$W_3Fe(As)(CO)_{10}Cp_3$, **10**, 218
$AsFe_2C_{10}H_6O_8$
$[Me_2As[Fe(CO)_4]_2]^-$, **7**, 55
$AsFe_3C_{10}HO_9$
$Fe_3(CH)(As)(CO)_9$, **7**, 273, 275
$AsFe_4C_{14}ClO_{14}$
$Fe_2(CO)_8As[Fe_2(CO)_6Cl]$, **7**, 56
$AsGa$
$GaAs$, **1**, 517
$AsGaC_{12}H_{27}$
$Bu^+GaAsBu_2{}^+$, **1**, 518
$AsGaSiC_{26}H_{48}$
$Cp_2{}^*GaAs(TMS)_2$, **1**, 518, 519
$AsInMn_4C_{41}H_{21}ClO_{17}$
$[AsPh_4][InCl[Mn(CO)_5][Mn_3\text{-}(CO)_{12}(H)]]$, **1**, 561
$AsInMn_6C_{48}H_{22}O_{24}$
$[AsPh_4][In[Mn_3(CO)_{12}(H)]_2]$, **1**, 561
$AsLiC_8H_{12}$
$(Me_4C_4As)Li$, **2**, 324
$AsLiC_{16}H_{20}O$
$Ph_2AsLi(OEt_2)$, **2**, 328
$AsLiC_{24}H_{34}O_6$
$LiAsPh_2{\cdot}3(dioxane)$, **1**, 11
$AsLiSi_4C_{14}H_{38}$
$[(TMS)_2CH]_2AsLi$, **2**, 328
$AsLiSi_4C_{47}H_{83}FO_2$
$LiAs(Pr^i_3C_6H_2Si)_{.2}(F)Si\text{-}(Pr^i_3Si){\cdot}2THF$, **1**, 11

$AsMnC_{28}H_{22}O_4$
$(AsPh_4)[MnH_2(CO)_4]$, **6**, 68
$AsMn_2C_{24}H_{31}O_4$
$[Mn(CO)_2Cp^*]_2AsH$, **6**, 114
$AsMoC_{10}H_{11}O_3$
$Mo(AsMe_2)(CO)_3Cp$, **5**, 428
$AsMoNi_2C_{37}H_{30}O_7$
$Ni_2Mo[As[Mo_3(As)(CO)_5Cp_3]](CO)_2Cp_3$, **9**, 173
$AsMoPt_2C_{27}H_{20}O_4P$
$MoPt_2(As)(CO)_4(PPh_3)Cp$, **10**, 218
$AsMoRh_3C_{25}H_{20}O_5$
$MoRh_3(As)(CO)_5Cp_4$, **10**, 218
$AsMo_2NiC_{19}H_{15}O_4$
$Mo_2Ni(As)(CO)_4Cp_3$, **10**, 218
$NiMo_2(As)(CO)_4Cp_3$, **9**, 173
$AsNaC_{16}H_{18}O_2$
$Ph_2AsNa(dioxane)$, **2**, 328
$AsOs_3C_9H_3O_9$
$Os_3H_2(AsH)(CO)_9$, **7**, 735
$AsOs_3C_{11}H_3O_{11}$
$Os_3H_2(AsH)(CO)_{11}$, **7**, 735
$AsOs_3C_{13}H_7O_{11}$
$Os_3H(AsMe_2)(CO)_{11}$, **7**, 735, 736
$AsOs_3C_{15}H_7O_9$
$Os_3H_2(AsPh)(CO)_9$, **7**, 703
$AsOs_4C_{18}H_7O_{12}$
$Os_4(H)_2(AsPh)(CO)_{12}$, **7**, 861
$AsOs_5C_{21}H_5O_{15}$
$Os_5(AsPh)(CO)_{15}$, **7**, 925
$AsPtC_{20}H_{17}Cl_2$
$PtCl_2(H_2CCHC_6H_4AsPh_2)$, **9**, 552
$AsPtC_{26}H_{22}F_6O_3$
$Pt[CH(CH_2OMe)C_6H_4AsPh_2](CF_3C(O)CH_2C(O)CF_3)$, **9**, 552
$AsPtC_{26}H_{26}O_2$
$Pt[CH_2C(CH_2)C_6H_4AsPh_2](acac)$, **9**, 552
$Pt[CHC(Me)C_6H_4AsPh_2](acac)$, **9**, 552
$AsPtC_{26}H_{27}O_2$
$[Pt(H_2CCMeC_6H_4AsPh_2)(acac)]^+$, **9**, 552
$AsPtC_{27}H_{29}ClP$
$PtClMe(PMe_2Ph)(AsPh_3)$, **9**, 438
$AsPtC_{45}H_{45}O_2P_2$
$[Pt(PPh_3)_2(C_3H_5)][As(OCH_2CHCH_2)_2]$, **9**, 574
$AsRhZrC_{46}H_{50}O_4S_2$
$[AsPh_4][Cp_2^*Zr(S)_2Rh(CO)_2]$, **10**, 47
$AsRu_2C_8H_6O_6$
$Ru_2(AsMe_2)(CO)_6$, **7**, 649
$AsRu_5C_{25}H_9O_{13}$
$Ru_5(C_6H_4)(AsPh)(CO)_{13}$, **7**, 924
$AsRu_{12}TlC_{58}H_{20}O_{32}$
$[AsPh_4][Tl[Ru_6C(CO)_{16}]_2$, **1**, 568
$AsSiC_{10}H_{17}$
$Ph(Me)AsTMS$, **2**, 29
$AsSi_2C_{12}H_{24}$
$PhAs(TMS)_2$, **2**, 29
$AsSi_2C_{25}H_{48}P$
$Bu^t_3C_6H_2PAsCH(TMS)_2$, **2**, 326
$AsSi_2C_{39}H_{67}$
$(Pr^i_3Ph)_2SiAsSiPr^i_3$, **2**, 30
$AsSi_2ZrC_{18}H_{32}Cl$
$Zr[As(TMS)_2](Cl)(C_5H_4Me)_2$, **4**, 518
$AsSi_3C_9H_{27}$
$(TMS)_3As$, **2**, 30
$AsSi_3C_{12}H_{33}Cl_2$
$(TMS\text{-}CH_2)_3AsCl_2$, **2**, 337
$As_2AlB_{11}C_{49}H_{54}$
$(AsPh_4)_2[B_{11}H_{11}AlMe]$, **1**, 258, 271
$As_2Au_2BFe_4C_{48}H_{30}O_{12}$
$Fe_4(CO)_{12}B(AuAsPh_3)_2$, **1**, 297, 299
$As_2Au_2BFe_4C_{48}H_{31}O_{12}$
$HFe_4(CO)_{12}B(AuAsPh_3)_2$, **7**, 18
$As_2B_8Co_2C_{10}H_{18}$
$[(Cp)Co]_2B_8H_8As_2$, **1**, 266, 267
$As_2B_9CH_{13}$
$B_9H_{10}As_2Me$, **1**, 266
$As_2B_9NiC_{16}H_{31}$
$(Me_2PhP)_2Ni(As_2B_9H_9)$, **1**, 266
$As_2B_{10}H_{10}$
$B_{10}H_{10}As_2$, **1**, 266
$As_2B_{10}SnC_{48}H_{52}Cl_2$
$(Ph_4As)_2B_{10}H_{12}SnCl_2$, **1**, 259, 260
$As_2B_{10}SnC_{50}H_{58}Cl_2$
$(Ph_4As)_2B_{10}H_{12}SnMeCl_2$, **1**, 259
$As_2B_{20}CoH_{20}$
$Co(B_{10}H_{10}As)_2^{3-}$, **1**, 266
$As_2Co_2Cr_3C_{20}O_{20}$
$Cr_3Co_2(As)_2(CO)_{20}$, **10**, 48
$As_2Co_2PtC_{33}H_{24}O_7$
$Co_2Pt(CO)_7(dpae)$, **10**, 39, 47
$As_2Cr_2Fe_3C_{19}O_{19}$
$Fe_3(CO)_9[AsCr(CO)_5]_2$, **7**, 56
$As_2FeC_9H_9NO_4$
$(CO)_4FeAs(Me)_2N(Me)AsMe_2$, **7**, 55
$As_2FeC_{38}H_{30}Cl_2O_2$
$Fe(CO)_2(AsPh_3)_2Cl_2$, **7**, 55, 78
$As_2Fe_2C_{10}H_{12}O_6$
$Fe_2(CO)_6(AsMe_2)_2$, **7**, 55
$As_2Fe_2C_{13}H_{15}NO_8$
$[(CO)_4FeAs(Me)_2]_2N(Me)$, **7**, 55
$As_2Fe_2C_{14}H_{18}N_2O_6S$
$Fe_2(CO)_6(Bu^tAsNSNAsBu^t)$, **7**, 56
$As_2Fe_6C_{22}O_{22}$
$[Fe_2(CO)_8As]_2Fe_2(CO)_6$, **7**, 56
$As_2InMn_5C_{68}H_{40}O_{20}$
$[AsPh_4]_2[In[Mn(CO)_4]_5]$, **1**, 561
$As_2IrC_{40}H_{36}Cl$
$Ir(Cl)(C_4H_6)(AsPh_3)_2$, **8**, 319, 320
$As_2IrRh_2C_{67}H_{58}Cl_2O_3P_4$
$[Rh_2IrCl_2(CO)_3[AsPh(CH_2PPh_2)_2]_2]^+$, **10**, 60
$As_2Ir_4C_{20}H_{16}O_{10}$
$Ir_4(C)_{10}(diars)$, **8**, 499
$As_2Ir_4C_{21}H_{16}O_{11}$
$Ir_4(CO)_{11}(diars)$, **8**, 495
$As_2LiC_{20}H_{43}O_2$
$Bu^t_2AsAsBu^tLi(THF)_2$, **2**, 328
$As_2Li_2C_{40}H_{60}O_4$
$[LiAsPh_2 \cdot 2Et_2O]_2$, **1**, 11
$As_2Mn_4C_{28}H_{20}O_8$
$[Mn(CO)_2Cp]_4As_2$, **6**, 114
$As_2Mn_4C_{48}H_{60}O_8$
$[Mn(CO)_2Cp^*]_4As_2$, **6**, 114
$As_2Mn_4In_2C_{42}H_{40}Cl_2O_{18}$
$[AsPh_4]_2[Mn_2(CO)_8[InCl[Mn(CO)_5]]_2$, **1**, 561
$As_2NiC_{55}H_{49}P_2$
$[Ni(dapm)_2Cp]^+$, **9**, 160
$As_2NiC_{57}H_{53}P_2$
$[Ni(arphos)_2Cp]^+$, **9**, 160
$As_2Ni_{10}C_{30}H_{10}O_{18}$
$[Ni_{10}(AsPh)_2(CO)_{18}]^{2-}$, **9**, 20
$As_2Os_3C_{20}H_{28}N_2O_8$
$Os_3H(NSNAsBu^t_2)(AsBu^t)(CO)_8$, **7**, 698
$As_2Pd_2C_{38}H_{32}Cl_2$
$[Pd[(AsPh_2CH_2)C_6H_4](Cl)]_2$, **9**, 264
$As_2PtC_{51}H_{42}O$
$Pt(AsPh_3)_2[PhHCC(O)CHPh]$, **9**, 578
$As_2Pt_{24}C_{78}H_{40}O_{30}$
$(AsPh_4)_2[Pt_{24}(CO)_{30}]$, **9**, 398
$As_2Pt_{26}C_{80}H_{40}O_{32}$
$(AsPh_4)_2[Pt_{26}(CO)_{32}]$, **9**, 398
$As_2Pt_{38}C_{92}H_{40}O_{44}$
$(AsPh_4)_2[Pt_{38}(CO)_{44}]$, **9**, 398
$As_2RhC_{37}H_{30}ClOP_4$
$RhCl(CO)P_4)(AsPh_3)_2$, **8**, 161
$As_2Ru_4C_{36}H_{38}O_{12}$
$[Ru_2(C_6H_4O_2)(CO)_4(AsPh_3)]_2$, **7**, 666
$As_2Si_4ZrC_{22}H_{46}$
$Zr[As(TMS)_2]_2Cp_2$, **4**, 518
$As_2Si_6C_{20}H_{54}$
$(TMS)_3CAsAsC(TMS)_3$, **2**, 326
$As_2Si_8C_{28}H_{76}$
$[(TMS)_2CH]_2AsAs[CH(TMS)_2]_2$, **2**, 325
$As_2WC_{15}H_{30}Br_2F_6O_6P_2$
$WBr_2(CO)[P(OMe)_3]_2[Me_2AsC(CF_3)=C(CF_3)AsMe_2]$, **5**, 262
$As_2W_2C_7I_2O_7$
$W(Co)_4(I)(As_2)WI(CO)_3$, **5**, 262
$As_2ZrC_{34}H_{30}$
$Zr(AsPh_2)_2Cp_2$, **4**, 518
$As_3C_3H_9S_3$
$(MeAsS)_3$, **2**, 334
$As_3C_{14}H_{31}$
$As(C_2H_4CH=CH_2)(C_3H_6AsMe_2)_2$, **9**, 122
$As_3GaC_{54}H_{66}$
$[(Mes)_2As]_3Ga$, **1**, 518, 519

$As_3Ga_3C_{24}H_{60}$
[$Me_2Ga(AsPr_2Pr^i{}_2)]_3$, **1**, 522
$As_3Ga_3Si_6C_{24}H_{66}$
$[Br_2Ga(As(CH_2TMS)_2]_3$, **1**, 522
$As_3In_3C_{12}H_{36}$
$[Me_2In(AsMe_2)]_3$, **1**, 522
$As_3Li_3Si_6C_{24}H_{60}$
$[LiAs(CH\text{-}TMS)_2]_3$, **1**, 11
$As_3NiC_{14}H_{31}Br_2$
$[Ni[As(C_2H_4CH{=}CH_2)(C_3H_6\text{-}AsMe_2)_2Br]Br]$, **9**, 122
$As_3Ni_9C_{33}H_{15}O_{15}$
$[Ni_9(AsPh)_3(CO)_{15}]^{2-}$, **9**, 19
$As_3Rh_6C_{31}H_{45}O_{11}$
$Rh_6(CO)_{11}(AsBu^t{}_2)_2(AsBu^t)$, **8**, 477
$As_3Ru_3C_{47}H_{37}O_7$
$Ru_3H_2(AsPh)(CO)_7(AsPh_3)_2$, **7**, 703
$As_3TiC_{28}H_{25}$
$Ti(As_3Ph_3)Cp_2$, **4**, 410
$As_4B_{18}CoH_{18}$
$[Co(As_2B_9H_9)_2]^-$, **1**, 266
$As_4C_4H_{12}O_4$
$(MeAsO)_4$, **2**, 334
$As_4C_4H_{12}S_4$
$(MeAsS)_4$, **2**, 334
$As_4C_{36}H_{44}O_4$
$(Bu^t{}_6H_2AsO)_4$, **2**, 334
$As_4Ir_4C_{36}H_{72}O_4$
$Ir_4(CO)_4(AsBu^t{}_2)_4$, **8**, 497
$As_4Li_2C_{32}H_{70}O_2$
$[LiAs(AsBu^t{}_2)Bu^t\cdot THF]_2$, **1**, 11
$As_5C_5H_{15}$
$(MeAs)_5$, **2**, 334
$As_5Mo_2C_{10}H_{10}$
$CpMo(As_5)MoCp$, **5**, 426
$As_6Co_4C_{72}H_{60}P_4$
$Co_4As_6(PPh_3)_4$, **8**, 443
$As_6Ga_2Si_6C_{32}H_{88}$
$[(TMS\text{-}CH_2)_2AsGa(As(CH_2\text{-}TMS)_2)]_2$, **1**, 520, 521
$As_7Ga_5Si_6C_{66}H_{102}$
$[(PhAsH)(TMS\text{-}CH_2)_2Ga\text{-}(PhAs)_6[(TMS\text{-}CH_2)\text{-}Ga]_4]$, **1**, 522, 523
$AuAs_4Ir_2C_{65}H_{58}ClOP_4$
$[IrCl(CO)[AsPh(CH_2P\text{-}Ph_2)_2]_2Au]^{2+}$, **10**, 68
$AuBC_8H_{24}P_2$
$AuMe_2[CH_2PMe_2BH_2PMe_2\text{-}CH_2]$, **3**, 22
$AuBC_{54}H_{45}F_4OP_3$
$[[Au(PPh_3)]_3O][BF_4]$, **10**, 258, 259
$AuBFe_4C_{30}H_{17}O_{12}P$
$HFe_4(CO)_{12}(AuPPh_3)BH$, **1**, 297, 299, 302
$AuB_2CoC_{35}H_{44}P$
$CpCo[Et_4Me(C_3B_2)]Au(PPh_3)$, **1**, 206, 207
$AuB_2FeC_{22}H_{20}O_4P$
$Fe(CO)_4(B_2H_5)Au(PPh_3)$, **10**, 62
$AuB_2OsC_{22}H_{20}O_4P$
$(PPh_3)Au(CO)_4Os(B_2H_5)$, **1**, 308
$AuB_2RuC_{63}H_{61}F_8P_4$
$[RhAu[HC(PPh_2)_3](cod)\text{-}(PPh_3)][BF_4]_2$, **10**, 40
$AuB_5C_{18}H_{23}P$
$Au(PPh_3)B_5H_8$, **1**, 326
$AuB_8C_5H_{22}NS_2$
$S_2CNEt_2AuB_8H_{12}$, **1**, 337
$AuB_9PtC_{32}H_{56}P_3$
$PtH(PEt_3)_2(C_2B_9H_{10})Au(PPh_3)$, **10**, 71
$AuB_9WC_{32}H_{37}O_2P$
$[C_2B_9H_9Me_2]W(CO)_2(\equiv CTol)Au(PPh_3)$, **10**, 8, 9
$AuB_{10}C_3H_{22}P$
$(AuPMe_3)B_{10}H_{13}$, **1**, 352–4
$AuB_{10}C_{15}H_{45}P$
$[[AuP(C_5H_{11})_3]B_{10}H_{12}]^-$, **1**, 352, 354
$AuB_{20}H_{24}$
$[(B_{10}H_{12})Au(B_{10}H_{12})]^-$, **1**, 352
$AuB_{20}H_{25}$
$[(B_{10}H_{12})Au(B_{10}H_{13})]^{2-}$, **1**, 352
AuCBrO
AuBr(CO), **3**, 48
AuCClO
AuCl(CO), **3**, 48
Au(CO)Cl, **10**, 305
$AuCFO_4S$
$AuOSO_2F(CO)$, **3**, 48
AuC_2H_6ClS
$AuCl(SMe_2)$, **10**, 305
AuC_2O_2
$[Au(CO)_2]^+$, **3**, 47, 48
AuC_4H_8ClS
$AuCl(SC_4H_8)$, **10**, 258, 303
$AuC_4H_{12}Cl_2$
$Au_2(Cl)_2Me_4$, **3**, 33
$AuC_4H_{12}P$
$AuMe(PMe_3)$, **3**, 4
$AuC_6H_9F_9P$
$Au(CF_3)_3(PMe_3)$, **3**, 40
$AuC_6H_{16}ClO_2S_2$
$Au[CH_2S(O)Me_2]_2Cl$, **3**, 21
$AuC_6H_{17}OS$
$AuMe_3[CH_2S(O)Me_2]$, **3**, 22
$AuC_7H_{14}O_2$
$AuMe_2(acac)$, **3**, 34
AuC_8H_5Cl
$[AuCl(CCPh)]^-$, **3**, 18
$AuC_8H_{14}Cl$
$AuCl(C_8H_{14})$, **3**, 44
$AuC_9H_{15}F_9P$
$Au(CF_3)_3(PEt_3)$, **3**, 40
$AuC_{10}H_8F_5S$
$Au(SC_4H_8)(C_6F_5)$, **10**, 313
$AuC_{14}H_{40}ClN_6P_2$
$Au[CH_2P(NMe_2)_3]_2Cl$, **3**, 21
$AuC_{17}H_8F_{10}O_2$
$Au(C_6F_5)_2(acac)$, **3**, 33
$AuC_{18}H_{15}ClP$
$AuCl(PPh_3)$, **3**, 18; **8**, 332
$AuC_{18}H_{15}NO_3P$
$AuPPh_3NO_3$, **8**, 332
$AuC_{19}H_{17}ClP$
$Au(CH_2Cl)(PPh_3)$, **3**, 4
$AuC_{20}H_{17}NP$
$Au(CN)(CH_2PPh_3)$, **3**, 22
$AuC_{22}H_8F_{15}S$
$Au(C_6F_5)_3(SC_4H_8)$, **3**, 41
$AuC_{22}H_{10}F_{15}O$
$Au(C_6F_5)_3(OEt_2)$, **3**, 41
$AuC_{22}H_{26}P$
$AuMe_3(CH_2PPh_3)$, **3**, 22
$AuC_{22}H_{31}B_{10}P$
$(Ph_3P)Au(Me_2C_2B_{10}H_{10})$, **1**, 419
$AuC_{23}H_{23}O_2P$
$Au(acac)(PPh_3)$, **3**, 6
$AuC_{24}H_{20}$
$[AuPh_4]^-$, **3**, 42
$AuC_{26}H_{20}P$
$Au(CCPh)(PPh_3)$, **3**, 12, 18
$AuC_{28}H_{28}P_2$
$Au_2[(CH_2)_2PPh_2]_2$, **3**, 27
$AuC_{31}H_{28}O_2$
$Au(C_2Ph_4)(acac)$, **3**, 35, 37
$AuC_{37}H_{30}ClP_2$
$AuCl(Ph_3PCPPh_3)$, **3**, 22
$AuC_{38}H_{34}ClO_4P_2$
$[Au(CH_2PPh_3)_2]ClO_4$, **3**, 23
$AuC_{40}H_{34}Cl$
$Au[C(CH_2)_2PPh_3]_2Cl$, **3**, 21
$AuC_{46}H_{46}O_4P_2$
$[Ph_3PNPPh_3][Au(acac)_2]$, **3**, 18
$AuC_{60}H_{30}F_{20}NP_2$
$[Au(C_6F_5)_4][N(PPh_3)_2]$, **3**, 43
$AuCo_3RuC_{30}H_{15}O_{12}P$
$RuCo_3(AuPPh_3)(CO)_{12}$, **10**, 48
$AuCo_3RuC_{48}H_{30}O_{12}P_2$
$[Au(PPh_3)_2][RuCo_3(CO)_{12}]$, **10**, 48
$AuCrC_{23}H_{16}O_5P$
$(Ph_3P)Au(H)Cr(CO)_5$, **5**, 223
$AuFeC_{42}H_{39}O_6P_2$
$Fe(CO)_3[P(OMe)_3][Au(PPh_3)_2]$, **10**, 54
$AuFe_2C_{26}H_{15}O_8P$
$[Fe_2(CO)_8(AuPPh_3)]^-$, **7**, 81
$AuFe_3C_{32}H_{25}NO_9P$
$AuFe_3(HC{=}NBu^t)(CO)_9(PPh_3)$, **10**, 261
$AuFe_3Ir_2C_{30}H_{15}O_{12}P$
$[Fe_2Ir_2(CO)_{12}(AuPPh_3)]^-$, **7**, 83
$AuFe_4C_{67}H_{45}NO_{13}P_3$
$[N(PPh_3)_2][AuFe_4(CO)_{13}\text{-}(PPh_3)]$, **10**, 299
$AuHgC_{26}H_{24}P_2S_2$
$[AuHg[CH_2P(S)Ph_2]_2]^{2+}$, **10**, 71
$[Au(SPPh_2CH_2)_2Hg]^+$, **10**, 71
$AuHg_2IrC_{51}H_{44}Cl_6OP_4$
$IrCl(CO)(dppm)\text{-}(dppmAuCl)Hg(Cl)_2\text{-}HgCl_2$, **10**, 68
$AuIrC_{51}H_{44}ClOP_4$
$[IrCl(CO)(dppm)_2Au]^+$, **10**, 68
$AuIrC_{54}H_{50}N_2P_4$
$[Ir(MeNC)_2(dppm)_2Au]^{2+}$, **10**, 68

$AuIrC_{61}H_{59}P_4$
$[Ir(H)(C_2H_4)(triphos)Au(PPh_3)]^+$, **10**, 63
$AuIrC_{64}H_{55}N_2P_3$
$AuIr(H)_2(bipy)(PPh_3)_3^{2+}$, **8**, 404
$AuIrC_{66}H_{58}N_2P_6$
$[Ir(CN)_2[Au[PPh(CH_2PPh_2)_2]_2]^+$, **10**, 68
$AuMnC_{23}H_{15}O_5P$
$Mn(CO)_5[Au(PPh_3)]$, **6**, 12
$AuMnC_{38}H_{31}ClO_9P_2$
$[Mn[PPh_2Au(PTol_3)](CO)_5]ClO_4$, **6**, 88
$AuMnSiC_{39}H_{35}O_2P$
$Mn(SiMePh_2)(CO)_2Cp'Au(PPh_3)$, **10**, 48
$AuMo_2C_{16}H_{10}O_6$
$[Au[Mo(CO)_3Cp]_2]^-$, **5**, 408
$AuNiOs_3C_{32}H_{22}O_9P$
$[NiOs_3(CO)_9Cp(H)_2[Au(PPh_3)]]^-$, **9**, 21
$AuOsC_{29}H_{33}NP$
$[Os(CNMe)(PMe_3)(C_6H_6)Au(PPh_3)]^+$, **10**, 54
$AuOsC_{37}H_{32}ClINOP_2$
$OsAu(CH_2)ClI(NO)(PPh_3)_2$, **7**, 338
$AuOsRu_3C_{35}H_{34}F_3O_9P_4S$
$[AuRu_3[(Me_2P)_3CH](CO)_9(PPh_3)][OsSCF_3]$, **10**, 299
$AuOs_3C_{16}H_{16}O_{10}P$
$Os_3H(AuPEt_3)(CO)_{10}$, **7**, 720
$AuOs_3C_{19}H_{20}O_{10}P$
$Os_3(AuPEt_3)(C_3H_5)(CO)_{10}$, **7**, 696, 788
$AuOs_3C_{26}H_{16}O_{10}P$
$Os_3(C_2Ph)(AuPMe_2Ph)(CO)_{10}$, **7**, 694
$AuOs_3C_{28}H_{16}O_{10}P$
$[HOs_3Au(CO)_{10}(PPh_3)]$, **10**, 59
$AuPtC_8H_{15}Cl_2S$
$PtCl(SMe_2)(CCBu^t)(AuCl)$, **9**, 571
$AuPtC_{31}H_{31}ClN_2P$
$PtClMe_2(CH_2AuPPh_3)(bipy)$, **3**, 4
$AuPtC_{31}H_{31}Cl_2P_2$
$PtCl(CCBu^t)(dppm)AuCl$, **9**, 571
$AuPtC_{56}H_{53}ClP_4$
$[PtCl(CCBu^t)(dppm)_2(Au)]^+$, **9**, 571
$AuPtC_{62}H_{62}P_4$
$[Pt(CCBu^t)_2(dppm)_2(Au)]^+$, **9**, 571
$AuRe_3C_{27}H_{18}O_9P$
$[AuRe_3(H)_3(CO)_9(PPh_3)]$, **10**, 304
$AuRe_3C_{28}H_{18}O_{10}P$
$[AuRe_3(H)_3(CO)_{10}(PPh_3)]^-$, **10**, 304
$AuRh_3RuC_{30}H_{15}O_{12}P$
$RuRh_3(AuPPh_3)(CO)_{12}$, **7**, 685
$AuRuC_{73}H_{62}F_6OP_5$
$[RuAu(H)_2(CO)(PPh_3)_4](PF_6)$, **10**, 41
$AuRu_2C_{35}H_{44}N_2O_{13}P_5$
$[Ru_2Au(PPh_3)][(CO)_5(etmdp)_2]^+$, **7**, 627
$AuRu_3C_{19}H_{15}IO_{12}P$
$Ru_4(C)(AuPEt_3)(I)(CO)_{12}$, **7**, 852
$AuRu_3C_{58}H_{42}O_7P_3$
$Ru_3(AuPPh_3)(CO)_7(dppm)(CCPh)$, **7**, 774
$AuRu_4C_{30}H_{18}O_{12}P$
$H_3Ru_4[Au(PPh_3)](CO)_{12}$, **10**, 42
$AuRu_4C_{31}H_{16}O_{12}P$
$(H)Ru_4(C)(AuPPh_3)(CO)_{12}$, **7**, 852
$AuRu_5C_{35}H_{18}O_{15}P$
$Ru_5(C)(MeCO)(AuPPh_3)(CO)_{14}$, **7**, 920
$AuSb_2C_2F_{11}O$
$[Au(CO)_2][Sb_2F_{11}]$, **3**, 48
$AuUC_2F_6O_2$
$[Au(CO)_2][UF_6]$, **3**, 48
$AuVC_{24}H_{15}O_6P$
$[V(CO)_6][AuPPh_3]$, **5**, 9
$Au_2AgAs_4C_{86}H_{44}ClF_{30}P_4$
$Au(C_6F_5)_3(dppm)Ag(OClO_3)(dppm)Au(C_6F_5)_3$, **10**, 71
$Au_2Ag_2C_{38}H_{34}Cl_2O_8P_2$
$[Ag(OClO_3)(AuCH_2PPh_3)]_2$, **10**, 71
$Au_2Ag_2C_{68}H_{50}P_2$
$Au_2Ag_2(CCPh)_4(PPh_3)_2$, **3**, 75
$Au_2As_2BFe_4C_{48}H_{30}O_{12}$
$Fe_4(CO)_{12}B(AuAsPh_3)_2$, **1**, 297, 299
$Au_2As_2BFe_4C_{48}H_{31}O_{12}$
$HFe_4(CO)_{12}B(AuAsPh_3)_2$, **7**, 18
$Au_2BC_{32}H_{57}N_2P_4$
$[Au_2(dppm)_2(S_2CNEt_2)][BH_3CN]$, **1**, 288, 290
$Au_2BC_{35}H_{35}F_4P_2$
$[MeC_6H_4[Au_2(Ph_2PCH_2CH_2)_2]][BF_4]$, **3**, 12
$Au_2BC_{43}H_{37}F_4P_2$
$[MeC_6H_4(AuPPh_3)_2][BF_4]$, **3**, 12
$Au_2BFeC_{46}H_{39}F_4P_2$
$[[(Cp)Fe(C_5H_4)](AuPPh_3)_2][BF_4]$, **3**, 12
$Au_2BFe_4C_{24}H_{31}O_{12}P_2$
$HFe_4(CO)_{12}B[AuPEt_3]_2$, **7**, 18
$Au_2BFe_4C_{48}H_{31}O_{12}P_2$
$Fe_4(CO)_{12}(AuPPh_3)_2BH$, **1**, 297, 299, 302
$Au_2B_3Fe_8C_{30}H_{22}O_{24}P_2$
$[H[Fe_4(CO)_{12}BH]_2Au][Au(PMe_3)_2BH]$, **1**, 297, 300
$Au_2B_8C_{10}H_{30}N_2S_4$
$(S_2CNEt_2)_2Au_2B_8H_{10}$, **1**, 337
$Au_2B_8C_{36}H_{76}P_2$
$[(C_6H_{11})_3PAu]_2B_8H_{10}$, **1**, 337
$Au_2B_{20}C_6H_{30}P_2$
$[B_{10}H_{12}AuPMe_3]_2$, **1**, 352, 353
$Au_2B_{20}C_6H_{42}P_2$
$[(Me_3P)_2Au][(B_{10}H_{12})Au(B_{10}H_{12})]$, **1**, 352, 354
Au_2CCl_4O
$Au_2Cl_4(CO)$, **3**, 48
$Au_2C_4Br_2F_{12}$
$Au_2(Br)_2(CF_3)_4$, **3**, 33
$Au_2C_4F_{12}I_2$
$Au_2(I)_2(CF_3)_4$, **3**, 33
$Au_2C_9H_{24}P_2$
$(Me_3P)_2C(AuMe)_2$, **3**, 13
$Au_2C_{28}H_{28}Br_3IP_2$
$Au_2(Br)[(CH_2)_2PPh_2]_2[IBr_2]$, **3**, 27
$Au_2C_{28}H_{28}P_2$
$Au_2[(CH_2)_2PPh]_2$, **3**, 30
$Au_2C_{30}H_{32}P_2$
$[Ph_2P(CH_2)_3Au]_2$, **3**, 4
$Au_2C_{39}H_{30}N_2P_2$
$(NC)_2C(AuPPh_3)_2$, **3**, 13
$Au_2C_{42}H_{34}P_2$
$Au_2(CCPh)_2(Ph_2PCH_2CH_2PPh_2)$, **3**, 12
$Au_2C_{56}H_{40}Cl_2$
$Au_2(Cl)_2(C_4Ph_4)_2$, **3**, 34
$Au_2CrRu_5C_{29}H_{30}O_{17}P_2$
$Au_2Ru_5Cr(CO)_{17}(PEt_3)_2$, **10**, 278, 288
$Au_2CrRu_5C_{30}H_{30}O_{17}P_2$
$Au_2Ru_5CrC(CO)_{17}(PEt_3)_2$, **10**, 278
$Au_2Cu_4C_{32}H_{40}Br_2N_4$
$Au_2Cu_4Br_2(C_6H_4NMe_2)_4$, **3**, 122
$Au_2FeC_{40}H_{30}O_4P_2$
$Au_2Fe(CO)_4(PPh_3)_2$, **10**, 256
$Fe(CO)_4(AuPPh_3)_2$, **7**, 81
$Au_2FeC_{40}H_{66}O_4P_2$
$Fe(CO)_4[AuP(cy)_3]_2$, **7**, 81
$Au_2Fe_2Si_2C_{62}H_{62}O_{12}P_4$
$[Fe(CO)_3[Si(OMe)_3](dppm)Au]_2$, **10**, 62
$Au_2Fe_4C_{25}H_{30}O_{12}P_2$
$Fe_4(CO)_{12}C[AuPEt_3]_2$, **7**, 27
$Au_2Fe_4C_{41}H_{22}O_{16}P_2$
$[Fe_2(CO)_8][AuPh_2P(CH_2)PPh_2Au)Fe_2(CO)_8]^{2-}$, **7**, 81
$Au_2Fe_5C_{27}H_{30}O_{14}P_2$
$Fe_5(CO)_{14}C[AuPEt_3]_2$, **7**, 26
$Au_2H_6Cl_2P_2$
$[AuCl(PH_3)]_2$, **3**, 3
$Au_2Li_2C_{36}H_{48}N_4$
$Au_2Li_2[C_6H_4(CH_2NMe_2)]_4$, **3**, 13, 101
$Au_2Os_3C_{46}H_{30}O_{10}P_2$
$Au_2Os_3(CO)_{10}(PPh_3)_2$, **10**, 305
$Au_2Os_4C_{48}H_{32}O_{12}P_2$
$Au_2Os_4(H)_2(CO)_{12}(PPh_3)_2$, **10**, 278
$Au_2Os_5C_{51}H_{30}O_{14}P_2$
$Os_5(C)[Au(PPh_3)]_2(CO)_{14}$, **7**, 920
$Au_2Os_8C_{58}H_{30}O_{22}P_2$
$Au_2Os_8(CO)_{22}(PPh_3)_2$, **10**, 289

$Au_2PtC_{16}H_{30}Cl_2S_2$
$Pt(CCBu^t)_2(SMe_2)_2(AuCl)_2$, **9**, 571
$Au_2PtC_{52}H_{48}Cl_2P_4S_4$
$Pt[(SPPh_2CH_2)_2AuCl]_2$, **10**, 71
$Au_2PtC_{52}H_{48}P_4S_4$
$Pt[(SPPh_2CH_2)_2Au]_2$, **10**, 71
$Au_2Ru_4C_{29}H_{22}O_{12}P_2$
$Ru_4(C)(AuPMe_2Ph)_2(CO)_{12}$, **7**, 852
$Au_2Ru_4C_{48}H_{32}O_{12}P_2$
$Au_2Ru_4(H)_2(CO)_{12}(PPh_3)_2$, **10**, 289, 292
$Au_2Ru_4C_{49}H_{30}O_{13}$
$Ru_4Au_2(CO)_{13}(PPh_3)_2$, **7**, 845
$Au_2Ru_6C_{29}H_{30}O_{16}P_2$
$Au_2Ru_6C(CO)_{16}(PEt_3)_2$, **10**, 278, 288
$Au_2Si_4C_{19}H_{40}N$
$Au_2[C_5H_4N[C(TMS)_2]_2]$, **3**, 12
$Au_2Tl_2C_{26}H_{24}P_2S_2$
$[AuTl(Ph_2PCH_2S)]_2$, **1**, 569
$Au_2Tl_2C_{68}H_{56}P_4$
$[AuTl(C_5H_4PPh_2)_2]_2$, **1**, 531
$Au_3BC_9H_{27}F_4OP_3$
$[(Me_3PAu)_3O][BF_4]$, **1**, 300
$Au_3BC_{57}H_{45}F_4N_2P_3$
$[(CN)_2C(AuPPh_3)_3][BF_4]$, **3**, 12
$Au_3BC_{83}H_{65}F_4P_3$
$[Ph_4C_5(AuPPh_3)_3][BF_4]$, **3**, 12
$Au_3BFeC_{66}H_{45}P_3$
$Fe(CO)_{12}B(AuPPh_3)_3$, **1**, 294
$Au_3BFe_4C_{66}H_{45}O_{12}P_3$
$Fe_4(CO)_{12}B(AuPPh_3)_3$, **1**, 297, 299
$Au_3BSi_2C_{61}H_{63}F_4P_3$
$(Ph_3PAu)_3[C(TMS)_2][BF_4]$, **3**, 15
$Au_3B_{20}C_{12}H_{61}P_4$
$[(Me_3P)_2Au]_2[(B_{10}H_{12})Au-(B_{10}H_{13})]$, **1**, 352, 354
$Au_3C_{12}H_{15}N_6$
$[Au[C(NMe)CHCHN]]_3$, **3**, 11
$Au_3C_{58}H_{54}ClP_4$
$[Me_3PC(AuPPh_3)_3]Cl$, **3**, 15
$Au_3CoRu_3C_{66}H_{45}O_{12}P_3$
$Au_3CoRu_3(CO)_{12}(PPh_3)_3$, **10**, 292
$Au_3Cu_2C_{48}H_{30}$
$[Au_3Cu_2(CCPh)_6]^-$, **3**, 109
$Au_3FeC_{58}H_{99}F_6O_4P_3$
$[Fe(CO)_4[AuP(cy)_3]_3][PF_6]$, **7**, 81
$Au_3IrC_{65}H_{58}Cl_3OP_6$
$[IrCl(CO)[PPh(CH_2PPh_2)_2]_2Au-(AuCl)_2]^+$, **10**, 68
$Au_3Nb_3SiC_8H_{19}$
$Au_3Nb_3(H)_6(C_5H_4TMS)$, **10**, 317
$Au_3ReC_{78}H_{82}P_6$
$Au_3ReH_4(PMe_2Ph)_3(PPh_3)_3$, **10**, 256
$Au_3RhC_{91}H_{76}F_6OP_6$
$[Au_3Rh(H)(CO)(PPh_3)_5][PF_6]$, **10**, 302
$Au_3Ru_3C_{65}H_{48}O_{10}P_3$
$Au_3Ru_3(COMe)(CO)_9(PPh_3)_3$, **10**, 292
$Au_3Ru_3C_{68}H_{54}O_8P_3$
$Au_3Ru_3(CO)_8(C_2Bu^t)(PPh_3)_3$, **10**, 292
$Au_3Ru_3C_{74}H_{60}O_8P_3$
$Au_3Ru_3(C_{12}H_{15})(CO)_8(PPh_3)_3$, **10**, 292
$Au_3Ru_4C_{66}H_{46}O_{12}P_3$
$Au_3Ru_4(H)(CO)_{12}(PPh_3)_3$, **10**, 292
$Ru_4Au_3(H)(CO)_{12}(PPh_3)_3$, **7**, 847
$Au_4C_{24}H_8F_{12}$
$[Au(C_6F_3H_2)]_4$, **3**, 13
$Au_4C_{76}H_{68}ClO_{16}P_4$
$Au_2Au_2(CH_2PPh_3)_4(OClO_3)_4$, **10**, 317
$Au_4Fe_2C_{60}H_{24}O_8P_4$
$Au_4Fe_2[Ph_2P(CH_2)_2PPh_2]_2-(CO)_8$, **10**, 298
$Au_4Fe_6C_{96}H_{64}O_{18}P_4$
$[(Ph_3PAu)_2Fe_3(CO)_9P]_2C_6H_4$, **7**, 50
$Au_5BOs_{10}C_{97}H_{60}F_4O_{24}P_4$
$[Au_5Os_{10}C(CO)_{24}(PPh_3)_4][BF_4]$, **10**, 295
$Au_5C_{45}H_{55}$
$[Au(C_6H_2Me_3)]_5$, **3**, 13
$Au_6B_2PtC_{174}H_{145}P_7$
$[Au_6Pt(PPh_3)_7][BPh_4]_2$, **10**, 256
$Au_6B_{20}C_{24}H_{84}P_4$
$(H_{12}B_{10}Au)(AuPEt_3)_4(AuB_{10}-H_{12})$, **1**, 352, 353
$Au_6C_{126}H_{126}P_6$
$[AuP(C_6H_4Me)_3]_6^{2+}$, **3**, 17
$Au_6PtC_{127}H_{105}OP_7$
$[Au_6Pt(CO)(PPh_3)_7]^{2+}$, **10**, 298
$Au_8C_{131}H_{114}N_3O_6P_7$
$[Au_8(PPh_3)_7(CNBu^t)](NO_3)_2$, **3**, 43
$Au_8HgPtC_{126}H_{105}Cl_3P_7$
$Pt(AuPPh_3)_7AuClHgCl_2^+$, **10**, 344
$Au_8HgPtC_{126}H_{105}N_3O_9P_7$
$Pt(AuPPh_3)_7Au(NO_3)Hg-(NO_3)_2^+$, **10**, 344
$Au_8Hg_2PtC_{144}H_{120}Cl_2P_8$
$Pt(AuPPh_3)_8(HgCl)_2^{2+}$, **10**, 47
$Au_8Hg_2PtC_{144}H_{120}N_2O_6P_8$
$Pt(AuPPh_3)_8Hg_2(NO_3)_2^{2+}$, **10**, 344
$Au_8Hg_2PtC_{144}H_{120}N_4O_{12}P_8$
$[Au_8Hg_2Pt(PPh_3)_8(ONO_2)_2]-[NO_3]_2$, **10**, 256
$Au_8Hg_2PtC_{144}H_{120}P_8$
$Pt(AuPPh_3)_8Hg_2^{4+}$, **10**, 343
$Au_8PtC_{136}H_{123}N_4O_6P_7$
$[PtAu_8(PPh_3)_7(CNBu^t)_2](NO_3)_2$, **3**, 43
$Au_8PtC_{144}H_{120}P_8$
$[Pt(AuPPh_3)_8]^{2+}$, **10**, 37
$Au_9C_{118}H_{108}N_5O_9P_6$
$[Au_9(PPh_3)_6(CNBu^t)](NO_3)_3$, **3**, 43
$Au_{10}C_{114}H_{75}F_{20}P_5$
$Au_{10}(C_6F_5)_4(PPh_3)_5$, **3**, 43
$Au_{13}Ag_{12}SbC_{180}H_{150}Br_8F_6$
$[Ag_{12}Au_{13}(PPh_3)_{10}Br_8][SbF_6]$, **10**, 256
$Au_{18}Ag_{20}C_{360}H_{576}Cl_{14}P_{12}$
$Ag_{20}Au_{18}[P(C_6H_4Me_4)_3]_{12}Cl_{14}$, **10**, 256

B

$BAgF_4$
  $Ag[BF_4]$, **10**, 2
$BAlTi_2C_{20}H_{28}$
  $(Cp_2TiH_2)_2AlH_2BH_2$, **1**, 290–2
$BAs_2Au_2Fe_4C_{48}H_{30}O_{12}$
  $Fe_4(CO)_{12}B(AuAsPh_3)_2$, **1**, 297, 299
$BAs_2Au_2Fe_4C_{48}H_{31}O_{12}$
  $HFe_4(CO)_{12}B(AuAsPh_3)_2$, **7**, 18
$BAuC_8H_{24}P_2$
  $AuMe_2[CH_2PMe_2BH_2PMe_2\text{-}CH_2]$, **3**, 22
$BAuC_{54}H_{45}F_4OP_3$
  $[[Au(PPh_3)]_3O][BF_4]$, **10**, 258, 259
$BAu_2C_{32}H_{57}N_2P_4$
  $[Au_2(dppm)_2(S_2CNEt_2)]\text{-}[BH_3CN]$, **1**, 288, 290
$BAu_2C_{35}H_{35}F_4P_2$
  $[MeC_6H_4[Au_2(Ph_2PCH_2\text{-}CH_2)_2]][BF_4]$, **3**, 12
$BAu_2C_{43}H_{37}F_4P_2$
  $[MeC_6H_4(AuPPh_3)_2][BF_4]$, **3**, 12
$BAu_2FeC_{46}H_{39}F_4P_2$
  $[[(Cp)Fe(C_5H_4)](AuPPh_3)_2]\text{-}[BF_4]$, **3**, 12
$BAu_2Fe_4C_{24}H_{31}O_{12}P_2$
  $HFe_4(CO)_{12}B[AuPEt_3]_2$, **7**, 18
$BAu_2Fe_4C_{48}H_{31}O_{12}P_2$
  $Fe_4(CO)_{12}(AuPPh_3)_2BH$, **1**, 297, 299, 302
$BAu_3C_9H_{27}F_4OP_3$
  $[(Me_3PAu)_3O][BF_4]$, **1**, 300
$BAu_3C_{57}H_{45}F_4N_2P_3$
  $[(CN)_2C(AuPPh_3)_3][BF_4]$, **3**, 12
$BAu_3C_{83}H_{65}F_4P_3$
  $[Ph_4C_5(AuPPh_3)_3][BF_4]$, **3**, 12
$BAu_3FeC_{66}H_{45}P_3$
  $Fe(CO)_{12}B(AuPPh_3)_3$, **1**, 294
$BAu_3Fe_4C_{66}H_{45}O_{12}P_3$
  $Fe_4(CO)_{12}B(AuPPh_3)_3$, **1**, 297, 299
$BAu_3Si_2C_{61}H_{63}F_4P_3$
  $(Ph_3PAu)_3[C(TMS)_2][BF_4]$, **3**, 15
$BAu_5Os_{10}C_{97}H_{60}F_4O_{24}P_4$
  $[Au_5Os_{10}C(CO)_{24}(PPh_3)_4][BF_4]$, **10**, 295
$BBeC_{13}H_{19}N$
  $CpBe[MeBu^t(NBC_3H_2)]$, **1**, 212, 213
$BBeC_{13}H_{20}N$
  $(BuMeH_2azaborolyl)BeCp$, **1**, 46
$BBeC_{21}H_{40}N_6$
  $(Bupz)_3BH_3BeH$, **1**, 50
BCH_3
  $CH_2{=}BH$, **1**, 134, 201
  MeB, **1**, 144
BCH_4
  $[CH_2{=}BH_2]^-$, **1**, 201
BCH_4N
  $CH_2{=}B{=}NH_2$, **1**, 201
BCH_5
  $MeBH_2$, **11**, 196, 197
BC_2H_3
  $CH{=}CHBH$, **1**, 131, 133
$BC_2H_3Cl_2$
  $CH_2{=}CHBCl_2$, **11**, 211, 212
$BC_2H_3S_2$
  $BS_2C_2H_3$, **1**, 179, 180
$BC_2H_5N_2$
  $NCHCH_2NHBH$, **1**, 161
$BC_2H_5O_2$
  $(OH)(BO_2C_2H_4)$, **1**, 173
$BC_2H_7Cl_2S$
  $HBCl_2{\cdot}SMe_2$, **11**, 198
$BC_2H_7N_2$
  diazaborolidine, **1**, 161
BC_2H_8
  $H_2B{\cdot}C_2H_6$, **11**, 196, 197
BC_2H_8ClS
  $H_2BCl{\cdot}SMe_2$, **11**, 196, 197
BC_2H_9S
  $Me_2S{\cdot}BH_3$, **11**, 193, 194
$BC_3H_4Br_3$
  $CH_2{=}CBrCH_2BBr_2$, **11**, 202
BC_3H_5
  $BHCH{=}CHCH_2$, **1**, 131, 132
$BC_3H_5N_2$
  $NCHCHCHNBH_2$, **1**, 132
  dioxaborole, **1**, 172
  $pzBH_2$, **1**, 140
$BC_3H_6N_2$
  $pzBH_3$, **1**, 167
BC_3H_7
  C_3H_6BH, **1**, 131
$BC_3H_7S_2$
  $BS_2C_3H_7$, **1**, 181
$BC_3H_9N_2$
  $BHNHCH_2CH_2CH_2NH$, **1**, 164
$BC_4H_6F_6N$
  $(F_3C)_2BNMe_2$, **11**, 223
BC_4H_7
  $BHCH_2CH{=}CHCH_2$, **1**, 150
$BC_4H_{10}F_3O$
  $BF_3{\cdot}OEt_2$, **12**, 73, 104
$BC_4H_{11}O$
  $THF{\cdot}BH_3$, **11**, 193, 194
BC_5H_5
  BCHCHCHCHCH, **1**, 152
  borabenzene, **1**, 209
  C_5BH_5, **1**, 131, 132
BC_5H_6
  BHCHCHCHCHCH, **1**, 209
BC_5H_9
  $Me(BCHCHCH_2CH_2)$, **1**, 202
BC_5H_{11}
  C_5BH_{11}, **1**, 136
$BC_6H_5O_2$
  $C_6H_4O_2BH$, **1**, 174; **11**, 198
BC_6H_7
  $BH(CH)_5CH$, **1**, 132
  BC_6H_7, **1**, 153, 154
$BC_6H_{11}Br_2$
  $CyBBr_2$, **11**, 205
$BC_6H_{11}O$
  $(OCH_2CH_2CH_2B)CH_2\text{-}CH{=}CH_2$, **11**, 236
BC_6H_{13}
  $Me_2(C_4BH_7)$, **11**, 247, 248
$BC_6H_{13}OS$
  $Pr^i(BOSC_3H_6)$, **1**, 185
$BC_6H_{13}O_2$
  $Me_2CC(Me_2)OB(H)O$, **11**, 198
  $Pr^i(BO_2C_3H_6)$, **1**, 175
$BC_6H_{14}Cl$
  $MeCH(Me)C(Me_2)BHCl$, **11**, 198
BC_6H_{15}
  $MeCH(Me)C(Me)_2BH_2$, **11**, 194, 196, 197
  $(thex)BH_2$, **11**, 196, 197
$BC_7H_{12}F_3O_3S$
  $Me_2(C_4BH_6)OTf$, **11**, 240, 241, 257, 258
BC_7H_{15}
  $Me_2(C_5BH_9)$, **11**, 196
BC_8H_9Cl
  $MeClB(CH_2Br)C_6H_4$, **1**, 156
$BC_8H_{14}Cl$
  Cl-9-BBN, **11**, 240
BC_8H_{15}
  9-BBN, **1**, 138, 168
  (BBN-H), **11**, 198
  borabicyclononane, **1**, 143
  C_8BH_{15}, **11**, 196
BC_8H_{16}
  $B(CH_2)_7CH_2$, **1**, 146
$BC_8H_{16}N$
  $Bu^tNB(Me)CHCH_2$, **1**, 212
  $MeBCHCHCH_2N(Bu^t)$, **1**, 212
$BC_8H_{17}O_2S$
  $(C_4SH_7)B(OEt)_2$, **11**, 248
$BC_8H_{17}O_3$
  $(C_4OH_7)B(OEt)_2$, **11**, 248
$BC_8H_{18}NPS_2$
  $Bu^t(Me_3P)(BNSC)(S)$, **1**, 184
BC_9H_7
  boranaphthalene, **1**, 152
BC_9H_{11}
  borepine, **1**, 210
$BC_9H_{14}F_3O_3S$

TfO-9-BBN, **11**, 240, 241
$BC_9H_{15}F_6N_2O$
Bu^{t}Me(CF_3)$_2$(BN_2C(O)), **1**, 165
$BC_9H_{16}ClO_2$
CH_2=CHCClH-($BOCMe_2CMe_2O$), **11**, 235, 236, 253, 254
BC_9H_{17}
borabicyclodecane, **1**, 143
(CH_2=$CHCH_2$)($BO_2C_2Me_4$), **11**, 253, 254
$BC_9H_{21}N_2O_2$
$MeOCH_2OCH$=$CHCH_2B$-(NMe_2)$_2$, **11**, 235, 236
$BC_9H_{21}N_4$
Pr$^i_3$(N_4B), **1**, 162, 163
BC_9H_{25}
(Pr$^i_3C_6H_2$)BH_2, **11**, 196, 197
(tripp)BH_2, **11**, 196, 197
$BC_{10}H_{12}NO_4$
Me(NO_2)Ph($BO_2C_3H_4$), **1**, 175
$BC_{10}H_{15}$
boraadamantane, **1**, 145
$BC_{10}H_{17}O_4$
OC(Me)$_2$C(Me)$_2$OBCH=$CHCO_2Me$, **11**, 205, 206
$BC_{10}H_{19}$
$MeC_9H_{14}BH_2$, **11**, 196, 197, 247, 248
$BC_{10}H_{19}O_4$
MeO_2CC(=CH_2)CH_2-($BOCMe_2CMe_2O$), **11**, 235, 236
MeO_2CCH=CHB(OPr)$_2$, **11**, 211, 212
$BC_{10}H_{20}ClS$
(C_8BH_{14})SMe_2Cl, **11**, 196, 197
$BC_{10}H_{23}$
[MeCH(Me)CH(Me)]$_2$BH, **11**, 194, 198
$BC_{11}H_8N$
(B_6N_7-phenalene), **1**, 158
$BC_{11}H_{17}NOP$
Ph($NOPC_5H_9$)BH_3, **11**, 261
$BC_{11}H_{19}N_2O$
boraoxaadamantane, **1**, 145
$BC_{11}H_{19}O_2$
(OMe)$_2$borabicyclononene, **1**, 154
$BC_{11}H_{21}$
Et(C_9H_{14})BH_2, **11**, 247, 248
$BC_{11}H_{24}$
CH_2=CHC(=CH_2)CH_2BPr_2, **11**, 236
$BC_{12}H_8Cl$
Me(dibenzoborole), **1**, 156
$BC_{12}H_{22}Br$
Cy_2BBr, **11**, 205
$BC_{12}H_{22}Cl$
Cy_2BCl, **11**, 240, 241
$BC_{12}H_{22}I$
Cy_2BI, **11**, 240, 241
$BC_{12}H_{23}$
(C_5H_7)CH_2BPr_2, **11**, 236
Cy_2BH, **11**, 198
$BC_{12}H_{25}O_2$
BuCH=CHB(OPr^i)$_2$, **1**, 83
(C_5H_{11})$_2$($BO_2C_2H_3$), **1**, 173
$BC_{12}H_{27}NP$
Bu$^t_3$(NPB), **1**, 182
$BC_{12}H_{28}ClN_2$
ClB(NPr_2)$_2$, **11**, 203
$BC_{13}H_{11}N_2O$
Ph_2(BN_2OC), **1**, 182
$BC_{13}H_{18}NO$
Ph($BNOC_7H_{13}$), **1**, 182
$BC_{13}H_{19}$
Me(boratricyclotridecane), **1**, 144
$BC_{13}H_{20}NO_4S$
[(Mes)SO_2]Pr^i(O)($BNOC_2H_2$), **11**, 262
$BC_{13}H_{21}$
Pr^i(boraadamantane)C_3H_7, **1**, 145, 147
$BC_{13}H_{22}F_3SO_3$
Cy_2BOTf, **11**, 240, 241
$BC_{13}H_{23}O_6$
(CH_2=$CHCH_2$)($BO_2C_2H_2$)(CO_2Pr^i)$_2$, **11**, 253, 254
$BC_{14}H_{14}O_2P$
Ph_2($BPO_2C_2H_4$), **1**, 183
$BC_{14}H_{19}NP$
(NEt_2)Ph(BPC_4H_4), **1**, 169, 170
$BC_{14}H_{23}$
Me_5C_5B(OH_2)$_4$, **1**, 135
$BC_{14}H_{24}N_2S$
Bu$^t_2$Ph(BN_2S), **1**, 184
$BC_{14}H_{25}O_6$
(MeCH=$CHCH_2$)($BO_2C_2H_2$)-(CO_2Pr^i)$_2$, **11**, 252
$BC_{14}H_{25}S_2$
($C_4S_2H_7$)CH_2(C_9H_{14})BH_2, **11**, 194, 247, 248
$BC_{14}H_{32}NP_2$
(Bu^{t}P)$_2$(BNPr$^i_2$), **1**, 169
$BC_{15}H_{14}BrO_2$
CH_2=$CBrCH_2B$(OPh)$_2$, **11**, 202
$BC_{15}H_{18}NOP$
MePh(C_2NOPH_2)$PhBH_3$, **11**, 246
$BC_{15}H_{20}OP$
Me(OH)$CHCH_2PPh_2$·BH_3, **11**, 246
$BC_{15}H_{24}NO_2$
[C_4N(Bz)H_7]B(OEt)$_2$, **11**, 248
$BC_{16}H_{16}Cl$
Ph_2(C_4H_6B)Cl, **11**, 240, 241
$BC_{16}H_{19}PS_3$
(biph)(Me_3P)(BS_3CH), **1**, 180
$BC_{18}H_{23}$
(Mes)$_2$BH, **11**, 198
$BC_{18}H_{32}$
($C_{10}H_{17}$)(BC_8H_{14}), **11**, 260, 261
$BC_{19}H_3NO_4S$
Et(Pr$^i_3C_6H_2SO_2$)(O)-($BNOC_2H_2$), **11**, 262
$BC_{20}H_{34}Cl$
($C_{10}H_{17}$)$_2$BCl, **11**, 260, 261
Opc_2BCl, **11**, 240, 241
$BC_{20}H_{35}$
($C_{10}H_{17}$)$_2$BH, **11**, 198, 247, 248
$BC_{21}H_{34}F_3O_3S$
Ipc_2BOTf, **11**, 240, 241
$BC_{22}H_{25}N_2O_4S$
($PhSO_2$)Bu(O)($BNOC_2H$)CH_2-(C_8NH_6), **11**, 263
$BC_{22}H_{26}N$
($C_{13}H_8$=B($NCMe_2CH_2CH_2$-CH_2CMe_2), **1**, 201
$BC_{22}H_{38}Cl$
($C_{11}H_{19}$)$_2$BCl, **11**, 260, 261
$BC_{23}H_{25}N_2O_4$
CH_2=$CHCH_2$($BO_2C_2H_2$)C(O)-N(Bz)CH_2CH_2N(Bz)CO, **11**, 253, 254
$BC_{24}H_{40}$
Ipc_2BCH_2CH=CHMe, **11**, 252
$BC_{25}H_{40}$
CH_2=CHC(=CH_2)CH_2BIpc_2, **11**, 252, 253
$BC_{26}H_{17}O_3$
($C_{10}H_6$)$_2O_2BOPh$, **11**, 262, 263
$BC_{26}H_{20}$
Ph_4(BC_2), **1**, 148
$BC_{26}H_{43}$
(C_6H_9)$BIpc_2$, **11**, 247, 248
$BC_{26}H_{43}O_2$
Ipc_2BCH_2CH=CHOMOM, **11**, 252
$BC_{26}H_{56}N_{45}$
(NPr$^i_2$)$_3$(Pr^i)$_2BN_2C_2$, **1**, 161
$BC_{28}H_{29}$
(fluorenyl)($MesEt_2$)-methyleneborane, **1**, 149, 150
$BC_{28}H_{49}$
(longifolyl)$_2$BH, **11**, 247, 248
$BC_{30}H_{18}BrF_{12}N_2O_4S$
Ph_2[(CF_3)$_2C_6H_3SO_2$]$_2$-($BN_2C_2H_2$)Br, **11**, 240, 241, 257, 258
$BC_{30}H_{44}O$
$PhOCH_2C$(=CH_2)CH_2B(Icr)$_2$, **11**, 252, 253
$BC_{31}H_{29}N_2O_4S_2$
($TolSO_2$)$_2Ph_2$($BN_2C_2H_2$)-CH=C=CH_2, **11**, 252, 253
$BC_{32}H_{42}OP$
BU_3Ph_3($BOPC_2$), **1**, 184
$BC_{35}H_{47}N$
($C_{10}H_{16}$)$_2BCH_2CH$=$CHNPh_2$, **11**, 252
$BC_{39}H_{47}O_2P$
[(OH)C_6H_4]Bu_3Ph_3($BPOC_3$), **1**, 179
$BC_{48}H_{35}$
Ph(BC_6), **1**, 154
$BC_{54}H_{47}$
Ph(Tol)$_6$(BC_6), **1**, 154
$BCoC_4H_2O_4$
(CO)$_4CoBH_2$, **1**, 290–2
$BCoC_{10}H_{14}N$

CoCp[(CH)$_3$N(Me)B(Me)], **8**, 86
$BCoC_{10}H_{15}NO_2$
(CO)$_2$Co[Bu^{t}Me(C$_3$H$_3$BN)], **1**, 212
$BCoC_{13}H_{19}N$
CpCo[MeBu^t(BC$_3$H$_2$N)], **1**, 212
$BCoC_{13}H_{20}N$
CpCo[MeBu^t(BC$_3$H$_3$N)], **1**, 212
$BCoC_{15}H_{14}$
Co(C$_4$H$_4$BPh)Cp, **1**, 204
$BCoC_{15}H_{15}N_3$
(terpy)Co(BH$_4$), **1**, 285
$BCoC_{15}H_{27}NSi$
CpCo(C$_{10}$H$_{22}$BNSI), **1**, 213
$BCoC_{20}H_{23}F_4O_2$
[Cp*Co(cot)Rh(CO)$_2$]BF$_4$, **8**, 245
$BCoC_{30}H_{38}P_2$
Co(BPh$_4$)(PMe$_3$)$_2$, **8**, 90
$BCoC_{36}H_{56}P_4$
[Co(PMe$_3$)$_4$][BPh$_4$], **8**, 90
$BCoC_{36}H_{58}P_4$
[CoH$_2$(PMe$_3$)$_4$]BPh$_4$, **8**, 32
$BCoC_{37}H_{54}NP_3$
[Co(MeCN)(C$_2$H$_4$)(PMe$_3$)$_3$]-BPh$_4$, **8**, 72
$BCoC_{39}H_{30}$
CoCp(C$_4$BPh$_5$), **8**, 86
$BCoC_{52}H_{46}O_2P_2$
(CO)$_2$(dppm)Co(dppm)BH$_2$, **1**, 291, 292
$BCo_2C_{14}H_{18}BO_6$
(CO)$_6$Bu$^t_2$(BNCo$_2$), **1**, 211
$BCo_2C_{20}H_{23}$
Co$_2$(CO)$_6$(CBC$_8$H$_4$)(CBu^t), **1**, 139
$BCo_3C_{27}H_{25}P$
(CpCo)$_3$PBPh$_2$, **1**, 297, 298
$BCo_6C_{60}H_{110}P_6S_8$
[Co$_6$S$_8$(PEt$_3$)$_6$][BPh$_4$], **8**, 453
$BCrC_5H_5F_4NO_2$
Cr(BF$_4$)(NO)$_2$Cp, **5**, 419
$BCrC_{12}H_{37}P_4$
Cr(BH$_4$)H(dmpe)$_2$, **1**, 284
$BCrC_{13}H_{13}O_3$
(CO)$_3$Cr[Me(C$_9$H$_{10}$B)], **1**, 210, 211
$BCrC_{14}H_{21}NO_3$
(CO)$_3$Cr[Bu$^t_2$(BNC$_3$H$_3$)], **1**, 212
$BCrC_{14}H_{22}NO_4Si$
(CO)$_4$Cr[Et$_2$Me$_4$(C$_2$SiNB)], **1**, 213
$BCrC_{20}H_{30}F_4NOS$
[Cr(CO)(NO)(CS)(C$_6$Et$_6$)]-[BF$_4$], **5**, 500
$BCrC_{20}H_{30}F_4NO_3$
[Cr(CO)$_2$(NO)(C$_6$Et$_6$)][BF$_4$], **5**, 500
$BCrC_{41}H_{34}NO_5P_2$
[PPN][(CO)$_5$Cr(H)(BH$_3$)], **1**, 283, 284
$BCrFeC_{25}H_{21}O_3$
CpFe(C$_4$H$_4$BPh)Cr(CO)$_3$Bz, **7**, 195
$BCuC_8H_{12}F_4N_4$
[Cu(NCMe)$_4$][BF$_4$], **10**, 257
$BCuC_{13}H_{16}P$
(triphos)Cu(BH$_4$), **1**, 286
$BCuC_{30}H_{27}N_2P$
Ph$_3$PCu(phen)(BH$_4$), **1**, 286
$BCuC_{36}H_{37}P_3$
(PHPh$_2$)$_3$Cu(BH$_4$), **1**, 286
$BCuC_{36}H_{70}P_2$
(PCy$_3$)$_2$Cu(BH$_4$), **1**, 286
$BCuC_{44}H_{39}P$
Ph$_3$PCu[(Me$_2$)(Ph$_2$phen)](BH$_4$), **1**, 286
$BCuRu_2C_{38}H_{34}Cl_2F_4N_2O_4$
[CuRu$_2$(Ph$_2$PPyOMe)$_2$(Cl$_2$)-(CO)$_2$][BF$_4$], **10**, 298
$BCu_2C_{72}H_{64}ClO_4P_4$
[[Cu(PPh$_3$)$_2$]$_2$(BH$_4$)][ClO$_4$], **1**, 286
$BCu_3C_{91}H_{76}F_4P_6$
[Cu$_3$(CCPh)$_2$(Ph$_2$PCH$_2$PPh$_2$)$_3$]-[BF$_4$], **3**, 113
$BFeC_7H_6F_4O_4$
[(CH$_2$=CHMe)Fe(CO)$_4$][BF$_4$], **12**, 967
$BFeC_7H_8O_2$
[FeH(CO)$_2$(C$_4$H$_4$BMe)]$^-$, **7**, 165
$BFeC_8H_7O_3$
Fe(CO)$_3$(C$_4$H$_4$BMe), **7**, 165
$BFeC_{10}H_{10}$
borataferrocene, **1**, 204
$BFeC_{10}H_{13}$
FeH(C$_4$H$_4$BMe)Cp, **7**, 195
$BFeC_{11}H_{13}F_4O_3$
[Cp(CO)$_2$Fe(THF)][BF$_4$], **12**, 395
$BFeC_{12}H_{13}F_4O_4$
Fe(CO)$_3$[Me$_2$(OMe)C$_6$H$_4$]BF$_4$, **7**, 157
$BFeC_{12}H_{37}P_4$
(dmpe)$_2$Fe(H)(BH$_4$), **1**, 284, 285
$BFeC_{13}H_9O_3$
Fe(CO)$_3$(C$_4$H$_4$BPh), **1**, 204; **7**, 138
$BFeC_{13}H_{22}NO_3Si$
(CO)$_3$Fe(C$_{10}$H$_{22}$BNSI), **1**, 213
$BFeC_{18}H_{26}$
Fe(CpBMe)(C$_6$Me$_6$), **1**, 209
$BFeC_{28}Cl_4H_{29}NO_3P$
[(Ph$_3$P)(CO)$_3$Fe≡CNPr$^i_2$]$^+$-[BCl$_4$]$^-$, **7**, 120
$BFeC_{41}H_{41}P_3$
(triphos)Fe(H)(BH$_4$), **1**, 284, 285
$BFeRu_3C_{12}H_3O_{12}$
H[FeRu$_3$(CO)$_{12}$BH$_2$], **1**, 297, 301
$BFeRu_3C_{48}H_{31}NO_{12}P_2$
[PPN][FeRu$_3$(CO)$_{12}$BH], **1**, 297, 301
$BFe_2C_{11}H_9F_4O_6P_2S$
[Fe$_2$(CO)$_6$(PMe$_2$)$_2$(SMe)][BF$_4$], **7**, 44
$BFe_2C_{20}H_{10}F_4O_8P$
[Fe$_2$(CO)$_8$PPh$_2$][BF$_4$], **7**, 43
$BFe_2C_{20}H_{19}$
CpFe(C$_4$H$_4$BPh)FeCp, **7**, 195
$BFe_3C_9H_4O_9$
[(H)Fe$_3$(CO)$_9$(BH$_3$)]$^-$, **1**, 296, 297, 299, 302
$BFe_3C_9H_5O_9$
(H)Fe$_3$(CO)$_9$(BH$_4$), **1**, 294, 296, 297, 299; **7**, 16, 17
$BFe_3C_{10}H_3O_{10}$
(H)(CO)Fe$_3$(CO)$_9$BH$_2$, **1**, 296, 297, 299
$BFe_3C_{10}H_7O_9$
(H)Fe$_3$(CO)$_9$(BH$_3$Me), **1**, 296, 297
$BFe_3C_{12}H_5O_9$
(MeC$_2$BH$_2$)Fe$_3$(CO)$_9$, **1**, 201, 202
$BFe_3C_{46}H_{32}NO_9P_2$
[PPN][(CO)Fe$_3$(CO)$_9$BH$_2$], **1**, 296, 297, 299
$BFe_3Rh_3C_{15}H_2O_{15}$
H$_2$Fe$_3$Rh$_3$(CO)$_{15}$B, **1**, 294, 304
$BFe_4C_{12}H_2O_{12}$
[HFe$_4$(CO)$_{12}$BH]$^-$, **1**, 296, 297, 302; **7**, 17, 18
$BFe_4C_{12}H_3O_{12}$
HFe$_4$(CO)$_{12}$BH$_2$, **1**, 294, 297, 298, 302
$BFe_4C_{12}O_{12}$
[Fe$_4$(CO)$_{12}$B]$^{3-}$, **1**, 298
$BFe_4C_{13}HCl_2$
(H)Fe$_4$(CO)$_{12}$CBCl$_2$, **1**, 297, 300
$BFe_4C_{13}H_3O_{12}$
(H)Fe$_4$(CO)$_{12}$CBH$_2$, **1**, 297, 300
HFe$_4$(CO)$_{12}$CBH$_2$, **7**, 26
$BFe_4C_{13}H_3O_{13}$
HFe$_4$(CO)$_{12}$CBH(OH), **7**, 26
$BFe_4C_{48}H_{32}NOP_2$
[PPN][HFe$_4$(CO)$_{12}$BH], **1**, 297, 298, 302
$BFe_4C_{64}H_{54}NO_{10}P_2$
[PPN][HFe$_4$(CO)$_{10}$(PPhMe$_2$)$_2$-BH], **1**, 297, 298
$BFe_4Rh_2C_{16}O_{16}$
[Fe$_4$Rh$_2$(B)(CO)$_{16}$]$^-$, **10**, 215
$BFe_4Rh_2C_{52}H_{30}NO_{16}P_2$
[PPN][Fe$_4$Rh$_2$(CO)$_{16}$B], **1**, 304
$BGaC_{23}H_{34}P$
Bu$^t_2$GaP(Ph)B(Bu$^t_6$H$_2$), **1**, 519
$BHfC_{39}H_{39}N_2$
[HfMe(MeCN)$_2$Cp$_2$][BPh$_4$], **4**, 602
$BInC_{14}H_{30}F_4O$
[Pr$^i_2$In(THF)$_2$][BF$_4$], **1**, 514
$BInMoC_6H_6F_4N_2O_2$
[Mo(MeCN)$_2$(CO)$_2$In][BF$_4$], **12**, 639
$BIrC_5H_6N_6O_3$
Ir(CO)(HBpz$_3$)(CO$_2$H)(H), **8**, 351

$BIrC_{11}H_{10}N_6O_2$
$IR(HBpz_3)(CO)_2$, **8**, 351
$BIrC_{11}H_{11}F_4N_6O_2$
$[Ir(CO)_2(HBpz_3)(H)]BF_4$, **8**, 351
$BIrC_{12}H_{27}ClNSi$
$IrCl(Et)(C_{10}H_{22}BNSI)$, **1**, 213
$BIrC_{13}H_{19}N_7O_2$
$Ir(CO)(HBpz_3)(CONHPr)(H)$, **8**, 351
$BIrC_{13}H_{30}ClF_7OP_2S$
$[Ir(CO)(Cl)(F)(SF_2)(PEt_3)_2)^+]$-$[BF_4^-]$, **8**, 360
$BIrC_{15}H_{32}ClO_2P_3$
$Ir(H)(BO_2C_6H_4)(Cl)(PMe_3)_3$, **1**, 293, 295, 296
$BIrC_{17}H_{43}P_3$
$IrH_2(PMe_3)_3(BC_8H_{14})$, **1**, 293, 295
$BIrC_{44}H_{44}F_4P_2$
$[Ir(cod)(H)_2(PPh_3)_2]BF_4$, **8**, 376
$BIrC_{58}H_{80}P_2$
$[Ir(cod)(dcpe)]BPh_4$, **8**, 367
$BIrC_{76}H_{68}OP_4$
$[Ir(CO)(dppe)_2]BPh_4$, **8**, 362
$BIr_2C_{20}H_{35}$
$[Cp^*IrH]_2(H)(BH_4)$, **1**, 285
$BLiC_4H_{11}N$
$(pyrrolidine)BH_3Li$, **11**, 244
$BLiC_8H_{16}O_2$
$\overline{OC(Me)_2C(Me)_2OB}CH(Li)Me$, **11**, 208, 209
$BLiC_{18}H_{33}$
$(C_{10}H_{16})(BC_8H_{15})Li$, **11**, 260, 261
$BLiC_{26}H_{40}O$
$[(BzO)CH_2CH_2](C_9H_{14})(BC_8$-$H_{15})Li$, **11**, 260, 261
$BMgC_{21}H_{37}ClN_6$
$[HB(Bu^tpz)_3]MgCl$, **1**, 104
$BMgC_{30}H_{56}N_{12}$
$Mg[HB(Me_2pz)_3]_2$, **1**, 104
$BMn_2C_{18}H_{13}O_6$
$(EtC_4H_3BPh)[Mn(CO)_3]_2$, **1**, 202
$BMn_2C_{31}H_{27}O_6P_2$
$Mn_2(H)(BH_4)(CO)_6(dppm)$, **1**, 284
$BMoC_8H_5F_4O_3$
$Mo(BF_4)(CO)_3Cp$, **5**, 404
$BMoC_9H_7ClO_3$
$(CO)_3Mo[(C_6H_7B)Cl]$, **1**, 210, 211
$BMoC_9H_8O_3$
$(CO)_3Mo(C_6H_7BH)$, **1**, 210, 211
$BMoC_{10}H_{11}F_4O_2$
$Mo(BF_4)(CO)_2(CH_2CHMe)Cp$, **5**, 406
$BMoC_{11}H_{11}F_4N_2O_2$
$[Mo(CO)_2(NCMe)_2Cp][BF_4]$, **5**, 406
$BMoC_{14}H_{22}NO_4Si$
$(CO)_4Mo(C_{10}H_{22}BNSi)$, **1**, 213
$BMoC_{15}H_{12}O_3$
$(CO)_3Mo[C_6H_7B)Ph]$, **1**, 210, 211
$BMoC_{15}H_{13}F_4N_2O_2$
$Mo(CO)_2(NCMe)_2(Ind)BF_4$, **5**, 202
$BMoC_{16}H_{20}F_3O_3$
$[CH_2{=}CHCH{=}CHCH_2CH$-$(OBF_3)Pr^i]MoCp(CO)_2$, **12**, 971
$BMoC_{17}H_{12}F_4O_2$
$Mo(CCPh)(CO)_2(C_7H_7)(BF_4)$, **5**, 166
$BMoC_{22}H_{16}F_4N_4O_2$
$[Mo(CO)_2(bipy)_2]BF_4$, **5**, 237
$BMoSi_2C_{19}H_{31}NP$
$CpMo(CO)_2[P(BH_3)(Ph)$-$[N(TMS)_2]]$, **1**, 288, 290
$BMo_3C_{14}H_{27}F_4O_{16}$
$[Mo_3(O)(CMe)(O_2CMe)_6$-$(OH_2)_3]BF_4$, **5**, 322
$BNaC_3H_{10}O_3$
$NaHB(OMe)_3$, **11**, 424
$BNaC_{24}H_{20}$
$NaBPh_4$, **12**, 149
$BNaH_4$
$NaBH_4$, **12**, 135
$BNbC_{12}H_{13}Cl_2F_4N$
$[Cp_2NbCl_2]BF_4{\cdot}NCMe$, **5**, 126
$BNbC_{18}H_{29}N$
$Cp_2[Bu^t_2(NbBN)]$, **1**, 211
$BNbC_{20}H_{21}ClF_4O$
$[Cp_2Nb[C(O)CHEtPh]Cl]BF_4$, **5**, 98
$BNbC_{28}H_{26}$
$[Ph_2B(Ph)_2]Nb(MeCCMe)$, **5**, 132
$BNbC_{34}H_{30}F_4P_2$
$[Cp_2Nb(PPh_2)_2]BF_4$, **5**, 127
$BNbSi_2C_{46}H_{40}F_4$
$[Cp_2Nb(SiPh_3)_2]BF_4$, **5**, 127
$BNiC_9H_{12}N_6$
$Ni(C_3H_5)(HBpz_3)$, **9**, 140
$BNiC_{11}H_{30}F_4OP_3$
$[Ni(COMe)(PMe_3)_3][BF_4]$, **9**, 59
$BNiC_{12}H_{20}N_2$
$Ni(C_4H_4BNMe_2)_2$, **1**, 204
$BNiC_{13}H_{35}F_4P_4$
$[NiMe(dmpe)_2][BF_4]$, **9**, 3, 56
$BNiC_{13}H_{37}F_4P_4$
$[NiMe(PMe_3)_2(dmpe)][BF_4]$, **9**, 3
$BNiC_{13}H_{39}F_4P_4$
$[Ni(Me)(PMe_3)_4][BF_4]$, **9**, 59, 61
$BNiC_{14}H_{37}F_4OP_4$
$[Ni[C(O)Me](PMe_3)_2(dmpe)]$-$[BF_4]$, **9**, 3
$BNiC_{15}H_{13}O_2$
$Ni(C_3H_5)[(O_2C_6H_4)_2B]$, **9**, 33
$BNiC_{15}H_{13}O_4$
$Ni(C_3H_5)[B(C_6H_4O_2)_2]$, **9**, 124
$BNiC_{15}H_{24}Cl_5F_4OP_2$
$[Ni(C_6Cl_5)(PMe_3)_2[C(OMe)$-$Me]][BF_4]$, **9**, 13
$BNiC_{16}H_{15}O_2$
$Ni(C_4H_7)[(O_2C_6H_4)_2B]$, **9**, 33
$BNiC_{17}H_{29}F_4N_2P_2$
$[Ni(Me)(PMe_3)_2(bipy)][BF_4]$, **9**, 56
$BNiC_{19}H_{23}$
$Ni(CpBPh)(C_8H_{13})$, **9**, 176
$BNiC_{20}H_{44}F_4P_2$
$[Ni(C_2H_4)(H)(PBu^t_2C_2H_4$-$PBuSt_2)][BF_4]$, **9**, 113
$BNiC_{33}H_{45}F_4P_4$
$[Ni(Me)(dppe)(PMe_3)_2][BF_4]$, **9**, 59
$BNiC_{34}H_{45}F_4OP_4$
$[Ni[C(O)Me](PMe_3)_2(dppe)]$-$[BF_4]$, **9**, 3
$BNiC_{69}H_{63}P_2$
$[Ni(CHMePPh_3)_2Cp][BPh_4]$, **9**, 36
$BNiSiC_{13}H_{38}F_4P_3$
$[Ni(CH_2TMS)(PMe_3)_3][BF_4]$, **9**, 56
$BNiTiC_{38}H_{38}F_4P_2S_2$
$[TiCp_2(SCH_2CH_2PPh_2)_2Ni]BF_4$, **10**, 27
$BNi_2C_{15}H_{17}$
$Ni_2(C_4H_4BMe)Cp_2$, **9**, 179
$BOsC_7H_5F_4O_4$
$[Os(CO)_4(C_3H_5)]BF_4$, **7**, 459
$BOsC_{12}H_{13}F_4$
$[Os(C_6H_6)(C_6H_7)]BF_4$, **7**, 552
$BOs_2C_{13}H_{12}F_4O_8P$
$[Os_2(CH_2CHPMe_3)(CO)_8]BF_4$, **7**, 653
$BOs_3C_{10}H_3O_{10}$
$(H)_3(CO)_9Os_3BCO$, **1**, 296, 297
$BOs_3C_{10}H_5O_9$
$(H)_3(CO)_9Os_3BCH_2$, **1**, 296, 297, 301
$Os_3H_3(BCH_2)(CO)_9$, **7**, 707
$Os_3H(BH_2CH_2)(CO)_9$, **7**, 707
$BOs_3C_{10}H_9O_{10}$
$(H)_3(CO)_9Os_3BCO$, **1**, 296, 297
$BOs_3C_{14}H_5F_4O_{11}$
$[Os_3(CO)_{11}(C_3H_5)][BF_4]$, **7**, 787, 788
$BOs_3C_{15}H_5F_4O_{10}$
$[Os_3(CO)_{10}CP][BF_4]$, **7**, 794, 795
$BPdC_8H_{12}F_8N_4$
$[Pd(MeCN)_4][BF_4]_2$, **12**, 1225, 1226
$BPdC_{45}H_{23}F_{24}N_2O_2$
$[Pd(bipy)(CO)(COMe)]$-$[B(CF_3)_2C_6H_3)_4]$, **12**, 1225
$BPdC_9H_{21}F_4N_2$
$[Pd(allyl)(TMEDA)][BF_4]$, **9**, 353
$BPdC_{15}H_{25}F_4N_2$
$[Pd(PhC_3H_4)(TMEDA)][BF_4]$, **9**, 327
$BPdC_{19}H_{38}F_4P$
$[PdCp(CH_2{=}CHPh)(PEt_3)]$-$[BF_4]$, **9**, 299

$BPdC_{27}H_{29}F_4NOP$
$[Pd(MeC_3H_4)[CH(COMe)PPh_3](CNMe)]BF_4$, **9**, 228
$BPdC_{29}H_{32}ClF_4NOP$
$[Pd(ClC_3H_4)(CNBu^t)Ph_3PCHCOMe)][BF_4]$, **9**, 350
$BPdC_{30}H_{35}F_4NOP$
$[Pd(MeC_3H_4)[CH(COMe)PPh_3](CNBu^t)]BF_4$, **9**, 215, 230
$BPdC_{31}H_{27}ClF_4P$
$[PdCp(PPh_3)(ClC_6H_4CH{=}CH_2)][BF_4]$, **9**, 301
$BPdC_{32}H_{30}F_4OP$
$[PdCp(PPh_3)][(MeO)C_6H_4\text{-}CH{=}CH_2][BF_4]$, **9**, 301
$BPdC_{39}H_{35}F_4P_2$
$[Pd(C_3H_5)(PPh_3)_2][BF_4]$, **9**, 355
$BPd_2RhC_{26}H_{30}F_4S_4$
$Rh(pyS)_4[Pd(C_3H_5)]_2[BF_4]$, **9**, 354
$BPtC_{18}H_{21}$
$Pt(cod)(C_4H_4BPh)$, **9**, 582
$BPtC_{33}H_{39}P_2$
$Pt[B(Me)CH_2CH(Me)CH(Me)CH_2](dppe)$, **9**, 581
$BPtC_{42}H_{37}$
$Pt(cod)(BC_4Ph_5)$, **9**, 582
$BPtC_{44}H_{35}F_4O_2P_2$
$[Pt(COPh)(CO)(PPh_3)_2]BF_4$, **9**, 462
$BPt_2C_{34}H_{56}ClF_4P_3$
$[(Et_3P)ClPt(C{=}CHPh)Pt(C{\equiv}CPh)(PEt_3)_2]BF_4$, **9**, 489
$BPt_2C_{40}H_{66}F_4P_4$
$[(Et_3P)_2Pt(C{=}CHPh)Pt(C{\equiv}CPh)(PEt_3)_2]BF_4$, **9**, 489
$BReC_5F_4O_5$
$Re(CO)_5(FBF_3)$, **6**, 177
$BReC_{20}H_{32}F_4$
$[ReCp^*{}_2H_2][BF_4]$, **6**, 252
$BReC_{24}H_{22}Cl_2F_4NOP$
$[Re(ClCH_2Cl)(NO)(PPh_3)Cp][BF_4]$, **6**, 177
$BReC_{24}H_{30}F_5NP_2$
$[Re[N(H)Ph]FMe_2(PMe_3)_2][BF_4]$, **6**, 234
$BReC_{25}H_{37}F_4OP_3$
$[Re(H)_2(H_2)(CO)(PMe_2Ph)_3][BF_4]$, **6**, 181
$[Re(H)_4(CO)(PMe_2Ph)_3][BF_4]$, **6**, 181
$BReC_{29}H_{32}Cl_2F_4NOP$
$[Re(Cl)(CH_2Cl)(NO)(PPh_3)Cp^*][BF_4]$, **6**, 179
$BReC_{42}H_{40}F_4P_2$
$[Re(C_6H_7)H_3(PPh_3)_2][BF_4]$, **6**, 252
$BReC_{55}H_{53}F_5P_4$
$[ReF(CCH_2Me)(dppe)_2][BF_4]$, **6**, 203, 215
$BRe_2C_{12}HF_4O_{10}$
$[(CO)_5Re(CCH)Re(CO)_5][BF_4]$, **6**, 194
$BRe_2C_{12}H_3F_4O_{10}$
$[Re(CO)_5(CHCH_2)Re(CO)_5][BF_4]$, **6**, 205, 215
$BRhC_{14}H_{20}N_4$
$Rh(H_2Bpz_2)(cod)$, **8**, 253
$BRhC_{15}H_{27}NSi$
$CpRh(C_{10}H_{22}BNSi)$, **1**, 213
$BRhC_{16}H_{23}$
$[Cp^*Rh(CpBMe)]^+$, **1**, 209
$BRhC_{17}H_{22}N_6$
$Rh(HBpz_3)(cod)$, **8**, 253
$BRhC_{17}H_{22}N_6O_2$
$Tp^*Rh(CO)_2$, **8**, 216
$BRhC_{18}H_{26}N_6O$
$Tp^*Rh(CO)(C_2H_4)$, **8**, 230
$BRhC_{20}H_{30}F_4$
$Rh(C_6Me_6)(cod)BF_4$, **8**, 275
$BRhC_{22}H_{24}N_8O_2$
$(pzBpz_3)Rh(duroquinone)$, **8**, 270
$BRhC_{24}H_{47}ClO_2P_2$
$RhHCl(BO_2C_6H_4)(PPr^i{}_3)_2$, **1**, 292, 295, 296
$BRhC_{42}H_{35}ClO_2P_2$
$RhHCl(BO_2C_6H_4)(PPh_3)_2$, **1**, 292, 295
$BRhTiC_{68}H_{66}P_2$
$[TiCp_2(CH_2PPh_2)_2Rh(C_8H_{12})][BPh_4]$, **4**, 371
$BRh_2C_{25}H_{31}F_4$
$[CpRh(cot)Rh(C_6Me_6)]BF_4$, **8**, 232
$BRuC_7H_5F_4O_4$
$[Ru(CO)_4(C_3H_5)]BF_4$, **7**, 459
$BRuC_9H_7F_4O_3$
$[Ru(CO)_3(C_6H_7)]BF_4$, **7**, 455
$BRuC_9H_8F_4NO_2$
$[Ru(NCMe)(CO)_2Cp]BF_4$, **7**, 483
$BRuC_9H_9F_4S$
$[Ru(C_4H_4S)Cp]BF_4$, **7**, 523, 538
$BRuC_9H_{32}P_3$
$RuH(BH_4)(PMe_3)_3$, **7**, 381
$BRuC_{10}H_{11}F_4N_2O$
$[Ru(NCMe)_2(CO)Cp]BF_4$, **7**, 483
$BRuC_{11}H_{17}F_4$
$[Ru(C_2H_4)_3Cp]BF_4$, **7**, 505
$BRuC_{12}H_{20}ClF_4O$
$[RuCl(H_2O)(C_{12}H_{18})]BF_4$, **7**, 462
$BRuC_{13}H_{15}F_4O_3$
$[Ru(CO)_3(Cp^*)]BF_4$, **7**, 482
$BRuC_{14}H_{13}F_4O$
$[Ru(CO)(cot)Cp]BF_4$, **7**, 484
$BRuC_{14}H_{17}F_4O$
$[Ru(CO)(cod)Cp]BF_4$, **7**, 484
$BRuC_{15}H_{23}F_4O$
$[Ru(CO)(C_7H_{11})(C_7H_{12})]BF_4$, **7**, 466
$BRuC_{16}H_{13}F_4O_2$
$[Ru(CH_2CCHPh)(CO)_2Cp]BF_4$, **7**, 390
$BRuC_{16}H_{23}$
$(Cp^*)Ru(CpBMe)$, **1**, 209
$BRuC_{18}H_{20}P$
$(Ph_3P)Ru(H)(BH_4)$, **1**, 284, 285
$BRuC_{21}H_{26}P$
$(ttp)Ru(H)(BH_4)$, **1**, 284, 285
$BRuC_{23}H_{35}F_4$
$[RuCp(C_6Et_6)][BF_4]$, **7**, 531
$BRuC_{24}H_{22}F_4O_2P$
$[Ru(CO)(PPh_3)(H_2O)Cp]BF_4$, **7**, 493
$BRuC_{26}H_{13}O_{12}$
$HRu(CO)_{12}B(H)C(Ph)CHPh$, **1**, 301, 303
$BRuC_{26}H_{24}F_4OP$
$[RuMe(CO)(PPh_3)(C_6H_6)]BF_4$, **7**, 571
$BRuC_{26}H_{26}F_4O_2P$
$[Ru(CO)(PPh_3)(EtOH)Cp]BF_4$, **7**, 493
$BRuC_{28}H_{28}F_4O_2P$
$[Ru(CO)(PPh_3)(THF)Cp]BF_4$, **7**, 493
$BRuC_{32}H_{36}F_4P$
$[RuH(CH_2CHC_6H_4PPh_2)(C_6Me_6)]BF_4$, **7**, 571
$BRuC_{34}H_{25}F_4O$
$[Ru(CO)(C_4Ph_4)Cp]BF_4$, **7**, 474
$BRuC_{38}H_{32}F_5O_3P_2$
$[RuF(OH_2)(CO)_2(PPh_3)_2]BF_4$, **7**, 308
$BRuC_{38}H_{33}F_4O_3P_2$
$[RuH(CO)_2(OH_2)(PPh_3)_2]BF_4$, **7**, 314, 315
$BRuC_{40}H_{33}F_4O_3P_2S$
$[Ru(SMe)(CO)_3(PPh_3)_2]BF_4$, **7**, 317
$BRuC_{41}H_{55}P_2$
$[RuMe(PMe_3)_2[C_5Me_4(CH_2)]]BPh_4$, **7**, 502
$BRuC_{46}H_{39}F_4P_2S$
$[Ru(C_5H_4CH_2C_4H_3S)(PPh_3)_2]BF_4$, **7**, 523
$BRuC_{47}H_{35}F_4$
$[RuCp(C_6Ph_6)]BF_4$, **7**, 531
$BRu_2C_{15}H_{13}F_4O_3$
$[Ru_2Cp_2(CMe)(CO)_3]BF_4$, **7**, 665
$BRu_2C_{23}H_{32}ClF_4O_2$
$[Ru_2Cp^*{}_2(Cl)(CH_2)(CO)_2]BF_4$, **7**, 668
$BRu_2C_{27}H_{38}F_3O_3S$
$(Cp^*)_2Ru_2(CpBMe)(OTf)$, **1**, 209
$BRu_2C_{42}H_{44}F_4N_3$
$[Ru_2(NHPh)_3(C_6Me_6)_2]BF_4$, **7**, 559
$BRu_3C_9H_4O_9$
$[Ru_3(CO)_9BH_4]^-$, **1**, 296, 297, 300, 301
$BRu_3C_9H_5O_9$
$Ru_3BH_5(CO)_9$, **7**, 708
$Ru_3(CO)_9BH_5$, **1**, 296, 297, 300, 301
$BRu_3C_{31}H_{46}Cl_2F_4$
$[Ru_3Cp^*{}_3(CH)(Cl)_2]BF_4$, **7**, 668
$BRu_3WC_{16}H_7O_{11}$

H(CpW)[Ru$_3$(CO)$_{11}$BH], **1**, 297, 301
WRu$_3$(H)(BH)(CO)$_{11}$Cp, **10**, 216, 217
$BRu_3WC_{30}H_{17}O_{11}$
H(CpW)Ru$_3$(CO)$_{11}$BC(Ph)-CPhH, **1**, 301, 303
WRu$_3$(H)[BC(Ph)=CPhH]-(CO)$_{11}$Cp, **10**, 216, 217
$BRu_4C_{12}HO_{12}$
[(H)Ru$_4$(B)(CO)$_{12}$]$^{2-}$, **7**, 851
$BRu_4C_{12}H_3O_{12}$
HRu$_4$(CO)$_{12}$BH$_2$, **1**, 297, 300
(H)Ru$_4$(H)$_2$(B)(CO)$_{12}$, **7**, 850
$BRu_4C_{21}H_{10}O_{12}$
Ru$_4$(H)[BC(Ph)C(H)(Me)]-(CO)$_{12}$, **7**, 892
$BRu_4C_{26}H_{13}O_{12}$
(H)Ru$_4$(H)[BC(Ph)C(H)(Ph)]-(CO)$_{12}$, **7**, 892
$BRu_4C_{48}H_{32}NO_{12}P_2$
[PPN][HRu$_4$(CO)$_{12}$BH], **1**, 304
$BRu_6C_{17}HO_{17}$, **1**, 294
HRu$_6$B(CO)$_{17}$, **1**, 303, 304
$BRu_6C_{54}H_{32}NO_{18}P_2$
[PPN][Ru$_6$(H$_2$)(CO)$_{18}$B], **1**, 303, 304
$BScC_{10}H_{14}$
Cp$_2$Sc(BH$_4$), **1**, 278; **4**, 113
$BScSi_4C_{22}H_{46}$
[C$_5$H$_3$(TMS)$_2$]$_2$Sc[BH$_4$], **1**, 278
[(TMS)$_2$C$_5$H$_3$]$_2$Sc(BH$_4$), **4**, 49
$BSiC_8H_{16}Cl_3$
Bu^{t}CH=C(SiMe$_2$Cl)BCl$_2$, **1**, 157
$BSiC_{10}H_{22}Cl_2N$
Bu^{t}CH=C(SiMe$_2$Cl)B(Cl)-NMe$_2$, **1**, 157
$BSiC_{10}H_{22}N$
Et$_2$Me$_4$[BNSiC$_2$], **1**, 213
$BSiC_{10}H_{22}P$
Et$_2$Me$_4$[BPSiC$_2$], **1**, 213
$BSiC_{10}H_{22}S$
Et$_2$Me$_4$[BSSiC$_2$], **1**, 213
$BSiC_{10}H_{22}Se$
Et$_2$Me$_4$[BSeSiC$_2$], **1**, 213
$BSiC_{11}H_{23}S$
Et$_2$Me$_3$(BS$_2$SiC$_4$H$_4$), **1**, 180
$BSiC_{18}H_{15}$
Ph$_3$SiB, **2**, 22
$BSiC_{19}H_{37}S$
(BBN)CH$_2$CH=C(TMS)SPh, **11**, 200
$BSiC_{21}H_{33}O_6$
PhMe$_2$SiCH=CHCH$_2$(BO$_2$-C$_2$H$_2$)(CO$_2$Pr^i)$_2$, **11**, 252
$BSiC_{27}H_9F_{20}$
(TMS)B(C$_6$F$_5$)$_4$, **2**, 4
$BSiUC_{19}H_{33}Cl_2N_6$
[HB(Me$_2$Pz)$_3$]UCl$_2$(CH$_2$TMS), **4**, 192
$BSiZrC_{43}H_{45}F_4$
[Zr[C(TMS)CMe$_2$]Cp'$_2$][B-(C$_6$H$_4$F)$_4$], **4**, 606
$BSiZrC_{45}H_{55}O$
[Zr(CH$_2$CH$_2$TMS)(THF)Cp'$_2$][BPh$_4$], **4**, 598
$BSiZrC_{52}H_{69}O$
[Zr(CH$_2$TMS)(THF)Cp*$_2$][BPh$_4$], **4**, 595
$BSi_2TiC_{41}H_{29}F_{20}$
[TiMe[C$_5$H$_4$(TMS)]$_2$]-[B(C$_6$F$_5$)$_4$], **4**, 395
$BSi_2UC_{22}H_{41}Cl_2N_6$
[HB(Me$_2$Pz)$_3$]UCl$_2$[CH-(TMS)$_2$], **4**, 192
$BSi_3C_{26}H_{33}$
(TMS)$_2$(Ph$_3$Si)(BC$_2$H$_2$), **1**, 148
$BSi_4C_{33}H_{27}F_{20}$
(TMS)$_3$SiB(C$_6$F$_5$)$_4$, **2**, 4
$BSi_6ThC_{18}H_{58}N_3$
[(TMS)$_2$N]$_3$Th(BH$_4$), **4**, 134
$BTaC_{11}H_{13}F_4$
[TaMeCp$_2$][BF$_4$], **4**, 593
$BTaC_{12}H_{16}F_4$
[Cp$_2$TaMe$_2$][BF$_4$], **5**, 70
$BTaC_{12}H_{33}O_3P_3$
Ta(BH$_4$)(CO)$_3$(PMe$_3$)$_3$, **5**, 81
$BTaC_{12}H_{42}P_4$
TaH$_2$(BH$_4$)(PMe$_3$)$_4$, **1**, 280; **5**, 81
$BTaC_{15}H_{24}$
Cp*CpTa(BH$_4$), **5**, 81
$BTcC_{12}H_{10}N_6O_3$
Tc(HBpz$_3$)(CO)$_3$, **6**, 156
$BTcC_{18}H_{22}N_6O_3$
Tc[HB(Me$_2$C$_3$HN$_2$)$_3$](CO)$_3$, **6**, 156
$BTcC_{29}H_{25}N_6O_2P$
Tc(HBpz$_3$)(CO)$_2$(PPh$_3$), **6**, 156
$BThC_{45}H_{33}F_{20}$
[Cp*$_2$ThMe][B(C$_6$F$_5$)$_4$], **4**, 191
$BTiC_{11}H_{14}$
CH$_2$(C$_5$H$_4$)$_2$Ti(BH$_4$), **1**, 279, 280
$BTiC_{13}H_{25}$
Ti(C$_5$H$_3$Bu$_2{}^t$)(H)$_2$BH$_2$, **4**, 261
$BTiC_{20}H_{22}N_2O_3$
(salen)Ti[BH$_4$]$_2$·(THF), **1**, 279, 280
$BTiC_{35}H_{33}$
[TiMeCp$_2$][BPh$_4$], **4**, 395
$BTiC_{44}H_{73}O_2$
[Ti(OH)Cp*$_2$(H$_2$O)][BPh$_4$], **4**, 402
$BTiC_{45}H_{53}$
[TiMeCp*$_2$][BPh$_4$], **4**, 367
$BTiC_{49}H_{61}O$
[TiMeCp*$_2$(THF)][BPh$_4$], **4**, 368, 394
$BTiC_{49}H_{61}S$
[TiMeCp*$_2$(THT)][BPh$_4$], **4**, 395
$BTiC_{57}H_{47}N_2$
[Ti[N=C(Me)Ph](C$_9$H$_7$)$_2$-(PhCN)][BPh$_4$], **4**, 366
$BTiC_{57}H_{62}NO$
[Ti(OH)(HN=CPh$_2$)Cp*$_2$]-[BPh$_4$], **4**, 399
$BUC_{15}H_{19}$
Cp$_3$U(BH$_4$), **4**, 158
$BUC_{18}H_{21}N_2$
Cp$_3$U(NCBH$_3$)(MeCN), **4**, 159
$BUC_{39}H_{35}$
Cp$_3$UBPh$_4$, **4**, 159
$BVC_6H_{22}P_2$
V(BH$_4$)(PMe$_3$)$_2$, **1**, 283
$BVC_{18}H_{18}F_4N$
Cp$_2$V[C(Ph)NMe]BF$_4$, **5**, 35
$BWC_{15}H_{15}IN_4O_4$
WI(CO)(MeC$_2$Me)Tp', **5**, 367
$BWC_{15}H_{15}N_4O_4$
[W(CO)(MeC$_2$Me)Tp']$^+$, **5**, 367
$BWC_{19}H_{30}F_4N_3$
[W(CnEt)(CNEt)$_2$Cp*]BF$_4$, **5**, 202, 203
$BWC_{27}H_{27}N_4O_4$
[W(CO)(HCCPh)$_2$Tp']$^+$, **5**, 370
$BWC_{58}H_{52}F_5P_4$
[WF(CHCH$_2$Ph)(dppe)$_2$]BF$_4$, **5**, 311
$BW_2C_{18}H_{19}O_4$
W$_2$(MeCB(H)Et)(CO)$_4$Cp$_2$, **1**, 292–4
$BYC_{14}H_{22}O$
Cp$_2$Y(BH$_4$)(THF), **4**, 48
$BZnC_{25}H_{53}N_6$
[HB(Bu^{t}pz)$_3$]ZnBu^t, **3**, 182
$BZrC_{11}H_{17}$
Cp$_2$Zr(BH$_4$)Me, **4**, 554
Cp$_2$Zr(H)BH$_3$Me, **1**, 288, 290
$BZrC_{16}H_{22}Cl_2N_5$
ZrCl$_2$Cp[H$_2$B(pz)$_2$](NCBu^t), **4**, 548
$BZrC_{24}H_{33}N_2$
Zr(NBu^{t}BPhNBu^t)Cp$_2$, **4**, 508
$BZrC_{25}H_{15}F_{15}N$
Zr[C(NBu^t)Me][MeB(C$_6$F$_5$)$_3$], **4**, 610
$BZrC_{30}H_{16}F_{15}$
ZrMe[MeB(C$_6$F$_5$)$_3$]Cp$_2$, **4**, 593
$BZrC_{34}H_{20}F_{15}O_2$
[Zr[C(O)Me](CO)Cp'$_2$][MeB-(C$_6$F$_3$)$_3$], **4**, 610
$BZrC_{34}H_{24}F_{15}$
ZrMe[MeB(C$_6$F$_5$)$_3$](Me$_2$C$_5$H$_3$)$_2$, **4**, 600
$BZrC_{35}H_{25}F_{15}N$
[Zr[C(NBu^t)(Me)]Cp$_2$][BMe-(C$_6$F$_3$)$_3$], **4**, 621
$BZrC_{37}H_{40}N$
[Zr(Bz)(MeCN)Cp$_2$][BPh$_4$], **4**, 596
$BZrC_{38}H_{32}F_{15}$
[Zr(H)Cp*$_2$][HB(C$_6$F$_5$)$_3$], **4**, 612
$BZrC_{38}H_{36}NO$
[Zr[C(O)Me](MeCN)Cp$_2$]-[BPh$_4$], **4**, 610
$BZrC_{38}H_{39}O$
[Zr(H)(THF)Cp$_2$][BPh$_4$], **4**, 594
$BZrC_{39}H_{33}D_6N_2$
[ZrMe(CD$_3$CN)$_2$Cp$_2$][BPh$_4$], **4**, 593
$BZrC_{39}H_{34}F_{15}$

[Zr(H)Cp*$_2$][MeB(C_6F_5)$_3$], **4**, 603, 612
$BZrC_{39}H_{41}O$
[ZrMe(THF)Cp$_2$][BPh$_4$], **4**, 595, 601, 603
$BZrC_{40}H_{41}O_2$
[Zr[C(O)Me](THF)Cp$_2$][BPh$_4$], **4**, 610
$BZrC_{40}H_{43}O$
[Zr(H)(THF)Cp'$_2$][BPh$_4$], **4**, 611
$BZrC_{40}H_{49}P_2$
[Zr(H)(PMe$_3$)$_2$Cp$_2$][BPh$_4$], **4**, 612
$BZrC_{40}H_{51}O_2$
[Zr(OBu)(THF)Cp'$_2$][BPh$_4$], **4**, 613
$BZrC_{41}H_{48}P$
[ZrEt(PMe$_3$)Cp'$_2$][BPh$_4$], **4**, 597, 607
$BZrC_{42}H_{47}OS$
[Zr(SBu^t)(THF)Cp$_2$][BPh$_4$], **4**, 613
$BZrC_{42}H_{47}O_2$
[Zr(OBu^t)(THF)Cp$_2$][BPh$_4$], **4**, 613
$BZrC_{42}H_{53}P_2$
[Zr(H)(PMe$_3$)$_2$Cp'$_2$][BPh$_4$], **4**, 603, 612
$BZrC_{43}H_{52}P$
[Zr(CH$_2$CHMe$_2$)(PMe$_3$)-(C_5H_4Me)$_2$][BPh$_4$], **4**, 598
$BZrC_{44}H_{43}O$
[Zr(Ph)(THF)Cp$_2$][BPh$_4$], **4**, 606
$BZrC_{44}H_{47}O_2$
[Zr[C(Me)C(Me)C(O)Me]-(THF)Cp$_2$][BPh$_4$], **4**, 606
[Zr[C(O)CMeCMe$_2$](THF)-Cp$_2$][BPh$_4$], **4**, 606
$BZrC_{44}H_{51}O$
[Zr(CH$_2$CHMe$_2$)(THF)Cp'$_2$]-[BPh$_4$], **4**, 598
$BZrC_{44}H_{57}O_2$
[ZrMe$_2$(THF)$_2$Cp*][BPh$_4$], **4**, 615
$BZrC_{45}H_{41}$
Zr(Bz)$_3$(Ph)BPh$_3$, **4**, 616
$BZrC_{45}H_{45}O$
[Zr(Bz)(THF)Cp$_2$][BPh$_4$], **4**, 596
$BZrC_{45}H_{49}O_2$
[Zr[C(O)CHCMe(Pr)]-(THF)Cp$_2$][BPh$_4$], **4**, 606
$BZrC_{45}H_{51}N_2$
[Zr[C(NBu^t)Me](CNBu^t)Cp$_2$]-[BPh$_4$], **4**, 610
$BZrC_{46}H_{28}F_{15}$
Zr(Bz)$_3$[BzB(C_6F_5)$_3$], **4**, 616
$BZrC_{46}H_{65}OP_2$
[ZrMe$_2$(THF)(dmpe)Cp*]-[BPh$_4$], **4**, 615
$BZrC_{47}H_{55}O$
[Zr[CMeC(Me)(Pr)](THF)-Cp'$_2$][BPh$_4$], **4**, 607
$BZrC_{48}H_{31}F_{24}N_3$
[Zr(MeCN)$_3$Cp$_2$][B[(CF$_3$)$_2$-C_6H_3]$_4$], **4**, 614
$BZrC_{48}H_{51}F_4$
[Zr(CH$_2$CCMe)Cp*$_2$]-[B(C_6H_4F)$_4$], **4**, 594
$BZrC_{48}H_{66}F_4$
[Zr(C_5Me_4CH$_2$CH$_2$-CHCHCH$_2$)Cp*]-[B(C_6H_4F)$_4$], **4**, 604
$BZrC_{49}H_{31}F_{20}P$
ZrMe(PMe$_3$)[C_5H_4CMe$_2$-(fluorenyl)][B(C_6F_5)$_4$], **4**, 593, 596
$BZrC_{49}H_{34}F_{15}$
[Zr(Bz)$_2$[CH$_2$CH(Me)Bz]][B-(Bz)(C_6F_5)$_3$], **4**, 617
$BZrC_{51}H_{57}F_4$
[Zr[CH$_2$C(Me)C(Me)C(Me)-CH(Me)](C_5H_4Bu^t)$_2$]-[B(C_6H_4F)$_4$], **4**, 599
$BZrC_{52}H_{53}O$
[Zr(Bz)$_2$(THF)Cp*][BPh$_4$], **4**, 615
$BZrC_{53}H_{56}OP$
[Zr(Bz)[O(CH$_2$)$_4$PMe$_2$Ph]Cp$_2$]-[BPh$_4$], **4**, 601
$BZrC_{58}H_{52}F_{15}O_3$
[Zr(Bz)$_3$(THF)$_3$][B(Bz)(C_6F_5)$_3$], **4**, 616
$BZrC_{65}H_{39}F_{20}$
[(In)CH$_2$]$_2$ZrMe[CH$_2$CH$_2$-(C_9H_7)$_2$][B(C_6F_5)$_4$], **4**, 570
$BZrSiC_{43}H_{45}F_4$
[Zr[C(TMS)CMe$_2$]Cp'$_2$]-[B(C_6H_4F)$_4$], **4**, 606
$B_2PdC_{36}H_{34}F_8N_2P_2$
Pd(dppe)(py)$_2$(BF$_4$)$_2$, **12**, 789
$B_2AgRu_8C_{60}H_{34}NO_{24}P_2$
[PPN][HRu$_4$(CO)$_{12}$BH]$_2$Ag, **1**, 304
$B_2Ag_4Ru_8C_{60}H_{34}Cl_2NO_{24}P_2$
[PPN][[HRu$_4$(CO)$_{12}$BH]$_2$Ag$_4$-(Cl)]Cl, **1**, 304
$B_2AuCoC_{35}H_{44}P$
CpCo[Et$_4$Me(C_3B_2)]Au(PPh$_3$), **1**, 206, 207
$B_2AuFeC_{22}H_{20}O_4P$
Fe(CO)$_4$(B_2H_5)Au(PPh$_3$), **10**, 62
$B_2AuOsC_{22}H_{20}O_4P$
(PPh$_3$)Au(CO)$_4$Os(B_2H_5), **1**, 308
$B_2AuRuC_{63}H_{61}F_8P_4$
[RhAu[HC(PPh$_2$)$_3$](cod)-(PPh$_3$)][BF$_4$]$_2$, **10**, 40
$B_2Au_6PtC_{174}H_{145}P_7$
[Au$_6$Pt(PPh$_3$)$_7$][BPh$_4$]$_2$, **10**, 256
$B_2BeC_{16}H_{30}N_2$
(BuMeH$_2$azaborolyl)$_2$Be, **1**, 46
$B_2BeC_{20}H_{29}N_2$
[MeBu^t(NBC$_3$H$_2$)]Be[MeBu^t-(NMC$_3$H$_3$)], **1**, 212, 213
$B_2C_2H_4$
$C_2B_2H_4$, **1**, 131, 150
$B_2C_2H_4N_2$
$B_2N_2C_2H_4$, **1**, 166
$B_2C_2H_5S$
H$_2$BS=BHCH=CH, **1**, 214
$B_2C_2H_6$
$C_2B_2H_6$, **1**, 150
$B_2C_2H_6N_2$
$C_2B_2N_2H_6$, **1**, 164
$B_2C_2H_6S$
H$_2$BSBH$_2$CH=CH, **1**, 214
$B_2C_2H_7P$
H$_2$BPHBH$_2$CH=CH, **1**, 214
$B_2C_2H_{10}$
(MeBH$_2$)$_2$, **11**, 206
$B_2C_2H_{14}$
(MeBH$_4$)$_2$, **1**, 38
$B_2C_3H_6$
CH$_2$B(H)C(H)C(H)B(H), **1**, 152
$B_2C_3H_6Cl_2S$
Cl$_2$($B_2SC_3H_6$), **1**, 181
$B_2C_4H_4$
$B_2C_4H_4$, **1**, 153
diborafulvene, **1**, 199, 208, 214
$B_2C_4H_6$
BHBHCHCHCHCH, **1**, 209
$B_2C_4H_6$, **1**, 153
$B_2C_4H_8$
BHCHCHBHCH$_2$CH$_2$, **1**, 209, 210
diboracyclohexene, **1**, 145
$B_2C_4H_{12}N_2S$
Me$_4$(B_2N_2S), **1**, 185
$B_2C_4H_{12}N_2S_2$
[(Me$_2$N)BS]$_2$, **1**, 179
$B_2C_4H_{14}$
(Me$_2$BH)$_2$, **11**, 206
$B_2C_5H_8$
($B_2C_5H_8$), **1**, 154
$B_2C_5H_{10}N_2S_2$
MeB(pz)$_2$(S_2)BMe, **1**, 185
$B_2C_5H_{13}Cl_3N_2$
NMe$_2$CH$_2$CH$_2$N(Me)B(Cl)B-Cl$_2$, **1**, 165
$B_2C_6H_4N_2$
diazaboranaphthalene, **1**, 164, 165
$B_2C_6H_{16}N_2S_2$
Et$_2$Me$_2$($B_2N_2S_2$), **1**, 185
$B_2C_6H_{18}O_2P_2$
(Me$_3$P)$_2$(B_2O_2), **1**, 171, 172
$B_2C_6H_{18}P_2S_2$
[(Me$_3$P)BS]$_2$, **1**, 179
$B_2C_6H_{18}P_2Se_2$
[(Me$_3$P)BSe]$_2$, **1**, 179
$B_2C_7H_{17}N_2P$
(NMe$_2$)$_2$Me(PB$_2$C$_2$H$_2$), **1**, 169
$B_2C_8H_{14}N$
nido-$C_4B_2H_4$(Me)(NCHMe$_2$), **1**, 231
$B_2C_{10}H_{14}$
($C_6H_4B_2$CH)Me$_3$, **1**, 206, 207
$B_2C_{10}H_{20}$

$Me_6(B_2C_4H_2)$, **1**, 153
$B_2C_{10}H_{27}NP_2$
[(TMP)B]$_2$NBu^t, **1**, 159
$B_2C_{11}H_{25}N$
(Bu^{t}B)(Pr^{i}B)BBu^t, **1**, 159
$B_2C_{12}H_{12}O_3$
$(MeO)_2(C_{10}H_6B_2O)$, **1**, 177
$B_2C_{12}H_{25}N$
$(C_9H_{14})(CH_2CH_2NMeBH)BH_3$, **11**, 261
$B_2C_{12}H_{28}N_2S_2$
[(Pr$^i_2$N)BS]$_2$, **1**, 179
$B_2C_{14}H_{12}O_2S_2$
$[(MeO)C_6H_3(BOS)]_2$, **1**, 185
$B_2C_{14}H_{28}O_5$
(Bu^t)$_2$Et$_2$(B$_2$O$_5$C$_2$), **1**, 179
$B_2C_{14}H_{30}N_2$
(Pr$^i_2$N)$_2$(B$_2$C$_2$H$_2$), **1**, 150
$B_2C_{14}H_{36}N_2P_2$
[(Me$_3$P)BNBu^t]$_2$, **1**, 159
$B_2C_{16}H_{18}$
$Pr_2BC(=CH_2)C(=CH_2)BPr_2$, **11**, 236
$B_2C_{16}H_{34}$
$Pr_2BCH_2C(=CH_2)CH_2BPr_2$, **11**, 236
$B_2C_{17}H_{20}N_6$
$[Me_2C_6H_4N_2B]_2N=C=N$, **1**, 162
$B_2C_{18}H_{22}$
$Pr_2BCH_2C(=CH_2)C(=CH_2)$-$CH_2BPr_2$, **11**, 236
$B_2C_{20}H_{35}N_2$
(Pr$^i_2$N=)$_2$Ph[BPBCH=CH, **1**, 214
$B_2C_{26}H_{40}P_2$
B(xylyl)=P(Bu^t)B(xylyl)=PBu^t, **1**, 211, 212
$B_2C_{28}H_{42}$
Mes$_2$(Bu^t)$_2$(B$_2$C$_2$H$_2$), **1**, 150
$B_2C_{32}H_{34}N_2O_6$
[(OPr^i)(C$_6$H$_4$O$_2$B)]$_2$-PhHC=NN=CHPh, **1**, 178, 179
$B_2C_{32}H_{40}O_4P_2$
$[Ph_2P(O)OBEt_2]_2$, **1**, 177
$B_2C_{36}H_{58}O_2$
(Bu$^t_3$C$_6$C$_6$H$_2$)$_2$(B$_2$O$_2$), **1**, 171
$B_2C_{60}H_{36}O_6$
$[(C_{10}H_6)_2O_2]_3B_2$, **11**, 262
$B_2Cl_2C_{26}H_{92}$
[(Bu$^t_2$C$_5$H$_3$)$_2$Cl[(H)$_2$B(H)$_2$]]$_2$, **1**, 277
$B_2Cl_2C_{52}H_{92}$
[(C$_5$H$_3$Bu$^t_2$)$_2$Cl[(H)$_2$BH$_2$]]$_2$, **1**, 286, 287
$B_2CoAuC_{35}H_{44}P$
CpCo[Et$_4$Me(C$_3$B$_2$)]Au(PPh$_3$), **1**, 206, 207
$B_2CoC_{14}H_{23}$
CpCo[Et$_2$Me$_2$(C$_3$B$_2$)], **1**, 205, 206
$B_2CoC_{17}H_{29}$
CpCo[Et$_4$Me(C$_3$B$_2$)], **1**, 206, 207
$B_2CoC_{21}H_{40}N_2$
CpCoPr$^i_2$N=BHCH=C(Me)-CH$_2$B=NPr$^i_2$, **1**, 205
$B_2CoC_{44}H_{47}F_8NP_4$
[Co(PP$_3$)(MeCN)](BF$_4$)$_2$, **8**, 20
$B_2CoFeC_{18}H_{25}$
CpFe(C$_3$B$_2$Me$_5$)CoCp, **7**, 195
$B_2CoTlC_{17}H_{23}$
CpTl[Me$_2$C$_2$(BMe)$_2$CH]CoCp, **1**, 206, 207
$B_2Co_2C_{22}H_{25}P$
Cp$_2$Co$_2$(PPh$_2$)B$_2$H$_5$, **1**, 309, 310
$B_2Co_2C_{52}H_{56}P_4$
(dppe)$_2$Co$_2$(BH$_4$)$_2$, **1**, 285
$B_2Co_4C_{20}H_{24}$
Co$_4$(H)$_2$(B$_2$H$_2$)Cp$_4$, **8**, 449
(H)$_2$[CpCo]$_4$B$_4$H$_2$, **1**, 312
$B_2Co_4C_{26}H_{27}P$
Co$_4$(PPh)(B$_2$H$_2$)Cp$_4$, **8**, 448
Ph[CpCo]$_4$PB$_2$H$_2$, **1**, 312
$B_2Co_5C_{14}HO_{14}$
Co$_5$(CO)$_{14}$B$_2$H, **8**, 450
$B_2CrC_{10}H_{22}O_4P_2$
Cr(CO)$_4$(B$_2$H$_4$·2PMe$_3$), **1**, 311
$B_2CrC_{11}H_{22}O_5P_2$
Cr(CO)$_5$(B$_2$H$_4$·2PMe$_3$), **1**, 311
$B_2CrC_{14}H_{18}O_4$
Cr(CO)$_4$[Me$_2$C$_2$(BMe)$_2$-C=CMe$_2$], **1**, 208
$B_2CrC_{20}H_{36}N_2O_4$
(CO)$_4$Cr[Bu$_4$(B$_2$N$_2$)], **1**, 211
$B_2CrC_{22}H_{36}N_2O_2$
Cr(CO)$_2$(C$_4$H$_4$BNPr$^i_2$)$_2$, **1**, 204
$B_2CrC_{22}H_{37}O_4P$
Cr(CO)$_4$[Et$_2$C$_2$(BMe)$_2$CCMe$_2$P-Et$_3$], **1**, 208
$B_2CrC_{30}H_{40}O_4P_2$
(CO)$_4$Cr[xylyl$_2$(B$_2$P$_2$)], **1**, 211
$B_2CuC_{37}H_{40}NP_2$
(PPh$_3$)$_2$Cu(H$_3$BCNB$_3$H$_7$), **1**, 317
$B_2Cu_4Ru_8C_{60}H_{32}Cl_2NO_{24}P_2$
[PPN][[HRu$_4$(CO)$_{12}$B]$_2$Cu$_4$-(Cl)]Cl, **1**, 304, 305
$B_2FeC_7H_{10}O_2$
CpFe(CO)$_2$(B$_2$H$_5$), **1**, 307–9
$B_2FeC_{12}H_{18}F_8N_6$
[Fe(CNMe)$_6$][BF$_4$]$_2$, **7**, 12
$B_2FeC_{13}H_{18}N_2$
H$_2$B(Fcpz)BH$_2$, **1**, 168
$B_2FeC_{13}H_{18}O_3$
Fe(CO)$_3$[Me$_2$C$_2$(BMe)$_2$-C=CMe2], **1**, 208
$B_2FeC_{14}H_{22}S$
(benzene)Fe[Et$_2$Me$_2$(BSBC$_2$)], **1**, 214
$B_2FeC_{15}H_{16}O_3$
(CO)$_3$[Me$_4$(C$_7$B$_2$)], **1**, 208
$B_2FeC_{15}H_{23}$
CpFe[Et$_2$Me(C$_4$H$_2$B$_2$], **1**, 210
$B_2FeC_{17}H_{14}O_3$
(CO)$_3$Fe(Me$_2$diboraanthracene), **1**, 210
$B_2FeC_{17}H_{20}O_3$
(CO)$_3$[Et$_2$Me$_2$(C$_8$H$_4$B$_2$)], **1**, 210
$B_2FeC_{18}H_{41}N_2O_2P$
(CO)$_2$(PMe$_3$)Fe[(Pr$^i_2$N)$_2$C$_4$B$_2$], **1**, 209, 210
$B_2FeC_{19}H_{24}$
Fe(PhMe)[Et$_4$Me(C$_3$B$_2$)], **1**, 206
$B_2FeC_{19}H_{32}N_2O_3$
(CO)$_3$Fe[(Pr$^i_2$N)$_2$C$_4$B$_2$], **1**, 209, 210
$B_2FeC_{23}H_{35}N_2O_3P$
(CO)$_3$Fe[(Pr$^i_2$N)$_2$Ph(C$_2$H$_2$B$_2$P)], **1**, 214
$B_2FeC_{29}H_{40}O_3P_2$
(CO)$_3$Fe(xylyl$_2$Bu$^t_2$)P$_2$B$_2$, **1**, 211
$B_2FeC_{30}H_{40}O_4P_2$
(CO)$_4$Fe(xylyl$_2$Bu$^t_2$)P$_2$B$_2$, **1**, 211
$B_2Fe_2C_6H_4O_6$
$[Fe_2(CO)_6B_2H_4]^{2-}$, **7**, 16
$B_2Fe_2C_6H_5O_6$
$[B_2H_5Fe_2(CO)_6]^-$, **1**, 306, 307
$[Fe_2(CO)_6B_2H_5]^-$, **7**, 16
$B_2Fe_2C_6H_6O_6$
B$_2$H$_6$Fe$_2$(CO)$_6$, **1**, 298, 306, 307
Fe$_2$(CO)$_6$B$_2$H$_6$, **7**, 16
$B_2Fe_2C_{24}H_{30}$
(FeCp)$_2$[Et$_2$Me$_2$(C$_8$H$_4$B$_2$)], **1**, 210
$B_2H_4N_2$
BHNHBHNH, **1**, 211
B_2HgF_8
Hg(BF$_4$)$_2$, **11**, 412, 413, 416, 419, 421
$B_2Li_2Fe_2C_6H_4O_6$
Li$_2$[B$_2$H$_4$Fe$_2$(CO)$_6$], **1**, 307
$B_2MnC_{21}H_{36}N_2O$
Mn(CO)(C$_4$H$_4$BNPr$^i_2$)$_2$, **1**, 204
$B_2MoC_{10}H_{16}$
MoCp$_2$H(B$_2$H$_5$), **1**, 308, 321
$B_2MoC_{14}H_{11}F_8N_3O_2$
[Mo(CO)$_2$(NCMe)(bipy)]-[BF$_4$]$_2$, **5**, 237
$B_2MoC_{14}H_{18}O_4$
Mo(CO)$_4$[Me$_2$C$_2$(BMe)$_2$-C=CMe$_2$], **1**, 208
$B_2Mo_2C_{44}H_{32}F_8N_8O_4$
[Mo$_2$(Co)$_4$(bipy)$_4$][BF$_4$]$_2$, **5**, 237
$B_2NiC_8H_{10}$
Ni(C$_4$H$_4$BH)$_2$, **1**, 202
$B_2NiC_8H_{22}O_2P_2$
Ni(CO)$_2$(B$_2$H$_4$·2PMe$_3$), **1**, 311
$B_2NiC_{10}H_{14}$
Ni(C$_4$H$_4$BMe)$_2$, **1**, 204
$B_2NiC_{13}H_{21}S$
CpNi[Et$_2$Me$_2$(BSBC$_2$)], **1**, 214
$B_2NiC_{16}H_{20}N_2$
Ni(C$_3$H$_3$NBu^{t}BMe)$_2$, **9**, 145
$B_2NiC_{20}H_{18}$
Ni(C$_4$H$_4$BPh)$_2$, **1**, 204
$B_2NiC_{20}H_{44}N_2Si_2$
Ni(C$_{10}$H$_{22}$BNSi)$_2$, **1**, 213
$B_2NiC_{22}H_{20}$
Ni(CpBPh)$_2$, **9**, 176
$B_2NiC_{22}H_{32}$

(cod)Ni[$Et_2Me_2C_8H_4B_2$, **1**, 210
$B_2Ni_2C_{14}H_{16}O_2$
$Ni_2(CO)_2(CpBMe)_2$, **9**, 176
$B_2Ni_2C_{19}H_{25}$
$Cp_2Ni_2[Et_2MeC_4H_2B_2]$, **1**, 210
$B_2Ni_3C_{30}H_{38}$
$Ni[Me_3(C_7B_2)(NiCp)]_2$, **1**, 206
$B_2OsC_4H_6O_4$
$H(CO)_4Os(B_2H_5)$, **1**, 308, 309
$B_2OsC_7H_{14}O_5$
$Me_3O(CO)_4Os(B_2H_5)$, **1**, 308
$B_2Os_2C_{30}H_{68}P_2$
$[Os[BH_4]H_3P(C_5H_9)_3]_2$, **1**, 284, 285
$B_2Os_3C_{16}H_7O_{12}$
$(H)_3Os_3(CO)_9B(COBO_2C_6H_4)$, **1**, 296, 297
$B_2Os_3C_{18}H_{20}ClO_{10}$
$Os_3H_2[(C_8H_{17}BO)C{=}BHCl]$-$(CO)_9$, **7**, 703
$B_2PdC_{20}H_{18}$
$Pd(C_4H_4BPh)_2$, **1**, 204
$B_2Pd_2C_{24}H_{28}N_{12}$
$(C_3H_5)Pd(pz)_3B_2(pz)_3Pd(C_3H_5)$, **9**, 350
$B_2PtC_{20}H_{18}$
$Pt(C_4H_4BPh)_2$, **9**, 582
$B_2PtC_{20}H_{36}$
(cod)Pt[$Et_4Me(C_3B_2)$], **1**, 206
$B_2PtC_{20}H_{44}P_2$
$Pt[Me_2BC(Me)CHCHC(Me)B$-$Me_2](Et_2PCH_2CH_2PEt_2)$, **9**, 574
$B_2ReC_{24}H_{46}P_3$
$Re(BH_4)_2H(PMe_2Ph)_3$, **1**, 284
$B_2RhC_{18}H_{27}N_2$
$(Cp^*)Rh[C_4(BNMe_2)_2]$, **1**, 209
$B_2Rh_2C_{22}H_{44}ClN_2Si_2$
$[RhCl(C_{10}H_{22}BNSi)]$, **1**, 213
$B_2Rh_2C_{24}H_{40}N_2$
$(cod)_2Rh_2[(Me_2N)_2C_2H_4B_2]$, **1**, 209
$B_2Rh_2C_{32}H_{44}F_8$
$[(C_6Me_6)Rh(cot)Rh(C_6Me_6)]$ $(BF_4)_2$, **8**, 245
$B_2Rh_2C_{133}H_{124}N_2P_6$
$[Rh_2(BuNC)_2(dppm)_3][BPh_4]_2$, **8**, 167
$B_2Rh_4C_{40}H_{64}F_8$
$[Rh_4H_4Cp^*_4][BF_4]_2$, **8**, 470
$B_2RuC_7H_{10}O_2$
$Ru(B_2H_5)(CO)_2Cp$, **7**, 487
$B_2RuC_{10}H_{21}F_8O_3$
$[RuCp^*(H_2O)_3][BF_4]_2$, **7**, 668
$B_2RuC_{13}H_{31}P$
$Cp^*Ru(PMe_3)(B_2H_7)$, **1**, 304, 305
$B_2RuC_{15}H_{30}F_8P_2S$
$[Ru[CCH(SMe_2)](PMe_3)_2Cp]$-$(BF_4)_2$, **7**, 517
$B_2RuC_{20}H_{34}N_2$
$(C_6Me_6)Ru[C_4H_4B_2(NMe_2)_2]$, **1**, 209
$B_2RuC_{24}H_{20}F_8$
$[Ru(PhPh)_2](BF_4)_2$, **7**, 550
$B_2RuC_{38}H_{30}F_8O_2P_2$
$Ru(BF_4)_2(CO)_2(PPh_3)_2$, **7**, 308, 309
$B_2Ru_2C_{50}H_{50}P$
$[[(triphos)(H)Ru]_2(BH_4)]$-$[BPh_4]$, **1**, 284, 285
$B_2Ru_3C_9H_6O_9$
$HRu_3(CO)_9B_2H_5$, **1**, 308
$B_2Ru_3C_{26}H_{21}O_8P$
$HRu_3(CO)_8(PPh_3)B_2H_5$, **1**, 308, 309
$B_2Ru_3C_{45}H_{35}NP_2$
$[PPN][Ru_3(CO)_9B_2H_5]$, **1**, 303
$B_2Ru_9C_{23}O_{23}$
$Ru_9(CO)_{23}B_2$, **1**, 294
$B_2Si_2C_{30}H_{49}$
$(Dur)_2(TMS)_2Ph(B_2C_2)C_8H_{13}$, **1**, 150
$B_2Si_2UC_{14}H_{32}$
$[(TMS)_2C_8H_6]U(BH_4)_2$, **4**, 185
$B_2Si_4C_{24}H_{54}$
$(TMS)_2CB(Bu^t)C{=}C{=}C$-$(TMS)_2]BBu^t$, **1**, 135
$B_2Si_5C_{20}H_{46}Cl_2N_2$
$[(TMS)_2Me$-agaborole$]SiCl_2$, **1**, 161
$B_2Si_6C_{20}H_{54}O$
$[(TMS)_3C]_2(B_2O)$, **1**, 171
$B_2Si_6SnC_{26}H_{47}$
$[(TMS)_2]_2Sn{=}[CB(Bu^t)C$-$(TMS)_2BBu^t]$, **2**, 287
$B_2SnC_{17}H_{29}$
$CpSn[Et_4Me(C_3B_2)]$, **1**, 206
$B_2Tc_2C_{34}H_{44}N_{14}O_4$
$[Tc[HB(Me_2C_3HN_2)_3]$-$(CO)_2]_2N_2$, **6**, 156
$B_2ThC_2H_{12}$
$Th[BH_3Me]_2$, **1**, 290, 291
$B_2TiC_{12}H_{40}P_4$
$Ti(BH_4)_2(dmpe)_2$, **1**, 279, 280
$B_2TiC_{23}H_{38}O_2$
(bisphenoxo)$Ti(BH_4)_2$, **1**, 279, 280
$B_2TlCoC_{13}H_{20}$
$TlMe_5(C_2B_2C)CoCp$, **1**, 532
$B_2TlCoC_{17}H_{23}$
$CpTl[Me_2C_2(BMe)_2CH]CoCp$, **1**, 206, 207
$B_2UC_{10}H_{18}$
$Cp_2U(BH_4)_2$, **4**, 146
$B_2UC_{10}H_{21}O$
$[(C_8H_8)U(BH_4)_2(OEt)]^-$, **4**, 185
$B_2UC_{18}H_{22}$
$(C_9H_7)_2U(BH_4)_2$, **4**, 163
$B_2UC_{18}H_{32}P_2$
$(C_5Me_4P)_2U(BH_4)_2$, **4**, 178
$B_2VC_{12}H_{40}P_4$
$V(BH_4)_2(dmpe)_2$, **1**, 280, 283
$B_2V_2C_{50}H_{52}P_4$
$[VCl(dppm)(BH_4)]_2$, **1**, 280, 283
$B_2V_2Zn_2C_{52}H_{64}P_4$
$V_2Zn_2H_4[BH_4]_2(PMePh_2)_4$, **1**, 280, 282, 283
$B_2YC_{10}H_{18}$
$[Cp_2Y(BH_4)_2]^-$, **4**, 48
$B_2ZnC_6H_{22}Cl_2P_2$
$ZnCl_2(B_2H_4{\cdot}2PMe_3)$, **1**, 311
$B_2ZrC_{10}H_{18}$
$Cp_2Zr(BH_4)_2$, **4**, 554
$B_2ZrC_{12}H_{42}P_4$
$ZrH_2(BH_4)_2(dmpe)_2$, **1**, 279, 280
$B_2ZrC_{27}H_{46}O_3$
(bisphenoxo)$Zr(THF)(BH_4)_2$, **1**, 279, 280
$B_2ZrC_{63}H_{59}N_2$
$[ZrMe(MeCN)_2Cp_2][BPh_4]_2$, **4**, 614
$B_2ZrC_{64}H_{59}N_3$
$[Zr(MeCN)_3Cp_2][BPh_4]_2$, **4**, 614
$B_2ZrC_{72}H_{65}N_3$
$[Zr(MeCN)(Me_2C_{10}H_6N_2)Cp_2]$-$[BPh_4]_2$, **4**, 614
$B_2Zr_2C_{38}H_{54}O_2$
$[Cp_2ZrOCH_2B(C_8H_{14})H]_2$, **1**, 292, 293
$B_3Au_2Fe_8C_{30}H_{22}O_{24}P_2$
$[H[Fe_4(CO)_{12}BH]_2Au][Au$-$(PMe_3)_2BH]$, **1**, 297, 300
$B_3C_2H_5$
$C_2B_3H_5$, **1**, 198, 208
$B_3C_8H_{23}N_2$
$BH_2N(Bu^t)BHN(Bu^t)BH_2$, **1**, 160
$B_3C_9H_{15}O_2$
$Me_3Ph(B_3NO_2)$, **1**, 183
$B_3C_9H_{21}N_3P_2$
$P_2(Pr^iNB)_3$, **1**, 264
$B_3C_{10}H_{21}$
closo-$C_2B_3Me_2Et_3$, **1**, 235
$B_3C_{12}H_{12}N_3$
$(C_4H_4BN)_3$, **1**, 132
$B_3C_{12}H_{23}N_4O_2$
$EtB(pz)_2(O_2BEt)BEt$, **1**, 183
$B_3C_{39}H_{60}P_3$
$B(xylyl)P(Bu^t){=}B(xylyl)P$-$(Bu^t){=}B(xylyl)PBu^t$, **1**, 211, 212
$B_3Co_3C_{16}H_{18}O$
$Co_3B_3H_3(CO)Cp_3$, **8**, 432
$(CpCo)_3(CO)B_3H_3$, **1**, 318
$B_3CrC_{15}H_{30}N_3O_3$
$Cr(CO)_3(B_3N_3Et_6)$, **5**, 499
$B_3CrC_{42}H_{60}O_3P_3$
$(CO)_3Cr[xyxlyl_3Bu^t_3(P_3B_3)]$, **1**, 211
$B_3FeC_6H_7O_6$
$[Fe(CO)_3]_2B_3H_7$, **1**, 317, 318
$B_3FeC_{18}H_{22}OP$
$CpFe(CO)B_3H_7(PPh_2)$, **1**, 313, 316, 317
$B_3FeC_{42}H_{36}NO_6P_2$
$[PPN][[Fe(CO)_3]_2B_3H_6]$, **1**, 317, 318
$B_3FePdC_{34}H_{35}P_2$
$(dppf)PdB_3H_7$, **1**, 317
$B_3FePtC_{34}H_{35}P_2$
$[(Ph_2PC_5H_4)_2Fe]PtB_3H_7$, **1**, 317
$B_3Fe_2C_6H_7O_6$
$(CO)_6Fe_2B_3H_7$, **1**, 313, 316

B_3GaH_{10}
$H_2GaB_3H_8$, **1**, 258
$B_3H_3O_3$
boroxane, **1**, 171
$B_3H_6N_3$
borazine, **1**, 158, 160
$B_3HoC_3H_{18}$
$Ho[BH_3Me]_3$, **1**, 288, 290
$B_3IrC_{36}H_{40}P_2$
$(PPh_3)_2H_2IrB_3H_8$, **1**, 313, 314, 316
$B_3IrC_{37}H_{38}OP_2$
$(IrB_3H_7)(CO)H(PPh_3)_2$, **1**, 313, 316
$B_3NbC_{10}H_{18}$
$NbCp_2B_3H_8$, **1**, 313, 315
$B_3NiC_{18}H_{32}$
$[Ni(cod)(C_3Et_2HB_3Me_3)]^+$, **9**, 181
$B_3NiC_{23}H_{28}P$
$[CpNi(PPh_3)][B_3H_8]$, **1**, 316
$B_3O_9C_{30}H_6O_{30}$
$[(H)_2Os_3(CO)_9(C)]_3[B_3O_3]$, **1**, 296, 297
$B_3OsC_{37}H_{39}OP_2$
$(CO)H(PPh_3)_2OsB_3H_8$, **1**, 313, 315
$B_3Os_9C_{30}H_6O_{30}$
$[(H)_2Os_3(CO)_9(C)]_3[B_3O_3]$, **1**, 296, 297
$B_3PdC_{25}H_{29}P_2$
$[Ph_2P(CH_2)PPh_2]Pd(B_3H_7)$, **1**, 316
$B_3PtC_{16}H_{29}P_2$
$(Me_2PhP)_2PtB_3H_7$, **1**, 313, 316, 317
$B_3RuC_{37}H_{38}ClOP_2$
$RuCl(B_3H_8)(Co)(PPh_3)_2$, **1**, 315
$B_3RuC_{37}H_{39}OP_2$
$(CO)(PPh_3)_2HRuB_3H_8$, **1**, 313–15
$B_3Si_2C_{14}H_{29}N_4S_2$
$Me_8Ph(B_3N_4S_2Si_2)$, **1**, 185
$B_3TiC_4H_{22}O_2$
$Ti(BH_4)_3(DME)$, **1**, 279, 280
$B_3TiC_6H_{30}P_3$
$Ti(BH_4)_3(PMe_3)_2$, **1**, 279, 280
$B_3UC_5H_{17}$
$[CpU(BH_4)_3]^-$, **4**, 138, 144
$B_3UC_6H_{19}$
$(MeC_5H_4)U(BH_4)_3$, **4**, 144
$B_3UC_8H_{20}$
$[(C_8H_8)U(BH_4)_3]^-$, **4**, 185
$B_3UC_9H_{24}$
$(C_6H_3Me_3)U(BH_4)_3$, **4**, 180
$B_3UC_{10}H_{27}$
$[Cp^*U(BH_4)_3]^-$, **4**, 138
$B_3UC_{12}H_{44}P_4$
$(dmpe)_2U(BH_4)_3$, **1**, 286, 288, 289
$B_3UC_{34}H_{40}P_2$
$U(BH_4)_3(Ph_2PPy)_2$, **1**, 286, 289
$B_3VC_6H_{30}P_2$
$V(BH_4)_3(PMe_3)_2$, **1**, 282, 283
$B_3WC_3H_{20}P$
$WH_3(PMe_3)B_3H_8$, **1**, 313, 315
$B_3YbC_3H_{18}$
$Yb[BH_3Me]_3$, **1**, 288, 290
$B_3ZrC_5H_{17}$
$Cpzr(BH_4)_3$, **1**, 279, 280; **4**, 554
$B_3ZrC_6H_{29}P_2$
$ZrH(BH_4)_3(dmpe)$, **1**, 279, 280
$B_3ZrC_{10}H_{27}$
$Cp^*Zr(BH_4)_3$, **4**, 555
$B_4AlC_{10}H_{30}$
$(Et_3N)(AlH)(EtC)_2B_4H_4$, **1**, 258, 259
$B_4C_2H_6$
closo-$C_2B_4H_6$, **1**, 218, 223
$C_2B_4H_6$, **1**, 219, 228, 232
$B_4C_2H_6F$
[F-*nido*-$C_2B_4H_6]^-$, **1**, 235, 236
$B_4C_2H_7$
[*nido*-$C_2B_4H_7]^-$, **1**, 229
$B_4C_2H_8$
$C_2B_4H_8$, **1**, 232
$B_4C_3H_{12}$
hypho-$C_3B_4H_{12}$, **1**, 227
$B_4C_5H_{16}$
nido-$C_2B_4H_7[TMS]$, **1**, 219
$B_4C_6H_{16}$
nido-$C_2B_4H_6Et_2$, **1**, 223
$Et_2C_2B_4H_6$, **1**, 207, 208
$B_4C_{10}H_{20}$
nido-$C_4B_4H_4Me_2Et_2$, **1**, 228
Me_2-*nido*-$Et_2C_4B_4H_4$, **1**, 231
$B_4C_{10}H_{28}N_4$
$(Me_2N)_4(C_2B_4H_4)$, **1**, 137
$B_4C_{16}H_{36}$
$(Bu^t)_4B_4$, **1**, 245
$B_4C_{28}H_{60}N_4$
$(NPr^i_2)_4(B_4C_4H_4)$, **1**, 154
$B_4Cl_4P_2$
$P_2B_4Cl_4$, **1**, 264
$B_4CoC_5H_{13}$
$CpCoB_4H_8$, **1**, 320
$B_4CoC_{10}H_{23}$
$Cp^*CoB_4H_8$, **1**, 320
$B_4Co_2C_{20}H_{32}N_4O_4$
$Co_2(CO)_4[C_4H_4B_2(NMe_2)_2]_2$, **1**, 209
$B_4Co_2C_{20}H_{36}$
$Cp^*_2Co_2B_4H_6$, **1**, 321
$B_4Co_2NiC_{26}H_{40}$
$CpCo(C_3B_2Me_5)Ni(C_3B_2Me_5)$-CoCp, **9**, 178
$B_4Co_3C_{30}H_{49}$
$Co_3B_4H_4Cp^*_3$, **8**, 432
$(Cp^*Co)_3B_4H_4$, **1**, 322
$B_4CuC_{12}H_{44}ClP_4$
$[Cu(B_2H_4{\cdot}2PMe_3)_2]Cl$, **1**, 311
$B_4FeC_3H_8O_3$
$[Fe(CO)_3]B_4H_8$, **1**, 320
$B_4FeC_{17}H_{32}N_4O$
$Fe(CO)[C_4H_4B_2(NMe_2)_2]_2$, **1**, 209
$B_4FeC_{18}H_{21}OP$
$CpFe(CO)B_4H_6(PPh_2)$, **1**, 317, 321
$B_4FeNiC_{25}H_{37}$
$(C_9H_8)Fe(Et_2C_2B_4H_4)NiCp^*$, **9**, 179
$B_4Fe_2C_{12}H_{46}P_4$
$[Fe(PMe_3)_2]_2(H)B_4H_9$, **1**, 319
$B_4Fe_4C_{12}H_8O_{12}$
$[B_2H_4Fe_2(CO)_6]_2$, **7**, 16, 17
$B_4H_8Fe_4(CO)_{12}$, **1**, 307
$B_4GaSi_2C_{12}H_{33}$
$Bu^tGa(TMS)_2C_2B_4H_4$, **1**, 258
$B_4GaSi_2C_{20}H_{37}$
(bipyrimidine)$Bu^tGa(TMS)_2$-$C_2B_4H_4$, **1**, 258
$B_4GeSi_2C_8H_{22}$
$Ge(TMS)_2C_2B_4H_4$, **1**, 260
$B_4InSi_2C_{11}H_{29}$
$Me_2CHIn[(TMS)_2C_2B_4H_4]$, **1**, 532
$B_4InSi_2C_{11}H_{39}$
$(Me_2CH)In(TMS)_2C_2B_4H_{14}$, **1**, 258
$B_4IrC_7H_{27}OP_2$
$(CO)(PMe_3)_2Ir(B_4H_9)$, **1**, 319, 320
$B_4IrC_{17}H_{31}OP_2$
$(CO)(PMe_2Ph)_2Ir(B_4H_9)$, **1**, 319
$Ir(B_4H_9)(CO)(PMe_2Ph)_2$, **8**, 339
$B_4Ir_2C_{57}H_{45}O_3P_2$
$(CO)_3(PPh_3)(Ph_2PC_6H_4)_2$-$(Ir_2B_4H_2)$, **1**, 322
$B_4MoC_{10}H_{16}$
$CpMo(C_5H_4)(B_4H_7)$, **1**, 321, 322
B_4MoI
$MoIB_4$, **12**, 893
$B_4NbC_{10}H_{19}$
$Cp_2NbB_4H_9$, **1**, 319
$B_4NiC_{16}H_{32}N_4$
$Ni[C_4H_4B_2(NMe_2)_2]_2$, **1**, 209
$B_4NiC_{28}H_{31}ClP_2$
$NiCl(dppe)C_2B_4H_7$, **9**, 181
$B_4NiC_{28}H_{40}$
$Ni[Et_2Me_2(C_8H_4B_2)]_2$, **1**, 210
$B_4NiC_{42}H_{43}Br$
$NiBr(PPh_3)(Et_2C_2B_4H_3PPh_3)$, **9**, 181
$B_4NiPtC_{29}H_{51}$
$CpNi(MeCB_2C_2Et_4)Pt$-$(MeCB_2C_2Et_4)$, **9**, 179
$B_4Ni_2C_{20}H_{36}$
$[Ni[Me_2C_2(BMe)_2C{=}CMe_2]]_2$, **1**, 208
$B_4OsC_{37}H_{38}OP_2$
$(OsB_4H_8)(CO)(PPh_3)_2$, **1**, 320
$B_4Pt_2Ni_2C_{34}H_{56}$
$[Pt(MeCB_2C_2Et_4)NiCp]_2$, **9**, 179
$B_4RuC_{10}H_{25}$
$Cp^*RuHB_4H_9$, **1**, 319, 320
$B_4RuC_{12}H_{26}$
$(C_6Me_6)RuB_4H_8$, **1**, 320
$B_4RuC_{37}H_{38}OP_2$
$(CO)(PPh_3)_2RuB_4H_8$, **1**, 320, 321, 323
$B_4SiC_6H_{18}N_2$

$(TMS)_2N_2B_4$, **1**, 261
$B_4SiC_8H_{22}$
$(TMS)_2$-*closo*-$C_2B_4H_4$, **1**, 231
$B_4SiSnC_6H_{16}$
$Sn(TMS)MeC_2B_4H_4$, **1**, 260
$B_4Si_2C_{26}H_{51}P$
$[(Bu^t)_3C_6H_2]P(TMS)_2C_2B_4H_4$, **1**, 265
$B_4Ta_2C_{10}H_{22}$
$Cp_2Ta_2(B_2H_6)_2$, **1**, 306
$B_4ThC_4H_{24}$
$Th(BH_3Me)_4$, **1**, 290, 291
$B_4TlC_{18}H_{24}F_{20}O_4Te_4$
$[(C_6H_3Me_3)_2Tl][B(OTeF_5)_4]$, **1**, 536
$B_4UC_4H_{24}$
$U(BH_3Me)_4$, **1**, 290, 291
$B_4UC_{10}H_{40}P_2$
$(dmpe)U(BH_3Me)_4$, **1**, 290, 291
$B_4UC_{16}H_{56}P_2$
$(dmpe)_2U(MeBH_3)_4$, **1**, 290, 291
$B_4V_2C_{12}H_{54}OP_4$
$[V(BH_4)_2(PMe_3)_2]_2O$, **1**, 282, 283
$B_4ZrC_4H_{24}$
$Zr(BH_3Me)_4$, **1**, 287, 290
$B_4Zr_2C_{12}H_{54}P_4$
$Zr_2H_4(BH_4)_4(dmpe)_2$, **1**, 279, 280
$B_5AuC_{18}H_{23}P$
$Au(PPh_3)B_5H_8$, **1**, 326
$B_5BeC_5H_{13}$
$(CpBe)B_5H_8$, **1**, 48, 326
B_5CH_7
CB_5H_7, **1**, 219
$B_5C_2H_6Br$
$BrC_2B_5H_6$, **1**, 236
$B_5C_2H_6F$
$FC_2B_5H_6$, **1**, 227
$B_5C_2H_7$
closo-$C_2B_5H_7$, **1**, 233, 234
$B_5C_4H_{20}N$
$(Me_3N)CB_5H_{11}$, **1**, 230
$B_5C_6H_{15}$
Et_2-*closo*-$C_2B_5H_5$, **1**, 228, 232
$B_5C_6H_{16}$
[*nido*-$C_2B_5H_6Et_2]^-$, **1**, 235
$B_5C_7H_{17}$
$MeEt_2$-*closo*-$C_2B_5H_4$, **1**, 228
$B_5C_8H_6F_{12}O_2$
$[HOC(CF_3)_2C(CF_3)_2O]C_2B_5H_6$, **1**, 237
$B_5C_{12}B_{16}$
$PhEt_2$-*closo*-C_2B_5H, **1**, 228
$B_5C_{12}H_{18}P$
$(Ph_2P)B_5H_8$, **1**, 264
$B_5C_{20}H_{47}N_3$
$(Pr^i{}_2N)_3$-*closo*-$C_2B_5H_5$, **1**, 238
$B_5CoC_4H_8O_4$
$[Co(CO)_4]B_5H_8$, **1**, 326
$B_5Co_2C_{20}H_{37}$
$Cp^*{}_2Co_2B_5H_7$, **1**, 327
$B_5Co_2C_{20}H_{39}$
$Cp^*{}_2Co_2B_5H_9$, **1**, 324, 325
$B_5CuC_{26}H_{32}P_2$
$Cu[Ph_2PCH_2]_2B_5H_8$, **1**, 326
$B_5FeC_3H_5O_3$
$[B_5H_5Fe(CO)_3]^{2-}$, **1**, 327
$B_5FeC_3H_9O_3$
$(CO)_3FeB_5H_9$, **1**, 323
$B_5FeC_7H_{13}O_2$
$[CpFe_9(CO)_2]B_5H_8$, **1**, 326
$B_5FeC_{19}H_{22}O_2P$
$Cp(CO)_2FeB_5H_7(PPh_2)$, **1**, 317, 321, 327
$B_5MnC_3H_{10}O_3$
$(CO)_3MnB_5H_{10}$, **1**, 323
$B_5NiC_{22}H_{36}$
$Ni(C_3Et_2HB_2Me_2)(C_3Et_2C_3$-$B_3Me_3)$, **9**, 181
$B_5O_3C_{10}H_{10}O_9$
$(H)_2Os_3(CO)_9(CB_5H_8)$, **1**, 296, 297
$B_5OsC_{37}H_{39}OP_2$
$(Ph_3P)_2(CO)OsB_5H_9$, **1**, 323
$B_5OsPtC_{45}H_{49}ClOP_3$
$(CO)(PPh_3)_2HOs(PMe_2Ph)$-$ClPtB_5H_7$, **1**, 324
$B_5OsPtC_{64}H_{52}OP_3$
$(CO)(PPh_3)(PPh_2)Os(PPh_3)$-Pt- B_5H_7Ph, **1**, 324
$B_5RuC_{12}H_{27}$
$(C_6Me_6Ru)B_5H_9$, **1**, 324
$B_5RuC_{37}H_{39}OP_2$
$(CO)(PPh_3)_2RuB_5H_9$, **1**, 323
$B_5SiC_8H_{26}OP$
$(TMS$-$O)Bu^tC{=}PB_5H_8$, **1**, 264
$B_5Si_2C_7H_{26}P$
$(TMS)_2C{=}PB_5H_8$, **1**, 263
$B_5Si_2C_7H_{27}ClP$
$(TMS)_2CHP(Cl)B_5H_8$, **1**, 264
B_5ZnCH_{11}
$(MeZn)B_5H_8$, **1**, 326
$B_5Zr_2C_6H_{41}P_2$
$Zr_2H_3(BH_4)_5(PMe_3)_2$, **1**, 279–81
B_6CH_{12}
MeB_6H_9, **1**, 245
B_6CH_{14}
MeB_6H_{11}, **1**, 247
$B_6C_2H_9$
[*nido*-$C_2B_6H_9]^-$, **1**, 223, 230
$B_6C_2H_{10}$
nido-$C_2B_6H_{10}$, **1**, 223
$B_6C_2H_{11}$
[*arachno*-$C_2B_6H_{11}]^-$, **1**, 230
$B_6C_2H_{12}$
$C_2B_6H_{12}$, **1**, 228, 229
$B_6C_4H_4Cl_6$
$C_4H_4B_6Cl_6$, **1**, 147
$B_6C_4H_8$
$(C_2B_3H_4)_2$, **1**, 231
$B_6C_4H_{10}$
nido-$C_4B_6H_1$, **1**, 228
$B_6C_{10}H_{22}$
$C_4H_4B_6Me_6$, **1**, 147
$B_6C_{20}H_{42}$
$Me_4C_4B_6Et_6$, **1**, 229
$B_6CuFeC_{40}H_{39}O_4P_2$
$Cu(PPh_3)_2B_6H_9Fe(CO)_4$, **1**, 327, 328
$B_6FeC_3H_6O_3$
$[B_6H_6Fe(CO)_3]^{2-}$, **1**, 328
$B_6FeKC_4H_9O_4$
$K[Fe(CO)_4B_6H_9]$, **1**, 327
$B_6H_9S_2$
$[S_2B_6H_9]^-$, **1**, 269, 270
$B_6NiC_{20}H_{40}$
$Ni(Me_3Et_2C_3B_3H)_2$, **9**, 181
$B_6NiC_{21}H_{47}N$
$[Bu_4N][CpNiB_6H_6]$, **1**, 328, 330
$B_6U_2C_7H_{31}$
$[(BH_4)_3U(C_7H_7)U(BH_4)_3]^-$, **4**, 177
$B_6Zr_3H_{66}P_4$
$Zr_3H_6(BH_4)_6(PMe_3)_4$, **1**, 279, 280
$B_7C_2H_{11}$
nido-$C_2B_7H_{11}$, **1**, 230
$B_7C_2H_{13}$
arachno-$C_2B_7H_{13}$, **1**, 223, 230
$B_7C_2H_{13}S$
$C_2B_7H_{12}(SH)$, **1**, 237
$B_7C_4H_{11}$
nido-$C_4B_7H_{11}$, **1**, 228, 229
$C_4B_7H_{11}$, **1**, 229
$B_7C_4H_{13}$
Me-*nido*-$C_3B_7H_{10}$, **1**, 228
$B_7C_4H_{14}$
arachno-$(CH_2)C_3B_7H_{12}$, **1**, 228
$B_7C_5H_{14}N$
arachno-$(NCCH_2)C_3B_7H_{12}$, **1**, 228
$B_7C_6H_{17}$
$(Me)_3$-*nido*-$C_3B_7H_8$, **1**, 228
$B_7C_8H_{21}$
$Me_4C_4B_7H_9$, **1**, 231
$B_7Co_2C_{10}H_{22}S_2$
$Cp_2{}^*Co_2S_2B_7H_7$, **1**, 269
$B_7H_9S_2$
$S_2B_7H_9$, **1**, 269
$B_7NiC_{13}H_{28}$
$(C_4H_3Me_3)NiC_3B_7H_7Me_3$, **9**, 181
$B_7Ni_2C_{11}H_{18}$
$Cp_2Ni_2CB_7H_8$, **9**, 181
$B_7WC_{20}H_{48}NO_4$
$[Bu_4N][(CO)_4WB_7H_{12}]$, **1**, 329, 331
$B_8AlC_8H_{25}O$
$AlEt(OEt_2)C_2B_8H_{10}$, **1**, 259
$B_8As_2Co_2C_{10}H_{18}$
$[(Cp)Co]_2B_8H_8As_2$, **1**, 266, 267
$B_8AuC_5H_{22}NS_2$
$S_2CNEt_2AuB_8H_{12}$, **1**, 337
$B_8Au_2C_{10}H_{30}N_2S_4$
$(S_2CNEt_2)_2Au_2B_8H_{10}$, **1**, 337
$B_8Au_2C_{36}H_{76}P_2$
$[(C_6H_{11})_3PAu]_2B_8H_{10}$, **1**, 337
$B_8CH_{12}S$
CSB_8H_{12}, **1**, 425
$B_8CH_{13}N$
CNB_8H_{13}, **1**, 262
B_8CH_{14}

arachno-CB_8H_{14}, **1**, 230
$B_8C_2H_{10}$
$C_2B_8H_{10}$, **1**, 226
$B_8C_4H_{10}$
$(C_2B_4H_5)_2$, **1**, 231
$B_8C_4H_{14}$
closo-$C_2B_8H_8Me_2$, **1**, 227
$C_2B_8H_8Me_2$, **1**, 241
$B_8C_4H_{16}$
$C_2B_8H_{10}Me_2$, **1**, 227
$B_8C_4H_{21}N$
(Me_3N)-*arachno*-CB_8H_{12}, **1**, 230
$B_8C_6H_{21}N$
$Bu^tC_2NB_8H_{12}$, **1**, 262
$B_8C_{32}H_{36}$
$(Bz)_4C_4B_8H_8$, **1**, 220, 229
$B_8CoC_5H_{13}Se_2$
$(Cp)Co(B_8H_8Se_2)$, **1**, 270
$B_8Co_2C_{10}H_{22}$
$Cp_2Co_2B_8H_{12}$, **1**, 321, 335
$B_8Co_2C_{20}H_{42}$
$(Cp^*Co)_2B_8H_{12}$, **1**, 335
$B_8FeC_9H_{21}O$
$(Mes)Fe(OB_8H_{10})$, **1**, 334
$B_8FeC_9H_{22}O$
$(Bu^t{}_6H_3)FeOB_8H_{10}$, **1**, 267
$B_8H_{13}N$
$B_8H_{12}NH$, **1**, 260
$B_8IrC_6H_{26}ClP_2$
$H(PMe_3)_2Ir(B_8H_7Cl)$, **1**, 331, 337
$B_8IrC_6H_{27}P_2$
$(Me_3P)_2HIrB_8H_8$, **1**, 337
$B_8IrC_7H_{28}ClOP_2$
$(Me_3P)_2(CO)IrB_8H_{10}Cl$, **1**, 331, 336
$B_8IrC_7H_{29}OP_2$
$(Me_3P)_2(CO)IrB_8H_{11}$, **1**, 336
$B_8IrC_7H_{30}ClOP_2$
$(Me_3P)_2(CO)HIrB_8H_{11}Cl$, **1**, 332, 333
$B_8IrC_7H_{31}OP_2$
$(Me_3P)_2(CO)HIrB_8H_{12}$, **1**, 332, 333, 336
$B_8IrC_{10}H_{38}OP_3$
$(Me_3P)_3(CO)IrB_8H_{11}$, **1**, 336
$B_8IrC_{12}H_{30}N$
$Cp^*(MeNC)IrB_8H_{12}$, **1**, 332, 339
$B_8IrC_{55}H_{55}P_3$
$(PPh_3)_3IrCB_8H_{10}$, **8**, 322, 323
$B_8IrPtC_{13}H_{47}OP_4$
$(Me_3P)_2(CO)HIrB_8H_{10}Pt$-$(PMe_3)_2$, **1**, 332, 333
$B_8N_{12}N_2$
$N_2B_8H_{12}$, **1**, 262
$B_8NiC_6H_{14}$
$[(Cp)Ni]CB_8H_9$, **9**, 181
$B_8PdC_{16}H_{34}P_2$
$(Me_2PhP)_2PdB_8H_{12}$, **1**, 337
$B_8PdC_{36}H_{42}P_2$
$(PPh_3)_2PdB_8H_{12}$, **1**, 337
$B_8PdPtC_{32}H_{54}P_4$
$(PhMe_2P)_4PtPdB_8H_{10}$, **1**, 337
$B_8PtC_{16}H_{33}ClP_2$
$Cl(PhMe_2P)_2PtB_8H_{11}$, **1**, 337
$B_8PtC_{16}H_{34}P_2$
$(PMe_2Ph)_2PtB_8H_{12}$, **1**, 336, 364
$B_8PtC_{17}H_{36}OP_2$
$(OMe)(PhMe_2P)_2PtB_8H_{11}$, **1**, 337
$B_8PtC_{36}H_{42}P_2$
$(PPh_3)_2PtB_8H_{12}$, **1**, 337
$B_8Pt_2C_{16}H_{36}P_2$
$Pt_2(PMe_2Ph)_2(B_2H_5)(B_6H_9)$, **1**, 310, 327, 337
$B_8Pt_2C_{32}H_{54}P_4$
$(PhMe_2P)_4Pt_2B_8H_{10}$, **1**, 337
$B_8ReC_{24}H_{46}P_3$
$(PMe_2Ph)_3H_2ReB_8H_{11}$, **1**, 333, 336
$B_8RhC_{13}H_{27}$
$Cp^*Me[RhC_2B_8H_9]$, **1**, 339
$B_8RhC_{18}H_{36}$
$Cp^*[RhB_8H_{10}](PMe_2Ph)$, **1**, 341
$B_8RhC_{18}H_{36}P$
$Cp^*RhB_8H_{10}(PMe_2Ph)$, **1**, 336
$B_8RhC_{34}H_{35}P_2$
$(PPh_2)(PHPh_2)(Cp^*Rh)B_8H_9$, **1**, 332, 333
$B_8RuC_6H_{20}$
$(C_6H_6)RuB_8H_{14}$, **1**, 332, 333
$B_8SiC_5H_{22}$
arachno-$C_2B_8H_{13}(TMS)$, **1**, 228, 229
$B_8Si_2C_{10}H_{28}$
nido-$C_4B_8H_{10}(TMS)_2$, **1**, 229
$B_8Si_2SnC_{12}H_{32}$
$Sn[(TMS)MeC_2B_4H_4]_2$, **1**, 260
$B_8Si_4C_{16}H_{44}$
$(TMS)_4Si(SiC_2B_4H_4)_2$, **1**, 259
$B_8Si_4GeC_{16}H_{44}$
$[(TMS)_2C_2B_4H_4]_2Ge$, **1**, 260
$B_8U_3C_{27}H_{79}O_5$
$[U(BH_4)_2(THF)_5][(BH_4)_3U$-$(C_7H_7)U(BH_4)_3]$, **4**, 177
$B_9AlC_4H_{15}Cl_4P$
$ClP(MeC)_2B_9H_9AlCl_3$, **1**, 265
$B_9AsC_4H_{15}Cl$
$ClAs(MeC)_2B_9H_9$, **1**, 265
$B_9As_2CH_{13}$
$B_9H_{10}As_2Me$, **1**, 266
$B_9As_2NiC_{16}H_{31}$
$(Me_2PhP)_2Ni(As_2B_9H_9)$, **1**, 266
$B_9AuPtC_{32}H_{56}P_3$
$PtH(PEt_3)_2(C_2B_9H_{10})Au(PPh_3)$, **10**, 71
$B_9AuWC_{32}H_{37}O_2P$
$[C_2B_9H_9Me_2]W(CO)_2$-$(\equiv CTol)Au(PPh_3)$, **10**, 8, 9
$B_9C_2H_{12}$
$[$*nido*-$C_2B_9H_{12}]^-$, **1**, 230
$B_9C_2H_{12}S_2$
$MeCNS_2B_7H_9$, **1**, 269, 270
$B_9C_2H_{13}N$
$(NH_2)C_2B_9H_{11}$, **1**, 230
$B_9C_4H_{15}ClP$
$ClP(MeC)_2B_9H_9$, **1**, 265
$B_9C_4H_{20}N$
$(Me_3N)CB_9H_{11}$, **1**, 240
$B_9C_5H_{22}N$
$(Me_3N)(Me)$-*nido*-CB_9H_{10}, **1**, 229
$B_9C_5H_{24}N$
$(Me_3N)Me$-*arachno*-CB_9H_{12}, **1**, 230
$B_9C_6H_{25}NS$
$[Me_4N][C_2B_9H_{12}]^-$, **1**, 230
$B_9C_{11}H_{22}$
$[$*nido*-$C_2B_9H_9Me_2(Bz)]^-$, **1**, 238
$B_9C_{16}H_{50}N$
$[Bu_4N][B_9H_{14}]$, **1**, 329
$B_9CoC_4H_{19}$
$[EtC_2B_4H_4]Co[B_5H_{10}]$, **1**, 323
$B_9CoC_5H_{15}N$
$(Cp)CoNHB_9H_9$, **1**, 263
$B_9CoRhC_{30}H_{48}O_2P$
$Co(CO)(C_4Me_4)Rh(CO)(PPh_3)$-$(C_2B_9H_{11})$, **10**, 64
$B_9Co_2WC_{18}H_{22}O_6$
$[[C_2B_9H_9Me_2]WCo_2(CO)_6$-$(\equiv CTol)]^-$, **10**, 8, 9
$B_9CuC_{36}H_{44}P_2$
$Cu(PPh_3)_2(B_9H_{14})$, **1**, 339
$B_9FeC_7H_{19}S$
$(PhMe)FeSB_9H_{11}$, **1**, 268
$B_9FeC_7H_{21}$
$(PhMe)[FeB_9H_{13}]$, **1**, 341, 342
$B_9FeC_9H_{25}$
$(C_6Me_3H_3)[FeB_9H_{13}]$, **1**, 341, 342
$B_9FeWC_{17}H_{22}O_5$
$[W(CO)_2(C_2B_9H_9Me_2)$-$(CTol)Fe(CO)_3]^-$, **10**, 43
$B_9FeWC_{18}H_{22}O_6$
$[W(CO)_2(C_2B_9H_8Me_2)(CHTol)$-$(CO)Fe(CO)_3]^-$, **10**, 43
$B_9GaC_5H_{19}$
$MeEtGaC_2B_9H_{11}$, **1**, 532
$B_9GeC_{23}H_{33}Cl$
$Ph_3PMe[ClGe(B_9H_9C_2Me_2)]$, **1**, 260
B_9H_9P
$[PB_9H_9]^-$, **1**, 264
$B_9H_9Se_2$
$B_9H_9Se_2$, **1**, 270
$B_9H_{10}N$
NHB_9H_9, **1**, 418
$B_9H_{10}P_2$
$[P_2B_9H_{10}]^-$, **1**, 264
$B_9H_{12}N$
$B_9H_{11}NH$, **1**, 260
$B_9H_{13}N$
$[B_9H_{12}NH]^-$, **1**, 260
$B_9IrC_6H_{28}P_2$
$(H)(PMe_3)_2[IrB_9H_9]$, **1**, 347
$B_9IrC_6H_{32}P_2$
$(H)(PMe_3)_2[IrB_9H_{13}]$, **1**, 347
$B_9IrC_8H_{25}Cl$
$(C_8H_{12})Cl[IrB_9H_{13}]$, **1**, 342
$B_9IrC_{10}H_{24}S$
$(Cp^*)IrSB_9H_9$, **1**, 269
$B_9IrC_{10}H_{25}ClS$
$(Cp^*)IrSB_9H_{10}Cl$, **1**, 269

$B_9IrC_{10}H_{26}ClN$
  $Cl(Cp*)IrNHB_9H_{10}$, **1**, 262, 263
$B_9IrC_{10}H_{28}$
  $Cp*[IrB_9H_{13}]$, **1**, 339, 341, 342
$B_9IrC_{14}H_{32}N_2$
  $Cp*(MeNC)[IrB_9H_{11}](MeNC)$, **1**, 339
$B_9IrC_{26}H_{38}P_2$
  $(H)(PPh_3)(Ph_2PC_6H_4)[IrB_9H_8]$, **1**, 347
$B_9IrC_{26}H_{42}P_2$
  $(PPh_3)(Ph_2PC_6H_4)H[IrB_9H_{12}]$, **1**, 347
$B_9IrC_{36}H_{38}P_2$
  $(PPh_3)[(Ph_2P)C_6H_4](H)IrB_9H_8$, **1**, 345, 346
$B_9IrC_{36}H_{42}P_2$
  $(H)(PPh_3)(PPh_2C_6H_4[IrB_9H_{12}]$, **1**, 342–5
$B_9IrC_{36}H_{43}P$
  $(PPh_3)_2[IrB_9H_{13}]$, **1**, 342
$B_9IrC_{36}H_{43}P_3$
  $(PPh_3)(PH_3)_2(Ph_2PC_6H_4)H$-$[IrB_9H_7]$, **1**, 347, 348
$B_9IrC_{36}H_{44}P$
  $(H)(PPh_3)_2[IrB_9H_{13}]$, **1**, 347
$B_9IrC_{36}H_{55}$
  $(PPh_3)_3[IrB_9H_{10}]$, **1**, 343, 344
$B_9IrC_{38}H_{39}P_2$
  $(OEt)(PPh_3Cl(Ph_2P)C_6H_4$-$[IrB_9H_5]$, **1**, 343
$B_9IrC_{54}H_{53}P_2$
  $(PPh_3)_2(Ph_2P)C_6H_4[IrB_9H_9]$, **1**, 343, 345
$B_9IrC_{54}H_{54}P_2$
  $(H)(PPh_3)_2(PPh_2C_6H_4[IrB_9$-$H_{10}])$, **1**, 343
$B_9IrC_{54}H_{55}P_3$
  $(PPh_3)_3IrB_9H_{10}$, **8**, 322, 323
$B_9IrRhC_{20}H_{41}$
  $Cp*_2(H)[RhIrB_9H_{10}]$, **1**, 342
$B_9MoC_{10}H_{17}$
  $CpMo(C_3H_3)C_2B_9H_9$, **1**, 321
$B_9MoWC_{21}H_{29}O_2$
  $MoW(CO)_2(C_7H_7)(CTol)(C_2$-$B_9H_9Me_2)$, **10**, 33
$B_9NiC_{12}H_{19}N_2$
  $Ni(C_2B_9H_{11})(bipy)$, **9**, 182
$B_9NiC_{16}H_{29}Cl_2P_2$
  $(PMe_2Ph)_2Cl_2[NiB_9H_7]$, **1**, 349
$B_9OsC_{12}H_{31}$
  $(C_6Me_6)[OsB_9H_{12}]$, **1**, 341, 342
$B_9PdSb_2C_{16}H_{31}$
  $(Me_2PhP)_2PdSb_2B_9H_9$, **1**, 267
$B_9PtC_{25}H_{44}OP_3$
  $(PMe_2Ph)_3Pt_2B_9H_8(OMe)$, **1**, 349
$B_9PtRhC_{33}H_{57}OP_3$
  $Rh(PPh_3)(C_2B_9H_{11})(H)(CO)Pt$-$(PEt_3)_2$, **10**, 64
$B_9PtWC_{18}H_{45}O_2P_2$
  $W(CO)_2(Me_2C_2B_9H_9)Pt(PEt_3)_2$, **10**, 32
$B_9RhC_{10}H_{28}$
  $Cp*[RhB_9H_{13}]$, **1**, 332, 338, 339, 341, 342, 347
$B_9RhC_{12}H_{31}$
  $(C_6Me_6)[RuB_9H_{13}]$, **1**, 347
$B_9RhC_{14}H_{32}N_2$
  $Cp*(MeNC)_2(RhB_9H_{11})$, **1**, 338, 339
$B_9RhC_{15}H_{36}NO$
  $(C_6Me_6)(OMe)RuNMeC(H)$-B_9H_{11}, **1**, 338
$B_9RhC_{18}H_{37}$
  $Cp*[RhB_9H_{11}](PMe_2Ph)$, **1**, 341
$B_9RhC_{18}H_{37}P$
  $Cp*[RhB_9H_{11}](PMe_2Ph)$, **1**, 347
$B_9RhC_{26}H_{40}N_2$
  $(MeC_6H_4NC)_2(Cp*RhB_9H_{11})$, **1**, 338
$B_9RhC_{26}H_{44}P_2$
  $Cp*[RhB_9H_7](PMe_2Ph)_2$, **1**, 341, 347
$B_9RhC_{36}H_{40}P_2S$
  $(Ph_3P)_2RhSB_9H_{10}$, **1**, 268
$B_9RhWC_{50}H_{52}O_2P_2$
  $[C_2B_9H_9Me_2]W(CO)_2$-$(\equiv CTol)Rh(PPh_3)_2$, **10**, 8, 9
$B_9Rh_2C_{16}H_{29}P$
  $Ph(Cp*)Rh_2PB_9H_9$, **1**, 265
$B_9Rh_2C_{20}H_{38}ClS$
  $S[(Cp*)Rh]_2ClB_9H_8$, **1**, 268
$B_9Rh_2C_{54}H_{52}Cl_2P_3S$
  $(PPh_3)_2Cl_2Ph_2PC_6H_4Rh_2SB_9H_8$, **1**, 268
$B_9RuC_{10}H_{23}$
  $(MeC_6H_4Pr^i)[RuB_9H_9]$, **1**, 346
$B_9RuC_{10}H_{24}ClS$
  $(cymene)RuSB_9H_{10}Cl$, **1**, 269
$B_9RuC_{10}H_{24}N$
  $(MeC_6H_4Pr^i)RuNHB_9H_9$, **1**, 263
$B_9RuC_{12}H_{27}$
  $(C_6Me_6)RuB_9H_9$, **1**, 338, 344, 347
$B_9RuC_{12}H_{31}$
  $(C_6Me_6)[RuB_9H_{13}]$, **1**, 338, 340–2, 344, 347
$B_9RuC_{13}H_{33}O$
  $(C_6Me_6)(OMe)RuB_9H_{12}$, **1**, 338
$B_9RuC_{14}H_{31}O$
  $(C_6Me_6)[RuB_9H_8](OEt)$, **1**, 346
$B_9RuC_{15}H_{37}N$
  $(C_6Me_6)Ru(NMe_2CH_2)$, **1**, 338
$B_9RuC_{20}H_{40}P$
  $(C_6Me_6)[RuB_9H_{11}](PMe_2Ph)$, **1**, 340, 341
$B_9RuC_{22}H_{45}NP$
  $(NMe_2)(PMe_2Ph)(C_6Me_6)$-$RuB_9H_{10}$, **1**, 338
$B_9RuC_{30}H_{44}P$
  $(C_6Me_6)[RuB_9H_{11}](PPh_3)$, **1**, 340
$B_9RuC_{54}H_{53}Cl$
  $(PPh_3)HCl[RuB_9H_7](PPh_3)_2$, **1**, 346
$B_9RuC_{54}H_{53}P_3$
  $(PPh_3)HCl[RuB_9H_7](PPh_3)_2$, **1**, 345, 346
$B_9Ru_2C_{20}H_{37}S$
  $[(cymene)Ru]_2SB_9H_9$, **1**, 269
$B_9ScSi_2C_{19}H_{45}$
  $[Cp*Sc(C_2B_9H_{11})CH(TMS)_2]^-$, **4**, 70
$B_9SiC_{10}H_{26}Br_5$
  $(Pr^i{}_3Si)Br_5CB_9H_5$, **2**, 5
$B_9SiC_{12}H_{29}Br_5N$
  $(Pr^i{}_3Si)(NCMe)Br_5CB_9H_5$, **2**, 5
$B_9SnC_6H_{21}$
  $Me_2Sn[(MeC)_2B_9H_9]$, **1**, 260
$B_9SnC_{23}H_{33}Cl$
  $Ph_3PMe[ClSn(B_9H_9C_2Me_2)]$, **1**, 260
$B_9TlC_2H_{11}$
  [*closo*-$TlC_2B_9H_{11}]^-$, **1**, 517
  $TlC_2B_9H_{11}$, **1**, 258
$B_9Tl_2C_4H_{15}$
  $Tl[Tl(MeC)_2B_9H_9]$, **1**, 258
$B_9Tl_3C_2H_{12}O$
  $Tl_3(C_2B_9H_{12})O$, **1**, 230
$B_9ZrC_{15}H_{29}$
  $ZrMeCp'_2(C_2B_9H_{12})$, **4**, 592
$B_9ZrC_{17}H_{37}O$
  $Cp*(C_2B_9H_{11})Zr(Me)(THF)$, **4**, 572
$B_9ZrC_{25}H_{49}$
  $ZrMe(C_5Me_4Et)_2(C_2B_9H_{12})$, **4**, 600
$B_{10}AsCH_{15}$
  $B_{10}H_{12}AsMe$, **1**, 266
$B_{10}AsC_4H_{24}N$
  $NMe_4[B_{10}H_{12}As]$, **1**, 265
$B_{10}AsC_6H_{16}$
  $[PhAsB_{10}H_{11}]^-$, **1**, 266
$B_{10}AsC_6H_{17}$
  $PhAsB_{10}H_{12}$, **1**, 266
$B_{10}AsH_{10}P$
  $PAsB_{10}H_{10}$, **1**, 264
$B_{10}AsH_{12}$
  $[B_{10}H_{12}As]^-$, **1**, 266
$B_{10}As_2H_{10}$
  $B_{10}H_{10}As_2$, **1**, 266
$B_{10}As_2SnC_{48}H_{52}Cl_2$
  $(Ph_4As)_2B_{10}H_{12}SnCl_2$, **1**, 259, 260
$B_{10}As_2SnC_{50}H_{58}Cl_2$
  $(Ph_4As)_2B_{10}H_{12}SnMeCl_2$, **1**, 259
$B_{10}AuC_3H_{22}P$
  $(AuPMe_3)B_{10}H_{13}$, **1**, 352–4
$B_{10}AuC_{15}H_{45}P$
  $[[AuP(C_5H_{11})_3]B_{10}H_{12}]^-$, **1**, 352, 354
$B_{10}BiH_{10}P$
  $PBiB_{10}H_{10}$, **1**, 264
$B_{10}Bi_2H_{10}$
  $Bi_2B_{10}H_{10}$, **1**, 267
$B_{10}CH_{15}P$
  $MePB_{10}H_{12}$, **1**, 264
$B_{10}C_2H_{11}$
  $C_2B_{10}H_{11}$, **1**, 226
$B_{10}C_2H_{12}$

$C_2B_{10}H_{12}$, **1**, 223, 226, 227, 233,
$B_{10}C_2H_{20}$
$(MeB_5H_7)_2$, **1**, 246
$B_{10}C_3H_{12}$
CH-*closo*-$C_2B_{10}H_{11}$, **1**, 219
$B_{10}C_3H_{13}$
$[C_2B_{10}H_{11}]$=CH_2, **1**, 227
$B_{10}C_5H_{19}N$
$(OC)(Me_3N)CB_{10}H_{10}$, **1**, 239
$B_{10}C_7H_{16}N$
$C_2B_9H_{11}BNCp$, **1**, 230
$B_{10}C_8H_{16}$
$C_2B_{10}H_{11}Ph$, **1**, 241
$B_{10}C_{12}H_{23}P$
$(PPh_2)B_{10}H_{13}$, **1**, 263
$B_{10}CaC_{10}H_{24}N_4$
$(MeCN)_4CaB_{10}C_2H_{12}$, **1**, 116
$B_{10}CoC_6H_{18}P$
$(Cp)MeCoPB_{10}H_{10}$, **1**, 265
$B_{10}CoC_{13}H_{28}NO_3$
$[BzNMe_3][(CO)_3CoB_{10}H_{12}]$, **1**, 351
$B_{10}CoWC_{24}H_{37}O_2$
$W(C_2B_{10}H_9Me_2)(CO)_2$(CTol)-CoCp*, **10**, 43
$B_{10}Co_2C_{13}H_{28}O_5S_2$
$(CO)_4(CO)(SEt_2)_2Co_2B_{10}H_8$, **1**, 361
$B_{10}FeC_7H_{18}S$
$(PhMe)FeSB_{10}H_{10}$, **1**, 268
$B_{10}FeC_7H_{20}O_2$
$CpFe(CO)_2(B_5H_7)(B_5H_8)$, **1**, 358, 361
$B_{10}FeWC_{17}H_{23}O_5$
$W(CO)_2(C_2B_{10}H_{10}Me_2)$-(CTol)$Fe(CO)_3$, **10**, 43
$B_{10}GaC_6H_{11}F_6O_4$
$C_2B_{10}H_{11}Ga(O_2CCF_3)_2$, **1**, 227
$B_{10}GeC_2H_{18}$
$Me_2GeB_{10}H_{12}$, **1**, 259
$B_{10}H_{10}P_2$
$P_2B_{10}H_{10}$, **1**, 264
$B_{10}H_{12}P$
$[PB_{10}H_{12}]^-$, **1**, 264
$B_{10}H_{12}Te$
$B_{10}H_{12}Te$, **1**, 270
$B_{10}H_{13}N$
$NHB_{10}H_{12}$, **1**, 261
$B_{10}H_{14}N_4$
$B_{10}H_{12}(N_3)NH_2$, **1**, 261
$B_{10}InC_6H_{19}Cl_2O$
$C_2B_{10}H_{11}InCl_2(THF)C_4H_8O$, **1**, 227
$B_{10}IrC_{29}H_{33}NOP$
$Ir(PhC_2B_{10}H_{10})$(CO)(MeCN)-(PPh_3), **8**, 351
$B_{10}IrC_{33}H_{35}NOP$
$Ir(PhC_2B_{10}H_{10})$(CO)(PhCN)-(PPh_3), **8**, 320
$B_{10}LiH_{15}$
$Li[(B_5H_7)(B_5H_8)]$, **1**, 358
$B_{10}MgC_{14}H_{34}O_4$
$Mg(MeC_2B_{10}H_{10})_2(dioxane)_2$, **1**, 88
$B_{10}Mg_2C_4H_{14}Br_2$
$C_2B_{10}H_{10}(CH_2MgBr)_2$, **1**, 242
$B_{10}OsC_{40}H_{46}OP_2$
$(OEt)_2(PPh_3)(Ph_2P\overline{C_6H_4)Os}$-$B_{10}H_7]$, **1**, 355
$B_{10}PdC_{18}H_{25}PSe$
$(PPh_3)PdSeB_{10}H_{10}$, **1**, 270
$B_{10}PdC_{36}H_{39}ClP_2Te$
$Cl(PPh_3)PdTeB_{10}H_9(PPh_3)$, **1**, 271
$B_{10}PtC_{12}H_{40}P_2Te$
$(Et_3P)_2TePtB_{10}H_{10}$, **1**, 428
$B_{10}PtC_{13}H_{32}NPS_2$
$(SCSNEt_2)(PMe_2Ph)PtB_{10}H_{11}$, **1**, 352
$B_{10}PtC_{16}H_{32}P_2$
$(PMe_2Ph)_2PtB_{10}H_{10}$, **1**, 351
$B_{10}PtC_{16}H_{33}ClP_2$
$Cl[PMe_2Ph]_2PtB_{10}H_{11}$, **1**, 351
$B_{10}PtC_{16}H_{33}P_2$
$(PMe_2Ph)_2B_{10}H_{12}$, **1**, 362
$B_{10}PtC_{16}H_{34}P_2$
$[(PMe_2Ph)_2PtB_{10}H_{12}]$, **1**, 351
$B_{10}PtC_{18}H_{39}ClP_2$
$Cl[PMe_2Ph]_2(SMe_2)PtB_{10}H_{11}$, **1**, 351
$B_{10}PtC_{22}H_{37}P_2$
$Ph(PMe_2Ph)_2PtB_{10}H_{10}$, **1**, 265
$B_{10}PtC_{36}H_{40}P_2Se$
$(PPh_3)_2SePtB_{10}H_{10}$, **1**, 271
$B_{10}PtC_{36}H_{57}P_3Se$
$[PtH(PPh_3)_3][SeB_{10}H_{11}]$, **1**, 271
$B_{10}Pt_2C_{32}H_{53}ClP_4$
$(PMe_2Ph)_4Pt_2B_{10}H_9Cl$, **1**, 351, 362, 365
$B_{10}RhC_{10}H_{25}$
$Cp^*RhB_{10}H_{10}$, **1**, 358
$B_{10}RhC_{10}H_{28}Cl$
$Cp^*RhB_{10}H_{13}Cl$, **1**, 342, 351
$B_{10}RhC_{11}H_{27}O$
$Cp^*RhB_{10}H_9(OMe)$, **1**, 358
$B_{10}RhC_{14}H_{34}N$
$Cp^*(NEt_2)RhB_{10}H_9$, **1**, 358
$B_{10}RhC_{16}H_{40}NO$
$(Cp^*)Rh(NEt_3)OB_{10}H_{10}$, **1**, 267
$B_{10}RhC_{17}H_{34}ClOP_2$
$(PMe_2Ph)_2(H)Cl(OMe)$-$RhB_{10}H_8$, **1**, 357
$B_{10}RhC_{18}H_{35}ClOP$
$Cp^*[RhOB_{10}H_9Cl(PMe_2Ph)]$, **1**, 267, 351
$B_{10}RhC_{18}H_{37}O_2P_2$
$(PMe_2Ph)_2(H)(OMe)_2RhB_{10}H_8$, **1**, 357
$B_{10}RhC_{19}H_{39}Cl_3P$
$Cp^*[RhB_{10}H_{11}Cl(PMe_2Ph)]$-$CH_2Cl_2$, **1**, 351
$B_{10}RhC_{36}H_{42}NP$
$[(Ph_3P)_2RhH]HNB_{10}H_{10}$, **1**, 262
$B_{10}RuC_{10}H_{24}$
$(MePr^iC_6H_4)RuB_{10}H_{10}$, **1**, 357–9
$B_{10}RuC_{12}H_{31}$
$(C_6Me_6)[RuB_{10}H_{13}]^-$, **1**, 350, 351
$B_{10}RuC_{22}H_{29}O_5P$
$(PPh_3)(MeCO_2)_2RuB_{10}H_7(OH)$, **1**, 357
$B_{10}RuC_{40}H_{48}O_2P_2$
$(PPh_3)_2RuB_{10}H_8(OEt)_2$, **1**, 355
$B_{10}RuC_{40}H_{48}P_2$
$(PPh_3)_2(OEt)_2RuB_{10}H_8$, **1**, 353, 355
$B_{10}Ru_2C_{28}H_{51}Cl_2O_2P_3$
$(PMe_3)_2ClRuClRu(PPh_3)B_{10}H_8$-$(OEt)_2$, **1**, 355
$B_{10}Ru_2C_{28}H_{58}O_2$
$[(C_6Me_6)_2Ru_2H_4]RuB_{10}H_8$-$(OEt)_2$, **7**, 796, 797
$B_{10}Ru_2C_{30}H_{56}O_4$
$[(C_6Me_6)_2RuH_2(CO)_2]RuB_{10}H_8$-$(OEt)_2$, **7**, 796, 797
$B_{10}Ru_2C_{42}H_{48}O_6P_2$
$(PPh_3)_2Ru(MeCO_2)_3(H)_2$-$RuB_{10}H_7$, **1**, 357
$B_{10}Ru_3C_{28}H_{58}O_2$
$(C_6Me_6)_2Ru_2H_4[RuB_{10}H_8$-$(OEt)_2]$, **1**, 357, 360
$B_{10}Ru_3C_{29}H_{58}Cl_2O_2$
$Ru_3H_2(C_6Me_6)_2(CH_2Cl_2)$-$[B_{10}H_8(OEt)_2]$, **7**, 710, 723
$B_{10}Ru_3C_{30}H_{56}O_4$
$(C_6Me_6)_2Ru_2H_2(CO)_2RuB_{10}H_8$-$(OEt)_2$, **1**, 358, 360
$B_{10}SbH_{10}P$
$PSbB_{10}H_{10}$, **1**, 264
$B_{10}Sb_2H_{10}$
$Sb_2B_{10}H_{10}$, **1**, 267
$B_{10}Si_2C_2H_{18}$
$Me_2Si_2B_{10}H_{12}$, **1**, 259
$B_{10}Si_2C_{10}H_{20}S$
$(Me_2S)[(TMS)_2CH]CB_{10}H$, **1**, 227
$B_{10}SrC_8H_{21}N_3$
$(MeCN)_3SrC_2B_{10}H_{12}$, **1**, 116
$B_{10}TlC_6H_{11}F_6O_4$
$C_2B_{10}H_{11}Tl(O_2CCF_3)_2$, **1**, 227
$B_{10}YbC_{16}H_{38}N_7$
$[(MeCN)_6Yb][(H_2)B_{10}H_{12}]\cdot$2MeCN, **1**, 349, 350
$B_{10}ZnC_8H_{36}N_2$
$[NMe_4]_2[ZnB_{10}H_{12}]$, **1**, 352
$B_{11}AgIrC_{38}H_{42}ClOP_2$
$IrCl(CO)(PPh_3)_2Ag(B_{11}CH_{12})$, **10**, 63
$B_{11}AlAs_2C_{49}H_{54}$
$(AsPh_4)_2[B_{11}H_{11}AlMe]$, **1**, 258, 271
$B_{11}AlNa_2CH_{14}$
$Na_2[B_{11}H_{11}AlMe]$, **1**, 258
$B_{11}AsH_{11}$
$[B_{11}H_{11}As]^-$, **1**, 266
$B_{11}BiH_{11}$
$[B_{11}H_{11}Bi]^-$, **1**, 267
$B_{11}CH_{14}N$
$MeNB_{11}H_{11}$, **1**, 261
$B_{11}CH_{14}P$
$MePB_{11}H_{11}$, **1**, 264
$B_{11}C_3H_{19}P$
$(Me_3N)PB_{11}H_{10}$, **1**, 264
$B_{11}C_3H_{21}N$

$CB_{11}H_{11}[NH(Me)_2]$, **1**, 228

$B_{11}C_4H_{20}N$

(Me_3N)-*closo*-$[CB_{11}H_{11}]$, **1**, 228

$B_{11}C_6H_{16}P$

$PhPB_{11}H_{11}$, **1**, 264

$B_{11}C_6H_{27}N$

$Et_3NH[NB_{11}H_{11}]$, **1**, 261

$B_{11}C_{10}H_{35}N_2$

$[Et_3NMe][(Me_2N)CB_{11}H_{11}]$, **1**, 228

$B_{11}C_{10}H_{35}N_2S$

$[Et_3NMe][(Me_2N)(HS)$-*closo*-$CB_{11}H_{10}$, **1**, 229

$B_{11}C_{11}H_{27}N$

$(Me_3N)(RhCh_2)CB_{11}H_{11}$, **1**, 228

$B_{11}C_{20}H_{32}N_2OP$

$[N(PPh_3)][MeNB_{11}H_{11}(OMe)]$, **1**, 261

$B_{11}CsRhC_{37}H_{43}OP_2$

$Cs[Rh(CO)(PPh_3)_2]B_{11}H_{13}$, **1**, 363

$B_{11}FeC_3H_{11}O_3$

$[B_{11}H_{11}Fe(CO)_3]^{2-}$, **1**, 363

$B_{11}GeH_{11}$

$[B_{11}H_{11}Ge]^{2-}$, **1**, 259

$B_{11}H_{11}P$

$[PB_{11}H_{11}]^-$, **1**, 264

$B_{11}H_{12}N$

$HNB_{11}H_{11}$, **1**, 261

$B_{11}NaReC_5H_{14}O_5$

$Na[Re(CO)_5]B_{11}H_{14}$, **1**, 363

$B_{11}Na_2H_{13}$

$Na_2[B_{11}H_{13}]$, **1**, 258

$B_{11}PbH_{11}$

$[B_{11}H_{11}Pb]^{2-}$, **1**, 259

$B_{11}PdC_{36}H_{41}F_4OP_2Te$

$[(H_2O)(PPh_3)PdTeB_{10}H_9$-$(PPh_3)]BF_4$, **1**, 271

$B_{11}ReC_5H_{14}O_5$

$[Re(CO)_5]B_{11}H_{14}$, **1**, 363

$B_{11}RhC_{37}H_{44}OP_2$

$[Rh(CO)(PPh_3)_2]B_{11}H_{14}$, **1**, 363

$B_{11}SbH_{11}$

$[B_{11}H_{11}Sb]^-$, **1**, 267

$B_{11}SnC_{21}H_{34}Cl_2$

$[MeSnB_{11}H_{11}]PPh_3Me\cdot CH_2Cl_2$, **1**, 260

$B_{11}SnH_{11}$

$[B_{11}H_{11}Sn]^{2-}$, **1**, 259

$B_{11}ZrC_{12}H_{30}$

$ZrMe(CB_{11}H_{12})Cp^*$, **4**, 615

$B_{11}ZrC_{13}H_{33}$

$ZrMe_2(CB_{11}H_{12})Cp^*$, **4**, 615

$B_{11}ZrC_{14}H_{29}$

$ZrMe(CB_{11}H_{12})Cp'_2$, **4**, 591, 600

$B_{11}ZrC_{16}H_{38}O$

$[ZrMe(THF)Cp^*][CB_{11}H_{12}]$, **4**, 615

$B_{11}ZrC_{18}H_{29}$

$Zr(Bz)(CB_{11}H_{12})Cp_2$, **4**, 591, 600

$B_{11}ZrC_{18}H_{37}O$

$[ZrMe(THF)Cp_{3'}][CB_{11}H_{12}]$, **4**, 591

$B_{11}ZrC_{22}H_{36}N$

$[Zr(Bz)(MeCN)Cp_{2'}][CB_{11}H_{12}]$, **4**, 591

$B_{12}C_2H_{10}O_2$

$(OC)_2B_{12}H_{10}$, **1**, 245

$B_{12}Pt_2C_{16}H_{40}P_2$

$Pt_2(PMe_2Ph)_2(B_6H_9)_2$, **1**, 327, 328, 337, 363

$B_{12}Pt_3C_{32}H_{62}P_4$

$Pt_2(B_6H_9)(B_6H_8)[PtH(PMe_2$-$Ph)_2](PMe_2Ph)_2$, **1**, 327

$B_{12}RuC_{24}H_{45}P_3$

$(PMe_2Ph)_3RuB_{12}H_{12}$, **1**, 363

$B_{12}RuC_{42}H_{56}ClNP_2$

$P(PPh_3)_2ClRuB_{12}H_{11}(NEt_3)$, **1**, 363

$B_{13}FeC_3H_{13}O_3$

$[B_{13}H_{13}Fe(CO)_3]^{2-}$, **1**, 363

$B_{14}Pt_3C_{32}H_{60}P_4$

$(PMe_2Ph)_4Pt_3B_{14}H_{16}$, **1**, 327, 329, 364

$B_{16}AlC_4H_{16}$

$Al(C_2B_6H_8)_2$, **1**, 259

$B_{16}AlC_4H_{20}$

$Al(C_2B_8H_0)_2$, **1**, 259

$B_{16}PtC_{16}H_{40}P_2$

$(PMe_2Ph)PtB_{16}H_{18}(PMe_2Ph)$, **1**, 327, 330, 364

$B_{16}Pt_2C_{31}H_{55}P_2$

$(PMe_2Ph)_2Pt_2B_{16}H_{15}(C_6H_4Me)$-$(PMe_2Ph)$, **1**, 364

$B_{16}Pt_2C_{31}H_{55}P_3$

$(PMe_2Ph)_2Pt_2B_{16}H_{15}(C_6H_4Me)$-$(PMe_2Ph)$, **1**, 327, 329

$B_{17}CoC_5H_{26}$

$CpCoB_{17}H_{21}$, **1**, 365

$B_{17}Rh_2C_{20}H_{49}$

$(Cp^*Rh)_2B_{17}H_{19}$, **1**, 365, 366

$B_{18}Al_4C_8H_{32}$

$Al[(H)_2(AlEt_2)AlC_2B_9H_9]$-$[AlC_2B_9H_{11}]$, **1**, 258

$B_{18}As_4CoH_{18}$

$[Co(As_2B_9H_9)_2]^-$, **1**, 266

$B_{18}C_4H_{22}$

$(nido$-$C_2B_8H_{11})C_2B_{10}H_{11}$, **1**, 231

$B_{18}C_{10}H_{40}N_2S_4$

$[Me_3NH]_2[C_2B_9H_{10}(S)_2]_2$, **1**, 240

$B_{18}C_{16}H_{48}N_2S_2$

$[Me_4N]_2[(CH_2S)_2C_2B_9H_{10}]_2$, **1**, 230

$B_{18}CoThC_{24}H_{53}$

$[Cp^*_2ThH][Co(B_9C_2H_{11})_2]$, **4**, 176

$B_{18}FeTh_2C_{46}H_{88}$

$[Cp^*_2ThMe]_2[Fe(B_9C_2H_{11})_2]$, **4**, 174

$B_{18}Ga_3C_4H_{22}$

$Ga(GaC_2B_9H_{11})_2$, **1**, 259

$B_{18}GeC_{46}H_{66}$

$(Ph_3PMe)_2[Ge(B_9H_9C_2Me_2)_2]$, **1**, 260

$B_{18}NiC_4H_{22}$

$Ni(C_2B_9H_{11})_2^-$, **9**, 181

$B_{18}NiC_6H_{26}O_2$

$Ni[(MeO)C_2B_9H_{10})_2]$, **9**, 181

$B_{18}Ni_3C_4H_{22}$

$Ni(C_2B_9H_{11}Ni)_2$, **9**, 181

$B_{18}PtC_8H_{31}P$

$(PMe_2Ph)PtB_{18}H_{20}$, **1**, 365, 366

$B_{18}Rh_2C_{16}H_{50}S_2$

$Cp^*[RhB_9(SMe_2)H_{10}]RhB_9H_7$-$(SMe_2)_2$, **1**, 347

$B_{18}Rh_2C_{16}H_{50}S_3$

$Cp^*RhB_9(SMe_2)H_{10}RhB_9H_7$-$(SMe_2)_2$, **1**, 341

$B_{18}Rh_2C_{20}H_{54}O$

$O[Cp^*[RhB_9H_{12}]]_2$, **1**, 342

$B_{18}SiC_{14}H_{32}N$

$(C_2B_9H_{11})(C_2B_9H_{11})Si$-$(C_5H_5N)_2$, **1**, 259

$B_{18}Si_3C_4H_{22}$

$Si(SiC_2B_9H_{11})_2$, **1**, 259

$B_{18}Zr_2C_{25}H_{54}$

$[Cp^*(C_2B_9H_{11})Zr]_2(CH_2)$, **4**, 572

$B_{20}AsC_6H_{26}Cl$

$(C_2B_{10}H_{10}Me)_2AsCl$, **1**, 241

$B_{20}As_2CoH_{20}$

$Co(B_{10}H_{10}As)_2^{3-}$, **1**, 266

$B_{20}AuH_{24}$

$[(B_{10}H_{12})Au(B_{10}H_{12})]^-$, **1**, 352

$B_{20}AuH_{25}$

$[(B_{10}H_{12})Au(B_{10}H_{13})]^{2-}$, **1**, 352

$B_{20}Au_2C_6H_{30}P_2$

$[B_{10}H_{12}AuPMe_3]_2$, **1**, 352, 353

$B_{20}Au_2C_6H_{42}P_2$

$[(Me_3P)_2Au][(B_{10}H_{12})Au$-$(B_{10}H_{12})]$, **1**, 352, 354

$B_{20}Au_3C_{12}H_{61}P_4$

$[(Me_3P)_2Au]_2[(B_{10}H_{12})Au$-$(B_{10}H_{13})]$, **1**, 352, 354

$B_{20}Au_6C_{24}H_{84}P_4$

$(H_{12}B_{10}Au)(AuPEt_3)_4$-$(AuB_{10}H_{12})$, **1**, 352, 353

$B_{20}C_8H_{28}ClO_4P$

$[C_2B_{10}H_{10}(OH_2O)_2]_2PCl$, **1**, 241

$B_{20}C_{80}H_{83}P_4$

$[(Ph_3PMe)_2]_2(C_2B_{10}H_{11})_2$, **1**, 221

$B_{20}CuC_6H_{34}N_5$

$[Et_3NH][Cu(B_{10}H_9N_2)_2]$, **1**, 362

$B_{20}FeC_{30}H_{62}N$

$[CpFeC_5H_4CH_2N(Me)_2Et]_2$-$[B_{20}H_{18}]$, **1**, 365

$B_{20}MgC_{14}H_{42}O_4$

$(C_2B_{10}H_{10}Me)_2Mg2(C_4H_8O_2)$, **1**, 243

$B_{20}PdC_{22}H_{59}N_3$

$[BzNMe_3]_2[(B_{10}H_{12})_2Pd]\cdot MeCN$, **1**, 351

$B_{20}Pd_2C_{12}H_{32}Br_{20}N_2$

$[Pd(C_3H_5)(NCMe_2)]_2[B_{10}Br_{10}]$, **9**, 354

$B_{20}PtC_{20}H_{56}N_2$

$[BzNMe_3]_2[(B_{10}H_{12})_2Pt]$, **1**, 351

$B_{24}C_{12}N_{24}$

$C_{12}B_{24}N_{24}$, **1**, 132
$B_{30}C_9H_{39}O_4P$
$(C_2B_{10}H_{11}CH_2)_3PO$, **1**, 227
$B_{30}Hg_3C_6H_{30}$
$(B_{10}H_{10}C_2Hg)_3$, **3**, 165
$B_{40}Hg_4C_8H_{40}Cl$
$(B_{10}H_{10}C_2Hg)_4Cl^-$, **3**, 164
$B_{40}Hg_4C_{32}H_{56}I$
$(PhB_{10}H_9C_2Hg)_4I^-$, **3**, 165
BaC_2H_6
$BaMe_2$, **1**, 110
$BaC_{10}H_{10}$
$BaCp_2$, **1**, 112, 114
$BaC_{18}H_{22}$
$Ba(CMe_2Ph)_2$, **1**, 110
$BaC_{18}H_{26}$
$Ba(C_5H_4Bu^t)_2$, **1**, 112
$BaC_{20}H_{30}$
$BaCp_2$*, **1**, 113
$BaC_{26}H_{30}N_4$
Ba(fluorenyl)$_2(NH_3)_4$, **1**, 115
$BaC_{26}H_{30}O_2$
$Ba(C_9H_7)_2(THF)_2$, **1**, 115
$BaC_{28}H_{24}$
$Ba(CPh_2CH_2CH_2CPh_2)$, **1**, 111
$BaC_{28}H_{46}O_2$
$BaCp_2$*$(THF)_2$, **1**, 112
$BaC_{32}H_{42}O_4$
$Ba(C{\equiv}CPh)_2(THF)_4$, **1**, 111
$BaC_{38}H_{30}$
$Ba(CPh_3)_2$, **1**, 110, 111
$BaC_{42}H_{38}$
$Ba(Ph_3C_3H_4)_2$, **1**, 112
$BaC_{44}H_{38}O_2$
Ba(fluorenyl)$_2(Et_2O)_2$, **1**, 115
$BaC_{46}H_{44}$
$Ba[CPh(CH_2CMe_2Ph)C_6H_4$-$C_6H_4CPh(CH_2CMe_2Ph)]$, **1**, 111
$BaC_{46}H_{46}O_2$
$Ba(CPh_3)_2(THF)_2$, **1**, 111
$BaC_{74}H_{66}$
$Ba[CPh(CH_2CMe_2Ph)C_6H_4$-$C_6H_4(CPh{=}CH_2)]_2$, **1**, 110
$BaC_{76}H_{60}P_4$
Ba[(fluorenyl)$PPh_2CHPPh_2]_2$, **1**, 116
$BaLiC_{25}H_{50}O_2$
Li[BaCp*$CH(TMS)_2](THF)_2$, **1**, 119
$BaLiSi_2C_{35}H_{65}O_2$
$Li[Cp_2$*$BaCH(TMS)_2](THF)_2$, **1**, 115
$BaSiC_6H_{12}O_6$
$Ba[Si(OCH_2CH_2O)_3]$, **2**, 4
$BaSi_4C_{22}H_{42}$
$Ba[C_5H_3(TMS)_2]_2$, **1**, 112
$Ba_2Co_{25}S_{27}$
$Ba_2Co_{25}S_{27}$, **8**, 455
$Ba_2KSi_5C_{98}H_{95}O_9$
$KBa_2(OSiPh_3)_5(DME)_2$, **2**, 34
$Ba_2Si_4C_{48}H_{108}O_4$
$Ba_2(OSiBu^t_3)_4$, **2**, 34
$Ba_2Ta_2C_{20}H_{36}Br_2$
Cp*$_2Ta_2(Br)_2(B_2H_6)$, **1**, 306
$Ba_3Si_6C_{108}H_{90}O_6$
$Ba_3(OSiPh_3)_6$, **2**, 34
$BeAlC_{21}H_{42}P$
Cp*$BePBu^t_2{\cdot}AlMe_3$, **1**, 43
$BeAl_2C_{16}H_{24}$
$Me_4(MeC{\equiv}C)_4Al_2Be$, **1**, 39
$BeBC_{13}H_{19}N$
$CpBe[MeBu^t(NBC_3H_2)]$, **1**, 212, 213
$BeBC_{13}H_{20}N$
(BuMeH_2azaborolyl)BeCp, **1**, 46
$BeBC_{21}H_{40}N_6$
$(Bupz)_3BH_3BeH$, **1**, 50
$BeB_2C_{16}H_{30}N_2$
(BuMeH_2azaborolyl)$_2$Be, **1**, 46
$BeB_2C_{20}H_{29}N_2$
$[MeBu^t(NBC_3H_2)]Be[MeBu^t$-$(NMC_3H_3)]$, **1**, 212, 213
$BeB_5C_5H_{13}$
$(CpBe)B_5H_8$, **1**, 48, 326
$BeCH_3Br$
MeBeBr, **1**, 53
BeC_2
$BeC{\equiv}C$, **1**, 39
BeC_2H_6
EtBeH, **1**, 50
BeC_4H_{10}
Et_2Be, **1**, 54
BeC_5H_5Cl
CpBeCl, **1**, 48
$BeC_5H_{11}Cl$
(MeBu)BeCl, **1**, 52
BeC_6H_6
$(MeC{\equiv}C)_2Be$, **1**, 39
$BeC_6H_{15}NO$
(Me_2N)(MeEtNCO)BeH, **1**, 50
$BeC_{10}H_{10}$
Cp_2Be, **1**, 38, 44
$BeC_{10}H_{15}Cl$
Cp*BeCl, **1**, 46, 48
$BeC_{10}H_{22}$
$(MeBu)_2Be$, **1**, 52
$BeC_{12}H_{20}O$
Cp*BeOEt, **1**, 43
$BeC_{14}H_{30}N_2$
$(MeBe(C{\equiv}CMe)NMe_3)_2$, **1**, 39
$BeC_{15}H_{20}$
CpMeCp*, **1**, 46
$BeC_{18}H_{33}P$
Cp*$BePBu^t_2$, **1**, 42, 43
$BeLiC_{12}H_{27}$
$LiBeBu^t_3$, **1**, 42
$BeYbC_{31}H_{33}$
Cp_2*Yb(Me)BeCp*, **1**, 48
$BeYbC_{31}H_{48}$
Cp*$_2$Yb(Me)BeCp*, **4**, 67
$Be_2C_4H_{12}$
$(Me_2Be)_2$, **1**, 40
$Be_2C_{10}H_{26}N_4$
$(HBeN(Me)C_2H_4NMe_2)_2$, **1**, 51
$BiB_{10}H_{10}P$
$PBiB_{10}H_{10}$, **1**, 264
$BiB_{11}H_{11}$
$[B_{11}H_{11}Bi]^-$, **1**, 267
$BiC_5H_5Cl_2$
$CpBiCl_2$, **2**, 329
$BiC_{12}H_{10}Br_2$
$[Ph_2BiBr_2]^-$, **2**, 332
$BiC_{12}H_{10}I_2$
$[Ph_2BiI_2]^-$, **2**, 332
$BiC_{12}H_{27}$
Bu_3Bi, **11**, 502
$BiC_{16}H_{10}F_6O_4$
$[Ph_2Bi(OCOCF_3)_2]^-$, **2**, 333
$BiC_{19}H_{15}O_3$
Ph_3BiCO_3, **11**, 504, 505
$BiC_{22}H_{21}O_4$
$Ph_3Bi(OAc)_2$, **11**, 504, 510
$BiC_{24}H_{46}N_6O_2P_2$
$[Ph_2Bi(HMPA)_2]^+$, **2**, 332
$BiC_{26}H_{20}F_3O_2$
$Ph_4Bi(OCOCF_3)$, **11**, 509
$BiC_{26}H_{25}O_2$
$Ph_3Bi{=}C_6(O)_2Me_2H_4$, **11**, 506
$BiC_{30}H_{25}$
Ph_5Bi, **11**, 505
$BiCl_3$
$BiCl_3$, **11**, 502
$BiCoC_{16}H_{10}O_4$
$CoBiPh_2(CO)_4$, **8**, 96
$BiCoC_{33}H_{25}O_3P$
$CoBiPh_29Co)_3PPh_3$, **8**, 96
$BiCoFe_2C_{10}O_{10}$
$[BiFe_2Co(CO)_{10}]^-$, **7**, 60
$BiCo_3C_9O_9$
$Co_3(CO)_9(Bi)$, **8**, 429
$BiCo_3C_{12}O_{12}$
$Co_3(CO)_{12}(Bi)$, **8**, 429
$BiCrFe_3C_{27}O_{27}$
$[Bi[Fe(CO)_{14}][Fe_2(CO)_8]$-$[Cr(CO)_5]]^-$, **7**, 60
$BiFeC_{16}O_{16}$
$[Bi[Fe(CO)_4]_4]^-$, **7**, 60
$BiFe_3C_9H_3O_9$
$H_3BiFe_3(CO)_9$, **7**, 59
$BiFe_3C_{10}HO_9$
$Fe_3(CO)_9Bi(CH)$, **7**, 59
$BiFe_3C_{10}O_{10}$
$[BiFe_3(CO)_{10}]^-$, **7**, 58
$BiFe_3C_{11}H_3O_{10}$
$BiFe_3(CO)_9(COMe)$, **7**, 58
$BiFe_3C_{12}ClO_{12}$
$[ClBi[Fe(CO)_4]_3]^{2-}$, **7**, 60
$BiFe_4C_{16}O_{16}$
$[Bi[Fe(CO)_4]_4]^{3-}$, **7**, 58
$BiRu_3C_9H_3O_9$
$Ru_3H_3(Bi)(CO)_9$, **7**, 701
$BiSi_3C_{12}H_{33}$
$(TMS{-}CH_2)_3Bi$, **2**, 331
$BiSi_6C_{21}H_{57}$
$[(TMS)_2CH]_3Bi$, **2**, 331
$Bi_2B_{10}H_{10}$
$Bi_2B_{10}H_{10}$, **1**, 267
$Bi_2C_{12}H_{10}Cl_6$
$[Ph_2Bi_2Cl_6]^{2-}$, **2**, 334
$Bi_2C_{12}H_{28}$
$Pr_2BiBiPr_2$, **2**, 325
$Bi_2C_{24}H_{20}$
$Ph_2BiBiPh_2$, **2**, 326
$Bi_2C_{25}H_{22}$
$(Ph_2Bi)_2CH_2$, **11**, 503

$Bi_2C_{36}H_{30}Cl_2O$
$(Ph_3BiCl)_2O$, **11**, 504
$Bi_2Co_4C_{11}O_{11}$
$[Co_4(CO)_{11}Bi_2]^-$, **8**, 450
$Bi_2FeC_{28}H_{20}O_4$
$Fe(CO)_4(BiPh_2)_2$, **7**, 60
$Bi_2Fe_2C_{32}O_8$
$[Fe(CO)_4(BiPh_2)]_2$, **7**, 60
$Bi_2Fe_3C_9O_9$
$Bi_2Fe_3(CO)_9$, **7**, 59
$Bi_2Fe_4C_{13}O_{13}$
$[Fe_3(CO)_9Bi[BiFe(CO)_4]]^{2-}$, **7**, 59
$Bi_2Mo_2C_{14}H_{10}$
$Mo_2(CO)_4(Bi_2)Cp_2$, **5**, 427
$Bi_2Si_4C_{12}H_{36}$
$(TMS)_4Bi_2$, **2**, 31
$Bi_2Sm_2C_{40}H_{60}$
$(Bi_2)[SmCp^*_2]_2$, **4**, 83
$Bi_4Fe_4C_{13}O_{13}$
$[Bi_4Fe_4(CO)_{13}]^{2-}$, **7**, 60
$BkC_{10}H_{10}Cl$
Cp_2BkCl, **4**, 138
$BrCrC_{40}H_{34}NO_4P_2$
$[PPN][(H)_2BH_2Cr(CO)_4]$, **1**, 283, 284

C

$CaB_{10}C_{10}H_{24}N_4$
$(MeCN)_4CaB_{10}C_2H_{12}$, **1**, 116
$CaCH_3I$
MeCaI, **1**, 109
CaC_2H_3I
CH_2=CHCaI, **1**, 109
CaC_2H_6
$CaMe_2$, **1**, 110
CaC_6H_5I
PhCaI, **1**, 109
CaC_6H_6
PhCaH, **1**, 118
CaC_6H_9N
$(C_5H_4Me)Ca(NH_2)$, **1**, 118
CaC_8H_8
Ca(cot), **1**, 117
$CaC_{10}H_{10}$
$CaCp_2$, **1**, 114
$CaC_{12}H_{10}$
$CaPh_2$, **1**, 109, 110
$CaC_{12}H_{14}$
$Ca(C_5H_4Me)_2$, **1**, 118
$CaC_{16}H_{15}Br$
$(MePh_2C_3H_2)CaBr$, **1**, 108, 109
$CaC_{16}H_{24}O_2$
$Ca(C_5H_4Me)_2(DME)$, **1**, 112
$CaC_{18}H_{26}$
$Ca(C_5H_4Bu^t)_2$, **1**, 112
$CaC_{18}H_{26}O_2$
$CaCp_2(THF)_2$, **1**, 112, 119
$CaC_{18}H_{31}IO_2$
$Cp^*CaI(THF)_2$, **1**, 109, 110
$CaC_{19}H_{33}IO_2$
$(C_5Me_4Et)CaI(THF)_2$, **1**, 109, 110, 118
$CaC_{20}H_{30}$
$CaCp^*_2$, **1**, 112
$CaC_{20}H_{30}O_2$
$Ca(C_5H_4Me)_2(THF)_2$, **1**, 112
$CaC_{24}H_{40}O$
$CaCp_2^*(OEt_2)$, **1**, 112
$CaC_{25}H_{45}IO_2$
$(C_5HPr^i_4)Ca(I)(THF)_2$, **1**, 109
$CaC_{26}H_{42}O$
$(C_5Me_4Et)_2Ca(THF)$, **1**, 109, 112
$CaC_{26}H_{43}N_3O$
$(C_5Me_4Et)CaI(THF)_2$, **1**, 118
$(C_5Me_4Et)Ca[OC_6H_2(CH_2NMe_2)_3]$, **1**, 118
$CaC_{26}H_{44}N_2$
$Ca[C_5Me_4(CH_2CH_2NMe_2)]_2$, **1**, 112, 114
$CaC_{26}H_{45}P$
$Cp_2^*Ca(PEt_3)$, **1**, 114
$CaC_{28}H_{46}$
$Ca(C_5H_2Pr^i_3)_2$, **1**, 112, 113
$CaC_{28}H_{46}O_2$
$Cp^*_2Ca(THF)_2$, **1**, 109, 110
$CaC_{30}H_{38}N_2$
$Cp_2^*Ca(bipy)$, **1**, 114
$CaC_{30}H_{42}O_4$
$Ca(THF)_4$(anthracene), **1**, 117
$CaC_{30}H_{50}O$
$Ca(C_5H_3Bu^t_2)_2(THF)$, **1**, 112
$CaC_{32}H_{54}O$
$Ca(C_5H_2Pr^i_3)_2(THF)$, **1**, 112
$CaC_{34}H_{58}$
$Ca(C_5HPr^i_4)_2$, **1**, 112, 113
$CaC_{40}H_{52}N_2$
$Cp^*_2Ca(xylylisocyanide)_2$, **1**, 114
$CaC_{52}H_{68}O_6$
$Ca(THF)_6(anthracene)_2$, **1**, 117
$CaHfC_{40}H_{60}Cl_2$
$Cp^*_2CaCl(HfClCp^*_2)$, **1**, 115
$CaKC_{24}H_{36}O_4$
$[K(DME)]_2[Ca(cot)_2]$, **1**, 117
$CaK_2C_{32}H_{48}O_4$
$[K(THF)_2]_2Ca(cot)_2$, **1**, 119
$CaSiC_9H_{14}$
TMS-C_6H_4CaH, **1**, 118
$CaSiC_{30}H_{48}$
Cp^*_2Ca(TMS-C≡CC≡C-TMS), **1**, 114
$CaSi_2C_7H_{19}Br$
$[(TMS)_2CH]CaBr$, **1**, 109
$CaSi_2C_{20}H_{34}O$
$Ca[C_5H_4(TMS)]_2(THF)$, **1**, 112
$CaSi_2C_{28}H_{57}NO_3$
$Cp^*Ca[N(TMS)_2](THF)_3$, **1**, 118
$CaSi_2C_{29}H_{58}O_3$
$Cp^*Ca[CH(TMS)_2](THF)_3$, **1**, 115
$CaSi_4C_{14}H_{38}$
$Ca[CH(TMS)_2]_2$, **1**, 109
$CaSi_4C_{22}H_{42}$
$Ca[C_5H_3(TMS)_2]_2$, **1**, 112, 118
$CaSi_4C_{26}H_{62}O_3$
$Ca[CH(TMS)_2]_2(THF)_3$, **1**, 110
$CaYC_{40}H_{60}Cl$
$Cp^*_2Ca(Cl)YCp^*_2$, **1**, 115
$CaZrC_{40}H_{60}Cl_2$
$Cp^*_2CaClZrClCp^*_2$, **1**, 120
$CaZrC_{42}H_{60}O_2$
$Cp^*_2Ca(OC)_2ZrCp^*_2$, **1**, 115, 120
$Ca_2C_{36}H_{62}I_2O_4$
$[Cp^*CaI(THF)_2]_2$, **1**, 109
$Ca_2C_{48}H_{72}N_2$
$(Cp^*_2Ca)_2(N_2C_4Me_4)$, **1**, 114
$Ca_2Cr_2C_{64}H_{82}O_3$
$[Cp^*_2Ca(CO)_2Cr(CO)(Mes)]_2$, **1**, 115
$Ca_2Si_{16}C_{88}H_{168}O_2$
$[[(TMS)_2C_5H_3]_4CaO]_2$, **1**, 119
CdC_2F_6
$Cd(CF_3)_2$, **3**, 176, 200; **11**, 183
CdC_2H_6
$CdMe_2$, **3**, 200
CdC_7H_5BrN
Br(CNPh)Cd, **3**, 199
$CdC_{10}H_{10}$
Cp_2Cd, **3**, 197
$CdC_{10}H_{22}$
$(C_5H_{11})_2$, **3**, 200
$CdC_{12}F_{26}$
$Cd(C_6F_{13})_2$, **3**, 196
$CdC_{16}F_{34}$
$Cd(C_8F_{17})_2$, **3**, 196
$CdCo_2C_8O_8$
$Cd[Co(CO)_4]_2$, **10**, 48
$CdFeSiC_{23}H_{23}BrNO_6P$
$Fe(CO)_3[Si(OMe)_3](Ph_2PPy)$-CdBr, **10**, 338
$CdFeSiC_{24}H_{24}ClO_3P$
$Fe(CO)_3(PPh_3)(TMS)CdCl$, **10**, 338
$CdFeSiC_{26}H_{26}ClO_7P_2$
$Fe(CO)_3[Si(OMe)_3](Ph_2PCH_2$-C(O)Ph)CdCl, **10**, 338
$CdFeSiC_{31}H_{31}ClO_6P_2$
$Fe(CO)_3[Si(OMe)_3](PhP_2CH_2P$-$Ph_2)$CdCl, **10**, 338
$CdFeSiC_{46}H_{43}BrO_8P_2$
$Fe(CO)_3[Si(OMe)_3](Ph_2PCH_2$-C(O)Ph)CdBr($Ph_2P$-$CH_2$C(O)Ph), **10**, 339
$CdFe_2Si_2C_{18}H_{36}O_6P_2$
$[Fe(TMS)(CO)_3(PMe_3)]_2Cd$, **10**, 54
$CdFe_2Si_2C_{48}H_{50}N_2O_{12}P_2$
$[Fe(CO)_3[Si(OMe)_3](Ph_2P$-$CH_2Py)]_2Cd$, **10**, 339
$CdFe_2Si_2C_{64}H_{60}O_{14}P_2$
$[(CO)_3[Si(OMe)](PhFeC_6H_4C$-$(O)CH_2PPh_2)]_2Cd$, **10**, 339
$CdLiSi_3C_{33}H_{49}Br_2O_2$
$Li(THF)_2CdBr_2[C(SiMe_2Ph)_3]$, **3**, 198
$CdMn_2Si_2C_{42}H_{40}O_4$
$[(C_5H_4Me)Mn(SiMePh_2)$-$(CO)_2]_2Cd$, **6**, 144
$CdMn_2Si_2C_{42}H_{42}O_4$
$[MeCp(CO)_2SiMePh_2Mn]_2Cd$, **10**, 337
$CdSi_2C_8H_{22}$
$[(TMS)CH_2]_2Cd$, **3**, 196
$CdSi_2F_6$
$Cd(SiF_3)_2$, **3**, 176, 200
$CdSi_3C_{10}H_{27}Br$
$CdBr[C(TMS)_3]$, **3**, 198

$CdSi_3C_{10}H_{27}Cl$
  $CdCl[C(TMS)_3]$, **3**, 198
$CdSi_3C_{10}H_{28}O$
  $Cd(OH)[C(TMS)_3]$, **3**, 198
$CdSi_4C_{13}H_{36}Br_3O$
  $(Br)_3[O\text{-}TMS]Cd[C(TMS)_3]$, **3**, 198
$CdSi_6C_{20}H_{54}O_2$
  $Cd[C(TMS)_2(SiMe_2OMe)]_2$, **3**, 179
$CdSi_6C_{22}H_{54}$
  $[(TMS)_2(CH_2CHMe_2Si)C]_2Cd$, **3**, 196
$Cd_2Fe_2Si_2C_{52}H_{52}Cl_2O_{14}P_2$
  $[[Ph_2PCH_2C(O)Ph]Fe(CO)_3\text{-}(MeO)_2SiOMeCdCl]_2$, **10**, 333
$Cd_3Mn_2C_{26}H_{38}$
  $(Cp_2Mn)_2(CdMe_2)_3$, **6**, 146
$Cd_3Si_9C_{30}H_{81}Br_4$
  $[[Cd[C(TMS)_3]]_3(Br)_4]^-$, **3**, 198
$CeC_5H_5Cl_3$
  $CpCeCl_3$, **4**, 35
$CeC_{10}H_{15}Cl_3$
  $[Cp^*CeCl_3]^-$, **4**, 68
$CeC_{12}H_{21}Cl_3$
  $[(C_5Me_4Pr)CeCl_3]^-$, **4**, 68
$CeC_{16}H_{16}$
  $[Ce(C_8H_8)_2]^-$, **4**, 181
  $Ce(C_8H_8)_2$, **4**, 102, 104, 183
$CeC_{18}H_{20}$
  $Ce(MeC_8H_7)_2$, **4**, 104
$CeC_{18}H_{22}O$
  $Cp_3Ce(OPr^i)$, **4**, 58
$CeC_{19}H_{23}O$
  $Cp_3Ce(THF)$, **4**, 55, 58
$CeC_{20}H_{25}Cl_2N_2$
  $Cp^*CeCl_2(py)_2$, **4**, 68
$CeC_{20}H_{29}O_6$
  $[CpCe(acac)_3]^-$, **4**, 34
$CeC_{27}H_{39}$
  $Ce(Bu^tC_5H_4)_3$, **4**, 54
$CeC_{28}H_{46}S_2$
  $[Cp^*_2Ce(THT)_2]^-$, **4**, 74
$CeC_{30}H_{72}N_6$
  $[Ce[(CH_2)_3NMe_2]_6]^{3-}$, **4**, 20
$CeC_{31}H_{29}O$
  $(C_9H_7)_3Ce(THF)$, **4**, 58
$CeC_{32}H_{26}N$
  $(C_9H_7)_3Ce(py)$, **4**, 63
$CeC_{32}H_{46}S_2$
  $[(MeC_5H_4)_2Ce(SBu^t)]_2$, **4**, 40
$CeLiC_{21}H_{42}O_6$
  $CeMe_6[Li(acac)]_3$, **4**, 21
$CeLiC_{32}H_{58}Cl_2N_2$
  $(Bu^t_2C_5H_3)_2Ce(Cl)_2Li\text{-}(TMEDA)$, **4**, 37
$CeNaC_{18}H_{26}N_2O_8$
  $Cp_2Ce(NO_3)_2Na(THF)_2$, **4**, 36
$CeSi_2C_{27}H_{49}$
  $Cp^*_2CeCH(TMS)_2$, **4**, 84
$CeSi_4C_{22}H_{51}N_2$
  $Cp^*Ce[N(TMS)_2]_2$, **4**, 72
$CeSi_6C_{33}H_{63}$
  $Ce[(TMS)_2C_5H_3]_3$, **4**, 54
$CeSi_6C_{34}H_{64}$
  $Ce[(TMS)_3C_8H_5]_2$, **4**, 104
$Ce_2B_2C_{26}H_{92}$
  $[(Bu^t_2C_5H_3)_2Ce[(H)_2B(H)_2]]_2$, **1**, 277
$Ce_2B_2C_{52}H_{92}$
  $[(C_5H_3Bu^t_2)_2Ce[(H)_2BH_2]]2$, **1**, 286, 287
$Ce_2C_{24}H_{24}$
  $(C_8H_8)_3Ce_2$, **4**, 103
$Ce_2C_{32}H_{48}Cl_2O_4$
  $[(C_8H_8)Ce(Cl)(THF)_2]_2$, **4**, 99
$CoAsFeMoWC_{19}H_{16}O_7S$
  $FeCoMoW(S)(AsMe_2)Cp_2\text{-}(CO)_7$, **10**, 37
$CoAsMoFeWC_{19}H_{16}O_7S$
  $MoWFeCo(S)(AsMe_2)\text{-}(CO)_7Cp_2$, **10**, 244
$CoAs_2B_{20}H_{20}$
  $Co(B_{10}H_{10}As)_2^{3-}$, **1**, 266
$CoAs_4B_{18}H_{18}$
  $[Co(As_2B_9H_9)_2]^-$, **1**, 266
$CoAuB_2C_{35}H_{44}P$
  $CpCo[Et_4Me(C_3B_2)]Au(PPh_3)$, **1**, 206, 207
$CoAu_3Ru_3C_{66}H_{45}O_{12}P_3$
  $Au_3CoRu_3(CO)_{12}(PPh_3)_3$, **10**, 292
$CoBC_4H_2O_4$
  $(CO)_4CoBH_2$, **1**, 290–2
$CoBC_{10}H_{14}N$
  $CoCp[(CH)_3N(Me)B(Me)]$, **8**, 86
$CoBC_{10}H_{15}NO_2$
  $(CO)_2Co[MeBu^t(C_3H_3BN)]$, **1**, 212
$CoBC_{13}H_{20}N$
  $CpCo[MeBu^t(BC_3H_3N)]$, **1**, 212
$CoBC_{15}H_{14}$
  $Co(C_4H_4BPh)Cp$, **1**, 204
$CoBC_{15}H_{15}N_3$
  $(terpy)Co(BH_4)$, **1**, 285
$CoBC_{15}H_{27}NSi$
  $CpCo(C_{10}H_{22}BNSi)$, **1**, 213
$CoBC_{20}H_{23}F_4O_2$
  $[Cp^*Co(cot)Rh(CO)_2]BF_4$, **8**, 245
$CoBC_{30}H_{38}P_2$
  $Co(BPh_4)(PMe_3)_2$, **8**, 90
$CoBC_{36}H_{56}P_4$
  $[Co(PMe_3)_4][BPh_4]$, **8**, 90
$CoBC_{36}H_{58}P_4$
  $[CoH_2(PMe_3)_4]BPh_4$, **8**, 32
$CoBC_{39}H_{30}$
  $CoCp(C_4BPh_5)$, **8**, 86
$CoBC_{52}H_{46}O_2P_2$
  $(CO)_2(dppm)Co(dppm)BH_2$, **1**, 291, 292
$CoB_2C_{14}H_{23}$
  $CpCo[Et_2Me_2(C_3B_2)]$, **1**, 205, 206
$CoB_2C_{21}H_{40}N_2$
  $CpCoPr^i_2N{=}BHCH{=}C(Me)C\text{-}H_2BBPr^i_2$, **1**, 205
$CoB_2C_{44}H_{47}F_8NP_4$
  $[Co(PP_3)(MeCN)](BF_4)_2$, **8**, 20
$CoB_2FeC_{18}H_{25}$
  $CpFe(C_3B_2Me_5)CoCp$, **7**, 195
$CoB_2TlC_{13}H_{20}$
  $TlMe_5(C_2B_2C)CoCp$, **1**, 532
$CoB_4C_5H_{13}$
  $CpCoB_4H_8$, **1**, 320
$CoB_4C_{10}H_{23}$
  $Cp^*CoB_4H_8$, **1**, 320
$CoB_5C_4H_8O_4$
  $[Co(CO)_4]B_5H_8$, **1**, 326
$CoB_8C_5H_{13}Se_2$
  $(Cp)Co(B_8H_8Se_2)$, **1**, 270
$CoB_9C_4H_{19}$
  $[EtC_2B_4H_4]Co[B_5H_{10}]$, **1**, 323
$CoB_9C_5H_{15}N$
  $(Cp)CoNHB_9H_9$, **1**, 263
$CoB_9C_6H_{17}N$
  $(Cp)CoNMeB_9H_9$, **1**, 263
$CoB_9RhC_{30}H_{48}O_2P$
  $Co(CO)(C_4Me_4)Rh(CO)(PPh_3)\text{-}(C_2B_9H_{11})$, **10**, 64
$CoB_{10}C_6H_{18}P$
  $(Cp)MeCoPB_{10}H_{10}$, **1**, 265
$CoB_{10}C_{13}H_{28}NO_3$
  $[BzNMe_3][(CO)_3CoB_{10}H_{12}]$, **1**, 351
$CoB_{10}WC_{24}H_{37}O_2$
  $W(C_2B_{10}H_9Me_2)(CO)_2(CTol)\text{-}CoCp^*$, **10**, 43
$CoB_{17}C_5H_{26}$
  $CpCoB_{17}H_{21}$, **1**, 365
$CoB_{18}ThC_{24}H_{53}$
  $[Cp^*_2ThH][Co(B_9C_2H_{11})_2]$, **4**, 176
$CoBiC_{16}H_{10}O_4$
  $CoBiPh_2(CO)_4$, **8**, 96
$CoBiC_{33}H_{25}O_3P$
  $CoBiPh_2(CO)_3PPh_3$, **8**, 96
$CoBiFe_2C_{10}O_{10}$
  $[BiFe_2Co(CO)_{10}]^-$, **7**, 60
$CoCH_3$
  $[CoMe]^+$, **8**, 35
  $CoMe$, **8**, 27
$CoCH_4$
  $CoCH_4$, **8**, 76
  $MeCoH$, **8**, 76
$CoC_2H_2NO_3$
  $Co(CO)_2(NO)(H_2)$, **8**, 76
$CoC_2H_7D_2O_2$
  $CoMe_2(DO)(DOH)$, **8**, 62
$CoC_2H_9D_2O_2$
  $HOCH_2CH_2Co(DH)_2(H_2O)$, **8**, 65
CoC_3HO_3
  $CoH(CO)_3$, **8**, 27, 32
$CoC_3H_2O_3$
  $Co(H_2)(CO)_3^-$, **8**, 5
$CoC_3H_3O_3$
  $Co(H_3)(CO)_3$, **8**, 5, 76
CoC_3O_3
  $[Co(CO)_3]^{3-}$, **8**, 4
CoC_4HO_4
  $CoH(CO)_4$, **8**, 26, 28–30, 32, 421
$CoC_4H_{14}D_2O_3P$
  $CoMe(DH)_2P(OMe)_3$, **8**, 52

$CoC_4H_{14}D_2P$
$CoMe(DH)_2PMe_3$, **8**, 60
$CoC_4H_{16}B_4H_{29}$
$Cp^*Co(Et_2C_2B_4H_4)$, **1**, 376
$CoC_4H_{20}B_{18}I$
$Co(C_2B_9H_{10})_2I$, **1**, 414
$CoC_4H_{22}B_{18}$
$Co(C_2B_9H_{11})_2^-$, **1**, 376
CoC_4O_4
$[Co(CO)_4]^-$, **8**, 5, 27, 29, 30, 505
CoC_5ClN_5
$[Co(CN)_5Cl]^{3-}$, **8**, 8
CoC_5HN_5
$[Co(CN)_5H]^{3-}$, **8**, 12
CoC_5HO_5
$Co(COH)(CO)_4$, **8**, 28
CoC_5HO_{17}
$Co(CHO)(CO_4)_4$, **8**, 28
$CoC_5H_3O_4$
$CoC(O)Me(CO)_3$, **8**, 34
$CoMe(CO)_4$, **8**, 27, 35
CoC_5H_5NO
$[CoCpNO]^-$, **8**, 14
$CoC_5H_5NO_2$
$CoCp(NO)_2$, **8**, 13
$CoC_5H_5N_2S_2$
$CoCp(S_2N_2)$, **8**, 26
$CoC_5H_7D_2N$
$[Co(DH)_2py]^-$, **8**, 47
$CoC_5H_8D_2N$
$CoH(DH)_2py$, **8**, 48
CoC_5N_5
$[Co(CN)_5]^{2-}$, **8**, 11
$CoC_6H_2NO_4$
$Co(CH_2CN)(CO)_4$, **8**, 8
$CoC_6H_3O_5$
$Co[C(O)Me](CO)_4$, **8**, 4, 34
$CoC_6H_3O_6$
$Co(MeOCO)(CO)_4$, **8**, 37
$CoC_6H_5I_2O$
$CoCpI_2(CO)$, **8**, 23, 67
CoC_6H_5O
$CoCp(CO)$, **8**, 85
$CoC_6H_5O_3$
$(allyl)Co(CO)_3$, **12**, 651, 654
$Co(C_3H_5)(CO)_3$, **8**, 78
$CoC_6H_6O_3$
$(CH_2{=}CHMe)Co(CO)_3$, **12**, 975
CoC_6H_8
$\overline{CH(Me)CHCHCH(Me)Co}$, **12**, 757
$CoC_6H_{18}O_7P_2$
$Co[P(O)(OMe)_2P(OMe)_4]$, **8**, 93
CoC_6N_5O
$[Co(CN)_5CO]^{2-}$, **8**, 8
CoC_6N_6
$[Co(CN)_6]^{3-}$, **8**, 11
$CoC_7H_4O_4$
$Co(CO)_3(C_4H_4O)$, **8**, 37
$CoC_7H_5F_3IO$
$CoCp(CF_3)I(CO)$, **8**, 23
$CoC_7H_5NO_5$
$Co(CH_2CN)(COOMe)(CO)_3$, **8**, 8
$CoC_7H_5O_2$
$CoCp(CO)_2$, **8**, 67, 70, 83, 429
$CpCo(CO)_2$, **12**, 744, 747, 749, 750
$CoC_7H_5O_3$
$Co(CO)_3(CH_2{=}CHCH{=}CH)$, **12**, 654
$CoC_7H_5O_6$
$EtOCOCo(CO)_4$, **8**, 34
$CoC_7H_7Cl_2O$
$CoCpCl(CH_2Cl)(CO)$, **8**, 67
$CoC_7H_{11}NO_7$
$[Co(Me)(nta)(H_2O)]^-$, **8**, 62
$CoC_7H_{18}NOP_2$
$Co(NO)(CO)(PMe_3)_2$, **8**, 14
$CoC_8H_5F_6O$
$CoCp(CF_3)_2(CO)$, **8**, 67
$CoC_8H_5O_3$
$CoCp(CO)_3$, **8**, 85
CoC_8H_7BrNO
$CoCp'Br(CN)(CO)$, **8**, 11
$CoC_8H_7O_2$
$CoCp'(CO)_2$, **8**, 15, 83
$CoC_8H_7O_5$
$Co(CO)_4(MeCH_2CH_2CO)$, **8**, 41
$CoC_8H_7O_6$
$EtOCOCH_2Co(CO)_4$, **8**, 34
$CoC_8H_{13}D_2O$
$Ph(Me)(H)CCo(DH)_2(H_2O)$, **8**, 64
$CoC_8H_{16}P$
$CoCp(H)_2P(Me)_3$, **8**, 69
CoC_9HO_9
$Co_3H(CO)_9$, **8**, 421
CoC_9H_9
$CpCo(C_4H_4)$, **12**, 745, 761
$CoC_9H_{10}O_4$
$[MeC(O)CH_2CH{=}CHMe]Co(CO)_3$, **12**, 975
CoC_9H_{11}
$CoCp(C_4H_6)$, **8**, 79
$CoC_9H_{11}NO_3$
$Co(CO)_3N(Me){=}CHCH{=}CMe_2$, **8**, 91
$CoC_9H_{12}I$
$Co(MeC_3H_4)CpI$, **8**, 22
CoC_9H_{13}
$CpCo(C_2H_4)_2$, **8**, 73, 429, 435, 512; **12**, 757
$CoC_9H_{15}D_2NO_2$
$Co(MeOCOEt)(DH)_2py$, **8**, 61
$CoC_9H_{16}I_2O_3P$
$CoCp'(I)_2P(OMe)_3$, **8**, 22
$CoC_9H_{19}O_6P_2$
$CoMe(CO)_2[(MeO)_2PCH_2CH_2P(OMe)_2]$, **8**, 19
$CoC_9H_{21}B_9S$
$(Cp)Co(C_2B_9H_{10}(Me_2S))$, **1**, 414, 416
$CoC_9H_{21}O_8P_2$
$CoMe(CO)_2[P(OMe)_3]_2$, **8**, 19
$CoC_9H_{27}BrP$
$CoBr(PMe_3)_3$, **8**, 40
$CoC_9H_{27}BrP_3$
$CoBr(PMe_3)_3$, **8**, 22, 74, 90
$CoC_{10}F_5O_4$
$Co(C_6F_5)(CO)_4$, **8**, 41
$CoC_{10}HO_{10}$
$Co_3(CO)_9(COH)$, **8**, 421
$CoC_{10}H_{10}$
$CoCp_2$, **8**, 88
$CoC_{10}H_{11}$
$CpCo(pyran)$, **12**, 764
$CoC_{10}H_{15}$
$CoCp'(C_2H_4)_2$, **8**, 73
$CoC_{10}H_{16}NO_4$
$Et_3NH^+[Co(CO)_4]^-$, **8**, 29
$CoC_{10}H_{18}P$
$CoCp(PMe_3)(C_2H_4)$, **8**, 73
$CoC_{10}H_{19}O_3P$
$[CoCp[P(OMe)_3Et]]^+$, **8**, 76
$CoC_{10}H_{20}P$
$CoCp(Me)_2P(Me)_3$, **8**, 69
$CoC_{11}H_7O_2$
$Co(C_9H_7)(CO)_2$, **8**, 85
$CoC_{11}H_7O_4$
$CoBz(CO)_4$, **8**, 38
$CoC_{11}H_{11}O_3P$
$Co(CO)_3(PMe_2Ph)^-$, **8**, 7
$CoC_{11}H_{13}$
$CoCp(C_6H_8)$, **8**, 82
$CpCoCy$, **12**, 759, 760
$CoC_{11}H_{14}$
$Co(PhMe)(C_2Me_2)$, **8**, 89
$CoC_{11}H_{15}O_2$
$Co(acac)Cp'$, **8**, 67
$CoC_{11}H_{21}P$
$CoCp'(C_2H_4)H[P(OMe)_3]^+$, **8**, 72
$CoC_{11}H_{23}O_9P_3$
$[CpCo[(OMe)_2PO]_3]^-$, **8**, 16
$CoC_{11}H_{24}O_6P_2$
$[CoHCp[P(OMe)_3]_2]^+$, **8**, 30
$CoC_{11}H_{32}P_3$
$CoH(C_2H_4)(PMe_3)_3$, **8**, 34
$CoC_{12}H_{11}O_2$
$CoCp'(O_2C_6H_4)$, **8**, 24
$CoC_{12}H_{13}P_2$
$CoCp'(dmpe)$, **8**, 19
$CoC_{12}H_{14}D_2N$
$Co(Bz)(DH)_2py$, **8**, 64
$CoC_{12}H_{16}OP$
$CoCp[CpP(O)Me_2]$, **8**, 89
$CoC_{12}H_{17}$
$Cp'Co(H)_3CoCp'$, **8**, 93
$CoC_{12}H_{23}O_9P_3$
$CoCp[P(O)(OMe)_2]_3^-$, **8**, 21
$CoC_{12}H_{30}N_3O_5P_2$
$Co(NO)(NO_2)_2(PEt_3)_2$, **8**, 14
$CoC_{12}H_{32}O_9P_3$
$Co(allyl)[P(OMe)_3]_3$, **8**, 78
$CoC_{12}H_{36}O_{12}P_4$
$Co[P(OMe)_3]_4$, **8**, 93
$CoC_{13}H_{17}$
$CpCo(cod)$, **12**, 754
$CoC_{14}H_{10}O_4$
$Co(OOCPh)_2$, **12**, 426
$CoC_{14}H_{11}O_2$
$Co(BzC_5H_4)(CO)_2$, **8**, 85

$CoC_{14}H_{15}$
  CoCp'(cot), **8**, 82
$CoC_{14}H_{17}$
  $CoCp(Tol)[CH_2CH_2(H)]$, **8**, 72
$CoC_{14}H_{19}$
  CoCp'(cod), **8**, 82
$CoC_{14}H_{19}O_5$
  $[OC(O)C(Bu^tCHMeCHEt]Co(CO)_3$, **12**, 976
$CoC_{14}H_{22}O_2P_2$
  $CpCo(PMe_2)_2(SO_2)CoCp$, **8**, 97
$CoC_{14}H_{23}$
  $Co(C_2H_4)_2Cp^*$, **8**, 432
$CoC_{14}H_{25}$
  $CoCp^*(Me)_2(C_2H_4)$, **8**, 70
$CoC_{14}H_{30}B_9$
  $(C_6Me_6H)Co(C_2B_9H_{11})$, **1**, 376
$CoC_{14}H_{42}O_7P_5$
  $Co[P(O)(OMe)_2](POMe_3)_4$, **8**, 16
$CoC_{15}F_{10}O_3$
  $Co(C_6F_5)_2(CO)_3$, **8**, 8
$CoC_{15}H_{18}$
  $Co(C_8H_8Me)Cp'$, **8**, 82
$CoC_{15}H_{23}P_2$
  $CoCp(Bu^t{}_2C_2P_2)$, **8**, 24
$CoC_{15}H_{25}$
  $Co(C_5Me_4Et)(C_2H_4)_2$, **8**, 73
$CoC_{15}H_{42}D_2P$
  $Co(Bz)(DH)_2PCy_3$, **8**, 60
$CoC_{16}H_{10}O_4P$
  $CoPPh_2(CO)_4$, **8**, 96
$CoC_{16}H_{13}O_4P$
  $[Co(CO)_3(PPh_2OMe)]^-$, **8**, 5
$CoC_{16}H_{30}N_4$
  $CoMe_2(tim)$, **8**, 62
$CoC_{16}H_{34}O_9P_3$
  $Co(Bz)[P(OMe)_3]_3$, **8**, 78
$CoC_{16}H_{35}P_3$
  $Co(PMe_3)_3(nbd)$, **8**, 73
$CoC_{17}H_{29}B_2$
  $CpCo[Et_4Me(C_3B_2)]$, **1**, 206, 207
$CoC_{17}H_{35}O_9$
  $[CpCo[P(O)(OEt)_2]_3]^-$, **8**, 83
$CoC_{17}H_{35}P_2$
  $CoCp(PEt_3)_2$, **8**, 83
$CoC_{17}H_{36}P_3$
  $CoPh(C_2H_4)(PMe_3)_3$, **8**, 72
$CoC_{17}H_{40}O_9P_3$
  $Co(cycloacetenyl)[P(OMe)_3]_3$, **8**, 78
$CoC_{18}H_{14}$
  $Co(C_9H_7)_2$, **8**, 85
$CoC_{18}H_{15}INO_2P$
  $CoI(NO)_2(PPh_3)$, **8**, 13
$CoC_{18}H_{21}$
  $Co_2Cp'_3$, **8**, 93
$CoC_{18}H_{27}O_3$
  $Co(C_3Bu^t{}_3)(CO)_3$, **8**, 78
$CoC_{18}H_{53}P_2$
  $Co(C_3H_5(Pr^i{}_2P(CH_2)_3PPr_2{}^i))$, **8**, 93
$CoC_{19}H_8F_{10}$
  $Co(C_6F_5)_2(PhMe)$, **8**, 8, 41
$CoC_{19}H_{22}O_3P_2$
  $[Co(CO)_3(PMe_2Ph)_2]^+$, **8**, 6
$CoC_{19}H_{41}P_2$
  $Co(C_5Me_2H_5)(PEt_3)_2$, **8**, 86
$CoC_{20}H_{15}NO_3P$
  $PPh_3Co(NO)(CO)_2$, **12**, 353
$CoC_{20}H_{16}N_4$
  $[Co(bipy)_2]^{2+}$, **12**, 1189
$CoC_{20}H_{19}N_4O$
  $[CoH(bipy)_2(H_2O)]^+$, **8**, 30
$CoC_{20}H_{20}F_{10}O$
  $Co(C_6F_5)_2(ether)_2$, **8**, 41
$CoC_{20}H_{24}PS_2$
  $CoCp(SPh)_2(PMe_3)$, **8**, 25
$CoC_{20}H_{26}$
  $Co(C_6Me_6)(cot)$, **8**, 82
$CoC_{20}H_{28}N_4O_4$
  $Co(cqd)_2)$, **12**, 416
$CoC_{21}H_{15}O_3P$
  $Co(CO)_3(PPh_3)^-$, **8**, 7
$CoC_{21}H_{16}O_3P$
  $CoH(CO)_3PPh_3$, **8**, 29
$CoC_{22}H_{19}O_3$
  $Co(C_3Ph_2Bu^t)(CO)_3$, **8**, 78
$CoC_{22}H_{27}D_2N_2$
  $C_6(adamantyl)(DH)_2[(Me_2N)-C_5H_4C_5N]$, **8**, 50
$CoC_{22}H_{27}N_4O_2$
  CoMe(Mesalen)(en), **8**, 64
$CoC_{23}H_{20}O_2P$
  $Co(allyl)(CO)_2(PPh_3)$, **9**, 347
$CoC_{24}H_{24}N_4O_4$
  $Co[[(C_5H_4N)NH]_2C_8H_5](OAc)-(OOBu^t)$, **12**, 1189
$CoC_{24}H_{36}$
  $Co(C_6Me_6)_2$, **8**, 90
$CoC_{25}H_{40}NP_3$
  $[Co(MeCN)(PhC{\equiv}CPh)(PMe_3)_3]^+$, **8**, 75
$CoC_{26}H_{23}O_2$
  $[(OH)MePh_2(O)C_8H_4]CoCp$, **12**, 744
$CoC_{27}H_{24}P$
  $Cp(PPh_3)CoCHCHCHCH$, **12**, 759, 764, 766
$CoC_{27}H_{54}O_3P_2$
  $[Co(CO)_3(PBu_3)_2]^+$, **8**, 91
$CoC_{28}H_{28}$
  $Co(nbd)_4$, **8**, 71
$CoC_{29}H_{27}O_4P_2$
  $Co(COMe)(CO)_2(PPh_2H)-(PPh_2OMe)$, **8**, 5
$CoC_{31}H_{40}Cl_2P_4$
  $CoCl_2(PMe_3)_2(dppm)$, **8**, 18
$CoC_{32}H_{24}O_6P_2$
  $Co_2(Co)_6(dppe)$, **8**, 93
$CoC_{32}H_{29}OP_2$
  $[CoCp(CO)(dppe)]^{2+}$, **8**, 8
$CoC_{32}H_{31}P_2$
  CoCp'(dppe), **8**, 19
$CoC_{32}H_{42}$
  $Co(C_9(Me)_7)_2$, **8**, 88
$CoC_{36}H_{30}ClN_2O_2P_2$
  $[(OC)Co(CO)(PPh_2py)_2CoCl]$, **8**, 18
$CoC_{36}H_{30}Cl_2P_2$
  $Co(PPh_3)_2Cl_2$, **12**, 171
$CoC_{36}H_{32}NO$
  $PhC(Me)N(Me)P(Ph)_2Co(Cp)-C(O)CH_2CH_2CH_2$, **12**, 378
$CoC_{36}H_{85}O_{12}P_4$
  $CoH[P(OPr^i)_3]_4$, **8**, 16
$CoC_{37}H_{35}P$
  $CoCp(PhC{\equiv}CPh)PPh_3$, **8**, 74
$CoC_{39}H_{30}O_6P$
  $(CO_2Me)_3Ph-(CoCCCHCHCp)PPh_3$, **12**, 759, 760
$CoC_{41}H_{33}INO_4P_2$
  $PPN^+[Co(COMe)I(CO)_3]^-$, **8**, 37
$CoC_{41}H_{35}P_2$
  $CoCp(PPh_3)_2$, **8**, 15
$CoC_{42}H_{40}N$
  $CoH(np_3)$, **8**, 35
$CoC_{43}H_{39}NS$
  $[Co(np_3)CS]^+$, **8**, 9
$CoC_{51}H_{44}OP_4$
  $Co(CO)(dppm)_2{}^+$, **8**, 8, 18
$CoC_{52}H_{44}ClO_2P_4$
  $CoCl(dppm)_2(CO)_2$, **8**, 18
$CoC_{52}H_{44}O_2P_4$
  $Co(dppm)_2(CO)_2$, **8**, 18
$CoC_{52}H_{49}P_4$
  $CoH(dppe)_2$, **8**, 30
$CoC_{54}H_{45}ClP_3$
  $ClCo(PPh_3)_3$, **12**, 752
$CoC_{54}H_{46}N_2P_3$
  $CoH(N_2)(PPh_3)_3$, **8**, 21, 23, 34
$CoC_{55}H_{46}OP_3$
  $CoH(CO)(PPh_3)_3$, **8**, 30
$CoC_{57}H_{50}P_4$
  $[CoHCp(Ph_2PC_2H_2PPh_2)_2]^+$, **8**, 30
$CoC_{60}H_{50}OP_3$
  $Co(OPh)(PPh_3)_3$, **8**, 21
$CoC_{72}H_{168}O_{24}P_8$
  $Co_2[P(OPr^i)_3]_8$, **8**, 16
$CoCrC_{16}H_{36}O_5P_3$
  $Cr(CO)_3(PMe_3)(PBu^t{}_2)(CO)Co-(CO)(PMe_3)$, **10**, 44
$CoCrFeC_{13}H_5O_8S$
  $CrFeCo(S)(CO)_8Cp$, **10**, 244
$CoCr_3C_{18}H_{15}O_4S_3$
  $Cr_3Co(O)(S)_3(CO)_3Cp_3$, **10**, 228, 233, 237
$CoFeC_{11}H_5O_6$
  $FeCp(CO)_2Co(CO)_4$, **10**, 54
$CoFeC_{13}H_7O_8S_2$
  $Fe_2Co(C_5H_4CO_2Me)(CO)_6S_2$, **7**, 66
$CoFeC_{16}H_{17}O_3$
  $Fe(CO)_3(CMeCMeCMeCMe)-CoCp$, **10**, 60
$CoFeC_{20}H_{17}O_{11}$
  $FeCp[C_4(Co_2Me)_4]Co(CO)_3$, **10**, 60
$CoFeCrC_{13}H_5O_8S$
  $FeCoCr(S)Cp(CO)_8$, **10**, 37
$CoFeIrRuC_{16}H_9O_{10}P$
  $FeRuCoIr(H)(PMe)(CO)_{10}Cp$, **10**, 244

$CoFeMnC_{11}H_3O_{10}P$
$MnFeCo(Co)_{10}(PMe)$, **10**, 15
$CoFeMoC_{13}H_5O_8S$
$MoFeCo(S)Cp(CO)_8$, **10**, 49
$CoFeNiC_{15}H_{14}O_6P$
$FeCoNiCp[PBu^t](CO)_6$, **9**, 20, 173
$CoFeNiC_{16}H_{14}O_8P$
$CoFeNi(Me_3P)Cp(CO)_8$, **9**, 21
$CoFePd_2C_{56}H_{44}NO_7P_4$
$FeCoPd_2(CO)_6(NO)(dppm)_2$, **10**, 243
$CoFeWC_{12}H_{26}O_8P_3$
$WFeCo(PMe)(CO)_8(PMe_3)_2Cp$, **10**, 222
$CoFeWC_{14}H_8O_8P$
$WFeCo(PMe)(CO)_8Cp$, **10**, 37, 226, 244, 245
$CoFeWC_{19}H_{22}O_8P_2$
$W(CO)Cp(CO)(dppm)[PMeFe(CO)_4]Co(CO)_2$, **10**, 44
$CoFe_2C_8NO_9S$
$Fe_2Co(CO)_8(NO)(S)$, **7**, 70
$CoFe_2C_9HO_9S$
$Fe_2Co(H)(S)(CO)_9$, **10**, 237
$HFe_2Co(CO)_9(S)$, **7**, 70
$CoFe_2C_{11}HO_{10}$
$Fe_2Co(CO)_{10}(CH)$, **7**, 23
$CoFe_2C_{11}O_{10}$
$[Fe_2Co(CCO)(CO)_9]^-$, **7**, 271, 272
$CoFe_2C_{12}H_5O_6SeTe$
$Fe_2(CO)_6[SeTeCo(CO)Cp]$, **7**, 75
$CoFe_2C_{12}H_5O_7Te_2$
$Fe_2(CO)_6Te_2Co(CO)Cp$, **7**, 77
$CoFe_2C_{14}H_5O_9$
$Fe_2CoCp(CO)_9$, **10**, 41
$CoFe_2C_{15}H_{16}O_8P_2$
$[Fe_2Co(CO)_8(dmpe)C]^-$, **7**, 22
$CoFe_2C_{16}H_{14}O_7P$
$Fe_2Co(PBu^t)(CO)_7Cp$, **10**, 222
$CoFe_2C_{30}H_{25}N_4O_5P_2$
$Fe_2Co(PPh_2)_2(CO)(NO)_4Cp$, **10**, 223
$CoFe_2NiC_{14}H_8O_8P$
$Fe_2CoNiCp(PMe)(CO)_8$, **9**, 20
$CoFe_2NiC_{15}H_{11}O_8P_2$
$NiCoFe_2(PMe)_2(CO)_8Cp$, **9**, 173
$CoFe_3C_{13}O_{13}$
$[Fe_3Co(CO)_{13}]^-$, **10**, 19, 20
$CoFe_3C_{20}H_{20}NO_{12}$
$[Et_4N][Fe_3Co(CO)_{12}]$, **10**, 45
$CoFe_3C_{21}H_{20}NO_{13}$
$[Et_4N][Fe_3Co(CO)_{13}]$, **10**, 47
$CoFe_3GeC_{14}O_{14}$
$[Fe_3Co(Ge)(CO)_{14}]^-$, **10**, 215
$[GeCoFe_3(CO)_{14}]^-$, **7**, 30
$CoFe_4C_{15}O_{14}$
$[CoFe_4(CO)_{14}C]^-$, **7**, 25
$CoGaC_7H_{11}NO_4$
$GaH_2(NMe_3)[Co(CO)_4]$, **1**, 556
$CoGaC_{18}H_{30}O_5$
$Ga(CH_2Bu^t)_2(THF)[Co(CO)_4]$, **1**, 556
$CoGaSi_2C_{16}H_{30}O_5$
$Ga(CH_2TMS)_2(THF)[Co(CO)_4]$, **1**, 556
CoH_2
$[Co(H_2)]^-$, **8**, 75
$CoHgOs_3C_{19}H_9O_{13}$
$(C_2Bu^t)(CO)_9Os_3[HgCo(CO)_4]$, **10**, 340
$CoInC_{14}H_{24}N_2O_4$
$In(CH_2CH_2CH_2NMe_2)_2[Co(CO)_4]$, **1**, 564
$CoInC_{22}H_{15}Cl_2O_4P$
$InCl_2(PPh_3)[Co(CO)_4]$, **1**, 565
$CoIrC_{18}H_{22}O_2$
$Cp(CO)Co(CH_2)IrCp^*(CO)$, **8**, 387
$CoLiC_{11}H_{31}P_3$
$Li[Co(C_2H_4)(PMe_3)_3]$, **1**, 26
$CoMgC_{34}H_{86}N_4O_4P_6$
$Mg(THF)_4[N_2Co(PMe_3)_3]_2$, **1**, 104
$CoMnC_9O_9$
$Mn(CO)_5Co(CO)_4$, **10**, 48
$CoMnC_{11}H_6O_7$
$Mn(CO)_4(MeC_2Me)Co(CO)_3$, **10**, 52
$CoMnC_{11}H_8O_5P$
$Mn(CO)_2Cp(PHMe)Co(CO)_3$, **10**, 53
$CoMnC_{55}H_{44}O_5P_4$
$Mn(CO)_4(dppm)_2Co(CO)$, **10**, 52
$CoMn_2C_{26}H_{28}O_4PS$
$Mn(CO)_2Cp'(CO)[CSMn(CO)_2Cp']Co(PMe_3)Cp$, **10**, 52
$CoMoC_{10}H_{17}B_9$
$CpCoMo(C_3H_3)(C_2B_9H_9)$, **1**, 414
$CoMoC_{13}H_8O_5$
$Co(CH_2CCH)(CO)_3Mo(CO)_2(Cp)$, **12**, 687
$CoMoC_{14}H_6O_5$
$Mo(CO)_2Cp(MeC_2Me)Co(CO)_3$, **10**, 43
$CoMoC_{16}H_{13}O_3$
$MoCp(CO)_3Co(C_4Me_4)$, **10**, 43
$CoMoC_{17}H_{17}O_4$
$Mo(CO)_2Cp(C_4Me_4)Co(CO)_2$, **10**, 43
$CoMoC_{19}H_{13}O_5$
$CoMo(CO)_5Cp(MeC_2Ph)$, **10**, 13
$CoMoNiC_{17}H_{13}O_5$
$NiCoMo(CMe)(CO)_5Cp_2$, **9**, 173
$CoMoPd_2C_{56}H_{44}O_6P_4$
$MoCoPd_2(CO)_6(dppm)_2$, **10**, 243
$CoMoRhC_{27}H_{30}O_7$
$MoCoRh(CO)_7Cp^*_2$, **10**, 244
$CoMoRuC_{15}H_9O_8$
$HMoRuCo(CMe)Cp(CO)_8$, **10**, 41
$CoMoW_2C_{27}H_{21}O_9$
$MoW_2Co(CO)_9Cp'_3$, **10**, 244
$CoNaC_4O_4$
$NaCo(CO)_4$, **12**, 354, 974, 975
$CoNiC_{25}H_{29}PS_2$
$Co(PMe_3)Cp(SPh)_2NiCp$, **10**, 64
$CoNiRuC_{19}H_{17}O_6$
$CpRuCoNi(CO)_6(C_4Me_4)$, **10**, 166
$CoNiW_2C_{24}H_{19}O_7$
$NiCoW_2(CO)_7CpCp'_2$, **9**, 173
$CoNi_2C_{17}H_{15}O_2$
$Ni_2Co(CO)_2Cp_3$, **9**, 172, 173
$CoNi_2C_{22}H_{25}O_2$
$[CoNi_2(CO)_2Cp_2Cp^*]^-$, **10**, 211
$Ni_2Co(CO)_2Cp_2Cp^*$, **9**, 172
$CoOs_3C_{14}H_8O_9$
$Os_3Co(H)_3(CO)_9Cp$, **10**, 210
$CoOs_3C_{14}H_9O_9$
$Os_3Co(H)_4(CO)_9Cp$, **10**, 210
$CoOs_3C_{15}H_7O_{10}$
$Os_3Co(Cp)(CO)_{10}(H)_2$, **10**, 7
$CoPb_2C_{39}H_{30}O_3$
$(Ph_3Pb)_2Co(CO)_3$, **2**, 312
$CoPd_2C_{26}H_{49}O_4P_2$
$CoPd_2(CO)_4(MeC_3H_4)(PPr^i_3)_2$, **10**, 157
$CoRhC_7O_7$
$Co(CO)_4Rh(CO)_3$, **10**, 62
$CoRh(CO)_7$, **8**, 30, 505, 508; **10**, 45
$CoRhC_8O_8$
$CoRh(CO)_8$, **8**, 506
$CoRhC_{11}H_{10}O_5$
$Co(CO)_3(BuC_2H)Rh(CO)_2$, **10**, 64
$CoRhC_{11}H_{18}O_5P_2$
$Co(CO)_4Rh(Co)(PMe_3)_2$, **10**, 62
$CoRhC_{17}H_{15}O_3$
$[Co(CO)_3(C_7H_7)Rh(nbd)]^+$, **10**, 60
$CoRhC_{17}H_{17}$
$CpCo(C_7H_7)RhCp$, **10**, 39
$CoRhC_{17}H_{30}O_5P_2$
$CoRh(CO)_5(PEt_3)_2$, **8**, 508
$CoRhC_{19}H_{38}O_3P_2$
$(CO)_2Co(H)(PBu^t_2)Rh(CO)(HPBu^t_2)$, **10**, 45
$CoRhC_{20}H_{10}O_6$
$CoRh(CO)_6(PhCCPh)$, **8**, 508
$CoRhC_{20}H_{30}S_8$
$CoRhCp^*_2(S_8)$, **10**, 64
$CoRhC_{23}H_{28}$
$[CoCp(C_8H_8)RhCp^*]^+$, **10**, 64
$CoRhC_{24}H_{15}O_6P$
$CoRh(CO)_6(PPh_3)$, **10**, 45
$CoRhC_{28}H_{40}O_2$
$CoCp^*[CHCMe_2CMeC(O)](CO)RhCp^*$, **10**, 64
$CoRhC_{53}H_{45}O_3P_4$
$[Co(CO)_2(H)(dppm)_2Rh(CO)]^+$, **10**, 64
$CoRhSiC_{14}H_{11}O_6$
$Co(CO)_4Rh(SiMe_2Ph)(CO)_2$, **10**, 62
$CoRh_2Ru_2C_{19}H_{14}O_8P$

$Ru_2CoRh_2(H)(PMe)(CO)_8Cp_2$, **10**, 243
$CoRh_6C_{15}NO_{15}$
$[Rh_6CoN(CO)_{15}]^{2-}$, **8**, 513
$CoRuC_{19}H_{10}O_7$
$Ru(CO)_4(PPh_2)Co(CO)_3$, **10**, 61
$CoRuC_{45}H_{35}O_4P_2$
$CpRu(PPh_3)_2Co(CO)_4$, **10**, 47
$CoRuC_{53}H_{40}O_5P_3$
$RuCo(PPh_2)(CO)_5(PPh_3)_2$, **7**, 366
$CoRuWC_{15}H_9O_8$
$HWRuCo(CMe)Cp(CO)_8$, **10**, 41
$CoRuWC_{19}H_{10}O_8P$
$WRuCoCp(CO)_8(PPh)$, **10**, 19, 20
$CoRu_2C_{15}H_6NO_9$
$Ru_2Co(H)(NPh)(CO)_9$, **10**, 218
$CoRu_3C_{12}H_2O_{12}$
$[Ru_3Co(H)_2(CO)_{12}]^-$, **10**, 203
$CoRu_3C_{12}H_3O_{12}$
$H_3Ru_3Co(CO)_{12}$, **10**, 43, 58
$CoRu_3C_{12}NO_{12}$
$Ru_3Co(N)(CO)_{12}$, **10**, 197
$CoRu_3C_{13}O_{13}$
$[Ru_3Co(CO)_{13}]^-$, **10**, 19, 20, 48
$CoRu_3C_{14}H_7O_{12}S$
$Ru_3Co(H)(SMe_2)(CO)_{12}$, **10**, 234
$CoRu_3C_{21}H_{20}NO_{13}$
$[Et_4N][Ru_3Co(CO)_{13}]$, **10**, 51
$CoRu_5C_{43}H_{25}O_{12}P_2$
$CpCoRu_5(CO)_{12}(C_2Ph)(PPh)(PPh_2)$, **10**, 146
$CoSiC_7H_9O_4$
$[(TMS)Co(CO)_4]^+$, **12**, 975
$CoSmC_{24}H_{40}I_2O_9$
$[SmI_2(THF)_5][Co(CO)_4]$, **4**, 108
$CoSn_2C_{39}H_{30}O_3$
$Co(SnPh_3)_2(CO)_3^-$, **8**, 5
$CoTiC_{16}H_{27}O_7$
$Ti(OBu^t)_3Co(CO)_4$, **10**, 27
$CoTiC_{20}H_{20}Cl_2$
$[CoCp_2][TiCl_2Cp_2]$, **4**, 248
$CoTiC_{22}H_{26}O_6$
$[TiCp_2(THF)_2][Co(CO)_4]$, **4**, 259
$CoTiC_{24}H_{30}O_4$
$TiCp*_2[Co(CO)_4]$, **4**, 227
$CoTlB_2C_{17}H_{23}$
$CpTl[Me_2C_2(BMe)_2CH]CoCp$, **1**, 206, 207
$CoU_3C_{51}H_{45}N_6$
$[Cp_3U]_3Co(CN)_6$, **4**, 190
$CoVC_{14}H_{10}O_4$
$Cp_2V[Co(CO)_4]$, **5**, 45
$CoWC_{13}H_8O_5$
$Co(CH_2CCH)(CO)_3W(CO)_2(Cp)$, **12**, 687
$CoWC_{37}H_{28}O_7P$
$W(CO)_4[CTolCMeCMeC(O)]Co(CO)_2(PPh_3)$, **10**, 44
$CoWC_{39}H_{31}O_4P_2$
$W(CO)_2Cp(PPh_2)(H)Co(CO)_2(PPh_3)$, **10**, 44
$CoWZnC_{16}H_5O_{11}$
$[Co(CO)_4]_2[WCp(CO)_3]Zn^-$, **10**, 340
$CoW_3C_{27}H_{21}O_9$
$W_3Co(CO)_9(CpMe)_3$, **10**, 7, 210
$Co_2AsMoC_{17}H_{21}O_6P$
$MoCo_2(PBu^t)(AsMe_2)(CO)_6Cp$, **10**, 222
$Co_2AsMoC_{18}H_{15}O_3$
$MoCo_2(As)(CO)_3Cp_3$, **10**, 218
$Co_2As_2B_8C_{10}H_{18}$
$[(Cp)Co]_2B_8H_8As_2$, **1**, 266, 267
$Co_2As_2Cr_3C_{20}O_{20}$
$Cr_3Co_2(As)_2(CO)_{20}$, **10**, 218
$Co_2As_2PtC_{33}H_{24}O_7$
$Co_2Pt(CO)_7(dpae)$, **10**, 39, 47
$Co_2BC_{14}H_{18}BO_6$
$(CO)_6Bu^t_2(BNCo_2)$, **1**, 211
$Co_2BC_{20}H_{23}$
$Co_2(CO)_6(CBC_8H_{14})(CBu^t)$, **1**, 139
$Co_2B_2C_{22}H_{25}P$
$Cp_2Co_2(PPh_2)B_2H_5$, **1**, 309, 310
$Co_2B_2C_{52}H_{56}P_4$
$(dppe)_2Co_2[BH_4]_2$, **1**, 285
$Co_2B_4C_{20}H_{32}N_4O_4$
$Co_2(CO)_4]C_4H_4B_2(NMe_2)_2]_2$, **1**, 209
$Co_2B_4C_{20}H_{36}$
$Cp*_2Co_2B_4H_6$, **1**, 321
$Co_2B_4NiC_{26}H_{40}$
$CpCo(C_3B_2Me_5)Ni(C_3B_2Me_5)CoCp$, **9**, 178
$Co_2B_5C_{20}H_{37}$
$Cp*_2Co_2B_5H_7$, **1**, 327
$Co_2B_5C_{20}H_{39}$
$Cp*_2Co_2B_5H_9$, **1**, 324, 325
$Co_2B_7C_{10}H_{22}S_2$
$Cp_2*Co_2S_2B_7H_7$, **1**, 269
$Co_2B_8C_{10}H_{22}$
$Cp_2Co_2B_8H_{12}$, **1**, 321, 335
$Co_2B_8C_{20}H_{42}$
$(Cp*Co)_2B_8H_{12}$, **1**, 335
$Co_2B_9WC_{18}H_{22}O_6$
$[[C_2B_9H_9Me_2]WCo_2(CO)_6(\equiv CTol)]^-$, **10**, 8, 9
$Co_2B_{10}C_{13}H_{28}O_5S_2$
$(CO)_4(CO)(SEt_2)_2Co_2B_{10}H_8$, **1**, 361
$Co_2C_5H_9F_{12}N_3O_2$
$[MeN(PF_2)_2]_3Co_2(CO)_2$, **8**, 101
$Co_2C_7H_{21}F_{17}N_5P_8$
$Me_2NPF_3Co[NMe(PF_2)_2]_3CoPF_2NMe_2$, **8**, 19
$Co_2C_7O_7$
$Co_2(CO)_7$, **8**, 30
$Co_2C_8H_2O_6$
$Co_2(CO)_6(HCCH)$, **12**, 704
$Co_2C_8H_3O_5$
$Co_2(CO)_5(CH_2CCH)$, **12**, 686
$Co_2C_8O_8$
$Co_2(CO)_8$, **8**, 6, 30; **11**, 407; **12**, 354, 704, 747
$Co_2C_9H_3O_6$
$[(HCCCH_2)Co_2(CO)_6]^-$, **12**, 690
$Co_2C_9H_4O_6$
$Co_2(CO)_6(HCCMe)$, **12**, 707
$Co_2C_{10}H_{10}NO_2$
$CpCo(NO)_2CoCp$, **8**, 13, 14
$Co_2C_{10}N_{10}$
$[Co_2(CN)_{10}]^{6-}$, **8**, 12
$Co_2C_{11}H_7O_6$
$Co_2(HCCCMe_2)(CO)_6$, **12**, 687
$Co_2C_{12}H_{10}O_2$
$[CoCp(CO)]_2^-$, **8**, 67, 73
$Co_2C_{12}H_{12}O$
$CpCo(CO)(CH_2)CoCp$, **8**, 94
$Co_2C_{12}H_{14}Cl$
$Cp'CoClCoCp'$, **8**, 22
$Co_2C_{13}H_8O_6$
$Co_2(CO)_6(nbd)$, **8**, 79
$Co_2C_{13}H_{10}I_4O_2$
$I_2(CO)Co[(C_5H_4)_2CH_2]Co(CO)I_2$, **8**, 22
$Co_2C_{13}H_{12}O_2$
$Cp(CO)Co(CH_2)Co(CO)Cp$, **8**, 94, 95
$Co_2C_{13}H_{28}O_3P_4$
$Co_2(CO)_3(dmpm)_2$, **8**, 18
$Co_2C_{14}H_6O_6$
$Co_2(CO)_6(PhCCH)$, **12**, 718
$Co_2C_{14}H_{10}O_4$
$[CoCp_2][Co(CO)_4]$, **8**, 89
$Co_2C_{14}H_{14}O_2$
$[CoCp'(CO)]_2$, **8**, 93
$Co_2C_{14}H_{16}O_2$
$[CoCpMeCO]_2$, **8**, 67, 95
$Co_2C_{14}H_{22}P_2$
$CpCo(PMe_2)_2CoCp$, **8**, 97
$Co_2C_{14}H_{30}N_2O_{12}P_4$
$[MeN[P(OMe)_2]_2]_2Co_2(CO)_4$, **8**, 101
$Co_2C_{14}H_{36}N_4P$
$(Me_3P)(N_2)Co(PBu^t_2)Co(N_2)(PMe_3)$, **8**, 97
$Co_2C_{16}H_{18}O_6$
$Bu^t_2C_2[Co(CO)_3]_2$, **8**, 98
$(C_8H_{17}CCH)Co_2(CO)_6$, **12**, 692, 693
$Co_2C_{18}H_{16}O_4$
$Co_2(CO)_4(nbd)_2$, **8**, 79
$Co_2C_{18}H_{28}N_2P_2$
$[Cp(MeNC)Co(PMe_2)_2Co(MeNC)Cp]^{2+}$, **8**, 97
$Co_2C_{20}H_{36}O_4P_2$
$(CO)_2Co(PBu^t_2)_2Co(CO)_2$, **8**, 96
$Co_2C_{22}H_{30}O_2$
$Cp*Co(CO)_2CoCp*$, **10**, 63
$Co_2C_{23}H_{18}O_8$
$(BzOCH_2CCH_2OBz)Co_2(CO)_6$, **12**, 693
$Co_2C_{23}H_{38}B_8$
$CH_2[(C_5H_4)Co(Et_2C_2B_4H_4)]_2$, **1**, 387
$Co_2C_{24}H_{72}O_{24}P_8$

$Co_2[P(OMe)_3]_8$, **8**, 92

$Co_2C_{25}H_{28}Cl_2P$
$CpClCo(PMe_2)(CH_2PPh_2)$-$CoClCp$, **8**, 97

$Co_2C_{25}H_{28}P_2$
$CpCo(PMe_2)(CH_2PPh_2)CoCp$, **8**, 97

$Co_2C_{28}H_{72}B_5P_2$
$[(Et_3P)_3CoH_2](Et_3P)Co$-$(Me_2C_2B_5H_4)$, **1**, 395

$Co_2C_{30}H_{20}O_6$
$(CO)_3Co(PPh_2)_2Co(CO)_3$, **8**, 96

$Co_2C_{30}H_{27}O_8P$
$Co_2(CO)_5(PhCCH)$(glyphos), **12**, 718

$Co_2C_{30}H_{39}O_2P$
$Cp(CO)Co[P(Bu^t{}_3C_6H_2)Co$-$(CO)Cp$, **8**, 97

$Co_2C_{30}H_{54}O_6P_2$
$Co_2(CO)_6(PBu^t{}_3)_2$, **8**, 8

$Co_2C_{32}H_{56}B_6$
$Cp^*{}_2Co_2(Et_4C_4B_6H_6)$, **1**, 377

$Co_2C_{33}H_{36}IO_3P_4$
$[(Co)CoI(CO)$(dmpm)-(dppm)$Co(CO)]^+$, **8**, 18

$Co_2C_{36}H_{44}$
$Co_2(C_9H_{11})_4$, **8**, 93

$Co_2C_{41}H_{33}O_4P_5$
$(CO)_2Co(H)(PPh_2)$(dppm)Co-$(CO)_2$, **8**, 18

$Co_2C_{53}H_{44}IO_3P_4$
$[Co_2(CO)_3(dppm)_2I]^+$, **8**, 98

$Co_2C_{53}H_{44}O_3P_4$
$Co_2(CO)_3(dppm)_2$, **8**, 98

$Co_2C_{58}H_{68}P_7$
$Me_3PCo(PMe_2)(dppm)_2$-$CoPMe_3$, **8**, 18

$Co_2C_{62}P_6$
$[Co(PMe_3)_3]_2$(nbd), **8**, 73

$Co_2CdC_8O_8$
$Cd[Co(CO)_4]_2$, **10**, 48

$Co_2Cr_2C_{14}H_{10}O_4S_3$
$Cr_2Co_2(S)_3(CO)_4Cp_2$, **10**, 232, 233, 237

$Co_2Cr_2C_{20}H_{20}S_3$
$Cr_2Co_2(S)_3Cp_4$, **10**, 232, 233

$Co_2Cr_2Fe_4C_{24}H_{14}O_{12}S_7$
$Cr_2Fe_4Co_2(S)_7(CO)_{12}Cp'_2$, **10**, 239

$Co_2CuC_2O_8$
$[Co(CO)_4CuCo(CO)_4]^-$, **10**, 63

$Co_2FeC_9HNO_9$
$FeCo_2(NH)(CO)_9$, **10**, 197, 218

$Co_2FeC_9O_9S$
$FeCo_2(S)(CO)_9$, **10**, 237

$Co_2FeC_9O_9Se$
$FeCo_2(CO)_9Se$, **7**, 74

$Co_2FeC_9O_9Te$
$FeCo_2(CO)_9Te$, **7**, 77

$Co_2FeC_{10}H_3O_9P$
$FeCo_2(PMe)(CO)_9$, **10**, 226

$Co_2FeC_{15}H_5O_9P$
$FeCo_2(PPh)(CO)_9$, **10**, 227

$Co_2FeC_{15}H_6O_9P$
$[FeCo_2(PPhH)(CO)_9]^-$, **10**, 227

$Co_2FeC_{16}H_{11}O_8PS$
$FeCo_2(S)(CO)_8(PMe_2Ph)$, **10**, 244

$Co_2FeC_{21}H_{11}O_9P_2$
$FeCo_2(PPh)(PPhH)(CO)_9$, **10**, 222

$Co_2Fe_2C_{23}H_{10}O_{11}P_2$
$Fe_2Co_2(CO)_{11}(PPh)_2$, **10**, 45

$Co_2Fe_2Ge_2C_{17}H_6O_{15}$
$Fe_2(CO)_7[GeMe[Co(CO)_4]]_2$, **7**, 30

$Co_2Fe_4HgC_{28}H_{16}O_{16}$
$[(CoMe)Cp(CO)_7CoFe_2]_2(Hg)$, **10**, 332, 339

$Co_2GaC_{42}H_{30}ClO_6P_2$
$GaCl[Co(CO)_3(PPh_3)]_3$, **1**, 555

$Co_2GeC_{17}H_6O_{15}$
$(CO)_3Co[Ge(Co(CO)_4Me]_2Co$-$(CO)_4$, **8**, 96

$Co_2GeMo_2C_{21}H_{10}O_{11}$
$Mo_2Co_2(Ge)(CO)_{11}Cp_2$, **10**, 215

$Co_2HgC_8O_8$
$[(CO)_4Co]_2Hg$, **10**, 330

$Co_2HgC_{12}H_8O_4$
$Co(C_4H_4)(CO)_2HgCo(C_4H_4)$, **10**, 63

$Co_2InC_{16}H_{19}O_9$
$In(CH_2TMS)(THF)[Co(CO)_4]_2$, **1**, 564

$Co_2InC_{42}H_{30}ClO_6P_2$
$InCl[Co(CO)_3(PPh_3)]_2$, **1**, 565

$Co_2InC_{60}H_{45}ClO_6P_3$
$InCl(PPh_3)[Co(CO)_3(PPh_3)]_2$, **1**, 565

$Co_2IrC_{22}H_{25}O_2$
$Co_2Ir(CO)_2Cp_2Cp^*$, **10**, 209, 211

$Co_2IrC_{22}H_{40}O_2$
$Co_2Ir(CO)_2Cp_2Cp^*$, **8**, 512

$Co_2Ir_2C_8F_{12}O_8P_4$
$Co_2Ir_2(CO)_8(PF_3)_4$, **8**, 512

$Co_2Ir_2C_{12}O_{12}$
$Co_2Ir_2(CO)_{12}$, **8**, 505, 510; **10**, 43

$Co_2Ir_2C_{18}H_{27}O_{18}P_3$
$Co_2Ir_2(CO)_9[P(OMe)_3]_3$, **8**, 511

$Co_2K_2C_{16}H_{44}P_4$
$[K[Co(C_2H_4)(PMe_3)]_2]_2$, **1**, 26

$Co_2MnC_{11}O_{10}$
$[MnCo_2(CO)_9(CCO)]^-$, **10**, 132

$Co_2MoC_{15}H_5F_6O_9P$
$[CpMoCo_2(CO)_8(CCO)][PF_6]$, **10**, 109

$Co_2MoC_{23}H_{22}O_{10}$
$Cp^*MoCo_2(CO)_8(CCo_2Pr^i)$, **10**, 112

$Co_2MoNiC_{19}H_{12}O_8$
$NiCo_2Mo(CO)_8CpCp'$, **9**, 173

$Co_2Mo_2C_{14}H_{14}O_2S_4$
$Mo_2Co_2(S)_4(CO)_2Cp'_2$, **10**, 236, 237

$Co_2Mo_2C_{15}H_{12}O_4S_3$
$Cp(C_5H_4Me)_2Mo_2Co_2(CO)_4S_3$, **10**, 51

$Co_2Mo_2C_{16}H_{14}O_4S_3$
$Mo_2Co_2(S)_3(CO)_4Cp'_2$, **10**, 236, 237

$Co_2Mo_2C_{31}H_{32}O_7P_2$
$Mo_2Co_2[C_6H_4(PBu^t)_2]$-$(CO)_7Cp_2$, **10**, 224

$Co_2Mo_2C_{34}H_{40}O_4$
$Mo_2Co_2(CO)_4Cp_2Cp^*{}_2$, **10**, 209

$Co_2Mo_3C_{38}H_{46}N_2O_8$
$Mo_3Co_2(N)(NH)(CO)_8Cp_3^*$, **10**, 218

$Co_2NiC_{13}H_8O_6$
$NiCo_2(CMe)(CO)_6Cp$, **9**, 173; **10**, 156

$Co_2Ni_{10}C_{21}O_{20}$
$[Ni_{10}Co_2C(CO)_{20}]^{2-}$, **10**, 156

$Co_2PdC_{31}H_{24}O_7P_2$
$Co_2Pd(CO)_7$(dppe), **10**, 48

$Co_2PtC_{15}H_{12}O_7$
$PtCo_2(CO)_7$(cod), **9**, 563

$Co_2PtC_{20}H_{14}NO_8$
$Co_2Pt(CO)_8(picoline)_2$, **10**, 211

$Co_2PtC_{22}H_{22}N_2O_8$
$Pt[Co(CO)_4]_2(CNCy)_2$, **10**, 39, 46, 49

$Co_2PtC_{33}H_{22}O_7P_2$
$Co_2Pt(CO)_7$(dppe), **10**, 46, 47

$Co_2PtC_{33}H_{24}O_7P_2$
$PtCo_2(CO)_7$(dppe), **10**, 39

$Co_2Pt_2C_{14}H_{12}O_6$
$Pt_2Co_2(CO)_6$(cod), **9**, 563

$Co_2Pt_2C_{22}H_{22}O_8$
$Pt_2Co_2(CO)_8$(cod)(EtCCEt), **9**, 563

$Co_2Pt_2C_{43}H_{30}O_7P_2$
$Co_2Pt_2(CO)_7(PPh_3)_2$, **10**, 46, 47

$Co_2Pt_2C_{44}H_{30}O_8P_2$
$Co_2Pt_2(CO)_8(PPh_3)_2$, **10**, 37, 40, 49, 60

$Co_2Pt_3C_{50}H_{44}O_6$
$Pt_3Co_2(CO)_6(cod)_2(PhCCPh)_2$, **9**, 563

$Co_2ReC_{22}H_{15}O_7$
$ReCo_2(CO)_7(C_7H_8)$(CTol), **10**, 128

$Co_2RhC_{31}H_{51}B_5$
$CpCo(Et_2C_2B_3H_3)RhCp^*$, **1**, 388

$Co_2Rh_2C_8F_{12}O_8P_4$
$Co_2Rh_2(CO)_8(PF_3)_4$, **8**, 511

$Co_2Rh_2C_{10}F_6O_{10}P_2$
$Co_2Rh_2(CO)_{10}(PF_3)_2$, **8**, 511

$Co_2Rh_2C_{12}O_{12}$
$Co_2Rh_2(CO)_{12}$, **8**, 505, 506, 508; **10**, 7, 40, 43, 45, 46, 48, 50, 58, 60

$Co_2Rh_2C_{24}H_{10}O_{10}$
$Co_2Rh_2(CO)_{10}(PhCCPh)$, **8**, 508; **10**, 36

$Co_2Rh_2C_{27}H_{30}O_7$
$Co_2Rh_2(CO)_7Cp^*{}_2$, **8**, 512

$Co_2Rh_2C_{29}H_{15}O_{11}P$
$Co_2Rh_2(CO)_{11}(PPh_3)$, **8**, 511; **10**, 191

$Co_2Rh_2RuC_{18}H_{10}O_8S$
$RuCo_2Rh_2(S)(CO)_8Cp_2$, **10**, 243

$Co_2Rh_4C_{14}O_{13}$
$[Co_2Rh_4C(CO)_{13}]^{2-}$, **8**, 512
$Co_2Rh_4C_{16}O_{16}$
$Co_2Rh_4(CO)_{16}$, **8**, 510, 512
$Co_2RuC_9O_9S$
$RuCo_2(S)(CO)_9$, **10**, 236, 237
$Co_2RuC_{11}O_{11}$
$RuCo_2(CO)_{11}$, **10**, 43, 52, 58, 197
$Co_2RuC_{13}H_6O_9$
$RuCo_2(CO)_9(MeC_2Me)$, **10**, 142
$Co_2RuC_{23}H_{10}O_9$
$Co_2Ru(CO)_9(PhCCPh)$, **9**, 98
$Co_2Ru_2C_{12}H_2O_{12}$
$H_2Ru_2Co_2(CO)_{12}$, **10**, 35
$Ru_2Co_2(H)_2(CO)_{12}$, **10**, 197
$Co_2Ru_2C_{13}O_{13}$
$Ru_2Co_2(CO)_{13}$, **10**, 35, 197
$Co_2Ru_2C_{15}H_6O_{11}$
$Ru_2Co_2(C_2Me_2)(CO)_{11}$, **10**, 35
$Co_2SiC_{12}H_{11}O_6$
$(TMS\text{-}CCCH_2)Co_2(CO)_6$, **12**, 693
$Co_2TiC_{28}H_{30}O_8$
$TiCp^*{}_2[(OC)Co(CO)_3]_2$, **4**, 227, 405
$Co_2V_2C_{12}H_{14}N_2O_2S_4$
$(MeCp)_2V_2S_4[Co(NO)]_2$, **5**, 32
$Co_2WC_{14}H_6O_8$
$WCo_2(CH)Cp(CO)_8$, **10**, 49
$Co_2WC_{21}H_{12}O_8$
$CpWCo_2(CO)_8(\equiv CTol)$, **10**, 7, 8
$Co_2ZnC_8O_8$
$Co(CO)_4ZnCo(CO)_4$, **10**, 63
$Zn[Co(CO)_4]_2$, **10**, 48
$Co_3Ag_3C_{108}H_{108}N_{12}$
$Ag_3Co_3(CNC_6H_3Me_2)_{12}$, **10**, 310
$Co_3AuRuC_{30}H_{15}O_{12}P$
$RuCo_3(AuPPh_3)(CO)_{12}$, **10**, 48
$Co_3AuRuC_{48}H_{30}O_{12}P_2$
$[Au(PPh_3)_2][RuCo_3(CO)_{12}]$, **10**, 48
$Co_3BC_{27}H_{25}P$
$(CpCo)_3PBPh_2$, **1**, 297, 298
$Co_3B_3C_{16}H_{18}O$
$Co_3B_3H_3(CO)Cp_3$, **8**, 432
$(CpCo)_3(CO)B_3H_3$, **1**, 318
$Co_3B_4C_{30}H_{49}$
$Co_3B_4H_4Cp^*{}_3$, **8**, 432
$(Cp^*Co)_3B_4H_4$, **1**, 322
$Co_3BiC_9O_9$
$Co_3(CO)_9(Bi)$, **8**, 429
$Co_3BiC_{12}O_{12}$
$Co_3(CO)_{12}(Bi)$, **8**, 429
$Co_3C_5H_8NO$
$[Co_3(CH)[N(O)CPr)]]^+$, **8**, 433
$Co_3C_9H_5NO_7S_2$
$Co_3(CO)_7(S)(MeCSNH_2)$, **8**, 429
$Co_3C_9O_9S$
$SCo_3(CO)_9$, **8**, 428
$Co_3C_{10}O_{10}$
$[Co_3(CO)_{10}]^-$, **8**, 6, 421
$Co_3C_{11}H_3O_9$
$Co_3(CO)_9(CMe)$, **8**, 421, 423–5
$Co_3C_{11}O_{11}$
$[Co_3(CO)_9(CCO_2)]^-$, **8**, 424
$Co_3C_{12}H_2F_3O_9$
$Co_3(CO)_9(CCH_2CF_3)$, **10**, 139
$Co_3C_{12}H_3O_{10}$
$[Co_3(CO)_9[CC(O)Me]]^+$, **8**, 440
$Co_3C_{12}H_6NO_8$
$Co_3(CO)_8(CMe)(NCMe)$, **8**, 425
$Co_3C_{15}H_{15}N_2O_2$
$Co_3(NO)_2Cp_3$, **8**, 435
$Co_3C_{15}H_{15}S_2$
$Co_3(S)_2Cp_3$, **8**, 436
$Co_3C_{16}H_5O_9$
$Co_3(CO)_9(CPh)$, **8**, 423
$Co_3C_{16}H_{13}O_4$
$Co_3(CO)_4(CMe)Cp_2$, **8**, 432
$Co_3C_{16}H_{15}S_2$
$Co_3(S)(CS)Cp_3$, **8**, 436
$Co_3C_{17}H_{15}O_2$, **8**, 430
$Co_3(CO)_2Cp_3$, **8**, 429
$Co_3C_{17}H_{18}$
$[Co_3H(CH_2)Cp_3]^+$, **8**, 532
$Co_3C_{17}H_{18}NS$
$Co_3(S)(CNMe)Cp_3$, **8**, 436
$Co_3C_{18}H_{15}O_3$
$Co_3(CO)_3Cp_3$, **8**, 431
$Co_3C_{18}H_{18}O$
$[Co_3(CH)(HCCOH)Cp_3]^+$, **8**, 433
$Co_3C_{19}H_{21}$
$Co_3(CMe)_2Cp_3$, **8**, 432
$Co_3C_{20}H_{18}O_2$
$[Co_3(CO)_2(C_6H_6)_3]^+$, **8**, 438
$Co_3C_{20}H_{21}O$
$Co_3(MeCCMe)(CO)Cp_3$, **8**, 433
$Co_3C_{20}H_{23}$
$Co_3(CPr)(CH)Cp_3$, **8**, 433
$Co_3C_{21}H_{21}$
$Co_3(C_6H_6)Cp_3$, **8**, 437
$Co_3C_{21}H_{21}O_3$
$Co_3(CO)_3(MeCp)_3$, **8**, 430
$Co_3C_{21}H_{25}$
$Co_3((CH)CBu^t)Cp_3$, **8**, 433
$Co_3C_{21}H_{55}O_3P_6$
$Co_3H(CO)_3(PMe_3)_6$, **8**, 422
$Co_3C_{22}H_{15}O_5$
$Co_3(CO)_5(CPh)Cp_2$, **8**, 423
$Co_3C_{23}H_{30}NO$
$Co_3(CO)(NH)(MeCp)_2Cp^*$, **8**, 436
$Co_3C_{24}H_{25}$
$Co_3(C_9H_{10})Cp_3$, **8**, 437
$Co_3C_{25}H_{30}NO_6S$
$Co_3(S)(NBu^t)(C_5H_4CO_2Me)_3$, **8**, 536
$Co_3C_{28}H_{26}$
$Co_3(CPh)(Tol)_3$, **8**, 438
$Co_3C_{29}H_{25}$
$Co_3(CPh)_2Cp_3$, **8**, 434
$Co_3C_{33}H_{45}O_3$
$Co_3(CO)_3Cp^*{}_3$, **8**, 431
$Co_3C_{34}H_{25}O_7P_2$
$Co_3(CO)_7(dmmp)(CMe)$, **8**, 426
$Co_3C_{45}H_{34}O_6P_3$
$Co_3(CO)_6(tripod)(CMe)$, **8**, 426
$Co_3C_{54}H_{93}P_3S_6$
$Co_3(C_6H_4S_2)_3(PBu_3)_3$, **8**, 429
$Co_3C_{60}H_{30}O_4P_2$
$[Co(CO)_2PPh_3]_3$, **8**, 422
$Co_3C_{62}H_{47}O_5P_4$
$Co_3(CO)_5(CPh)(dppm)_2$, **8**, 426
$Co_3FeC_{12}HO_{12}$
$HFeCo_3(CO)_{12}$, **10**, 47, 49, 52, 58
$Co_3FeC_{12}O_{12}$
$[FeCo_3(CO)_{12}]^-$, **10**, 47, 52, 197
$FeCo_3(CO)_{12}$, **10**, 341
$Co_3FeC_{15}H_5O_9S$
$FeCo_3(CS)(CO)_9Cp$, **10**, 209
$Co_3FeC_{16}H_{12}NO_{12}$
$[Me_4N][FeCo_3(CO)_{12}]$, **10**, 49
$Co_3FeC_{20}H_{10}O_9$
$Co_3(CO)_9(CFc)$, **8**, 426
$Co_3FeC_{20}H_{20}NO_{12}$
$[Et_4N][FeCo_3(CO)_{12}]$, **10**, 45
$Co_3FeC_{22}H_{16}NO_{12}$
$[(PhCH_2)Me_3N][FeCo_3(CO)_{12}]$, **10**, 45
$Co_3FeC_{27}H_{26}$
$Co_3(CH)(CFc)Cp_3$, **8**, 435
$Co_3FeC_{29}H_{15}O_{11}P$
$[FeCo_3(CO)_{11}(PPh_3)]^-$, **10**, 191, 192
$Co_3Fe_2C_{27}H_{35}$
$Co_3(CFc)_2Cp_3$, **8**, 433, 435
$Co_3Fe_3C_{20}O_{18}$
$[Fe_2Co(CO)_9CCFeCo_2(CO)_9]^{2-}$, **7**, 22
$Co_3GaC_{63}H_{45}O_9P_3$
$Ga[Co(CO)_3(PPh_3)]_3$, **1**, 556
$Co_3GeC_{10}H_3O_9$
$Co_3(CO)_9(GeMe)$, **8**, 428
$Co_3HgC_{12}O_{12}$
$Hg(Co(CO)_4)_3$, **8**, 6
$Co_3InC_{20}H_{20}INO_{12}$
$I[NEt_4][InI[Co(CO)_4]_3]$, **1**, 565
$Co_3InC_{44}H_{30}Cl_2O_{11}P_2$
$Co(CO)_3(PPh_3)_2[InCl_2[Co\text{-}(CO)_4]_2]$, **1**, 565
$Co_3InC_{63}H_{45}O_9P_3$
$In[Co(CO)_3(PPh_3)]_3$, **1**, 565
$Co_3MnC_{14}O_{14}$
$Co_3(CO)_9[CMn(CO)_5]$, **10**, 127
$Co_3Mo_2C_{30}H_{30}NO_{10}$
$Mo_2Co_3(N)(CO)_{10}Cp^*{}_2$, **10**, 218
$Co_3NaC_{39}H_{117}O_{39}P_{13}$
$[Co[P(OMe)_3]_5{}^+][Co[P\text{-}(OMe)_3]_4Na[(MeO)_3P]_4\text{-}Co]$, **8**, 92
$Co_3Ni_7C_{17}O_{15}$
$[Co_3Ni_7C_2(CO)_{15}]^{3-}$, **9**, 20
$Co_3Ni_7C_{18}O_{16}$
$[Co_3Ni_7(CO)_{16}C_2]^{2-}$, **10**, 156
$Co_3Ni_7C_{18}O_{16}$

$[Co_3Ni_7C_2(CO)_{16}]^{2-}$, **9**, 20
$Co_3Ni_9C_{21}O_{20}$
$[Co_3Ni_9C(CO)_{20}]^{3-}$, **9**, 20; **10**, 15, 156
$Co_3OsC_{20}H_{20}NO_{12}$
$[Et_4N][OsCo_3(CO)_{12}]$, **10**, 48
$Co_3ReC_{18}H_{18}N_2O_{10}$
$ReCo_3(NBu^t)_2(CO)_{10}$, **10**, 218
$Co_3RhC_{12}O_{12}$
$Co_3Rh(CO)_{12}$, **8**, 505, 510, 511; **10**, 40, 43, 46, 48, 50
$Co_3Rh_9C_{30}O_{30}$
$[Co_3Rh_9(CO)_{30}]^{2-}$, **10**, 42
$Co_3RuC_{11}NO_{12}$
$RuCo_3(CO)_{11}(NO)$, **10**, 197
$Co_3RuC_{12}HO_{12}$
$HRuCo_3(CO)_{12}$, **10**, 43, 45, 48, 52, 58, 203
$Co_3RuC_{12}O_{12}$
$[RuCo_3(CO)_{12}]^-$, **10**, 52, 191, 192, 203
$Co_3RuC_{19}H_{19}OP$
$[RuCo_3(CO)(PPh_2)(CCBu^t)]^-$, **10**, 145
$Co_3RuC_{20}H_{20}NO_{12}$
$[Et_4N][RuCo_3(CO)_{12}]$, **10**, 51
$Co_3RuC_{30}H_{30}NO_{10}$
$[Et_4][RuCo_3(CO)_{10}(C_2Ph_2)]$, **10**, 48
$Co_3SiC_{19}H_{24}NO$
$Co_3(CO)(N\text{-}TMS)Cp_3$, **8**, 435
$Co_3Si_2Yb_2C_{60}H_{86}O_4$
$[Cp^*{}_2Yb]_2(CO)_4Co_3(TMS\text{-}C_5H_4)_2$, **4**, 108
$Co_3TiC_{20}H_{10}O_{10}$
$TiCp_2[OCCo_3(CO)_9]$, **4**, 258
$Co_3TiC_{22}H_{14}O_{10}$
$TiCp'_2[OCCo_3(CO)_9]$, **4**, 258
$Co_3VC_{20}H_{10}O_{10}$
$Cp_2V[OCCo_3(CO)_9]$, **5**, 45
$Co_3ZnC_{12}O_{12}$
$[Co(CO)_4]_3Zn^-$, **10**, 340
$Co_4As_6C_{72}H_{60}P_4$
$Co_4As_6(PPh_3)_4$, **8**, 443
$Co_4B_2C_{20}H_{24}$
$Co_4(H)_2(B_2H_2)Cp_4$, **8**, 449
$(H)_2[CpCo]_4B_4H_2$, **1**, 312
$Co_4B_2C_{26}H_{27}P$
$Co_4(PPh)(B_2H_2)Cp_4$, **8**, 448
$Ph[CpCo]_4PB_2H_2$, **1**, 312
$Co_4Bi_2C_{11}O_{11}$
$[Co_4(CO)_{11}Bi_2]^-$, **8**, 450
$Co_4C_{11}O_{11}Te_2$
$Co_4(CO)_{11}Te_2$, **7**, 77
$Co_4C_{12}H_2O_{10}$
$Co_4(CO)_{10}(HCCH)$, **8**, 449
$Co_4C_{12}O_{12}$
$Co_4(CO)_{12}$, **8**, 421, 439, 440; **10**, 49
$Co_4C_{13}H_3O_{12}$
$[Co_4(CO)_{11}(COMe)]^-$, **8**, 440
$Co_4C_{16}H_{18}O_{16}P_2$
$Co_4(CO)_{10}[P(OMe)_3]_2$, **8**, 441
$Co_4C_{17}H_7O_{11}PS$
$Co_4(CO)_{10}[PPh(OMe)](S)$, **8**, 447
$Co_4C_{18}H_6O_{10}S$
$Co_4(CO)_{10}(S)(HCCPh)$, **8**, 448
$Co_4C_{18}H_{25}F_{20}N_5OP_{12}$
$Co_4(CO)(PPh)_2[(F_2P)_2NMe]_5$, **8**, 447
$Co_4C_{18}H_{28}O_8P_4$
$Co_4(CO)_8(dmpm)_2$, **8**, 441
$Co_4C_{19}H_{10}F_6O_5$
$Co_4(CO)_5(CF_3CCCF_3)Cp_2$, **8**, 450
$Co_4C_{19}H_{22}F_{16}N_4O_3P_{10}$
$Co_4(CO)_3(PPh)_2[(F_2P)_2NMe]_4$, **8**, 447
$Co_4C_{20}H_{20}S_2$
$Co_4(S)_2Cp_4$, **8**, 449
$Co_4C_{20}H_{26}O_8P_4$
$Co_4(CO)_8(PPh)_2(dmpe)$, **8**, 447
$Co_4C_{21}H_{11}O_9P_2$
$Co_4(CO)_9(H)(PPh)_2$, **8**, 445
$Co_4C_{22}H_{10}O_{10}P_2$
$Co_4(CO)_{10}(PPh)_2$, **8**, 423, 445, 446
$Co_4C_{22}H_{20}O_2$
$Co_4(CO)_2Cp_4$, **8**, 444
$Co_4C_{22}H_{24}$
$Co_4(H)(CMe)Cp_4$, **8**, 444
$Co_4C_{23}H_{12}O_9P_2$
$Co_4(CO)_9(C{=}CH_2)(PPh)_2$, **8**, 445
$Co_4C_{24}H_{17}O_4P$
$Co_4(CO)_4(HCCH)(PPh_3)$, **8**, 449
$Co_4C_{24}H_{24}O$
$Co(\overline{C_2CH_2OCH_2})Cp_4$, **8**, 449
$Co_4C_{29}H_{14}O_9$
$Co_4(CO)_9$(triptycene), **8**, 444
$Co_4C_{29}H_{15}O_{11}P$
$Co_4(CO)_{11}(PPh_3)$, **8**, 441
$Co_4C_{46}H_{31}O_9P_3$
$Co_4(CO)_9$(tripod), **8**, 442
$Co_4C_{48}H_{40}O_8P_4$
$Co_4(CO)_8(tripod)[P(OMe)_3]$, **8**, 443
$Co_4C_{58}H_{44}O_8P_4$
$Co_4(CO)_8(dppm)_2$, **8**, 441
$Co_4C_{72}H_{60}P_4Se_4$
$Co_4(Se)_4(PPh_3)_4$, **8**, 444
$Co_4C_{96}H_{80}P_8$
$Co_4(PPh)_4(PPh_3)_4$, **8**, 443
$Co_4Cu_4C_{16}O_{16}$
$[Co(CO)_4Cu]_4$, **10**, 63
$Co_4Fe_2Ge_2C_{22}O_{22}$
$[Co_2(CO)_7]GeFe_2(CO)_8Ge[Co_2(CO)_7]$, **7**, 30
$Co_4GeC_{11}H_3O_{10}$
$Co_4(CO)_{10}(GeMe)$, **8**, 428
$Co_4Ge_2C_{13}H_6O_{11}$
$Co_4(CO)_{11}(GeMe)_2$, **8**, 448
$Co_4Ge_2C_{16}H_6O_{14}$
$Co_2[Ge(Me)[Co(CO)_4]]_2(CO)_6$, **8**, 448
$Co_4HgRuC_{16}O_{16}$
$RuCo_3(CO)_{12}(Hg)Co(CO)_4$, **10**, 341
$Co_4Sb_2C_{11}O_{11}$
$[Co_4(CO)_{11}Sb_2]^-$, **8**, 450
$Co_4Si_2C_{13}H_6O_{11}$
$Co_4(CO)_{11}(SiMe)_2$, **8**, 448
$Co_4Zn_2C_{14}O_{14}$
$[ZnCo(CO)_4]_2Co_2(CO)_6$, **10**, 333
$Co_4Zn_2C_{15}O_{15}$
$Co_2[ZnCo(CO)_4]_2(CO)_7]$, **10**, 49
$Co_5B_2C_{14}HO_{14}$
$Co_5(CO)_{14}B_2H$, **8**, 450
$Co_5C_{17}H_{18}O_{11}P_3$
$Co_5(CO)_{11}(PMe_2)_3$, **8**, 450
$Co_5Fe_2Ge_2C_{22}O_{22}$
$[Fe_2Co_5(Ge)_2(CO)_{22}]^-$, **10**, 215
$[Ge_2Co_5Fe_2(CO)_{22}]^-$, **7**, 30
$Co_5GeC_{16}O_{16}$
$[Co_5(CO)_{16}Ge]^-$, **8**, 428
$Co_5Ge_2C_{16}H_3O_{15}$
$Co_4(CO)_{11}(GeMe)[GeCo(CO)_4]$, **8**, 448
$Co_6BC_{60}H_{110}P_6S_8$
$[Co_6S_8(PEt_3)_6][BPh_4]$, **8**, 453
$Co_6C_6O_6S_8$
$Co_6(S)_8(CO)_6$, **8**, 453
$Co_6C_{13}NO_{13}$
$[Co_6N(CO)_{13}]^-$, **8**, 453, 454
$Co_6C_{13}O_{12}S_2$
$Co_6C(CO)_{12}(S)_2$, **8**, 452
$Co_6C_{14}O_{13}$
$[Co_6C(CO)_{13}]^{2-}$, **8**, 451
$Co_6C_{14}O_{14}$
$[Co_6(CO)_{14}]^{4-}$, **8**, 451
$Co_6C_{15}HO_{15}$
$[Co_6H(CO)_{15}]^-$, **8**, 473
$Co_6C_{15}NO_{15}$
$[Co_6N(CO)_{15}]^-$, **8**, 452, 454
$Co_6C_{15}O_{14}$
$[Co_6C(CO)_{14}]^-$, **8**, 451
$Co_6C_{15}O_{15}$
$[Co_6(CO)_{15}]^{2-}$, **8**, 451
$Co_6C_{16}O_{14}S$
$Co_6(C_2)(CO)_{14}(S)$, **8**, 452
$Co_6C_{16}O_{15}$
$[Co_6(C)(CO)_{15}]^{2-}$, **8**, 451
$Co_6C_{16}O_{16}$
$Co_6(CO)_{16}$, **8**, 451
$Co_6C_{16}O_{16}P$
$[Co_6P(CO)_{16}]^-$, **8**, 453
$Co_6C_{51}H_{31}NO_{15}P_2$
$[PPN][Co_6H(CO)_{15}]$, **8**, 451
$Co_6C_{108}H_{90}P_6S_8$
$Co_6S_8(PPh_3)_6$, **8**, 453
$Co_6Ge_2C_{19}O_{19}$
$Co_4(CO)_{11}[GeCo(CO)_4]_2$, **8**, 448
$Co_6HgRu_2C_{24}O_{24}$
$[RuCo_3(CO)_{12}]_2(Hg)$, **10**, 341
$Co_6Hg_9C_{18}O_{18}$
$Hg_9Co_6(CO)_{18}$, **10**, 341
$Co_6Ni_2C_{18}O_{16}$
$[Co_6Ni_2(CO)_{16}C_2]^{2-}$, **10**, 156
$Co_6Ni_6C_{18}O_{16}$
$[Co_6Ni_2C_2(CO)_{16}]^{2-}$, **9**, 20
$Co_6TiC_{30}H_{10}O_{20}$

$TiCp_2[OCCo_3(CO)_9]_2$, **4**, 227, 405
$Co_6ZnC_{38}H_{16}O_{22}$
$Zn[(CO)_9Co_3CC_6H_4(CH_2)_2\text{-}CO_2]_2$, **3**, 187
$Co_7C_5NO_{15}$
$[Co_7N(CO)_{15}]^{2-}$, **8**, 454
$Co_7C_{108}H_{90}BrP_6S_6$
$Co_7(S)_6(PPh_3)_6Br$, **8**, 454
$Co_7GeC_{20}O_{20}$
$[Co_3(CO)_9GeCo_4(CO)_{11}]^-$, **8**, 428
$Co_7Ge_2C_{21}O_{21}$
$[Co(CO)_3[GeCo_3(CO)_9]_2]^-$, **8**, 426, 428
$Co_8C_{19}O_{18}$
$[Co_8C(CO)_{18}]^{2-}$, **8**, 455
$Co_8C_{48}H_{40}S_7$
$[Co_8S_6(SPh)_8]^{4-}$, **8**, 455
$Co_9C_{108}H_{90}P_6Se_{11}$
$Co_9Se_{11}(PPh_3)_6$, **8**, 456
$Co_9SiC_{21}O_{21}$
$[Co_9Si(CO)_{21}]^{2-}$, **8**, 455
$Co_{11}C_{24}O_{22}$
$[Co_{11}C_2(CO)_{22}]^{3-}$, **8**, 456, 484
$Co_{13}C_{25}O_{24}$
$[Co_{13}C(CO)_{24}]^{3-}$, **8**, 456
$Co_{13}C_{26}O_{24}$
$[Co_{13}C_2(CO)_{24}]^{4-}$, **8**, 456
$Co_{14}C_{26}N_3O_{26}$
$[Co_{14}(CO)_{26}N_3]^{3-}$, **8**, 457
$Co_{18}Zn_4C_{66}O_{67}$
$Zn_4O[(CO)_9Co_3CCO_2]_6$, **3**, 187
$Co_{25}Ba_2S_{27}$
$Ba_2Co_{25}S_{27}$, **8**, 455
$CrAsC_{29}H_{20}O_5$
$Cr(CO)_5(AsPh)_4$, **5**, 260
$CrAsMnC_{15}O_{15}$
$As[Mn(CO)_5][Cr(CO)_5]_2$, **6**, 86
$CrAuC_{23}H_{16}O_5P$
$(Ph_3P)Au(H)Cr(CO)_5$, **5**, 223
$CrAu_2Ru_5C_{29}H_{30}O_{17}P_2$
$Au_2Ru_5Cr(CO)_{17}(PEt_3)_2$, **10**, 278, 288
$CrAu_2Ru_5C_{30}H_{30}O_{17}P_2$
$Au_2Ru_5CrC(CO)_{17}(PEt_3)_2$, **10**, 278
$CrBC_5H_5F_4NO_2$
$Cr(BF_4)(NO)_2Cp$, **5**, 419
$CrBC_{12}H_{37}P_4$
$Cr(BH_4)H(dmpe)_2$, **1**, 284
$CrBC_{13}H_{13}O_3$
$(CO)_3Cr[Me(C_9H_{10}B)]$, **1**, 210, 211
$CrBC_{14}H_{21}NO_3$
$(CO)_3Cr[Bu^t_2(BNC_3H_3)]$, **1**, 212
$CrBC_{14}H_{22}NO_4Si$
$(CO)_4Cr[Et_2Me_4(C_2SiNB)]$, **1**, 213
$CrBC_{20}H_{30}F_4NOS$
$[Cr(CO)(NO)(CS)(C_6Et_6)]\text{-}[BF_4]$, **5**, 500
$CrBC_{20}H_{30}F_4NO_3$
$[Cr(CO)_2(NO)(C_6Et_6)][BF_4]$, **5**, 500
$CrBC_{40}H_{34}NO_4P_2$
$[PPN][(H)_2BH_2Cr(CO)_4]$, **1**, 283, 284
$CrBC_{41}H_{34}NO_5P_2$
$[PPN][(CO)_5Cr(H)(BH_3)]$, **1**, 283, 284
$CrBFeC_{25}H_{21}O_3$
$CpFe(C_4H_4BPh)Cr(CO)_3Bz$, **7**, 195
$CrB_2C_{10}H_{22}O_4P_2$
$Cr(CO)_4(B_2H_4{\cdot}2PMe_3)$, **1**, 311
$CrB_2C_{11}H_{22}O_5P_2$
$Cr(CO)_5(B_2H_4{\cdot}2PMe_3)$, **1**, 311
$CrB_2C_{14}H_{18}O_4$
$Cr(CO)_4[Me_2C_2(BMe)_2\text{-}C{=}CMe_2]$, **1**, 208
$CrB_2C_{20}H_{36}N_2O_4$
$(CO)_4Cr[Bu_4(B_2N_2)]$, **1**, 211
$CrB_2C_{22}H_{36}N_2O_2$
$Cr(CO)_2(C_4H_4BNPr^i_2)_2$, **1**, 204
$CrB_2C_{22}H_{37}O_4P$
$Cr(CO)_4[Et_2C_2(BMe)_2CCMe_2P\text{-}Et_3]$, **1**, 208
$CrB_2C_{30}H_{40}O_4P_2$
$(CO)_4Cr[xylyl_2(B_2P_2)]$, **1**, 211
$CrB_3C_{15}H_{30}N_3O_3$
$Cr(CO)_3(B_3N_3Et_6)$, **5**, 499
$CrB_3C_{42}H_{60}O_3P_3$
$(CO)_3Cr[xylyl_3Bu^t_3(P_3B_3)]$, **1**, 211
$CrBiFe_3C_{27}O_{27}$
$[Bi[Fe(CO)_{14}][Fe_2(CO)_8][Cr\text{-}(CO)_5]]^-$, **7**, 60
$CrC_2H_{12}ClN_2O_4$
$[Cr(CN)_2(NH_3)_4]ClO_4$, **5**, 305
$CrC_2H_{12}N_6$
$[Cr(CN)_2(NH_3)_4]^+$, **5**, 307
$CrC_4Cl_2O_4$
$CrCl_2(CO)_4$, **5**, 218
CrC_5FO_5
$CrF(CO)_5$, **5**, 218
CrC_5HO_5
$CrH(CO)_5$, **5**, 221
$CrC_5H_5ClNO_2$
$Cr(Cl)(NO)_2Cp$, **5**, 419
$CrC_5N_2O_5$
$Cr(CO)_5(N_2)$, **5**, 235
CrC_5O_5
$[Cr(CO)_5]^{2-}$, **5**, 160, **12**, 556
$[Cr(CO)_5]^-$, **5**, 159
CrC_6HO_6
$[Cr(CHO)(CO)_5]^{-\cdot}$, **5**, 161
$CrC_6H_3ClO_4$
$CrCl(CO)_4(CMe)$, **5**, 192
$CrC_6H_6F_9P_3$
$Cr(PF_3)_3(C_6H_6)$, **5**, 516
$CrC_6H_7ClNO_2$
$Cr(CH_2Cl)(NO)_2Cp$, **5**, 419
CrC_6NO_5
$Cr(CN)(CO)_5$, **5**, 161
CrC_6N_6
$[Cr(CN)_6]^{3-}$, **5**, 305
CrC_6O_5S
$Cr(CO)_5(CS)$, **5**, 167
CrC_6O_5Se
$Cr(CO)_5(CSe)$, **5**, 167
CrC_6O_6
$Cr(CO)_6$, **5**, 156, 332; **12**, 981
$CrC_7Cl_3NO_5$
$Cr(CNCCl_3)(CO)_5$, **5**, 164
$CrC_7F_3NO_5$
$Cr(CO)_5(CNCF_3)$, **5**, 162
$CrC_7HF_2NO_5$
$Cr(CO)_5(CNCF_2H)$, **5**, 162
$CrC_7H_2FNO_5$
$Cr(CO)_5(CNCFH_2)$, **5**, 162
$CrC_7H_4O_3S$
$(C_4H_4S)Cr(CO)_3$, **12**, 1030
$CrC_7H_4O_5$
$Cr(CO)_5(C_2H_4)$, **5**, 332
$CrC_7N_2O_5$
$Cr(CO)_5(CNCN)$, **5**, 162
$CrC_8F_3NO_5$
$Cr(CNCFCF_2)(CO)_5$, **5**, 164
$CrC_8HCl_2NO_5$
$Cr[CNC(Cl)CH(Cl)](CO)_5$, **5**, 164
$CrC_8H_5NO_3$
$Cr(CO)_3(py)$, **5**, 531
$CrC_8H_5O_3$
$[Cr(CO)_3Cp]^-$, **5**, 404
$CrC_8H_6O_3$
$CrH(CO)_3Cp$, **5**, 406
$CrC_8H_6O_4$
$Cr(CO)_4(C_4H_6)$, **5**, 347
$CrC_8H_6O_6$
$(CO)_5Cr{=}C(OMe)Me$, **12**, 550, 560, 566
$Cr(CO)_5[C(OMe)Me]$, **5**, 173
$CrC_8H_8O_4$
$Cr(CO)_4(C_2H_4)_2$, **5**, 332
$CrC_8H_{16}Cl_2O_2$
$CrCl_2(THF)_2$, **5**, 288
$CrC_9H_4O_7$
$Cr(CO)_5(HCCCO_2Me)$, **5**, 363
$CrC_9H_5ClO_3$
$(ClPh)Cr(CO)_3$, **12**, 980, 983, 987, 1018, 1022
$Cr(CO)_3(PhCl)$, **5**, 490
$CrC_9H_5DO_3$
$(DPh)Cr(CO)_3$, **12**, 1019
$CrC_9H_5FO_3$
$Cr(CO)_3(PhF)$, **5**, 509, 512
$FPhCr(CO)_3$, **12**, 983, 988, 989, 1018, 1021
$CrC_9H_5IO_3$
$(IPh)Cr(CO)_3$, **12**, 1019
$CrC_9H_6F_4NO_5P$
$Cr(CO)_5[(F_3C)P{=}CF(NMe_2)]$, **5**, 247
$CrC_9H_6O_2S$
$Cr(CO)_2(CS)(C_6H_6)$, **5**, 493, 508
$CrC_9H_6O_3$
$(C_6H_6)Cr(CO)_3$, **12**, 995, 1010, 1011, 1018, 1040
$[Cr(CO)_3(C_6H_6)]^{2-}$, **5**, 495
$Cr(CO)_3(C_6H_6)$, **5**, 495, 497
$CrC_9H_7O_3$
$(chd)Cr(CO)_3$, **12**, 995
$CrC_9H_8O_3$
$CrMe(CO)_3Cp$, **5**, 410, 413

$CrC_9H_9O_2$
$Cr(H_2)(CO)_2)(Tol)$, **5**, 224
$CrC_9H_{12}O_4P$
$[Cr(CO)_4(PMe_3)(CMe)]^+$, **5**, 195
CrC_9H_{15}
$Cr(C_3H_5)_3$, **5**, 335
$CrC_9H_{15}O_2P$
$Cr(CO)_2(PMe_3)(C_4H_6)$, **5**, 347
$CrC_9H_{15}O_6P$
$(CH_2{=}CHMe)Cr[P(OMe)_3]-(CO)_3$, **12**, 976
$CrC_{10}H_5FO_4$
$Cr(CO)_3(FC_6H_4CHO)$, **12**, 1047, 1048
$CrC_{10}H_6O_4$
$Cr(CO)_3(C_6H_5CHO)$, **12**, 1044
$CrC_{10}H_7O_3$
$[Cr(CO)_3(C_7H_7)]^+$, **5**, 358, 359
$CrC_{10}H_8O_2$
$Cr(CO)_2(C_2H_2)(C_6H_6)$, **5**, 508
$CrC_{10}H_8O_3$
$Cr(CO)_3(C_7H_8)$, **5**, 355
$Cr(CO)_3(PhMe)$, **5**, 518, 520
$(PhMe)Cr(CO)_3$, **12**, 1020, 1054, 1055
$CrC_{10}H_8O_4$
$Cr(CO)_3(C_6H_5CH_2OH)$, **12**, 1025, 1049
$Cr(CO)_3(BzOH)$, **5**, 517
$[(MeO)C_6H_5]Cr(CO)_3$, **12**, 996, 1001, 1007, 1022, 1034–6
$CrC_{10}H_9NO_3$
$Cr(CO)_3(EtPy)$, **12**, 1063, 1064
$Cr(CO)_3(NH_2C_6H_4Me)$, **12**, 1067
$CrC_{10}H_{10}NO_5$
$[Cr(CO)_5(CNEt_2)]^+$, **5**, 196
$CrC_{10}H_{10}N_2$
$Cr(py)_2$, **5**, 531
$CrC_{10}H_{11}NO_5$
$Cr(CO)_5[C[\overline{N(CH_2)_3C}H_2]Me]$, **5**, 190
$CrC_{10}H_{12}O_6$
$Cr(CO)_3(MeOC_6H_4CH_2CO_2H)$, **12**, 1060
$CrC_{10}H_{15}O_3P$
$Cr(CO)_3(PMe_3)(C_4H_6)$, **5**, 347
$CrC_{11}H_5F_3O_4$
$Cr(CO)_3(CF_3C_6H_4CHO)$, **12**, 1045
$CrC_{11}H_6O_4$
$Cr(CO)_3$(benzocyclobutanone), **12**, 1042, 1043
$CrC_{11}H_6O_5$
$Cr(CO)_3(C_6H_4(CHO)_2)$, **12**, 1045
$CrC_{11}H_7NO_3$
(indole)$Cr(CO)_3$, **12**, 981, 998, 1002
$CrC_{11}H_8O_3$
$Cr(CO)_3(C_6H_5CH{=}CH_2)$, **12**, 1044, 1064, 1065
$Cr(CO)_3(C_8H_8)$, **5**, 357, 484
$CrC_{11}H_8O_4$
$Cr(CO)_3(C_6H_5C(O)Me)$, **12**, 1040, 1067
$Cr(CO)_3$(indananone), **12**, 1065, 1066
$Cr(CO)_3$(OH-cyclobutabenzene), **12**, 1067
$Cr(CO)_3((OH)CH{=}C_6H_4{=}CH_2$, **12**, 1067
$Cr(CO)_3$(phthalan), **12**, 1057, 1058
$Cr(CO)_3$(tolualdehyde), **12**, 1045
$Cr(CO)_3(MeC_6H_4CHO)$, **5**, 505, 524
$Cr(CO)_3(PhCOMe)$, **5**, 505
$CrC_{11}H_8O_5$
$Cr(CO)_3$(acetylcresyl), **12**, 1042, 1043
$Cr(CO)_3((MeO)C_6H_4CHO)$, **12**, 1046–8
$CrC_{11}H_9FO_3$
$(Me_2FC_6H_3)Cr(CO)_3$, **12**, 990, 991
$CrC_{11}H_{10}O$
$Cr(CO)Cp_2$, **5**, 442
$CrC_{11}H_{10}O_3$
$Cr(CO)_3(EtC_6H_5)$, **12**, 1055, 1060
$Cr(CO)_3(MeC_6H_4Me)$, **5**, 482
$CrC_{11}H_{10}O_4$
$Cr(CO)_3(BzOMe)$, **12**, 1056
$Cr(CO)_3(C_6H_5CH(OH)Me)$, **12**, 1039, 1040, 1049
$Cr(CO)_3(BzOMe)$, **5**, 520
$Cr(CO)_3(PhCHOHMe)$, **5**, 505
$[(MeO)MeC_6H_4]Cr(CO)_3$, **12**, 1009
$CrC_{11}H_{11}N$
$Cr(py)(C_6H_6)$, **5**, 531
$CrC_{11}H_{11}NO_3$
$Cr(CO)_3(PhNMe_2)$, **5**, 514
$CrC_{11}H_{15}$
$Cr(C_3H_5)_2Cp$, **5**, 335
$CrC_{11}H_{20}ClO_5$
$Cr(CH_2Cl)(acac)_2(OH_2)$, **5**, 301
$CrC_{11}N_4O_5$
$Cr(CO)_5(TCNE)$, **5**, 332
$CrC_{12}H_6O_4$
$Cr(CO)_3$(isochroman), **12**, 1059, 1060
$CrC_{12}H_6O_5$
$(CO)_5Cr{=}CHPh$, **12**, 391
$CrC_{12}H_8O_3$
$Cr(CO)_3$(indene), **12**, 1065
$Cr(CO)_3(C_9H_8)$, **5**, 501
$CrC_{12}H_8O_4$
$Cr(CO)_3$(indanone), **12**, 1042, 1067
$CrC_{12}H_{10}Cl_2$
$[Cr(PhCl)_2]^+$, **5**, 477
$CrC_{12}H_{10}O_2$
$Cr(CO)_2Cp_2$, **5**, 442
$CrC_{12}H_{10}O_3$
$Cr(CO)_3$(indane), **12**, 1057, 1058
$Cr(CO)_3(BzCHCH_2)$, **5**, 521
$CrC_{12}H_{10}O_4$
$Cr(CO)_3$(chroman), **12**, 1059
$Cr(CO)_3$(indanol), **12**, 1042, 1051, 1053
$CrC_{12}H_{10}O_5$
$Cr(CO)_3$(acetylanisyl), **12**, 1042, 1043
$Cr(CO)_3(C_6H_5CH_2CO_2Me)$, **12**, 1055, 1056
$CrC_{12}H_{11}NO_3$
[Me(indoline)]$Cr(CO)_3$, **12**, 998
$CrC_{12}H_{11}NO_4$
$Cr(CO)_3((Me_2N)C_6H_4CHO)$, **12**, 1047
$CrC_{12}H_{11}O_4$
$Cr(CO)_3(MeC_6H_5C(O)Me)$, **12**, 1065, 1066
$CrC_{12}H_{12}$
$Cr(C_6H_6)_2$, **5**, 475
$CrCp(C_7H_7)$, **5**, 361
$CrC_{12}H_{12}O_2$
$(C_6H_6)Cr(CO)_2(CCMe_2)$, **5**, 508
$CrC_{12}H_{12}O_5$
$Cr(CO)_3(MeC_6H_5CH_2COMe)$, **12**, 1061, 1062
$Cr(CO)_3(MeOC_6H_4CH_2OMe)$, **5**, 519
$CrC_{12}H_{12}O_6$
$[[(MeO)CH_2O](HOCH_2)Ph]Cr-(CO)_3$, **12**, 1026
$CrC_{12}H_{14}O_3$
$Cr(CO)_2(THF)(C_6H_6)$, **5**, 507
$CrC_{12}H_{15}NO_3$
$(MeCHPhNMe_2)Cr(CO)_3$, **12**, 1025
$CrC_{12}H_{16}Cl_2$
$[CrMe(Cl)Cp]_2$, **5**, 439
$CrC_{12}H_{16}O_4$
$Cr(H_2)(CO)_4$(cyclooctene), **5**, 224
$CrC_{12}H_{16}O_5$
$Cr(CO)_3(MeOC_6H_4CH_2O-CH_2CH{=}CHMe)$, **12**, 1061, 1063
$CrC_{12}H_{22}O_2P_2$
$Cr(CO)_2(dmpe)(C_4H_6)$, **5**, 347
$CrC_{13}H_8O_3$
$Cr(CO)_3(C_{10}H_8)$, **5**, 495, 503, 516
$CrC_{13}H_8O_6$
$(CO)_5Cr{=}C(OMe)Ph$, **12**, 390
$CrC_{13}H_{10}O_3$
$(C_{10}H_{10})Cr(CO)_3$, **12**, 981, 999, 1028, 1064, 1065
$CrC_{13}H_{10}O_4$
$Cr(CO)_3$(methylindanone), **12**, 1042
$Cr(CO)_3$(tetralone), **12**, 1040–2, 1065–7
$CrC_{13}H_{11}NO_2$
$Cr(CO)_2(py)(C_6H_6)$, **5**, 495
$CrC_{13}H_{11}O_5$

$Cr(CO)_3(C_6H_5C(Me)OC$-(O)Me), **12**, 1051
$CrC_{13}H_{12}O_3S$
$Cr(CO)_3(C_6H_5C(SEt)=CH_2)$, **12**, 1064
$CrC_{13}H_{12}O_4$
$Cr(CO)_3(BzOCH_2CH=CH_2)$, **12**, 1056, 1057
$Cr(CO)_3$(MeO-indane), **12**, 1057
$Cr(CO)_3$(tetralol), **12**, 1051
$CrC_{13}H_{12}O_5$
$Cr(CO)_3$(MeO-chroman), **12**, 1059
$CrC_{13}H_{12}O_6$
$Cr(CO)_5(\overline{C_8H_{12}O})$, **12**, 518, 519
$CrC_{13}H_{13}$
CrCp(cot), **5**, 357
$CrC_{13}H_{13}NO_3$
$Cr(CO)_3$(Me-isoquinoline), **12**, 1059, 1060
$Cr(CO)_3$(Me-quinoline), **12**, 1059
$(MeC_9NH_{10})Cr(CO)_3$, **12**, 997, 998
$CrC_{13}H_{14}O_4$
$Cr(CO)_3(MeC_6H_4CH(OH)$-$CH_2Me)$, **12**, 1067
$CrC_{13}H_{14}O_4Si$
$Cr(CO)_3((TMS)C_6H_4CHO))$, **12**, 1048
$CrC_{13}H_{14}O_5$
$Cr(CO)_3[MeC_6H_4CH(OMe)_2]$, **12**, 1053, 1054
$Cr(CO)_3[C_6H_4(CH_2OMe)_2]$, **5**, 519
$Cr(CO)_5(C_8H_{14})$, **5**, 240
$Cr(CO)_5$(coe), **5**, 332
$CrC_{13}H_{14}O_6$
$Cr(CO)_3((MeO)C_6H_4CH$-$(OMe)_2)$, **12**, 1053
$CrC_{13}H_{15}NO_3$
$Cr(CO)_3(PhCHMeNMe_2)$, **5**, 510
$CrC_{13}H_{15}NO_4$
$Cr(CO)_3((Me_2N)C_6H_4CH$-(OH)Me), **12**, 1047
$CrC_{13}H_{17}O$
$Cr(CO)(Me_2pdyl)Cp$, **5**, 353
$CrC_{13}H_{20}ClNO_5$
$[NEt_4][CrCl(CO)_5]$, **5**, 217
$CrC_{13}H_{22}Br_2N_3$
$CrBr_2(py)(NBu^t)_2$, **5**, 291
$CrC_{13}H_{23}BrN_2$
$CrBr(NBu^t)_2Cp$, **5**, 291
$CrC_{13}H_{30}ClP_2$
$CrCl(PMe_3)_2(Me_2pdyl)$, **5**, 352
$CrC_{14}H_{12}O_4$
$Cr(CO)_3$(Me-tetralone), **12**, 1040, 1041
$CrC_{14}H_{12}O_5$
$Cr(CO)_3$(MeO-tetralone), **12**, 1041, 1065
$Cr(CO)_3(MeO)C_6H_4C(O)$-CH=CHMe, **12**, 1065, 1066
$[(MeO)C_{10}H_{11}O]Cr(CO)_3$, **12**, 1030, 1031
$CrC_{14}H_{14}O_4$
$Cr(CO)_3$(MeO-tetralin), **12**, 1057
$CrC_{14}H_{14}O_5$
$Cr(CO)_3(\overline{CHOCH(Me)C}H$-(Me)O), **12**, 1052, 1054
$Cr(CO)_3((MeO)C_6H_4)CH$-(OAc)Me, **12**, 1052, 1054
$CrC_{14}H_{14}O_6$
$Cr(CO)_3((MeO)C_6H_4CH$-(OAc)Me), **12**, 1052, 1053
$CrC_{14}H_{15}NO_3$
$Cr(CO)_3((Me_2N)$indane), **12**, 1057
$Cr(CO)_3$(Me-benzazepine), **12**, 1054, 1060, 1061
$CrC_{14}H_{17}NO_3$
$Cr(CO)_3(C_6H_5CH_2CH(Me_2$-N)Me), **12**, 1056
$CrC_{15}H_{10}O_3$
$Cr(CO)_3(PhOPh)$, **5**, 512
$CrC_{15}H_{10}O_4$
$(Ph_2O)Cr(CO)_3$, **12**, 990
$CrC_{15}H_{13}O_5$
$[(MeO)C_{10}H_{11}O(Me)]Cr(CO)_3$, **12**, 1030, 1031
$CrC_{15}H_{14}O_5$
$Cr(CO)_3$(acetoxytetralone), **12**, 1042
$[(MeO)_2C_{10}H_8]Cr(CO)_3$, **12**, 999, 1011, 1029
$CrC_{15}H_{14}O_6$
$(CO)_5Cr=C(OMe)CH=CH$-(C_6H_9), **12**, 567
$CrC_{15}H_{16}O_4$
$Cr(CO)_3C(OH)(Me)C(MeCr$-$(CO)_3(C_6H_5CH(OH)$-$Me)CH=CH_2$, **12**, 1049, 1050
$CrC_{15}H_{16}O_5$
$Cr(CO)_3(C_6H_5C(OAc)CH$-(Me)Me, **12**, 1051
$Cr(CO)_3((MeO)_2$tetraline), **12**, 1063, 1064
$[(MeO)_2C_{10}H_{10}]Cr(CO)_3$, **12**, 1027
$CrC_{15}H_{18}O_3$
$Cr(CO)_3(C_6Me_6)$, **5**, 482
$CrC_{15}H_{22}$
$Cr(C_5H_7)Cp^*$, **5**, 353
$CrC_{15}H_{24}O_2P$
$Cr(CO)_2(PMe_3)Cp^*$, **5**, 400
$CrC_{16}H_{12}O_3$
$Cr(CO)_3(BzPh)$, **5**, 518
$CrC_{16}H_{14}O_3$
$Cr(CO)_3(C_6H_5$-bicyclo[2.2.1]-nonane), **12**, 1065
$CrC_{16}H_{14}O_7$
$Cr(CO)_3((MeO)$-(ethylenedioxy)-tetralone), **12**, 1041, 1042
$CrC_{16}H_{16}O_4$
$Cr(CO)_3(C_6H_5C(OAc)CH$-$(Me)CH=CH_2$, **12**, 1051
$CrC_{16}H_{17}NO_3$
$Cr(CO)_3$(Me-benzisoindoline), **12**, 1057, 1058
$CrC_{16}H_{20}O_2$
$Cr(CO)_2(HCCH)(C_6Me_6)$, **5**, 363
$CrC_{16}H_{29}NO_3$
$Cr(NO)(OPr^i)_2Cp^*$, **5**, 419
$CrC_{16}H_{38}P_4$
$Cr(dmpe)_2(C_4H_6)$, **5**, 349
$CrC_{16}H_{40}P_2$
$Cr(dmpe)_2(C_2H_4)_2$, **5**, 333
$CrC_{17}H_{10}N_2O_5$
$Cr(CO)_5(HNC_6H_4NPh)$, **5**, 227
$CrC_{17}H_{10}O_3$
(phenanthrene)$Cr(CO)_3$, **12**, 1028, 1029
$CrC_{17}H_{12}NO_3$
$Cr(CO)_3(MeC_6H_4CNPh)$, **5**, 528
$CrC_{17}H_{13}NO_3$
$Cr(CO)_3(MeC_6H_4CH=NC_6H_5)$, **12**, 1048, 1049, 1062
$CrC_{17}H_{15}NOS_2$
$Cr(NO)(SPh)_2Cp$, **5**, 419
$CrC_{18}H_{11}NO_5$
$Cr(CO)_5[C(NPh_2)H]$, **5**, 185, 186
$CrC_{18}H_{12}O_4$
$Cr(CO)_3(PhCHCHCOPh)$, **5**, 524
$CrC_{18}H_{15}NO_4$
$Cr(CO)_3PhC(O)C_6H_4(NMe_2)$, **12**, 1047
$CrC_{18}H_{15}N_3O_3$
$Cr(CO)_3(pyridine)_3$, **12**, 986
$CrC_{18}H_{23}NO_6$
$Cr(CO)_3((MeO)_2C_6H_3CH_2N$-$(Me)CH_2CH(OH)C_6H_5$, **12**, 1049, 1050
$CrC_{18}H_{24}O_5Si$
$Cr(CO)_3((MeO)_2(TMS)$-tetralone), **12**, 1051
$CrC_{18}H_{37}NO_6P$
$[(CH_2=CHCH_2Me)Cr[P$-$(OMe)_3](CO)_3][NEt_4]$, **12**, 976
$CrC_{18}H_{42}P_2$
$Cr(dmpe)_2(C_3H_5)_2$, **5**, 335
$CrC_{19}H_{15}NO_2$
$Cr(CO)(NO)(CPh_2)Cp$, **5**, 169, 418
$CrC_{19}H_{18}N_2O_5$
$Cr(CO)_4[Me_2Ph(C_4H_5)(\overline{C(O)N}$ - $\overline{C_2H_2N}$)], **12**, 532
$CrC_{19}H_{19}O_5P$
$Cr(CO)_2[P(OMe)_3](PhCCPh)$, **5**, 363
$CrC_{19}H_{25}NO_6$

$Cr(CO)_3((MeO)_2C_6H_3CH_2N$-(Me)$CH_2$(Me)CH(OH)-$C_6H_5$), **12**, 1049, 1050

$CrC_{20}H_{15}NO_5$
$Cr(CO)_3$(indoxyl), **12**, 1042, 1043

$CrC_{20}H_{16}$
$Cr(C_{10}H_8)_2$, **5**, 472, 473

$CrC_{20}H_{17}NO_5$
$Cr(CO)_3$- (benzoyldimethylindolinol), **12**, 1051, 1053

$CrC_{20}H_{26}O_7$
[PhCHOCH[C(MeO)Me]CH-[C(MeO)Me]O]$Cr(CO)_3$, **12**, 1026

$CrC_{20}H_{28}O_8$
$Cr(CO)_4(coe)_2$, **5**, 332

$CrC_{20}H_{44}$
$Cr(CH_2Bu^t)_4$, **5**, 283

$CrC_{21}H_{28}O_4$
$Cr(CO)_3(C_6Et_5COMe)$, **5**, 500

$CrC_{21}H_{30}O_2S$
$Cr(CO)_2(CS)(C_6Et_6)$, **5**, 499

$CrC_{21}H_{30}O_3$
$Cr(CO)_3(C_6Et_6)$, **5**, 499

$CrC_{22}H_{26}O_5$
$Cr(CO)_4[[(OEt)PhC](C_9H_{16})]$, **12**, 535, 536

$CrC_{23}H_{27}NO_7$
$Cr(CO)_3$(canadine), **12**, 1060, 1061

$CrC_{24}H_{30}O_8$
$Cr(CO)_3(MeO)C_6H_4CH(Me)C$-$H_2CH$=CHC(Me)CH-$(CO_2Et)_2$, **12**, 1052, 1054

$CrC_{24}H_{40}P_2$
$Cr(Mes)_2(PMe_3)_2$, **5**, 304

$CrC_{25}H_{20}O_2P$
$Cr(CO)_2(PPh_3)Cp$, **5**, 400

$CrC_{30}H_{35}ClNO_4P$
$[NEt_4][CrCl(CO)_4(PPh_3)]$, **5**, 220

$CrC_{30}H_{54}F_{12}N_6P_2$
$[Cr(CNBu^t)_6](PF_6)_2$, **5**, 310

$CrC_{31}H_{29}P$
$Cr(PMe_3)(PhCCPh)_2$, **5**, 363

$CrC_{31}H_{41}O$
$Cr(Mes)_3(THF)$, **5**, 286

$CrC_{35}H_{63}F_{12}N_7P_2$
$[Cr(CNBu^t)_7][(PF_6)_2]$, **5**, 309

$CrC_{39}H_{68}O_3P_2$
$Cr(H_2)(CO)_3(PCy_3)_2$, **5**, 223

$CrC_{42}H_{30}N_6$
$Cr(CNPh)_6$, **5**, 309

$CrC_{44}H_{30}O_2$
$Cr(CO)_2(Ph_2C_2)(C_4Ph_4)$, **5**, 351

$CrC_{45}H_{30}O_3$
$Cr(CO)_3(C_6Ph_6)$, **5**, 498

$CrC_{53}H_{44}O_3P_4$
$Cr(CO)_3(dppm)_2$, **5**, 246

$CrC_{78}H_{108}N_6$
$Cr(CNC_6H_4Pr^i_2)_6$, **5**, 309

$CrCoC_{16}H_{36}O_5P_3$
$Cr(CO)_3(PMe_3)(PBu^t_2)(CO)Co$-$(CO)(PMe_3)$, **10**, 44

$CrCoFeC_{13}H_5O_8S$
$CrFeCo(S)(CO)_8Cp$, **10**, 244

$CrFeC_9HO_9$
$[Cr(CO)_5(H)Fe(CO)_4]^-$, **10**, 42

$CrFeC_9O_9$
$Cr(CO)_5Fe(CO)_4$, **10**, 30

$CrFeC_{11}H_{12}O_7$
$Cr(CO)_4(PMe_2)_2Fe(CO)_3$, **10**, 43

$CrFeC_{13}H_5O_7$
$Cr(CO)_3[PhFe(CO)_4]^-$, **7**, 112

$CrFeC_{13}H_{10}NO_3$
$[Cr(NO)Cp(CO)_2Fe(CO)Cp]^-$, **10**, 42

$CrFeC_{16}H_{13}O_3$
$CrCp(C_8H_8)Fe(CO)_3$, **10**, 42

$CrFeC_{18}H_{18}$
$CrCp(C_8H_8)FeCp$, **10**, 42

$CrFeCoC_{13}H_5O_8S$
$FeCoCr(S)Cp(CO)_8$, **10**, 37

$CrFe_3C_{14}O_{13}$
$[CrFe_3(C)(CO)_{13}]^{2-}$, **10**, 90

$CrFe_3C_{14}O_{14}$
$[CrFe_3(CO)_{14}]^{2-}$, **10**, 191, 192

$CrHgMoC_{49}H_{35}NO_8P_2$
$(PPN)[(CO)_5CrHgMoCp$-$(CO)_3]$, **10**, 336

$CrIrC_{35}H_{35}O_4P_2$
$(CO)_4Cr(PPh_2)_2Ir(cod)Me$, **8**, 372

$CrLiC_9H_4ClO_3$
$(ClLiC_6H_4)Cr(CO)_3$, **12**, 1022

$CrLiC_9H_4FO_3$
$(FLiC_6H_4)Cr(CO)_3$, **12**, 1022, 1036

$CrLiC_9H_5O_3$
$(LiPh)Cr(CO)_3$, **12**, 1018

$CrLiC_{10}H_6FO_4$
$[FLi(MeO)C_6H_3]Cr(CO)_3$, **12**, 1036

$CrLiC_{16}H_{44}N_4$
$(TMEDA)Li(Me)_2Cr(Me)_2Li$-(TMEDA), **5**, 323

$CrLi_2C_{16}H_{46}N_4$
$[Li(TMEDA)]_2[CrMe_4]$, **5**, 288

$CrMgC_9H_5BrO_3$
$(PhMgBr)Cr(CO)_3$, **12**, 1018

$CrMnC_{25}H_{33}O_3$
$Cp^*Mn(CO)_3Cr(C_6Me_6)$, **6**, 116

$CrMoC_{22}H_{14}O_6$
$Cr(CO)_3(azulene)Mo(CO)_3$-$(C_6H_6)$, **10**, 29

$CrNaC_5HO_5$
$Na[CrH(CO)_5]$, **5**, 223

$CrNiC_{14}H_{10}O_4$
$Cp_2NiCr(CO)_4$, **9**, 95
$Cr(CO)_3CpNi(CO)Cp$, **10**, 45

$CrNiC_{18}H_{36}ClO_4P_3$
$Cr(CO)_4(PMe_3)(PBu^t_2)NiCl$-$(PMe_3)$, **10**, 45

$CrOsC_2H_6NO_4$
$[Os(N)Me_2(CrO_4)]^-$, **10**, 50

$CrOsC_{12}H_9O_{12}P$
$Cr(CO)_5Os(CO)_4[P(OMe)_3]$, **10**, 29

$CrOsC_{18}H_{18}N_2O_8$
$Cr(CO)_5Os(CO)_3(Bu^tNC)_2$, **10**, 30

$CrOsC_{31}H_{20}O_7P_2$
$Cr(CO)_4(PPh_2)_2Os(CO)_3$, **10**, 43

$CrPdC_{32}H_{50}O_2P_2$
$Cr(CO)_2Cp(PCy_2)Pd(CO)$-$(PHCy_2)$, **10**, 47

$CrPdC_{49}H_{36}INO_5P_2$
$PdI[(CH_2C_6H_4NC)Cr(CO)_5]$-$(PPh_3)_2$, **9**, 235

$CrPtC_{26}H_{23}O_6P$
$Cr(CO)_6(PPh_2)Pt(C_8H_{13})$, **10**, 47

$CrPt_2C_{30}H_{54}O_3P_2$
$CpCrPt_2(CO)_3(MeC_3H_4)$-$(PPr^i_3)_2$, **10**, 124

$CrReC_{16}H_5O_9$
$Cr(\equiv CPh)(CO)_4Re(CO)_5$, **10**, 30

$CrRhC_{10}H_8O_3$
$Cr(CO)_3Rh(nbd)$, **5**, 502

$CrRhC_{20}H_{19}O_3$
$Cr(CO)_3(C_9H_7)Rh(cod)$, **5**, 502

$CrRhC_{54}H_{44}ClO_4P_4$
$Cr(CO)_3(dppm)_2Rh(CO)Cl$, **10**, 44

$CrRhC_{55}H_{44}O_5P_4$
$[Cr(CO)_3(dppm)_2Rh(CO)_2]^+$, **10**, 44, 45

$CrRhC_{57}H_{47}O_2P_4$
$Cr(CO)Cp(dppm)(CO)Rh$-(dppm), **10**, 44

$CrRuC_{18}H_{15}NO_2$
$Cr(CO)_2(NO)(C_5H_4CH_2CpRu$-$(C_5H_4))$, **7**, 610

$CrRu_3C_{21}H_{13}O_{11}P$
$Ru_3(CO)_8[C_6H_4Cr(CO)_3](PBu^t)$, **10**, 17

$CrRu_5C_{18}O_{17}$
$[CrRu_5C(CO)_{17}]^{2-}$, **10**, 91

$CrSiC_{11}H_{14}O_3$
$Cr(CO)_3(PhTMS)$, **5**, 488

$CrSiC_{13}H_{14}O_4$
$Cr(CO)_3(TMS$-$C_6H_4CHO)$, **5**, 524

$CrSiC_{16}H_{18}O_3$
$[(TMS)C_{10}H_9]Cr(CO)_3$, **12**, 1028

$CrSiC_{17}H_{21}NO_4$
$[[(TMS)CH_2CH_2OCH_2]$-indole$]Cr(CO)_3$, **12**, 1029

$CrSiC_{17}H_{22}O_4$
$[[CH_2$=CHCH(OMe)CH-(TMS)]Ph$]Cr(CO)_3$, **12**, 1035

$CrSiC_{19}H_{28}O_5$
$[(Pr^i_3SiO)(MeO)C_6H_4]Cr(CO)_3$, **12**, 1024

$CrSiC_{22}H_{31}NO_3$
$[(Me_2Bu^tSi)Me(indoline)]Cr$-$(CO)_3$, **12**, 1025

$CrSiC_{22}H_{36}O_2$
$Cr[C{=}C(TMS)_2](CO)_2(C_6Me_6)$, **5**, 363
$CrSiC_{25}H_{25}NO_7$
$Cr(CO)_5[Ph_2(TMS)(\overline{C_2HNO})$-(OEt)C], **12**, 528, 529
$CrSiC_{26}H_{30}O_2$
$Cr(CO)_2(HSiPh_2H)(C_6Me_6)$, **5**, 507
$CrSi_2C_{14}H_{36}N_2O_2$
$Cr(O\text{-}TMS)_2(NBu^t)_2$, **5**, 291
$CrSi_2C_{30}H_{36}O_2$
$Cr(CO)_2[CC(TMS)_2](C_6Me_6)$, **5**, 508
$CrTeC_{17}H_{14}O_7$
$Cr(CO)_5[Te(C_5H_4OMe)_2]$, **5**, 262
$CrWC_{16}H_{13}O_3$
$CrCp(C_8H_8)W(CO)_3$, **10**, 33
$CrZrC_{40}H_{34}O_4P_2$
$Cp_2Zr(CH_2PPh_2)_2Cr(CO)_4$, **4**, 576
$Cr_2Al_2C_{48}H_{138}P_{16}$
$[CrH(dmpe)_4]_2[(H)_2AlH$-$(H)_2AlH(H)_2]$, **1**, 548
$Cr_2AsMnC_{14}O_{14}$
$As[Cr_2(CO)_9][Mn(CO)_5]$, **6**, 86
$Cr_2As_2Fe_3C_{19}O_{19}$
$Fe_3(CO)_9[AsCr(CO)_5]_2$, **7**, 56
$Cr_2C_{10}HO_{10}$
$Cr_2(H)(CO)_{10}$, **5**, 221
$Cr_2C_{10}H_{10}Cl_4$
$[CrCl_2Cp]_2$, **5**, 438
$Cr_2C_{10}H_{10}N_2O_4$
$[Cr(NO)_2Cp]_2$, **5**, 419
$Cr_2C_{10}H_{10}P_5$
$CpCr(P_5)CrCp$, **5**, 426
$Cr_2C_{10}O_{10}$
$[Cr_2(CO)_{10}]^{2-}$, **5**, 159
$Cr_2C_{12}H_{10}I_2N_2O_4$
$[CrI(CO)(NO)Cp]_2$, **5**, 419
$Cr_2C_{12}H_{20}$
$Cr_2(C_3H_5)_4$, **5**, 336
$Cr_2C_{14}H_{10}O_4S$
$Cr_2(CO)_4(S)Cp_2$, **5**, 433
$Cr_2C_{14}H_{10}O_4S_2$
$Cr_2(CO)_4(S_2)Cp_2$, **5**, 433
$Cr_2C_{14}H_{10}O_4Se$
$Cr_2(CO)_4(Se)Cp_2$, **5**, 434
$Cr_2C_{14}H_{10}O_4Se_2$
$Cr_2(CO)_4(Se_2)Cp_2$, **5**, 434
$Cr_2C_{15}H_{10}O_5S_2$
$Cr_2(CO)_5(S_2)Cp_2$, **5**, 433
$Cr_2C_{15}H_{10}O_5Se_2$
$Cr_2(CO)_5(Se_2)Cp_2$, **5**, 434
$Cr_2C_{15}H_{12}O_3$
$Cr_2(CO)_3(C_6H_6)_2$, **5**, 508, 509
$Cr_2C_{16}H_{10}O_6$
$Cr_2(CO)_6Cp_2$, **5**, 400
$Cr_2C_{16}H_{10}O_6Te$
$Cr_2(CO)_6(Te)Cp_2$, **5**, 434
$Cr_2C_{18}H_{18}$
$Cr_2(C_8H_8)Cp_2$, **5**, 450
$Cr_2C_{18}H_{28}O_2S_4$
$Cr_2(S_4)(OBu^t)_2Cp_2$, **5**, 434
$Cr_2C_{18}H_{28}S_3$
$Cr_2(S)(SBu^t)_2Cp_2$, **5**, 434
$Cr_2C_{20}H_{30}Cl_2$
$[CrClCp^*]_2$, **5**, 439
$Cr_2C_{20}H_{30}O_4$
$Cr_2(O)_4Cp^*_2$, **5**, 431
$Cr_2C_{20}H_{30}S_5$
$Cr_2(S)(S_2)_2Cp^*_2$, **5**, 433
$Cr_2C_{22}H_{20}$
$(C_6H_6)Cr(C_{10}H_8)Cr(C_6H_6)$, **5**, 473
$Cr_2C_{22}H_{20}N_2O_2$
$Cr_2(O)_2(NPh)_2Cp_2$, **5**, 420
$Cr_2C_{24}H_{22}$
$(C_6H_6)Cr(PhPh)Cr(C_6H_6)$, **5**, 474
$Cr_2C_{24}H_{40}N_4$
$Cr_2(C_6H_4NMe_2)_4$, **5**, 322
$Cr_2C_{26}H_{30}O_6$
$Cr_2(CO)_6Cp^*_2$, **5**, 400
$Cr_2C_{27}H_{33}$
(Mes)Cr(Mes)Cr(Mes), **5**, 474, 475
$Cr_2C_{30}H_{38}N_2O_2$
$Cr_2(OBu^t)_2(NPh)_2Cp_2$, **5**, 420
$Cr_2C_{40}H_{60}N_2O_4$
$[Cr(CO)_2(C_6Et_6)]_2(N_2)$, **5**, 499, 506
$Cr_2Ca_2C_{64}H_{82}O_3$
$[Cp^*_2Ca(CO)_2Cr(CO)(Mes)]_2$, **1**, 115
$Cr_2Co_2C_{14}H_{10}O_4S_3$
$Cr_2Co_2(S)_3(CO)_4Cp_2$, **10**, 232, 233, 237
$Cr_2Co_2C_{20}H_{20}S_3$
$Cr_2Co_2(S)_3Cp_4$, **10**, 232, 233
$Cr_2Co_2Fe_4C_{24}H_{14}O_{12}S_7$
$Cr_2Fe_4Co_2(S)_7(CO)_{12}Cp'_2$, **10**, 239
$Cr_2FeC_{17}H_{19}O_3S_3$
$Cp_2Cr_2Fe(CO)_3(SBu^t)(S)_2$, **7**, 71, 72
$Cr_2Fe_3C_{17}O_{16}$
$[Cr_2Fe_3C(CO)_{16}]^{2-}$, **7**, 21, 22; **10**, 90
$Cr_2HgC_{18}H_{10}O_6$
$Ph_2Hg[Cr(CO)_3]_2$, **12**, 1019
$Hg[PhCr(CO)_3]_2$, **5**, 490
$Cr_2InC_{16}H_{10}ClO_6$
$InCl[(CO)_3Cp]_2$, **1**, 558
$Cr_2InC_{21}H_{15}ClNO_6$
$InCl(py)[Cr(CO)_3Cp]_2$, **1**, 559
$Cr_2InNaC_{24}H_{26}Cl_2O_8$
$[Na(THF)_2][InCl_2[Cr$-$(CO)_3Cp]_2$, **1**, 558
$Cr_2Ni_2C_{20}H_{20}S_3$
$Cr_2Ni_2(S)_3Cp_4$, **10**, 232
$Cr_2PdC_{14}H_{32}Cl_2S_3$
$Cr_2Pd(S)(SBu^t)_2Cl_2Cp'_2$, **10**, 228
$Cr_2PdC_{30}H_{20}N_2O_6$
$[Cp(CO)_3Cr]Pd(PhCN)_2[Cr$-$(CO)_3Cp]$, **9**, 204
$Cr_2Re_2TiC_{36}H_{14}O_{18}$
$Re(CO)_5Cr(CO)_4(CTol)Ti$-$(CTol)Cr(CO)_4Re(CO)_5$, **10**, 45
$Cr_2RhC_{16}H_{19}O_2S_3$
$Cr_2Rh(S)_2(SBu^t)(CO)_2Cp_2$, **10**, 228
$Cr_2Ru_3C_{27}H_9O_{15}P$
$Ru_3(CO)_9[P[C_6H_5Cr(CO)_3]]$-$[C_6H_4Cr(CO)_3]$, **10**, 16
$Cr_2TiC_{10}H_{14}O_9$
$[TiCp_2(H_2O)_2][Cr_2O_7]$, **4**, 401
$Cr_2WC_{17}H_5O_{12}P$
$Cr_2W(P)(CO)_{12}Cp$, **10**, 218
$Cr_3As_2Co_2C_{20}O_{20}$
$Cr_3Co_2(As)_2(CO)_{20}$, **10**, 218
$Cr_3C_{39}H_{35}N_4$
$Cr_3(NPh)_4Cp_3$, **5**, 420
$Cr_3CoC_{18}H_{15}O_4S_3$
$Cr_3Co(O)(S)_3(CO)_3Cp_3$, **10**, 228, 233, 237
$Cr_3InC_{24}H_{15}O_9$
$In[Cr(CO)_3Cp]_3$, **1**, 557
$Cr_3TlC_{24}H_{15}O_9$
$Tl[Cr(CO)_3Cp]_3$, **1**, 567
$Cr_4C_{20}H_{20}S_4$
$Cr_4(S_4)Cp_4$, **5**, 433
$Cr_4(S)_4Cp_4$, **5**, 436
$Cr_4C_{40}H_{64}$
$[Cr(H)Cp^*]_4$, **5**, 439
$Cr_4Fe_2C_{24}H_{38}S_6Se_2$
$Cr_4Fe_2(S)_4(Se)_2(SBu^t)_2Cp_4$, **10**, 234, 235
$Cr_4MnC_{28}H_{38}S_6$
$Cr_4Mn(S)_4(SBu^t)_2Cp_4$, **10**, 234, 235
$Cr_6Fe_3C_{33}H_{30}O_3S_9$
$Cr_6Fe_3(S)_9(CO)_3Cp_6$, **10**, 234, 235
$Cr_6Rh_2C_{38}H_{48}S_9$
$Cr_6Rh_2(S)_7(SBu^t)_2Cp_6$, **10**, 234, 235
$CsB_{11}RhC_{37}H_{43}OP_2$
$Cs[Rh(CO)(PPh_3)_2]B_{11}H_{13}$, **1**, 363
$CsCH_3$
CsMe, **1**, 20
$CsFO_4S$
$CsSO_4F$, **11**, 218
$CuBC_8H_{12}F_4N_4$
$[Cu(NCMe)_4][BF_4]$, **10**, 257
$CuBC_{13}H_{16}P$
$(triphos)Cu(BH_4)$, **1**, 286
$CuBC_{30}H_{27}N_2P$
$Ph_3PCu(phen)(BH_4)$, **1**, 286
$CuBC_{36}H_{37}P_3$
$(PHPh_2)_3Cu(BH_4)$, **1**, 286
$CuBC_{36}H_{70}P_2$
$(PCy_3)_2Cu(BH_4)$, **1**, 286
$CuBC_{44}H_{39}P$
$Ph_3PCu[(Me_2)(Ph_2phen)](BH_4)$, **1**, 286
$CuBRu_2C_{38}H_{34}Cl_2F_4N_2O_4$
$[CuRu_2(Ph_2PPyOMe)_2(Cl_2)$-$(CO)_2][BF_4]$, **10**, 298
$CuB_2C_{27}H_{40}NP_2$
$(PPh_3)_2Cu(H_3BCNB_3H_7)$, **1**, 317
$CuB_4C_{12}H_{44}ClP_4$
$[Cu(B_2H_4{\cdot}2PMe_3)_2]Cl$, **1**, 311

$CuB_5C_{26}H_{32}P_2$
$Cu[Ph_2PCH_2]_2B_5H_8$, **1**, 326
$CuB_6FeC_{40}H_{39}O_4P_2$
$Cu(PPh_3)_2B_6H_9Fe(Co)_4$, **1**, 327, 328
$CuB_9C_{36}H_{44}P_2$
$Cu(PPh_3)_2(B_9H_{14})$, **1**, 339
$CuB_{20}C_6H_{34}N_5$
$[Et_3NH][Cu(B_{10}H_9N_2)_2]$, **1**, 362
$CuCF_3O_3S$
CuOTf, **12**, 415, 426
$CuCH_3$
CuMe, **3**, 63
$CuC_2F_6O_6S_2$
$Cu(OTf)_2$, **12**, 403
CuC_2F_5
$CuCF_2CF_3$, **3**, 64
CuC_2H_6BrS
$CuBr{\cdot}SMe_2$, **12**, 83
CuC_4F_{12}
$[Cu(CF_3)_4]^-$, **3**, 64
CuC_4H_9
BuCu, **12**, 119
Bu^tCu, **12**, 119
$CuC_5H_6F_6NS_2$
$Cu(CF_3)_2(S_2CNMe_2)$, **3**, 64
CuC_5H_8N
$CuCCCH_2NMe_2$, **3**, 88
CuC_6F_5
$Cu(C_6F_5)$, **3**, 58, 76
CuC_6H_5
PhCu, **3**, 76; **12**, 119
$CuC_7H_4F_3$
$Cu(C_6H_4CF_3)$, **3**, 58
$CuC_7H_{10}F_6NS_2$
$Cu(CF_3)_2(S_2CNEt_2)$, **3**, 117
CuC_8F_9
$CuCF_2CF_2C_6F_5$, **3**, 64
$CuC_8H_{10}N$
$Cu(C_6H_4NMe_2)$, **3**, 66
$CuC_8H_{12}F_6N_4P$
$[Cu(NCMe)_4][PF_6]$, **10**, 258, 311
$CuC_9H_{21}ClO_3P$
$CuCl[P(OPr^i)_3]$, **12**, 403
$CuC_{10}H_{14}N$
$Cu[C_6H_3CH_2(NMe_2)(Me)]$, **3**, 58, 122
$CuC_{10}H_{16}O_4$
$Cu(acac)_2$, **12**, 403, 426
$CuC_{14}H_{23}O$
Cu(THF)Cp*, **10**, 313
$CuC_{16}H_{24}N_3O_2$
Cu(semicorrin), **12**, 407
$CuC_{17}H_{26}P$
$Cu(CCPh)(PPr^i_3)$, **3**, 68
$CuC_{18}H_{15}IP$
$CuI{\cdot}PPh_3$, **12**, 64
$CuC_{20}H_8F_{24}O_8$
$Cu(CF_3COCH_2COCF_3)_4$, **12**, 443
$CuC_{20}H_{35}S$
$Cu(C_6H_2Bu^t_3)(DMS)$, **3**, 84
$CuC_{24}H_{17}$
$Cu(C_6H_2Ph_3)$, **3**, 82
$CuC_{28}H_{24}N_2O_2$
$(benzylsalicylaldiminato)_2Cu$, **12**, 400
$CuC_{30}H_{41}P$
$[Cu(CCBu^t)_2(PCy_3)]^-$, **3**, 100
$CuC_{44}H_{33}P$
$[Cu(fluorenyl)_2(PPh_3)]^-$, **3**, 64
$CuC_{46}H_{45}P_3$
$CuPh[MeC(CH_2PPh_2)_3]$, **3**, 110
$CuC_{47}H_{45}P_2$
$Cu[C(PPh_3)_2]Cp^*$, **3**, 115
$CuC_{55}H_{48}P_3$
$CuMe(PPh_3)_3$, **3**, 110
CuCl
CuCl, **12**, 426
$CuCl_2$
$CuCl_2$, **12**, 142, 143
$CuCl_2O_8$
$Cu(ClO_4)_2$, **12**, 1189
$CuCo_2C_2O_8$
$[Co(CO)_4CuCo(CO)_4]^-$, **10**, 63
$CuFeC_{40}H_{30}O_4P_2$
$Fe(CO)_4(CuPPh_3)_2$, **7**, 79, 80; **10**, 54
$CuFeC_{56}H_{63}P_4$
$Fe(PEtPh_2)_3(H)_3Cu(PEtPh_2)$, **10**, 61
$CuFe_2C_{26}H_{15}O_8P$
$[Fe_2(CO)_8(CuPPh_3)]^-$, **7**, 79
$CuFe_3C_{11}IO_{10}$
$[Fe_3CuI(CO)_9(CCO)]^{2-}$, **7**, 22
$Fe_3CuI(CO)_9(CCO)$, **7**, 271, 272
$CuFe_4C_{31}H_{15}O_{13}P$
$[Fe_4(CO)_{13}(CuPPh_3)]^-$, **7**, 80
$CuFe_4C_{66}H_{45}NO_{12}P_3$
$[N(PPh_3)_2][CuFe_4(CO)_{12}$-$(PPh_3)]$, **10**, 299
$CuIrC_{19}H_{33}ClOP_4$
$Ir(C_2Ph)(CO)(dmpm)_2CuCl$, **10**, 68
$CuLiC_2H_3N$
MeCu(CN)Li, **12**, 85
$CuLiC_2H_6$
Me_2CuLi, **12**, 673
$CuLiC_4H_9I$
$BuCu{\cdot}LiI$, **12**, 83
$CuLiC_4H_{10}$
Et_2CuLi, **12**, 107
$CuLiC_5H_3NS$
LiCu(CN)(thienyl), **12**, 64, 79, 107
$CuLiC_8H_{18}$
Bu^t_2CuLi, **12**, 107
$CuLiC_{12}H_{10}$
Ph_2CuLi, **12**, 107
$CuLiC_{25}H_{42}O_3P$
$LiCu(PBu^t_2)Me(THF)_3$, **3**, 103
CuLiClI
$CuI{\cdot}2LiCl$, **12**, 83
$CuLiSiC_9H_{11}N$
$PhMe_2SiCu(CN)Li$, **12**, 96
$CuLiSi_6C_{28}H_{86}O_4$
$[Cu[C(TMS)_3]_2][Li(THF)_4]$, **3**, 106
$CuLiSn_2C_{24}H_{54}$
$(Bu_3Sn)_2CuLi$, **2**, 231
$CuLi_2CCl_2N$
$CuCN{\cdot}2LiCl$, **12**, 67, 72
$CuLi_2C_3H_6N$
$Me_2Cu(CN)Li_2$, **12**, 77, 85–7, 89
$CuLi_2C_3H_9$
Li_2CuMe_3, **3**, 65
$CuLi_2C_3H_{10}NO_2$
$Li_2CuMe_2(CN)(H_2O)_2$, **3**, 127
$CuLi_2C_8H_{18}I$
$Bu_2CuLi{\cdot}LiI$, **12**, 83
$CuLi_2C_9H_{10}N$
$(dioxenyl)_2Cu(CN)Li_2$, **12**, 76
$CuLi_2C_9H_{12}NS$
$Bu(thienyl)Cu(CN)Li_2$, **12**, 92
$CuLi_2C_9H_{18}N$
$Bu_2Cu(CN)Li_2$, **12**, 100
$CuLi_2Si_2C_4H_9N$
$(TMS)_2Cu(CN)Li_2$, **12**, 99
$CuLi_2Si_2C_{17}H_{22}N$
$(PhMe_2Si)_2Cu(CN)Li_2$, **12**, 97
$CuLi_2SnC_{14}H_{30}N$
$Bu_3Sn(Me)CuCNLi_2$, **2**, 231
$CuMoPdC_{43}H_{37}O_2P_3$
$CuMoPd(PPh_2)(CO)_2(PPh_2$-$H)_2Cp$, **10**, 311
$CuNb_2C_{20}H_{26}F_6P$
$[[Cp_2NbH(H)_2]_2Cu]PF_6$, **5**, 76
$CuOs_{10}C_{79}H_{45}NO_{24}P_3$
$[N(PPh_3)_2][CuOs_{10}C(CO)_{24}(PPh_3)]$, **10**, 299
$CuPtWC_{25}H_{27}O_2$
$CuPtW(CC_6H_4Me)(CO)_2$-$CpCp^*$, **3**, 116
$CuReC_{10}H_{10}Cl_2$
$[ReCp_2][CuCl_2]$, **6**, 184
$CuRh_9C_{25}H_9N_3O_{19}$
$CuRh_9(CO)_{19}(NCMe)_3$, **10**, 317
$CuRuC_{15}H_{24}ClO_7P_2$
$[Cl[OP(OMe)_2]_2Ru$-$(\eta$-$C_6Me_6)]Cu(CO)$, **7**, 564
$CuRuC_{18}H_{36}ClP$
$RuH(PCy_3)(H)_2CuCl$, **10**, 61
$CuRuC_{50}H_{46}ClP_4$
$Ru(dppm)_2(H)_2CuCl$, **10**, 62
$CuRuC_{66}H_{48}O_{12}P_3$
$[Cu(PPh_3)_3][Ru(H_3)(CO)_{12}]$, **10**, 302
$CuRu_2C_{21}H_{40}N_4O_{13}P_4$
$[Ru_2[Cu(MeCN)_2](CO)_5$-$(etmdp)_2]^+$, **7**, 627
$CuRu_3C_{83}H_{60}IN_2O_{10}P_4$
$[PPN]_2[Ru_3CuI(CO)_9(CCO)]$, **7**, 755, 757
$CuRu_4C_{30}H_{18}O_{12}P$
$CuRu_4(H)_3(CO)_{12}(PPh_3)$, **10**, 41, 302
$CuSi_2C_{21}H_{43}N_3O_2$
Cu(azasemicorrin), **12**, 407
$CuSnC_{12}H_{27}$
Bu_3SnCu, **12**, 101
$CuSnC_5H_{15}S$
$Me_3SnCu{\cdot}Me_2S$, **2**, 231
$Cu_2Ag_2C_{32}H_{20}$
$Ag_2Cu_2(CCPh)_4$, **3**, 109

$Cu_2AsRu_4C_{38}H_{26}O_{12}P$
$Cu_2Ru_4(H)_2[Ph_2As(CH_2)_2PPh_2](CO)_{12}$, **10**, 299
$Cu_2Au_3C_{48}H_{30}$
$[Au_3Cu_2(CCPh)_6]^-$, **3**, 109
$Cu_2BC_{72}H_{64}ClO_4P_4$
$[[Cu(PPh_3)_2]_2(BH_4)][ClO_4]$, **1**, 286
$Cu_2C_8H_{20}P_2$
$Cu_2[(CH_2)_2PMe_2]_2$, **3**, 79
$Me_2P(CH_2CuCH_2)_2PMe_2$, **3**, 76
$Cu_2C_8H_{22}P_2$
$[Cu(CH_2PMe_3)]_2$, **3**, 115
$Cu_2C_{50}H_{40}S$
$Cu_2(C_6H_2Ph_3)_2(DMS)$, **3**, 84
$Cu_2C_{66}H_{78}P_3$
$Cu_2(CCBu^t)_2(PCy_3)(PPh_3)_2$, **3**, 64
$Cu_2C_{66}H_{81}P_3$
$Cu_2(CCBu^t)_2(PCy_3)(PPh_3)_2$, **3**, 100
$Cu_2C_{70}H_{70}P_4$
$[Cu(Mes)_2][Cu(dppe)_2]$, **3**, 105
$Cu_2C_{76}H_{68}P_6$
$Cu_2(PPh_2)_2(dppe)_2$, **3**, 68
$Cu_2LiC_3H_9$
$LiCu_2Me_3$, **3**, 65
$Cu_2Li_2C_4H_{12}$
$Li_2Cu_2Me_4$, **3**, 126
$[Me_2CuLi]_2$, **12**, 121
$Cu_2Li_2C_{28}H_{28}$
$Li_2Cu_2(Tol)_4$, **3**, 65
$Cu_2Li_2C_{30}H_{38}S_3$
$Li_2Cu_2(Ph)_4(DMS)_3$, **3**, 64, 101, 121, 124
$Cu_2Li_2C_{32}H_{40}O_2$
$Li_2Cu_2(Ph)_4(Et_2O)_2$, **3**, 64, 101, 124
$[[Li(Et_2O)](CuPh_2)]_2$, **1**, 26, 28
$Cu_2Li_2C_{36}H_{48}N_4$
$Li_2Cu_2(C_6H_4CH_2NMe_2)_4$, **3**, 58, 64, 101, 121, 124
$Cu_2Li_2C_{36}H_{48}O_2$
$Li_2Cu_2(Tol)_4(Et_2O)_2$, **3**, 64, 127
$Cu_2Li_2C_{40}H_{56}N_4$
$Li_2Cu_2[C_6H_4CH(Me)NMe_2]_4$, **3**, 101
$Cu_2Li_2Si_4C_{20}H_{56}S_2$
$Li_2Cu_2(CH_2TMS)_4(DMS)_2$, **3**, 64, 101, 120
$Cu_2Li_3C_{30}H_{25}$
$[Li_3CuPh_3][CuPh_2]$, **3**, 65
$Cu_2Li_3C_{38}H_{49}S_4$
$Li_3Cu_2Ph_5(DMS)_4$, **3**, 101, 127
$Li_3(CuPh_2)(CuPh_3)(DMS)_4$, **3**, 101
$Cu_2Re_4C_{64}H_{102}P_8$
$Cu_2Re_4H_{14}(PMe_2Ph)_8$, **10**, 256
$Cu_2Rh_{12}C_{30}O_{30}$
$Cu_2[Rh_{12}(CO)_{30}]$, **10**, 42
$Cu_2Ru_4C_{30}H_{44}O_{12}P_2$
$Cu_2Ru_4(H)_2(CO)_{12}(PPr^i_3)_2$, **10**, 278
$Cu_2Ru_4C_{33}H_{23}O_{12}P$
$Cu_2Ru_4(H)_2(CO)_{12}[P(Bz)_2(Bz)]$, **10**, 300
$Cu_2Ru_4C_{38}H_{26}O_{12}P_2$
$Cu_2Ru_4(H)_2[Ph_2P(CH)_2)_2PPh_2](CO)_{12}$, **10**, 299
$Cu_2Ru_4C_{48}H_{32}O_{12}P_2$
$Cu_2Ru_4H_2(CO)_{12}(PPh_3)_2$, **10**, 289
$H_2Ru_4[Cu(PPh_3)]_2(CO)_{12}$, **10**, 41
$Cu_2Ru_4C_{49}H_{32}O_{13}P_2$
$Cu_2Ru_4(CO)_{13}(PPh_3)_2$, **10**, 289
$Cu_2Ru_4C_{54}H_{44}O_{12}P_2$
$Cu_2Ru_4(H)_2(CO)_{12}[P(Bz)_2(Bz)]_2$, **10**, 289
$Cu_2Ru_4(H)_2(CO)_{12}[P(Bz)_3]_2$, **10**, 289
$Cu_2Si_2C_{12}H_{23}N_2$
$Cu_2[C(py)(TMS)_2]_2$, **3**, 77, 88
$Cu_3BC_{91}H_{76}F_4P_6$
$[Cu_3(CCPh)_2(Ph_2PCH_2PPh_2)_3][BF_4]$, **3**, 113
$Cu_3C_{23}H_{21}O_4$
$Cu_3Mes(O_2CPh)_2$, **3**, 93
$Cu_3C_{24}H_{38}BrN_4$
$Cu_3Br[C_6H_4CH_2N(Me)CH_2CH_2NMe_2]_2$, **3**, 121
$Cu_3C_{30}H_{56}N_6$
$[Cu_3Ph_2(PMDTA)_2]^+$, **3**, 115
$Cu_3C_{45}H_{50}N_2PS_2$
$Cu_3Mes[SC_6H_4(CH_2NMe_2)]_2(PPh_3)$, **3**, 93, 99
$Cu_3C_{75}H_{63}P_6$
$Cu_3(Ph_2PCHPPh_2)_3$, **3**, 68, 80
$Cu_3Fe_3C_{12}O_{12}$
$[Cu_3Fe_3(CO)_{12}]^{3-}$, **7**, 79, 80
$Cu_3Fe_3(CO)_{12}$, **10**, 310
$Cu_3Li_2C_5H_{15}$
$Li_2Cu_3Me_5$, **3**, 65
$Cu_3Li_2C_{36}H_{30}$
$[Li_2Cu_3Ph_6]^-$, **3**, 109
$Cu_4Au_2C_{32}H_{40}Br_2N_4$
$Au_2Cu_4Br_2(C_6H_4NMe_2)_4$, **3**, 122
$Cu_4B_2Ru_8C_{60}H_{32}Cl_2NO_{24}P_2$
$[PPN][[HRu_4(CO)_{12}B]_2Cu_4(Cl)]Cl$, **1**, 304, 305
$Cu_4C_{24}F_{20}$
$Cu_4(C_6F_5)_4$, **3**, 65
$Cu_4C_{24}H_{38}Br_2N_4$
$Cu_4Br_2[C_6H_3(CH_2NMe_2)_2]_2$, **3**, 91
$Cu_4C_{28}H_{32}S_2$
$Cu_4Ph_4(DMS)_2$, **3**, 84, 120, 121
$Cu_4C_{32}H_{40}S_2$
$Cu_4(C_6H_4Me)_4(DMS)_2$, **3**, 84, 129
$Cu_4C_{32}H_{48}N_4$
$Cu_4(C_6H_4CH_2NMe_2)_4$, **3**, 66
$Cu_4C_{36}H_{46}N_2S_2$
$Cu_4Mes_2(SC_6H_4CH_2NMe_2)_2$, **3**, 93
$Cu_4C_{36}H_{48}N_2$
$[Cu(C_6H_4CH_2NMe_2)]_4$, **3**, 125
$Cu_4C_{36}H_{48}N_4$
$Cu_4(C_6H_4CH_2NMe_2)_4$, **3**, 87, 121
$Cu_4C_{40}H_{52}S$
$Cu_4Mes_4(THT)$, **3**, 84
$Cu_4C_{40}H_{56}N_4$
$Cu_4[C_6H_3CH_2NMe_2(Me)]_4$, **3**, 87, 117, 125
$Cu_4C_{44}H_{60}S_2$
$Cu_4Mes_4(THT)_2$, **3**, 121
$Cu_4C_{48}H_{48}N_4$
$Cu_4(C_{10}H_6NMe_2)_4$, **3**, 88, 120
$Cu_4C_{60}H_{92}$
$Cu_4(C_6H_2Pr^i_3)_4$, **3**, 82, 124
$Cu_4C_{100}H_{80}N_4P_4$
$Cu_4(CCPh)_4[PPh_2(py)]_4$, **3**, 97
$Cu_4C_{108}H_{80}P_4$
$Cu_4(CCPh)_4(PPh_3)_4$, **3**, 97
$Cu_4Co_4C_{16}O_{16}$
$[Co(CO)_4Cu]_4$, **10**, 63
$Cu_4LiC_{36}H_{30}$
$[LiCu_4Ph_6]^-$, **3**, 109
$Cu_4Li_5C_{62}H_{69}S_4$
$Li_5Cu_4Ph_9(DMS)_4$, **3**, 101
$Li_5(CuPh_2)_3(CuPh_3)(DMS)_4$, **3**, 101
$Cu_4MgC_{40}H_{40}O$
$MgCu_4Ph_6(Et_2O)$, **3**, 102
$Cu_4Mg_2C_{72}H_{92}N_4S_4$
$[Cu_4Mes_4][Mg(SC_6H_4CH_2NMe_2)_2]_2$, **3**, 104
$Cu_4Mg_2C_{76}H_{100}N_4S_4$
$[Cu_4Mes_4][Mg[SC_6H_4CH(Me)NMe_2]_2]_2$, **3**, 128
$Cu_4Si_4C_{16}H_{44}$
$Cu_4(CH_2TMS)_4$, **3**, 58, 76, 77, 125
$Cu_4Si_4C_{36}H_{60}N_4$
$Cu_4[CH(py)(TMS)]_4$, **3**, 77
$Cu_5C_{16}H_{14}Br_4$
$[Cu_5(C_6H_4CHCH_2)_2Br_4]^-$, **3**, 109
$Cu_5C_{32}H_{28}Br_2$
$[Cu_5(C_6H_4CHCH_2)_4Br_2]^-$, **3**, 109
$Cu_5C_{36}H_{30}$
$[Cu_5Ph_6]^-$, **3**, 108
$Cu_5C_{36}H_{66}Br_3N_8$
$Cu_5Br_3[C_6H_3[CH_2N(Me)CH_2CH_2NMe_2]_2]_2$, **3**, 91
$Cu_5C_{45}H_{55}$
Cu_5Mes_5, **3**, 63, 69, 79, 82
$Cu_5Fe_4C_{16}O_{16}$
$[Cu_5Fe_4(CO)_{16}]^{3-}$, **7**, 79, 80
$Cu_5LiC_{52}H_{62}O_4$
$[Li(THF)_4][Cu_5Ph_6]$, **3**, 65
$Cu_6C_{12}H_{12}Br_2O_8$
$Cu_6Br_2(MeOxl)_4$, **3**, 89, 91
$Cu_6C_{16}H_{20}B_2N_2$
$Cu_6Br_2(C_6H_4NMe_2)_2$, **3**, 120
$Cu_6C_{32}H_{40}Br_2N_4$
$Cu_6Br_2(C_6H_4NMe_2)_4$, **3**, 58, 66, 89, 122, 124, 125
$Cu_6C_{34}H_{34}N_2$
$Cu_6(C_6H_4NMe_2)_2(CCC_6H_4Me)_2$, **3**, 120
$Cu_6C_{48}H_{33}N_4S_4$
$[Cu_3(CCBu^t)[SC_6H_4(CH_2NMe_2)]_2]_2$, **3**, 99

$Cu_6C_{50}H_{54}N_4$
$Cu_6(CCC_6H_4Me)_2(C_6H_4N$-
$Me_2)_4$, **3**, 89, 97
$Cu_6C_{52}H_{74}N_4S_4$
$[Cu_3(CCBu^t)[SC_6H_4[CH$-
$(Me)NMe_2]]_2]_2$, **3**, 99
$Cu_6Fe_4C_{16}O_{16}$
$[Cu_6Fe_4(CO)_{16}]^{2-}$, **7**, 79, 80
$Cu_6Li_4C_{72}H_{60}$
$[Li_2Cu_3Ph_6]^-{}_2$, **3**, 101
$Cu_7Ag_6C_{112}H_{70}$
$[Ag_6Cu_7(CCPh)_{14}]^-$, **3**, 75, 109
$Cu_8C_{56}H_{56}O_8$
$Cu_8(C_6H_4OMe)_8$, **3**, 88
$Cu_{10}C_{54}H_{66}O_2$
$Cu_{10}Mes_6(O)_2$, **3**, 95
$Cu_{24}C_{144}H_{216}$
$Cu_{24}(CCBu^t)_{24}$, **3**, 120

D

$DyC_{24}H_{33}$
$Dy(Pr^iC_5H_4)_3$, **4**, 130

E

ErC_9H_{18}
$Er(MeCHCH_2)_3$, **4**, 14
$ErC_{10}H_{10}I$
Cp_2ErI, **4**, 36
$ErC_{15}H_{15}$
$ErCp_3$, **4**, 53
$ErC_{24}H_{33}$
$Er(Pr^iC_5H_4)_3$, **4**, 130
$ErC_{27}H_{39}$
$Er(BuC_5H_4)_3$, **4**, 130
$ErLiC_{18}H_{32}N_2$
$Cp_2Er(Me)_2Li(TMEDA)$, **4**, 43
$ErMo_3C_{26}H_{29}O_{14}$
$(EtOH)(H_2O)_4Er[CpMo(CO)_3]_3$, **4**, 13
$Er_2C_{20}H_{24}N_2$
$[Cp_2Er(NH_2)]_2$, **4**, 41
$Er_2C_{34}H_{48}N_2$
$[(MeC_5H_4)_2Er(HCNBu^t)]_2$, **4**, 47
$Er_2K_2C_{48}H_{64}O_4$
$(C_8H_8)Er(C_8H_8)K(C_8H_8)Er(C_8H_8)K(THF)_4$, **4**, 104
EuC_2H_4
$Eu(C_2H_4)$, **4**, 1
EuC_8H_8
$Eu(C_8H_8)$, **4**, 98
$EuC_{10}H_{10}$
$EuCp_2$, **4**, 28
$EuC_{20}H_{30}$
$EuCp^*_2$, **4**, 64, 66
$EuC_{24}H_{38}O$
$Cp^*_2Eu(THF)$, **4**, 65
$EuC_{28}H_{48}O_2$
$Cp^*_2Eu(THF)(Et_2O)$, **4**, 65
$EuC_{36}H_{66}O_4$
$[Cp^*Eu(OBu^t)_2]_2$, **4**, 70
$Eu_2C_{36}H_{62}I_2O_4$
$[Cp^*Eu(I)(THF)_2]_2$, **4**, 64
$Eu_2C_{72}H_{102}O_4$
$[Cp^*_2Eu(CCPh)(THF)_2]_2$, **4**, 90

F

$FeAsC_7H_8O_5$
$(CO)_4FeAs(Me)(OEt)$, **7**, 55
$FeAsC_{25}H_{21}O_4$
$Fe(CO)_4As(Tol)_3$, **7**, 55
$FeAsCoMoWC_{19}H_{16}O_7S$
$MoWFeCo(S)(AsMe_2)(CO)_7Cp_2$, **10**, 37, 244
$FeAsMnC_{11}H_6O_9$
$(CO)_5Mn(AsMe_2)Fe(CO)_4$, **10**, 39
$FeAsMoC_{14}H_{11}O_7$
$Cp(CO)_3Mo(AsMe_2)Fe(CO)_4$, **10**, 39
$FeAsMo_3C_{25}H_{15}O_{10}$
$Mo_3Fe(As)(CO)_{10}Cp_3$, **10**, 218
$FeAsW_3C_{25}H_{15}O_{10}$
$W_3Fe(As)(CO)_{10}Cp_3$, **10**, 218
$FeAs_2C_9H_9NO_4$
$(CO)_4FeAs(Me)_2N(Me)AsMe_2$, **7**, 55
$FeAs_2C_{38}H_{30}Cl_2O_2$
$Fe(CO)_2(AsPh_3)_2Cl_2$, **7**, 55, 78
$FeAuB_2C_{22}H_{20}O_4P$
$Fe(CO)_4(B_2H_5)Au(PPh_3)$, **10**, 62
$FeAuC_{42}H_{39}O_6P_2$
$Fe(CO)_3[P(OMe)_3][Au(PPh_3)_2]$, **10**, 54
$FeAu_2BC_{46}H_{39}F_4P_2$
$[[(Cp)Fe(C_5H_4)](AuPPh_3)_2]$-$[BF_4]$, **3**, 12
$FeAu_2C_{40}H_{30}O_4P_2$
$Au_2Fe(CO)_4(PPh_3)_2$, **10**, 256
$Fe(CO)_4(AuPPh_3)_2$, **7**, 81
$FeAu_2C_{40}H_{66}O_4P_2$
$Fe(CO)_4[AuP(cy)_3]_2$, **7**, 81
$FeAu_3BC_{66}H_{45}O_{12}P_3$
$Fe(CO)_{12}B(AuPPh_3)_3$, **1**, 294
$FeAu_3C_{58}H_{99}F_6O_4P_3$
$[Fe(CO)_4[AuP(cy)_3]_3][PF_6]$, **7**, 81
$FeBC_7H_6F_4O_4$
$[(CH_2{=}CHMe)Fe(CO)_4][BF_4]$, **12**, 967
$FeBC_7H_8O_2$
$[FeH(CO)_2(C_4H_4BMe)]^-$, **7**, 165
$FeBC_8H_7O_3$
$Fe(CO)_3(C_4H_4BMe)$, **7**, 165
$FeBC_{10}H_{10}$
borataferrocene, **1**, 204
$FeBC_{10}H_{13}$
$FeH(C_4H_4BMe)Cp$, **7**, 195
$FeBC_{11}H_{13}F_4O_3$
$[Cp(CO)_2Fe(THF)][BF_4]$, **12**, 395
$FeBC_{12}H_3P_4$
$(dmpe)_2Fe(H)(BH_4)$, **1**, 284, 285
$FeBC_{12}H_{13}F_4O_4$
$Fe(CO)_3[Me_2(OMe)C_6H_4]BF_4$, **7**, 157
$FeBC_{13}H_9O_3$
$Fe(CO)_3(C_4H_4BPh)$, **1**, 204; **7**, 138
$FeBC_{13}H_{22}NO_3Si$
$(CO)_3Fe(C_{10}H_{22}BNSi)$, **1**, 213
$FeBC_{18}H_{26}$
$Fe(CpBMe)(C_6Me_6)$, **1**, 209
$FeBC_{28}Cl_4H_{29}NO_3P$
$[(Ph_3P)(CO)_3Fe{\equiv}CNPr^i_2]^+$-$[BCl_4]^-$, **7**, 120
$FeBC_{41}H_{41}P_3$
$(triphos)Fe(H)(BH_4)$, **1**, 284, 285
$FeBCrC_{25}H_{21}O_3$
$CpFe(C_4H_4BPh)Cr(CO)_3Bz$, **7**, 195
$FeBRu_3C_{12}H_3O_{12}$
$H[FeRu_3(CO)_{12}BH_2]$, **1**, 297, 301
$FeBRu_3C_{48}H_{31}NO_{12}P_2$
$[PPN][FeRu_3(CO)_{12}BH]$, **1**, 297, 301
$FeB_2C_7H_{10}O_2$
$CpFe(CO)_2(B_2H_5)$, **1**, 307–9
$FeB_2C_{12}H_{18}F_8N_6$
$[Fe(CNMe)_6][BF_4]_2$, **7**, 12
$FeB_2C_{13}H_{18}O_3$
$Fe(CO)_3[Me_2C_2(BMe)_2$-$C{=}CMe_2]$, **1**, 208
$FeB_2C_{14}H_{22}S$
$(benzene)Fe[Et_2Me_2(BSBC_2)]$, **1**, 214
$FeB_2C_{15}H_{16}O_3$
$(CO)_3[Me_4(C_7B_2)]$, **1**, 208
$FeB_2C_{15}H_{23}$
$CpFe[Et_2Me(C_4H_2B_2)]$, **1**, 210
$FeB_2C_{17}H_{14}O_3$
$(CO)_3Fe(Me_2$diboraanthracene), **1**, 210
$FeB_2C_{17}H_{20}O_3$
$(CO)_3[Et_2Me_2(C_8H_4B_2)]$, **1**, 210
$FeB_2C_{18}H_{41}N_2O_2P$
$(CO)_2(PMe_3)Fe[(Pr^i_2N)_2C_4B_2]$, **1**, 209, 210
$FeB_2C_{19}H_{24}$
$Fe(PhMe)[Et_4Me(C_3B_2)]$, **1**, 206
$FeB_2C_{19}H_{32}N_2O_3$
$(CO)_3Fe[(Pr^i_2N)_2C_4B_2]$, **1**, 209, 210
$FeB_2C_{23}H_{35}N_2O_3P$
$(CO)_3Fe[(Pr^i_2N)_2Ph(C_2H_2B_2P)]$, **1**, 214
$FeB_2C_{29}H_{40}O_3P_2$
$(CO)_3Fe[xylyl_2Bu^t_2(P_2B_2)$, **1**, 211
$FeB_2C_{30}H_{40}O_4P_2$
$(CO)_4Fe[xylyl_2Bu^t_2(P_2B_2)$, **1**, 211
$FeB_2CoC_{18}H_{25}$
$CpFe(C_3B_2Me_5)CoCp$, **7**, 195
$FeB_3C_6H_7O_6$
$[Fe(CO)_3]_2B_3H_7$, **1**, 317, 318
$FeB_3C_{18}H_{22}OP$
$CpFe(CO)B_3H_7(PPh_2)$, **1**, 313, 316, 317
$FeB_3C_{42}H_{36}NO_6P_2$
$[PPN][[Fe(CO)_3]_2B_3H_6]$, **1**, 317, 318
$FeB_3PdC_{34}H_{35}P_2$
$(dppf)PdB_3H_7$, **1**, 317
$FeB_3PtC_{34}H_{35}P_2$
$[(Ph_2PC_5H_4)_2Fe]PtB_3H_7$, **1**, 317
$FeB_4C_3H_8O_3$
$[Fe(CO)_3]B_4H_8$, **1**, 320
$FeB_4C_{17}H_{32}N_4O$
$[Fe(CO)][C_4H_4B_2(NMe_2)_2]_2$, **1**, 209
$FeB_4C_{18}H_{21}OP$
$CpFe(CO)B_4H_6(PPh_2)$, **1**, 317, 321
$FeB_4NiC_{25}H_{37}$
$(C_9H_8)Fe(Et_2C_2B_4H_4)NiCp^*$, **9**, 179
$FeB_5C_3H_5O_3$
$[B_5H_5Fe(CO)_3]^{2-}$, **1**, 327
$FeB_5C_3H_9O_3$
$(CO)_3FeB_5H_9$, **1**, 323
$FeB_5C_7H_{13}O_2$
$[CpFe_9(CO)_2]B_5H_8$, **1**, 326
$FeB_5C_{19}H_{22}O_2P$
$Cp(CO)_2FeB_5H_7(PPh_2)$, **1**, 327
$FeB_6C_3H_6O_3$
$[B_6H_6Fe(CO)_3]^{2-}$, **1**, 328
$FeB_6CuC_{40}H_{39}O_4P_2$
$Cu(PPh_3)_2B_6H_9Fe(Co)_4$, **1**, 327, 328
$FeB_6KC_4H_9O_4$
$K[Fe(CO)_4B_6H_9]$, **1**, 327
$FeB_8C_9H_{21}O$
$(Mes)Fe(OB_8H_{10})$, **1**, 267, 334
$FeB_9C_7H_{19}S$
$(PhMe)FeSB_9H_{11}$, **1**, 268
$FeB_9C_7H_{21}$
$(PhMe)[FeB_9H_{13}]$, **1**, 341, 342
$FeB_9C_9H_{25}$
$(C_6Me_3H_3)[FeB_9H_{13}]$, **1**, 341, 342
$FeB_9WC_{17}H_{22}O_5$
$[W(CO)_2(C_2B_9H_9Me_2)(CTol)$-$Fe(CO)_3]^-$, **10**, 43

$FeB_9WC_{18}H_{22}O_6$
$[W(CO)_2(C_2B_9H_8Me_2)(CHTol)(CO)Fe(CO)_3]^-$, **10**, 43
$FeB_{10}C_7H_{18}S$
$(PhMe)FeSB_{10}H_{10}$, **1**, 268
$FeB_{10}C_7H_{20}O_2$
$CpFe(CO)_2(B_5H_7)(B_5H_8)$, **1**, 358, 361
$FeB_{10}WC_{17}H_{23}O_5$
$W(CO)_2(C_2B_{10}H_{10}Me_2)(CTol)Fe(CO)_3$, **10**, 43
$FeB_{11}C_3H_{11}O_3$
$[B_{11}H_{11}Fe(CO)_3]^{2-}$, **1**, 363
$FeB_{13}C_3H_{13}O_3$
$[B_{13}H_{13}Fe(CO)_3]^{2-}$, **1**, 363
$FeB_{18}Th_2C_{46}H_{88}$
$[Cp^*{}_2ThMe]_2[Fe(B_9C_2H_{11})_2]$, **4**, 174
$FeB_{20}C_{30}H_{62}N$
$[CpFeC_5H_4CH_2N(Me)_2Et]_2[B_{20}H_{18}]$, **1**, 365
$FeBiC_{16}O_{16}$
$[Bi[Fe(CO)_4]_4]^-$, **7**, 60
$FeBi_2C_{28}H_{20}O_4$
$Fe(CO)_4(BiPh_2)_2$, **7**, 60
$FeC_{26}H_{23}O_2P$
$Cp(CO)(PPh_3)FeCOMe$, **12**, 377
$FeCH_4$
HFeMe, **7**, 108
$FeCN_3O_3$
$Fe(CO)(NO)_2(N_2)$, **7**, 31
$FeC_2N_2O_4$
$Fe(CO)_2(NO)_2$, **7**, 32
FeC_2O_2
$[Fe(CO)_2]^-$, **7**, 9
$FeC_2O_2Se_8$
$[Fe(CO)_2(Se_4)_2]^{2-}$, **7**, 74
$FeC_3H_8O_3P_2$
$Fe(CO)(PH_3)_2[C(O)H]_2$, **7**, 40
$HFe(CO)_2(PH_3)_2[C(O)H]$, **7**, 40
FeC_3NO_4
$[Fe(CO)_3(NO)]^-$, **7**, 32
FeC_3O_3
$Fe(CO)_3$, **12**, 745
$[Fe(CO)_3]^-$, **7**, 9
FeC_4HIO_4
$HFe(CO)_4I$, **7**, 79
FeC_4HO_4
$[HFe(CO)_4]^-$, **7**, 7, 11; **12**, 355
$FeC_4H_2O_4$
$H_2Fe(CO)_4$, **7**, 6, 86; **12**, 351
FeC_4O_4
$[Fe(CO)_4]^{2-}$, **7**, 5, 8
$FeC_4O_4Br_2$
$Fe(CO)_4Br_2$, **7**, 78
FeC_5HO_5
$[Fe(CO)_4[C(O)H]]^-$, **7**, 7
$[HFe(CO)_5]^+$, **7**, 6
$FeC_5H_2O_5$
$(CO)_4Fe{=}CHOH$, **7**, 118
$FeC_5H_3O_4Se$
$Fe(CO)_4SeMe$, **7**, 74
$FeC_5H_4N_4O$
$Fe(CO)(N_2)_2(cbd)$, **7**, 137
$FeC_5H_4O_3$
$Fe(CO)_3(C_2H_4)$, **7**, 122
$FeC_5H_7O_5S$
$Fe(CO)_2CH_2CH_2SO_3Me$, **7**, 175
FeC_5O_4S
$Fe(CO)_4(CS)$, **7**, 16
FeC_5O_5
$Fe(CO)_5$, **7**, 3, 8; **12**, 353, 426
FeC_5O_6
$[Fe(CO)_4(CO_2)]^{2-}$, **7**, 85
$FeC_6F_4O_4$
$Fe(CO)_4(C_2F_4)$, **7**, 121
$FeC_6F_6O_4$
$Fe(CO)_4(CF_3)_2$, **7**, 78
$FeC_6H_4N_2O_2$
$Fe(CO)_2(N_2)(cbd)$, **7**, 137
$FeC_6H_4O_3$
$(CH_2{=}C{=}CH_2)Fe(CO)_3$, **12**, 355
$FeC_6H_4O_4$
$Fe(CO)_4(C_2H_4)$, **7**, 121
$FeC_6H_5ClO_3$
$FeCl(CO)_3(C_3H_5)$, **7**, 127, 129
$FeC_6H_5O_3$
$Fe(CO)_3(C_3H_5)$, **7**, 129
FeC_6H_6
$Fe(C_6H_6)^+$, **7**, 106
$FeC_6H_6O_2$
$Fe(CO)_2(H_2)(cbd)$, **7**, 137
$FeC_6H_7NO_3$
$Fe(CO)_2(NO)(CH_2CHCHMe)$, **7**, 127
$FeC_6H_{11}O_2PS$
$Fe(CO)_2(PMe_3)_2(CH_2S)$, **7**, 124
$FeC_7H_3O_7$
$[Fe(CO)_5(CO_2Me)]^+$, **7**, 115
$FeC_7H_4O_3$
$Fe(CO)_3(\overline{CH{=}CHCH{=}CH})$, **12**, 624
$Fe(CO)_3(cbd)$, **7**, 136, 144
$FeC_7H_4O_5$
$Fe(CO)_3(OCHCH{=}CHCHO)$, **7**, 147
$FeC_7H_4O_5S$
$Fe(CO)_3(C_4H_4SO_2)$, **7**, 138
$FeC_7H_5ClO_2$
FpCl, **7**, 239
$FeC_7H_5O_3$
$Fe(CO)_3(CH_2{=}CHCH{=}CH)$, **12**, 649, 650
$FeC_7H_5O_4$
$[Fe(CO)_4(C_3H_5)]^+$, **7**, 127
$FeC_7H_6O_2$
$CpFe(CO)_2H$, **7**, 108
$FeC_7H_6O_3$
$Fe(CO)_3(CH_2{=}CHCH{=}CH_2)$, **12**, 629
$Fe(CO)_3$(butadiene), **7**, 142, 145, 154
$Fe(CO)_3[C(CH_2)_3]$, **7**, 134, 239
$FeC_7H_6O_6$
$Fe(CO)_4(CO_2Me)Me$, **7**, 114
$FeC_7H_7O_3$
$Fe(CO)_3(MeCH_2CHCH)$, **7**, 133
$FeC_7H_9O_3$
$[(MeCH{=}CH_2CH_2)Fe(CO)_3]^-$, **12**, 968
$FeC_7H_{10}B_4O_3$
$(CO)_3Fe(Me_2C_2B_4H_4)$, **1**, 389
$FeC_7H_{10}IO_2$
$FeI(CO)(=\overline{COCH_2CH_2C}Me_2)$, **7**, 176
$FeC_7H_{15}B_9$
$(Cp)Fe(C_2B_9H_{10})$, **1**, 415
$FeC_8H_4O_8$
$Fe(CO)_3[C_4(OH)_4CO]$, **7**, 138
$FeC_8H_5F_3O_3$
$Fe(O_2CCF_3)(CO)Cp$, **7**, 239
$FeC_8H_5O_3$
$Fe(CO)_3Cp$, **7**, 154
$FeC_8H_6O_4$
$Fe(CO)_3(CH_2{=}C(CHO)CH{=}CH_2)$, **12**, 625, 631, 632
$Fe(CO)_3[(CH_2)_2C(CHCHO)]$, **7**, 135
$FeC_8H_6O_6$
$Fe(CH_2{=}CHCO_2Me)(CO)_4$, **1**, 84
$FeC_8H_6O_8$
$Fe(CO)_4(CO_2Me)_2$, **7**, 115
$FeC_8H_7BrO_3$
$FeBr(CO)_3(C_5H_7)$, **7**, 127
$FeC_8H_7O_2$
$[Cp(CO)_2Fe{=}CH_2]^+$, **12**, 394
$FeC_8H_8O_3$
$Fe(CO)_3(CH_2{=}CHC(Me){=}CH_2)$, **12**, 631
$Fe(CO)_3(CH_2{=}C(Me)CH{=}CH_2)$, **7**, 127, 135
$FeC_8H_8O_4$
$Fe(CO)_3(CH_2{=}C(CH_2OH)CH{=}CH_2)$, **12**, 633
$Fe(CO)_3(H_2C{=}CHCH{=}CHCH_2OH)$, **7**, 156
$FeC_8H_8O_5$
$(CO)_4Fe{=}C(OEt)Me$, **7**, 116
$FeC_8H_{10}NO_4$
$(MeC(O)CH_2CH{=}CHMe)Fe(CO)_2(NO)$, **12**, 967
$FeC_8H_{12}O_2$
$Fe(CO)_2(C_2H_4)_3$, **7**, 121
$FeC_8H_{17}B_9$
$(C_6H_6)Fe(C_2B_9H_{11})$, **1**, 377
$FeC_8H_{18}O_{10}P_2S$
$Fe(CO)_2[P(OMe)_3]_2SO_2$, **7**, 41
$FeC_9H_5NO_4$
$Fe(CO)_4(py)$, **7**, 33
$FeC_9H_7NO_3$
$Fe(CO)_3$(azepine), **7**, 142
$FeC_9H_7O_3$
$Fe(CO)_3(chd)$, **12**, 650, 653
$FeC_9H_8O_3$
$Fe(CO)_3(chd)$, **7**, 141, 142, 154
$FeC_9H_8O_4$
$Fe(CO)_3(CH_2{=}CHCH{=}CHC(O)Me)$, **12**, 631
$Fe(CO)_3(CH_2{=}C(Me)CH{=}CHCHO)$, **12**, 631

Fe(CO)$_3$(MeCH=CH-CH=CHCHO), **12**, 628
FeC$_9$H$_8$O$_5$
Fe(CO)$_3$(CH$_2$=CH-CH=CHCOOMe), **12**, 629
FeC$_9$H$_8$O$_7$
Fe(CO)$_4$(COMe)(CO$_2$Et), **7**, 114
FeC$_9$H$_8$O$_8$
Fe(CO)$_4$(CO$_2$Me)(CH$_2$CO$_2$Me), **7**, 114
FeC$_9$H$_9$ClO$_2$
Fe(CO)$_2$Cp'CH$_2$Cl, **7**, 172
FeC$_9$H$_9$IO$_2$
FeI(CO)$_2$(C$_7$H$_9$), **7**, 162
FeC$_9$H$_9$O$_2$
[Cp(CO)$_2$Fe=CHMe]$^+$, **12**, 395
[Fe(CO)$_2$Cp(C$_2$H$_4$)]$^+$, **7**, 182
FeC$_9$H$_{10}$O
Fe(CO)(allyl)Cp, **7**, 184
FeC$_9$H$_{10}$O$_2$
Fe(CO)$_2$Cp'Me, **7**, 177
FeH(CO)$_2$(C$_7$H$_9$), **7**, 144, 179
FeC$_9$H$_{10}$O$_2$S
Fe(CO)$_2$CpCH$_2$SMe, **7**, 183
FeC$_9$H$_{10}$O$_3$
CpFe(CO)$_2$CH$_2$OMe, **12**, 388
FeC$_9$H$_{10}$O$_4$
Fe(CO)$_3$(CH$_2$=CHC(C(OH)-Me)=CH$_2$), **12**, 633
FeC$_9$H$_{11}$IO$_2$
FeI(CO)$_2$(C$_7$H$_{11}$), **7**, 162
FeC$_9$H$_{11}$O$_3$
Fe(CO)$_3$[CH$_2$C(CHMe$_2$)CH$_2$], **7**, 129
Fe(CO)$_3$[CH$_2$C(Me)CMe$_2$], **7**, 129
FeC$_9$H$_{13}$O$_3$
(CH$_2$=CHMe)Fe(CO)$_3$Pr^i, **12**, 968
FeC$_9$H$_{14}$OS$_3$
Fe(CO)[C(SMe)$_3$]Cp, **7**, 183
FeC$_9$H$_{16}$O
Fe(CO)(C$_2$H$_4$)$_4$, **7**, 121
FeC$_9$H$_{20}$O$_9$P$_2$
Fe(CO)$_2$[P(OMe)$_3$]$_2$(CH$_2$O), **7**, 124
FeC$_{10}$F$_{15}$N$_5$
Fe(CNCF$_3$)$_5$, **7**, 12
FeC$_{10}$H$_5$Cl$_2$O$_4$P
Fe(CO)$_4$PPhCl$_2$, **7**, 38
FeC$_{10}$H$_5$O$_4$S
[Fe(CO)$_4$(SPh)]$^-$, **7**, 61
FeC$_{10}$H$_6$O$_4$
Fe(CO)$_3$(tropone), **12**, 625, 626
FeC$_{10}$H$_7$O$_4$P
Fe(CO)$_4$PPhH$_2$, **7**, 38
FeC$_{10}$H$_8$O$_3$
Fe(CO)$_3$(chpt), **7**, 150, 151
Fe(CO)$_3$(nbd), **7**, 142, 154
FeC$_{10}$H$_8$S$_3$
Fe(C$_5$H$_4$SSSC$_5$H$_4$), **7**, 189
FeC$_{10}$H$_8$Se$_3$
Fe(C$_5$H$_4$SeSeSeC$_5$H$_4$), **7**, 189
FeC$_{10}$H$_9$O$_3$
Fe(CO)$_3$(Mechd), **12**, 656
FeC$_{10}$H$_{10}$
FeCp$_2$, **7**, 185
FeC$_{10}$H$_{10}$O$_3$
Fe(CO)$_2$Cp'(COMe), **7**, 171
FeC$_{10}$H$_{10}$O$_4$
Fe(CO)$_2$Cp'CO$_2$Me, **7**, 168
Fe(CO)$_4$(C$_6$H$_{10}$), **7**, 121
FeC$_{10}$H$_{10}$O$_5$
Fe(CO)$_3$(CH$_2$=CH-CH=CHCH$_2$COOMe), **12**, 629, 630
FeC$_{10}$H$_{10}$O$_8$
Fe(CO)$_4$(CH$_2$CO$_2$Me)$_2$, **7**, 114
FeC$_{10}$H$_{11}$BrO$_3$
FeBr(CO)$_3$(C$_7$H$_{11}$), **7**, 162
FeC$_{10}$H$_{11}$ClO$_3$
FeCl(CO)$_3$(C$_7$H$_{11}$), **7**, 128
FeC$_{10}$H$_{11}$IO$_2$
FeI(CO)(=COCH$_2$CH$_2$CH$_2$)Cp, **7**, 176
FeC$_{10}$H$_{11}$O
Fe(CO)(C$_3$H$_5$)(C$_6$H$_6$), **7**, 199
FeC$_{10}$H$_{11}$O$_2$
[Fe(CO)$_2$Cp(CH$_2$=CHMe)]$^+$, **7**, 177
FeC$_{10}$H$_{12}$
Fe(C$_5$H$_7$)Cp, **7**, 184
FeC$_{10}$H$_{12}$O
Fe(CO)(CH$_2$CHCHMe)Cp, **7**, 184
FeC$_{10}$H$_{12}$O$_3$
Fe(CO)$_2$CpCH(OMe)Me, **7**, 176
FeC$_{10}$H$_{12}$O$_4$
Fe(CO)$_3$(MeCH=CH-CH=CHC(OH)Me), **12**, 633
FeC$_{10}$H$_{13}$O$_3$
Cp(CO)$_2$FeCH$_2$OMe$_2$, **12**, 431
FeC$_{10}$H$_{14}$
(C$_5$H$_7$)$_2$Fe, **7**, 187
FeC$_{10}$H$_{14}$OS$_3$
Fe(SMe)(CO)[C(SMe)$_2$]Cp, **7**, 183
FeC$_{10}$H$_{14}$O$_2$
Fe(CO)$_2$Me(C$_7$H$_{11}$), **7**, 162
FeC$_{10}$H$_{20}$NOP
Fe(PMe$_3$)(MeNO)MeCp, **7**, 171
FeC$_{10}$H$_{21}$IO$_3$P$_2$
Fe(CO)$_2$(PMe$_3$)$_2$[C(O)Me]I, **7**, 40, 78
FeC$_{10}$H$_{21}$P
[Fe(PMe$_3$)(Me$_2$C$_5$H$_6$)]$^+$, **7**, 161
FeC$_{11}$F$_{27}$O$_2$P$_3$
Fe(CO)$_2$[P(CF$_3$)$_3$]$_3$, **7**, 39
FeC$_{11}$H$_8$O$_3$
Fe(CO)$_3$C$_6$H$_4$(CH$_2$)$_2$, **7**, 138
Fe(CO)$_3$(cot), **7**, 149, 153, 154
FeC$_{11}$H$_8$O$_4$S
Fe(CO)$_4$(MeSPh), **7**, 61
FeC$_{11}$H$_9$Cl$_2$
[(Cl$_2$C$_6$H$_4$)FeCp]$^+$, **12**, 992, 993
FeC$_{11}$H$_9$NO$_2$
(C$_4$H$_4$N)Fe(CO)$_2$Cp, **7**, 195
FeC$_{11}$H$_9$O$_4$
Fe(CO)$_3$[MeC(O)(C$_6$H$_6$)], **7**, 143
FeC$_{11}$H$_{10}$NO$_2$
[(NO$_2$C$_6$H$_5$)FeCp]$^+$, **12**, 993
FeC$_{11}$H$_{10}$O
Fe(CO)Cp$_2$, **7**, 184
FeC$_{11}$H$_{11}$
[(C$_6$H$_6$)FeCp]$^+$, **12**, 1004
FeC$_{11}$H$_{11}$O$_3$
(cod)Fe(CO)$_3$, **12**, 673
[Fe(CO)$_2$Cp[(OMe)C$_3$H$_3$]]$^+$, **12**, 524, 526
FeC$_{11}$H$_{12}$
Fe(C$_6$H$_7$)Cp, **7**, 187
FeC$_{11}$H$_{12}$O
Fe(CO)(C$_5$H$_7$)Cp, **7**, 184
FeC$_{11}$H$_{12}$O$_2$
(CH$_2$)$_2$CHCH$_2$Fe(CO)$_2$Cp, **7**, 167
Fe(CO)$_2$CpCH$_2$C(Me)=CH$_2$, **7**, 167
FeC$_{11}$H$_{12}$O$_3$
Fe(CO)$_3$(cod), **7**, 155
Fe(CO)(CH$_2$CHCHCO$_2$)Cp, **7**, 184
Fe(COMe)(CO)$_2$(C$_7$H$_9$), **7**, 162
FeC$_{11}$H$_{12}$O$_6$
(CO)$_4$[(EtO)$_2$C$_3$H$_2$]Fe, **7**, 116
FeC$_{11}$H$_{13}$O$_2$
[Fe(CO)$_2$Cp'CMe$_2$]$^+$, **7**, 177
FeC$_{11}$H$_{14}$O$_3$
MeOCH$_2$CH$_2$Fe(CO)$_2$Cp', **7**, 168
FeC$_{11}$H$_{17}$OP
Fe(CO)(CH$_2$PEt$_2$)Cp, **7**, 184
Fe(PMe$_3$)(CO)(PhMe), **7**, 198
FeC$_{11}$H$_{17}$O$_2$
[Fe(CO)(OMe$_2$)(CH$_2$=CH-Me)Cp]$^+$, **7**, 182
FeC$_{11}$H$_{20}$O$_9$P$_2$
(CO)$_2$[P(OMe)$_3$]$_2$-Fe=C=CHCHO, **7**, 117, 118
FeC$_{11}$H$_{24}$P$_2$
CpFeH(PMe$_3$)$_2$, **7**, 200
FeC$_{12}$H$_7$O$_3$
Fe(CO)$_3$(C$_9$H$_7$), **7**, 129
FeC$_{12}$H$_8$Cl$_3$O$_4$
Fe(CO)$_3$[(OMe)(CCl$_3$)MePh], **7**, 163
FeC$_{12}$H$_8$O$_3$
FeH(CO)$_3$(C$_9$H$_7$), **7**, 164
FeC$_{12}$H$_8$O$_4$
Fe(CO)$_3$(PhCH=CHCHO), **12**, 634
FeC$_{12}$H$_{10}$O$_4$
Fe(CO)$_3$(C(Ac)CH=CH-CH=CHCH=CH), **12**, 630
Fe(picolinicacid)$_2$, **12**, 1189
FeC$_{12}$H$_{10}$O$_5$
(CO)$_4$Fe=C(OEt)Ph, **12**, 392
FeC$_{12}$H$_{12}$
Fe(C$_6$H$_6$)$_2$, **7**, 108

$FeC_{12}H_{12}O_4$
$Fe(CO)_3(MeCOC_7H_9)$, **7**, 162
$FeC_{12}H_{12}O_5$
$(CO)_4Fe{=}C(OEt)C{\equiv}CPr$, **7**, 116
$Fe(CO)_3(MeCH{=}CH\text{-}CH{=}CHCH{=}CHCOOMe)$, **12**, 628
$FeC_{12}H_{14}$
$(C_6H_6)(C_6H_8)Fe$, **7**, 188, 199
$[Fe(Cp')_2]^+$, **7**, 192
$FeC_{12}H_{14}O$
$Fe(CO)(C_6H_9)Cp$, **7**, 184
$FeC_{12}H_{14}O_2$
$Fe(CO)_2Cp''C(Me){=}CH_2$, **7**, 183
$FeC_{12}H_{14}O_6$
$(CO)_4Fe{=}C(COBu^t)OEt$, **7**, 116
$FeC_{12}H_{15}O$
$[Fe(CO)(C_6H_{10})Cp]^+$, **7**, 185
$FeC_{12}H_{16}$
$Fe(C_6H_8)_2$, **7**, 155, 188
$FeCp(Me_2Cp)$, **7**, 188
$FeC_{12}H_{16}O_2$
$FeH(CO)_2Cp^*$, **7**, 163
$FeC_{12}H_{18}P_2$
$Fe(Me_2PC_4H_3)_2$, **7**, 196
$FeC_{12}H_{19}O_6P$
$Fe(CO)[P(OMe)_3](COCH_2O\text{-}Me)Cp$, **7**, 171
$FeC_{12}H_{22}O_{10}P_2$
$(CO)_2[P(OMe)_3]_2\text{-}Fe{=}C{=}CHCO_2Me$, **7**, 119
$FeC_{12}H_{32}P_4$
$Fe(dmpm)_2(C_2H_4)$, **7**, 115
$FeC_{12}O_4$
$Fe(CO)_3(MeC(O)CHCH{=}CH\text{-}CH{=}CHCH{=}CH)$, **12**, 629, 630
$FeC_{13}H_{10}O_3$
$Fe(CO)_3Me(C_9H_7)$, **7**, 171
$FeC_{13}H_{10}O_4$
$Fe(bda)(CO)_3$, **7**, 147
$[PhCH{=}CHC(O)Me]Fe(CO)_3$, **7**, 147
$FeC_{13}H_{11}O_5$
$Fe(CO)_3[(C_3O_2H_4)C_7H_7]$, **7**, 152
$FeC_{13}H_{12}O_5$
$Fe(CO)_3(CH_2{=}CHCH{=}CH\text{-}CH{=}CHCH{=}CHCOOMe)$, **12**, 628
$Fe(CO)_3(MeCH{=}CHCH{=}CH\text{-}(CHCH_2CH{=}CHC(O)O)$, **12**, 626
$FeC_{13}H_{14}I_3$
$Fe(C_5H_4CH_2CH_2CH_2C_5H_4)I_3$, **7**, 190
$FeC_{13}H_{14}O_6$
$Fe(CO)_3(MeC(OH)C{=}C\text{-}C{=}CC{=}CCOOMe)$, **12**, 626
$FeC_{13}H_{15}O_2$
$Fe(CO)_2Cp''(MeCCMe)$, **7**, 183
$FeC_{13}H_{16}F_6NO_2P$
$Fe(CO)_2(CNBu^t)(C_6H_7)PF_6$, **7**, 157
$FeC_{13}H_{16}O_3$
$Fe(CO)_3(limonene)$, **7**, 154
$FeC_{13}H_{18}O_3P_2$
$Fe(CO)_3[Bu^t_2(P_2C_2)]$, **7**, 137
$FeC_{13}H_{18}O_4Si$
$Fe(CO)_3(MeCH{=}CHCH{=}CHC\text{-}(TMS\text{-}O){=}CH_2)$, **12**, 630
$FeC_{13}H_{23}O_2P$
$Fe(COMe)(CO)(PMe_3)(C_7H_{11})$, **7**, 162
$FeC_{13}H_{36}P_4$
$FeH(dmpe)_2Me$, **7**, 111
$FeC_{14}H_{11}O_2$
$[Cp(CO)_2Fe{=}CHPh]^+$, **12**, 393, 395, 396
$FeC_{14}H_{13}IN_2O$
$FeI(CO)(bipy)(C_3H_5)$, **7**, 129
$FeC_{14}H_{14}$
$Fe(Tol)_2$, **7**, 109
$FeC_{14}H_{15}O$
$[Fe(C_5H_4COEt)(C_6H_6)]^+$, **7**, 200
$FeC_{14}H_{16}O_7$
$Fe(CO)_3(MeOOCCH{=}CH\text{-}CH{=}CHC(OH)CH_2CH_2\text{-}CH_2CHO)$, **12**, 633
$FeC_{14}H_{20}N$
$[Fe(C_4Me_4NMe)Cp]^+$, **7**, 195
$FeC_{14}H_{22}$
$Fe(Me_2Cp)_2$, **7**, 187
$FeC_{14}H_{24}N_4O_5$
$[C(NMe_2)_3]^+[Fe(CO)_4CON\text{-}Me_2]^-$, **7**, 112
$FeC_{14}H_{27}P$
$Fe(butadiene)_2(PEt_3)$, **7**, 142
$FeC_{14}H_{30}O_8P_2$
$[Fe(CO)_2[P(OEt)_3]_2]^{2-}$, **7**, 39
$FeC_{14}H_{32}P_3$
$[Fe(PMe_3)_3Cp]^+$, **7**, 200
$FeC_{14}H_{35}P_4$
$FeH(dmpe)_2(CH{=}CH_2)$, **7**, 121
$FeC_{14}H_{36}P_4$
$Fe(dmpe)_2(C_2H_4)$, **7**, 121
$FeC_{15}H_6F_4O_3$
$Fe(CO)_3(C_{12}H_6F_4)$, **7**, 154
$FeC_{15}H_{10}F_{20}OP_2$
$Fe(CO)(dfepe)(butadiene)$, **7**, 141
$FeC_{15}H_{10}O_4$
$PhCo(CO)Fe(CO)_2Cp$, **7**, 168
$FeC_{15}H_{11}NO_4$
$Fe(CO)_2CpCH{=}CHC_6H_4NO_2$, **7**, 191
$FeC_{15}H_{11}N_3O_4$
$Fe(CO)_2CpCH{=}NN{=}CH\text{-}C_6H_4NO_2$, **7**, 191
$FeC_{15}H_{11}O_3$
$Fe(CO)_3(BzC_5H_4)$, **7**, 164
$FeC_{15}H_{13}IO_2$
$FeI(CO)_2(C_5H_4CHMePh)$, **7**, 166
$FeC_{15}H_{13}NO_3$
$Fe(CO)_3Cp_2(CNMe)$, **7**, 243
$FeC_{15}H_{14}INO_2$
$Fe(CO)_2CpCH{=}CHC_5H_4NMeI$, **7**, 191
$FeC_{15}H_{17}O_2$
$[Fe(CO)_2Cp'(C_7H_{10})]^+$, **7**, 183
$FeC_{15}H_{20}O_8P$
$[Fe(CO)[P(OMe)_3](MeO_2C\text{-}CCCO_2Me)Cp]^+$, **7**, 183
$FeC_{15}H_{23}O_2$
$Fe(acac)Cp^*$, **7**, 165
$FeC_{15}H_{24}O_6$
$Fe(acac)_3$, **12**, 179
$FeC_{15}H_{30}O_3P_2$
$Fe(CO)_3(PEt_3)_2]$, **7**, 38
$FeC_{16}H_{10}Cl_2O_4P_2$
$Fe(CO)_4[P(Cl)PhP(CO)Ph]$, **7**, 36
$FeC_{16}H_{11}NO_3$
$Fe(CO)_3(C_7H_6NPh)$, **7**, 152
$FeC_{16}H_{13}NO_3S$
$Fe(CO)_3Cp_2[C(CN)SMe]$, **7**, 243
$FeC_{16}H_{14}N_2O_9$
$Fe(CO)_3(MeCH{=}CHCH{=}CHC\text{-}(Me)(Odnb))$, **12**, 633
$FeC_{16}H_{14}O_9$
$Fe(CO)_3(MeOOCCH{=}CHCH{=}CH\text{-}CH{=}CC(O))C(Me)_2OC\text{-}(O))$, **12**, 632
$FeC_{16}H_{15}NO_3$
$Fe(CO)_3(MeCH{=}CHCH{=}CHCH{=}NBz)$, **12**, 632
$FeC_{16}H_{17}NO_3S$
$Fe(CO)_3[PhCHCHC(S)NEt_2]$, **7**, 125
$FeC_{16}H_{20}O_5$
$Fe(CO)_2Cp'COCH_2CO_2Bu^t$, **7**, 173
$FeC_{16}H_{22}$
$Fe(Me_2Ph)_2$, **7**, 188
$FeC_{16}H_{25}B_{18}N_2$
$Fe(C_2B_9H_{10}NEt_3)_2$, **1**, 412
$FeC_{16}H_{28}NO_3$
$Fe(CO)_2(PEt_3)(tropone)$, **7**, 141
$FeC_{16}H_{32}P_2$
$Fe(PMe_3)_2(C_5H_7)_2$, **7**, 128
$FeC_{16}H_{34}P_2$
$FeH(PMe_3)_2Cp^*$, **7**, 163
$FeC_{16}H_{37}P_3$
$Fe(PMe_3)_3[(CH_2)_2C(CHCMe{=}CH_2)]$, **7**, 134
$FeC_{16}H_{38}P_2$
$Fe(PEt_3)_2(C_2H_4)_2$, **7**, 121
$FeC_{17}H_{11}NO_3$
$Fe(CO)_2(NO)(C_3Ph_2H)$, **7**, 127
$FeC_{17}H_{13}NO_3$
$Fe(CO)_3(C_7H_6NTol)$, **7**, 152
$FeC_{17}H_{16}NO_3S$
$Fe(CO)_3Cp_2[C(CN)SMe_2]^+$, **7**, 243
$FeC_{17}H_{16}N_2$
$Fe(bipy)(PhMe)$, **7**, 198
$FeC_{17}H_{20}O$
$Fe(CO)[CH_2C(Me)C(Me)\text{-}CH_2]_2$, **7**, 155

$FeC_{17}H_{25}N_3$
$Fe[Pr^iN{=}CHCH{=}NPr^i][NC(Me_2C_6H_3)]$, **7**, 116
$FeC_{17}H_{28}O_7$
$Fe(CO)_3$(cyclooctene)(diethylfumarate), **7**, 122
$FeC_{17}H_{31}P$
$Fe(C_7H_{11})_2(PMe_3)$, **7**, 161
$FeC_{17}H_{32}O_3P_2$
$Fe(CO)_3$(dippe), **7**, 114
$FeC_{17}H_{35}O_9P_3$
$Fe[P(OMe)_3]_3[C_6H_4(CH_2)_2]$, **7**, 138
$FeC_{17}H_{37}F_6P_2$
$[Fe(PEt_3)_2(C_5H_7)]^+PF_6^-$, **7**, 162
$FeC_{17}H_{38}F_3O_3P_3S$
$[Fe(PMe_3)_3(C_7H_{11})]^+OTf^-$, **7**, 162
$FeC_{17}H_{41}P_4$
$Fe(dmpe)_2(CMe{=}CMe_2)^+$, **7**, 110
$FeC_{17}H_{44}P_4$
$FeH(dmpe)_2(C_5H_{11})$, **7**, 111
$FeC_{18}H_{12}F_{20}P_2$
$Fe(PhCH{=}CH_2)(CF_3CF_2)_2PCH_2CH_2P(CF_2CF_3)_2$, **7**, 198
$FeC_{18}H_{13}NO_3$
$Fe(CO)_3(PhCH{=}CHCH{=}NPh$, **7**, 154
$FeC_{18}H_{16}O_4$
$Fe(CO)_2Cp'COC({=}CH_2)OBz$, **7**, 173
$FeC_{18}H_{18}$
$Fe(C_9H_9)_2$, **7**, 187
$FeC_{18}H_{19}NO_3$
$Fe(CO)_3[PhCH{=}CHC(Me){=}C{=}NBu^t]$, 7, 139, 199
$FeC_{18}H_{33}P$
$Fe(PEt_3)$(butadiene)(cod), **7**, 155
$FeC_{18}H_{34}P_2$
$Fe(PMe_3)_2(chd)_2$, **7**, 133
$FeC_{18}H_{41}P_3$
$Fe(PMe_2Ph)_3(C_2H_4)_2$, **7**, 121
$FeC_{19}H_{10}F_5N_3O_3P_3$
$Fe(CO)_3[C_4Ph_2(P_3N_3F_5)]$, **7**, 136
$FeC_{19}H_{32}O_3$
$Fe(CO)_3(cyclooctene)_2$, **7**, 121, 122
$FeC_{19}H_{36}N_2O_4$
$[Bu_4N][Fe(CO)_3(NO)]$, **12**, 354
$FeC_{19}H_{36}O_4P_2$
$Fe(CO)_2(PEt_3)_2({=}C{=}CHCO_2Et)$, **7**, 123
$FeC_{20}H_{16}$
$Fe(naphthalene)_2$, **7**, 155
$FeC_{20}H_{19}N$
$(C_6H_6)[(Bz)CN(C_6H_6)]Fe$, **7**, 199
$FeC_{20}H_{20}O_5$
$Fe(CO)_3[PhCH{=}CHC(Me){=}C{=}CHCO_2Bu^t]$, **7**, 139
$FeC_{20}H_{22}O_5$
$Fe(CO)_3(C_5H_{11}C{\equiv}CCH_2C{\equiv}CCH_2CH{=}CHCH{=}CHCOOMe)$, **12**, 626
$FeC_{20}H_{24}$
Fe(cod)(Bz)Cp, **7**, 114
$FeC_{20}H_{28}O_5$
$Fe(CO)_3(MeC(Me){=}CHCH_2CH_2CH(Me)CH_2CH{=}CHC(Me){=}CHCOOEt)$, **12**, 626
$FeC_{20}H_{30}$
$Cp^*{}_2Fe$, **7**, 186
$FeC_{20}H_{32}$
$Fe(Bu^tC_6H_7)_2$, **7**, 188
$FeC_{20}H_{37}ClP_4$
$FeCl(dmpe)_2(C{\equiv}CPh)$, **7**, 111
$FeC_{20}H_{38}P_2$
$Fe(depe)(C_5H_7)_2$, **7**, 128
$FeC_{21}H_{10}N_4O_2$
$Fe(CO)_2CpC[{=}C(CN)_2]CPhC(CN)_2$, **7**, 174
$FeC_{21}H_{15}O_3Se_3$
$[Fe(CO)_3(SePh)_3]^-$, **7**, 74
$FeC_{21}H_{15}O_3Te_3$
$[Fe(CO)_3(TePh)_3]^-$, **7**, 76
$FeC_{21}H_{24}S$
$FeCp^*(C_5H_4SPh)$, **7**, 61
$FeC_{21}H_{36}O_3P_2$
$Fe(CO)_2(PEt_3)_2(PhCHO)$, **7**, 124
$FeC_{22}H_{15}O_4P$
$Fe(CO)_4(PPh_3)$, **12**, 673
$FeC_{22}H_{19}O_3P$
$Fe(CO)_2Cp'COCH_2PPh_2$, **7**, 173
$FeC_{22}H_{20}O_3S$
$Fe(CO)_3[CS(CPh_2)(CHBu^t)]$, **7**, 135
$FeC_{22}H_{28}N_2$
$(Mes)Fe[N{=}C(Me)(Mes)](NCMe)$, **7**, 110
$FeC_{22}H_{28}O_9P_2$
$Fe(CO)_2[P(OMe)_3]_2(OCCPh_2)$, **7**, 124
$FeC_{22}H_{36}O_2P_2$
$Fe(CO)_2(PEt_3)_2(PhC{\equiv}CH)$, **7**, 123
$FeC_{22}H_{43}BrO_3P_2$
$FeBr(CO)_2(dippe)(COCH_2Bu^t)$, **7**, 111
$FeC_{22}H_{51}P_3$
$Fe(butadiene)(PEt_3)_3$, **7**, 142
$FeC_{23}H_{17}O_4P$
$Ph_3PCH_2Fe(CO)_4$, **7**, 112
$FeC_{23}H_{19}O_3P$
$Fe(CO)_3(PPh_3)(C_2H_4)$, **7**, 121
$FeC_{23}H_{29}$
$[FeCp(Me_8C_{10})]^+$, **7**, 200
$FeC_{23}H_{37}O_9P_3$
$Fe[P(OMe)_3]_3(PhC{\equiv}CPh)$, **7**, 123
$FeC_{24}H_{18}O_5$
$[(OH)MePh_2(O)C_8H4]Fe(CO)_3$, **12**, 744, 745
$FeC_{24}H_{21}OP$
$FeH(PPh_3)(CO)Cp$, **7**, 163
$FeC_{24}H_{21}O_2PS_4$
$Fe(CO)_2(PPh_3)[S_2C_2(SMe)_2]$, **7**, 16, 68
$FeC_{24}H_{24}O_2PS_4$
$Fe(CO)_2(PPh_3)[SC(SMe)C(SMe)S]$, **7**, 62
$FeC_{24}H_{29}NO$
$Fe(NO)(Bz)_2Cp^*$, **7**, 168
$FeC_{24}H_{32}P_2$
$Fe(PMe_3)_2(C_5H_4{=}CPh_2)$, **7**, 149
$FeC_{24}H_{36}$
$[Fe(C_6Me_6)_2]^+$, **7**, 203
$FeC_{24}H_{37}O_3P$
$Fe(CO)_2(PBu_3)(bda)$, **7**, 147
$FeC_{25}H_{21}O_2P$
$Fe(CO)_2(PPh_3)(C_5H_6)$, **7**, 163
$FeC_{25}H_{22}O_2P$
$(CH_2{=}CHCH{=}CHCH_2)Fe(CO)_2(PPh_3)$, **12**, 667
$FeC_{25}H_{23}O_2P$
$Fe(CO)_2(PPh_3)$(isoprene), **7**, 127
$FeC_{25}H_{23}O_3P$
$C_4H_8Fe(CO)_3(PPh_3)$, **7**, 116
$FeC_{25}H_{45}P_5$
$Fe(C_2Bu^t{}_2P_3)(C_3Bu^t{}_3P_2)$, **7**, 197
$FeC_{26}H_{10}D_5O$
$Fe(CO)(C_6D_5)(PhCCPh)Cp$, **7**, 170
$FeC_{26}H_{22}O_2P$
$(chd)Fe(CO)_2(PPh_3)$, **12**, 658
$FeC_{26}H_{23}Cl_2OP$
$FeCl_2[(HO)Ph_2CCH_2PPh_2]$, **7**, 40
$FeC_{26}H_{23}NO_3P$
$[Fe(NO)(PPh_3)(COCOMe)Cp]^+$, **7**, 171
$FeC_{26}H_{24}OP$
$[Cp(CO)(PPh_3)Fe{=}CHMe]^{+}$' **12**, 394
$FeC_{26}H_{25}O_3P$
$[MeCH{=}C(Me)C(Me)(O)]Fe(CO)_2(PPh_3)$, **12**, 968
$FeC_{26}H_{48}O_2$
$Fe(CO)_2(cyclooctene)_3$, **7**, 121
$FeC_{27}H_{22}N_3O_3P$
$Fe(N_2C_6H_4COMe)(CO)(NO)(PPh_3)$, **7**, 32, 40
$FeC_{27}H_{23}O_4P$
$Fe(CO)_3(P(OPh)_3)\overline{C(O)CH_2}CH{=}CHCH{=}CHCH_2$, **12**, 629, 630
$FeC_{27}H_{23}O_6P$
$(CO)_3(P(OPh)_3)(\overline{C(O)CH_2}CH{=}CHCH{=}CCH_2)$, **12**, 630
$FeC_{27}H_{27}O_4P$
$Fe(CO)Me[P(OPh)_3](C_7H_9)$, **7**, 162
$FeC_{28}H_{20}N_2O_2$
$Fe(NO)_2(C_4Ph_4)$, **7**, 136
$FeC_{28}H_{21}O_2P$
$Fe(CO)_2(PPh_3)(C_8H_6)$, **7**, 137
$FeC_{28}H_{30}NOP$
$Fe(CO)(Ph_2PNMeCHMePh)MeCp$, **7**, 172

$FeC_{28}H_{30}P_2$
$Fe(dppe)Me_2$, **7**, 168
$FeC_{28}H_{46}P_2$
$Fe(dippe)(Bz)_2$, **7**, 114
$FeC_{29}H_{22}O_4P_2$
$Fe(CO)_4(Ph_2P(CH_2)PPh_2)$, **7**, 37
$FeC_{29}H_{29}F_3NO_4P$
$(Ph_3P)(CO)_2Fe(OCOCF_3)(\equiv CNPr^i_2)$, **7**, 120
$FeC_{29}H_{29}IO_2P_2$
$FeI(PPh_2Me)_2(CO)_2Me$, **7**, 111
$FeC_{29}H_{31}$
$Fe(Me_2Ph_2C_6H_4)$(xylyl), **7**, 204
$FeC_{29}H_{36}NO_3P$
$[Et_4N][HFe(CO)_3(PPh_3)]$, **7**, 39
$FeC_{30}H_{24}O_6P_2$
$Fe(CO)_2(Ph_2PCH_2CO_2)_2$, **7**, 40
$FeC_{30}H_{28}NOP$
$Fe(py)(PPh_3)(COMe)Cp$, **7**, 171
$FeC_{30}H_{28}P_2$
FeH(dppm)Cp, **7**, 163, 164
$FeC_{30}H_{36}$
$Fe(PhBu^t{}_6H_4)_2$, **7**, 188
$FeC_{30}H_{36}NO_7P$
$Et_4N^+[Fe(CO)_3[P(OPh)_3]COH]^-$, **7**, 112
$FeC_{31}H_{41}O_4$
$Fe(CO)_3$(ergosterol), **12**, 626
$FeC_{32}H_{23}O_{10}P$
$Fe_3(MeCCOEt)(CO)_9(PPh_3)$, **7**, 277
$FeC_{32}H_{26}O_3P_2$
$Fe(CO)_3(Ph_2PCH_2Fe(CO)_3(Ph_2PCH_2C_6H_4)_2)_2C_3H_4$, **7**, 38
$FeC_{32}H_{29}OP_2$
$[Fe(CO)(dppe)Cp]^+$, **6**, 6
$FeC_{32}H_{30}OP_2$
FeH(dppe)(CO)Cp, **7**, 163
$FeC_{32}H_{30}P_2S_2$
Fe(dppm)(SCMeS)Cp, **7**, 185
$FeC_{32}H_{32}P_2$
Fe(dppe)MeCp, **7**, 168
$FeC_{32}H_{42}$
$[Fe(C_9Me_7)_2]^+$, **7**, 191
$FeC_{32}H_{50}N_4$
$Fe(C_4Me_4NHNC_4Me_4)_2$, **7**, 195
$FeC_{33}H_{29}O_5P$
$Fe(CO)_2[P(OMe)_3](C_4Ph_4)$, **7**, 137
$FeC_{34}H_{33}P_2$
$[Fe(dppe)(=C=CHMe)Cp]^+$, **7**, 179
$FeC_{35}H_{35}P_2$
$[Fe(dppe)(=C=CMe_2)Cp]^+$, **7**, 179
$FeC_{36}H_{30}N_2O_2P_2$
$Fe(NO)_2(PPh_3)_2$, **7**, 40
$FeC_{36}H_{36}P_2$
$Fe(dppm)(CCBu^t)Cp$, **7**, 169
$FeC_{36}H_{37}O_4PS$
$Ph_3P(CO)_3Fe=C(OEt)CH(SPh)CH_2Bu$, **7**, 116
$FeC_{37}H_{30}N_2O_3P$
$Fe(CO)(NO)_2(PPh_3)_2$, **7**, 40
$FeC_{37}H_{31}NOP_2$
$HFe(CO)(NO)(PPh_3)_2$, **7**, 32
$FeC_{37}H_{40}OP_2$
Fe(dppe)(CO)(Cp*H), **7**, 164
$FeC_{38}H_{30}IO_2P_2$
$HFe(CO)_2I(PPh_3)_2$, **7**, 40
$FeC_{38}H_{36}P_2$
Fe(dppe)(Bz)Cp, **7**, 114
$FeC_{39}H_{30}O_2P_2S$
$Fe(CO)_2(CS)(PPh_3)_2$, **7**, 16
$FeC_{39}H_{33}O_4P$
$Fe(CO)[P(OPh)_3](CPhCPhMe)Cp$, **7**, 169
$FeC_{39}H_{34}BrOP$
$FeBr(PMe_3)(CO)(C_5Ph_5)$, **7**, 166
$FeC_{40}H_{35}O_2P$
$Fe(PMe_3)(CO)(CHO)(C_5Ph_5)$, **7**, 166
$FeC_{40}H_{59}O_3P_2$
$Fe(CO)_3[(Bu_3C_6H_2)PCHP(Bu_3C_6H_2)]^-$, **7**, 130
$FeC_{41}H_{30}NO_2P$
$Fe(CO)(NO)(PPh_3)(Ph_3C_4O)$, **7**, 127
$FeC_{41}H_{31}P_2$
$FeH(PPh_3)_2Cp$, **7**, 164
$FeC_{41}H_{32}O_3P_2$
$Fe(CO)_3(Ph_2PCH_2C_6H_4)_2$, **7**, 38
$FeC_{41}H_{33}O_3P_3$
$Fe(CO)_3(Ph_2P)_2CHCH_2PPh_2$, **7**, 38
$FeC_{42}H_{60}O_{10}P_4$
$[Fe(CO)_2[PPh(OEt)_2]_4]^+$, **7**, 79
$FeC_{43}H_{35}NO_6P_2$
$[PPN]^+[Fe(CO)_4CH_2CO_2Me]^-$, **7**, 112
$FeC_{43}H_{37}ClO_4P_2$
$FeCl(CO)_2(PPh_3)_2[C\equiv CCH(OMe)_2]$, **7**, 111
$FeC_{45}H_{30}O_6$
$Fe(dbm)_3$, **12**, 168
$FeC_{48}H_{37}O_2P$
$Fe(CO)(PPh_3)(COMe)(C_{27}H_{19})$, **7**, 172
$FeC_{51}H_{44}I_2OP_4$
$FeI_2(CO)(dppm)_2$, **7**, 79
$FeC_{52}H_{40}O_2P_2$
$Fe(CO)_2(PPh_3)_2(PhC\equiv CPh)$, **7**, 123
$FeC_{52}H_{64}N_4S_4$
$Fe(CNC_6H_{11})_4(SPh)_4$, **7**, 61
$FeC_{53}H_{46}N_4P_2$
$[Fe(CNPh)_2[C[N(H)Ph]_2](dppe)]^{2+}$, **7**, 14
$FeC_{54}H_{58}O_2$
$Fe(THF)_2[CPh=C(Ph)(Mes)]_2$, **7**, 110
$FeC_{57}H_{46}NO_6P_3$
$[PPN][HFe(CO)_3[P(OPh)_3]]$, **7**, 39
$FeC_{70}H_{50}Br$
$Fe(C_5Ph_5)_2Br$, **7**, 166
$FeC_{74}H_{62}$
$Fe[CPh=CPhCPh=C(Ph)(Mes)]_2$, **7**, 110
$FeC_{80}H_{70}$
$[(Bz)_5C_5]_2Fe$, **7**, 186
$FeCdSiC_{23}H_{23}BrNO_6P$
$Fe(CO)_3[Si(OMe)_3](Ph_2Ppy)CdBr$, **10**, 338
$FeCdSiC_{24}H_{24}ClO_3P$
$Fe(CO)_3(PPh_3)(TMS)CdCl$, **10**, 338
$FeCdSiC_{26}H_{26}ClO_7P_2$
$Fe(CO)_3[Si(OMe)_3](Ph_2PCH_2C(O)Ph)CdCl$, **10**, 338
$FeCdSiC_{31}H_{31}ClO_6P_2$
$Fe(CO)_3[Si(OMe)_3](Ph_2PCH_2PPh_2)CdCl$, **10**, 338
$FeCdSiC_{46}H_{43}BrO_8P_2$
$Fe(CO)_3[Si(OMe)_3](Ph_2PCH_2C(O)Ph)CdBr(Ph_2PCH_2C(O)Ph)$, **10**, 339
$FeCl_2O_8$
$Fe(ClO_4)_2$, **12**, 1189
$FeCoC_8O_8$
$[FeCo(CO)_8]^-$, **10**, 19, 20
$FeCoC_{11}H_5O_6$
$FeCp(CO)_2Co(CO)_4$, **10**, 54
$FeCoC_{13}H_7O_8S_2$
$Fe_2Co(C_5H_4CO_2Me)(CO)_6S_2$, **7**, 66
$FeCoC_{16}H_{17}O_3$
$Fe(CO)_3(CMeCMeCMeCMe)CoCp$, **10**, 60
$FeCoC_{20}H_{17}O_{11}$
$FeCp[C_4(Co_2Me)_4]Co(CO)_3$, **10**, 60
$FeCoCrC_{13}H_5O_8S$
$FeCoCr(S)Cp(CO)_8$, **10**, 37, 244
$FeCoIrRuC_{16}H_9O_{10}P$
$FeRuCoIr(H)(PMe)(CO)_{10}Cp$, **10**, 244
$FeCoMnC_{11}H_3O_{10}P$
$MnFeCo(Co)_{10}(PMe)$, **10**, 15
$FeCoMoC_{13}H_5O_8S$
$MoFeCo(S)Cp(CO)_8$, **10**, 49
$FeCoNiC_{15}H_{14}O_6P$
$FeCoNiCp[PBu^t](CO)_6$, **9**, 20, 173
$FeCoNiC_{16}H_{14}O_8P$
$CoFeNi(Me_3P)Cp(CO)_8$, **9**, 21
$FeCoPd_2C_{56}H_{44}NO_7P_4$
$FeCoPd_2(CO)_6(NO)(dppm)_2$, **10**, 243
$FeCoWC_{12}H_{26}O_8P_3$
$WFeCo(PMe)(CO)_8(PMe_3)_2Cp$, **10**, 222
$FeCoWC_{14}H_8O_8P$
$WFeCo(PMe)(CO)_8Cp$, **10**, 37, 244
$FeCoWC_{19}H_{22}O_8P_2$
$W(CO)Cp(CO)(dppm)[PMeFe(CO)_4]Co(CO)_2$, **10**, 44
$FeCo_2C_9HNO_9$
$FeCo_2(NH)(CO)_9$, **10**, 197, 218

$FeCo_2C_9O_9S$
$FeCo_2(S)(CO)_9$, **10**, 237
$FeCo_2C_9O_9Se$
$FeCo_2(CO)_9Se$, **7**, 74
$FeCo_2C_9O_9Te$
$FeCo_2(CO)_9Te$, **7**, 77
$FeCo_2C_{10}H_3O_9P$
$FeCo_2(PMe)(CO)_9$, **10**, 226
$FeCo_2C_{15}H_5O_9P$
$FeCo_2(PPh)(CO)_9$, **10**, 227
$FeCo_2C_{15}H_6O_9P$
$[FeCo_2(PPhH)(CO)_9]^-$, **10**, 227
$FeCo_2C_{16}H_{11}O_8PS$
$FeCo_2(S)(CO)_8(PMe_2Ph)$, **10**, 244
$FeCo_2C_{21}H_{11}O_9P_2$
$FeCo_2(PPh)(PPhH)(CO)_9$, **10**, 222
$FeCo_3C_{12}HO_{12}$
$HFeCo_3(CO)_{12}$, **10**, 47, 49, 52, 58
$FeCo_3C_{12}O_{12}$
$[FeCo_3(CO)_{12}]^-$, **10**, 47, 52, 197
$FeCo_3(CO)_{12}$, **10**, 341
$FeCo_3C_{15}H_5O_9S$
$FeCo_3(CS)(CO)_9Cp$, **10**, 209
$FeCo_3C_{16}H_{12}NO_{12}$
$[Me_4N][FeCo_3(CO)_{12}]$, **10**, 49
$FeCo_3C_{20}H_{10}O_9$
$Co_3(CO)_9(CFc)$, **8**, 426
$FeCo_3C_{20}H_{20}NO_{12}$
$[Et_4N][FeCo_3(CO)_{12}]$, **10**, 45
$FeCo_3C_{22}H_{16}NO_{12}$
$[(PhCH_2)Me_3N][FeCo_3(CO)_{12}]$, **10**, 45
$FeCo_3C_{27}H_{26}$
$Co_3(CH)(CFc)Cp_3$, **8**, 435
$FeCo_3C_{29}H_{15}O_{11}P$
$[FeCo_3(CO)_{11}(PPh_3)]^-$, **10**, 191, 192
$FeCrC_9HO_9$
$[Cr(CO)_5(H)Fe(CO)_4]^-$, **10**, 42
$FeCrC_9O_9$
$Cr(CO)_5Fe(CO)_4$, **10**, 30
$FeCrC_{11}H_{12}O_7$
$Cr(CO)_4(PMe_2)_2Fe(CO)_3$, **10**, 43
$FeCrC_{13}H_5O_7$
$Cr(CO)_3[PhFe(CO)_4]^-$, **7**, 112
$FeCrC_{13}H_{10}NO_3$
$[Cr(NO)Cp(CO)_2Fe(CO)Cp]^-$, **10**, 42
$FeCrC_{16}H_{13}O_3$
$CrCp(C_8H_8)Fe(CO)_3$, **10**, 42
$FeCrC_{18}H_{18}$
$CrCp(C_8H_8)FeCp$, **10**, 42
$FeCr_2C_{17}H_{19}O_3S_3$
$Cp_2Cr_2Fe(CO)_3(SBu^t)(S)_2$, **7**, 71, 72
$FeCuC_{40}H_{30}O_4P_2$
$Fe(CO)_4(CuPPh_3)_2$, **7**, 79, 80; **10**, 54
$FeCuC_{56}H_{63}P_4$
$Fe(PEtPh_2)_3(H)_3Cu(PEtPh_2)$, **10**, 61
$FeGaC_{11}H_{15}I_2O_3$
$GaI_2(Et_2O)[Fe(CO)_2Cp)]$, **1**, 555
$FeGaC_{16}H_{26}NO_2$
$GaEt(CH_2CH_2CH_2NEt_2)[Fe(CO)_2Cp]$, **1**, 556
$FeGaC_{24}H_{24}NO_2$
$Ga(C_5H_4Me)_2(py)[Fe(CO)_2Cp)]$, **1**, 555
$FeGeSiC_9H_{14}O_4$
$(CO)_4Fe(Me_2SiCH_2GeMe_2)$, **7**, 30
$FeHCl_4$
$HFeCl_4$, **2**, 121
$FeHfC_{22}H_{31}O_4P_2$
$Cp_2HHf(PEt_2)_2Fe(CO)_4$, **4**, 516
$FeHgC_{10}H_9Cl$
$CpFeC_5H_4HgCl$, **12**, 134, 135
$FeHgC_4O_3S$
$HgFe(CO)_3(CS)$, **7**, 16
$FeHgC_7H_5ClO_5S$
$Fe(CO)_4(HgCl)[C(S)OEt]$, **7**, 16
$FeHgMnSi_2C_{54}H_{43}O_5$
$Fe(CO)_3PPh_3SiMePh_2$-$HgMnCpSiPh_3(CO)_2$, **10**, 339
$FeHgMn_2Si_2C_{60}H_{51}O_5$
$MeCpMn(CO)_2SiPh_3MnHgFe(CO)_3PPh_3SiMePh_2$, **10**, 337
$FeHgRuC_{20}H_{18}$
$[Cp(C_5H_4)FeCp(C_5H_4)Ru(Hg)]^{2+}$, **7**, 618
$FeHg_2C_4Br_2O_4$
$Fe(CO)_4(HgBr)_2$, **10**, 334
$FeHg_2C_4Cl_2O_4$
$Fe(CO)_4(HgCl)_2$, **10**, 324
$FeHg_2C_6N_2O_4S_2$
$Fe(CO)_4(HgSCN)_2$, **7**, 82
$FeHg_2C_{30}H_{16}N_6O_4S_2$
$Fe(CO)_4(HgSCN)_2(phen)_2$, **7**, 82
$FeHg_2C_{33}H_{25}O_3P$
$(CO)_3(PPh_3)Fe(HgPh)_2$, **10**, 339
$FeInC_4Br_2O_4$
$[InBr_2[Fe(CO)_4]]^-$, **7**, 19
$FeInC_{17}H_{29}N_2O_2$
$In(CH_2CH_2CH_2NMe_2)_2[Fe(CO)_2Cp]$, **1**, 562
$FeInSiC_{26}H_{26}Cl_2O_7P$
$Fe(CO)_3[Si(OMe)_3][InCl_2]$-$[PPh_2CH_2C(O)Ph]$, **1**, 564
$FeIn_2Si_6C_{22}H_{54}Cl_2O_4$
$In_2[C(TMS)_3]_2(Cl)_2[Fe(CO)_4]$, **1**, 564
$FeIrC_{19}H_{23}O_3$
$Fe(CO)_2(CMeCHCHCMe)$-$(CO)IrCp^*$, **10**, 60
$FeIr_3C_{12}O_{12}$
$[FeIr_3(CO)_{12}]^-$, **7**, 83; **10**, 191, 192
$FeIr_4C_{15}O_{15}$
$[FeIr_4(CO)_{15}]^{2-}$, **7**, 83; **10**, 192
$FeIr_4Na_2C_{15}O_{15}$
$Na_2[FeIr_4(CO)_{15}]$, **10**, 43
$FeIr_5C_{15}HO_{15}$
$[FeIr_5H(CO)_{15}]^{2-}$, **7**, 83; **10**, 197
$FeIr_5C_{15}H_2O_{15}$
$[H_2FeIr_5(CO)_{15}]^-$, **7**, 83
$FeIr_5C_{15}O_{15}$
$[FeIr_5(CO)_{15}]^{3-}$, **7**, 83
$FeIr_5C_{16}O_{16}$
$[FeIr_5(CO)_{16}]^-$, **10**, 192
$FeK_2C_4O_4$
$K_2Fe(CO)_4$, **12**, 352
$FeK_3C_6N_6$
$K_3Fe(CN)_6$, **12**, 1145
$FeLiC_{27}H_{33}$
$Li[Fe(Mes)_3]$, **7**, 110
$FeLi_2C_4H_{12}$
Li_2FeMe_4, **7**, 110
$FeLi_2C_{40}H_{30}N_2$
$Li_2[Fe(NPh_2)_2(C{\equiv}CPh)_2]$, **7**, 114
$FeLi_2C_{48}H_{48}O_2$
$[Li(OEt_2)]_2[Fe(nap)_4]$, **7**, 110
$FeLi_3C_{30}H_{21}$
$Li_3[Fe(nap)_3]$, **7**, 110
$FeLi_3C_{30}H_{21}N_2$
$Li_3Fe(N_2)(nap)_3$, **7**, 110
$FeLi_4C_{40}H_{60}O_4$
$[Li(OEt_2)]_4[FePh_4]$, **7**, 110
$FeLi_4C_{40}H_{62}O_4$
$[FeH_2Ph_4][Li(OEt_2)]_4$, **7**, 110
$FeLi_5C_{56}H_{73}O_5$
$[Li(THF)]_5[Ph_3Fe(H)_3FePh_3]$, **7**, 110
$FeMnC_{10}H_3O_9$
$Mn(CO)_4[MeC(O)]Fe(CO)_4$, **10**, 52
$FeMnC_{12}H_5O_6S_2$
$Mn[S_2CFe(CO)_2Cp](CO)_4$, **6**, 89
$FeMnC_{12}H_5O_7$
$Mn(CO)_5Fe(CO)_2Cp$, **10**, 48
$FeMnC_{15}H_9O_8$
$Mn(CO)_2Cp[C{=}CH(CO_2Me)]Fe(CO)_4$, **10**, 52
$FeMnC_{20}H_{10}O_8P$
$FeMn(CO)_8(PPh_2)$, **7**, 78
$Mn(CO)_4(PPh_2)Fe(CO)_4$, **6**, 87
$FeMnC_{21}H_{27}O_3$
$Cp^*Mn(CO)_3Fe(C_4Me_4)$, **6**, 116
$FeMnC_{22}H_{16}O_5P$
$Mn(CO)_4(H)(PPh_2)Fe(CO)Cp$, **6**, 88
$FeMnC_{24}H_{15}O_6S_3$
$FeMn(SPh)_3(CO)_6$, **7**, 61, 62
$FeMnC_{33}H_{22}BrO_8P_2$
$(CO)_4Fe(dppm)Mn(CO)_4Br$, **7**, 83
$FeMoC_4H_6N_2O_2S_2$
$[MoFeS_4(SCN)_2(OMe)_2]^{2-}$, **10**, 37
$FeMoC_{10}H_5BrO_5Te_2$
$CpMoFe(CO)_5Te_2Br$, **7**, 77
$FeMoC_{12}H_{15}I_2N$

$MoCp_2(H)_2FeI_2(MeCN)$, **10**, 43
$FeMoC_{14}H_{10}O_5$
$Mo(CO)_2Cp(CH_2C{=}CMe)Fe(CO)_3$, **10**, 42
$FeMoC_{15}H_{15}NO_5S_2Te_2$
$CpMoFe(CO)_5Te_2S_2CNEt_2$, **7**, 77
$FeMoC_{16}H_{10}O_4S_2$
$Mo(CO)_4(SPh)_2Fe$, **10**, 43
$FeMoC_{19}H_{14}O_3$
$Mo(C_6H_6)(azulene)Fe(CO)_3$, **10**, 42
$FeMoC_{33}H_{22}O_8P_2$
$(CO)_4Fe(dppm)Mo(CO)_4$, **7**, 83
$MoFe(CO)_8(dppm)$, **10**, 17
$FeMoC_{34}H_{22}O_9P_2$
$(CO)_4Fe(dppm)Mo(CO)_5$, **7**, 83
$FeMoC_{44}H_{54}N_4O_4S_2$
$(CyNC)_4Fe(SPh)_2Mo(CO)_4$, **7**, 61
$FeMoNiC_{27}H_{22}O_7$
$Cp_2FeMoNi(CO)_5(PhC_2Co_2Pr^i)$, **10**, 162
$FeMoSbC_{10}H_5F_6O_5Te_2$
$[CpMoFe(CO)_5Te_2][SbF_6]$, **7**, 77
$FeMo_2C_2H_6OS_8$
$[Mo_2FeS_8O(OMe)_2]^{3-}$, **10**, 37
$FeMo_2C_{17}H_{10}O_7Te_2$
$FeMoCp(CO)_5Te_2[Mo(CO)_2Cp]$, **7**, 77
$Mo_2Fe(Te)_2(CO)_7Cp_2$, **10**, 228
$FeMo_2C_{27}H_{30}O_8$
$Cp^*{}_2Mo_2Fe(CO)_7(O)$, **7**, 60
FeN_5O_2
$Fe(NO)_2(N_2)_2$, **7**, 31
$FeNaC_4O_4$, **12**, 351
$NaFe(CO)_4$, **12**, 351
$FeNaC_6H_6O_3$
$[(CH_2{=}CHMe)Fe(CO)_3]Na$, **12**, 967, 968
$FeNaC_7H_5O_2$
$NaFeCp(CO)_2$, **12**, 617
$FeNbC_{10}H_{10}Cl_4$
$Cp_2Nb(Cl)_2FeCl_2$, **5**, 126, 127
$FeNiC_{17}H_{36}ClO_3P_3$
$Fe(CO)_3(PMe_3)(PBu^t{}_2)NiCl(PMe_3)$, **10**, 61
$FeNiC_{24}H_{40}B_4$
$Cp^*{}_2NiFe(Me_2C_2B_4H_4)$, **1**, 397
$FeNiC_{30}H_{25}O_5P$
$FeNi(CO)_3(PPh_3)[C(CO_2Me){=}CH_2]Cp$, **10**, 61
$FeNiC_{34}H_{30}Cl_2P_2$
$Ni(dppf)Cl_2$, **12**, 170
$FeNi_2C_{17}H_{19}O_3P$
$Ni_2Fe(PBu^t)(CO)_3Cp_2$, **9**, 173
$FeOs_3C_{13}HO_{13}$
$[HFeOs_3(CO)_{13}]^-$, **7**, 83
$FeOs_3C_{13}H_2O_{13}$
$H_2FeOs_3(CO)_{13}$, **10**, 52
$FeOs_3C_{22}H_{19}O_8P$
$Os_3(CO)_8[C_5H_3PEt_2Fe(C_5H_4)(H)_3]$, **10**, 17
$FeOs_3C_{23}H_{23}O_8P$
$Os_3H_2(Pr^i{}_2PC_4H_3FeC_5H_4)(CO)_8$, **7**, 709
$FePb_2C_{16}H_{28}$
$(Me_3PbCp)_2Fe$, **2**, 306
$FePdC_{16}H_{10}ClO_4P$
$(CO)_4Fe(PPh_2)Pd(Cl)$, **10**, 41
$FePdC_{28}H_{23}PS_2$
$Fe(C_5H_4S)_2Pd(PPh_3)$, **10**, 61
$FePdC_{31}H_{31}IN_3P_2$
$(MeNC)_3Fe(dppm)PdI$, **7**, 79
$FePdC_{34}H_{28}Cl_2P_2$
$[Fe(C_5H_4PPh_2)_2]PdCl_2$, **1**, 317
$FePdC_{34}H_{30}Cl_2P_2$
$Pd(dppf)Cl_2$, **12**, 167, 174–6, 180, 181, 186, 214
$FePdC_{52}H_{44}I_2O_2P_4$
$Fe(I)(CO)_2(dppm)_2PdI$, **7**, 79
$FePdC_{57}H_{53}I_2NO_2P_4$
$[(Bu^tNC)(CO)_2Fe(dppm)_2PdI]I$, **7**, 79
$FePdPtC_{54}H_{44}O_4P_4$
$FePdPt(CO)_4(dppm)_2$, **7**, 84; **10**, 239, 242
$FePdSiSnC_{49}H_{46}O_6P_2$
$Fe(CO)_3[Si(OMe)_2OMe](dppm)Pd(SnPh_3)$, **10**, 61
$FePdSi_4C_{37}H_{49}Cl_3O_6P_2$
$PdFe(CO)_3(dppm)[Si(OTMS)_3Cl_3]$, **10**, 17, 18
$FePd_2C_{54}H_{44}O_4P_2$
$Pd_2Fe(dppm)_2(CO)_4$, **10**, 19
$FePd_2C_{54}H_{44}O_4P_4$
$Pd_2Fe(CO)_4(dppm)_2$, **7**, 84; **9**, 205
$FePtC_{27}H_{24}O_3$
$PtFe(CO)_3(PhCCHCHCPh)(cod)$, **9**, 564
$FePtC_{28}H_{22}Br_2O_4P_2$
$(CO)_3Fe(dppm)(CO)PtBr_2$, **7**, 84
$FePtC_{28}H_{22}I_2O_3P_2$
$(CO)_3Fe(dppm)(I)PtI$, **7**, 84
$FePtC_{34}H_{37}O_3P_3$
$Fe(CO)_2[CH(OH)C(=CH_2)CH_2](dppm)Pt(PMe_3)$, **10**, 61
$FePtC_{35}H_{28}O_8P_2$
$(CO)_4Fe(dppm)Pt(MeO_2C{-}CCCO_2Me)$, **9**, 572
$FePtC_{47}H_{37}O_4P_3$
$PtFe(CO)_3(CO)(dppm)(PPh_3)$, **10**, 17
$FePtC_{48}H_{39}O_3P_3$
$Fe(CO)_2[C_2H_2C(O)](dppm)Pt(PPh_3)$, **10**, 61
$FePtC_{49}H_{41}O_3P_3$
$Fe(CO)_3[C(CH_2)CH_2](dppm)Pt(PPh_3)$, **9**, 563; **10**, 61
$FePtC_{49}H_{42}O_3P_3$
$Fe(CO)_3(CMe{=}CH_2)(dppm)Pt(PPh_3)$, **9**, 563; **10**, 61
$FePt_5C_{33}H_{60}O_9P_4$
$FePt_5(CO)_9(PEt_3)_4$, **7**, 84
$FePt_5C_{59}H_{45}O_5P_3$
$FePt_5(CO)_5(PPh_3)_3$, **10**, 212
$FeReC_{29}H_{20}O_7P$
$(CO)_4Re(CO_2)Fe(CO)(PPh_3)Cp$, **6**, 200
$FeReC_{30}H_{20}O_8P$
$(CO)_5Re(CO_2)Fe(CO)(PPh_3)Cp$, **6**, 200
$FeRhC_{18}H_{16}O_3$
$[Fe(CO)_3(C_8H_8)Rh(nbd)]^+$, **10**, 60
$FeRhC_{18}H_{19}O_3$
$Fe(CO)_3(C_7H_7)Rh(cod)$, **10**, 60
$FeRhC_{21}H_{37}O_5P_2$
$Fe(CO)_4(PBu^t{}_2)Rh(CO)(PBu^t{}_2H)$, **7**, 44
$FeRhC_{23}H_{40}O_3P_2$
$Fe(CO)_2(PBu^t{}_2H)(PBu^t)(CO)Rh(cod)$, **7**, 44
$FeRhC_{30}H_{22}ClO_5P_2$
$(CO)_4Fe(dppm)RhCl(CO)$, **7**, 83
$FeRhC_{38}H_{30}ClN_2O_4P_2$
$FeRh(Ph_2Ppy)_2(CO)_4Cl$, **10**, 47
$FeRhSiC_{32}H_{31}O_7P_2$
$Fe(CO)_3(dppm)[Si(OMe)_3]Rh(CO)$, **10**, 61
$FeRhWC_{27}H_{19}O_5$
$CpWRhFe(CO)_5(C_9H_7)(\equiv CTol)$, **10**, 8
$FeRh_2C_{15}O_{14}$
$Rh_2Fe_3(C)(CO)_{14}$, **10**, 134
$FeRh_2C_{26}H_{30}O_6$
$Cp^*{}_2Rh_2Fe(CO)_6$, **10**, 7
$FeRh_2C_{29}H_{55}O_5P_3$
$Rh_2Fe(CO)_3(PBu^t{}_2H)(CO)_2(PBu^t{}_2)_2$, **7**, 53
$FeRh_4C_{15}O_{15}$
$[FeRh_4(CO)_{15}]^{2-}$, **7**, 83; **10**, 47, 191, 193
$FeRh_5C_{16}O_{16}$
$[FeRh_5(CO)_{16}]^-$, **7**, 83; **10**, 193
$FeRuC_8O_8$
$[Fe(CO)_4Ru(CO)_4]^{2-}$, **10**, 54
$FeRuC_{12}H_8O_6$
$Fe(CO)_3(CHCMeCMeCH)Ru(CO)_3$, **10**, 54
$FeRuC_{12}H_{13}NO_5$
$Fe(CO)_2(CMeCHCHNPr^i)(H)Ru(CO)_3$, **10**, 54
$FeRuC_{13}HO_{13}$
$[HFeRu(CO)_{13}]^-$, **10**, 52
$FeRuC_{13}H_2O_{13}$
$H_2FeRu(CO)_{13}$, **10**, 52
$FeRuC_{13}H_{13}NO_6$
$Fe(CO)_3(CHMeCHCHNPr^i)Ru(CO)_3$, **10**, 54
$FeRuC_{14}H_{10}O_4$
$Fe(CO)Cp(CO)_2Ru(CO)Cp$, **10**, 54
$FeRuC_{14}H_{20}N_2O_5$
$Fe(CO)_3(H)(NPr^i{=}CH{-}CH{=}NPr^i)RuMe(CO)_2$, **10**, 54
$FeRuC_{18}H_{13}NO_6$

$Fe(CO)_3(CPh{=}CHCHN{=}CMe_2)Ru(CO)_3$, **10**, 54
$FeRuC_{20}H_{18}$
$FeCp(C_5H_4)RuCp(C_5H_4)$, **7**, 616
$FeRuC_{22}H_{24}O_6$
$Fe(CO)_3(C_{16}H_{24})Ru(CO)_3$, **10**, 54, 60
$FeRuC_{24}H_{16}O_5$
$FeRu(CO)(CPh{=}CHPh)(CO)_4Cp$, **10**, 54
$FeRuC_{25}H_{35}$
$[Cp^*RuCp^*FeCp]^+$, **7**, 617
$FeRuC_{26}H_{42}P_2$
$Ru[C_5H_4FeCp](PMe_3)_2Cp^*$, **7**, 399
$FeRuC_{30}H_{20}O_6$
$(CO)_3Fe(PPh_2)_2Ru(CO)_3$, **10**, 45
$FeRuC_{34}H_{20}O_6$
$Fe(CO)_3(CPhCPhCPhCPh)Ru(CO)_3$, **10**, 55
$FeRu(CO)_6(C_2Ph_2)_2$, **10**, 130
$FeRu_2C_{12}H_{10}NO_{12}P$
$FeRu_2(NH)(CO)_9[P(OMe)_3]$, **10**, 218
$FeRu_2C_{12}O_{12}$
$FeRu_2(CO)_{12}$, **10**, 37, 43, 196
$FeRu_2C_{46}H_{30}O_{10}P_2$
$Ru_2Fe(CO)_{10}(PPh_3)_2$, **7**, 685
$FeRu_3C_{12}HNO_{12}$
$FeRu_3(H)(N)(CO)_{12}$, **10**, 225
$FeRu_3C_{12}NO_{12}$
$[FeRu_3(N)(CO)_{12}]^-$, **7**, 32; **10**, 225
$FeRu_3C_{13}H_2O_{13}$
$H_2FeRu_3(CO)_{13}$, **10**, 37, 41, 43, 61
$FeRu_3C_{16}H_{18}NO_{10}P_2$
$[FeRu_3(N)(CO)_{10}[P(OMe)_3]_2]^-$, **10**, 218
$FeRu_3C_{24}H_{24}O_8P_2$
$Ru_3H_2(PC_5H_4FeC_5H_4PPr^i{}_2)(CO)_8$, **7**, 709
$FeRu_3C_{27}H_{30}O_8P_2$
$Ru_3H_2(CH_2CHMePC_5H_4FeC_5H_4PPr^i{}_2)(CO)_8$, **7**, 709
$FeRu_3C_{37}H_{20}O_{13}P_2$
$FeRu_3(CO)_{13}(PPh_2)_2$, **7**, 55
$FeRu_4C_{27}H_{14}O_{11}P$
$Ru_4[PCpFeCp](C_6H_4)(CO)_{11}$, **7**, 881
$FeRu_4C_{39}H_{22}O_{13}P_2$
$FeRu_4(C)(dppm)(CO)_{13}$, **7**, 27; **10**, 129
$FeRu_5C_{39}H_{20}O_{14}$
$FeRu_5(C)(PPh_2)_2(CO)_{14}$, **10**, 129
$FeRu_5C_{39}H_{20}O_{14}P_2$
$Ru_5Fe(CO)_{14}(PPh_2)_2C$, **7**, 27
$FeRu_5C_{41}H_{28}O_{13}P_2S_2$
$FeRu_5(CCH_2)(SMe)_2(PPh_2)_2(CO)_{13}$, **10**, 130, 131
$FeSbC_{38}H_{31}F_6O_2P_2$
$[HFe(CO)_2(PPh_3)_2][SbF_6]$, **7**, 40
$FeSiC_9H_{12}O_5$
$(CO)_4Fe{=}C(O\text{-}TMS)Me$, **7**, 116
$FeSiC_{12}H_{20}O_2$
$FeH(CO)_2(SiEt_3)(cbd)$, **7**, 144
$FeSiC_{13}H_{13}NO_4$
$Fe(CO)_4[Si(H)C_6H_4CH_2NMe_2]$, **7**, 28
$FeSiC_{14}H_{18}NO_4$
$[Ph[(TMS)O]C{=}CHMe]Fe(CO)_2(NO)$, **12**, 967
$FeSiC_{15}H_{16}O_4$
$Fe(CO)_3[PhCH{=}CHC(TMS)(O)]$, **7**, 147
$FeSiC_{16}H_{33}P_3$
$Fe[MeSi(CH_2PMe_2)_3](C_6H_6)$, **7**, 156
$Fe(tmps)(C_6H_6)$, **7**, 149
$FeSiC_{24}H_{37}P_3$
$Fe(MeSi(CH_2PMe_2)_3](PhC{\equiv}CPh)$, **7**, 123
$Fe(tmps)(PhCCPh)$, **7**, 149
$FeSiC_{28}H_{22}O_3$
$Fe(CO)_3(Ph_3SiC_7H_7)$, **7**, 129
$FeSiC_{39}H_{31}O_3P$
$HFe(CO)_3(PPh_3)(SiPh_3)$, **7**, 27
$FeSi_2C_{11}H_{18}O_4P$
$[(CO)_4FePC(TMS)_2]^-$, **7**, 37
$FeSi_2C_{11}H_{20}O$
$Fe(CO)(TMS)({=}SiMe_2)Cp$, **7**, 178
$FeSi_2C_{12}H_{18}O_4$
$Fe(CO)_4(TMS\text{-}C{\equiv}C\text{-}TMS)$, **7**, 123
$FeSi_6C_{20}H_{58}$
$[(TMS)_3CH_2]_2Fe$, **7**, 186
$FeSnC_{16}H_{26}$
$(Me_3Sn)_2(ferrocene)$, **2**, 249
$FeSn_2C_{42}H_{36}O$
$FeH(CO)(SnPh_3)_2Cp$, **7**, 169
$FeThC_{22}H_{20}O_2$
$Fe(CO)_2CpThCp_3$, **4**, 188; **10**, 54
$FeTiC_{13}H_{23}N_3O_2$
$Ti(NMe_2)_3Fe(CO)_2Cp$, **10**, 27
$FeTi_2C_{38}H_{34}Cl_4N_2O$
$[[TiCp_2(PhCH{=}CHCN)]_2(O)][FeCl_4]$, **4**, 398
$FeTlC_3NO_4$
$Tl[Fe(CO)_3(NO)]$, **1**, 568
$FeVC_{15}H_{15}Br_3$
$[Cp_2Fe][CpVBr_3]$, **5**, 30
$FeVC_{23}H_{26}N$
$Cp_2V[C_5H_3(CH_2NMe_2)FeCp]$, **5**, 44
$FeV_2C_{12}H_{17}N_2O_2S_4$
$(MeCp)_2V_2S_4Fe(NO)_2$, **5**, 32
$FeV_2C_{15}H_{14}O_3S_4$
$(MeCp)_2V_2S_4Fe(CO)_3$, **5**, 32
$V_2Fe(S)_2(S_2)(CO)_3Cp'_2$, **10**, 232
$FeVa_2C_{15}H_{14}O_3S_3$
$(C_5H_4Me)_2Va_2Fe(CO)_3(S)_3$, **7**, 71, 72
$FeWC_8O_8Se_2$
$[FeW(CO)_8Se_2]^{2+}$, **7**, 74
$FeWC_{14}H_{10}O_5$
$W(CO)_2Cp(MeC{=}C{=}CH_2)Fe(CO)_3$, **10**, 42
$FeWC_{16}H_{14}O_3$
$CpW(CO)_2(C_3H_4)Fe(CO)Cp$, **10**, 15
$W(CO)_2Cp(CH_2{=}C{=}CH_2)Fe(CO)Cp$, **10**, 42
$FeWC_{19}H_{10}O_5$
$Cp(CO)_2FeCCCCW(CO)_3Cp$, **5**, 165
$FeWC_{19}H_{12}O_6$
$CpWFe(CO)_6(CTol)$, **10**, 91
$FeWC_{22}H_{18}O_5$
$CpWFe(CO)_5(CTolCMeCMe)$, **10**, 96
$FeWC_{57}H_{44}O_7P_4$
$(dppm)(CO)_3W(dppm)Fe(CO)_4$, **7**, 83
$FeW_2C_{17}H_{10}O_7S$
$Cp_2W_2Fe(CO)_7S$, **7**, 66
$FeYbC_{10}H_9N_3O_4$
$(MeCN)_3YbFe(CO)_4$, **7**, 85
$FeZrC_{17}H_{15}ClO_2$
$Cp_2ZrCl[Fe(CO)_2Cp]$, **4**, 535
$FeZrC_{28}H_{37}O$
$Zr(Cp_2Fe)(OH)(C_5H_4Bu^t)_2$, **4**, 527
$FeZrC_{34}H_{30}N_2O_2P_2$
$Cp_2Zr(PPh_2)Fe(NO)_2$, **4**, 517
$FeZrC_{35}H_{33}O_3P$
$Zr(OBu^t)Cp[C_5H_4(PPh_2)]Fe(CO)_2Cp$, **10**, 27
$Fe_2AsC_{10}H_6O_8$
$[Me_2As[Fe(CO)_4]_2]^-$, **7**, 55
$Fe_2As_2C_{10}H_{12}O_6$
$Fe_2(CO)_6(AsMe_2)_2$, **7**, 55
$Fe_2As_2C_{13}H_{15}NO_8$
$[(CO)_4FeAs(Me)_2]_2N(Me)$, **7**, 55
$Fe_2As_2C_{14}H_{18}N_2O_6S$
$Fe_2(CO)_6(Bu^tAsNSNAsBu^t)$, **7**, 56
$Fe_2AuC_{26}H_{15}O_8P$
$[Fe_2(CO)_8(AuPPh_3)]^-$, **7**, 81
$Fe_2Au_2Si_2C_{62}H_{62}O_{12}P_4$
$[Fe(CO)_3[Si(OMe)_3](dppm)Au]_2$, **10**, 62
$Fe_2Au_4C_{60}H_{24}O_8P_4$
$Au_4Fe_2[Ph_2P(CH_2)_2PPh_2]_2(CO)_8$, **10**, 298
$Fe_2BC_{11}H_9F_4O_6P_2S$
$[Fe_2(CO)_6(PMe_2)_2(SMe)][BF_4]$, **7**, 44
$Fe_2BC_{20}H_{10}F_4O_8P$
$[Fe_2(CO)_8PPh_2][BF_4]$, **7**, 43
$Fe_2BC_{20}H_{19}$
$CpFe(C_4H_4BPh)FeCp$, **7**, 195
$Fe_2B_2C_6H_4O_6$
$[Fe_2(CO)_6B_2H_4]^{2-}$, **7**, 16
$Fe_2B_2C_6H_5O_6$
$[Fe_2(CO)_6B_2H_5]^-$, **1**, 306, 307; **7**, 16
$Fe_2B_2C_6H_6O_6$

$Fe_2(CO)_6B_2H_6$, **1**, 298, 306, 307; **7**, 16
$Fe_2B_2C_{24}H_{30}$
$(FeCp)_2[Et_2Me_2(C_8H_4B_2)]$, **1**, 210
$Fe_2B_2Li_2C_6H_4O_6$
$Li_2[B_2H_4Fe_2(CO)_6]$, **1**, 307
$Fe_2B_3C_6H_7O_6$
$(CO)_6Fe_2B_3H_7$, **1**, 313, 316
$Fe_2B_4C_{12}H_{46}P_4$
$[Fe(PMe_3)_2]_2(H)B_4H_9$, **1**, 319
$Fe_2BiCoC_{10}O_{10}$
$[BiFe_2Co(CO)_{10}]^-$, **7**, 60
$Fe_2Bi_2C_{32}O_8$
$[Fe(CO)_4(BiPh_2)]_2$, **7**, 60
$Fe_2C_6H_2O_6S_2$
$Fe_2(CO)_6(SH)_2$, **7**, 66
$Fe_2C_6H_3O_6S$
$Fe_2(CO)_6(H)_2(SH)$, **7**, 64
$Fe_2C_6H_3O_6Se$
$(H)_2Fe_2(CO)_6(SeH)$, **7**, 74
$Fe_2C_6H_3S_2$
$Fe_2(CO)_6(H)(SH)_2$, **7**, 64
$Fe_2C_6H_4O_6P_2$
$Fe_2(CO)_6(PH_2)_2$, **7**, 44
$Fe_2C_6O_6S_2$
$[Fe_2(CO)_6S_2]^{2-}$, **7**, 65, 66
$Fe_2C_6O_6Se_2$
$Fe_2(CO)_6Se_2$, **7**, 74
$Fe_2C_6O_6Te$
$Fe_2(CO)_6Te$, **7**, 76
$Fe_2C_6O_6Te_3$
$Fe_2(CO)_6(Te)(Te_2)$, **7**, 76
$Fe_2C_6O_6Te_4$
$Fe_2(CO)_6(Te_2)_2$, **7**, 76
$Fe_2C_7H_2O_6Te_2$
$Fe_2(CO)_6(Te_2CH_2)$, **7**, 76
$Fe_2C_8HClO_7$
$Fe_2(CO)_6(Cl)(CHCO)$, **7**, 243
$Fe_2C_8HO_8$
$[HFe_2(CO)_8]^-$, **7**, 8
$Fe_2C_8HO_8S$
$Fe_2(CO)_8(SH)$, **7**, 64
$Fe_2C_8H_6IO_6P$
$Fe_2(CO)_6I(PMe_2)$, **7**, 44, 78
$Fe_2C_8H_6O_6S_2$
$Fe_2(CO)_6(SMe)_2$, **7**, 65, 66
$Fe_2C_8H_6O_6Te_2$
$Fe_2(CO)_6(TeMe)_2$, **7**, 76
$Fe_2C_8I_2O_8$
$Fe_2(CO)_8I_2$, **7**, 79
$Fe_2C_8O_8$
$[Fe_2(CO)_8]^{2-}$, **7**, 5, 8
$Fe_2C_9H_2O_8$
$Fe_2(CO)_8(CH_2)$, **7**, 239
$Fe_2C_9H_5NO_3$
$Fe_2(CN)(CO)_3Cp^{2-}$, **7**, 233
$Fe_2C_9H_8Br_2NO_6P_2$
$Fe_2(CO)_6[P(NPr^i)Br][P(H)Br]$, **7**, 45
$Fe_2C_9H_8O_6S_2$
$Fe_2(CO)_6(SMe)(H_2CSMe)$, **7**, 65
$Fe_2C_9O_9$
$Fe_2(CO)_9$, **7**, 4, 8, 264; **12**, 350, 745
$Fe_2C_{10}F_2O_8$
$Fe_2(CO)_8(CCF_2)$, **7**, 245
$Fe_2C_{10}H_4O_9$
$Fe_2(CO)_8(CHOMe)$, **7**, 242, 243
$Fe_2C_{10}H_7NO_7P_2$
$[Fe_2(CO)_6[P(NPr^i)C(O)P]]^-$, **7**, 45
$Fe_2C_{10}H_8O_6$
$Fe_2(CO)_6(CH_2CH_2)_2$, **7**, 236
$Fe_2C_{10}H_9O_6S_3$
$[Fe_2(CO)_6(S)(SSBu^t)]^-$, **7**, 66
$Fe_2C_{10}H_{10}N_2O_2$
$Fe_2(NO)_2Cp_2$, **7**, 232, 241
$Fe_2C_{10}H_{24}N_6O_4$
$[Fe(en)_3][Fe(CO)_4]$, **7**, 33
$Fe_2C_{11}H_4O_{10}$
$Fe_2(CO)_8(CHO_2CMe)$, **7**, 242
$Fe_2C_{11}H_5NO_9$
$Fe_2(CO)_8[CH(NHCOMe)]$, **7**, 242
$Fe_2C_{11}H_9NO_6S$
$Fe_2(CO)_6[SC(NMe_2)CMe]$, **7**, 70
$Fe_2C_{11}H_9O_7S$
$[Fe_2(SBu^i)(CO)_7]^-$, **6**, 221
$Fe_2C_{11}H_{12}O_6S_2$
$Fe_2(CO)_6(SMe)(SBu^t)$, **7**, 65
$Fe_2C_{11}H_{13}BrNO_7P_2$
$Fe_2(CO)_6[P(NPr^i)OEt]$-$[P(H)Br]$, **7**, 45
$Fe_2C_{12}H_8O_5$
$Fe_2(CO)_5Cp(CHCH_2)$, **7**, 247
$Fe_2C_{12}H_8O_8$
$[Fe(CO)_3(CH_2COCH_2)]_2$, **7**, 128
$Fe_2C_{12}H_{12}$
$Fe_2(C_6H_6)_2$, **7**, 192
$Fe_2C_{12}H_{13}N_2O_2$
$Fe_2(NO)_2Cp_2(CH_2CH)$, **7**, 241
$Fe_2C_{12}H_{18}B_6$
$Cp_2Fe_2(C_2B_6H_8)$, **1**, 398
$Fe_2C_{12}O_{12}$
$Fe_3(CO)_{12}$, **7**, 264
$Fe_2C_{13}H_{16}N_2O_7P_2$
$Fe_2(CO)_6[P(NPr^i)C(OH)(H)P$-$(NPr^i)]$, **7**, 45
$Fe_2C_{14}H_6O_6Se_2$
$Fe_2(CO)_6[Se_2C(H)C(Ph)]$, **7**, 74
$Fe_2C_{14}H_6O_6Te_2$
$Fe_2(CO)_6Te_2C(H)C(Ph)$, **7**, 77
$Fe_2C_{14}H_{10}IO_7P$
$Fe_2(CO)_6[P(C_6H_4OMe)Me](I)$, **7**, 79
$Fe_2C_{14}H_{10}O_3S$
$Fe_2(CO)_3(CS)Cp_2$, **7**, 233
$Fe_2C_{14}H_{10}O_4$
$Fe_2Cp_2(CO)_4$, **7**, 163, 232
$Fe_2C_{14}H_{11}N_2O_6P_2$
$HN[PPhFe(CO)(NO)_2]_2$, **7**, 40
$Fe_2C_{14}H_{11}O_3$
$Fe_2(CO)_3Cp_2(CH)^+$, **7**, 242–4
$Fe_2C_{14}H_{12}O_3$
$Fe_2(CO)_3Cp_2(CH_2)$, **7**, 237, 239, 242
$Fe_2C_{14}H_{12}O_4$
$Fe_2(CO)_3Cp_2(CHOH)$, **7**, 242
$Fe_2C_{14}H_{14}O_2$
$Fe_2(CO)_2Cp_2(CHMe)$, **7**, 240
$Fe_2C_{14}H_{16}N_2O_6$
$Fe_2(CO)_6(Pr^iNCHCHNPr^i)$, **7**, 235
$Fe_2C_{15}H_{12}O_4$
$CpFe(CO)_3Fe(CO)(C_6H_7)$, **7**, 163
$Fe_2C_{15}H_{11}NO_6SSe$
$Fe_2(CO)_6(Me_2NCS)(SePh)$, **7**, 74
$Fe_2C_{15}H_{12}O_3$
$Fe_2(CO)_3Cp_2(CCH_2)$, **7**, 242, 245
$Fe_2C_{15}H_{13}O_3$
$Fe_2(CO)_3Cp_2(CHCH_2)^+$, **7**, 247
$Fe_2(CO)_3Cp_2(CMe)^+$, **7**, 242, 243, 246
$Fe_2C_{15}H_{13}O_3S$
$Fe_2(CO)_3Cp_2[C(SMe)]^+$, **7**, 242
$Fe_2C_{15}H_{14}O_3$
$Fe_2(CO)_3Cp_2(CH_2CH_2)$, **7**, 236
$Fe_2(CO)_3Cp_2(CHMe)$, **7**, 240, 242
$Fe_2C_{15}H_{21}ClN_2O_6P_3$
$Fe_2(CO)_6[P(NPr^i)P(Cl)$ $(NPr^i_2)P]$, **7**, 45
$Fe_2C_{16}H_{12}O_5$
$Fe(CO)_2CpCH_2C(O)Fe$ $(CO)_2Cp$, **7**, 176
$Fe_2C_{16}H_{13}N_2O_6$
$Fe_2(CO)_6(NC_6H_4NBu^t)$, **7**, 33
$Fe_2C_{16}H_{13}O_5$
$[Fe(CO)_2CpC(Me)OFe$ $(CO)_2Cp]^+$, **7**, 176
$Fe_2C_{16}H_{14}O_4$
$[Fe(CO)_2(C_6H_7)]_2$, **7**, 163
$Fe_2C_{17}H_{15}O_3$
$[Fe(CO)[Fe(CO)_2CpCH{=}CH$-$OH{=}CH_2)]Cp]^+$, **7**, 185
$Fe_2C_{17}H_{17}O_4$
$Fe_2(CO)_3Cp_2[CHCMeOMe]^+$, **7**, 248
$Fe_2C_{17}H_{28}O_5P_2$
$Fe(CO)_2(PMe_3)_2(C_3H_5)Fe$-$(CO)_3(C_3H_5)$, **7**, 129
$Fe_2C_{18}H_{10}O_6S_2$
$Fe_2(CO)_6(SPh)_2$, **7**, 61
$Fe_2C_{18}H_{12}O_6$
$Fe_2(CO)_6(C_{12}H_{12})$, **7**, 159
$Fe_2C_{18}H_{14}O_6$
$Fe(CO)_3(C_6H_7C_6H_7)Fe(CO)_3$, **7**, 163
$Fe_2C_{18}H_{18}O_4$
$[Fe(CO)_2(C_7H_9)]_2$, **7**, 162
$Fe_2C_{18}H_{22}O_4$
$[Fe(CO)_2(C_7H_{11})]_2$, **7**, 162
$Fe_2C_{18}H_{30}B_8$
$Cp_2Fe_2(Me_4C_4B_8H_8)$, **1**, 398
$Fe_2C_{19}H_{10}NO_7$
$Fe_2(CO)_6(PhNC(O)C_6H_4NH)$, **7**, 33
$Fe_2C_{19}H_{18}O_6$
$Fe(CO)_2Cp'COOCH_2CH_2Fe$-$(CO)_2Cp'$, **7**, 168

$Fe_2C_{19}H_{20}NO_3$
$Fe_2(CO)_3Cp_2(CHCNBu^t)^+$, **7**, 243
$Fe_2C_{19}H_{28}N_2O_7P_2$
$Fe_2(CO)_6[P(NPr^i_2)COP(NPr^i_2)]$, **7**, 45
$Fe_2C_{19}H_{30}N_2P_2$
$(Pr^i_2NPHCHPNPr^i_2)Fe_2(CO)_6$, **7**, 45
$Fe_2C_{20}H_{20}O_3$
$Fe_2(CO)_3Cp_2[CH(CCBu^t)]$, **7**, 244
$Fe_2(CO)_3Cp_2[CH(CHCMe-CMeCH_2)]$, **7**, 244
$Fe_2C_{21}H_{20}O_6$
$Fe(CO)_2Cp'CO(CH_2)_3COFe-(CO)_2Cp'$, **7**, 173
$Fe_2C_{21}H_{30}NO_7P$
$[Et_4N][Fe_2(CO)_7(PPh_2)]$, **7**, 42
$Fe_2C_{21}H_{36}O_5P_2$
$Fe_2(CO)_5(PBu^t_2)_2$, **7**, 47
$Fe_2C_{22}H_{22}$
$CpFe(C_6H_6C_6H_6)FeCp$, **7**, 188
$Fe_2C_{23}H_{30}O_3$
$Fe_2(CO)_3Cp*_2$, **7**, 234
$Fe_2C_{24}H_{15}O_6PS$
$Fe_2(CO)_6(PPh_2)(SPh)$, **7**, 44, 64
$Fe_2C_{24}H_{15}O_6P_3$
$Fe_2(CO)_6[P(Ph)]_3$, **7**, 44
$Fe_2C_{24}H_{22}$
$(CpFe)_2(C_{14}H_{12})$, **7**, 193
$Fe_2C_{24}H_{30}O_4$
$Fp*_2$, **7**, 232
$Fe_2C_{24}H_{31}O_3$
$Fe_2(CO)_3Cp*_2(CH)^+$, **7**, 241
$Fe_2C_{25}H_{24}O_5P$
$Fe_2(CO)_5(PPh_2)(CH_2C-Bu^tCHCH_2)$, **7**, 248
$Fe_2C_{25}H_{28}O_5P_2$
$Fe_2(CO)_5(PPh_2)(PBu_2)$, **7**, 47
$Fe_2C_{25}H_{32}O_3$
$Fe_2(CO)_3Cp*_2(CCH_2)$, **7**, 245
$Fe_2C_{25}H_{35}$
$[CpFe(Cp*)FeCp*]^+$, **7**, 194
$Fe_2C_{26}H_{15}O_6P$
$Fe_2(CO)_6(PPh_2)(CCPh)$, **7**, 43
$Fe_2C_{26}H_{20}O_7P$
$Fe_2(CO)_6[C(O)CBu^tCC(H)P-Ph_2]$, **7**, 41
$Fe_2C_{26}H_{28}O_6P_2$
$Fe_2(CO)_6(PPh_2)(PBu^t_2)$, **7**, 47
$Fe_2C_{26}H_{30}O_4$
$Fe(CO)_2Cp*CCCp*(CO)_2Fe$, **7**, 169
$Fe_2C_{26}H_{31}O_4$
$Fe(CO)_2Cp*(HCCFp*)$, **7**, 245
$Fe_2C_{28}H_{22}O_4$
$(C_5H_4C_5H_4)[Fe(CO)_2Bz]_2$, **7**, 166
$Fe_2C_{28}H_{58}B_8$
$[C_5Me_4)FeH(Et_2C_2B_4H_4)]_2-C_6H_4$, **1**, 388, 390
$Fe_2C_{30}H_{20}O_6P_4$
$Fe_2(CO)_6[P(Ph)]_4$, **7**, 44
$Fe_2C_{30}H_{21}O_6P_2$
$[Fe_2(CO)_5[C(O)H](PPh_2)_2]^-$, **7**, 47
$[Fe_2(CO)_5(CO)(PPh_2)(PPh_2H)]^-$, **7**, 47
$Fe_2C_{30}H_{27}NO_6P$
$Fe_2(H)(CO)_6(PPh_2)(CHCPhNEt_2)^+$, **7**, 239
$Fe_2C_{30}H_{38}$
$Cp*Fe(C_5H_4C_5H_4)FeCp*$, **7**, 192
$Fe_2C_{35}H_{28}O_8P_2$
$Fe_2(CO)_6(CHCO_2Et)(dppm)$, **7**, 238
$Fe_2C_{36}H_{27}O_5P_3$
$Fe_2(CO)_5(PPh_2)(Ph_2PCH_2PPh)$, **7**, 43
$Fe_2C_{36}H_{46}O_8P_2$
$Fe_2(CO)_6(POC_6H_2Bu^t_2Me)_2$, **7**, 47, 48
$Fe_2C_{36}H_{58}B_8$
$[(C_5Me_4)FeH(Et_2C_2B_4H_4)]_2-C_6H_4$, **1**, 387
$Fe_2C_{42}H_{30}O_6P_4$
$Fe_2(CO)_6(PhPPPh_2)_2$, **7**, 44
$Fe_2C_{42}H_{33}O_5P_3$
$HFe_2(CO)_5(dppm)(PPh_2)$, **7**, 43
$Fe_2C_{43}H_{30}N_6O_{13}$
$[Fe(py)_6][Fe_4(CO)_{13}]$, **7**, 33
$Fe_2C_{44}H_{40}O_4P_2S_2$
$Fe_2(CO)_4(PPh_3)_2(SEt)_2$, **7**, 49
$Fe_2C_{45}H_{49}N_2O_6P_2$
$Fe_2(CO)_4(PPh_2)(dppm)[HNNC-(H)CO_2Et]$, **7**, 43
$Fe_2C_{48}H_{42}N_6O_4P_4$
$Fe_2(NO)_4[Ph_2PN(H)PPh_2]_2$, **7**, 40
$Fe_2C_{49}H_{40}O_7P_2$
$[PPh_4][Fe_2(CO)_6[Ph_2PCHC-(Bu^t)C(O)]]$, **7**, 41
$Fe_2C_{53}H_{32}O_4P_4$
$Fe_2(CO)_4(dppm)(PPh_2)_2$, **7**, 43
$Fe_2C_{54}H_{40}O_6P_3$
$Ph_2P[Fe(CO)_3PPh_3]_2$, **7**, 43
$Fe_2C_{63}H_{45}N_9$
$Fe_2(CNPh)_9$, **7**, 12
$Fe_2CdSi_2C_{18}H_{36}O_6P_2$
$[Fe(TMS)(CO)_3(PMe_3)]_2Cd$, **10**, 54
$Fe_2CdSi_2C_{48}H_{50}N_2O_{12}P_2$
$[Fe(CO)_3[Si(OMe)_3](Ph_2P-CH_2Py)]_2Cd$, **10**, 339
$Fe_2CdSi_2C_{64}H_{60}O_{14}P_2$
$[(CO)_3[Si(OMe)](PhFeC_6H_4C-(O)CH_2PPh_2)]_2Cd$, **10**, 339
$Fe_2Cd_2Si_2C_{52}H_{52}Cl_2O_{14}P_2$
$[[Ph_2PCH_2C(O)Ph]Fe(CO)_3-(MeO)_2SiOMeCdCl]_2$, **10**, 333
$Fe_2CoC_8NO_9S$
$Fe_2Co(CO)_8(NO)(S)$, **7**, 70
$Fe_2CoC_9HO_9S$
$Fe_2Co(H)(S)(CO)_9$, **7**, 70; **10**, 237
$Fe_2CoC_{11}HO_{10}$
$Fe_2Co(CO)_{10}(CH)$, **7**, 23
$Fe_2CoC_{11}O_{10}$
$[Fe_2Co(CCO)(CO)_9]^-$, **7**, 22, 271, 272
$Fe_2CoC_{12}H_5O_6SeTe$
$Fe_2(CO)_6[SeTeCo(CO)Cp]$, **7**, 75
$Fe_2CoC_{12}H_5O_7Te_2$
$Fe_2(CO)_6Te_2Co(CO)Cp$, **7**, 77
$Fe_2CoC_{14}H_5O_9$
$Fe_2CoCp(CO)_9$, **10**, 41
$Fe_2CoC_{15}H_{16}O_8P_2$
$[Fe_2Co(CO)_8(dmpe)C]^-$, **7**, 22
$Fe_2CoC_{16}H_{14}O_7P$
$Fe_2Co(PBu^t)(CO)_7Cp$, **10**, 222
$Fe_2CoC_{30}H_{25}N_4O_5P_2$
$Fe_2Co(PPh_2)_2(CO)(NO)_4Cp$, **10**, 223
$Fe_2CoNiC_{14}H_8O_8P$
$Fe_2CoNiCp(PMe)(CO)_8$, **9**, 20
$Fe_2CoNiC_{15}H_{11}O_8P_2$
$NiCoFe_2(PMe)_2(CO)_8Cp$, **9**, 173
$Fe_2Co_2C_{23}H_{10}O_{11}P_2$
$Fe_2Co_2(CO)_{11}(PPh)_2$, **10**, 45
$Fe_2Co_2Ge_2C_{17}H_6O_{15}$
$Fe_2(CO)_7[GeMe[Co(CO)_4]]_2$, **7**, 30
$Fe_2Co_3C_{27}H_{35}$
$Co_3(CFc)_2Cp_3$, **8**, 433, 435
$Fe_2Co_4Ge_2C_{22}O_{22}$
$[Co_2(CO)_7]GeFe_2(CO)_8Ge[Co_2-(CO)_7]$, **7**, 30
$Fe_2Co_5Ge_2C_{22}O_{22}$
$[Fe_2Co_5(Ge)_2(CO)_{22}]^-$, **7**, 30; **10**, 215
$Fe_2Cr_2C_{17}O_{17}P_2$
$Fe_2(CO)_7[PCr(CO)_5]_2$, **7**, 46
$Fe_2Cr_4C_{24}H_{38}S_6Se_2$
$Cr_4Fe_2(S)_4(Se)_2(SBu^t)_2Cp_4$, **10**, 234, 235
$Fe_2CuC_{26}H_{15}O_8P$
$[Fe_2(CO)_8(CuPPh_3)]^-$, **7**, 79
$Fe_2GaC_{14}H_{10}ClO_4$
$GaCl[Fe(CO)_2Cp]_2$, **1**, 555
$Fe_2Ga_2C_{20}H_{22}O_{10}$
$Fe_2(CO)_8[Ga(C_2H_3)(THF)]_2$, **1**, 555
$Fe_2Ga_2C_{24}H_{32}$
$[CpFe(CpGaMe_2)]_2$, **1**, 509
$Fe_2GeC_6O_6Se_2$
$[Fe_2(CO)_6Se_2]Ge$, **7**, 74
$Fe_2GeSiC_{13}H_{14}O_8$
$Fe_2(CO)_8(Me_2SiCH_2GeMe_2)$, **7**, 29
$Fe_2Ge_2C_8H_4O_8$
$[Fe(CO)_4(GeH_2)]_2$, **7**, 30
$Fe_2Ge_2MnC_{32}H_{54}O_{14}$
$(CO)_4FeGe(OBu^t)_3Mn(OBu^t)_3GeFe(CO)_4$, **7**, 29
$Fe_2H_2Cl_8$
$H[FeCl_4HFeCl_4]$, **2**, 121
$Fe_2Hf_2C_{28}H_{20}O_8$
$[Cp_2Hf[Fe(CO)_4]]_2$, **4**, 535, 536
$Fe_2HgSi_2C_{38}H_{44}O_6P_2$

$[Fe(SiPh_2Me)(CO)_3(PMe_3)]_2$-Hg, **10**, 54

$Fe_2InC_8BrO_8$
$[InBr[Fe(CO)_4]_2]^{2-}$, **7**, 19

$Fe_2InC_{14}H_{10}ClO_4$
$InCl[Fe(CO)_2Cp]_2$, **1**, 562

$Fe_2InC_{14}H_{10}IO_4$
$InI[Fe(CO)_2Cp]_2$, **1**, 562

$Fe_2InC_{18}H_8N_2O_8$
$In(bipy)[Fe(CO)_4]_2$, **1**, 563

$Fe_2InC_{19}H_{15}ClNO_4$
$InCl(py)[Fe(CO)_2Cp]_2$, **1**, 562

$Fe_2In_2C_8Br_2Cl_2O_8$
$[Fe_2(CO)_8[InClBr]_2]^{2-}$, **1**, 563

$Fe_2In_2C_{16}H_{16}Cl_2O_{18}$
$Fe_2(CO)_8[InCl(THF)]_2$, **1**, 563

$Fe_2IrC_{43}H_{30}ClO_7P_2Te_2$
$Fe_2(CO)_6Te_2[IrCl(CO)(PPh_3)_2]$, **7**, 77

$Fe_2Ir_3C_{14}O_{14}$
$[Fe_2Ir_3(CO)_{14}]^-$, **7**, 83

$Fe_2LiC_7HO_6S_2$
$Fe_2(CO)_6S_2CHLi$, **7**, 68

$Fe_2LiC_{18}H_{10}O_6P$
$Li[Fe_2(CO)_6(PPh_2)]$, **7**, 43

$Fe_2MnC_{14}H_8O_8P$
$CpMnFe_2(PMe)(CO)_8$, **10**, 35

$Fe_2MnC_{18}H_{15}O_8Te$
$Fe_2MnCp^*(CO)_8(Te)$, **7**, 77

$Fe_2MnC_{19}H_{16}O_8P$
$MnFe_2(PCy)(CO)_8Cp$, **10**, 222

$Fe_2MnSbC_{37}H_{25}O_8P$
$MnFe_2(PPh)(CO)_8(SbPh_3)Cp$, **10**, 222

$Fe_2Mn_2C_{14}O_{14}$
$Fe_2Mn(CO)_9[SMn(CO)_5]$, **7**, 72

$Fe_2Mn_2C_{20}H_{10}O_{10}P_2$
$Fe_2(CO)_6[PMn(CO)_2Cp]_2$, **7**, 46

$Fe_2MoC_{13}H_6O_8S$
$HFe_2Mo(CO)_8Cp(S)$, **7**, 70

$Fe_2Mo_2C_{17}H_{10}O_7Te_2$
$Cp_2Mo_2Fe_2(CO)_7Te_2$, **7**, 77

$Fe_2Mo_2C_{31}H_{30}O_{11}$
$Cp^*_2Mo_2Fe_2(CO)_{11}$, **7**, 83; **10**, 209, 211

$Fe_2N_4O_4S_2$
$[Fe_2(NO)_4S_2]^{2-}$, **7**, 66

$Fe_2NiC_{13}H_5O_8$
$CpNiFe_2(CO)_8$, **7**, 84

$Fe_2NiC_{17}H_{14}O_6$
$CpNiFe_2(CO)_6(C{\equiv}CBu^t)$, **10**, 102

$Fe_2Ni_2C_{22}H_{20}O_6$
$Cp_2Ni_2Fe_2(CO)_6(C_2Et_2)$, **10**, 151

$Fe_2Ni_2C_{36}H_{38}O_6P_2$
$[CpNi(PMe_3)_2][CpNiFe_2(CO)_6$-$(C_2Ph_2)]$, **10**, 150

$Fe_2OsC_{12}O_{12}$
$Fe_2Os(CO)_{12}$, **7**, 83; **10**, 191

$Fe_2Pd_2C_{32}H_{20}Cl_2O_4P_2$
$[(OC)_4Fe(PPh_2)Pd(Cl)]_2$, **10**, 38

$Fe_2Pd_2C_{32}H_{20}Cl_2O_8P_2$
$[(CO)_4Fe(PPh_2)Pd(Cl)]_2$, **10**, 47

$Fe_2PtC_{16}H_{12}O_8$
$PtFe_2(CO)_8(cod)$, **9**, 564

$Fe_2PtC_{17}H_{22}N_4O_8$
$Pt[Fe(Co)_3NO]_2(CNCy)_2$, **10**, 49

$Fe_2PtC_{20}H_{10}N_4O_8$
$Pt[Fe(CO)_3(NO)]_2(NCPh)_2$, **7**, 32

$Fe_2PtC_{24}H_{15}O_6PSeTe$
$Fe_2(CO)_6[SeTePt(PPh_3)_2]$, **7**, 75

$Fe_2PtC_{33}H_{22}O_8P_2$
$Fe_2Pt(dppm)(CO)_8$, **7**, 84

$Fe_2PtC_{42}H_{30}O_6P_2Te_2$
$Fe_2(CO)_6Te_2Pt(PPh_3)_2$, **7**, 77

$Fe_2PtC_{44}H_{30}O_8P_2$
$Fe_2Pt(CO)_8(PPh_3)_2$, **10**, 61

$Fe_2PtC_{50}H_{36}O_6P_2Se_2$
$(CO)_6Fe_2[SeC(Ph)C(H)Se]Pt$-$(PPh_3)_2$, **9**, 564

$Fe_2Pt_2C_{44}H_{31}O_8P_2$
$[Fe_2Pt_2(H)(CO)_8(PPh_3)_2]^-$, **7**, 84

$Fe_2Pt_2C_{44}H_{32}O_8P_2$
$H_2Fe_2Pt_2(CO)_8(PPh_3)_2$, **7**, 84

$Fe_2Pt_5C_{28}H_{24}O_{12}$
$Fe_2Pt_5(CO)_{12}(cod)_2$, **9**, 564; **10**, 212, 213

$Fe_2Pt_5C_{43}H_{32}O_7$
$Pt_5Fe_2(CO)_7(PhCCPh)_2(cod)$, **9**, 564

$Fe_2RhC_{11}H_5O_6Te_2$
$CpRhFe_2(CO)_6Te_2$, **7**, 77

$Fe_2RhC_{12}H_5O_7Te_2$
$CpRhFe_2(CO)_7Te_2$, **7**, 77

$Fe_2RhC_{42}H_{30}O_6P_3$
$Fe_2Rh(PPh_2)_3(CO)_6$, **10**, 227

$Fe_2RhC_{44}H_{30}O_8P_3$
$Fe_2Rh(PPh_2)_3(CO)_8$, **10**, 226, 227

$Fe_2RhC_{60}H_{45}O_6P_4$
$Fe_2Rh(PPh_2)_3(CO)_6(PPh_3)$, **10**, 226, 227

$Fe_2Rh_2C_{30}H_{40}O_8P_4$
$Fe_2Rh_2(CO)_6(CO)_2(PPh_2)_4$, **7**, 44

$Fe_2Rh_2C_{58}H_{44}O_8P_4$
$Fe_2Rh_2(dppm)_2(CO)_8$, **7**, 83

$Fe_2Rh_2C_{60}H_{44}O_8P_4$
$Rh_2Fe_2(CO)_8[(Ph_2P)_2C{=}CH_2]_2$, **10**, 6

$Fe_2Rh_4C_{16}O_{16}$
$[Fe_2Rh_4(CO)_{16}]^{2-}$, **7**, 83

$Fe_2RuC_{12}O_{12}$
$Fe_2Ru(CO)_{12}$, **7**, 83; **10**, 37, 43, 196

$Fe_2RuC_{23}H_{10}O_9$
$Fe_2Ru(CO)_9(C_2Ph_2)$, **10**, 130

$Fe_2RuC_{36}H_{20}O_8$
$Fe_2Ru(CO)_8(C_2Ph_2)_2$, **10**, 130

$Fe_2Ru_2C_{11}O_{11}SeTe$
$Fe_2Ru_2(CO)_{11}SeTe$, **7**, 75

$Fe_2Ru_2C_{11}O_{11}Te_2$
$Fe_2Ru_2(Te)_2(CO)_{11}$, **10**, 11, 12, 232, 233

$Fe_2Ru_2Co_{20}H_{10}O_8$
$Fe_2Ru(Cp)(CO)_6[CCRu$-$(CO)_2Cp]$, **7**, 22

$Fe_2Ru_3C_{14}O_{15}$
$[Fe_2Ru_3(CO)_{14}O]^{2-}$, **7**, 60, 61; **10**, 232, 233

$Fe_2Ru_3C_{17}O_{17}Te_2$
$Fe_2(CO)_6Te_2Ru_3(CO)_{11}$, **7**, 76

$Fe_2Ru_6C_{41}H_{30}O_{17}$
$Fe_2Ru_6(C_2)_2(CO)_{17}Cp^*_2$, **7**, 937

$Fe_2Sb_2C_{37}H_{25}O_7$
$Ph_2Sb[Fe(CO)_4][Fe(CO)_3$-$SbPh_3]$, **7**, 57

$Fe_2SiC_{10}H_6O_8$
$Me_2SiFe_2(CO)_8$, **7**, 29

$Fe_2Si_2C_{44}H_{30}O_8$
$[Fe(CO)_4SiPh_3]_2$, **7**, 28

$Fe_2Si_2Zn_2C_{68}H_{74}Cl_2O_{12}P_4$
$[Fe(CO)_3(Si(OEt)_3(Ph_2PCH_2$-$PPh_2)ZnCl]_2$, **10**, 338

$Fe_2Si_4C_{18}H_{36}O_6P_2$
$Fe_2(CO)_6[PC(TMS)_2]_2$, **7**, 44

$Fe_2Sn_2W_2C_{18}Cl_2O_{18}$
$Fe_2(CO)_8[Sn[W(CO)_5]Cl]_2$, **7**, 31

$Fe_2TiC_{16}H_{10}O_6Te_2$
$TiCp_2(Te)_2Fe_2(CO)_6$, **4**, 416

$Fe_2TlC_8O_8$
$[Tl[Fe(CO)_4]_2]^-$, **7**, 20

$Fe_2VC_{30}H_{28}$
$Cp_2V(C_{10}H_9Fe)_2$, **5**, 44

$Fe_2V_2C_{12}H_{14}N_2O_2S_4$
$(MeCp)_2V_2S_4[Fe(NO)]_2$, **5**, 32

$Fe_2WC_{10}O_{10}Te_2$
$Fe_2W(CO)_{10}Te_2$, **7**, 77

$Fe_2WC_{17}H_{11}O_9PS_2$
$Fe_2W(CO)_9(PMe_2Ph)S_2$, **7**, 72

$Fe_2W_2C_{15}O_{15}S_2$
$W_2Fe_2(S)_2(CO)_{15}$, **10**, 228, 233

$Fe_2W_2C_{16}H_{11}NO_7S_2$
$Fe_2W_2(CO)_5(py)(S)_2(OPr^i)_2$, **7**, 66

$Fe_2W_2C_{22}H_{11}O_{14}PS_2$
$Fe_2W(CO)_9(PMe_2Ph)(S)[SW$-$(CO)_5]$, **7**, 72

$Fe_2ZrC_{34}H_{40}O_4$
$Fe_2(CO)_2Cp_2(C_2O_2ZrCp^*_2)$, **7**, 232

$Fe_3AsC_{10}HO_9$
$Fe_3(CH)(As)(CO)_9$, **7**, 273, 275

$Fe_3As_2Cr_2C_{19}O_{19}$
$Fe_3(CO)_9[AsCr(CO)_5]_2$, **7**, 56

$Fe_3AuC_{32}H_{25}NO_9P$
$AuFe_3(HC{=}NBu^t)(CO)_9(PPh_3)$, **10**, 261

$Fe_3AuIr_2C_{30}H_{15}O_{12}P$
$[Fe_2Ir_2(CO)_{12}(AuPPh_3)]^-$, **7**, 83

$Fe_3BC_9H_4O_9$
$[(H)Fe_3(CO)_9(BH_3)]^-$, **1**, 296, 297, 299, 302

$Fe_3BC_9H_5O_9$
$(H)Fe_3(CO)_9(BH_4)$, **1**, 294, 296, 297, 299; **7**, 16, 17

$Fe_3BC_{10}H_3O_{10}$
$(H)(CO)Fe_3(CO)_9BH_2$, **1**, 296, 297, 299

$Fe_3BC_{10}H_7O_9$
$(H)Fe_3(CO)_9(BH_3Me)$, **1**, 296, 297

$Fe_3BC_{12}H_5O_9$
$(MeC_2BH_2)Fe_3(CO)_9$, **1**, 201, 202
$Fe_3BC_{46}H_{32}NO_9P_2$
$[PPN][(CO)Fe_3(CO)_9BH_2]$, **1**, 296, 297, 299
$Fe_3BRh_3C_{15}H_2O_{15}$
$H_2Fe_3Rh_3(CO)_{15}B$, **1**, 294, 304
$Fe_3BiC_9H_3O_9$
$H_3BiFe_3(CO)_9$, **7**, 59
$Fe_3BiC_{10}HO_9$
$Fe_3(CO)_9Bi(CH)$, **7**, 59
$Fe_3BiC_{11}H_3O_{10}$
$BiFe_3(CO)_9(COMe)$, **7**, 58
$Fe_3BiC_{12}ClO_{12}$
$[ClBi[Fe(CO)_4]_3]^{2-}$, **7**, 60
$Fe_3Bi_2C_9O_9$
$Bi_2Fe_3(CO)_9$, **7**, 59
$Fe_3C_8HN_2O_8$
$[Fe_3(CO)_8(NO)(NH)]^-$, **7**, 34
$Fe_3C_9H_2O_9$
$H_2Fe_3(CO)_9$, **7**, 70
$Fe_3C_9H_2O_9P$
$[HFe_3(CO)_9(PH)]^-$, **7**, 50
$Fe_3C_9O_9S$
$[Fe_3(CO)_9S]^{2-}$, **7**, 70
$Fe_3C_9O_9S_2$
$Fe_3(CO)_9S_2$, **7**, 61, 70, 71
$Fe_3C_9O_9SeTe$
$Fe_3(CO)_9(Se)(Te)$, **7**, 75
$Fe_3C_9O_9Te_2$
$Fe_3(CO)_9Te_2$, **7**, 76–8
$Fe_3C_9O_{10}$
$[Fe_3(CO)_9(O)]^{2-}$, **7**, 60, 61
$Fe_3C_9O_{12}S_2$
$[Fe_3(CO)_7(CCO)(SO_2)_2]^{2-}$, **7**, 23
$Fe_3C_{10}HNO_{10}$
$Fe_3(CO)_{10}(NH)$, **7**, 35
$Fe_3C_{10}H_3O_9$
$[Fe_3(H)(HCH)(CO)_9]^-$, **7**, 275, 276
$Fe_3C_{10}H_4O_9$
$Fe_3(CH_4)(CO)_9$, **7**, 276
$Fe_3C_{10}H_7NO_8S_2$
$Fe_3(CO)_8(NMe_2H)S_2$, **7**, 71
$Fe_3C_{11}HO_{10}$
$Fe_3(CO)_{16}(CH)$, **7**, 22
$Fe_3C_{11}HO_{11}$
$[HFe_3(CO)_{11}]^-$, **7**, 7, 11, 262, 263
$Fe_3C_{11}H_2O_8S_4$
$Fe_3(CO)_8(CS_2C_2H_2)S_2$, **7**, 71
$Fe_3C_{11}H_4O_9$
$Fe_3(H)(CO)_9(CMe)$, **7**, 262, 263
$Fe_3C_{11}H_5NO_9$
$Fe_3(MeC{=}NH)(H)(CO)_9$, **10**, 261
$Fe_3C_{11}O_{10}$
$[Fe_3(CCO)(CO)_9]^{2-}$, **7**, 21, 270, 271
$Fe_3C_{11}O_{11}$
$[Fe_3(CO)_{11}]^{2-}$, **7**, 6, 8, 270
$Fe_3C_{12}BO_9$, **1**, 393
$Fe_3C_{12}O_{12}$
$Fe_3(CO)_{12}$, **7**, 4, 8, 263; **12**, 430
$Fe_3C_{13}H_3O_{11}$
$[Fe_3(CO)_9(CCOC(O)Me)]^-$, **7**, 22, 23
$Fe_3C_{13}H_4O_{11}$
$Fe_3(CO)_9[C(H)COC(O)Me]$, **7**, 22
$Fe_3C_{14}H_7O_{12}$
$[Fe_3(H)(CHCO_2Et)(CO)_{10}]^-$, **7**, 282
$Fe_3C_{14}H_9NO_9S$
$Fe_3(CO)_9(S)(MeC_2NMe_2)$, **7**, 70
$Fe_3C_{15}H_6NO_9$
$[HFe_3(CO)_9NPh]^-$, **7**, 34
$Fe_3C_{15}H_7O_9P$
$H_2Fe_3(CO)_9(PPh)$, **7**, 52
$Fe_3C_{15}H_8O_{11}$
$Fe_3(MeCCOEt)(CO)_{10}$, **7**, 277
$Fe_3C_{15}H_{12}NO_{13}P$
$Fe_3(CO)_{10}[P(OMe)_3](NCMe)$, **7**, 33
$Fe_3C_{16}H_9NO_{11}$
$Fe_3(CO)_{11}(CNBu^t)$, **7**, 13
$Fe_3C_{16}H_{18}O_{16}P_2$
$Fe_3(CO)_{10}[P(OMe)_3]_2$, **7**, 49
$Fe_3C_{17}H_{15}N_3O_{11}P_3$
$Fe_2(CO)_6[(Et_2N)_2PP]Fe(CO)_4]P[C(O)NEt]$, **7**, 44
$Fe_3C_{19}H_7NO_{11}$
$Fe_3(CO)_{11}(NCC_6H_4Me)$, **7**, 33
$Fe_3C_{20}H_{13}O_9$
$Fe_2(CO)_8[C(Fc)(OMe)]$, **7**, 238
$Fe_3C_{20}H_{18}N_2O_{10}$
$Fe_3(CO)_{10}(CNBu^t)_2$, **7**, 13
$Fe_3C_{21}H_{10}O_9P_2$
$Fe_3(CO)_9(PPh)_2$, **7**, 52
$Fe_3C_{21}H_{11}O_9P_2$
$Fe_3(CO)_9(PPh)(PPhH)$, **7**, 53
$Fe_3C_{21}H_{33}O_7P_3$
$HFe_3(CO)_7(PMeBu^t)(Bu^tPCH_2PBu^t)$, **7**, 49
$Fe_3C_{22}H_{30}N_3O_{11}P_2$
$Fe_2(CO)_6[(Et_2N)_2P[Fe(CO)_4]PC(O)NEt_2]$, **7**, 45
$Fe_3C_{23}H_{19}O_{10}P$
$Fe_3(CO)_9[Bu^tPC(Ph)C(OEt)]$, **7**, 52
$Fe_3C_{25}H_{22}O_6$
$HC[CH_2Fe(CO)_2Cp)]_3$, **7**, 167
$Fe_3C_{26}H_{15}N_2O_9P$
$Fe_3(CO)_8(NO)(NPPh_3)$, **7**, 34
$Fe_3C_{26}H_{15}O_8PS_2$
$Fe_3(CO)_8(PPh_3)S_2$, **7**, 64
$Fe_3C_{26}H_{15}O_8PSeTe$
$Fe_3(CO)_8(PPh_3)(Se)(Te)$, **7**, 75
$Fe_3C_{26}H_{20}O_8S$
$Fe_3(CO)_8[H(Ph)CCPh)(SBu^t)$, **7**, 70
$Fe_3C_{27}H_{15}N_2O_9P$
$Fe_3(CO)_9(NO)(NPPh_3)$, **7**, 32
$Fe_3C_{31}H_{18}N_6O_8$
$(Fe(CO)_2CpH)_2[Fe(CO)_2CpH[C_2(CN)_3O]_2]$, **7**, 191
$Fe_3C_{32}H_{22}O_7P_2S_2$
$Fe_3(CO)_7(dppm)S_2$, **7**, 71
$Fe_3C_{32}H_{22}O_8P_2$
$H_2Fe_3(CO)_8(PPh_2)_2$, **7**, 49
$Fe_3C_{33}H_{22}O_7P_2$
$Fe_3(CO)_6(CO)(PPh_2)_2(CCH_2)$, **7**, 41
$Fe_3C_{42}H_{30}O_6S_6$
$(CO)_3Fe(SPh)_3Fe(SPh)_3Fe(CO)_3$, **7**, 61
$Fe_3C_{43}H_{30}O_7P_2S_2$
$Fe_3(CO)_7(PPh_3)_2S_2$, **7**, 64
$Fe_3C_{43}H_{40}N_2O_7P_2$
$Fe_3(CO)_7(PhCCPh)[Et_2NPC(Ph)C(Ph)PNEt_2]$, **7**, 53, 54
$Fe_3C_{53}H_{36}O_9P_2$
$[PPh_4][Fe_3(CO)_9(PhPCPhCPhH)]$, **7**, 49, 50
$Fe_3CoC_{13}O_{13}$
$[Fe_3Co(CO)_{13}]^-$, **10**, 19, 20
$Fe_3CoC_{20}H_{20}NO_{12}$
$[Et_4N][Fe_3Co(CO)_{12}]$, **10**, 45
$Fe_3CoC_{21}H_{20}NO_{13}$
$[Et_4N][Fe_3Co(CO)_{13}]$, **10**, 47
$Fe_3CoGeC_{14}O_{14}$
$[Fe_3Co(Ge)(CO)_{14}]^-$, **7**, 30; **10**, 215
$Fe_3Co_3C_{20}O_{18}$
$[Fe_2Co(CO)_9CCFeCo_2(CO)_9]^{2-}$, **7**, 22
$Fe_3CrC_{14}O_{13}$
$[CrFe_3(C)(CO)_{13}]^{2-}$, **10**, 90
$Fe_3CrC_{14}O_{14}$
$[CrFe_3(CO)_{14}]^{2-}$, **10**, 191, 192
$Fe_3Cr_2C_{17}O_{16}$
$[Cr_2Fe_3C(CO)_{16}]^{2-}$, **7**, 21, 22; **10**, 90
$Fe_3Cr_6C_{33}H_{30}O_3S_9$
$Cr_6Fe_3(S)_9(CO)_3Cp_6$, **10**, 234, 235
$Fe_3CuC_{11}IO_{10}$
$[Fe_3CuI(CO)_9(CCO)]^{2-}$, **7**, 22, 271, 272
$Fe_3Cu_3C_{12}O_{12}$
$[Cu_3Fe_3(CO)_{12}]^{3-}$, **7**, 79, 80
$Cu_3Fe_3(CO)_{12}$, **10**, 310
$Fe_3GaC_{21}H_{15}O_6$
$Ga[Fe(CO)_2Cp]_3$, **1**, 555
$Fe_3GeMnC_{19}H_5O_{14}$
$GeFe_3Mn(CO)_{14}Cp$, **7**, 30
$Fe_3GeMnC_{20}H_7O_{14}$
$MnFe_3(Ge)(CO)_{14}Cp'$, **10**, 215
$Fe_3Ge_2Re_2C_{19}O_{19}$
$Fe_3(CO)_9[GeRe(CO)_5]_2$, **7**, 30
$Fe_3InC_{12}O_{12}$
$[In[Fe(CO)_4]_3]^{3-}$, **7**, 19
$Fe_3InC_{21}H_{15}O_6$
$In[Fe(CO)_2Cp]_3$, **1**, 562
$Fe_3InC_{36}H_{30}NO_{12}$
$[NMe_3(Bz)]_3[In[Fe(CO)_4]_3]$, **1**, 563
$Fe_3Li_6C_{42}H_{56}N_4$
$[(C_5H_4Li)_2Fe]_3(TMEDA)_2$, **7**, 189
$Fe_3MnC_{12}O_{13}$

$[Fe_3Mn(CO)_{12}O]^-$, **7**, 60, 61; **10**, 232, 233, 235
$Fe_3MnC_{14}HO_{13}$
$MnFe_3(CO)_{13}(CH)$, **7**, 25
$Fe_3MnC_{16}H_5O_{11}P_2$
$Fe_3(CO)_9(P)[PMn(CO)_2Cp]$, **7**, 53
$Fe_3NiC_{12}HO_{12}$
$[Fe_3Ni(H)(CO)_{12}]^-$, **9**, 20; **10**, 203
$Fe_3NiC_{12}O_{12}$
$[Fe_3Ni(CO)_{12}]^{2-}$, **7**, 84; **9**, 20
$Fe_3Ni_3C_{14}O_{13}$
$[Fe_3Ni_3(CO)_{13}C]^{2-}$, **7**, 22; **9**, 20; **10**, 146
$Fe_3PtC_{29}H_{16}O_{11}P$
$[HFe_3Pt(CO)_{11}(PPh_3)]^-$, **7**, 84
$Fe_3Pt_3C_{15}O_{15}$
$Fe_3Pt_3(CO)_{15}$, **7**, 84; **9**, 564; **10**, 5, 192
$Fe_3RhC_{12}HO_{12}$
$[HFe_3Rh(CO)_{12}]^{2-}$, **7**, 83
$Fe_3RhC_{13}HO_{12}$
$RhFe_3(CO)_{12}(CH)$, **7**, 25
$Fe_3RhC_{13}H_3O_{11}$
$HFe_3Rh(CO)_{11}(C{=}CH_2)$, **10**, 41
$Fe_3RhC_{13}O_{12}$
$[RhFe_3(CO)_{12}C]^-$, **7**, 21
$Fe_3RhC_{19}H_7O_{11}$
$HFe_3Rh(CO)_{11}(C{=}CHPh)$, **10**, 45
$Fe_3RhC_{30}H_{25}O_8P_2$
$Fe_3Rh(PPh)_2(CO)_8Cp^*$, **10**, 226
$Fe_3RhC_{31}H_{25}O_9P_2$
$Cp^*RhFe_3(CO)_9(PPh)_2$, **7**, 55
$Fe_3Rh_2C_{15}O_{14}$
$Rh_2Fe_3(CO)_{14}C$, **7**, 22
$Fe_3Rh_2C_{17}H_5O_{13}$
$Fe_3Rh_2(CO)_{13}[C(Me)CCH_2]$, **10**, 141, 142
$Fe_3Rh_2C_{41}H_{25}O_{13}P$
$[Ph_4P][Fe_3Rh_2(CO)_{10}(CO)_3(MeC{=}C{=}CH_2)]$, **10**, 41
$Fe_3Rh_3C_{40}H_{20}O_{15}P$
$[Ph_4P][Fe_3Rh_3C(CO)_{15}]$, **10**, 45
$Fe_3RuC_{11}O_{11}SeTe$
$Fe_3Ru(CO)_{11}SeTe$, **7**, 75
$Fe_3Ru_5C_{43}H_{20}O_{17}$
$Fe_3Ru_5(C)_2(PPh_2)_2(CO)_{17}$, **10**, 129
$Fe_3SbC_{10}HO_9$
$Fe_3(CO)_9(Sb)(CH)$, **7**, 58
$Fe_3SbC_{10}O_{10}$
$SbFe_2(CO)_{10}$, **7**, 58
$Fe_3SbC_{12}ClO_{12}$
$[ClSb[Fe(CO)_4]_3]^{2-}$, **7**, 58
$Fe_3SbC_{13}H_{10}O_9$
$[HFe_3(CO)_9(SbBu^t)]^-$, **7**, 58
$Fe_3SbCrC_{17}O_{17}$
$[Sb[Fe(CO)_4][Fe_2(CO)_8][Cr(CO)_5]]^-$, **7**, 58
$Fe_3SiC_{18}H_{22}O_9P$
$[HFe_3(CO)_9[PSiPr^i_3]]^-$, **7**, 50
$Fe_3SnC_{12}O_{12}$
$[Sn[Fe(CO)_4]_3]^{2-}$, **7**, 31
$Fe_3SnC_{18}H_{10}O_8$
$Fe(CO)VSn[Fe(CO)_2Cp]_2$, **7**, 31
$Fe_3WC_{14}O_{14}$
$[WFe_3(CO)_{14}]^{2-}$, **10**, 194
$Fe_3WC_{15}BrO_{15}P$
$WFe_3(P)(Br)(CO)_{15}$, **10**, 218
$Fe_3WC_{19}H_{11}O_{11}PS_2$
$Fe_3W(CO)_{11}(PMe_2Ph)S_2$, **7**, 72
$Fe_3Yb_2C_{51}H_{60}O_{11}$
$[Cp^*{}_2Yb]_2[Fe_3(CO)_{11}]$, **4**, 108
$Fe_3ZnC_{37}H_{51}NO_6$
$(Bu_4N)[Zn[FeCp(CO)_2]_3]$, **10**, 338
$Fe_4AsC_{14}ClO_{14}$
$Fe_2(CO)_8As[Fe_2(CO)_6Cl]$, **7**, 56
$Fe_4As_2Au_2BC_{48}H_{30}O_{12}$
$Fe_4(CO)_{12}B(AuAsPh_3)_2$, **1**, 297, 299
$Fe_4As_2Au_2BC_{48}H_{31}O_{12}$
$HFe_4(CO)_{12}B(AuAsPh_3)_2$, **7**, 18
$Fe_4AuBC_{30}H_{17}O_{12}P$
$HFe_4(CO)_{12}(AuPPh_3)BH$, **1**, 297, 299, 302
$Fe_4AuC_{67}H_{45}NO_{13}P_3$
$[N(PPh_3)_2][AuFe_4(CO)_{13}(PPh_3)]$, **10**, 299
$Fe_4Au_2BC_{24}H_{31}O_{12}P_2$
$HFe_4(CO)_{12}B[AuPEt_3]_2$, **7**, 18
$Fe_4Au_2BC_{48}H_{31}O_{12}P_2$
$Fe_4(CO)_{12}(AuPPh_3)_2BH$, **1**, 297, 299, 302
$Fe_4Au_2C_{25}H_{30}O_{12}P_2$
$Fe_4(CO)_{12}C[AuPEt_3]_2$, **7**, 27
$Fe_4Au_2C_{41}H_{22}O_{16}P_2$
$[Fe_2(CO)_8][AuPh_2P(CH_2)PPh_2Au)Fe_2(CO)_8]^{2-}$, **7**, 81
$Fe_4Au_3BC_{66}H_{45}O_{12}P_3$
$Fe_4(CO)_{12}B(AuPPh_3)_3$, **1**, 297, 299
$Fe_4BC_{12}H_2O_{12}$
$[HFe_4(CO)_{12}BH]^-$, **1**, 296, 297, 302; **7**, 17, 18
$Fe_4BC_{12}H_3O_{12}$
$HFe_4(CO)_{12}BH_2$, **1**, 294, 297, 298, 302
$Fe_4BC_{12}O_{12}$
$[Fe_4(CO)_{12}B]^{3-}$, **1**, 298
$Fe_4BC_{13}HCl_2$
$(H)Fe_4(CO)_{12}CBCl_2$, **1**, 297, 300
$Fe_4BC_{13}H_3O_{12}$
$HFe_4(CO)_{12}CBH_2$, **1**, 297, 300; **7**, 26
$Fe_4BC_{13}H_3O_{13}$
$HFe_4(CO)_{12}CBH(OH)$, **7**, 26
$Fe_4BC_{48}H_{32}NO_{12}P_2$
$[PPN][HFe_4(CO)_{12}BH]$, **1**, 297, 298, 302
$Fe_4BC_{62}H_{54}NO_{10}P_2$
$[PPN][HFe_4(CO)_{10}(PPhMe_2)_2BH]$, **1**, 297, 298
$Fe_4BRh_2C_{16}O_{16}$
$[Fe_4Rh_2(B)(CO)_{16}]^-$, **10**, 215
$Fe_4BRh_2C_{52}H_{30}NO_{16}P_2$
$[PPN][Fe_4Rh_2(CO)_{16}B]$, **1**, 304
$Fe_4B_4C_{12}H_8O_{12}$
$[B_4H_4fe_4(CO)_{12}]$, **1**, 307
$[B_2H_4Fe_2(CO)_6]_2$, **7**, 16, 17
$Fe_4BiC_{16}O_{16}$
$[Bi[Fe(CO)_4]_4]^{3-}$, **7**, 58
$Fe_4Bi_2C_{13}O_{13}$
$[Fe_3(CO)_9Bi[BiFe(CO)_4]]^{2-}$, **7**, 59
$Fe_4Bi_4C_{13}O_{13}$
$[Bi_4Fe_4(CO)_{13}]^{2-}$, **7**, 60
$Fe_4C_{11}N_2O_{12}$
$Fe_4(CO)_{11}(NO)N$, **7**, 35
$Fe_4C_{11}O_{11}S_2$
$Fe_4(CO)_{11}(S)_2$, **7**, 72
$Fe_4C_{12}HNO_{12}$
$HFe_4(CO)_{12}N$, **7**, 34
$Fe_4C_{12}NO_{12}$
$[Fe_4(CO)_{12}M]^-$, **7**, 35
$Fe_4C_{12}O_{12}S_3$
$[[SFe_2(CO)_6]_2S]^{2-}$, **7**, 66
$Fe_4C_{12}O_{12}Se_4$
$[[Fe_2(CO)_6(Se)]_2Se_2]^{2-}$, **7**, 74
$Fe_4Se_4(CO)_{12}$, **7**, 74
$Fe_4C_{12}O_{12}Te_4$
$Fe_4(CO)_{12}Te_4$, **7**, 77
$Fe_4C_{13}HO_{13}$
$[Fe_4(H)(CO)_{13}]^-$, **7**, 7, 276
$Fe_4C_{13}H_2O_{12}$
$HFe_4(CO)_{12}CH$, **7**, 12
$Fe_4C_{13}O_{13}$
$[Fe_4(CO)_{13}]^{2-}$, **7**, 6, 24, 268
$Fe_4C_{14}ClO_{14}P$
$Fe_2(CO)_8(P)Fe_2(CO)_6Cl$, **7**, 44
$Fe_4C_{14}H_3O_{12}$
$[Fe_4(CO)_{12}CMe]^-$, **7**, 25
$Fe_4C_{14}H_3O_{14}$
$[Fe_4(CO)_{12}C(O)OMe]^-$, **7**, 23, 24
$Fe_4C_{14}O_{13}$
$Fe_4(CO)_{13}C$, **7**, 23
$Fe_4C_{15}H_3O_{13}$
$[Fe_4(CO)_{12}CC(O)Me]^-$, **7**, 25
$Fe_4C_{15}H_8O_{12}Te_4$
$[(CO)_6Fe_2(TeMe)Te]_2(CH_2)$, **7**, 77
$Fe_4C_{20}H_6O_{12}Se_4$
$Fe_2(CO)_6Se_2C(H)C(Ph)Se_2Fe_2(CO)_6$, **7**, 74
$Fe_4C_{20}H_{20}S_4$
$Fe_4(Cp)_4(S)_4$, **7**, 284
$Fe_4C_{20}H_{20}S_5$
$Fe_4Cp_4(S)_3(S_2)$, **7**, 285, 286
$Fe_4C_{20}H_{20}S_6$
$Fe_4Cp_4(S)_2(S_2)_2$, **7**, 285, 286
$Fe_4C_{20}H_{20}Se_4$
$Fe_4Cp_4(Se)_4$, **7**, 285
$Fe_4C_{20}H_{30}S_4$
$Fe_4Cp^*{}_2S_4$, **7**, 286
$Fe_4C_{21}H_{19}O_{13}P$
$Fe_4(CO)_{11}(PBu^t)(COEt)_2$, **7**, 52
$Fe_4C_{24}H_{20}O_4$
$[CpFe(CO)]_4$, **7**, 185
$Fe_4C_{28}H_{20}N_4O_8$
$Fe_4(CO)_8(py)_4$, **7**, 33
$Fe_4C_{36}H_{20}O_{12}P_4$

$Fe_2(CO)_6[PhPP[Fe_2(CO)_6$-$(PPh_2)]PPh]$, **7**, 44
$Fe_4C_{40}H_{60}S_4$
$Fe_4Cp^*{}_4S_4$, **7**, 286
$Fe_4C_{40}H_{60}S_5$
$Fe_4S_5Cp^*{}_4$, **7**, 286
$Fe_4CoC_{15}O_{14}$
$[CoFe_4(CO)_{14}C]^-$, **7**, 25
$Fe_4Co_2Cr_2C_{24}H_{14}O_{12}S_7$
$Cr_2Fe_4Co_2(S)_7(CO)_{12}Cp'_2$, **10**, 239
$Fe_4Co_2HgC_{28}H_{16}O_{16}$
$[(COMe)Cp(CO)_7CoFe_2]_2(Hg)$, **10**, 332, 339
$Fe_4CuC_{31}H_{15}O_{13}P$
$[Fe_4(CO)_{13}(CuPPh_3)]^-$, **7**, 80
$Fe_4CuC_{66}H_{45}NO_{12}P_3$
$[N(PPh_3)_2][CuFe_4(CO)_{12}(PPh_3)$, **10**, 299
$Fe_4Cu_5C_{16}O_{16}$
$[Cu_5Fe_4(CO)_{16}]^{3-}$, **7**, 79, 80
$Fe_4Cu_6C_{16}O_{16}$
$[Cu_6Fe_4(CO)_{16}]^{2-}$, **7**, 79, 80
$Fe_4GeC_{14}O_{14}$
$[Fe_3(CO)_{10}[GeFe(CO)_4]]^{2-}$, **7**, 30
$Fe_4GeC_{16}O_{16}$
$Ge[Fe_2(CO)_8]_2$, **7**, 30
$Fe_4HgSn_2C_{16}Cl_4O_{16}$
$[(CO)_4FeSnCl_2Fe(CO)_4]_2Hg^{2-}$, **10**, 339
$Fe_4Hg_4C_{16}O_{16}$
$[HgFe(CO)_4]_4$, **10**, 334
$Fe_4InC_{28}H_{20}O_8S_4$
$InS_4[Fe(CO)_2Cp]_4$, **1**, 562
$Fe_4InC_{28}H_{20}O_8Se_4$
$InSe_4[Fe(CO)_2Cp]_4$, **1**, 562
$Fe_4Mo_2C_{20}O_{19}$
$Mo_2Fe_4C(CO)_{19}$, **10**, 91
$Fe_4PbC_{12}N_4O_{16}$
$Ph[Fe(CO)_3(NO)]_4$, **7**, 31
$Fe_4Pt_6C_{22}O_{22}$
$[Fe_4Pt_6(CO)_{22}]^{2-}$, **7**, 84
$Fe_4RhC_{29}H_{25}O_{12}P_2$
$Fe_4Rh(PPh)_2(CO)_{12}Cp$, **10**, 222, 223
$Fe_4RhC_{39}H_{20}O_{14}P$
$[Ph_4P][Fe_4RhC(CO)_{14}]$, **10**, 45
$Fe_4Rh_2C_{17}O_{16}$
$Fe_4Rh_2C(CO)_{16}$, **10**, 45
$Fe_4Ru_2C_{22}O_{22}$
$[Fe_2Ru(CO)_{11}]_2{}^{2-}$, **7**, 83
$Fe_4SbC_{13}HO_{13}$
$[(H)Fe_3(CO)_9[SbFe(CO)_4]]^{2-}$, **7**, 58
$Fe_4SbC_{13}H_2O_{13}$
$[(H)_2Fe_3(CO)_9[SbFe(CO)_4]]^-$, **7**, 58
$Fe_4SbC_{16}O_{16}$
$[Sb[Fe_2(CO)_8][Fe(CO)_4]_2]$, **7**, 58
$[Sb[Fe(CO)_4]_4]^{3-}$, **7**, 58
$Fe_4SnC_{16}O_{16}$
$[Sn[Fe_2(CO)_8][Fe(CO)_4]_2]^{2-}$, **7**, 31
$SnFe_4(CO)_{16}$, **7**, 31
$Fe_4Tl_2C_{32}H_{40}N_2O_{16}$
$[NEt_4]_2[[Fe(CO)_4]_2[TlFe$-$(CO)_4]_2]$, **1**, 567
$Fe_4B_4C_{12}H_8O_{12}$
$B_4H_8Fe_4(CO)_{12}$, **1**, 307
$Fe_5Au_2C_{27}H_{30}O_{14}P_2$
$Fe_5(CO)_{14}C[AuPEt_3]_2$, **7**, 26
$Fe_5C_{13}Br_2O_{12}$
$Fe_5(CO)_{12}(Br)_2C$, **7**, 79
$Fe_5C_{14}NO_{14}$
$[Fe_5(CO)_{13}(NO)C]^-$, **7**, 26
$Fe_5C_{14}O_{14}Te_4$
$[Fe_5(CO)_{14}Te_4]^{2-}$, **7**, 77
$Fe_5C_{14}O_{15}S$
$[Fe_5(CO)_{13}(SO_2)C]^{2-}$, **7**, 26
$Fe_5C_{15}O_{14}$
$[Fe_5C(CO)_{14}]^{2-}$, **7**, 26, 267
$Fe_5C_{16}H_3O_{14}$
$[Fe_5C(COMe)(CO)_{13}]^-$, **7**, 267
$Fe_5C_{16}O_{15}$
$Fe_5C(CO)_{15}$, **7**, 267
$Fe_5C_{20}H_{16}O_{13}P_2$
$Fe_5(CO)_{13}(dmpe)C$, **7**, 26
$Fe_5C_{37}H_{33}O_{12}P_3$
$Fe_5(CO)_{12}(PMe_2Ph)_3C$, **7**, 26
$Fe_5Ge_2C_{23}H_{10}O_{13}$
$Fe_3(CO)_9[GeFe(CO)_2Cp]_2$, **7**, 30
$Fe_5MnC_{23}H_5O_{18}P_2$
$Fe_3(CO)_8(PMn(CO)_2Cp)[PFe_2$-$(CO)_8]$, **7**, 53
$Fe_5MoC_6I_3O_6S_6$
$MoFe_5(S)_6(CO)_6I_3$, **10**, 237
$Fe_5RhC_{21}H_{12}NO_{16}$
$[Me_4N][Fe_5RhC(CO)_{16}]$, **10**, 50
$Fe_5RhC_{41}H_{20}O_{16}P$
$[Ph_4P][Fe_5RhC(CO)_{16}]$, **10**, 45
$Fe_5Sb_2C_{17}O_{17}$
$[Fe_3(CO)_9[SbFe(CO)_4]_2]^{2-}$, **7**, 58
$Fe_5Si_2C_{23}H_{10}O_{13}$
$Fe_3(CO)_9[SiFeCp(CO)_2]_2$, **7**, 29
$Fe_6As_2C_{22}O_{22}$
$[Fe_2(CO)_8As]_2Fe_2(CO)_6$, **7**, 56
$Fe_6Au_4C_{96}H_{64}O_{18}P_4$
$[(Ph_3PAu)_2Fe_3(CO)_9P]_2C_6H_4$, **7**, 50
$Fe_6C_{12}O_{12}S_6$
$[Fe_6S_6(CO)_{12}]^{2-}$, **7**, 68, 69
$Fe_6C_{16}O_{17}S$
$[Fe_6(CO)_{15}(SO_2)C]^{2-}$, **7**, 27
$Fe_6C_{17}O_{16}$
$[Fe_6C(CO)_{16}]^{2-}$, **7**, 27, 267
$Fe_6C_{24}H_8O_{18}P_2$
$[H_2Fe_3(CO)_9P]_2C_6H_4$, **7**, 50
$Fe_6HgC_{22}H_6O_{20}$
$[Fe_3(COCH_3)(CO)_9]_2(Hg)$, **10**, 339
$Fe_6In_2C_{24}O_{24}$
$[In_2Fe_6(CO)_{24}]^{2-}$, **7**, 19
$Fe_6In_2C_{32}H_{20}NO_{24}$
$[NEt_4][In_2Fe_6(CO)_{24}]$, **1**, 563
$Fe_6MoC_{16}O_{16}S_6$
$[MoFe_6(S)_6(CO)_{16}]^{2-}$, **10**, 232, 233
$Fe_6Sb_2C_{20}O_{20}$
$[Fe_4(CO)_{12}[SbFe(CO)_4]_2]^{2-}$, **7**, 58
$Fe_6Tl_2C_{24}O_{24}$
$[Tl_2Fe_6(CO)_{24}]^{2-}$, **7**, 20
$Fe_6Tl_2C_{40}H_{40}O_{24}$
$[NEt_4]_2[Tl_2Fe_6(CO)_{24}]$, **1**, 568
$Fe_7Hg_3C_{24}H_6O_{24}$
$Hg[Fe(CO)_4[HgFe_3(OMe)$-$(CO)_9]_2]$, **10**, 339
$Fe_7Mo_2C_{22}O_{22}S_8$
$Mo_2Fe_7(S)_8(CO)_{22}$, **10**, 232, 233
$Fe_8Ag_{13}C_{32}O_{33}$
$[Ag_{13}Fe_8(CO)_{32}]^{4-}$, **7**, 80
$Fe_8Au_2B_3C_{30}H_{22}O_{24}P_2$
$[H[Fe_4(CO)_{12}BH]_2Au]$-$[Au(PMe_3)_2BH]$, **1**, 297, 300
$Fe_8C_{24}O_{24}Te_6$
$[Fe_8(CO)_{24}Te_6]^{2-}$, **7**, 77, 78
$Fe_8H_8O_3$
$Fe(CO)_3(isoprene)$, **7**, 145
$Fe_8Hg_3C_{32}H_6O_{30}$
$Hg[Fe(CO)_4(Hg)Fe_3(CO)_{10}$-$(COCH_3)]_2$, **10**, 334, 335
$Fe_8Tl_4C_{30}O_{30}$
$[Tl_4Fe_8(CO)_{30}]^{4-}$, **7**, 20
$Fe_8Tl_4C_{62}H_{80}N_4O_{30}$
$[NEt_4]_4[Tl_4Fe_8(CO)_{30}]$, **1**, 568
$Fe_{10}Tl_6C_{40}H_{40}O_{24}$
$[NEt_4]_2[Tl_2Fe_6(CO)_{24}]$, **1**, 568

G

GaAs
GaAs, **1**, 517
$GaAsC_{12}H_{27}$
$Bu^tGaAsBu^t_2$, **1**, 518
$GaAsSiC_{26}H_{48}$
$Cp^*_2GaAs(TMS)_2$, **1**, 518, 519
$GaAs_3C_{54}H_{66}$
$[(Mes)_2As]_3Ga$, **1**, 518, 519
$GaBC_{23}H_{34}P$
$Bu^t_2GaP(Ph)B(Bu^t_6H_2)$, **1**, 519
GaB_3H_{10}
$H_2GaB_3H_8$, **1**, 258
$GaB_4Si_2C_{12}H_{33}$
$Bu^tGa(TMS)_2C_2B_4H_4$, **1**, 258
$GaB_4Si_2C_{20}H_{37}$
(bipyrimidine)$Bu^tGa(TMS)_2$-$C_2B_4H_4$, **1**, 258
$GaB_9C_5H_{19}$
$MeEtGaC_2B_9H_{11}$, **1**, 532
$GaB_{10}C_6H_{11}F_6O_4$
$C_2B_{10}H_{11}Ga(O_2CCF_3)_2$, **1**, 227
GaC_2H_7
$GaMe_2H$, **1**, 515
GaC_3
GaC_3, **1**, 506
$GaC_4H_{10}NO_2$
$NH_2CH_2CO_2GaMe_2$, **1**, 525
GaC_5H_5
GaCp, **1**, 529
$GaC_6H_{14}I_4S_2$
$[I_2Ga(SPr^i)]_2$, **1**, 525
GaC_7H_{11}
Me_2GaCp, **1**, 529
$GaC_8H_{12}NO$
$NH_2C_6H_4OGaMe_2$, **1**, 525
$GaC_8H_{13}NO$
$C_5H_3N(CH_2NH_2)(O)GaMe_2$, **1**, 525
$GaC_9H_{18}N$
$Ga[(CH_2)_3]_3N$, **1**, 505, 506
$GaC_{10}H_{15}$
GaCp*, **1**, 504
$GaC_{10}H_{28}BrN_2$
$[Me_2Ga(Bu^tNH_2)_2]Br$, **1**, 514
$GaC_{11}H_{18}N$
$Me_2GaC_6H_4CH_2NMe_2$, **1**, 506
$GaC_{12}H_{27}$
Bu^t_3Ga, **1**, 505
$GaC_{15}H_{13}N_2O_4$
$[CH{=}NOHC_6H_3O]_2GaMe$, **1**, 525
$GaC_{15}H_{15}$
$Ga(Cp)_3$, **1**, 529, 530
$GaC_{15}H_{33}$
$(Bu^tCH_2)_3Ga$, **1**, 505
$GaC_{16}H_{38}S_2$
$[Bu^t_2Ga(SH)]_2$, **1**, 525, 526
$GaC_{18}H_{15}$
Ph_3Ga, **1**, 504
$GaC_{18}H_{23}P$
$Bu^tCH_2GaCH_2PPh_2$, **1**, 508
$GaC_{27}H_{33}$
$Ga(Mes)_3$, **1**, 507
$GaC_{40}H_{67}S_2$
$GaBu[S(Bu^t_3C_6H_2)]_2$, **1**, 525
$GaC_{54}H_{87}S_3$
$Ga[S(Bu^t_3C_6H_2)]_3$, **1**, 525
$GaC_{54}H_{90}P_3$
$Ga[PH(Bu^t_3C_6H_2]_3$, **1**, 519
$GaCoC_7H_{11}NO_4$
$GaH_2(NMe_3)[Co(CO)_4]$, **1**, 556
$GaCoC_{18}H_{30}O_5$
$Ga(CH_2Bu^t)_2(THF)[Co(CO)_4]$, **1**, 556
$GaCoSi_2C_{16}H_{30}O_5$
$Ga(CH_2TMS)_2(THF)[Co(CO)_4]$, **1**, 556
$GaCo_2C_{42}H_{30}ClO_6P_2$
$GaCl[Co(CO)_3(PPh_3)]_3$, **1**, 555
$GaCo_3C_{63}H_{45}O_9P_3$
$Ga[Co(CO)_3(PPh_3)]_3$, **1**, 556
$GaFeC_{11}H_{15}I_2O_3$
$GaI_2(Et_2O)[Fe(CO)_2Cp)]$, **1**, 555
$GaFeC_{16}H_{26}NO_2$
$GaEt(CH_2CH_2CH_2NEt_2)[Fe(CO)_2Cp]$, **1**, 556
$GaFeC_{24}H_{24}NO_2$
$Ga(C_5H_4Me)_2(py)[Fe(CO)_2Cp)]$, **1**, 555
$GaFe_2C_{14}H_{10}ClO_4$
$GaCl[Fe(CO)_2Cp]_2$, **1**, 555
$GaFe_3C_{21}H_{15}O_6$
$Ga[Fe(CO)_2Cp]_3$, **1**, 555
GaH_3
GaH_3, **1**, 515
$GaKSi_3C_{12}H_{34}$
$KGaH(CH_2TMS)_3$, **1**, 516
$GaMnC_8H_{11}NO_5$
$GaH_2(NMe_3)[Mn(CO)_5]$, **1**, 556
$GaMn_3NaC_{15}ClO_{15}$
$Na[GaCl[Mn(CO)_5]_3$, **1**, 553
$GaMoC_{12}H_{11}Cl_2O_3$
$[(CO)_3Mo(Mes)]GaCl_2$, **1**, 508
$GaMoC_{13}H_{15}I_2O_3$
$GaI_2(THF)[Mo(CO)_2(C_7H_7)]$, **1**, 553
$GaMoC_{13}H_{17}I_2O_4$
$GaI_2(Et_2O)[Mo(CO)_3(C_5H_4Me)]$, **1**, 553
$GaMoC_{15}H_{14}IO_3$
$GaI[Mo(CO)_3(C_5H_4Me)_2]$, **1**, 553
$GaMoC_{30}H_{33}O_3$
$[(CO)_3Mo(Mes)]Ga(Mes)_2$, **1**, 507
$GaMo_2C_{33}H_{33}O_3$
$[(CO)_3Mo(Mes)]_2Ga(Mes)$, **1**, 507, 508
$GaMo_3C_{24}H_{15}O_9$
$Ga[Mo(CO)_3Cp]_3$, **1**, 553
$GaNiC_{11}H_{17}N_4$
$Ni(C_3H_5)(GaMe_2pz_2)$, **9**, 140
$GaNiC_{19}H_{35}O$
$Ga(CH_2Bu^t)_2(THF)[NiCp]$, **1**, 557
$GaPtC_{41}H_{81}P_2$
$Pt[(C_6H_{11})_2PCH_2CH_2P\text{-}(C_6H_{11})_2](CH_2Bu^t)[Ga\text{-}(CH_2Bu^t)_2]$, **1**, 557
$GaRbC_8H_{21}O_2$
$Rb[Me_3GaOC(Me)OGaMe_3]$, **1**, 510
$GaRe_2C_{42}H_{30}I_3O_6P_2$
$Re_2(CO)_6(PPh_3)_2(I)(GaI_2)$, **1**, 554
$GaRe_3C_{62}H_{45}I_2O_8P_3$
$Re_2(CO)_4(PPh_3)_2(I)_2[GaRe\text{-}(CO)_4(PPh_3)]$, **1**, 554
$GaRe_4C_{82}H_{60}Cl_3O_{10}P_4$
$Re_3(CO)_6(PPh_3)_3(Cl)_3[GaRe\text{-}(CO)_4(PPh_3)]$, **1**, 554
$GaSiC_{12}H_{33}$
$Ga(CH_2TMS)_3$, **1**, 514
$GaSiC_{22}H_{24}P$
$Bu^tGaPSiPh_3$, **1**, 524
$GaSiC_{30}H_{42}N$
$Bu^t_2GaNBu^t(SiPh_3)$, **1**, 520
$GaSiC_{44}H_{62}P$
$Bu^t_2GaP(Bu^t_3C_6H_2)(SiPh_3)$, **1**, 519, 520
$GaSi_6TeC_{23}H_{65}$
$[(TMS)_2CH]_2GaTeSi(TMS)_3$, **1**, 525
$GaSi_8C_{23}H_{65}Te$
$[(TMS)_2CH]_2GaTeSi(TMS)_3$, **1**, 526
$GaSi_8TeC_{23}H_{65}$
$[(TMS)_2CH]_2GaTeSi(TMS)_3$, **1**, 525
$GaTlSe_2$
$TlGaSe_2$, **1**, 524
$Ga_2As_6Si_6C_{32}H_{88}$
$[(TMS\text{-}CH_2)_2AsGa(As(CH_2\text{-}TMS)_2)]_2$, **1**, 520, 521
Ga_2Br_4
$Br_2Ga\text{-}GaBr_2$, **1**, 537
$Ga_2C_6H_{18}$
$(Me_3Ga)_2$, **1**, 504
$Ga_2C_{10}H_{13}Cl_4$
[(durene)Ga]$[GaCl_4]$, **1**, 535
$Ga_2C_{12}H_{18}Br_4$

$[(C_6Me_6)Ga][GaBr_4]$, **1**, 535
$Ga_2C_{12}H_{18}Cl_4$
$[(C_6Me_6)Ga][GaCl_4]$, **1**, 535
$Ga_2C_{17}H_{22}Cl_4$
$[(PhMe)(C_6H_2Me_4)Ga][GaCl_4]$, **1**, 535
$Ga_2C_{18}H_{24}Cl_4$
$[(C_6H_3Me_3)_2Ga][GaCl_4]$, **1**, 534, 535
$Ga_2C_{20}H_{48}P_2$
$[Ga(PBu^t_2)Me_2]_2$, **1**, 520
$Ga_2C_{24}H_{24}Br_4$
[(paracyclophane)Ga]$GaBr_4$, **1**, 536, 537
$Ga_2C_{24}H_{30}O_2$
$[Cp_2Ga(OEt)]_2$, **1**, 525
$Ga_2C_{24}H_{54}O_2$
$[Bu^t_2Ga(OBu^t)]_2$, **1**, 525
$Ga_2C_{24}H_{54}P_4$
$Ga_2P_4Bu^t_6$, **1**, 522
$Ga_2C_{24}H_{54}S_4$
$[Bu^t_2Ga(SSBu^t)]_2$, **1**, 527
$Ga_2C_{24}H_{54}Se_2$
$[Bu^t_2Ga(SeBu^t)]_2$, **1**, 527
$Ga_2C_{24}H_{66}$
$[(TMS\text{-}CH_2)_3Ga]_2$, **1**, 505
$Ga_2C_{28}H_{30}S_2$
$[Ph_2Ga(SEt)]_2$, **1**, 525
$Ga_2C_{32}H_{40}NS_6$
$[Et_4N][(PhS)_2Ga(S)]_2$, **1**, 525
$Ga_2C_{32}H_{54}Te_2$
$[(Bu^tCH_2)_2GaTePh]_2$, **1**, 525
$Ga_2C_{69}H_{103}P$
$(Bu^t_6H_2)P[Ga(Pr^i_3C_6H_2)_2]_2$, **1**, 519
$Ga_2Fe_2C_{20}H_{22}O_{10}$
$Fe_2(CO)_8[Ga(C_2H_3)(THF)]_2$, **1**, 555
$Ga_2Fe_2C_{24}H_{32}$
$[CpFe(CpGaMe_2)]_2$, **1**, 509
$Ga_2LiC_{20}H_{45}P_2$
$Bu^tGa(Bu^t_2Ga)(PBu^t_2Li)$, **1**, 523, 524
$Ga_2Re_4C_{18}O_{18}$
$Re_2(CO)_8[GaRe(CO)_5]_2$, **1**, 554
$Ga_2Si_8C_{28}H_{72}$
$[[(TMS)_2HC]_2Ga]_2$, **1**, 537
$Ga_3As_3C_{24}H_{60}$
$[Me_2Ga(AsPr^i_2)]_3$, **1**, 522
$Ga_3As_3Si_6C_{24}H_{66}$
$[Br_2Ga(As(CH_2TMS)_2]_3$, **1**, 522
$Ga_3B_{18}C_4H_{22}$
$Ga(GaC_2B_9H_{11})_2$, **1**, 259
$Ga_3C_6H_{24}N_3$
$[Me_2Ga(NH_2)]_3$, **1**, 521
$Ga_3C_{20}H_{45}Cl_2P_2$
$(Bu^tGaPBu^t)_2GaCl_2Bu^t$, **1**, 522
$Ga_3C_{24}H_{57}O_3$
$[Bu^t_2Ga(OH)]_3$, **1**, 522, 527
$Ga_3C_{24}H_{60}N_3$
$[Bu^t_2Ga(NH_2)]_3$, **1**, 521
$Ga_3C_{24}H_{60}P_3$
$[Bu^t_2Ga(PH_2)]_3$, **1**, 522
$Ga_3C_{28}H_{63}P_2$
$Bu^tGa(PBu^t)_2(GaBu^t_2)_2$, **1**, 522–4
$Ga_3C_{51}H_{63}Cl_3$
$[(Bu^tCH_2)ClGa(PPh_2)]_3$, **1**, 522
$Ga_3C_{90}H_{84}P_3$
$[(Ph_3C_6H_2)GaP(C_6H_{11})]_3$, **1**, 522
$Ga_3Sb_3C_{24}H_{54}Cl_6$
$[Cl_2Ga(SbBu^t_2)]_3$, **1**, 522
$Ga_3Sb_3C_{30}H_{72}$
$[Me_2Ga(SbBu^t_2)]_3$, **1**, 522
$Ga_3Si_8C_{24}H_{72}$
$[[(TMS)_2HC]_2Ga]_2$, **1**, 538
$Ga_4C_{16}H_{36}S_4$
$[Bu^tGa(S)]_4$, **1**, 527
$Ga_4C_{16}H_{36}Se_4$
$[Bu^tGa(Se)]_4$, **1**, 527
$Ga_4C_{48}H_{108}S_6$
$[Bu^t_2Ga(SSBu^t)(SBu^t)GaBu^t_2]_2$, **1**, 527
$Ga_4C_{95}H_{145}P_5$
$Ga_4(Pr^i_3C_6H_2)_3$(P-(adamantyl))$_4$PH-(adamantyl), **1**, 522
$Ga_4Re_8C_{32}O_{32}$
$Re_4(CO)_{12}[GaRe(CO)_5]_4$, **1**, 554
$Ga_4Tl_4C_{54}H_{72}Br_{16}$
$(C_6H_3Me_3)_6Tl_4(GaBr_4)_4$, **1**, 536
$Ga_5As_7Si_6C_{66}H_{102}$
$[(PhAsH)(TMS\text{-}CH_2)_2Ga\text{-}(PhAs)_6[(TMS\text{-}CH_2)\text{-}Ga]_4]$, **1**, 522, 523
$Ga_5C_{20}H_{50}Cl_7O_5$
$[(Et_2O)_2ClGa]Ga[GaCl_2\text{-}(Et_2O)]_3$, **1**, 539
$GdC_4H_{22}B_9$
$Gd(C_2B_9H_{11})_2$, **1**, 417
$GdC_{14}H_{14}ClO_2$
$Gd(C_6H_4OMe)_2Cl$, **4**, 21
$GdC_{21}H_{31}Cl_2O_3$
$(C_9H_7)GdCl_2(THF)_3$, **4**, 63
$GdC_{21}H_{33}$
$Gd(C_7H_{11})_3$, **4**, 96
$GdC_{22}H_{37}Cl_2O_4$
$PhGdCl_2(THF)_4$, **4**, 22, 128
$GdC_{36}H_{60}$
$Gd(Bu^t_3C_6H_3)_2$, **4**, 4
$GdC_{40}H_{40}Cl_4$
$[Cp_2Gd(Cl)]_4$, **4**, 36
$Gd_2C_{26}N_2O_2$
$[Cp_2Gd(ONCMe_2)]_2$, **4**, 41
$GeB_4Si_2C_8H_{22}$
$Ge(TMS)_2C_2B_4H_4$, **1**, 260
$GeB_8Si_4C_{16}H_{44}$
$[(TMS)_2C_2B_4H_4]_2Ge$, **1**, 260
$GeB_9C_{23}H_{33}Cl$
$Ph_3PMe[ClGe(B_9H_9C_2Me_2)]$, **1**, 260
$GeB_{10}C_2H_{18}$
$Me_2GeB_{10}H_{12}$, **1**, 259
$GeB_{11}H_{11}$
$[B_{11}H_{11}Ge]^{2-}$, **1**, 259
$GeB_{18}C_{46}H_{66}$
$(Ph_3PMe)_2[Ge(B_9H_9C_2Me_2)_2]$, **1**, 260
$GeBr_2$
$GeBr_2$, **2**, 190
$GeCH_4$
$H_2Ge{=}CH_2$, **2**, 193
GeC_2H_6
Me_2Ge, **2**, 147, 166, 173, 191
$GeC_2H_6Cl_2$
Me_2GeCl_2, **2**, 145
$GeC_2H_6N_3$
Me_2GeN_3, **2**, 191
GeC_2H_6O
Me_2GeO, **2**, 204
GeC_2H_6S
$Me_2Ge{=}S$, **2**, 198
GeC_3H_{10}
Me_3GeH, **2**, 157
GeC_4H_{12}
Me_4Ge, **2**, 188
$GeC_5H_{14}O$
$Et_2GeH(OMe)$, **2**, 193
GeC_6H_{14}
$Me_2Ge(CH_2)_4$, **2**, 190
GeC_7H_{16}
$Me_2Ge(CH_2)_5$, **2**, 190
GeC_8H_{18}
$Me_2Ge(CH_2)_6$, **2**, 190
$GeC_{10}H_{16}ClNO_2$
$[(ClEt_2Ge)NH](MeOOC)\text{-}[C_4H_2S]$, **2**, 148
$GeC_{16}H_{34}N_2$
$Et_2[Ge(CH_2)_4C][(CH_2)_2N]\text{-}(CH_2)_3NMe_2$, **2**, 207
$GeC_{18}H_{15}$
Ph_3Ge, **2**, 189
$GeC_{18}H_{15}N_3$
Ph_3GeN_3, **2**, 164
$GeC_{18}H_{22}$
$(Mes)_2Ge$, **2**, 192
$GeC_{18}H_{25}N$
Et_3GeNPh_2, **2**, 166
$GeC_{20}H_{30}$
Cp^*_2Ge, **2**, 149
$GeC_{24}H_{20}$
$GePh_4$, **1**, 83
$GeC_{27}H_{18}N_3$
$(Ph_3Ge)(CN)_3(C_6H_3)$, **2**, 189
$GeC_{27}H_{35}N$
$(Mes)_3GeNH_2$, **2**, 158, 163
$GeC_{28}H_{44}$
$[Bu^t_3C_6H_2]Cp^*Ge$, **2**, 192
$GeC_{35}H_{25}Cl_3$
$Ph_5C_5GeCl_3$, **2**, 194
$GeC_{36}H_{58}$
$[Bu^t_3C_6H_2]_2Ge$, **2**, 192
$GeC_{36}H_{58}S$
$[Bu_{3'}C_6H_2]_2GeS$, **2**, 168
$GeC_{36}H_{58}Se_2$
$[(Bu^t_3C_6H_2)Se]_2Ge$, **2**, 191
$GeC_{50}H_{42}P_4$
$Ge[CH(PPh_2)_2]_2$, **2**, 192
$GeCl_2$
$GeCl_2$, **2**, 190
$GeCoFe_3C_{14}O_{14}$
$[GeCoFe_3(CO)_{14}]^-$, **7**, 30; **10**, 215
$GeCo_2Mo_2C_{21}H_{10}O_{11}$

$Mo_2Co_2(Ge)(CO)_{11}Cp_2$, **10**, 215
$GeCo_3C_{10}H_3O_9$
$Co_3(CO)_9(GeMe)$, **8**, 428
$GeCo_4C_{11}H_3O_{10}$
$Co_4(CO)_{10}(GeMe)$, **8**, 428
$GeCo_5C_{16}O_{16}$
$[Co_5(CO)_{16}Ge]^-$, **8**, 428
$GeCo_7C_{20}O_{20}$
$[Co_3(CO)_9GeCo_4(CO)_{11}]^-$, **8**, 428
GeF_2
GeF_2, **2**, 145, 190
$GeFeSiC_9H_{14}O_4$
$(CO)_4Fe(Me_2SiCH_2GeMe_2)$, **7**, 30
$GeFe_2C_6O_6Se_2$
$[Fe_2(CO)_6Se_2]Ge$, **7**, 74
$GeFe_2SiC_{13}H_{14}O_8$
$Fe_2(CO)_8(Me_2SiCH_2GeMe_2)$, **7**, 29
$GeFe_3MnC_{19}H_5O_{14}$
$GeFe_3Mn(CO)_{14}Cp$, **7**, 30
$GeFe_3MnC_{20}H_7O_{14}$
$MnFe_3(Ge)(CO)_{14}Cp'$, **10**, 215
$GeFe_4C_{14}O_{14}$
$[Fe_3(CO)_{10}[GeFe(CO)_4]]^{2-}$, **7**, 30
$GeFe_4C_{16}O_{16}$
$Ge[Fe_2(CO)_8]_2$, **7**, 30
$GeHCl_3$
Cl_3GeH, **2**, 151
GeH_4
GeH_4, **2**, 151
GeI_2
GeI_2, **2**, 190
$GeKH_3$
H_3GeK, **2**, 175
$GeLiC_6H_{15}$
Et_3GeLi, **2**, 176
$GeLiSiC_7H_{19}$
$(TMS)Et_2GeLi$, **2**, 176
$GeLi_2C_{12}H_{10}$
Ph_2GeLi_2, **2**, 175
$GeMnC_8H_{11}O_2$
$MeCpMn(CO)_2GeH_3^-$, **10**, 337
$GeMn_2C_{14}H_{10}Cl_2O_4$
$Mn_2(GeCl_2)(CO)_4Cp_2$, **6**, 114
$GeMn_2C_{14}H_{10}O_4$
$[Mn(CO)_2Cp]_2Ge$, **6**, 114
$GeMoC_8H_5IO_3$
$Mo(GeI)(CO)_3Cp$, **5**, 409
$GeNi_2C_{22}H_{10}F_{10}$
$Ni_2Ge(C_6F_5)_2Cp_2$, **9**, 168
$GeOs_2C_8Cl_4O_8$
$Os_2Cl(GeCl_3)(CO)_8$, **7**, 302
$GeSiC_{10}H_{24}N_2$
$Bu^t{}_2Me_2[\overline{SiNGeN}]$, **2**, 190
$GeSiC_{21}H_{24}$
Ph_3GeTMS, **2**, 23
$Ph_3SiGeMe_3$, **2**, 23
$GeSiC_{25}H_{24}$
$MePhHSiGePh_3$, **2**, 178
$GeSiH_4$
$H_2Ge{=}SiH_2$, **2**, 193
$GeSiSnC_{11}H_{30}$
$Bu^tSiMe_2GeMe_2SnMe_3$, **2**, 285
$GeSiSnC_{25}H_{36}$
$Bu^tSiMe_2GeMe_2SnPh_3$, **2**, 285
$GeSi_2C_{56}H_{70}$
$Ge[CH_2Si(Mes)_3]_2$, **2**, 192
$GeSi_3C_{20}H_{42}$
$(TMS)_3CGeCp^*$, **2**, 192
$GeSi_4C_{12}H_{36}N_2$
$Ge[N(TMS)_2]_2$, **2**, 193
$GeSi_4C_{12}H_{38}N_2O_2$
$[(TMS)_2N]_2Ge(OH)_2$, **2**, 163
$GeSi_5C_{17}H_{46}$
$(TMS)_3CGeCH(TMS)_2$, **2**, 192
$GeSi_6C_{42}H_{85}S$
$[[(TMS)_2CH]_3Ph][Pr^i{}_3C_6H_2]$-Ge=S, **2**, 205
$GeSnC_{21}H_{24}$
$Ph_3SnGeMe_3$, **2**, 284
$GeSnC_{48}H_{40}$
$Ph_3SnGePh_2SnPh_3$, **2**, 284
$GeTiC_4H_{11}Cl_3$
$Me_3GeCH_2TiCl_3$, **12**, 579
$GeTiC_{12}H_{10}Cl_2F_6$
$TiCl[Ge(CF_3)_2Cl]Cp_2$, **4**, 397
$GeTiC_{29}H_{28}$
$TiMe(GePh_3)Cp_2$, **4**, 397
$GeTiC_{30}H_{28}O$
$Ti(COMe)(GePh_3)Cp_2$, **4**, 397
$Ge_2C_4H_{12}$
$Me_2Ge{=}GeMe_2$, **2**, 195
$Ge_2C_4H_{12}Cl_2$
$Me_4Ge_2Cl_2$, **2**, 194
$Ge_2C_{12}H_{22}O_5$
$(MeO_2CCH{=}CHGeMe_2)_2O$, **2**, 196
$Ge_2C_{12}H_{30}O$
$(Et_3Ge)_2O$, **2**, 166
$Ge_2C_{18}H_{22}$
$(Mes)_2Ge$, **2**, 169
$Ge_2C_{24}H_{20}$
Ph_4Ge_2, **2**, 156
$Ge_2C_{36}H_{30}$
Ge_2Ph_6, **1**, 83
$Ge_2C_{36}H_{44}$
$(Mes)_2Ge{=}Ge(Mes)_2$, **2**, 192
$Ge_2C_{36}H_{44}S_2$
$[(Mes)_2GeS]_2$, **2**, 170
$Ge_2C_{40}H_{52}$
$[Et_2C_6H_3]_4Ge_2$, **2**, 169
$Ge_2C_{54}H_{66}$
$(Mes)_3GeGe(Mes)_3$, **2**, 187
$Ge_2C_{56}H_{44}S_2$
$[(Mes)_2GeS]_2$, **2**, 198
$Ge_2Co_2Fe_2C_{17}H_6O_{15}$
$Fe_2(CO)_7[GeMe[Co(CO)_4]]_2$, **7**, 30
$Ge_2Co_4C_{13}H_6O_{11}$
$Co_4Ge_2C_{13}H_6O_{11}$, **8**, 448
$Ge_2Co_4C_{16}H_6O_{14}$
$Co_2[Ge(Me)[Co(CO)_4]]_2(CO)_6$, **8**, 448
$Ge_2Co_4Fe_2C_{22}O_{22}$
$[Co_2(CO)_7]GeFe_2(CO)_8Ge[Co_2$-$(CO)_7]$, **7**, 30
$Ge_2Co_5C_{16}H_3O_{15}$
$Co_4(CO)_{11}(GeMe)[GeCo$-$(CO)_4]$, **8**, 448
$Ge_2Co_5Fe_2C_{22}O_{22}$
$[Fe_2Co_5(Ge)_2(CO)_{22}]^-$, **7**, 30; **10**, 215
$Ge_2Co_6C_{19}O_{19}$
$Co_4(CO)_{11}[GeCo(CO)_4]_2$, **8**, 448
$Ge_2Co_7C_{21}O_{21}$
$[Co(CO)_3[GeCo_3(CO)_9]_2]^-$, **8**, 426, 428
$Ge_2Fe_2C_8H_4O_8$
$[Fe(CO)_4(GeH_2)]_2$, **7**, 30
$Ge_2Fe_2MnC_{32}H_{54}O_{14}$
$(CO)_4FeGe(OBu^t)_3Mn$-$(OBu^t)_3GeFe(CO)_4$, **7**, 29
$Ge_2Fe_3Re_2C_{19}O_{19}$
$Fe_3(CO)_9[GeRe(CO)_5]_2$, **7**, 30
$Ge_2Fe_5C_{23}H_{10}O_{13}$
$Fe_3(CO)_9[GeFe(CO)_2Cp]_2$, **7**, 30
Ge_2H_4
$H_2Ge{=}GeH_2$, **2**, 193
$Ge_2HgC_{42}H_{30}F_{18}P_2$
$(CF_3)_3Ge(PPh_3)_2HgGe(CF_3)_3$, **10**, 343
$Ge_2HgMnC_{54}H_{26}F_{30}O_2$
$[Mn(THF)_2][HgCp_2[Ge$-$(C_6F_5)_3]_2]$, **6**, 138
$Ge_2HgPtC_{72}H_{30}F_{30}P_2$
$Pt(PPh_3)_2HgGe(C_6F_5)_3Ge$-$(C_6F_5)_3$, **10**, 342
$Ge_2HgPt_2C_{78}H_{60}F_{18}P_4$
$[(CF_3)_3Ge(PPh_3)_2Pt]_2Hg$, **10**, 343
$Ge_2RuC_7H_6Cl_6O$
$Ru(GeCl_3)_2(CO)(C_6H_6)$, **7**, 586
$Ge_2Si_8C_{28}H_{56}$
$[[(TMS)_2CH]_2Ge]_2$, **2**, 192
$Ge_3C_9H_{27}N$
$(Me_3Ge)_3N$, **2**, 158
$Ge_3C_{48}H_{40}$
Ge_3Ph_8, **1**, 83
$Ge_3C_{54}H_{66}$
$[(Mes)_2Ge]_3$, **2**, 169, 192
Ge_3H_9P
$(H_3Ge)_3P$, **2**, 158
$Ge_4C_{48}H_{40}$
Ph_8Ge_4, **2**, 156
$Ge_4C_{60}H_{50}$
Ge_4Ph_{10}, **1**, 83
$Ge_4HgNi_2C_{82}H_{70}$
$[(Ph_3GeHg)(Ph_3Ge)CpNi]_2Hg$, **10**, 334, 335
$Ge_5C_{10}H_{30}$
$[GeMe_2]_5$, **2**, 191
$Ge_6C_{12}H_{36}$
$[GeMe_2]_6$, **2**, 192

H

$HfBC_{39}H_{39}N_2$
[HfMe(MeCN)$_2$Cp$_2$][BPh$_4$], **4**, 602
HfC_6O_6
[Hf(CO)$_6$]$^{2-}$, **4**, 436
$HfC_9H_5O_4$
[Hf(Cp)(CO)$_4$]$^-$, **4**, 436
$HfC_9H_{15}Cl_3O_2$
HfCl$_3$Cp(DME), **4**, 545
$HfC_{10}H_{10}BrNO_3$
HfBr(NO$_3$)Cp$_2$, **4**, 537
$HfC_{10}H_{10}Cl_2O_4$
[Hf(Cl)Cp$_2$][ClO$_4$], **4**, 614, 621
$HfC_{10}H_{10}Cl_2O_8$
[HfCp$_2$][ClO$_4$]$_2$, **4**, 621
$HfC_{13}H_{16}Cl_2$
HfCl$_2$Cp[C$_5$H$_2$(Me)$_3$], **4**, 484
$HfC_{13}H_{24}$
HfMe$_3$Cp*, **4**, 546
$HfC_{14}H_{10}N_2S_2$
Hf[S$_2$C$_2$(CN)$_2$]Cp$_2$, **4**, 530
$HfC_{14}H_{16}$
Hf(C$_4$H$_6$)Cp$_2$, **4**, 446
Hf(C$_7$H$_8$)$_2$, **4**, 434
$HfC_{15}H_{23}P$
Hf(PMe$_3$)(C$_2$H$_4$)Cp$_2$, **4**, 442
$HfC_{15}H_{27}O_4P_3$
Hf(CO)$_4$[MeC(CH$_2$PMe$_2$)$_3$], **4**, 436
$HfC_{16}H_{28}P_2$
Hf(PMe$_3$)$_2$Cp$_2$, **4**, 441
$HfC_{17}H_{16}S_2$
Hf(S$_2$C$_6$H$_3$Me)Cp$_2$, **4**, 530
$HfC_{17}H_{25}P$
Hf(PhMe)$_2$(PMe$_3$), **4**, 433, 434
$HfC_{18}H_{24}$
Hf(C$_8$H$_8$)Cp*H, **4**, 474, 628
$HfC_{18}H_{26}Cl_2$
HfCl$_2$[C$_5$H$_2$(Me)$_3$]Cp*, **4**, 484
$HfC_{18}H_{28}$
Cp*Hf(MeC$_3$H$_4$)(C$_4$H$_6$), **4**, 565
$HfC_{18}H_{30}P_2$
Hf(PEt$_2$)$_2$Cp$_2$, **4**, 514
Hf(PHBu^t)$_2$Cp$_2$, **4**, 514
$HfC_{19}H_{27}P$
Hf(C$_6$H$_8$)(PMe$_3$)Cp$_2$, **4**, 443
$HfC_{20}H_{14}O_2$
Hf(CO)$_2$(C$_9$H$_7$)$_2$, **4**, 440, 566
$HfC_{20}H_{32}O$
HfH(OH)Cp*$_2$, **4**, 509, 527
$HfC_{20}H_{38}N_3S_6$
CpHf(Me$_2$dtc)$_3$, **4**, 548
$HfC_{21}H_{25}N$
Hf(C$_8$H$_8$)$_2$(CNBu^t), **4**, 628
$HfC_{21}H_{34}$
HfH(Me)Cp*$_2$, **4**, 505
$HfC_{21}H_{34}O$
HfH(OMe)Cp*$_2$, **4**, 521
$HfC_{21}H_{35}N$
HfH(NHMe)Cp*$_2$, **4**, 508
$HfC_{22}H_{30}O_2$
Hf(CO)$_2$Cp*$_2$, **4**, 440, 450
$HfC_{22}H_{36}$
Cp*Hf(MeC$_3$H$_4$)$_3$, **4**, 565
$HfC_{26}H_{36}O$
HfH(OPh)Cp*$_2$, **4**, 521, 527
Hf(OH)(Ph)Cp*$_2$, **4**, 521, 527
$HfC_{26}H_{37}P$
HfH(PHPh)Cp*$_2$, **4**, 514
$HfC_{26}H_{40}O_2$
Hf[OCHC(CH$_2$CHMe$_2$)O]-Cp*$_2$, **4**, 524
$HfC_{27}H_{42}O$
Hf(OCHCHCBu^{t}H)Cp*$_2$, **4**, 523
$HfC_{28}H_{15}F_{15}O_3$
Hf(OC$_6$F$_5$)$_3$Cp*, **4**, 546
$HfC_{28}H_{42}O_2$
Hf(OCHCHCBu^{t}CHO)Cp*$_2$, **4**, 523
$HfC_{30}H_{36}Se_2$
Hf(SePh)$_2$(C$_5$H$_4$Bu^t)$_2$, **4**, 529
$HfC_{33}H_{41}O_2P$
HfH(O$_2$CPPh$_2$)Cp*$_2$, **4**, 516
$HfC_{34}H_{41}O_4P$
Hf[OC(O)H][OC(O)PPh$_2$]Cp*$_2$, **4**, 516
$HfC_{34}H_{54}P_2$
Hf(PCy$_2$)$_2$Cp$_2$, **4**, 514
$HfC_{34}H_{58}N_2$
Hf(C$_5$H$_2$NBu$_3$^{t})$_2$, **4**, 434
$HfC_{36}H_{60}$
Hf(C$_6$H$_3$Bu$_3$^{t})$_2$, **4**, 433
$HfC_{42}H_{44}N_4$
Hf[N(Me$_2$C$_6$H$_3$)-(methylpyridylvinyl)]$_2$-Cp$_2$, **4**, 508
$HfCaC_{40}H_{60}Cl_2$
Cp$_2$*CaCl(HfClCp$_2$*), **1**, 115
$HfFeC_{22}H_{31}O_4P_2$
Cp$_2$HHf(PEt$_2$)$_2$Fe(CO)$_4$, **4**, 516
$HfMoC_{22}H_{30}O_4P_2$
HfCp$_2$(PEt$_2$)$_2$Mo(CO)$_4$, **10**, 27
$HfSiC_{21}H_{37}Cl$
HfCl(SiH$_2$Ph)CpCp*, **4**, 504
$HfSi_2C_{11}H_{21}Cl_3$
HfCl$_3$[(TMS)$_2$C$_5$H$_3$], **4**, 546
$HfSi_2C_{16}H_{26}Cl_2$
HfCl$_2$(C$_5$H$_4$TMS)$_2$, **4**, 484
$HfSi_4C_{22}H_{46}P_2$
Hf[P(TMS)$_2$]$_2$Cp$_2$, **4**, 514
$HfSi_4C_{24}H_{47}Cl$
HfCl[Si(TMS)$_3$]CpCp*, **4**, 504
$HfSi_4C_{24}H_{47}Cl_2N$
HfCl$_2$[Si(TMS)$_3$](py)Cp*, **4**, 545
$HfSi_4C_{28}H_{48}$
Hf[C$_8$H$_6$(TMS)$_2$]$_2$, **4**, 627
$Hf_2C_{36}H_{48}P_2$
[Hf(PMe$_3$)Cp$_2$]$_2$Cp$_2$, **4**, 441
$Hf_2C_{50}H_{52}NP_4$
[Cp$_2$Hf(PBzCH$_2$CH$_2$PPh)]$_2$N, **4**, 517
$Hf_2Fe_2C_{28}H_{20}O_8$
[Cp$_2$Hf[Fe(CO)$_4$]]$_2$, **4**, 535, 536
$Hf_2Si_2C_{36}H_{48}Cl_2$
[Cp*Cp(Cl)Hf]$_2$[(SiH$_2$)$_2$C$_6$H$_4$], **4**, 504
$HgAuC_{26}H_{24}P_2S_2$
[AuHg[CH$_2$P(S)Ph$_2$]$_2$]$^{2+}$, **10**, 71
[Au(SPPh$_2$CH$_2$)$_2$Hg]$^+$, **10**, 71
$HgAu_8PtC_{126}H_{105}Cl_3P_7$
Pt(AuPPh$_3$)$_7$AuClHgCl$_2$$^+$, **10**, 344
$HgAu_8PtC_{126}H_{105}N_3O_9P_7$
Pt(AuPPh$_3$)$_7$Au(NO$_3$)Hg-(NO$_3$)$_2$$^+$, **10**, 344
HgB_2F_8
Hg(BF$_4$)$_2$, **11**, 412, 413, 416, 419, 421
$HgCH_2Br_2$
BrCH$_2$HgBr, **3**, 153
$HgCH_2Cl_2$
ClCH$_2$HgCl, **3**, 153
$HgCH_3Cl$
MeHgCl, **3**, 137
$HgCH_4$
MeHgH, **3**, 166
HgC_2F_6
(CF$_3$)$_2$Hg, **3**, 143
HgC_2H_5D
EtHgD, **3**, 166
HgC_2H_6
EtHgH, **3**, 166
Me$_2$Hg, **3**, 155
$HgC_2H_6O_6S_2$
Hg(MeSO$_3$)$_2$, **11**, 411, 429
$HgC_2N_2O_2$
Hg(CNO)$_2$, **3**, 155
$HgC_2N_2S_2$
Hg(SCN)$_2$, **11**, 400
$HgC_2N_6O_{12}$
Hg[C(NO$_2$)$_3$]$_2$, **11**, 392, 427
HgC_2O_2
[Hg(CO)$_2$]$^{2+}$, **3**, 154
HgC_3H_5IO
MeCOCH$_2$HgI, **3**, 139
$HgC_4F_6O_4$
Hg(O$_2$CCF$_3$)$_2$, **11**, 411, 413, 414, 417, 418, 425
HgC_4H_6
Hg(CH$_2$=CH)$_2$, **11**, 400

$HgC_4H_6O_2$
$Hg(OAc)_2$, **12**, 135, 136, 142, 143
$HgC_4H_6O_4$
$Hg(OAc)_2$, **11**, 392, 399, 411, 417, 418, 420, 425
HgC_6DF_5
C_6F_5HgD, **3**, 166
HgC_6HF_5
C_6F_5HgH, **3**, 166
HgC_6H_5D
PhHgD, **3**, 166
HgC_6H_6
PhHgH, **3**, 166
$HgC_6H_{10}O_2$
$(MeCOCH_2)_2Hg$, **3**, 139
HgC_7H_6NO
$HgC_6H_4CONH_2$, **3**, 137
$HgC_7H_{13}BrO$
$Bu^tCOCHMeHgBr$, **3**, 140
$HgC_8F_{14}O_4$
$Hg(O_2CCF_2CF_2CF_3)_2$, **11**, 413
HgC_8H_8O
PhHgOAc, **12**, 141
$HgC_8H_8O_2$
PhHgOAc, **12**, 133
HgC_8H_{18}
Bu^t_2Hg, **11**, 427
$HgC_{10}H_{10}$
Cp_2Hg, **3**, 160
$HgC_{10}H_{10}O_4$
$AcOC_6H_4HgOAc$, **3**, 160
$HgC_{10}H_{11}F_6NO_6S_2$
$Hg(O_3SCF_3)_2 \cdot PhNMe_2$, **11**, 427
$HgC_{10}H_{15}Cl$
Cp*HgCl, **3**, 160
$HgC_{11}H_5Cl_5$
Cl_5C_5HgPh, **3**, 161
$HgC_{11}H_8ClN$
$[(C_5H_4N)C_6H_4]HgCl$, **3**, 160
$HgC_{12}F_8$
$(C_6F_4C_6F_4)Hg$, **3**, 138
$HgC_{12}H_8$
$(C_6H_4C_6H_4)Hg$, **3**, 138
$HgC_{12}H_9ClN_2$
$ClHgC_6H_4NNPh$, **3**, 160
$HgC_{12}H_{10}$
Ph_2Hg, **11**, 410
$HgC_{14}H_{16}O_4S_2$
$Hg(O_2SPhMe)_2$, **11**, 429
$HgC_{14}H_{16}O_6S_2$
$Hg(O_3SPhMe)_2$, **11**, 429
$HgC_{14}H_{26}O_2$
$(Bu^tCOCHMe)_2Hg$, **3**, 140
$HgC_{16}H_{18}F_6O_4$
$Me_6C_6Hg(TFA)_2$, **3**, 141
$HgC_{16}H_{20}N_2$
$[(Me_2N)C_6H_4]_2Hg$, **3**, 159
$HgC_{17}H_{11}ClN_2O_2$
$ClHgC_{10}H_6NCHC_6H_4NO_2$, **3**, 160
$HgC_{20}H_{30}$
$Cp*_2Hg$, **3**, 160
$HgC_{25}H_{18}F_6O_6S$
$C_6H_4(HgOCOCF_3)_2SC(C_6H_4O$-$Me)_2$, **3**, 163
$HgCl_2O_8$
$Hg(ClO_4)_2$, **11**, 414
$HgCoOs_3C_{19}H_9O_{13}$
$(C_2Bu^t)(CO)_9Os_3[HgCo(CO)_4]$, **10**, 340
$HgCo_2C_8O_8$
$[(CO)_4Co]_2Hg$, **10**, 330
$HgCo_2C_{12}H_8O_4$
$Co(C_4H_4)(CO)_2HgCo(C_4H_4)$, **10**, 63
$HgCo_2Fe_4C_{28}H_{16}O_{16}$
$[(COMe)Cp(CO)_7CoFe_2]_2(Hg)$, **10**, 332, 339
$HgCo_3C_{12}O_{12}$
$Hg(Co(CO)_4)_3$, **8**, 6
$HgCo_4RuC_{16}O_{16}$
$RuCo_3(CO)_{12}(Hg)Co(CO)_4$, **10**, 341
$HgCo_6Ru_2C_{24}O_{24}$
$[RuCo_3(CO)_{12}]_2(Hg)$, **10**, 341
$HgCrMoC_{49}H_{35}NO_8P_2$
$(PPN)[(CO)_5CrHgMoCp(CO)_3]$, **10**, 336
$HgCr_2C_{18}H_{10}O_6$
$Hg[PhCr(CO)_3]_2$, **5**, 490
$Ph_2Hg[Cr(CO)_3]_2$, **12**, 1019
$HgFeC_4O_3S$
$HgFe(CO)_3(CS)$, **7**, 16
$HgFeC_7H_5ClO_5S$
$Fe(CO)_4(HgCl)[C(S)OEt]$, **7**, 16
$HgFeC_{10}H_9Cl$
$CpFeC_5H_4HgCl$, **12**, 134, 135
$HgFeMnSi_2C_{54}H_{43}O_5$
$Fe(CO)_3PPh_3SiMePh_2Hg$-$MnCpSiPh_3(CO)_2$, **10**, 339
$HgFeMn_2Si_2C_{60}H_{51}O_5$
$MeCpMn(CO)_2SiPh_3MnHgFe$-$(CO)_3PPh_3SiMePh_2$, **10**, 337
$HgFeRuC_{20}H_{18}$
$[Cp(C_5H_4)FeCp(C_5H_4)Ru$-$(Hg)]^{2+}$, **7**, 618
$HgFe_2Si_2C_{38}H_{44}O_6P_2$
$[Fe(SiPh_2Me)(CO)_3$-$(PMe_3)]_2Hg$, **10**, 54
$HgFe_4Sn_2C_{16}Cl_4O_{16}$
$[(CO)_4FeSnCl_2Fe(CO)_4]_2Hg^{2-}$, **10**, 339
$HgFe_6C_{22}H_6O_{20}$
$[Fe_3(COMe)(CO)_9]_2(Hg)$, **10**, 339
$HgGe_2C_{42}H_{30}F_{18}P_2$
$(CF_3)_3Ge(PPh_3)_2HgGe(CF_3)_3$, **10**, 343
$HgGe_2MnC_{54}H_{26}F_{30}O_2$
$[Mn(THF)_2][HgCp_2[Ge$-$(C_6F_5)_3]_2]$, **6**, 138
$HgGe_2PtC_{72}H_{30}F_{30}P_2$
$Pt(PPh_3)_2HgGe(C_6F_5)_3Ge$-$(C_6F_5)_3$, **10**, 342
$HgGe_2Pt_2C_{78}H_{60}F_{18}P_4$
$[(CF_3)_3Ge(PPh_3)_2Pt]_2Hg$, **10**, 343
$HgGe_4Ni_2C_{82}H_{70}$
$[(Ph_3GeHg)(Ph_3Ge)CpNi]_2Hg$, **10**, 334, 335
$HgMnPtC_{29}H_{30}Cl_5O_5P_2$
$[(PPh_3)_2C_6Cl_5Pt]HgMn(CO)_5$, **10**, 343
$HgMnSiC_{21}H_{21}BrO_2$
$MeCp(CO)_2SiMePh_2MnHgBr$, **10**, 337
$HgMn_2C_{20}H_{10}ClO_8P$
$(PPh_2)(CO)_8Mn_2(HgCl)$, **10**, 337
$HgMn_2C_{26}H_{15}O_8P$
$(PPh_2)(CO)_8Mn_2(HgPh)$, **10**, 337
$HgMn_2Si_2C_{42}H_{42}O_4$
$[MeCp(CO)_2SiMePh_2Mn]_2Hg$, **10**, 337
$HgMn_4C_{40}H_{20}O_{16}P_2$
$[(PPh_2)(CO)_8Mn_2]_2(Hg)$, **10**, 337
$HgMoC_8H_5ClO_3$
$[Cp(CO)_3Mo]HgCl$, **10**, 336
$HgMoC_8H_5O_3$
$HgMoCp(CO)_3$, **10**, 334
$HgMoC_{14}H_5Cl_5O_3$
$(C_6Cl_5)HgMoCp(CO)_3$, **10**, 336
$HgMoPtC_{50}H_{35}Cl_5O_3P_2$
$(PPh_3)_2PtHgMoCp(CO)_3$-(C_6Cl_5), **10**, 336
$Pt(C_6Cl_5)(PPh_3)_2[HgMo$-$(CO)_3Cp]$, **9**, 467
$HgMoRu_3C_{23}H_{14}O_{12}$
$(C_2Bu^t)(CO)_9Ru_3(Hg)MoCp$-$(CO)_3$, **10**, 339
$Ru_3(CO)_9(CCBu^t)[Hg[Mo$-$(CO)_3Cp]]$, **5**, 409
$HgMo_2C_{16}H_{10}O_6$
$[Cp(CO)_3Mo]_2Hg$, **10**, 336
$HgMo_2C_{38}H_{32}Cl_2O_3P_2$
$ClHgMoCp(CO)(PPh_2CH_2P$-$Ph_2)MoCp(CO)_2Cl$, **10**, 336
$HgMo_2C_{50}H_{40}O_4P_2$
$(CpMo(CO)_2PPh_3)_2Hg$, **10**, 330
HgN_2O_6
$Hg(NO_3)_2$, **11**, 417, 424, 429
$HgNi_4C_{70}H_{74}N_{10}P_4$
$[Ni_2(CNMe)(CNMe)_4(dppm)]_2$-$(Hg)^{2+}$, **10**, 334
$HgNi_5C_{70}H_{74}Cl_4N_5P_4$
$[Ni_2(CNMe)(CNMe)_4(Ph_2P$-$CH_2PPh_2)]_2(Hg)NiCl_4$, **10**, 342
$HgNi_5C_{70}H_{74}Cl_4N_{10}P_4$
$[Ni_4Hg(CNMe)_{10}(dppm)_2]$-$[NiCl_4]$, **9**, 22
$HgOsC_{40}H_{35}ClIN_3O_3P_2$
$OsCl(HgI)[C(N_2)CO_2Et](NO)$-$(PPh_3)_2$, **7**, 389
$HgOs_2C_{18}H_2O_{18}S_2$
$[(H)(S)(CO)_9Os]_2Hg$, **10**, 340
$HgOs_3C_{15}H_9IO_9$
$(C_2Bu^t)(CO)_9Os_3(HgI)$, **10**, 340
$HgOs_6C_{18}H_2O_{18}S_2$
$[(H)(S)(CO)_9Os_3]_2(Hg)$, **10**, 332

$HgOs_{10}C_{25}ClO_{24}$
$Os_{10}(CO)_{24}C(HgCl)^-$, **10**, 332
$HgOs_{10}C_{27}F_3O_{26}$
$Os_{10}C(CO)_{24}(HgO_2CCF_3)^-$, **10**, 340
$HgOs_{18}C_{44}HO_{42}$
$H[Os_9C(CO)_{21}]_2(Hg)^{3-}$, **10**, 329
$HgOs_{20}C_{50}O_{48}$
$[Os_{10}(CO)_{24}C]_2(Hg)^{2-}$, **10**, 332
$HgPdC_{66}H_{54}Cl_2P_4$
$[(PhC{\equiv}C)_2Pd](dppm)_2(HgCl_2)$, **9**, 230
$HgPd_6C_{42}H_{90}O_6P_6$
$Pd_6Hg(CO)_6(PEt_3)_6$, **9**, 207
$HgPtC_{10}H_{11}BrNO_2$
$(Me_2NCH_2C_6H_3)PtHg(O_2CBr)$, **10**, 342, 343
$HgPtC_{22}H_{30}N_2O_4$
$(Me_2NCH_2C_6H_4)_2PtHg(O_2CMe)_2$, **10**, 342, 343
$HgPt_3C_{76}H_{66}OP_6$
$(CO)(PPh_2CH_2PPh_2)_3Pt_3(Hg)^{2+}$, **10**, 343
$HgPt_6C_{54}H_{54}N_6$
$(Hg)[Pt_3(Me_2C_6H_3NC)_3(Me_2C_6H_3NC)_3]_2$, **10**, 343
$HgPt_6C_{108}H_{108}N_{12}$
$(Hg)[Pt_3(Me_2C_6H_3NC)_6]_2$, **10**, 327, 329
$HgReC_{24}H_{45}Cl_2P_2$
$ReH_7(PPhPr^i{}_2)_2HgCl_2$, **10**, 48
$HgRe_7C_{22}ClO_{21}$
$Re_7C(CO)_{21}(HgCl)^{2-}$, **10**, 338
$HgRh_2C_{54}H_{51}ClN_2O_{10}P_4$
$Rh_2(Co)_2(H)[(PhO)_2PN(Et)P(OPh)_2]_2(HgCl)$, **10**, 341
$HgRuC_{25}H_{30}O_{10}$
$Cp^*Ru[C_5(HgOAc)_5]$, **3**, 162
$HgRuC_{33}H_{32}Cl_2OP_2$
$Ru[C(CCPh)C(HgCl)Ph]Cl(CO)(PMe_2Ph)_2$, **7**, 411
$HgRuC_{37}H_{32}Cl_3NOP_2$
$Ru(CH_2Cl)(HgCl)Cl(NO)(PPh_3)_2$, **7**, 339
$HgRuC_{40}H_{30}F_3O_3P_2$
$[Ru(HgCF_3)(CO)_3(PPh_3)_2]^+$, **7**, 347
$HgRuC_{40}H_{30}F_6O_2P_2$
$Ru(CF_3)(HgCF_3)(CO)_2(PPh_3)_2$, **7**, 347, 389
$HgRuC_{44}H_{39}Cl_2NOP_2$
$Ru(CH_2HgC_6H_4Me)Cl_2(NO)(PPh_3)_2$, **7**, 339
$HgRuC_{45}H_{35}F_3O_2P_2$
$RuPh(HgCF_3)(CO)_2(PPh_3)_2$, **7**, 397
$HgRu_2C_{17}H_{34}ClN_2O_{13}P_4$
$[Ru_2(HgCl)(CO)_5(etmdp)_2]^+$, **7**, 627
$HgRu_2C_{26}H_{78}P_8$
$Hg[RuMe(PMe_3)_4]_2$, **7**, 381
$HgRu_3C_{15}H_9BrO_9$
$(C_2Bu^t)(CO)_9Ru_3(HgBr)$, **10**, 332, 339
$HgRu_3C_{15}H_9IO_9$
$(C_2Bu^t)(CO)_9Ru_3(HgI)$, **10**, 330–2
$HgRu_3C_{15}H_9O_9$
$C_2Bu^t(CO)_9Ru_3(HgBr)$, **10**, 330
$HgRu_6C_{30}H_{18}O_{18}$
$[(C_2Bu^t)(CO)_9Ru_3]_2[Hg]$, **10**, 327, 330, 332, 339
$HgRu_6C_{47}H_{33}O_{17}P$
$(PPh_3)(C_2Bu^t)(CO)_8Ru_3(Hg)Ru_3(CO)_9(C_2Bu^t)$, **10**, 332
$HgSi_2C_6H_{18}$
$(TMS)_2Hg$, **2**, 3; **3**, 154
$Hg_2AuIrC_{51}H_{44}Cl_6OP_4$
$IrCl(CO)(dppm)(dppmAuCl)Hg(Cl)_2HgCl_2$, **10**, 68
$Hg_2Au_8PtC_{144}H_{120}Cl_2P_8$
$Pt(AuPPh_3)_8(HgCl)_2{}^{2+}$, **10**, 344
$Hg_2Au_8PtC_{144}H_{120}N_2O_6P_8$
$Pt(AuPPh_3)_8Hg_2(NO_3)_2{}^{2+}$, **10**, 344
$Hg_2Au_8PtC_{144}H_{120}N_4O_{12}P_8$
$[Au_8Hg_2Pt(PPh_3)_8(ONO_2)_2][NO_3]_2$, **10**, 256
$Hg_2Au_8PtC_{144}H_{120}P_8$
$Pt(AuPPh_3)_8Hg_2{}^{4+}$, **10**, 343
$Hg_2C_3H_4Cl_2O$
$(ClHg)_2CMeCHO$, **3**, 157
$Hg_2C_4H_6Cl_2O$
$(ClHg)_2CEtCHO$, **3**, 157
$Hg_2C_6H_4Cl_2$
$C_6H_4(HgCl)_2$, **3**, 163
$Hg_2C_7H_5NO$
$Hg_2C_6H_3CONH_2$, **3**, 137
$Hg_2C_{10}H_6Cl_2$
$C_{10}H_6(HgCl)_2$, **3**, 163
$Hg_2C_{12}H_{11}O$
$(PhHg)_2OH^+$, **3**, 160
$Hg_2FeC_4Br_2O_4$
$Fe(CO)_4(HgBr)_2$, **10**, 334
$Hg_2FeC_4Cl_2O_4$
$Fe(CO)_4(HgCl)_2$, **10**, 324
$Hg_2FeC_6N_2O_4S_2$
$Fe(CO)_4(HgSCN)_2$, **7**, 82
$Hg_2FeC_{30}H_{16}N_6O_4S_2$
$Fe(CO)_4(HgSCN)_2(phen)_2$, **7**, 82
$Hg_2FeC_{33}H_{25}O_3P$
$(CO)_3(PPh_3)Fe(HgPh)_2$, **10**, 339
$Hg_2MnC_{20}H_{17}O_2$
$(C_5H_4Me)Mn(HgPh)_2(CO)_2$, **6**, 143
$Hg_2NbC_{10}H_{11}Cl_2$
$Cp_2NbH(HgCl)_2$, **10**, 335
$Hg_2Ni_2C_{80}H_{88}I_4N_8$
$[Ni(CNXy)_4HgI_2]_2$, **9**, 11
$Hg_2Os_5C_{21}HF_9O_{21}$
$HOs_5(CO)_{15}(O_2CCF_3)(HgO_2CCF_3)_2$, **10**, 327, 328, 340
$Hg_2Os_9C_{22}O_{21}$
$[Os_9C(CO)_{21}](Hg_2)^{2-}$, **10**, 340
$Hg_2Pd_4C_{28}H_{60}Br_2O_4P_4$
$Pd_4Hg_2Br_2(CO)_4(PEt_3)_4$, **9**, 207
$Hg_2Pt_2C_{39}H_{104}O_6P_3$
$[(CO)_3(PPhPr^i{}_2)_3Pt]_2(Hg_2)$, **10**, 343
$Hg_2Pt_3C_{75}H_{66}P_6$
$(PPh_2CH_2PPh_2)_3Pt_3(Hg)_2{}^{2+}$, **10**, 343
$Hg_2RuC_4I_2O_4$
$(HgI)_2Ru(CO)_4$, **10**, 327
$Hg_2Ru_4C_{19}H_9O_{13}$
$(C_2Bu^t)(CO)_9Ru_3(Hg)_2Ru(CO)_4$, **10**, 340
$Hg_2Ru_7C_{34}H_{18}O_{22}$
$Ru(CO)_4[HgRu_3(CO)_9(C_2Bu^t)]_2$, **10**, 334, 335
$Hg_2Ru_{10}C_{30}Cl_4O_{28}$
$[Ru_5C(CO)_{14}(Cl)(HgCl)]_2$, **10**, 327, 328
$Hg_3B_{30}C_6H_{30}$
$(B_{10}H_{10}C_2Hg)_3$, **3**, 165
$Hg_3C_2HN_3O_{11}$
$(O_3NHg)_3CCO_2H$, **3**, 159
$Hg_3C_3H_9O$
$(MeHg)_3O^+$, **3**, 155
$Hg_3C_3H_9S$
$(MeHg)_3S^+$, **3**, 155
$Hg_3C_{18}Cl_{12}$
$(C_6Cl_4Hg)_3$, **3**, 138
$Hg_3C_{18}F_{12}$
$(C_6F_4Hg)_3$, **3**, 138, 163
$Hg_3C_{18}H_{12}$
$(C_6H_4Hg)_3$, **3**, 138
$Hg_3Fe_7C_{24}H_6O_{24}$
$Hg[Fe(CO)_4[HgFe_3(OMe)(CO)_9]_2]$, **10**, 339
$Hg_3Fe_8C_{32}H_6O_{30}$
$Hg[Fe(CO)_4(Hg)Fe_3(CO)_{10}(COMe)]_2$, **10**, 334, 335
$Hg_3NbC_{25}H_{40}N_3S_6$
$Cp_2Nb[HgS_2CNEt_2]_3$, **10**, 335
$Hg_3Os_9C_{33}O_{33}$
$[Os_3(CO)_{11}Hg]_3$, **10**, 334, 335, 340
$Hg_3Os_{18}C_{44}O_{42}$
$[Os_9C(CO)_{21}]_2(Hg_3)$, **10**, 331
$Hg_3Ru_9C_{33}O_{33}$
$[Ru_3Hg(CO)_{11}]_3$, **7**, 720
$Hg_3Ru_{18}C_{44}O_{42}$
$[Ru_9C(CO)_{21}]_2(Hg_3)^{2-}$, **10**, 340
$[Ru_{18}Hg_3(C)_2(CO)_{42}]^{2-}$, **7**, 949
$Hg_4B_{40}C_8H_{40}Cl$
$(B_{10}H_{10}C_2Hg)_4Cl^-$, **3**, 164
$Hg_4B_{40}C_{32}H_{56}I$
$(PhB_{10}H_9C_2Hg)_4I^-$, **3**, 165
$Hg_4Fe_4C_{16}O_{16}$
$[HgFe(CO)_4]_4$, **10**, 334
$Hg_4Mn_4C_{32}H_{32}O_8$
$[MeCp(CO)_2MnHg]_4$, **10**, 337, 338
$Hg_4Mo_8C_{32}H_{20}O_{12}$
$(CpMo(CO)_3HgMo)_4$, **10**, 327, 329
$Hg_5C_{15}H_5Cl_2F_{30}$
$[(CF_3)_2CHHg]_5Cl_2{}^{2-}$, **3**, 164
$Hg_5C_{15}H_5F_{30}$
$[(CF_3)_2CHg]_5$, **3**, 164
$Hg_6Rh_4C_{36}H_{108}P_{12}$
$Hg_6Rh_4(PMe_3)_{12}$, **10**, 327, 329

$Hg_9Co_6C_{18}O_{18}$
$Hg_9Co_6(CO)_{18}$, **10**, 341
$HoB_3C_3H_{18}$
$Ho[BH_3Me]_3$, **1**, 288, 290
HoC_6H_{18}
$[HoMe_6]^-$, **4**, 20
$HoC_{18}H_{23}O$
$(MeC_5H_4)_3Ho(H_2O)$, **4**, 56
$HoC_{36}H_{60}$
$Ho(Bu^t{}_3C_6H_3)_2$, **4**, 7

I

$InAsMn_4C_{41}H_{21}ClO_{17}$
$[AsPh_4][InCl[Mn(CO)_5][Mn_3(CO)_{12}(H)]]$, **1**, 561

$InAsMn_6C_{48}H_{22}O_{24}$
$[AsPh_4][In[Mn_3(CO)_{12}(H)]_2]$, **1**, 561

$InAs_2Mn_5C_{68}H_{40}O_{20}$
$[AsPh_4]_2[In[Mn(CO)_4]_5]$, **1**, 561

$InBC_{14}H_{30}F_4O$
$[Pr^i_2In(THF)_2][BF_4]$, **1**, 514

$InBMoC_6H_6F_4N_2O_2$
$[Mo(MeCN)_2(CO)_2In][BF_4]$, **12**, 639

$InB_4Si_2C_{11}H_{29}$
$Me_2CHIn[(TMS)_2C_2B_4H_4]$, **1**, 532

$InB_4Si_2C_{11}H_{39}$
$(Me_2CH)In(TMS)_2C_2B_4H_{14}$, **1**, 258

InC_5H_5
$InCp$, **1**, 529

InC_6H_7
InC_5H_4Me, **1**, 530

$InC_7H_{18}Br_3$
$Br_3InCH_2(TMEDA)$, **1**, 514

InC_9H_{13}
$In(C_5H_4Bu^t)$, **1**, 530

$InC_9H_{21}N_2$
$Me_3InN(CH_2CH_2)_3N$, **1**, 504

$InC_{10}H_{15}$
$In(Cp^*)$, **1**, 530, 539

$InC_{11}H_{18}N$
$Me_2InC_6H_4CH_2NMe_2$, **1**, 506

$InC_{12}H_{33}$
$(TMS\text{-}CH_2)_3In$, **1**, 505

$InC_{17}H_{14}P$
$In(C_5H_4PPh_2)$, **1**, 531

$InC_{18}H_{15}Se_3$
$In(SePh)_3$, **1**, 525

$InC_{19}H_{17}Br_3P$
$Br_3InCH_2PPh_3$, **1**, 514

$InC_{21}H_{21}$
$(Bz)_3In$, **1**, 506

$InC_{24}H_{54}O_4$
$[Bu^t_2In(O_2Bu^t)]_2$, **1**, 526

$InC_{24}H_{54}P_3$
$In(PBu^t_2)_3$, **1**, 518

$InC_{27}H_6F_{27}$
$In[(CF_3)_3C_6H_2]_3In$, **1**, 507, 508

$InC_{27}H_6F_{27}S_3$
$In[S(CF_3)_3C_6H_2]_3$, **1**, 525

$InC_{27}H_{33}$
$In(Mes)_3$, **1**, 507

$InC_{30}H_{70}O_{11}$
$(INOPr^i)_5(OPr^i)_5(O)$, **1**, 527

$InC_{31}H_{45}ClN$
$[NMe_4][InCl(Mes)_3]$, **1**, 511

$InC_{40}H_{35}$
$InC_5(Bz)_5$, **1**, 530, 539

$InC_{49}H_{44}O_9P_4$
$[InMe(OH)(O_2PPh_2)_4]$, **1**, 526, 527

$InCoC_{14}H_{24}N_2O_4$
$In(CH_2CH_2CH_2NMe_2)_2[Co(CO)_4]$, **1**, 564

$InCoC_{22}H_{15}Cl_2O_4P$
$InCl_2(PPh_3)[Co(CO)_4]$, **1**, 565

$InCo_2C_{16}H_{19}O_9$
$In(CH_2TMS)(THF)[Co(CO)_4]_2$, **1**, 564

$InCo_2C_{42}H_{30}ClO_6P_2$
$InCl[Co(CO)_3(PPh_3)]_2$, **1**, 565

$InCo_2C_{60}H_{45}ClO_6P_3$
$InCl(PPh_3)[Co(CO)_3(PPh_3)]_2$, **1**, 565

$InCo_3C_{20}H_{20}INO_{12}$
$I[NEt_4][In[Co(CO)_4]_3]$, **1**, 565

$InCo_3C_{44}H_{30}Cl_2O_{11}P_2$
$Co(CO)_3(PPh_3)_2[InCl_2[Co(CO)_4]_2]$, **1**, 565

$InCo_3C_{63}H_{45}O_9P_3$
$In[Co(CO)_3(PPh_3)]_3$, **1**, 565

$InCr_2C_{16}H_{10}ClO_6$
$InCl[CO)_3Cp]_2$, **1**, 558

$InCr_2C_{21}H_{15}ClNO_6$
$InCl(py)[Cr(CO)_3Cp]_2$, **1**, 559

$InCr_2NaC_{24}H_{26}Cl_2O_8$
$[Na(THF)_2][InCl_2[Cr(CO)_3Cp]_2]$, **1**, 558

$InCr_3C_{24}H_{15}O_9$
$In[Cr(CO)_3Cp]_3$, **1**, 557

$InFeC_4Br_2O_4$
$[InBr_2[Fe(CO)_4]]^-$, **7**, 19

$InFeC_{17}H_{29}N_2O_2$
$In(CH_2CH_2CH_2NMe_2)_2[Fe(CO)_2Cp]$, **1**, 562

$InFeSiC_{26}H_{26}Cl_2O_7P$
$Fe(CO)_3[Si(OMe)_3][InCl_2]\text{-}[PPh_2CH_2C(O)Ph]$, **1**, 564

$InFe_2C_8BrO_8$
$[InBr[Fe(CO)_4]_2]^{2-}$, **7**, 19

$InFe_2C_{14}H_{10}ClO_4$
$InCl[Fe(CO)_2Cp]_2$, **1**, 562

$InFe_2C_{14}H_{10}IO_4$
$InI[Fe(CO)_2Cp]_2$, **1**, 562

$InFe_2C_{18}H_8N_2O_8$
$In(bipy)[Fe(CO)_4]_2$, **1**, 563

$InFe_2C_{19}H_{15}ClNO_4$
$InCl(py)[Fe(CO)_2Cp]_2$, **1**, 562

$InFe_3C_{12}O_{12}$
$[In[Fe(CO)_4]_3]^{3-}$, **7**, 19

$InFe_3C_{21}H_{15}O_6$
$In[Fe(CO)_2Cp]_3$, **1**, 562

$InFe_3C_{36}H_{30}NO_{12}$
$[NMe_3(Bz)]_3[In[Fe(CO)_4]_3]$, **1**, 563

$InFe_4C_{28}H_{20}O_8S_4$
$InS_4[Fe(CO)_2Cp]_4$, **1**, 562

$InFe_4C_{28}H_{20}O_8Se_4$
$InSe_4[Fe(CO)_2Cp]_4$, **1**, 562

$InFe_6C_{32}H_{20}NO_{24}$
$[NEt_4][In_2Fe_6(CO)_{24}]$, **1**, 563

$InIrC_{12}H_{36}BrP_3$
$IrMe_2(InMeBr)(PMe_3)_3$, **1**, 565

$InIrC_{13}H_{39}P_3$
$IrMe_2(InMe_2)(PMe_3)_3$, **1**, 565

$InIrC_{15}H_{43}P_3$
$IrH(Et)(InEt_2)(PMe_3)_3$, **1**, 565

$InKC_{15}H_{34}$
$KHIn(CH_2Bu^t)_3$, **1**, 516

$InKSi_3C_{12}H_{34}$
$KInH(CH_2TMS)_3$, **1**, 516

$InKSi_4C_{16}H_{44}$
$KIn(CH_2TMS)_4$, **1**, 516

$InMnC_{15}H_{24}N_2O_5$
$In(CH_2CH_2CH_2NMe_2)_2[Mn(CO)_5]$, **1**, 561

$InMoC_4H_3O_3$
$MoMe(CO)_3In$, **12**, 639

$InMoNaC_{16}H_{21}Cl_3O_5$
$[Na(THF)_2][InCl_3[Mo(CO)_3Cp]]$, **1**, 557

$InMo_2C_{20}H_{18}ClO_7$
$InCl(THF)[Mo(CO)_3Cp]_2$, **1**, 559

$InMo_2C_{20}H_{22}Cl_2NO_6$
$[NMe_4][InCl_2[Mo(CO)_3Cp]_2]$, **1**, 559

$InMo_2NaC_{24}H_{26}Cl_2O_8$
$[Na(THF)_2][InCl_2[Mo(CO)_3Cp]_2]$, **1**, 557

$InMo_2SnC_{22}H_{27}O_{13}$
$(CO)_5MoSn(OBu^t)_3InMo(CO)_5$, **1**, 560

$InMo_3C_{24}H_{15}O_9$
$In[Mo(CO)_3Cp]_3$, **1**, 557

$InNaC_4H_{12}$
$Na(InEt_2H_2)$, **1**, 515

$InNiC_{16}H_{27}N_2O$
$In(CH_2CH_2CH_2NMe_2)_2[Ni(CO)Cp]$, **1**, 566

$InPtSi_3C_{38}H_{81}P_2$
$Pt(CH_2TMS)[In(CH_2TMS)_2]\text{-}[Cy_2PCH_2CH_2PCy_2]$, **1**, 566

$InRe_3C_{14}I_2O_{14}$
$In[I_2Re(CO)_4][Re(CO)_5]_2$, **1**, 560

$InRe_4C_{14}I_3O_{14}$
$In[Re_2(CO)_6(I)_2][Re_2(CO)_8(I)]$, **1**, 560

$InRhC_5H_4O$

InRh(cyclopentadienone), **8**, 242
$InRhC_6H_{10}$
InRh(hexadiene), **8**, 271
$InRhC_{28}H_{20}$
$InRh(Ph_4C_4)$, **8**, 242
$InRhZrC_{15}H_{23}IP_2$
$[Cp_2Zr(PEt)_2Rh(In)(Me)]I$, **8**, 198
$InRh_2C_{13}H_{13}$
(CpRh)(cot)(RhIn), **8**, 245
$InRu_2C_{14}H_{10}ClO_4$
$InCl[Ru(CO)_2Cp]_2$, **1**, 562
$InRu_2C_{14}H_{10}Cl_2O_4$
$InCl_2[Ru(CO)_2Cp]_2$, **1**, 562
$InRu_3C_{21}H_{15}O_6$
$In[Ru(CO)_2Cp]_3$, **1**, 562
$InSiC_8H_{13}$
$In(C_5H_4TMS)$, **1**, 530
$InSiC_{12}H_{33}$
$In(CH_2TMS)_3$, **1**, 514
$InSi_6C_{18}H_{57}$
$[(TMS)_2HC]_3In$, **1**, 537
$InSnC_{12}H_{27}O_3$
$In(OBu^t)_3Sn$, **1**, 525
$In_2As_2Mn_4C_{42}H_{40}Cl_2O_{18}$
$[AsPh_4]_2[Mn_2(CO)_8[InCl[Mn(CO)_5]]_2]$, **1**, 561
$In_2C_6H_{30}$
$(Me_3In)_2$, **1**, 504
$In_2C_{18}H_4F_{18}$
$[In[(CF_3)_3C_6H_2]]_2$, **1**, 525
$In_2C_{18}H_{24}Br_4$
$[(C_6H_3Me_3)_2In][InBr_4]$, **1**, 535
$In_2C_{20}H_{46}O_2$
$[Bu^t_2In(OEt)]_2$, **1**, 525
$In_2C_{24}H_{54}O_4$
$[Bu^t_2In(O_2Bu^t)]_2$, **1**, 525, 526
$In_2C_{30}H_{67}$
$(Bu^tCH_2)_3InHIn(CH_2Bu^t)_3$, **1**, 516
$In_2C_{36}H_8F_{36}$
$[[(CF_3)_3C_6H_2]_2In]_2$, **1**, 538
$In_2C_{46}H_{34}P_2$
$[(Bu^tCH_2)_2InCH_2PPh_2]_2$, **1**, 508, 509
$In_2FeSi_6C_{22}H_{54}Cl_2O_4$
$In_2[C(TMS)_3]_2(Cl)_2[Fe(CO)_4]$, **1**, 564
$In_2Fe_2C_8Br_2Cl_2O_8$
$[Fe_2(CO)_8[InClBr]_2]^{2-}$, **1**, 563
$In_2Fe_2C_{16}H_{16}Cl_2O_{18}$
$Fe_2(CO)_8[InCl(THF)]_2$, **1**, 563
$In_2Fe_6C_{24}O_{24}$
$[In_2Fe_6(CO)_{24}]^{2-}$, **7**, 19
$In_2KC_{30}H_{67}$
$K[H[In(CH_2Bu^t)_3]_2]$, **1**, 516, 517
$In_2LiSi_6C_{28}H_{75}$
$(TMS)_3CIn(H)_2Li(THF)_2(H)In(H)_2C(TMS)_3$, **1**, 515, 516
$In_2Mo_2C_6O_6$
$[InMo(CO)_3]_2$, **12**, 639
$In_2Re_2C_{26}H_{24}I_4O_6P_2$
$Re_2(CO)_4[InI_2(H_2O)]_2(PPh_2)_2$, **1**, 560
$In_2Rh_2C_8H_8$
(InRh)(cot)(RhIn), **8**, 245
$In_2Si_8C_{28}H_{72}$
$[[(TMS)_2HC]_2In]_2$, **1**, 537
$In_3As_3C_{12}H_{36}$
$[Me_2In(AsMe_2)]_3$, **1**, 522
$In_3Sb_3C_{30}H_{72}$
$[Me_2In(SbBu^t_2)]_3$, **1**, 522
$In_4Mo_4C_{32}H_{20}O_{12}S_4$
$In_4S_4[Mo(CO)_3Cp]_4$, **1**, 559
$In_4Mo_4C_{32}H_{20}O_{12}Te_4$
$In_4Te_4[Mo(CO)_3Cp]_4$, **1**, 559
$In_4Si_{12}C_{40}H_{114}O_{10}$
$[O(TMS)_3CIn]_4(OH)_6$, **1**, 526, 527
$IrAs_2C_{40}H_{36}Cl$
$Ir(Cl)(C_4H_6)(AsPh_3)_2$, **8**, 319, 320
$IrAs_2Rh_2C_{67}H_{58}Cl_2O_3P_4$
$[Rh_2IrCl_2(CO)_3[AsPh(CH_2PPh_2)_2]_2]^+$, **10**, 64
$IrAuC_{51}H_{44}ClOP_4$
$[IrCl(CO)(dppm)_2Au]^+$, **10**, 68
$IrAuC_{54}H_{50}N_2P_4$
$[Ir(MeNC)_2(dppm)_2Au]^{2+}$, **10**, 68
$IrAuC_{61}H_{59}P_4$
$[Ir(H)(C_2H_4)(triphos)Au(PPh_3)]^+$, **10**, 63
$IrAuC_{64}H_{55}N_2P_3$
$AuIr(H)_2(bipy)(PPh_3)_3^{2+}$, **8**, 404
$IrAuC_{66}H_{58}N_2P_6$
$[Ir(CN)_2[Au[PPh(CH_2PPh_2)_2]_2]^+$, **10**, 68
$IrAuHg_2C_{51}H_{44}Cl_6OP_4$
$IrCl(CO)(dppm)(dppmAuCl)Hg(Cl)_2HgCl_2$, **10**, 68
$IrAu_3C_{65}H_{58}Cl_3OP_6$
$[IrCl(CO)[PPh(CH_2PPh_2)_2]_2Au(AuCl)_2]^+$, **10**, 68
$IrBC_5H_6N_6O_3$
$Ir(CO)(HBpz_3)(COOH)(H)$, **8**, 351
$IrBC_{11}H_{10}N_6O_2$
$IR(HBpz_3)(CO)_2$, **8**, 351
$IrBC_{11}H_{11}F_4N_6O_2$
$[Ir(CO)_2(HBpz_3)(H)]BF_4$, **8**, 351
$IrBC_{12}H_{27}ClNSi$
$IrCl(Et)(C_{10}H_{22}BNSi)$, **1**, 213
$IrBC_{13}H_{19}N_7O_2$
$Ir(CO)(HBpz_3)(CONHPr)(H)$, **8**, 351
$IrBC_{13}H_{30}ClF_7OP_2S$
$[Ir(CO)(Cl)(F)(SF_2)(PEt_3)_2]^+[BF_4]^-$, **8**, 360
$IrBC_{15}H_{32}ClO_2P_3$
$Ir(H)(BO_2C_6H_4)(Cl)(PMe_3)_3$, **1**, 293, 295, 296
$IrBC_{17}H_{43}P_3$
$IrH_2(PMe_3)_3(BC_8H_{14})$, **1**, 293, 295
$IrBC_{44}H_{44}F_4P_2$
$[Ir(cod)(H)_2(PPh_3)_2]BF_4$, **8**, 376
$IrBC_{58}H_{80}P_2$
$[Ir(cod)(dcpe)]BPh_4$, **8**, 367
$IrBC_{76}H_{68}OP_4$
$[Ir(CO)(dppe)_2]BPh_4$, **8**, 362
$IrB_3C_{36}H_{40}P_2$
$(PPh_3)_2H_2IrB_3H_8$, **1**, 313, 314, 316
$IrB_3C_{37}H_{38}OP_2$
$(IrB_3H_7)(CO)H(PPh_3)_2$, **1**, 313, 316
$IrB_4C_7H_{27}OP_2$
$(CO)(PMe_3)_2Ir(B_4H_9)$, **1**, 319, 320
$IrB_4C_{17}H_{31}OP_2$
$Ir(B_4H_9)(CO)(PMe_2Ph)_2$, **1**, 319; **8**, 339
$IrB_8C_6H_{26}ClP_2$
$H(PMe_3)_2Ir(B_8H_7Cl)$, **1**, 331, 337
$IrB_8C_6H_{27}P_2$
$(Me_3P)_2HIrB_8H_8$, **1**, 337
$IrB_8C_7H_{28}ClOP_2$
$(Me_3P)_2(CO)IrB_8H_{10}Cl$, **1**, 331, 336
$IrB_8C_7H_{29}OP_2$
$(Me_3P)_2(CO)IrB_8H_{11}$, **1**, 336
$IrB_8C_7H_{30}ClOP_2$
$(Me_3P)_2(CO)HIrB_8H_{11}Cl$, **1**, 332, 333
$IrB_8C_7H_{31}OP_2$
$(Me_3P)_2(CO)HIrB_8H_{12}$, **1**, 332, 333, 336
$IrB_8C_{10}H_{38}OP_3$
$(Me_3P)_3(CO)IrB_8H_{11}$, **1**, 336
$IrB_8C_{12}H_{30}N$
$Cp^*(MeNC)IrB_8H_{12}$, **1**, 332, 339
$IrB_8C_{55}H_{55}P_3$
$(PPh_3)_3IrCB_8H_{10}$, **8**, 322, 323
$IrB_8PtC_{13}H_{47}OP_4$
$(Me_3P)_2(CO)HIrB_8H_{10}Pt(PMe_3)_2$, **1**, 332, 333
$IrB_9C_6H_{28}P_2$
$(H)(PMe_3)_2[IrB_9H_9]$, **1**, 347
$IrB_9C_6H_{32}P_2$
$(H)(PMe_3)_2[IrB_9H_{13}]$, **1**, 347
$IrB_9C_8H_{25}Cl$
$(C_8H_{12})Cl[IrB_9H_{13}]$, **1**, 342
$IrB_9C_{10}H_{24}S$
$(Cp^*)IrSB_9H_9$, **1**, 269
$IrB_9C_{10}H_{25}ClS$
$(Cp^*)IrSB_9H_{10}Cl$, **1**, 269
$IrB_9C_{10}H_{26}ClN$
$Cl(Cp^*)IrNHB_9H_{10}$, **1**, 262, 263
$IrB_9C_{10}H_{28}$
$Cp^*[IrB_9H_{13}]$, **1**, 339, 341, 342
$IrB_9C_{14}H_{32}N_2$
$Cp^*(MeNC)[IrB_9H_{11}](MeNC)$, **1**, 339
$IrB_9C_{26}H_{38}P_2$
$(H)(PPh_3)(Ph_2PC_6H_4)[IrB_9H_8]$, **1**, 347
$IrB_9C_{26}H_{42}P_2$

$(PPh_3)(Ph_2PC_6H_4)H[IrB_9H_{12}]$, **1**, 347
$IrB_9C_{36}H_{38}P_2$
$(PPh_3)[(Ph_2P)C_6H_4](H)IrB_9H_8$, **1**, 345, 346
$IrB_9C_{36}H_{42}P_2$
$(H)(PPh_3)(PPh_2C_6H_4[IrB_9H_{12}]$, **1**, 342–5
$IrB_9C_{36}H_{43}P$
$(PPh_3)_2[IrB_9H_{13}]$, **1**, 342
$IrB_9C_{36}H_{43}P_3$
$(PPh_3)(PH_3)_2(Ph_2PC_6H_4)H$-$[IrB_9H_7]$, **1**, 347, 348
$IrB_9C_{36}H_{44}P$
$(H)(PPh_3)_2[IrB_9H_{13}]$, **1**, 347
$IrB_9C_{36}H_{55}$
$(PPh_3)_3[IrB_9H_{10}]$, **1**, 343, 344
$IrB_9C_{38}H_{39}P_2$
$(OEt)(PPh_3Cl(Ph_2P)C_6H_4$-$[IrB_9H_5])$, **1**, 343
$IrB_9C_{54}H_{53}P_2$
$(PPh_3)_2(Ph_2P)C_6H_4[IrB_9H_9]$, **1**, 343, 345
$IrB_9C_{54}H_{54}P_2$
$(H)(PPh_3)_2(PPh_2C_6H_4[IrB_9$-$H_{10}])$, **1**, 343
$IrB_9C_{54}H_{55}P_3$
$(PPh_3)_3IrB_9H_{10}$, **8**, 322, 323
$IrB_9RhC_{20}H_{41}$
$Cp*_2(H)[RhIrB_9H_{10}]$, **1**, 342
$IrB_{10}C_{29}H_{33}NOP$
$Ir(PhC_2B_{10}H_{10})(CO)(MeCN)$-$(PPh_3)$, **8**, 351
$IrB_{10}C_{33}H_{35}NOP$
$Ir(PhC_2B_{10}H_{10})(CO)(PhCN)$-$(PPh_3)$, **8**, 320
$IrC_2Br_2O_2$
$Ir(CO)_2Br_2$, **8**, 320
$IrC_2O_{11}P_3$
$(CO)_2Ir(P_3O_9)^{2-}$, **8**, 323
IrC_3ClO_3
$Ir(CO)_3Cl$, **8**, 304
IrC_3O_3
$Ir(CO)_3^{3-}$, **8**, 332
IrC_4O_4
$[Ir(CO)_4]^-$, **8**, 332, 501
IrC_6H_7O
$CpIr(CO)(H)_2$, **8**, 385
$IrC_7H_5O_2$
$CpIr(CO)_2$, **8**, 377, 387
$IrC_7H_8O_4$
$Ir(CO)_2(acac)$, **8**, 491
$IrC_7H_{18}ClOP_2$
$Ir(CO)(Cl)(PMe_3)_2$, **8**, 305, 359
$IrC_8H_{12}Cl_2$
$Ir(Cl)_2(cod)$, **8**, 341
$IrC_8H_{12}O_9P_3$
$Ir(cod)(P_3O_9)^{2-}$, **8**, 395
$IrC_8H_{18}ClO_2P_2$
$Ir(CO)_2Cl(PMe_3)_2$, **8**, 305
$IrC_9H_{10}O$
CpIr(allyl)(CO), **8**, 381
IrC_9H_{11}
CpIr(butadiene), **8**, 380
IrC_9H_{13}
$CpIr(C_2H_4)_2$, **8**, 379, 380
IrC_9H_{15}
$Ir(C_3H_5)_3$, **8**, 304
$IrC_{10}H_{10}D$
CpIr(CpD), **8**, 380
$IrC_{10}H_{11}$
$CpIr(C_5H_6)$, **8**, 380, 381
$IrC_{10}H_{18}P$
$CpIr(C_2H_4)PMe_3$, **8**, 378
$IrC_{10}H_{19}$
$Cp*IrH_4$, **8**, 385, 405
$IrC_{11}H_5F_6$
$CpIr(C_6F_6)$, **8**, 380
$IrC_{11}H_7O_2$
$(In)Ir(CO)_2$, **8**, 391
$IrC_{11}H_8ClN_2O$
Ir(bipy)(CO)Cl, **8**, 317
$IrC_{11}H_{15}O_2$
$Cp'Ir(C_2H_4)_2$, **8**, 379
$IrC_{12}H_2P_4Se_2$
$Ir(dmpe)_2Se_2^+$, **8**, 361
$IrC_{12}H_{12}P_4$
$Ir(dmpe)_2^+$, **8**, 361
$IrC_{12}H_{15}O_2$
$Cp*Ir(CO)_2$, **8**, 377, 387, 388, 512
$IrC_{12}H_{16}Cl$
$IrCl(chd)_2$, **8**, 376
$IrC_{12}H_{18}ClN_2$
IrCl(cod)(Meimidazole), **8**, 374
$IrC_{12}H_{35}P_2$
$IrH_5(PEt_3)_2$, **8**, 406
$IrC_{12}H_{36}ClP_4$
$IrCl(PMe_3)_4$, **8**, 354, 355
$IrC_{12}H_{36}F_6P_5$
$[Ir(PMe_3)_4^+][PD_6^-]$, **8**, 348, 349
$IrC_{12}H_{36}P_4$
$Ir(PMe_3)_4^+$, **8**, 304
$IrC_{12}H_{37}P_4$
$IrH(PMe_3)_4$, **8**, 403
$IrC_{12}H_{38}P_4$
$H_2Ir(PMe_3)_4^+$, **8**, 304
$IrC_{12}H_{39}O_2P_4$
$[Ir(H)_2(PMe_3)_4]^+[HCO_2]^-$, **8**, 402
$IrC_{13}H_{14}N_6O_3$
$Ir(CO)(CO_2Me)(H)(tpzm)$, **8**, 351
$IrC_{13}H_{15}$
$(In)Ir(C_2H_4)_2$, **8**, 338, 339, 391, 392
$IrC_{13}H_{18}Cl_2N_2$
$Cp*IrCl_2(pyrazole)$, **8**, 389, 390
$IrC_{13}H_{20}O_2$
Ir(acac)(cod), **8**, 367
$IrC_{13}H_{20}P$
$(In)Ir(H)(Me)(PMe_3)$, **8**, 391, 392
$IrC_{13}H_{24}Cl_2P$
$Cp*IrCl_2(PMe_3)$, **8**, 383
$IrC_{13}H_{24}I_2P$
$Cp*IrI_2(PMe_3)$, **8**, 395
$IrC_{13}H_{25}ClPS$
$Cp*Ir(Cl)(SH)(PMe_3)$, **8**, 389, 390
$IrC_{13}H_{26}P$
$Cp*Ir(H)_2(PMe_3)$, **8**, 385
$IrC_{13}H_{30}ClOP_2$
$Ir(CO)(Cl)(PEt_3)_2$, **8**, 309, 359
$IrC_{13}H_{30}Cl_2F_2OP_3$
$Ir(CO)Cl_2(PEt_3)_2(PF_2)$, **8**, 309
$IrC_{13}H_{36}OP_4$
$Ir(CO)(PMe_3)_4^+$, **8**, 349
$IrC_{14}H_{19}$
Cp'Ir(cod), **8**, 380
$IrC_{14}H_{24}N$
$Cp*Ir(NBu^t)$, **8**, 387
$IrC_{14}H_{26}I_2P$
$CpIrI_2(PPr^i_3)$, **8**, 383
$IrC_{14}H_{27}ClP$
$Cp*Ir(Cl)(Me)(PMe_3)$, **8**, 383
$IrC_{14}H_{27}OS$
$Cp*Ir(DMSO)Me_2$, **8**, 386
$IrC_{14}H_{28}P$
$CpIrH_2(PPr^i_3)$, **8**, 384, 385
$IrC_{14}H_{29}P$
$CpIrH_3(PPr^i_3)^+$, **8**, 385
$IrC_{14}H_{33}O_2P_3$
$Ir(CO)_2(PH_3)(PEt_3)_2^+$, **8**, 339
$IrC_{14}H_{42}O_{15}P_5$
$Ir[P(O)(OMe)_2][P(OMe)_3]_4$, **8**, 338, 339
$IrC_{15}H_{14}O_2$
$(In)Ir(CO)_2(C_4H_7)$, **8**, 391
$IrC_{15}H_{16}O_3P$
$(In)Ir(CO)_3(PMe_3)$, **8**, 391
$IrC_{15}H_{20}O_2P$
$(In)Ir(CO)_2(H)(Me)(PMe_3)$, **8**, 391
$IrC_{15}H_{23}Cl$
Cp*Ir(acac)Cl, **8**, 389
$IrC_{15}H_{24}$
$Cp*Ir(allyl)(C_2H_4)$, **8**, 395
$IrC_{15}H_{24}F_6O_6PS_2$
$Cp*Ir(PMe_3)(OSO_2CF_3)_2$, **8**, 383
$IrC_{15}H_{30}O_3P_2$
$Ir(CO)_3(PEt_3)_2^+$, **8**, 339
$IrC_{15}H_{53}ClP_3$
$Ir(Cl)(H)(Ph)(PMe_3)_3$, **8**, 355
$IrC_{16}H_{22}Cl_4P_2$
$IrCl_4(PMe_2Ph)_2$, **8**, 405
$IrC_{16}H_{24}$
$Ir(cod)_2^+$, **8**, 329
$IrC_{16}H_{24}Cl$
$IrCl(cod)_2$, **8**, 345
$IrC_{16}H_{33}$
$Cp*Ir(C_6Me_6)^{2+}$, **8**, 393
$IrC_{16}H_{38}ClP_2$
$IrCl(C_2H_4)_2(PEt_3)_2$, **8**, 348
$IrC_{17}H_{18}N$
$Ir(cod)[(Me_2NCH_2)C_6H_4]$, **8**, 370
$IrC_{17}H_{22}ClOP_2$
$Ir(CO)Cl(PMe_2Ph)_2$, **8**, 306
$IrC_{17}H_{27}O$
$Cp*Ir(CO)(C_6H_{11})(H)$, **8**, 377
$IrC_{17}H_{35}NP$
$Cp*Ir(H)(CH_2CMe_2NH_2)PMe_3$, **8**, 378
$IrC_{17}H_{38}ClOP_2$

$Ir(CO)Cl(PBu_2H)_2$, **8**, 305
$IrC_{17}H_{39}ClP_3$
$[Ir(cod)(PMe_3)_3]Cl$, **8**, 370
$IrC_{17}H_{41}ClP_3$
$Ir(Cl)(C_8H_{14})(PMe_3)_3$, **8**, 349, 350
$IrC_{17}H_{43}ClP_3$
$Ir(C_8H_{14})(Cl)(PMe_3)_3$, **8**, 398, 400
$IrC_{18}H_{20}ClO_4$
$[Ir(azulene)(cod)]ClO_4$, **8**, 393
$IrC_{18}H_{21}O$
$(In)Ir(CO)(C_8H_{14})$, **8**, 391
$IrC_{18}H_{22}F_6O_4$
$Ir(cod)(THF)_2^+$, **8**, 368
$IrC_{18}H_{29}P$
$Cp^*Ir(allyl)PMe_3^+$, **8**, 395, 396
$IrC_{18}H_{45}ClP_3$
$IrCl(PEt_3)_3$, **8**, 400
$IrC_{18}H_{46}ClP_2$
$Ir(Cl)H_4(PPr^i_3)_2$, **8**, 405
$IrC_{19}H_{25}N_6$
$Cp^*Ir(pz)_2(Hpz)$, **8**, 390
$IrC_{19}H_{33}O_3$
$Cp^*Ir(acetone)_3^{2+}$, **8**, 380, 381
$IrC_{19}H_{36}P$
$Cp^*Ir(Cy)(H)PMe_3$, **8**, 378
$IrC_{19}H_{42}ClOP_2$
$Ir(CO)Cl(PPr^i_3)_2$, **8**, 305
$IrC_{19}H_{42}OP_2$
$Ir(CO)(PBu_2Me)_2Cl$, **8**, 306
$IrC_{20}H_{12}Cl_{10}$
$Ir(cod)(C_6Cl_5)_2$, **8**, 366, 367
$IrC_{20}H_{16}Cl_2N_4$
$Ir(bipy)_2Cl_2^+$, **8**, 358
$IrC_{20}H_{23}ClN_2$
$Cp^*Ir(bipy)Cl^+$, **8**, 386
$IrC_{20}H_{23}N_2$
$Cp^*Ir(bipy)$, **8**, 386
$IrC_{20}H_{24}N_2$
$Cp^*Ir(bipy)H^+$, **8**, 386
$IrC_{20}H_{25}Cl_2N_2O$
$[Cp^*Ir(bipy)(H_2O)]Cl_2$, **8**, 386
$IrC_{20}H_{36}N_4$
$Ir(CNBu^t)_4^+$, **8**, 350
$IrC_{20}H_{46}ClP_2$
$Ir(Cl)(C_2H_4)(PPr^i_3)_2$, **8**, 354
$IrC_{20}H_{46}F_3OP_2$
$Ir(OCH_2CF_3)(H)_2(PPr^i_3)_2$, **8**, 406, 407
$IrC_{20}H_{49}ClP_4$
$HIr(Cl)(Et_2PCH_2CH_2PEt_2)_2^+$, **8**, 304
$IrC_{21}H_{28}OP_2$
$Ir(C_4H_6)(CO)(PMe_2Ph)_2$, **8**, 339
$IrC_{21}H_{46}ClOP_2$
$Ir(CO)(Cl)(H)(PPr^i_3)_2$-$(CH{=}CH_2)$, **8**, 355
$IrC_{21}H_{49}P_2$
$IrH_4Bu^t_2PCH_2CH_2CHCH_2$-$CH_2PBu^t_2$, **8**, 405
$IrC_{22}H_{18}$
$[CpIr(C_5H_3Ph_2)]^+$, **8**, 380, 381
$IrC_{22}H_{26}Cl_2P$
$Cp^*IrCl_2(PHPh_2)$, **8**, 388, 389
$IrC_{22}H_{36}P$
$Cp^*Ir(Pr)(Ph)(PMe_3)$, **8**, 378
$IrC_{23}H_{41}ClOP$
$IrCl(cod)(Cy_2PCH_2CH_2OMe)$, **8**, 368, 369
$IrC_{24}H_{34}ClP_3$
$IrCl_2(H)(PMe_2Ph)_3$, **8**, 357
$IrC_{24}H_{34}Cl_2P_3$
$IrCl_2(H)(PMe_2Ph)_3$, **8**, 347
$IrC_{24}H_{36}P_3$
$IrH_3(PMe_2Ph)_3$, **8**, 405
$IrC_{24}H_{38}P_3$
$Ir(H)_3(PMe_2Ph)_3$, **8**, 402, 403
$IrC_{25}H_{51}ClNP_2$
$Ir(Cl)(H)(py)(PPr^i_3)_2$-$(CH{=}CH_2)$, **8**, 354
$IrC_{25}H_{54}ClOP_2$
$Ir(CO)Cl(PBu^t_3)_2$, **8**, 305
$IrC_{26}H_{26}$
$Ir(cod)(C_5H_4CPh_2)^+$, **8**, 367
$IrC_{26}H_{38}NP_3$
$Ir(MeCN)(H)_2(PMe_2Ph)_3^+$, **8**, 402, 403
$IrC_{26}H_{39}P_3$
$Ir(C_2H_4)(H)_2(PMe_2Ph)_3^+$, **8**, 405
$IrC_{26}H_{40}ClO_3P_3$
$Ir(Cl)(H)[P(O)(OMe)_2](PMe_2$-$Ph)_3$, **8**, 350
$IrC_{26}H_{42}S_2$
$Ir(Mes)_2(SEt_2)_2$, **8**, 320
$IrC_{27}H_{24}BrOP_2$
$IrBr(CO)(dppe)$, **8**, 308, 333
$IrC_{27}H_{24}ClOP_2$
$Ir(CO)Cl(dppe)$, **8**, 314
$IrC_{27}H_{24}IOP_2$
$Ir(CO)(dppe)I$, **8**, 404
$IrC_{27}H_{26}ClOP_2$
$Ir(CO)(Cl)(PMePh_2)_2$, **8**, 361
$IrC_{27}H_{27}OP_2$
$Ir(CO)(dppe)H_3$, **8**, 404
$IrC_{27}H_{33}$
$Ir(Mes)_3$, **8**, 304
$IrC_{27}H_{42}P_3$
$IrMe_3(PMe_2Ph)_3$, **8**, 355, 356
$IrC_{28}H_{24}NOP_2$
$Ir(CO)(CN)(dppe)$, **8**, 320
$IrC_{28}H_{30}Cl_2P$
$Cp^*IrCl_2PPh_3$, **8**, 383
$IrC_{28}H_{37}O_2P_3$
$Ir(CO)[C(O)Me](Me)(PMe_2$-$Ph)_3$, **8**, 355
$IrC_{28}H_{39}P_3$
$Ir(MeC{\equiv}CMe)(PMe_2Ph)_3^+$, **8**, 405
$IrC_{28}H_{40}N_2P_3$
$Ir(MeCN)_2(H)(PMe_2Ph)_3^{2+}$, **8**, 402
$IrC_{28}H_{41}P_3$
$Ir(C_2H_4)_2(PMe_2Ph)_3^+$, **8**, 355
$IrC_{28}H_{48}O_2P_2$
$Ir(CO)_2(dcpe)^+$, **8**, 367
$IrC_{29}H_{31}OP_2$
$Ir(CO)(dppe)(Et)(H)_2$, **8**, 354, 355
$IrC_{29}H_{36}Np$
$Ir(cod)(CH_2pyMe)PEtPh_2$, **8**, 373
$IrC_{30}H_{36}OP$
$Cp^*Ir(OEt)(H)(PPh_3)$, **8**, 383
$IrC_{31}H_{50}NP$
$Ir(cod)(py)(PCy_3)^+$, **8**, 305, 366
$IrC_{32}H_{45}P_4$
$Ir(H)(PMe_2Ph)_4$, **8**, 403
$IrC_{34}H_{36}OP_2$
$Ir(cod)(Ph_2POCH_2CH_2PPh_2)^+$, **8**, 370, 371
$IrC_{35}H_{60}OP_2$
$Ir(CO)(cod)(dcpe)^+$, **8**, 367
$IrC_{36}H_{30}ClN_2P_2$
$Ir(Cl)(N_2)(PPh_3)_2$, **8**, 307, 311
$IrC_{36}H_{30}ClP_2$
$Ir(Cl)(PPh_3)_2$, **8**, 306
$IrC_{36}H_{35}OP_2$
$Ir(CO)(dppe)(Mes)$, **8**, 314
$IrC_{36}H_{35}P_2$
$Ir(H)_5(PPh_3)_2$, **8**, 402
$IrC_{36}H_{44}$
$Ir(Mes)_4$, **8**, 304
$IrC_{36}H_{47}N_2P_4S_2$
$[Ir(H)_2(PMe_2Ph)_4^+][HS(S)CC$-$(CN)_2^-]$, **8**, 403
$IrC_{36}H_{66}ClO_2P_2S$
$Ir(Cl)(SO_2)(PCy_3)_2$, **8**, 320, 322
$IrC_{36}H_{71}P_2$
$IrH_5(PCy_3)_2$, **8**, 406, 407
$IrC_{36}H_{72}P_2$
$IrH_2(H_2)_2(PCy_3)_2^+$, **8**, 406, 407
$IrC_{37}H_{30}ClNO_2P_2$
$Ir(CO)(Cl)(NO)(PPh_3)_2^+$, **8**, 339
$IrC_{37}H_{30}ClOP_2$
$Ir(CO)Cl(PPh_3)_2$, **8**, 306
$IrC_{37}H_{30}ClOP_2$
$Ir(CO)(Cl)(PPh_3)_2$, **1**, 316; **8**, 304, 305, 312, 320, 332, 333, 347, 350, 359
$IrC_{37}H_{30}ClO_3P_2S$
$Ir(CO)(Cl)(SO_2)(PPh_3)_2$, **8**, 362
$IrC_{37}H_{30}ClO_5P_2$
$Ir(CO)(ClO_4)(PPh_3)_2$, **8**, 315
$IrC_{37}H_{30}ClP_2Se$
$Ir(Cl)(CSe)(PPh_3)_2$, **8**, 320
$IrC_{37}H_{30}Cl_2OP_2$
$Ir(CO)Cl_3(PPh_3)_2$, **8**, 362
$IrC_{37}H_{30}IOP_2$
$Ir(CO)I(PPh_3)_2$, **8**, 311
$IrC_{37}H_{30}NO_2P_2$
$Ir(CO)(NO)(PPh_3)_2$, **8**, 318
$IrC_{37}H_{32}ClOP_2$
$Ir(CO)(Cl)(H)_2(PPh_3)_2$, **8**, 402
$IrC_{38}H_{26}F_9O_2P_2$
$IrH_2(O_2CCF_3)[P(C_6H_4F)_3]_2$, **8**, 354
$IrC_{38}H_{30}ClOP_2Se_2$
$Ir(CO)(Cl)(CSe_2)(PPh_3)_2$, **8**, 320
$IrC_{38}H_{30}F_3O_4P_2S$
$Ir(CO)(PPh_3)_2(OSO_2CF_3)$, **8**, 315, 333
$IrC_{38}H_{30}O_2P_2$

Ir(CO)$_2$(PPh$_3$)$_2$, **8**, 340
$IrC_{38}H_{31}O_2P_2$
Ir(CO)$_2$(H)(PPh$_3$)$_2$, **8**, 340
$IrC_{38}H_{32}F_3O_5P_2S$
[Ir(H$_2$O)(CO)(PPh$_3$)$_2$]$^+$[CF$_3$-SO$_3$]$^-$, **8**, 315
$IrC_{38}H_{32}IOP_2$
Ir(CO)(I)(CH$_2$)(PPh$_3$)$_2$, **8**, 311, 312, 345
$IrC_{38}H_{33}OP_2$
Ir(CO)(Me)(PPh$_3$)$_2$, **8**, 311, 312, 319
$IrC_{38}H_{33}O_2P_2$
Ir(CO)(OMe)(PPh$_3$)$_2$, **8**, 319
$IrC_{38}H_{33}O_3P_2$
Ir(Me)(CO)(O$_2$)(PPh$_3$)$_2$, **8**, 318
$IrC_{38}H_{34}ClO_2P_2$
Ir(Cl)(C$_2$H$_4$)(O$_2$)(PPh$_3$)$_2$, **8**, 318
$IrC_{38}H_{38}B_7P_2$
(CO)(Ph$_3$P)$_2$Ir(CB$_7$H$_8$), **1**, 379
$IrC_{39}H_{30}O_3P_2$
Ir(CO)$_3$(PPh$_3$)$_2$$^+$, **8**, 316
$IrC_{39}H_{32}ClINOP_2$
Ir(CO)Cl(CH$_2$CN)I(PPh$_3$)$_2$, **8**, 312
$IrC_{39}H_{32}NOP_2$
Ir(CO)(CH$_2$CN)(PPh$_3$)$_2$, **8**, 312
$IrC_{39}H_{33}F_6NOP_3$
[Ir(MeCN)(CO)(PPh$_3$)$_2$][PF$_6$], **8**, 317, 320
$IrC_{40}H_{30}NO_4P_2$
(PPN)Ir(CO)$_4$, **8**, 355, 357
$IrC_{40}H_{32}N_3P_2$
Ir(dppm)(H)(CNBu^t)$_3$$^{2+}$, **8**, 350
$IrC_{40}H_{34}P_2S$
Ir(C$_4$H$_4$S)(PPh$_3$)$_2$$^+$, **8**, 381
$IrC_{40}H_{36}O_2P_2$
Ir(CO)$_2$(Me)$_2$(PPh$_3$)$_2$$^+$, **8**, 355, 356
$IrC_{41}H_{39}ClO_2P_3$
Ir(Cl)(O$_2$)(triphos), **8**, 318, 319
$IrC_{41}H_{39}P_6$
Ir(P$_3$)(triphos), **8**, 364
$IrC_{43}H_{30}F_5O_2P_2$
Ir(CO)(OC$_6$F$_5$)(PPh$_3$)$_2$, **8**, 316
$IrC_{43}H_{42}ClOP_2$
Ir(CO)Cl[P(Tol)$_3$]$_2$, **8**, 316
$IrC_{43}H_{43}ClP_3$
IrCl(C$_2$H$_4$)(triphos), **8**, 398
$IrC_{43}H_{43}O_2P_2$
Ir(CO)(OH)[P(Tol)$_3$]$_2$, **8**, 316
$IrC_{44}H_{42}P_2$
Ir(cod)(PPh$_3$)$_2$$^+$, **8**, 358
$IrC_{44}H_{45}OP_2$
Ir(CO)Me[P(Tol)$_3$]$_2$, **8**, 313
$IrC_{44}H_{45}O_2P_2$
Ir(CO)(OMe)[P(Tol)$_3$]$_2$, **8**, 316
$IrC_{44}H_{48}O_2P_2$
Ir(H)$_2$(THF)$_2$(PPh$_3$)$_2$$^+$, **8**, 402
$IrC_{45}H_{38}P_2$
(In)Ir(H)(PPh$_3$)$_2$$^+$, **8**, 391
$IrC_{45}H_{39}OP_2$
Ir(CO)(H)(styrene)(PPh$_3$)$_2$, **8**, 404
$IrC_{46}H_{40}N_2P_2$
Ir(bipy)(H)$_2$(PPh$_3$)$_2$$^+$, **8**, 404
$IrC_{46}H_{41}OP_2$
Ir(CO)(C$_6$H$_2$Me$_3$)(PPh$_3$)$_2$, **8**, 398
$IrC_{46}H_{44}N_2O_4P_2$
Ir$_2$(pz)$_2$(CO)$_2$(PPh$_2$OCH$_2$CH$_2$-py)$_2$, **8**, 330
$IrC_{47}H_{43}OP_2$
Ir(CO)(C$_6$H$_3$Et$_2$)(PPh$_3$)$_2$, **8**, 312
$IrC_{49}H_{51}P_3$
Ir(cod)(tripod)$^+$, **8**, 370
$IrC_{51}H_{40}ClN_2O_2P_2$
Ir[N$_2$C(C(O)Ph)$_2$]Cl(PPh$_3$)$_2$, **8**, 311
$IrC_{51}H_{44}ClOP_4$
[Ir(dppm)$_2$(CO)]Cl, **8**, 326, 327
$IrC_{52}H_{44}I_2O_2P_4$
Ir$_2$(CO)I$_2$(CO)(dppm)$_2$, **8**, 324
$IrC_{52}H_{47}P_2$
Ir(C≡CPh)(cod)(PPh$_3$)$_2$, **8**, 370
$IrC_{52}H_{48}P_4$
Ir(dppe)$_2$$^+$, **8**, 332
$IrC_{52}H_{49}P_4$
Ir(dppe)$_2$H, **8**, 332, 403
$IrC_{52}H_{50}P_4$
Ir(dppe)$_2$H$_2$$^+$, **8**, 403
$IrC_{53}H_{48}OP_4$
Ir(CO)(dppe)$_2$$^+$, **8**, 364
$IrC_{53}H_{64}OP_2$
Ir(CO)(CH$_2$Bu^t)$_2$[P(Tol)$_3$]$_2$, **8**, 319, 320
$IrC_{54}H_{33}NO_{11}P_2$
[Ir(MeCN)(CO)(PPh$_3$)$_2$][C$_5$-(CO$_2$Me)$_5$], **8**, 392, 393
$IrC_{54}H_{47}ClP_3$
Ir(Cl)(H)$_2$(PPh$_3$)$_3$, **8**, 402
$IrC_{54}H_{48}P_3$
IrH$_3$(PPh$_3$)$_3$, **8**, 404
$IrC_{55}H_{45}ClOP_3$
Ir(CO)(Cl)(PPh$_3$)$_3$$^+$, **8**, 339
$IrC_{55}H_{46}OP_3$
Ir(CO)(H)(PPh$_3$)$_3$, **8**, 316, 347, 364, 402
$IrC_{70}H_7O$
(In)Ir(CO)(C$_{60}$), **8**, 391
$IrCoC_{18}H_{22}O_2$
Cp(CO)Co(CH$_2$)IrCp*(CO), **8**, 387
$IrCoFeRuC_{16}H_9O_{10}P$
FeRuCoIr(H)(PMe)(CO)$_{10}$Cp, **10**, 244
$IrCo_2C_{22}H_{25}O_2$
Co$_2$Ir(CO)$_2$Cp$_2$Cp*, **10**, 209, 211
$IrCo_2C_{22}H_{40}O_2$
Co$_2$Ir(CO)$_2$Cp$_2$Cp*, **8**, 512
$IrCrC_{35}H_{35}O_4P_2$
(CO)$_4$Cr(PPh$_2$)$_2$Ir(cod)Me, **8**, 372
$IrCuC_{19}H_{33}ClOP_4$
Ir(C$_2$Ph)(CO)(dmpm)$_2$CuCl, **10**, 68
$IrFeC_{19}H_{23}O_3$
Fe(CO)$_2$(CMeCHCHCMe)-(CO)IrCp*, **10**, 60
$IrFe_2C_{43}H_{30}ClO_7P_2Te_2$
Fe$_2$(CO)$_6$Te$_2$[IrCl(CO)(PPh$_3$)$_2$], **7**, 77
$IrH_{12}O_2P_4$
Ir(O$_2$)(PH$_3$)$_4$, **8**, 364
$IrInC_{12}H_{36}BrP_3$
IrMe$_2$(InMeBr)(PMe$_3$)$_3$, **1**, 565
$IrInC_{13}H_{39}P_3$
IrMe$_2$(InMe$_2$)(PMe$_3$)$_3$, **1**, 565
$IrInC_{15}H_{43}P_3$
IrH(Et)(InEt$_2$)(PMe$_3$)$_3$, **1**, 565
$IrLiC_{19}H_{41}N_3$
Cp*IrH$_3$Li(pmdeta), **8**, 385
$IrLiSiC_{22}H_{49}N_3$
Cp*IrH$_2$(TMS)Li(pmdeta), **8**, 385
$IrMnC_{24}H_{37}O_2P$
Cp'(CO)$_2$Mn(PBu$^t_2$)Ir(cod), **8**, 372, 373
$IrMoC_{13}H_8O_6$
Mo(CO)$_3$(CHCMeCHC-MeCH)Ir- (CO)$_3$, **10**, 45
IrP_4
IrPP$_3$$^+$, **8**, 341
$IrPtC_{24}H_{64}P_4$
[IrH$_2$(PEt$_3$)$_2$(H)$_2$Pt(PEt$_3$)$_2$]$^+$, **10**, 64
IrH(PEt$_3$)$_3$(H)$_2$PtH(PEt$_3$), **10**, 64
$IrPtC_{39}H_{46}NS_4$
Ir(cod)(SPPh$_2$)$_2$Pt(S$_2$CNPr$^i_2$), **10**, 64
$IrPtC_{41}H_{34}O_3P_2$
[Ir(CO)(PPh$_3$)$_2$(H)-(CH=CH$_2$)Pt(CO)$_2$]$^+$, **10**, 64
$IrPtC_{75}H_{63}OP_4$
(PPh$_3$)$_2$(CO)Ir(CH=CH$_2$)Pt-(PPh$_3$)$_2$$^+$, **8**, 333
$IrPtC_{75}H_{64}OP_4$
(Ph$_3$P)$_2$Pt(H)(H$_2$CCH)Ir(CO)-(PPh$_3$)$_2$, **9**, 549
[PtIr(PPh$_3$)$_4$(CO)(C$_2$H$_3$)(H)]$^+$, **10**, 16
$IrReC_{37}H_{30}O_5P_2$
Ir(CO)(ReO$_4$)(PPh$_3$)$_2$, **8**, 315
$IrReC_{52}H_{88}O_4P_4$
Re(CO)$_2$(PCy$_2$)$_3$Ir(CO)$_2$(PCy$_2$), **10**, 53
$IrReC_{52}H_{88}P_4$
(PCy$_2$)(CO)$_2$Re(PCy$_2$)$_2$Ir(CO)$_2$-(PCy$_2$), **8**, 372
$IrReC_{54}H_{106}P_6$
(PCy$_2$)$_2$Re(PCy$_2$)$_2$Ir(PMe$_3$)$_2$, **8**, 372
$IrRhC_{24}H_{19}ClNO_2$
RhCp(CO)(NC$_5$H$_4$PPh$_2$)IrCl-(CO), **10**, 64
$IrRhC_{27}H_{36}N_6$
Cp*Ir(pz)$_3$Rh(cod), **8**, 257
$IrRhC_{27}H_{37}N_6$
[Cp*Ir(pzH)(pz)$_2$Rh(cod)]$^+$, **8**, 257
$IrRhC_{30}H_{27}ClF_4N_4$
Cp*IrCl(pz)$_2$Rh(tfb), **8**, 257
$IrRh_2C_{33}H_{45}O_3$
Rh$_2$Ir(CO)$_3$Cp*$_3$, **8**, 512

$IrRh_2C_{36}H_{40}F_6P$
$Rh_2Cp^*{}_2Ir(PPh_2)(CF_3CCCF_3)$, **8**, 512
$IrRh_3C_{12}O_{12}$
$Rh_3Ir(CO)_{12}$, **8**, 505, 510
$IrRuC_{33}H_{31}N_4OP$
$H(CO)(PPh_3)Ru$-(biimidazolato)Ir(cod), **10**, 39
$IrRuC_{51}H_{46}N_4OP_2$
$(H)(CO)(PPh_3)_2Ru(bim)Ir(cod)$, **8**, 374
$IrRuC_{51}H_{49}N_4OP_2$
$(H)(CO)(PPh_3)_2Ru(pz)_2Ir(cod)$, **8**, 374
$IrRuC_{51}H_{54}N_4OP_2$
$H(CO)(PPh_3)_2Ru$-(biimidazolato)Ir(cod), **8**, 305
$IrRuC_{58}H_{58}ClP_4$
(dppm)(H)Ru(H)(dppm)(Cl)(Ir-(cod), **8**, 327
$IrRu_3C_{50}H_{36}O_8P_3$
$Ru_3Ir(H)_2(PPh_2)(PPhC_6H_4)$-$(CO)_8(PPh_3)$, **10**, 223
$IrSiC_{13}H_{27}$
$Cp^*IrH_3(TMS)$, **8**, 385
$IrSiC_{28}H_{32}NP_2$
$Ir(H_2)[N(SiMe_2)(CH_2PPh_2)_2]$, **8**, 345
$IrSiC_{35}H_{37}OP_2$
$Ir(PPh_2CH_2CH_2SiMe_2)(PPh_3)$-$(CO)(H)_2$, **8**, 345
$IrSi_2C_{22}H_{47}$
$Cp^*Ir(H)_2(SiEt_3)_2$, **8**, 385
$IrSi_2C_{31}H_{39}INP_2$
$Ir(I)(Me)[N(SiMe_2CH_2PPh_2)_2]$, **8**, 342, 345
$IrSi_2C_{34}H_{42}NP_2$
$Ir[C(CH_2)_3][N(SiMe_2CH_2P$-$Ph_2)_2]$, **8**, 342, 344
$IrSnC_{15}H_{29}$
$Cp^*Ir(C_2H_4)(H)(SnMe_3)$, **8**, 402
$IrSnC_{26}H_{29}Cl_3P$
$Ir(cod)(H)_2(SnCl_3)(PPh_3)$, **8**, 402
$IrTaC_{14}H_{14}O_2$
$Cp_2Ta(CH_2)_2Ir(CO)_2$, **8**, 304; **10**, 28, 41, 49
$IrV_2C_{30}H_{29}ClPS_4$
$(MeCp)_2V_2S_4IrCl(PPh_3)$, **5**, 32
$IrWC_{17}H_{15}O_7$
$W(CO)_5Ir(CO)_2Cp^*$, **10**, 29
$IrWC_{47}H_{36}O_5P_3$
$WIr(CO)_5(H)(PPh_3)(PPh_2)_2$, **8**, 333
$Ir_2As_4AuC_{65}H_{58}ClOP_4$
$[IrCl(CO)[AsPh(CH_2PPh_2)_2]_2$-$Au]^{2+}$, **10**, 68
$Ir_2AuFe_3C_{30}H_{15}O_{12}P$
$[Fe_2Ir_2(CO)_{12}(AuPPh_3)]^-$, **7**, 83
$Ir_2BC_{20}H_{35}$
$[Cp^*IrH]_2(H)(BH_4)$, **1**, 285
$Ir_2B_4C_{57}H_{45}O_3P_2$
$(CO)_3(PPh_3)(Ph_2P\overline{C_6H_4)_2(Ir_2B}_4$-$H_2)$, **1**, 322
$Ir_2C_7H_3N_2O_4$
$Ir_2(CO)_4(pz)$, **8**, 330
$Ir_2C_8H_{16}Cl_2$
$[Ir(C_2H_4)_2Cl]_2$, **8**, 379, 380
$Ir_2C_{10}H_{10}Cl_4$
$Cp_2Ir_2Cl_4$, **8**, 385
$Ir_2C_{12}H_{18}O_4S_2$
$Ir_2(CO)_4(SBu^t)_2$, **8**, 330
$Ir_2C_{13}H_{20}I_2O_4S_2$
$Ir_2(CO)_4(I)_2(SBu^t)_2(CH_2)$, **8**, 330, 331
$Ir_2C_{16}H_{24}Cl_2$
$[IrCl(cod)]_2$, **8**, 364–6
$Ir_2C_{16}H_{24}I_4$
$Ir_2(cod)_2I_4$, **8**, 366, 367
$Ir_2C_{16}H_{28}Cl_2$
$[Ir(Cl)(C_8H_{14})]_2$, **8**, 317
$Ir_2C_{16}H_{36}O_2P_2S_2$
$Ir_2(CO)_2(PMe_3)_2(SBu^t)_2$, **8**, 330
$Ir_2C_{18}H_{30}O$
$[Ir(OMe)(cod)]_2$, **8**, 370
$Ir_2C_{18}H_{30}O_2$
$[Ir(cod)(OMe)]_2$, **8**, 366, 367, 375
$Ir_2C_{20}H_{30}Cl_4$
$Cp^*{}_2Ir_2Cl_4$, **8**, 304, 385, 386
$Ir_2C_{20}H_{32}Cl_2$
$Cp^*{}_2Ir_2Cl_2(H)_2$, **8**, 378, 379
$Ir_2C_{20}H_{33}$
$Cp^*{}_2Ir_2(H)_3{}^+$, **8**, 385, 405
$Ir_2C_{20}H_{33}O_3$
$[Cp^*Ir]_2(OH)_3$, **8**, 389, 390
$Ir_2C_{22}H_{30}N_4$
$[Ir(cod)]_2(pz)_2$, **8**, 395
$Ir_2C_{27}H_{22}Cl_2O_2P_2$
$Ir_2(CO)_2(Cl)_2(dppm)$, **8**, 326
$Ir_2C_{28}H_{22}Cl_2O_3P_2$
$Ir_2(Cl)_2(CO)_3(dppm)$, **8**, 324
$Ir_2C_{32}H_{32}Cl_2$
$[IrCl(cot)_2]_2$, **8**, 334
$Ir_2C_{32}H_{56}Cl$
$[IrCl(C_8H_{14})_2]_2$, **8**, 324
$Ir_2C_{32}H_{56}Cl_2$
$[(C_8H_{14})_2IrCl]_2$, **8**, 348, 380, 381
$Ir_2C_{34}H_{72}O_8P_2S_2$
$Ir_2(CO)_2[P(OBu^t)_3]_2(SBu^t)_2$, **8**, 330
$Ir_2C_{36}N_2O_3P_2$
$Ir_2(NO)_2(O)(PPh_3)_2$, **8**, 331
$Ir_2C_{40}H_{44}P_2$
$[Ir(cod)(PPh_2)]_2$, **8**, 372
$Ir_2C_{42}H_{30}O_6P_2$
$Ir_2(CO)_6(PPh_3)_2$, **8**, 304
$Ir_2C_{52}H_{44}Cl_2O_2P_4$
$Ir_2Cl_2(CO)_2(dppm)_2$, **8**, 324
$Ir_2C_{52}H_{44}O_2P_4S$
$Ir_2(CO)_2(dppm)_2(S)$, **8**, 324
$Ir_2C_{53}H_{44}O_3P_4$
$Ir_2(CO)_3(dppm)_2$, **8**, 324, 326
$Ir_2C_{54}H_{44}Cl_2O_4P_4$
$[Ir_2(Cl)(CO)_4(dppm)_2]Cl$, **8**, 324
$Ir_2C_{54}H_{57}P_4$
$[Ir_2H_5(dppp)_2]^+$, **8**, 487
$Ir_2C_{56}H_{52}Cl_2O_2P_4$
$Ir_2(CO)_2(Cl)_2(dppp)_2$, **8**, 326, 327
$Ir_2C_{94}H_{44}Cl_2O_2P_4$
$C_{60}[Ir(CO)Cl(PMe_2Ph)_2]_2$, **8**, 306
$Ir_2Co_2C_8F_{12}O_8P_4$
$Co_2Ir_2(CO)_8(PF_3)_4$, **8**, 512
$Ir_2Co_2C_{12}O_{12}$
$Co_2Ir_2(CO)_{12}$, **8**, 505, 510; **10**, 43
$Ir_2Co_2C_{18}H_{27}O_{18}P_3$
$Co_2Ir_2(CO)_9[P(OMe)_3]_3$, **8**, 511
$Ir_2Nb_4W_8C_4HO_{42}$
$[(CO)_2Ir]_2H(Nb_2W_4O_{19})_2{}^{5-}$, **8**, 323
$Ir_2Pt_2C_{61}H_{45}O_7P_3$
$Ir_2Pt_2(CO)_7(PPh_3)_3$, **10**, 41
$Ir_2Rh_2C_{12}O_{12}$
$Rh_2Ir_2(CO)_{12}$, **8**, 505, 510
$Ir_2Rh_2C_{27}H_{30}O_7$
$Rh_2Ir_2(CO)_7Cp^*{}_2$, **8**, 512
$Ir_2Ru_4C_{22}H_2O_{22}$
$[Re_4Ir_2(H)_2(CO)_{22}]^{2-}$, **10**, 199
$Ir_3C_6O_6S$
$[Ir_3(S)(CO)_6]^-$, **8**, 488
$Ir_3C_{18}H_{15}O_3$
$Cp_3Ir_3(CO)_3$, **8**, 490
$Ir_3C_{24}H_{36}IO_2$
$Ir_3(cod)_3(I)(O)_2$, **8**, 366, 489
$Ir_3C_{24}H_{37}S_2$
$Ir_3H(S)_2(cod)_3$, **8**, 489
$Ir_3C_{27}H_{15}O_7P_2$
$Ir_3(CO)_7(PhPC{=}CPPh_2)$, **8**, 489
$Ir_3C_{29}H_{54}O_5P_3$
$Ir_3(CO)_5(PBu^t{}_2)_3$, **8**, 487
$Ir_3C_{32}H_{51}P_2S_2$
$Ir_3H(S)_2(cod)_2(PBu^t{}_2H)_2$, **8**, 489
$Ir_3C_{33}H_{64}P_3S_2$
$Ir_3H(S)_2(cod)_3(PMe_3)_3$, **8**, 489
$Ir_3C_{43}H_{32}O_6P_3$
$Ir_3(Ph)(PPh)(dppm)(CO)_6$, **8**, 490
$Ir_3C_{51}H_{48}N_2O_5P_3$
$Ir_3(PPh_2)_3(CO)_5(CNBu^t)_2$, **8**, 487
$Ir_3C_{59}H_{111}N_3P_3$
$[Ir_3H_7(PCy_3)_3(py)_3]^{2+}$, **8**, 487
$Ir_3C_{75}H_{60}O_3P_5$
$Ir_3(PPh_2)_3(CO)_3(PPh_3)_2$, **8**, 487
$Ir_3C_{81}H_{85}P_6$
$[Ir_3H_7(dppp)_3]^{2+}$, **8**, 487
$Ir_3C_{82}H_{83}OP_6$
$[Ir_3H_5(CO)(dppp)_3]^{2+}$, **8**, 487
$Ir_3FeC_{12}O_{12}$
$[FeIr_3(CO)_{12}]^-$, **7**, 83; **10**, 191, 192
$Ir_3Fe_2C_{14}O_{14}$
$[Fe_2Ir_3(CO)_{14}]^-$, **7**, 83
$Ir_3Rh_9C_{30}O_{30}$
$[Rh_9Ir_3(CO)_{30}]^{2-}$, **10**, 42
$Ir_3RuC_{12}O_{12}$
$[RuIr_3(CO)_{12}]^-$, **10**, 191, 192
$Ir_3W_{12}C_{108}H_{96}O_{40}P_7$

$[(Ph_3P)_2IrH_2]_3PW_{12}O_{40}$, **8**, 404
$Ir_3W_{12}C_{132}H_{126}O_{40}P_7$
$[(PPh_3)_2Ir(cod))]_3PW_{12}O_{40}$, **8**, 323
$Ir_4As_2C_{20}H_{16}O_{10}$
$Ir_4(CO)_{10}(diars)$, **8**, 499
$Ir_4As_2C_{21}H_{16}O_{11}$
$Ir_4(CO)_{11}(diars)$, **8**, 495
$Ir_4As_4C_{36}H_{72}O_4$
$Ir_4(CO)_4(AsBu^t{}_2)_4$, **8**, 497
$Ir_4C_{10}H_2O_{10}$
$[Ir_4H_2(CO)_{10}]^{2-}$, **8**, 492
$Ir_4C_{11}BrO_{11}$
$[Ir_4(CO)_{11}Br]^-$, **8**, 492, 495, 497
$Ir_4C_{11}HO_{11}$
$[Ir_4H(CO)_{11}]^-$, **8**, 492, 499
$Ir_4C_{11}H_2O_{11}$
$Ir_4H_2(CO)_{11}$, **8**, 492
$Ir_4C_{11}IO_{11}$
$[Ir_4(CO)_{11}I]^-$, **8**, 497
$Ir_4C_{11}O_{11}$
$[Ir_4(CO)_{11}]^{2-}$, **8**, 492
$Ir_4C_{11}O_{13}S$
$Ir_4(CO)_{11}(SO_2)$, **8**, 498
$Ir_4C_{12}H_6O_9S_2$
$Ir_4(CO)_9(trithiane)$, **8**, 497
$Ir_4C_{12}NO_{11}S$
$[Ir_4(CO)_{11}(SCN)]^-$, **8**, 489, 493
$Ir_4C_{12}O_{12}$
$Ir_4(CO)_{12}$, **8**, 490–2, 495, 497
$Ir_4C_{13}H_3NO_{11}$
$Ir_4(CO)_{11}(CNMe)$, **8**, 494
$Ir_4C_{13}H_3O_{13}$
$[Ir_4(CO)_{11}(CO_2Me)]^-$, **8**, 492, 494
$Ir_4C_{13}H_4O_{11}$
$Ir_4(CO)_{11}(C_2H_4)$, **8**, 499
$Ir_4C_{14}H_5O_{13}$
$[Ir_4(CO)_{11}(CH_2CO_2Me)]^-$, **8**, 493
$Ir_4C_{16}H_9NO_{11}$
$Ir_4(CO)_{11}(CNBu^t)$, **8**, 493
$Ir_4C_{16}H_{10}O_{14}$
$[Ir_4(CO)_{10}(CH_2CO_2Me)_2]^{2-}$, **8**, 493
$Ir_4C_{17}H_7O_{11}P$
$Ir_4(CO)_{11}(PH_2Ph)$, **8**, 500
$Ir_4C_{18}H_{10}O_{11}$
$Ir_4(CO)_{11}(norbornene)$, **8**, 498
$Ir_4C_{20}H_{32}O_8P_2$
$Ir_4(CO)_8(dmpe)_2$, **8**, 495
$Ir_4C_{20}H_{36}O_8P_4$
$Ir_4(CO)_8(PMe_3)_4$, **8**, 495
$Ir_4C_{22}H_{11}O_{10}P$
$Ir_4H(CO)_{10}(PPh_2)$, **8**, 497
$Ir_4C_{23}H_{11}O_{11}P$
$Ir_4(CO)_{11}(PHPh_2)$, **8**, 497; **10**, 261
$Ir_4C_{26}H_{45}O_{26}P_5$
$Ir_4(CO)_{11}[P(OMe)_3]_5$, **8**, 503
$Ir_4C_{34}H_{23}O_9P$
$Ir_4(CO)_9(PPh_3)(nbd)$, **8**, 499
$Ir_4C_{35}H_{56}$
$Ir_4(H)_4(cod)_4(HCCMe)$, **8**, 499
$Ir_4C_{39}H_{26}O_9P_2$
$Ir_4(CO)_9(H)(PPh_2)(PPh_3)$, **8**, 497
$Ir_4C_{39}H_{31}O_7P$
$Ir_4(CO)_7(PPh_3)(nbd)_2$, **8**, 499
$Ir_4C_{46}H_{30}O_{10}P_2$
$Ir_4(CO)_{10}(PPh_3)_2$, **8**, 494
$Ir_4C_{46}H_{31}O_9P_3$
$Ir_4(CO)_9[HC(PPh_2)_3]$, **8**, 495
$Ir_4(CO)_9(tripod)$, **8**, 497
$Ir_4C_{54}H_{88}O_6P_4$
$Ir_4(CO)_6(PCy_2)_4$, **8**, 497
$Ir_4C_{63}H_{45}O_9P_3$
$Ir_4(CO)_9(PPh_3)_3$, **8**, 495
$Ir_4C_{84}H_{65}O_6P_5$
$Ir_4(CO)_6(PPh_3)_4(PPh)$, **8**, 495
$Ir_4FeC_{15}O_{15}$
$[FeIr_4(CO)_{15}]^{2-}$, **7**, 83; **10**, 192
$Ir_4FeNa_2C_{15}O_{15}$
$Na_2[FeIr_4(CO)_{15}]$, **10**, 43
$Ir_4RuC_{15}O_{15}$
$[RuIr_4(CO)_{15}]^{2-}$, **10**, 192
$Ir_4SiC_{22}H_{33}O_9P_3$
$Ir_4(CO)_9[MeSi(PEt_2)_3]$, **8**, 495
$Ir_5C_{26}H_{20}O_{10}P$
$Ir_5(CO)_{10}(PPh)Cp^*$, **8**, 500
$Ir_5FeC_{15}HO_{15}$
$[FeIr_5H(CO)_{15}]^{2-}$, **7**, 83; **10**, 197
$Ir_5FeC_{15}H_2O_{15}$
$[H_2FeIr_5(CO)_{15}]^-$, **7**, 83
$Ir_5FeC_{15}O_{15}$
$[FeIr_5(CO)_{15}]^{3-}$, **7**, 83
$Ir_5FeC_{16}O_{16}$
$[FeIr_5(CO)_{16}]^-$, **10**, 192
$Ir_6C_{16}O_{16}$
$Ir_6(CO)_{16}$, **8**, 500
$Ir_6C_{14}ClO_{14}$
$[Ir_6(CO)_{14}Cl]^-$, **8**, 502
$Ir_6C_{15}ClO_{15}$
$[Ir_6(CO)_{15}Cl]^-$, **8**, 503
$Ir_6C_{15}NO_{16}$
$[Ir_6(CO)_{15}(NO)]^-$, **8**, 502
$Ir_6C_{15}O_{15}$
$[Ir_6(CO)_{15}]^{2-}$, **8**, 501
$Ir_6C_{16}NO_{15}S$
$[Ir_6(CO)_{15}(SCN)]^-$, **8**, 489
$Ir_6C_{16}O_{16}$
$Ir_6(CO)_{16}$, **8**, 491
$Ir_6C_{17}H_3O_{17}$
$[Ir_6(CO)_{15}(CO_2Me)]^-$, **8**, 501, 503
$Ir_6C_{18}H_5O_{16}$
$[Ir_6(CO)_{15}(COEt)]^-$, **8**, 502, 503
$Ir_6C_{19}H_6O_{19}$
$[Ir_6(CO)_{15}(CO_2Me)_2]^-$, **8**, 501
$Ir_6C_{28}H_{10}O_{14}$
$Ir_6(CO)_{14}(PhCCPh)$, **8**, 503
$Ir_6C_{40}H_{20}O_{12}$
$Ir_6(CO)_{12}(PhCCPh)_2$, **8**, 503
$Ir_6C_{84}H_{60}O_{16}P_4$
$Ir_6(CO)_{12}[P(OPh_3)]_4$, **8**, 503
$Ir_6Rh_6C_{30}O_{30}$
$[Rh_6Ir_6(CO)_{30}]^{2-}$, **10**, 42
$Ir_8C_{20}O_{20}$
$[Ir_8(CO)_{20}]^{2-}$, **8**, 492
$Ir_8C_{22}O_{22}$
$[Ir_8(CO)_{22}]^{2-}$, **8**, 499
$Ir_9C_{20}O_{20}$
$[Ir_9(CO)_{20}]^{3-}$, **8**, 504
$Ir_{12}C_{26}O_{26}$
$[Ir_{12}(CO)_{26}]^{2-}$, **8**, 504
$Ir_{19}H_{14}F_4O_2$
Ir(acac)(tfb), **8**, 389, 390

K

$KAlC_{12}H_{28}$
  $K(AlBu^i_3H)$, **1**, 447
$KAlC_{16}H_{36}$
  $[Bu^i_4Al]^-K^+$, **1**, 446
$KB_2FeC_4H_5O_4$
  $K[Fe(CO)_4(B_2H_5)]$, **1**, 531
$KB_6FeC_4H_9O_4$
  $K[Fe(CO)_4B_6H_9]$, **1**, 327
$KBa_2Si_5C_{98}H_{95}O_9$
  $KBa_2(OSiPh_3)_5(DME)_2$, **2**, 34
KCH_3
  MeK, **1**, 20
$KC_{52}H_{59}O_3$
  $K[C_5(Bz)_5{\cdot}(THF)_3]$, **1**, 22–4
$KCaC_{24}H_{36}O_4$
  $[K(DME)]_2[Ca(cot)_2]$, **1**, 117
$KGaSi_3C_{12}H_{34}$
  $KGaH(CH_2TMS)_3$, **1**, 516
$KGeH_3$
  H_3GeK, **2**, 175
$KInC_{15}H_{34}$
  $KHIn(CH_2Bu^t)_3$, **1**, 516
$KInSi_3C_{12}H_{34}$
  $KInH(CH_2TMS)_3$, **1**, 516
$KInSi_4C_{16}H_{44}$
  $KIn(CH_2TMS)_4$, **1**, 516
$KIn_2C_{30}H_{67}$
  $K[H[In(CH_2Bu^t)_3]_2]$, **1**, 516, 517
$KLuSi_6C_{21}H_{57}Cl$
  $[(TMS)_2CH]_3Lu(Cl)K$, **4**, 19
$KLuSi_6C_{25}H_{67}ClO$
  $[(TMS)_2CH]_3Lu(Cl)K(Et_2O)$, **4**, 19
$KNiC_6H_5O$
  $K[NiCp(CO)]$, **9**, 5
$KReC_{12}H_{24}P_2$
  $[Re(PMe_3)_2(C_6H_6)][K]$, **6**, 210
$KRuC_8H_5N_2O$
  $K[Ru(CN)_2(CO)Cp]$, **7**, 482
$KRu_3C_{13}H_6NO_{11}$
  $K[Ru_3[C(O)NMe_2](CO)_{10}]$, **7**, 748
KSeCN
  KSeCN, **11**, 528
$KSiC_3H_9$
  TMS-K, **2**, 99
$KSiC_{18}H_{15}O$
  Ph_3SiOK, **2**, 34
$KSi_2C_{26}H_{26}N$
  $[(TMS)_2NK{\cdot}Toluene]_2$, **1**, 6
$KSnC_{15}H_{33}$
  $[Bu^tCH_2]_3SnK$, **2**, 283
$KSnC_{18}H_{15}$
  Ph_3SnK, **2**, 283
$KSnC_{30}H_{45}$
  $(Me_2CH)_3SnK{\cdot}3Toluene$, **1**, 15, 16
$KUC_{16}H_{22}$
  $(MeC_7H_8)U(MeC_7H_8)K$, **4**, 181
$KVC_8H_8O_3Ge$
  $[CpV(CO)_3(GeH_3)]K$, **5**, 21
$KVC_{13}H_{17}$
  $[CpV(C_8H_{12})]K$, **5**, 24
$K_2Al_{12}C_{48}H_{108}$
  $[Bu^iAl]_{12}K_2$, **1**, 446
$K_2C_{38}H_{50}N_4$
  $[K(fluorenyl){\cdot}TMEDA]_2$, **1**, 23, 25
$K_2CaC_{32}H_{48}O_4$
  $[K(THF)_2]_2Ca(cot)_2$, **1**, 119
$K_2Co_2C_{16}H_{44}P_4$
  $[K[Co(C_2H_4)(PMe_3)_2]_2]$, **1**, 26
$K_2FeC_4O_4$
  $K_2Fe(CO)_4$, **12**, 352
$K_2Os_2C_{48}H_{72}P_6$
  $[K[OsH_3(PMe_2Ph)_3]]_2$, **1**, 26, 28
$K_2Os_3C_{11}O_{11}$
  $K_2[Os_3(CO)_{11}]$, **7**, 719
$K_2Ru_2C_{108}H_{92}P_2$
  $K_2[(Ph_3P)_2Ph_2PC_6H_4RuH_2]_2$, **1**, 26, 27
$K_2YbC_{20}H_{36}O_2$
  $[K(DME)]_2Yb(cot)_2$, **1**, 25–7
$K_2Zr_2C_{32}H_{42}P_2$
  $K_2[ZrCp_2(PC_6H_{11})]_2$, **4**, 468
$K_3FeC_6N_6$
  $K_3Fe(CN)_6$, **12**, 1145
$K_3Mn_3C_{12}O_{12}$
  $K_3[Mn_3(CO)_{12}]$, **6**, 12
$K_3Rh_7C_{16}O_{16}$
  $K_3[Rh_7(CO)_{16}]$, **8**, 476
$K_4NiC_4N_4$
  $K_4[Ni(CN)_4]$, **9**, 14
$K_4Si_4C_{12}H_{36}O_4$
  $[KO\text{-}TMS]_4$, **1**, 12
$K_4Si_4C_{56}H_{68}O_4$
  $[C_6H_6KOSiMe_2Ph]_4$, **2**, 34
$K_5WC_7N_7$
  $K_5[W(CN)_7]$, **5**, 305
$K_8Si_8C_{156}H_{150}O_{14}$
  $K_8(OSiPh_3)_8(DME)_3$, **2**, 34

L

$LaC_9H_{11}O$
$BzLa(H)(CHCH_2)$, **4**, 21
$LaC_{15}H_{15}$
$LaCp_3$, **4**, 52
$LaC_{18}H_{21}$
$La(MeC_5H_4)_3$, **4**, 53
$LaC_{18}H_{22}NO$
$Cp_3La(DMF)$, **4**, 56
$LaC_{21}H_{25}N_2$
$Cp_3La(EtCN)_2$, **4**, 57
$LaC_{22}H_{26}N$
$Cp_3La(CNC_6H_{11})$, **4**, 57
$LaC_{32}H_{44}Cl_2O_4$
$[[C_5H_4(CH_2)_2OMe]_2La(Cl)]_2$, **4**, 38
$LaC_{43}H_{41}OP_2$
$CpLa(C_5H_4PPh_2)_2(THF)$, **4**, 54
$LaNaC_{24}H_{28}O$
$Na[La(Cp)_4](THF)$, **4**, 58
$LaSi_2C_{29}H_{58}O_3$
$[Cp^*La[CH(TMS)_2(THF)_3]]^-$, **4**, 72
$LaSi_4C_{24}H_{53}$
$Cp^*La[CH(TMS)_2]_2$, **4**, 72
$LaSi_6C_{21}H_{57}$
$La[CH(TMS)_2]_3$, **4**, 23
$La_2C_{40}H_{62}I_4O_6$
$(PhCHCHCHCHPh)[LaI_2-(THF)_3]_2$, **4**, 26
$La_2C_{48}H_{74}O$
$Cp^*_2(THF)La(CH_2CHCHCH_2)-LaCp^*_2$, **4**, 89
$LiAgSi_6C_{36}H_{86}O_4$
$[Ag[C(TMS)_3]_2][Li(THF)_4]$, **3**, 106
$LiAlC_9H_{22}$
$LiAlBu^i_2MeH$, **11**, 293, 302
$LiAlC_{24}H_{20}$
$LiAlPh_4$, **11**, 293
$LiAlH_4$
$LiAlH_4$, **11**, 279; **12**, 135, 1073
$LiAlNiC_{24}H_{52}N_4$
$[Li(TMEDA)_2][Ni(cdt)(AlH_4)]$, **1**, 551
$LiAlNi_2C_{48}H_{84}N_4$
$[Li(TMEDA)_2][Ni(cdt)]_2-[(H)_2AlBu^i_2]$, **1**, 551
$LiAlSi_4C_{12}H_{36}$
$(TMS)_4AlLi$, **2**, 22
$LiAsC_8H_{12}$
$(Me_4C_4As)Li$, **2**, 324
$LiAsC_{16}H_{20}O$
$Ph_2AsLi(OEt_2)$, **2**, 328
$LiAsC_{24}H_{34}O_6$
$LiAsPh_2\cdot 3(dioxane)$, **1**, 11
$LiAsSi_4C_{14}H_{38}$
$[(TMS)_2CH]_2AsLi$, **2**, 328
$LiAsSi_4C_{47}H_{83}FO_2$
$LiAs(Pr^i_3C_6H_2Si)_2(F)Si-(Pr^i_3Si)\cdot 2THF$, **1**, 11
$LiAs_2C_{20}H_{43}O_2$
$Bu^t_2AsAsBu^tLi(THF)_2$, **2**, 328
$LiBC_4H_{11}N$
$(pyrrolidine)BH_3Li$, **11**, 244
$LiBC_8H_{16}O_2$
$OC(Me)_2C(Me)_2OBCH(Li)Me$, **11**, 208, 209
$LiBC_{18}H_{33}$
$(C_{10}H_{16})-(BC_8H_{15})Li$, **11**, 260, 261
$LiBC_{26}H_{40}O$
$[(BzO)CH_2CH_2](C_9H_{14})-(BC_8H_{15})Li$, **11**, 260, 261
$LiB_{10}H_{15}$
$Li[(B_5H_7)(B_5H_8)]$, **1**, 358
$LiBaC_{25}H_{50}O_2$
$Li[BaCp^*CH(TMS)_2](THF)_2$, **1**, 119
$LiBaSi_2C_{35}H_{65}O_2$
$Li[Cp_2^*BaCH(TMS)_2](THF)_2$, **1**, 115
$LiBeC_{12}H_{27}$
$LiBeBu^t_3$, **1**, 42
$LiCH_3$
MeLi, **11**, 3
LiCN
LiCN, **12**, 225
$LiC_2H_{11}N_4$
$(LiNH)(NH_2)CH_2CH_2(NH_2)_2$, **11**, 5
LiC_4H_4N
$(C_4H_4N)Li$, **11**, 5
LiC_4H_6N
$Li[(Me)_2CC{\equiv}N]$, **12**, 988
LiC_4H_8NO
$CH_2NLiCH_2CH_2OCH_2$, **11**, 5
LiC_4H_9
BuLi, **11**, 3
Bu^sLi, **11**, 3
Bu^tLi, **11**, 3
$LiC_4H_{10}N$
$(Et_2N)Li$, **11**, 5
LiC_4H_{12}
$[MeLi]_4$, **1**, 19, 20
$LiC_5H_9S_2$
$LiMeCSCH_2CH_2CH_2S$, **12**, 988
$LiC_5H_{10}N$
$(C_5H_{10}N)Li$, **11**, 5
$LiC_5H_{10}N_4$
$(CH_2N)_2CH_2CH_2(NLi)(NCH_2)$, **11**, 5
LiC_6H_5
PhLi, **11**, 3
LiC_6H_9
$LiCCBu^t$, **1**, 21
$LiC_6H_{14}N$
$(Pr^i_2N)Li$, **1**, 5; **11**, 4
$LiC_9H_{11}NO$
$Li[PhC(O)NMe_2]$, **11**, 5
$LiC_9H_{18}N$
$[(C_6H_{11})Pr^iN]Li$, **11**, 5
$(Me_4C_5H_6N)Li$, **11**, 5
$LiC_{10}H_7$
(nap)Li, **11**, 4
$LiC_{10}H_{10}$
$[LiCp_2]^-$, **1**, 22, 23
$LiC_{12}H_{26}N$
$[(C_8H_{17})Bu^tN]Li$, **11**, 5
$LiC_{12}H_{29}O_2$
$Bu^tLi\cdot 2Et_2O$, **1**, 19, 20
$LiC_{15}H_{23}N_2$
Li(indenyl)·TMEDA, **1**, 24
$LiC_{19}H_{23}O$
$PhCHCHC(Ph)HLi\cdot Et_2O$, **1**, 19, 22, 24
$LiCdSi_3C_{33}H_{49}Br_2O_2$
$Li(THF)_2CdBr_2[C(SiMe_2Ph)_3]$, **3**, 198
$LiCeC_{21}H_{42}O_6$
$CeMe_6[Li(acac)]_3$, **4**, 21
$LiCeC_{32}H_{58}Cl_2N_2$
$(Bu^t_2C_5H_3)_2Ce(Cl)_2Li-(TMEDA)$, **4**, 37
$LiCoC_{11}H_{31}P_3$
$Li[Co(C_2H_4)(PMe_3)_3]$, **1**, 26
$LiCrC_9H_4ClO_3$
$(ClLiC_6H_4)Cr(CO)_3$, **12**, 1024
$LiCrC_9H_4FO_3$
$(FLiC_6H_4)Cr(CO)_3$, **12**, 1022, 1036
$LiCrC_9H_5O_3$
$(LiPh)Cr(CO)_3$, **12**, 1018
$LiCrC_{10}H_6FO_4$
$[FLi(MeO)C_6H_3]Cr(CO)_3$, **12**, 1036
$LiCrC_{16}H_{44}N_4$
$(TMEDA)Li(Me)_2Cr(Me)_2-Li(TMEDA)$, **5**, 323
$LiCuC_2H_3N$
MeCu(CN)Li, **12**, 85
$LiCuC_2H_6$
Me_2CuLi, **12**, 100
$LiCuC_4H_9I$
BuCu·LiI, **12**, 83
$LiCuC_4H_{10}$
Et_2CuLi, **12**, 107
$LiCuC_5H_3NS$
LiCu(CN)(thienyl), **12**, 64, 79, 107
$LiCuC_8H_{18}$
Bu^t_2CuLi, **12**, 107

$LiCuC_{12}H_{10}$
Ph_2CuLi, **12**, 107
$LiCuC_{25}H_{42}O_3P$
$LiCu(PBu^t_2)Me(THF)_3$, **3**, 103
LiCuClI
CuI·2LiCl, **12**, 83
$LiCuSiC_9H_{11}N$
$PhMe_2SiCu(CN)Li$, **12**, 96
$LiCuSi_6C_{28}H_{86}O_4$
$[Cu[C(TMS)_3]_2][Li(THF)_4]$, **3**, 106
$LiCuSn_2C_{24}H_{54}$
$(Bu_3Sn)_2CuLi$, **2**, 231
$LiCu_2C_3H_9$
$LiCu_2Me_3$, **3**, 65
$LiCu_4C_{36}H_{30}$
$[LiCu_4Ph_6]^-$, **3**, 109
$LiCu_5C_{52}H_{62}O_4$
$[Li(THF)_4][Cu_5Ph_6]$, **3**, 65
$LiErC_{18}H_{32}N_2$
$Cp_2Er(Me)_2Li(TMEDA)$, **4**, 43
$LiFeC_{27}H_{33}$
$Li[Fe(Mes)_3]$, **7**, 110
$LiFe_2C_7HO_6S_2$
$Fe_2(CO)_6S_2CHLi$, **7**, 68
$LiFe_2C_{18}H_{10}O_6P$
$Li[Fe_2(CO)_6(PPh_2)]$, **7**, 43
$LiGa_2C_{20}H_{45}P_2$
$Bu^tGa(Bu^t_2Ga)(PBu^t_2Li)$, **1**, 523, 524
$LiGeC_6H_{15}$
Et_3GeLi, **2**, 176
$LiGeSiC_7H_{19}$
$(TMS)Et_2GeLi$, **2**, 176
$LiIn_2Si_6C_{28}H_{75}$
$(TMS)_3CIn(H)_2Li(THF)_2(H)In(H)_2C(TMS)_3$, **1**, 515, 516
$LiIrC_{19}H_{41}N_3$
$Cp^*IrH_3Li(pmdeta)$, **8**, 385
$LiIrSiC_{22}H_{49}N_3$
$Cp^*IrH_2(TMS)Li(pmdeta)$, **8**, 385
$LiLuC_{24}H_{49}N_2S_2$
$Cp^*Lu(SBu^t)_2Li(TMEDA)$, **4**, 73
$LiLuC_{40}H_{46}N_2P_2$
$Cp_2Lu(PPh_2)_2Li(TMEDA)$, **4**, 42
$LiMgSi_3C_{21}H_{51}Br_2$
$Li(THF)_2(Br)_2Mg[C(TMS)_3(THF)]$, **1**, 85
$LiMnC_8Cl_4O_3$
$Mn(CO)_3(C_5Cl_4Li)$, **6**, 110
$LiMnC_{14}H_{18}O_2$
$Li[Mn(C_6Me_6)(CO)_2]$, **6**, 70
$LiMn_2C_{10}HO_{10}$
$Li[(CO)_5MnMn(CO)_4(CHO)]$, **6**, 26
$LiNaC_{20}H_{48}N_{12}$
$[(Me_2N)_2CN]_4LiNa$, **1**, 4, 5
$LiNiC_{13}H_{32}N_3$
$Li(PMPTA)(Me)Ni(C_2H_4)_2$, **9**, 110
$LiNiC_{14}H_{34}N_3$
$(C_2H_4)_2Ni(Me)Li(PMDTA)$, **9**, 34
$LiNiC_{15}H_{35}NO$
$Ni(C_2H_4)_2[C(O)NMe_2]-(PMPTA)Li$, **9**, 112
$LiNiC_{17}H_{43}N_4$
$[Li(TMEDA)_2][NiMe(C_2H_4)_2]$, **9**, 34, 110
$LiNiC_{18}H_{45}N_4$
$[Li(TMEDA)_2][NiEt(C_2H_4)_2]$, **9**, 34
$LiNiC_{23}H_{43}N_4$
$Ni(C_2H_4)_2[=CPh(NBu^t)]Li-(PMTA)$, **9**, 111
$LiNiC_{24}H_{59}N_6$
$[Li(PMPTA)_2][Ni_2(H)(C_2H_4)_4]$, **9**, 111
$LiPbC_{18}H_{15}$
Ph_3PbLi, **2**, 312
$LiPbC_{27}H_{38}N_3$
$Ph_3PbLi(pmdeta)$, **2**, 311
$LiPdCl_3$
$LiPdCl_3$, **11**, 404; **12**, 133, 135, 136
$LiReSiC_{25}H_{20}O_2$
$[Re(H)(CO)_2(C_5H_4SiPh_3)][Li]$, **6**, 211
$LiRhC_{13}H_{25}P$
$Li[Cp^*RhH(PMe_3)]$, **8**, 198
$LiRhC_{42}H_{39}OP_3$
Li[Rh(CO)(triphos)], **8**, 190
$LiRhSi_2C_{22}H_{50}N_2$
$(cod)Rh(CH_2TMS)_2Li-(TMEDA)$, **8**, 152
$LiRuC_{14}H_{23}$
$Li[Ru(C_2H_4)_2Cp^*]$, **7**, 505
$LiSbC_{21}H_{45}N_3$
$LiSb(C_6H_{11})_2$·PMDETA, **1**, 11
$LiSbSi_2C_6H_{18}$
$(TMS)_2SbLi$, **2**, 30
$LiSiC_3H_9$
TMSLi, **12**, 96
$LiSiC_7H_{13}$
$(TMS)C{\equiv}CCH_2CH_2Li$, **12**, 79
$LiSiC_8H_{11}$
$PhMe_2SiLi$, **12**, 96
$LiSiC_3H_9$
(TMS)SeLi, **2**, 36
$LiSiC_4H_9N_2$
$Li(TMS)CN_2$, **11**, 341
$LiSiC_8H_{11}$
$(Me_2PhSi)Li$, **2**, 94
$LiSiC_{10}H_{16}N$
[Bz(TMS)N]Li, **11**, 5
$LiSiC_{17}H_{22}N$
$[Bu^t(SiMePh_2)N]Li$, **11**, 5
$LiSiC_{18}H_{15}$
Ph_3SiLi, **2**, 94
$LiSiVC_{16}H_{38}O_3$
$[(TMS-CH_2)V(OBu^t)_3]Li$, **5**, 13
$LiSi_2C_6H_{18}N$
$(TMS)_2NLi$, **1**, 5; **11**, 5
$LiSi_2C_6H_{18}N_2$
$[SiMe_2NMe]_2Li$, **11**, 5
$LiSi_2C_6H_{18}P$
$(TMS)_2PLi$, **2**, 28
$LiSi_2VC_{26}H_{60}N_2$
$[(Bu^t_3SiN)_2VMe_2]Li$, **5**, 16
$LiSi_4C_{21}H_{51}O_3$
$(TMS)_3SiLi(THF)_3$, **2**, 22
$LiSi_4UC_{30}H_{58}Cl_2O_2$
$[(TMS)_2C_5H_3]_2U(Cl)_2Li(THF)_2$, **4**, 138
$LiSi_4VC_{24}H_{60}O_4$
$LiV(O\text{-}TBDMS)_4$, **5**, 14
$LiSnC_3H_9$
Me_3SnLi, **2**, 228, 284; **11**, 363
$LiSnC_4H_{11}$
$(Me_3Sn)CH_2Li$, **11**, 365
$LiSnC_5H_{15}$
Me_5SnLi, **2**, 225
$LiSnC_6H_9$
$(CH_2{=}CH)_3SnLi$, **2**, 228
$LiSnC_{12}H_{12}O$
$(furyl)_3SnLi$, **2**, 283
$LiSnC_{12}H_{27}$
Bu_3SnLi, **11**, 363
$LiSnC_{18}H_{15}$
Ph_3SnLi, **2**, 283
$LiTlC_{19}H_{33}N_3$
CpTlCpLi·PMDETA, **1**, 15
$LiU_2C_{30}H_{36}Cl_5O_2$
$[CH_2(C_5H_4)_2]_2U_2(Cl)_5Li(THF)_2$, **4**, 161
$LiVC_{16}H_{36}O_4$
$Li[Bu^tVO(Bu^nO)_3]$, **5**, 15
$LiVC_{91}H_{95}O_4P_6$
$[[Ph_2PCHPPh_2]_3V][Li(THF)_4]$, **5**, 12
$LiYC_{36}H_{56}O$
$Cp^*_2Y(CCBu^t)_2Li(THF)$, **4**, 90
$LiZnC_3H_9$
$LiMe_3Zn$, **11**, 182
$LiZnC_{12}H_{27}$
Bu_3ZnLi, **11**, 182
$LiZrC_3H_9$
Me_3ZnLi, **12**, 90
$Li_2Ag_2C_{32}H_{40}O_2$
$Li_2Ag_2(Ph)_4(OEt_2)_2$, **3**, 125
$Li_2Ag_2C_{36}H_{48}N_4$
$Li_2Ag_2(C_6H_4CH_2NMe_2)_4$, **3**, 58, 124, 125
$Li_2Ag_3C_{36}H_{30}$
$[Li_2Ag_3Ph_6]^-$, **3**, 109
$Li_2As_2C_{40}H_{60}O_4$
$[LiAsPh_2{\cdot}2Et_2O]_2$, **1**, 11
$Li_2As_4C_{32}H_{70}O_2$
$[LiAs(AsBu^t_2)Bu^t{\cdot}THF]_2$, **1**, 11
$Li_2Au_2C_{36}H_{48}N_4$
$Au_2Li_2[C_6H_4(CH_2NMe_2)]_4$, **3**, 13, 101
$Li_2B_2Fe_2C_6H_4O_6$
$Li_2[B_2H_4Fe_2(CO)_6]$, **1**, 307
Li_2CH_2
Li_2CH_2, **1**, 21
$Li_2C_{28}H_{41}ClN_2O_4$
$[(C_6H_4)_2(NH)NLi{\cdot}LiCl{\cdot}4THF]$, **1**, 10
$Li_2C_{30}H_{46}N_4$
$[LiCCPh{\cdot}TMPDA]_2$, **1**, 19, 21
$Li_2C_{30}H_{72}N_8O_2P_2$
$[Bu^t_2CNLi{\cdot}HMP]_2$, **1**, 3, 4

$Li_2C_{36}H_{48}N_2O_2$
$[(Bz)_2NLi\cdot Et_2O]_2$, **1**, 8
$Li_2C_{128}H_{140}NO_7$
$[Li_2N(C_{10}H_7)_{10}\cdot 6Et_2O]\cdot Et_2O$, **1**, 17
$Li_2CrC_{16}H_{46}N_4$
$[Li(TMEDA)]_2[CrMe_4]$, **5**, 288
Li_2CuCCl_2N, **12**, 72
$CuCN\cdot 2LiCl$, **12**, 67
$Li_2CuC_3H_6N$
$Me_2Cu(CN)Li_2$, **12**, 77, 85–7, 89
$Li_2CuC_3H_9$
Li_2CuMe_3, **3**, 65
$Li_2CuC_3H_{10}NO_2$
$Li_2CuMe_2(CN)(H_2O)_2$, **3**, 127
$Li_2CuC_8H_{18}I$
$Bu_2CuLi\cdot LiI$, **12**, 83
$Li_2CuC_9H_{10}N$
$(dioxenyl)_2Cu(CN)Li_2$, **12**, 76
$Li_2CuC_9H_{12}NS$
$Bu(thienyl)Cu(CN)Li_2$, **12**, 92
$Li_2CuC_9H_{18}N$
$Bu_2Cu(CN)Li_2$, **12**, 100
$Li_2CuSi_2C_4H_9N$
$(TMS)_2Cu(CN)Li_2$, **12**, 99
$Li_2CuSi_2C_{17}H_{22}N$
$(PhMe_2Si)_2Cu(CN)Li_2$, **12**, 97
$Li_2CuSnC_{14}H_{30}N$
$Bu_3Sn(Me)CuCNLi_2$, **2**, 231
$Li_2Cu_2C_4H_{12}$
$[Me_2CuLi]_2$, **12**, 121
$Li_2Cu_2Me_4$, **3**, 126
$Li_2Cu_2C_{28}H_{28}$
$Li_2Cu_2(Tol)_4$, **3**, 65
$Li_2Cu_2C_{30}H_{38}S_3$
$Li_2Cu_2Ph_4(DMS)_3$, **3**, 64, 101, 121, 124
$Li_2Cu_2C_{32}H_{40}O_2$
$Li_2Cu_2Ph_4(Et_2O)_2$, **3**, 64, 101, 124
$[[Li(Et_2O)](CuPh_2)]_2$, **1**, 26, 28
$Li_2Cu_2C_{36}H_{48}N_4$
$Li_2Cu_2(C_6H_4CH_2NMe_2)_4$, **3**, 58, 64, 101, 121, 124
$Li_2Cu_2C_{36}H_{48}O_2$
$Li_2Cu_2(Tol)_4(Et_2O)_2$, **3**, 64, 127
$Li_2Cu_2C_{40}H_{56}N_4$
$Li_2Cu_2[C_6H_4(CH(Me)NMe_2]_4$, **3**, 101
$Li_2Cu_2Si_4C_{20}H_{56}S_2$
$Li_2Cu_2(CH_2TMS)_4(DMS)_2$, **3**, 64, 101, 120
$Li_2Cu_3C_5H_{15}$
$Li_2Cu_3Me_5$, **3**, 65
$Li_2Cu_3C_{36}H_{30}$
$[Li_2Cu_3Ph_6]^-$, **3**, 109
$Li_2FeC_4H_{12}$
Li_2FeMe_4, **7**, 110
$Li_2FeC_{40}H_{30}N_2$
$Li_2[Fe(NPh_2)_2(C\equiv CPh)_2]$, **7**, 114
$Li_2FeC_{48}H_{48}O_2$
$[Li(OEt_2)]_2[Fe(nap)_4]$, **7**, 110
$Li_2GeC_{12}H_{10}$
Ph_2GeLi_2, **2**, 175
$Li_2MgC_4H_{12}$
Li_2MgMe_4, **1**, 105
$Li_2MgC_{16}H_{44}N_4$
$[Li(TMEDA)]_2MgMe_4$, **1**, 105
$Li_2MgC_{28}H_{28}$
Li_2MgBz_4, **1**, 105
$Li_2MgC_{40}H_{60}N_4$
$[Li(TMEDA)]_2MgBz_4$, **1**, 105
$Li_2MgC_{44}H_{30}N_4$
$[Li(TMEDA)]_2Mg(C\equiv CPh)_4$, **1**, 105
$Li_2Mg_2C_{18}H_{50}N_4$
$[Li(TMEDA)MgMe_3]_2$, **1**, 105
$Li_2Mg_2C_{48}H_{62}N_4$
$[Li(TMEDA)]_2Mg_2Ph_6$, **1**, 105
$Li_2Mg_2C_{48}H_{86}N_2$
$Li_2[Mg(TMEDA)(C\equiv CBu^t)_3]_2$, **1**, 105, 106
$Li_2NiC_{18}H_{40}O_4$
$Ni(C_2H_4)(C_4H_8Li_2)(O_2C_4H_8)(OEt_2)_2$, **9**, 35
$Li_2NiC_{28}H_{68}N_6$
$(C_2H_4)_2Ni(C_3H_7)_2[Li(PMDTA)]_2$, **9**, 34
Li_2PdCl_4
Li_2PdCl_4, **11**, 399, 401, 402, 404, 407; **12**, 133, 135, 136
$Li_2Ru_3C_{14}H_6O_{12}$
$Li_2[Ru_3[C(O)Me]_2(CO)_{10}]$, **7**, 748
$Li_2Si_2C_6H_{18}$
$[(TMS)Li]^+{}_2$, **2**, 22
$Li_2Si_2ZrC_{64}H_{65}Cl_2O_2P_2$
$[[LiCl]_2(THF)_2][ZrCp_2](PHSiPh_3)(PSiPh_3)(C_5H_4C_5H_4)$, **4**, 467
$Li_2TaC_{14}H_{25}O_2S_3$
$Li_2(DME)Cp^*TaS_3$, **5**, 120
$Li_2TaC_{16}H_{31}N_2Se_3$
$Cp^*TaSe_3Li_2(TMEDA)$, **5**, 120
$Li_2TaC_{22}H_{47}N_4S_3$
$Li_2(TMEDA)_2Cp^*TaS_3$, **5**, 120
$Li_2Ta_2C_{32}H_{54}O_4S_8$
$[Li_2(THF)_3][Cp^*Ta(S_2)_2]_2(O)$, **5**, 120
$Li_2WC_5O_5$
$Li_2[W(CO)_5]$, **5**, 160
$Li_2ZrSi_2C_{64}H_{65}Cl_2O_2P_2$
$[[Li(Cl)]_2(THF)_2][ZrCp_2(PHSiPh_3)(PSiPh_3)(C_5H_4C_5H_4)]$, **4**, 469
$Li_2Zr_2C_{56}H_{78}O_6P_2$
$[Li(THF)_3]_2[ZrCp_2(PPh)]_2$, **4**, 467
$Li_3As_3Si_6C_{24}H_{60}$
$[LiAs(CH_2TMS)_2]_3$, **1**, 11
$Li_3C_{12}H_{28}ClN_2$
$[(Me_2CH)_2NLi]_2\cdot LiCl$, **1**, 10
$Li_3Cu_2C_{30}H_{25}$
$[Li_3CuPh_3][CuPh_2]$, **3**, 65
$Li_3Cu_2C_{38}H_{49}S_4$
$Li_3Cu_2Ph_5(DMS)_4$, **3**, 101, 127
$Li_3(CuPh_2)(CuPh_3)(DMS)_4$, **3**, 101
$Li_3FeC_{30}H_{21}$
$Li_3[Fe(nap)_3]$, **7**, 110
$Li_3FeC_{30}H_{21}N_2$
$Li_3Fe(N_2)(nap)_3$, **7**, 110
$Li_3RuC_{24}H_{66}N_6$
$[RuMe_6][Li(TMEDA)]_3$, **7**, 377
$Li_3TaC_{22}H_{39}ClO_3Se_3$
$Cp^*TaSe_3Li_3Cl(THF)_3$, **5**, 120
$Li_4C_4H_{12}$
$[MeLi]_4$, **1**, 30
$Li_4C_{16}H_{36}$
$[Bu^tLi]_4$, **1**, 19, 20
$Li_4C_{22}H_{50}Br_2O_4$
$[(C_3H_5)Li\cdot Et_2O]_2\cdot(LiBr\cdot Et_2O)_2$, **1**, 29
$Li_4C_{24}H_{38}I_2N_4O_2$
$[(Me_2NCH_2)_2\cdot C_6H_3OLi]_2(LiI)_2$, **1**, 13, 14
$Li_4C_{28}H_{64}N_8$
$[(H_2C(CH_2)_3NLi)_2TMEDA]_2$, **1**, 6, 7
$Li_4C_{32}H_{44}S_4$
$Li_4Ph_4(DMS)_4$, **3**, 121
$Li_4C_{36}H_{45}BrO_3$
$(PhLi\cdot Et_2O)_3\cdot LiBr$, **1**, 29, 30
$Li_4C_{36}H_{48}N_4$
$Li_4(C_6H_4CH_2NMe_2)_4$, **3**, 121
$Li_4C_{40}H_4$
$[LiCCBu^t\cdot THF]_4$, **1**, 19, 21
$Li_4C_{40}H_{52}N_8$
$[H_2C(CH)_4NLi\cdot(H_2C(CH)_4NH)]_4$, **1**, 6
$Li_4C_{40}H_{60}O_4$
$[PhLi\cdot Et_2O]_4$, **1**, 29
$Li_4C_{43}H_{87}N_{13}O_7P_4S$
$(PhOLi)_3\cdot LiNCS\cdot 4HMP$, **1**, 13, 14
$Li_4C_{44}H_{52}N_4$
$(LiCCPh)_4(TMEDA)_2$, **1**, 19, 21, 22
$Li_4C_{72}H_{56}N_8$
$[Ph_2CNLi\cdot py]_4$, **1**, 3, 4
$Li_4FeC_{40}H_{60}O_4$
$[Li(OEt_2)]_4[FePh_4]$, **7**, 110
$Li_4FeC_{40}H_{62}O_4$
$[FeH_2Ph_4][Li(OEt_2)]_4$, **7**, 110
$Li_4Na_2C_{66}H_{90}N_6$
$[Bu^t(Ph)CN]_6Li_4Na_2$, **1**, 4, 5
$Li_4Na_4C_{36}H_{40}O_4$
$[(Me_2NaCH_2\cdot C_6H_2)OLi]_4$, **1**, 13
$Li_4Si_4C_{42}H_{86}O._8S_2$
$[LiPhSC(TMS)_2O_2Li\cdot 2Et_2O]_2$, **1**, 17, 18
$Li_5Cu_4C_{62}H_{69}S_4$
$Li_5Cu_4Ph_9(DMS)_4$, **3**, 101
$Li_5(CuPh_2)_3(CuPh_3)(DMS)_4$, **3**, 101
$Li_5FeC_{56}H_{73}O_5$
$[Li(THF)]_5[Ph_3Fe(H)_3FePh_3]$, **7**, 110
$Li_5NbC_{56}H_{63}O_5$
$Nb(C_6H_4)_2Ph_3(LiPh\cdot THF)(Li\cdot THF)_4$, **5**, 97
$Li_5SmC_{32}H_{72}O_8$
$Li_5Sm(OBu^t)_8$, **4**, 20
$Li_6C_{24}H_{54}$
$[BuLi]_6$, **1**, 19, 20

$Li_6C_{42}H_{94}N_{12}$
[(H$_2$C(CH$_2$)$_3$NLi)$_3$PMDETA]$_2$, **1**, 6, 8
$Li_6Fe_3C_{42}H_{56}N_4$
[(C$_5$H$_4$Li)$_2$Fe]$_3$(TMEDA)$_2$, **7**, 189
$Li_6Si_6C_{18}H_{54}$
[TMSLi]$_6$, **1**, 15
$Li_6Ta_2C_{50}H_{110}Cl_2N_{10}S_6$
[Li$_3$(TMEDA)$_2$Cp*TaS$_3$Cl]$_2$-(TMEDA), **5**, 120
$Li_8C_{32}H_{72}O_4$
[BuLi·Bu^{t}OLi]$_4$, **1**, 13
$Li_8C_{68}H_{88}N_8$
[LiPh$_2$CH$_2$N(Li)CH$_2$CH$_2$N-Me$_2$]$_4$, **1**, 17, 18
$Li_{11}C_{84}H_{78}N_9OP$
[(Ph$_2$CNLi)$_6$Li$_5$·HMP]$^-$, **1**, 13
$Li_{12}C_{88}H_{140}O_4$
(LiCCBu^t)$_{12}$·4THF, **1**, 19, 21, 22
$Li_{12}Si_6C_{100}H_{164}O_{23}S_6$
[(TMS-CSO$_2$Ph)Li$_2$]$_6$Li$_2$O·10THF, **1**, 17
$Li_{24}Si_{12}C_{84}H_{168}N_{12}O_6$
[(TMS-CCN)Li$_2$]$_{12}$·6Et$_2$O, **1**, 17
$Li_{24}Si_{24}C_{126}H_{290}N_{12}O_6$
[(TMS)$_2$C(Li$_2$)CN]$_{12}$·6Et$_2$O·C$_6$H$_{14}$, **1**, 17
LnC_6H_{10}
Ln(C$_6$H$_{10}$), **4**, 2
$LnC_{12}H_{18}$
Ln(C$_4$H$_6$)$_3$, **4**, 2
$LnC_{12}H_{20}$
Ln(Me$_2$C$_4$H$_4$)$_2$, **4**, 2
$Ln_2C_{18}H_{30}$
Ln$_2$(C$_6$H$_{10}$)$_3$, **4**, 2
LuC_6H_{18}
[LuMe$_6$]$^-$, **4**, 20
$LuC_{13}H_{19}O$
(C$_8$H$_8$)LuMe(THF), **4**, 102
$LuC_{15}H_{15}$
LuCp$_3$, **4**, 53
$LuC_{15}H_{19}O$
Cp$_2$Lu[C(O)Bu^t], **4**, 40
$LuC_{17}H_{29}Cl_2O_2$
(Bu^{t}C$_5$H$_4$)LuCl$_2$(THF)$_2$, **4**, 32
$LuC_{18}H_{23}$
(C$_8$H$_8$)LuCp*, **4**, 106
$LuC_{18}H_{27}$
Lu(CCBu^t)$_3$, **4**, 27
$LuC_{19}H_{23}O_2$
CpLu(C$_{10}$H$_8$)(DME), **4**, 34
$LuC_{19}H_{29}F_6O_9S_2$
CpLu(O$_3$SCF$_3$)$_2$(THF)$_3$, **4**, 33
$LuC_{19}H_{35}ClNO_2$
CpLu[CH$_2$CH(Me)CH$_2$NMe$_2$]-(Cl)(THF)$_2$, **4**, 34
$LuC_{20}H_{22}$
[Cp$_2$Lu(H)]$_2$, **4**, 47
$LuC_{20}H_{26}NO$
Cp$_2$Lu(NC$_4$H$_2$Me$_2$)(THF), **4**, 41
$LuC_{22}H_{35}O$
Cp*$_2$LuOEt, **4**, 78
$LuC_{24}H_{37}$
Cp*$_2$Lu[CH$_2$C(Me)CH$_2$], **4**, 88
$LuC_{24}H_{40}P$
Cp*$_2$Lu[(CH$_2$)$_2$PMe$_2$], **4**, 91
$LuC_{26}H_{28}OP$
Cp$_2$LuPPh$_2$(THF), **4**, 42
$LuC_{26}H_{35}$
Cp*$_2$LuPh, **4**, 94
$LuC_{27}H_{31}O_2$
CpLu(C$_{14}$H$_{10}$)(THF)$_2$, **4**, 34
$LuC_{28}H_{47}O_2$
Cp*$_2$Lu(OBu^t)(THF), **4**, 78
$LuC_{30}H_{34}$
[(H)[Cp$_2$Lu(H)]$_3$]$^-$, **4**, 48
$LuC_{30}H_{66}P_3$
Lu[(CH$_2$)$_2$PBu$^t_2$]$_3$, **4**, 17
$LuC_{31}H_{45}O_2$
Cp*$_2$LuOBz(THF), **4**, 48
$LuC_{32}H_{36}$
[Lu(C$_6$H$_3$Me$_2$)$_4$]$^-$, **4**, 21
$LuC_{36}H_{60}$
Lu(Bu$^t_3$C$_6$H$_3$)$_2$, **4**, 6
$LuC_{40}H_{44}O_2$
[(Ph$_2$C$_4$H$_4$)$_2$Lu(THF)$_2$]$^{2-}$, **4**, 21
$LuC_{48}H_{43}$
(C$_8$H$_8$)Lu[C$_5$(Bz)$_5$], **4**, 106
$LuKSi_6C_{21}H_{57}Cl$
[(TMS)$_2$CH]$_3$Lu(Cl)K, **4**, 19
$LuKSi_6C_{25}H_{67}ClO$
[(TMS)$_2$CH]$_3$Lu(Cl)K(Et$_2$O), **4**, 19
$LuLiC_{24}H_{49}N_2S_2$
Cp*Lu(SBu^t)$_2$Li(TMEDA), **4**, 73
$LuLiC_{40}H_{46}N_2P_2$
Cp$_2$Lu(PPh$_2$)$_2$Li(TMEDA), **4**, 42
$LuRe_2C_{46}H_{69}OP_4$
Cp$_2$Lu(THF)Re$_2$H$_7$(PMe$_2$Ph)$_4$, **4**, 109
$LuRuC_{21}H_{23}O_3$
Cp$_2$(THF)LuRu(CO)$_2$Cp, **4**, 107; **10**, 54
$LuRuC_{27}H_{35}O_2$
Cp*$_2$LuRu(CO)$_2$Cp, **4**, 107
$LuRuSi_4C_{33}H_{55}O_3$
[(TMS)$_2$C$_5$H$_3$]$_2$(THF)LuRu-(CO)$_2$Cp, **4**, 107
$LuSiC_{18}H_{29}O$
Cp$_2$LuCH$_2$TMS(THF), **4**, 43
$LuSiC_{24}H_{41}$
Cp*$_2$LuCH$_2$TMS, **4**, 94
$LuSi_2C_{17}H_{34}Cl_2$
Cp*Lu[CH(TMS)$_2$]Cl$_2$, **4**, 73
$LuSi_2C_{32}H_{49}O$
(C$_5$Me$_3$Ph$_2$)Lu(CH$_2$TMS)$_2$-(THF), **4**, 73
$LuSi_3C_{21}H_{45}Cl$
Cp*Lu[CH(TMS)$_2$](CH$_2$TMS)-Cl, **4**, 73
$Lu_2C_{28}H_{38}O_2$
[Cp$_2$Lu(H)(THF)]$_2$, **4**, 47
$Lu_2C_{30}H_{31}$
[Cp$_3$Lu(H)LuCp$_3$]$^-$, **4**, 56
$Lu_2C_{36}H_{54}$
[(Bu^{t}C$_5$H$_4$)$_2$Lu(H)]$_2$, **4**, 47
$Lu_2C_{40}H_{62}$
[Cp*$_2$Lu(H)]$_2$, **4**, 93, 117
$Lu_2C_{42}H_{66}$
[Cp*$_2$LuMe]$_2$, **4**, 67, 84, 115, 117
$Lu_2C_{52}H_{56}O_2P_2$
[Cp$_2$Lu[O(CH$_2$)$_4$PPh$_2$]]$_2$, **4**, 42

M

$MgAl_2C_{16}H_{24}$
$(Me_2Al)_2Mg(C{\equiv}CMe)_4$, **1**, 107
$MgBC_{21}H_{37}ClN_6$
$[HB(Bu^tpz)_3]MgCl$, **1**, 104
$MgBC_{30}H_{56}N_{12}$
$Mg[HB(Me_2pz)_3]_2$, **1**, 104
$MgB_{10}C_{14}H_{34}O_4$
$Mg(MeC_2B_{10}H_{10})_2(dioxane)_2$, **1**, 88
$MgB_{20}C_{14}H_{42}O_4$
$(C_2B_{10}H_{10}Me)_2Mg_2(C_4H_8O_2)$, **1**, 243
$MgCH_3Br$
MeMgBr, **1**, 75; **12**, 644
$MgCH_4$
MeMgH, **1**, 69
MgC_2Br_2Cl
$CF_3CClBrMgBr$, **1**, 70
MgC_2ClF_2
$CF_3CFClMgF$, **1**, 70
MgC_2H_5Br
EtMgBr, **1**, 65, 66, 87
MgC_2H_5Cl
EtMgCl, **1**, 67
MgC_2H_5ClS
$MeSCH_2MgCl$, **1**, 70
MgC_2H_5I
EtMgI, **1**, 87
MgC_2H_6
$MgMe_2$, **1**, 73, 88, 95
MgC_3H_5Br
$(C_3H_5)MgBr$, **1**, 61
MgC_3H_5Cl
$CH_2{=}CHCH_2MgCl$, **1**, 58, 67
MgC_3H_6O
$Mg[(CH_2)_3]O$, **1**, 92
MgC_3H_7Br
Pr^iMgBr, **1**, 67
MgC_4H_7Cl
$MeCH{=}CHCH_2MgCl$, **1**, 67
MgC_4H_9Br
BuMgBr, **1**, 67, 69, 76
Bu^tMgBr, **1**, 78
MgC_4H_9Cl
BuMgCl, **1**, 63, 66, 67, 69
Bu^tMgCl, **1**, 66
MgC_4H_{10}
$MgEt_2$, **1**, 66, 86, 87, 95, 96
MgC_5H_4BrN
(py)MgBr, **1**, 65
MgC_5H_6
CpMgH, **1**, 101
MgC_5H_7Br
$Me_2C{=}C{=}CHMgBr$, **1**, 70
MgC_5H_{10}
$\overline{MgCH_2CH_2CH_2CH_2CH_2}$, **1**, 92
$Mg(CH_2)_5$, **1**, 85
MgC_6H_5Br
PhMgBr, **1**, 64, 65; **12**, 154
MgC_6H_5Cl
PhMgCl, **1**, 65
MgC_6H_5F
PhMgF, **1**, 59
$MgC_6H_{12}Br$
$[Me(C_4H_7)CH_2]MgBr$, **1**, 79
MgC_6H_{14}
$MgPr_2$, **1**, 87
$MgC_6H_{15}N$
Et_2NMgEt, **1**, 87
$MgC_7H_5Br_3O$
$MeOC_6H_2Br_2MgBr$, **1**, 65
MgC_7H_6
$\overline{Mg(C_6H_4)CH_2}$, **1**, 67, 68, 71, 76
$MgC_7H_6Br_2O$
$MeOC_6H_3BrMgBr$, **1**, 65
MgC_7H_7Cl
BzMgCl, **1**, 72
MgC_7H_{10}
CpMgEt, **1**, 87
$MgC_7H_{10}BrN$
$NC(CH_2)_3CH{=}CHCH_2MgBr$, **1**, 77
$MgC_7H_{13}Br$
$(C_7H_{13})MgBr$, **1**, 60
MgC_8H_8
Mg(cot), **1**, 98
$MgC_8H_{13}Cl$
$(ClMgCH_2)$(norbornene), **1**, 79
MgC_8H_{14}
$(CH_2{=}CMeCH_2)_2Mg$, **1**, 93
$(CH_2)_2(C_6H_{10})Mg$, **1**, 98
MgC_8H_{18}
$BuMgBu^s$, **1**, 85, 96
$MgBu_2$, **1**, 85, 86, 95
$MgBu^i_2$, **1**, 69
MgC_9H_6BrN
(quinolyl)MgBr, **1**, 66
MgC_9H_9Cl
$(CH_2{=}CH)C_6H_4CH_2MgCl$, **1**, 72
$MgC_9H_{10}BrF_{12}O_3$
$EtO_2C(CF_2)_4OCF_2CF_2MgBr$, **1**, 70
$MgC_9H_{11}BrO$
$PhO(CH_2)_3MgBr$, **1**, 70
MgC_9H_{12}
$Cp(CH_2{=}CMeCH_2)Mg$, **1**, 93
$MgC_9H_{14}O$
CpMgH(THF), **1**, 101
$MgC_9H_{17}BrN$
(tmph)MgBr, **1**, 64
$MgC_{10}H_{10}$
Cp_2Mg, **1**, 86, 93, 94
$MgC_{10}H_{12}$
$Mg(CH_2CH_2)_2C_6H_4$, **1**, 92
$MgC_{10}H_{12}Cl$
$PhPr^iCHMgCl$, **1**, 77
$MgC_{10}H_{14}$
CpMg(neopentyl), **1**, 94
$MgC_{10}H_{14}BrO_2$
$(MeOCH_2)_2C_6H_4MgBr$, **1**, 76
$[MeO(CH_2CH_2O)CH_2]$Ph-MgBr, **1**, 74
$MgC_{10}H_{14}O_2$
$[(MeOCH_2)_2C_6H_4]Mg$, **1**, 76
$MgC_{10}H_{15}Cl$
Cp*MgCl, **1**, 84
$MgC_{10}H_{15}N$
Ph(Et)NMgEt, **1**, 87
$MgC_{10}H_{16}Br$
(adamantane)MgBr, **1**, 60
$MgC_{10}H_{17}Cl$
$(Me_2C{=}CH(CH_2)_2C$-$Me{=}CHCH_2)MgCl$, **1**, 77
$MgC_{10}H_{22}$
$Mg(neopentyl)_2$, **1**, 87, 90, 94, 96
$MgC_{10}H_{22}O_2$
$(MeOCH_2CH_2CH_2CH_2)_2Mg$, **1**, 90
$MgC_{11}H_{21}NO_2S$
$EtMg(NCS)(THF)_2$, **1**, 104
$MgC_{12}H_{10}$
$MgPh_2$, **1**, 64, 65, 86, 88, 96
$MgC_{12}H_{13}O_3$
PhMg(OOCMe)(THF), **1**, 102
$MgC_{12}H_{22}O_2$
$Mg(butadiene)(THF)_2$, **1**, 98
$MgC_{13}H_{27}BrO_3$
$MeMgBr(THF)_3$, **1**, 74
$MgC_{14}H_{23}ClO$
Cp*MgCl(THF), **1**, 63
$MgC_{14}H_{30}O_2$
$Mg(CH_2Bu^t)_2(dioxane)$, **1**, 88
$MgC_{15}H_{14}O$
$PhCH{=}CHCH_2MgOPh$, **1**, 101
$MgC_{15}H_{26}O_3$
$CpMg(OEt)(THF)_2$, **1**, 102
$MgC_{15}H_{29}BrN_2$
$CpMgBr(Et_2NCH_2CH_2NEt_2)$, **1**, 74
$MgC_{16}H_{18}O_4$
$Mg[C_6H_3(OMe)_2]_2$, **1**, 90
$MgC_{16}H_{29}N_2Br$
EtMgBr(isosparteine), **1**, 73
EtMgBr(sparteine), **1**, 73
$MgC_{16}H_{34}$
$Mg(C_8H_{17})_2$, **1**, 86
$MgC_{18}H_{26}$
$Mg(Bu^tC_5H_4)_2$, **1**, 91
$MgC_{18}H_{33}N_2Cl$
Bu^tMgCl(sparteine), **1**, 73

$MgC_{19}H_{22}N_2O_3$
CpMg(OOCH$_2$CH$_2$OEt)(py)$_2$, **1**, 102
$MgC_{20}H_{24}$
Bu(Pr-fluorenyl)Mg, **1**, 93
$MgC_{20}H_{26}O_2$
MgPh$_2$(THF)$_2$, **1**, 88
$MgC_{20}H_{30}$
Cp*$_2$Mg, **1**, 93, 94, 445
$MgC_{22}H_{26}O_2$
Mg(Me-anthracene)(THF)$_2$, **1**, 99
$MgC_{22}H_{30}O_2$
Mg(Tol)$_2$(THF)$_2$, **1**, 88
$MgC_{23}H_{35}N_2Br$
EtMgBr(benzylsparteine), **1**, 73
$MgC_{24}H_{50}N_2$
Mg(C≡CBu^t)$_2$(TMEDA)$_2$, **1**, 91
$MgC_{26}H_{34}O_3$
Mg(anthracene)(THF)$_3$, **1**, 59, 63, 72, 99–101
$MgC_{26}H_{40}O_3$
Mg(C$_6$H$_4$Bu^t)$_2$[MeO(CH$_2$C-H$_2$O)$_2$Me, **1**, 89
$MgC_{27}H_{36}O_3$
Mg(Me-anthracene)(THF)$_3$, **1**, 99
$MgC_{30}H_{34}N_2O_2$
Mg[N(py)Ph]$_2$(THF)$_2$, **1**, 103
$MgC_{30}H_{38}N_2$
Cp*$_2$Mg(bipy), **1**, 93
$MgC_{30}H_{48}O_5$
Mg(C$_6$H$_4$Bu^t)$_2$[MeO(CH$_2$C-H$_2$O)$_4$Me], **1**, 89
$MgC_{36}H_{44}O_6$
[Mg(anisyl)$_2$(THF)]$_2$, **1**, 90
$MgC_{38}H_{42}O_3$
Mg(C$_{14}$H$_8$Ph$_2$)(THF)$_3$, **1**, 99
$MgC_{56}H_{68}O_6$
Mg(THF)$_6$(fluoranthrene)$_2$, **1**, 99
$MgC_{76}H_{84}O_6$
Mg(THF)$_6$(C$_{14}$H$_8$Ph$_2$)$_2$, **1**, 99
$MgCoC_{34}H_{86}N_4O_4P_6$
Mg(THF)$_4$[N$_2$Co(PMe$_3$)$_3$]$_2$, **1**, 104
$MgCrC_9H_5BrO_3$
(PhMgBr)Cr(CO)$_3$, **12**, 1018
$MgCu_4C_{40}H_{40}O$
MgCu$_4$Ph$_6$(Et$_2$O), **3**, 102
MgEtCl
EtMgCl, **1**, 87
MgH_2
MgH$_2$, **1**, 59, 86, 100
$MgLiSi_3C_{21}H_{51}Br_2$
Li(THF)$_2$(Br)$_2$Mg[C(TMS)$_3$-(THF)], **1**, 85
$MgLi_2C_4H_{12}$
Li$_2$MgMe$_4$, **1**, 105
$MgLi_2C_{16}H_{44}N_4$
[Li(TMEDA)]$_2$MgMe$_4$, **1**, 105
$MgLi_2C_{28}H_{28}$
Li$_2$MgBz$_4$, **1**, 105
$MgLi_2C_{40}H_{60}N_4$
[Li(TMEDA)]$_2$MgBz$_4$, **1**, 105
$MgLi_2C_{44}H_{30}N_4$
[Li(TMEDA)]$_2$Mg(C≡CPh)$_4$, **1**, 105
$MgMo_2C_{24}H_{56}O_6$
[Mg(THF)$_4$][MoMe$_4$(O)]$_2$, **5**, 289
$MgNiC_{12}H_{30}N_2$
(C$_2$H$_4$)$_2$Ni(Me)Mg(Me)(TMEDA), **9**, 34, 110
$MgNiC_{20}H_{28}O$
(C$_2$H$_4$)$_2$Ni(Ph)MgPh(OEt$_2$), **9**, 35
$MgNi_2C_{24}H_{36}O$
[(C$_2$H$_4$)$_2$Ni]$_2$[Ph$_2$Mg(OEt$_2$)], **9**, 35
$MgSbC_{12}H_{10}$
Ph$_2$SbMg, **11**, 491
$MgSiC_5H_{14}$
TMS-MgEt, **1**, 87
$MgSiC_6H_{15}ClO$
(Pr^{i}O)Me$_2$SiCH$_2$MgCl, **1**, 70
$MgSi_2C_{11}H_{29}ClO$
MgCl[CH(TMS)$_2$](OEt$_2$), **1**, 72
$MgSi_2C_{20}H_{54}$
Mg[C(TMS)$_3$]$_2$, **1**, 72, 85
$MgSi_2C_{28}H_{42}O_2$
Mg[C$_{14}$H$_8$(TMS)$_2$](THF)$_2$, **1**, 100
$MgSi_3C_{10}H_{27}Br$
(TMS)$_3$CMgBr, **1**, 70
$MgSi_3C_{20}H_{46}BrN_2$
[C$_5$H$_3$(TMS)$_3$]MgBr(TMEDA), **1**, 74
$MgSi_3C_{21}H_{48}N_2$
MgMe[C$_5$H$_2$(TMS)$_3$]-(TMEDA), **1**, 86
$MgSi_4C_{14}H_{38}$
Mg[CH(TMS)$_2$]$_2$, **1**, 86, 87, 92
$MgSi_4C_{18}H_{48}O$
Mg[CH(TMS)$_2$]$_2$(OEt$_2$), **1**, 88
$MgSi_4C_{22}H_{44}N_2$
Mg[C(TMS)$_2$(py)]$_2$, **1**, 90
$MgSi_6C_{20}H_{54}$
Mg[C(TMS)$_3$]$_2$, **1**, 87, 88, 92
$MgSi_6C_{28}H_{58}$
Mg[(TMS)$_3$C$_5$H$_2$]$_2$, **1**, 92
$MgSi_6C_{34}H_{74}N_2$
Mg[C$_5$H$_2$(TMS)$_3$]$_2$(TMEDA), **1**, 86
$MgThC_{27}H_{42}Cl_2O$
(C$_8$H$_8$)Cp*Th(Cl)$_2$Mg(CH$_2$-Bu^t)- (THF), **4**, 186
$MgU_2C_{32}H_{28}Cl_2O_8$
U$_2$(CO$_2$Bz)$_4$MgCl$_2$, **4**, 133
$Mg_2B_{10}C_4H_{14}Br_2$
C$_2$B$_{10}$H$_{10}$(CH$_2$MgBr)$_2$, **1**, 242
$Mg_2CH_2Br_2$
CH$_2$(MgBr)$_2$, **1**, 59, 71, 83
Mg_2CH_3Br
MeMg$_2$Br, **1**, 59
Mg_2CH_6
MeMg$_2$H$_3$, **1**, 101
$Mg_2C_3H_4Br_2$
(BrMg)$_2$(C$_3$H$_4$), **1**, 59
$Mg_2C_4H_8Br_2$
(CH$_2$)$_4$(MgBr)$_2$, **1**, 71, 76
$Mg_2C_4H_8Br_2O$
O[(CH$_2$)$_2$MgBr]$_2$, **1**, 71
$Mg_2C_6H_4Cl_2$
(ClMg)$_2$C$_6$H$_4$, **1**, 76
$Mg_2C_7H_6Br_2$
(BrMg)C$_6$H$_4$CH$_2$MgBr, **1**, 67, 71, 85
$Mg_2C_7H_7Br_2$
[(BrMgCH$_2$)Ph]MgBr, **1**, 76
$Mg_2C_8H_8Cl_2$
(ClMgCH$_2$)$_2$C$_6$H$_4$, **1**, 85
$Mg_2C_8H_{16}$
[Mg(CH$_2$)$_4$]$_2$, **1**, 76, 92
$Mg_2C_9H_{10}Br_2$
BrMg(CH$_2$)$_3$C$_6$H$_4$MgBr, **1**, 67, 71
$Mg_2C_{10}H_{12}Br_2$
(BrMgCH$_2$CH$_2$)$_2$C$_6$H$_4$, **1**, 71
$Mg_2C_{10}H_{14}O_2$
(MeOMgCH$_2$)$_2$C$_6$H$_4$, **1**, 102
$Mg_2C_{10}H_{20}$
Mg(CH$_2$)$_5$Mg(CH$_2$)$_5$, **1**, 92
$Mg_2C_{12}H_{30}$
[Mg$_2$Et$_4$(Et)$_2$]$^{2-}$, **1**, 90
$Mg_2C_{12}H_{32}N_4$
[MeMgNMe(CH$_2$CH$_2$NMe$_2$)]$_2$, **1**, 103
$Mg_2C_{14}H_{12}Cl_2$
(ClMgCH$_2$)$_2$(biphenyl), **1**, 71
$Mg_2C_{14}H_{34}N_2$
[MgEt(CH$_2$CH$_2$CH$_2$NMe$_2$)]$_2$, **1**, 90
$Mg_2C_{15}H_{15}Br_2P$
(BrMgC$_6$H$_4$CH$_2$)$_2$PMe, **1**, 68, 70, 71, 76
$Mg_2C_{16}H_{16}Br_2$
(BrMgC$_6$H$_4$)CH$_2$CH$_2$O-CH$_2$CH$_2$(C$_6$H$_4$MgBr), **1**, 76
$Mg_2C_{16}H_{16}Br_2O$
[(BrMg)C$_6$H$_4$)]CH$_2$CH$_2$O-CH$_2$CH$_2$[(BrMg)C$_6$H$_4$], **1**, 85
$Mg_2C_{18}H_{30}Cl_2O_2$
[CpMgCl(OEt$_2$)]$_2$, **1**, 74
$Mg_2C_{23}H_{40}Cl_2O_2$
[Cp*MgCl(OEt$_2$)]$_2$, **1**, 74
$Mg_2C_{24}H_{48}Cl_3O_6$
[Mg$_2$Cl$_3$(THF)$_6$]$^+$, **1**, 99
$Mg_2C_{28}H_{46}Cl_2$
[Cp*MgCl(THF)]$_2$, **1**, 109
$Mg_2Cu_4C_{72}H_{92}N_4S_4$
[Cu$_4$Mes$_4$][Mg(SC$_6$H$_4$CH$_2$N-Me$_2$)$_2$]$_2$, **3**, 104
$Mg_2Cu_4C_{76}H_{100}N_4S_4$
[Cu$_4$Mes$_4$][Mg[SC$_6$H$_4$CH-(Me)NMe$_2$]$_2$]$_2$, **3**, 128
$Mg_2Li_2C_{18}H_{50}N_4$
[Li(TMEDA)MgMe$_3$]$_2$, **1**, 105
$Mg_2Li_2C_{48}H_{62}N_4$
[Li(TMEDA)]$_2$Mg$_2$Ph$_6$, **1**, 105
$Mg_2Li_2C_{48}H_{86}N_2$
Li$_2$[Mg(TMEDA)(C≡Bu^t)$_3$]$_2$, **1**, 105, 106
Mg_2MeCl

MeMgMgCl, **1**, 78
$Mg_2SiC_4H_{10}Br_2$
(TMS)CH(MgBr)$_2$, **1**, 71
$Mg_2Si_2C_{23}H_{50}BrO_4$
(TMS)$_2$C[MgBr(THF)$_2$]$_2$, **1**, 73
$Mg_2Si_8C_{32}H_{84}O_2$
[Mg[CH(TMS)$_2$]$_2$]$_2$(dioxane), **1**, 88
$Mg_2U_2C_{32}H_{28}Cl_4O_{16}$
[U(CO$_2$Bz)$_4$MgCl$_2$]$_2$, **4**, 133
$Mg_2U_2C_{58}H_{56}Cl_4O_2$
[U(Bz)$_3$(COBz)MgCl$_2$]$_2$, **4**, 133
$Mg_2U_2C_{58}H_{56}Cl_4O_4$
[U(Bz)$_3$(CO$_2$Bz)MgCl$_2$]$_2$, **4**, 133
$Mg_2U_2C_{76}H_{88}Cl_2O_2P_4$
[Cp*U[(CH$_2$)PPh$_2$(CH$_2$)]]$_2$Mg-[CH$_2$PMePh$_2$]$_2$(O)$_2$(Cl)$_2$, **4**, 166
$Mg_3C_{48}H_{72}O_6$
[Mg[(CH$_2$)C$_6$H$_4$(CH$_2$)]-(THF)$_2$]$_3$, **1**, 76
$Mg_3C_{48}H_{81}NO_2$
[(xylidene)(THF)$_2$Mg]$_3$, **1**, 88, 89
$Mg_4C_{24}H_{16}$
[MgC$_6$H$_4$]$_4$, **1**, 85
$Mg_4C_{28}H_{20}O_4$
[CpMg(OEt)]$_4$, **1**, 102
$Mg_4C_{28}H_{40}O_4$
[CpMg(OEt)]$_4$, **1**, 101, 102
$Mg_4C_{30}H_{66}Br_2$
[Mg(neopentyl)$_2$]$_2$[Mg-(neopentyl)Br]$_2$, **1**, 88
$Mg_4C_{40}H_{48}O_4$
[(C$_6$H$_4$)(THF)Mg]$_4$, **1**, 91
$Mg_4C_{56}H_{56}O_4$
[(C$_{10}$H$_6$)(THF)Mg]$_4$, **1**, 91
$Mg_4C_{72}H_{72}O_4$
[(PhC=CPh)(THF)Mg]$_4$, **1**, 91, 97
$MnAsC_{28}H_{22}O_4$
(AsPh$_4$)[MnH$_2$(CO)$_4$], **6**, 68
$MnAsCrC_{15}O_{15}$
As[Mn(CO)$_5$][Cr(CO)$_5$]$_2$, **6**, 86
$MnAsCr_2C_{14}O_{14}$
As[Cr$_2$(CO)$_9$][Mn(CO)$_5$], **6**, 86
$MnAsFeC_{11}H_6O_9$
(CO)$_5$Mn(AsMe$_2$)Fe(CO)$_4$, **10**, 39
$MnAuC_{23}H_{15}O_5P$
Mn(CO)$_5$[Au(PPh$_3$)], **6**, 12
$MnAuC_{38}H_{31}ClO_9P_2$
[Mn[PPh$_2$Au(PTol$_3$)](CO)$_5$]-ClO$_4$, **6**, 88
$MnAuSiC_{39}H_{35}O_2P$
Mn(SiMePh$_2$)(CO)$_2$Cp'Au-(PPh$_3$), **10**, 48
$MnBC_6F_6O_5$
[Mn(CF$_2$)(CO)$_5$]BF$_4$, **6**, 42
$MnB_2C_{21}H_{36}N_2O$
Mn(CO)(C$_4$H$_4$BNPr^i$_2$)$_2$, **1**, 204
$MnB_5C_3H_{10}O_3$
(CO)$_3$MnB$_5$H$_{10}$, **1**, 323
$MnC_9H_7ClO_3$
[(MeClC$_6$H$_4$)Mn(CO)$_3$]$^+$, **12**, 994
$MnC_9H_7O_3$
(chd)Mn(CO)$_3$, **12**, 675, 1006
$MnC_{10}H_9O_3$
(cycloheptadienyl)Mn(CO)$_3$, **12**, 679
$MnC_{11}H_{12}O_4$
[Me(OMe)C$_6$H$_6$]Mn(CO)$_3$, **12**, 1006
$MnCN_3O_4$
Mn(CO)(NO)$_3$, **6**, 123
$MnC_2HF_9O_2P_3$
MnH(CO)$_2$(PF$_3$)$_3$, **6**, 72
$MnC_3HF_6O_3P_2$
MnH(CO)$_3$(PF$_3$)$_2$, **6**, 72
MnC_3O_3
[Mn(CO)$_3$]$^-$, **6**, 47
MnC_4FO_7S
Mn(O$_3$SF)(CO)$_4$, **6**, 84
$MnC_4H_4O_4P$
MnH(CO)$_4$(PH$_3$), **6**, 73
MnC_4H_{12}
[MnMe$_4$]$^{2-}$, **6**, 128, 139
MnC_4NO_5
Mn(CO)$_4$(NO), **6**, 123
MnC_5ClO_5
MnCl(CO)$_5$, **6**, 14
MnC_5ClO_9
Mn(OClO$_3$)(CO)$_5$, **6**, 84
$MnC_5F_5O_6Te$
Mn(OTeF$_5$)(CO)$_5$, **6**, 41, 84
$MnC_5H_5O_2$
CpMnO$_2$, **6**, 146
$MnC_5H_9Cl_2N$
MnCl$_2$(CNBu^t), **6**, 133
$MnC_5H_9I_2N$
MnI$_2$(CNBu^t), **6**, 133
MnC_5H_{15}
[MnMe$_5$]$^{2-}$, **6**, 139, 145
MnC_5IO_5
MnI(CO)$_5$, **6**, 1, 6
MnC_5O_5
[Mn(CO)$_5$]$^-$, **6**, 10, 22; **7**, 277
Mn(CO)$_5$, **6**, 10
MnC_5O_7
Mn(CO)$_5$(O$_2$), **6**, 10
$MnC_6Br_3O_5$
Mn(CBr$_3$)(CO)$_5$, **6**, 42
$MnC_6Cl_3O_5$
Mn(CCl$_3$)(CO)$_5$, **6**, 42
$MnC_6F_3O_5$
Mn(CF$_3$)(CO)$_5$, **6**, 42
$MnC_6F_3O_9S$
Mn(OTf)(CO)$_5$, **6**, 23
MnC_6HO_6
Mn(CHO)(CO)$_5$, **6**, 26
MnC_6H_{18}
[MnMe$_6$]$^{2-}$, **6**, 145
$MnC_6I_3O_5$
Mn(CI$_3$)(CO)$_5$, **6**, 42
MnC_6O_6
[Mn(CO)$_6$]$^+$, **6**, 2, 13
$MnC_7Cl_2F_3O_5$
Mn(CCl$_2$CF$_3$)(CO)$_5$, **6**, 42
$MnC_7H_3O_6$
Mn(COMe)(CO)$_5$, **6**, 22
$MnC_7H_4O_5$
[Mn(CO)$_5$(CH$_2$CH$_2$)]$^+$, **6**, 28
$MnC_7H_5N_2O_2$
Mn(CO)$_2$(N$_2$)Cp, **6**, 113, 115
$MnC_7H_5O_2$
[Mn(CO)$_2$Cp]$^{2-}$, **6**, 68, 115
$MnC_7H_5O_4$
Mn(C$_3$H$_5$)(CO)$_4$, **6**, 44
$MnC_7H_6O_2$
[MnHCp(CO)$_2$]$^-$, **6**, 68
$MnC_7H_7O_2$
MnCp(H)$_2$(CO)$_2$, **6**, 68, 73, 115
$MnC_8H_5O_3$
CpMn(CO)$_3$, **6**, 136
$MnC_8H_5O_5$
Mn(C$_3$H$_5$)(CO)$_5$, **6**, 47
$MnC_8H_5O_7$
Mn[C(O)OEt](CO)$_5$, **6**, 36
$MnC_8H_6ClO_2$
(C$_6$H$_6$)Mn(CO)$_2$Cl, **6**, 34
$MnC_8H_7O_3$
Mn(C$_5$H$_7$)(CO)$_3$, **6**, 98, 102
MnC_8H_{14}
CpMnMe$_3$, **6**, 146
MnC_8H_{16}
Mn(C$_4$H$_8$)$_2$, **6**, 139
MnC_8H_{20}
[MnEt$_4$]$^{2-}$, **6**, 129, 138
$MnC_9H_4NO_7$
Mn(NCOCH$_2$CH$_2$CO)(CO)$_5$, **6**, 84
$MnC_9H_7O_7$
Mn[CH$_2$C(O)OEt](CO)$_5$, **6**, 36
$MnC_9H_8O_3$
Mn(chd)(CO)$_3$, **6**, 94, 98
$MnC_9H_9O_3$
Mn(C$_6$H$_9$)(CO)$_3$, **6**, 100
$MnC_9H_{11}Cl_2OP$
Mn(Cl)$_2$(CO)(PPhMe$_2$), **6**, 15
$MnC_9H_{11}O_2$
Mn(Me)$_2$(CO)$_2$Cp, **6**, 113
$MnC_{10}H_7O_3$
Mn(C$_7$H$_7$)(CO)$_3$, **6**, 96, 101
$MnC_{10}H_9O_3$
Mn(C$_7$H$_9$)(CO)$_3$, **6**, 98
$MnC_{10}H_9O_5$
Mn(CHMeCHCHMe)(CO)$_5$, **6**, 35
$MnC_{10}H_{10}$
Cp$_2$Mn, **6**, 134
$MnC_{10}H_{10}O_2$
Mn(allyl)Cp(CO)$_2$, **6**, 99
$MnC_{10}H_{24}N_2$
Mn(CH$_2$CH$_2$CH$_2$NMe$_2$)$_2$, **6**, 132
$MnC_{10}H_{28}P_2$
MnMe$_4$(dmpe), **6**, 145
$MnC_{10}H_{30}P_2$
MnMe$_4$(PMe$_3$)$_2$, **6**, 145
$MnC_{11}H_4O_8P$
Mn[PO(O$_2$C$_6$H$_4$)](CO)$_5$, **6**, 85
$MnC_{11}H_{13}O_3$
MnCp'(CO)$_2$[C(Me)OMe], **6**, 59

$MnC_{11}H_{21}ClN_2$
CpMnCl(TMEDA), **6**, 134
$MnC_{11}H_{23}IP_2$
$CpMnI(PMe_3)_2$, **6**, 134
$MnC_{11}H_{24}O_{12}P_3$
$Mn(CO)_3[P(OMe)_3]_2[PO(OMe)_2]$, **6**, 85
$MnC_{11}H_{27}N_2$
$MnMe(CH_2CH_2CH_2NMe_2)_2$, **6**, 141
$MnC_{12}H_5O_6$
$Mn[C(O)Ph](CO)_5$, **6**, 25
$MnC_{12}H_7O_5$
$MnBz(CO)_5$, **6**, 35
$MnC_{12}H_{12}O_4P$
$MnH(CO)_4(PMe_2Ph)$, **6**, 37
$MnC_{12}H_{13}O_3$
$Mn(CO)_3(C_5H_4Bu^t)$, **6**, 110
$Mn(CO)_3(C_5Me_4H)$, **6**, 110
$MnC_{12}H_{13}O_4$
$MnCp(CO)_2[CCHCH(OMe)_2]$, **6**, 63
$MnC_{12}H_{14}$
$(C_5H_4Me)_2Mn$, **6**, 133, 135
$MnC_{12}H_{18}NO_4$
$Mn(CO)_3(tppo)$, **6**, 10
$MnC_{12}H_{23}IP_2$
$(C_5H_4Me)MnI(dmpe)$, **6**, 134
$MnC_{12}H_{25}ClP_2$
$(C_5H_4Me)MnCl(PMe_3)_2$, **6**, 134
$MnC_{12}H_{25}P_2$
$MnH(C_6H_6)(PMe_3)_2$, **6**, 68
$MnC_{13}H_5O_5$
$Mn(CCPh)(CO)_5$, **6**, 57
$MnC_{13}H_7O_6$
$Mn[C(O)C_7H_7](CO)_5$, **6**, 47
$MnC_{13}H_8N_2O_3$
$[Mn(CO)_3(bipy)]^-$, **6**, 11
$MnC_{13}H_9O_6$
$Mn(CH_2C_6H_4OMe)(CO)_5$, **6**, 37
$MnC_{13}H_{14}$
$[C_5H_4(CH_2)_3C_5H_4]Mn$, **6**, 136
$MnC_{13}H_{15}O_3$
$Mn(CO)_3Cp^*$, **6**, 110
$MnC_{13}H_{17}ClNO_2$
$MnCp^*(CHCHCl)(CO)(NO)$, **6**, 57, 123
$MnC_{13}H_{19}P$
$Cp_2Mn(PMe_3)$, **6**, 137
$MnC_{13}H_{20}Br_2NP$
$MnBr_2(CNBu^t)(PMe_2Ph)$, **6**, 133
$MnC_{13}H_{24}O_3P_2$
$Mn(CO)_3(depe)$, **6**, 8
$MnC_{14}H_7O_5$
$Mn(CH_2CCPh)(CO)_5$, **6**, 43
$MnC_{14}H_{10}ClO_2$
$MnCp(CO)_2[C(Cl)Ph]$, **6**, 63
$MnC_{14}H_{13}NO_2$
$Mn(CO)_2(NHTol)Cp$, **6**, 112
$MnC_{14}H_{14}NO_2$
$Mn(CO)_2(NH_2Tol)Cp$, **6**, 112
$MnC_{14}H_{15}N_2O_2$
$Mn(CO)_2(H_2NC_6H_4NH_2)(MeCp)$, **6**, 112
$MnC_{14}H_{17}O_2$
$MnCp^*(CO)_2(HCCH)$, **6**, 57
$MnC_{14}H_{18}$
$(C_5H_4Et)_2Mn$, **6**, 135
$MnC_{14}H_{18}O$
$Cp_2Mn(THF)$, **6**, 138
$MnC_{14}H_{32}O_9P_3$
$CpMn[P(OMe)_3]_3$, **6**, 137
$MnC_{14}H_{34}P_2$
$MnBu^t_2(dmpe)$, **6**, 131
$MnC_{14}H_{37}P_4$
$MnH(CH_2{=}CH_2)(dmpe)_2$, **6**, 132
$MnC_{14}H_{38}P_4$
$Mn(CH_2CH_4)(dmpe)_2$, **6**, 132
$[MnMe_2(dmpe)_2]^-$, **6**, 141
$MnMe_2(dmpe)_2$, **6**, 132
$MnC_{15}H_{11}O_2$
$Mn(Cp)(CO)_2(CCHPh)$, **6**, 66, 67
$MnC_{15}H_{12}NO_4$
$Mn(Cp)(CO)(NO)[C(O)C(O)C_6H_4Me]$, **6**, 68
$MnC_{15}H_{18}O_3$
$Mn(C_6Me_6)(CO)_3$, **6**, 94
$MnC_{15}H_{29}Cl_2P_2$
$Mn(C_6H_2Me_3)Cl_2(PMe_3)_2$, **6**, 140
$MnC_{15}H_{31}O_2P_2S_2$
$Mn(S_2CH)(CO)_2(PEt_3)_2$, **6**, 89
$MnC_{16}H_9O_3$
$Mn(CO)_3(C_{13}H_9)$, **6**, 111
$MnC_{16}H_{14}O_3$
$[MnCp'(CO)_2[C(O)Tol]]^-$, **6**, 121
$MnC_{16}H_{15}F_9N_3$
$MnCp^*(CNCF_3)_3$, **6**, 120
$MnC_{16}H_{21}O_5P$
$CpMn(CO)_2[P(OPr^i)_3]$, **7**, 53
$MnC_{16}H_{24}O_2S$
$Cp^*Mn(CO)_2(SBu^t)$, **6**, 134
$MnC_{16}H_{26}N_2$
$Cp_2Mn(TMEDA)$, **6**, 136
$MnC_{16}H_{27}O_3P$
$Me_4P[MnCp(CO)_2C(O)CHPMe_3]$, **6**, 23
$MnC_{16}H_{33}O_3P_4$
$[Mn(CO)_3(depe)(PMe_3)]^+$, **6**, 6
$MnC_{17}H_8O_9P$
$Mn[P(O_2C_6H_4)_2](CO)_5$, **6**, 85
$MnC_{17}H_{22}O_5$
$[Mn(CO)_3(DBCat)]^-$, **6**, 11
$MnC_{17}H_{40}P_4$
$[MnH(C_5H_7)(dmpe)_2]^+$, **6**, 144
$MnC_{18}H_{15}O_{13}$
$Mn(CO)_3[C_5(CO_2Me)_5]$, **6**, 110
$MnC_{18}H_{17}Cl_2N$
$[C_5H_4(CH_2)_3C_5H_4]Mn(Cl_2C_5H_3N)$, **6**, 136
$MnC_{18}H_{20}I_2N_2$
$[Mn[C_6H_4(CH_2NMe_2)]_2I_2]^{2-}$, **6**, 129
$MnC_{18}H_{24}$
$Mn(C_8H_{12})_2$, **6**, 103
$MnC_{18}H_{24}N_2$
$Mn[C_6H_4(CH_2NMe_2)]_2$, **6**, 132
$MnC_{18}H_{26}$
$(C_5Me_4H)_2Mn$, **6**, 134
$MnC_{18}H_{27}F_5N_3$
$Mn(C_6F_5)(NBu^t)_3$, **6**, 146
$MnC_{19}H_{15}F_5NO_2$
$MnCp^*(CO)_2(CNC_6F_5)$, **6**, 120
$MnC_{19}H_{16}O_5P$
$Mn(H)(CO)_4[PPh_2CH_2CH_2CHO]$, **6**, 33
$MnC_{19}H_{27}N_2$
$MnMe(C_6H_4CH_2NMe_2)_2$, **6**, 140
$MnC_{20}H_{20}NO_2$
$Mn(Cp)(CO)_2(Bu^tNCCCHPh)$, **6**, 66
$MnC_{20}H_{30}$
$[Cp^*_2Mn]^-$, **6**, 144
$MnC_{20}H_{30}F_6P$
$[Cp^*_2Mn][PF_6]$, **6**, 144
$MnC_{20}H_{32}N_3$
$CpMn(CNBu^t)_3$, **6**, 136
$MnC_{20}H_{37}NOP_4$
$[Mn(CO)(NCPh)(dmpe)_2]^-$, **6**, 141
$MnC_{20}H_{40}P_4$
$Mn[(CH_2)_2C_6H_4](dmpe)_2$, **6**, 132
$MnC_{20}H_{45}BrO_2P_3S_2$
$[Mn(S_2CPEt_3)(CO)_2(PEt_3)_2]Br$, **6**, 89
$MnC_{22}H_{15}O_2$
$MnCp(CO)_2(CCCPh_2)$, **6**, 64
$MnC_{22}H_{15}O_4P$
$[Mn(CO)_4(PPh_3)]^-$, **6**, 10
$MnC_{22}H_{16}ClO_2$
$[MnCp(CO)_2(CCHCPh_2)]Cl$, **6**, 64
$MnC_{22}H_{30}O_4$
$MnBz_2(diox)_2$, **6**, 132
$MnC_{22}H_{37}O_3P_2Te$
$Mn(TeBz)(CO)_3(PEt_3)_2$, **6**, 90
$MnC_{23}H_{16}O_5P$
$Mn(CHO)(PPh_3)(CO)_4$, **6**, 26
$MnC_{23}H_{17}IO_4P$
$Mn(I)(CH_2PPh_3)(CO)_4$, **6**, 42
$MnC_{23}H_{34}O_4P$
$MnBz(CO)_4(PBu_3)$, **6**, 35
$MnC_{23}H_{49}N_2$
$Mn[C(=NBu^t)CH_2Bu^t]_2(PMe_3)$, **6**, 130
$MnC_{24}H_{20}O_5P$
$Mn(CH_2OMe)(CO)_4(Ph_3P)$, **6**, 42
$MnC_{24}H_{36}O_4P$
$Mn(CH_2CH_2Ph)(CO)_4(PBu_3)$, **6**, 35
$MnC_{24}H_{48}N_4$
$[Mn(CH_2NC_5H_{10})_4]^{2-}$, **6**, 128
$MnC_{24}H_{52}$
$[Mn(CH_2CH_2Bu^t)_4]^{2-}$, **6**, 129
$MnC_{26}H_{44}P_2$
$Mn(CH_2CMe_2Ph)_2(PMe_3)_2$, **6**, 131
$MnC_{27}H_{33}$
$[Mn(C_6H_2Me_3)_3]^-$, **6**, 129
$MnC_{27}H_{54}O_3P_2$

$Mn(CO)_3(PBu_3)_2$, **6**, 9
$MnC_{28}H_{33}I_2N_2P$
$MnI_2(PPh_3)(CNBu^t)_2$, **6**, 133
$MnC_{28}H_{44}$
$Mn(C_7H_{11})_4$, **6**, 145, 146
$MnC_{29}H_{22}IO_4P$
$Mn(CH_2I)(CO)_4(Ph_4P)$, **6**, 42
$MnC_{30}H_{54}N_6$
$[Mn(CNBu^t)_6]^+$, **6**, 120
$MnC_{30}H_{38}O_4P$
$Mn(CCPh)(CO)_4(PCY_3)$, **6**, 58
$MnC_{30}H_{43}P_3$
$MnPh_2(PCy)_3$, **6**, 130
$MnC_{32}H_{30}OP_2$
$[Mn(H)(CO)(dppe)Cp]^+$, **6**, 115
$MnC_{33}H_{29}O_3P_2$
$Mn(CCCH_2OMe)(CO)_3(dppe)$, **6**, 58
$MnC_{35}H_{72}N_2O_{12}P_3$
$(Bu_4N)_2[Mn(CO)_3(P_3O_9)]$, **6**, 84
$MnC_{38}H_{30}N_3O_4P_2$
$[N(PPh_3)_2][Mn(CO)_2(NO)_2]$, **6**, 123
$MnC_{40}H_{31}O_4P_2$
$Mn(CHO)(PPh_3)_2(CO)_3$, **6**, 26, 59
$MnC_{40}H_{31}O_{10}P_2$
$Mn(CHO)[P(OPh)_3]_2(CO)_3$, **6**, 26
$MnC_{40}H_{32}ClO_3P_2$
$Mn(PPh_3)_2(CO)_3(CH_2Cl)$, **6**, 27
$MnC_{43}H_{35}O_3$
$Mn(CO)_3[C_5Bz_5]$, **6**, 110
$MnC_{52}H_{44}NOP_4$
$Mn(CN)(CO)(dppm)_2$, **6**, 16
$MnC_{55}H_{49}F_6NOP_4$
$[Mn(CNallyl)(CO)dppm)_2]$-$[PF_6]$, **6**, 120
$MnC_{56}H_{38}NO_8P_4$
$[PPN][Mn(phth)_2(CO)_4]$, **6**, 84
$MnC_{57}H_{36}O_{12}P_3$
$Mn(CHO)[P(OPh)_3]_3(CO)_2$, **6**, 26
$MnC_{72}H_{84}F_{12}N_6O_6P_2$
$[Mn(Bu^tDiNC)_3][PF_6]_2$, **6**, 133
$MnCoC_9O_9$
$Mn(CO)_5Co(CO)_4$, **10**, 48
$MnCoC_{11}H_6O_7$
$Mn(CO)_4(MeC_2Me)Co(CO)_3$, **10**, 52
$MnCoC_{11}H_8O_5P$
$Mn(CO)_2Cp(PHMe)Co(CO)_3$, **10**, 53
$MnCoC_{55}H_{44}O_5P_4$
$Mn(CO)_4(dppm)_2Co(CO)$, **10**, 52
$MnCoFeC_{11}H_3O_{10}P$
$MnFeCo(Co)_{10}(PMe)$, **10**, 15
$MnCo_2C_{11}O_{10}$
$[MnCo_2(CO)_9(CCO)]^-$, **10**, 132
$MnCo_3C_{14}O_{14}$
$Co_3(CO)_9[CMn(CO)_5]$, **10**, 127
$MnCrC_{25}H_{33}O_3$
$Cp^*Mn(CO)_3Cr(C_6Me_6)$, **6**, 116
$MnCr_4C_{28}H_{38}S_6$
$Cr_4Mn(S)_4(SBu^t)_2Cp_4$, **10**, 234, 235
$MnCr_4C_{32}H_{46}S_6$
$[(C_5H_4Me)_2Cr_2(SBu^t)(S)_2]_2Mn$, **6**, 138
$MnFeC_{10}H_3O_9$
$Mn(CO)_4[MeC(O)]Fe(CO)_4$, **10**, 52
$MnFeC_{12}H_5O_6S_2$
$Mn[S_2CFe(CO)_2Cp](CO)_4$, **6**, 89
$MnFeC_{12}H_5O_7$
$Mn(CO)_5Fe(CO)_2Cp$, **10**, 48
$MnFeC_{15}H_9O_8$
$Mn(CO)_2Cp[C{=}CH(Co_2Me)]$-$Fe(CO)_4$, **10**, 52
$MnFeC_{20}H_{10}O_8P$
$FeMn(CO)_8(PPh_2)$, **7**, 78
$MnFeC_{21}H_{27}O_3$
$Cp^*Mn(CO)_3Fe(C_4Me_4)$, **6**, 116
$MnFeC_{22}H_{16}O_5P$
$Mn(CO)_4(H)(PPh_2)Fe(CO)Cp$, **6**, 88
$MnFeC_{24}H_{15}O_6S_3$
$FeMn(SPh)_3(CO)_6$, **7**, 61, 62
$MnFeC_{33}H_{22}BrO_8P_2$
$(CO)_4Fe(dppm)Mn(CO)_4Br$, **7**, 83
$MnFeHgSi_2C_{54}H_{43}O_5$
$Fe(CO)_3PPh_3SiMePh_2Hg$-$MnCpSiPh_3(CO)_2$, **10**, 339
$MnFe_2C_{14}H_8O_8P$
$CpMnFe_2(PMe)(CO)_8$, **10**, 35
$MnFe_2C_{18}H_{15}O_8Te$
$Fe_2MnCp^*(CO)_8(Te)$, **7**, 77
$MnFe_2C_{19}H_{16}O_8P$
$MnFe_2(PCy)(CO)_8Cp$, **10**, 222
$MnFe_2Ge_2MnC_{32}H_{54}O_{14}$
$(CO)_4FeGe(OBu^t)_3Mn$-$(OBu^t)_3GeFe(CO)_4$, **7**, 29
$MnFe_2SbC_{37}H_{25}O_8P$
$MnFe_2(PPh)(CO)_8(SbPh_3)Cp$, **10**, 222
$MnFe_3C_{12}O_{13}$
$[Fe_3Mn(CO)_{12}O]^-$, **7**, 60, 61; **10**, 232, 233, 235
$MnFe_3C_{14}HO_{13}$
$MnFe_3(CO)_{13}(CH)$, **7**, 25
$MnFe_3C_{16}H_5O_{11}P_2$
$Fe_3(CO)_9(P)[PMn(CO)_2Cp]$, **7**, 53
$MnFe_3GeC_{19}H_5O_{14}$
$GeFe_3Mn(CO)_{14}Cp$, **7**, 30
$MnFe_3GeC_{20}H_7O_{14}$
$MnFe_3(Ge)(CO)_{14}Cp'$, **10**, 215
$MnFe_5C_{23}H_5O_{18}P_2$
$Fe_3(CO)_8(PMn(CO)_2Cp)[PFe_2$-$(CO)_8]$, **7**, 53
$MnGaC_8H_{11}NO_5$
$GaH_2(NMe_3)[Mn(CO)_5]$, **1**, 556
$MnGeC_8H_{11}O_2$
$MeCpMn(CO)_2GeH_3^-$, **10**, 337
$MnHgGe_2C_{54}H_{26}F_{30}O_2$
$[Mn(THF)_2][HgCp_2[Ge$-$(C_6F_5)_3]_2]$, **6**, 138
$MnHgPtC_{29}H_{30}Cl_5O_5P_2$
$[(PPh_3)_2C_6Cl_5Pt]HgMn(CO)_5$, **10**, 343
$MnHgSiC_{21}H_{21}BrO_2$
$MeCp(CO)_2SiMePh_2MnHgBr$, **10**, 337
$MnHg_2C_{20}H_{17}O_2$
$(C_5H_4Me)Mn(HgPh)_2(CO)_2$, **6**, 143
$MnInC_{15}H_{24}N_2O_5$
$In(CH_2CH_2CH_2NMe_2)_2[Mn$-$(CO)_5]$, **1**, 561
$MnIrC_{24}H_{37}O_2P$
$Cp'(CO)_2Mn(PBu^t_2)Ir(cod)$, **8**, 372, 373
$MnLiC_8Cl_4O_3$
$Mn(CO)_3(C_5Cl_4Li)$, **6**, 110
$MnLiC_{14}H_{18}O_2$
$Li[Mn(C_6Me_6)(CO)_2]$, **6**, 70
$MnMoC_{13}H_5O_8$
$Mo(CO)_3CpMn(CO)_5$, **10**, 30
$MnMoC_{23}H_{15}IO_6P$
$CpMo(CO)_2(I)(PPh_2)Mn(CO)_4$, **6**, 88
$MnMoC_{23}H_{16}O_6P$
$Mn(CO)_4(H)(PPh_2)Mo$-$(CO)_2Cp$, **6**, 88; **10**, 33
$MnMoC_{24}H_{14}O_7P$
$Mo(CO)_3(C_5H_4PPh_2)Mn(CO)_4$, **10**, 33
$MnMoC_{25}H_{20}O_6P$
$Mn(CO)_4(H)(PTol_2)Mo$-$(CO)_2Cp$, **6**, 88
$MnMoC_{30}H_{27}O_4P$
$MnMo[CH(Me)CHCH_2](PPh_2)$-$(CO)_4Cp_2$, **10**, 261
$MnNaC_5O_5$
$Na[Mn(CO)_5]$, **6**, 1
$MnNa_2C_4HO_4$
$Na_2[MnH(CO)_4]$, **6**, 68
$MnNa_3C_4O_4$
$Na_3[Mn(CO)_4]$, **6**, 68
$MnNbC_{16}H_{10}O_6$
$Cp_2Nb(CO)_2Mn(CO)_4$, **5**, 65
$MnNb_2C_{27}H_{21}O_7$
$[[Cp_2(CO)Nb]_2(H)][Mn(CO)_5]$, **5**, 81
$MnNb_2W_4C_{51}H_{108}N_3O_{22}$
$(Bu_4N)_3[Mn(CO)_3(Nb_2W_4O_{19})]$, **6**, 84
MnO_2
MnO_2, **12**, 1183
$MnPbC_{25}H_{21}O_2$
$MnHCp(CO)_2(PbPh_3)$, **6**, 68
$MnPdC_{28}H_{22}BrO_3P_2$
$MnPdBr(CO)_3(dppm)_2$, **10**, 47
$MnPdC_{28}H_{22}IO_3P_2$
$MnPdI(CO)_3(dppm)_2$, **10**, 47
$MnPdC_{40}H_{68}O_4P_3$
$(Cy_2PH)_2Pd(PCy_2)Mn(CO)_4$, **6**, 88
$MnPtC_{30}H_{22}ClO_5P_2$

$Mn(CO)_5PtCl(dppm)$, **10**, 48
$MnPtC_{52}H_{40}O_4P_3$
$MnPt(PPh_2)(CO)_4(PPh_3)_2$, **10**, 226
$MnReC_{10}O_{10}$
$Mn(CO)_5Re(CO)_5$, **10**, 48
$MnRe(CO)_{10}$, **6**, 4; **10**, 4
$MnRhC_{54}H_{44}O_4P_4S$
$MnRh(S)(dppm)_2(CO)_4$, **6**, 89
$MnRhC_{54}H_{44}O_6P_4S$
$MnRh(SO_2)(dppm)_2(CO)_4$, **6**, 89
$MnRuC_{18}H_{14}NO_6$
$MnRu(CO)_6[PhC=C(H)C(H)=NPr^i]$, **10**, 39
$MnRuC_{22}H_{11}O_5P$
$Ru(CO)Cp(H)(PPh_2)Mn(CO)_4$, **10**, 52
$MnRuC_{22}H_{16}O_5P$
$CpRu(CO)(H)(PPh_2)Mn(CO)_4$, **6**, 87, 88
$MnRuC_{25}H_{22}O_4P$
$Mn(CO)_3(CMe=CHMe)-(PPh_2)Ru(CO)Cp$, **10**, 52
$MnRuC_{54}H_{40}O_6P_3$
$Mn(PPh_3)(CO)_3(PPh_2)Ru(CO)_3-(PPh_3)$, **6**, 87
$MnRuC_{59}H_{44}N_2O_4P_2$
$[Ru[CCPh(N_2C_6H_4Me)(CO)_4]-(PPh_3)_2Cp]^+$, **7**, 352
$MnRu_3C_{50}H_{30}NO_{13}P$
$[PPN][MnRu_3C(CO)_{13}]$, **7**, 755
$MnSbC_{14}H_{10}Cl_6O_2$
$[MnCp(CO)_2(CPh)]SbCl_6$, **6**, 63
$MnSiC_7H_6Cl_3O_2$
$CpMnH(SiCl_3)(CO)_2$, **6**, 143
$MnSiC_8H_8Cl_3O_2$
$(C_5H_4Me)MnH(SiCl_3)(CO)_2$, **6**, 143
$MnSiC_8H_9Cl_3O_2$
$(C_5H_4Me)MnH(SiCl_3)(CO)_2$, **6**, 143
$MnSiC_8H_9O_5$
$Mn(TMS)(CO)_5$, **6**, 33
$MnSiC_{10}H_{11}O_3$
$Mn(CO)_3(C_5H_4SiHMe_2)$, **6**, 110
$MnSiC_{13}H_{11}O_5$
$Mn(SiMe_2Ph)(CO)_5$, **6**, 25
$MnSiC_{13}H_{16}Cl_5O_2$
$Mn(H)(SiEt_3)(CO)_2(C_5Cl_5)$, **6**, 116
$MnSiC_{15}H_{15}O_6$
$Mn[CH(OSiMe_2Ph)Me](CO)_5$, **6**, 25
$MnSiC_{25}H_{21}O_2$
$Mn(H)(SiPh_3)Cp(CO)_2$, **6**, 73
$MnSiZnC_{21}H_{21}ClO_2$
$MeCp(CO)_2SiMePh_2MnZnCl$, **10**, 337
$MnSi_2C_{14}H_{38}P_2$
$Mn(CH_2TMS)_2(dmpe)$, **6**, 131
$MnSi_4C_{14}H_{38}$
$Mn[CH(TMS)_2]_2$, **6**, 130
$MnSi_4C_{16}H_{20}$
$Mn(CH_2TMS)_4$, **6**, 145
$MnSi_4C_{16}H_{44}$
$[Mn(CH_2TMS)_4]^{2-}$, **6**, 128
$Mn(CH_2TMS)_4$, **6**, 138
$MnSi_4C_{18}H_{46}O$
$Mn[CH(TMS)_2]_2(THF)$, **6**, 132
$MnSi_4C_{48}H_{102}N_4O_6$
$Mn_4(CH_2Bu^t)_2(NCO)_2[OC-(CH_2Bu^t)NTMS]_4$, **6**, 130
$MnSi_6C_{20}H_{54}$
$Mn[C(TMS)_3]_2$, **6**, 130
$MnSnC_{23}H_{15}O_4S_2$
$Mn(S_2CSnPh_3)(CO)_4$, **6**, 89
$MnSnC_{26}H_{23}O_2$
$(C_5H_4Me)Mn(CO)_2H(SnPh_3)$, **6**, 143
$MnSnC_{27}H_{33}O_3P_2S$
$Mn(SSnPh_3)(CO)_3(PMe_3)_2$, **6**, 89
$MnSnC_{30}H_{36}NO_4$
$(NEt_4)[MnH(CO)_4(SnPh_3)]$, **6**, 68
$MnSn_2C_{25}H_{20}Cl_8O_2P$
$[CpMn(SnCl_3)(CO)_2(PPh_3)]-[SnCl_5]$, **6**, 144
$MnTaC_{17}H_{16}O_2$
$TaCp_2(H)Mn(CO)_2Cp$, **10**, 28
$MnTiC_{25}H_{22}ClO_3$
$(CO)_2CpMn=C(Tol)OTiCp_2Cl$, **12**, 567
$MnTiC_{33}H_{32}Cl_2O_2P$
$TiCl_2Cp[C_5Me_4PPh_2MnCp-(CO)_2]$, **4**, 356
$MnUC_{30}H_{32}OP$
$MnCp[C(OUCp_3)CHPPhMe_2]$, **6**, 56
$MnVC_{14}H_{10}O_4P_3$
$CpMn(CO)P_3(CO)_3VCp$, **5**, 23
$MnYbC_{20}H_{28}I_2O_6$
$[(C_5H_4)Mn(CO)_3]YbI_2(THF)_3$, **4**, 111
$MnYbC_{25}H_{30}O_5$
$Cp^*{}_2YbMn(CO)_5$, **4**, 108; **10**, 48
$Mn_2Ag_2C_{43}H_{39}ClO_{10}P_3$
$Ag_2Mn_2(H)_2(Ph_2PCH_2PPh_2)-(CO)_6(PPh_3)(OClO_3)$, **10**, 317
$Mn_2Al_2C_{22}H_{36}N_6O_{10}$
$Mn_2(CO)_9[C(NMe_2)OAl_2-(NMe_2)_5]$, **6**, 8
$Mn_2Al_2C_{24}H_{72}P_8$
$[Mn(dmpe)_2]_2[(H)_2AlH(H)_2-AlH(H)_2]$, **1**, 550
$Mn_2AsC_{24}H_{31}O_4$
$[Mn(CO)_2Cp^*]_2AsH$, **6**, 114
$Mn_2BC_{18}H_{13}O_6$
$(EtC_4H_3BPh)[Mn(CO)_3]_2$, **1**, 202
$Mn_2BC_{31}H_{27}O_6P_2$
$Mn_2(H)(BH_4)(CO)_6(dppm)$, **1**, 284
$Mn_2C_6O_6Se_4$
$[Mn_2(Se_2)_2(CO)_6]^{2-}$, **6**, 90
$Mn_2C_6O_6Se_8$
$[Mn_2(Se_4)_2(CO)_6]^{2-}$, **6**, 90
$Mn_2C_8H_4O_8P_2$
$Mn_2(PH_2)_2(CO)_8$, **6**, 86
$Mn_2C_8H_{24}$
Mn_2Me_8, **6**, 139
$Mn_2C_9F_3O_9P$
$Mn_2(CO)_9(PF_3)$, **6**, 8
$Mn_2C_{10}F_6O_8S_2$
$[Mn_2(SCF_3)_2(CO)_8]$, **6**, 89
$Mn_2C_{10}H_3O_9PS$
$Mn[SPMe[Mn(CO)_5]](CO)_4$, **6**, 87
$Mn_2C_{10}O_9S$
$Mn_2(CO)_9(CS)$, **6**, 7
$Mn_2C_{10}O_{10}$
$Mn_2(CO)_{10}$, **6**, 1, 2
$Mn_2C_{12}H_4O_{10}$
$(CO)_5MnCH_2CH_2Mn(CO)_5$, **6**, 28
$Mn_2C_{12}O_{12}$
$Mn(CO)_5(COCO)Mn(CO)_5$, **6**, 24
$Mn_2C_{13}H_8O_9S$
$Mn_2(CO)_9(C_4SH_8)$, **6**, 8
$Mn_2C_{14}H_8O_{10}$
$(CO)_5Mn(CH_2)_4Mn(CO)_5$, **6**, 22, 41
$Mn_2C_{14}H_{10}O_9S$
$Mn_2(CO)_9(\overline{SCH_2CMe_2CH_2})$, **6**, 8, 9
$Mn_2C_{16}H_8O_6$
$(CO)_3Mn(C_5H_4C_5H_4)Mn(CO)_3$, **6**, 110, 111
$Mn_2C_{17}H_{21}$
$Cp_2Mn_2(Me_2C_5H_5)$, **6**, 135
$Mn_2C_{18}H_{10}Br_2O_6Se_2$
$Mn_2(Br)_2(PhSeSePh)(CO)_6$, **6**, 15
$Mn_2C_{18}H_{10}O_{10}$
$Mn(CO)_5Mn[C(OEt)Ph](CO)_4$, **6**, 67
$Mn_2C_{20}H_{11}O_8P$
$Mn_2(H)(PPh_2)(CO)_8$, **6**, 87
$Mn_2C_{20}H_{12}O_8P_2$
$Mn_2(PHPh)_2(CO)_8$, **6**, 86
$Mn_2C_{20}H_{30}Cl_2O_2$
$[MnCl(THF)Cp']_2$, **6**, 15
$Mn_2C_{23}H_{30}O_3$
$Mn_2(CO)_3Cp^*{}_2$, **6**, 116
$Mn_2C_{24}H_{30}O_4P$
$[[Mn(CO)_2Cp^*]_2P]^+$, **6**, 114
$Mn_2C_{24}H_{31}O_4P$
$[Mn(CO)_2Cp^*]_2PH$, **6**, 114
$Mn_2C_{26}H_8O_{10}$
$(CO)_5Mn(CCC_6H_4C_6H_4-CC)Mn(CO)_5$, **6**, 57
$Mn_2C_{26}H_{15}O_8PS_2$
$Mn_2(PPhSPh)(SPh)(CO)_8$, **6**, 86
$Mn_2C_{26}H_{16}N_4O_6$
$[Mn(CO)_3(bipy)]_2$, **6**, 9
$Mn_2C_{26}H_{46}Cl_2N_4S_2$
$Mn_2(CH_2SPh)_2(Cl)_2(TMEDA)_2$, **6**, 130
$Mn_2C_{26}H_{58}P_2$
$Mn_2(CH_2Bu^t)_2(PBu^t{}_2)_2$, **6**, 130

$Mn_2C_{27}H_{15}O_9P$
$Mn_2(CO)_9(PPh_3)$, **6**, 12
$Mn_2C_{30}H_{60}O_6P_4Te_2$
$Mn_2(Te_2)(CO)_6(PEt_3)_4$, **6**, 90
$Mn_2C_{31}H_{20}O_7P_2$
$Mn_2(PPh_2)_2(CO)_7$, **6**, 86
$Mn_2C_{33}H_{22}IO_8P_2$
$[Mn_2I(CO)_8(dppm)]^+$, **6**, 15
$Mn_2C_{36}H_{28}N_2O_6$
$Mn_2(CO)_6[TolNC(Bz)C(Bz)N\text{-}Tol]$, **6**, 85
$Mn_2C_{36}H_{30}$
$[Mn_2Ph_6]^{2-}$, **6**, 129
$Mn_2C_{36}H_{48}N_4$
$[Mn[CH_2C_6H_4(NMe_2)]_2]_2$, **6**, 132
$Mn_2C_{36}H_{58}Br_4N_4P_2$
$[Mn(CNBu^t)_4(PMe_2Ph)_2]\text{-}[MnBr_4]$, **6**, 133
$Mn_2C_{38}H_{28}Cl_2N_4O_{14}$
$[Mn_2(salen)_2(CO)_2][ClO_4]_2$, **6**, 144
$Mn_2C_{40}H_{52}$
$Mn_2(CH_2CMe_2Ph)_4$, **6**, 130
$[Mn(CH_2CMe_2Ph)_2]_2$, **6**, 130
$Mn_2C_{44}H_{30}O_8P_2$
$Mn_2(CO)_8(PPh_3)_2$, **6**, 7
$Mn_2C_{44}H_{30}O_9P_2S$
$Mn_2(SO)(CO)_8(PPh_3)_2$, **6**, 89
$Mn_2C_{44}H_{30}O_{10}P_2S$
$Mn_2(SO_2)(CO)_8(PPh_3)_2$, **6**, 89
$Mn_2C_{44}H_{68}O_6P_2S_4$
$Mn_2(S_2CHPCy_3)_2(CO)_6$, **6**, 89
$Mn_2C_{56}H_{44}O_6P_4$
$Mn_2(CO)_6(dppm)_2$, **6**, 9
$Mn_2C_{56}H_{45}O_6P_4$
$[Mn_2(H)(CO)_6(dppm)_2]^+$, **6**, 9
$Mn_2C_{56}H_{84}O_2$
$Mn_2(CH_2CMe_2Ph)_2(OC_6H_2\text{-}Bu^t_3)_2$, **6**, 130
$Mn_2C_{57}H_{48}NO_5P_4$
$[Mn_2(H)(CO)_5(NCMe)\text{-}(dppm)_2]^+$, **6**, 9
$Mn_2CdSi_2C_{42}H_{40}O_4$
$[(C_5H_4Me)Mn(SiMePh_2)\text{-}(CO)_2]_2Cd$, **6**, 144
$Mn_2CdSi_2C_{42}H_{42}O_4$
$[MeCp(CO)_2SiMePh_2Mn]_2Cd$, **10**, 337
$Mn_2Cd_3C_{26}H_{38}$
$(Cp_2Mn)_2(CdMe_2)_3$, **6**, 146
$Mn_2CoC_{26}H_{28}O_4PS$
$Mn(CO)_2Cp'(CO)[CSMn\text{-}(CO)_2Cp']Co(PMe_3)Cp$, **10**, 52
$Mn_2FeHgSi_2C_{60}H_{51}O_5$
$MeCpMn(CO)_2SiPh_3MnHgFe\text{-}(CO)_3PPh_3SiMePh_2$, **10**, 337
$Mn_2Fe_2C_{14}O_{14}$
$Fe_2Mn(CO)_9[SMn(CO)_5]$, **7**, 72
$Mn_2Fe_2C_{20}H_{10}O_{10}P_2$
$Fe_2(CO)_6[PMn(CO)_2Cp]_2$, **7**, 46
$Mn_2GeC_{14}H_{10}Cl_2O_4$
$Mn_2(GeCl_2)(CO)_4Cp_2$, **6**, 114
$Mn_2GeC_{14}H_{10}O_4$
$[Mn(CO)_2Cp]_2Ge$, **6**, 114
$Mn_2HgC_{20}H_{10}ClO_8P$
$(PPh_2)(CO)_8Mn_2(HgCl)$, **10**, 337
$Mn_2HgC_{26}H_{15}O_8P$
$(PPh_2)(CO)_8Mn_2(HgPh)$, **10**, 337
$Mn_2HgSi_2C_{42}H_{42}O_4$
$[MeCp(CO)_2SiMePh_2Mn]_2Hg$, **10**, 337
$Mn_2LiC_{10}HO_{10}$
$Li[(CO)_5MnMn(CO)_4(CHO)]$, **6**, 26
$Mn_2PtC_{28}H_{30}N_2O_4$
$[(C_5H_4Me)Mn(CO)_2(H)]_2Pt\text{-}(picoline)_2$, **6**, 143
$Mn_2Pt(H)_2(CO)_4(picoline)_2Cp'_2$, **10**, 211
$Mn_2PtC_{33}H_{20}O_9P_2$
$(CO)Pt[(PPh_2)Mn(CO)_4]_2$, **6**, 88
$Mn_2Pt(PPh_2)_2(CO)_9$, **10**, 226
$Mn_2PtC_{44}H_{31}O_8P_3$
$Mn_2Pt(H)(PPh_2)_3(CO)_8$, **10**, 225
$Mn_2Pt(PPh_2)_2(CO)_8(PPh_2H)$, **10**, 225
$Mn_2PtC_{50}H_{35}O_8P_3$
$Mn_2Pt(PPh_2)_2(CO)_8(PPh_3)$, **10**, 226
$Mn_2Pt_2C_{56}H_{42}O_8P_4$
$Mn(CO)_4(PPh_2)(H)Pt(PPh_2)_2\text{-}Pt(PPh_2)(H)Mn(CO)_4$, **10**, 53
$Mn_2ReC_{14}HO_{14}$
$Mn(CO)_5Mn(CO)_4(H)Re(CO)_5$, **10**, 48
$Mn_2SbC_{13}H_7Cl_2O_7$
$Mn(CO)_5(SbCl_2)Mn(CO)_2\text{-}(C_5H_4Me)$, **6**, 86
$Mn_2Si_2SnC_{16}H_{14}Cl_8O_4$
$Cl_2Sn[Mn(SiCl_3)(CO)_2(C_5H_4\text{-}Me)]_2$, **6**, 143
$Mn_2Si_4C_{22}H_{62}P_2$
$Mn_2(CH_2TMS)_4(PMe_3)_2$, **6**, 130
$Mn_2SnC_{24}H_{30}O_4$
$[Mn(CO)_2Cp^*]_2Sn$, **6**, 114
$Mn_2Sn_2C_{44}H_{30}O_8S_2$
$Mn_2(SSnPh_3)_2(CO)_8$, **6**, 89
$Mn_2TeC_{24}H_{30}O_4$
$[Mn(CO)_2Cp^*]_2Te$, **6**, 115
$Mn_2TeC_{26}H_{40}O_4$
$Mn_2(TePh)_2(CO)_4Cp_2$, **6**, 116
$Mn_2TiC_{22}H_{34}O_{10}$
$Mn_2(CO)_9(\overline{CCH_2CH_2TiCp^*_2O})$, **6**, 8
$Mn_3C_{12}H_6O_{12}P_3$
$Mn_3(PH_2)_3(CO)_{12}$, **6**, 86
$Mn_3C_{40}H_{36}Cl_4N_8O_{14}$
$[Mn_3O_4(bipy)_4(H_2O)_2](ClO_4)_4$, **12**, 1189
$Mn_3C_{52}H_{80}Br_6N_4O_2P_4$
$[Mn(CNBu^t)_4(PMe_2Ph)_2]\text{-}[MnBr_3(OPMe_2Ph)]_2$, **6**, 133
$Mn_3C_{54}H_{66}$
$[Mn(C_6H_2Me_3)_2]_3$, **6**, 130
$Mn_3GaNaC_{15}ClO_{15}$
$Na[GaCl[Mn(CO)_5]_3]$, **1**, 553
$Mn_3K_3C_{12}O_{12}$
$K_3[Mn_3(CO)_{12}]$, **6**, 12
$Mn_3SbC_{18}H_7ClO_{12}$
$Mn_2(CO)_{10}(SbCl)Mn(CO)_2\text{-}(C_5H_4Me)$, **6**, 86
$Mn_3Si_2C_{38}H_{84}N_2O_2$
$Mn_3(CH_2Bu^t)_4[OC(CH_2Bu^t)N\text{-}TMS]_2$, **6**, 130
$Mn_3Si_9C_{34}H_{89}Cl_4O$
$Mn_3[C(TMS)_3]_3Cl_4(THF)$, **6**, 129
$Mn_3TeC_{21}H_{15}O_6$
$[Mn(CO)_2Cp]_3Te$, **6**, 115
$Mn_4AsInC_{41}H_{21}ClO_{17}$
$[AsPh_4][InCl[Mn(CO)_5][Mn_3\text{-}(CO)_{12}(H)]]$, **1**, 561
$Mn_4As_2C_{28}H_{20}O_8$
$[Mn(CO)_2Cp]_4As_2$, **6**, 114
$Mn_4As_2C_{48}H_{60}O_8$
$[Mn(CO)_2Cp^*]_4As_2$, **6**, 114
$Mn_4As_2In_2C_{42}H_{40}Cl_2O_{18}$
$[AsPh_4]_2[Mn_2(CO)_8[InCl[Mn\text{-}(CO)_5]]_2]$, **1**, 561
$Mn_4C_{16}H_{20}N_4O_{16}$
$[Mn_4(OH)_4(CO)_{12}]\cdot 2en$, **6**, 85
$Mn_4C_{40}H_{88}$
$[Mn(CH_2Bu^t)_2]_4$, **6**, 130
$Mn_4HgC_{40}H_{20}O_{16}P_2$
$[(PPh_2)(CO)_8Mn_2]_2(Hg)$, **10**, 337
$Mn_4Hg_4C_{32}H_{32}O_8$
$[MeCp(CO)_2MnHg]_4$, **10**, 337, 338
$Mn_4Pd_4C_{40}H_{28}N_4O_{12}$
$[Mn(CO)_2Cp'(NC)Pd(CO)]_4$, **10**, 53
$[Pd(CO)(NC)Mn(CO)_2(C_5H_4\text{-}Me)]_4$, **9**, 207
$Mn_5As_2InC_{68}H_{40}O_{20}$
$[AsPh_4]_2[In[Mn(CO)_4]_5]$, **1**, 561
$Mn_6AsInC_{48}H_{22}O_{24}$
$[AsPh_4][In[Mn_3(CO)_{12}(H)]_2]$, **1**, 561
$MoAl_2C_{10}H_{24}Cl_2NO_4$
$Mo(NO)_2(OPr^iAlCl_2Et)_2$, **1**, 552
$MoAsC_{10}H_{11}O_3$
$Mo(AsMe_2)(CO)_3Cp$, **5**, 428
$MoAsCoFeWC_{19}H_{16}O_7S$
$MoWFeCo(S)(AsMe_2)\text{-}(CO)_7Cp_2$, **10**, 37, 244
$MoAsCo_2C_{17}H_{21}O_6P$
$MoCo_2(PBu^t)(AsMe_2)(CO)_6Cp$, **10**, 222
$MoAsCo_2C_{18}H_{15}O_3$
$MoCo_2(As)(CO)_3Cp_3$, **10**, 218
$MoAsFeC_{14}H_{11}O_7$

$Cp(CO)_3Mo(AsMe_2)Fe(CO)_4$, **10**, 39

$MoAsNi_2C_{37}H_{30}O_7$
$Ni_2Mo[As[Mo_3(As)(CO)_5Cp_3]]-(CO)_2Cp_3$, **9**, 173

$MoAsPt_2C_{27}H_{20}O_4P$
$MoPt_2(As)(CO)_4(PPh_3)Cp$, **10**, 218

$MoAsRh_3C_{25}H_{20}O_5$
$MoRh_3(As)(CO)_5Cp_4$, **10**, 218

$MoBC_8H_5F_4O_3$
$Mo(BF_4)(CO)_3Cp$, **5**, 404

$MoBC_9H_7ClO_3$
$(CO)_3Mo[(C_6H_7B)Cl]$, **1**, 210, 211

$MoBC_9H_8O_3$
$(CO)_3Mo(C_6H_7BH)$, **1**, 210, 211

$MoBC_{10}H_{11}F_4O_2$
$Mo(BF_4)(CO)_2(CH_2CHMe)Cp$, **5**, 406

$MoBC_{11}H_{11}F_4N_2O_2$
$[Mo(CO)_2(NCMe)_2Cp][BF_4]$, **5**, 406

$MoBC_{14}H_{22}NO_4Si$
$(CO)_4Mo(C_{10}H_{22}BNSi)$, **1**, 213

$MoBC_{15}H_{12}O_3$
$(CO)_3Mo[(C_6H_7B)Ph]$, **1**, 210, 211

$MoBC_{15}H_{13}F_4N_2O_2$
$Mo(CO)_2(NCMe)_2(Ind)BF_4$, **5**, 202

$MoBC_{16}H_{20}F_3O_3$
$[CH_2=CHCH=CHCH_2CH-(OBF_3)Pr^{i}]MoCp(CO)_2$, **12**, 971

$MoBC_{22}H_{16}F_4N_4O_2$
$[Mo(CO)_2(bipy)_2]BF_4$, **5**, 237

$MoBInC_6H_6F_4N_2O_2$
$[Mo(MeCN)_2(CO)_2In][BF_4]$, **12**, 639

$MoBSi_2C_{19}H_{31}NP$
$CpMo(CO)_2[P(BH_3)(Ph)[N-(TMS)_2]]$, **1**, 288, 290

$MoB_2C_{10}H_{16}$
$MoCp_2H(B_2H_5)$, **1**, 321

$MoB_2C_{14}H_{11}F_8N_3O_2$
$[Mo(CO)_2(NCMe)(bipy)]-[BF_4]_2$, **5**, 237

$MoB_2C_{14}H_{18}O_4$
$Mo(CO)_4[Me_2C_2(BMe)_2-C=CMe_2]$, **1**, 208

$MoB_4C_{10}H_{16}$
$CpMo(C_5H_4)(B_4H_7)$, **1**, 321, 322

MoB_4I
$MoIB_4$, **12**, 893

$MoB_9C_{10}H_{17}$
$CpMo(C_3H_3)C_2B_9H_9$, **1**, 321

$MoB_9WC_{21}H_{29}O_2$
$MoW(CO)_2(C_7H_7)(CTol)(C_2-B_9H_9Me_2)$, **10**, 33

$MoCH_2ClO$
$MoOCl(=CH_2)$, **12**, 595

$MoC_2H_5Cl_2$
$MoCl_2Me(=CH_2)$, **12**, 595

$MoC_5H_5ClNO_2$
$Mo(Cl)(NO)_2Cp$, **5**, 419

$MoC_5H_5Cl_2$
$MoCl_2Cp$, **5**, 441

$MoC_5H_5Cl_3$
$MoCl_3Cp$, **5**, 441

$MoC_6H_6Br_2N_2O_2$
$MoBr_2(CO)_2(NCMe)_2$, **5**, 220

$MoC_6H_{15}Cl_2O_3$
$(EtO)_3MoCl_2$, **12**, 595

$MoC_6H_{18}N_3O_6P$
$MoO_5(HMPA)$, **11**, 413

MoC_6O_6
$Mo(CO)_6$, **5**, 156; **12**, 430, 638

$MoC_7Br_2N_2O_3$
$MoBr_2(CO)_3(NCMe)_2$, **5**, 218

$MoC_7H_5Cl_3O_2$
$MoCl_3(CO)_2Cp$, **5**, 407

$MoC_7H_5O_2$
$Mo(CO)_2Cp$, **12**, 641, 644

$MoC_7H_5O_2P_3$
$Mo(CO)_2(P_3)Cp$, **5**, 424

$MoC_7H_9ClNO_2$
$(CH_2=CHMe)Mo(CO)_2-(MeCN)Cl$, **12**, 969

$MoC_7H_{17}O_4$
$(EtO)_3MoO(=CH_2)$, **12**, 595

$MoC_8H_5O_3$
$[CpMo(CO)_3]^{+}$, **12**, 639

$MoC_8H_8O_4$
$Mo(CO)_4(C_2H_4)_2$, **5**, 332

$MoC_8H_{16}Cl_3O_3$
$MoOCl_3(THF)_2$, **12**, 595

MoC_8N_8
$[Mo(CN)_8]^{4-}$, **5**, 305

$MoC_9H_5ClF_6$
$MoCl(F_3CCCCF_3)Cp$, **5**, 369

$MoC_9H_5F_3O_3$
$Mo(CF_3)(CO)_3Cp$, **5**, 413

$MoC_9H_7IO_2$
$MoI(CO)_2(C_7H_7)$, **5**, 359

$MoC_9H_7OP_3$
$[Mo(CO)_3(CH_2)Cp]^{+}$, **5**, 168

$MoC_9H_8O_3$
$MoMe(CO)_3Cp$, **5**, 404, 414

$MoC_9H_9N_3O_3$
$Mo(CO)_3(MeCN)_3$, **12**, 638, 893, 969

$MoC_9H_9O_3$
$[Mo(CO)_2(MeCHO)Cp]^{-}$, **5**, 414

$MoC_9H_{10}I_2N$
$MoI_2(NCMe)(C_7H_7)$, **5**, 360

$MoC_9H_{10}NO_2$
$(allyl)Mo(NO)(CO)Cp$, **12**, 643
$[Mo(CO)(NO)(C_3H_5)Cp]^{+}$, **5**, 342

$MoC_9H_{10}O_2$
$MoH(CO)_2(C_2H_4)Cp$, **5**, 334, 412

$MoC_9H_{11}ClN_2O_2$
$(allyl)Mo(CO)_2(MeCN)_2Cl$, **12**, 638
$MoCl(CO)_2(NCMe)_2(C_3H_5)$, **5**, 337

$MoC_9H_{11}Cl_2$
$MoCl_2(C_4H_6)Cp$, **5**, 350

$MoC_9H_{11}NO$
$Mo(NO)(C_4H_6)Cp$, **5**, 349

$MoC_9H_{11}NO_2$
$CH_2=CHMeMo(CO)(NO)Cp$, **12**, 970

$MoC_9H_{11}O_5P$
$Mo(CO)_4(Me_2POCH_2-CH=CH_2)$, **5**, 258

$MoC_9H_{13}ClNO$
$[CH_2=C(Me)Me]MoCP(NO)Cl$, **12**, 971

$MoC_9H_{24}N_2OP_2S_4$
$Mo(S_2CNMe_2)(NO)(PMe_3)[SC-(PMe_3)S]$, **5**, 303

$MoC_9H_{27}ClNOP_3S_2$
$MoCl(NO)(PMe_3)_2[SC-(PMe_3)S]$, **5**, 303

$MoC_{10}H_8O_3$
$Mo(CO)_3(C_5H_4CH_2CH_2)$, **5**, 410
$Mo(CO)_3(C_7H_8)$, **5**, 354

$MoC_{10}H_{10}Cl_2$
$MoCl_2Cp_2$, **5**, 445

$MoC_{10}H_{10}O$
$Mo(O)Cp_2$, **5**, 445, 448

$MoC_{10}H_{10}O_2$
$(allyl)Mo(CO)_2Cp$, **12**, 638, 640, 642, 643
$Mo(CO)_2(C_3H_5)Cp$, **5**, 338, 339, 406
$MoMe(CO)_2(C_7H_7)$, **5**, 359

$MoC_{10}H_{10}O_3$
$Mo(CO)_3(H_2)(nbd)$, **5**, 224, 347

$MoC_{10}H_{10}S$
$Mo(S)Cp_2$, **5**, 449

$MoC_{10}H_{10}S_2$
$Mo(S_2)Cp_2$, **5**, 449

$MoC_{10}H_{10}S_4$
$Mo(S_4)Cp_2$, **5**, 449

$MoC_{10}H_{11}Cl$
$MoHClCp_2$, **5**, 445

$MoC_{10}H_{11}O_2$
$(CH_2=CHMe)Mo(CO)_2Cp$, **12**, 968–70, 973

$MoC_{10}H_{12}$
MoH_2Cp_2, **5**, 443

$MoC_{10}H_{12}IO_3$
$MoI(CO)_3(Me_2pdyl)$, **5**, 352

$MoC_{10}H_{12}O_3$
$Mo(CO)_3(Me_2pdyl)$, **5**, 352

$MoC_{10}H_{13}NO_3$
$Mo(CO)(NO)[C(OEt)Me]Cp$, **5**, 418

$MoC_{10}H_{15}ClO_2$
$Mo(Cl)(O)_2Cp^{*}$, **5**, 431

$MoC_{10}H_{15}Cl_4$
$MoCl_4Cp^{*}$, **5**, 440

$MoC_{10}H_{16}O_6$
$MoO_2(acac)_2$, **12**, 815

$MoC_{11}H_8O_3$
$Mo(CH_2CCH)(CO)_3Cp$, **5**, 415

$MoC_{11}H_9O_4$
$[Mo(CO)_4(C_7H_9)]^{+}$, **5**, 354

$MoC_{11}H_{10}NO_2$

[Mo(NCMe)(CO)$_2$(C$_7$H$_7$)]$^+$, **5**, 360
$MoC_{11}H_{10}O_2$
Mo(CO$_2$)Cp$_2$, **5**, 445
$MoC_{11}H_{10}O_3$
Mo(CO$_3$)Cp$_2$, **5**, 445
$MoC_{11}H_{11}NO_5P$
Mo(CO)$_4$(Me$_2$POpy), **5**, 258
$MoC_{11}H_{12}O_2$
Mo(CO)$_2$(MeC$_3$H$_4$)Cp, **5**, 339
$MoC_{11}H_{12}O_3$
Mo(CO)$_3$(Me$_2$chd), **5**, 354
$MoC_{11}H_{15}$
Mo(C$_3$H$_5$)$_2$Cp, **5**, 336
$MoC_{11}H_{24}P_2$
MoH$_3$(dmpe)Cp, **5**, 406
$MoC_{12}H_8N_2O_4$
Mo(CO)$_4$(bipy), **12**, 890
$MoC_{12}H_9O_3$
[Mo(CO)$_2$(C$_4$H$_4$CO)Cp]$^+$, **5**, 352
$MoC_{12}H_{12}$
Mo(C$_7$H$_7$)Cp, **5**, 361, 450
$MoC_{12}H_{12}O_2$
(isoprene)MO(CO)$_2$Cp, **12**, 655
Mo(CO)$_2$(C$_5$H$_7$)Cp, **5**, 353
$MoC_{12}H_{13}ClN$
[MoCl(NCMe)Cp$_2$]$^+$, **5**, 445, 446
$MoC_{12}H_{13}N$
Mo(MeCN)Cp$_2$, **5**, 446
$MoC_{12}H_{13}O_3$
[CH$_2$=CHCH$_2$C(O)Me]Mo-(CO)$_2$Cp, **12**, 975
$MoC_{12}H_{15}NO_2S_2$
Mo(CO)$_2$(S$_2$CNEt$_2$)Cp, **5**, 438
$MoC_{12}H_{15}O$
[Mo(CO)(C$_2$H$_4$)(MeC$_2$-Me)Cp]$^+$, **5**, 367
$MoC_{12}H_{16}O_2$
Mo(CO)$_2$(MeC$_4$H$_5$)$_2$, **5**, 347
$MoC_{12}H_{18}$
Mo(CH$_2$CHCHCH$_2$)$_3$, **5**, 346
$MoC_{12}H_{20}$
Mo(C$_3$H$_5$)$_4$, **5**, 336
$MoC_{12}H_{20}N_2O_2S_4$
Mo(CO)$_2$(S$_2$CNEt$_2$)$_2$, **5**, 269, 271
$MoC_{12}H_{24}ClP$
MoCl(PMe$_3$)(C$_3$H$_5$)$_3$, **5**, 336
$MoC_{12}H_{24}O_8P_2$
Mo(CO)$_2$[P(OMe)$_3$]$_2$(C$_4$H$_6$), **5**, 347
$MoC_{12}H_{27}O_2P_3S_2$
Mo(CO)$_2$(PMe$_3$)$_2$(S$_2$CPMe$_3$), **5**, 269
$MoC_{13}H_8O_6$
(CO)$_5$Mo=C(OMe)Ph, **12**, 390
$MoC_{13}H_{10}N_2O_2$
Mo(CO)$_2$(NNPh)Cp, **5**, 423
$MoC_{13}H_{12}O_2$
(chd)Mo(CO)$_2$Cp, **12**, 638, 970
$MoC_{13}H_{12}S_2$
Mo[C(S)SCHCH]Cp$_2$, **5**, 450
$MoC_{13}H_{13}ClN_2O_2$
MoCl(CO)$_2$(NCMe)$_2$(C$_5$H$_7$), **5**, 352
$MoC_{13}H_{13}O_2$
Mo(CO)$_2$(chd)Cp, **5**, 339
$MoC_{13}H_{13}O_3$
(C$_6$H$_8$O)Mo(CO)$_2$Cp, **12**, 974
$MoC_{13}H_{17}Cl$
MoCl(Me$_2$C$_2$)$_2$Cp, **5**, 350
$MoC_{13}H_{20}Br_2N_4O_2$
MoBr$_2$(CO)$_2$[CH$_2$(Me$_2$pz)$_2$], **5**, 220
$MoC_{13}H_{23}BrO_2P_2$
MoBr(CO)$_2$(dmpe)(C$_5$H$_7$), **5**, 353
$MoC_{13}H_{23}IP_2$
MoI(dmpe)(C$_7$H$_7$), **5**, 361
$MoC_{13}H_{23}O_2P_2$
[Mo(CO)$_2$(dmpe)(C$_5$H$_7$)]$^+$, **5**, 353
$MoC_{13}H_{25}IP_2$
MoI(PMe$_3$)$_2$(C$_7$H$_7$), **5**, 361
$MoC_{14}H_5F_{12}IO$
MoI(CO)[C$_4$(CF$_3$)$_4$]Cp, **5**, 351
$MoC_{14}H_{14}$
Mo(Tol)(C$_7$H$_7$), **5**, 361
$MoC_{14}H_{14}O_2$
(cycloheptadiene)Mo(CO)$_2$Cp, **12**, 638
[Mo(CO)$_2$(C$_7$H$_9$)Cp, **5**, 354
$MoC_{14}H_{15}F_6O$
Mo=CHC(Me)$_2$Ph(OCMe-(CF$_3$)$_2$), **12**, 1221
$MoC_{14}H_{16}$
Mo(C$_7$H$_8$)$_2$, **5**, 354
Mo(C$_7$H$_9$)(C$_7$H$_7$), **5**, 354
Mo(PhMe)$_2$, **5**, 479
$MoC_{14}H_{16}NO_2$
[Mo(CNBu^t)(CO)$_2$(C$_7$H$_7$)]$^+$, **5**, 360
$MoC_{14}H_{16}N_2$[Mo(CNMe)$_2$Cp$_2$]$^{2+}$, **5**, 446
$MoC_{14}H_{17}O$
[Mo(CO)(MeCCMe)$_2$Cp]$^+$, **5**, 370
$MoC_{14}H_{18}S_4$
Mo(SCH$_2$CH$_2$S)$_2$Cp$_2$, **5**, 435
$MoC_{14}H_{19}O_2$
[MeCH(Me)CH=CHCH$_2$-Me]Mo(CO)$_2$Cp, **12**, 970
$MoC_{14}H_{27}NO_3$
[Bu^{t}NH$_3$][MoO$_3$Cp*], **5**, 431
$MoC_{14}H_{30}P_2$
Mo(PMe$_3$)$_2$(C$_4$H$_6$)$_2$, **5**, 348
$MoC_{14}H_{32}ClP_3$
MoCl(PMe$_3$)$_3$Cp, **5**, 442
$MoC_{15}H_5F_{11}O$
Mo(C$_6$F$_5$)(O)(F$_3$CCCCF$_3$)Cp, **5**, 371
$MoC_{15}H_5F_{12}IO_2$
MoI(CO)[C$_4$(CF$_3$)$_4$CO]Cp, **5**, 352
$MoC_{15}H_7F_5O_2$
Mo(C$_6$F$_5$)(CO)$_2$(C$_7$H$_7$), **5**, 360
$MoC_{15}H_{11}N_3O_3$
Mo(CO)$_3$(bipy)(MeCN), **12**, 890, 893
$MoC_{15}H_{12}O_2$
Mo(≡CC$_6$H$_4$Me)(CO)$_2$(Cp), **1**, 411
$MoC_{15}H_{12}O_3$
Mo(CH$_2$Ph)(CO)$_3$Cp, **5**, 412, 413
$MoC_{15}H_{13}ClN_2O_2$
MoCl(CO)$_2$(bipy)(C$_3$H$_5$), **5**, 338
$MoC_{15}H_{16}O_2$
(cod)Mo(CO)$_2$Cp, **12**, 638, 649
$MoC_{15}H_{18}N_3O_2$
[Mo(CO)$_2$(NCMe)$_3$(C$_7$H$_9$)]$^+$, **5**, 354
$MoC_{15}H_{18}O_2$
(cyclooctenyl)Mo(CO)$_2$Cp, **12**, 649
$MoC_{15}H_{19}F_6O_{10}P_2$
Mo[OC(O)CF$_3$][HOC(O)CF$_3$]-[P(OMe)$_3$]$_2$Cp, **5**, 320
$MoC_{15}H_{23}NO_2$
Mo(CO)(NO)(C$_4$H$_8$)Cp*, **5**, 334
$MoC_{15}H_{25}BrN_2O_2$
MoBr(CO)$_2$(Bu^{t}NCHCHNBu^t)-(C$_3$H$_5$), **5**, 338
$MoC_{15}H_{33}O_6P_2$
Mo[P(OMe)$_3$]$_2$(MeC$_2$Me)Cp, **5**, 367
$MoC_{15}H_{35}P_3$
Mo(Me)(PMe$_3$)$_3$Cp, **5**, 442
$MoC_{16}H_{10}O_3$
Mo(CCPh)(CO)$_3$Cp, **5**, 417
$MoC_{16}H_{16}O_2$
Mo(CO)$_2$(nbd)$_2$, **5**, 347
$MoC_{16}H_{19}O_4$
[OCH(Me)CH=CHCH$_2$CH-(MeOCH$_2$)]Mo(CO)$_2$Cp, **12**, 972
$MoC_{16}H_{20}O_2$
Mo(CO)$_2$(CH$_2$CHCCH$_2$)Cp*, **5**, 344
$MoC_{16}H_{44}P_4$
Mo(PMe$_3$)$_4$(C$_2$H$_4$)$_2$, **5**, 333
$MoC_{17}H_{12}O_2$
Mo(CCPh)(CO)$_2$(C$_7$H$_7$), **5**, 166
$MoC_{17}H_{15}BrN_2O_2$
MoBr(CO)$_2$(bipy)(C$_5$H$_7$), **5**, 353
$MoC_{17}H_{18}OS_2$
Mo[S(Me$_2$C$_6$H$_3$)]$_2$(CO), **5**, 530, 531
$MoC_{17}H_{24}O_2$
Mo(COMe)(CO)(C$_4$H$_6$)Cp*, **5**, 349
$MoC_{18}H_{12}N_4O_2$
[(CH)$_4$Mecyclohexenyl]Mo-(CO)$_2$Cp, **12**, 972
$MoC_{18}H_{14}N_2O_6$
Mo(CO)$_2$(bipy)(CO$_2$Me-CCCO$_2$Me), **5**, 363
$MoC_{18}H_{18}O_3$
Mo[C(OEt)Ph](CO)$_2$(C$_7$H$_8$), **5**, 354

$MoC_{18}H_{22}O_2$
Mo(Mes)$_2$(O)$_2$, **5**, 289, 296
$MoC_{18}H_{25}N_2O$
[Mo(CNBu^t)$_2$(CO)(C$_7$H$_7$)]$^+$, **5**, 360
$MoC_{18}H_{28}ClP_2$
[MoCl(PMe$_3$)$_2$(≡CPh)Cp]$^+$, **5**, 321
$MoC_{18}H_{28}O_6P_2$
Mo[P(OMe)$_3$]$_2$(≡CPh)Cp, **5**, 321
$MoC_{19}H_{15}O_2P$
[Mo(CO)$_2$(PPh$_2$)Cp]$^{2-}$, **5**, 428
$MoC_{19}H_{18}O_4$
Mo[C(OEt)Ph](CO)$_3$(nbd), **5**, 347
$MoC_{19}H_{21}O_3$
[Ph(OH)CHCH$_2$CH=C(Me)-Me]MoCp(CO)$_2$, **12**, 971
$MoC_{19}H_{21}P$
Mo(PMes)Cp$_2$, **5**, 448
$MoC_{19}H_{31}ClNO$
[CH$_2$C(Me)Me]Mo(neo-menthylCp)(NO)Cl, **12**, 971
$MoC_{19}H_{41}N$
Mo(CH$_2$Bu^t)$_2$(=CHBu^t)(NBu^t), **5**, 313
$MoC_{20}H_{16}$
Mo(C$_{10}$H$_8$)$_2$, **5**, 479, 529
$MoC_{20}H_{24}O_4$
Mo$_2$(CO)$_4$(Me$_2$chd)$_2$, **5**, 354
$MoC_{20}H_{39}P$
Mo(PEt$_3$)(Me$_2$pdyl)$_2$, **5**, 353
$MoC_{20}H_{40}FN_3O_2S_4$
[NEt$_4$][MoF(CO)$_2$(S$_2$CNEt$_2$)$_2$], **5**, 218
$MoC_{21}H_{27}N$
Mo(NBu^t)(C$_5$H$_4$Pr^i)(C$_9$H$_7$), **5**, 421
$MoC_{21}H_{30}O_3$
Mo(CO)$_3$(C$_6$Et$_6$), **5**, 495, 499
$MoC_{21}H_{50}OP_2$
MoH$_2$(CO)(Et$_2$PCH$_2$CH$_2$PEt$_2$)$_2$, **5**, 224
$MoC_{22}H_{16}N_4O_2$
Mo(CO)$_2$(bipy)$_2$, **5**, 236, 237
$MoC_{22}H_{26}B_{10}O_4P$
(CO)$_3$(Ph$_3$P)Me(HO)(CB$_{10}$-H$_{10}$)$^-$, **1**, 380
$MoC_{22}H_{36}$
MoMe$_2$Cp*$_2$, **5**, 446
$MoC_{22}H_{36}N_4O_2$
Mo(CO)$_2$(Bu^{t}NC)$_4$, **12**, 893
$MoC_{22}H_{40}N_3O_4S_4$
[NEt$_4$][Mo(CO)$_4$(S$_2$CNEt$_2$)$_2$], **5**, 271
$MoC_{24}H_{22}$
Mo(PhCHCHPh)Cp$_2$, **5**, 444
$MoC_{24}H_{23}F_{12}NO_2$
Mo(CHCMe$_2$Ph)(NPh)[OCMe-(CF$_3$)$_2$]$_2$, **12**, 1091
$MoC_{25}H_{21}N_2O_3P$
Mo(CO)$_3$(NCMe)$_2$PPh$_3$, **5**, 234
$MoC_{25}H_{22}S_3$
[Mo(C$_7$H$_7$)(SPh)$_3$]$^+$, **5**, 362
$MoC_{26}H_{24}N_4P_2$
Mo(N$_2$)$_2$(dppe), **12**, 893
$MoC_{27}H_{20}O_3P_2S_4$
Mo(CO)$_3$(S$_2$PPh$_2$)$_2$, **5**, 269
$MoC_{27}H_{23}O_2P$
Mo(CO)(PPh$_3$)[CH$_2$CHC-(O)]Cp, **5**, 344
$MoC_{27}H_{23}O_3P$
Mo(COMe)(CO)$_2$(PPh$_3$)Cp, **5**, 413, 414
$MoC_{27}H_{29}P$
Mo(PPh$_3$)(MeC$_2$Me)Cp, **5**, 367
$MoC_{27}H_{40}O_4$
Mo(CO)(EtCCEt)$_2$[(OMe)Et$_2$-(O)(C$_9$H$_7$O)], **12**, 481, 482
$MoC_{28}H_{25}P$
Mo(PPh$_3$)Cp$_2$, **5**, 448
$MoC_{28}H_{29}OP$
Mo(CO)(PPh$_3$)(MeC$_2$Me)Cp, **5**, 367
$MoC_{28}H_{44}$
Mo(norbornyl)$_4$, **5**, 282
$MoC_{29}H_{22}O_4P_2$
Mo(CO)$_4$(dppm), **5**, 234
$MoC_{30}H_{44}ClF_6N_6$
[MoCl(CNBu^t)$_4$(bipy)](PF$_6$), **5**, 310
$MoC_{30}H_{44}NO_2$
Mo(CO)$_2$(CNCy)$_4$, **5**, 162
$MoC_{30}H_{56}IN_6$
[MoI(CNBu^t)$_4$-(Bu^{t}HNCCNHBu^t)]$^+$, **5**, 372
$MoC_{30}H_{56}N_6$
[Mo[C(H)NHBu^t](CNBu^t)$_5$]$^{2+}$, **5**, 310
$MoC_{31}H_{29}ClO_2P_2$
MoCl(CO)$_2$(dppe)(C$_3$H$_5$), **5**, 338
$MoC_{31}H_{36}F_6O_2$
(Me(F$_3$C)$_2$CO)$_2$Mo[(Pr^i$_2$-C$_6$H$_3$)CH=][Me$_2$PhC-CH=], **12**, 596
$MoC_{33}H_{32}P_2$
MoH(dppe)(C$_7$H$_7$), **5**, 361
$MoC_{34}H_{32}N_2S_4$
Mo(S$_2$CNMe$_2$)$_2$(PhCCPh)$_2$, **5**, 366
$MoC_{34}H_{33}OP_2$
[Mo(CO)(dppe)(C$_2$H$_4$)Cp]$^+$, **5**, 334
$MoC_{34}H_{37}NOP_2S_2$
Mo(CO)(S$_2$CNEt$_2$)(dppm)-(C$_3$H$_5$), **5**, 337
$MoC_{35}H_{31}IO_2P_2$
MoI(CO)$_2$(dppe)(C$_7$H$_7$), **5**, 360
$MoC_{35}H_{37}P_2$
MoH(C$_2$H$_4$)(dppe)(PhMe), **5**, 529
$MoC_{35}H_{63}N_7$
[Mo(CNBu^t)$_7$]$^{2+}$, **5**, 311
$MoC_{36}H_{72}N_4S_4$
Mo(SBu^t)$_4$(CNBu^t)$_4$, **5**, 310
$MoC_{39}H_{35}O$
[Mo(CO)(PhCCPh)$_2$Cp*]$^+$, **5**, 370
$MoC_{40}H_{34}F_6P_3$
[Mo(C≡CPh)(dppe)(C$_7$H$_7$)]PF$_6$, **5**, 311
$MoC_{41}H_{34}P_2$
Mo(C≡CPh)(dppe)(C$_7$H$_7$), **5**, 311
$MoC_{41}H_{36}P_2$
Mo(CCPh)(dppe)(C$_7$H$_7$), **5**, 361
$MoC_{41}H_{37}P_2$
[Mo(CCHPh)(dppe)(C$_7$H$_7$)]$^+$, **5**, 361
$MoC_{42}H_{30}IN_6$
[MoI(CNPh)$_6$]$^+$, **5**, 310
$MoC_{44}H_{38}P_2$
Mo(dppe)(C$_9$H$_7$)$_2$, **5**, 448
$MoC_{49}H_{77}F_{12}N_7P_2$
[Mo(CNCy)$_7$](PF$_6$)$_2$, **5**, 309
$MoC_{52}H_{52}P_4$
Mo(PPh$_2$Me)$_3$(PhPMePh), **5**, 529
$MoC_{53}H_{44}O_3P_4$
Mo(CO)$_3$(dppm)$_2$, **5**, 227
$MoC_{53}H_{48}ClNP_4$
MoCl(CN)(dppe)$_2$, **5**, 318
$MoC_{53}H_{50}ClNP_4$
MoCl(≡CNH$_2$)(dppe)$_2$, **5**, 318
$MoC_{53}H_{50}OP_4$
Mo(H$_2$)(CO)(dppe)$_2$, **5**, 224
$MoC_{54}H_{44}N_2O_2P_4S_2$
Mo(NCS)$_2$(CO)$_2$(dppm)$_2$, **5**, 227
$MoC_{54}H_{48}F_7O_2P_5$
[MoF(CO)$_2$(dppe)$_2$][PF$_6$], **5**, 218
$MoC_{54}H_{54}N_6$
Mo(CNC$_6$H$_3$Me$_2$)$_6$, **5**, 309
$MoC_{55}H_{54}P_4$
MoH(dppe)$_2$(C$_3$H$_5$), **5**, 342
$MoC_{56}H_{56}P_4$
Mo(dppe)$_2$(C$_2$H$_4$)$_2$, **5**, 333
$MoC_{58}H_{52}O_4P_4$
MoH$_2$(C≡CCO$_2$Me)$_2$(dppe)$_2$, **5**, 311
$MoC_{60}H_{77}F_{12}N_5P_2$
[Mo(CNCy)$_5$(dppm)](PF$_6$)$_2$, **5**, 310
$MoCl_2O_2$
MoO$_2$Cl$_2$, **12**, 595
$MoCl_3O$
MoOCl$_3$, **12**, 1090
$MoCoC_{10}H_{17}B_9$
CpCoMo(C$_3$H$_3$)(C$_2$B$_9$H$_9$), **1**, 414
$MoCoC_{13}H_8O_5$
Co(CH$_2$CCH)(CO)$_3$Mo(CO)$_2$-(Cp), **12**, 687
$MoCoC_{14}H_6O_5$
Mo(CO)$_2$Cp(MeC$_2$Me)Co-(CO)$_3$, **10**, 43
$MoCoC_{16}H_{13}O_3$
MoCp(CO)$_3$Co(C$_4$Me$_4$), **10**, 43
$MoCoC_{17}H_{17}O_4$

$Mo(CO)_2Cp(C_4Me_4)Co(CO)_2$, **10**, 43
$MoCoC_{19}H_{13}O_5$
$CoMo(CO)_5Cp(MeC_2Ph)$, **10**, 13
$MoCoFeC_{13}H_5O_8S$
$MoFeCo(S)Cp(CO)_8$, **10**, 49
$MoCoNiC_{17}H_{13}O_5$
$NiCoMo(CMe)(CO)_5Cp_2$, **9**, 173
$MoCoPd_2C_{56}H_{44}O_6P_4$
$MoCoPd_2(CO)_6(dppm)_2$, **10**, 243
$MoCoRhC_{27}H_{30}O_7$
$MoCoRh(CO)_7Cp^*_2$, **10**, 244
$MoCoRuC_{15}H_9O_8$
$HMoRuCo(CMe)Cp(CO)_8$, **10**, 41
$MoCoW_2C_{27}H_{21}O_9$
$MoW_2Co(CO)_9Cp'_3$, **10**, 244
$MoCo_2C_{15}H_5F_6O_9P$
$[CpMoCo_2(CO)_8(CCO)][PF_6]$, **10**, 109
$MoCo_2C_{23}H_{22}O_{10}$
$Cp^*MoCo_2(CO)_8(CCO_2Pr^i)$, **10**, 112
$MoCo_2NiC_{19}H_{12}O_8$
$NiCo_2Mo(CO)_8CpCp'$, **9**, 173
$MoCrC_{22}H_{14}O_6$
$Cr(CO)_3(azulene)Mo(CO)_3(C_6H_6)$, **10**, 29
$MoCrHgC_{49}H_{35}NO_8P_2$
$(PPN)[(CO)_5CrHgMoCp(CO)_3]$, **10**, 336
$MoCuPdC_{43}H_{37}O_2P_3$
$CuMoPd(PPh_2)(CO)_2(PPh_2H)_2Cp$, **10**, 311
$MoFeC_4H_6N_2O_2S_2$
$[MoFeS_4(SCN)_2(OMe)_2]^{2-}$, **10**, 37
$MoFeC_{10}H_5BrO_5Te_2$
$CpMoFe(CO)_5Te_2Br$, **7**, 77
$MoFeC_{12}H_{15}I_2N$
$MoCp_2(H)_2FeI_2(MeCN)$, **10**, 43
$MoFeC_{14}H_{10}O_5$
$Mo(CO)_2Cp(CH_2C{=}CMe)Fe(CO)_3$, **10**, 42
$MoFeC_{15}H_{15}NO_5S_2Te_2$
$CpMoFe(CO)_5Te_2S_2CNEt_2$, **7**, 77
$MoFeC_{16}H_{10}O_4S_2$
$Mo(CO)_4(SPh)_2Fe$, **10**, 43
$MoFeC_{19}H_{14}O_3$
$Mo(C_6H_6)(azulene)Fe(CO)_3$, **10**, 42
$MoFeC_{33}H_{22}O_8P_2$
$(CO)_4Fe(dppm)Mo(CO)_4$, **7**, 83
$MoFe(CO)_8(dppm)$, **10**, 17
$MoFeC_{34}H_{22}O_9P_2$
$(CO)_4Fe(dppm)Mo(CO)_5$, **7**, 83
$MoFeC_{44}H_{54}N_4O_4S_2$
$(CyNC)_4Fe(SPh)_2Mo(CO)_4$, **7**, 61
$MoFeNiC_{27}H_{22}O_7$
$Cp_2FeMoNi(CO)_5(PhC_2Co_2Pr^i)$, **10**, 162
$MoFeSbC_{10}H_5F_6O_5Te_2$
$[CpMoFe(CO)_5Te_2][SbF_6]$, **7**, 77
$MoFe_2C_{13}H_6O_8S$
$HFe_2Mo(CO)_8Cp(S)$, **7**, 70
$MoFe_5C_6I_3O_6S_6$
$MoFe_5(S)_6(CO)_6I_3$, **10**, 237
$MoFe_6C_{16}O_{16}S_6$
$[MoFe_6(S)_6(CO)_{16}]^{2-}$, **10**, 232, 233
$MoGaC_{12}H_{11}Cl_2O_3$
$[(CO)_3Mo(Mes)]GaCl_2$, **1**, 508
$MoGaC_{13}H_{15}I_2O_3$
$GaI_2(THF)[Mo(CO)_2(C_7H_7)]$, **1**, 553
$MoGaC_{13}H_{17}I_2O_4$
$GaI_2(Et_2O)[Mo(CO)_3(C_5H_4Me)]$, **1**, 553
$MoGaC_{15}H_{14}IO_3$
$GaI[Mo(CO)_3(C_5H_4Me)_2]$, **1**, 553
$MoGaC_{30}H_{33}O_3$
$[(CO)_3Mo(Mes)]Ga(Mes)_2$, **1**, 507
$MoGeC_8H_5IO_3$
$Mo(GeI)(CO)_3Cp$, **5**, 409
$MoHfC_{22}H_{30}O_4P_2$
$HfCp_2(PEt_2)_2Mo(CO)_4$, **10**, 27
$MoHgC_8H_5ClO_3$
$[Cp(CO)_3Mo]HgCl$, **10**, 336
$MoHgC_8H_5O_3$
$HgMoCp(CO)_3$, **10**, 334
$MoHgC_{14}H_5Cl_5O_3$
$(C_6Cl_5)HgMoCp(CO)_3$, **10**, 336
$MoHgPtC_{50}H_{35}Cl_5O_3P_2$
$(PPh_3)_2PtHgMoCp(CO)_3(C_6Cl_5)$, **10**, 336
$MoHgRu_3C_{23}H_{14}O_{12}$
$(C_2Bu^t)(CO)_9Ru_3(Hg)MoCp(CO)_3$, **5**, 409; **10**, 339
$MoInC_4H_3O_3$
$MoMe(CO)_3In$, **12**, 639
$MoInNaC_{16}H_{21}Cl_3O_5$
$[Na(THF)_2][InCl_3[Mo(CO)_3Cp]]$, **1**, 557
$MoIrC_{13}H_8O_6$
$Mo(CO)_3(CHCMeCHCMeCH)Ir(CO)_3$, **10**, 45
$MoMnC_{13}H_5O_8$
$Mo(CO)_3CpMn(CO)_5$, **10**, 30
$MoMnC_{23}H_{15}IO_6P$
$CpMo(CO)_2(I)(PPh_2)Mn(CO)_4$, **6**, 88
$MoMnC_{23}H_{16}O_6P$
$CpMo(CO)_2(H)(PPh_2)Mn(CO)_4$, **6**, 88; **10**, 33
$MoMnC_{24}H_{14}O_7P$
$Mo(CO)_3(C_5H_4PPh_2)Mn(CO)_4$, **10**, 33
$MoMnC_{25}H_{20}O_6P$
$Mn(CO)_4(H)(PTol_2)Mo(CO)_2Cp$, **6**, 88
$MoMnC_{30}H_{27}O_4P$
$MnMo[CH(Me)CHCH_2](PPh_2)(CO)_4Cp_2$, **10**, 261
$MoNaC_8H_5O_3$
$Na[CpMo(CO)_3]$, **12**, 638, 970
$MoNaO_4$
$NaMoO_4$, **11**, 413
$MoNbC_{16}H_{11}O_6$
$Cp_2(CO)Nb(H)Mo(CO)_5$, **5**, 81
$MoNbC_{26}H_{20}F_6O_4PS_2$
$[Cp_2Nb(SPh)_2Mo(CO)_4]PF_6$, **5**, 128
$MoNbC_{26}H_{20}O_4S_2$
$[NbCp_2(SPh)_2Mo(CO)_4]^+$, **10**, 28
$MoNbC_{29}H_{29}O_4S_2$
$Nb(PhC_2Ph)Cp(SPr^i)_2Mo(CO)_4$, **10**, 28
$MoNiC_{15}H_{12}O_4$
$Mo(CO)_3CpNi(CO)Cp'$, **9**, 144, 169; **10**, 32, 45
$MoNiC_{17}H_{18}O_2$
$NiMo(CO)_2(MeC_2Me)CpCp'$, **9**, 136
$MoNiC_{17}H_{21}S_2$
$MoCp_2(SMe)_2NiCp$, **10**, 45
$MoNiC_{18}H_{20}O_2$
$NiCp(C_3H_2Me_2)(CO)Mo(CO)Cp'$, **9**, 144
$MoNiC_{18}H_{20}O_3$
$Cp^*NiMo(CO)_3Cp$, **9**, 95, 169; **10**, 32
$MoNiC_{19}H_{20}O_4$
$Ni(CO)Cp^*Mo(CO)_3Cp$, **9**, 173; **10**, 45
$NiCp^*(CO)_2Mo(CO)_2Cp$, **9**, 169
$MoNiC_{19}H_{22}O_3$
$Mo(CO)_2Cp(CO)(CH_2)NiCp^*$, **9**, 19, 169; **10**, 45
$MoNiC_{23}H_{21}F_8N_2O_3P_5$
$Mo(CO)_3[MeN(PF_2)_2]_2Ni(PPh_3)$, **10**, 45
$MoNiC_{24}H_{22}O_4$
$NiMo(CO)_2Cp(PhC_2CO_2Pr^i)Cp$, **9**, 136
$MoNiC_{30}H_{32}O_2S_2$
$NiCp^*(SPh)(CO)Mo(SPh)(CO)Cp'$, **9**, 169
$MoNiC_{34}H_{27}O_4$
$NiMo(CO)_2Cp(PhC_2Ph)(CO)_2CpCp'$, **9**, 136
$MoNiC_{41}H_{38}O_2$
$NiCp^*(PhC_2H)Ni(C_4H_2Ph_2)Mo(CO)_2Cp$, **9**, 169
$MoNiC_{46}H_{59}O_4P_3$
$Mo(CO)_4(PCy_2)_2Ni(PPh_3)$, **10**, 45
$MoNi_3C_{23}H_{20}O_3$
$Ni_3Mo(CO)_3Cp_4$, **9**, 173
$MoOsC_{12}H_9O_{12}P$
$Mo(CO)_5Os(CO)_4P(OMe)_3$, **10**, 30
$MoOs_3WC_{34}H_{20}O_8$
$Cp_2MoWOs_3(CO)_8(C)(CPh)(C_2Ph)$, **10**, 105
$MoPbC_{20}H_{18}O_7$
$Pb[Mo(CO)_3Cp]_2(THF)$, **5**, 409
$MoPdC_{55}H_{44}N_2O_3P_4$

Mo(CN)(dppm)$_2$Pd(CN)(CO)$_3$, **10**, 47
$MoPdPtC_{57}H_{49}ClO_2P_2$
MoPdPt(CO)$_2$Cl(dppm)$_2$Cp, **10**, 238
$MoPd_2C_{31}H_{52}O_3P_2$
CpMoPd$_2$(CO)$_3$Cp(PPr$^i_3$)$_2$, **10**, 124
$MoPtC_{23}H_{21}F_8N_2O_3P_5$
Pt(PPh$_3$)[F$_2$PN(Me)PF$_2$]$_2$Mo-(CO)$_3$, **10**, 17, 18
$MoPtC_{57}H_{50}O_2P_4$
Mo(CO)$_2$Cp(dppm)PtH(dppm), **10**, 47
$MoPtHgC_{50}H_{35}Cl_5O_3P_2$
Pt(C$_6$Cl$_5$)(PPh$_3$)$_2$[HgMo-(CO)$_3$Cp], **9**, 467
$MoReC_{15}H_{11}N_2O_6$
Mo(CO)$_3$CpRe(CO)$_3$(CNMe)$_2$, **10**, 30
$MoRhC_{18}H_{14}O_2$
[Mo(C$_6$H$_6$)(azulene)Rh(CO)$_2$]$^+$, **10**, 43
$MoRhC_{18}H_{27}O_4P_2$
Mo(CO)$_4$(PMe$_2$)$_2$RhCp*, **10**, 44
$MoRhC_{26}H_{20}O_3P$
CpMoRh(CO)$_3$(PPh$_3$), **10**, 2
$MoRhC_{44}H_{35}O_3P_2$
CpMoRh(CO)$_2$(CO)(PPh$_3$)$_2$, **10**, 39
$MoRhC_{53}H_{44}ClO_5P_4S$
Mo(CO)$_3$(SO$_2$)(dppm)$_2$RhCl, **10**, 44
$MoRh_2C_{27}H_{30}O_7$
MoRh$_2$(CO)$_7$Cp*$_2$, **10**, 210
$MoRh_3C_{65}H_{50}O_6P_3$
MoRh$_3$(CO)$_6$(PPh$_3$)$_3$Cp, **10**, 211
$MoRuC_{15}H_8O_5$
Mo(CO)$_3$(C$_5$H$_4$C$_5$H$_4$)Ru(CO)$_2$, **10**, 42
$MoRuC_{31}H_{20}O_7P_2$
(CO)$_4$Mo(PPh$_2$)$_2$Ru(CO)$_3$, **10**, 39
$MoRu_2C_{21}H_{10}O_8$
CpMoRu$_2$(CO)$_8$(C≡CPh), **10**, 20, 21
$MoRu_2C_{46}H_{40}P_2S_4$
[Ru(PPh$_3$)Cp]$_2$(MoS$_4$), **7**, 524
$MoSbC_{39}H_{36}I_2O_3P$
MoI$_2$(CO)$_3$(PPh$_2$Cy)(SbPh$_3$), **5**, 262
$MoSiC_{18}H_{24}O_3$
[(TMS-O)cod]Mo(CO)$_2$Cp, **12**, 974
$MoSi_2C_{14}H_{22}$
Mo(SiMe$_2$SiMe$_2$)Cp$_2$, **5**, 447
$MoSi_2C_{18}H_{29}$
[MoH(TMSC$_2$TMS)Cp$_2$]$^+$, **5**, 368
$MoSnC_{30}H_{54}Cl_3N_6$
Mo(SnCl$_3$)(CNBu^t)$_6$, **5**, 310
$MoTiC_{22}H_{23}O_4$
Cp$_2$Ti(THF)(CO)Mo(CO)$_2$Cp, **5**, 399
$MoTlC_8H_5O_3$
Tl[Mo(CO)$_3$Cp], **1**, 567
$MoWC_{14}H_9O_6$
CpMo(CO)$_2$(C$_3$H$_4$C=OCpW-(CO)$_3$), **10**, 15
$MoWC_{14}H_{10}O_4$
Mo(CO)$_2$CpW(CO)$_2$Cp, **10**, 29
MoW(CO)$_4$Cp$_2$, **10**, 32
$MoWC_{17}H_{14}O_4$
CpMo(CO)$_2$(C$_3$H$_4$)W(CO)$_2$Cp, **10**, 15
$MoWC_{18}H_{16}O_4$
Mo(CO)Cp(CO)(MeC$_2$Me)W-(CO)$_2$Cp, **10**, 33
$MoWC_{20}H_{13}O_7$
Mo(CO)$_5$[CH(Tol)]W(CO)$_2$Cp, **10**, 33
$MoWC_{24}H_{20}O_5$
MoW(CO)$_4$(CTolCOMe)Cp$_2$, **5**, 415
$MoWC_{28}H_{30}O_3$
Mo(CO)$_2$Cp(CTolCMeCMe)-(MeC$_2$Me)$_2$W(CO), **10**, 33
$MoWC_{35}H_{30}NO_6P_2$
[(CO)$_4$Mo(PPh$_2$)$_2$W(CO)$_2$-(CNEt$_2$0]$^-$, **5**, 197
$MoW_2C_{19}H_{15}NO_5$
MoW$_2$(N)(O)(CO)$_4$Cp$_3$, **10**, 218
$MoW_2C_{37}H_{25}O_4$
W(C$_4$Ph$_4$)Mo(CO)$_2$W(CO)$_2$Cp, **10**, 33
$MoYbC_{46}H_{44}O_6$
Yb(THF)$_2$[(C$_5$H$_4$Ph$_2$)$_2$Mo-(CO)$_4$], **4**, 110
$MoZrC_{19}H_{18}O_3$
ZrMeCp$_2$[(CO)Mo(CO)$_2$Cp], **4**, 535, 593
$MoZrC_{40}H_{30}O_4P_2$
ZrCp$_2$(PPh$_2$)$_2$Mo(CO)$_4$, **10**, 27
$Mo_2AsNiC_{19}H_{15}O_4$
Mo$_2$Ni(As)(CO)$_4$Cp$_3$, **9**, 173; **10**, 218
$Mo_2As_5C_{10}H_{10}$
CpMo(As$_5$)MoCp, **5**, 426
$Mo_2B_2C_{44}H_{32}F_8N_8O_4$
[Mo$_2$(Co)$_4$(bipy)$_4$][BF$_4$]$_2$, **5**, 237
$Mo_2Bi_2C_{14}H_{10}$
Mo$_2$(CO)$_4$(Bi$_2$)Cp$_2$, **5**, 427
$Mo_2C_4H_{16}O_2S_2$
Mo$_2$(CO)$_2$(SMe)$_2$Cp$_2$, **9**, 173
$Mo_2C_8Cl_4O_8$
Mo$_2$Cl$_4$(CO)$_8$, **5**, 218
$Mo_2C_8H_{12}O_8$
Mo$_2$(OAc)$_4$, **12**, 426
$Mo_2C_{10}HO_{10}$
Mo$_2$(H)(CO)$_{10}$, **5**, 222
$Mo_2C_{10}H_{11}S_4$
Mo$_2$(S)$_2$(SH)$_2$Cp$_2$, **5**, 435
$Mo_2C_{11}H_{12}S_4$
Mo$_2$(S)$_2$(S$_2$CH$_2$)Cp$_2$, **5**, 435
$Mo_2C_{14}H_{10}O_4P_2$
Mo$_2$(CO)$_4$(P$_2$)Cp$_2$, **5**, 426
$Mo_2C_{14}H_{16}O_2S_2$
Mo$_2$(CO)$_2$(SMe)$_2$Cp$_2$, **5**, 437
$Mo_2C_{14}O_4$
FvMo$_2$(μ-MeCCCH$_2$)(CO)$_4$$^-$, **12**, 687
$Mo_2C_{15}H_{19}O_2S_3$
[Mo$_2$(CO)$_2$(SMe)$_3$Cp$_2$]$^+$, **5**, 437
$Mo_2C_{16}H_8O_6$
Mo$_2$(CO)$_6$(C$_{10}$H$_8$), **5**, 380
$Mo_2C_{16}H_{15}NO_3$
Mo$_2$(O)$_3$(NPh)Cp$_2$, **5**, 421
$Mo_2C_{16}H_{17}O_4P$
Mo$_2$(H)(PMe$_2$)(CO)$_4$Cp$_2$, **5**, 431
$Mo_2C_{16}H_{22}Cl_4$
Mo$_2$(Cl)$_4$(C$_5$H$_4$Pr^i)$_2$, **5**, 440
$Mo_2C_{17}H_{12}O_5$
Mo(CO)$_2$(C$_7$H$_7$)Mo(CO)$_3$Cp, **5**, 362
$Mo_2C_{17}H_{13}O_4$
[(HCCCH$_2$)Mo$_2$(CO)$_4$Cp$_2$$^-$, **12**, 690
$Mo_2C_{17}H_{14}O_4$
Mo$_2$(CO)$_4$(CH$_2$CCH$_2$)Cp$_2$, **5**, 380
$Mo_2C_{18}H_{14}O_3$
Mo$_2$(CO)$_3$(C$_5$H$_4$)Cp$_2$, **5**, 443
$Mo_2C_{18}H_{19}O_4P$
Mo$_2$(CO)$_4$(PCBu^t)Cp$_2$, **5**, 426
$Mo_2C_{18}H_{30}Cl_2$
[MoCl(C$_3$H$_5$)$_3$]$_2$, **5**, 336
$Mo_2C_{19}H_{16}O_4$
Mo$_2$(CO)$_4$(CCCMe$_2$)Cp$_2$, **5**, 172
$Mo_2C_{19}H_{19}O_4P$
Mo$_2$(CO)$_4$(PCBu^t)Cp$_2$, **5**, 399
$Mo_2C_{20}H_{18}$
Mo$_2$(C$_5$H$_4$)$_2$Cp$_2$, **5**, 443
$Mo_2C_{20}H_{20}$
Cp$_2$Mo(H)(C$_5$H$_4$)MoCp, **5**, 443
$Mo_2C_{20}H_{30}O_4$
Mo$_2$(O)$_4$Cp*$_2$, **5**, 432
$Mo_2C_{20}H_{30}O_5$
[Mo(O)$_2$Cp*]$_2$(O), **5**, 432, 433
$Mo_2C_{20}H_{30}P_6$
Cp*Mo(P$_6$)MoCp*, **5**, 426
$Mo_2C_{20}H_{30}S_4$
Mo$_2$S$_4$Cp*$_2$, **5**, 435
$Mo_2C_{20}H_{30}S_{12}$
Mo$_2$(S$_2$)$_6$Cp*$_2$, **5**, 435
$Mo_2C_{22}H_{20}N_2O_2$
Mo$_2$(O)$_2$(NPh)$_2$Cp$_2$, **5**, 423
$Mo_2C_{28}H_{20}Cl_8$
[MoCl$_3$(PhCCPh)]$_2$Cl$_2$, **5**, 371
$Mo_2C_{28}H_{25}N_3$
Mo$_2$(NPh)$_3$Cp$_2$, **5**, 423
$Mo_2C_{28}H_{72}P_4$
Mo$_2$Me$_4$(PEt$_3$)$_4$, **5**, 322
$Mo_2C_{30}H_{20}O_4$
Mo$_2$(CO)$_4$(PhCCCCPh)Cp$_2$, **5**, 417
$Mo_2C_{30}H_{26}ClN_4O_4$
[Mo(CO)$_2$(bipy)(C$_3$H$_5$)]$_2$Cl, **5**, 338
$Mo_2C_{34}H_{34}O_6$

$Mo_2(CO)_4Cp_2$(17β-ethynylestradiol), **12**, 687
$Mo_2C_{35}H_{30}OP_2$
$Mo_2(CO)(PPh_2)_2Cp_2$, **5**, 430
$Mo_2C_{35}H_{30}O_2P_2$
$Mo_2(CO)(O)(PPh_2)_2Cp_2$, **5**, 430
$Mo_2C_{36}H_{30}O_2P_2$
$Mo_2(CO)_2(PPh_2)_2Cp_2$, **5**, 430
$Mo_2C_{81}H_{68}P_4$
(C_7H_7)(dppe)Mo[=C=C(Ph)C-(Ph)=C=C]Mo(dppe)-(C_7H_7), **5**, 311
$Mo_2Co_2C_{14}H_{14}O_2S_4$
$Mo_2Co_2(S)_4(CO)_2Cp'_2$, **10**, 236, 237
$Mo_2Co_2C_{15}H_{12}O_4S_3$
$Cp(C_5H_4Me)_2Mo_2Co_2(CO)_4S_3$, **10**, 51
$Mo_2Co_2C_{16}H_{14}O_4S_3$
$Mo_2Co_2(S)_3(CO)_4Cp'_2$, **10**, 236, 237
$Mo_2Co_2C_{31}H_{32}O_7P_2$
$Mo_2Co_2[C_6H_4(PBu^t)_2]$-$(CO)_7Cp_2$, **10**, 224
$Mo_2Co_2C_{34}H_{40}O_4$
$Mo_2Co_2(CO)_4Cp_2Cp^*_2$, **10**, 209
$Mo_2Co_2GeC_{21}H_{10}O_{11}$
$Mo_2Co_2(Ge)(CO)_{11}Cp_2$, **10**, 215
$Mo_2Co_3C_{30}H_{30}NO_{10}$
$Mo_2Co_3(N)(CO)_{10}Cp^*_2$, **10**, 218
$Mo_2FeC_2H_6OS_8$
$[Mo_2FeS_8O(OMe)_2]^{3-}$, **10**, 37
$Mo_2FeC_{17}H_{10}O_7Te_2$
$FeMoCp(CO)_5Te_2$[Mo-$(CO)_2Cp$], **7**, 77
$Mo_2Fe(Te)_2(CO)_7Cp_2$, **10**, 228
$Mo_2FeC_{27}H_{30}O_8$
$Cp^*_2Mo_2Fe(CO)_7(O)$, **7**, 60
$Mo_2Fe_2C_{17}H_{10}O_7Te_2$
$Cp_2Mo_2Fe_2(CO)_7Te_2$, **7**, 77
$Mo_2Fe_2C_{31}H_{30}O_{11}$
$Cp^*_2Mo_2Fe_2(CO)_{11}$, **7**, 83; **10**, 209, 211
$Mo_2Fe_4C_{20}O_{19}$
$Mo_2Fe_4C(CO)_{19}$, **10**, 91
$Mo_2Fe_7C_{22}O_{22}S_8$
$Mo_2Fe_7(S)_8(CO)_{22}$, **10**, 232, 233
$Mo_2GaC_{33}H_{33}O_3$
$[(CO)_3Mo(Mes)]_2$Ga(Mes), **1**, 507, 508
$Mo_2HgC_{16}H_{10}O_6$
$[Cp(CO)_3Mo]_2Hg$, **10**, 336
$Mo_2HgC_{38}H_{32}Cl_2O_3P_2$
ClHgMoCp(CO)$(PPh_2CH_2P$-$Ph_2)MoCp(CO)_2Cl$, **10**, 336
$Mo_2HgC_{50}H_{40}O_4P_2$
$(CpMo(CO)_2PPh_3)_2Hg$, **10**, 330
$Mo_2In_2C_6O_6$
$[InMo(CO)_3]_2$, **12**, 639
$Mo_2InC_{20}H_{18}ClO_7$
InCl(THF)$[Mo(CO)_3Cp]_2$, **1**, 559
$Mo_2InC_{20}H_{22}Cl_2NO_6$
$[NMe_4][InCl_2[Mo(CO)_3Cp]_2]$, **1**, 559
$Mo_2InNaC_{24}H_{26}Cl_2O_8$
$[Na(THF)_2][InCl_2$[Mo-$(CO)_3Cp]_2]$, **1**, 557
$Mo_2InSnC_{22}H_{27}O_{13}$
$(CO)_5MoSn(OBu^t)_3InMo(CO)_5$, **1**, 560
$Mo_2MgC_{24}H_{56}O_6$
$[Mg(THF)_4][MoMe_4(O)]_2$, **5**, 289
$Mo_2NiPtC_{30}H_{29}O_4$
$PtNiMo_2(CC_6H_4Me)(CO)_4$-(cod)$Cp_2$, **9**, 564
$Mo_2Ni_2C_{21}H_{20}OS_2$
$Mo_2Ni_2(S)_2(CO)Cp_4$, **10**, 234, 235
$Mo_2Ni_2C_{29}H_{25}O_2$
$Ni_2Mo_2(CPh)(CO)_2Cp_4$, **9**, 173
$Mo_2Ni_2C_{34}H_{40}O_4$
$Ni_2Mo_2(CO)_4Cp_2Cp^*_2$, **9**, 173
$Mo_2Pd_2C_{52}H_{40}O_6P_2$
$Mo_2Pd_2(CO)_6Cp_2(PPh_3)_2$, **10**, 49
$Mo_2Pd_2C_{54}H_{44}O_6P_2$
$Mo_2Pd_2(CO)_6(PPh_2Bz)_2Cp_2$, **9**, 204; **10**, 204
$Mo_2PtC_{30}H_{32}N_2O_6$
$Pt[Mo(CO)_3Cp]_2(CNCy)_2$, **10**, 37
$Mo_2Pt_2C_{62}H_{51}O_4P_3$
$Mo(CO)_2Cp(PPh_2)(H)Pt$-$(PPh_2)_2Pt(PPh_2)Mo$-$(CO)_2Cp$, **10**, 47
$Mo_2Pt_2C_{62}H_{52}O_4P_4$
$Mo_2Pt_2(H)_2(PPh_2)_4(CO)_4Cp_2$, **10**, 223
$Mo_2Pt_2C_{72}H_{113}O_6P_3$
$Mo_2Pt_2(CO)_6(PCy_3)_3(C_5H_4$-$Me)_2$, **10**, 204
$Mo_2Pt_4WC_{47}H_{49}O_6$
$Pt_4WMo_2(CMe)(CC_6H_4Me)$-$(CO)_6(cod)_2Cp_3$, **9**, 564
$Mo_2Pt_5W_3C_{46}H_{39}O_{10}$
$Pt_5W_3Mo_2(CMe)(CC_6H_4Me)$-$(CC_6H_4Me)(CO)_{10}$-(cod)$Cp_2$, **9**, 564
$Mo_2RuC_{17}H_{10}O_7S$
$Mo_2Ru(S)(CO)_7Cp_2$, **10**, 49, 236
$Mo_2RuC_{17}O_7S$
$RuMo_2Cp_2(CO)_7(S)$, **10**, 20
$Mo_2Ru_3C_{22}H_{10}O_{12}S$
$Ru_3Mo_2(CO)_{12}Cp_2(S)$, **10**, 20
$Mo_2Ru_4C_{24}H_{10}O_{14}S$
$Mo_2Ru_4(S)(CO)_{14}Cp_2$, **10**, 234–6
$Mo_2Ru_5C_{25}H_{10}O_{15}S$
$Mo_2Ru_5(S)(CO)_{15}Cp_2$, **10**, 234
$Mo_2Si_2C_{22}H_{28}O_4$
$Mo_2(CO)_4(TMSC_2TMS)Cp_2$, **5**, 376
$Mo_2Si_4C_{20}H_{56}N_2$
$Mo_2(CH_2TMS)_4(NMe_2)_2$, **5**, 321
$Mo_2Ti_2C_{24}H_{28}O_8$
$[TiMeO_4(C_5H_4Me)_2]_2$, **4**, 405
$Mo_2Ti_2C_{36}H_{30}O_6$
$[Cp_2Ti(CO)Mo(CO)_2Cp]_2$, **5**, 399, 402
$Mo_2Ti_2C_{54}H_{70}O_4$
$Mo_2(CO)_2(COTiCp^*_2)_2Cp_2$, **5**, 399, 402
$Mo_3AsFeC_{25}H_{15}O_{10}$
$Mo_3Fe(As)(CO)_{10}Cp_3$, **10**, 218
$Mo_3BC_{14}H_{27}F_4O_{16}$
$[Mo_3(O)(CMe)(O_2CMe)_6$-$(OH_2)_3]BF_4$, **5**, 322
$Mo_3Co_2C_{38}H_{46}N_2O_8$
$Mo_3Co_2(N)(NH)(CO)_8Cp^*_3$, **10**, 218
$Mo_3ErC_{26}H_{29}O_{14}$
$(EtOH)(H_2O)_4Er$[CpMo-$(CO)_3]_3$, **4**, 13
$Mo_3GaC_{24}H_{15}O_9$
$Ga[Mo(CO)_3Cp]_3$, **1**, 553
$Mo_3InC_{24}H_{15}O_9$
$In[Mo(CO)_3Cp]_3$, **1**, 557
$Mo_3TlC_{24}H_{15}O_9$
$Tl[Mo(CO)_3Cp]_3$, **1**, 567
$Mo_4C_{12}F_8O_{12}$
$[MoF_2(CO)_3]_4$, **5**, 218
$Mo_4In_4C_{32}H_{20}O_{12}S_4$
$In_4S_4[Mo(CO)_3Cp]_4$, **1**, 559
$Mo_4In_4C_{32}H_{20}O_{12}Te_4$
$In_4Te_4[Mo(CO)_3Cp]_4$, **1**, 559
$Mo_4Na_2Pd_4C_{32}H_{20}O_{12}$
$Na_2[Pd_4[Mo(CO)_3Cp]_4]$, **9**, 202; **10**, 51
$Mo_5TiC_5H_5O_{18}$
$[TiCp(Mo_5O_{18})]^{3-}$, **4**, 308
$Mo_8Hg_4C_{32}H_{20}O_{12}$
$(CpMo(CO)_3HgMo)_4$, **10**, 327, 329
$MuC_{46}H_{37}NO_3P_2$
$Mn(CN)(CO)_2[P(OPh)_3]$-(dppm), **7**, 315

N

$NaAlSi_4C_{12}H_{36}$
$(TMS)_4AlNa$, **2**, 23
$NaAsC_{16}H_{18}O_2$
Ph_2AsNa(dioxane), **2**, 328
$NaBC_3H_{10}O_3$
$NaHB(OMe)_3$, **11**, 424
$NaBC_{24}H_{20}$
$NaBPh_4$, **12**, 149
$NaBH_4$
$NaBH_4$, **12**, 135
$NaB_{11}ReC_5H_{14}O_5$
$Na[Re(CO)_5]B_{11}H_{14}$, **1**, 363
$NaCH_3$
MeNa, **1**, 20
$NaC_5H_6O_4$
$NaCH(CO_2Me)_2$, **12**, 644, 652, 673
$NaC_7H_7O_2$
$Na[MeOC(O)C_5H_4]$, **8**, 85
$NaC_{18}H_{32}N_3$
$Na(PhC_3H_4)\cdot$PMDETA, **1**, 23, 24
$NaC_{22}H_{32}N_3$
Na(fluorenyl)·PMDETA, **1**, 23, 25
$NaC_{25}H_{30}N_3$
NaCp*·3py, **1**, 22, 23
$NaCeC_{18}H_{26}N_2O_8$
$Cp_2Ce(NO_3)_2Na(THF)_2$, **4**, 36
$NaCoC_4O_4$
$[Co(CO)_4]Na$, **12**, 354, 974, 975
$NaCo_3C_{39}H_{117}O_{39}P_{13}$
$[Co[P(OMe)_3]_5{}^+][Co[P(OMe)_3]_4Na[(MeO)_3P]_4Co]$, **8**, 92
$NaCrC_5HO_5$
$Na[CrH(CO)_5]$, **5**, 223
$NaFeC_4O_4$
$NaFe(CO)_4$, **12**, 351
$NaFeC_6H_6O_3$
$[(CH_2{=}CHMe)Fe(CO)_3]Na$, **12**, 967, 968
$NaFeC_7H_5O_2$
$NaFeCp(CO)_2$, **12**, 617
$NaGaMn_3C_{15}ClO_{15}$
$Na[GaCl[Mn(CO)_5]_3]$, **1**, 553
$NaInC_4H_{12}$
$Na(InEt_2H_2)$, **1**, 515
$NaInCr_2C_{24}H_{26}Cl_2O_8$
$[Na(THF)_2][InCl_2[Cr(CO)_3Cp]_2]$, **1**, 558
$NaInMoC_{16}H_{21}Cl_3O_5$
$[Na(THF)_2][InCl_3[Mo(CO)_3Cp]]$, **1**, 557
$NaInMo_2C_{24}H_{26}Cl_2O_8$
$[Na(THF)_2][InCl_2[Mo(CO)_3Cp]_2]$, **1**, 557
$NaLaC_{24}H_{28}O$
$Na[La(Cp)_4](THF)$, **4**, 58
$NaLiC_{20}H_{48}N_{12}$
$[(Me_2N)_2CN]_4LiNa$, **1**, 4, 5
$NaMnC_4H_2O_4$
$Na[MnH_2(CO)_4]$, **6**, 68
$NaMnC_5O_5$
$Na[Mn(CO)_5]$, **6**, 1
$NaMoC_8H_5O_3$
$Na[CpMo(CO)_3]$, **12**, 638, 970
$NaMoO_4$
$NaMoO_4$, **11**, 413
$NaNiC_{20}H_{49}N_4$
$[Na(TMEDA)_2][Ni_2(H)(C_2H_4)_4]$, **9**, 111
$NaNi_2C_{20}H_{41}N_4$
$[Na(TMEDA)_2][Ni_2(C_2H_4)_4(H)]$, **1**, 26, 28
NaPdCO
$[PdCONa]^+$, **9**, 195
$NaReC_7H_5NO_3$
$[Re(CN)(C_3H_5)(CO)_3]Na$, **6**, 209
$NaRhC_{11}H_{23}O_3P_2$
$[CpRhMe(PMe_3)[P(ONa)(OMe)_2]]^+$, **8**, 195
$NaRu_2C_4O_4Se_{16}$
$[Na[Ru(CO)_2(Se_4)_2]_2]^{3-}$, **7**, 317
NaSeCN
NaSeCN, **11**, 528
$NaTc_3C_{13}H_{12}O_{13}$
$Na[Tc_3(OMe)_4(CO)_9]$, **6**, 154
$NaTeC_4H_3OS$
$(C_4H_3S)TeONa$, **11**, 583
NaTeH
NaHTe, **11**, 572
$NaVC_6O_6$
$NaV(CO)_6$, **5**, 2
$NaVC_8H_6O_3$
$[CpV(CO)_3H]Na$, **5**, 24
$Na_2AlB_{11}CH_{14}$
$Na_2[B_{11}H_{11}AlMe]$, **1**, 258
$Na_2B_{11}H_{13}$
$Na_2[B_{11}H_{13}]$, **1**, 258
$Na_2FeIr_4C_{15}O_{15}$
$Na_2[FeIr_4(CO)_{15}]$, **10**, 43
$Na_2Li_4C_{66}H_{90}N_6$
$[Bu^t(Ph)CN]_6Li_4Na_2$, **1**, 4, 5
$Na_2MnC_4HO_4$
$Na_2[MnH(CO)_4]$, **6**, 68
$Na_2Mo_4Pd_4C_{32}H_{20}O_{12}$
$Na_2[Pd_4[Mo(CO)_3Cp]_4]$, **9**, 202; **10**, 51
$Na_2OsC_4O_4$
$Na_2[Os(CO)_4]$, **7**, 300
$Na_2Pd_2C_{14}H_{22}Cl_2O_8$
$Na_2[Pd[CH(CO_2Et)_2]_2(Cl)]_2$, **9**, 241
$Na_2Pt_{30}C_{60}O_{60}$
$Na_2[Pt_{30}(CO)_{60}]$, **9**, 397
$Na_2Rh_2C_{44}H_{36}Cl_2O_{18}S_6$
$[P(SO_3NaC_6H_4)_3]_2Rh(Cl)_2Rh$(cod), **8**, 263
$Na_2VC_5HO_5$
$Na_2[HV(CO)_5]$, **5**, 3
Na_2WO_4
Na_2WO_4, **11**, 413; **12**, 1178
$Na_3MnC_4O_4$
$Na_3[Mn(CO)_4]$, **6**, 68
$Na_3VC_5O_5$
$Na_3[V(CO)_5]$, **5**, 3
$Na_4C_4H_{12}$
$[MeNa]_4$, **1**, 20
$Na_4C_{80}H_{108}N_8$
$[Na(fluorenyl)\cdot TMPDA]_4$, **1**, 23, 25
$Na_4Li_4C_{36}H_{40}O_4$
$[(Me_2NaCH_2\cdot C_6H_2)OLi]_4$, **1**, 13
$Na_4Ni_2C_{48}H_{68}O_5$
$[Na_4(THF)_5][Ph_2Ni(C_2H_4)]_2$, **1**, 26, 27
$NbBC_{12}H_{13}Cl_2F_4N$
$[Cp_2NbCl_2]BF_4\cdot NCMe$, **5**, 126
$NbBC_{18}H_{29}N$
$Cp_2[Bu^t_2(NbBN)]$, **1**, 211
$NbBC_{20}H_{21}ClF_4O$
$[Cp_2Nb[C(O)CHEtPh]Cl]BF_4$, **5**, 98
$NbBC_{28}H_{26}$
$[Ph_2B(Ph)_2]Nb(MeCCMe)$, **5**, 132
$NbBC_{34}H_{30}F_4P_2$
$[Cp_2Nb(PPh_2)_2]BF_4$, **5**, 127
$NbBSi_2C_{46}H_{40}F_4$
$[Cp_2Nb(SiPh_3)_2]BF_4$, **5**, 127
$NbB_3C_{10}H_{18}$
$NbCp_2B_3H_8$, **1**, 313, 315
$NbB_4C_{10}H_{19}$
$Cp_2NbB_4H_9$, **1**, 319
$NbC_4H_{10}Cl_3O_2$
$NbCl_3(DME)$, **12**, 344
$NbC_8H_6Cl_3$
$Ph[\overline{C{=}CHNb}]Cl_3$, **12**, 337, 338
$NbC_8H_{16}Cl_4$
$NbCl_4(THF)_2$, **12**, 344
$NbC_8H_{16}Cl_4O_2$
$NbCl_4(THP)_2$, **12**, 1088
$NbC_{12}H_{22}$
$(C_{10}H_{21})[\overline{C{=}CHNb}]$, **12**, 336, 337
$NbC_4H_{10}Cl_3O_2$
$NbCl_3(DME)$, **5**, 140, 142
$NbC_8H_7O_3$
$CpNbH_2(CO)_3$, **5**, 80
$NbC_8H_{16}Cl_4O_2$
$NbCl_4(THF)_2$, **5**, 143

$NbC_9H_{11}Cl_2$
$CpNb(C_4H_6)Cl_2$, **5**, 140
$NbC_{10}H_{10}ClO$
$Cp_2Nb(O)Cl$, **5**, 126, 127
$NbC_{10}H_{13}$
Cp_2NbH_3, **5**, 76
$NbC_{12}H_{12}$
$(C_7H_7)NbCp$, **5**, 132
$NbC_{12}H_{12}Br$
$(C_7H_7)NbCpBr$, **5**, 132
$NbC_{12}H_{16}$
Cp_2NbMe_2, **5**, 70
$NbC_{12}H_{16}O_2P$
$(C_7H_7)Nb(CO)_2(PMe_3)$, **5**, 132
$NbC_{12}H_{19}$
$Nb(CH_2CHCHMe)(C_4H_6)_2$, **5**, 110
$NbC_{12}H_{21}Cl_3NOS$
$Cp^*Nb[NS(O)Me_2]Cl_3$, **5**, 119
$NbC_{12}H_{34}Cl_2P_4$
$NbH_2Cl_2(dmpe)_2$, **5**, 73
$NbC_{12}H_{41}P_4$
$NbH_5(PMe_3)_4$, **5**, 75
$NbC_{13}H_{17}$
$Cp_2Nb(CHMe_2)$, **5**, 92
$CpNb(C_4H_6)_2$, **5**, 112, 140
$NbC_{13}H_{19}BrP$
$Cp_2NbBr(PMe_3)$, **5**, 127
$NbC_{13}H_{19}Cl_2$
$CpNb(C_4Me_4H_2)Cl_2$, **5**, 112
$NbC_{13}H_{25}IP_2$
$(C_7H_7)Nb(PMe_3)_2I$, **5**, 132
$NbC_{13}H_{26}Cl_2P_2$
$(C_7H_8)Nb(PMe_3)_2Cl_2$, **5**, 132
$NbC_{14}H_{13}O_2$
$CpNb(C_7H_8)(CO)_2$, **5**, 64
$NbC_{14}H_{16}$
$(PhMe)_2Nb$, **5**, 129
$NbC_{14}H_{17}$
$Cp_2Nb(CH_2CHCHMe)$, **5**, 110
$Cp_2Nb(MeCCMe)H$, **5**, 97
$NbC_{14}H_{22}P$
$(C_7H_7)Nb(C_4H_6)(PMe_3)$, **5**, 132
$NbC_{14}H_{23}OP_2$
$(C_7H_7)Nb(dmpe)(CO)$, **5**, 132
$NbC_{14}H_{29}P_2$
$Nb(CH_2CHCHMe)(C_4H_6)$-(dmpe), **5**, 110
$NbC_{14}H_{32}Cl_2P_3$
$CpNbCl_2(PMe_3)_3$, **5**, 119
$NbC_{15}H_{13}O_2$
$CpNb(CO)_2(C_8H_8)$, **5**, 64
$NbC_{15}H_{19}$
$Cp_2Nb(MeCHCHCHMe)$, **5**, 110
$NbC_{15}H_{19}O_2$
$NbCp_2(O_2CBu^t)$, **4**, 258
$NbC_{16}H_{21}$
$Cp_2Nb(MeCCMe)(CH_2Me)$, **5**, 96
$NbC_{16}H_{25}Cl_2$
$Cp^*Nb(EtCCEt)Cl_2$, **5**, 96
$NbC_{16}H_{37}ClO_2P$
$Nb(CHBu^t)(OBu^t)_2Cl(PMe_3)$, **5**, 138
$NbC_{17}H_{16}NS$
$Cp_2Nb(SCHNPh)$, **5**, 110
$NbC_{17}H_{24}P$
$(C_7H_7)Nb(C_7H_8)PMe_3$, **5**, 132
$NbC_{17}H_{26}NS_2$
$Cp_2NbMe_2(S_2CNEt_2)$, **5**, 70
$NbC_{18}H_{10}F_{12}$
$Cp_2Nb(CCF_3CCF_3CCF_3CCF_3)$, **5**, 108
$NbC_{18}H_{15}$
$(C_9H_7)_2NbH$, **5**, 110
$NbC_{18}H_{16}NOS$
$Cp_2Nb(SCHNPh)(CO)$, **5**, 110
$NbC_{18}H_{24}$
$(C_6H_3Me_3)_2Nb$, **5**, 129
$NbC_{20}H_{30}Cl_2$
$Cp^*_2NbCl_2$, **5**, 126
$NbC_{20}H_{31}S_3$
$Cp^*_2NbS_3H$, **5**, 128
$Cp^*_2Nb(SH)(S_2)$, **5**, 128
$NbC_{20}H_{35}P$
$(C_7H_{10})_2Nb(PEt_3)$, **5**, 127
$NbC_{21}H_{19}$
$(C_9H_7)_2Nb(CH_2CHCH_2)$, **5**, 110
$NbC_{22}H_{18}Cl_2$
$Cp_2Nb(SC_6H_4Cl)_2$, **5**, 128
$NbC_{22}H_{20}F_6PS_2$
$[Cp_2Nb(SPh)_2]PF_6$, **5**, 128
$NbC_{22}H_{21}Cl_2$
$Cp'Nb(TolCCTol)Cl_2$, **5**, 96
$NbC_{22}H_{34}Cl_2$
$(C_5Me_4Et)_2NbCl_2$, **5**, 121
$NbC_{24}H_{24}$
$[Nb(C_8H_8)_3]^-$, **5**, 110
$NbC_{24}H_{25}$
$(C_9H_7)_2Nb(MeCHCHCHEt)$, **5**, 110
$NbC_{25}H_{41}ClN_4$
$[CpNb(CNBu^t)_4Cl]^+$, **5**, 66
$NbC_{28}H_{30}Cl_3NP$
$Cp^*Nb(NPPh_3)Cl_3$, **5**, 119
$NbC_{34}H_{30}P_2$
$Cp_2Nb(PPh_2)_2$, **5**, 127
$NbC_{48}H_{35}O$
$CpNb(C_4Ph_4)(PhCCPh)(CO)$, **5**, 113
$NbC_{54}H_{41}O_3$
$Nb(OC_6H_3PhC_6H_7)(OC_6H_3$-$Ph_2)_2$, **5**, 113, 142
$NbC_{60}H_{53}O_2$
$Nb(CH_2C_6H_4Me)_3(OC_6H_3Ph_2)_2$, **5**, 142
$NbC_{72}H_{53}O_3$
$Nb(OC_6H_2Ph_2C_6H_7)(OC_6H_2$-$Ph_3)_2$, **5**, 113
$NbCl_5$
$NbCl_5$, **12**, 1087
$NbFeC_{10}H_{10}Cl_4$
$Cp_2Nb(Cl)_2FeCl_2$, **5**, 126, 127
$NbHg_3C_{25}H_{40}N_3S_6$
$Cp_2Nb[HgS_2CNEt_2]_3$, **10**, 335
$NbLi_5C_{56}H_{63}O_5$
$Nb(C_6H_4)_2Ph_3(LiPh{\cdot}THF)$-$(Li{\cdot}THF)_4$, **5**, 97
$NbMnC_{16}H_{10}O_6$
$Cp_2Nb(CO)_2Mn(CO)_4$, **5**, 65
$NbMoC_{16}H_{11}O_6$
$Cp_2(CO)Nb(H)Mo(CO)_5$, **5**, 81
$NbMoC_{26}H_{20}F_6O_4PS_2$
$[Cp_2Nb(SPh)_2Mo(CO)_4]PF_6$, **5**, 128
$NbMoC_{26}H_{20}O_4S_2$
$[NbCp_2(SPh)_2Mo(CO)_4]^+$, **10**, 28
$NbMoC_{29}H_{29}O_4S_2$
$Nb(PhC_2Ph)Cp(SPr^i)_2Mo(CO)_4$, **10**, 28
$NbNiC_{14}H_{20}Cl_2S_2$
$NbCl_2Cp(SEt)_2NiCp$, **10**, 28
$NbNiC_{25}H_{25}Cl_2S_2$
$NiCp(SEt)_2NbCl_2Cp_2$, **9**, 169
$NbSiC_{14}H_{21}Cl$
$Cp_2Nb(CH_2TMS)Cl$, **5**, 70
$NbSiC_{23}H_{53}O_2P_4$
$Nb(COSiPr^i{}_3)(dmpe)_2(CO)$, **5**, 65
$NbSi_2C_{16}H_{26}Br$
$(C_5H_4TMS)_2NbBr$, **5**, 126
$NbSi_2C_{16}H_{26}Cl$
$(C_5H_4TMS)_2NbCl$, **5**, 127
$NbSi_2C_{16}H_{26}ClO$
$(C_5H_4TMS)_2Nb(O)Cl$, **5**, 127, 128
$NbSi_2C_{16}H_{26}ClO_2$
$(C_5H_4TMS)_2Nb(O_2)Cl$, **5**, 127
$NbSi_2C_{16}H_{26}I$
$(C_5H_4TMS)_2NbI$, **5**, 126
$NbSi_2C_{16}H_{29}$
$(C_5H_4TMS)_2NbH_3$, **5**, 76, 79
$NbSi_2C_{19}H_{35}ClF_6P_2$
$[[C_5H_4(TMS)]_2NbCl[P$-$(OMe)_3]]PF_6$, **5**, 127
$NbSi_2C_{19}H_{35}ClO_3$
$(C_5H_4TMS)_2NbCl[P(OMe)_3]$, **5**, 127
$NbSi_2C_{22}H_{33}O$
$Cp_2Nb(TMS$-CC-$TMS)$-$[C(COMe)CH_2]$, **5**, 97
$NbSi_2C_{23}H_{31}ClNO$
$(C_5H_4TMS)_2NbCl(OCNPh)$, **5**, 101
$NbSi_2C_{31}H_{39}ClN$
$(C_5H_4TMS)_2Nb(PhNCC$-$MePh)Cl$, **5**, 101
$NbSi_2C_{46}H_{40}$
$Cp_2Nb(SiPh_3)_2$, **5**, 127
$NbSi_4C_{22}H_{42}ClO$
$[C_5H_3(TMS)_2]_2Nb(O)Cl$, **5**, 128
$NbSi_4C_{22}H_{42}ClO_2$
$[C_5H_3(TMS)_2]_2NbCl(O_2)$, **5**, 127
$NbSi_4C_{22}H_{45}$
$[C_5H_3(TMS)_2]_2NbH_3$, **5**, 76, 79
$NbSi_4C_{42}ClO$
$[C_5H_3(TMS)_2]_2Nb(O)Cl$, **5**, 127
$NbSnC_{14}H_{19}O$
$Cp_2Nb(CO)(SnMe_3)$, **5**, 128
$NbSnC_{15}H_{22}N$
$Cp_2Nb(NCMe)(SnMe_3)$, **5**, 101
$NbTlC_{22}H_{10}BrClF_{10}$
$Cp_2Nb(Cl)(Br)Tl(C_6F_5)_2$, **5**, 126, 127

$NbZnC_{15}H_{17}$
Cp_2NbH_2ZnCp, **10**, 335
$NbZn_2C_{20}H_{21}$
$Cp_2NbH(ZnCp)_2$, **5**, 81; **10**, 335
$NbZrC_{31}H_{43}O$
$Cp_2(H)NbCHOZr(H)Cp^*_2$, **5**, 133
$NbZrC_{32}H_{43}O_2$
$Cp_2(CO)NbCH_2OZr(H)Cp^*_2$, **5**, 133
$Nb_2C_{14}H_{10}Cl_2O_4$
$[CpNbCl(CO)_2]_2$, **5**, 65
$Nb_2C_{14}H_{10}Cl_4O_4$
$[CpNbCl(Cl)(CO)_2]_2$, **5**, 140
$Nb_2C_{14}H_{14}Cl_4O_4$
$[Cp'NbCl_2(CO)_2]_2$, **5**, 64
$Nb_2C_{16}H_{20}Br_3NO_8$
$[NEt_4][Nb_2(Br)_3(CO)_8]$, **5**, 140
$Nb_2C_{17}H_{24}P$
$(C_7H_7)Nb(C_7H_8)Nb(PMe_3)$, **5**, 132
$Nb_2C_{17}H_{25}F_6P_2$
$[(C_7H_7)Nb(C_7H_9)Nb(PMe_3)]PF_6$, **5**, 132
$Nb_2C_{18}H_{22}Cl_2$
$[CpNb(C_4H_6)Cl]_2$, **5**, 140
$Nb_2C_{20}H_{20}Br_3$
$Cp_4Nb_2Br_3$, **5**, 126
$Nb_2C_{20}H_{20}Br_4O$
$[Cp_2NbBr_2]_2(O)$, **5**, 127
$Nb_2C_{20}H_{20}Cl_3$
$[Cp_2NbCl]_2(Cl)$, **5**, 126
$Nb_2C_{20}H_{20}Cl_3O_4$
$[Cp_4Nb_2Cl_2]ClO_4$, **5**, 126
$Nb_2C_{20}H_{30}Cl_4$
$[Cp^*NbCl_2]_2$, **5**, 114
$Nb_2C_{20}H_{30}Cl_6O$
$[Cp^*NbCl_3]_2(O)$, **5**, 119
$Nb_2C_{20}H_{32}Cl_4O_3$
$[Cp^*NbCl_2(OH)]_2(O)$, **5**, 119
$Nb_2C_{22}H_{34}P_6$
$[(C_5Me_4Et)Nb]_2(P_6)$, **5**, 121
$Nb_2C_{24}H_{36}Br_4$
$[(C_6Me_6)Nb(Br)_2]_2$, **5**, 129
$Nb_2C_{26}H_{42}P_6$
$[(C_5H_3Bu^t{}_2)NB]_2(P_6)$, **5**, 121
$Nb_2C_{30}H_{54}Cl_6N_6$
$Nb_2Cl_6(CNBu^t)_6$, **5**, 66
$Nb_2C_{36}H_{36}Cl_6S_2$
$[Nb(PhCCPh)Cl_2(Cl)(THT)]_2$, **5**, 93
$Nb_2C_{68}H_{60}OP_4$
$[Cp_2Nb(PPh_2)_2]_2(O)$, **5**, 127
$Nb_2CuC_{20}H_{26}F_6P$
$[[Cp_2NbH(H)]_2Cu]PF_6$, **5**, 76
$Nb_2MnC_{27}H_{21}O_7$
$[[Cp_2(CO)Nb]_2(H)][Mn(CO)_5]$, **5**, 81
$Nb_2MnW_4C_{51}H_{108}N_3O_{22}$
$(Bu_4N)_3[Mn(CO)_3(Nb_2W_4O_{19})]$, **6**, 84
$Nb_2RhW_4C_{10}H_{15}O_{19}$
$[Cp^*Rh(Nb_2W_4O_{19})]^{2-}$, **8**, 178
$Nb_2Si_4C_{92}H_{30}O$
$[Cp_2Nb(SiPh_3)_2]_2(O)$, **5**, 127
$Nb_2Sn_2C_{26}H_{38}O$
$[Cp_2Nb(SnMe_3)]_2(O)$, **5**, 128
$Nb_2UW_{10}C_{95}H_{195}N_5O_{38}$
$[NBu_4]_5[Cp_3U(NbW_5O_{19})_2]$, **4**, 189
$Nb_2W_4O_{19}$
$[Nb_2W_4O_{19}]^{4-}$, **8**, 178, 249
$Nb_2ZnC_{20}H_{24}$
$[Cp_2NbH_2]_2Zn$, **5**, 81; **10**, 335
$Nb_3Ag_3C_{30}H_{48}O_{12}P_6$
$Ag_3Nb_3(CO)_{12}[Me_2P(CH_2)_2PMe_2]_3$, **10**, 310
$Nb_3Au_3SiC_8H_{19}$
$Au_3Nb_3(H)_6(C_5H_4TMS)$, **10**, 317
$Nb_3C_{25}H_{45}Cl_8N_5$
$Nb_3Cl_8(CNBu^t)_5$, **5**, 66, 67
$Nb_3C_{30}H_{45}Cl_4O_3$
$[Cp^*NbCl(O)]_3(O)(Cl)$, **5**, 120
$Nb_3C_{33}H_{51}S_7$
$(C_5Me_4Et)_3Nb_3S_7$, **5**, 121
$Nb_3C_{33}H_{51}S_{12}$
$(C_5Me_4Et)_3Nb_3S_{12}$, **5**, 121
$Nb_3C_{36}H_{54}Br_7$
$[(C_6Me_6)_3Nb_3Br_6]Br$, **5**, 129
$Nb_3C_{36}H_{54}Cl_7$
$[(C_6Me_6)_3Nb_3Cl_6]Cl$, **5**, 129
$Nb_3SiW_9O_{40}$
$SiNb_3W_9O_{40}{}^{7-}$, **8**, 178
$Nb_4C_{56}H_{40}Cl_8O$
$[Nb_4OCl_8(C_4Ph_4)_2]^{2-}$, **5**, 112, 113
$Nb_4C_{56}H_{40}Cl_{12}$
$[(PhCCPh)NbCl_3]_4$, **5**, 140
$Nb_4Ir_2W_8C_4HO_{42}$
$[(CO)_2Ir]_2H(Nb_2W_4O_{19})_2{}^{5-}$, **8**, 323
$Nb_6SnC_{96}H_{87}Cl_{19}$
$(Bu_3Sn)[Nb_6(C_4Ph_4)_3Cl_{19}]$, **5**, 113
$NdC_{11}H_{10}N$
Cp_2NdCN, **4**, 36
$NdC_{18}H_{21}$
$Nd(MeC_5H_4)_3$, **4**, 53
$NdC_{18}H_{28}ClN_2$
$(C_5H_4CH_2CH_2NMe_2)_2NdCl$, **4**, 38
$NdC_{21}H_{20}$
$(Cp_3NdPh)^-$, **4**, 57
$NdC_{21}H_{33}$
$Nd(C_7H_{11})_3$, **4**, 95
$NdC_{23}H_{36}NS_2$
$Cp^*_2Nd(S_2CNMe_2)$, **4**, 68
$NdC_{24}H_{33}O_3$
$Nd(C_5H_4CH_2CH_2OMe)_3$, **4**, 55
$NdC_{24}H_{38}ClO$
$Cp^*_2NdCl(THF)$, **4**, 68
$NdC_{33}H_{45}$
$Nd[MeC_5H_2(CH_2)_5]_3$, **4**, 54
$NdC_{33}H_{50}OSe$
$Cp^*_2Nd(SeMes)(THF)$, **4**, 68
$NdC_{36}H_{39}O_3$
$Cp_2Nd(OC_6H_3Ph_2)(THF)_2$, **4**, 39
$NdC_{42}H_{40}$
$Nd[(C_5H_4)CMe_2(C_{14}H_{10})]_2$, **4**, 45
$NdSi_2C_{27}H_{49}$
$Cp^*_2NdCH(TMS)_2$, **4**, 85, 130
$NdSi_3C_{27}H_{49}$
$[Me_2Si(C_5Me_4)_2]NdCH(TMS)_2$, **4**, 61
$NdSi_4C_{22}H_{51}N_2$
$Cp^*Nd[N(TMS)_2]_2$, **4**, 68
$Nd_2C_{38}H_{58}$
$[(Bu^tC_5H_4)_2Nd(Me)]_2$, **4**, 44
$Nd_6C_{50}H_{76}O_2$
$Nd_6Cl_{12}(C_7H_{11})_6(THF)_2$, **4**, 119
$Nd_{12}Al_6C_{62}H_{154}Cl_{12}O_2$
$[Nd_6Al_3(Cl)_6(Et)_{14}(OPr^i)]_2$, **4**, 23
$NiAlC_7H_{18}$
$[Ni(HAlMe_3)(C_2H_4)_2]^-$, **9**, 110
$NiAlC_{12}H_{20}$
$Ni(cdt)[AlH_4]$, **1**, 551
$NiAlC_{21}H_{38}N$
$NiHAlMe_2(cdt)(C_7H_{13}N)$, **9**, 111
$NiAlLiC_{24}H_{52}N_4$
$[Li(TMEDA)_2][Ni(cdt)(AlH_4)]$, **1**, 551
$NiAsMo_2C_{19}H_{15}O_4$
$Mo_2Ni(As)(CO)_4Cp_3$, **10**, 218
$NiAs_2B_9C_{16}H_{31}$
$(Me_2PhP)_2Ni(As_2B_9H_9)$, **1**, 266
$NiAs_2C_{55}H_{49}P_2$
$[Ni(dapm)_2Cp]^+$, **9**, 160
$NiAs_2C_{57}H_{53}P_2$
$[Ni(arphos)_2Cp]^+$, **9**, 160
$NiAs_3C_{14}H_{31}Br_2$
$[Ni[As(C_2H_4CH{=}CH_2)(C_3H_6AsMe_2)_2Br]Br]$, **9**, 122
$NiAuOs_3C_{32}H_{22}O_9P$
$[NiOs_3(CO)_9Cp(H)_2[Au(PPh_3)]]^-$, **9**, 21
$NiBC_9H_{12}N_6$
$Ni(C_3H_5)(HBpz_3)$, **9**, 140
$NiBC_{11}H_{30}F_4OP_3$
$[Ni(COMe)(PMe_3)_3][BF_4]$, **9**, 59
$NiBC_{12}H_{20}N_2$
$Ni(C_4H_4BNMe_2)_2$, **1**, 204
$NiBC_{13}H_{35}F_4P_4$
$[NiMe(dmpe)_2][BF_4]$, **9**, 3, 56
$NiBC_{13}H_{39}F_4P_4$
$[Ni(Me)(PMe_3)_4][BF_4]$, **9**, 59, 61
$NiBC_{14}H_{37}F_4OP_4$
$[Ni[C(O)Me](PMe_3)_2(dmpe)][BF_4]$, **9**, 3
$NiBC_{15}H_{13}O_2$
$Ni(C_3H_5)[(O_2C_6H_4)_2B]$, **9**, 33
$NiBC_{15}H_{13}O_4$
$Ni(C_3H_5)[B(C_6H_4O_2)_2]$, **9**, 125
$NiBC_{15}H_{24}Cl_5F_4OP_2$
$[Ni(C_6Cl_5)(PMe_3)_2[C(OMe)Me][BF_4]$, **9**, 13
$NiBC_{16}H_{15}O_2$
$Ni(C_4H_7)[(O_2C_6H_4)_2B]$, **9**, 33
$NiBC_{17}H_{29}F_4N_2P_2$

[Ni(Me)(PMe$_3$)$_2$(bipy)][BF$_4$], **9**, 56
$NiBC_{19}H_{23}$
Ni(CpBPh)(C$_8$H$_{13}$), **9**, 176
$NiBC_{20}H_{44}F_4P_2$
[Ni(C$_2$H$_4$)(H)(PBu^t$_2$C$_2$H$_4$P-BuSt$_2$)][BF$_4$], **9**, 113
$NiBC_{33}H_{45}F_4P_4$
[Ni(Me)(dppe)(PMe$_3$)$_2$][BF$_4$], **9**, 59
$NiBC_{34}H_{45}F_4OP_4$
[Ni[C(O)Me](PMe$_3$)$_2$(dppe)]-[BF$_4$], **9**, 3
$NiBC_{69}H_{63}P_2$
[Ni(CHMePPh$_3$)$_2$Cp][BPh$_4$], **9**, 36
$NiBSiC_{13}H_{38}F_4P_3$
[Ni(CH$_2$TMS)(PMe$_3$)$_3$][BF$_4$], **9**, 56
$NiBTiC_{38}H_{38}F_4P_2S_2$
[TiCp$_2$(SCH$_2$CH$_2$PPh$_2$)$_2$Ni]BF$_4$, **10**, 27
$NiB_2C_8H_{22}O_2P_2$
Ni(CO)$_2$(B$_2$H$_4$·2PMe$_3$), **1**, 311
$NiB_2C_{10}H_{14}$
Ni(C$_4$H$_4$BMe)$_2$, **1**, 204
$NiB_2C_{13}H_{21}S$
CpNi[Et$_2$Me$_2$(BSBC$_2$)], **1**, 214
$NiB_2C_{16}H_{20}N_2$
Ni(C$_3$H$_3$NBu^{t}BMe)$_2$, **9**, 145
$NiB_2C_{20}H_{18}$
Ni(C$_4$H$_4$BPh)$_2$, **1**, 204
$NiB_2C_{20}H_{44}N_2Si_2$
Ni(C$_{10}$H$_{22}$BNSi)$_2$, **1**, 213
$NiB_2C_{22}H_{20}$
Ni(CpBPh)$_2$, **9**, 176
$NiB_2C_{22}H_{32}$
(cod)Ni[Et$_2$Me$_2$C$_8$H$_4$B$_2$], **1**, 210
$NiB_3C_{18}H_{32}$
[Ni(cod)(C$_3$Et$_2$HB$_3$Me$_3$)]$^+$, **9**, 181
$NiB_3C_{23}H_{28}P$
[CpNi(PPh$_3$)][B$_3$H$_8$], **1**, 316
$NiB_4C_{16}H_{32}N_4$
Ni[C$_4$H$_4$B$_2$(NMe$_2$)$_2$]$_2$, **1**, 209
$NiB_4C_{28}H_{31}ClP_2$
NiCl(dppe)C$_2$B$_4$H$_7$, **9**, 181
$NiB_4C_{28}H_{40}$
Ni[Et$_2$Me$_2$(C$_8$H$_4$B$_2$)]$_2$, **1**, 210
$NiB_4C_{42}H_{43}Br$
NiBr(PPh$_3$)(Et$_2$C$_2$B$_4$H$_3$PPh$_3$), **9**, 181
$NiB_4Co_2C_{26}H_{40}$
CpCo(C$_3$B$_2$Me$_5$)Ni(C$_3$B$_2$Me$_5$)-CoCp, **9**, 178
$NiB_4PtC_{29}H_{51}$
CpNi(MeCB$_2$C$_2$Et$_4$)Pt-(MeCB$_2$C$_2$Et$_4$), **9**, 179
$NiB_5C_{22}H_{36}$
Ni(C$_3$Et$_2$HB$_2$Me$_2$)(C$_3$Et$_2$C$_3$-B$_3$Me$_3$), **9**, 181
$NiB_6C_{20}H_{40}$
Ni(Me$_3$Et$_2$C$_3$B$_3$H)$_2$, **9**, 181
$NiB_6C_{21}H_{47}N$
[Bu$_4$N][CpNiB$_6$H$_6$], **1**, 328, 330
$NiB_7C_{13}H_{28}$
(C$_4$H$_3$Me$_3$)NiC$_3$B$_7$H$_7$Me$_3$, **9**, 181
$NiB_8C_6H_{14}$
[(Cp)Ni]CB$_8$H$_9$, **9**, 181
$NiB_9C_{12}H_{19}N_2$
Ni(C$_2$B$_9$H$_{11}$)(bipy), **9**, 182
$NiB_9C_{16}H_{29}Cl_2P_2$
(PMe$_2$Ph)$_2$Cl$_2$[NiB$_9$H$_7$], **1**, 349
$NiB_{18}C_4H_{22}$
Ni(C$_2$B$_9$H$_{11}$)$_2$$^-$, **9**, 181
$NiB_{18}C_6H_{26}O_2$
[Ni(MeOC$_2$B$_9$H$_{10}$)$_2$], **9**, 181
$NiCH_4$
NiH(Me), **9**, 32
$NiC_2H_2O_2X$
CHCHC(O)ONiX, **12**, 759
$NiC_2H_3F_{10}P_3$
Ni(C$_2$H$_3$F)(PF$_3$)$_3$, **9**, 119
NiC_3H_6
NiCH$_2$CH$_2$CH$_2$, **9**, 30
Ni(CH=CH$_2$)(Me), **9**, 30
Ni(CH$_2$)(C$_2$H$_4$), **9**, 30
$NiC_3H_6Cl_2$
(CH$_2$=CHMe)NiCl$_2$, **12**, 960
NiC_3H_8
NiH(Me)(C$_2$H$_4$), **9**, 32
$NiC_4H_{14}P_2$
Ni(CH$_2$CH$_2$CH$_2$CH$_2$)(PH$_3$)$_2$, **9**, 73
$NiC_4N_2O_3$
[Me$_4$N][Ni(CO)$_3$(CN)], **9**, 14
NiC_4O_4
Ni(CO)$_4$, **9**, 1; **11**, 407; **12**, 359, 369, 370, 960, 966
$NiC_5H_8Br_2$
(CH$_2$=C(Me)CH=CH$_2$)NiBr$_2$, **12**, 961
NiC_5O_5
Ni(CO)$_5$, **12**, 353
NiC_6H_{12}
Ni(C$_2$H$_4$)$_3$, **9**, 110
NiC_6H_5O
[NiCp(CO)]$^-$, **9**, 4
$NiC_6H_8O_4S$
Ni[CH$_2$S(O)Me$_2$](CO)$_3$, **9**, 33
$NiC_6H_9Br_2$
(cyclohexenyl)NiBr$_2$, **12**, 962
NiC_6H_{10}
Ni(C$_3$H$_5$)$_2$, **9**, 137
$NiC_6H_{10}Cl_2$
Ni$_2$Cl$_2$(C$_3$H$_5$)$_2$, **9**, 137
$NiC_6H_{10}I_2$
Ni$_2$I$_2$(C$_3$H$_5$)$_2$, **9**, 137
NiC_6H_{12}
Ni(C$_2$H$_4$)$_3$, **9**, 32, 108, 111, 113
NiC_6H_{13}
Ni(Et)(C$_2$H$_4$)$_2$, **9**, 34
$NiC_6H_{16}Cl_2P_2$
Ni(dmpe)Cl$_2$, **12**, 174
$NiC_7H_{11}O_3P$
Ni(CH$_2$PMe$_3$)(CO)$_3$, **9**, 36
$NiC_7H_{12}O$
Ni(Me)(MeCHCHCHMe)(CO), **9**, 65
$NiC_7H_{13}P$
Ni(C$_2$H$_2$)$_2$(PMe$_3$), **9**, 133
$NiC_7H_{16}Br_2O$
MeCH=CHCH$_2$O(TMS)NiBr$_2$, **12**, 960
$NiC_7H_{16}OS$
Ni[CH$_2$S(O)Me$_2$](C$_2$H$_4$)$_2$, **9**, 32
$NiC_7H_{17}P$
Ni(C$_2$H$_4$)$_2$(PMe$_3$), **9**, 112, 113
$NiC_7H_{21}ClP_2$
NiClMe(PMe$_3$)$_2$, **9**, 61
$NiC_8F_{12}N_4$
Ni(CNCF$_3$)$_4$, **9**, 10
NiC_8H_{10}
Ni(C$_3$H$_5$)Cp, **9**, 141
$NiC_8H_{12}Cl_2$
NiCl$_2$(C$_4$Me$_4$), **8**, 80
$NiC_8H_{12}O_2$
Ni(Me)(MeCHCHCHMe)(CO)$_2$, **9**, 65
$NiC_8H_{19}P$
Ni(CH$_2$PMe$_3$)(C$_2$H$_4$)$_2$, **9**, 36, 112
$NiC_8H_{20}P_2$
Ni(C$_2$H$_2$)(PMe$_3$)$_2$, **9**, 132, 133
$NiC_8H_{22}N_2$
Ni(Me)$_2$(TMEDA), **9**, 63
NiC_9H_{14}
Ni(Et)(C$_2$H$_4$)Cp, **9**, 71, 113
NiMe(propene)Cp, **9**, 72
$NiC_9H_{17}P_2S_2$
[Ni[P(S)Me$_2$]$_2$Cp]$^-$, **9**, 163
$NiC_9H_{21}F_3O_2P_2$
Ni(Me)[OC(O)(CF$_3$)](PMe$_3$)$_2$, **9**, 39
$NiC_9H_{22}N_2O$
Ni(C$_2$H$_4$)(H$_2$C=O)(TMEDA), **9**, 128
$NiC_9H_{23}BrO_2P_2$
NiBr[C(O)CH$_2$OMe](PMe$_3$)$_2$, **9**, 58
$NiC_{10}H_{10}$
[NiCp$_2$]$^+$, **9**, 154
[NiCp$_2$]$^{2-}$, **9**, 152
[NiCp$_2$]$^-$, **9**, 152
NiCp$_2$, **9**, 154
$NiC_{10}H_{10}Br_2N_2$
NiBr$_2$(bipy), **12**, 966
$NiC_{10}H_{11}$
[Ni(C$_5$H$_6$)Cp]$^+$, **9**, 109
$NiC_{10}H_{14}$
NiMe[CH$_2$=C(CH$_2$)$_2$]Cp, **9**, 119
Ni(Me)(MeCCMe)Cp, **9**, 72
$NiC_{10}H_{15}BrO_2$
NiBr(Ph)(glyme), **9**, 30
$NiC_{10}H_{16}O_4$
Ni(acac)$_2$, **1**, 68; **12**, 156, 165, 170, 175, 176, 430
$NiC_{10}H_{19}OP$
Ni(CH$_2$OMe)(PMe$_3$)Cp, **9**, 57
$NiC_{10}H_{22}P_2$
Ni(C$_4$H$_6$)(PM$_2$C$_2$H$_4$PMe$_2$), **9**, 117

$NiC_{11}H_{14}$
Ni[CH$_2$C(Me)CH(CH=CH$_2$)Cp], **9**, 140
NiEt(C$_2$H$_4$)(C$_5$H$_4$Et), **9**, 113
$NiC_{11}H_{14}O_4$
Ni(CH$_2$CH=CH$_2$)(COBu)-(CO)$_3$, **9**, 33
$NiC_{11}H_{15}BrO$
NiBr(CO)Cp*, **9**, 167
$NiC_{11}H_{15}IO$
NiI(CO)Cp*, **9**, 169
$NiC_{11}H_{15}N_2O_2$
NiCp(CNCH$_2$CH$_2$OH)$_2$, **9**, 13
$NiC_{11}H_{15}O$
[Ni(CO)Cp*]$^-$, **9**, 167
$NiC_{11}H_{18}$
Ni(C$_2$H$_4$Et)(C$_2$H$_4$)Cp, **9**, 113
NiEt(C$_2$H$_4$)(C$_5$H$_4$Et), **9**, 113
$NiC_{11}H_{18}ClP$
NiCl(CH$_2$C$_6$H$_4$Me)(PMe$_3$), **9**, 146
$NiC_{11}H_{21}N$
Ni(C$_2$H$_4$)$_2$(C$_7$H$_{13}$N), **9**, 113
$NiC_{11}H_{22}N_2O_2$
Ni(C$_3$H$_4$CH$_2$CO$_2$)(TMEDA), **9**, 117
$NiC_{12}H_8Cl_3N_2$
Ni(CClCCl$_2$)(bipy), **9**, 61
$NiC_{12}H_8N_2O_2$
Ni(bipy)(CO)$_2$, **12**, 752
$NiC_{12}H_{13}$
[Ni(nbd)Cp]$^+$, **9**, 121
$NiC_{12}H_{14}N_2$
Ni(Me)$_2$(bipy), **9**, 61, 63
$NiC_{12}H_{18}$
Ni(C$_{12}$H$_{18}$), **9**, 117
Ni(cdt), **9**, 108, 110, 111
$NiC_{12}H_{19}$
[Ni(C$_3$H$_4$Me)(cod)]$^+$, **9**, 142
$NiC_{12}H_{20}Cl_2O_3$
[CH$_2$=CHCH$_2$OCH(Cy)-(CO$_2$Me)]NiCl$_2$, **12**, 963
$NiC_{12}H_{21}O_6$
(CO)$_3$Ni[P(OPr^i)$_3$], **12**, 747
$NiC_{12}H_{23}ClP_2$
NiCl(Ph)(PMe$_3$)$_2$, **9**, 51
$NiC_{12}H_{26}N_3$
NiMe[P(NMe$_2$)$_3$]Cp, **9**, 161
$NiC_{12}H_{26}O_2P$
Ni(Me)(acac)(PEt$_3$), **9**, 37
$NiC_{12}H_{27}P$
Ni(CH$_2$Bu^t)[CH$_2$C(Me)CH$_2$]-(PMe$_3$), **9**, 65
$NiC_{13}H_{14}$
Ni(Ph)(C$_2$H$_4$)Cp, **9**, 71
$NiC_{13}H_{21}P_3$
Ni(P$_3$)(CH$_3$Bu$^t_2$), **9**, 168
$NiC_{13}H_{22}$
NiMe(CH$_2$=CH$_2$)Cp*, **9**, 113
$NiC_{13}H_{22}N_2$
Ni(dbn)(CH$_2$=CHC$_2$H$_4$-CH=CH$_2$), **9**, 124
$NiC_{13}H_{24}Br_2P_2$
NiBr(CH$_2$C$_6$H$_4$Br)(PMe$_3$)$_2$, **9**, 47
$NiC_{13}H_{27}BrP_2$
NiBr(C$_6$H$_4$Me)(PMe$_3$)$_2$, **9**, 56
$NiC_{13}H_{32}N_3O$
Ni[C(O)NEt$_2$](NHEt$_2$)$_2$, **9**, 59
$NiC_{13}H_{39}P_4$
[Ni(Me)(PMe$_3$)$_4$]$^+$, **9**, 62
$NiC_{14}H_{14}N_2$
Ni(CH$_2$=CHCH=CH$_2$)(bipy), **9**, 121
$NiC_{14}H_{14}O_2$
Ni$_2$(CO)$_2$Cp'$_2$, **9**, 168
Ni(C$_5$H$_4$COMe)$_2$, **9**, 154
$NiC_{14}H_{16}$
Ni$_2$(C$_4$H$_6$)Cp$_2$, **9**, 114
$NiC_{14}H_{16}N_2$
Ni(CH$_2$CH$_2$CH$_2$CH$_2$)(bipy), **9**, 80
$NiC_{14}H_{18}$
Ni(C$_5$H$_4$Et)$_2$, **9**, 113
$NiC_{14}H_{18}N_2$
Ni(Et)$_2$(bipy), **9**, 61
$NiC_{14}H_{19}N_3$
Ni(Et)(NMe$_2$)(bipy), **9**, 62
$NiC_{14}H_{20}O_2$
Ni(CH$_2$=CHCH=CHCO$_2$Me)-(cod), **9**, 121
$NiC_{14}H_{21}Br$
NiBr[(CH$_2$=CH(CH$_2$)$_3$C$_5$Me$_4$)], **9**, 117
$NiC_{14}H_{22}$
Ni(CH=CH$_2$)(C$_2$H$_4$)Cp*, **9**, 114
$NiC_{14}H_{22}Cl_3P_2$
Ni[C$_2$Cl$_2$(C$_6$H$_4$Cl)](PMe$_3$)$_2$, **9**, 61
$NiC_{14}H_{23}$
[Ni(C$_3$H$_2$Me$_3$)(cod)]$^+$, **9**, 142
$NiC_{14}H_{24}N_4$
Ni(dbn)$_2$, **9**, 124
$NiC_{14}H_{25}BrOP_2$
NiBr[C(O)C$_6$H$_4$Me](PMe$_3$)$_2$, **9**, 56
$NiC_{14}H_{25}Br_2O_2P_2$
Ni[C$_6$H(OMe)$_2$Br$_2$](PMe$_3$)$_2$, **9**, 58
$NiC_{14}H_{27}O_2P_2$
Ni[C$_6$H$_3$(OMe)$_2$](PMe$_3$)$_2$, **9**, 58
$NiC_{14}H_{28}N_2$
Ni(C$_4$H$_6$)$_2$(TMEDA), **9**, 117
$NiC_{14}H_{30}P_2$
Ni(PMe$_3$)$_2$(cod), **9**, 47
$NiC_{14}H_{32}N_4$
Ni(C$_{10}$H$_{20}$N$_4$Me$_4$), **9**, 91
$NiC_{14}H_{36}P_2$
Ni(Me)$_2$(PEt$_3$)$_2$, **9**, 61
$NiC_{15}H_{17}NO_3$
Ni(CO)$_3$[=CPh(NBu^{t}Me)], **9**, 112
$NiC_{15}H_{19}$
NiCp(C$_3$H$_3$Me$_2$)Cp, **9**, 119
$NiC_{15}H_{20}$
NiCpCp*, **9**, 114
$NiC_{15}H_{22}N$
Ni(C$_2$H$_4$)$_2$Ph[=C(NBu^t)]$^-$, **9**, 111
$NiC_{15}H_{23}Cl_5OP_2$
Ni(C$_6$Cl$_5$)(PMe$_3$)$_2$-[C(OMe)=CH$_2$], **9**, 13
$NiC_{15}H_{23}N$
Ni(C$_2$H$_4$)$_2$[=CPh(NHBu^t)], **9**, 112
$NiC_{15}H_{24}N_2$
Ni(cod)(dbn), **9**, 124
$NiC_{15}H_{27}ClNP$
Ni(CH$_2$C$_6$H$_4$NMe$_2$)Cl(PEt$_3$), **9**, 37
$NiC_{15}H_{27}F_3OP_2$
Ni(Me)[OCH(CF$_3$)Ph](PMe$_3$)$_2$, **9**, 54
$NiC_{15}H_{29}ClP_2$
NiCl(C$_6$H$_2$Me$_3$)(PMe$_3$)$_2$, **9**, 58
$NiC_{15}H_{39}F_3O_2P_4$
[Ni(Me)(PMe$_3$)$_4$][CF$_3$COO], **9**, 39
$NiC_{16}H_{10}F_{10}O_2$
Ni(C$_6$F$_5$)$_2$(glyme), **9**, 30
$NiC_{16}H_{10}N_2$
[Ni(PhC$_2$Ph)(CN)$_2$]$^{2-}$, **9**, 133
$NiC_{16}H_{18}$
Ni(C$_6$H$_8$)Cp$_2$, **9**, 144
$NiC_{16}H_{18}N_2$
NiEt$_2$(phen), **9**, 55, 63
$NiC_{16}H_{24}$
Ni(cod)$_2$, **9**, 47, 108, 109; **12**, 752, 907, 910, 917, 918, 960, 961
$NiC_{16}H_{25}N$
Ni(C$_2$H$_4$)$_2$[=CPh(NMeBu^t)], **9**, 111
$NiC_{16}H_{29}P$
Ni(CH$_2$PMe$_3$)(cdt), **9**, 36
$NiC_{16}H_{30}P_2$
Ni[CH$_2$C(Me)$_2$C$_6$H$_4$](PMe$_3$)$_2$, **9**, 56
$NiC_{16}H_{31}ClP_2$
NiCl(CH$_2$CMe$_2$Ph)(PMe$_3$)$_2$, **9**, 75
$NiC_{16}H_{36}IN_3O$
NiI(NHEt$_2$)$_2$[C$_3$H$_4$C(O)NEt$_2$], **9**, 143
$NiC_{16}H_{36}P_2$
Ni(C$_2$H$_4$)(PPr$^i_2$C$_2$H$_4$PPr$^i_2$), **9**, 113
$NiC_{17}H_5F_{10}$
[Ni(C$_6$F$_5$)$_2$(Cp)]$^-$, **9**, 33
$NiC_{17}H_{14}$
Ni(C$_6$H$_4$Ph)(Cp), **9**, 72
$NiC_{17}H_{25}P$
Ni(C$_6$H$_2$Me$_3$)(PMe$_3$)Cp, **9**, 57
$NiC_{17}H_{28}O_2P$
Ni(Me)(PhC(O)CH$_2$C(O)Me)(PEt$_3$), **9**, 37
$NiC_{17}H_{30}P_2S_2$
Ni[SC(S)CH$_2$CMe$_2$C$_6$H$_4$](PMe$_3$)$_2$, **9**, 129
$NiC_{17}H_{39}N_4O$
Ni[(CH$_2$)$_3$OH](C$_{10}$H$_{20}$N$_4$Me$_4$), **9**, 91
$NiC_{18}H_{15}Cl_2P$
Ni(PPh$_3$)Cl$_2$, **12**, 170
$NiC_{18}H_{18}F_{10}O_6P_2$
Ni(C$_6$F$_5$)$_2$[P(OMe)$_3$]$_2$, **9**, 62

$NiC_{18}H_{20}N_2$
Ni(cod)(bipy), **9**, 119
$NiC_{18}H_{26}$
$Ni(C_5H_4Bu^t)_2$, **9**, 154
$NiC_{18}H_{32}$
$Ni(C_3Me_5H_2)Cp^*$, **9**, 144
$NiC_{18}H_{34}P_2$
$Ni(H_2C_4)[Pr^i_2P(CH_2)_2PPr^i_2]$, **9**, 134
$NiC_{18}H_{36}P_2S$
$Ni(Ph)(SH)(PEt_3)_2$, **9**, 62
$NiC_{19}H_7F_{10}$
$Ni(C_6F_5)_2(Tol)$, **9**, 33
$NiC_{19}H_8F_{10}$
$Ni(C_6F_5)_2(nbd)$, **9**, 122
$NiC_{19}H_{24}N_5S$
$Ni[C_6H_3(CH_2NMe_2)_2](NCS)_2$-(py), **9**, 86
$NiC_{19}H_{28}P_2S$
NiPh[S(tolyl)](dmpe), **9**, 61
$NiC_{19}H_{37}N_2P$
$Ni[C(NBu^t)CH_2Bu^t](NC_4H_2$-$Me_2)(PMe_3)$, **9**, 13
$NiC_{19}H_{38}P_2$
$Ni(Me)(Ph)(PEt_3)_2$, **9**, 62
$NiC_{20}H_{12}F_{10}$
$Ni(C_6F_5)_2(cod)$, **9**, 122
$NiC_{20}H_{21}P$
$Ni(CH_2CH_2CH_2PPh_2)Cp$, **9**, 70
$NiC_{20}H_{26}O_2P$
$Ni(acac)(Me)(PEtPh_2)$, **9**, 37
$NiC_{20}H_{28}B_9BrP_2$
$(Ph_3P)BrNi(MePCB_9N_{10})$, **1**, 380
$NiC_{20}H_{30}$
$[NiCp^*_2]^+$, **9**, 156
$[NiCp^*_2]^{2+}$, **9**, 156
$NiCp^*_2$, **9**, 155
$NiC_{20}H_{31}$
$[Ni(C_5Me_5H)Cp^*]^+$, **9**, 166
$NiC_{20}H_{36}NP$
$Ni(C_2H_4)(PEt_3)[=CPh$-$(NBu^tMe)]$, **9**, 112
$NiC_{20}H_{36}N_4$
$Ni(CNBu^t)_4$, **9**, 11
$NiC_{20}H_{48}P_4$
$Ni_2(C_2H_4)_4(dmpe)_2$, **9**, 113
$NiC_{21}H_{15}O_2P$
$Ni(CO)_3PPh_3$, **12**, 918
$NiC_{21}H_{15}O_3P$
$Ni(CO)_3PPh_3$, **12**, 916, 917
$NiC_{21}H_{31}$
$[Ni(cod)_2(C_5H_7)]^+$, **9**, 124
$NiC_{21}H_{36}$
$Ni(norbornene)_3$, **9**, 119
$NiC_{22}H_{20}P$
$Ni[PPh_2C_6H_4(CH=CH_2)_2]$, **9**, 122
$NiC_{22}H_{25}O_2P$
$Ni(C_8H_{13})[PPh_2CH_2C(O)O]$, **9**, 137, 140
$NiC_{22}H_{44}P_2$
$Ni(cod)[Pr^i_2P(CH_2)_2PPr^i_2]$, **9**, 134
$NiC_{23}H_{18}N_2O$
$Ni(Ph_2C=O)(bipy)$, **9**, 127
$NiC_{24}H_{12}$
Ni(TBC), **9**, 134
$NiC_{24}H_{15}ClF_5O_4P$
$Ni(C_6F_5)(PPh_3)(ClO_4)$, **9**, 40
$NiC_{24}H_{23}P$
$NiMe(PPh_3Cp)$, **9**, 163
$NiC_{24}H_{36}$
$Ni_2(C_4H_6)Cp^*_2$, **9**, 114
$NiC_{24}H_{40}P_2$
$Ni(C_{10}H_8)(PPr^i_2C_2H_4PPr^i_2)$, **9**, 123, 174
$NiC_{24}H_{46}Cl_2P_2$
$NiCl(C_6H_4Cl)(PPr^i_3)_2$, **9**, 95
$NiC_{24}H_{46}P_2$
$Ni(C_6H_4)(PPr^i_3)_2$, **9**, 95
$NiC_{24}H_{54}Br_2P_2$
$(Bu_3P)_2NiBr_2$, **12**, 156, 157, 747
$NiC_{25}H_{21}O_3P$
$Ni(CO)_3[P(Mes)=CPh_2]$, **9**, 126
$NiC_{25}H_{22}Cl_2P_2$
$Ni(dppm)Cl_2$, **12**, 173
$NiC_{25}H_{37}ClOP_2$
$NiCl[C(Ph)C(H)(CO[CH_2C$-$Me_2Ph]](PMe_3)_2$, **9**, 78
$NiC_{25}H_{40}ClP$
$NiCl(Bz)(PCy_3)$, **9**, 146
$NiC_{25}H_{42}ClF_6NP_2$
$Ni(Cl)[C(CF_3)C(CF_3)CH_2$-$C_6H_4NMe_2](PEt_3)_2$, **9**, 37
$NiC_{25}H_{42}P_2S$
$NiPh[S(tolyl)](PEt_3)_2$, **9**, 61
$NiC_{25}H_{46}O_2P$
$Ni(acac)(Et)(PCy_3)$, **9**, 37
$NiC_{26}H_{10}F_{10}N_2$
$Ni(C_6F_5)_2(PhCN)_2$, **9**, 33
$NiC_{26}H_{24}Cl_2P_2$
$Ni(dppe)Cl_2$, **12**, 168
$NiC_{26}H_{24}P_2$
Ni(dppe), **9**, 47
$NiC_{26}H_{27}Cl_5OP_2$
$Ni(C_6Cl_5)(C=CHCH_2CH_2O)$-$(PMe_2Ph)_2$, **9**, 62
$NiC_{26}H_{28}PS$
$Cp(Ph_3P)NiCH_2SMe_2$, **12**, 431
$[Cp(PPh_3)NiCH_2SMe_2]^+$, **12**, 393
$NiC_{26}H_{30}NO_3P$
$Ni(Et)(CN)[P(OTol)_3](C_2H_4)$, **9**, 37, 123
$NiC_{26}H_{45}P$
$Ni(PCy_3)(cod)$, **9**, 47
$NiC_{27}H_{18}N_2O_3$
Ni[CPhCPhC(O)C(O)](CO)-(bipy), **9**, 122, 149
$NiC_{27}H_{26}$
$NiPh(PPh_3)(C_3H_6)$, **9**, 143
$NiC_{27}H_{26}Cl_2P_2$
$Ni(dppp)Cl_2$, **12**, 165, 167, 170, 171, 174
$NiC_{27}H_{38}O_4P_2$
$Ni[C_6H_3(OMe)_2]_2(PMe_3)$-$(PMe_2Ph)$, **9**, 56
$NiC_{27}H_{47}NOP_2$
$Ni(P{\equiv}CNPr^i_2)(CO)(PCy_3)$, **9**, 136
$NiC_{28}H_{22}F_5P$
$Ni(C_6F_5)[CH_2C(Me)CH_2]$-$(PPh_3)$, **9**, 65
$NiC_{28}H_{30}B_4P_2$
$(dppe)Ni(C_2B_4H_6)$, **1**, 391
$NiC_{28}H_{30}P_2$
$Ni(Me)_2(dppe)$, **9**, 61
$NiC_{28}H_{40}P_2$
$Ni(CCPh)_2(PEt_3)_2$, **9**, 98
$NiC_{29}H_{25}P$
$Ni(Ph)(PPh_3)Cp$, **9**, 66
$NiC_{30}H_{24}O_6$
$Ni[(MeO)_6TBC]$, **9**, 135
$NiC_{30}H_{27}O_2S$
$Ni(CH_2SO_2Ph)(PPh_3)Cp$, **9**, 163
$NiC_{30}H_{31}NOP_2$
$Ni[CH_2CH_2CH_2C(O)NH]$-(dppe), **9**, 85
$NiC_{30}H_{32}OP_2$
$Ni(Ph)(Ph_2PCHCPhO)(CH_2P$-$Me_3)$, **9**, 36
$NiC_{30}H_{32}P_2$
Me(allyl)(dppe)Ni, **12**, 965
$NiC_{30}H_{36}P_2$
$Ni(C_6H_4Me)_2(PMe_2Ph)_2$, **9**, 64
$NiC_{30}H_{45}P_5$
$Ni_3(P)(P_4)Cp^*_3$, **9**, 168
$NiC_{30}H_{46}N_2$
$Ni(C_2H_4)_2[DAD(C_6H_3Pr^i_2)_2]$, **9**, 113
$NiC_{30}H_{54}P_2$
$Ni(C_4H_6)(dcpe)$, **9**, 117
$NiC_{31}H_{25}P$
$Ni(CCPh)(PPh_3)Cp$, **9**, 66, 72
$NiC_{31}H_{29}O_2PS$
$Ni[CH(Me)SO_2Ph](PPh_3)Cp$, **9**, 72
$NiC_{31}H_{42}OP_2$
$Ni(Ph)(Ph_2PCHCMeO)(CH_2P$-$Pr^i_3)$, **9**, 36
$NiC_{31}H_{48}OP_2$
$Ni(Ph_2C=C=O)(PBu^t_2CH_2P$-$Bu^t_2)$, **9**, 128
$NiC_{32}H_{26}Cl_6P_2$
$NiCl(C_6Cl_5)(PMePh_2)_2$, **9**, 56
$NiC_{32}H_{29}N_2P$
$Ni(bipy)[P(xyl)=CPh_2]$, **9**, 126
$NiC_{32}H_{29}P_2$
Ni(Ph)(dppe), **9**, 63
$NiC_{32}H_{36}OP$
$Ni(Ph)[OC(Ph)CHPPh_2](PEt_3)$, **9**, 59
$NiC_{32}H_{38}Br_2O_4P_2$
$Ni[C_6H_2Br(OMe)_2]_2(PMe_2Ph)_2$, **9**, 56
$NiC_{32}H_{39}NP_2$
$Ni(P=NBu^t)(PPh_3)Cp^*$, **9**, 166
$NiC_{32}H_{40}Se_2$
$Ni(SePh)_2Cp^*_2$, **9**, 169, 170
$NiC_{32}H_{41}O_2P$
$Ni(C_{14}H_8O_2)(PCy_3)$, **9**, 123
$NiC_{32}H_{52}P_2$
$Ni(C_6H_4)(dcpe)$, **9**, 133, 134

$NiC_{32}H_{56}O_2P_2$
Ni(CH$_2$=CHCH=CHCO$_2$Me)(dcpe), **9**, 121
$NiC_{34}H_{26}O$
Ni(C$_4$Ph$_4$HCO)Cp, **9**, 145
$NiC_{34}H_{35}ClF_6P_4$
[Ni(triphos)Cl]PF$_6$, **12**, 176
$NiC_{36}H_{22}F_{10}P_2$
Ni(C$_6$F$_5$)$_2$(PPh$_2$H)$_2$, **9**, 30
$NiC_{36}H_{30}Cl_2P_2$
Ni(PPh$_3$)$_2$Cl$_2$, **12**, 171, 964
$NiC_{37}H_{25}N_4P$
Ni[C[=C(CN)$_2$]CPh+C(CN)$_2$](PPh$_3$Cp), **9**, 161
$NiC_{37}H_{33}P_2$
NiMe(PPh$_3$)$_2$, **9**, 130
$NiC_{37}H_{66}O_2P_2$
Ni(CO$_2$)(PCy$_3$)$_2$, **9**, 128
$NiC_{37}H_{68}O_2P_2$
NiH(O$_2$CH)(PCy$_3$)$_2$, **9**, 57
$NiC_{38}H_{30}Cl_4P_2$
NiCl(CClCCl$_2$)(PPh$_3$)$_2$, **9**, 61
$NiC_{38}H_{30}O_2P_2$
(PPh$_3$)$_2$Ni(CO)$_2$, **12**, 365, 377
$NiC_{38}H_{31}NS_2$
Ni(S$_2$CNMe$_2$)(C$_5$Ph$_5$), **9**, 154
$NiC_{38}H_{32}OP_2$
Ni(CH$_2$=C=O)(PPh$_3$)$_2$, **9**, 9
$NiC_{38}H_{32}P_2$
Ni(C$_2$H$_2$)(PPh$_3$)$_2$, **9**, 133
$NiC_{38}H_{34}BrP_2$
NiBr(Ph)$_2$(dppe), **9**, 63
$NiC_{38}H_{34}P_2$
Ni(C$_2$H$_4$)(PPh$_3$)$_2$, **9**, 47, 118, 119
$NiC_{38}H_{36}P_2$
Ni(Me)$_2$(PPh$_3$)$_2$, **9**, 62
$NiC_{38}H_{64}P_2$
Ni(C$_{14}$H$_{10}$)(PBu$_3$)$_2$, **9**, 123
$NiC_{38}H_{68}P_2$
Ni(C$_2$H$_2$)(PCy$_3$)$_2$, **9**, 132
$NiC_{38}H_{70}P_2$
Ni(PCy$_3$)$_2$(C$_2$H$_4$), **9**, 37
$NiC_{39}H_{30}O_2P_2$
(C$_3$O$_2$)Ni(PPh$_3$)$_2$, **9**, 9, 122, 148
$NiC_{39}H_{35}ClO_2P_2$
Ni[C(O)OEt]Cl(PPh$_3$)$_2$, **9**, 4
$NiC_{40}H_{32}O_2P_2$
Ni[Ph$_2$PCH=C(O)Ph]$_2$, **9**, 8
$NiC_{40}H_{36}P_2$
Ni[CH$_2$=C(CH$_2$)$_2$](PPh$_3$)$_2$, **9**, 119
$NiC_{40}H_{39}O_4P_2$
Ni[MeC(O)P(O)(OMe)$_2$](PPh$_3$)$_2$, **9**, 128
$NiC_{40}H_{47}O_2PS$
Ni[CH(C$_{10}$H$_{21}$)SO$_2$Ph](PPh$_3$)Cp, **9**, 72
$NiC_{41}H_{34}Br_2NP_2$
NiBr(BrC$_5$H$_4$N)(PPh$_3$)$_2$, **9**, 65
$NiC_{41}H_{35}P_2$
[Ni(PPh$_3$)$_2$Cp]$^+$, **9**, 153
$NiC_{41}H_{38}P_2$
Ni(CH$_2$=C=CHEt)(PPh$_3$)$_2$, **9**, 121, 144
$NiC_{41}H_{42}OP_2$
Ni(Me)(OBu^t)(PPh$_3$)$_2$, **9**, 62
$NiC_{42}H_{30}BrCl_5P_2$
NiBr(C$_6$Cl$_5$)(PPh$_3$)$_2$, **9**, 62
$NiC_{42}H_{30}Cl_6P_2$
NiCl(C$_6$Cl$_5$)(PPh$_3$)$_2$, **9**, 55, 56
$NiC_{42}H_{33}BrCl_2P_2$
NiBr(C$_6$H$_3$Cl$_2$)(PPh$_3$)$_2$, **9**, 60
$NiC_{42}H_{34}Br_2P_2$
NiBr(C$_6$H$_4$Br)(PPh$_3$)$_2$, **9**, 65
$NiC_{42}H_{35}ClP_2$
NiCl(Ph)(PPh$_3$)$_2$, **9**, 55
$NiC_{42}H_{38}O_2P_2$
Ni(CH$_2$=CHCH=CHCO$_2$Me)(PPh$_3$)$_2$, **9**, 121
$NiC_{42}H_{39}P_3S_2$
Ni(CS$_2$)(triphos), **9**, 129
$NiC_{42}H_{42}NP_3$
Ni[N(C$_2$H$_4$PPh$_2$)$_3$], **9**, 129
$NiC_{42}H_{66}Cl_6P_2$
NiCl(C$_6$Cl$_5$)(PCy$_3$)$_2$, **9**, 63
$NiC_{42}H_{72}P_2$
NiH(Ph)(PCy$_3$)$_2$, **9**, 57
$NiC_{43}H_{30}F_5OP_2$
Ni(C$_6$F$_5$)(CO)(PPh$_3$)$_2$, **9**, 62
$NiC_{43}H_{37}ClP_2$
NiCl(Bz)(PPh$_3$)$_2$, **9**, 146
$NiC_{43}H_{38}OP_2$
Ni(PhMeC=C=O)(PPh$_3$)$_2$, **9**, 128
$NiC_{43}H_{40}P_2S$
Ni(CH$_2$SMe)(PPh$_3$)$_2$Cp, **9**, 57
$NiC_{43}H_{42}NP_3S_2$
Ni(CS$_2$)[N(C$_2$H$_4$PPh$_2$)$_3$], **9**, 129
$NiC_{43}H_{48}Cl_5NP_2S$
Ni(NCS)(C$_6$Cl$_5$)(PCy$_3$)(PPh$_3$), **9**, 56
$NiC_{43}H_{72}OP_2$
Ni(PhCHO)(PCy$_3$)$_2$, **9**, 127
$NiC_{43}H_{72}O_2P_2$
Ni(Ph)(O$_2$CH)(PCy$_3$)$_2$, **9**, 57
$NiC_{44}H_{36}OP_2$
Ni(Ph)(Ph$_2$PCHCPhO)(PPh$_3$), **9**, 36
Ni(CPhH=C=O)(PPh$_3$)$_2$, **9**, 128
$NiC_{44}H_{39}F_6P_2$
[Ni[Ph$_3$PCH(CH$_2$)CHPPh$_3$]Cp][PF$_6$], **9**, 36
$NiC_{44}H_{40}O_4P_2$
Ni(CO$_2$MeCH$_2$C$_2$CH$_2$CO$_2$Me)(PPh$_3$)$_2$, **9**, 130
$NiC_{44}H_{42}O_2P_2$
Ni(CO)$_2$[P(Mes)=CPh$_2$]$_2$, **9**, 126
$NiC_{44}H_{45}NP_2S_2$
[Ph$_2$PC$_2$H$_4$N(C$_2$H$_4$PPh$_2$)$_2$NiCS$_2$Me]$^+$, **9**, 129
$NiC_{44}H_{54}P_2$
Ni(C$_8$H$_4$Ph$_4$)(PEt$_3$)$_2$, **9**, 122, 123
$NiC_{45}H_{41}BrP_2$
NiBr(C$_6$H$_2$Me$_3$)(PPh$_3$)$_2$, **9**, 60
$NiC_{45}H_{47}NO_6P_2$
Ni(CN)(Et)[P(OC$_6$H$_4$Me)$_3$]$_2$, **9**, 54
$NiC_{46}H_{40}OP_2$
Ni(Ph(Et)C=C=O)(PPh$_3$)$_2$, **9**, 128
$NiC_{48}H_{44}NP_3S$
Ni(PhNCS)(triphos), **9**, 129
$NiC_{48}H_{50}F_{12}N_4O_4P_4$
[Ni*CNCH$_2$CH$_2$OH)$_4$(PPh$_3$)$_2$][PF$_6$]$_2$, **9**, 11
$NiC_{50}H_{42}P_2$
Ni(CHPh=CHPh)(PPh$_3$)$_2$, **9**, 119
$NiC_{50}H_{47}P_3$
Ni(PPh$_3$)$_2$[PPh$_2$C$_6$H$_4$(CH=CH$_2$)], **9**, 122
$NiC_{51}H_{43}NP_2$
Ni(MAD-Ph)(PPh$_3$)$_2$, **9**, 118
$NiC_{54}H_{42}N_2O_4P_2$
Ni[Ph$_2$PC[C(O)Ph][=C(O)NHPh]]$_2$, **9**, 8
$NiC_{55}H_{41}OP$
Ni[Ph$_2$PCH=C(O)Ph](C$_5$Ph$_5$), **9**, 153
$NiC_{55}H_{49}BrP_4$
NiBr(dppm)$_2$Cp, **9**, 160
$NiC_{55}H_{49}ClP_4$
NiCl(dppm)$_2$Cp, **9**, 160
$NiC_{57}H_{45}ClO_4P_2$
[Ni(PPh$_3$)$_2$(C$_3$Ph$_3$)][ClO$_4$], **9**, 295
$NiC_{57}H_{45}F_6P_3$
[Ni(PPh$_3$)$_2$(C$_3$Ph$_3$)][PF$_6$], **9**, 296
$NiC_{57}H_{45}P_2$
[Ni(C$_3$Ph$_3$)(PPh$_3$)$_2$]$^+$, **9**, 146
$NiC_{58}H_{42}$
[Ni(C$_5$HPh$_4$)$_2$]$^+$, **9**, 154
$NiC_{62}H_{54}P_3$
[Ni(triphos)(C$_3$Ph$_3$)]$^+$, **9**, 146
$NiC_{72}H_{60}N_4P_4S_4$
[Ni(PPh$_3$)$_2$]$_2$(SNSN)$_2$, **9**, 146
$NiC_{72}H_{60}P_4$
Ni(PPh$_3$)$_4$, **12**, 167, 172
$NiC_{108}H_{90}Cl_2S_6$
Ni$_8$Cl$_2$(PPh$_3$)$_6$S$_6$, **9**, 20
$NiCl_2$
NiCl$_2$, **1**, 69, 70
$NiCoC_{25}H_{29}PS_2$
Co(PMe$_3$)Cp(SPh)$_2$NiCp, **10**, 64
$NiCoFeC_{15}H_{14}O_6P$
FeCoNiCp[PBu^t](CO)$_6$, **9**, 20, 173
$NiCoFeC_{16}H_{14}O_8P$
CoFeNi(Me$_3$P)Cp(CO)$_8$, **9**, 21
$NiCoFe_2C_{14}H_8O_8P$
Fe$_2$CoNiCp(PMe)(CO)$_8$, **9**, 20
$NiCoFe_2C_{15}H_{11}O_8P_2$
NiCoFe$_2$(PMe)$_2$(CO)$_8$Cp, **9**, 173
$NiCoMoC_{17}H_{13}O_5$
NiCoMo(CMe)(CO)$_5$Cp$_2$, **9**, 173
$NiCoRuC_{19}H_{17}O_6$
CpRuCoNi(CO)$_6$(C$_4$Me$_4$), **10**, 166
$NiCoW_2C_{24}H_{19}O_7$
NiCoW$_2$(CO)$_7$CpCp'$_2$, **9**, 173

$NiCo_2C_{13}H_8O_6$
CpNiCo$_2$(CO)$_6$(CMe), **9**, 173; **10**, 156
$NiCo_2MoC_{19}H_{12}O_8$
NiCo$_2$Mo(CO)$_8$CpCp', **9**, 173
$NiCrC_{14}H_{10}O_4$
CpNiCr(CO)$_4$, **9**, 95
Cr(CO)$_3$CpNi(CO)Cp, **10**, 45
$NiCrC_{18}H_{36}ClO_4P_3$
Cr(CO)$_4$(PMe$_3$)(PBu$^t_2$)NiCl-(PMe$_3$), **10**, 45
$NiFeB_4C_{25}H_{37}$
(C$_9$H$_8$)Fe(Et$_2$C$_2$B$_4$H$_4$)NiCp*, **9**, 179
$NiFeC_{17}H_{36}ClO_3P_3$
Fe(CO)$_3$(PMe$_3$)(PBu$^t_2$)NiCl-(PMe$_3$), **10**, 61
$NiFeC_{24}H_{40}B_4$
Cp*$_2$NiFe(Me$_2$C$_2$B$_4$H$_4$), **1**, 397
$NiFeC_{30}H_{25}O_5P$
FeNi(CO)$_3$(PPh$_3$)[C(Co$_2$Me)=CH$_2$]Cp, **10**, 61
$NiFeC_{34}H_{30}Cl_2P_2$
Ni(dppf)Cl$_2$, **12**, 170
$NiFeMoC_{27}H_{22}O_7$
Cp$_2$FeMoNi(CO)$_5$(PhC$_2$Co$_2$Pr^i), **10**, 162
$NiFe_2C_{13}H_5O_8$
CpNiFe$_2$(CO)$_8$, **7**, 84
$NiFe_2C_{17}H_{14}O_6$
CpNiFe$_2$(CO)$_6$(C≡CBu^t), **10**, 102
$NiFe_3C_{12}HO_{12}$
[Fe$_3$Ni(H)(CO)$_{12}$]$^-$, **9**, 20; **10**, 203
$NiFe_3C_{12}O_{12}$
[Fe$_3$Ni(CO)$_{12}$]$^{2-}$, **7**, 84; **9**, 20
$NiGaC_{11}H_{17}N_4$
Ni(C$_3$H$_5$)(GaMe$_2$pz$_2$), **9**, 140
$NiGaC_{19}H_{35}O$
Ga(CH$_2$Bu^t)$_2$(THF)[NiCp], **1**, 557
$NiInC_{16}H_{27}N_2O$
In(CH$_2$CH$_2$CH$_2$NMe$_2$)$_2$[Ni-(CO)Cp], **1**, 566
$NiKC_6H_5O$
K[NiCp(CO)], **9**, 5
$NiLiC_{13}H_{32}N_3$
Li(PMPTA)(Me)Ni(C$_2$H$_4$)$_2$, **9**, 110
$NiLiC_{14}H_{34}N_3$
(C$_2$H$_4$)$_2$Ni(Me)Li(PMDTA), **9**, 34
$NiLiC_{15}H_{35}NO$
Ni(C$_2$H$_4$)$_2$[C(O)NMe$_2$]-(PMPTA)Li, **9**, 112
$NiLiC_{17}H_{43}N_4$
[Li(TMEDA)$_2$][NiMe(C$_2$H$_4$)$_2$], **9**, 34, 110
$NiLiC_{23}H_{43}N_4$
Ni(C$_2$H$_4$)$_2$[=CPh(NBu^t)]Li-(PMTA), **9**, 111
$NiLiC_{24}H_{59}N_6$
[Li(PMPTA)$_2$][Ni$_2$(H)(C$_2$H$_4$)$_4$], **9**, 111
$NiLi_2C_{18}H_{40}O_4$
Ni(C$_2$H$_4$)(C$_4$H$_8$Li$_2$)(O$_2$C$_4$H$_8$)(OEt$_2$)$_2$, **9**, 35
$NiLi_2C_{28}H_{68}N_6$
(C$_2$H$_4$)$_2$Ni(C$_3$H$_7$)$_2$[Li-(PMDTA)]$_2$, **9**, 34
$NiMgC_{12}H_{30}N_2$
(C$_2$H$_4$)$_2$Ni(Me)Mg(Me)-(TMEDA), **9**, 34, 110
$NiMgC_{20}H_{28}O$
(C$_2$H$_4$)$_2$Ni(Ph)MgPh(OEt$_2$), **9**, 35
$NiMoC_{15}H_{12}O_4$
Mo(CO)$_3$CpNi(CO)Cp', **9**, 144, 169; **10**, 32, 45
$NiMoC_{17}H_{18}O_2$
NiMo(CO)$_2$(MeC$_2$Me)CpCp', **9**, 136
$NiMoC_{17}H_{21}S_2$
MoCp$_2$(SMe)$_2$NiCp, **10**, 45
$NiMoC_{18}H_{20}O_2$
NiCp(C$_3$H$_2$Me$_2$)(CO)Mo-(CO)Cp', **9**, 144
$NiMoC_{18}H_{20}O_3$
Cp*NiMo(CO)$_3$Cp, **9**, 95, 169; **10**, 32
$NiMoC_{19}H_{20}O_4$
Ni(CO)Cp*Mo(CO)$_3$Cp, **9**, 173
NiCp*(CO)$_2$Mo(CO)$_2$Cp, **9**, 169
$NiMoC_{19}H_{22}O_3$
Mo(CO)$_2$Cp(CO)(CH$_2$)NiCp*, **9**, 19, 169; **10**, 45
$NiMoC_{23}H_{21}F_8N_2O_3P_5$
Mo(CO)$_3$[MeN(PF$_2$)$_2$]$_2$Ni-(PPh$_3$), **10**, 45
$NiMoC_{24}H_{22}O_4$
NiMo(CO)$_2$Cp(PhC$_2$CO$_2$-Pr^i)Cp, **9**, 136
$NiMoC_{30}H_{32}O_2S_2$
NiCp*(SPh)(CO)Mo(SPh)-(CO)Cp', **9**, 169
$NiMoC_{34}H_{27}O_4$
NiMo(CO)$_2$Cp(PhC$_2$Ph)-(CO)$_2$CpCp', **9**, 136
$NiMoC_{41}H_{38}O_2$
NiCp*(PhC$_2$H)Ni(C$_4$H$_2$Ph$_2$)-Mo(CO)$_2$Cp, **9**, 169
$NiMoC_{46}H_{59}O_4P_3$
Mo(CO)$_4$(PCy$_2$)$_2$Ni(PPh$_3$), **10**, 45
$NiMo_2AsC_{19}H_{15}O_4$
NiMo$_2$(As)(CO)$_4$Cp$_3$, **9**, 173
$NiMo_2PtC_{30}H_{29}O_4$
PtNiMo$_2$(CC$_6$H$_4$Me)(CO)$_4$-(cod)Cp$_2$, **9**, 564
$NiNaC_{20}H_{49}N_4$
[Na(TMEDA)$_2$][Ni$_2$(H)-(C$_2$H$_4$)$_4$], **9**, 111
$NiNbC_{14}H_{20}Cl_2S_2$
NbCl$_2$Cp(SEt)$_2$NiCp, **10**, 28
$NiNbC_{25}H_{25}Cl_2S_2$
NiCp(SEt)$_2$NbCl$_2$Cp$_2$, **9**, 169
$NiOs_3C_{14}H_7O_9$
[NiOs$_3$(CO)$_9$Cp(H)$_2$]$^-$, **9**, 20
$NiOs_3C_{14}H_8O_9$
NiOs$_3$(CO)$_9$Cp(H)$_3$, **9**, 20; **10**, 38, 40, 41, 340
$NiPdC_{54}H_{50}ClN_2P_4$
[Ni(MeNC)(dppm)$_2$(CNMe)-PtCl]$^+$, **10**, 69
$NiPdC_{56}H_{53}N_{34}$
[Ni(MeNC)$_3$(dppm)$_2$Pd]$^{2+}$, **10**, 69
$NiRh_6C_{16}O_{16}$
[NiRh$_6$(CO)$_{16}$]$^{2-}$, **10**, 193–5
$NiRuC_{28}H_{23}PS_2$
Ru(C$_5$H$_4$S)$_2$Ni(PPh$_3$), **10**, 61
$NiRu_3C_{14}H_8O_9$
Ru$_3$NiCp(H)$_3$(CO)$_9$, **9**, 21; **10**, 38, 40, 41
$NiSiC_7H_{16}Br_2$
[CH$_2$=C(Me)CH$_2$(TMS)]NiBr$_2$, **12**, 962, 963
$NiSiC_9H_{20}Cl_2O$
[CH$_2$=CHCH$_2$(OSiBu^{t}Me$_2$)]-NiCl$_2$, **12**, 962
$NiSiC_{10}H_{27}ClN_2P_2$
NiCl[C(N$_2$)TMS](PMe$_3$)$_2$, **9**, 19, 94
$NiSiC_{11}H_{27}ClP_2$
NiCl(CC-TMS)(PMe$_3$)$_2$, **9**, 60
$NiSiC_{11}H_{29}ClOP_2$
Ni(COCH$_2$TMS)Cl(PMe$_3$)$_2$, **9**, 3
$NiSiC_{16}H_{26}$
Ni(cod)[C$_4$H$_2$Me$_2$(SiMe$_2$)], **9**, 124
$NiSiC_{26}H_{30}$
Ni(cod)[CPh=CHCH=CPh-(SiMe$_2$)], **9**, 124
$NiSiC_{30}H_{42}OP_2$
Ni(C$_8$H$_{13}$)[P(Mes)C(O-TMS)P-(Mes)], **9**, 145
$NiSiC_{47}H_{44}P_2$
Ni[PhC$_2$(TMS)](PPh$_3$)$_2$, **9**, 133
$NiSi_2C_3Cl_6O_3$
Ni(SiCl$_3$)$_2$(CO)$_3$, **9**, 2
$NiSi_2C_6H_6F_6$
Ni(SiF$_3$)$_2$(C$_6$H$_6$), **9**, 174
$NiSi_2C_7H_8Cl_6$
Ni(SiCl$_3$)$_2$(Tol), **9**, 2
$NiSi_2C_7H_8F_6$
Ni(SiF$_3$)$_2$(TolH), **9**, 174
$NiSi_2C_{11}H_{31}NP_2$
NiMe[N(SiMe$_2$CH$_2$PMe$_2$)$_2$], **9**, 58
$NiSi_2C_{14}H_{40}P_2$
Ni(CH$_2$TMS)$_2$(PMe$_3$)$_2$, **9**, 56
$NiSi_2C_{18}H_{32}N_2$
Ni(CH$_2$TMS)$_2$(py)$_2$, **9**, 56
$NiSi_2C_{32}H_{58}P_2$
Ni(CH$_2$SiMe$_2$Ph)$_2$(Pr$^i_2$PCH$_2$-CH$_2$PPr$^i_2$), **9**, 56
$NiSi_2C_{44}H_{48}P_2$
Ni(TMS)C$_2$(TMS)(PPh$_3$)$_2$, **9**, 130
$NiSi_3C_{35}H_{71}P_2$
Ni[Si(TMS)(Mes)=C=CPh-(TMS)](PEt$_3$)$_2$, **9**, 125
$NiSi_4C_{13}H_{36}ClOP$

Ni(CO)[P(Cl)=C(TMS)$_2$]$_2$, **9**, 126, 127
$NiSi_4C_{16}H_{12}N_4$
Ni(CN-TMS)$_4$, **9**, 10
$NiSi_4C_{18}H_{51}ClP_2$
NiCl(PEt$_3$)[C(TMS)$_2$=P=C(TMS)$_2$], **9**, 127
$NiSi_4C_{20}H_{55}P_3$
Ni[P[CH(TMS)$_2$=C(TMS)$_2$]-(PMe$_3$)$_2$], **9**, 126
$NiSi_4SnC_{18}H_{46}$
Ni(C$_2$H$_4$)$_2$[=Sn[CH(TMS)$_2$]$_2$], **9**, 112
$NiSi_4SnC_{52}H_{70}P_2$
Ni(PPh$_3$)$_2$[HC$_2$SnH[CH-(TMS)$_2$]$_2$], **9**, 133
$NiSi_4Sn_4C_{40}H_{96}N_8$
Ni[Sn(NBu^t)$_2$(SiMe$_2$)]$_4$, **9**, 14
$NiSnC_{29}H_{25}O$
Ni(SnPh$_3$)(CO)Cp*, **9**, 167
$NiSn_2C_{16}H_{26}$
Ni(C$_5$H$_4$SnMe$_3$)$_2$, **9**, 154
$NiTaC_{40}H_{57}N_2O$
Ta(TMEDA)(CH$_2$)(Me)-(OCMe$_2$CH$_2$PPh$_2$)$_2$Ni, **10**, 28
$NiThC_{22}H_{34}O_2P_2$
Cp*$_2$Th(PH$_2$)$_2$Ni(CO)$_2$, **4**, 188
$NiThC_{36}H_{30}O_2P_2$
Ni(CO)$_2$(PPh$_2$)$_2$ThCp$_2$, **10**, 71
$NiWC_{15}H_{12}O_4$
NiCp(CO)W(C$_5$H$_4$Me)(CO)$_3$, **9**, 15
$NiWC_{16}H_{20}O_2$
W(CO)$_2$Cp'(CH$_2$CCMe$_2$)NiCp, **10**, 45
$NiWC_{18}H_{19}O_2$
W(CO)$_2$Cp'(CHCMeCMe)-NiCp, **10**, 45
$NiWC_{21}H_{26}O_2$
W(CO)$_2$Cp(CH$_2$C(CH$_2$)$_2$NiCp*, **10**, 45
$NiWC_{22}H_{28}O_2$
NiCp*[C(Me)C(Me)CH$_2$]-(CO)W(CO)Cp, **9**, 19
$NiWC_{30}H_{32}O_2S_2$
NiCp*(SPh)(CO)W(SPh)-(CO)Cp', **9**, 169
$NiW_2C_{28}H_{36}O_4$
Cp*$_2$W$_2$Ni(CO)$_4$(CMe)$_2$, **10**, 119
$NiW_2C_{30}H_{24}O_4$
W(CO)$_2$Cp(CTol)NiW-(CO)$_2$Cp(CTol), **10**, 45
$NiW_2W_2C_{40}H_{44}O_4$
NiW$_2$(CC$_6$H$_4$Me)$_2$(CO)$_4$Cp*$_2$, **9**, 23
$Ni_2AlLiC_{48}H_{84}N_4$
[Li(TMEDA)$_2$][[Ni(cdt)]$_2$-[(H)$_2$AlBu$^i_2$]], **1**, 551
$Ni_2AsMoC_{37}H_{30}O_7$
Ni$_2$Mo[As[Mo$_3$(As)(CO)$_5$Cp$_3$]](CO)$_2$Cp$_3$, **9**, 173
$Ni_2BC_{15}H_{17}$
Ni$_2$(C$_4$H$_4$BMe)Cp$_2$, **9**, 179
$Ni_2B_2C_{14}H_{16}O_2$
Ni$_2$(CO)$_2$(CpBMe)$_2$, **9**, 176
$Ni_2B_2C_{19}H_{25}$
Cp$_2$Ni$_2$[Et$_2$MeC$_4$H$_2$B$_2$], **1**, 210
$Ni_2B_4C_{20}H_{36}$
[Ni[Me$_2$C$_2$(BMe)$_2$C=CMe$_2$]]$_2$, **1**, 208
$Ni_2B_4Pt_2C_{34}H_{56}$
[Pt(MeCB$_2$C$_2$Et$_4$)NiCp]$_2$, **9**, 179
$Ni_2B_7C_{11}H_{18}$
Cp$_2$Ni$_2$CB$_7$H$_8$, **9**, 181
$Ni_2C_3H_5Cl_2$
[(allyl)NiCl]$_2$, **12**, 747
$Ni_2C_8Cl_8O_4$
[Ni(C$_3$Cl$_3$)Cl(CO)$_2$]$_2$, **9**, 145, 146
$Ni_2C_{10}H_{10}F_6O_4$
[(CH$_2$=CHCH$_2$)Ni(OOCCF$_3$)]$_2$, **12**, 1225
$Ni_2C_{10}H_{14}P_2S_2$
CpNi[P(S)H$_2$]$_2$NiCp, **9**, 163
$Ni_2C_{10}H_{30}O_2P_2$
[Ni(Me)(OMe)(PMe$_3$)]$_2$, **9**, 37
$Ni_2C_{11}H_{18}B_7$
Cp$_2$Ni$_2$(CB$_7$H$_8$), **1**, 379
$Ni_2C_{12}H_{10}O_2$
Ni$_2$(CO)$_2$Cp$_2$, **9**, 4, 15, 144, 159, 167, 173
$Ni_2C_{12}H_{20}Br_4$
(cyclohexenyl)NiBr$_2$, **12**, 961, 962
$Ni_2C_{12}H_{24}$
[Ni(Me)(MeCHCHCHMe)]$_2$, **9**, 65
$Ni_2C_{12}H_{24}P_2$
Ni$_2$(C$_2$H$_2$)$_3$(PMe$_3$)$_2$, **9**, 133
$Ni_2C_{12}H_{33}O_3P_3$
Ni$_2$Me$_2$(CO$_3$)(PMe$_3$)$_3$, **9**, 39
$Ni_2C_{13}H_{24}Br_2P_2$
NiBr[CH$_2$C$_6$H$_4$[NiBr(PMe$_3$)$_2$]], **9**, 146
$Ni_2C_{14}H_{10}F_6N_2$
Ni$_2$(CF$_3$NC)$_2$Cp$_2$, **9**, 10, 168
$Ni_2C_{14}H_{14}O_2$
Ni$_2$(CO)$_2$Cp'$_2$, **9**, 167
$Ni_2C_{14}H_{16}$
Ni$_2$(MeC$_2$Me)Cp$_2$, **9**, 135
$Ni_2C_{14}H_{22}P_2S_2$
CpNi[P(S)Me$_2$]$_2$NiCp, **9**, 163
$Ni_2C_{14}H_{32}P_2$
Ni$_2$(C$_2$H$_4$)$_4$(dmpe), **9**, 113
$Ni_2C_{14}H_{38}P_4$
Ni$_2$(C$_2$H$_2$)(PMe$_3$)$_4$, **9**, 133
$Ni_2C_{15}H_{19}P$
Ni$_2$(P≡CBu^t)Cp$_2$, **9**, 136
$Ni_2C_{15}H_{35}ClP_2$
Ni$_2$(C$_3$H$_5$)Cl(PEt$_3$)$_2$, **9**, 143
$Ni_2C_{16}H_{24}F_{12}N_8P_2$
[Ni$_2$(CNMe)$_8$][PF$_6$]$_2$, **9**, 17
$Ni_2C_{17}H_{51}O_2P_5$
[Ni(Me)(PMe$_3$)$_4$][(O$_2$)Ni(Me)-(PMe$_3$)], **9**, 39
$Ni_2C_{18}H_{16}N_2$
Ni$_2$(C$_4$H$_4$)(DAD-Ph$_2$), **9**, 118
$Ni_2C_{19}H_{42}BrP_4$
Ni$_2$(CH$_2$C$_6$H$_4$)Br(PMe$_3$)$_4$, **9**, 12
$Ni_2C_{19}H_{42}Br_2P_4$
(Me$_3$P)$_2$BrNi(CH$_2$C$_6$H$_4$)NiBr(PMe$_3$)$_2$, **9**, 47
$Ni_2C_{20}H_{16}N_2O_2$
Ni$_2$(C$_4$H$_4$)(CO)$_2$(DAD-Ph$_2$), **9**, 118
$Ni_2C_{20}H_{18}$
Ni$_2$(C$_5$H$_4$C$_5$H$_4$)Cp$_2$, **9**, 156
$Ni_2C_{21}H_{42}Br_2O_2P_4$
(Me$_3$P)$_2$BrNiC(O)CH$_2$C$_6$H$_4$C-(O)NiBr(PMe$_3$)$_2$, **9**, 58
$Ni_2C_{22}H_{20}S_2$
Ni$_2$(SPh)$_2$Cp$_2$, **9**, 154
$Ni_2C_{22}H_{30}O_2$
Ni$_2$(CO)$_2$Cp*$_2$, **9**, 168
$Ni_2C_{22}H_{30}S_4$
Ni$_2$(C$_2$S$_4$)Cp*$_2$, **9**, 154
$Ni_2C_{22}H_{38}O_2P_2$
[Ni(CH$_2$C$_6$H$_4$Me)(PMe$_3$)-(OH)]$_2$, **9**, 39
$Ni_2C_{22}H_{48}P_2$
Ni$_2$(C$_2$H$_4$)$_4$(PPr$^i_2$C$_2$H$_4$PPr$^i_2$), **9**, 113
$Ni_2C_{24}Cl_2F_{20}$
[Ni(Cl)(C$_6$F$_5$)$_2$]$_2$, **9**, 33
$Ni_2C_{24}H_{20}$
Ni$_2$(PhC$_2$Ph)Cp$_2$, **9**, 95, 136
$Ni_2C_{24}H_{28}N_2$
Ni$_2$(MeCN)$_2$(C$_3$H$_3$PhMe)$_2$, **9**, 143
$Ni_2C_{24}H_{50}N_4$
Ni$_2$(C$_4$H$_6$)$_3$(TMEDA)$_2$, **9**, 117
$Ni_2C_{26}H_{25}NO$
(CpNi)$_2$[PhCC(O)NHCH-(Me)Ph], **9**, 94
$Ni_2C_{26}H_{46}O_2P_2$
[Ni(CH$_2$CMe$_2$Ph)(OH)-(PMe$_3$)]$_2$, **9**, 39
$Ni_2C_{27}H_{25}NO$
Ni$_2$Cp$_2$[PhC$_2$C(O)N-HCHMePh], **9**, 136
$Ni_2C_{28}H_{27}NO$
Ni$_2$CpCp'(PhC$_2$C(O)N-HCHMePh), **9**, 136
$Ni_2C_{30}H_2F_{20}N_4$
[Ni(C$_6$F$_5$)$_2$(NCCHCN)]$_2^{2-}$, **9**, 33
$Ni_2C_{32}H_{34}N_4$
Ni$_2$(C$_4$H$_6$)$_2$(bipy)$_2$, **9**, 118
$Ni_2C_{34}H_{70}P_4$
Ni$_2$(C$_4$H$_4$)($_2$H$_2$)(PPr$^i_2$C$_2$H$_4$P-Pr$^i_2$)$_2$, **9**, 113
$Ni_2C_{36}H_{54}$
Ni$_2$(C$_8$H$_{12}$)$_2$Cp*$_2$, **9**, 120
$Ni_2C_{36}H_{60}Br_2P_4$
Ni$_2$Br$_2$(CPh=CHCH=CPh)-(PEt$_2$C$_2$H$_4$PEt$_2$)$_2$, **9**, 118
$Ni_2C_{36}H_{60}F_{12}P_6$
[[(depe)Ni]$_2$[HCC(Ph)C-(Ph)CH]][PF$_6$]$_2$, **9**, 94
$Ni_2C_{36}H_{60}P_4$
Ni$_2$(CH=CPhCPh=CH)(PEt$_2$-C$_2$H$_4$PEt$_2$)$_2$, **9**, 118
$Ni_2C_{38}H_{38}F_{20}NO_2$

$[NBu_4][Ni(OH)(C_6F_5)_2]_2$, **9**, 33, 34

$Ni_2C_{38}H_{54}$
$Ni_2Cp^*{}_2(C_5Me_4C_5Me_4)$, **9**, 154, 156

$Ni_2C_{40}H_{40}Cl_2F_{20}N_2$
$[NEt_4]_2[Ni(Cl)(C_6F_5)_2]_2$, **9**, 33, 34

$Ni_2C_{40}H_{72}Br_2P_4$
$Ni_2Br_2(CPh{=}CHCH{=}CPh)$-$(PEt_3)_4$, **9**, 118

$Ni_2C_{42}H_{64}Br_2P_4$
$[(PPh_3)_2BrNi]_2(C_6H_4)$, **9**, 37

$Ni_2C_{44}H_{24}Cl_2P_2$
$[(PCy_3)NiCl]_2(CH_2C_6H_4CH_2)$, **9**, 37

$Ni_2C_{44}H_{30}Br_2O_2$
$[NiBr(CO)(C_3Ph_3)]_2$, **9**, 147

$Ni_2C_{44}H_{36}Cl_2P_2$
$Ni_2Cl_2(PhC_2H)(PPh_3)_2$, **9**, 133

$Ni_2C_{44}H_{74}Br_2P_2$
$[(PCY_3)NiBr]_2(CH_2C_6H_4CH_2)$, **9**, 37

$Ni_2C_{45}H_{43}NO_9$
$[(C_5H_4CO_2Me)_2Ni]_2[PhCCC$-$(O)NHCH(Me)Ph]$, **9**, 94

$Ni_2C_{46}H_{41}Br_2P_2$
$Ni_2Br_2(PPh_3)_2(C_3H_3PhMe)$, **9**, 143

$Ni_2C_{48}H_{40}Br_2P_2$
$[(PPh_3)NiBr]_2[CH_2(C_{10}H_6)$-$CH_2]$, **9**, 37

$Ni_2C_{51}H_{44}Cl_2OP_4$
$Ni_2(CO)Cl_2(dppm)_2$, **9**, 16

$Ni_2C_{53}H_{44}O_3P_4$
$Ni_2(CO)_3(dppm)_2$, **9**, 16

$Ni_2C_{54}H_{44}O_4P_4$
$Ni_2(CO)_4(dppm)_2$, **9**, 16

$Ni_2C_{54}H_{46}N_2P_4S_2$
$Ni_2(C{=}CH_2)(SCN)_2(dppm)_2$, **9**, 16

$Ni_2C_{54}H_{82}O_2$
$Ni_2(CO)_2(C_{10}H_5Bu^t{}_4)_2$, **9**, 168

$Ni_2C_{54}H_{84}O_2$
$Ni_2(CO)_2(C_{10}H_6Bu^t{}_4)_2$, **9**, 168

$Ni_2C_{56}H_{72}F_{22}N_2$
$[NBu_4]_2[Ni(F)(C_6F_5)_2]_2$, **9**, 33, 34

$Ni_2C_{56}H_{103}O_3P_3$
$(Cy_3P)(Me)Ni(O_2CO)NiH$-$(PCy_3)_2$, **9**, 57

$Ni_2C_{56}H_{103}O_3P_4$
$(Cy_3P)(Me)Ni(O_2CO)NiH$-$(PCy_3)_2$, **9**, 39

$Ni_2C_{57}H_{49}Cl_2NP_4$
$Ni_2(CNPh)Cl_2(dppm)_2$, **9**, 16, 17

$Ni_2C_{58}H_{56}F_{12}N_4P_{16}$
$[Ni_2(CNMe)_4(dppm)_2][PF_6]_2$, **9**, 17

$Ni_2C_{58}H_{100}P_4$
$Ni_2(C_6H_4)(dcpe)_2$, **9**, 134

$Ni_2C_{60}H_{60}N_2P_2$
$Ni_2(MAD\text{-}Pr^i)_2(PPh_3)_2$, **9**, 118

$Ni_2C_{62}H_{70}Cl_2P_2$
$Ni_2Cl_2(PPh_3)_2(C_{10}H_4Bu^t{}_4)$, **9**, 168

$Ni_2C_{62}H_{118}Cl_2P_4$
$Ni_2Cl_2(PhC_2Ph)(PBu_3)_4$, **9**, 133

$Ni_2C_{63}H_{45}Br_3$
$[C_3Ph_3][Ni_2Br_3(C_3Ph_3)_2]$, **9**, 147

$Ni_2C_{66}H_{56}N_2P_2$
$Ni_2(MAD\text{-}Ph)_2(PPh_3)_2$, **9**, 118

$Ni_2C_{68}H_{42}P_2$
$Ni_2(P_2)(C_5HPh_4)_2$, **9**, 136

$Ni_2C_{86}H_{72}N_2$
$Ni_2[N(C_6H_3Ph_2){=}C(Me)C$-$(Me){=}N(C_6H_3Ph_2)]_2$-$(C_6H_8)$, **9**, 144

$Ni_2CoC_{17}H_{15}O_2$
$Ni_2Co(CO)_2Cp_3$, **9**, 172, 173

$Ni_2CoC_{22}H_{25}O_2$
$[CoNi_2(CO)_2Cp_2Cp^*]^-$, **10**, 211
$Ni_2Co(CO)_2Cp_2Cp^*$, **9**, 172

$Ni_2Co_6C_{18}O_{16}$
$[Co_6Ni_2(CO)_{16}C_2]^{2-}$, **10**, 156

$Ni_2Cr_2C_{20}H_{20}S_3$
$Cr_2Ni_2(S)_3Cp_4$, **10**, 232

$Ni_2FeC_{17}H_{19}O_3P$
$Ni_2Fe(PBu^t)(CO)_3Cp_2$, **9**, 173

$Ni_2Fe_2C_{22}H_{20}O_6$
$Cp_2Ni_2Fe_2(CO)_6(C_2Et_2)$, **10**, 151

$Ni_2Fe_2C_{36}H_{38}O_6P_2$
$[CpNi(PMe_3)_2][CpNiFe_2(CO)_6$-$(C_2Ph_2)]$, **10**, 150

$Ni_2GeC_{22}H_{10}F_{10}$
$Ni_2Ge(C_6F_5)_2Cp_2$, **9**, 168

$Ni_2Ge_4HgC_{82}H_{70}$
$[(Ph_3GeHg)(Ph_3Ge)CpNi]_2Hg$, **10**, 334, 335

$Ni_2Hg_2C_{80}H_{88}I_4N_8$
$[Ni(CNXy)_4HgI_2]_2$, **9**, 11

$Ni_2MgC_{24}H_{36}O$
$[(C_2H_4)_2Ni]_2[Ph_2Mg(OEt_2)]$, **9**, 35

$Ni_2Mo_2C_{21}H_{20}OS_2$
$Mo_2Ni_2(S)_2(CO)Cp_4$, **10**, 234, 235

$Ni_2Mo_2C_{29}H_{25}O_2$
$Ni_2Mo_2(CPh)(CO)_2Cp_4$, **9**, 173

$Ni_2Mo_2C_{34}H_{40}O_4$
$Ni_2Mo_2(CO)_4Cp_2Cp^*{}_2$, **9**, 173

$Ni_2NaC_{20}H_{41}N_4$
$[Na(TMEDA)_2][Ni_2(C_2H_4)_4$-$(H)]$, **1**, 26, 28

$Ni_2Na_4C_{48}H_{68}O_5$
$[Na_4(THF)_5][Ph_2Ni(C_2H_4)]_2$, **1**, 26, 27

$Ni_2SiC_{80}H_{72}P_4$
$[Ni(PPh_3)_2]_2(CH_2{=}CH)_2Si$-$(CH{=}CH_2)_2$, **9**, 119

$Ni_2Si_2C_{14}H_{30}Br_2$
$Ni_2Br_2[C_3H_4[CH_2(TMS)]]_2$, **9**, 143

$Ni_2Si_2C_{14}H_{30}Cl_2O_2$
$Ni_2Cl_2[C_3H_3Me(O\text{-}TMS)]_2$, **9**, 143

$Ni_2Si_2C_{30}H_{46}Cl_2O_2P_4$
$[Ni(CO)[P[PCl(Bu^t)]{=}C(TMS)Ph]]_2$, **9**, 127

$Ni_2Si_2C_{44}H_{48}O_2P_2$
$[Ni(CO)[P(PPh_2){=}C(TMS)Ph]]_2$, **9**, 127

$Ni_2WC_{30}H_{32}O_2$
$Ni_2W(CC_6H_4Me)(CO)_2Cp_2Cp^*$, **9**, 23

$Ni_2Zn_4C_{30}H_{30}$
$Ni_2Zn_4Cp_6$, **9**, 173; **10**, 342

$Ni_2Zn_4C_{40}H_{50}$
$Ni_2Zn_4Cp_4Cp^*{}_2$, **9**, 173

$Ni_3B_2C_{30}H_{38}$
$Ni[Me_3(C_7B_2)(NiCp)]_2$, **1**, 206

$Ni_3B_{18}C_4H_{22}$
$Ni(C_2B_9H_{11}Ni)_2$, **9**, 181

$Ni_3C_{16}H_{16}$
$(NiCp)_3CH$, **9**, 72

$Ni_3C_{17}H_8F_{18}$
$Ni_3(CF_3C_2CF_3)_3(C_8H_8)$, **9**, 22

$Ni_3C_{17}H_{15}O_2$
$[Ni_3(CO)_2Cp_3]^-$, **9**, 173

$Ni_3C_{18}H_{18}O_2$
$(CpNi)_3(CCO_2Me)$, **9**, 94
$Ni_3(CCO_2Me)Cp_3$, **9**, 171

$Ni_3C_{20}H_{21}O_2$
$Ni_3(CO)_2Cp'_3$, **9**, 173

$Ni_3C_{22}H_{20}$
$(CpNi)_3(CPh)$, **9**, 94

$Ni_3C_{27}H_{44}$
$Ni_3(H)_2(CpBu^t)_3$, **9**, 171

$Ni_3C_{30}H_{45}P_5$
$Ni_3(P)(P_4)Cp^*{}_3$, **9**, 168

$Ni_3C_{32}H_{45}O_2$
$Ni_3(CO)_2Cp^*{}_3$, **9**, 173

$Ni_3C_{38}H_{56}O_2P_2$
$Ni_3(CH_2C_6H_4Me)_4(PMe_3)_2$-$(OH)_2$, **9**, 39

$Ni_3C_{39}H_{65}$
$Ni_3(H)_2(C_5H_3Bu^t{}_2)_3$, **9**, 172

$Ni_3C_{45}H_{75}P_3$
$Ni_3(C_6H_4)_3(PPr^i{}_3)_3$, **9**, 134

$Ni_3C_{75}H_{66}I_2$
$Ni_3(dppm)_3I_2$, **9**, 21

$Ni_3C_{76}H_{66}I_2O$
$[Ni_3(dppm)_3(CO)I]I$, **9**, 21

$Ni_3Fe_3C_{14}O_{13}$
$[Ni_3Fe_3(CO)_{13}C]^{2-}$, **7**, 22; **9**, 20; **10**, 146

$Ni_3MoC_{23}H_{20}O_3$
$Ni_3Mo(CO)_3Cp_4$, **9**, 173

$Ni_3Os_3C_{24}H_{15}O_9$
$Os_3Ni_3Cp_3(Co)_9$, **10**, 38

$Ni_3Pt_3C_{12}O_{12}$
$[Ni_3Pt_3(CO)_{12}]^{2-}$, **9**, 399

$Ni_3Ru_3C_{14}O_{13}$
$[Ni_3Ru_3(CO)_{13}C]^{2-}$, **10**, 146

$Ni_4C_{16}F_{18}O_4$
$Ni_4(CF_3C_2CF_3)_3(CO)_4$, **9**, 22

$Ni_4C_{20}H_{16}$
$[Ni_2(C_5H_4)(C_5H_4)]_2$, **9**, 151

$Ni_4C_{35}H_{63}N_7$
$Ni_4(CNBu^t)_7$, **9**, 22

$Ni_4C_{36}H_{57}$
$Ni_4H(CpBu^t)_4$, **9**, 172

$Ni_4C_{80}H_{144}N_4P_4$

$Ni_4(MeCN)_4(PCy_3)_4$, **9**, 135
$Ni_4HgC_{70}H_{74}N_{10}P_4$
$[Ni_2(CNMe)_5(dppm)]_2(Hg)^{2+}$, **10**, 334
$Ni_4Zn_8C_{180}H_{190}$
$[Zn_4Ni_2Cp_4Cp^*_5]_2$, **10**, 342
$Ni_5C_{12}O_{12}$
$[Ni_5(CO)_{12}]^{2-}$, **9**, 19
$Ni_5HgC_{70}H_{74}Cl_4N_5P_4$
$[Ni_2(CNMe)_5(Ph_2PCH_2PPh_2)]_2$-(Hg)$NiCl_4$, **10**, 342
$Ni_5HgC_{70}H_{74}Cl_4N_{10}P_4$
$[Ni_4Hg(CNMe)_{10}(dppm)_2]$-$[NiCl_4]$, **9**, 22
$Ni_5Si_4C_{20}H_{38}ClO_6P_2$
$[Ni_5Cl(CO)_6[(TMS)_2CHP=PCH(TMS)_2]]^{2-}$, **9**, 20
$Ni_6C_{12}O_{12}$
$[Ni_6(CO)_{12}]^{2-}$, **9**, 19
$Ni_6C_{18}O_{16}$
$[Ni_6(CO)_{16}C_2]^{2-}$, **9**, 20
$Ni_6C_{62}H_{90}S_6$
$(Ni_3Sp^*_3)_2(C_2S_6)$, **9**, 154
$Ni_6Co_6C_{18}O_{16}$
$[Co_6Ni_2C_2(CO)_{16}]^{2-}$, **9**, 20
$Ni_7Co_3C_{17}O_{15}$
$[Co_3Ni_7C_2(CO)_{15}]^{3-}$, **9**, 20
$Ni_7Co_3C_{18}O_{16}$
$[Co_3Ni_7C_2(CO)_{16}]^{2-}$, **9**, 20; **10**, 156
$Ni_8C_{17}O_{16}$
$[Ni_8(CO)_{16}C]^{2-}$, **9**, 20
$Ni_8C_{108}H_{90}Cl_4P_{10}$
$Ni_8Cl_4(PPh_3)_4(PPh)_6$, **9**, 20
$Ni_8C_{112}H_{90}O_4P_{10}$
$Ni_8(CO)_4(PPh_3)_4(PPh)_6$, **9**, 20
$Ni_9As_3C_{33}H_{15}O_{15}$
$[Ni_9(AsPh)_3(CO)_{15}]^{2-}$, **9**, 19
$Ni_9C_{18}O_{17}$
$[Ni_9(CO)_{17}C]^{2-}$, **9**, 20
$Ni_9C_{18}O_{18}$
$[Ni_9(CO)_{18}]^{2-}$, **9**, 19
$Ni_9Co_3C_{21}O_{20}$
$[Ni_9Co_3C(CO)_{20}]^{2-}$, **10**, 156
$[Ni_9Co_3C(CO)_{20}]^{3-}$, **9**, 20; **10**, 15, 156
$Ni_9Pt_3C_{21}H_2O_{21}$
$[Ni_9Pt_3(H)_2(CO)_{21}]^{2-}$, **10**, 200, 201
$Ni_9Pt_3C_{21}H_4O_{21}$
$Ni_9Pt_3(H_4)(CO)_{21}$, **10**, 213
$Ni_{10}As_2C_{30}H_{10}O_{18}$
$[Ni_{10}(AsPh)_2(CO)_{18}]^{2-}$, **9**, 20
$Ni_{10}C_{18}O_{16}$
$[Ni_{10}(CO)_{16}C_2]^{2-}$, **9**, 20
$Ni_{10}C_{19}O_{18}$
$[Ni_{10}(CO)_{18}C]^{2-}$, **9**, 20
$Ni_{10}Co_2C_{21}O_{20}$
$[Ni_{10}Co_2C(CO)_{20}]^{2-}$, **10**, 156
$Ni_{12}C_{21}H_2O_{21}$
$[H_2Ni_{12}(CO)_{21}]^{2-}$, **9**, 19
$[H_2Ni_{12}(CO)_{21}]^{3-}$, **9**, 19
$Ni_{12}C_{21}O_{21}$
$[Ni_{12}(CO)_{21}]^{4-}$, **9**, 19
$Ni_{16}C_{27}O_{23}$
$[Ni_{16}(CO)_{23}(C_2)_2]^{2-}$, **9**, 20
$Ni_{34}C_{42}HO_{38}$
$[HNi_{34}(CO)_{38}C_4]^{2-}$, **9**, 20
$Ni_{34}C_{42}O_{38}$
$[Ni_{34}(CO)_{38}C_4]^{3-}$, **9**, 20
$Ni_{35}C_{43}O_{39}$
$[Ni_{35}(CO)_{39}C_4]^{6-}$, **9**, 20
$Ni_{38}C_{48}HO_{42}$
$[HNi_{38}(CO)_{42}C_6]^{5-}$, **9**, 20
$Ni_{38}Pt_6C_{48}HO_{48}$
$[Ni_{38}Pt_6(CO)_{48}H]^{5-}$, **9**, 21
$Ni_{38}Pt_6C_{48}H_2O_{48}$
$[Ni_{38}Pt_6(CO)_{48}H_2]^{2-}$, **9**, 21
$Ni_{38}Pt_6C_{48}H_3O_{48}$
$[Ni_{38}Pt_6(H)_3(CO)_{48}]^{3-}$, **10**, 200, 201
$NpC_{15}H_{15}Cl$
Cp_3NpCl, **4**, 151
$NpC_{16}H_{16}$
$Np(C_8H_8)_2$, **4**, 181
$NpC_{20}H_{30}Cl_2$
$Cp^*_2NpCl_2$, **4**, 165
$NpC_{24}H_{28}$
$Np(MeC_5H_4)_4$, **4**, 160
$NpC_{27}H_{39}O_3$
$Cp_3Np(THF)_3$, **4**, 140
$NpSi_6C_{21}H_{57}$
$Np[CH(TMS)_2]_3$, **4**, 132
$Np_3WC_5H_9$
$WNp_3(\equiv CBu^t)$, **5**, 315

O

$OsAg_3Rh_3C_{178}H_{171}F_3P_9S$
$Ag_3Rh_3(H)_9[MeC(CH_2PPh_3)_3]_3$-$[OsSCF_3]$, **10**, 256
$OsAuB_2C_{22}H_{20}O_4P$
$(PPh_3)Au(CO)_4Os(B_2H_5)$, **1**, 308
$OsAuC_{29}H_{33}NP$
$[Os(CNMe)(PMe_3)(C_6H_6)Au$-$(PPh_3)]^+$, **10**, 54
$OsAuC_{37}H_{32}ClINOP_2$
$OsAu(CH_2)ClI(NO)(PPh_3)_2$, **7**, 338
$OsAuRu_3C_{35}H_{34}F_3O_9P_4S$
$[AuRu_3[(Me_2P)_3CH](CO)_9$-$(PPh_3)][OsSCF_3]$, **10**, 299
$OsBC_7H_5F_4O_4$
$[Os(CO)_4(C_3H_5)]BF_4$, **7**, 459
$OsBC_{12}H_{13}F_4$
$[Os(C_6H_6)(C_6H_7)]BF_4$, **7**, 552
$OsB_2C_4H_6O_4$
$H(CO)_4Os(B_2H_5)$, **1**, 308, 309
$OsB_2C_7H_{14}O_5$
$Me_3O(CO)_4Os(B_2H_5)$, **1**, 308
$OsB_3C_{37}H_{39}OP_2$
$(CO)H(PPh_3)_2OsB_3H_8$, **1**, 313, 315
$OsB_4C_{37}H_{38}OP_2$
$(OsB_4H_8)(CO)(PPh_3)_2$, **1**, 320
$OsB_5C_{37}H_{39}OP_2$
$(Ph_3P)_2(CO)OsB_5H_9$, **1**, 323
$OsB_5PtC_{45}H_{49}ClOP_3$
$(CO)(PPh_3)_2HOs(PMe_2Ph)$-$ClPtB_5H_7$, **1**, 324
$OsB_9C_{12}H_{31}$
$(C_6Me_6)[OsB_9H_{12}]$, **1**, 341, 342
$OsB_{10}C_{40}H_{46}OP_2$
$[(OEt)_2(PPh_3)(Ph_2PC_6H_4)$-$OsB_{10}H_7]$, **1**, 355
$OsC_3F_5O_3$
$Os(CO)_3(F)_5$, **7**, 869
$OsC_3H_6Cl_4O_2$
$[Os(CO)Cl_4(EtOH)]^-$, **7**, 319
OsC_4HO_4
$OsH(CO)_4$, **7**, 363
$OsC_4H_2O_4$
$OsH_2(CO)_4$, **7**, 300, 362
$OsC_4I_2O_4$
$OsI_2(CO)_4$, **7**, 307, 308
$OsC_5Cl_2O_5$
$[OsCl(CO)_5]Cl$, **7**, 307
$OsC_5H_4O_4$
$OsHMe(CO)_4$, **7**, 378, 380
OsC_5O_5
$Os(CO)_5$, **7**, 301
$OsC_6H_4O_4$
$Os(CO)_4(C_2H_4)$, **7**, 442
$OsC_6H_6O_4$
$OsMe_2(CO)_4$, **7**, 378, 380
$OsC_6H_{20}F_9N_5O_9S_3$
$[Os(NH_3)_5(C_3H_5)](OTf)_3$, **7**, 459
$OsC_6H_{25}N_5$
$[Os(NH_3)_5(C_6H_{10})]^{3+}$, **7**, 445
$OsC_7F_6O_7$
$Os(CF_3CO_2)(CO)_3$, **7**, 643
$OsC_7H_5IO_2$
$OsI(CO)_2Cp$, **7**, 479
$OsC_7H_6O_2$
$OsH(CO)_2Cp$, **7**, 307, 479
$OsC_8H_8O_2$
$OsMe(CO)_2Cp$, **7**, 480
$OsC_8H_{12}Br_2N_4$
$OsBr_2(CNMe)_4$, **7**, 328
$OsC_8H_{12}O_2$
$Os(CO)_2(C_2H_4)_3$, **7**, 442
$OsC_8H_{20}N_4$
$[Os(en)_2(C_2H_2)_2]^{2+}$, **7**, 447
$OsC_8H_{22}N_4$
$[Os(en)_2(C_2H_4)(C_2H_2)]^{2+}$, **7**, 447
$OsC_8H_{24}ClN_6$
$OsCl(N_2H_4)_3(cod)$, **7**, 451
$OsC_8H_{25}F_6N_5O_6S_2$
$[Os(NH_3)_5(C_6H_{10})](OTf)_2$, **7**, 445
$OsC_8H_{28}N_8$
$[Os(N_2H_4)_4(cod)]^{2+}$, **7**, 451
$OsC_8I_2O_8$
$OsI_2(CO)_8$, **7**, 301
$OsC_9H_9NO_4$
$Os(CO)_4(CNBu^t)$, **7**, 331
$OsC_9H_{10}O$
$Os(CO)(C_2H_4)(C_6H_6)$, **7**, 569
$OsC_9H_{11}O$
$[OsH(C_2H_4)(CO)(C_6H_6)]^+$, **7**, 339
$OsC_9H_{13}HO_4$
$[Os(O)_2OCHMeCHMeO]py$, **12**, 1138
$OsC_9H_{27}P_3$
$OsH(CH_2PMe_2)(PMe_3)_2$, **7**, 384
$OsC_{10}H_4O_9$
$H_3Os(CH)(CO)_9$, **1**, 294
$OsC_{10}H_{10}$
Cp_2Os, **7**, 604
$OsC_{10}H_{10}ClF_6P$
$[OsClCp_2][PF_6]$, **7**, 609
$OsC_{10}H_{11}Cl_2O$
$OsCl_2(CO)(Mes)$, **1**, 84
$OsC_{10}H_{14}F_9N_3O_{12}S_3$
$[Os(CPh)(NH_3)_3](OTf)_3$, **7**, 358
$OsC_{10}H_{14}O$
$OsH_2(CO)(C_6H_3Me_3)$, **7**, 568, 569
$OsC_{10}H_{16}ClN_2O_5$
$OsCl(O_2CMe)_2(CO)(py)_2$, **7**, 319
$OsC_{11}H_{12}O_2$
$Os(CO)_2(C_6H_3Me_3)$, **7**, 568, 569
$OsC_{11}H_{14}F_6N_3P$
$[Os(NCMe)_3Cp]PF_6$, **7**, 495
$OsC_{11}H_{16}O$
$OsH(Me)(CO)(C_6H_3Me_3)$, **7**, 387
$OsC_{11}H_{18}O$
$OsH_3(CO)Cp^*$, **7**, 480
$OsC_{11}H_{25}NO_2$
$Os(N)Me_3(THF)_2$, **7**, 374
$OsC_{11}H_{26}OP_2$
$Os(CO)(PMe_3)_2(C_2H_4)_2$, **7**, 442
$OsC_{12}H_{12}$
$Os(C_6H_6)_2$, **7**, 552
$OsC_{12}H_{14}$
$Os(C_6H_6)(C_6H_8)$, **7**, 552
$OsC_{12}H_{15}IO_2$
$OsI(CO)_2Cp^*$, **7**, 479
$OsC_{12}H_{16}O_2$
$OsH(CO)_2Cp^*$, **7**, 479
$OsC_{12}H_{17}ClO$
$OsCl(Me)(CO)(cym)$, **7**, 568
$OsC_{12}H_{18}N_6$
$[Os(CNMe)_6]^{2+}$, **7**, 328
$OsC_{12}H_{20}Cl_2OS$
$OSCl_2(cym)(Me_2SO)$, **7**, 564
$OsC_{12}H_{23}P$
$OsH_2(PMe_3)(C_6H_3Me_3)$, **7**, 573
$OsC_{12}H_{24}F_6IP_3$
$[OsI(PMe_3)_2(C_6H_6)]PF_6$, **7**, 562
$OsC_{12}H_{25}P_2$
$[Os(PMe_3)_2(C_6H_7)]^-$, **7**, 572
$OsC_{12}H_{31}OP_3$
$Os(CO)(C_2H_4)(PMe_3)_3$, **7**, 306
$OsC_{12}H_{36}P_4$
$OsH(CH_2PMe_2)(PMe_3)_3$, **7**, 385
$Os(PMe_3)_4$, **7**, 385
$OsC_{12}H_{38}P_4$
$OsH_2(PMe_3)_4$, **7**, 391
$OsC_{13}H_{10}O_3$
$Os(CO)_3(C_{10}H_{10})$, **7**, 479
$OsC_{13}H_{18}N_2O_3$
$Os(CO)_3(CNBu^t)_2$, **7**, 331
$OsC_{13}H_{18}O$
$Os(CO)(C_6Me_6)$, **7**, 569
$OsC_{13}H_{18}O_3$
$Os(CH_2OH)(CO)_2Cp^*$, **7**, 483
$OsC_{13}H_{20}O$
$OsH_2(CO)(C_6Me_6)$, **7**, 569
$OsC_{13}H_{25}P$
$OsH(Me)(PMe_3)(C_6H_3Me_3)$, **7**, 573
$OsC_{13}H_{40}P_4$
$OsHMe(PMe_3)_4$, **7**, 385

$OsC_{14}H_{22}$
$Os(C_7H_{11})_2$, **7**, 465
$OsC_{14}H_{28}F_6N_4O_6S_2$
$[Os(C_4Me_4)(en)_2](O_3SCF_3)_2$, **7**, 342
$OsC_{14}H_{36}ClN_6$
$[OsCl(NH_2NMe_2)_3(cod)]^+$, **7**, 451
$OsC_{14}H_{42}P_4$
$OsMe_2(PMe_3)_4$, **7**, 382, 384
$OsC_{15}H_{21}P$
$OsH(Ph)(PMe_3)(C_6H_6)$, **7**, 395
$OsC_{15}H_{28}IP$
$OsH(I)(PMeBu^t_2)(C_6H_6)$, **7**, 586
$OsC_{15}H_{44}ClP_5$
$OsCl(CH_2PMe_2)(PMe_3)_4$, **7**, 383, 384
$OsC_{16}H_{16}ClO$
$OsPhCl(CO)(Mes)$, **1**, 84
$OsC_{16}H_{17}BrO$
$OsBr(Ph)(CO)(C_6H_3Me_3)$, **7**, 399
$OsC_{16}H_{21}O$
$Os(CO)(C_6H_{10})(Mes)$, **1**, 84
$OsC_{16}H_{22}O$
$Os(C_6H_{10})(CO)(C_6H_3Me_3)$, **7**, 568
$OsC_{16}H_{27}F_6IOP_2$
$[OsI(CO)(PMeBu^t_2)(C_6H_6)]PF_6$, **7**, 580, 590
$OsC_{16}H_{29}P$
$Os(Me)(CH_2CHMePPr^i_2)$-(C_6H_6), **7**, 428
$OsC_{16}H_{30}P$
$[OsH(Me)(CH_2CHMePPr^i_2)$-$(C_6H_6)]^+$, **7**, 428
$OsC_{16}H_{31}P$
$OsH(Me)(PPr^i_3)(C_6H_6)$, **7**, 573
$OsC_{17}H_{29}F_6IP_2$
$[OsI(CCH_2)(PMeBu^t_2)(C_6H_6)]$-$PF_6$, **7**, 580, 590
$OsC_{17}H_{48}P_4$
$OsH(CH_2Bu^t)(PMe_3)_4$, **7**, 384, 395
$OsC_{18}H_{12}F_{10}O_3$
$Os(O)(C_6F_5)_2(OCMe_2CMe_2O)$, **7**, 394, 395
$OsC_{18}H_{22}O_2$
$OsO_2(Mes)_2$, **7**, 291, 393, 394
$OsC_{18}H_{32}O_2$
$Os(OBu^t)_2(cym)$, **7**, 559
$OsC_{18}H_{40}Cl_2O_2P_2$
$Os(CHOCH_2CH_2PPr^i_2)Cl_2$-$(Pr_2SiPCH_2CH_2OMe)$, **7**, 345
$OsC_{19}H_{33}ClN_3O_4$
$OsCl(O_2CMe)_2(CNBu^t)_3$, **7**, 329
$OsC_{19}H_{42}P_4$
$Os(CH_2C_6H_4)(PMe_3)_4$, **7**, 382, 384, 425
$OsC_{19}H_{45}OP_2$
$[OsH(H_2)(CO)(PPr^i_3)_2]^+$, **7**, 366
$OsC_{19}H_{46}OP_2$
$OsH_4(CO)(PPr^i_3)_2$, **7**, 366, 371
$OsC_{19}H_{47}O_2P_2$
$[OsH(H_2)(CO)(OH_2)(PPr^i_3)_2]^+$, **7**, 366
$OsC_{20}H_{30}$
Cp^*_2Os, **7**, 604
$OsC_{20}H_{36}Br_2N_4$
$OsBr_2(CNBu^t)_4$, **7**, 328
$OsC_{20}H_{42}$
$Os(CBu^t)(CH_2Bu^t)_3$, **7**, 341
$Os(CHBu^t)_2(CH_2Bu^t)_2$, **7**, 340
$OsC_{20}H_{42}Cl_2O_2P_2$
$OsCl_2(CO)_2(PPr^i_3)_2$, **7**, 367
$OsC_{20}H_{46}Cl_2N_2P_2$
$Os(CHNMeCH_2CH_2PPr^i_2)Cl_2$-$(Pr^i_2PCH_2CH_2NMe_2)$, **7**, 346
$OsC_{21}H_{38}OP_2$
$OsH_2(CH_2CHPh)(CO)(PPr^i_2)_2$, **7**, 412
$OsC_{22}HIO_{22}$
$Os_8(H)I(CO)_{22}$, **7**, 948
$OsC_{22}H_{10}F_{10}N_2O_2$
$Os(O)_2(C_6F_5)_2(py)_2$, **7**, 395
$OsC_{22}H_{16}F_3N_4O$
$[Os(CF_3)(CO)(bipy)_2]^+$, **7**, 389
$OsC_{22}H_{19}O_3P$
$OsHMe(CO)_3(PPh_3)$, **7**, 378, 380
$OsC_{22}H_{22}O$
$OsPh_2(CO)(C_6H_3Me_3)$, **7**, 399
$OsC_{22}H_{31}NO$
$Os(O)(NBu^t)(C_6H_2Me_3)_2$, **7**, 393
$OsC_{22}H_{34}INP$
$Os[C(py)CH_2]I(PMeBu^t_2)$-$(C_6H_6)]^+$, **7**, 407
$OsC_{23}H_{21}O_3P$
$OsMe_2(CO)_3(PPh_3)$, **7**, 378
$OsC_{24}H_{42}N_4O_4$
$Os(O_2CMe)_2(CNBu^t)_4$, **7**, 329
$OsC_{24}H_{44}$
$OsCy_4$, **7**, 377
$OsC_{24}H_{46}N_6$
$Os(NH_2NCMe_2)_2(CNBu^t)_2$-$(cod)$, **7**, 451
$OsC_{24}H_{49}O_3P_2$
$Os[CH_2CH_2C(O)Me](CO)_2$-$(PPr^i_3)_2]^+$, **7**, 392
$OsC_{25}H_{22}BrP$
$OsBr(CH_2CHC_6H_4PPh_2)Cp$, **7**, 505
$OsC_{25}H_{30}O$
$Os(C_6H_{11})_2(CO)(C_6Me_6)$, **7**, 568
$OsC_{26}H_{41}F_3N_2O_4S$
$Os(CBu^t)(CH_2Bu^t)_2(py)_2(OTf)$, **7**, 359
$OsC_{26}H_{48}ClNOP_2$
$OsH(CCPh)Cl(NO)(PPr^i_3)_2$, **7**, 412
$OsC_{27}H_{27}P$
$OsH(C_6H_4PPh_2)(C_6H_3Me_3)$, **7**, 573, 574
$OsC_{27}H_{28}ClP$
$OsH(Cl)(PPh_3)(C_6H_3Me_3)$, **7**, 574
$OsC_{27}H_{31}NO_2$
$Os(C_6H_2Me_3)_2(O)_2(CNC_6H_3$-$Me_2)$, **7**, 393, 394
$OsC_{27}H_{33}O_2P$
$Os(CO)_2(PMe_2Ph)(Cp^*Bz)$, **7**, 482
$OsC_{27}H_{52}OP_2$
$OsH_2(H_2CCHPh)(CO)(PPr^i_3)_2$, **7**, 371
$OsC_{28}H_{28}$
$Os(C_6H_4Me)_4$, **7**, 393, 394, 571
$OsC_{28}H_{31}P$
$OsH(Me)(PPh_3)(C_6H_3Me_3)$, **7**, 574
$OsC_{28}H_{48}O_2P_2$
$OsH(CCPh)(CO)_2(PPr^i_3)_2$, **7**, 413
$OsC_{28}H_{65}OP_3$
$OsH_2(CO)(PPr^i_3)_3$, **7**, 366
$OsC_{30}H_{41}N_3O$
$Os(C_6H_4Me)_2(CO)(CNBu^t)_3$, **7**, 393, 394
$OsC_{31}H_{43}F_6NP$
$[Os(NCPh_2)(PPr^i_3)(C_6H_3Me_3)]$-$PF_6$, **7**, 586
$OsC_{32}H_{30}N_6$
$[OsCp^*_2][C_6(CN)_6]$, **7**, 608
$OsC_{32}H_{32}NOPS_3$
$Os(SCC_6H_4Me)(S_2CNEt_2)$-$(CO)(PPh_3)$, **7**, 422
$OsC_{37}H_{30}ClF_2NOP_2$
$Os(CF_2)Cl(NO)(PPh_3)_2$, **7**, 389
$OsC_{37}H_{30}ClNOP_2S_2$
$OsCl(NO)(CS_2)(PPh_3)_2$, **7**, 320, 323
$OsC_{37}H_{30}INOP_2S$
$OsI(NO)(CS)(PPh_3)_2$, **7**, 324
$OsC_{37}H_{30}I_5NOP_2S$
$[OsI_2(NO)(CS)(PPh_3)_2]I_3$, **7**, 324, 325
$OsC_{37}H_{31}Cl_4OP_3$
$OsHCl(CO)(PPh_3)_2(PCl_3)$, **7**, 311
$OsC_{37}H_{32}ClNO_2P_2S$
$Os(CH_2SO)Cl(NO)(PPh_3)_2$, **7**, 338
$OsC_{37}H_{32}ClNO_3P_2S$
$Os[CH_2S(O)O]Cl(NO)(PPh_3)_2$, **7**, 338
$OsC_{37}H_{32}Cl_3NOP_2$
$Os(CH_2Cl)Cl_2(NO)(PPh_3)_2$, **7**, 387
$Os(CH_2PPh_3)Cl_3(NO)(PPh_3)$, **7**, 338, 387
$OsC_{37}H_{33}Cl_2NOP_2$
$Os(Me)Cl_2(NO)(PPh_3)_2$, **7**, 337
$OsC_{37}H_{33}Cl_2NO_2P_2$
$Os(CH_2OH)Cl_2(NO)(PPh_3)_2$, **7**, 387, 388
$OsC_{37}H_{33}OP_3S$
$[OsH(CS)(OH_2)(PPh_3)_3]^+$, **7**, 325
$OsC_{37}H_{57}O_3P_2$

[Os[C(Bz)OCMeO](CCPh)-(CO)(PPr$^i_3$)$_2$]$^+$, **7**, 416, 417
$OsC_{37}H_{67}ClOP_2$
OsClH(CO)(PCy$_3$)$_2$, **7**, 366, 368
$OsC_{38}H_{30}ClNO_6P_2S$
[Os(NO)(CO)(CS)(PPh$_3$)$_2$]-ClO$_4$, **7**, 324
$OsC_{38}H_{30}Cl_2O_2P_2$
OsCl$_2$(CO)$_2$(PPh$_3$)$_2$, **7**, 308, 309
$OsC_{38}H_{30}Cl_4P_2S$
OsCl$_2$(CCl$_2$)(CS)(PPh$_3$)$_2$, **7**, 321
$OsC_{38}H_{30}O_2P_2$
OsH(C$_6$H$_4$PPh$_2$)(CO)$_2$(PPh$_3$), **7**, 417, 428
$OsC_{38}H_{32}O_2P_2$
OsH$_2$(CO)$_2$(PPh$_3$)$_2$, **7**, 304, 366
$OsC_{39}H_{30}Cl_2O_2P_2$
Os(CHClC$_6$H$_4$PPh$_2$)Cl(CO)$_2$-(PPh$_3$), **7**, 429
$OsC_{39}H_{30}O_2P_2Te$
Os(CTe)(CO)$_2$(PPh$_3$)$_2$, **7**, 358
$OsC_{39}H_{32}O_2P_2Se$
Os(CH$_2$Se)(CO)$_2$(PPh$_3$)$_2$, **7**, 388
$OsC_{39}H_{34}O_3P_2$
Os(CH$_2$OH)H(CO)$_2$(PPh$_3$)$_2$, **7**, 387
$OsC_{40}H_{32}O_3P_2Se_2$
Os[CH$_2$OC(Se)Se](CO)$_2$-(PPh$_3$)$_2$, **7**, 388
$OsC_{40}H_{33}O_2P_2Te$
[Os(CTeMe)(CO)$_2$(PPh$_3$)$_2$]$^+$, **7**, 325
$OsC_{40}H_{34}OP_2S$
Os(CO)(CS)(PPh$_3$)$_2$(C$_2$H$_4$), **7**, 443
$OsC_{40}H_{34}O_2P_2$
Os(CO)$_2$(PPh$_3$)$_2$(C$_2$H$_4$), **7**, 443
$OsC_{40}H_{35}ClINO_3P_2$
Os[CH(CO$_2$Et)C$_6$H$_4$PPh$_2$]ClI-(NO)(PPh$_3$), **7**, 361
$OsC_{40}H_{35}ClNO_3P_2$
[Os(CCO$_2$Et)Cl(NO)(PPh$_3$)$_2$]$^+$, **7**, 361
$OsC_{40}H_{35}ClO_3P_2$
Os(CH$_2$OMe)Cl(CO)$_2$(PPh$_3$)$_2$, **7**, 387, 388
$OsC_{40}H_{36}Cl_2O_2P_2$
Os[C(OH)Et]Cl$_2$(CO)(PPh$_3$)$_2$, **7**, 343
$OsC_{40}H_{37}Cl_2NO_3P_2$
OsCl$_2$(CH$_2$CO$_2$Et)(NO)(PPh$_3$)$_2$, **7**, 389, 390
$OsC_{40}H_{37}Cl_2N_3O_3P_2$
OsCl$_2$(N$_2$)[N(OH)CHCO$_2$Et]-(PPh$_3$)$_2$, **7**, 337, 376
$OsC_{40}H_{42}P_2$
OsMe$_4$(PPh$_3$)$_2$, **7**, 377, 379
$OsC_{41}H_{35}ClO_3P_2$
Os[C(O)Et]Cl(CO)$_2$(PPh$_3$)$_2$, **7**, 343
$OsC_{42}H_{32}O_5P_2$
Os(CO)$_2$(PPh$_3$)$_2$(C$_4$H$_2$O$_3$), **7**, 444
$OsC_{42}H_{34}NOP_2S$
Os(SC$_5$H$_4$N)(CO)(PPh$_3$)$_2$, **7**, 370
$OsC_{42}H_{34}OP_2S$
Os(SCCHCHCHCH)(CO)-(PPh$_3$)$_2$, **7**, 326
$OsC_{42}H_{36}NOPS$
[Os(H$_2$)(SC$_5$H$_4$N)(CO)-(PPh$_3$)$_2$]$^+$, **7**, 370
$OsC_{43}H_{36}F_6O_5P_2S$
Os(O$_2$CCF$_3$)$_2$(EtOH)(CS)-(PPh$_3$)$_2$, **7**, 326
$OsC_{43}H_{39}OP_3$
OsH$_2$(CO)(PPh$_3$)$_2$(PH$_2$Ph), **7**, 365
$OsC_{44}H_{35}ClO_2P_2$
OsCl(Ph)(CO)$_2$(PPh$_3$)$_2$, **7**, 357
$OsC_{44}H_{35}ClO_3P_2$
Os(CPh)(OOCO)Cl(PPh$_3$)$_2$, **7**, 357
$OsC_{44}H_{36}ClNOP_2$
OsCl(NO)(PPh$_3$)$_2$(HCCPh), **7**, 412
$OsC_{44}H_{38}ClO_6P_3$
[OsH(CO)$_2$(PPh$_3$)$_2$(PH$_2$Ph)]-ClO$_4$, **7**, 366
$OsC_{44}H_{39}ClN_2O_3P_2S_2$
Os[CH$_2$S(NSO$_2$C$_6$H$_4$Me)O]Cl-(NO)(PPh$_3$)$_2$, **7**, 338
$OsC_{45}H_{37}ClO_2P_2$
Os(C$_6$H$_4$Me)Cl(CO)$_2$(PPh$_3$)$_2$, **7**, 326, 357
$OsC_{45}H_{38}O_2P_2$
OsH(C$_6$H$_4$Me)(CO)$_2$(PPh$_3$)$_2$, **7**, 397
$OsC_{45}H_{40}ClO_2P_2$
[Os(CHC$_6$H$_4$Me)Cl(OH$_2$)(CO)-(PPh$_3$)$_2$]$^+$, **7**, 360
$OsC_{45}H_{40}OP$
[Os(CCHBu^t)(CO)(PPh$_3$)Cp*]$^+$, **7**, 352
$OsC_{45}H_{41}ClNO_5P_3$
[OsH(CO)(NCMe)(PPh$_3$)$_2$-(PH$_2$Ph)]ClO$_4$, **7**, 365
$OsC_{45}H_{42}Cl_2NOP_2$
[Os(CC$_6$H$_4$NMe$_2$)Cl$_2$(OH$_2$)-(PPh$_3$)$_2$]$^+$, **7**, 357
$OsC_{45}H_{45}NO_2P_2S_2$
Os[C(O)Et](S$_2$CNEt$_2$)(CO)-(PPh$_3$)$_2$, **7**, 418
$OsC_{45}H_{45}N_5$
Os(CNC$_6$H$_3$Me$_2$)$_5$, **7**, 327
$OsC_{46}H_{37}O_2P_2$
[Os(CC$_6$H$_4$Me)(CO)$_2$(PPh$_3$)$_2$]$^+$, **7**, 360
$OsC_{46}H_{38}O_2P_2$
Os(CCCpMe)(CO)$_2$(PPh$_3$)$_2$, **7**, 361
$OsC_{46}H_{41}ClN_2O_4PS_2$
Os[C(CH$_2$)CH$_2$S(NSO$_2$C$_6$H$_4$-Me)O]Cl(NO)(PPh$_3$)$_2$, **7**, 407
$OsC_{46}H_{46}ClOP_3$
OsHCl(CO)(PPh$_3$)$_3$, **7**, 459
$OsC_{47}H_{37}F_3O_3P_2S$
Os(SCC$_6$H$_4$Me)(O$_2$CCF$_3$)(CO)-(PPh$_3$)$_2$, **7**, 422
$OsC_{50}H_{41}O_2P_3$
Os(CO)$_2$(PPh$_3$)$_2$(PHPh$_2$), **7**, 366
$OsC_{51}H_{40}P_2S$
Os(CS)(PPh$_3$)$_2$(PhC$_2$Ph), **7**, 444
$OsC_{51}H_{41}ClP_2S$
Os(CPhCHPh)Cl(CS)(PPh$_3$)$_2$, **7**, 325, 400
$OsC_{52}H_{40}O_2P_2$
Os(CO)$_2$(PPh$_3$)$_2$(PhC$_2$Ph), **7**, 444
$OsC_{52}H_{41}ClOP_2S$
Os(SCCPhCHPh)Cl(CO)-(PPh$_3$)$_2$, **7**, 326
$OsC_{53}H_{40}O_2P_2S$
Os[C(S)CPhCPh](CO)$_2$(PPh$_3$)$_2$, **7**, 326
$OsC_{53}H_{47}ClN_2P_2S$
Os(SCC$_6$H$_4$NMe$_2$)Cl(CN-C$_6$H$_4$Me)(PPh$_3$)$_2$, **7**, 361
$OsC_{53}H_{47}Cl_2N_2P_2$
[Os(CC$_6$H$_4$NMe$_2$)Cl$_2$(CN-C$_6$H$_4$Me)(PPh$_3$)$_2$]$^+$, **7**, 357
$OsC_{53}H_{97}OP_4$
[OsH(CO)(depe)$_2$]$^+$, **7**, 308, 310
$OsC_{54}H_{43}O_2P_2S$
[Os[C(SMe)CPhCPh](CO)$_2$-(PPh$_3$)$_2$]$^+$, **7**, 326
$OsC_{54}H_{45}Cl_2P_3$
OsCl$_2$(PPh$_3$)$_3$, **7**, 363
$OsC_{55}H_{45}ClNO_5P_3S$
[Os(NO)(CS)(PPh$_3$)$_3$]ClO$_4$, **7**, 324
$OsC_{55}H_{45}Cl_2OP_3$
OsCl$_2$(CO)(PPh$_3$)$_3$, **7**, 308
$OsC_{56}H_{46}OP_3S$
[OsH(CO)(CS)(PPh$_3$)$_3$]$^+$, **7**, 325
$OsC_{59}H_{48}ClNP_2S$
Os(CPhCHPh)Cl(CS)(CN-C$_6$H$_4$Me)(PPh$_3$)$_2$, **7**, 326
$OsC_{62}H_{53}OP_3$
OsH(C$_6$H$_4$Me)(CO)(PPh$_3$)$_3$, **7**, 397
$OsCo_3C_{20}H_{20}NO_{12}$
[Et$_4$N][OsCo$_3$(CO)$_{12}$], **10**, 48
$OsCrC_2H_6NO_4$
[Os(N)Me$_2$(CrO$_4$)]$^-$, **10**, 50
$OsCrC_{12}H_9O_{12}P$
Cr(CO)$_5$Os(CO)$_4$[P(OMe)$_3$], **10**, 29
$OsCrC_{18}H_{18}N_2O_8$
Cr(CO)$_5$Os(CO)$_3$(Bu^{t}NC)$_2$, **10**, 30
$OsCrC_{31}H_{20}O_7P_2$
Cr(CO)$_4$(PPh$_2$)$_2$Os(CO)$_3$, **10**, 43
$OsFe_2C_{12}O_{12}$
Fe$_2$Os(CO)$_{12}$, **7**, 83; **10**, 191
$OsHgC_{40}H_{35}ClIN_3O_3P_2$

$OsCl(HgI)[C(N_2)CO_2Et](NO)(PPh_3)_2$, **7**, 389
$OsMoC_{12}H_9O_{12}P$
$Mo(CO)_5Os(CO)_4P(OMe)_3$, **10**, 30
$OsNa_2C_4O_4$
$Na_2[Os(CO)_4]$, **7**, 300
OsO_4
OsO_4, **12**, 1138
$OsPtB_5C_{64}H_{52}OP_3$
$(CO)(PPh_3)(PPh_2)Os(PPh_3)Pt-B_5H_7Ph$, **1**, 324
$OsRe_2C_{18}H_8O_{14}$
$Os[CH_2CH_2Re(CO)_5]_2(CO)_4$, **7**, 392, 393
$OsRh_2C_{37}H_{35}O_2P_3$
$OsRh_2(H)_2(CO_2)(cod)_2(PMe_2Ph)_3$, **10**, 215
$OsRh_3C_{12}O_{12}$
$[OsRh_3(CO)_{12}]^-$, **10**, 192
$OsRh_4C_{15}O_{15}$
$[OsRh_4(CO)_{15}]^{2-}$, **10**, 192
$OsRuC_{11}H_2O_9$
$RuOs[HCCHC(O)](CO)_8$, **7**, 652
$OsRuC_{12}F_6O_8$
$RuOs(hfb)(CO)_8$, **7**, 652
$OsRuC_{14}H_{10}O_8$
$Os(CO)_4(CBu^tCH)Ru(CO)_4$, **10**, 54, 55
$OsSbC_{40}H_{33}Cl_2F_7NOP_2$
$[OsCl(CClF)(CO)(NCMe)(PPh_3)_2]SbF_6$, **7**, 347
$OsSbC_{54}H_{49}F_6O_2P_4$
$[Os(CHO)(CO)(dppe)_2]SbF_6$, **7**, 418
$OsSiC_{10}H_{29}Cl_2P_2$
$Os(CH_2TMS)Cl_2(PMe_3)_2$, **7**, 377, 379
$OsSiC_{16}H_{22}P_4$
$Os(CH_2SiMe_2CH_2)(PMe_3)_4$, **7**, 425
$OsSiC_{16}H_{46}P_4$
$Os[(CH_2)_2SiMe_2](PMe_3)_4$, **7**, 382, 384
$OsSiC_{58}H_{56}ClP_3$
$OsHCl(CH_2TMS)(C_6H_4PPh_2)(PPh_3)_2$, **7**, 377, 379
$OsSiWC_{13}H_{38}P_3S_4$
$[Os(CH_2TMS)(PMe_3)_3(S)_2WS_2]^-$, **7**, 375
$OsSi_2C_{10}H_{22}N_3S_2$
$[Os(N)(NCS)_2(CH_2TMS)_2]^-$, **7**, 375
$OsSi_2C_{10}H_{28}O$
$Os(O)Me_2(CH_2TMS)_2$, **7**, 373
$OsSi_2C_{13}H_{27}N$
$OsN(CH_2TMS)_2Cp$, **7**, 504
$OsSi_2C_{14}H_{40}ClNP_2$
$Os(N)(CH_2TMS)_2Cl(PMe_3)_2$, **7**, 375
$OsSi_2C_{21}H_{49}N$
$Os(CBu^t)(CH_2Bu^t)_2[N(TMS)_2]$, **7**, 361
$OsSi_2WC_9H_{25}NS_4$
$Os(N)(CH_2TMS)_2(S)(SMe)WS_2$, **7**, 375
$OsSi_2WC_{10}H_{22}N_3S_4$
$[Os(N)(CH_2TMS)_2(S)_2W(SCN)_2]^-$, **7**, 375
$OsSi_3C_{12}H_{33}N$
$Os(N)(CH_2TMS)_3$, **7**, 376
$OsSi_3C_{13}H_{36}ClN$
$Os(NMe)Cl(CH_2TMS)_3$, **7**, 376
$OsSi_4C_{16}H_{44}O$
$Os(O)(CH_2TMS)_4$, **7**, 372
$Os_2AgSi_4C_{26}H_{54}N_2$
$[Ag[NOs(CH_2TMS)_2Cp]_2]^+$, **7**, 375
$Os_2B_2C_{30}H_{68}P_2$
$[Os[BH_4]H_3P(C_5H_9)_3]_2$, **1**, 284, 285
$Os_2C_6Cl_4O_6$
$Os_2(Cl)_4(CO)_6$, **7**, 307, 308
$Os_2C_6I_2O_6$
$Os_2(I)_2(CO)_6$, **7**, 647
$Os_2C_8Cl_2O_8$
$Os_2Cl_2(CO)_8$, **7**, 307, 635
$Os_2C_8H_2O_8$
$(CO)_4HOsOsH(CO)_4$, **7**, 362, 378
$Os_2C_8H_{16}O_8$
$[Os(O)_2OCHMeCHMeO]_2$, **12**, 1138
$Os_2C_8I_2O_8$
$Os_2I_2(CO)_8$, **7**, 634
$Os_2C_8O_8$
$[Os_2(CO)_8]^{2-}$, **7**, 301, 625, 626, 632
$Os_2C_9H_4O_8$
$Os_2H(Me)(CO)_8$, **7**, 378, 652
$Os_2C_9H_6ClO_9$
$[Os_2[OC(Me)O]_2Cl(CO)_5]^-$, **7**, 643
$Os_2C_9H_6O_6S$
$Os_2(SCH_2CH_2CH)(H)(CO)_6$, **7**, 647, 648
$Os_2C_{10}H_3O_8$
$[Os_2(C_2H_3)(CO)_8]^+$, **7**, 653
$Os_2C_{10}H_4O_8$
$Os_2(CH_2CH_2)(CO)_8$, **7**, 652
$Os_2C_{10}H_6O_8$
$Os_2Me_2(CO)_8$, **7**, 652
$Os_2C_{10}H_6O_{10}$
$Os_2[OC(Me)O]_2(CO)_6$, **7**, 643
$Os_2C_{10}O_9$
$Os_2[CH_2C(O)](CO)_8$, **7**, 653
$Os_2C_{11}H_4O_9$
$Os_2[CH_2C(O)CH_2](CO)_8$, **7**, 653
$Os_2[CH(Me)C(O)](CO)_8$, **7**, 652
$Os_2C_{11}H_6O_8$
$Os_2(CH_2CH_2CH_2)(CO)_8$, **7**, 652
$Os_2C_{11}H_{10}O_6S$
$Os_2[S(CH_2)_3CMe](H)(CO)_6$, **7**, 647, 648
$Os_2C_{12}F_6O_8$
$Os_2(hfb)(CO)_8$, **7**, 652
$Os_2C_{13}H_{14}O_{10}$
$Os_2[OC(Me)O]_2(CO)_5(THF)$, **7**, 643
$Os_2C_{14}H_{10}O_4$
$Os_2Cp_2(CO)_4$, **7**, 655
$Os_2C_{16}H_8N_2O_8$
$Os_2(NC_5H_4O)_2(CO)_6$, **7**, 644
$Os_2C_{16}H_{10}N_4O_6$
$Os_2(NC_5H_4NH)_2(CO)_6$, **7**, 644
$Os_2C_{18}H_{10}IO_6P$
$Os_2(PPh_2)(I)(CO)_6$, **7**, 650
$Os_2C_{18}H_{25}$
$[Os_2(H)_3(Mes)_2]^+$, **7**, 673, 674
$Os_2C_{18}H_{26}$
$Os_2(H)_2H_2(Mes)_2$, **7**, 673, 674
$Os_2C_{18}H_{27}Cl$
$[Os_2(H)_3(C_6H_3Me_3)_2]Cl$, **7**, 567
$Os_2C_{18}H_{28}$
$[OsH(C_6H_3Me_3)]_2(H)_2$, **7**, 567
$Os_2C_{20}H_{20}$
$[Os_2Cp_4]^{2+}$, **7**, 609
$Os_2C_{20}H_{20}F_{12}P_2$
$[Os_2Cp_4][PF_6]_2$, **7**, 609, 610
$Os_2C_{21}H_{26}O_6S_3$
$Os_2[OC(SCH_2CMe_2CH_2)_3](CO)_5$, **7**, 648
$Os_2C_{21}H_{32}O$
$Os_2Cp^*_2(H)_2(CO)$, **7**, 656
$Os_2C_{22}H_{14}O_7S$
$Os_2[SC(H)(C_6H_4)CH_2Bz](CO)_7$, **7**, 647, 648
$Os_2C_{22}H_{32}O_2$
$Os_2Cp^*_2(H)_2(CO)_2$, **7**, 656
$Os_2C_{24}H_{30}O_4$
$Os_2Cp^*_2(CO)_4$, **7**, 655
$Os_2C_{26}H_{54}$
$Os_2(C_3H_5)_2(CH_2Bu^t)_4$, **7**, 377, 674
$Os_2C_{39}H_{32}O_3P_2$
$Os[OsH(CHO)(CO)_2(PPh_3)_2]$, **7**, 417, 418
$Os_2C_{40}H_{34}Cl_2O_4P_2$
$Os_2[OC(Me)P]_2(Ph_2PC_6H_4)_2Cl_2$, **7**, 674
$Os_2C_{42}H_{38}Cl_2O_4P_2$
$Os_2[OC(Et)O]_2(Ph_2PC_6H_4)_2Cl_2$, **7**, 674
$Os_2C_{48}H_{36}O_{10}P_2$
$Os_2(DMAD)(CO)_6(PPh_3)_2$, **7**, 652
$Os_2C_{58}H_{51}O_6P_4$
$[Os_2[OC(Me)O](CO)_4(dppe)_2]^+$, **7**, 643
$Os_2GeC_8Cl_4O_8$
$Os_2Cl(GeCl_3)(CO)_8$, **7**, 302
$Os_2HgC_{18}H_2O_{18}S_2$
$[(H)(S)(CO)_9Os]_2Hg$, **10**, 340
$Os_2K_2C_{48}H_{72}P_6$
$[K[OsH_3(PMe_2Ph)_3]]_2$, **1**, 26, 28
$Os_2RhC_{19}H_{15}O_9$
$Os_2RhCp^*(CO)_9$, **7**, 653
$Os_2Si_2C_{36}H_{36}O_6P_2$
$Os_2(Ph_2PCH_2SiMe_2)_2(CO)_6$, **7**, 633
$Os_2Si_4C_{26}H_{52}N_4S_2$

$Os_2(SC_5H_4N)_2(N)_2(CH_2TMS)_4$, **7**, 375
$Os_2WC_{20}H_{11}O_{12}PS_2$
$WOs_3(S)_2(CO)_{12}(PMe_2Ph)$, **10**, 228
$Os_2WC_{21}H_{10}O_8$
$CpWOs_2(CO)_8(C{\equiv}CPh)$, **10**, 21
$Os_2WC_{22}H_{13}NO_7$
$CpWOs_2(CO)_7(NCMe)(C_2Ph)$, **10**, 102
$Os_2WC_{22}H_{18}N_2O_{12}$
$W(CO)_5Os(CO)_3(Bu^tNC)Os(CO)_4(Bu^tNC)$, **10**, 30
$Os_2WC_{22}H_{18}O_{18}P_2$
$W(CO)_5Os(CO)_4Os(CO)_3[P(OCH_2)_3CMe]_2$, **10**, 30
$Os_3AlRuC_{12}H_3O_{12}$
$[Al][H_3RuOs_3(CO)_{12}]$, **10**, 52
$Os_3AsC_9H_3O_9$
$Os_3H_2(AsH)(CO)_9$, **7**, 735
$Os_3AsC_{11}H_3O_{11}$
$Os_3H_2(AsH)(CO)_{11}$, **7**, 735
$Os_3AsC_{13}H_7O_{11}$
$Os_3H(AsMe_2)(CO)_{11}$, **7**, 735, 736
$Os_3AsC_{15}H_7O_9$
$Os_3H_2(AsPh)(CO)_9$, **7**, 703
$Os_3As_2C_{20}H_{28}N_2O_8$
$Os_3H(NSNAsBu^t{}_2)(AsBu^t)(CO)_8$, **7**, 698
$Os_3AuC_{16}H_{16}O_{10}P$
$Os_3H(AuPEt_3)(CO)_{10}$, **7**, 720
$Os_3AuC_{19}H_{20}O_{10}P$
$Os_3(AuPEt_3)(C_3H_5)(CO)_{10}$, **7**, 696, 787, 788
$Os_3AuC_{26}H_{16}O_{10}P$
$Os_3(C_2Ph)(AuPMe_2Ph)(CO)_{10}$, **7**, 694
$Os_3AuC_{28}H_{16}O_{10}P$
$[HOs_3Au(CO)_{10}(PPh_3)]$, **10**, 59
$Os_3AuNiC_{32}H_{22}O_9P$
$[NiOs_3(CO)_9Cp(H)_2[Au(PPh_3)]]^-$, **9**, 21
$Os_3Au_2C_{46}H_{30}O_{10}P_2$
$Au_2Os_3(CO)_{10}(PPh_3)_2$, **10**, 305
$Os_3BC_{10}H_3O_{10}$
$(H)_3(CO)_9Os_3BCO$, **1**, 296, 297
$Os_3BC_{10}H_5O_9$
$(H)_3(CO)_9Os_3BCH_2$, **1**, 296, 297, 301
$Os_3H_3(BCH_2)(CO)_9$, **7**, 707
$Os_3H(BH_2CH_2)(CO)_9$, **7**, 707
$Os_3BC_{10}H_9O_{10}$
$(H)_3(CO)_9Os_3BCO$, **1**, 296, 297
$Os_3BC_{14}H_5F_4O_{11}$
$[Os_3(CO)_{11}(C_3H_5)][BF_4]$, **7**, 787, 788
$Os_3BC_{15}H_5F_4O_{10}$
$[Os_3(CO)_{10}Cp][BF_4]$, **7**, 794, 795
$Os_3B_2C_{16}H_7O_{12}$
$(H)_3Os_3(CO)_9B(COBO_2C_6H_4)$, **1**, 296, 297
$Os_3B_2C_{18}H_{20}ClO_{10}$
$Os_3H_2[(C_8H_{17}BO)C{=}BHCl](CO)_9$, **7**, 703
$Os_3B_5C_{10}H_{10}O_9$
$(H)_2Os_3(CO)_9(CB_5H_8)$, **1**, 296, 297
$Os_3C_9H_2O_9S$
$Os_3H_2(S)(CO)_9$, **7**, 726, 727, 866
$Os_3C_9H_2O_{10}$
$Os_3H_2(O)(CO)_9$, **7**, 723
$Os_3C_9O_8S_3$
$Os_3(S)_2(CO)_8(CS)$, **7**, 726
$Os_3C_9O_9Se_2$
$Os_3Se_2(CO)_9$, **7**, 726
$Os_3C_9O_9Te_2$
$Os_3(Te)_2(CO)_9$, **7**, 726
$Os_3C_{10}Cl_2O_{10}$
$Os_3Cl_2(CO)_{10}$, **7**, 307, 723
$Os_3C_{10}HO_{10}$
$Os_3(H)(CO)_{10}$, **7**, 923
$Os_3C_{10}H_2ClO_{10}$
$[Os_3H_2Cl(CO)_{10}]^-$, **7**, 722
$Os_3C_{10}H_2O_9S$
$Os_3(SCH_2)(CO)_9$, **7**, 697
$Os_3C_{10}H_2O_{10}$
$Os_3H_2(CO)_{10}$, **7**, 687, 689, 691, 694, 719
$Os_3C_{10}H_2O_{11}$
$Os_3H(OH)(CO)_{10}$, **7**, 723
$Os_3C_{10}H_3NO_{10}$
$Os_3H(NH_2)(CO)_{10}$, **7**, 716
$Os_3C_{10}H_3O_{10}$
$[Os_3H_3(CO)_{10}]^-$, **7**, 688, 719
$Os_3C_{10}H_3O_{10}P$
$Os_3H(PH_2)(CO)_{10}$, **7**, 735
$Os_3C_{10}H_4O_9$
$Os_3(H)_3(CO)_9(CH)$, **7**, 757
$Os_3C_{10}O_{10}S$
$Os_3(S)(CO)_{10}$, **7**, 727
$Os_3C_{11}HDO_{11}$
$Os_3H(cod)(CO)_{10}$, **7**, 720
$Os_3C_{11}HO_{11}$
$[Os_3H(CO)_{11}]^-$, **7**, 688, 689, 719
$Os_3C_{11}H_2ClO_{10}$
$[Os_3(CH_2)(Cl)(CO)_{10}]^-$, **7**, 722
$Os_3C_{11}H_2F_6O_{15}S_2$
$Os_3H_2(CF_3SO_3)_2(CO)_9$, **7**, 725
$Os_3C_{11}H_2O_9$
$Os_3H(CO)_9(CCH)$, **7**, 775
$Os_3C_{11}H_2O_{10}$
$H_2Os_3(CO)_9(CCO)$, **7**, 748
$Os_3C_{11}H_2O_{10}S$
$Os_3(CH_2S)(CO)_{10}$, **7**, 705, 750
$Os_3C_{11}H_2O_{10}S_2$
$Os_3(S)(CH_2S)(CO)_{10}$, **7**, 702
$Os_3C_{11}H_2O_{11}$
$Os_3H_2(CO)_{11}$, **7**, 688
$Os_3C_{11}H_2O_{12}$
$Os_3H(HCO_2)(CO)_{10}$, **7**, 724
$Os_3C_{11}H_3NO_{10}$
$H_2Os_3(CO)_{10}(C{=}NH)$, **7**, 751
$HOs_3(CO)_{10}(CNH_2)$, **7**, 751
$Os_3C_{11}H_3O_{11}$
$[Os_3(OMe)(CO)_{10}]^-$, **7**, 748
$Os_3C_{11}H_3O_{11}P$
$Os_3(CO)_{11}(PH_3)$, **7**, 735
$Os_3C_{11}H_4O_9$
$Os_3H_2(C{=}CH_2)(CO)_9$, **1**, 296; **7**, 765
$Os_3C_{11}H_4O_{10}$
$Os_3H(Me)(CO)_{10}$, **7**, 691, 696
$Os_3C_{11}H_6O_{10}$
$Os_3(H)_3(COMe)(CO)_9$, **7**, 930
$Os_3C_{11}H_9NO_8S$
$Os_3(H)_2(CO)_8[C(H)NMe_2](S)$, **7**, 752
$Os_3C_{11}O_{10}$
$[Os_3(CO)_9(CCO)]^{2-}$, **7**, 755
$Os_3C_{12}Cl_2O_{12}$
$Os_3Cl_2(CO)_{12}$, **7**, 723
$Os_3C_{12}HO_{12}$
$[Os_3[C(O)H](CO)_{11}]^-$, **7**, 748
$[Os_3H(CO)_{12}]^+$, **7**, 719
$Os_3C_{12}H_2F_3NO_{10}$
$Os_3H(N{=}CCHCF_3)(CO)_{10}$, **7**, 733
$Os_3C_{12}H_2O_{11}$
$Os_3(CH_2)(CO)_{11}$, **7**, 692, 748, 749
$Os_3(CO)_{10}(CH_2CO)$, **7**, 748
$Os_3(CO)_{10}[HCC(O)H]$, **7**, 749
$Os_3C_{12}H_2O_{11}S$
$Os_3(CO)_{11}(SCH_2)$, **7**, 750
$Os_3C_{12}H_2O_{12}$
$H_2Os_3(CO)_{12}$, **7**, 911
$Os_3C_{12}H_3N_2O_{10}$
$Os_3(CO)_{10}(NCMe)_2$, **7**, 841
$Os_3C_{12}H_3O_{12}$
$[Os_3[OC(OMe)](CO)_{10}]^-$, **7**, 748
$Os_3C_{12}H_4O_{10}$
$Os_3H(CO)_{10}(COMe)$, **7**, 760
$Os_3C_{12}H_4O_{11}$
$Os_3H(OCH{=}CH_2)(CO)_{10}$, **7**, 723
$Os_3(OCMe)(OH)(CO)_{10}$, **7**, 751
$Os_3C_{12}H_4O_{12}$
$Os_3H(MeCO_2)(CO)_{10}$, **7**, 724
$Os_3C_{12}H_5F_3O_{14}S$
$Os_3H_2(MeCO_2)(TfO)(CO)_9$, **7**, 725
$Os_3C_{12}H_5NO_9$
$Os_3(CH_2)(CO)_9(MeCN)$, **7**, 749
$Os_3C_{12}H_5O_{11}$
$H_2Os_3(CO)_9[CC(O)OMe]$, **7**, 748
$Os_3C_{12}H_7NO_9$
$Os_3H_2(CHCH{=}NMe)(CO)_9$, **7**, 704
$Os_3C_{12}H_7N_2O_6$
$Os_3(CO)_6[MeC(CH_2NC)_2]$, **7**, 711
$Os_3C_{12}H_8O_9S$
$Os_3H(CO)_9(SMe)(CH_2CH_2)$, **7**, 766
$Os_3C_{12}IO_{12}$
$[Os_3I(CO)_{12}]^+$, **7**, 723
$Os_3C_{12}I_2O_{12}$
$Os_3(CO)_{12}(I)_2$, **7**, 838
$Os_3C_{12}O_{12}$

$Os_3(CO)_{12}$, **7**, 452, 684, 685, 689, 838, 909, 918, 948

$Os_3C_{13}HF_6NO_{12}$
$Os_3H[(CF_3)_2NO](CO)_{11}$, **7**, 725

$Os_3C_{13}H_2NO_{12}$
$[Os_3(CH_2)(NCO)(CO)_{11}]^-$, **7**, 750

$Os_3C_{13}H_3NO_{11}$
$Os_3(CO)_{11}(MeCN)$, **7**, 684, 717

$Os_3C_{13}H_3O_{12}$
$[Os_3[C(O)Me](CO)_{11}]^-$, **7**, 748

$Os_3C_{13}H_4O_{12}$
$HOs_3(CO)_9[CHCH_2C(O)OC(O)]$, **7**, 749

$Os_3C_{13}H_5NO_9$
$Os_3H_2(C_4H_3N)(CO)_9$, **7**, 704

$Os_3C_{13}H_6NO_{11}$
$[Os_3[C(O)NMe_2](CO)_{10}]^-$, **7**, 748

$Os_3C_{13}H_6O_{10}S$
$Os_3(SCH_2CHCH_2)(H)(CO)_{10}$, **7**, 647

$Os_3C_{13}H_6O_{11}$
$Os_3[C(O)Me](OCMe)(CO)_9$, **7**, 748

$Os_3C_{13}H_7NO_{10}$
$HOs_3(CO)_{10}(CNMe_2)$, **7**, 752
$Os_3H(CH_2CH{=}NMe)(CO)_{10}$, **7**, 696

$Os_3C_{13}H_7NO_{11}$
$Os_3H(Me_2C{=}NO)(CO)_{10}$, **7**, 697, 725, 726

$Os_3C_{13}H_8O_9$
$Os_3H_2(CO)_9(MeCCMe)$, **7**, 780, 781

$Os_3C_{13}H_8O_9S$
$Os_3H(C_4H_7S)(CO)_9$, **7**, 699

$Os_3C_{13}H_8O_{12}$
$HOs_3(CO)_{10}[C(H)MeO(OMe)]$, **7**, 748

$Os_3C_{13}H_9NO_{10}$
$Os_3H(CO)_{10}[N(H)Pr]$, **7**, 751

$Os_3C_{13}H_{11}NO_{10}$
$Os_3H(CO)_9(OMe)[C(H)NMe_2]$, **7**, 752

$Os_3C_{13}H_{13}NO_8S$
$Os_3(H)_2(CO)_8[C(Et)NMe_2](S)$, **7**, 752, 753

$Os_3C_{13}H_{18}O$
$Os(CH_2CHMe)(CO)(C_6H_3Me_3)$, **7**, 386

$Os_3C_{13}NO_{11}$
$Os_3(OH)(Me_2C{=}N)(CO)_{10}$, **7**, 697

$Os_3C_{14}F_6N_2O_{10}$
$Os_3(CO)_{10}(CNCF_3)_2$, **7**, 754

$Os_3C_{14}H_2F_6O_{10}$
$Os_3H(CF_3CCHCF_3)(CO)_{10}$, **7**, 701

$Os_3C_{14}H_2O_{13}$
$Os_3(CH_2CO)(CO)_{12}$, **7**, 750; **10**, 148

$Os_3C_{14}H_4O_{11}$
$Os_3H(C_4H_3O)(CO)_{10}$, **7**, 692, 694

$Os_3C_{14}H_4O_{13}$
$Os_3(CO)_{11}(\overline{COCH_2CH_2O})$, **7**, 752

$Os_3C_{14}H_5NO_{10}$
$Os_3H(C_5H_4NO)(CO)_9$, **7**, 733

$Os_3C_{14}H_6NO_{12}$
$[Os_3[C(O)NMe_2](CO)_{11}]^-$, **7**, 748

$Os_3C_{14}H_6N_2O_{10}$
$Os_3(CO)_{10}(NCMe)_2$, **7**, 911, 932

$Os_3C_{14}H_6N_2O_{14}$
$Os_3(MeO_2CN_2CO_2Me)(CO)_{10}$, **7**, 698

$Os_3C_{14}H_6O_9P$
$H_2Os_3(CO)_9(PC_5H_4)$, **7**, 754, 755

$Os_3C_{14}H_6O_{10}$
$Os_3(CO)_{10}(C_4H_6)$, **7**, 697, 714

$Os_3C_{14}H_6O_{11}$
$HOs_3(CO)_{10}(HCCHCMeO)$, **7**, 749
$Os_3(CO)_{10}(HCCCH_2CH_2OH)$, **7**, 782

$Os_3C_{14}H_6O_{11}S$
$Os_3(CO)_{11}(\overline{SCH_2CH_2CH_2})$, **7**, 728

$Os_3C_{14}H_7O_9P$
$Os_3H_2(PCp)(CO)_9$, **7**, 704

$Os_3C_{14}H_8O_{10}$
$Os_3H(CO)_9[CCC(Me)_2OH]$, **7**, 776

$Os_3C_{14}H_8O_{11}$
$HOs_3(CO)_{10}(CH{=}CHOEt)$, **7**, 749

$Os_3C_{14}H_9NO_9S$
$Os_3(CO)_9(MeC_2NMe_2)(S)$, **7**, 752

$Os_3C_{14}H_9NO_{10}$
$Os_3H(\overline{NCH_2CH_2CH_2CH_2})(CO)_{10}$, **7**, 731

$Os_3C_{14}H_9NO_{12}$
$Os_3H(EtMeNCO_2)(CO)_{110}$, **7**, 724

$Os_3C_{14}H_{10}ClO_9S$
$Os_3(Cl)(\overline{SCH_2CMe_2CH_2})(CO)_9$, **7**, 722

$Os_3C_{14}H_{10}O_{11}$
$Os_3H[C(Me)OMe](OCMe)(CO)_9$, **7**, 751
$Os_3H(OBu^i)(CO)_{10}$, **7**, 719

$Os_3C_{15}F_6N_2O_{11}$
$Os_3(CO)_{11}(CNCF_3)_2$, **7**, 754

$Os_3C_{15}H_2F_6O_{11}$
$Os_3(CO)_{11}[CF_3(H)CC(H)CF_3]$, **7**, 767

$Os_3C_{15}H_5ClNO_9$
$[Os_3Cl(NPh)(CO)_9]^-$, **7**, 731

$Os_3C_{15}H_5ClO_{10}$
$Os_3Cl(CO)_{10}Cp$, **7**, 794, 795

$Os_3C_{15}H_5NO_{10}$
$HOs_3(CO)_{10}(py)$, **7**, 716, 752

$Os_3C_{15}H_6N_2O_{15}$
$Os_3(CO)_{11}(MeO_2CN_2CO_2Me)$, **7**, 700

$Os_3C_{15}H_6O_9$
$Os_3H_2(CO)_9(C_6H_4)$, **7**, 784

$Os_3C_{15}H_6O_{10}$
$Os_3(CO)_{10}(C_5H_6)$, **7**, 793
$Os_3H_2(C_6H_4O)(CO)_9$, **7**, 704
$Os_3H(CO)_{10}Cp$, **7**, 794, 795

$Os_3C_{15}H_7NO_9$
$H_2Os_3(CO)_9(C_6H_4NH)$, **7**, 704, 752
$Os_3H_2(PhN)(CO)_9$, **7**, 731

$Os_3C_{15}H_7NO_{10}$
$Os_3H(C_4H_3NMe)(CO)_{10}$, **7**, 692, 751

$Os_3C_{15}H_7O_9$
$[Os_3(CO)_9(C_6H_7)]^-$, **7**, 796

$Os_3C_{15}H_8O_9$
$Os_3(C_6H_8)(CO)_9$, **7**, 709
$Os_3H(C_6H_7)(CO)_9$, **7**, 705, 793

$Os_3C_{15}H_9NO_{10}$
$Os_3(CO)_{10}(MeCCNMe_2)$, **7**, 696, 781

$Os_3C_{15}H_{10}O_{10}S$
$Os_3(CO)_{10}$(thiacyclohexane), **7**, 751

$Os_3C_{15}H_{11}ClO_{10}S$
$Os_3H(SCH_2CMe_2CH_2Cl)(CO)_{10}$, **7**, 728

$Os_3C_{15}H_{11}NO_{10}$
$Os_3H(EtCNMeCH_2)(CO)_{10}$, **7**, 694

$Os_3C_{15}H_{11}NO_{11}$
$Os_3(CH_2NMe{=}CHCCO_2Me)(CO)_9$, **7**, 708
$Os_3H(MeO_2CHCCNMe_2)(CO)_9$, **7**, 705

$Os_3C_{15}H_{13}NO_9$
$H_2Os_3(CO)_9(HC_2NEt_2)$, **7**, 701, 752

$Os_3C_{15}H_{13}NO_{10}$
$(OMe)Os_3(CO)_9[(Me)CHCNMe_2]$, **7**, 706, 730, 752

$Os_3C_{15}H_{15}NO_{10}$
$H(OMe)Os_3(CO)_9[C(Et)NMe_2]$, **7**, 703, 752

$Os_3C_{16}H_6NO_9S$
$HOs_3(S)(HC{=}NPh)(CO)_9$, **7**, 943

$Os_3C_{16}H_6O_{10}$
$Os_3H(C_7H_5O)(CO)_9$, **7**, 708

$Os_3C_{16}H_6O_{10}P$
$[Os_3H(PPh)(CO)_{10}]^-$, **7**, 736

$Os_3C_{16}H_6O_{10}S$
$Os_3(H)(SPh)(CO)_{10}$, **7**, 922

$Os_3C_{16}H_7N_3O_{10}$
$Os_3H(NHNNPh)(CO)_{10}$, **7**, 696, 731

$Os_3C_{16}H_8O_9$
$Os_3(CO)_9(PhMe)$, **7**, 795
$Os_3(H)_3(CO)_9(CPh)$, **7**, 757

$Os_3C_{16}H_8O_{10}$
$Os_3(CO)_{10}(C_6H_8)$, **7**, 793

$Os_3C_{16}H_8O_{14}$
$Os_3(CO)_{10}(\overline{COCH_2CH_2O})_2$, **7**, 752

$Os_3C_{16}H_9NO_8$
$Os_3(CO)_8(MeCN)(C_6H_6)$, **7**, 795

$Os_3C_{16}H_{11}NO_{10}$
$Os_3H(N{=}C_6H_{10})(CO)_{10}$, **7**, 704
$Os_3C_{16}H_{11}NO_{11}$
$Os_3(CO)_{11}[C(Et)NMe_2]$, **7**, 752, 754
$Os_3C_{16}H_{11}O_{12}P$
$Os_3H(Ph_2PO_2)(CO)_{10}$, **7**, 725
$Os_3C_{16}H_{12}N_2O_{11}$
$Os_3H_2(Bu^tCH{=}NNCO)(CO)_{10}$, **7**, 732
$Os_3C_{16}H_{12}O_9$
$HOs_3(CO)_8[C(O)C(=CHMe)CHCHEt]$, **7**, 749
$Os_3C_{16}H_{12}O_{10}$
$Os_3(CO)_{10}(allene)_2$, **7**, 690
$Os_3C_{16}H_{13}NO_{10}$
$Os_3H[CHC(Me)NHCHMe_2](CO)_{10}$, **7**, 750
$Os_3(OMe)(MeCHCNMe_2)(CO)_{10}$, **7**, 694
$Os_3C_{16}H_{13}NO_{11}$
$Os_3(CO)_9[H_2CN(Me)_2CH_2CC(OMe)O]$, **7**, 751
$Os_3C_{16}H_{18}Br_2O_{16}P_2$
$Os_3Br_2(CO)_{10}[P(OMe)_3]_2$, **7**, 723
$Os_3C_{17}H_5N_3O_{11}$
$Os_3(PhN_3CO)(CO)_{10}$, **7**, 697
$Os_3C_{17}H_6FNO_{10}S$
$Os_3H(SCH{=}NC_6H_4F)(CO)_{10}$, **7**, 728
$Os_3C_{17}H_9NO_{12}$
$Os_3H(Pr^iNC)(C_2HCO_2)(CO)_{10}$, **7**, 725
$Os_3(Pr^iNHC)(C_2HCO_2)(CO)_{10}$, **7**, 724
$Os_3C_{17}H_9N_3O_9$
$Os_3(CO)_9[MeC(CH_2NC)_3]$, **7**, 711, 713, 716
$Os_3C_{17}H_9O_{10}P$
$Os_3H(PMePh)(CO)_{10}$, **7**, 696
$Os_3C_{17}H_{10}O_9$
$Os_3(CO)_9(H)_3(CTol)$, **10**, 94
$Os_3C_{17}H_{11}F_6N_3O_{15}S_2$
$[Os_3(H)_2(CO)_9(MeCN)_3][CF_3SO_3]_2$, **7**, 686
$Os_3C_{17}H_{11}NO_{11}$
$Os_3(CO)_{11}(NH{=}C_6H_{10})$, **7**, 704, 711
$Os_3C_{17}H_{12}O_9$
$Os_3(CO)_9[MeCC(Me)C(Me)CMe]$, **7**, 787
$Os_3C_{17}H_{13}O_8PS_2$
$Os_3(CO)_8(PMe_2Ph)(S)(SCH_2)$, **7**, 750
$Os_3C_{17}H_{13}O_9P$
$Os_3H_2(CO)_9(PMe_2Ph)$, **7**, 688, 690
$Os_3C_{17}H_{15}NO_{10}$
$Os_3H[CHCHN(Et)Pr](CO)_{10}$, **7**, 751
$Os_3C_{17}H_{18}N_2O_7S$
$Os_3(S)(MeC_2NMe_2)_2(CO)_7$, **7**, 696
$Os_3C_{18}H_8O_{10}$
$Os_3(CO)_{10}(H)(CTol)$, **10**, 94
$Os_3C_{18}H_{12}O_{12}$
$Os_3(CO)_9[MeOCH_2CC(H)COC(H)CCH_2OMe]$, **7**, 789, 790
$Os_3C_{18}H_{13}NO_9S$
$Os_3H(CO)_9[C(H)NMe_2](SPh)$, **7**, 752
$Os_3C_{18}H_{13}O_9PS_2$
$Os_3(CO)_9(PMe_2Ph)(S)(SCH_2)$, **7**, 690, 749
$Os_3C_{18}H_{16}N_2O_{10}$
$Os_3(CO)_{10}(Pr^iN{=}CHCH{=}NPr^i)$, **7**, 699
$Os_3C_{18}H_{16}O_{10}S_2$
$Os_3(CO)_{10}(\overline{SCH_2CH_2CH_2C}H_2)_2$, **7**, 728
$Os_3C_{18}H_{18}NO_{10}P$
$Os_3(CO)_{10}(PEt_3)(MeCN)$, **7**, 687
$Os_3C_{19}H_8O_{10}$
$Os_3H_2(C_{10}H_6O)(CO)_9$, **7**, 697
$Os_3C_{19}H_{10}O_9S$
$Os_3H(CO)_9(SPh)(\overline{CCCH_2}CH_2)$, **7**, 783
$Os_3C_{19}H_{12}O_9P_2$
$Os_3H(C_{10}H_{11}P_2)(CO)_9$, **7**, 708
$Os_3C_{19}H_{12}O_{11}$
$HOs_3(CO)_9(CPh)[C(OMe)_2]$, **7**, 748
$Os_3C_{19}H_{15}NO_{12}$
$Os_3(CNMe_2)[CH_2C(CO_2Me)C(CO_2Me)CH](CO)_8$, **7**, 710
$Os_3C_{20}H_9NO_{10}$
$Os_3H(C_9H_5NMe)(CO)_{10}$, **7**, 751
$Os_3C_{20}H_{13}F_3NO_{10}P$
$Os_3H[NCH(CF_3)(PMe_2Ph)](CO)_{10}$, **7**, 694
$Os_3C_{21}H_4O_{20}S_2$
$[Os_3H(CO)_{10}]_2(SCH_2S)$, **7**, 728
$Os_3C_{21}H_{11}O_9P$
$H_2Os_3(CO)_9[P(Ph)(C_6H_4)]$, **7**, 704, 754
$Os_3C_{21}H_{13}O_9P$
$Os_3H(PC_4H_2Me_2)(C_6H_4)(CO)_9$, **7**, 703
$Os_3(PhPCHCMe{=}CMeCH)(CO)_9$, **7**, 709
$Os_3C_{21}H_{17}O_9P$
$Os_3H(CO)_9(MeCCCH_2PMe_2Ph)$, **7**, 791
$Os_3C_{21}H_{33}N_3O_6$
$[Os(O)_2Me_2(py)]_3$, **7**, 373
$Os_3C_{22}H_{10}O_{10}Se_2$
$Os_3(SePh)_2(CO)_{10}$, **7**, 690
$Os_3C_{22}H_{11}O_{10}P$
$Os_3H(PPh_2)(CO)_{10}$, **7**, 696
$Os_3C_{22}H_{11}O_{12}P$
$Os_3H(Ph_2PO_2)(CO)_{10}$, **7**, 696
$Os_3C_{22}H_{13}O_9P$
$Os_3H(C_6H_4PMePh)(CO)_9$, **7**, 707
$Os_3C_{22}H_{22}O_{10}Te$
$Os_3(CO)_{10}(TeCy_2)$, **7**, 728
$Os_3C_{23}H_8O_{20}S_2$
$[Os_3H(CO)_{10}]_2(SCH_2CH_2CH_2S)$, **7**, 728
$Os_3C_{23}H_{12}F_3O_{11}P$
$Os_3H_2(CF_3CO_2)(CO)_9(PPh_2)$, **7**, 725
$Os_3C_{23}H_{12}N_2O_{10}$
$Os_3H(NHN{=}CPh_2)(CO)_{10}$, **7**, 696
$Os_3C_{24}H_{12}O_{10}$
$Os_3(H)(OH)(C{=}C{=}CPh_2)(CO)_9$, **7**, 723
$Os_3C_{24}H_{54}O_{24}P_6$
$Os_3(CO)_6[P(OMe)_3]_6$, **7**, 711
$Os_3C_{24}H_{59}O_{24}P_6$
$Os_3(CO)_6[P(OMe)_3]_6$, **7**, 713
$Os_3C_{25}H_{10}O_9$
$Os_3(C_2Ph)_2(CO)_9$, **7**, 694, 695
$Os_3C_{25}H_{12}O_{10}S$
$Os_3(CO)_{10}[C(Ph)CH_2C(Ph)S]$, **7**, 750
$Os_3C_{26}H_{12}O_{10}$
$Os_3(CO)_9(HC_2Ph)_2(CO)$, **7**, 749
$Os_3C_{26}H_{22}O_{10}P_2$
$Os_3(CO)_{10}(PMe_2Ph)_2$, **7**, 714, 716, 720
$Os_3C_{26}H_{28}O_{10}$
$Os_3(CO)_{10}(coe)_2$, **7**, 841
$Os_3C_{27}H_{13}NO_{10}P$
$Os_3(CO)_{10}[(Ph_2P)C_5H_3N]$, **7**, 714, 718
$Os_3C_{27}H_{17}O_8PS$
$Os_3H(CHS)(CO)_8(PPh_3)$, **7**, 706
$Os_3C_{27}H_{17}O_{10}P$
$Os_3(CO)_{10}(Ph_2PCCPr^i)$, **7**, 690
$Os_3C_{27}H_{19}NO_8$
$Os_3H(PhCCPhCNMeCEt)(CO)_8$, **7**, 709
$Os_3C_{28}H_{17}O_{10}P$
$Os_3H_2(CO)_{10}(PPh_3)$, **7**, 688
$Os_3C_{28}H_{36}O_{13}S_3$
$Os_3[(SCH_2CMe_2CH_2)_3][OC(O)Me]_2(CO)_9$, **7**, 648
$Os_3C_{29}H_{19}O_{11}P$
$Os_3H(MeCO_2)(CO)_9(PPh_3)$, **7**, 724
$Os_3C_{29}H_{21}O_8$
$Os_3(CEt)(CPhCPhCHCEt)(CO)_8$, **7**, 710
$Os_3C_{30}H_{47}O_6P_3$
$Os_3H_2(PBu^t_2)_2(CO)_6(PHBu_2^t)$, **7**, 736
$Os_3C_{32}H_{19}O_9P$
$Os_3H(C_5H_3PPh_3)(CO)_9$, **7**, 705
$Os_3C_{32}H_{22}O_{10}PS$
$HOs_3(CO)_{10}(DHT\text{-}H)(PPh_3)$, **7**, 751
$Os_3C_{33}H_{19}O_{10}P$
$Os_3H(C_5H_3PPh_3)(CO)_{10}$, **7**, 693
$Os_3C_{33}H_{19}O_{10}S$
$Os_3H(MeC_4H_2S)(CO)_{10}$, **7**, 692
$Os_3C_{33}H_{22}O_8P_2$
$Os_3H(C_6H_4PBzPPh_2)(CO)_8$, **7**, 689
$Os_3H(CO)_8(Ph_2PCH_2PPhC_6H_4)$, **7**, 716

$Os_3C_{33}H_{24}NO_9P$
$Os_3(PPh_2)(CO)_9(PhCCCNBu^t)$, **7**, 775
$Os_3C_{33}H_{24}O_8P_2$
$Os_3H(Ph_2PCH_2PPhC_6H_6)(CO)_8$, **7**, 701
$Os_3C_{33}H_{25}O_9P$
$Os_3(CO)_9(PPh_3)(EtCCEt)$, **7**, 780, 781
$Os_3C_{33}H_{33}O_9P_3$
$Os_3(CO)_9(PMe_2Ph)_3$, **7**, 715
$Os_3C_{34}H_{22}O_9P_2$
$Os_3H(C_6H_4PBzPPh_2)(CO)_9$, **7**, 707
$Os_3C_{35}H_{22}O_9S_2$
$Os_3H_2(SCPh_2CO)(CO)_8$-$(Ph_2C{=}S)$, **7**, 697
$Os_3C_{35}H_{22}O_{10}P_2$
$Os_3(CO)_{10}(dppm)$, **7**, 716, 717
$Os_3C_{36}H_{20}O_8$
$Os_3H(CO)_8[Ph_2C_2(Ph)C_2$-$(C_6H_4)]$, **7**, 787
$Os_3C_{36}H_{24}O_{10}P_2$
$Os_3(CO)_{10}(dppe)$, **7**, 715
$Os_3C_{36}H_{25}O_{10}P_2$
$[Os_3(H)(CO)_{10}(dppe)]^+$, **7**, 720
$Os_3C_{36}H_{31}O_{11}P_5$
$Os_3(CO)_8(dppm)[P(OMe)_3]$, **7**, 307
$Os_3C_{44}H_{30}O_8P_2$
$HOs_3(CO)_8(PPh_3)(PPh_2C_6H_4)$, **7**, 755
$Os_3C_{44}H_{31}O_8P_2$
$Os_3H(Ph_2PNPh)(CO)_8(Ph_2P$-$NHPh)$, **7**, 707
$Os_3C_{45}H_{30}O_9P_2$
$HOs_3(CO)_9(PPh_3)(PPh_2C_6H_4)$, **7**, 755, 756
$Os_3C_{48}H_{32}N_2O_{11}P_2$
$[PPN][Os_3(CH_2)(NCO)(CO)_{10}]$, **7**, 750
$Os_3CoC_{14}H_8O_9$
$Os_3Co(H)_3(CO)_9Cp$, **10**, 210
$Os_3CoC_{14}H_9O_9$
$Os_3Co(H)_4(CO)_9Cp$, **10**, 210
$Os_3CoC_{15}H_7O_{10}$
$Os_3Co(Cp)(CO)_{10}(H)_2$, **10**, 7
$Os_3CoHgC_{19}H_9O_{13}$
$(C_2Bu^t)(CO)_9Os_3[HgCo(CO)_4]$, **10**, 340
$Os_3FeC_{13}HO_{13}$
$[HFeOs_3(CO)_{13}]^-$, **7**, 83
$Os_3FeC_{13}H_2O_{13}$
$H_2FeOs_3(CO)_{13}$, **10**, 52
$Os_3FeC_{22}H_{19}O_8P$
$Os_3(CO)_8[C_5H_3PEt_2Fe(C_5H_4)$-$(H)_3]$, **10**, 17
$Os_3FeC_{23}H_{23}O_8P$
$Os_3H_2(Pr^i{}_2PC_4H_3FeC_5H_4)$-$(CO)_8$, **7**, 709
$Os_3HgC_{15}H_9IO_9$
$(C_2Bu^t)(CO)_9Os_3(HgI)$, **10**, 340
$Os_3K_2C_{11}O_{11}$
$K_2[Os_3(CO)_{11}]$, **7**, 719
$Os_3MoWC_{34}H_{20}O_8$
$Cp_2MoWOs_3(CO)_8(C)(CPh)$-$(C_2Ph)$, **10**, 105
$Os_3NiC_{14}H_7O_9$
$[NiOs_3(CO)_9Cp(H)_2]^-$, **9**, 20
$Os_3NiC_{14}H_8O_9$
$(H)_3CpNiOs_3(CO)_8(CO)$, **10**, 340
$NiOs_3(CO)_9Cp(H)_3$, **9**, 20; **10**, 38, 40, 41
$Os_3Ni_3C_{24}H_{15}O_9$
$Os_3Ni_3Cp_3(Co)_9$, **10**, 38
$Os_3PtC_{25}H_{22}O_9P_2S_2$
$Os_3Pt(S)_2(CO)_9(PMe_2Ph)_2$, **10**, 235
$Os_3PtC_{28}H_{35}O_{10}P$
$Os_3Pt(H)_2(CO)_{10}(PCy_3)$, **10**, 35
$Os_3PtC_{28}H_{36}O_{10}P$
$[Os_3Pt(H)_3(CO)_{10}(PCy_3)]^+$, **10**, 202
$Os_3PtC_{32}H_{33}O_8P_3S$
$Os_3Pt(S)(CO)_8(PMe_2Ph)_3$, **10**, 235
$Os_3PtC_{38}H_{51}O_9PS_2$
$Os_3Pt(S)_2(CO)_9(PCy_3)(C{=}$$C_6H_6Me_4)$, **10**, 235
$Os_3PtSn_2C_{32}H_{45}Cl_4O_{11}P$
$Os_3PtSn(H)_2Cl(SnCl_3)(CO)_{10}$-$(OEt_2)(PCy_3)$, **10**, 217, 218
$Os_3Pt_2C_{14}O_{14}$
$Os_3Pt_2(CO)_{14}$, **10**, 192, 193
$Os_3Pt_2C_{25}H_{26}O_9$
$Os_3Pt_2(H)_2(CO)_9(cod)_2$, **10**, 212
$Os_3Pt_2C_{26}H_{24}O_{10}$
$Os_3Pt_2(CO)_{10}(cod)_2$, **10**, 5, 212
$Os_3RhC_{15}H_{10}O_{12}$
$H_2Os_3Rh(CO)_{10}(acac)$, **10**, 60
$Os_3RuC_{12}H_3O_{12}$
$[H_3RuOs_3(CO)_{12}]^-$, **10**, 60
$Os_3Ru_6C_{28}HO_{28}$
$[HOs_3(CO)_{10}(O_2C)Ru_6(C)$-$(CO)_{16}]^-$, **7**, 937
$Os_3SiC_{15}H_{10}O_{11}$
$Os_3(CH\text{-}TMS)(CO)_{11}$, **7**, 750, 755, 757
$Os_3SiC_{15}H_{16}O_{10}$
$Os_3H[Si(OEt)_3](CO)_9$, **7**, 708
$Os_3SiC_{16}H_{16}O_{11}$
$Os_3H[O(SiEt_3)](CO)_{10}$, **7**, 724
$Os_3H[Si(OEt)_3](CO)_{10}$, **7**, 708
$Os_3SiC_{16}H_{16}O_{13}$
$Os_3H[Si(OEt)_3](CO)_{10}$, **7**, 696
$Os_3SiC_{18}H_{10}NO_{13}$
$Os_3H[Si(OEt)_3](CO)_{10}(MeCN)$, **7**, 726
$Os_3SiC_{18}H_{19}NO_{13}$
$Os_3H[Si(OEt)_3](MeCN)(CO)_{10}$, **7**, 708
$Os_3Si_2C_{12}Cl_6O_{12}$
$Os_3(CO)_{12}(SiCl_3)_2$, **7**, 838
$Os_3Si_2SnC_{21}H_{23}NO_{10}$
$Os_3H[SnC(TMS)_2(C_5H_4N)]$-$(CO)_{10}$, **7**, 696
$Os_3Si_2SnC_{22}H_{26}O_{13}$
$Os_3[SnCH(TMS)_2](MeO_2C$-$CCH_2CO_2Me)(CO)_9$, **7**, 698
$Os_3SnC_{12}H_2Cl_2O_{11}$
$Os_3(SnCl_2)(CH_2)(CO)_{11}$, **7**, 750
$Os_3SnSi_4C_{24}H_{38}O_{10}$
$Os_3H_2[[CH(TMS)_2]SnOC[CH$-$(TMS)_2]](CO)_9$, **7**, 705
$Os_3WC_{14}HO_{14}$
$[WOs_3(H)(CO)_{14}]^-$, **10**, 191, 203
$Os_3WC_{14}H_6O_{11}$
$WOs_3(H)(O)_2(CO)_9Cp$, **10**, 235
$Os_3WC_{16}H_6O_{11}$
$WOs_3(H)(CO)_{11}Cp$, **10**, 208
$Os_3WC_{16}H_6O_{11}S$
$Os_3W(CO)_{11}Cp(H)(S)$, **10**, 15
$Os_3WC_{20}H_{11}O_{12}PS_2$
$WOs_3(S)_2(CO)_{12}(PMe_2Ph)$, **10**, 228, 236
$Os_3WC_{21}H_{20}O_{10}$
$Cp^*WOs_3(CO)_9(H)(O)(CHMe)$, **10**, 96
$Os_3WC_{24}H_{10}O_{11}$
$CpWOs_3(CO)_{11}(C{\equiv}CPh)$, **10**, 21, 104
$Os_3WC_{27}H_{22}O_{11}PS_2$
$WOs_3(S)_2(CO)_{11}(PMe_2Ph)_2$, **10**, 236, 237
$Os_3WC_{28}H_{22}O_{12}P_2S_2$
$WOs_3(S)_2(CO)_{12}(PMe_2Ph)_2$, **10**, 236, 237
$Os_4AsC_{18}H_7O_{12}$
$Os_4(H)_2(AsPh)(CO)_{12}$, **7**, 861
$Os_4AuC_{19}H_{16}O_{13}P$
$Os_4Au(H)(CO)_{13}(PEt_3)$, **7**, 850
$Os_4Au_2C_{48}H_{32}O_{12}P_2$
$Au_2Os_4(H)_2(CO)_{12}(PPh_3)_2$, **10**, 278
$Os_4C_{11}H_3NO_{12}$
$Os_4(H)_3(CO)_{11}(NO)$, **7**, 846, 861
$Os_4C_{11}H_4O_{11}$
$[Os_4(H)_4(CO)_{11}]^{2-}$, **7**, 849
$Os_4C_{12}H_2IO_{12}$
$[Os_4(H)_2(CO)_{12}I]^-$, **7**, 870
$Os_4C_{12}H_2O_{12}S$
$Os_4(H)_2(S)(CO)_{12}$, **7**, 866
$Os_4C_{12}H_2O_{12}S_2$
$Os_4(H)_2(S)_2(CO)_{12}$, **7**, 866
$Os_4C_{12}H_3IO_{12}$
$Os_4(H)_3(I)(CO)_{12}$, **7**, 869
$Os_4C_{12}H_3NO_{13}$
$Os_4(H)_3(NO)(CO)_{12}$, **7**, 853
$Os_4C_{12}H_3O_{12}$
$[Os_4(H)_3(CO)_{12}]^-$, **7**, 849
$Os_4C_{12}H_4O_{12}$
$Os_4(H)_4(CO)_{12}$, **7**, 862
$Os_4C_{12}H_4O_{13}$
$Os_4(H)_3(OH)(CO)_{12}$, **7**, 862
$Os_4C_{12}H_5O_{16}P$
$Os_4(H)_3(OPO_3H_2)(CO)$, **7**, 862
$Os_4C_{12}NO_{12}$
$[Os_4(N)(CO)_{12}]^-$, **7**, 854
$Os_4C_{12}O_{12}S$
$Os_4(S)(CO)_{12}$, **7**, 865

$Os_4C_{12}O_{12}S_2$
$Os_4(S)_2(CO)_{12}$, **7**, 865, 866
$Os_4C_{12}O_{16}$
$Os_4O_4(CO)_{12}$, **7**, 723, 864
$Os_4C_{13}H_2O_{13}$
$Os_4(H)_2(CO)_{13}$, **7**, 870
$Os_4C_{13}H_3BrO_{13}$
$Os_4(H)_2(Br)(CO)_{13}(H)$, **7**, 870
$Os_4C_{13}H_3NO_{12}$
$Os_4(NMe)(CO)_{12}$, **7**, 853
$Os_4C_{13}O_{12}S_2$
$Os_4(S)(CS)(CO)_{12}$, **7**, 901
$Os_4C_{13}O_{13}S$
$Os_4(S)(CO)_{13}$, **7**, 865
$Os_4C_{13}O_{13}S_2$
$Os_4(S)_2(CO)_{13}$, **7**, 866
$Os_4C_{14}H_4NO_{13}$
$[Os_4(H)[NC(O)Me](CO)_{12}]^-$, **7**, 854
$Os_4C_{14}H_7NO_{12}S$
$Os_4(S)(CO)_{12}(NMe_2H)$, **7**, 866
$Os_4C_{14}H_{13}O_{14}P$
$Os_4(H)_4(CO)_{11}[P(OMe)_3]$, **7**, 846
$Os_4C_{14}O_{14}$
$Os_4(CO)_{14}$, **7**, 840
$Os_4C_{15}F_3O_{15}P$
$Os_4(CO)_{15}(PF_3)$, **7**, 843
$Os_4C_{15}H_2O_{15}$
$Os_4(H)(CO)_{15}(H)$, **7**, 848
$Os_4C_{15}H_4O_{13}S$
$Os_4(S)[C_2(CO_2Me)(H)](CO)_{11}$, **7**, 877
$Os_4C_{15}H_6O_{11}$
$Os_4(H)(C_3H_2Me)(CO)_{11}$, **7**, 879
$Os_4C_{15}H_6O_{13}S$
$Os_4(H)_2(S)[C_2(CO_2Me)(H)](CO)_{11}$, **7**, 878
$Os_4C_{15}O_{15}$
$Os_4(CO)_{15}$, **7**, 838, 841
$Os_4C_{16}F_6O_{16}$
$Os_4(O_2CCF_3)_2(CO)_{12}$, **7**, 862
$Os_4C_{16}H_6O_{12}$
$Os_4(C_4H_6)(CO)_{12}$, **7**, 879
$Os_4C_{16}H_6O_{14}S$
$Os_4(H)_2(S)[C_2(CO_2Me)(H)](CO)_{12}$, **7**, 878
$Os_4C_{16}H_7N_2O_{12}$
$[Os_4(H)(CO)_{12}(NCMe)_2]^+$, **7**, 850
$Os_4C_{16}H_8O_{10}$
$Os_4(H)_2(C_6H_6)(CO)_{10}$, **7**, 887
$Os_4C_{16}H_9O_{13}P$
$Os_4(CO)_{13}(PMe_3)$, **7**, 841
$Os_4C_{16}H_{11}O_{13}P$
$Os_4(H)_2(CO)_{13}(PMe_3)$, **7**, 848
$Os_4C_{16}O_{16}$
$Os_4(CO)_{16}$, **7**, 838, 842
$Os_4C_{17}H_9O_{14}P$
$Os_4(CO)_{14}(PMe_3)$, **7**, 842
$Os_4C_{17}H_{10}O_{11}$
$Os_4(H)_2(C_6H_8)(CO)_{11}$, **7**, 880
$Os_4C_{17}H_{11}ClO_{12}S$
$Os_4(H)(SCH_2CMe_2CH_2Cl)(CO)_{12}$, **7**, 864
$Os_4C_{17}H_{12}O_{11}$
$Os_4(H)_3(C_6H_9)(CO)_{11}$, **7**, 886
$Os_4C_{18}H_9O_{15}P$
$Os_4(CO)_{15}(PMe_3)$, **7**, 302
$Os_4C_{18}H_{10}ClO_{13}S$
$[Os_4(SCH_2CMe_2CH_2Cl)(CO)_{13}]^-$, **7**, 864
$Os_4C_{18}H_{11}ClO_{13}S$
$Os_4(H)(SCH_2CMe_2CH_2Cl)(CO)_{13}$, **7**, 864
$Os_4C_{18}H_{13}O_{12}P$
$Os_4(H)_2(PCy)(CO)_{12}$, **7**, 858
$Os_4C_{19}H_{10}O_{11}$
$Os_4(H)_3(CHCHPh)(CO)_{11}$, **7**, 873
$Os_4C_{19}H_{16}O_9$
$[C_5Me_3(CH_2)_2]Os_4(H)_3(CO)_9$, **7**, 888
$Os_4C_{20}H_6O_{12}S$
$Os_4(S)[C_2(Ph)H](CO)_{12}$, **7**, 876
$Os_4C_{20}H_{16}O_{10}$
$(C_5Me_4CH_2)Os_4(H)_2(CO)_{10}$, **7**, 887
$Os_4C_{20}H_{18}N_2O_{11}S$
$Os_4(S)[C(Me)C(NMe_2)C(NMe_2)C(Me)](CO)_{11}$, **7**, 891
$Os_4C_{21}H_6O_{12}$
$Os_4(C_9H_6)(CO)_{12}$, **7**, 872
$Os_4C_{21}H_{12}O_{11}S$
$Os_4(S)[C(CH_2)_2][(CH_2)_2CCHC(CO_2Me)](CO)_{11}$, **7**, 889
$Os_4C_{21}H_{14}O_{11}$
$(C_5Me_4CH_2)Os_4(CO)_{11}$, **7**, 887
$Os_4C_{21}H_{18}N_2O_{11}S$
$Os_4(S)[C_2(NMe_2)(Me)][C_2(NMe_2)(Me)](CO)_{11}$, **7**, 879, 891
$Os_4C_{22}H_{16}O_{12}$
$Cp^*Os_4(H)(CO)_{12}$, **7**, 884
$Os_4C_{23}H_{16}O_{13}$
$Cp^*Os_4(H)(CO)_{13}$, **7**, 884
$Os_4C_{24}H_{18}N_2O_{14}$
$(CO)_3(Bu^tCN)_2OsOs_3(CO)_{11}$, **7**, 843
$Os_4C_{24}H_{27}N_3O_9S$
$Os_4(S)[C(Me)C(NMe_2)C(Me)C(NMe_2)][C(Me)C(NMe_2)](CO)_9$, **7**, 891
$Os_4C_{25}H_{12}O_{13}S$
$Os_4(S)[C(CO_2Me)C(CO_2Me)C(H)C(Ph)](CO)_{11}$, **7**, 890
$Os_4C_{25}H_{12}O_{15}S$
$Os_4[SC_2(H)(Ph)][C_2(CO_2Me)_2](CO)_{11}$, **7**, 890
$Os_4(S)[C(CO_2Me)C(CO_2Me)C(H)C(Ph)](CO)_{11}$, **7**, 890
$Os_4C_{25}H_{14}O_9$
$Os_4(C_6H_6)(C_6H_8)(CO)_9$, **7**, 887
$Os_4C_{27}H_{14}O_9$
$Os_4(H)(C_9H_6)(CO)_9(C_9H_7)$, **7**, 872
$Os_4C_{37}H_{22}O_9$
$Os_4(H)_2(C_2Ph_2)_2(CO)_9$, **7**, 873
$Os_4PtC_{15}H_2O_{15}$
$Os_4Pt(H)_2(CO)_{15}$, **10**, 200
$Os_4PtC_{31}H_{15}O_{13}PS$
$Os_4Pt(S)(CO)_{13}(PPh_3)$, **10**, 232, 233
$Os_4PtC_{31}H_{15}O_{13}S$
$Os_4Pt(CO)_{13}(PPh_3)(S)$, **10**, 20, 21
$Os_4Pt_2C_{18}O_{18}$
$Os_4Pt_2(CO)_{18}$, **10**, 9, 10, 197
$Os_4Pt_2C_{27}H_{24}O_{11}$
$Os_4Pt_2(CO)_{11}(cod)_2$, **10**, 212
$Os_4Pt_2C_{28}H_{24}O_{12}$
$Os_4Pt_2(CO)_{12}(cod)_2$, **10**, 212, 213
$Os_4SiC_{17}V_{10}O_{14}$
$Os_4(H)(CO)_{14}(TMS)$, **7**, 842
$Os_4Sn_4C_{22}H_{24}O_{14}$
$Os_4(SnMe_2)_4(CO)_{14}$, **7**, 852
$Os_4Sn_4C_{22}H_{24}O_{16}$
$Os_4(O)_2(SnMe_2)_4(CO)_{14}$, **7**, 852
$Os_5AsC_{21}H_5O_{15}$
$Os_5(AsPh)(CO)_{15}$, **7**, 925
$Os_5Au_2C_{51}H_{30}O_{14}P_2$
$Os_5(C)[Au(PPh_3)]_2(CO)_{14}$, **7**, 920
$Os_5C_{14}H_2O_{14}S_2$
$Os_5(H)_2(S)_2(CO)_{14}$, **7**, 922
$Os_5C_{15}HO_{15}$
$[Os_5(H)(CO)_{15}]^-$, **7**, 911, 918
$Os_5C_{15}H_2O_{15}$
$Os_5(H)_2(CO)_{15}$, **7**, 911, 918
$Os_5C_{15}IO_{15}$
$[Os_5I(CO)_{15}]^-$, **7**, 918
$Os_5C_{15}O_{15}$
$[Os_5(CO)_{15}]^{2-}$, **7**, 911, 917, 940
$Os_5C_{15}O_{15}S$
$Os_5(S)(CO)_{15}$, **7**, 922
$Os_5C_{16}HO_{15}$
$H_2Os_5(C)(CO)_{15}$, **7**, 920
$Os_5C_{16}H_2O_{15}$
$H_2Os_5(C)(CO)_{15}$, **7**, 920
$Os_5C_{16}H_2O_{16}$
$Os_5(H)_2(CO)_{16}$, **7**, 918
$Os_5C_{16}H_3O_{16}P$
$Os_5[P(OMe)](CO)_{15}$, **7**, 919, 924, 925
$Os_5C_{16}H_5NO_{14}$
$H_2Os_5(CO)_{14}(NCMe)$, **7**, 918
$Os_5C_{16}I_2O_{15}$
$Os_5(C)I_2(CO)_{15}$, **7**, 919, 920
$Os_5C_{16}O_{16}$
$Os_5(CO)_{16}$, **7**, 909, 917
$Os_5C_{17}H_3NO_{15}$
$Os_5(CO)_{15}(NCMe)$, **7**, 918
$Os_5C_{17}H_7O_{17}P$
$Os_5(H)(C)[OP(OMe)_2](CO)_{14}$, **7**, 919, 920
$Os_5C_{17}H_{10}O_{18}P_2$
$Os_5(H)(C)[OP(OMe)OP(OMe)_2](CO)_{13}$, **7**, 919, 920
$Os_5C_{17}O_{17}$
$Os_5(CO)_{17}$, **7**, 918
$Os_5C_{18}H_9O_{15}P$
$Os_5(CO)_{15}(PMe_3)$, **7**, 841
$Os_5C_{18}H_9O_{18}P$

$Os_5(CO)_{15}[P(OMe)_3]$, **7**, 917, 918
$Os_5C_{18}H_{10}O_{12}$
$Os_5(H)_4(C_6H_6)(CO)_{12}$, **7**, 849
$Os_5C_{18}O_{18}$
$Os_5(CO)_{18}$, **7**, 911, 917
$Os_5C_{19}H_7NO_{14}$
$Os_5(H)_2(CO)_{14}(py)$, **7**, 918, 923
$Os_5C_{19}H_{16}O_{19}P_2$
$Os_5(H)(C)[OP(OMe)_2](CO)_{13}$-$[P(OMe)_3]$, **7**, 919, 920
$Os_5C_{19}O_{19}$
$Os_5(CO)_{19}$, **7**, 911, 917
$Os_5C_{20}H_5NO_{14}$
$Os_5(H)(C)(NC_5H_4)(CO)_{14}$, **7**, 920, 923
$Os_5C_{20}H_5NO_{18}$
$H_2Os_5(CO)_{18}(NCMe)$, **7**, 918
$Os_5C_{20}H_7NO_{15}$
$H_2Os_5(CO)_{15}(py)$, **7**, 923
$Os_5C_{20}H_{17}O_{14}P$
$Os_5(H)_2(CO)_{14}(PEt_3)$, **7**, 918
$Os_5C_{21}H_5O_{15}P$
$Os_5(PPh)(CO)_{15}$, **7**, 925, 938
$Os_5C_{21}H_6O_{15}Se$
$Os_5(H)(SePh)(CO)_{15}$, **7**, 922, 923
$Os_5C_{22}H_{26}O_{16}P_2$
$Os_5(H)_2(CO)_{13}(PEt_3)[P(OMe)_3]$, **7**, 918
$Os_5C_{23}H_7O_{15}$
$Os_5(H)_2(C_2Ph)(CO)_{15}$, **7**, 930, 932
$Os_5C_{25}H_{10}N_2O_{13}$
$Os_5(H)(PhNC_6H_4N)(CO)_{13}$, **7**, 909, 923
$Os_5C_{29}H_{12}O_{13}$
$Os_5(PhC_2H)_2(CO)_{13}$, **7**, 930
$Os_5C_{41}H_{20}O_{13}$
$Os_5(PhC_2Ph)_2(CO)_{13}$, **7**, 930, 932
$Os_5C_{41}H_{22}O_{13}$
$Os_5(H)_2(PhC_2Ph)[PhC_2(H)$-$C_6H_4](CO)_{13}$, **7**, 930, 932
$Os_5C_{42}H_{24}O_{15}P_2$
$Os_5(C)(CO)_{15}(dppe)$, **7**, 919
$Os_5Hg_2C_{21}HF_9O_{21}$
$HOs_5(CO)_{15}(O_2CCF_3)(HgO_2C$-$CF_3)_2$, **10**, 327, 328, 340
$Os_5PtC_{15}H_6O_{15}S$
$Os_5Pt(H)_6(S)(CO)_{15}$, **10**, 202, 235
$Os_5PtC_{16}H_6O_{16}$
$Os_5Pt(H)_6(CO)_{16}$, **10**, 200, 201
$Os_5PtC_{18}H_4O_{18}$
$Os_5Pt(H)_4(CO)_{18}$, **10**, 200, 201
$Os_5PtC_{51}H_{30}O_{15}S$
$Os_5Pt(CO)_{15}(PPh_3)_2(S)$, **10**, 20
$Os_5Pt_2C_{17}H_6O_{17}$
$Os_5Pt_2(H)_6(CO)_{17}$, **10**, 200, 201
$Os_5RhC_{21}H_6O_{15}$
$Os_5Rh(CO)_{15}(C_6H_6)$, **10**, 5, 6
$Os_6AgC_{22}H_2O_{22}$
$[Os_3H(CO)_{11}]_2Ag$, **7**, 720
$Os_6C_{16}O_{16}S_2$
$Os_6(S)_2(CO)_{16}$, **7**, 727, 941, 942
$Os_6C_{17}HO_{17}$
$[HOs_6(CO)_{17}]^-$, **7**, 933
$Os_6C_{17}H_2O_{17}S_2$
$Os_6(H)_2(S)_2(CO)_{17}$, **7**, 942, 943
$Os_6C_{17}O_{17}S$
$Os_6(S)(CO)_{17}$, **7**, 726–8, 942
$Os_6C_{17}O_{17}S_2$
$Os_6(S)_2(CO)_{17}$, **7**, 941, 942
$Os_6C_{18}HO_{18}$
$[Os_6(H)(CO)_{18}]^-$, **7**, 933
$Os_6C_{18}HO_{18}P$
$Os_6(H)(P)(CO)_{18}$, **7**, 939
$Os_6C_{18}H_2O_{18}$
$Os_6(H)_2(CO)_{18}$, **7**, 933
$Os_6C_{18}H_2O_{18}S_2$
$Os_6(H)_2(S)_2(CO)_{18}$, **7**, 942, 943
$Os_6C_{18}H_2O_{18}Se$
$Os_6(H)(SeH)(CO)_{18}$, **7**, 944
$Os_6C_{18}H_2O_{19}S_2$
$Os_6(H)(S)_2(OH)(CO)_{18}$, **7**, 943
$Os_6C_{18}H_4O_{16}S_3$
$Os_6(S)[S(CH_2)_2S](CO)_{16}$, **7**, 909, 943
$Os_6C_{18}NO_{18}$
$[Os_6(NCO)(CO)_{17}]^-$, **7**, 940
$Os_6C_{18}O_{17}$
$Os_6(C)(CO)_{17}$, **7**, 934
$Os_6C_{18}O_{18}$
$[Os_6(CO)_{18}]^{2-}$, **7**, 911, 928, 930, 932, 933, 945
$Os_6C_{18}O_{18}P$
$[Os_6(P)(CO)_{18}]^-$, **7**, 939
$Os_6C_{19}H_2O_{19}$
$H_2Os_6(CO)_{19}$, **7**, 933
$Os_6C_{19}H_3NO_{17}$
$Os_6(CO)_{17}(NCMe)$, **7**, 930
$Os_6C_{19}H_4O_{17}S_2$
$Os_6[S(CH_2)_2S](CO)_{17}$, **7**, 942, 943
$Os_6C_{19}H_9NO_{16}S_2$
$Os_6(H)_2(S)_2[C(H)NMe_2](CO)_{16}$, **7**, 943
$Os_6C_{19}O_{19}S$
$Os_6(S)(CO)_{19}$, **7**, 727, 942
$Os_6C_{19}O_{20}$
$Os_6(O)(CO)_{19}$, **7**, 938
$Os_6C_{20}H_3NO_{18}$
$Os_6(NCMe)(CO)_{18}$, **7**, 940
$Os_6C_{20}H_6O_{16}$
$Os_6(CMe)_2(CO)_{16}$, **7**, 936, 944, 945
$Os_6C_{20}O_{20}S_2$
$Os_6(S)_2(CO)_{20}$, **7**, 728
$Os_6C_{21}H_2O_{21}P$
$(CO)_{10}(H)Os_3(PH)Os_3(CO)_{11}$, **7**, 939
$Os_6C_{21}H_6O_{16}$
$Os_6(C)(MeC_2Me)(CO)_{16}$, **7**, 936, 944
$Os_6C_{21}H_9NO_{15}$
$Os_6(CO)_{15}(MeC_2Me)(NCMe)$, **7**, 944
$Os_6C_{21}O_{21}$
$Os_6(CO)_{21}$, **7**, 932
$Os_6C_{22}F_6O_{22}$
$Os_6(O_2CCF_3)_2(CO)_{18}$, **7**, 862, 938
$Os_6C_{22}H_2O_{22}$
$Os_6(H)_2(CO)_{22}$, **7**, 933
$Os_6C_{22}H_2O_{24}$
$[Os_3H(CO)_{10}]_2(C_2O_4)$, **7**, 724
$Os_6C_{22}H_3NO_{20}$
$Os_6(CO)_{20}(NCMe)$, **7**, 932
$Os_6C_{22}H_6O_{20}$
$[Os_3(H)_3(CO)_9(CCO)]_2$, **7**, 757
$Os_6C_{22}H_{12}O_{16}S_4$
$Os_6[S(CH_2CH_2S)_3](CO)_{16}$, **7**, 909, 942, 943
$Os_6C_{23}H_6N_2O_{19}$
$Os_6(CO)_{19}(NCMe)_2$, **7**, 932
$Os_6C_{23}H_{11}NO_{15}S_2$
$Os_6(S)(SPh)(CNMe_2)(CO)_{15}$, **7**, 944
$Os_6C_{24}H_7O_{18}P$
$Os_6(H)_2(PPh)(CO)_{18}$, **7**, 925, 938, 939
$Os_6C_{24}H_{11}NO_{15}S_2$
$Os_6(CO)_{15}(S)(SPh)(CNMe_2)$, **7**, 752
$Os_6C_{24}H_{11}N_{15}O_{15}S_2$
$Os_6(H)_2(SMe)_2CNMe_2)_2(CO)_{12}$, **7**, 944
$Os_6C_{24}H_{12}O_{16}$
$Os_6(H)(C_8H_{11})(CO)_{16}$, **7**, 945
$Os_6C_{26}H_{18}N_2O_{16}$
$Os_6(CNBu^t)_2(CO)_{16}$, **7**, 932
$Os_6C_{27}H_{10}N_2O_{17}$
$Os_6(CO)_{17}(py)_2$, **7**, 940
$Os_6C_{28}H_{18}N_2O_{14}S_2$
$Os_6(H)_6(S)_2(HC{=}NPh)_2(CO)_{14}$, **7**, 943
$Os_6C_{29}H_{11}O_{17}P$
$Os_6(CO)_{17}(PHPh_2)$, **7**, 938
$Os_6C_{29}H_{12}N_2O_{15}S_2$
$Os_6(S)_2(HC{=}NPh)_2(CO)_{15}$, **7**, 909, 943
$Os_6C_{29}H_{16}N_2O_{15}S_2$
$Os_6(H)_4(S)_2(HC{=}NPh)_2(CO)_{15}$, **7**, 943
$Os_6C_{29}H_{36}O_{29}P_4$
$Os_6(CO)_{17}[P(OMe)_3]_4$, **7**, 931, 932
$Os_6C_{30}H_{10}O_{16}$
$Os_6(CPh)_2(CO)_{16}$, **7**, 944, 945
$Os_6C_{34}H_{14}N_2O_{18}$
$Os_6(CNC_6H_4Me)_2(CO)_{18}$, **7**, 932
$Os_6HgC_{18}H_2O_{18}S_2$
$[(H)(S)(CO)_9Os_3]_2(Hg)$, **10**, 332
$Os_6PtC_{18}O_{18}$
$Os_6Pt(H)_8(CO)_{18}$, **10**, 200, 201
$Os_6PtC_{26}H_{15}NO_{17}$
$Os_6Pt(NCMe)(CO)_{17}(cod)$, **10**, 225, 226
$Os_6Pt_2C_{32}H_{24}O_{16}$
$Pt_2Os_6(CO)_{16}(cod)_2$, **10**, 5
$Os_6Pt_2C_{52}H_{30}O_{16}P_2S_4$
$Os_6Pt_2(H)_2(S)_4(CO)_{16}(PPh_2$-$C_6H_4)_2$, **10**, 232, 233

$Os_6Pt_4C_{38}H_{24}O_{22}$
  $Cp^*W_2Ru_4(CO)_{12}(PPh)_2$, **10**, 10
$Os_7C_{19}O_{19}S$
  $Os_7(S)(CO)_{19}$, **7**, 946
$Os_7C_{20}HO_{19}$
  $[Os_7(H)(C)(CO)_{19}]^-$, **7**, 946
$Os_7C_{20}H_2O_{19}$
  $Os_7(H)_2(C)(CO)_{19}$, **7**, 946
$Os_7C_{20}H_2O_{20}$
  $H_2Os_7(CO)_{20}$, **7**, 945, 946
$Os_7C_{20}O_{20}$
  $[Os_7(CO)_{20}]^{2-}$, **7**, 945, 947
$Os_7C_{20}O_{20}S_2$
  $Os_7(S)_2(CO)_{20}$, **7**, 946
$Os_7C_{21}HO_{21}$
  $H_2Os_7(CO)_{21}$, **7**, 945, 946
$Os_7C_{21}H_5NO_{19}$
  $H_2Os_7(CO)_{19}(NCMe)$, **7**, 947
$Os_7C_{21}O_{21}$
  $Os_7(CO)_{21}$, **7**, 945
$Os_7C_{22}H_2O_{22}$
  $Os_7(H)_2(CO)_{22}$, **7**, 945, 946
$Os_7C_{23}H_6O_{19}$
  $Os_7(MeC_2Me)(CO)_{19}$, **7**, 947, 948
$Os_7C_{23}H_8O_{19}$
  $H_2Os_7(CO)_{19}(MeC_2Me)$, **7**, 944
$Os_7C_{23}H_9O_{23}P$
  $Os_7(CO)_{20}[P(OMe)_3]$, **7**, 945
$Os_7C_{28}H_8N_2O_{24}$
  $H_2Os_7(CO)_{24}(NCMe)_2$, **7**, 945
$Os_7C_{32}H_{10}O_{18}$
  $Os_7(CPh)_2(CO)_{18}$, **7**, 947, 948
$Os_7C_{43}H_{20}O_{15}$
  $Os_7(CPh)_4(CO)_{15}$, **7**, 947, 948
$Os_7Pt_2C_{23}H_8O_{23}$
  $Os_7Pt_2(H)_8(CO)_{23}$, **10**, 200, 201
$Os_8Au_2C_{58}H_{30}O_{22}P_2$
  $Au_2Os_8(CO)_{22}(PPh_3)_2$, **10**, 289
$Os_8C_{22}HO_{22}$
  $[Os_8(H)(CO)_{22}]^-$, **7**, 948
$Os_8C_{22}H_2O_{22}$
  $H_2Os_8(CO)_{22}$, **7**, 948
$Os_8C_{22}O_{22}$
  $[Os_8(CO)_{22}]^{2-}$, **7**, 947, 948
$Os_8C_{23}O_{22}$
  $Os_8(C)(CO)_{22}$, **7**, 948, 949
$Os_8C_{23}O_{23}$
  $Os_8(CO)_{23}$, **7**, 947
$Os_8C_{25}H_{11}O_{25}P$
  $H_2Os_8(CO)_{22}[P(OMe)_3]$, **7**, 948
$Os_8C_{26}HO_{27}$
  $[HOs_3(CO)_{10}(O_2C)Os_5(CO)_{15}]^-$, **7**, 918
$Os_8C_{32}H_8N_2O_{28}$
  $H_2Os_8(CO)_{28}(NCMe)_2$, **7**, 948
$Os_9B_3C_{30}H_6O_{30}$
  $[(H)_2Os_3(CO)_9(C)]_3[B_3O_3]$, **1**, 296, 297
$Os_9C_{24}HO_{24}$
  $[Os_9(H)(CO)_{24}]^-$, **7**, 949, 950
$Os_9C_{24}O_{24}$
  $[Os_9(CO)_{24}]^{2-}$, **7**, 950
$Os_9C_{25}H_5O_{21}$
  $[Os_9[CHC(Me)CH](CO)_{21}]^-$, **7**, 909, 950
$Os_9C_{28}HO_{29}$
  $[HOs_3(CO)_{10}(O_2C)Os_6(CO)_{17}]^-$, **7**, 749, 938
$Os_9C_{31}HO_{32}$
  $[HOs_3(CO)_{10}(O_2C)Os_6(CO)_{20}]^-$, **7**, 938
$Os_9C_{36}H_8N_2O_{32}$
  $H_2Os_9(CO)_{32}(NCMe)_2$, **7**, 950
$Os_9Hg_2C_{22}O_{21}$
  $[Os_9C(CO)_{21}](Hg_2)^{2-}$, **10**, 340
$Os_9Hg_3C_{33}O_{33}$
  $[Os_3(CO)_{11}Hg]_3$, **10**, 334, 335, 340
$Os_{10}Au_5BC_{97}H_{60}F_4O_{24}P_4$
  $[Au_5Os_{10}C(CO)_{24}(PPh_3)_4][BF_4]$, **10**, 295
$Os_{10}C_{23}INO_{23}$
  $[Os_{10}(C)I(CO)_{22}(NO)]^{2-}$, **7**, 951, 952
$Os_{10}C_{23}O_{23}S_2$
  $Os_{10}(S)_2(CO)_{23}$, **7**, 953
$Os_{10}C_{24}H_4O_{24}$
  $[H_4Os_{10}(CO)_{24}]^-$, **7**, 950, 951
$Os_{10}C_{24}NO_{24}$
  $[Os_{10}(C)(CO)_{23}(NO)]^-$, **7**, 952
$Os_{10}C_{25}H_2O_{24}$
  $H_2Os_{10}(C)(CO)_{24}$, **7**, 951
$Os_{10}C_{25}NO_{25}$
  $[Os_{10}(C)(NO)(CO)_{24}]^-$, **7**, 952
$Os_{10}C_{25}O_{24}$
  $[Os_{10}(C)(CO)_{24}]^{2-}$, **7**, 908, 909, 923, 950, 951, 953
$Os_{10}C_{26}O_{26}$
  $[Os_{10}(CO)_{26}]^{2-}$, **7**, 950
$Os_{10}C_{27}H_9I_2O_{26}P$
  $Os_{10}(C)(I)_2(CO)_{23}[P(OMe)_3]$, **7**, 952
$Os_{10}C_{31}HO_{32}$
  $HOs_3(CO)_{10}(O_2C)Os_7(CO)_{20}$, **7**, 945
$Os_{10}C_{34}H_{36}O_{33}P_4$
  $Os_{10}(C)(CO)_{21}[P(OMe)_3]_4$, **7**, 952
$Os_{10}C_{40}O_{40}$
  $[Os_{20}(CO)_{40}]^{2-}$, **7**, 953, 954
$Os_{10}CuC_{79}H_{45}NO_{24}P_3$
  $[N(PPh_3)_2][CuOs_{10}C(CO)_{24}$-$(PPh_3)]$, **10**, 299
$Os_{10}HgC_{25}ClO_{24}$
  $Os_{10}(CO)_{24}C(HgCl)^-$, **10**, 332
$Os_{10}HgC_{27}F_3O_{26}$
  $Os_{10}C(CO)_{24}(HgO_2CCF_3)^-$, **10**, 340
$Os_{11}C_{28}IO_{27}$
  $[Os_{11}C(CO)_{27}I]^-$, **7**, 953
$Os_{11}C_{28}I_2O_{27}$
  $Os_{11}C(CO)_{27}I_2$, **7**, 953
$Os_{11}C_{28}O_{27}$
  $[Os_{11}(C)(CO)_{27}]^{2-}$, **7**, 953, 954
$Os_{11}C_{30}O_{33}$
  $Os_{11}(O)_3(CO)_{30}$, **7**, 909, 953, 954
$Os_{17}C_{36}O_{36}$
  $[Os_{17}(CO)_{36}]^{2-}$, **7**, 909, 953, 954
$Os_{18}HgC_{44}HO_{42}$
  $H[Os_9C(CO)_{21}]_2(Hg)^{3-}$, **10**, 329
$Os_{18}Hg_3C_{44}O_{42}$
  $[Os_9C(CO)_{21}]_2(Hg_3)$, **10**, 331
$Os_{20}C_{40}O_{40}$
  $[Os_{20}(CO)_{40}]^{2-}$, **7**, 909, 954
$Os_{20}HgC_{50}O_{48}$
  $[Os_{10}(CO)_{24}C]_2(Hg)^{2-}$, **10**, 332

P

$PbB_{11}H_{11}$
$[B_{11}H_{11}Pb]^{2-}$, **1**, 259
PbC_2H_6
Me_2Pb, **2**, 305
PbC_3H_9
Me_3Pb^+, **2**, 305
PbC_4F_{12}
$Pb(CF_3)_4$, **2**, 307
$PbC_4H_3F_9$
$MePb(CF_3)_3$, **2**, 307
$PbC_4H_6F_6$
$Me_2Pb(CF_3)_2$, **2**, 307
PbC_4H_{12}
Me_4Pb, **12**, 154
$PbC_4H_{12}S$
Me_3PbSMe, **2**, 309
PbC_5H_5Cl
CpPbCl, **2**, 315
$PbC_5H_{10}O_4$
$Me_3Pb(ox)$, **2**, 309
$PbC_8H_{12}O_8$
$Pb(OAc)_4$, **11**, 403
PbC_8H_{14}
Me_3PbCp, **2**, 306, 313
PbC_8H_{20}
Et_4Pb, **2**, 317; **11**, 462
$PbC_8H_{24}O_4$
$Ph(OAc)_4$, **12**, 141
$PbC_{10}H_{10}$
Cp_2Pb, **2**, 305, 307, 313, 315
$PbC_{10}H_{12}O_6S$
$(C_4H_3S)Pb(O_2CMe)_3$, **11**, 467
$PbC_{10}H_{12}O_7$
$(C_4H_3O)Pb(O_2CMe)_3$, **11**, 467
$PbC_{11}H_{20}$
Et_3PbCp, **2**, 313
$PbC_{12}H_{14}O_6$
$PhPb(O_2CMe)_3$, **11**, 467
$PbC_{13}H_{13}F_3O_6$
$[(CF_3)C_6H_4]Pb(O_2CMe)_3$, **11**, 482
$PbC_{13}H_{16}O_7$
$[(MeO)C_6H_4]Pb(O_2CMe)_3$, **11**, 465, 466, 469, 471
$PbC_{14}H_{10}F_6O_6S_2$
$Ph_2Pb(OTf)_2$, **2**, 309
$PbC_{14}H_{14}O_6$
$PhC{\equiv}CPb(O_2CMe)_3$, **11**, 469
$PbC_{14}H_{16}O_6$
$PhCH{=}CHPb(O_2CMe)_3$, **11**, 471
$PbC_{14}H_{18}O_8$
$[(MeO)_2C_6H_3]Pb(O_2CMe)_3$, **11**, 466, 482
$PbC_{14}H_{32}O$
$Bu_3(MeOCH_2)Pb$, **11**, 463
$PbC_{15}H_{18}O_7$
$(CH_2{=}CHCH_2O)C_6H_4Pb(O_2C\text{-}Me)_3$, **11**, 480
$PbC_{15}H_{20}O_9$
$[(MeO)_3C_6H_2]Pb(O_2CMe)_3$, **11**, 466
$PbC_{16}H_{36}$
Bu_4Pb, **2**, 317; **11**, 462
$PbC_{17}H_{15}$
$CpPh_2Pb$, **2**, 313
$PbC_{18}H_{15}$
Ph_3Pb, **2**, 306, 307
$PbC_{21}H_{20}$
$(CH_2{=}CH{-}CH_2)Ph_3Pb$, **11**, 464
$PbC_{22}H_{20}$
Ph_2PbCp_2, **2**, 307
$PbC_{23}H_{20}$
Ph_3PbCp, **2**, 307, 313
$PbC_{24}H_{20}S$
Ph_3PbSPh, **2**, 309
$PbC_{24}H_{22}$
$(CpCH_2)Ph_3Pb$, **2**, 308
$PbC_{28}H_{28}$
$Pb(Tol)_4$, **2**, 306
$PbC_{36}H_{30}ClOP$
$Ph_3PbCl{\cdot}Ph_3PO$, **2**, 308
$PbC_{50}H_{42}P_4$
$Pb[(Ph_2P)_2CH]_2$, **2**, 315
$PbFe_4C_{12}N_4O_{16}$
$Ph[Fe(CO)_3(NO)]_4$, **7**, 31
$PbLiC_{18}H_{15}$
Ph_3PbLi, **2**, 312
$PbLiC_{27}H_{38}N_3$
$Ph_3PbLi(pmdeta)$, **2**, 311
$PbMnC_{25}H_{21}O_2$
$MnHCp(CO)_2(PbPh_3)$, **6**, 68
$PbMoC_{20}H_{18}O_7$
$Pb[Mo(CO)_3Cp]_2(THF)$, **5**, 409
$PbPtC_{39}H_{39}ClP_2$
$PtCl(PbMe_3)(PPh_3)_2$, **9**, 436
$PbPt_2C_{80}H_{72}N_2$
$[Bu_4N]_2Pb[Pt(C_6F_5)_4]_2$, **9**, 474
$PbSiC_{27}H_{42}$
$Ph_3PbSi(TMS)_3$, **2**, 312
$PbSi_4C_{14}H_{38}$
$Pb[CH(TMS)_2]_2$, **2**, 313
$PbSi_4C_{62}H_{76}P_4$
$Pb[(Ph_2P)_2C(TMS)_2]_2$, **2**, 315
$PbSnC_{24}H_{40}O_6$
$[(Bu_3Sn)C_6H_4]Pb(O_2CMe)_3$, **11**, 467
$PbSnC_{42}H_{42}$
$(Tol)_3PbSn(Tol)_3$, **2**, 312
$PbVC_{24}H_{15}O_6$
$Ph_3PbV(CO)_6$, **2**, 312; **5**, 6
$PbWC_{26}H_{20}O_3$
$CpW(Co)_3PbPh_3$, **2**, 313
$Pb_2C_6H_{18}$
Pb_2Me_6, **2**, 305
$Pb_2C_{11}H_{22}$
$C_5H_4(PbMe_3)_2$, **2**, 306
$Pb_2C_{18}H_{22}O_{12}$
$C_6H_4[(Pb(O_2CMe)_3]_2$, **11**, 467
$Pb_2C_{36}H_{30}$
$Ph_3PbPbPh_3$, **2**, 312; **11**, 464
$Pb_2C_{36}H_{66}$
Cy_6Pb_2, **2**, 312
$Pb_2C_{42}H_{42}$
$Pb_2(Tol)_6$, **2**, 306
$Pb_2C_{46}H_{43}$
$Ph(Tol)_3PbPbPh_2(Tol)$, **2**, 312
$Pb_2CoC_{39}H_{30}O_3$
$(Ph_3Pb)_2Co(CO)_3$, **2**, 312
$Pb_2FeC_{16}H_{28}$
$(Me_3PbCp)_2Fe$, **2**, 306
$Pb_2Si_6C_{18}H_{54}$
$(TMS)_6Pb_2$, **2**, 23
$Pb_3C_{36}H_{30}S_3$
$(Ph_2PbS)_3$, **2**, 310
$PdBC_8H_{12}F_8N_4$
$[Pd(MeCN)_4][BF_4]_2$, **12**, 1225, 1226
$PdBC_9H_{21}F_4N_2$
$[Pd(allyl)(TMEDA)][BF_4]$, **9**, 353
$PdBC_{15}H_{25}F_4N_2$
$[Pd(PhC_3H_4)(TMEDA)][BF_4]$, **9**, 327
$PdBC_{19}H_{38}F_4P$
$[PdCp(CH_2{=}CHPh)(PEt_3)]\text{-}[BF_4]$, **9**, 299
$PdBC_{27}H_{29}F_4NOP$
$[Pd(MeC_3H_4)[CH(COMe)PPh_3](CNMe)]BF_4$, **9**, 228
$PdBC_{29}H_{32}ClF_4NOP$
$[Pd(ClC_3H_4)(CNBu^t)Ph_3P\text{-}CHCOMe)][BF_4]$, **9**, 350
$PdBC_{30}H_{35}F_4NOP$
$[Pd(MeC_3H_4)[CH(COMe)\text{-}PPh_3](CNBu^t)]BF_4$, **9**, 215, 230
$PdBC_{31}H_{27}ClF_4P$
$[PdCp(PPh_3)(ClC_6H_4CH{=}CH_2)][BF_4]$, **9**, 301
$PdBC_{32}F_4O_4P$
$[PdCp(PPh_3)][(MeO)C_6H_4\text{-}CH{=}CH_2][BF_4]$, **9**, 301
$PdBC_{39}H_{35}F_4P_2$
$[Pd(C_3H_5)(PPh_3)_2][BF_4]$, **9**, 355
$PdBC_{45}H_{23}F_{24}N_2O_2$
$[Pd(bipy)(CO)(COMe)]\text{-}[B((CF_3)_2C_6H_3)_4]$, **12**, 1225
$PdB_2C_{20}H_{18}$
$Pd(C_4H_4BPh)_2$, **1**, 204
$PdB_2C_{36}H_{34}F_8N_2P_2$
$Pd(dppe)(py)_2(BF_4)_2$, **12**, 789

$PdB_3C_{25}H_{29}P_2$
$[Ph_2P(CH_2)PPh_2]Pd(B_3H_7)$, **1**, 316
$PdB_3FeC_{34}H_{35}P_2$
$(dppf)PdB_3H_7$, **1**, 317
$PdB_8C_{16}H_{34}P_2$
$(Me_2PhP)_2PdB_8H_{12}$, **1**, 337
$PdB_8C_{36}H_{42}P_2$
$(PPh_3)_2PdB_8H_{12}$, **1**, 337
$PdB_8PtC_{32}H_{54}P_4$
$(PhMe_2P)_4PtPdB_8H_{10}$, **1**, 337
$PdB_9Sb_2C_{16}H_{31}$
$(Me_2PhP)_2PdSb_2B_9H_9$, **1**, 267
$PdB_{10}C_{18}H_{25}PSe$
$(PPh_3)PdSeB_{10}H_{10}$, **1**, 270
$PdB_{10}C_{36}H_{39}ClP_2Te$
$Cl(PPh_3)PdTeB_{10}H_9(PPh_3)$, **1**, 271
$PdB_{11}C_{36}H_{41}F_4OP_2Te$
$[(H_2O)(PPh_3)PdTeB_{10}H_9\text{-}(PPh_3)]BF_4$, **1**, 271
$PdB_{20}C_{22}H_{59}N_3$
$[BzNMe_3]_2[(B_{10}H_{12})_2Pd]\cdot MeCN$, **1**, 351
$PdCCl_3O$
$[PdCl_3(CO)]^-$, **9**, 196
$PdCH_5OP$
$Pd(H)_2(CO)(PH_3)$, **9**, 208
$PdH(COH)(PH_3)$, **9**, 208
PdCO
Pd(CO), **9**, 194
PdC_2F_6
$Pd(CF_3)_2$, **9**, 233
$PdC_2H_4Cl_3$
$[PdCl_3(C_2H_4)]^-$, **9**, 294
$[PdCl_3(CHMe]^-$, **9**, 295
$PdC_2H_{10}OP_2$
$Pd(PH_3)_2[H_2CC(O)CH_2]$, **9**, 579
PdC_2O_2
$[Pd(CO)_2]^+$, **9**, 195
PdC_3ClO_3
$[PdCl(CO)_3]^-$, **9**, 196
PdC_4H_4
PdCHCHCHCH, **12**, 760
$PdC_4H_6Cl_2N_2$
$PdCl_2(MeCN)_2$, **12**, 141, 154, 204, 222, 603, 615, 616
$PdCl_2(CNMe)_2$, **9**, 215
$PdC_4H_6O_2$
$Pd(OAc)_2$, **12**, 142, 259
$PdC_4H_6O_4$
$Pd(OAc)_2$, **11**, 399, 420, 491; **12**, 47, 133, 134, 142, 259, 300, 425, 1226
$PdC_4H_9P_2$
(trimethylenemethane)Pd-$(PH_3)_2$, **12**, 934
$PdC_4H_{12}Cl_2P_2S_2$
$Pd_2Cl_2(MeSCCSMe)(PH_3)_2$, **9**, 319
PdC_4O_4
$Pd(CO)_4$, **9**, 194
$PdC_5H_{15}P$
$PdMe_2(PMe_3)$, **9**, 240
$PdC_6Cl_4O_4$
$[PdCl_2(C_6Cl_2O_4)]^{2-}$, **9**, 270
$PdC_6H_9ClN_2$
$PdCl(C_3H_5)(pz)$, **9**, 350
$PdC_6H_9ClN_3$
$[PdCl(CNMe)_3]^+$, **9**, 210
$PdC_6H_9N_3$
$[Pd(CNMe)_3]^+$, **9**, 210
PdC_6H_{10}
$Pd(C_3H_5)_2$, **9**, 303, 324
$PdC_6H_{18}Cl_2S_2$
$Pd_2Cl_2(Me)_2(SMe_2)_2$, **9**, 306
$PdC_7H_8Cl_2$
$PdCl_2(nbd)$, **9**, 297
$PdC_7H_{12}O$
$Pd[(MeO)C_6H_9]$, **9**, 340
$PdC_7H_{16}ClP$
$PdCl(MeC_3H_4)(PMe_3)$, **9**, 344
$PdC_7H_{21}ClN_2$
$[PdCl(CH_2CH_2CHMeNMe_2)]\text{-}(NHMe_2)$, **9**, 265
$PdC_8H_9Cl_3NO$
$[PdCl_3[C(NHPh)(OMe)]]^-$, **9**, 217
$PdC_8H_{12}ClNO$
$Pd[CH_2CHCH_2C(O)CH_2CH_2]\text{-}(NCMe)Cl$, **9**, 238
$PdC_8H_{12}Cl_2$
$PdCl_2(cod)$, **9**, 249, 296, 310; **12**, 154, 155, 806
PdC_8H_{14}
$Pd(MeC_3H_4)_2$, **9**, 324
$PdC_8H_{18}S_3$
$PdMe_2$(trithiacyclononane), **9**, 261
$PdC_8H_{21}IP_2S_2$
$PdI(S_2CMe)(PMe_3)_2$, **9**, 271
$PdC_8H_{22}N_2$
$PdMe_2(TMEDA)$, **9**, 240
$PdC_8H_{22}N_2O_4S_2$
$Pd(SO_2Me)_2(TMEDA)$, **9**, 271
$PdC_8H_{27}B_9N_2$
$(NMe_2CH_2)_2Pd(C_2B_9H_{11})$, **1**, 398
$PdC_9H_5N_3$
$[PdPh(CN)_3]^{2-}$, **9**, 237
$PdC_9H_{14}Cl_2$
$PdCl(CH_2Cl)(cod)$, **9**, 259
$PdC_9H_{19}Cl_2N$
$Pd(CH_2{=}CHCMe_2CH_2NMe_2)\text{-}Cl(CH_2Cl)$, **9**, 238
$PdC_9H_{21}ClN$
$PdClMe(CH_2{=}CH_2CMe_2CH_2\text{-}NMe_2)$, **9**, 259
$PdC_9H_{22}NPS_2$
$PdPr(S_2CNMe_2)(PMe_3)$, **9**, 261
$PdC_{10}H_6BrF_5N_2$
$Pd(C_6F_5)Br(NCMe)_2$, **9**, 339
$PdC_{10}H_6ClF_5N_2$
$PdCl(C_6F_5)(CNMe)_2$, **9**, 212
$PdC_{10}H_6N_6S_4$
$[Pd[S_2C_2(CN)_2]_2](CNMe)_2$, **9**, 215
$PdC_{10}H_{10}Cl_2$
$PdCl_2[(CH_2{=}CH)_2C_6H_4)]$, **9**, 311, 313
$PdCl_2Cp_2$, **9**, 249, 310, 311
$PdC_{10}H_{14}O_4$
$Pd(acac)_2$, **12**, 154
$PdC_{10}H_{15}Cl_2$
$PdCl_2[(CH_2{=}CH)_2C_6H_9]$, **9**, 249
$PdC_{10}H_{18}N_2$
$Pd(CNBu^t)_2$, **9**, 211
$PdC_{10}H_{23}BrOP_2$
$PdBr(COCH_2CH{=}CH_2)\text{-}(PMe_3)_2$, **9**, 255
$PdC_{10}H_{24}ClO_2P_2$
$Pd(CH_2{=}CHCH_2COO)Cl\text{-}(PMe_3)_2$, **9**, 359
$PdC_{10}H_{24}NPS_2$
$PdMe(S_2CNMe_2)(PEt_3)$, **9**, 239
$PdC_{10}H_{26}Cl_2N_2$
$PdCl_2(CH_2CH_2NEt_2H)(NEt_2H)$, **9**, 249
$PdC_{10}H_{28}P_2$
$PdEt_2(PMe_3)_2$, **9**, 270
$PdC_{11}H_{11}IN_2$
PdIMe(bipy), **9**, 234
$PdC_{11}H_{19}BrN_2$
$PdBr(CHMeCH_2CH_2NMe_2)\text{-}(py)$, **9**, 237
$PdC_{11}H_{24}ClN_2O$
$PdCl(COCH_2CH_2NEt_2)\text{-}(NEt_2H)$, **9**, 256
$PdC_{11}H_{25}ClP_2$
$PdCl(C{\equiv}CPr)(PMe_3)_2$, **9**, 230
$PdC_{11}H_{25}P_2$
$Pd(C{\equiv}CPr)(PMe_3)_2$, **9**, 230
$PdC_{12}Br_4F_{10}$
$[Pd(C_6F_5)Br(Br)]_2{}^{2-}$, **9**, 262
$PdC_{12}H_{14}N_2$
$PdMe_2$(bipy), **9**, 267
$PdC_{12}H_{21}IN_2$
PdIPh(TMEDA), **9**, 239
$PdC_{12}H_{25}N_2PS_2$
$Pd(CH_2CH_2CN)(S_2CNMe_2)\text{-}(PEt_3)$, **9**, 238
$Pd[CH(CN)Me](S_2CNMe_2)\text{-}(PEt_3)$, **9**, 250
$PdC_{12}H_{27}BrP_2$
$PdBr(C{\equiv}CBu^t)(PMe_3)_2$, **9**, 231
$PdC_{12}H_{28}NPS_2$
$PdPr(S_2CNMe_2)(PEt_3)$, **9**, 240
$PdC_{12}H_{32}N_8$
$[Pd[C(NHMe)_2]_4]^{2+}$, **9**, 217
$PdC_{13}H_{11}N_2O_2$
$[Pd(COMe)(bipy)(CO)]^+$, **9**, 256
$PdC_{13}H_{12}Cl_2N_2O$
$PdCl_2[MeC_5H_4NCHC(O)\text{-}(C_5H_4N)]$, **9**, 229
$PdC_{13}H_{16}NP_2$
$[PdMe(dmpe)(NH_2Ph)]^+$, **9**, 268
$PdC_{13}H_{17}IN_2$
$PdIMe_3$(bipy), **9**, 276
$PdC_{13}H_{24}N_2$
PdMePh(TMEDA), **9**, 239
$PdC_{13}H_{26}P_2S$
$Pd(SPh)Me(PMe_3)_2$, **9**, 271
$PdC_{13}H_{30}BrF_3P_2$
$PdBr(CF_3)(PEt_3)_2$, **9**, 238

$PdC_{13}H_{30}NPS_2$
$PdBu^{i}(S_2CNMe_2)(PEt_3)$, **9**, 238
$PdC_{13}H_{33}BrP_2$
$PdBrMe(PEt_3)_2$, **9**, 240
$PdC_{14}F_{10}O_2$
$Pd(C_6F_5)_2(CO)_2$, **9**, 196
$PdC_{14}H_{10}Cl_2N_2$
$PdCl_2(PhCN)_2$, **12**, 151, 152, 396, 403
$PdC_{14}H_{12}ClF_5$
$Pd(C_6F_5)Cl$(cod), **9**, 308
$PdC_{14}H_{16}O_2$
Pd(quinone)(cod), **9**, 307
$PdC_{14}H_{18}N_2$
$PdEt_2$(bipy), **9**, 239
$PdC_{14}H_{23}O_2P_2$
$Pd(COPh)(CO)(PMe_3)_2$, **9**, 208
$PdC_{14}H_{25}BrSe_2$
$PdBrPh(SeEt_2)_2$, **9**, 240
$PdC_{14}H_{26}N_2$
PdMeBz(TMEDA), **9**, 240
$PdC_{14}H_{28}P_2S$
$Pd(SPh)Et(PMe_3)_2$, **9**, 269
$PdC_{14}H_{30}F_6P_2$
$Pd(CF_3)_2(PEt_3)_2$, **9**, 238
$PdC_{14}H_{30}NPS_2$
$PdPr[S_2CN(CH_2)_4](PEt_3)$, **9**, 239
$PdC_{14}H_{36}P_2$
$PdMe_2(PEt_3)_2$, **9**, 239
$PdC_{14}H_{40}Cl_2N_6P_2$
$PdCl_2[CH_2P(NMe_2)_3]_2$, **9**, 227
$PdC_{15}H_{15}ClN_2$
$PdClMe(Me_2phen)$, **9**, 271
$PdC_{15}H_{15}IN_2$
$PdIMe(Me_2phen)$, **9**, 234
$PdC_{15}H_{28}O_2P_2$
$PdBz(CO_2Me)(PMe_3)_2$, **9**, 238
$PdC_{15}H_{29}O_4P_4$
$Pd(COPh)[P(O)(OMe)_2](PMe_3)_2$, **9**, 237
$PdC_{16}H_8F_{10}O_2$
$Pd(C_6F_5)_2$(dioxane), **9**, 262
$PdC_{16}H_{13}BrN_2$
PdBrPh(bipy), **9**, 236
$PdC_{16}H_{13}IN_2$
PdIPh(bipy), **9**, 233
$PdC_{16}H_{15}ClN_2O$
$PdCl(COMe)(Me_2phen)$, **9**, 256
$PdC_{16}H_{16}N_2$
$PdMe_2(NCPh)_2$, **9**, 262
$PdC_{16}H_{24}$
$Pd(cod)_2$, **9**, 302
$PdC_{16}H_{25}ClO_4$
$[Pd(C_8H_{13})(cod)][ClO_4]$, **9**, 308
$PdC_{16}H_{29}P$
$PdCp^*(CH_2CH{=}CH_2)(PMe_3)$, **9**, 260
$PdC_{16}H_{32}P_2$
$Pd(C{\equiv}CH)_2(PEt_3)_2$, **9**, 230
$Pd(C{\equiv}CPr)_2(PMe_3)_2$, **9**, 230
$PdC_{17}H_{13}ClF_6N_2$
$PdCl[[OC(Me)NH]C_6H_4C(CF_3){=}C(CF_3)](py)$, **9**, 254
$PdC_{17}H_{14}N_2$
$PdCp[(PhN_2)C_6H_4]$, **9**, 264
$PdC_{17}H_{16}N_2$
PdMePh(bipy), **9**, 271
$PdC_{17}H_{33}NP_2$
$PdMe[(Pr^{i}N{=}CH)C_6H_4](PMe_3)_2$, **9**, 239
$PdC_{18}H_{15}Cl_2P$
$Pd(PPh_3)Cl_2$, **12**, 180
$PdC_{18}H_{19}IN_2$
$PdIMe_2Ph$(bipy), **9**, 274
$PdC_{18}H_{24}N_2$
$Pd[(NMe_2CH_2)C_6H_4]_2$, **9**, 268
$PdC_{18}H_{25}P$
$PdBz(C_5H_4CH_2CH{=}CH_2)(PMe_3)$, **9**, 260
$PdC_{18}H_{30}BrF_5P_2$
$PdBr(C_6F_5)(PEt_3)_2$, **9**, 233
$PdC_{18}H_{31}NO_2P_2$
$Pd(COPh)(CO\overline{N(CH_2)_3CH_2})(PMe_3)_2$, **9**, 208
$PdC_{18}H_{35}ClP_2$
$PdClPh(PEt_3)_2$, **9**, 238
$PdC_{18}H_{35}IP_2$
$PdIPh(PEt_3)_2$, **9**, 239
$PdC_{19}H_{15}ClN_2$
PdCl(Tol)(phen), **9**, 237
$PdC_{19}H_{15}OP$
$(PPh_3)Pd(CO)$, **12**, 242
$PdC_{19}H_{21}BrN_2$
$PdBrMe_2Bz$(bipy), **9**, 276
$PdC_{19}H_{29}ClOS_2$
$PdCl(C_{10}H_6OMe)(SEt_2)_2$, **9**, 240
$PdC_{19}H_{31}O_2$
$Pd(C_4CH_2(Me)Bu^{t}_2)$(acac), **9**, 316
$PdC_{19}H_{31}P$
$PdCp_2(PPr^{i}_3)$, **9**, 270
$PdC_{19}H_{37}ClP_2$
$PdClBz(PEt_3)_2$, **9**, 233, 260
$PdC_{19}H_{38}P_2$
$PdPhMe(PEt_3)_2$, **9**, 239
$PdC_{19}H_{46}P_2$
$PdH(Me)(PPr^{i}_3)_2$, **9**, 240
$PdC_{20}H_3F_{15}N$
$[Pd(C_6F_5)_3(CNMe)]^-$, **9**, 217
$PdC_{20}H_8$
$[Pd(C{\equiv}CC_6H_4C{\equiv}C)_2]^{2-}$, **9**, 231
$PdC_{20}H_{16}F_{10}S_2$
$Pd(C_6F_5)_2(THT)_2$, **9**, 240
$PdC_{20}H_{22}Cl_2N_2$
$PdCl_2(CNxylyl)_2$, **9**, 212, 217
$PdC_{20}H_{22}N_2O_8$
$Pd[CH(CO_2Me)_2]_2$(bipy), **9**, 241
$PdC_{20}H_{30}N_2$
$Pd(CH_2Bu^{t})_2$(bipy), **9**, 240, 269
$PdC_{20}H_{32}P_2$
$PdEt_2(PMe_2Ph)_2$, **9**, 267
$PdC_{20}H_{35}ClP_2$
$PdCl(C{\equiv}CPh)(PEt_3)_2$, **9**, 232
$PdC_{20}H_{38}BrP_2$
$PdBr(CH_2{=}CHPh)(PEt_3)_2$, **9**, 240
$PdC_{20}H_{45}P_2$
$Pd(Et)(Bu^{t}_2P(CH_2)_2PBu^{t}_2)$, **9**, 237
$PdC_{21}H_{13}F_{10}N$
$Pd(C_6F_5)_2(BzNMe_2)$, **9**, 382
$PdC_{21}H_{20}ClP$
$PdCl(C_3H_5)(PPh_3)$, **9**, 323
$PdC_{21}H_{22}IOPS$
$PdI[(CH_2)_2SOMe](PPh_3)$, **9**, 229
$PdC_{21}H_{33}IP_2$
$PdIMe(PEt_2Ph)_2$, **9**, 267
$PdC_{21}H_{39}P$
$PdCp(Bu^{t}C_3H_4)(PPr^{i}_3)$, **9**, 352
$PdC_{22}H_8F_{15}S$
$[Pd(C_6F_5)_3(THT)]^-$, **9**, 240, 270
$PdC_{22}H_{22}ClP$
$Pd(MeC_3H_4)Cl(PPh_3)$, **9**, 328
$PdC_{22}H_{23}NO_6$
$Pd[(MeO)C_6H_3CH{=}N(Tol)][C(CO_2Me){=}C(OMe)C(O)OMe]$, **9**, 252
$PdC_{22}H_{34}ClN_5$
$PdCl[[O{=}NN(Me)]C_6H_4C({=}NBu^{t})](CNBu^{t})_2$, **9**, 258
$PdC_{22}H_{36}P_2$
$PdMe_2(PEt_2Ph)_2$, **9**, 239, 267
$PdC_{22}H_{41}P$
$PdMe(C_3H_5)(PCy_3)$, **9**, 240
$PdC_{23}H_{19}O_3P$
$PdCp[P(OPh)_2(OC_6H_4)]$, **9**, 375
$PdC_{23}H_{24}ClNOP$
$PdCl[(NMe_2CH_2)C_6H_4](Ph_2PCH_2CO)$, **9**, 263
$PdC_{23}H_{24}Cl_2NO_3P$
$PdCl_2(CO_2Bu)(NO)(PPh_3)$, **9**, 257, 273
$PdC_{23}H_{24}Cl_2NP$
$PdCl_2(PPh_3)(CNBu^{t})$, **9**, 218
$PdC_{23}H_{29}BrP_2$
$PdBr(Tol)(PMe_2Ph)_2$, **9**, 240
$PdC_{23}H_{34}F_6O_2P_2$
$Pd[OCH(CF_3)Ph]Me(PMe_3)_2[HOCH(CF_3)Ph]$, **9**, 265
$PdC_{23}H_{41}OP$
$Pd(C_3H_5)(OAc)(PCy_3)$, **12**, 803
$PdC_{24}F_{20}$
$[Pd(C_6F_5)_4]^{2-}$, **9**, 240
$PdC_{24}H_{23}Cl_2OP$
$PdCl(ClC_3H_4)(Ph_3PCHCOMe)$, **9**, 350
$PdC_{24}H_{23}Cl_5P_2$
$Pd(C{\equiv}CH)(C_6Cl_5)(PMe_2Ph)_2$, **9**, 232
$PdC_{24}H_{24}ClOP$
$PdCl(C_3H_5)(Ph_3PCHCOMe)$, **9**, 350
$PdC_{24}H_{25}O_2P$
$Pd(CO_2Me)(MeC_3H_4)(PPh_3)$, **9**, 359
$PdC_{24}H_{25}P_2$
$Pd(C_3H_5)_2(PPh_3)$, **9**, 352
$PdC_{24}H_{34}N_2O_6$
$Pd[CH(COMe)(CO_2Et)]_2(Mepy)_2$, **9**, 241

PdC$_{24}$H$_{35}$NO$_4$P$_2$
PdPh[C(CO$_2$Me)=C(NHPh)-(CO$_2$Me)](PMe$_3$)$_2$, **9**, 252
PdC$_{24}$H$_{36}$
[Pd(C≡CBu^t)$_4$]$^{2-}$, **9**, 230
PdC$_{24}$H$_{40}$P$_2$
PdBu$_2$(PMe$_2$Ph)$_2$, **9**, 238
PdPh$_2$(PEt$_3$)$_2$, **9**, 239, 266
PdC$_{24}$H$_{43}$OP
Pd[(CH$_2$CHCH$_2$)$_2$O](PCy$_3$), **9**, 345
PdC$_{24}$H$_{44}$P$_2$
PdPh(C≡CBu^t)(PEt$_3$)$_2$, **9**, 230
PdC$_{25}$H$_{20}$F$_3$O$_2$P
Pd(O$_2$CCF$_3$)Cp(PPh$_3$), **9**, 267
PdC$_{25}$H$_{21}$ClNOP
PdCl[(COCH$_2$)py](PPh$_3$), **9**, 256
PdC$_{25}$H$_{22}$D$_2$P
PdCp(PPh$_3$)(CHD=CHD), **9**, 249
PdC$_{25}$H$_{27}$ClNO$_4$P
PdCl[(CHCO$_2$Et)$_2$PPh$_2$]py, **9**, 230
PdC$_{25}$H$_{28}$Cl$_5$OP$_2$
[Pd[C(OMe)Me](C$_6$Cl$_5$)-(PMe$_2$Ph)$_2$]$^+$, **9**, 218
PdC$_{25}$H$_{30}$NOP
Pd(C≡CBu)[PPh$_2$CH$_2$CH$_2$NC-(Me)=CHCMe(O)], **9**, 230
PdC$_{26}$H$_{26}$IN$_2$
PdI(MePh)C$_7$H$_{10}$(phen), **9**, 274
PdC$_{26}$H$_{28}$Cl$_5$OP$_2$
[Pd[CCH$_2$(CH$_2$)$_2$)](C$_6$Cl$_5$)-(PMe$_2$Ph)$_2$]$^+$, **9**, 218
PdC$_{26}$H$_{47}$OP
Pd[(CH$_2$CMeCH$_2$)$_2$O](PCy$_3$), **9**, 345
PdC$_{27}$H$_{38}$P$_2$
PdMePh(PEt$_2$Ph)$_2$, **9**, 239
PdC$_{28}$H$_{25}$P
PdCp$_2$(PPh$_3$), **9**, 238
PdC$_{28}$H$_{27}$P
PdPh(MeC$_3$H$_4$)(PPh$_3$), **9**, 238
PdC$_{28}$H$_{30}$P$_2$
PdMe$_2$(dppe), **9**, 239, 266
PdC$_{28}$H$_{32}$ClNO$_2$P
PdCl(ONBu$^t_2$)[CH(COPh)-(PPh$_2$Me)], **9**, 227
PdC$_{28}$H$_{32}$P$_2$
PdMe$_2$(PMePh$_2$)$_2$, **9**, 266
PdC$_{28}$H$_{40}$O$_4$
Pd(dbsq)$_2$, **9**, 348
PdC$_{28}$H$_{42}$F$_6$O$_4$P
Pd(tfacac)(OC(Me)=CHCOCF$_3$)(PCy$_3$), **9**, 261
PdC$_{29}$H$_{22}$F$_{10}$N$_3$
[Pd[C(=NTol)N(Tol)CHN-(Tol)](C$_6$F$_5$)$_2$]$^-$, **9**, 257
PdC$_{29}$H$_{31}$BrP$_2$
PdBr(CH=CHMe)(PMePh$_2$)$_2$, **9**, 235
PdC$_{29}$H$_{31}$OP
Pd(CH$_2$COBu^t)Cp(PPh$_3$), **9**, 240
PdC$_{29}$H$_{32}$O$_4$P
Pd(CH$_2$COCH$_2$COEt)(acac)-(PPh$_3$), **9**, 260
PdC$_{30}$H$_{24}$F$_{10}$N$_4$
Pd(C$_6$F$_5$)$_2$[C(NHMe)(NHBz)]$_2$, **9**, 217
PdC$_{30}$H$_{29}$O$_2$P
Pd(OOBu^t)(C$_3$H$_5$)(PPh$_3$), **9**, 352
PdC$_{30}$H$_{40}$F$_5$P$_3$
Pd(C$_6$F$_5$)(PPh$_2$)(PEt$_3$)$_2$, **9**, 237
PdC$_{31}$H$_{27}$ClF$_5$OPS
PdCl(C$_6$F$_5$)[CH(COMe)PPh$_3$]-(THT), **9**, 228
PdC$_{31}$H$_{28}$P
[PdCp(CH$_2$=CHPh)(PPh$_3$)]$^+$, **9**, 297
PdC$_{32}$H$_{24}$ClF$_5$P$_2$
PdCl(C$_6$F$_5$)(dppe), **9**, 267
PdC$_{32}$H$_{64}$N$_4$O$_4$P$_2$
Pd[C(=N$_2$)CO$_2$Et]$_2$(PBu$_3$)$_2$, **9**, 237
PdC$_{32}$H$_{72}$P$_2$
Pd(dippb)$_2$, **12**, 242
PdC$_{33}$H$_{28}$Cl$_2$NO$_3$
[Pd(MeNO$_2$C$_6$H$_3$)Cl$_2$(CO)]-[PPh$_3$Bz], **9**, 196
PdC$_{33}$H$_{32}$O$_2$P$_2$
Pd(C$_2$H$_4$)(diop), **9**, 302
PdC$_{33}$H$_{33}$ClO$_5$PS
[PdCp[CH$_2$(COPh)SMe$_2$]-(PPh$_3$)]ClO$_4$, **9**, 229
PdC$_{34}$H$_{28}$O$_2$
Pd(dba)$_2$, **12**, 260, 264, 906, 907, 910–13, 916–18
PdC$_{34}$H$_{33}$ClO$_2$P$_2$
PdCl(CH$_2$CH=CH$_2$)(diop), **9**, 235
PdC$_{35}$H$_{29}$N$_2$P
Pd(C$_6$H$_4$NNPh)Cp(PPh$_3$), **9**, 379
PdC$_{35}$H$_{44}$Cl$_4$N$_2$OP$_2$
Pd$_2$Cl$_4$(CO)(Ph$_2$PCh$_2$NEt$_2$)$_2$, **9**, 198
PdC$_{35}$H$_{47}$N$_2$P
Pd(C$_6$H$_4$NNPh)Cp(PCy$_3$), **9**, 379
PdC$_{36}$H$_{25}$ClO
PdCl(CO)(C$_5$Ph$_5$), **9**, 197
PdC$_{36}$H$_{30}$ClP
[PdCl(PPh$_3$)$_2$]$^+$, **10**, 47
PdC$_{36}$H$_{30}$Cl$_2$P$_2$
PdCl$_2$(PPh$_3$)$_2$, **11**, 408
Pd(PPh$_3$)$_2$Cl$_2$, **12**, 172, 201, 214, 222, 260
PdC$_{36}$H$_{30}$N$_2$O$_7$P$_2$
Pd(ONO$_2$)$_2$(PPh$_3$)(OPPh$_3$), **9**, 271
PdC$_{36}$H$_{30}$P$_2$
Pd(PPh$_3$)$_2$, **12**, 183, 201
PdC$_{36}$H$_{108}$O$_{12}$P$_4$
[(Pr^{i}O)$_3$P]$_4$Pd, **12**, 928
PdC$_{37}$H$_{33}$BrP$_2$
PdBrMe(PPh$_3$)$_2$, **9**, 238
PdC$_{37}$H$_{33}$IO$_2$P$_2$
PdIPh(diop), **9**, 235
PdC$_{38}$H$_{30}$Cl$_2$P$_2$
PdCl$_2$(C≡CPPh$_3$)(PPh$_3$), **9**, 230
PdC$_{38}$H$_{33}$ClNP$_2$
PdCl(PPh$_3$)$_2$(CNMe), **9**, 217
PdC$_{38}$H$_{33}$ClOP$_2$
PdCl(COMe)(PPh$_3$)$_2$, **9**, 237, 259
PdC$_{38}$H$_{33}$ClO$_2$P$_2$
PdCl(CO$_2$Me)(PPh$_3$)$_2$, **9**, 257
PdC$_{38}$H$_{38}$P$_2$
Pd(Ph)(C≡CBu^t)(dppe), **9**, 230
PdC$_{38}$H$_{68}$P$_2$
PdH(C≡CH)(PCy$_3$)$_2$, **9**, 230
PdC$_{39}$H$_{35}$ClOP$_2$
PdCl[CH$_2$C(O)Me](PPh$_3$)$_2$, **9**, 235
PdC$_{39}$H$_{35}$P$_2$
[Pd(C$_3$H$_5$)(PPh$_3$)$_2$]$^+$, **9**, 239
PdC$_{39}$H$_{37}$IOP$_2$S
[Pd[(CH$_2$)$_2$SOMe](PPh$_3$)$_2$]I, **9**, 229, 230
PdC$_{39}$H$_{38}$IO$_3$P$_3$
PdI[CH$_2$PO(OMe)$_2$](PPh$_3$)$_2$, **9**, 228
PdC$_{39}$H$_{39}$ClO$_8$P$_2$
PdCl[C(CO$_2$Me)=C(CO$_2$Me)C-(CO$_2$Me)=CMe-(CO$_2$Me)](dppe), **9**, 253
PdC$_{40}$H$_{20}$F$_{10}$
Pd(C$_6$F$_5$)$_2$(PhCCPh)$_2$, **9**, 317
PdC$_{40}$H$_{34}$P$_2$
Pd(CH$_2$=CHC$_6$H$_4$PPh$_2$)$_2$, **9**, 303
PdC$_{40}$H$_{35}$ClN$_2$O$_2$P$_2$
PdCl[C(=N$_2$)CO$_2$Et](PPh$_3$)$_2$, **9**, 238
PdC$_{40}$H$_{35}$ClP$_2$
PdCl(CH=CHCOOMe)(PPh$_3$)$_2$, **9**, 302
PdC$_{40}$H$_{66}$F$_6$P$_2$
Pd(PCy$_3$)$_2$(F$_3$CCCCF$_3$), **9**, 317
PdC$_{41}$H$_{35}$ClO$_4$P$_2$
[PdCp(PPh$_3$)$_2$][ClO$_4$], **9**, 306
PdC$_{41}$H$_{38}$ClNOP$_2$
PdCl[C(NHMe)=CHCOMe]-(PPh$_3$)$_2$, **9**, 257
PdC$_{41}$H$_{40}$P$_2$
Pd(CH$_2$CMe$_2$CH$_2$)(PPh$_3$)$_2$, **9**, 266, 267
PdC$_{41}$H$_{72}$F$_3$NO$_3$P$_2$
Pd(ONO$_2$)[C(Ph)=CH(CF$_3$)]-(PCy$_3$)$_2$, **9**, 252
PdC$_{41}$H$_{74}$O$_2$P$_2$
Pd(OAc)(C$_3$H$_5$)(PCy$_3$)$_2$, **9**, 345
PdC$_{42}$H$_{30}$ClF$_3$
PdCl(C$_6$F$_5$)(PPh$_3$)$_2$, **9**, 259
PdC$_{42}$H$_{36}$O$_4$P$_2$
Pd(PPh$_3$)$_2$(MeOOCCCCO-OMe), **9**, 317
PdC$_{42}$H$_{37}$Cl$_2$F$_6$PSb$_2$
[PdCp(SbPh$_3$)$_2$][PF$_6$]CH$_2$Cl$_2$, **9**, 377
PdC$_{42}$H$_{41}$BrOP$_2$

$PdBr[CH_2C(O)Bu^t](PPh_3)_2$, **9**, 235
$PdC_{42}H_{71}ClP_2$
$PdClPh(PCy_3)_2$, **9**, 234
$PdC_{42}H_{71}IP_2$
$PdIPh(PCy_3)_2$, **9**, 233
$PdC_{43}H_{30}F_5OP_2$
$[Pd(C_6F_5)(CO)(PPh_3)_2]^+$, **9**, 196
$PdC_{43}H_{30}F_8$
$Pd(PPh_3)_2(C_7F_8)$, **9**, 302
$PdC_{43}H_{35}ClOP_2$
$PdCl(COPh)(PPh_3)_2$, **9**, 233
$PdC_{43}H_{37}$
$Pd(C_5Ph_5)(cod)$, **9**, 272
$PdC_{43}H_{37}BrO_2P_2$
$PdBr[C{=}C(CMe{=}CH_2)CH_2C(O)O](PPh_3)_2$, **9**, 255
$PdC_{43}H_{37}ClP_2$
$BzPdCl(PPh_3)_2$, **12**, 201
$PdClBz(PPh_3)_2$, **9**, 265
$PdC_{43}H_{71}ClOP_2$
$PdCl(COPh)(PCy_3)_2$, **9**, 256
$PdC_{44}H_{37}BrP_2$
$PdBr(CH{=}CHPh)(PPh_3)_2$, **9**, 235
$PdC_{44}H_{39}P_3$
$Pd(PPh_3)(dppe)$, **9**, 345
$PdC_{46}H_{39}BrOP_2$
$PdBr(PhCH{=}CHCH_2CO)(PPh_3)_2$, **9**, 256
$PdC_{46}H_{46}BrNP_2$
$PdBr[C({=}CMe_2)CH{=}C{=}NBu^t](PPh_3)_2$, **9**, 258
$PdC_{46}H_{52}P_2$
$Pd(CH_2Bu^t)_2(PPh_3)_2$, **9**, 266
$PdC_{48}H_{30}F_{10}$
$Pd(C_6F_5)_2(PPh_3)_2$, **9**, 237
$PdC_{50}H_{50}ClOP_2$
$PdCl[Bu^t_2OC_6H_2](PPh_3)_2$, **9**, 272
$PdC_{51}H_{47}P_4$
$PdMe(dppm)_2$, **9**, 226
$PdC_{52}H_{51}P_4$
$[Pd(Me)_2(dppm)_2(H)]^+$, **9**, 226
$PdC_{55}H_{45}OP_3$
$Pd(CO)(PPh_3)_3$, **9**, 197
$PdC_{56}H_{38}P_2$
$[Pd(C{\equiv}CC_6H_4C{\equiv}C)_2(PPh_3)_2]^{2-}$, **9**, 230
$PdC_{56}H_{128}N_8O_8$
$[Pd[C(NHMe)[NHCH_2CH(OEt)_2]]_4]^{2+}$, **9**, 217
$PdC_{57}H_{45}ClO_4P_2$
$[Pd(PPh_3)_2(C_3Ph_3)][ClO_4]$, **9**, 296
$PdC_{57}H_{50}N_2O_3P_2$
$Pd(dba)[Ph_2PC_6H_4NHC(O)Me]_2$, **9**, 310
$PdC_{60}H_{50}P_3$
$[PdPh(PPh_3)_3]^+$, **9**, 237
$PdC_{63}H_{51}F_6P_3$
$[Pd(PPh_3)_2(C_3Ph_3)][PF_6]{\cdot}C_6H_6$, **9**, 296
$PdC_{66}H_{54}P_4$
$Pd(C{\equiv}CPh)_2(dppm)_2$, **9**, 230
$PdC_{72}H_{60}P_4$
$Pd(PPh_3)_4$, **11**, 168, 401; **12**, 167, 172, 174, 175, 180, 181, 183, 186, 201, 214, 222, 225, 260, 304, 306, 766, 906, 907, 911, 949, 950
$PdCl_2$
$PdCl_2$, **1**, 68; **12**, 425, 747
$PdCo_2C_{31}H_{24}O_7P_2$
$Co_2Pd(CO)_7(dppe)$, **10**, 48
$PdCrC_{32}H_{50}O_2P_2$
$Cr(CO)_2Cp(PCy_2)Pd(CO)(PHCy_2)$, **10**, 47
$PdCrC_{49}H_{36}INO_5P_2$
$PdI[(CH_2C_6H_4NC)Cr(CO)_5](PPh_3)_2$, **9**, 235
$PdCr_2C_{14}H_{32}Cl_2S_3$
$Cr_2Pd(S)(SBu^t)_2Cl_2Cp'_2$, **10**, 228
$PdCr_2C_{30}H_{20}N_2O_6$
$[Cp(CO)_3Cr]Pd(PhCN)_2[Cr(CO)_3Cp]$, **9**, 204
$PdCuMoC_{43}H_{37}O_2P_3$
$CuMoPd(PPh_2)(CO)_2(PPh_2H)_2Cp$, **10**, 311
$PdFeC_{34}H_{30}Cl_2P_2$
$Pd(dppf)Cl_2$, **12**, 165, 167, 174–6, 180, 181, 186, 214
$PdFeC_{16}H_{10}ClO_4P$
$(CO)_4Fe(PPh_2)Pd(Cl)$, **10**, 41
$PdFeC_{28}H_{23}PS_2$
$Fe(C_5H_4S)_2Pd(PPh_3)$, **10**, 61
$PdFeC_{31}H_{31}IN_3P_2$
$(MeNC)_3Fe(dppm)PdI$, **7**, 79
$PdFeC_{34}H_{28}Cl_2P_2$
$[Fe(C_5H_4PPh_2)_2]PdCl_2$, **1**, 317
$PdFeC_{52}H_{44}I_2O_2P_4$
$Fe(I)(CO)_2(dppm)_2PdI$, **7**, 79
$PdFeC_{57}H_{53}I_2NO_2P_4$
$[(Bu^tNC)(CO)_2Fe(dppm)_2PdI]I$, **7**, 79
$PdFePtC_{54}H_{44}O_4P_4$
$FePdPt(CO)_4(dppm)_2$, **7**, 84; **10**, 239, 242
$PdFeSiSnC_{49}H_{46}O_6P_2$
$Fe(CO)_3[Si(OMe)_2OMe](dppm)Pd(SnPh_3)$, **10**, 61
$PdFeSi_4C_{37}H_{49}Cl_3O_6P_2$
$PdFe(CO)_3(dppm)[Si(O\text{-}TMS)_3Cl_3]$, **10**, 17, 18
PdHCl
HPdCl, **12**, 606
$PdHgC_{66}H_{54}Cl_2P_4$
$[(PhC{\equiv}C)_2Pd](dppm)_2(HgCl_2)$, **9**, 230
$PdLiCl_3$
$LiPdCl_3$, **11**, 404; **12**, 133, 135, 136
$PdLi_2Cl_4$
Li_2PdCl_4, **11**, 399, 401, 402, 404, 407; **12**, 133, 135, 136
$PdMnC_{28}H_{22}BrO_3P_2$
$MnPdBr(CO)_3(dppm)_2$, **10**, 47
$PdMnC_{28}H_{22}IO_3P_2$
$MnPdI(CO)_3(dppm)_2$, **10**, 47
$PdMnC_{40}H_{68}O_4P_3$
$(Cy_2PH)_2Pd(PCy_2)Mn(CO)_4$, **6**, 88
$PdMoC_{55}H_{44}N_2O_3P_4$
$Mo(CN)(dppm)_2Pd(CN)(CO)_3$, **10**, 47
$PdMoPtC_{57}H_{49}ClO_2P_2$
$MoPdPt(CO)_2Cl(dppm)_2Cp$, **10**, 238
PdNaCO
$[PdCONa]^+$, **9**, 195
$PdNiC_{54}H_{50}ClN_2P_4$
$[Ni(MeNC)(dppm)_2(CNMe)PtCl]^+$, **10**, 69
$PdNiC_{56}H_{53}N_{34}$
$[Ni(MeNC)_3(dppm)_2Pd]^{2+}$, **10**, 69
$PdPtC_{12}H_{18}N_6$
$[PdPt(MeNC)_6]^{2+}$, **9**, 210
$PdPtC_{55}H_{45}ClOP_3$
$PdCl(PPh_3)(CO)PtCl(PPh_3)_2$, **9**, 197
$PtCl(PPh_3)(CO)PdCl(PPh_3)_2$, **9**, 197
$PdPtC_{56}H_{44}ClF_5P_4$
$[(C_6F_5)Pt](dppm)_2(PdCl)$, **9**, 226
$(ClPt)(dppm)_2[Pd(C_6F_5)]$, **9**, 226
$PdPtC_{64}H_{47}F_{10}NP_4$
$PdPt(C_6F_5)_2(dppm)_2(CNMe)$, **9**, 214
$PdRe_7C_{31}H_9O_{21}$
$[Re_7C(CO)_{21}Pd(C_9H_9)]^{2-}$, **10**, 128
$Re_7C(CO)_{21}Pd(PhC_3H_4)$, **9**, 350
$PdRhC_{28}H_{34}N_4S_4$
$[Rh(pyS)_4Pd(C_3H_4Me)_2]^+$, **10**, 64
$PdRhC_{32}H_{37}F_6N_3P_2$
$[Cp(CNBu^t)Rh(Ph_2PPy)PdCl(CNBu^t)]PF_6$, **9**, 215
$PdRhC_{33}H_{25}ClN_2OPS_4$
$RhCl(CO)(benzothiazole\text{-}2\text{-}thiolate)_2Pd(PPh_3)$, **10**, 64
$PdRhC_{48}H_{33}Cl_2N_4P_2$
$[RhCl(PhCN)_2(NC_5H_4PPh_2)_2PdCl]^+$, **10**, 64, 67
$PdRuC_{16}H_{22}Cl_2N_2$
$(cymene)ClRuCl(Pz)Pd(C_3H_5)$, **9**, 350
$PdRuC_{36}H_{28}Cl_2N_2O_4P_2$
$RuCl(CO)_2(PPh_2C_5H_4N)_2PdCl$, **10**, 61
$PdSiC_{20}H_{34}ClPSi$
$PdCl[CH(TMS)PMe_2Ph](cod)$, **9**, 230
$PdSiC_{43}H_{45}P_2$
$[Pd[(CH_2TMS)C_3H_4](PPh_3)_2]^+$, **9**, 374
$PdSi_2C_{14}H_{28}P_2$
$Pd(CH_2TMS)_2(dmpe)$, **9**, 240

$PdSnC_5H_7Cl_3O$
$Pd(SnCl_3)(MeC_3H_4)(CO)$, **9**, 359
$PdSnC_{12}H_{15}Cl_3$
$Pd(SnCl_3)(MeC_3H_4)$(styrene), **9**, 350
$PdSnC_{30}H_{31}ClP_2$
$Pd(C{\equiv}CH)(SnMe_2Cl)$(dppe), **9**, 231
$PdTiC_{20}H_{26}ClP$
$TiCp_2(CH_2)(Cl)PdMe(PMe_2Ph)$, **4**, 387
$Pd_2As_2C_{38}H_{32}Cl_2$
$[Pd[(AsPh_2CH_2)C_6H_4](Cl)]_2$, **9**, 264
$Pd_2BRhC_{26}H_{30}F_4S_4$
$Rh(pyS)_4[Pd(C_3H_5)]_2[BF_4]$, **9**, 354
$Pd_2B_2C_{24}H_{28}N_{12}$
$(C_3H_5)Pd(pz)_3B_2(pz)_3Pd(C_3H_5)$, **9**, 350
$Pd_2B_{20}C_{12}H_{32}Br_{20}N_2$
$[Pd(C_3H_5)(NCMe_2)]_2[B_{10}Br_{10}]$, **9**, 354
$Pd_2C_{80}H_{71}Cl_3O_3P_2$
$(dba)_3Pd_2(CHCl_3)$(dppb), **12**, 47, 48
$Pd_2C_2Cl_4O_2$
$Pd_2(CO)_2Cl_4$, **9**, 195
$Pd_2C_4H_8Cl_4$
$Pd_2Cl_4(C_2H_4)_2$, **9**, 296
$Pd_2C_6H_{10}Cl_2$
$Pd_2Cl_2(C_3H_5)_2$, **9**, 323, 344, 350
$Pd_2C_6H_{10}S$
$Pd_2(C_3H_5)_2S$, **9**, 352
$Pd_2C_6H_{18}Cl_2S_2$
$[PdMe(SMe_2)(Cl)]_2$, **9**, 263
$Pd_2C_8H_{12}Cl_2N_4$
$Pd_2Cl_2(CNMe)_4$, **9**, 210
$Pd_2C_8H_{16}$
$[Pd(C_3H_5)(Me)]_2$, **9**, 240
$Pd_2C_{10}H_{16}O_4$
$Pd_2(O_2CMe)_2(C_3H_5)_2$, **9**, 352
$Pd_2C_{10}H_{18}Cl_2$
$Pd_2Cl_2(Me_2C_3H_3)_2$, **9**, 323
$Pd_2C_{10}H_{24}Cl_2N_2$
$[Pd(CH_2CH_2CH_2NMe_2)(Cl)]_2$, **9**, 238
$Pd_2C_{11}H_{28}Cl_2OP_4$
$Pd_2Cl_2(CO)(dmpm)_2$, **9**, 197
$Pd_2C_{11}H_{30}O_3P_4$
$Pd_2(OH)_2(CO)(dmpm)_2$, **9**, 198
$Pd_2C_{12}Br_4F_{10}$
$[PdBr(C_6F_5)(Br)]_2^{2-}$, **9**, 264
$Pd_2C_{12}H_{18}Cl_2O_2$
$[Pd[(CH_2{=}CHCH_2CH_2C(O)-CH_2](Cl)]_2$, **9**, 238
$Pd_2C_{12}H_{18}N_6$
$[Pd_2(CNMe)_6]^{2+}$, **9**, 210, 211
$Pd_2C_{12}H_{18}O_4$
$Pd_2(OAc)_2(C_8H_{12})$, **9**, 336
$Pd_2C_{14}H_{12}Cl_4N_2O_4$
$[Pd[(Me)(NO_2)C_6H_3]Cl(Cl)]_2^{2-}$, **9**, 263
$Pd_2C_{14}H_{14}N_2O_8$
$[Pd[C(CN){=}CHOC(O)Me]-(O_2CMe)]_2$, **9**, 249
$Pd_2C_{14}H_{16}I_2N_2$
$[Pd[(NH_2CH_2)C_6H_4](I)]_2$, **9**, 244
$Pd_2C_{14}H_{26}Cl_4N_2$
$[Pd[CH{=}C(Cl)CMe_2NMe_2]-(Cl)]_2$, **9**, 252
$Pd_2C_{16}H_6Cl_2F_{10}N_2$
$Pd_2Cl_2(C_6F_5)_2(CNMe)_2$, **9**, 213
$Pd_2C_{16}H_{14}Br_2O_4S_2$
$[Pd(CH_2SO_2Tol)(Br)]_2^{2-}$, **9**, 240
$Pd_2C_{16}H_{16}Br_2N_2$
$[Pd[(MeN{=}CH)C_6H_4]Br]_2$, **9**, 236
$Pd_2C_{16}H_{16}Cl_2N_2O_2$
$[Pd[(HC{=}C(CH_2OH)C_5H_4N]-(Cl)]_2$, **9**, 245
$Pd_2C_{16}H_{18}Cl_2S_2$
$[Pd[(SMeCH_2)C_6H_4](Cl)]_2$, **9**, 255
$Pd_2C_{18}H_{20}Cl_2N_2O_2$
$[Pd[(NMe_2)C_6H_4C(O)](Cl)]_2$, **9**, 245
$Pd_2C_{20}H_{36}Cl_2N_4$
$Pd_2Cl_2(Bu^tNC)_4$, **9**, 210, 215
$Pd_2C_{22}H_{30}O_2$
$Pd_2(CO)_2Cp^*_2$, **9**, 200, 375
$Pd_2C_{24}H_{20}Br_2F_{10}$
$[Pd[(CH_2{=}CH(CH_2)_2CHCH_2-C_6F_5](Br)]_2$, **9**, 250
$Pd_2C_{24}H_{38}O_2$
$[(pinene)Pd(OAc)]_2$, **12**, 612
$Pd_2C_{24}H_{47}BrOP_2$
$Cp(PPr^i_3)Pd(CO)PdBr(PPr^i_3)$, **9**, 378
$Pd_2C_{26}H_{60}Cl_2P_4$
$[(Et_3P)_2ClPd]_2(C{\equiv}C)$, **9**, 232
$Pd_2C_{28}H_{24}Cl_2F_{10}$
$Pd_2Cl_2(C_6F_5C_8H_{12})_2$, **9**, 308
$Pd_2C_{28}H_{48}Cl_4$
$Pd_2Cl_4(C_4Me_2Bu^t_2)_2$, **9**, 315
$Pd_2C_{29}H_{22}N_4O_5$
$Pd_2(OAc)_2(CO)(phen)_2$, **9**, 202
$Pd_2C_{29}H_{23}N_4O_5$
$Pd_2H(OAc)_2(CO)(phen)_2$, **9**, 202
$Pd_2C_{30}H_{46}O_4S_2$
$[PdPh(SBu_2)]_2(C_2O_4)$, **9**, 260
$Pd_2C_{32}H_{34}N_3P_2$
$[[(MeNC)Pd]_2(dppm)-[CMe{=}NMe]]^{3+}$, **9**, 258
$Pd_2C_{36}F_{30}S_2$
$[Pd(C_6F_5)_2(SC_6F_5)]_2^{2-}$, **9**, 263
$Pd_2C_{36}H_{28}Cl_2N_2O_2P_2$
$Pd_2Cl_2(CO)_2[(Ph_2P)C_5H_4N]_2$, **9**, 196
$Pd_2C_{36}H_{32}I_2O_4$
$[Pd[[(P(OPh)_2O]C_6H_4](I)]_2$, **9**, 247
$Pd_2C_{36}H_{58}F_{12}O_4P_2$
$[Pd(hfacac)(PPr^i_3)(CH_2C-H{=}CHCH_2)]_2$, **9**, 260
$Pd_2C_{37}H_{30}Cl_2OP_2$
$(PdClPPh_3)_2CO$, **9**, 197
$Pd_2C_{38}H_{36}Cl_2P_2$
$[PdPh(PMePh_2)(Cl)]_2$, **9**, 238
$Pd_2C_{38}H_{36}N_2$
$Pd_2(indenyl)_2(CNxylyl)_2$, **9**, 215
$Pd_2C_{40}H_{60}O_4P_2$
$[Pd_2(PBu^t_2)(PhO)(PBu^t_2H)]\cdot 3PhOH$, **9**, 383
$Pd_2C_{41}H_{36}O_5P_2$
$Pd_2(OAc)_2(CO)(PPh_3)_2$, **9**, 202
$Pd_2C_{42}H_{38}Cl_2O_4P_2$
$PdCl_2[CH(CO_2Me)PPh_3]_2$, **9**, 227
$Pd_2C_{42}H_{40}Cl_6O_2P_2$
$[Ph_3PCH_2COMe]_2[Pd_2Cl_6]$, **9**, 227
$Pd_2C_{42}H_{40}P_2$
$Pd_2(C_3H_5)_2(PPh_3)_2$, **9**, 352
$Pd_2C_{48}H_{30}Cl_2F_{10}P_2$
$[Pd(C_6F_5)(PPh_3)(Cl)]_2$, **9**, 238
$Pd_2C_{50}H_{44}Cl_2P_2$
$[PdBz(PPh_3)(Cl)]_2$, **9**, 266
$Pd_2C_{50}H_{44}Cl_2P_4$
$[PdCl(dppm)]_2$, **9**, 260
$Pd_2C_{50}H_{44}I_4P_4$
$Pd_2I_4(dppm)_2$, **9**, 319
$Pd_2C_{51}H_{42}O_3$
$Pd_2(dba)_3$, **12**, 302, 912
$Pd_2C_{52}H_{42}Cl_2I_2P_4$
$Pd_2I_2(CCCl_2)(dppm)_2$, **9**, 319
$Pd_2C_{52}H_{44}Cl_4P_4$
$Pd_2Cl_2(CCCl_2)(dppm)_2$, **9**, 319
$Pd_2C_{52}H_{46}O_4P_2$
$[PdPh(PPh_3)(OAc)]_2$, **9**, 259
$Pd_2C_{52}H_{47}Cl_2NP_4$
$Pd_2Cl_2(dppm)_2(CNMe)$, **9**, 215
$Pd_2C_{52}H_{96}P_4$
$Pd_2(dcpe)_2$, **9**, 318
$Pd_2C_{54}H_{48}Cl_4P_4$
$Pd_2Cl_2(CCCl_2)(dppe)_2$, **9**, 319
$Pd_2C_{54}H_{50}Cl_2P_4S$
$Pd_2Cl_2[CC(SMe)Me](dppm)_2$, **9**, 319
$Pd_2C_{56}H_{44}Cl_6O_2P_4S_2$
$(ClPd)(dppm)_2(SO_2S)[Pd(C_6-Cl_5)]$, **9**, 226
$Pd_2C_{56}H_{53}N_3P_4$
$Pd_2(CNMe)_2(dppm)_2(CNMe)$, **9**, 215
$Pd_2C_{57}H_{54}O_5P_4$
$Pd_2(OAc)_2(CO)(dppe)_2$, **9**, 202
$Pd_2C_{60}H_{55}Cl_2NO_2P_4$
$Pd_2Cl_2(dppm)_2-(CNCpCHCH_2CO_2Me)$, **9**, 215
$Pd_2C_{62}H_{44}Cl_{10}P_4$
$[Pd(C_6Cl_5)]_2(dppm)_2$, **9**, 226
$Pd_2C_{64}H_{56}Cl_4O_8$
$Pd_2Cl_4[[(MeO)C_6H_4]_2CC]_4$, **9**, 309
$Pd_2C_{72}H_{60}O_4P_4$
$[PdPh(PPh_3)(Ph_2PO_2)]_2$, **9**, 237
$Pd_2C_{84}H_{60}$
$Pd_2(C_5Ph_5)_2(PhCCPh)$, **9**, 375, 378
$Pd_2C_{109}H_{102}O_5P_8$

[Pd_2(CO)(dppe)$_4$](OAc)$_2$, **9**, 202

$Pd_2CoC_{26}H_{49}O_4P_2$
CoPd$_2$(CO)$_4$(MeC$_3$H$_4$)(PPr$^i_3$)$_2$, **10**, 157

$Pd_2CoFeC_{56}H_{44}NO_7P_4$
FeCoPd$_2$(CO)$_6$(NO)(dppm)$_2$, **10**, 243

$Pd_2CoMoC_{56}H_{44}O_6P_4$
MoCoPd$_2$(CO)$_6$(dppm)$_2$, **10**, 243

$Pd_2FeC_{54}H_{44}O_4P_4$
Pd$_2$Fe(CO)$_4$(dppm)$_2$, **7**, 84; **9**, 205; **10**, 19

$Pd_2Fe_2C_{32}H_{20}Cl_2O_8P_2$
[(CO)$_4$Fe(PPh$_2$)Pd(Cl)]$_2$, **10**, 47

$Pd_2MoC_{31}H_{52}O_3P_2$
CpMoPd$_2$(CO)$_3$Cp(PPr$^i_3$)$_2$, **10**, 124

$Pd_2Mo_2C_{52}H_{40}O_6P_2$
Mo$_2$Pd$_2$(CO)$_6$Cp$_2$(PPh$_3$)$_2$, **10**, 49

$Pd_2Mo_2C_{54}H_{44}O_6P_2$
Mo$_2$Pd$_2$(CO)$_6$(PPh$_2$Bz)$_2$Cp$_2$, **9**, 204; **10**, 204

$Pd_2Na_2C_{14}H_{22}Cl_2O_8$
Na$_2$[Pd[CH(CO$_2$Et)$_2$]$_2$(Cl)]$_2$, **9**, 241

$Pd_2Pt_2C_{40}H_{24}Br_4F_{20}$
[(C$_6$F$_5$)$_2$Pt]$_2$(Br)$_4$[Pd(cod)]$_2$, **9**, 470

$Pd_2Rh_2C_{54}H_{64}Cl_6N_6P_2$
[RhCl$_2$(Bu^{t}NC)$_2$(NC$_5$H$_4$PPh$_2$)-PdCl]$_2$, **10**, 64

$Pd_2Sn_2C_{37}H_{30}Cl_6OP_2$
[Pd(SnCl$_3$)(PPh$_3$)]$_2$CO, **9**, 197

$Pd_3C_{12}H_{18}O_{12}$
[Pd(OAc)$_2$]$_3$, **12**, 396

$Pd_3C_{32}H_{45}O_2$
[Pd$_3$Cp*$_3$(CO)$_2$]$^+$, **9**, 375

$Pd_3C_{42}H_{66}N_6$
Pd$_3$(CNCy)$_6$, **9**, 209, 215

$Pd_3C_{57}H_{45}O_3P_3$
Pd$_3$(CO)$_3$(PPh$_3$)$_3$, **9**, 200

$Pd_3C_{60}H_{66}N_6$
Pd$_3$(CNxylyl)$_6$, **9**, 209, 212

$Pd_3C_{75}H_{60}O_3P_4$
Pd$_3$(CO)$_3$(PPh$_3$)$_4$, **9**, 201

$Pd_3C_{76}H_{66}OP_6$
[Pd$_3$(CO)(dppm)$_3$]$^{2+}$, **9**, 198

$Pd_3C_{82}H_{72}F_9O_6P_7$
[Pd$_3$(MeO$_2$CCCCO$_2$Me)(CF$_3$C-O$_2$)(dppm)$_3$][PF$_6$], **9**, 320

$Pd_3C_{85}H_{77}NP_6$
[Pd$_3$(CNxylyl)(dppm)$_3$]$^{2-}$, **9**, 209

$Pd_3Sn_2C_{40}H_{77}O_{10}P_3$
Pd$_3$Sn$_2$(acac)$_4$(CO)$_2$(PEt$_3$)$_3$, **9**, 207

$Pd_4C_{12}H_{12}O_{12}$
Pd$_4$(MeCOO)$_4$(CO)$_4$, **9**, 198, 201

$Pd_4C_{16}H_{20}Cl_2N_8$
[Pd$_4$(bim)$_2$(C$_3$H$_5$)$_4$]CH$_2$Cl$_2$, **9**, 353

$Pd_4C_{43}H_{32}N_6O_7$
[Pd$_4$H$_2$(CO)$_3$(phen)$_3$](OAc)$_2$, **9**, 202

$Pd_4C_{53}H_{108}O_5P_4$
Pd$_4$(CO)$_5$(PBu$_3$)$_4$, **9**, 201

$Pd_4C_{54}H_{40}N_8O_6$
[Pd$_4$H$_2$(CO)$_2$(phen)$_4$](OAc)$_2$, **9**, 202

$Pd_4C_{54}H_{108}O_6P_4$
Pd$_4$(CO)$_6$(PBu$_3$)$_4$, **9**, 201

$Pd_4C_{58}H_{44}N_8O_{10}$
Pd$_4$(OAc)$_4$(CO)$_2$(phen)$_4$, **9**, 202

$Pd_4C_{60}H_{52}O_8$
Pd$_4$(OAc)$_4$(CPh$_2$)$_4$, **9**, 219

$Pd_4C_{74}H_{60}I_2O_2P_4$
Pd$_4$I$_2$(CO)$_2$(PPh$_3$)$_4$, **9**, 201

$Pd_4C_{75}H_{60}I_2O_3P_4$
Pd$_4$I$_2$(CO)$_3$(PPh$_3$)$_4$, **9**, 201

$Pd_4C_{77}H_{60}O_5P_4$
Pd$_4$(CO)$_5$(PPh$_3$)$_4$, **9**, 198, 201

$Pd_4Hg_2C_{28}H_{60}Br_2O_4P_4$
Pd$_4$Hg$_2$Br$_2$(CO)$_4$(PEt$_3$)$_4$, **9**, 207

$Pd_4Mn_4C_{40}H_{28}N_4O_{12}$
[Mn(CO)$_2$Cp'(NC)Pd(CO)]$_4$, **10**, 53
[Pd(CO)(NC)Mn(CO)$_2$(C$_5$H$_4$-Me)]$_4$, **9**, 207

$Pd_4Mo_4Na_2C_{32}H_{20}O_{12}$
Na$_2$[Pd$_4$[MoCp(CO)$_3$]$_4$], **10**, 51

$Pd_5C_{42}H_{55}O_8P_5S_3$
Pd$_5$(CO)$_2$(SO$_2$)$_3$(PMe$_2$Ph)$_5$, **9**, 202

$Pd_5C_{70}H_{77}N_7O_6S_3$
Pd$_5$(SO$_2$)$_3$(CNxylyl)$_7$, **9**, 209

$Pd_5C_{92}H_{75}O_6P_5S_2$
Pd$_5$(CO)$_2$(SO$_2$)$_2$(PPh$_3$)$_5$, **9**, 202

$Pd_6C_{24}H_{42}Se_3$
Pd$_6$(Se)$_3$(MeC$_3$H$_4$)$_6$, **9**, 351

$Pd_6C_{25}H_{63}O_4P_7$
Pd$_6$(CO)$_4$(PMe$_3$)$_7$, **9**, 202

$Pd_6C_{76}H_{60}Cl_2O_4P_4$
Pd$_6$Cl$_2$(CO)$_4$(PPh$_3$)$_4$, **9**, 203

$Pd_6C_{128}H_{112}Cl_{12}O_{16}$
[Pd$_2$Cl$_3$[[(MeO)C$_6$H$_4$]$_2$CC]$_4$]$_2$-[Pd$_2$Cl$_6$], **9**, 309

$Pd_6C_{162}H_{144}N_2P_{12}$
[[Pd$_3$(dppm)$_3$]$_2$(CNC$_6$Me$_4$N-C)]$^{4+}$, **9**, 209

$Pd_6H_{10}N_2$
PdMe$_2$(pyridazine), **9**, 240

$Pd_6HgC_{42}H_{90}O_6P_6$
Pd$_6$Hg(CO)$_6$(PEt$_3$)$_6$, **9**, 207

$Pd_7C_{28}H_{63}O_7P_7$
Pd$_7$(CO)$_7$(PMe$_3$)$_7$, **9**, 203

$Pd_8C_{29}H_{63}O_8P_7$
Pd$_8$(CO)$_8$(PMe$_3$)$_7$, **9**, 203

$Pd_{10}C_{38}H_{60}O_{14}P_4$
Pd$_{10}$(CO)$_{14}$(PEt$_3$)$_4$, **9**, 204

$Pd_{10}C_{48}H_{90}O_{12}P_6$
Pd$_{10}$(CO)$_{12}$(PEt$_3$)$_6$, **9**, 204

$Pd_{10}C_{62}H_{108}O_{14}P_4$
Pd$_{10}$(CO)$_{14}$(PBu$_3$)$_4$, **9**, 204

$Pd_{10}C_{120}H_{90}O_{12}P_6$
Pd$_{10}$(CO)$_{12}$(PPh$_3$)$_6$, **9**, 201

$Pd_{16}C_{43}H_{90}O_7P_6$
Pd$_{16}$(CO)$_7$(PEt$_3$)$_6$, **9**, 204

$Pd_{23}C_{82}H_{150}O_{22}P_{10}$
Pd$_{23}$(CO)$_{22}$(PEt$_3$)$_{10}$, **9**, 204

$PrC_{15}H_{15}$
PrCp$_3$, **4**, 52

$PrC_{20}H_{20}$
[PrCp$_4$]$^-$, **4**, 58

$PrC_{27}H_{42}N_3$
Pr(C$_5$H$_4$CH$_2$CH$_2$NMe$_2$)$_3$, **4**, 55

$PtAgC_{15}H_{24}BrN_3$
Pt[C$_6$H$_3$(CH$_2$NMe$_2$)$_2$]-(NTolCHMe)AgBr, **10**, 71

$PtAgC_{53}H_{47}Cl_2P_4$
Pt(MeCC)(Cl)(dppm)$_2$(AgCl), **8**, 239

$PtAgC_{84}H_{60}Br_4Cl_{10}P_2$
[(Ph$_3$P)Ag(Br)PtBr(C$_6$Cl$_5$)-(PPh$_3$)]$_2$, **9**, 474

$PtAsC_{20}H_{17}Cl_2$
PtCl$_2$(H$_2$CCHC$_6$H$_4$AsPh$_2$), **9**, 552

$PtAsC_{26}H_{22}F_6O_3$
Pt[CH(CH$_2$OMe)C$_6$H$_4$AsPh$_2$]-(CF$_3$C(O)CH$_2$C(O)CF$_3$), **9**, 552

$PtAsC_{26}H_{26}O_2$
Pt[CH$_2$C(CH$_2$)C$_6$H$_4$AsPh$_2$]-(acac), **9**, 552
Pt[CHC(Me)C$_6$H$_4$AsPh$_2$](acac), **9**, 552

$PtAsC_{26}H_{27}O_2$
[Pt(H$_2$CCMeC$_6$H$_4$AsPh$_2$)-(acac)]$^+$, **9**, 552

$PtAsC_{27}H_{29}ClP$
PtClMe(PMe$_2$Ph)(AsPh$_3$), **9**, 438

$PtAsC_{45}H_{45}O_2P_2$
[Pt(PPh$_3$)$_2$(C$_3$H$_5$)][As(OCH$_2$-CHCH$_2$)$_2$], **9**, 574

$PtAs_2C_{51}H_{42}O$
Pt(AsPh$_3$)$_2$[PhHCC(O)CHPh], **9**, 578

$PtAs_2Co_2C_{33}H_{24}O_7$
PtCo$_2$(CO)$_7$(dpae), **10**, 39, 47

$PtAuB_9C_{32}H_{56}P_3$
PtH(PEt$_3$)$_2$(C$_2$B$_9$H$_{10}$)Au(PPh$_3$), **10**, 71

$PtAuC_8H_{15}Cl_2S$
PtCl(SMe$_2$)(CCBu^t)(AuCl), **9**, 571

$PtAuC_{31}H_{31}ClN_2P$
PtClMe$_2$(CH$_2$AuPPh$_3$)(bipy), **3**, 4

$PtAuC_{31}H_{31}Cl_2P_2$
PtCl(CCBu^t)(dppm)AuCl, **9**, 571

$PtAuC_{56}H_{53}ClP_4$
[PtCl(CCBu^t)(dppm)$_2$(Au)]$^+$, **9**, 571

$PtAuC_{62}H_{62}P_4$
[Pt(CCBu^t)$_2$(dppm)$_2$(Au)]$^+$, **9**, 571

$PtAu_2C_{16}H_{30}Cl_2S_2$
Pt(CCBu^t)$_2$(SMe$_2$)$_2$(AuCl)$_2$, **9**, 571

$PtAu_2C_{52}H_{48}Cl_2P_4S_4$

$Pt[(SPPh_2CH_2)_2AuCl]_2$, **10**, 71
$PtAu_2C_{52}H_{48}P_4S_4$
$Pt[(SPPh_2CH_2)_2Au]_2$, **10**, 71
$PtAu_6B_2C_{174}H_{145}P_7$
$[Au_6Pt(PPh_3)_7][BPh_4]_2$, **10**, 256
$PtAu_6C_{127}H_{105}OP_7$
$[Au_6Pt(CO)(PPh_3)_7]^{2+}$, **10**, 298
$PtAu_8C_{136}H_{123}N_4O_6P_7$
$[PtAu_8(PPh_3)_7(CNBu^t)_2](NO_3)_2$, **3**, 43
$PtAu_8C_{144}H_{120}P_8$
$[Pt(AuPPh_3)_8]^{2+}$, **10**, 37
$PtAu_8HgC_{126}H_{105}N_3O_9P_7$
$Pt(AuPPh_3)_7Au(NO_3)Hg(NO_3)_2^{+}$, **10**, 344
$PtAu_8Hg_2C_{144}H_{120}Cl_2P_8$
$Pt(AuPPh_3)_8(HgCl)_2^{2+}$, **10**, 344
$PtAu_8Hg_2C_{144}H_{120}N_2O_6P_8$
$Pt(AuPPh_3)_8Hg_2(NO_3)_2^{2+}$, **10**, 344
$PtAu_8Hg_2C_{144}H_{120}P_8$
$Pt(AuPPh_3)_8Hg_2^{4+}$, **10**, 343
$PtBC_{18}H_{21}$
$Pt(cod)(C_4H_4BPh)$, **9**, 582
$PtBC_{33}H_{39}P_2$
$Pt[B(Me)CH_2CH(Me)CH(Me)CH_2](dppe)$, **9**, 581
$PtBC_{42}H_{37}$
$Pt(cod)(BC_4Ph_5)$, **9**, 582
$PtBC_{44}H_{35}F_4O_2P_2$
$[Pt(COPh)(CO)(PPh_3)_2]BF_4$, **9**, 462
$PtB_2C_{20}H_{18}$
$Pt(C_4H_4BPh)_2$, **9**, 582
$PtB_2C_{20}H_{36}$
$(cod)Pt[Et_4Me(C_3B_2)]$, **1**, 206
$PtB_2C_{20}H_{44}P_2$
$Pt[Me_2BC(Me)CHCHC(Me)BMe_2](Et_2PCH_2CH_2PEt_2)$, **9**, 574
$PtB_3C_{16}H_{29}P_2$
$(Me_2PhP)_2PtB_3H_7$, **1**, 313, 316, 317
$PtB_3FeC_{34}H_{35}P_2$
$[(Ph_2PC_5H_4)_2Fe]PtB_3H_7$, **1**, 317
$PtB_4NiC_{29}H_{51}$
$CpNi(MeCB_2C_2Et_4)Pt(MeCB_2C_2Et_4)$, **9**, 179
$PtB_5OsC_{45}H_{49}ClOP_3$
$(CO)(PPh_3)_2HOs(PMe_2Ph)ClPtB_5H_7$, **1**, 324
$PtB_5OsC_{64}H_{52}OP_3$
$(CO)(PPh_3)(PPh_2)Os(PPh_3)PtB_5H_7Ph$, **1**, 324
$PtB_8C_{16}H_{33}ClP_2$
$Cl(PhMe_2P)_2PtB_8H_{11}$, **1**, 337
$PtB_8C_{16}H_{34}P_2$
$(PMe_2Ph)_2PtB_8H_{12}$, **1**, 336, 364
$PtB_8C_{17}H_{36}OP_2$
$(OMe)(PhMe_2P)_2PtB_8H_{11}$, **1**, 337
$PtB_8C_{36}H_{42}P_2$
$(PPh_3)_2PtB_8H_{12}$, **1**, 337
$PtB_8IrC_{13}H_{47}OP_4$
$(Me_3P)_2(CO)HIrB_8H_{10}Pt(PMe_3)_2$, **1**, 332, 333
$PtB_8PdC_{32}H_{54}P_4$
$(PhMe_2P)_4PtPdB_8H_{10}$, **1**, 337
$PtB_9C_{25}H_{44}OP_3$
$(PMe_2Ph)_3Pt_2B_9H_8(OMe)$, **1**, 349
$PtB_9RhC_{33}H_{57}OP_3$
$Rh(PPh_3)(C_2B_9H_{11})(H)(CO)Pt(PEt_3)_2$, **10**, 64
$PtB_9WC_{18}H_{45}O_2P_2$
$W(CO)_2(Me_2C_2B_9H_9)Pt(PEt_3)_2$, **10**, 32
$PtB_{10}C_{12}H_{40}P_2Te$
$(Et_3P)_2TePtB_{10}H_{10}$, **1**, 271
$PtB_{10}C_{13}H_{32}NPS_2$
$(SCSNEt_2)(PMe_2Ph)PtB_{10}H_{11}$, **1**, 352
$PtB_{10}C_{16}H_{32}P_2$
$(PMe_2Ph)_2PtB_{10}H_{10}$, **1**, 351
$PtB_{10}C_{16}H_{33}ClP_2$
$Cl[PMe_2Ph]_2PtB_{10}H_{11}$, **1**, 351
$PtB_{10}C_{16}H_{33}P_2$
$(PMe_2Ph)_2B_{10}H_{12}$, **1**, 362
$PtB_{10}C_{16}H_{34}P_2$
$[(PMe_2Ph)_2PtB_{10}H_{12}]$, **1**, 351
$PtB_{10}C_{18}H_{39}ClP_2$
$Cl[PMe_2Ph]_2(SMe_2)PtB_{10}H_{11}$, **1**, 351
$PtB_{10}C_{22}H_{37}P_2$
$Ph(PMe_2Ph)_2PtB_{10}H_{10}$, **1**, 265
$PtB_{10}C_{36}H_{40}P_2Se$
$(PPh_3)_2SePtB_{10}H_{10}$, **1**, 271
$PtB_{10}C_{36}H_{57}P_3Se$
$[PtH(PPh_3)_3][SeB_{10}H_{11}]$, **1**, 271
$PtB_{16}C_{16}H_{40}P_2$
$(PMe_2Ph)PtB_{16}H_{18}(PMe_2Ph)$, **1**, 327, 330, 364
$PtB_{18}C_8H_{31}P$
$(PMe_2Ph)PtB_{18}H_{20}$, **1**, 365, 366
$PtB_{20}C_{20}H_{56}N_2$
$[BzNMe_3]_2[(B_{10}H_{12})_2Pt]$, **1**, 351
$PtCCl_3O$
$[PtCl_3(CO)]^-$, **9**, 400, 406, 409, 411
$PtCCl_5O$
$[PtCl_5(CO)]^-$, **9**, 411
PtC_2ClF_3O
$PtCl(CF_3CO)$, **9**, 410
$PtC_2Cl_2O_2$
$PtCl_2(CO)_2$, **9**, 409, 422
$PtC_2H_2Cl_2O$
$PtCl_2(H_2CCO)$, **9**, 559
$PtC_2H_4Cl_3$
$[PtCl_3(C_2H_4)]^-$, **9**, 534, 544
$PtC_2H_4Cl_4$
$[PtCl_3CH_2{=}CH_2)]Cl$, **12**, 602
$PtC_2H_{10}P_2$
$Pt(C_2H_4)(PH_3)_2$, **9**, 550
$PtC_2I_2O_2$
$PtI_2(CO)_2$, **9**, 409
$PtC_4H_8Cl_2$
$PtCl_2(C_2H_4)_2$, **9**, 547
$PtC_4H_8O_2$
$Pt(O_2)(C_2H_4)_2$, **9**, 537
$PtC_4H_9ClNO_2$
$[PtCl(gly)(C_2H_4)]^+$, **9**, 541
$PtC_4H_9Cl_2NO_2$
$PtCl_2(gly)(C_2H_4)$, **9**, 545
$PtC_4H_{10}Cl_2OS$
$PtCl_2(DMSO)(C_2H_4)$, **9**, 541, 545
PtC_4O_4
$Pt(CO)_4$, **9**, 392, 400
$PtC_5H_8Cl_2N_2$
$PtCl_2(N_2C_3H_4)(C_2H_4)$, **9**, 545
$PtC_5H_{10}ClNO_2$
$PtCl[OC(O)CH_2CH_2NH_2](C_2H_4)$, **9**, 545
$PtC_5H_{12}OS$
$PtMe_2(CO)(SMe_2)$, **9**, 409, 411
$PtC_6H_8F_4$
$Pt(C_2H_4)_2(C_2F_4)$, **9**, 550
PtC_6H_{10}
$Pt(C_3H_5)_2$, **9**, 324, 575
PtC_6H_{12}
$Pt(C_2H_4)_3$, **9**, 536, 537, 550
$PtC_6H_{12}N_2$
$PtMe_2(CNMe)_2$, **9**, 419, 457
$PtC_6H_{13}P_2$
$[Pt(PH_3)_2(C_6H_7)]^+$, **9**, 576
$PtC_6H_{14}Cl_2$
$Pt_2(Cl)_2Me_2(C_2H_4)_2$, **9**, 561, 562
$PtC_6H_{15}Cl_2N$
$PtCl_2(NHMe_2)(C_4H_8)$, **9**, 544
$PtC_6H_{16}S$
$PtMe_2(SMe_2)(C_2H_4)$, **9**, 542, 562
$PtC_6H_{18}O_2S_2$
$PtMe_2(DMSO)_2$, **9**, 439, 459
$PtC_7H_8Cl_2$
$PtCl_2(nbd)$, **9**, 297
$PtC_7H_9Cl_2N$
$PtCl_2(py)(C_2H_4)$, **9**, 495, 544
$PtC_7H_{12}Cl_2S$
$PtCl_2(C_7H_{12}S)$, **9**, 553
$PtC_7H_{21}BrS_2$
$PtMe_3Br(SMe_2)_2$, **9**, 504
$PtC_8H_8Cl_2$
$PtCl_2(cod)$, **9**, 437
PtC_8H_{10}
$Pt(C_3H_5)(Cp)$, **9**, 410, 417, 581
$PtC_8H_{12}Cl_2$
$PtCl_2(cod)$, **9**, 296, 554
$PtC_8H_{12}F_6O_3$
$PtMe_3(CF_3COCHCOCF_3)(H_2O)$, **9**, 500
PtC_8H_{14}
$PtBu^tCp$, **9**, 581
$PtC_8H_{14}Cl_2$
$PtCl_2(C_8H_{14})$, **9**, 556
$PtC_8H_{20}ClN_5$
$PtCl(CH_2CH_2N_3)(TMEDA)$, **9**, 548
$PtC_8H_{20}Cl_2N_2$
$PtCl_2(TMEDA)(C_2H_4)$, **9**, 548
$PtC_8H_{20}N_4O_4$
$Pt(CH_2CH_2NO_2)(NO_2)(TMEDA)$, **9**, 548
$PtC_8H_{22}OP_2$
$Pt(CH_2OCH_2)(PMe_3)_2$, **9**, 481
$PtC_8H_{24}P_2$

$PtMe_2(PMe_3)_2$, **9**, 416
$PtC_9H_8F_6$
$Pt(CF_3)_2(nbd)$, **9**, 472
PtC_9H_{14}
$Pt(Me)_2(nbd)$, **9**, 472
PtC_9H_{16}
$PtMe_3(C_5H_4Me)$, **9**, 581
$PtC_9H_{18}Cl_2N$
$PtCl_2(NC_5H_8Me_2)(C_2H_4)$, **9**, 545
$PtC_{10}H_{12}O_3$
$Pt(C_4H_2O_3)(C_6H_{10})$, **9**, 536, 537
$PtC_{10}H_{16}O$
Pt(COMe)MeEtCp, **9**, 581
$PtC_{10}H_{21}ClN_2O_2$
$[Pt(CH_2CHN)(Bu^t)O]Cl[N-(O)Bu^t]$, **9**, 548
$PtC_{10}H_{21}ClN_4O_2$
$Pt[CH_2CH_2NHC(O)]Cl(NCO)-(TMEDA)$, **9**, 548
$PtC_{10}H_{22}F_6OP_2$
$PtMe[OCH(CF_3)_2](PMe_3)_2$, **9**, 461
$PtC_{10}H_{23}P_2$
$[Pt(H_2CCHCH_2CH_2)(dmpe)]^+$, **9**, 548, 552
$PtC_{10}H_{25}P$
$PtEtH(C_2H_4)(PEt_3)$, **9**, 455
$PtC_{11}H_{10}O$
$Pt(Cp)_2(CO)$, **9**, 410
$PtC_{11}H_{15}ClO$
$Pt(Cp^*)Cl(CO)$, **9**, 410
$PtC_{11}H_{17}$
$[Pt(C_3H_5)(cod)]^+$, **9**, 410
$PtC_{11}H_{20}Cl_2$
$PtCl_2(C_2H_4)(C_9H_{16})$, **9**, 554, 556
$PtC_{12}H_{10}N_4$
$Pt(TCNE)(C_3H_5)_2$, **9**, 576
$PtC_{12}H_{10}O_2$
$Pt(CO)_2Cp_2$, **9**, 580
$PtC_{12}H_{14}$
$Pt(C_2H_4)Cp_2$, **9**, 580
$PtC_{12}H_{15}$
$[Pt(C_2H_4)(C_5H_6)Cp]^+$, **9**, 580
$PtC_{12}H_{15}O_2$
$Pt(CO)_2Cp^*$, **9**, 580
$PtC_{12}H_{16}N$
$PtMe_2[(C_3H_3N_2)_3CH]$, **9**, 440
$PtC_{12}H_{17}ClNOP$
$Pt[CMeCHN(CH_2)O]Cl-(PMe_2Ph)$, **9**, 559
$PtC_{12}H_{23}Cl$
$PtClMe(C_2H_4)(C_9H_{16})$, **9**, 554
$PtC_{12}H_{27}P_2$
$[Pt(H_2CCHCMe_2CH_2)-(dmpe)]^+$, **9**, 582
$PtC_{12}H_{29}F_3P_2O_2S$
$Pt(CH_2Bu^t)(OTf)(PMe_3)_2$, **9**, 518
$PtC_{12}H_{31}ClP_2$
$PtHCl(PEt_3)_2$, **9**, 407
$PtC_{12}H_{35}D_5P_2$
$Pt(Et)(C_2D_5)(PEt_3)_2$, **9**, 455
$PtC_{13}H_{14}Cl_2N_2$
$Pt(CH_2CH_2CH_2)Cl_2(bipy)$, **9**, 505
$PtC_{13}H_{15}ClNO$
$Pt(C_8H_{10}O)(Cl)(py)$, **9**, 549, 552
$PtC_{13}H_{16}Cl_2N_2$
$PtCl_2(py)[CH_2CH_2NC_5H_4-(Me)]$, **9**, 548
$PtC_{13}H_{17}$
$[Pt(C_4Me_4)Cp]^+$, **9**, 579
$PtC_{13}H_{22}Cl_2$
$PtCl_2(Pr^iMe_5C_5)$, **9**, 561
$PtC_{13}H_{24}$
$PtMe_3Cp^*$, **9**, 581
$PtC_{13}H_{25}O_4$
$PtMe_3(acac)_2$, **9**, 500
$PtC_{13}H_{30}ClOP_2$
$[PtCl(CO)(PEt_3)_2]^+$, **9**, 407
$PtC_{13}H_{30}F_3IP_3$
$PtI(CF_3)(PEt_3)_2$, **9**, 467
$PtC_{13}H_{31}ClO_2P_2$
$PtCl(COOH)(PEt_3)_2$, **9**, 407
$PtC_{14}H_{15}Cl_2N$
$PtCl_2(MeC_5H_4N)(PhCH=CH_2)$, **9**, 302
$PtC_{14}H_{18}$
$[Pt(C_4Me_4)(C_6H_6)]^+$, **9**, 579, 582
$PtC_{14}H_{19}Cl_2N$
$PtCl_2(py)(C_9H_{14})$, **9**, 542, 544
$PtC_{14}H_{22}Cl_2N$
$PtCl_2(py)[H_2CCH(CH_2)_6CH_2$, **9**, 549
$PtC_{14}H_{26}O_6P_2$
$Pt[CHClP(O)(OMe)_2]_2(cod)$, **9**, 471
$PtC_{14}H_{34}P_2$
$Pt(C_2H_4)(PEt_3)_2$, **9**, 455, 535
$PtC_{14}H_{26}Cl_2O_6P_2$
$Pt[CH(Cl)P(O)(OMe)_2]_2(cod)$, **9**, 558
$PtC_{16}H_{14}Cl_2N_2O_2$
$PtCl_2(CN\text{-}p\text{-}anisyl)_2$, **9**, 425
$PtC_{16}H_{16}Cl_2$
$PtCl_2(H_2CCHPh)_2$, **9**, 542
$PtC_{16}H_{22}$
$PtMe(C_6H_4Me)(cod)$, **9**, 458
$PtC_{16}H_{22}O_2S_2$
$PtPh_2(DMSO)_2$, **9**, 459
$PtC_{16}H_{22}S_2$
$PtPh_2(SMe_2)_2$, **9**, 459
$PtC_{16}H_{24}$
$Pt(cod)_2$, **9**, 436, 553
$PtC_{16}H_{32}P_2$
$Pt(C{\equiv}CH)_2(PEt_3)_2$, **9**, 493
$PtC_{16}H_{40}P_2$
$PtEt_2(PEt_3)_2$, **9**, 455
$PtC_{16}H_{41}B_5P_2$
$(Et_3P)_2Pt(Me_2C_2B_5H_5)$, **1**, 395
$PtC_{17}H_{20}OS$
$PtPh_2(CO)(SEt_2)$, **9**, 407
$PtC_{17}H_{23}Br$
$PtBr(C_6H_2Me_3)(cod)$, **9**, 437
$PtC_{17}H_{25}ClP_2$
$PtMeCl(PMe_2Ph)_2$, **4**, 387
$PtC_{17}H_{31}ClS_2$
$PtCl(C_6H_2Me_3)(SEt_2)_2$, **9**, 437
$PtC_{17}H_{33}B_9P_2$
$(PMe_2Ph)_2Pt(CB_9H_{11})$, **1**, 379
$PtC_{17}H_{35}P$
$Pt(MeC_3H_4)(MeC_3H_4)(Pr^i_3P)$, **9**, 352
$PtC_{17}H_{40}P_2$
$Pt(CH_2CMe_2CH_2)(PEt_3)_2$, **9**, 456, 481
$PtC_{18}H_{15}F_5$
$Pt(C_6F_5)(H_2CCHPh)[H_2CC-(Me)CH_2]$, **9**, 542, 545, 574
$PtC_{18}H_{18}Cl_2N_2O_3$
$PtCl_2(CN\text{-}p\text{-}anisyl)(COC_2H_4N\text{-}p\text{-}anisyl)$, **9**, 425
$PtC_{18}H_{20}F_6$
$PtMe[C_7H_5(CF_3)_2](cod)$, **9**, 557
$PtC_{18}H_{24}Cl_2OP_2$
$PtCl_2(PMe_2Ph)_2(H_2CCO)$, **9**, 559
$PtC_{18}H_{26}$
$PtMe(C_6H_2Me_3)(cod)$, **9**, 458
$PtC_{18}H_{28}N$
$Pt(Cp)(C_3H_5)(CNBu^t)_2$, **9**, 417
$PtC_{18}H_{34}INO_2P_2$
$PtI(C_6H_4NO_2)(PEt_3)_2$, **9**, 434
$PtC_{18}H_{35}O_4P_2$
$Pt[CH(CHCHCO_2Me)CH_2C-H_2CH(CHCHCO_2Me)-(PMe_3)_2$, **9**, 554
$PtC_{18}H_{36}OP_2$
$PtPh(OH)(PEt_3)_2$, **9**, 462
$PtC_{18}H_{40}I_2P_4$
$[Pt(CH_2PMe_3)_2(PMe_3)_2]I_2$, **9**, 495
$PtC_{19}H_{15}Cl_2OP$
$PtCl_2(CO)(PPh_3)$, **9**, 392
$PtC_{19}H_{20}Cl_2N_2$
$Pt(CHBzCH_2)Cl_2(py)_2$, **9**, 505
$PtC_{19}H_{23}ClN_3$
$[PtCl(py)(MeC_5H_4N)[CH_2C-H_2NC_5H_4(Me)]]^+$, **9**, 548
$PtC_{20}H_{18}ClOP$
$PtClPh(CO)(PMePh_2)$, **9**, 406
$PtC_{20}H_{19}P$
$Pt(PPh_3)(C_2H_4)$, **9**, 535
$PtC_{20}H_{24}Cl_2N_2$
$Pt_2Cl_2Me_2(Me_2phen)-(MeCCMe)$, **9**, 571
$PtC_{20}H_{26}B_7P$
$(PPh_3)Pt(C_2B_7H_{11})$, **1**, 395
$PtC_{20}H_{28}P_2$
$Pt(PMe_3)_2(PhCCPh)$, **9**, 565, 566
$PtC_{20}H_{43}P_2$
$[Pt(PEt_3)_2(C_8H_{13})]^-$, **9**, 574
$PtC_{21}H_{21}I_2PS_2$
$PtI_2(PPh_3)[C(SMe)_2]$, **9**, 426
$PtC_{21}H_{22}Cl_2NP$
$PtCl_2(CHNMe_2)(PPh_3)$, **9**, 495
$PtC_{21}H_{30}$
$Pt(norbornene)_3$, **9**, 535
$PtC_{21}H_{37}P$
$[Pt(PEt_3)[C_3H_4(CO_2Me)_2]-(cod)]^+$, **9**, 576

$PtC_{22}H_{20}N_2$
$PtPh_2(py)_2$, **9**, 439
$PtC_{22}H_{20}OP$
$[Pt(C_3H_5)(PPh_3)(CO)]^+$, **9**, 410
$PtC_{22}H_{22}$
Pt(PhCCPh)(cod), **9**, 566
$PtC_{22}H_{26}$
$Pt(C_3H_4Ph)Cp(cod)$, **9**, 580
$PtC_{22}H_{37}F_4P$
$Pt(PCy_3)(C_2H_4)(C_2F_4)$, **9**, 537
$PtC_{22}H_{37}O_3P$
$Pt(COCH_2CH_2COO)(PCy_3)$, **9**, 436
$PtC_{22}H_{50}P_2$
$PtH(CH_2Bu^t)(Bu^t{}_2PCH_2PBu^t{}_2)$, **9**, 517
$PtC_{22}H_{52}P_2$
$Pt(CH_2Bu^t)_2(PEt_3)_2$, **9**, 456
$PtMe_2(CH_2PMeBu^t{}_2)_2$, **9**, 495
$PtC_{24}H_{18}F_6N_2$
$Pt(C_6H_4CF_3)_2(py)_2$, **9**, 454
$PtC_{24}H_{20}$
$Pt(C_7H_5Ph_2)Cp$, **9**, 580
$PtC_{24}H_{25}P$
$Pt(PPh_3)(C_3H_5)_2$, **9**, 573, 576
$PtC_{24}H_{26}P$
$[Pt(PPh_3)(C_2H_4)(H_2C\text{-}CMeCH_2)]^+$, **9**, 545
$PtC_{24}H_{27}I_2PS_2$
$[PtI(PPh_3)[C(SMe)_2]_2]I$, **9**, 426
$PtC_{24}H_{53}ClP_2$
$Pt(CH_2CMe_2PBu^t{}_2)Cl(PBu^t{}_3)$, **9**, 507
$PtC_{25}H_{24}P$
$[Pt(PPh_3)(C_2H_4)(Cp)]^+$, **9**, 297, 576
$PtC_{25}H_{28}Cl_7O_5P_2$
$[Pt(C_6Cl_6)[C(OMe)Me]\text{-}(PMe_2Ph)]ClO_4$, **9**, 422
$PtC_{25}H_{30}N_3$
$[PtMe(NCMe)(Me_2phen)\text{-}(C_8H_{12})]^+$, **9**, 554
$PtC_{26}H_{30}N_2P$
$PtMe_2(pz)[P(C_6H_4Me)_3]$, **9**, 439
$PtC_{26}H_{30}P$
$[Pt(PPh_3)(C_4H_8)(H_2C\text{-}CHCHMe)]^+$, **9**, 545
$[Pt(PPh_3)(C_4H_8)(H_2C\text{-}CMeCH_2)]^+$, **9**, 545, 547, 574
$PtC_{26}H_{33}N_2P$
$[Pt(C_3H_5)(CNBu^t)_2(PPh_3)]^+$, **9**, 417
$PtC_{26}H_{49}P$
$Pt[H_2CC(Me)CH_2](PCy_3)\text{-}(C_3H_4Me)$, **9**, 573, 576
$PtC_{27}H_{24}P_2S_2$
$Pt(CS_2)(dppe)$, **9**, 426
$PtC_{27}H_{27}P$
$Pt(PPh_3)(C_9H_{12})$, **9**, 538
$PtC_{27}H_{30}I_2P_2S_2$
$[PtI(dppe)[C(SMe)_2)]]I$, **9**, 426
$PtC_{27}H_{45}P$
$PtCp(PCy_3)(C_3H_4Me)$, **9**, 573
$PtC_{27}H_{39}I_2NP$
$[PtI(CN\text{-}p\text{-}Tol)_2(PBu^t{}_2Pr]I$, **9**, 416
$PtC_{28}H_{20}$
$Pt(PhCCPh)_2$, **9**, 566
$PtC_{28}H_{30}OP_2$
PtMe(OMe)(dppe), **9**, 461, 462
$PtC_{28}H_{30}O_2P_2$
$Pt(OMe)_2(dppe)$, **9**, 458, 461
$PtC_{28}H_{32}P_2$
$Pt(Me)_2(PMePh_2)_2$, **9**, 493
$PtC_{28}H_{42}$
$Pt[C_8H_{12}Cp^*]Cp^*$, **9**, 581
$PtC_{28}H_{48}P_2$
$Pt(CH_2C_6H_4Me)_2(PEt_3)_2$, **9**, 434, 437
$PtC_{29}H_{25}P$
$PtPh(PPh_3)Cp$, **9**, 580
$PtC_{29}H_{27}OP$
$PtPh(PPh_3)(OH)(C_5H_6)$, **9**, 580
$PtC_{29}H_{28}NP$
$[Pt(PPh_3)(NCPh)[H_2CCH\text{-}(Me)CH_2]]^+$, **9**, 547
$PtC_{29}H_{30}N_2OP$
$PtH(PPh_3)[(C(NH\text{-}p\text{-}anisyl)\text{-}NC_2H_4CH_2]^+$, **9**, 424
$PtC_{29}H_{32}OP_2$
$Pt(CH_2Me)(OMe)(dppe)$, **9**, 458
$PtC_{29}H_{44}N_2P_2$
$Pt(C_6H_4NNPh)Cp(PEt_3)_2$, **9**, 580
$PtC_{30}H_{30}F_4P_2$
$Pt(Me)(CF_2CF_2Me)(dppe)$, **9**, 472
$PtC_{30}H_{30}I_2NO_2P$
$PtI_2(PPh_3)[C(OCOPh)(NEt_2)]$, **9**, 425
$PtC_{30}H_{30}P$
$[Pt(MeC_3H_4)(CH_2{=}CHPh)\text{-}(PPh_3)]^+$, **9**, 297
$PtC_{30}H_{34}P_2$
$PtMePh(PMePh_2)_2$, **9**, 462
$PtC_{30}H_{56}Cl_2P_2$
$Pt[(CH_2)_3PCy_2]Cl[Cy_2P\text{-}(CH_2)_3Cl]$, **9**, 509
$PtC_{31}H_{29}P_2$
$[Pt(dppe)Cp]^+$, **9**, 580
$PtC_{32}H_{33}F_{10}OPS_2$
$Pt(C_6F_6)_2(CO)[(SC(S)P\text{-}(C_6H_{11})_3]$, **9**, 406
$PtC_{33}H_{31}P_2$
$PtMePh(PMePh_2)_2$, **9**, 458
$PtC_{34}H_{32}O_2P_2$
$Pt(diop)(C_3H_4)$, **9**, 558
$PtC_{34}H_{33}ClNOP_2$
$[PtCl(C_5H_4NCH_2COMe)\text{-}(dppe)]^+$, **9**, 466
$PtC_{34}H_{38}OP_2$
$Pt(C_6H_9)(OMe)(dppp)$, **9**, 487
$PtC_{34}H_{54}P_2$
Pt(dcpe)(HCCPh), **9**, 568
Pt(H)(CCPh)(dcpe), **9**, 568
$PtC_{34}H_{60}P_2$
$Pt[C_8H_{12}Cp^*{}_2](PMe_3)_2$, **9**, 581
$PtC_{34}H_{62}P_2$
$Pt(CH_2CH_2CH_2CH_2)[P\text{-}(C_5H_9)_3]_2$, **9**, 481
$PtC_{35}H_{59}N_2P_3$
$[Pt(C_6H_4NNPh)(PEt_3)_3]Cp$, **9**, 580
$PtC_{36}H_{30}Cl_2P_2$
$PtCl_2(PPh_3)_2$, **12**, 247
$PtC_{36}H_{31}Cl_2P$
$PtCl_2(PPh_3)(C_4Me_2Ph_2)$, **9**, 579
$PtC_{37}H_{30}BCL_2F_3OP_2$
$[PtCl(CO)(PPh_3)_2](BF_3Cl)$, **9**, 406
$PtC_{37}H_{30}ClOSb$
$[PtCl(CO)(SbPh_3)_2]^+$, **9**, 407
$PtC_{37}H_{30}IOP$
$[PtI(CO)(PPh_3)_2]^+$, **9**, 407
$PtC_{37}H_{30}OP_2S$
$Pt(COS)(PPh_3)_2$, **9**, 400
$PtC_{37}H_{31}BF_6P_2$
$[PtH(CF_2)(PPh_3)_2]BF_4$, **9**, 420
$PtC_{37}H_{32}Cl_2P_2$
$PtCl_2(CH_2PPh_3)(PPh_3)$, **9**, 496
$PtC_{37}H_{32}OP_2$
$Pt(PPh_3)_2(OCH_2)$, **9**, 560
$PtC_{37}H_{33}ClP_2$
$PtCl(Me)(PPh_3)_2$, **9**, 493
$PtC_{37}H_{37}P$
$PtPh(PPh_3)[C_8H_{12}Cp]$, **9**, 580
$PtC_{37}H_{41}B_7P_2$
$(Ph_3P)_2Pt(CB_7H_{11})$, **1**, 379
$PtC_{37}H_{44}P_3$
$[Pt(dppe)(PEt_3)Cp]^+$, **9**, 580
$PtC_{38}H_{30}$
$Pt(C_3H_5)(C_5Ph_5)$, **9**, 581
$PtC_{38}H_{30}Cl_2P_2$
$PtCl(CCCl)(PPh_3)_2$, **9**, 566
$PtC_{38}H_{30}O_2P_2$
$Pt(CO)_2(PPh_3)_2$, **9**, 392, 399
$PtC_{38}H_{31}ClP_2$
$PtCl(C{\equiv}CH)(PPh_3)_2$, **9**, 493
$PtC_{38}H_{31}FP_2$
$Pt(PPh_3)_2(FCCH)$, **9**, 566
$PtC_{38}H_{32}OP$
$Pt(PPh_3)_2(H_2CCO)$, **9**, 559
$PtC_{38}H_{33}ClOP_2$
$PtCl(COMe)(PPh_3)_2$, **9**, 423
$PtC_{38}H_{34}Cl_2P_2$
$Pt(CH_2Cl)_2)(PPh_3)_2$, **9**, 495
$PtC_{38}H_{34}P_2$
$Pt(PPh_3)_2(C_2H_4)$, **9**, 538, 540
$PtC_{38}H_{36}OP_2$
$PtMe(OMe)(PPh_3)_2$, **9**, 458
$PtC_{38}H_{70}Cl_2P_2$
$PtCl_2(CH_2CH_2PCy_3)(PCy_3)$, **9**, 495
$PtC_{39}H_{32}F_6O_4P_2S_2$
$PtH[CH(SO_2CF_3)_2](PPh_3)_2$, **9**, 435
$PtC_{39}H_{33}P_2$
$[Pt(PPh_3)_2(C_3H_3)]^+$, **9**, 577
$PtC_{39}H_{34}OP_2$
$Pt(PPh_3)_2[H_2CC(O)CH_2]$, **9**, 579
$PtC_{39}H_{35}O_2P_2$
$[PtH(COC_2H_4O)(PPh_3)_2]^+$, **9**, 426

$PtC_{39}H_{35}P_2$
$[Pt(C_3H_5)(PPh_3)_2]^-$, **9**, 575
$[Pt(PPh_3)_2(C_3H_5)]^+$, **9**, 574
$PtC_{40}H_{20}F_{10}$
$Pt(C_6F_5)_2(PhCCPh)_2$, **9**, 568
$PtC_{40}H_{34}P_2$
$Pt(C_2H_3C_6H_4PPh_2)_2$, **9**, 551
Pt(Ph)(CCPh)(dppe), **9**, 568
$PtC_{40}H_{35}F_3O_3P_2S$
$Pt(CMe{=}CH_2)(OTf)(PPh_3)_2$, **9**, 408, 487
$PtC_{40}H_{35}F_3O_7P_2S$
$[Pt(COCOOMe)(H_2O)(PPh_3)_2]$-OTf, **9**, 434
$PtC_{40}H_{36}Cl_{20}N$
$[Bu_4N][Pt(C_6Cl_5)_4]$, **9**, 496
$PtC_{40}H_{36}P_2$
$Pt(PPh_3)_2(C_4H_6)$, **9**, 560
$PtC_{40}H_{42}O_{10}P_2$
$Pt[C(CO_2Me){=}C(OMe)(CO_2$-Me)]$_2$(dppe), **9**, 486
$PtC_{40}H_{58}P_2$
Pt(dcpe)(PhCCPh), **9**, 565
$PtC_{40}H_{66}F_6P_2$
$Pt(PCy_3)_2(F_3CCCCF_3)$, **9**, 317
$PtC_{41}H_{33}Cl_2NP_2$
$PtCl(NC_5H_3Cl)(PPh_3)_2$, **9**, 434
$PtC_{41}H_{37}P_2$
$[Pt[H_2CC(Me)CCH_2](PPh_3)_2]^-$, **9**, 577
$PtC_{41}H_{39}P_3$
$Pt(PPh_3)_2(PCBu^t)$, **9**, 569
$PtC_{42}H_{30}N_4P_2$
$Pt(PPh_3)_2[C_2(CN)_4]$, **9**, 536
$PtC_{42}H_{36}O_4P_2$
$Pt(PPh_3)_2(MeO_2CCCCO_2Me)$, **9**, 568
$PtC_{42}H_{36}P_2$
$Pt(C{\equiv}CPh)_2(PMePh_2)_2$, **9**, 490
$PtC_{42}H_{38}O_3P_2$
Pt[CH(COMe)COCH(COMe)]-$(PPh_3)_2$, **9**, 481
$PtC_{42}H_{41}P_2$
$[Pt(PPh_3)_2(Me_2CCHCHMe)]^+$, **9**, 576
$PtC_{42}H_{50}P$
$Pt(CH_2Bu^t)_2(Ph_2CHBzPPh_2)$, **9**, 518
$PtC_{43}H_{30}ClF_{11}P_2$
$PtCl(C_7F_{11})(PPh_3)_2$, **9**, 467
$PtC_{43}H_{35}ClOP_2$
$PtCl(COPh)(PPh_3)_2$, **9**, 462
$PtC_{43}H_{38}O_5P_2$
$Pt(PPh_3)_2[MeO_2CCHC(O)C$-$HCO_2Me]$, **9**, 578
$PtC_{43}H_{40}P_2$
$Pt(PPh_3)_2$(norbornene), **9**, 535
$PtC_{43}H_{43}ClOP_2$
$PtCl(COC_6H_{13})(PPh_3)_2$, **9**, 443, 460
$PtC_{43}H_{44}Cl_2NP_3$
$[Pt[(Ph_2PCH_2CH_2)_3N](CH_2$-Cl)Cl, **9**, 467
$PtC_{44}H_{35}ClO_2P_2$
$PtCl(COCOPh)(PPh_3)_2$, **9**, 462
$PtC_{44}H_{39}BrOP_2$
$PtBr(CH_2CH_2OPh)(PPh_3)_2$, **9**, 434, 455
$PtC_{44}H_{40}O_6P_2$
$Pt(PPh_3)_2[C_2H(CO_2Me)_3]$, **9**, 536
$PtC_{45}H_{35}F_3O_3P_2S$
$Pt(OTf)(CCPh)(PPh_3)_2$, **9**, 577
$PtC_{45}H_{37}P_2$
$[Pt(PPh_3)_2(C_3H_2Ph)]^+$, **9**, 577
$PtC_{45}H_{42}F_3O_3P_2S$
$[Pt(PPh_3)_2(MeCCCBu^t)](OTf)$, **9**, 577
$PtC_{45}H_{44}P_2$
$Pt(PPh_3)_2(C_9H_{14})$, **9**, 536
$PtC_{46}H_{37}OP_2$
$[Pt(CH_2CCPh)(CO)(PPh_3)_2]^+$, **9**, 577
$PtC_{46}H_{44}O_2P_2$
$Pt(C{\equiv}CCMe_2OH)_2(PPh_3)_2$, **9**, 493
$PtC_{47}H_{40}O_2P_2$
$Pt(PPh_3)_2(PhCCCO_2Et)$, **9**, 566
$PtC_{47}H_{44}P_2$
$Pt(PPh_3)_2(C_{11}H_{14})$, **9**, 565, 566
$PtC_{48}H_{39}FP_2S$
$PtPh(SC_6H_4F)(PPh_3)_2$, **9**, 436
$PtC_{48}H_{40}P_2$
$PtPh_2(PPh_3)_2$, **9**, 437
$PtC_{48}H_{40}P_4$
$Pt(PPh_3)_2(P_2Ph_2)$, **9**, 560
$PtC_{48}H_{45}I_2NO_2P_2$
$[PtI(PPh_3)_2[(C(OCOPh)$-$(NEt_2)]]I$, **9**, 425
$PtC_{48}H_{46}P_3$
$[Pt(CPhCCH_2)(PMe_3)(PPh_3)_2]^+$, **9**, 577
$PtC_{49}H_{45}NO_3P_2$
$Pt(CONEt_2)(COCOPh)(PPh_3)_2$, **9**, 425
$PtC_{49}H_{51}P_3$
$Pt(MesPCPh_2)_2(PCBu^t)$, **9**, 570
$PtC_{50}H_{44}Cl_2P_4$
$Pt_2Cl_2(dppm)_2$, **9**, 432
$PtC_{52}H_{40}P_2$
$Pt(C{\equiv}CPh)_2(PPh_3)_2$, **9**, 493
$PtC_{53}H_{36}P_2$
$Pt(PPh_3)_2(C_{17}H_6)$, **9**, 566
$PtC_{53}H_{42}F_6O_4P_2S_2$
$Pt(CPh{=}CHPh)[CH(Tf)_2]$-$(PPh_3)_2$, **9**, 486
$PtC_{53}H_{44}F_6O_5PS_2$
$Pt(CPh{=}CHPh)(H_2O)(PPh_3)_2$-$[CH(Tf)_2]$, **9**, 486
$PtC_{54}H_{45}P_3$
$Pt(PPh_3)_3$, **9**, 435
$PtC_{55}H_{45}OP_3$
$Pt(CO)(PPh_3)_3$, **9**, 392, 400
$PtC_{55}H_{47}Cl_2P_3$
$[PtCl(CH_2PPh_3)(PPh_3)_2]Cl$, **9**, 496
$PtC_{56}H_{44}ClF_5P_4$
$ClPt(dppm)_2Pt(C_6F_5)$, **9**, 433
$PtC_{56}H_{47}P_3$
$Pt(C_2H_3C_6H_4PPh_2)(PPh_3)_2$, **9**, 551
$PtC_{56}H_{49}I_2P_3$
$[PtI(CH_2PPh_3)_2(PPh_3)]I$, **9**, 495
$PtC_{56}H_{50}N_2P_2$
$Pt(C{\equiv}CCH_2NHBz)_2(PPh_3)_2$, **9**, 493
$PtC_{56}H_{72}Cl_{20}N_2$
$[Bu_4N]_2[Pt(C_6Cl_5)_4]$, **9**, 496
$PtC_{58}H_{51}P_3$
$Pt[Ph_2CP(Mes)](PPh_3)_2$, **9**, 560
$PtC_{59}H_{34}BF_4O_2P_2$
$[PtH(COC_2H_4O)(PPh_3)_2]BF_4$, **9**, 420
$PtC_{62}H_{66}ClP_7$
$PtCl(P_3C_2Bu^t_2)(dppe)_2$, **9**, 581
$PtC_{64}H_{50}P_2$
$Pt_2(PPh_3)_2(PhCCPh)_2$, **9**, 570
$PtC_{64}H_{58}P_6$
$Pt(dppe)_2(P_2Ph_2)$, **9**, 560
$PtC_{96}H_{30}P_2$
$Pt(C_{60})(PPh_3)_2$, **9**, 561
$PtCo_2C_{15}H_{12}O_7$
$PtCo_2(CO)_7(cod)$, **9**, 563
$PtCo_2C_{20}H_{14}NO_8$
$Co_2Pt(CO)_8(picoline)_2$, **10**, 211
$PtCo_2C_{22}H_{22}N_2O_8$
$Pt[Co(CO)_4]_2(CNCy)_2$, **10**, 39, 46, 49
$PtCo_2C_{33}H_{24}O_7P_2$
$PtCo_2(CO)_7(dppe)$, **10**, 39, 46, 47
$PtCrC_{26}H_{23}O_6P$
$Cr(CO)_6(PPh_2)Pt(C_8H_{13})$, **10**, 47
$PtCuWC_{25}H_{27}O_2$
$CuPtW(CC_6H_4Me)(CO)_2$-CpCp*, **3**, 116
$PtFeC_{27}H_{24}O_3$
$PtFe(CO)_3(PhCCHCHCPh)$-(cod), **9**, 564
$PtFeC_{28}H_{22}Br_2O_4P_2$
$(CO)_3Fe(dppm)(CO)PtBr_2$, **7**, 84
$PtFeC_{28}H_{22}I_2O_3P_2$
$(CO)_3Fe(dppm)(I)PtI$, **7**, 84
$PtFeC_{34}H_{37}O_3P_3$
$Fe(CO)_2[CH(OH)C$-$({=}CH_2)CH_2](dppm)Pt$-(PMe_3), **10**, 61
$PtFeC_{35}H_{28}O_8P_2$
$(CO)_4Fe(dppm)Pt(MeO_2C$-$CCCO_2Me)$, **9**, 572
$PtFeC_{47}H_{37}O_4P_3$
$PtFe(CO)_3(CO)(dppm)(PPh_3)$, **10**, 17
$PtFeC_{48}H_{39}O_3P_3$
$Fe(CO)_2[C_2H_2C(O)](dppm)Pt$-$(PPh_3)$, **10**, 61
$PtFeC_{49}H_{41}O_3P_3$
$Fe(CO)_3[C(CH_2)CH_2]$-$(dppm)Pt(PPh_3)$, **9**, 563; **10**, 61
$PtFeC_{49}H_{42}O_3P_3$
$Fe(CO)_3(CMe{=}CH_2)(dppm)Pt$-$(PPh_3)$, **9**, 563; **10**, 61
$PtFePdC_{54}H_{44}O_4P_4$
$FePdPt(CO)_4(dppm)_2$, **7**, 84; **10**, 239, 242

$PtFe_2C_{16}H_{12}O_8$
$PtFe_2(CO)_8(cod)$, **9**, 564
$PtFe_2C_{17}H_{22}N_4O_8$
$Pt[Fe(Co)_3NO]_2(CNCy)_2$, **10**, 49
$PtFe_2C_{20}H_{10}N_4O_8$
$Pt[Fe(CO)_3(NO)]_2(NCPh)_2$, **7**, 32
$PtFe_2C_{24}H_{15}O_6PSeTe$
$Fe_2(CO)_6[SeTePt(PPh_3)_2]$, **7**, 75
$PtFe_2C_{33}H_{22}O_8P_2$
$Fe_2Pt(dppm)(CO)_8$, **7**, 84
$PtFe_2C_{37}H_{33}O_6P_3$
$PtFe_2(dppe)(CO)_6(PCBu^t)$, **9**, 570
$PtFe_2C_{42}H_{30}O_6P_2Te_2$
$Fe_2(CO)_6Te_2Pt(PPh_3)_2$, **7**, 77
$PtFe_2C_{44}H_{30}O_8P_2$
$Fe_2Pt(CO)_8(PPh_3)_2$, **10**, 61
$PtFe_2C_{50}H_{36}O_6P_2Se_2$
$(CO)_6Fe_2[SeC(Ph)C(H)Se]Pt-(PPh_3)_2$, **9**, 564
$PtFe_3C_{29}H_{16}O_{11}P$
$[HFe_3Pt(CO)_{11}(PPh_3)]^-$, **7**, 84
$PtGaC_{41}H_{81}P_2$
$Pt[(C_6H_{11})_2PCH_2CH_2P-(C_6H_{11})_2](CH_2Bu^t)-[Ga(CH_2Bu^t)_2]$, **1**, 557
$PtGe_2HgC_{72}H_{30}F_{30}P_2$
$Pt(PPh_3)_2HgGe(C_6F_5)_3-Ge(C_6F_5)_3$, **10**, 342
PtH_2Cl_6
H_2PtCl_6, **2**, 114
$PtHgC_{10}H_{11}BrNO_2$
$(Me_2NCH_2C_6H_3)PtHg(O_2CBr)$, **10**, 342, 343
$PtHgC_{22}H_{30}N_2O_4$
$(Me_2NCH_2C_6H_4)_2PtHg(O_2C-Me)_2$, **10**, 342, 343
$PtHgMnC_{29}H_{30}Cl_5O_5P_2$
$[(PPh_3)_2C_6Cl_5Pt]HgMn(CO)_5$, **10**, 343
$PtHgMoC_{50}H_{35}Cl_5O_3P_2$
$Pt(C_6Cl_5)(PPh_3)_2[HgMo-(CO)_3Cp]$, **9**, 467; **10**, 336
$PtInSi_3C_{38}H_{81}P_2$
$Pt(CH_2TMS)[In(CH_2TMS)_2]-[Cy_2PCH_2CH_2PCy_2]$, **1**, 566
$PtIrC_{24}H_{64}P_4$
$[IrH_2(PEt_3)_2(H)_2Pt(PEt_3)_2]^+$, **10**, 64
$IrH(PEt_3)_3(H)_2PtH(PEt_3)$, **10**, 64
$PtIrC_{39}H_{46}NS_4$
$Ir(cod)(SPPh_2)_2Pt(S_2CNPr^i_2)$, **10**, 64
$PtIrC_{41}H_{34}O_3P_2$
$[Ir(CO)(PPh_3)_2(H)-(CH{=}CH_2)Pt(CO)_2]^+$, **10**, 64
$PtIrC_{75}H_{63}OP_4$
$(PPh_3)_2(CO)Ir(CH{=}CH_2)Pt-(PPh_3)_2^+$, **8**, 333
$PtIrC_{75}H_{64}OP_4$
$(Ph_3P)_2Pt(H)(H_2CCH)Ir(CO)-(PPh_3)_2$, **9**, 549
$[PtIr(PPh_3)_4(CO)(C_2H_3)(H)]^+$, **10**, 16
$PtMnC_{30}H_{22}ClO_5P_2$
$Mn(CO)_5PtCl(dppm)$, **10**, 48
$PtMnC_{52}H_{40}O_4P_3$
$MnPt(PPh_2)(CO)_4(PPh_3)_2$, **10**, 226
$PtMn_2C_{28}H_{30}N_2O_4$
$[(C_5H_4Me)Mn(CO)_2(H)]_2Pt-(picoline)_2$, **6**, 143
$Mn_2Pt(H)_2(CO)_4(picoline)_2Cp'_2$, **10**, 211
$PtMn_2C_{33}H_{20}O_9P_2$
$(CO)Pt[(PPh_2)Mn(CO)_4]_2$, **6**, 88
$Mn_2Pt(PPh_2)_2(CO)_9$, **10**, 226
$PtMn_2C_{44}H_{31}O_8P_3$
$Mn_2Pt(H)(PPh_2)_3(CO)_8$, **10**, 225
$Mn_2Pt(PPh_2)_2(CO)_8(PPh_2H)$, **10**, 225
$PtMn_2C_{50}H_{35}O_8P_3$
$Mn_2Pt(PPh_2)_2(CO)_8(PPh_3)$, **10**, 226
$PtMoC_{23}H_{21}F_8N_2O_3P_5$
$Pt(PPh_3)[F_2PN(Me)PF_2]_2Mo-(CO)_3$, **10**, 17, 18
$PtMoC_{57}H_{50}O_2P_4$
$Mo(CO)_2Cp(dppm)PtH(dppm)$, **10**, 47
$PtMoPdC_{57}H_{49}ClO_2P_2$
$MoPdPt(CO)_2Cl(dppm)_2Cp$, **10**, 238
$PtMo_2C_{30}H_{32}N_2O_6$
$Pt[Mo(CO)_3Cp]_2(CNCy)_2$, **10**, 37
$PtMo_2NiC_{30}H_{29}O_4$
$PtNiMo_2(CC_6H_4Me)(CO)_4-(cod)Cp_2$, **9**, 564
$PtOs_3C_{25}H_{22}O_9P_2S_2$
$Os_3Pt(S)_2(CO)_9(PMe_2Ph)_2$, **10**, 235
$PtOs_3C_{28}H_{35}O_{10}P$
$Os_3Pt(H)_2(CO)_{10}(PCy_3)$, **10**, 35
$PtOs_3C_{28}H_{36}O_{10}P$
$[Os_3Pt(H)_3(CO)_{10}(PCy_3)]^+$, **10**, 202
$PtOs_3C_{32}H_{33}O_8P_3S$
$Os_3Pt(S)(CO)_8(PMe_2Ph)_3$, **10**, 235
$PtOs_3C_{38}H_{51}O_9PS_2$
$Os_3Pt(S)_2(CO)_9(PCy_3)(C{=}C_6H_6Me_4)$, **10**, 235
$PtOs_3Sn_2C_{32}H_{45}Cl_4O_{11}P$
$Os_3PtSn(H)_2Cl(SnCl_3)(CO)_{10}-(OEt_2)(PCy_3)$, **10**, 217, 218
$PtOs_4C_{15}H_2O_{15}$
$Os_4Pt(H)_2(CO)_{15}$, **10**, 200
$PtOs_4C_{31}H_{15}O_{13}PS$
$Os_4Pt(S)(CO)_{13}(PPh_3)$, **10**, 20, 21, 232, 233
$PtOs_5C_{15}H_6O_{15}S$
$Os_5Pt(H)_6(S)(CO)_{15}$, **10**, 202, 235
$PtOs_5C_{16}H_6O_{16}$
$Os_5Pt(H)_6(CO)_{16}$, **10**, 200, 201
$PtOs_5C_{18}H_4O_{18}$
$Os_5Pt(H)_4(CO)_{18}$, **10**, 200, 201
$PtOs_5C_{51}H_{30}O_{15}S$
$Os_5Pt(CO)_{15}(PPh_3)_2(S)$, **10**, 20
$PtOs_6C_{18}H_8O_{18}$
$Os_6Pt(H)_8(CO)_{18}$, **10**, 200, 201
$PtOs_6C_{26}H_{15}NO_{17}$
$Os_6Pt(NCMe)(CO)_{17}(cod)$, **10**, 225, 226
$PtPdC_{55}H_{45}ClOP_3$
$PdCl(PPh_3)(CO)PtCl(PPh_3)_2$, **9**, 197
$PtCl(PPh_3)(CO)PdCl(PPh_3)_2$, **9**, 197
$PtPdC_{56}H_{44}ClF_5P_4$
$[(C_6F_5)Pt](dppm)_2(PdCl)$, **9**, 226
$(ClPt)(dppm)_2[Pd(C_6F_5)]$, **9**, 226
$PtPdC_{64}H_{47}F_{10}NP_4$
$PdPt(C_6F_5)_2(dppm)_2(CNMe)$, **9**, 214
$PtReC_{43}H_{37}O_2P_2$
$Cp(CO)_2Re(H)Pt(H)(PPh_3)_2$, **10**, 35
$PtReC_{60}H_{51}Cl_2N_2OP_4$
$ReCl(N_2Tol)[PPh_2C({=}CH_2)P-Ph_2]_2(CO)PtCl$, **10**, 53
$PtRe_2C_{16}H_{14}O_8$
$PtRe_2(CO)_8(cod)(H)_2$, **10**, 7
$PtRe_2C_{44}H_{32}O_8P_2$
$Re_2Pt(H)_2(CO)_8(PPh_3)_2$, **10**, 201, 202
$PtRe_3C_{14}H_3O_{14}$
$Re_3Pt(H)_3(CO)_{14}$, **10**, 202
$PtRe_4C_{16}H_6O_{16}$
$Re_4Pt(H)_6(CO)_{16}$, **10**, 199
$PtRe_4C_{17}O_{17}$
$[Re_4Pt(CO)_{17}]^{2-}$, **10**, 192
$PtRhC_{52}H_{58}F_6O_6P_3S_2$
$[(Ph_3P)_2Pt(H)(H_2CCH)Rh-(PMe_3)Cp^*](OSO_2CF_3)_2$, **9**, 549
$PtRhC_{55}H_{46}OP_4$
$[Rh(CO)(dppm)_2Pt(C_2H)_2]^+$, **10**, 64
$PtRhC_{55}H_{60}P_3$
$[(Ph_3P)_2Pt(H)(PhCC)Rh-(PMe_3)Cp^*]^{2+}$, **9**, 568
$PtRhC_{56}H_{53}ClN_3P_4$
$[RhCl(MeCN)_2(dppm)_2Pt-(MeCN)]^+$, **10**, 64
$PtRhC_{57}H_{50}OP_4$
$[Rh(CO)(dppm)_2Pt(C_2Me)_2]^+$, **10**, 64
$PtRh_2C_{33}H_{32}F_6O$
$PtRh_2(CO)(F_3CCCCF_3)Cp_2-(cod)Cp_2$, **9**, 564
$PtRh_2C_{41}H_{46}O_3P$
$[Rh_2Pt(H)(CO)_3(PPh_3)Cp^*_2]^+$, **10**, 210
$PtRh_4C_{12}O_{12}$

$[Rh_4Pt(CO)_{12}]^{2-}$, **10**, 188
$PtRh_4C_{29}H_{20}F_{12}O$
$PtRh_4(CO)(F_3CCCF_3)_2Cp_4$, **9**, 572
$PtRh_4C_{30}H_{20}F_{12}O_2$
$PtRh_4(CO)_2(F_3CCCCF_3)_2Cp_4$, **9**, 564
$PtRh_4C_{84}H_{60}N_2O_{12}P_4$
$[PPN]_2[Rh_4Pt(CO)_{12}]$, **10**, 42
$PtRh_5C_{15}O_{15}$
$[Rh_5Pt(CO)_{15}]^-$, **10**, 60
$PtRh_5C_{51}H_{30}NO_{15}P_2$
$[PPN][Rh_5Pt(CO)_{15}]$, **10**, 42
$PtRh_6C_{16}O_{16}$
$[Rh_6Pt(CO)_{16}]^{2-}$, **10**, 193
$PtRh_8C_{19}O_{19}$
$[Rh_8Pt(CO)_{19}]^{2-}$, **10**, 193–5, 213
$PtRh_{10}C_{21}NO_{21}$
$[Rh_{10}Pt(N)(CO)_{21}]^{3-}$, **10**, 218
$PtRh_{12}C_{24}O_{24}$
$[Rh_{12}Pt(CO)_{24}]^{4-}$, **10**, 193, 194
$PtRu_2C_{46}H_{34}O_6P_2$
$PtRu_2(CO)_6(PhCCPh)(dppe)$, **9**, 572
$PtRu_2C_{62}H_{74}S_3$
$Ru_2Cp^*{}_2(SPr^i)_2[SPt(PPh_3)_2S]$, **7**, 671
$PtRu_3C_{18}H_{14}O_{10}$
$PtRu_3(CO)_{10}(H)_2(cod)$, **7**, 847; **9**, 564
$PtRu_3C_{51}H_{43}O_7P_3$
$PtRu_3(CO)_7(PPh_2)(CCBu^t)(dppe)$, **9**, 572
$PtRu_4C_{21}H_{12}O_{13}$
$Ru_4Pt(CO)_{13}(cod)$, **10**, 212
$PtRu_4C_{21}H_{14}O_{13}$
$Ru_4Pt(H)_2(CO)_{13}(cod)$, **7**, 847; **9**, 564; **10**, 212
$PtRu_5C_{23}H_{12}O_{14}$
$PtRu_5(CO)_{14}(C)(cod)$, **10**, 147, 148
$PtRu_5C_{42}H_{20}O_{13}$
$Ru_5Pt(C)(CO)_{13}(C_2Ph_2)_2$, **10**, 147, 148
$PtRu_6C_{30}H_{20}O_{18}$
$PtRu_6(CO)_{18}(H)_2(CCBu^t)_2$, **9**, 572
$PtRu_8C_{49}H_{26}O_{23}P_2$
$Ru_8Pt_2(H)_2(CO)_{23}(dppe)$, **10**, 199, 200
$PtSi_2C_{10}H_{20}O_4$
$Pt[[H_2CCHSi(Me)_2]_2O]-(C_2H_2O_3)$, **9**, 553
$PtSi_2C_{23}H_{40}P_2$
$Pt(CH_2SiMe_2CHTMS)(PMe_2-Ph)_2$, **9**, 478
$PtSi_2C_{26}H_{33}OP$
$Pt(PPh_3)[[CH_2CHSi(Me)_2]_2O]$, **9**, 553
$PtSi_2C_{32}H_{72}OP_2$
$Pt(PBu^t{}_3)_2[[H_2CCHSi-(Me)_2]_2O]$, **9**, 553
$PtSnC_{12}H_{15}Cl_3$
$Pt(SnCl_3)(C_2H_3Ph)[H_2CC-(Me)CH_2]$, **9**, 547, 576
$PtSnC_{38}H_{33}Cl_3OP_2$
$PtCl(C(OSnCl_2)Me)(PPh_3)_2$, **9**, 423
$Pt(SnCl_3)(COMe)(PPh_3)_2$, **9**, 423
$PtSnC_{39}H_{39}ClP_2$
$PtMe(SnClMe_2)(PPh_3)_2$, **9**, 436
$PtSn_2CCl_7O$
$[PtCl(SnCl_3)_2(CO)]^-$, **9**, 407
$PtSn_5Cl_{15}$
$[Pt(SnCl_3)_5]^{3-}$, **9**, 407
$PtTaC_{15}H_{24}P$
$TaCp_2(CH_2)_2PtH(PMe_3)$, **10**, 28
$PtThC_{47}H_{59}P_3$
$Pt(PMe_3)(PPh_2)_2ThCp^*{}_2$, **4**, 188; **10**, 71
$PtTiC_{20}H_{26}ClP$
$TiCp_2(CH_2)(Cl)PtMe(PMe_2Ph)$, **4**, 387
$PtTiC_{21}H_{29}P$
$TiCp_2(CH_2)MePtMe(PMe_2Ph)$, **4**, 387
$PtTl_2C_4N_4$
$Tl_2Pt(CN)_4$, **1**, 569
$PtWC_{17}H_{20}O_3$
$Cp(CO)_3WPtMe(cod)$, **9**, 563
$PtWC_{22}H_{23}O_3$
$Cp(CO)_3WPtPh(cod)$, **9**, 563
$PtWC_{57}H_{46}O_3P_4$
$W(CO)_3(C_2H)(dppm)_2Pt(C_2H)$, **10**, 47
$PtW_4C_{60}H_{112}O_{10}P_2$
$Pt[C_2W_2(OBu^t)_5]_2(PMe_2Ph)_2$, **9**, 494
$PtYbC_{44}H_{50}O_2P_2$
$(THF)_2Yb(C_5H_4PPh_2)_2PtMe_2$, **9**, 444
$PtZrC_{39}H_{48O2}P_2$
$Cp^*ZrMe(OBz_2P)_2PtMe_2$, **4**, 551
$PtZrC_{52}H_{45}P_3$
$Cp_2Zr(PPh_2)_2Pt(PPh_3)$, **4**, 518
$Pt_2AgC_{22}H_8Cl_5F_{10}S$
$Pt(C_6F_5)_2(C_6Cl_5)PtAg(SC_4H_8)$, **10**, 69
$Pt_2AgC_{56}H_{46}F_{30}NO$
$[Bu_4N][Pt_2Ag(C_6F_5)_2(C_6F_5)_4]-O{\cdot}Et_2$, **9**, 468
$Pt_2Ag_2C_{38}H_{20}F_5$
$[Pt_2Ag_2(C_6F_5)(CCPh)_4]^{2-}$, **3**, 75
$Pt_2Ag_2C_{48}H_{72}$
$Pt_2Ag_2(CCBu^t)_8$, **3**, 75
$Pt_2Ag_2C_{84}H_{60}Br_4Cl_{10}P_4$
$[PtCr(C_6Cl_5)(PPh_3)(Br)Ag-(PPh_3)]_2$, **10**, 71
$Pt_2Ag_2C_{96}H_{60}Cl_{24}P_4$
$[Pt(C_6Cl_5)_2(Cl)_2Ag(PPh_3)_2]_2$, **10**, 71
$Pt_2AsMoC_{27}H_{20}O_4P$
$MoPt_2(As)(CO)_4(PPh_3)Cp$, **10**, 218
$Pt_2BC_{34}H_{56}ClF_4P_3$
$[(Et_3P)ClPt(C{=}CHPh)Pt-(C{\equiv}CPh)(PEt_3)_2]BF_4$, **9**, 489
$Pt_2BC_{40}H_{66}F_4P_4$
$[(Et_3P)_2Pt(C{=}CHPh)Pt-(C{\equiv}CPh)(PEt_3)_2]BF_4$, **9**, 489
$Pt_2B_4Ni_2C_{34}H_{56}$
$[PtMeCB_2C_2Et_4)NiCp]_2$, **9**, 179
$Pt_2B_8C_{16}H_{36}P_2$
$Pt_2(PMe_2Ph)_2(B_2H_5)(B_6H_9)$, **1**, 310, 327, 337
$Pt_2B_8C_{32}H_{54}P_4$
$(PhMe_2P)_4Pt_2B_8H_{10}$, **1**, 337
$Pt_2B_{10}C_{32}H_{53}ClP_4$
$(PMe_2Ph)_4Pt_2B_{10}H_9Cl$, **1**, 351, 362, 365
$Pt_2B_{12}C_{16}H_{40}P_2$
$Pt_2(PMe_2Ph)_2(B_6H_9)_2$, **1**, 327, 328, 337, 363
$Pt_2B_{16}C_{31}H_{55}P_3$
$(PMe_2Ph)_2Pt_2B_{16}H_{15}(C_6H_4-Me)(PMe_2Ph)$, **1**, 327, 329, 364
$Pt_2C_2Br_4O_2$
$Pt_2Br_4(CO)_2$, **9**, 409
$Pt_2C_2Cl_4O_2$
$[Pt_2Cl_4(CO)_2]^{2-}$, **9**, 406, 409, 410
$Pt_2C_2I_4O_2$
$Pt_2I_4(CO)_2$, **9**, 410
$Pt_2C_4H_6Cl_2O_2$
$Pt_2Cl_2Me_2(CO)_2$, **9**, 410, 544
$Pt_2C_4H_8Cl_4$
$Pt_2Cl_4(C_2H_4)_2$, **9**, 547
$Pt_2C_6H_{14}Cl_2$
$Pt_2Cl_2Me_2(C_2H_4)_2$, **9**, 410
$Pt_2C_6H_{22}Br_2O_4$
$[PtMe_2Br(MeOH)(OH)]_2$, **9**, 506
$Pt_2C_7H_{11}ClN_2$
$Pt_2(Cl)(N_2C_3H_3)(C_2H_4)_2$, **9**, 545
$Pt_2C_8H_{24}S_2$
$Pt_2Me_4(SMe_2)_2$, **9**, 419, 437
$Pt_2C_{11}H_{10}O$
$Pt_2(CO)Cp_2$, **9**, 400
$Pt_2C_{12}H_{10}O_2$
$Pt_2(CO)_2(Cp)_2$, **9**, 400
$Pt_2C_{12}H_{10}O_3$
$Pt(CO)(Cp)_2$, **9**, 400
$Pt_2C_{12}H_{36}S_2$
$Pt_2Me_8(SMe_2)_2$, **9**, 419, 463
$Pt_2C_{14}H_{40}P_4$
$Pt_2Me_4(dmpm)_2$, **9**, 463
$Pt_2C_{16}H_{14}Cl_2N_2$
$Pt_2Cl_2(NC_6H_4CHCH_2)_2$, **9**, 552
$Pt_2C_{16}H_{22}Cl_2O_4$
$[PtCl(C_6H_8CO_2Me)]_2$, **9**, 574
$Pt_2C_{18}H_{28}Cl_4$
$Pt_2Cl_4(C_9H_{14})_2$, **9**, 549
$Pt_2C_{20}H_{20}$
$CpPt(CpCp)PtCp$, **9**, 580
$Pt_2C_{20}H_{28}$
$Pt_2(Bu^tCCBu^t)Cp_2$, **9**, 580
$Pt_2C_{20}H_{36}O_4$
$[Pt(OMe)(C_8H_{12}OMe)]_2$, **9**, 557
$Pt_2C_{20}H_{50}Cl_2N_5$
$[(TMEDA)ClPtCH_2CH_2N-(Et)_2CH_2CH_2PtCl-(TMEDA)]^+$, **9**, 548

$Pt_2C_{21}H_{30}O$
$Pt_2(CO)Cp^*{}_2$, **9**, 400
$Pt_2C_{22}H_{24}Cl_4O_2$
$[Pt(Cl)(CH_2CH_2COC_6H_4Cl)(C_2H_4)]_2$, **9**, 562
$Pt_2C_{22}H_{30}O_2$
$Pt_2(CO)_2Cp^*{}_2$, **9**, 580
$Pt_2C_{22}H_{56}Cl_2N_6$
$[(TMEDA)ClPt(CH_2)_2NMe_2(CH_2)_2NMe_2(CH_2)_2PtCl(TMEDA)]^{2+}$, **9**, 548
$Pt_2C_{24}H_{52}Cl_2P_2$
$Pt_2(CH_2CMe_2PBu^t{}_2)_2(Cl)_2$, **9**, 507
$Pt_2C_{25}H_{20}O$
$Pt_2[C(Ph)C(Ph)CO]Cp_2$, **9**, 580
$Pt_2C_{28}H_{71}B_5P_4$
$(Et_3P)_4Pt_2(Me_2C_2B_5H_5)$, **1**, 395
$Pt_2C_{36}F_{30}$
$[Pt_2(C_6F_5)_2(C_6F_5)_4]^{2-}$, **9**, 470
$Pt_2C_{36}H_{32}Cl_4$
$Pt_2Cl_4(C_4Me_2Ph_2)_2$, **9**, 579
$Pt_2C_{38}H_{34}O_2P_2$
$Pt_2Me_2(CO)_2(C_5H_4PPh_2)_2$, **9**, 580
$Pt_2C_{46}H_{38}P_2$
$Pt_2Ph_2(C_5H_4PPh_2)_2$, **9**, 580
$Pt_2C_{50}H_{44}BCl_2F_4NOP_4$
$[Pt_2Cl_2(NO)(dppm)_2]BF_4$, **9**, 401
$Pt_2C_{52}H_{46}P_4$
$Pt_2[C_6H_4PPh(CH_2)_2PPh_2]_2$, **9**, 512
$Pt_2C_{54}H_{50}N_2P_4$
$[Pt_2(CNMe)_2(dppm)_2]^{2+}$, **9**, 415
$Pt_2C_{55}H_{45}OP_3S$
$Pt_2(S)(CO)(PPh_3)_3$, **9**, 400, 414
$Pt_2C_{56}H_{53}N_3P_4$
$[Pt_2(CNMe)_3(dppm)_2]^{2+}$, **9**, 415
$Pt_2C_{59}H_{44}F_9P_4$
$Pt_2(CCCF_3)_2(HCCCF_3)(dppm)_2$, **9**, 571
$Pt_2C_{62}H_{62}P_4$
$[Pt_2Cl(CCBu^t)(dppm)_2(CCBu^t)]^+$, **9**, 571
$Pt_2C_{62}H_{80}F_{60}N_4$
$[Bu_4N]_2Pt_2(C_6F_5)_6(bipy)$, **9**, 468
$Pt_2C_{64}H_{56}ClP_4$
$[(Ph_2P)_2Pt(C_4H_6)PtCl(PPh_3)_2]^-$, **9**, 578
$Pt_2C_{66}H_{64}P_4$
$Me_2Pt(dppm)_2Pt(C_6H_4Me)_2$, **9**, 445
$Pt_2C_{68}H_{54}P_4$
$Pt_2(C{\equiv}CPh)_2(dppm)_2$, **9**, 432
$Pt_2C_{71}H_{63}P_3$
$Pt_2Me(CH_2PPh_3)(CH_2)(dppm)_2$, **9**, 495
$Pt_2C_{72}H_{61}P_4$
$[Pt_2Ph(H)(PPh_2)(PPh_3)_3]^+$, **9**, 444
$Pt_2C_{74}H_{59}P_4$
$[Pt_2(CCPh)_2(dppm)_2(CCPh)]^+$, **9**, 571
$Pt_2C_{76}H_{69}F_6P_7$
$[MePt(dppm)_2Pt(dppm)]PF_6$, **9**, 432, 433
$Pt_2C_{79}H_{60}F_8$
$Pt_2(PPh_3)_4(C_7F_8)$, **9**, 302
$Pt_2Co_2C_{14}H_{12}O_6$
$Pt_2Co_2(CO)_6(cod)$, **9**, 563
$Pt_2Co_2C_{22}H_{22}O_8$
$Pt_2Co_2(CO)_8(cod)(EtCCEt)$, **9**, 563
$Pt_2Co_2C_{43}H_{30}O_7P_2$
$Co_2Pt_2(CO)_7(PPh_3)_2$, **10**, 46, 47
$Pt_2Co_2C_{44}H_{30}O_8P_2$
$Co_2Pt_2(CO)_8(PPh_3)_2$, **10**, 37, 40, 49, 60
$Pt_2CrC_{30}H_{54}O_3P_2$
$CpCrPt_2(CO)_3(MeC_3H_4)(PPr^i{}_3)_2$, **10**, 124
$Pt_2Fe_2C_{44}H_{31}O_8P_2$
$[Fe_2Pt_2(H)(CO)_8(PPh_3)_2]^-$, **7**, 84
$Pt_2Fe_2C_{44}H_{32}O_8P_2$
$H_2Fe_2Pt_2(CO)_8(PPh_3)_2$, **7**, 84
$Pt_2Ge_2HgC_{78}H_{60}F_{18}P_4$
$[(CF_3)_3Ge(PPh_3)_2Pt]_2Hg$, **10**, 343
$Pt_2Hg_2C_{39}H_{104}O_6P_3$
$[(CO)_3(PPhPr^i{}_2)_3Pt]_2(Hg)_2$, **10**, 343
$Pt_2Ir_2C_{61}H_{45}O_7P_3$
$Ir_2Pt_2(CO)_7(PPh_3)_3$, **10**, 41
$Pt_2Mn_2C_{56}H_{42}O_8P_4$
$Mn(CO)_4(PPh_2)(H)Pt(PPh_2)_2Pt(PPh_2)(H)Mn(CO)_4$, **10**, 53
$Pt_2Mo_2C_{62}H_{51}O_4P_3$
$Mo(CO)_2Cp(PPh_2)(H)Pt(PPh_2)_2Pt(PPh_2)Mo(CO)_2Cp$, **10**, 47
$Pt_2Mo_2C_{62}H_{52}O_4P_4$
$Mo_2Pt_2(H)_2(PPh_2)_4(CO)_4Cp_2$, **10**, 223
$Pt_2Mo_2C_{72}H_{113}O_6P_3$
$Mo_2Pt_2(CO)_6(PCy_3)_3(C_5H_4Me)_2$, **10**, 204
$Pt_2Os_3C_{14}O_{14}$
$Os_3Pt_2(CO)_{14}$, **10**, 192, 193
$Pt_2Os_3C_{25}H_{26}O_9$
$Os_3Pt_2(H)_2(CO)_9(cod)_2$, **10**, 212
$Pt_2Os_3C_{26}H_{24}O_{10}$
$Os_3Pt_2(CO)_{10}(cod)_2$, **10**, 5, 212
$Pt_2Os_4C_{18}O_{18}$
$Os_4Pt_2(CO)_{18}$, **10**, 9, 10, 197
$Pt_2Os_4C_{27}H_{24}O_{11}$
$Os_4Pt_2(CO)_{11}(cod)_2$, **10**, 212
$Pt_2Os_4C_{28}H_{24}O_{12}$
$Os_4Pt_2(CO)_{12}(cod)_2$, **10**, 212, 213
$Pt_2Os_5C_{17}H_6O_{17}$
$Os_5Pt_2(H)_6(CO)_{17}$, **10**, 200, 201
$Pt_2Os_6C_{32}H_{24}O_{16}$
$Pt_2Os_6(CO)_{16}(cod)_2$, **10**, 5
$Pt_2Os_6C_{52}H_{30}O_{16}P_2S_4$
$Os_6Pt_2(H)_2(S)_4(CO)_{16}(PPh_2C_6H_4)_2$, **10**, 232, 233
$Pt_2Os_7C_{23}H_8O_{23}$
$Os_7Pt_2(H)_8(CO)_{23}$, **10**, 200, 201
$Pt_2PbC_{80}H_{72}N_2$
$[Bu_4N]_2Pb[Pt(C_6F_5)_4]_2$, **9**, 474
$Pt_2Pd_2C_{40}H_{24}Br_4F_{20}$
$[(C_6F_5)_2Pt]_2(Br)_4[Pd(cod)]_2$, **9**, 470
$Pt_2Rh_9C_{22}O_{22}$
$[Rh_9Pt_2(CO)_{22}]^{3-}$, **10**, 193, 194, 213
$Pt_2Rh_{11}C_{24}O_{24}$
$[Rh_{11}Pt_2(CO)_{24}]^{3-}$, **10**, 193
$Pt_2Ru_3C_{36}H_{10}O_8$
$Ru_3Pt_2(CO)_8(C_2Ph_2)_2$, **10**, 151
$Pt_2Ru_3C_{46}H_{20}O_{18}$
$Pt_2Ru_3(CO)_{18}(PhCCPh)(PhCCPh)$, **9**, 572
$Pt_2Ru_4C_{18}O_{18}$
$Ru_4Pt_2(CO)_{18}$, **10**, 5, 6, 192, 193, 197
$Pt_2Ru_4C_{27}H_{26}O_{11}$
$Pt_2Ru_4(CO)_{11}(H)_2(cod)_2$, **7**, 847; **9**, 564; **10**, 212
$Pt_2Ru_4C_{42}H_{20}O_{14}$
$Pt_2Ru_4(CO)_{14}(PhCCPh)_2$, **9**, 572; **10**, 151
$Pt_2Ru_8C_{23}H_2O_{23}$
$Ru_8Pt_2(H)_2(CO)_{23}$, **7**, 848; **10**, 11, 199, 200
$Pt_2Sb_6C_{56}H_{59}F_6P_4$
$[Pt_2Et_3(dppm)_2]SbF_6$, **9**, 455
$Pt_2Si_6C_{24}H_{54}O_3$
$Pt_2[[H_2CCHSi(Me)_2]_2O]_3$, **9**, 553
$Pt_2Sn_2C_{10}H_{14}Cl_6O_2$
$[Pt(SnCl_3)(CO)(C_3H_4Me)]_2$, **9**, 574, 576
$Pt_2Sn_2C_{20}H_{30}Cl_8F_6P_2$
$Pt_2(Cl)_2[C(CF_3){=}C(CF_3)(SnCl_3)]_2(PEt_3)_2$, **9**, 486
$Pt_2Sn_3C_{24}H_{40}Cl_{12}$
$[Pt(C_4Et_4)Cl_3][Pt(C_4Et_4)(SnCl_3)_3]$, **9**, 579
$Pt_2W_3C_{45}H_{36}O_6$
$W(CO)_2Cp(CTol)Pt(CTol)W(CO)_2CpPt(CTol)W(CO)_2Cp$, **10**, 46
$Pt_3B_{12}C_{32}H_{62}P_4$
$Pt_2(B_6H_9)(B_6H_8)[PtH(PMe_2Ph)_2](PMe_2Ph)_2$, **1**, 327
$Pt_3B_{14}C_{32}H_{60}P_4$
$(PMe_2Ph)_4Pt_3B_{14}H_{16}$, **1**, 327, 329, 364
$Pt_3C_{12}O_{12}$
$[Pt_3(CO)_6]_2{}^{2-}$, **9**, 397
$Pt_3C_{14}H_{27}N_2O_3PSi$
$Pt_3(CO)_3(Bu^t(TMS)NP{=}NBu^t)$, **9**, 393
$Pt_3C_{24}H_{42}N_6$
$Pt_3(CNPr^i)_6$, **9**, 412
$Pt_3C_{30}H_{33}O_6$
$Pt_3(CO)_3[(C(OMe)(p\text{-}Tol)]_3$, **9**, 393
$Pt_3C_{30}H_{36}N_9O_3P_3$
$Pt_3(CO)_3[P(C_2H_4CN)_3]_3$, **9**, 392
$Pt_3C_{30}H_{54}N_6$
$Pt_3(CNBu^t)_6$, **9**, 411
$Pt_3C_{37}H_{32}Cl_6O$

$[(C_4Me_2Ph_2)Pt(Cl)_3Pt(C_4Me_2$-$Ph_2)][PtCl_3(CO)]$, **9**, 579

$Pt_3C_{39}H_{81}O_3P_3$
$Pt_3(CO)_3(PBu^t_3)_3$, **9**, 392, 393

$Pt_3C_{45}H_{42}O_3P_3$
$Pt_3(CO)_3(PBu^t_2Ph)_3$, **9**, 393, 394

$Pt_3C_{45}H_{69}O_3P_3$
$Pt_3(CO)_3(PBu^t_2Ph)_3$, **9**, 395, 397

$Pt_3C_{54}H_{99}O_6P_3S_3$
$Pt_3(SO_2)_3[P(C_6H_{11})_3]_3$, **9**, 396, 413

$Pt_3C_{55}H_{99}O_5P_3S_2$
$Pt_3(CO)(SO_2)_2[P(C_6H_{11})_3]_3$, **9**, 396, 397

$Pt_3C_{57}H_{45}O_3P_3$
$Pt_3(CO)_3(PPh_3)_3$, **9**, 392–4

$Pt_3C_{57}H_{99}O_3P_3$
$Pt_3(CO)_3[P(C_6H_{11})_3]_3$, **9**, 392, 393, 395

$Pt_3C_{72}H_{60}O_2P_4$
$Pt_3(Ph)(PPh_2)(SO_2)(PPh_3)_3$, **9**, 560

$Pt_3C_{75}H_{60}O_3P_4$
$Pt_3(CO)_3(PPh_3)_4$, **9**, 392, 395–7

$Pt_3C_{75}H_{67}P_6$
$[Pt_3(H)(dppm)_3]^+$, **9**, 404

$Pt_3C_{76}H_{66}ClOP_6$
$[Pt_3Cl(CO)(dppm)_3]^+$, **9**, 404

$Pt_3C_{76}H_{66}NP_6S$
$[Pt_3(S)(CN)(dppm)_3]^+$, **9**, 404

$Pt_3C_{77}H_{66}NOP_6S$
$[Pt_3(SCN)(CO)(dppm)_3]^+$, **9**, 404

$Pt_3C_{77}H_{66}O_2P_6$
$Pt_3(CO)_2(dppm)_3$, **9**, 403

$Pt_3C_{77}H_{69}P_6$
$[Pt_3(H)(dppm)_3(CCH_2)]^+$, **9**, 563

$Pt_3C_{78}H_{69}NOP_6$
$Pt_3(CNMe)(CO)(dppm)_3$, **9**, 414

$Pt_3C_{79}H_{70}P_6$
$[Pt_3[(Ph_2P)_2CHCHCH_2]$-$(dppm)_2(C_2H_2)]^+$, **9**, 572

$Pt_3C_{79}H_{71}P_6$
$[Pt_3(CHCH_2)(dppm)_3(C_2H_2)]^+$, **9**, 572

$Pt_3C_{100}H_{80}P_4$
$Pt_3(PPh_3)_4(PhCCPh)_2$, **9**, 572

$Pt_3C_{102}H_{91}NP_8$
$[Pt_3(CNMe(dppm)_4]^{2+}$, **9**, 415

$Pt_3Co_2C_{50}H_{44}O_6$
$Pt_3Co_2(CO)_6(cod)_2(PhCCPh)_2$, **9**, 563

$Pt_3Fe_3C_{15}O_{15}$
$Fe_3Pt_3(CO)_{15}$, **7**, 84; **9**, 564; **10**, 5, 192

$Pt_3HgC_{76}H_{66}OP_6$
$[(CO)(PPh_2CH_2PPh_2)_3Pt_3$-$(Hg)]^{2+}$, **10**, 343

$Pt_3Hg_2C_{75}H_{66}P_6$
$[(PPh_2CH_2PPh_2)_3Pt_3(Hg)_2]^{2+}$, **10**, 343

$Pt_3Ir_3C_{36}H_{45}O_6$
$Ir_3Pt_3(CO)_6Cp^*_3$, **10**, 5, 6, 210

$Pt_3Ni_3C_{12}O_{12}$
$[Ni_3Pt_3(CO)_{12}]^{2-}$, **9**, 399

$Pt_3Ni_9C_{21}H_2O_{21}$
$[Ni_9Pt_3(H)_2(CO)_{21}]^{2-}$, **10**, 200, 201

$Pt_3Ni_9C_{21}H_4O_{21}$
$Ni_9Pt_3(H)_4(CO)_{21}$, **10**, 213

$Pt_3RhTlC_{65}H_{111}Cl_2O_3P_3$
$[TlPt_3(CO)_3(PCy_3)_3][RhCl_2$-$(C_8H_{12})]$, **1**, 569

$Pt_3Ru_6C_{21}H_4O_{21}$
$Ru_6Pt_3(H)_4(CO)_{21}$, **10**, 199

$Pt_3Ru_6C_{22}H_2O_{22}$
$Ru_6Pt_3(H)_2(CO)_{22}$, **10**, 199, 200

$Pt_3Ru_6C_{34}H_{12}O_{20}$
$Ru_6Pt_3(H)_2(PhC{\equiv}CPh)(CO)_{20}$, **10**, 213

$Pt_3Ru_6C_{42}H_{12}O_{20}$
$Ru_6Pt_3(CO)_{20}(H)_2(C_2Ph_2)$, **10**, 38, 39

$Pt_3Ru_6C_{56}H_{30}O_{14}$
$Pt_3Ru_6(CO)_{14}(PhCCPh)_3$, **9**, 572

$Pt_3Ru_6C_{60}H_{30}O_{18}$
$Pt_3Ru_6(CO)_{18}(PhCCPh)_3$, **9**, 572; **10**, 152

$Pt_3Ru_7C_{22}H_2O_{22}$
$Ru_7Pt_3(H)_2(CO)_{22}$, **10**, 199

$Pt_3SnC_{36}Cl_3F_{30}$
$Pt_3(Cl)_3(Sn)(C_6F_5)_6$, **9**, 475

$Pt_3Sn_3C_{28}H_{45}Cl_9$
$[Pt(C_3H_4Me)(cod)]_2[Pt(SnCl_3)_3$-$(C_3H_4Me)]$, **9**, 574, 576

$Pt_4Ag_2C_{80}H_{72}Cl_{44}$
$[Bu_4N]_2[Pt(C_6Cl_5)_2]_4(Cl)_4(Ag)_2$, **9**, 475

$Pt_4C_{20}H_{32}Cl_8D_8$
$[Pt(CHMeCHMeCD_2)Cl]_4$, **9**, 505

$Pt_4C_{24}H_{40}Cl_8D_8$
$[Pt(CHMeCMe_2CD_2)Cl_2]_4$, **9**, 505

$Pt_4C_{29}H_{60}O_5P_4$
$Pt_4(CO)_5(PEt_3)_4$, **9**, 392, 393

$Pt_4C_{33}H_{45}O_3$
$[Pt_4(CO)_3Cp^*_3]^+$, **9**, 580

$Pt_4C_{37}H_{44}O_5P_4$
$Pt_4(CO)_5(PMe_2Ph)_4$, **9**, 392–4, 396

$Pt_4C_{77}H_{60}O_5P_4$
$Pt_4(CO)_5(PPh_3)_4$, **9**, 395

$Pt_4C_{77}H_{66}O_3P_6$
$Pt_4(CO)_2(dppm)_2(Ph_2PCH_2P$-$(O)Ph_2)$, **9**, 393

$Pt_4C_{101}H_{88}OP_8$
$[Pt_4(H)(CO)(dppm)_4]^+$, **9**, 400

$Pt_4C_{101}H_{88}O_2P_4$
$[Pt_4(CO)(dppm)_3(Ph_2PCH_2$-$P(O)Ph_2)]^+$, **9**, 400

$Pt_4C_{102}H_{89}O_3P_8$
$[Pt_4(H)(CO)_2(dppm)_3(Ph_2P$-$CH_2P(O)Ph_2)]^+$, **9**, 400

$Pt_4Mo_2WC_{47}H_{49}O_6$
$Pt_4WMo_2(CMe)(CC_6H_4Me)$-$(CO)_6(cod)_2Cp_3$, **9**, 564

$Pt_4Os_6C_{38}H_{24}O_{22}$
$Cp^*W_2Ru_4(CO)_{12}(PPh)_2$, **10**, 10

$Pt_4Rh_{18}C_{35}O_{35}$
$[Rh_{18}Pt_4(CO)_{35}]^{4-}$, **10**, 188, 193, 194

$Pt_4Ru_5C_{28}H_{12}O_{20}$
$Ru_5Pt_4(CO)_{20}(cod)$, **10**, 213

$Pt_5C_{30}H_{60}O_6P_4$
$Pt_5(CO)_6(PEt_3)_4$, **9**, 395

$Pt_5C_{78}H_{60}O_6P_4$
$Pt_5(CO)_6(PPh_3)_4$, **9**, 392–5; **10**, 60

$Pt_5FeC_{33}H_{60}O_9P_4$
$FePt_5(CO)_9(PEt_3)_4$, **7**, 84

$Pt_5FeC_{59}H_{45}O_5P_3$
$FePt_5(CO)_5(PPh_3)_3$, **10**, 212

$Pt_5Fe_2C_{28}H_{24}O_{12}$
$Fe_2Pt_5(CO)_{12}(cod)_2$, **10**, 212, 213
$Pt_5Fe_2(CO)_{12}(cod)_2$, **9**, 564

$Pt_5Fe_2C_{43}H_{32}O_7$
$Pt_5Fe_2(CO)_7(PhCCPh)_2(cod)$, **9**, 564

$Pt_5Mo_2W_3C_{46}H_{39}O_{10}$
$Pt_5W_3Mo_2(CMe)(CC_6H_4Me)$-$(CC_6H_4Me)(CO)_{10}$-$(cod)Cp_2$, **9**, 564

$Pt_5Ru_5C_{34}H_{26}O_{18}$
$Pt_5Ru_5(CO)_{18}(H)_2(cod)_2$, **7**, 847; **9**, 564; **10**, 213

$Pt_6Fe_4C_{22}O_{22}$
$[Fe_4Pt_6(CO)_{22}]^{2-}$, **7**, 84

$Pt_6HgC_{54}H_{54}N_6$
$(Hg)[Pt_3(Me_2C_6H_3NC)_3(Me_2$-$C_6H_3NC)_3]_2$, **10**, 343

$Pt_6HgC_{108}H_{108}N_{12}$
$(Hg)[Pt_3(Me_2C_6H_3NC)_6]_2$, **10**, 327, 329

$Pt_6Ni_{38}C_{48}HO_{48}$
$[Ni_{38}Pt_6(CO)_{48}H]^{5-}$, **9**, 21

$Pt_6Ni_{38}C_{48}H_2O_{48}$
$[Ni_{38}Pt_6(CO)_{48}H_2]^{2-}$, **9**, 21

$Pt_6Ni_{38}C_{48}H_3O_{48}$
$[Ni_{38}Pt_6(H)_3(CO)_{48}]^{3-}$, **10**, 200, 201

$Pt_7C_{120}H_{132}N_{12}$
$Pt_7(CNxylyl)_{12}$, **9**, 412

$Pt_9C_{18}O_{18}$
$[Pt_3(CO)_6]_3{}^{2-}$, **9**, 397

$Pt_{12}C_{24}O_{24}$
$[Pt_3(CO)_6]_4{}^{2-}$, **9**, 397

$Pt_{15}C_{30}O_{30}$
$[Pt_3(CO)_6]_5{}^{2-}$, **9**, 399

$Pt_{17}C_{57}H_{120}O_9P_8$
$Pt_{17}(CO)_9(PEt_3)_8$, **9**, 394

$Pt_{17}C_{60}H_{120}O_{12}P_8$
$Pt_{17}(CO)_{12}(PEt_3)_8$, **9**, 393

$Pt_{24}As_2C_{78}H_{40}O_{30}$
$(AsPh_4)_2[Pt_{24}(CO)_{30}]$, **9**, 398

$Pt_{26}As_2C_{80}H_{40}O_{32}$
$(AsPh_4)_2[Pt_{26}(CO)_{32}]$, **9**, 398

$Pt_{33}H_{32}O_2P_2$
$Pt(diop)P(C_2H_4)$, **9**, 538

$Pt_{38}As_2C_{92}H_{40}O_{44}$
$(AsPh_4)_2[Pt_{38}(CO)_{44}]$, **9**, 398
$PuC_{16}H_{16}$
$[Pu(C_8H_8)_2]^-$, **4**, 181
$PuC_{20}H_{30}Cl_2$
$Cp^*{}_2PuCl_2$, **4**, 165
$PuSi_6C_{21}H_{57}$
$Pu[CH(TMS)_2]_3$, **4**, 132

R

$RbCH_3$
MeRb, **1**, 20
$RbGaC_8H_{21}O_2$
$Rb[Me_3GaOC(Me)OGaMe_3]$, **1**, 510
$ReAlC_{18}H_{34}P_2$
$ReH_4(PMe_2Ph)_2[(H)_2AlMe_2]$, **1**, 550
$ReAlC_{26}H_{43}P_3$
$ReH_2(PMe_2Ph)_3[(H)_2AlMe_2]$, **1**, 550
$ReAlC_{42}H_{44}Cl_2OP_2$
$Re[C(H)Me](OAlCl_2)Et_2$-$(PPh_3)_2$, **6**, 244
$ReAu_3C_{78}H_{82}P_6$
$Au_3ReH_4(PMe_2Ph)_3(PPh_3)_3$, **10**, 256
$ReBC_5F_4O_5$
$Re(CO)_5(FBF_3)$, **6**, 177
$ReBC_{20}H_{32}F_4$
$[ReCp_2{}^*H_2][BF_4]$, **6**, 252
$ReBC_{24}H_{22}Cl_2F_4NOP$
$[Re(ClCH_2Cl)(NO)(PPh_3)Cp]$-$[BF_4]$, **6**, 177
$ReBC_{24}H_{30}F_5NP_2$
$[Re[N(H)Ph]FMe_2(PMe_3)_2]$-$[BF_4]$, **6**, 234
$ReBC_{25}H_{37}F_4OP_3$
$[Re(H)_2(H_2)(CO)(PMe_2Ph)_3]$-$[BF_4]$, **6**, 181
$[Re(H)_4(CO)(PMe_2Ph)_3][BF_4]$, **6**, 181
$ReBC_{29}H_{32}Cl_2F_4NOP$
$[Re(Cl)(CH_2Cl)(NO)$-$(PPh_3)Cp^*][BF_4]$, **6**, 179
$ReBC_{42}H_{40}F_4P_2$
$[Re(C_6H_7)H_3(PPh_3)_2][BF_4]$, **6**, 252
$ReBC_{55}H_{53}F_5P_4$
$[ReF(CCH_2Me)(dppe)_2][BF_4]$, **6**, 203, 215
$ReB_2C_{24}H_{46}P_3$
$Re(BH_4)_2H(PMe_2Ph)_3$, **1**, 284
$ReB_8C_{24}H_{46}P_3$
$(PMe_2Ph)_3H_2ReB_8H_{11}$, **1**, 333, 336
$ReB_{11}C_5H_{14}O_5$
$[Re(CO)_5]B_{11}H_{14}$, **1**, 363
$ReB_{11}NaC_5H_{14}O_5$
$Na[Re(CO)_5]B_{11}H_{14}$, **1**, 363
ReC_2H_5O
$Re(CH)(CH_2)(OH)_2$, **6**, 244
ReC_5HO_5
$ReH(CO)_5$, **6**, 169, 193
$ReC_5H_5I_2O$
$ReCp(O)I_2$, **6**, 246
ReC_5O_5
$Re(CO)_5$, **6**, 169
$ReC_6H_6NO_2$
$Re(H)(CO)(NO)Cp$, **6**, 180
$ReC_6H_{18}NO$
$ReMe_6(NO)$, **6**, 235
$ReC_7H_3NO_5$
$[Re(CO)_5(NCMe)]^+$, **6**, 6
$ReC_7H_4O_5$
$[Re(CH_2CH_2)(CO)_5]^+$, **6**, 205
$ReC_7H_5Br_2O_2$
$Re(Br)_2(CO)_2Cp$, **6**, 180
$ReC_7H_6NO_3$
$Re(CHO)(NO)(CO)Cp$, **6**, 192
$ReC_7H_7O_2$
$CpRe(CO)_2(H)_2$, **6**, 180; **8**, 333
$ReC_7H_8NO_2$
$[ReMe(NO)(CO)Cp]^+$, **6**, 192
$ReC_7H_8NO_3$
$Re(CH_2OH)(NO)(CO)Cp$, **6**, 192
$ReC_8H_8F_6NO_2P$
$[Re(CO)(NO)(MeCN)Cp][PF_6]$, **6**, 183
$ReC_8H_8NO_3$
$Re(CH_2CHCH_2)(CO)_3(CNMe)$, **6**, 209
$ReC_8H_9NO_2$
$[Re(CH_2CH_2)(CO)(NO)Cp]^+$, **6**, 192
$ReC_9H_6O_5$
$[Re(CH_2CHCHCH_2)(CO)_5]^+$, **6**, 205
$ReC_9H_7O_2$
$Re(CO)_2(C_7H_7)$, **6**, 213
$ReC_9H_{16}IO_3P_2$
$ReI(CO)_3(dmpe)$, **6**, 177
$ReC_9H_{16}I_3O_3P_2$
$Re(I_3)(CO)_3(dmpe)$, **6**, 177
ReC_9H_{18}
$Re{=}CH(CH_2)_7Me$, **6**, 244
$ReC_{10}H_7O_3$
$Re(CO)_3(C_7H_7)$, **6**, 213
$ReC_{10}H_{10}Cl_3$
$[ReCp_2Cl_2]Cl$, **6**, 252
$ReC_{10}H_{11}$
$ReHCp_2$, **6**, 180
$ReC_{10}H_{11}O_2$
$Re(CO)_2(CH_2CHMe)Cp$, **6**, 185
$ReC_{10}H_{15}OS_4$
$ReCp^*(O)(S_4)$, **6**, 251
$ReC_{10}H_{15}S_7$
$ReCp^*(S_3)(S_4)$, **6**, 251
$ReC_{10}H_{18}O_2$
$Re{=}CH(CH_2)_7CO_2Me$, **6**, 244
$ReC_{10}H_{19}Cl_2$
$Re(CBu^t)[C(H)Bu^t]Cl_2$, **6**, 241
$ReC_{11}H_7O_4$
$Re(CO)_4(C_7H_7)$, **6**, 213
$ReC_{11}H_{11}Cl_2N_2O$
$ReOCl_2Me(bipy)$, **6**, 233
$ReC_{11}H_{11}O$
$Re(H)(CO)Cp_2$, **6**, 213
$ReC_{11}H_{13}O_2$
$Re(CO)_2[CH(Me)CHMe]Cp$, **6**, 205, 215
$ReC_{11}H_{18}ClO$
$ReCp^*(O)ClMe$, **6**, 247
$ReC_{12}H_{11}O_2$
$Re(H)(CO)_2Cp_2$, **6**, 213
$ReC_{12}H_{15}Cl_2O_2$
$Re(CO)_2Cl_2Cp^*$, **6**, 216
$ReC_{12}H_{15}N_2O_2$
$Re(CO)_2(N_2)Cp^*$, **6**, 216
$ReC_{12}H_{16}F_6P$
$[ReCp_2Me_2][PF_6]$, **6**, 252
$ReC_{12}H_{18}Cl$
$ReCl(MeCCMe)_3$, **6**, 222
$ReC_{12}H_{18}I$
$Re(I)(MeCCMe)_3$, **6**, 222
$ReC_{12}H_{19}O_3$
$Re[O(CH_2)_2O]Cp^*(O)$, **6**, 248
$ReC_{12}H_{21}Cl_2$
$ReCp^*Cl_2Me_2$, **6**, 248
$ReC_{12}H_{21}O_3$
$ReCp^*(O)(OMe)_2$, **6**, 248
$ReC_{12}H_{24}IP_2$
$Re(I)(PMe_3)_2(C_6H_6)$, **6**, 210
$ReC_{12}H_{28}ClOP_2$
$[ReO[C(H)C(H)PMe_3]_2Me_2]Cl$, **6**, 233
$ReC_{13}H_{11}OS_4$
$Re(O)Me[(S)_2C_6H_4]_2$, **6**, 238
$ReC_{13}H_{12}NO_3$
$Re(O)Me[(O)(HN)C_6H_4]_2$, **6**, 238
$ReC_{13}H_{20}Br_2$
$ReCp^*Br_2(CH_2CHCH_2)$, **6**, 247
$ReC_{13}H_{21}O_3$
$Re[O(CH_2)_3O]Cp^*(O)$, **6**, 248
$ReC_{13}H_{24}Cl_4P$
$ReCp^*Cl_4(PMe_3)$, **6**, 246
$ReC_{13}H_{26}NO_2P_2$
$Re(Me)(NO)(CO)(PMe_3)_2Cp$, **6**, 212
$ReC_{13}H_{29}Cl_2N$
$Re[N(H)Bu^t](CBu^t)Cl_2(CH_2$-$Bu^t)$, **6**, 241
$ReC_{13}H_{38}OP_4$
$[Re(H_2)(CO)(PMe_3)_4]^+$, **6**, 183
$[Re(H)_2(CO)(PMe_3)_4]^+$, **6**, 183
$ReC_{14}H_{10}Cl_5$
$ReCl_5(PhCCPh)$, **6**, 235
$ReC_{14}H_{10}Cl_7OP$
$ReCl_4(PhCCPh)(OPCl_3)$, **6**, 232
$ReC_{14}H_{23}O_3$

Re[OC(H)(Me)C(H)(Me)O]-Cp*(O), **6**, 248
$ReC_{14}H_{23}O_3P_2$
Re(CO)$_3$(PMe$_3$)$_2$Cp, **6**, 213
$ReC_{14}H_{24}F_3O_3S$
ReCp*Me$_3$(OTf), **6**, 249
$ReC_{14}H_{24}N_2OP$
Re(CO)(N$_2$)(PMe$_3$)Cp*, **6**, 187, 216
$ReC_{14}H_{27}$
ReCp*Me$_4$, **6**, 248
$ReC_{14}H_{32}P_3$
Re(PMe$_3$)$_3$Cp, **6**, 187, 188
$ReC_{15}H_{21}NO_3P$
[Re(CO)$_2$(NO)(C$_6$Me$_7$)][PF$_6$], **6**, 210
$ReC_{15}H_{28}O$
Re(CH$_2$CMe$_2$CH$_2$)Cp*O, **6**, 247
$ReC_{16}H_{20}Cl_2NS$
ReCp*Cl$_2$[S(NH)C$_6$H$_4$], **6**, 250
$ReC_{16}H_{21}$
ReCp*(C$_6$H$_6$), **6**, 210
$ReC_{16}H_{27}IO_4P$
ReI(CO)$_4$(PBu$_3$), **6**, 177
$ReC_{16}H_{27}N_2O$
Re(NBu^t)$_2$(O)(xyl), **6**, 240
$ReC_{16}H_{45}P_4S$
Re(H)$_2$(SCHCHEt)(PMe$_3$)$_4$, **6**, 220
$ReC_{17}H_{20}NO_3$
ReCp*(O)[O(NPh)CO], **6**, 248
$ReC_{17}H_{21}O$
Re(CH$_2$C$_6$H$_4$)Cp*(O), **6**, 247
$ReC_{17}H_{29}N_2O$
Re(NBu^t)$_2$(O)(Mes), **6**, 240
$ReC_{17}H_{32}NO_4P_2$
Re(NPh)Me(O$_2$CMe)$_2$(PMe$_3$)$_2$, **6**, 234
$ReC_{18}H_{19}O_2$
Re(Ph)(CO)$_2$(C$_5$Me$_4$CH$_2$), **6**, 211
$ReC_{20}H_{13}O_4$
Re(CO)$_3$[C$_5$H$_2$Ph$_2$(OH)], **6**, 245
$ReC_{20}H_{26}N_3O_2$
ReO$_2$(CH$_2$Bu^t)(py)$_3$, **6**, 234
$ReC_{20}H_{31}$
ReHCp*$_2$, **6**, 180
$ReC_{23}H_{24}P$
ReCpH$_4$(PPh$_3$), **6**, 250
$ReC_{24}H_{22}NOP$
[Re(CH$_2$)(NO)(PPh$_3$)Cp]$^+$, **6**, 192
$ReC_{24}H_{25}O_3$
ReCp*(O)(O$_2$CCPh$_2$), **6**, 248
$ReC_{24}H_{27}S_4$
ReCp*(S$_2$C$_7$H$_6$)$_2$, **6**, 249
$ReC_{24}H_{36}N_3O$
Re(NBu^t)$_2$(xyl)$_2$(NO), **6**, 237
$ReC_{25}H_{21}F_6NO_2P$
[Re(PhCHPh$_2$)(CO)(NO)Cp]-[PF$_6$], **6**, 210
$ReC_{25}H_{23}OS_4$
Re(O)Me(SPh)$_4$, **6**, 238
$ReC_{25}H_{30}N_3$
Re(Nxyl)$_3$Me, **6**, 239
$ReC_{25}H_{37}OP_3$
[ReH$_4$(CO)(PMe$_2$Ph)$_3$]$^+$, **6**, 235
$ReC_{26}H_{28}NOPS$
Re(CH$_2$SMe$_2$)(NO)(PPh$_3$)Cp, **6**, 200
$ReC_{26}H_{35}O_2P_3$
[Re(H$_2$)(CO)$_2$(PMe$_2$Ph)$_3$]$^+$, **6**, 183
[Re(H)$_2$(CO)$_2$(PMe$_2$Ph)$_3$]$^+$, **6**, 183
$ReC_{28}H_{28}$
Re(MeC$_6$H$_4$)$_4$, **6**, 232
$ReC_{29}H_{31}N_2S_2$
Re(Nxyl)$_2$Me(SPh)$_2$, **6**, 239
$ReC_{32}H_{39}P_2$
ReH$_4$(isopropylallyl)(dppe), **6**, 235
$ReC_{32}H_{48}P_4$
[Re(H)$_4$(PMe$_2$Ph)$_4$]$^+$, **6**, 183
$ReC_{33}H_{46}OP_4$
[Re(H)$_2$(CO)(PMe$_2$Ph)$_4$]$^+$, **6**, 183
$ReC_{37}H_{30}Cl_2NO_2P_2$
ReCl$_2$(CO)(NO)(PPh$_3$)$_2$, **6**, 177
$ReC_{38}H_{30}Cl_5P$
[PPh$_4$][ReCl$_5$(PhCCPh), **6**, 232
$ReC_{39}H_{30}ClO_3P_2$
ReCl(CO)$_3$(PPh$_3$)$_2$, **6**, 177
$ReC_{39}H_{30}FO_3P_2$
ReF(CO)$_3$(PPh$_3$)$_2$, **6**, 176
$ReC_{39}H_{66}O_3P_2$
Re(CO)$_3$[P(Cy)$_3$]$_2$, **6**, 169
$ReC_{39}NO_5P_2$
Re(NO$_2$)(CO)$_3$(PPh$_3$)$_2$, **6**, 177
$ReC_{40}H_{30}O_4P_2$
[Re(CO)$_4$(PPh$_3$)$_2$]$^+$, **6**, 176
$ReC_{41}H_{32}P_2$
[Re(H)$_2$(PPh$_3$)$_2$Cp]$^{\cdot+}$, **6**, 186
ReH$_2$(PPh$_3$)$_2$Cp, **6**, 186
$ReC_{41}H_{33}P_2$
[Re(H)$_3$(PPh$_3$)$_2$Cp]$^+$, **6**, 186
$ReC_{41}H_{39}P_2$
Re(C$_5$H$_6$)(H)$_3$(PPh$_3$)$_2$, **6**, 186; **8**, 333
$ReC_{42}H_{45}OP_2$
ReOEt$_3$(PPh$_3$)$_2$, **6**, 232
$ReC_{43}H_{39}NP_2$
[Re(H)(NCMe)(PPh$_3$)$_2$Cp]$^+$, **6**, 186
$ReC_{47}H_{39}NOP_3$
Re(PPh$_2$)(NO)(PPh$_3$)(C$_5$H$_4$P-Ph$_2$), **6**, 211
$ReC_{52}H_{48}ClN_3O_7P_4$
[ReCl(NO)(dppe)$_2$][NO$_3$]$_2$, **6**, 216
$ReC_{56}H_{46}O_2P_3$
ReH(CO)$_2$(PPh$_3$)$_3$, **6**, 177
$ReC_{61}H_{56}ClP_4$
ReCl(CH$_2$CCHPh)(dppe)$_2$, **6**, 203
$ReCo_2C_{22}H_{15}O_7$
ReCo$_2$(CO)$_7$(C$_7$H$_8$)(CTol), **10**, 128
$ReCo_3C_{18}H_{18}N_2O_{10}$
ReCo$_3$(NBu^t)$_2$(CO)$_{10}$, **10**, 218
$ReCrC_{16}H_5O_9$
Cr(≡CPh)(CO)$_4$Re(CO)$_5$, **10**, 30
$ReCuC_{10}H_{10}Cl_2$
[ReCp$_2$][CuCl$_2$], **6**, 184
$ReFeC_{29}H_{20}O_7P$
(CO)$_4$Re(CO$_2$)Fe(CO)-(PPh$_3$)Cp, **6**, 232
$ReFeC_{30}H_{20}O_8P$
(CO)$_5$Re(CO$_2$)Fe(CO)-(PPh$_3$)Cp, **6**, 232
$ReHgC_{24}H_{45}Cl_2P_2$
ReH$_7$(PPhPr$^i_2$)$_2$HgCl$_2$, **10**, 48
$ReIrC_{37}H_{30}O_5P_2$
Ir(CO)(ReO$_4$)(PPh$_3$)$_2$, **8**, 315
$ReIrC_{52}H_{88}O_4P_4$
Re(CO)$_2$(PCy$_2$)$_3$Ir(CO)$_2$(PCy$_2$), **10**, 53
$ReIrC_{52}H_{88}P_4$
(PCy$_2$)(CO)$_2$Re(PCy$_2$)$_2$Ir(CO)$_2$-(PCy$_2$), **8**, 372
$ReIrC_{54}H_{106}P_6$
(PCy$_2$)$_2$Re(PCy$_2$)$_2$Ir(PMe$_3$)$_2$, **8**, 372
$ReKC_{12}H_{24}P_2$
[Re(PMe$_3$)$_2$(C$_6$H$_6$)][K], **6**, 210
$ReLiSiC_{25}H_{20}O_2$
[Re(H)(CO)$_2$(C$_5$H$_4$SiPh$_3$)][Li], **6**, 211
$ReMnC_{10}O_{10}$
MnRe(CO)$_{10}$, **6**, 4; **10**, 48
$ReMn_2C_{14}HO_{14}$
Mn(CO)$_5$Mn(CO)$_4$(H)Re(CO)$_5$, **10**, 48
$ReMoC_{15}H_{11}N_2O_6$
Mo(CO)$_3$CpRe(CO)$_3$(CNMe)$_2$, **10**, 30
$ReNaC_7H_5NO_3$
[Re(CN)(C$_3$H$_5$)(CO)$_3$]Na, **6**, 209
$RePtC_{43}H_{37}O_2P_2$
Cp(CO)$_2$Re(H)Pt(H)(PPh$_3$)$_2$, **10**, 35
$RePtC_{60}H_{51}Cl_2N_2OP_4$
ReCl(N$_2$Tol)[PPh$_2$C(=CH$_2$)P-Ph$_2$]$_2$(CO)PtCl, **10**, 53
$ReRhC_{48}H_{96}P_4$
(PCy$_2$H)ReH$_5$(PCy$_2$)$_2$RhH-(PCy$_2$H), **10**, 39
$ReRhC_{54}H_{44}O_4P_4$
Re(CO)$_3$(dppm)$_2$Rh(CO), **10**, 52
$ReRhC_{56}H_{100}P_4$
Re(PCy$_2$)$_2$(PCy$_2$)$_2$Rh(cod), **10**, 53
$ReRuC_{14}H_9O_7$
(CO)$_5$ReCH$_2$CH$_2$Ru(CO)$_2$Cp, **7**, 480
$ReSbC_{10}H_{10}Cl_2F_6$
[ReCp$_2$Cl$_2$][SbF$_6$], **6**, 252
$ReSbC_{32}H_{36}ClF_6N_3P_2$
[ReCl(CNHMe)(CNMe)$_2$-(PMePh$_2$)$_2$][SbF$_6$], **6**, 203, 215
$ReSiC_{19}H_{30}ClN_2O$

$\overline{Re[NCH_2N}C_5H_4]Cp^*Cl(OTMS)$, **6**, 251

$ReSi_2C_{13}H_{34}OP$
$ReO[C(O)Me](CH_2TMS)_2(PMe_3)$, **6**, 233

$ReSi_2C_{16}H_{39}N_2$
$Re(NBu^t)_2[C(H)TMS](CH_2TMS)$, **6**, 240

$ReSi_2C_{16}H_{40}ClN_2$
$Re(NBu^t)_2Cl(CH_2TMS)_2$, **6**, 239

$ReWC_{17}H_7O_9$
$[W(CO)_5(CTol)Re(CO)_4]^-$, **10**, 33

$ReWC_{17}H_8O_9$
$[W(CO)_5(CHTol)Re(CO)_4]^-$, **10**, 33

$ReYC_{22}H_{33}OP$
$Re(PMe_2Ph)H_4Y(THF)Cp_2$, **10**, 53

$ReZrC_{22}H_{27}O$
$Cp_2MeZr(OCHMe)Re(Cp)_2$, **6**, 184

$ReZrC_{37}H_{31}O_3P_2$
$Cp_2Zr(PPh_2)_2Re(H)(CO)_3$, **4**, 517

$Re_2Al_2C_{48}H_{78}P_6$
$[ReH_2(PMe_2Ph)_3]_2[(H)_2AlH(H)_2AlH(H)_2]$, **1**, 550

$Re_2Al_2C_{56}H_{94}O_2P_6$
$[ReH_2(PMe_2Ph)_3]_2[(H)_2AlH(OBu)_2AlH(H)_2]$, **1**, 550

$Re_2BC_{12}HF_4O_{10}$
$[(CO)_5Re(CCH)Re(CO)_5][BF_4]$, **6**, 194

$Re_2BC_{12}H_3F_4O_{10}$
$[Re(CO)_5(CHCH_2)Re(CO)_5][BF_4]$, **6**, 205, 215

$Re_2C_3H_7O_3$
$Me_2(O)Re(CH)(O)Re(O)Me_2$, **6**, 244

$Re_2C_3H_8O_3$
$Me_2(O)Re(CH_2)(O)Re(O)Me_2$, **6**, 244

$Re_2C_4H_{12}O_4$
$[[Re(O)OMe_2]_2]^-$, **6**, 234

$Re_2C_6Cl_2O_6S_8$
$Re_2Cl_2(CO)_6(S_8)$, **6**, 221

$Re_2C_9H_2O_{11}$
$Re_2(CO)_9(OH)_2$, **6**, 169

$Re_2C_{10}H_2O_8$
$Re_2(H)(C{\equiv}CH)(CO)_8$, **6**, 172

$Re_2C_{10}O_{10}$
$Re_2(CO)_{10}$, **6**, 168, 171

$Re_2C_{12}H_6O_8$
$Re_2(CH_2CHCHCH_2)(CO)_8$, **6**, 171

$Re_2C_{13}H_8O_8$
$Re_2(CHCHCMe_2)(CO)_8$, **6**, 171

$Re_2C_{14}H_9NO_9$
$Re_2(CO)_9(CNBu^t)$, **6**, 171

$Re_2C_{14}H_{16}N_2O_4$
$[ReO_2Me_2(py)]_2(O)$, **6**, 239

$Re_2C_{15}H_{19}O_8P_3$
$(CO)_4Re(tmpm)Re(CO)_4$, **8**, 336

$Re_2C_{16}H_{18}O_8S_2$
$Re_2(SBu^i)_2(CO)_8$, **6**, 221

$Re_2C_{17}H_6O_9$
$Re_2(H)(C{\equiv}CPh)(CO)_9$, **6**, 172

$Re_2C_{17}H_9NO_7$
$Re_2(H)(C{\equiv}CPh)(CO)_7(NCMe)$, **6**, 172

$Re_2C_{20}H_{30}Cl_2O_3$
$Cp^*(O)Re(O)_2ReCl_2Cp^*$, **6**, 247

$Re_2C_{20}H_{30}Cl_4O_2$
$[ReCp^*Cl_2(O)]_2$, **6**, 247

$Re_2C_{20}H_{30}Cl_6$
$[ReCp^*Cl_2(Cl)]_2$, **6**, 250

$Re_2C_{20}H_{30}I_6$
$[ReCp^*I_2(I)]_2$, **6**, 250

$Re_2C_{20}H_{30}O_2S_2$
$[ReCp^*(O)(S)]_2$, **6**, 251

$Re_2C_{20}H_{32}Cl_4$
$[ReCp^*Cl_2(H)]_2$, **6**, 250

$Re_2C_{20}H_{36}$
$[ReCp^*H_2(H)]_2$, **6**, 251

$Re_2C_{20}H_{44}O_2S_2$
$[Re(S)O(CH_2Bu^t)_2]_2$, **6**, 236

$Re_2C_{20}H_{44}O_3S$
$(Bu^tCCH_2)_2(O)Re(O)(S)Re(O)(CH_2Bu^t)_2$, **6**, 236

$Re_2C_{22}H_{40}$
$[Re(C_5Me_4Et)H_2(H)]_2$, **6**, 251

$Re_2C_{24}H_{32}O_4$
$Re_2(H)_2(CO)_4Cp^*_2$, **6**, 181

$Re_2C_{26}H_{26}O_2S_4$
$[Re(O)Me(SPh)(SPh)]_2$, **6**, 233

$Re_2C_{27}H_{15}O_9P$
$Re_2(CO)_9(PPh_3)$, **6**, 171

$Re_2C_{28}H_{20}Cl_8$
$[Re(Cl)Cl_3(PhCCPh)]_2$, **6**, 232

$Re_2C_{44}H_{30}O_8P_2$
$Re_2(CO)_8(PPh_3)_2$, **6**, 171

$Re_2C_{54}H_{40}O_6P_2Se_8$
$[Re_2(Se_4)_2(CO)_6][PPh_4]_2$, **6**, 220

$Re_2C_{60}H_{48}N_2O_8S_2$
$Re_2[CN(CH_2SO_2C_6H_4Me)]_2$, **6**, 171

$Re_2C_{77}H_{76}NP_4$
$[Re_2(H)_4H_3(PPh_3)_4(CNBu^t)]^+$, **6**, 232

$Re_2C_{82}H_{83}N_2P_4$
$[Re_2(H_3)H_2(PPh_3)_4(CNBu^t)_2]^{2+}$, **6**, 232

$Re_2Cr_2TiC_{36}H_{14}O_{18}$
$Re(CO)_5Cr(CO)_4(CTol)Ti(CTol)Cr(CO)_4Re(CO)_5$, **10**, 45

$Re_2Fe_3Ge_2C_{19}O_{19}$
$Fe_3(CO)_9[GeRe(CO)_5]_2$, **7**, 30

$Re_2GaC_{42}H_{30}I_3O_6P_2$
$Re_2(CO)_6(PPh_3)_2(I)(GaI_2)$, **1**, 554

$Re_2In_2C_{26}H_{24}I_4O_6P_2$
$Re_2(CO)_4[InI_2(H_2O)]_2(PPh_2)_2$, **1**, 560

$Re_2LuC_{46}H_{69}OP_4$
$Cp_2Lu(THF)Re_2H_7(PMe_2Ph)_4$, **4**, 109

$Re_2OsC_{18}H_8O_{14}$
$Os[CH_2CH_2Re(CO)_5]_2(CO)_4$, **7**, 392, 393

$Re_2PtC_{16}H_{14}O_8$
$PtRe_2(CO)_8(cod)(H)_2$, **10**, 7

$Re_2PtC_{44}H_{32}O_8P_2$
$Re_2Pt(H)_2(CO)_8(PPh_3)_2$, **10**, 201, 202

$Re_2Si_6C_{24}H_{66}O_3$
$[ReO(CH_2TMS)_3]_2(O)$, **6**, 236

$Re_2YC_{46}H_{69}OP_4$
$Cp_2Y(THF)Re_2H_7(PMe_2Ph)_4$, **4**, 109; **10**, 53

$Re_3AuC_{27}H_{18}O_9P$
$[AuRe_3(H)_3(CO)_9(PPh_3)]$, **10**, 304

$Re_3AuC_{28}H_{18}O_{10}P$
$[AuRe_3(H)_3(CO)_{10}(PPh_3)]^-$, **10**, 304

$Re_3C_6I_6O_6$
$Re_3I_6(CO)_6$, **6**, 177

$Re_3C_{10}H_4O_{10}$
$[Re_3H_4(CO)_{10}]^-$, **6**, 173

$Re_3C_{12}H_2O_{12}$
$[Re_3H_2(CO)_{12}]^-$, **6**, 173

$Re_3GaC_{62}H_{45}I_2O_8P_3$
$Re_2(CO)_4(PPh_3)_2(I)_2[GaRe(CO)_4(PPh_3)]$, **1**, 554

$Re_3InC_{14}I_2O_{14}$
$In[I_2Re(CO)_4][Re(CO)_5]_2$, **1**, 560

$Re_3PtC_{14}H_3O_{14}$
$Re_3Pt(H)_3(CO)_{14}$, **10**, 202

$Re_3RuC_{26}H_{15}O_{11}S_3$
$Re_3Ru(S)(pyS)_2(CO)_{11}(py)$, **10**, 236

$Re_3TlC_{15}O_{15}$
$Tl[Re(CO)_5]_3$, **1**, 567

$Re_4C_6I_8O_6$
$Re_4I_8(CO)_6$, **6**, 177

$Re_4C_{12}H_4O_{16}$
$Re_4(OH)_4(CO)_{12}$, **6**, 169

$Re_4C_{12}H_5O_{12}$
$[Re_4H_5(CO)_{12}]^-$, **6**, 175

$Re_4C_{14}H_5O_{14}$
$[Re_4H_5(CO)_{14}]^-$, **6**, 175

$Re_4C_{16}H_{11}N_2O_{12}$
$[Re_4H_5(CO)_{12}(MeCN)_2]^-$, **6**, 175

$Re_4C_{20}H_{30}O_{11}$
$Cp^*(O)Re(O)_2Re(OReO_3)_2Cp^*$, **6**, 247

$Re_4C_{22}H_{34}O_{12}$
$Cp^*[\overline{O(CH_2)_2O]Re}(O)_2Re(OReO_3)_2Cp^*$, **6**, 247

$Re_4C_{30}H_{20}O_{12}P$
$[Re_4H_5(CO)_{12}(PPh_3)]^-$, **6**, 175

$Re_4Cu_2C_{64}H_{102}P_8$
$Cu_2Re_4H_{14}(PMe_2Ph)_8$, **10**, 256

$Re_4GaC_{82}H_{60}Cl_3O_{10}P_4$
$Re_3(CO)_6(PPh_3)_3(Cl)_3[GaRe(CO)_4(PPh_3)]$, **1**, 554

$Re_4Ga_2C_{18}O_{18}$
$Re_2(CO)_8[GaRe(CO)_5]_2$, **1**, 554

$Re_4InC_{14}I_3O_{14}$

$Re_4InC_{14}I_3O_{14}$
In[$Re_2(CO)_6(I)_2$][$Re_2(CO)_8(I)$], **1**, 560
$Re_4PtC_{16}H_6O_{16}$
$Re_4Pt(H)_6(CO)_{16}$, **10**, 199
$Re_4PtC_{17}O_{17}$
$[Re_4Pt(CO)_{17}]^{2-}$, **10**, 192
$Re_4Tl_2C_{18}O_{18}$
$Re_2(CO)_8[TlRe(CO)_5]_2$, **1**, 567
$Re_6C_{19}H_2O_{18}$
$[Re_6H_2(C)(CO)_{18}]^{2-}$, **6**, 176
$Re_6C_{20}O_{19}$
$[Re_6C(CO)_{19}]^{2-}$, **6**, 176
$Re_7C_{22}HO_{21}$
$[Re_7H(C)(CO)_{21}]^{2-}$, **6**, 176
$Re_7C_{22}H_2O_{21}$
$[Re_7H_2(C)(CO)_{21}]^{-}$, **6**, 176
$Re_7C_{22}O_{21}$
$[Re_7C(CO)_{21}]^{2-}$, **6**, 175
$[Re_7C(CO)_{21}]^{3-}$, **6**, 175
$Re_7C_{23}O_{22}$
$[Re_7C(CO)_{22}]^{-}$, **6**, 175
$Re_7C_{40}H_{15}O_{24}P$
$[Re_7C(CO)_{21}[P(OPh)_3]]^{-}$, **6**, 176
$Re_7HgC_{22}ClO_{21}$
$Re_7C(CO)_{21}(HgCl)^{2-}$, **10**, 338
$Re_7PdC_{31}H_9O_{21}$
$[Re_7C(CO)_{21}Pd(C_9H_9)]^{2-}$, **10**, 128
$Re_7C(CO)_{21}Pd(PhC_3H_4)$, **9**, 350
$Re_7RhC_{24}O_{23}$
$[Re_7C(CO)_{21}Rh(CO)_2]^{-}$, **10**, 127
$Re_8C_{25}O_{24}$
$[Re_8C(CO)_{24}]^{2-}$, **6**, 175
$Re_8Ga_4C_{32}O_{32}$
$Re_4(CO)_{12}[GaRe(CO)_5]_4$, **1**, 554
$RhAgC_{60}H_{62}N_2P_4$
$Rh(Bu^tNC)_2(dppm)_2Ag$, **10**, 68
$RhAlC_{14}H_{32}P_2$
$Rh(PMe_3)_2(AlMe_3)Cp$, **1**, 552
$RhAsZrC_{46}H_{50}O_4S_2$
$[AsPh_4][Cp_2*Zr(S)_2Rh(CO)_2]$, **10**, 47
$RhAs_2C_{37}H_{30}ClOP_4$
$(RhCl(CO)P_4)(AsPh_3)_2$, **8**, 161
$RhAu_3C_{91}H_{76}F_6OP_6$
$[Au_3Rh(H)(CO)(PPh_3)_5][PF_6]$, **10**, 302
$RhBC_{14}H_{20}N_4$
$Rh(H_2Bpz_2)(cod)$, **8**, 253
$RhBC_{15}H_{27}NSi$
$CpRh(C_{10}H_{22}BNSi)$, **1**, 213
$RhBC_{16}H_{23}$
$[(Cp*Rh(CpBMe)]^+$, **1**, 209
$RhBC_{17}H_{22}N_6$
$Rh(HBpz_3)(cod)$, **8**, 253
$RhBC_{17}H_{22}N_6O_2$
$Tp*Rh(CO)_2$, **8**, 216
$RhBC_{18}H_{26}N_6O$
$Tp*Rh(CO)(C_2H_4)$, **8**, 216
$RhBC_{20}H_{30}F_4$
$Rh(C_6Me_6)(cod)BF_4$, **8**, 275
$RhBC_{22}H_{24}N_8O_2$
$(pzBpz_3)Rh(duroquinone)$, **8**, 270
$RhBC_{24}H_{47}ClO_2P_2$
$RhHCl(BO_2C_6H_4)(PPr^i_3)_2$, **1**, 292, 295, 296
$RhBC_{42}H_{35}ClO_2P_2$
$RhHCl(BO_2C_6H_4)(PPh_3)_2$, **1**, 292, 295
$RhBPd_2C_{26}H_{30}F_4S_4$
$Rh(pyS)_4[Pd(C_3H_5)]_2[BF_4]$, **9**, 354
$RhBTiC_{68}H_{66}P_2$
$[TiCp_2(CH_2PPh_2)_2Rh(C_8H_{12})]$-$[BPh_4]$, **4**, 371
$RhB_2C_{18}H_{27}N_2$
$(Cp*)Rh[C_4(BNMe_2)_2]$, **1**, 209
$RhB_8C_{13}H_{27}$
$Cp*Me[RhC_2B_8H_9]$, **1**, 339
$RhB_8C_{18}H_{36}P$
$Cp*RhB_8H_{10}(PMe_2Ph)$, **1**, 336, 341
$RhB_8C_{34}H_{35}P_2$
$(PPh_2)(PHPh_2)(Cp*Rh)B_8H_9$, **1**, 332, 333
$RhB_9C_{10}H_{28}$
$Cp*[RhB_9H_{13}]$, **1**, 332, 338, 339, 341, 342, 347
$RhB_9C_{14}H_{32}N_2$
$Cp*(MeNC)_2(RhB_9H_{11})$, **1**, 338, 339
$RhB_9C_{18}H_{37}P$
$Cp*[RhB_9H_{11}](PMe_2Ph)$, **1**, 341, 347
$RhB_9C_{26}H_{40}N_2$
$(MeC_6H_4NC)_2(Cp*RhB_9H_{11})$, **1**, 338
$RhB_9C_{26}H_{44}P_2$
$Cp*[RhB_9H_7](PMe_2Ph)_2$, **1**, 341, 347
$RhB_9C_{36}H_{40}P_2S$
$(Ph_3P)_2RhSB_9H_{10}$, **1**, 268
$RhB_9CoC_{30}H_{48}O_2P$
$Co(CO)(C_4Me_4)Rh(CO)(PPh_3)$-$(C_2B_9H_{11})$, **10**, 64
$RhB_9IrC_{20}H_{41}$
$Cp_2*(H)[RhIrB_9H_{10}]$, **1**, 342
$RhB_9PtC_{33}H_{57}OP_3$
$Rh(PPh_3)(C_2B_9H_{11})(H)(CO)Pt$-$(PEt_3)_2$, **10**, 64
$RhB_9WC_{50}H_{52}O_2P_2$
$[C_2B_9H_9Me_2]W(CO)_2$ $(\equiv CTol)Rh(PPh_3)_2$, **10**, 8, 9
$RhB_{10}C_{10}H_{25}$
$Cp*RhB_{10}H_{10}$, **1**, 358
$RhB_{10}C_{10}H_{28}Cl$
$Cp*[RhB_{10}H_{13}Cl]$, **1**, 342, 351
$RhB_{10}C_{11}H_{27}O$
$Cp*RhB_{10}H_9(OMe)$, **1**, 358
$RhB_{10}C_{14}H_{34}N$
$Cp*(NEt_2)RhB_{10}H_9$, **1**, 358
$RhB_{10}C_{16}H_{40}NO$
$(Cp*)Rh(NEt_3)OB_{10}H_{10}$, **1**, 267
$RhB_{10}C_{17}H_{34}ClOP_2$
$(PMe_2Ph)_2(H)Cl(OMe)$-$RhB_{10}H_8$, **1**, 357
$RhB_{10}C_{18}H_{35}ClOP$
$Cp*[RhOB_{10}H_9Cl(PMe_2Ph)]$, **1**, 267, 351
$RhB_{10}C_{18}H_{37}O_2P_2$
$(PMe_2Ph)_2(H)(OMe)_2RhB_{10}H_8$, **1**, 357
$RhB_{10}C_{19}H_{39}Cl_3P$
$Cp*[RhB_{10}H_{11}Cl(PMe_2Ph)]$-$CH_2Cl_2$, **1**, 351
$RhB_{10}C_{36}H_{42}NP$
$[(Ph_3P)_2RhH]HNB_{10}H_{10}$, **1**, 262
$RhB_{11}C_{37}H_{44}OP_2$
$[Rh(CO)(PPh_3)_2]B_{11}H_{14}$, **1**, 363
$RhB_{11}CsC_{37}H_{43}OP_2$
$Cs[Rh(CO)(PPh_3)_2]B_{11}H_{13}$, **1**, 363
$RhC_4Cl_2O_4$
$[ClRh(CO)_2]_2$, **11**, 407
$RhC_4H_{12}Cl_4N$
$RhCl_3Me_4NCl$, **12**, 748
RhC_4O_4
$[Rh(CO)_4]^-$, **8**, 468
$Rh(CO)_4$, **8**, 2
$RhC_6H_5N_2O$
$CpRh(CO)(N_2)$, **8**, 233
$RhC_7H_9N_2$
$CpRh(C_2H_4)(N_2)$, **8**, 233
$RhC_8H_8INO_2$
$[RhI_3(py)(CO)(COMe)]^-$, **8**, 203
$RhC_{10}H_{10}$
$[Cp_2Rh]^+$, **8**, 170
$RhC_{10}H_{17}O_5S$
$Cp*Rh(SO_4)\cdot H_2O$, **8**, 177
$RhC_{10}H_{18}P$
$CpRhH(PMe_3)(C_2H_3)$, **8**, 232
$CpRh(PMe_3)(C_2H_4)$, **8**, 215
$RhC_{10}H_{21}O_3$
$[Cp*Rh(H_2O)_3]^{2+}$, **8**, 178
$RhC_{10}H_{30}ClIP_3$
$RhCl(I)Me(PMe_3)_3$, **8**, 193
$RhC_{11}H_{18}N$
$(C_5H_4NMe_2)Rh(C_2H_4)_2$, **12**, 755
$RhC_{11}H_{19}O_5$
$Cp*Rh(CO_3)\cdot 2H_2O$, **8**, 177
$RhC_{12}H_{15}O_2$
$Cp*Rh(CO)_2$, **8**, 182, 185
$RhC_{12}H_{19}$
$Rh(cod)(CHMeCHCH_2)$, **8**, 274
$RhC_{12}H_{25}ClP_2$
$[CpRh(CH_2Cl)(PMe_3)_2]^+$, **8**, 195
$RhC_{13}H_{13}$
$CpRh(cot)$, **8**, 278
$RhC_{13}H_{26}P$
$Cp*RhH_2(PMe_3)$, **8**, 176, 215
$RhC_{14}H_{18}F_6O_2$
$[Rh(C_6Me_6)(CO)_2]PF_6$, **8**, 272
$RhC_{14}H_{23}$
$Rh(C_2H_4)_2Cp*$, **8**, 462
$RhC_{14}H_{24}PS_2$
$Cp*Rh(CS_2)(PMe_3)$, **8**, 240
$RhC_{14}H_{27}OS$
$Cp*RhMe_2(DMSO)$, **8**, 190

$RhC_{14}H_{28}P$
Cp*RhHMe(PMe_3), **8**, 191
$RhC_{15}H_{31}B_9N_2$
Cp*Rh(NHMe)CB_9H_9(CNEt), **1**, 380
$RhC_{15}H_{33}P_2$
Rh(Me)(cod)$(PMe_3)_2$, **8**, 267
$RhC_{16}H_{15}IP$
Cp*RhI(cyclopropyl)(PMe_3), **8**, 277
$RhC_{16}H_{31}P_2$
Cp*Rh(C_2H_4)(Me_2PPMe_2), **8**, 234
$RhC_{16}H_{33}O_9P_3$
$[Cp*Rh[P(O)(OMe)_2]_3]^-$, **8**, 183
$RhC_{16}H_{36}B_9N_2$
$(EtNC)_2$Cp*RhB_9H_{11}, **1**, 397
$RhC_{17}H_{22}$
Cp*Rh(Tol), **8**, 272
$RhC_{17}H_{30}O_4P$
Cp*Rh$(OAc)_2$(PMe_3), **8**, 177
$RhC_{17}H_{32}P_3$
Rh(CCPh)$(PMe_3)_3$, **8**, 189
$RhC_{17}H_{34}P$
Cp*$RhEt_2$(PMe_3), **8**, 191
$RhC_{18}H_{15}F_8$
Cp*Rh(C_8F_8), **8**, 275
$RhC_{18}H_{18}F_6N_2PS_3$
Rh(CS_2)(CSMe)[(Tol)NC(H)N-(Tol)]PF_6, **8**, 163
$RhC_{18}H_{27}B_2N_2$
(Cp*)Rh[$C_4(BNMe_2)_2$], **1**, 209
$RhC_{19}H_{24}F_6P$
Cp*Rh(C_6F_6)(PMe_3), **8**, 200
Cp*RhF(PMe_3)(C_6F_5), **8**, 233
$RhC_{19}H_{39}OP_2$
$[Cp*Rh(PMe_3)_2(acetone)]^{2+}$, **8**, 183
$RhC_{20}H_{30}Cl_4$
$[Cp*Rh(Cl)Cl]_2$, **8**, 182
$RhC_{20}H_{30}F_6P$
$[Cp*_2Rh][PF_6]$, **8**, 186
$RhC_{22}H_{27}O_8$
$[C_5Me(CO_2Me)_4]$Rh(cod), **8**, 268
$RhC_{22}H_{33}$
Cp*Rh(C_6Me_6), **8**, 244
$RhC_{23}H_8F_{15}O_2$
$[(C_6F_5)_3Rh(acac)]^-$, **8**, 183
$RhC_{24}H_{48}O_6$
$[Rh(C_6H_6)[P(OPr^i)_3]_2]^+$, **8**, 273
$RhC_{27}H_{33}$
$Rh(Mes)_3$, **8**, 273
$RhC_{30}F_{25}$
$[Rh(C_6F_5)_5]^{2-}$, **8**, 187
$RhC_{31}H_{32}O_3P_2$
$[CpRh(O_2H)(H_2O)(dppe)]^+$, **8**, 170
$RhC_{32}H_{50}OP$
Rh(OPh)(cod)(PCy_3), **8**, 263
$RhC_{34}H_{50}P$
Rh(C_2Ph)(cod)(PCy_3), **8**, 267
$RhC_{36}H_{52}PSi_2$
Cp*Rh$(CH_2TMS)_2$(PPh_3), **8**, 190
$RhC_{37}H_{25}O_2$
(C_5Ph_5)Rh$(CO)_2$, **8**, 174
$RhC_{37}H_{30}ClOP_2$
Rh$(Ph_3P)_2$(CO)Cl, **12**, 375, 376
$RhC_{37}H_{33}IP_2$
MeRhI$(PPh_3)_2$, **11**, 402
$RhC_{37}H_{40}B_8P_2$
$(Ph_3P)_2$(H)Rh(SCB_8H_9), **1**, 380
$RhC_{37}H_{41}B_8P_2$
$(Ph_3P)_2$(H)Rh(SCB_8H_{10}), **1**, 380
$RhC_{38}H_{34}NO_4$
[(C_5Ph_5)Rh(NO)[$P(OMe)_3$], **8**, 175
$RhC_{38}H_{42}B_9P_2$
$(Ph_3P)_2$HRh($C_2B_9H_{11}$), **1**, 183, 415, 416
$RhC_{38}H_{43}B_9P_3$
$(Ph_3P)_2$Rh($MePCB_9H_{10}$), **1**, 380
$RhC_{38}H_{62}O_2$
Rh$(Pr^i_3C_6H_2)_2(THT)_2$, **8**, 187
$RhC_{39}H_{39}ClP_3$
$(MePh_2P)_3$RhCl, **12**, 374
$RhC_{42}H_{35}O_2$
(C_5Bz_5)Rh$(CO)_2$, **8**, 175
$RhC_{43}H_{37}$
(C_5Ph_5)Rh(cod), **8**, 268
$RhC_{43}H_{40}O_2P_3$
(triphos)Rh(HCO)(CO), **8**, 205
$RhC_{43}H_{43}ClP_3$
RhCl(triphos)(C_2H_4), **8**, 232
$RhC_{44}H_{28}IN_4$
IRh(mesotetraphenylporphy-rin), **12**, 416
$RhC_{44}H_{46}P_4$
$[(PP_3)Rh(C_2H_4)]^+$, **8**, 232
$RhC_{44}H_{48}P_3$
Rh(triphos)Me_3, **8**, 188
$RhC_{45}H_{47}P_3$
$[Rh(triphos)(C_2H_4)_2]^+$, **8**, 232
$RhC_{48}H_{51}NOP_3$
Rh(COMe)(triphos)(Bu^tNC), **8**, 189
$RhC_{51}H_{47}P_4$
RhMe$(dppm)_2$, **8**, 189
$RhC_{54}H_{45}Cl$
ClRh$(PPh_3)_3$, **11**, 402
$RhC_{54}H_{45}ClP_3$
ClRh$(PPh_3)_3$, **12**, 374, 748, 755
$RhC_{55}H_{46}OP_3$
RhH(CO)$(PPh_3)_3$, **8**, 144
$RhC_{72}H_{61}P_4$
HRh$(PPh_3)_4$, **12**, 906, 907, 911
$RhC_{85}H_{66}O_{10}P_6$
$Rh_6(CO)_{10}(dppm)_3$, **8**, 474
$RhCoC_7O_7$
CoRh$(CO)_7$, **8**, 30, 505, 508; **10**, 45, 62
$RhCoC_8O_8$
CoRh$(CO)_8$, **8**, 506
$RhCoC_{11}H_{10}O_5$
Co$(CO)_3$(BuC_2H)Rh$(CO)_2$, **10**, 64
$RhCoC_{11}H_{18}O_5P_2$
Co$(CO)_4$Rh(Co)$(PMe_3)_2$, **10**, 62
$RhCoC_{17}H_{15}O_3$
$[Co(CO)_3(C_7H_7)Rh(nbd)]^+$, **10**, 64
$RhCoC_{17}H_{17}$
CpCo(C_7H_7)RhCp, **10**, 13
$RhCoC_{17}H_{30}O_5P_2$
CoRh$(CO)_5(PEt_3)_2$, **8**, 508
$RhCoC_{19}H_{38}O_3P_2$
$(CO)_2$Co(H)(PBu^t_2)Rh(CO)-($HPBu^t_2$), **10**, 45
$RhCoC_{20}H_{10}O_6$
CoRh$(CO)_6$(PhCCPh), **8**, 508
$RhCoC_{20}H_{30}S_8$
CoRhCp*$_2$(S_8), **10**, 64
$RhCoC_{23}H_{28}$
$[CoCp(C_8H_8)RhCp*]^+$, **10**, 64
$RhCoC_{24}H_{15}O_6P$
CoRh$(CO)_6$(PPh_3), **10**, 45
$RhCoC_{28}H_{40}O_2$
CoCp*[$CHCMe_2$CMeC(O)]-(CO)RhCp*, **10**, 64
$RhCoC_{53}H_{45}O_3P_4$
$[Co(CO)_2(H)(dppm)_2Rh(CO)]^+$, **10**, 64
$RhCoMoC_{27}H_{30}O_7$
MoCoRh$(CO)_7$Cp*$_2$, **10**, 244
$RhCoSiC_{14}H_{11}O_6$
Co$(CO)_4$Rh($SiMe_2$Ph)$(CO)_2$, **10**, 62
$RhCo_2C_{31}H_{51}B_5$
CpCo($Et_2C_2B_3H_3$)RhCp*, **1**, 388
$RhCo_3C_{12}O_{12}$
$Co_3Rh(CO)_{12}$, **8**, 505, 510, 511; **10**, 40, 43, 46, 48, 50
$RhCrC_{10}H_8O_3$
Cr$(CO)_3$Rh(nbd), **5**, 502
$RhCrC_{20}H_{19}O_3$
Cr$(CO)_3$(C_9H_7)Rh(cod), **5**, 502
$RhCrC_{54}H_{44}ClO_4P_4$
Cr$(CO)_3(dppm)_2$Rh(CO)Cl, **10**, 44
$RhCrC_{55}H_{44}O_5P_4$
$[Cr(CO)_3(dppm)_2Rh(CO)_2]^+$, **10**, 44, 45
$RhCrC_{57}H_{47}O_2P_4$
Cr(CO)Cp(dppm)(CO)Rh (dppm), **10**, 44
$RhCr_2C_{16}H_{19}O_2S_3$
$Cr_2Rh(S)_2(SBu^t)(CO)_2Cp_2$, **10**, 228
$RhFeC_{18}H_{16}O_3$
$[Fe(CO)_3(C_8H_8)Rh(nbd)]^+$, **10**, 60
$RhFeC_{18}H_{19}O_3$
Fe$(CO)_3$(C_7H_7)Rh(cod), **10**, 60
$RhFeC_{21}H_{37}O_5P_2$
Fe$(CO)_4$(PBu^t_2)Rh(CO)-(PBu^t_2H), **7**, 44
$RhFeC_{23}H_{40}O_3P_2$
Fe$(CO)_2$(PBu^t_2H)(PBu^t)-(CO)Rh(cod), **7**, 44

$RhFeC_{30}H_{22}ClO_5P_2$
$(CO)_4Fe(dppm)RhCl(CO)$, **7**, 83
$RhFeC_{38}H_{30}ClN_2O_4P_2$
$FeRh(Ph_2Ppy)_2(CO)_4Cl$, **10**, 47
$RhFeSiC_{32}H_{31}O_7P_2$
$Fe(CO)_3(dppm)[Si(OMe)_3]Rh(CO)$, **10**, 61
$RhFeWC_{27}H_{19}O_5$
$CpWRhFe(CO)_5(C_9H_7)(\equiv CTol)$, **10**, 8
$RhFe_2C_{11}H_5O_6Te_2$
$CpRhFe_2(CO)_6Te_2$, **7**, 77
$RhFe_2C_{12}H_5O_7Te_2$
$CpRhFe_2(CO)_7Te_2$, **7**, 77
$RhFe_2C_{42}H_{30}O_6P_3$
$Fe_2Rh(PPh_2)_3(CO)_6$, **10**, 227
$RhFe_2C_{44}H_{30}O_8P_3$
$Fe_2Rh(PPh_2)_3(CO)_8$, **10**, 226, 227
$RhFe_2C_{60}H_{45}O_6P_4$
$Fe_2Rh(PPh_2)_3(CO)_6(PPh_3)$, **10**, 226, 227
$RhFe_3C_{12}HO_{12}$
$[HFe_3Rh(CO)_{12}]^{2-}$, **7**, 83
$RhFe_3C_{13}HO_{12}$
$RhFe_3(CO)_{12}(CH)$, **7**, 25
$RhFe_3C_{13}H_3O_{11}$
$HFe_3Rh(CO)_{11}(C{=}CH_2)$, **10**, 41
$RhFe_3C_{13}O_{12}$
$[RhFe_3(CO)_{12}C]^-$, **7**, 21
$RhFe_3C_{19}H_7O_{11}$
$HFe_3Rh(CO)_{11}(C{=}CHPh)$, **10**, 45
$RhFe_3C_{30}H_{25}O_8P_2$
$Fe_3Rh(PPh)_2(CO)_8Cp^*$, **10**, 226
$RhFe_3C_{31}H_{25}O_9P_2$
$Cp^*RhFe_3(CO)_9(PPh)_2$, **7**, 55
$RhFe_4C_{29}H_{25}O_{12}P_2$
$Fe_4Rh(PPh)_2(CO)_{12}Cp$, **10**, 222, 223
$RhFe_4C_{39}H_{20}O_{14}P$
$[Ph_4P][Fe_4RhC(CO)_{14}]$, **10**, 45
$RhFe_5C_{21}H_{12}NO_{16}$
$[Me_4N][Fe_5RhC(CO)_{16}]$, **10**, 50
$RhFe_5C_{41}H_{20}O_{16}P$
$[Ph_4P][Fe_5RhC(CO)_{16}]$, **10**, 45
$RhInC_5H_4O$
InRh(cyclopentadienone), **8**, 257
$RhInC_6H_{10}$
InRh(hexadiene), **8**, 271
$RhIrC_{24}H_{19}ClNO_2$
$RhCp(CO)(NC_5H_4PPh_2)IrCl(CO)$, **10**, 64
$RhIrC_{27}H_{36}N_6$
$Cp^*Ir(pz)_3Rh(cod)$, **8**, 257
$RhIrC_{27}H_{37}N_6$
$[Cp^*Ir(pzH)(pz)_2Rh(cod)]^+$, **8**, 257
$RhIrC_{30}H_{27}ClF_4N_4$
$Cp^*IrCl(pz)_2Rh(tfb)$, **8**, 257
$RhLiC_{13}H_{25}P$
$Li[Cp^*RhH(PMe_3)]$, **8**, 200
$RhLiC_{42}H_{39}OP_3$
Li[Rh(CO)(triphos)], **8**, 194
$RhLiSi_2C_{22}H_{50}N_2$
$(cod)Rh(CH_2TMS)_2Li(TMEDA)$, **8**, 267
$RhMnC_{54}H_{44}O_4P_4S$
$MnRh(S)(dppm)_2(CO)_4$, **6**, 89
$RhMnC_{54}H_{44}O_6P_4S$
$MnRh(SO_2)(dppm)_2(CO)_4$, **6**, 89
$RhMoC_{18}H_{14}O_2$
$[Mo(C_6H_6)(azulene)Rh(CO)_2]^+$, **10**, 43
$RhMoC_{18}H_{27}O_4P_2$
$Mo(CO)_4(PMe_2)_2RhCp^*$, **10**, 44
$RhMoC_{26}H_{20}O_3P$
$CpMoRh(CO)_3(PPh_3)$, **10**, 2
$RhMoC_{44}H_{35}O_3P_2$
$CpMoRh(CO)_2(CO)(PPh_3)_2$, **10**, 39
$RhMoC_{53}H_{44}ClO_5P_4S$
$Mo(CO)_3(SO_2)(dppm)_2RhCl$, **10**, 44
$RhNaC_{11}H_{23}O_3P_2$
$[CpRhMe(PMe_3)[P(ONa)(OMe)_2]]^+$, **8**, 195
$RhNb_2W_4C_{10}H_{15}O_{19}$
$[Cp^*Rh(Nb_2W_4O_{19})]^{2-}$, **8**, 178
$RhOs_2C_{19}H_{15}O_9$
$Os_2RhCp^*(CO)_9$, **7**, 653
$RhOs_3C_{15}H_{10}O_{12}$
$H_2Os_3Rh(CO)_{10}(acac)$, **10**, 60
$RhOs_5C_{21}H_6O_{15}$
$Os_5Rh(CO)_{15}(C_6H_6)$, **10**, 5, 6
$RhPdC_{28}H_{34}N_4S_4$
$[Rh(pyS)_4Pd(C_3H_4Me)_2]^+$, **10**, 64
$RhPdC_{32}H_{37}F_6N_3P_2$
$[Cp(CNBu^t)Rh(Ph_2PPy)PdCl(CNBu^t)]PF_6$, **9**, 215
$RhPdC_{33}H_{25}ClN_2OPS_4$
$RhCl(CO)(benzothiazole\text{-}2\text{-}thiolate)_2Pd(PPh_3)$, **10**, 64
$RhPdC_{48}H_{33}Cl_2N_4P_2$
$[RhCl(PhCN)_2(NC_5H_4PPh_2)_2PdCl]^+$, **10**, 64, 67
$RhPtC_{52}H_{58}F_6O_6P_3S_2$
$[(Ph_3P)_2Pt(H)(H_2CCH)Rh(PMe_3)Cp^*](OSO_2CF_3)_2$, **9**, 549
$RhPtC_{55}H_{46}OP_4$
$[Rh(CO)(dppm)_2Pt(C_2H)_2]^+$, **10**, 64
$RhPtC_{56}H_{53}ClN_3P_4$
$[RhCl(MeCN)_2(dppm)_2Pt(MeCN)]^+$, **10**, 64
$RhPtC_{57}H_{50}OP_4$
$[Rh(CO)(dppm)_2Pt(C_2Me)_2]^+$, **10**, 64
$RhPt_3TlC_{65}H_{111}Cl_2O_3P_3$
$[TlPt_3(CO)_3(PCy_3)_3][RhCl_2(C_8H_{12})]$, **1**, 569
$RhReC_{48}H_{96}P_4$
$(PCy_2H)ReH_5(PCy_2)_2RhH(PCy_2H)$, **10**, 39
$RhReC_{54}H_{44}O_4P_4$
$Re(CO)_3(dppm)_2Rh(CO)$, **10**, 52
$RhReC_{56}H_{100}P_4$
$Re(PCy_2)_2(PCy_2)_2Rh(cod)$, **10**, 53
$RhRe_7C_{24}O_{23}$
$[Re_7C(CO)_{21}Rh(CO)_2]^-$, **10**, 127
$RhRuC_{31}H_{22}O_6P_2$
$[RuRh(CO)_6(dppm)]^+$, **7**, 302
$RhRuC_{33}H_{31}N_4OP$
$H(CO)(PPh_3)Ru(biimidazolato)Rh(cod)$, **10**, 39
$RhRuC_{35}H_{29}O_3P_2$
$Ru(CO)_2(dppm)(C_7H_7)Rh(CO)$, **10**, 61
$RhRuC_{53}H_{44}ClO_3P_4$
$RuRh(CO)_3Cl(dppm)_2$, **10**, 45
$RhRuC_{58}H_{57}ClP_4$
$[Ru(dppm)_2(H)(Cl)Rh(cod)]^+$, **10**, 61
$RhRu_3C_{14}H_9O_9$
$Ru_3Rh(H)_4(CO)_9Cp$, **10**, 211
$RhRu_3C_{15}H_7O_{10}$
$Ru_3Rh(H)_2(CO)_{10}Cp$, **10**, 209–11
$RhRu_3C_{67}H_{51}O_7P_4$
$Ru_3Rh(H)_2(PPh_2)(PPhC_6H_4)(CO)_7(PPh_3)_2$, **10**, 223
$RhRu_5C_{25}H_{15}O_{14}$
$Cp^*RhRu_5C(CO)_{14}$, **10**, 134
$RhRu_5C_{28}H_{21}O_{11}$
$Cp^*(C_6H_6)RhRu_5C(CO)_{11}$, **10**, 134
$RhRu_5C_{28}H_{25}O_7$
$Cp^*Cp_2RhRu_5C(CO)_7$, **10**, 134
$RhSiC_{13}H_{25}$
$CpRh(C_2H_4)(SiEt_3)H$, **2**, 115
$CpRhEt(SiEt_3)$, **2**, 116
$RhSiC_{15}H_{29}$
$CpRh(C_2H_4)(SiEt_3)Et$, **2**, 116
$RhSiC_{16}H_{30}P$
$Rh(C_2TMS)(cod)(PMe_3)$, **8**, 267
$RhSiC_{23}H_{27}F_{12}O$
$[C_5(OSiEt_3)(CF_3)_4]Rh(cod)$, **8**, 153
$RhWC_{32}H_{25}O_2$
$W(CO)_2Cp(CTolCHCPh)Rh(C_9H_7)$, **10**, 44
$RhWC_{47}H_{36}O_5P_3$
$W(CO)_4(PPh_2)_2RhH(CO)(PPh_3)$, **10**, 44
$RhZrC_{37}H_{34}ClOP_2$
$Cp_2Zr(CH_2PPh_2)_2Rh(CO)Cl$, **4**, 576
$RhZrC_{39}H_{42}ClO_4P_2$
$Cp^*Zr(OBz_2P)_2(OMe)(Cl)(Rh(CO)$, **4**, 551
$RhZrC_{53}H_{46}OP_2$
$Cp_2Zr(PPh_2)_2RhH(CO)(PPh_3)$, **10**, 44
$RhZrInC_{15}H_{23}IP_2$
$[Cp_2Zr(PEt)_2Rh(In)(Me)]I$, **8**, 198
$Rh_2AgC_{38}H_{35}O_3P_2$

Ag(OCOMe)Rh_2Cp_2(CO)-(dppm), **10**, 68
$Rh_2As_2IrC_{67}H_{58}Cl_2O_3P_4$
[$Rh_2IrCl_2(CO)_3$[AsPh(CH_2P-$Ph_2)_2]_2]^+$, **10**, 64
$Rh_2BC_{25}H_{31}F_4$
[CpRh(cot)Rh(C_6Me_6)]BF_4, **8**, 245
$Rh_2BFe_4C_{16}O_{16}$
[Fe_4Rh_2(B)(CO)$_{16}]^-$, **10**, 215
$Rh_2BFe_4C_{52}H_{30}NO_{16}P_2$
[PPN][$Fe_4Rh_2(CO)_{16}$B], **1**, 304
$Rh_2B_2C_{22}H_{44}ClN_2Si_2$
[RhCl($C_{10}H_{22}$BNSi)], **1**, 213
$Rh_2B_2C_{24}H_{40}N_2$
(cod)$_2Rh_2$[($Me_2N)_2C_4B_2$], **1**, 209
$Rh_2B_2C_{32}H_{44}F_8$
[(C_6Me_6)Rh(cot)Rh(C_6Me_6)]-($BF_4)_2$, **8**, 245
$Rh_2B_2C_{133}H_{124}N_2P_6$
[Rh_2(BuNC)$_2$(dppm)$_3$][$BPh_4]_2$, **8**, 167
$Rh_2B_9C_{16}H_{29}P$
Ph(Cp*)$Rh_2PB_9H_9$, **1**, 265
$Rh_2B_9C_{20}H_{38}ClS$
S[(Cp*)Rh]$_2ClB_9H_8$, **1**, 268
$Rh_2B_9C_{54}H_{52}Cl_2P_3S$
($PPh_3)_2Cl_2Ph_2PC_6H_4Rh_2SB_9H_8$, **1**, 268
$Rh_2B_{17}C_{20}H_{49}$
(Cp*Rh)$_2B_{17}H_{19}$, **1**, 365, 366
$Rh_2B_{18}C_{16}H_{50}S_2$
Cp*[RhB_9(SMe_2)H_{10}]RhB_9H_7-($SMe_2)_2$, **1**, 347
$Rh_2B_{18}C_{16}H_{50}S_3$
Cp*RhB_9(SMe_2)$H_{10}RhB_9H_7$-($SMe_2)_2$, **1**, 341
$Rh_2B_{18}C_{20}H_{54}O$
O[Cp*[RhB_9H_{12}]]$_2$, **1**, 342
$Rh_2C_4Cl_2O_4$
[ClRh(CO)$_2$]$_2$, **11**, 402
Rh_2(CO)$_4$(Cl)$_2$, **8**, 2, 327, 459, 505; **10**, 43
$Rh_2C_8F_{12}O_8$
Rh_2(TFA)$_4$, **12**, 447
Rh_2($CF_3CO_2)_4$, **8**, 228
$Rh_2C_8H_{12}O_8$
Rh_2(OAc)$_4$, **12**, 397, 403, 429
$Rh_2C_8H_{16}Cl_2$
[Rh(Cl)($C_2H_4)_2]_2$, **8**, 459
$Rh_2C_8H_{16}N_4O_4$
Rh_2(acam)$_4$, **12**, 429
$Rh_2C_8O_8$
Rh_2(CO)$_8$, **8**, 466
$Rh_2C_{10}H_{10}N_2O_2$
Rh_2(NO)$_2Cp_2$, **8**, 464
$Rh_2C_{12}H_{14}Cl_4$
[(C_5MeH_4)$RhCl_2]_2$, **8**, 171
$Rh_2C_{12}H_{20}Cl_2$
[Rh(Cl)(C_6H_{10})]$_2$, **8**, 270
$Rh_2C_{12}H_{22}O_2$
[Rh($C_3H_5)_2$(OH)]$_2$, **8**, 274
$Rh_2C_{12}H_{28}Cl_2O_2P_4$
Rh_2(CO)$_2Cl_2$(dmpm)$_2$, **8**, 145
$Rh_2C_{14}H_{10}F_6O$
(CpRh)$_2$($CF_3C_2CF_3$)(CO), **8**, 280
$Rh_2C_{14}H_{16}Cl_2$
Rh_2Cl_2(nbd)$_2$, **9**, 297
$Rh_2C_{16}F_{28}O_8$
Rh_2[$CF_3CF_2CF_2CO_2]_4$, **12**, 411, 413
Rh_2(pfb)$_4$, **12**, 429, 447
$Rh_2C_{16}H_{24}Cl_2$
[Rh(Cl)(cod)]$_2$, **8**, 250, 263
$Rh_2C_{16}H_{24}N_6$
[Rh(N_3)(cod)]$_2$, **8**, 472
$Rh_2C_{16}H_{36}O_8S_2$
[Rh(SBu^t)[P(OMe)$_3$](CO)]$_2$, **8**, 122
$Rh_2C_{16}O_{24}Cl_2$
Rh_2Cl_2(cod)$_2$, **9**, 296
$Rh_2C_{18}H_{18}$
(CpRh)$_2$(cot), **8**, 245, 281
$Rh_2C_{18}H_{21}O_2$
CpRh(cot)Rh(acac), **8**, 245
$Rh_2C_{18}H_{26}O_4$
[Rh(HCO_2)(cod)]$_2$, **8**, 247
$Rh_2C_{18}H_{30}Cl_2O_6S_4$
[RhCl(CO)(EtSCH_2SEt)]$_2$-(DMAD), **8**, 237
$Rh_2C_{18}H_{30}O_2$
[Rh(OMe)(cod)]$_2$, **8**, 246
$Rh_2C_{20}H_{28}N_4O_{16}$
Rh_2[MEOX]$_4$, **12**, 409
$Rh_2C_{20}H_{30}Br_2Se$
Cp*$_2Rh_2$(Br)$_2$(Se), **8**, 176
$Rh_2C_{20}H_{30}Cl_4$
[Cp*Rh(Cl)Cl]$_2$, **8**, 176
$Rh_2C_{20}H_{33}ClO_3$
[Cp*Rh(OH)$_3$RhCp*]Cl, **8**, 464
[Rh_2(OH)$_3$Cp*$_2$]Cl, **8**, 470
$Rh_2C_{20}H_{36}Cl_2N_4$
[Cp*RhCl(pz)]$_2$, **8**, 180
$Rh_2C_{20}H_{56}O_2P_6$
Rh_2H_2(CO)$_2$($PMe_3)_6$, **8**, 139
$Rh_2C_{22}H_{30}O_2$
[Rh(CO)Cp*]$_2$, **8**, 462
$Rh_2C_{22}H_{36}Cl_2S_2$
[Cp*Rh(SMe)Cl]$_2$, **8**, 179
$Rh_2C_{23}H_{28}$
(Cp*Rh)(cot)(RhCp), **8**, 245
$Rh_2C_{24}H_{36}N_4O_{12}$
Rh_2(MEPY)$_4$, **12**, 429
$Rh_2C_{24}H_{44}N_4O_4$
Rh_2(cap)$_4$, **12**, 429, 447
$Rh_2C_{24}H_{44}N_4O_{12}$
Rh_2[methylpyrrolidone-carboxylate]$_4$, **12**, 409
$Rh_2C_{26}H_{30}N_4O_4$
[Rh(pz)(duroquinone)]$_2$, **8**, 270
$Rh_2C_{30}H_{63}O_3P_3$
Rh_2(CO)$_3$($PPr^i_3)_3$, **8**, 139
$Rh_2C_{30}H_{66}O$
Np_3RhORhNp_3, **8**, 186
$Rh_2C_{32}H_{28}O_8$
Rh_2[C_6H_5CH(OH)$CO_2]_4$, **12**, 411
$Rh_2C_{32}H_{60}O_8$
Rh_2(oct)$_4$, **12**, 405, 429
$Rh_2C_{38}H_{61}Cl_2NO$
[Rh(Cl)(coe)$_2]_2$PhNO, **8**, 229
$Rh_2C_{44}H_{48}Cl_3N_2P_2$
[Rh_2Cl_2(Cl)(Bu^tNC)$_2$(Ph_2P-py)$_2]^{2+}$, **8**, 167
$Rh_2C_{52}H_{44}Cl_2O_2P_4$
Rh_2Cl_2(CO)$_2$(dppm)$_2$, **8**, 149
$Rh_2C_{52}H_{46}O_2P_4$
Rh_2(H)$_2$(CO)$_2$(dppm)$_2$, **8**, 147
$Rh_2C_{54}H_{50}N_2O_{10}P_4$
Rh_2(CO)$_2$[(PhO)$_2$PN(Et)P-(OPh)$_2]_2$, **8**, 150
$Rh_2C_{70}H_{50}Br_4$
[(C_5Ph_5)$RhBr_2]_2$, **8**, 174
$Rh_2C_{70}H_{50}N_2O_2$
[(C_5Ph_5)Rh(NO)]$_2$, **8**, 175
$Rh_2C_{72}H_{50}O_2$
[(C_5Ph_5)Rh(CO)]$_2$, **8**, 174
$Rh_2C_{80}H_{60}O_8$
Rh_2($Ph_3CCO_2)_4$, **12**, 411
$Rh_2C_{82}H_{76}O_4P_4$
[Rh(CO)($PPh_3)_2$(THF)]$_2$, **8**, 139
$Rh_2C_{82}H_{80}O_6P_4$
[Rh(CO)($PPh_3)_2$(DME)]$_2$, **8**, 139
$Rh_2C_{84}H_{52}O_8$
Rh_2(trp)$_4$, **12**, 450
$Rh_2CoRu_2C_{19}H_{14}O_8P$
Ru_2CoRh_2(H)(PMe)(CO)$_8Cp_2$, **10**, 243
$Rh_2Co_2C_8F_{12}O_8P_4$
Co_2Rh_2(CO)$_8$($PF_3)_4$, **8**, 511
$Rh_2Co_2C_{10}F_6O_{10}P_2$
Co_2Rh_2(CO)$_{10}$($PF_3)_2$, **8**, 511
$Rh_2Co_2C_{12}O_{12}$
Co_2Rh_2(CO)$_{12}$, **8**, 471, 505, 506, 508; **10**, 7, 40, 43, 45, 46, 48, 50, 58, 60
$Rh_2Co_2C_{24}H_{10}O_{10}$
Co_2Rh_2(C_2Ph_2)(CO)$_{10}$, **10**, 36
$Rh_2Co_2C_{27}H_{30}O_7$
Co_2Rh_2(CO)$_7$Cp*$_2$, **8**, 512
$Rh_2Co_2C_{29}H_{15}O_{11}P$
Co_2Rh_2(CO)$_{11}$(PPh_3), **8**, 511; **10**, 191
$Rh_2Co_2RuC_{18}H_{10}O_8S$
$RuCo_2Rh_2$(S)(CO)$_8Cp_2$, **10**, 243
$Rh_2Cr_6C_{38}H_{48}S_9$
Cr_6Rh_2(S)$_7$($SBu^t)_2Cp_6$, **10**, 234, 235
$Rh_2FeC_{15}O_{14}$
Rh_2Fe_3(C)(CO)$_{14}$, **10**, 134
$Rh_2FeC_{26}H_{30}O_6$
Cp_2*Rh_2Fe(CO)$_6$, **10**, 7
$Rh_2FeC_{29}H_{55}O_5P_3$
Rh_2Fe(CO)$_3$(PBu^t_2H)(CO)$_2$-($PBu^t_2)_2$, **7**, 53
$Rh_2Fe_2C_{30}H_{40}O_8P_4$
Fe_2Rh_2(CO)$_6$(CO)$_2$($PPh_2)_4$, **7**, 44
$Rh_2Fe_2C_{58}H_{44}O_8P_4$
Fe_2Rh_2(dppm)$_2$(CO)$_8$, **7**, 83
$Rh_2Fe_2C_{60}H_{44}O_8P_4$

$Rh_2Fe_2(CO)_8[(Ph_2P)_2C{=}CH_2]_2$, **10**, 6
$Rh_2Fe_3C_{15}O_{14}$
$Rh_2Fe_3(CO)_{14}C$, **7**, 22
$Rh_2Fe_3C_{17}H_5O_{13}$
$Fe_3Rh_2(CO)_{13}[C(Me)CCH_2]$, **10**, 141, 142
$Rh_2Fe_3C_{41}H_{25}O_{13}P$
$[Ph_4P][Fe_3Rh_2(CO)_{10}(CO)_3(MeC{=}C{=}CH_2)]$, **10**, 41
$Rh_2Fe_4C_{17}O_{16}$
$Fe_4Rh_2C(CO)_{16}$, **10**, 45
$Rh_2HgC_{54}H_{51}ClN_2O_{10}P_4$
$Rh_2(Co)_2(H)[(PhO)_2PN(Et)P(OPh)_2]_2(HgCl)$, **10**, 341
$Rh_2InC_{13}H_{13}$
(CpRh)(cot)(RhIn), **8**, 245
$Rh_2In_2C_8H_8$
(InRh)(cot)(RhIn), **8**, 176
$Rh_2IrC_{33}H_{45}O_3$
$Rh_2Ir(CO)_3Cp^*_3$, **8**, 512
$Rh_2IrC_{36}H_{40}F_6P$
$Rh_2Cp^*_2Ir(PPh_2)(CF_3CCCF_3)$, **8**, 512
$Rh_2Ir_2C_{12}O_{12}$
$Rh_2Ir_2(CO)_{12}$, **8**, 505, 510
$Rh_2Ir_2C_{27}H_{30}O_7$
$Rh_2Ir_2(CO)_7Cp^*_2$, **8**, 512
$Rh_2MoC_{27}H_{30}O_7$
$MoRh_2(CO)_7Cp^*_2$, **10**, 210
$Rh_2Na_2C_{44}H_{36}Cl_2O_{18}S_6$
$[P(SO_3NaC_6H_4)_3]_2Rh(Cl)_2Rh(cod)$, **8**, 263
$Rh_2OsC_{37}H_{35}O_2P_3$
$OsRh_2(H)_2(CO_2)(cod)_2(PMe_2Ph)_3$, **10**, 215
$Rh_2Pd_2C_{54}H_{64}Cl_6N_6P_2$
$[RhCl_2(Bu^tNC)_2(NC_5H_4PPh_2)PdCl]_2$, **10**, 64
$Rh_2PtC_{33}H_{32}F_6O$
$PtRh_2(CO)(F_3CCCCF_3)Cp_2(cod)Cp_2$, **9**, 564
$Rh_2PtC_{41}H_{46}O_3P$
$[Rh_2Pt(H)(CO)_3(PPh_3)Cp^*_2]^+$, **10**, 210
$Rh_2RuC_{12}O_{12}$
$RuRh_2(CO)_{12}$, **10**, 42
$Rh_2Ru_2C_{12}H_2O_{12}$
$H_2Ru_2Rh_2(CO)_{12}$, **10**, 39, 48
$Rh_2Ru_3C_{25}H_{20}O_{13}P_2$
$Ru_3Rh_2(PPh)(CO)_{13}(PEt_3)$, **10**, 222
$Rh_2Zn_2C_{38}H_{44}P_2$
$[Rh(Pr^i)_2P(Pr^i)_2]_2(H)_2(ZnBz)_2$, **10**, 341
$Rh_3Ag_3OsC_{178}H_{171}F_3P_9S$
$Ag_3Rh_3(H)_9[MeC(CH_2PPh_3)_3]_3[OsSCF_3]$, **10**, 256
$Rh_3AsMoC_{25}H_{20}O_5$
$MoRh_3(As)(CO)_5Cp_4$, **10**, 218
$Rh_3AuRuC_{30}H_{15}O_{12}P$
$RuRh_3(AuPPh_3)(CO)_{12}$, **7**, 685
$Rh_3BFe_3C_{15}H_2O_{15}$
$H_2Fe_3Rh_3(CO)_{15}B$, **1**, 294, 304
$Rh_3C_6O_6S_2$
$[Rh_3(S)_2(CO)_6]^-$, **8**, 461
$Rh_3C_{15}H_{15}N_2O_2$
$Rh_3(NO)_2Cp_3$, **8**, 464
$Rh_3C_{18}H_{15}O_3$
$Rh_3(CO)_3Cp_3$, **8**, 462
$Rh_3C_{18}H_{16}O_2$
$[Rh_3(CO)_2(CH)Cp_3]^+$, **8**, 463
$Rh_3C_{18}H_{57}P_6$
$[RhH[P(OMe)_3]_2]_3$, **8**, 457
$Rh_3C_{18}H_{59}O_{18}P_6$
$Rh_3H_5[P(OMe)_3]_6$, **8**, 458
$Rh_3C_{19}H_{16}F_3O_6S$
$[Rh_3(CO)_3(H)Cp_3][OTf]$, **8**, 463
$Rh_3C_{20}H_{15}F_6O$
$Rh_3(CO)(CF_3CCCF_3)Cp_3$, **8**, 465
$Rh_3C_{20}H_{21}$
$Rh_3(H)Cp_4$, **8**, 466
$Rh_3C_{20}H_{39}O_4P_2S_2$
$Rh_3H(S)_2(CO)_4(PBu^t_2H)_2$, **8**, 461
$Rh_3C_{21}H_{21}$
$Rh_3Cp_3(C_6H_6)$, **8**, 466
$Rh_3C_{24}H_{37}S_2$
$Rh_3H(S)_2(cod)_3$, **8**, 461
$Rh_3C_{25}H_{52}P_3S_2$
$Rh_3H(S)_2(cod)_2(PMe_3)_3$, **8**, 461
$Rh_3C_{27}H_{54}O_3P_3$
$[Rh(PBu^t_2)(CO)]_3$, **8**, 460
$Rh_3C_{29}H_{54}O_5P_3$
$Rh_3(PBu^t_2)_3(CO)_5$, **8**, 461
$Rh_3C_{30}H_{21}O_3$
$Rh_3(CO)_3(indenyl)_3$, **8**, 465
$Rh_3C_{30}H_{48}O$
$[Rh_3(H)_3(O)Cp^*_3]^+$, **8**, 464
$Rh_3C_{30}H_{48}O_3$
$[(Cp^*Rh)_3(OH)_3]^+$, **8**, 179
$Rh_3C_{31}H_{45}OS$
$Rh_3(CO)(S)Cp^*_3$, **8**, 465
$Rh_3C_{32}H_{45}O_2$
$Rh_3(CO)_2Cp^*_3$, **8**, 462
$Rh_3C_{32}H_{47}$
$Rh_3(CH)_2Cp^*_3$, **8**, 464
$Rh_3C_{32}H_{47}O_2$
$Rh_3(H)_2(CO)_2Cp^*_3$, **8**, 462
$Rh_3C_{33}H_{45}O_3$
$Rh_3(CO)_3Cp^*_3$, **8**, 463
$Rh_3C_{33}H_{80}ClO_{11}P_4$
$[Rh_3H_3(Cl)[OP(OPr^i)_2][P(OPr^i)_3]_3]^+$, **8**, 459
$Rh_3C_{35}H_{42}F_6N_3O$
$[Rh_3(CO)(CF_3CCCF_3)(CNBu^t)_3Cp_3]^+$, **8**, 465
$Rh_3C_{40}H_{30}Cl_2O_4P_3$
$Rh_3(PPh_2)_3(Cl)_2(CO)_4$, **8**, 460
$Rh_3C_{41}H_{30}O_5P_3$
$Rh_3(PPh_2)_3(CO)_5$, **8**, 459
$Rh_3C_{43}H_{30}O_7P_3$
$Rh_3(PPh_2)_3(CO)_7$, **8**, 459
$Rh_3C_{51}H_{48}N_2O_5P_3$
$Rh_3(PPh_2)_3(CO)_5(Bu^tCN)_2$, **8**, 459
$Rh_3C_{54}H_{41}O_6P_4$
$Rh_3(PPh_2)_3(CO)_6(PPh_2H)$, **8**, 459
$Rh_3C_{56}H_{44}ClO_{10}P_4$
$[Rh_3(CO)_6(dppm)_2]ClO_4$, **8**, 457
$Rh_3C_{64}H_{52}O_3P_5$
$Rh_3(PPh_2)_3(CO)_3(dppm)$, **8**, 459
$Rh_3C_{75}H_{60}O_3P_5$
$Rh_3(PPh_2)_3(CO)_3(PPh_3)_2$, **8**, 459
$Rh_3Fe_3C_{40}H_{20}O_{15}P$
$[Ph_4P][Fe_3Rh_3C(CO)_{15}]$, **10**, 45
$Rh_3IrC_{12}O_{12}$
$Rh_3Ir(CO)_{12}$, **8**, 505, 510
$Rh_3MoC_{65}H_{50}O_6P_3$
$MoRh_3(CO)_6(PPh_3)_3Cp$, **10**, 211
$Rh_3OsC_{12}O_{12}$
$[OsRh_3(CO)_{12}]^-$, **10**, 192
$Rh_3RuC_{12}HO_{12}$
$HRuRh_3(CO)_{12}$, **10**, 48
$Rh_3RuC_{46}H_{31}O_{10}P_2$
$HRuRh_3(CO)_{10}(PPh_3)_2$, **10**, 48
$Rh_4B_2C_{40}H_{64}F_8$
$[Rh_4H_4Cp^*_4][BF_4]_2$, **8**, 470
$Rh_4C_{12}H_6O_9S_3$
$Rh_4(CO)_9(ttcy)$, **8**, 469
$Rh_4C_{12}O_{12}$
$Rh_4(CO)_{12}$, **8**, 461, 466, 467, 469, 470; **10**, 39
$Rh_4C_{14}O_{14}$
$Rh_4(CO)_{14}$, **8**, 467
$Rh_4C_{24}H_{68}O_{16}P_8$
$[RhH(dmope)]_4$, **8**, 469
$Rh_4C_{27}H_{33}O_{18}P_3$
$Rh_4(CO)_9[P(OCH_2)_3CEt]_3$, **8**, 469
$Rh_4C_{29}H_{15}O_{14}P$
$Rh_4(CO)_{11}[P(OPh)_3]$, **8**, 468
$Rh_4C_{32}H_{49}$
$[Rh(H)(cod)]_4$, **8**, 267, 469
$Rh_4C_{40}H_{64}$
$[Rh_4(H)_4Cp^*_4]^{2+}$, **8**, 464
$Rh_4C_{44}H_{58}P_2$
$Rh_4(cod)_4(PPh)_2$, **8**, 470
$Rh_4C_{46}H_{30}O_{10}P_2$
$Rh_4(CO)_{10}(PPh_3)_2$, **8**, 468
$Rh_4C_{46}H_{31}O_9P_3$
$Rh_4(CO)_9[HC(PPh_2)_3]$, **8**, 469
$Rh_4C_{46}H_{66}O_{10}P_2$
$Rh_4(CO)_{10}(PCy_3)_2$, **8**, 469
$Rh_4C_{51}H_{46}O_{11}P_4$
$Rh_4(CO)_8[P(OEt)_3][HC(PPh_2)_3]$, **8**, 468
$Rh_4C_{54}H_{40}O_6P_4$
$Rh_4(PPh_2)_4(CO)_6$, **8**, 459
$Rh_4C_{58}H_{99}O_4P_5$
$Rh_4(CO)_4(PCy_2)_4(PCy)$, **8**, 471
$Rh_4C_{63}H_{45}O_{18}P_3$
$Rh_4(CO)_9[P(OPh)_3]_3$, **8**, 468
$Rh_4C_{70}H_{81}N_3P_2$
$[Rh_4(tz)(cod)_4(PPh_3)_2]^+$, **8**, 260
$Rh_4C_{76}H_{66}O_{22}P_4S_3$
$Rh_4(CO)_4(SO_2)_3[P(OPh)_3]_4$, **8**, 469
$Rh_4C_{80}H_{60}O_{20}P_4$
$Rh_4(CO)_{89}[P(OPh)_3]_4$, **8**, 469
$Rh_4Co_2C_{16}O_{16}$

$Co_2Rh_4(CO)_{16}$, **8**, 512
$Rh_4Co_2C_{14}O_{13}$
$[Co_2Rh_4C(CO)_{13}]^{2-}$, **8**, 512
$Rh_4Co_2C_{16}O_{16}$
$Co_2Rh_4(CO)_{16}$, **8**, 510
$Rh_4FeC_{15}O_{15}$
$[FeRh_4(CO)_{15}]^{2-}$, **7**, 83; **10**, 47
$Rh_4Fe_2C_{16}O_{16}$
$[Fe_2Rh_4(CO)_{16}]^{2-}$, **7**, 83
$Rh_4Hg_6C_{36}H_{108}P_{12}$
$Hg_6Rh_4(PMe_3)_{12}$, **10**, 327, 329
$Rh_4OsC_{15}O_{15}$
$[OsRh_4(CO)_{15}]^{2-}$, **10**, 192
$Rh_4PtC_{12}O_{12}$
$[Rh_4Pt(CO)_{12}]^{2-}$, **10**, 188
$Rh_4PtC_{29}H_{20}F_{12}O$
$PtRh_4(CO)(F_3CCCF_3)_2Cp_4$, **9**, 572
$Rh_4PtC_{30}H_{20}F_{12}O_2$
$PtRh_4(CO)_2(F_3CCCCF_3)_2Cp_4$, **9**, 564
$Rh_4PtC_{84}H_{60}N_2O_{12}P_4$
$[PPN]_2[Rh_4Pt(CO)_{12}]$, **10**, 42
$Rh_5C_{15}O_{15}$
$[Rh_5(CO)_{15}]^-$, **8**, 471, 479
$Rh_5C_{16}H_2NO_{14}$
$[Rh_5(CO)_{14}(CH_2CN)]^{2-}$, **8**, 471
$Rh_5FeC_{16}O_{16}$
$[FeRh_5(CO)_{16}]^-$, **7**, 83; **10**, 193
$Rh_5PtC_{15}O_{15}$
$[Rh_5Pt(CO)_{15}]^-$, **10**, 60
$Rh_5PtC_{51}H_{30}NO_{15}P_2$
$[PPN][Rh_5Pt(CO)_{15}]$, **10**, 42
$Rh_5RuC_{52}H_{30}NO_{16}$
$[PPN][RuRh_5(CO)_{16}]$, **10**, 45
$Rh_5RuC_{52}H_{30}NO_{16}P_2$
$[PPN][RuRh_5(CO)_{16}]$, **10**, 48
$Rh_5FeC_{16}O_{16}$
$[FeRh_5(CO)_{16}]^-$, **7**, 83
$Rh_6As_3C_{31}H_{45}O_{11}$
$Rh_6(CO)_{11}(AsBu^t{}_2)_2(AsBu^t)$, **8**, 477
$Rh_6C_{14}O_{13}$
$[Rh_6C(CO)_{13}]^{2-}$, **8**, 474, 475
$Rh_6C_{14}O_{14}$
$[Rh_6(CO)_4]^{4-}$, **8**, 473
$Rh_6C_{15}HO_{15}$
$[Rh_6H(CO)_{15}]^-$, **8**, 473, 476
$Rh_6C_{15}NO_{15}$
$[Rh_6N(CO)_{15}]^-$, **8**, 476, 484
$Rh_6C_{15}O_{15}$
$[Rh_6(CO)_{15}]^{2-}$, **8**, 472, 473, 476, 479
$Rh_6C_{16}HO_{15}$
$[Rh_6H(CO)_{15}C]^-$, **8**, 476
$Rh_6C_{16}H_{20}O_8P$
$Rh_6(CO)_8(PBu^t{}_2)(H)_2$, **8**, 476
$Rh_6C_{16}O_{15}$
$[Rh_6C(CO)_{15}]^{2-}$, **8**, 474
$Rh_6C_{16}O_{16}$
$Rh_6(CO)_{16}$, **8**, 461, 467, 472–4, 508; **10**, 48; **12**, 403, 415, 428, 429
$Rh_6C_{18}H_5O_{15}$
$[Rh_6(CO)_{15}(C_3H_5)]^-$, **8**, 474
$Rh_6C_{20}H_{10}O_{14}$
$Rh_6(CO)_{14}(dmbd)$, **8**, 474
$Rh_6C_{33}O_{15}P$
$Rh_6(CO)_{15}(PPh_3)$, **8**, 474
$Rh_6C_{37}H_{22}O_{11}P_2$
$[Rh_6C(CO)_{11}(dppm)]^{2-}$, **8**, 476
$Rh_6C_{40}H_{24}O_{14}P_2$
$Rh_6(CO)_{14}(dppe)$, **8**, 474
$Rh_6C_{84}H_{60}O_{24}P_4$
$Rh_6(CO)_{12}[P(OPh)_3]_4$, **8**, 474
$Rh_6CoC_{15}NO_{15}$
$[Rh_6CoN(CO)_{15}]^{2-}$, **8**, 513
$Rh_6Ir_6C_{30}O_{30}$
$[Rh_6Ir_6(CO)_{30}]^{2-}$, **10**, 42
$Rh_6NiC_{16}O_{16}$
$[NiRh_6(CO)_{16}]^{2-}$, **10**, 193–5
$Rh_6PtC_{16}O_{16}$
$[Rh_6Pt(CO)_{16}]^{2-}$, **10**, 193
$Rh_7C_{15}NO_{15}$
$[Rh_7N(CO)_{15}]^{2-}$, **8**, 477
$Rh_7C_{16}IO_{16}$
$[Rh_7I(CO)_{16}]^{2-}$, **8**, 477
$Rh_7C_{16}O_{16}$
$[Rh_7(CO)_{16}]^{3-}$, **8**, 477
$Rh_7C_{17}O_{17}$
$[Rh_7(CO)_{17}]^{3-}$, **8**, 472
$Rh_7C_{18}O_{17}$
$[Rh_7C(CO)_{17}]^-$, **8**, 483
$Rh_7K_3C_{16}O_{16}$
$K_3[Rh_7(CO)_{16}]$, **8**, 476
$Rh_8C_{20}O_{19}$
$Rh_8C(CO)_{19}$, **8**, 478
$Rh_8C_{24}H_6N_2O_{19}$
$[Rh(CO)_2(MeCN)_2][Rh_7C(CO)_{17}]$, **8**, 483
$Rh_8C_{39}H_{54}O_9P_6$
$Rh_8(CO)_9[P(C_5H_9)]_6$, **8**, 478
$Rh_8PtC_{19}O_{19}$
$[Rh_8Pt(CO)_{19}]^{2-}$, **10**, 193–5, 213
$Rh_9C_{19}O_{19}$
$[Rh_9(CO)_{19}]^{3-}$, **8**, 479
$Rh_9C_{21}O_{21}P$
$[Rh_9P(CO)_{21}]^{2-}$, **8**, 485
$Rh_9Co_3C_{30}O_{30}$
$[Co_3Rh_9(CO)_{30}]^{2-}$, **10**, 42
$Rh_9CuC_{25}H_9N_3O_{19}$
$CuRh_9(CO)_{19}(NCMe)_3$, **10**, 317
$Rh_9Ir_3C_{30}O_{30}$
$[Rh_9Ir_3(CO)_{30}]^{2-}$, **10**, 42
$Rh_9Pt_2C_{22}O_{22}$
$[Rh_9Pt_2(CO)_{22}]^{3-}$, **10**, 193, 194, 213
$Rh_{10}C_{21}O_{21}$
$[Rh_{10}(CO)_{21}]^{2-}$, **8**, 479
$Rh_{10}C_{22}O_{22}P$
$[Rh_{10}P(CO)_{22}]^{3-}$, **8**, 485
$Rh_{10}PtC_{21}NO_{21}$
$[Rh_{10}Pt(N)(CO)_{21}]^{3-}$, **10**, 218
$Rh_{11}C_{23}O_{23}$
$[Rh_{11}(CO)_{23}]^{3-}$, **8**, 479
$Rh_{11}H_{23}O_3P_2$
$CpRhMe(PMe_3)[P(O)(OMe)_2]$, **8**, 195
$Rh_{11}Pt_2C_{24}O_{24}$
$[Rh_{11}Pt_2(CO)_{24}]^{3-}$, **10**, 193
$Rh_{12}C_{23}HN_2O_{23}$
$[Rh_{12}N_2H(CO)_{23}]^{3-}$, **8**, 484
$Rh_{12}C_{25}HO_{25}$
$[Rh_{12}H(CO)_{25}]^-$, **8**, 479
$Rh_{12}C_{25}H_2O_{25}$
$Rh_{12}H_2(CO)_{25}$, **8**, 479
$Rh_{12}C_{26}O_{24}$
$[Rh_{12}C_2(CO)_{24}]^{2-}$, **8**, 483
$Rh_{12}C_{27}O_{25}$
$Rh_{12}C_2(CO)_{25}$, **8**, 456, 483, 484
$Rh_{12}C_{30}O_{30}$
$[Rh_{12}(CO)_{30}]^{2-}$, **8**, 471, 473, 479
$Rh_{12}C_{56}H_{24}O_{30}P_2$
$[Rh_6(CO)_{15}]PPh_2CH_2CH_2PPh_2$-$[Rh_6(CO)_{15}]$, **8**, 474
$Rh_{12}Cu_2C_{30}O_{30}$
$Cu_2[Rh_{12}(CO)_{30}]$, **10**, 42
$Rh_{12}PtC_{24}O_{24}$
$[Rh_{12}Pt(CO)_{24}]^{4-}$, **10**, 193, 194
$Rh_{13}C_{24}H_3O_{24}$
$[Rh_{13}H_3(CO)_{24}]^{2-}$, **8**, 482
$Rh_{14}C_{25}HO_{25}$
$[Rh_{14}HO(CO)_{25}]^{3-}$, **8**, 482
$Rh_{14}C_{25}N_2O_{25}$
$[Rh_{14}N_2(CO)_{25}]^{2-}$, **8**, 484
$Rh_{14}C_{25}O_{25}$
$[Rh_{14}(CO)_{25}]^{4-}$, **8**, 480, 481
$Rh_{14}C_{26}O_{26}$
$[Rh_{14}(CO)_{26}]^{2-}$, **8**, 481
$Rh_{14}C_{35}O_{33}$
$[Rh_{14}C_2(CO)_{33}]^{2-}$, **8**, 478, 483
$Rh_{15}C_{27}O_{27}$
$[Rh_{15}(CO)_{27}]^{3-}$, **8**, 480, 481
$Rh_{15}C_{30}O_{28}$
$[Rh_{15}C_2(CO)_{28}]^-$, **8**, 483
$Rh_{15}C_{30}O_{30}$
$[Rh_{15}(CO)_{30}]^{3-}$, **8**, 481
$Rh_{17}C_{30}O_{30}$
$[Rh_{17}(CO)_{30}]^{3-}$, **8**, 481
$Rh_{17}C_{32}O_{32}S_2$
$[Rh_{17}S_2(CO)_{32}]^{3-}$, **8**, 486
$Rh_{18}Pt_4C_{35}O_{35}$
$[Rh_{18}Pt_4(CO)_{35}]^{4-}$, **10**, 188, 193, 194
$Rh_{22}C_{37}O_{37}$
$[Rh_{22}(CO)_{37}]^{4-}$, **8**, 481
$Rh_{55}C_{144}H_{324}Cl_{20}P_{12}$
$Rh_{55}(PBu^t{}_3)_{12}Cl_{20}$, **8**, 486
$RuAlC_{13}H_{28}P$
$Ru(PMe_3)Cp^*[AlH_4]$, **1**, 551
$RuAlOs_3C_{12}H_3O_{12}$
$[Al][H_3RuOs_3(CO)_{12}]$, **10**, 52
$RuAuB_2C_{63}H_{61}F_8P_4$
$[RhAu[HC(PPh_2)_3](cod)(PPh_3)][BF_4]_2$, **10**, 40
$RuAuC_{73}H_{62}F_6OP_5$
$[RuAu(H)_2(CO)(PPh_3)_4](PF_6)$, **10**, 41
$RuAuCo_3C_{30}H_{15}O_{12}P$
$RuCo_3(AuPPh_3)(CO)_{12}$, **10**, 48
$RuAuCo_3C_{48}H_{30}O_{12}P_2$
$[Au(PPh_3)_2][RuCo_3(CO)_{12}]$, **10**, 48
$RuAuRh_3C_{30}H_{15}O_{12}P$
$RuRh_3(AuPPh_3)(CO)_{12}$, **7**, 685

$RuBC_7H_5F_4O_4$
[$Ru(CO)_4(C_3H_5)$]BF_4, **7**, 459
$RuBC_9H_7F_4O_3$
[$Ru(CO)_3(C_6H_7)$]BF_4, **7**, 455
$RuBC_9H_8F_4NO_2$
[Ru(NCMe)$(CO)_2$Cp]BF_4, **7**, 483
$RuBC_9H_9F_4S$
[$Ru(C_4H_4S)$Cp]BF_4, **7**, 523, 538
$RuBC_9H_{32}P_3$
$RuH(BH_4)(PMe_3)_3$, **7**, 381
$RuBC_{10}H_{11}F_4N_2O$
[$Ru(NCMe)_2$(CO)Cp]BF_4, **7**, 483
$RuBC_{11}H_{17}F_4$
[$Ru(C_2H_4)_3$Cp]BF_4, **7**, 505
$RuBC_{12}H_{20}ClF_4O$
[$RuCl(H_2O)(C_{12}H_{18})$]BF_4, **7**, 462
$RuBC_{13}H_{15}F_4O_3$
[$Ru(CO)_3$(Cp*)]BF_4, **7**, 482
$RuBC_{14}H_{13}F_4O$
[Ru(CO)(cot)Cp]BF_4, **7**, 484
$RuBC_{14}H_{17}F_4O$
[Ru(CO)(cod)Cp]BF_4, **7**, 484
$RuBC_{15}H_{23}F_4O$
[$Ru(CO)(C_7H_{11})(C_7H_{12})$]$BF_4$, **7**, 466
$RuBC_{16}H_{13}F_4O_2$
[$Ru(CH_2CCHPh)(CO)_2$Cp]BF_4, **7**, 390
$RuBC_{16}H_{23}$
(Cp*)Ru(CpBMe), **1**, 209
$RuBC_{18}H_{20}P$
$(Ph_3P)Ru(H)(BH_4)$, **1**, 284, 285
$RuBC_{21}H_{26}P$
(ttp)Ru(H)(BH_4), **1**, 284, 285
$RuBC_{23}H_{35}F_4$
[$RuCp(C_6Et_6)$]BF_4, **7**, 531
$RuBC_{24}H_{22}F_4O_2P$
[$Ru(CO)(PPh_3)(H_2O)$Cp]BF_4, **7**, 493
$RuBC_{26}H_{13}O_{12}$
$HRu(CO)_{12}$B(H)C(Ph)CHPh, **1**, 301, 303
$RuBC_{26}H_{24}F_4OP$
[$RuMe(CO)(PPh_3)(C_6H_6)$]BF_4, **7**, 571
$RuBC_{26}H_{26}F_4O_2P$
[$Ru(CO)(PPh_3)$(EtOH)Cp]BF_4, **7**, 493
$RuBC_{28}H_{28}F_4O_2P$
[$Ru(CO)(PPh_3)$(THF)Cp]BF_4, **7**, 493
$RuBC_{32}H_{36}F_4P$
[$RuH(CH_2CHC_6H_4PPh_2)$-(C_6Me_6)]BF_4, **7**, 571
$RuBC_{34}H_{25}F_4O$
[$Ru(CO)(C_4Ph_4)$Cp]BF_4, **7**, 474
$RuBC_{38}H_{32}F_5O_3P_2$
[$RuF(OH_2)(CO)_2(PPh_3)_2$]BF_4, **7**, 308
$RuBC_{38}H_{33}F_4O_3P_2$
[$RuH(CO)_2(OH_2)(PPh_3)_2$]BF_4, **7**, 314, 315
$RuBC_{40}H_{33}F_4O_3P_2S$
[$Ru(SMe)(CO)_3(PPh_3)_2$]BF_4, **7**, 317
$RuBC_{41}H_{55}P_2$
[$RuMe(PMe_3)_2[C_5Me_4(CH_2)]$]-$BPh_4$, **7**, 502
$RuBC_{46}H_{39}F_4P_2S$
[$Ru(C_5H_4CH_2C_4H_3S)(PPh_3)_2$]-$BF_4$, **7**, 523
$RuBC_{47}H_{35}F_4$
[$RuCp(C_6Ph_6)$]BF_4, **7**, 531
$RuB_2C_7H_{10}O_2$
$Ru(B_2H_5)(CO)_2$Cp, **7**, 487
$RuB_2C_{10}H_{21}F_8O_3$
[$RuCp^*(H_2O)_3$][BF_4]$_2$, **7**, 668
$RuB_2C_{13}H_{31}P$
$Cp^*Ru(PMe_3)(B_2H_7)$, **1**, 304, 305
$RuB_2C_{15}H_{30}F_8P_2S$
[$Ru[CCH(SMe_2)](PMe_3)_2$Cp]-$(BF_4)_2$, **7**, 517
$RuB_2C_{20}H_{34}N_2$
$(C_6Me_6)Ru[C_4H_4B_2(NMe_2)_2]$, **1**, 209
$RuB_2C_{24}H_{20}F_8$
[$Ru(PhPh)_2$][BF_4]$_2$, **7**, 550
$RuB_2C_{38}H_{30}F_8O_2P_2$
$Ru(BF_4)_2(CO)_2(PPh_3)_2$, **7**, 308, 309
$RuB_3C_{37}H_{38}ClOP_2$
$RuCl(B_3H_8)(Co)(PPh_3)_2$, **1**, 315
$RuB_3C_{37}H_{39}OP_2$
$RuH(B_3H_8)(CO)(PPh_3)_2$, **1**, 313–15
$RuB_4C_{10}H_{25}$
$Cp^*RuHB_4H_9$, **1**, 319, 320
$RuB_4C_{12}H_{26}$
$(C_6Me_6)RuB_4H_8$, **1**, 320
$RuB_4C_{37}H_{38}OP_2$
$(CO)(PPh_3)_2RuB_4H_8$, **1**, 320, 321, 323
$RuB_5C_{12}H_{27}$
$(C_6Me_6Ru)B_5H_9$, **1**, 324
$RuB_5C_{37}H_{39}OP_2$
$(CO)(PPh_3)_2RuB_5H_9$, **1**, 323
$RuB_8C_6H_{20}$
$(C_6H_6)RuB_8H_{14}$, **1**, 332, 333
$RuB_9C_{10}H_{23}$
$(MeC_6H_4Pr^i)[RuB_9H_9]$, **1**, 346
$RuB_9C_{10}H_{24}ClS$
(cymene)$RuSB_9H_{10}Cl$, **1**, 269
$RuB_9C_{10}H_{24}N$
$(MeC_6H_4Pr^i)RuNHB_9H_9$, **1**, 263
$RuB_9C_{12}H_{27}$
$(C_6Me_6)[RuB_9H_9]$, **1**, 338, 344, 347
$RuB_9C_{12}H_{31}$
$(C_6Me_6)[RuB_9H_{13}]$, **1**, 338, 340–2, 344, 347
$RuB_9C_{13}H_{33}O$
$(C_6Me_6)(OMe)RuB_9H_{12}$, **1**, 338
$RuB_9C_{14}H_{31}O$
$(C_6Me_6)[RuB_9H_8](OEt)$, **1**, 346
$RuB_9C_{15}H_{36}NO$
(C_6Me_6)(OMe)RuNMeC(H)-B_9H_{11}, **1**, 338
$RuB_9C_{15}H_{37}N$
$(C_6Me_6)Ru(NMe_2CH_2)$, **1**, 338
$RuB_9C_{20}H_{40}P$
$(C_6Me_6)[RuB_9H_{11}](PMe_2Ph)$, **1**, 340, 341
$RuB_9C_{22}H_{45}NP$
$(NMe_2)(PMe_2Ph)(C_6Me_6)$-$RuB_9H_{10}$, **1**, 338
$RuB_9C_{30}H_{44}P$
$(C_6Me_6)[RuB_9H_{11}](PPh_3)$, **1**, 340
$RuB_9C_{54}H_{53}Cl$
$(PPh_3)HCl[RuB_9H_7](PPh_3)_2$, **1**, 346
$RuB_9C_{54}H_{53}P_3$
$(PPh_3)HCl[RuB_9H_7](PPh_3)_2$, **1**, 345, 346
$RuB_{10}C_{10}H_{24}$
$(MePr^iC_6H_4)RuB_{10}H_{10}$, **1**, 357–9
$RuB_{10}C_{12}H_{31}$
$(C_6Me_6)[RuB_{10}H_{13}]^-$, **1**, 350, 351
$RuB_{10}C_{22}H_{29}O_5P$
$(PPh_3)(MeCO_2)_2RuB_{10}H_7(OH)$, **1**, 357
$RuB_{10}C_{40}H_{48}O_2P_2$
$(PPh_3)_2RuB_{10}H_8(OEt)_2$, **1**, 353, 355
$RuB_{12}C_{24}H_{45}P_3$
$(PMe_2Ph)_3RuB_{12}H_{12}$, **1**, 363
$RuB_{12}C_{42}H_{56}ClNP_2$
$P(PPh_3)_2ClRuB_{12}H_{11}(NEt_3)$, **1**, 363
$RuC_2H_{14}O_5$
$[Ru(H_2O)_5(C_2H_4)]^{2+}$, **7**, 445
RuC_3H_9ClN
$[Ru(N)ClMe_3]^-$, **7**, 373
RuC_3NO_4
$[Ru(CO)_3(NO)]^-$, **7**, 300, 301
RuC_4HO_4
$[RuH(CO)_4]^-$, **7**, 363
$RuC_4H_2O_4$
$RuH_2(CO)_4$, **7**, 362
$RuC_4H_{16}O_4$
$[Ru(H_2O)_4(C_2H_4)_2)_2]^{2+}$, **7**, 445
RuC_5H_5
$[RuCp]^+$, **7**, 534
RuC_5O_5
$Ru(CO)_5$, **7**, 836, 934
$RuC_6H_4O_4$
$Ru(CO)_4(C_2H_4)$, **7**, 302, 303, 442
$RuC_6H_{11}O_6P$
$RuH_2(CO)_3[P(OMe)_3]$, **7**, 362, 417
$RuC_6H_{12}O_3$
$[Ru(H_2O)_3(C_6H_6)]^{2+}$, **7**, 558
$RuC_7H_6O_2$
$RuH(CO)_2$Cp, **7**, 387, 479, 480
$RuC_7H_{21}O_7P_2$
$RuMe[P(O)(OMe)_2][P(OMe)_4]$, **7**, 385
$RuC_8H_6O_3$

$Ru(CHO)(CO)_2Cp$, **7**, 483
$RuC_8H_6O_4$
$Ru(CO_2H)(CO)_2Cp$, **7**, 483
$RuC_8H_6O_6$
$Ru(CO)_4(CH_2CHCO_2Me)$, **7**, 302, 303
$RuC_8H_6O_8$
$Ru(CO)_4(OAc)_2$, **12**, 415
$RuC_8H_8O_3$
$Ru(CH_2OH)(CO)_2(Cp)$, **7**, 387, 388, 483
$RuC_8H_{10}O$
$RuH(CO)(C_2H_4)Cp$, **7**, 481
$RuC_8H_{12}Cl_2OS$
$RuCl_2(DMSO)(C_6H_6)$, **7**, 561
$RuC_8H_{15}N_2O_5$
$Ru(O_2CCH_2)_2NCH_2CH_2N(CH_2CH_2OH)(CH_2CO_2)$, **7**, 448
$RuC_8H_{16}B_8$
$(C_6H_6)Ru(C_2B_8H_{10})$, **1**, 397
$RuC_8H_{17}B_7$
$(C_6H_6)Ru(C_2B_7H_{11})$, **1**, 397, 398
$RuC_9H_{18}O_3P_2$
$Ru(CO)_3(PMe_3)_2$, **7**, 368
$RuC_{10}H_8O_2$
$Ru(C_2Me)(CO)_2Cp$, **7**, 485
$Ru(CHCCH_2)(CO)_2Cp$, **7**, 390, 391
$RuC_{10}H_9F_5N_3P_3$
$N_3P_3F_5Cp(C_5H_4)Ru$, **7**, 610
$RuC_{10}H_{10}$
Cp_2Ru, **7**, 604, 605, 620
$RuC_{10}H_{10}Br$
$[RuBrCp_2]^+$, **7**, 608
$RuC_{10}H_{10}Cl$
$[RuClCp_2]^+$, **7**, 608
$RuC_{10}H_{10}I$
$[RuICp_2]I$, **7**, 618
$RuC_{10}H_{12}ClN_2$
$[RuCl(NCMe)_2(C_6H_6)]^+$, **7**, 560
$RuC_{10}H_{15}$
$[RuCp^*]^+$, **7**, 477
$RuC_{10}H_{15}Cl_2$
Cp^*RuCl_2, **7**, 604
$RuC_{10}H_{18}ClO_6P_2$
$[RuCl[PO(OMe)_2]_2(C_6H_6)]^-$, **7**, 564
$RuC_{10}H_{21}O_3$
$[Ru(H_2O)_3Cp^*]^{2+}$, **7**, 496
$RuC_{10}H_{27}Cl_2O_{10}P_3$
$RuCl_2(CO)[P(OMe)_3]_3$, **7**, 308
$RuC_{10}H_{29}O_{10}P_3$
$RuH_2(CO)[P(OMe)_3]_3$, **7**, 362, 417
$RuC_{11}H_8F_6O_8$
$Ru(O_2CCF_3)_2(CO)_3(THF)$, **7**, 316
$RuC_{11}H_8O_3$
$Ru(CO)_3[(CH_2)_2C_6H_4]$, **7**, 455
$RuC_{11}H_9F_3N_3P$
$[Ru(CNMe)_3Cp]PF_6$, **7**, 483
$RuC_{11}H_{10}O$
$Cp(C_5H_4)RuCHO$, **7**, 610
$RuC_{11}H_{10}O_2$
$Cp(C_5H_4)RuCO_2H$, **7**, 619
$RuC_{11}H_{11}O_2$
$[Ru(CHC_3H_5)(CO)_2Cp]^+$, **7**, 485
$RuC_{11}H_{11}O_3$
$[Ru(CO)_3(C_8H_{11})]^+$, **7**, 455
$RuC_{11}H_{12}O_3$
$Ru(CO)_3(cod)$, **7**, 452
$RuC_{11}H_{14}F_6N_3P$
$[Ru(CNMe)_3Cp]PF_6$, **7**, 495
$[Ru(NCMe)_3Cp]PF_6$, **7**, 495
$RuC_{11}H_{14}O_2$
$Ru[C(OMe)CH_2CH_2CH_2](CO)Cp$, **7**, 486
$RuC_{11}H_{15}S$
$[Ru(C_7H_{11})(C_4H_4S)]^+$, **7**, 466
$RuC_{11}H_{16}O$
$RuH(CO)(PrCHCH_2)Cp$, **7**, 481
$RuC_{11}H_{17}Cl_2NO$
$RuCl(CH_2Cl)(NO)Cp^*$, **7**, 388, 389, 489
$RuC_{11}H_{20}P$
$[Ru(CH_2CH_2H)(PMe_3)(C_6H_6)]^+$, **7**, 572
$[RuH(C_2H_4)(PMe_3)(C_6H_6)]^+$, **7**, 572
$RuC_{11}H_{23}Cl_2F_6P_3$
$[RuCl_2(PMe_3)_2Cp]PF_6$, **7**, 496
$RuC_{11}H_{32}O_3P_4S$
$RuH(O_3SMe)(dmpm)_2$, **7**, 381
$RuC_{11}H_{32}P_4$
$RuH(Me)(dmpm)_2$, **7**, 381, 383
$RuC_{12}H_6N_2O_2S_2$
$Ru(SC_5H_3N)_2(CO)_2$, **7**, 314, 315
$RuC_{12}H_8Cl_2O_2$
$Ru(C_5H_4)_2(COCl)_2$, **7**, 613, 614
$RuC_{12}H_{10}$
$Cp(C_5H_4)RuC_2H$, **7**, 610
$RuC_{12}H_{12}NO$
$[Ru(NCMe)(C_5H_4O)Cp]^+$, **7**, 609
$RuC_{12}H_{12}N_2O_8$
$Ru(CO)_4(OAc)_2(MeCN)_2$, **12**, 415
$RuC_{12}H_{12}O_2$
$Cp_2Ru(CHO)_2$, **7**, 610
$RuC_{12}H_{12}O_4$
$Cp_2Ru(CO_2H)_2$, **7**, 610
$RuC_{12}H_{13}$
$[RuCp(PhMe)]^+$, **7**, 533
$RuC_{12}H_{13}F_6P$
$[Ru(chpt)Cp]PF_6$, **7**, 507
$RuC_{12}H_{14}$
$Ru(C_6H_6)(chd)$, **7**, 576
$RuC_{12}H_{14}Cl_2O_2$
$RuCl(CO)_2(C_5Me_4CH_2Cl)$, **7**, 478
$RuC_{12}H_{14}O$
$RuCp(C_5H_4)CH(OH)Me$, **7**, 612
$RuC_{12}H_{15}BrO_2$
$RuBr(CO)_2Cp^*$, **7**, 487
$RuC_{12}H_{15}ClO_2$
$RuCl(CO)_2Cp^*$, **7**, 478
$RuC_{12}H_{15}Cl_2O_2P$
$Ru(PCl_2)(CO)_2Cp^*$, **7**, 487
$RuC_{12}H_{15}F_{12}N_3P_2$
$[Ru(NCMe)_3(C_6H_6)](PF_6)_2$, **7**, 559
$RuC_{12}H_{15}I$
$RuI(C_6H_7)(C_6H_8)$, **7**, 466
$RuC_{12}H_{15}O_2$
$[Ru(CO)_2Cp^*]^-$, **7**, 479
$RuC_{12}H_{15}O_2P_3$
$Ru(CO)_2Cp^*(P_3)$, **7**, 487
$RuC_{12}H_{16}$
$Ru(C_7H_{11})Cp$, **7**, 511
$RuC_{12}H_{16}O_5S$
$Ru(SO_3H)(CO)_2Cp^*$, **7**, 488
$RuC_{12}H_{17}O_2$
$[Ru(H_2)(CO)_2Cp^*]^+$, **7**, 480
$RuC_{12}H_{17}O_3$
$[Ru(CO)_2(H_2O)Cp^*]^+$, **7**, 482
$RuC_{12}H_{18}Cl_2$
$RuCl_2(C_{12}H_{18})$, **7**, 462
$RuC_{12}H_{18}O_2$
$Ru(C_6H_9O)_2$, **7**, 465
$RuC_{12}H_{18}O_4$
$Ru(OAc)_2(cod)$, **7**, 452
$RuC_{12}H_{19}ClO_2$
$RuCl(O_2CMe)(C_{10}H_{16})$, **7**, 461
$RuC_{12}H_{19}Cl_2NO$
$Ru(CH_2Cl)_2(NO)Cp^*$, **7**, 388, 389, 489
$RuC_{12}H_{20}Cl_2OS$
$RuCl_2(SOMe_2)(C_5Me_4CH_2)$, **7**, 478
$RuC_{12}H_{20}Cl_4S$
$RuCl_3(SMe_2)(C_5Me_4CH_2Cl)$, **7**, 478
$RuC_{12}H_{24}ClF_6P_3$
$[RuCl(PMe_3)_2(C_6H_6)]PF_6$, **7**, 563
$RuC_{12}H_{34}P_4$
$Ru(CH_2PMe_2)_2(PMe_3)_2$, **7**, 391
$RuC_{12}H_{35}ClP_4$
$RuCl(CH_2PMe_2)(PMe_3)_3$, **7**, 391
$RuC_{12}H_{36}P_4$
$RuH(CH_2PMe_2)(PMe_3)_3$, **7**, 381, 428
$RuC_{12}H_{37}ClP_4$
$RuClH(PMe_3)_4$, **7**, 381, 391
$RuC_{12}H_{38}P_4$
$RuH_2(PMe_3)_4$, **7**, 381, 391
$RuC_{13}H_{13}F_6P$
$[Ru(cot)Cp]PF_6$, **7**, 507
$RuC_{13}H_{15}F_6P$
$[Ru(C_8H_{10})Cp]PF_6$, **7**, 507
$RuC_{13}H_{16}O$
$Ru(CO)(C_6H_7)(C_6H_9)$, **7**, 454
$Ru(CO)(chd)_2$, **7**, 454
$RuC_{13}H_{16}O_4$
$Ru(CO_2H)(CO)_2Cp^*$, **7**, 483
$RuC_{13}H_{17}Cl$
$RuCl(cod)Cp$, **7**, 490, 495, 530
$RuC_{13}H_{18}$
$RuH(cod)Cp$, **7**, 506
$RuC_{13}H_{18}O_2$
$RuMe(CO)_2Cp^*$, **7**, 480

$RuC_{13}H_{18}O_3$
Ru(CH_2OH)$(CO)_2$Cp*, **7**, 483
$RuC_{14}H_{11}O_2$
[Ru(C_7H_6)$(CO)_2$Cp]$^+$, **7**, 486
$RuC_{14}H_{12}O_2$
Ru(C_7H_7)$(CO)_2$Cp, **7**, 480
$RuC_{14}H_{17}F_3O_3S$
[Ru(C_8H_{12})Cp]OTf, **7**, 508
$RuC_{14}H_{18}F_6O_5$
Ru(O_2CCF_3)$_2$(H_2O)($C_{10}H_{16}$), **7**, 461
$RuC_{14}H_{18}O_2$
Ru$(CO)_2$(C_6Me_6), **7**, 306, 568
$RuC_{14}H_{18}O_3S$
Ru(OTs)(C_7H_{11}), **7**, 578
$RuC_{14}H_{19}F_6O_3P$
[Ru(CHOMe)$(CO)_2$Cp]PF_6, **7**, 485
$RuC_{14}H_{20}O_3$
Ru(CH_2OMe)$(CO)_2$Cp*, **7**, 387
$RuC_{14}H_{21}NO$
Ru(C_4H_6)(NO)Cp*, **7**, 489
$RuC_{14}H_{22}$
Ru(C_7H_{11})$_2$, **7**, 465
$RuC_{14}H_{26}P_2S$
Ru(C_2SMe)(PMe_3)$_2$Cp, **7**, 517
$RuC_{14}H_{27}B_7$
(C_6Me_6)Ru($C_2B_7H_9$), **1**, 397
$RuC_{14}H_{27}ClO_2S_2$
RuCl(DMSO)$_2$Cp*, **7**, 524
$RuC_{14}H_{29}P_2$
RuEt(PMe_3)$_2$(C_6H_6), **7**, 572
$RuC_{14}H_{33}B_9N$
(C_6Me_6)Ru(MeNCB_9H_{12}), **1**, 380
$RuC_{14}H_{34}O_4P_4$
Ru(O_2CMe)$_2$(dmpm)$_2$, **7**, 381
$RuC_{14}H_{36}ClP_3$
RuCl(PMe_3)$_3$(C_5H_9), **7**, 467
$RuC_{14}H_{36}P_3$
[Ru(PMe_3)$_3$(C_5H_9)]$^+$, **7**, 467
$RuC_{14}H_{36}P_4$
Ru(dmpe)$_2$(C_2H_4), **7**, 444
$RuC_{14}H_{37}F_6N_6P$
[RuH(NH_2NMe_2)$_3$(cod)]PF_6, **7**, 451
$RuC_{14}H_{40}P_4$
Ru(PMe_3)$_4$(C_2H_4), **7**, 444
$RuC_{15}H_{10}ClNP_3$
[Ru(CCCPh_2)Cl(NP_3)]$^+$, **7**, 353
$RuC_{15}H_{10}F_{20}P_2$
RuH(dfepe)Cp, **7**, 498
$RuC_{15}H_{15}Br_5$
RuCp*(C_5Br_5), **7**, 611
$RuC_{15}H_{15}F_5$
Ru(Cp*)(C_5F_5), **7**, 604
$RuC_{15}H_{18}O_3S$
Ru(OTs)(C_8H_{11}), **7**, 578
$RuC_{15}H_{19}$
[RuCp($C_5Me_4CH_2$)]$^+$, **7**, 612
$RuC_{15}H_{20}$
RuCpCp*, **7**, 604, 611, 612
$RuC_{15}H_{23}O_2$
Ru(acac)Cp*, **7**, 497
$RuC_{15}H_{24}F_3O_4PS$
Ru(OTf)(CO)(PMe_3)Cp*, **7**, 493
$RuC_{15}H_{24}O$
RuMe_2(CO)(C_6Me_6), **7**, 569
$RuC_{15}H_{26}O_3P_2$
RuH(OC_6H_4Me)$(CO)_2$(PMe_3)$_2$, **7**, 368
$RuC_{15}H_{29}F_6P_2S_2$
[Ru[CC(SMe)$_2$](PMe_3)$_2$Cp]PF_6, **7**, 517
$RuC_{15}H_{29}P$
RuH_2(PPr^i_3)(C_6H_6), **7**, 573
$RuC_{15}H_{29}P_2S$
[Ru(CCSMe_2)(PMe_3)$_2$Cp]$^+$, **7**, 416
$RuC_{15}H_{29}P_2S_2$
[Ru[CC(SMe)$_2$](PMe_3)$_2$Cp]$^+$, **7**, 353
$RuC_{15}H_{30}OP_2S$
Ru[C(OMe)CH(SMe)]-(PMe_3)$_2$Cp], **7**, 406
$RuC_{15}H_{42}P_6$
Ru(dmpm)$_3$, **7**, 428
$RuC_{16}H_{12}O_2$
Ru(CH_2CCPh)$(CO)_2$Cp, **7**, 390, 391, 481
$RuC_{16}H_{20}F_3NO_3S$
[RuCp*(CpN)]OTf, **7**, 538
$RuC_{16}H_{21}$
[Ru(C_6H_6)Cp*]$^+$, **7**, 507, 530
$RuC_{16}H_{22}$
Ru(C_6H_7)Cp*, **7**, 509
Ru(C_8H_{10})(cod), **7**, 557, 575
$RuC_{16}H_{24}$
Ru(C_6H_9)Cp*, **7**, 511
$RuC_{16}H_{24}ClN_3$
[Ru(NCMe)$_3$Cp*]Cl, **7**, 477
$RuC_{16}H_{24}N_3$
[Ru(NCMe)$_3$Cp*]$^+$, **7**, 530, 604
$RuC_{16}H_{24}O$
Ru(OC_6H_9)Cp*, **7**, 512
$RuC_{16}H_{24}OP_2$
Ru(OH)(PMe_3)$_2$Cp*, **7**, 493
$RuC_{16}H_{24}S_2$
[Ru(C_4Me_4S)$_2$]$^{2+}$, **7**, 556
$RuC_{16}H_{26}$
Ru(C_8H_{13})$_2$, **7**, 465
Ru(MeCHCH_2)(C_3H_5)Cp*, **7**, 509
$RuC_{16}H_{27}$
[Ru(CH_2CH_2H)(C_2H_4)-(C_6Me_6)]$^+$, **7**, 577
$RuC_{16}H_{27}O_4P$
Ru$(CO)_4$(PBu_3), **7**, 642
$RuC_{16}H_{32}S_3$
[Ru[MeSCC(MeS)$_2$](PMe_3)$_2$-Cp]$^{2+}$, **7**, 406
$RuC_{16}H_{33}ClP_2$
RuCl(PMe_3)$_2$Cp*, **7**, 494
$RuC_{16}H_{34}P_2S$
Ru(SH)(PMe_3)$_2$Cp*, **7**, 522
$RuC_{16}H_{34}P_4$
RuH(Ph)(dmpm)$_2$, **7**, 381, 383, 396
$RuC_{16}H_{37}Cl_2NOP_2$
RuCl_2(CHNMe_2)(CO)(PEt_3)$_2$, **7**, 345
$RuC_{16}H_{38}N_2P_2$
RuH_2(CNBu^t)$_2$(PMe_3)$_2$, **7**, 335
$RuC_{16}H_{48}IO_{15}P_5$
[RuMe[P(OMe)$_3$]$_5$]I, **7**, 385
$RuC_{17}H_{16}$
Ru($C_{10}H_8$)(nbd), **7**, 576
$RuC_{17}H_{19}Cl$
RuCl(cod)(C_9H_7), **7**, 506
$RuC_{17}H_{22}O_3S$
RuCp*(OTs), **7**, 531
$RuC_{17}H_{23}$
RuCp*(nbd), **7**, 491
$RuC_{17}H_{23}O_3$
[Ru$(CO)_2$(CH_2CHCOEt)Cp*]$^+$, **7**, 484
$RuC_{17}H_{24}O$
RuCp*(C_5H_4)CHMe(OH), **7**, 611
$RuC_{17}H_{26}OP_2$
Ru($C_{10}H_8$)(CO)(PMe_3)$_2$, **7**, 656
$RuC_{17}H_{28}$
RuPr(C_4H_6)Cp*, **7**, 510
$RuC_{17}H_{35}O_2P_3$
RuH(OC_6H_4Me)(CO)(PMe_3)$_3$, **7**, 368, 370
$RuC_{17}H_{36}O_7P_2$
Ru(OMe)[P(OMe)$_3$]$_2$Cp*, **7**, 497
$RuC_{17}H_{36}P_2S_2$
[Ru[CC(SEt_2)SMe](PMe_3)$_2$-Cp]$^{2+}$, **7**, 416
$RuC_{17}H_{38}P_4$
RuH($C_6H_4PMe_2$)(PMe_3)$_3$, **7**, 428
$RuC_{18}H_{14}$
Ru(C_9H_7)$_2$, **7**, 618
$RuC_{18}H_{14}ClNOP_3$
[Ru[CCC(OMe)CHCPh_2]Cl-(NP_3)]$^+$, **7**, 353
$RuC_{18}H_{20}$
Ru($C_{10}H_8$)(cod), **7**, 454, 557, 576
$RuC_{18}H_{20}N_6$
[Ru(bipy)(CNMe)$_4$]$^{2+}$, **7**, 331
$RuC_{18}H_{22}$
Ru(C_8H_{10})(cod), **7**, 578
$RuC_{18}H_{23}Cl_3O_4S_2$
[RuCl_2(THT)$_2$Cp*]ClO_4, **7**, 523
$RuC_{18}H_{24}$
Ru(C_8H_9)Cp*, **7**, 509
$RuC_{18}H_{24}O_2P_2$
Ru$H_2$$(CO)_2$($PMe_2$Ph)$_2$, **7**, 368
$RuC_{18}H_{24}S_2$
Ru(C_6Me_4S)(C_4Me_4S), **7**, 556
$RuC_{18}H_{25}$
[Ru(C_8H_{10})Cp*]$^+$, **7**, 507
$RuC_{18}H_{26}$
Ru($C_5Me_4CH_2$)(cod), **7**, 508
Ru(C_8H_{11})Cp*, **7**, 507
$RuC_{18}H_{27}N$
RuCp*(C_4Me_4N), **7**, 604
$RuC_{18}H_{28}$
RuH(cod)Cp*, **7**, 507

$RuC_{18}H_{33}ClF_6OP_2$
[RuCl[C(OMe)Me](PMe$_3$)-(C_6Me_6)]PF_6, **7**, 579
[RuCl(PMe$_3$)[C(OCH$_2$CH$_2$-CH$_2$)](C$_6$Me$_6$)]PF$_6$, **7**, 581
$RuC_{18}H_{36}P_2$
Ru(PMe$_3$)$_2$(C$_6$Me$_6$), **7**, 572
$RuC_{18}H_{39}O_9P_3$
Ru[PO(OMe)$_2$][P(OMe)$_3$]$_2$Cp*, **7**, 497
$RuC_{18}H_{40}P_4$
Ru(C$_6$H$_4$)(PMe$_3$)$_4$, **7**, 377, 396, 425
$RuC_{18}H_{46}OP_4$
Ru[CHC(Bu^t)O](PMe$_3$)$_4$, **7**, 427
$RuC_{18}H_{46}P_4$
RuPh$_2$(PMe$_3$)$_4$, **7**, 396
$RuC_{19}H_{16}O$
Cp(C$_5$H$_4$)RuCH=CHCOPh, **7**, 610
$RuC_{19}H_{19}F_6N_2OP$
[Ru(EtOH)(phen)Cp]PF$_6$, **7**, 495
$RuC_{19}H_{26}Cl_2OP_2$
RuCl$_2$(PMe$_2$Ph)$_2$(CO)(C$_2$H$_4$), **7**, 448
$RuC_{19}H_{28}O_3$
Ru(CO)$_3$(C$_8$H$_{14}$)$_2$, **7**, 442
$RuC_{19}H_{30}NP$
Ru(NHPh)(PMe$_3$)Cp*, **7**, 478
$RuC_{19}H_{30}O$
Ru(CO)(CH$_2$CHBu^t)(C$_6$Me$_6$), **7**, 569
$RuC_{19}H_{32}P_2$
Ru(C$_2$C$_6$H$_9$)(PMe$_3$)$_2$Cp, **7**, 517
$RuC_{19}H_{33}F_6P_3$
[Ru(CCCPh$_2$)(PMe$_3$)$_2$Cp]PF$_6$, **7**, 517
$RuC_{19}H_{41}NO_2P_4$
Ru(C$_6$H$_4$NHCO$_2$)(PMe$_3$)$_4$, **7**, 425
Ru(NPhCO$_2$)(PMe$_3$)$_4$, **7**, 425
$RuC_{19}H_{42}OP_2$
Ru(CO)(PBu$^t_2$Me)$_2$, **7**, 459
$RuC_{19}H_{42}OP_4$
Ru(CH$_2$OC$_6$H$_4$)(PMe$_3$)$_4$, **7**, 396
$RuC_{19}H_{44}P_4$
RuH(Bz)(PMe$_3$)$_4$, **7**, 381
$RuC_{19}H_{45}ClOP_2$
RuH(H$_2$)Cl(CO)(PPr$^i_3$)$_2$, **7**, 371
$RuC_{19}H_{46}OP_2$
RuH$_2$(H$_2$)(CO)(PPr$^i_3$)$_2$, **7**, 371
$RuC_{19}H_{46}O_3P_4$
Ru[CHC(Bu^t)OCO$_2$](PMe$_3$)$_4$, **7**, 427
$RuC_{20}H_{17}N_2$
Cp(C$_5$H$_4$)Ru(bipy), **7**, 611
$RuC_{20}H_{18}F_6OP_2$
[Ru(CO)(Ph$_2$PCHCH$_2$)Cp]PF$_6$, **7**, 509
$RuC_{20}H_{18}F_{12}O_2P_2$
[RuCp(C$_5$H$_4$O)]$_2$[PF$_6$]$_2$, **7**, 609
$RuC_{20}H_{26}O_2P_2$
Ru(C$_2$H$_4$)(CO)$_2$(PMe$_2$Ph)$_2$, **7**, 368
$RuC_{20}H_{27}ClOP_2$
RuCl(CO)(PMe$_2$Ph)$_2$(C$_3$H$_5$), **7**, 459
$RuC_{20}H_{27}O_2P$
Ru(CHO)(CO)(PMe$_2$Ph)Cp*, **7**, 483
$RuC_{20}H_{28}$
[RuCp*C$_5$Me$_3$(CH$_2$)$_2$]$^{2+}$, **7**, 613
$RuC_{20}H_{29}Cl_2P$
RuCl$_2$(PMe$_2$Ph)(C$_6$Me$_6$), **7**, 571
$RuC_{20}H_{30}$
Cp*$_2$Ru, **7**, 604
Ru(C$_6$Me$_6$)(cod), **7**, 577
[RuCp*$_2$]$^+$, **7**, 608, 609, 612
$RuC_{20}H_{30}Br$
[RuBrCp*$_2$]$^+$, **7**, 608
$RuC_{20}H_{33}ClN_2$
RuCl(CNBu^t)$_2$Cp*, **7**, 495
$RuC_{20}H_{36}Br_2N_4$
RuBr$_2$(CNBu^t)$_4$, **7**, 336
$RuC_{20}H_{37}ClO_2P$
[Ru[C(OMe)CH$_2$CH$_2$OMe]Cl-(PMe$_3$)(C$_6$Me$_6$)]$^+$, **7**, 344
$RuC_{20}H_{37}P$
Ru(PEt$_3$)(cod)(C$_6$H$_{10}$), **7**, 460
$RuC_{20}H_{42}OP_4$
Ru[C(O)C$_6$H$_4$CH$_2$)(PMe$_3$)$_4$], **7**, 425
$RuC_{20}H_{42}O_2P_4$
Ru(CH$_2$C$_6$H$_4$CO$_2$)(PMe$_3$)$_4$, **7**, 425
$RuC_{20}H_{44}O_2P_2$
RuH$_2$(CO)$_2$(PPr$^i_3$)$_2$, **7**, 366
$RuC_{21}H_{15}O_3P$
Ru(CO)$_3$(PPh$_3$), **7**, 302
$RuC_{21}H_{16}BrN_4O$
[RuBr(CO)(bipy)$_2$]$^+$, **7**, 312
$RuC_{21}H_{16}Cl_2N_4O$
[RuCl(CO)(bipy)$_2$]Cl, **7**, 312
$RuC_{21}H_{17}Br$
Ru(CPhCHCHCPh)BrCp, **7**, 425
$RuC_{21}H_{17}Cl$
RuCl(C$_4$H$_2$Ph$_2$)Cp, **7**, 342
$RuC_{21}H_{17}ON_4$
[RuH(CO)(bipy)$_2$]$^+$, **7**, 368
$RuC_{21}H_{24}O$
RuCp*(C$_5$H$_4$)(O)Ph, **7**, 612
$RuC_{21}H_{39}P$
RuH(CH$_2$CHMePPr$^i_2$)(C$_6$Me$_6$), **7**, 574
$RuC_{21}H_{42}OP_4$
Ru(OCMeBz)(dmpe)$_2$, **7**, 427
$RuC_{21}H_{43}F_3O_4P_2S$
[Ru(OEt$_2$)(PMe$_3$)$_2$Cp*]OTf, **7**, 493
$RuC_{21}H_{47}OP_4$
Ru(OCMeBz)(PMe$_3$)$_4$, **7**, 427
$RuC_{22}H_{19}NO_2P$
[RuH(CO)$_2$(NCMe)(PPh$_3$)]$^+$, **7**, 366
$RuC_{22}H_{25}NO$
RuCp*(NO)Ph$_2$, **7**, 667
$RuC_{22}H_{34}O_{10}S_2$
[Ru(H$_2$O)$_4$(cod)](OTs)$_2$, **7**, 452
$RuC_{22}H_{36}F_6N_4O_4$
Ru(O$_2$CF$_3$)$_2$(CNBu^t)$_4$, **7**, 329
$RuC_{22}H_{38}P_4$
RuPh$_2$(dmpm)$_2$, **7**, 396
$RuC_{22}H_{42}ClP$
RuCl(PBu$^t_3$)Cp*, **7**, 496
$RuC_{22}H_{42}N_4O_4$
Ru(O$_2$Me)$_2$(CNBu^t)$_4$, **7**, 328, 329
$RuC_{22}H_{46}O_2P_4$
Ru[C(O)CH$_2$CPhMeO](PMe$_3$)$_4$, **7**, 427
$RuC_{23}H_{19}F_6P$
[RuCp(C$_5$H$_4$)CPh$_2$][PF$_6$], **7**, 612
$RuC_{23}H_{19}N_4$
[Ru(bipy)$_2$(CNMe)(CN)]$^+$, **7**, 331
$RuC_{23}H_{21}O_3P$
Ru[C(O)Et](OH)(CO)$_2$(PPh$_3$), **7**, 419
$RuC_{23}H_{22}$
Ru(C$_6$H$_3$Me$_3$)(C$_{14}$H$_{10}$), **7**, 550
$RuC_{23}H_{23}P$
RuH$_3$(PPh$_3$)Cp, **7**, 500
$RuC_{23}H_{24}N_2O$
Ru(C$_6$H$_4$NNPh)(CO)Cp*, **7**, 503
RuCp*(C$_6$H$_4$NNPh)(CO), **7**, 656
$RuC_{23}H_{28}Cl_2O_5P_2$
Ru[C(CO$_2$Me)CClC-(OMe)O]Cl(CO)(PMe$_2$-Ph)$_2$, **7**, 404
$RuC_{23}H_{28}P_2$
Ru(CCH)(PMe$_2$Ph)$_2$Cp, **7**, 415
$RuC_{23}H_{35}Cl$
RuCl(PPh$_3$)$_2$Cp, **7**, 509
$RuC_{23}H_{41}O_2P_3$
RuMe(PMe$_2$)(OC$_6$H$_4$Me)$_2$-(PMe$_3$)$_2$, **7**, 429
$RuC_{24}H_{18}$
[Ru(C$_6$Me$_6$)$_2$]$^{2+}$, **7**, 550
$RuC_{24}H_{20}ClOP$
RuCl(CO)(PPh$_3$)Cp, **7**, 491
$RuC_{24}H_{22}OP$
[Ru(H$_2$)(CO)(PPh$_3$)Cp]$^+$, **7**, 499
$RuC_{24}H_{24}N_4$
[Ru(bipy)$_2$(CNMe)$_2$]$^{2+}$, **7**, 331
$RuC_{24}H_{30}$
Ru(C$_6$H$_6$)(C$_{10}$Me$_8$), **7**, 550
$RuC_{24}H_{34}ClNO_2P_2$
Ru[C(O)Me]Cl(CO)(CNBu^t)-(PMe$_2$Ph)$_2$, **7**, 334
$RuC_{24}H_{36}$
[Ru(C$_6$Me$_6$)$_2$]$^{2+}$, **7**, 551
$RuC_{24}H_{38}F_3OP$
RuH$_2$(OCH$_2$CF$_3$)(PPr$^i_2$Ph)Cp*, **7**, 500
$RuC_{24}H_{38}P_2$
Ru(C$_2$Ph)(PMe$_3$)$_2$Cp*, **7**, 514
$RuC_{24}H_{42}N_4O_4$
Ru(OAc)$_2$(CNBu^t)$_4$, **7**, 329

$RuC_{24}H_{43}NOP_2$
$Ru(Ph)(NOEt)(PMe_3)_2Cp^+$, **7**, 488
$RuC_{24}H_{44}$
$RuCy_4$, **7**, 377
$RuC_{24}H_{46}P_4$
$Ru[C_{10}H_6(CH_2)_2](PMe_3)_4$, **7**, 382
$RuC_{25}H_{21}O_3P$
$Ru(CO_2H)(CO)(PPh_3)Cp$, **7**, 483
$RuC_{25}H_{23}O_3PS_2$
$Ru[S(SO_2Me)](CO)(PPh_3)Cp$, **7**, 525
$RuC_{25}H_{23}P$
$RuH(CH_2CHC_6H_4PPh_2)Cp$, **7**, 503, 505
$RuC_{25}H_{46}OP_4$
$Ru(OCHPhC_6H_4)(PMe_3)_4$, **7**, 425
$RuC_{25}H_{46}O_2P_4$
$Ru(OC_6H_3MePMe_2)(OC_6H_4Me)(PMe_3)_3$, **7**, 377, 429
$RuC_{26}H_{20}F_{12}O_4P_2S_2$
$[[Ru(CO)_2Cp]_2(PhSSPh)](PF_6)_2$, **7**, 522
$RuC_{26}H_{23}N_2OPS_2$
$Ru(SC_5H_4N)_2(CO)(PPh_3)$, **7**, 314, 315
$RuC_{26}H_{23}PS_2$
$Ru(S_2CCHMeC_6H_4PPh_2)Cp$, **7**, 524
$RuC_{26}H_{25}P$
$Ru(PPh_3)(allyl)Cp$, **7**, 509
$RuC_{26}H_{26}Cl_2O_2$
$RuCl_2(CO)_2(PhC_4Me_2H_2)_2$, **7**, 311
$RuC_{26}H_{26}O_6S_2$
$[Ru(C_6H_6)_2](OTs)_2$, **7**, 550
$RuC_{26}H_{33}F_6P_3$
$[Ru(CCCPh_2)(PMe_3)_2Cp]PF_6$, **7**, 353, 354, 517
$RuC_{26}H_{44}O_4$
$Ru(acac)_2(C_8H_{14})_2$, **7**, 448
$RuC_{26}H_{50}O_2P_4$
$Ru(OC_6H_4Me)_2(PMe_3)_4$, **7**, 427
$RuC_{26}H_{52}O_2$
$Ru(CO)_2(C_8H_{14})_3$, **7**, 442
$RuC_{27}H_{22}F_6N_3P$
$Ru[NHC(CF_3)NC(CF_3)NH](PPh_3)Cp$, **7**, 495
$RuC_{27}H_{27}P$
$Ru(PPh_3)(MeC_3H_4)Cp$, **7**, 502
$RuC_{27}H_{42}O_{12}P_4$
$Ru[C_6H_4OP(OPh)_2]H[P(OMe)_3]_3$, **7**, 385
$RuC_{27}H_{42}P_4$
$RuH(C_6H_4PPh_2)(PMe_3)_3$, **7**, 391, 395
$RuC_{28}H_{22}O_3P_2$
$Ru(CO)_3(dppm)$, **7**, 302
$RuC_{28}H_{23}N_4O$
$[Ru(Bz)(CO)(bipy)_2]^+$, **7**, 386
$RuC_{28}H_{27}O_2P$
$Ru(CO)_2(PPh_3)(cod)$, **7**, 452
$Ru[C(OMe)CHC(O)Me](PPh_3)Cp$, **7**, 406
$RuC_{28}H_{30}Cl_2P$
$[RuCl_2(PPh_3)Cp^*]^+$, **7**, 496
$RuC_{28}H_{37}P$
$RuH(CH_2CMe_2PPh_2)(C_6Me_6)$, **7**, 574
$RuC_{28}H_{43}NP_2$
$Ru(NPh_2)(PMe_3)_2Cp^*$, **7**, 494
$RuC_{28}H_{51}P$
$RuH_3(PCy_3)Cp^*$, **7**, 500
$RuC_{29}H_{22}O_4P_2$
$Ru(CO)_4(dppm)$, **7**, 302
$RuC_{29}H_{27}O_4P$
$Ru[C(CHCO_2Me)OCMeO](PPh_3)Cp$, **7**, 407, 416
$RuC_{29}H_{43}N_2O_2P_2$
$[Ru[C(O)Me](CO)(CNBu^t)_2(PMe_2Ph)_2]^+$, **7**, 419
$RuC_{29}H_{50}O_3P_2$
$Ru(CCPh)(O_2CMe)(CO)(PPr^i_3)_2$, **7**, 416, 417
$RuC_{29}H_{51}ClO_3P_2$
$Ru[C(CHPh)OCMeO]Cl(CO)(PPr^i_3)_2$, **7**, 416, 417
$RuC_{29}H_{51}OP$
$Ru(OMe)(PCy_3)Cp^*$, **7**, 496
$RuC_{30}H_{23}N_2O_2PS_2$
$Ru(SC_5H_4N)_2(CO)_2(PPh_3)$, **7**, 314, 315
$RuC_{30}H_{32}P$
$[Ru(C_6H_4PPh_2)(C_6Me_6)]^+$, **7**, 571
$RuC_{30}H_{33}F_{12}P$
$[Ru(C_6Me_6)(PhPPh_2)](PF_6)_2$, **7**, 550
$RuC_{30}H_{33}P$
$RuH(C_6H_4PPh_2)(C_6Me_6)$, **7**, 574
$RuC_{30}H_{37}O_2P_3$
$[(Mes)PP(Mes)P]Ru(CO)_2Cp^*$, **7**, 487
$RuC_{30}H_{42}S_2$
$Ru_3(S)_2(cym)_3$, **7**, 565
$RuC_{30}H_{44}O_2P_2$
$Bu^t_3C_6H_2PPRu(CO)_2Cp^*$, **7**, 487
$RuC_{31}H_{23}N_6O_2$
$[Ru[C_6H_3(NO_2)C_5H_4N](bipy)_2]^+$, **7**, 430
$RuC_{31}H_{27}O_3PS_2$
$Ru[S(SO_2Tol)](CO)(PPh_3)Cp$, **7**, 526
$RuC_{31}H_{29}F_6O_2P_3S$
$[Ru(CH_2SO_2)(dppm)Cp]PF_6$, **7**, 525
$RuC_{31}H_{36}Cl_3P$
$RuCl_3[P(Tol)_3]Cp^*$, **7**, 496
$RuC_{31}H_{40}ClO_5P_3$
$Ru[C(CO_2Me)CHC(OMe)O]Cl(CO)(PMe_2Ph)_3$, **7**, 405
$RuC_{31}H_{55}NP_2$
$Ru(NH_2)(Cy_2PCH_2CH_2PCy_2)Cp$, **7**, 493
$RuC_{32}H_{20}O_4$
$Ru(CO)_3(C_4Ph_4CO)$, **7**, 457, 657
$RuC_{32}H_{25}PS_2$
$Ru(S_2CCCPh)(PPh_3)Cp$, **7**, 416
$RuC_{32}H_{29}OP_2$
$[Ru(CO)(dppe)Cp]^+$, **7**, 501
$RuC_{32}H_{30}O_6P_2$
$Ru[C(O)OMe]_2(CO)_2(dppe)$, **7**, 422
$RuC_{32}H_{31}P_2$
$[Ru(CH_2)(dppe)Cp]^+$, **7**, 519
$RuC_{32}H_{32}O_2P_2$
$Ru(CO)_2(PMe_2Ph)_2(PhC_2Ph)$, **7**, 444
$RuC_{32}H_{34}O_3P_2S$
$Ru(SPh)(PPh_3)[P(OMe)_3]Cp$, **7**, 526
$RuC_{32}H_{41}P_3$
$Ru(PMe_2Ph)_3[C_6H_4(CH_2)_2]$, **7**, 456
$RuC_{32}H_{46}P_3$
$[RuH(PMe_2Ph)_3(cod)]^+$, **7**, 460
$[Ru(PMe_2Ph)_3(C_8H_{13})]^+$, **7**, 460
$RuC_{32}H_{54}P_2$
$Ru(Cy_2PCH_2CH_2PCy_2)(C_6H_6)$, **7**, 572
$RuC_{32}H_{55}NOP_2$
$Ru(CONH_2)(Cy_2PCH_2CH_2PCy_2)Cp$, **7**, 493
$RuC_{32}H_{55}NO_2P_2$
$Ru(OCONH_2)(Cy_2PCH_2CH_2PCy_2)Cp$, **7**, 493
$RuC_{32}H_{57}O_4P_2$
$[Ru[C(CHPh)OCMeO](OCMe_2)(CO)(PPr^i_3)_2]^+$, **7**, 416, 417
$RuC_{33}H_{54}ClF_3O_4P_2S$
$[RuCl(Cy_2PCH_2CH_2PCy_2)(C_6H_6)]OTf$, **7**, 562
$RuC_{34}H_{35}F_6OP_3$
$[Ru[C(OMe)Me](dppe)Cp]PF_6$, **7**, 343
$RuC_{34}H_{40}O_2$
$RuCp^*_2$, **7**, 612
$RuC_{35}H_{28}ClNO_4P_2$
$Ru[C(O)N(C_6H_4Cl)O](CO)_2(dppe)$, **7**, 424
$RuC_{35}H_{29}N_2P$
$Ru(C_6H_4NNPh)(PPh_3)Cp$, **7**, 503
$RuC_{36}H_{44}ClP_3$
$RuCl(CH_2C_{10}H_6Me)(PMe_2Ph)_3$, **7**, 382
$RuC_{36}H_{52}O_2P_2$
$Ru(CCPh)_2(CO)_2(PPr^i_3)_2$, **7**, 412
$RuC_{37}H_{25}BrO_2$
$RuBr(CO)_2(C_5Ph_5)$, **7**, 479
$RuC_{37}H_{32}Cl_2OP_2S$
$RuCl_2(CS)(OH_2)(PPh_3)_2$, **7**, 320
$RuC_{37}H_{39}Cl_2O$
$RuCl_2(CO)(PhC_4Me_2H_2)_3$, **7**, 311
$RuC_{37}H_{50}N_4O_2$

$[Ru(OH_2)(CO)[(Bu^t)_2bipy]_2]^{2+}$, **7**, 312

$RuC_{37}H_{56}O_3P_2$
$Ru[C(CHPh)OCMeO](CCPh)(CO)(PPr^i_3)_2$, **7**, 416, 417

$RuC_{38}H_{28}O_2$
$RuMe(CO)_2(C_5Ph_5)$, **7**, 480

$RuC_{38}H_{30}Cl_2F_2OP_2$
$RuCl_2(CF_2)(CO)(PPh_3)_2$, **7**, 347

$RuC_{38}H_{30}NO_3P_2$
$[Ru(CO)_2(NO)(PPh_3)_2]^+$, **7**, 303

$RuC_{38}H_{31}Cl_2N_4O_9P$
$[Ru(O)(bipy)_2(PPh_3)](ClO_4)_2$, **12**, 1182

$RuC_{38}H_{32}O_2P_2S_2$
$Ru(SH)_2(CO)_2(PPh_3)_2$, **7**, 317

$RuC_{38}H_{34}ClNOP_2$
$RuCl(NO)(PPh_3)_2(CH_2CH_2)$, **7**, 338

$RuC_{38}H_{44}N_4$
$Ru(dad)_2(C_2H_4)$, **7**, 444

$RuC_{38}H_{68}N_6$
$[Ru[C(NHBu^t)C_8H_{13}](CNBu^t)_5]^{2+}$, **7**, 335

$RuC_{39}H_{24}BrO_4$
$RuBr(CO)_2[C_5Ph_4[C_6H_4(CO)_2]]$, **7**, 479

$RuC_{39}H_{30}F_2O_2P_2$
$Ru(CF_2)(CO)_2(PPh_3)_2$, **7**, 348

$RuC_{39}H_{30}O_2P_2Se$
$Ru(CO)_2(CSe)(PPh_3)_2$, **7**, 325

$RuC_{39}H_{30}O_3P_2S_2$
$Ru(S_2CO)(CO)_2(PPh_3)_2$, **7**, 317

$RuC_{39}H_{31}ClF_2O_2P_2$
$Ru(CF_2H)Cl(CO)_2(PPh_3)_2$, **7**, 347

$RuC_{39}H_{31}F_3O_3P_2$
$RuH(O_2CCF_3)(CO)(PPh_3)_2$, **7**, 400

$RuC_{39}H_{33}IO_2P_2$
$Ru(OCMe)I(CO)(PPh_3)_2$, **7**, 386, 418

$RuC_{39}H_{33}NO_5P_2$
$Ru(NO_2)(O_2CMe)(CO)(PPh_3)_2$, **7**, 365, 392

$RuC_{39}H_{34}Cl_2O_2P_2$
$RuCl_2[CH(OMe)](CO)(PPh_3)_2$, **7**, 348

$RuC_{39}H_{34}INOP_2$
$RuHI(CO)(NCMe)(PPh_3)_2$, **7**, 366

$RuC_{39}H_{34}O_3P_2$
$RuH(O_2CMe)(CO)(PPh_3)_2$, **7**, 365, 366
$RuMe(O_2CH)(CO)(PPh_3)_2$, **7**, 388

$RuC_{39}H_{66}O_3P_2$
$Ru(CO)_3(PCy_3)_2$, **7**, 303

$RuC_{40}H_{30}F_4O_2P_2$
$Ru(CO)_2(PPh_3)_2(C_2F_4)$, **7**, 444

$RuC_{40}H_{33}NO_2P_2$
$Ru(CNMe)(CO)_2(PPh_3)_2$, **7**, 333

$RuC_{40}H_{34}Cl_2N_3O_8P$
$[Ru(bipy)_2(MeCn)(PPh_3)](ClO_4)_2$, **12**, 1182

$RuC_{40}H_{34}O_2P_2$
$Ru(PPh_3)_2(C_3H_4COO)$, **7**, 460

$RuC_{40}H_{36}Cl_2N_2OP_2$
$RuCl_2[C(NHCH_2)_2](CO)(PPh_3)_2$, **7**, 333, 345
$RuCl_2(CNNMe_2)(CO)(PPh_3)_2$, **7**, 333

$RuC_{40}H_{36}Cl_2O_2P_2$
$RuCl_2[C(OH)Et](CO)(PPh_3)_2$, **7**, 343, 418

$RuC_{40}H_{37}Cl_2NOP_2$
$RuCl_2(CHNMe_2)(CO)(PPh_3)_2$, **7**, 345

$RuC_{40}H_{38}ClNOP_2$
$Ru(CH_2CH_2CH_2NH_2)Cl(CO)(PPh_3)_2$, **7**, 393

$RuC_{40}H_{44}Cl_2N_4$
$[RuCl_2(CNC_6H_2Me_3)_4]^+$, **7**, 336, 376

$RuC_{40}H_{46}B_9ClP_2$
$(Ph_3P)_2ClHRu(Me_2C_2B_9H_9)$, **1**, 416
$(Ph_3P)_2ClRu(Me_2C_2B_9H_{10})$, **1**, 416

$RuC_{40}H_{74}P_2$
$RuH(C_6H_{10}PCy_2)(CH_2CH_2)_2(PCy_3)$, **7**, 428

$RuC_{41}H_{34}O_6P_2$
$Ru[(C_6H_4O)P(OPh)_2][P(OPh)_3]Cp$, **7**, 503

$RuC_{41}H_{34}P_2$
$Ru(C_6H_4PPh_2)(PPh_3)Cp$, **7**, 502

$RuC_{41}H_{35}ClP_2$
$RuCl(PPh_3)_2Cp$, **7**, 490

$RuC_{41}H_{35}P_2S_2$
$[Ru(PPh_3)_2Cp(S_2)]^{2+}$, **7**, 522

$RuC_{41}H_{36}N_2O_2P_2$
$Ru(CO)_2(CNCH_2CH_2NH_2)(PPh_3)_2$, **7**, 333

$RuC_{41}H_{36}P_2$
$RuH(PPh_3)_2Cp$, **7**, 501

$RuC_{41}H_{36}P_2S$
$Ru(SH)(PPh_3)_2Cp$, **7**, 522

$RuC_{41}H_{36}P_2Se$
$Ru(SeH)(PPh_3)_2Cp$, **7**, 522

$RuC_{41}H_{37}N_2OP_2$
$[RuH(NCMe)_2(CO)(PPh_3)_2]^+$, **7**, 364

$RuC_{41}H_{37}P_2S$
$[Ru(H_2S)(PPh_3)_2Cp]^+$, **7**, 522

$RuC_{41}H_{39}ClP_2$
$RuCl(PPh_3)_2(C_5H_9)$, **7**, 467

$RuC_{42}H_{32}O_5P_2$
$Ru(CO)_2(C_4H_2O_3)(PPh_3)_2$, **7**, 304

$RuC_{42}H_{33}ClNO_2P_2$
$Ru(C_4H_3N)Cl(CO)_2(PPh_3)_2$, **7**, 397

$RuC_{42}H_{34}Cl_3NOP_2$
$RuCl_2[CCl(C_4H_4N)](CO)(PPh_3)_2$, **7**, 347

$RuC_{42}H_{36}NP_2$
$[Ru(CNH)(PPh_3)_2(Cp)]^+$, **7**, 333

$RuC_{42}H_{42}OP_3$
$[RuH(H_2)(CO)(triphos)]^+$, **7**, 370

$RuC_{43}H_{35}ClI_2P_2$
$Ru(CPh)ClI_2(PPh_3)_2$, **7**, 357

$RuC_{43}H_{35}ClOP_2$
$RuClPh(CO)(PPh_3)_2$, **7**, 293

$RuC_{43}H_{36}N_2OP_2$
$Ru[C_6H_4(NH)_2](CO)(PPh_3)_2$, **7**, 312

$RuC_{43}H_{40}O_2P_3$
$RuH(CO)_2(triphos)$, **7**, 306

$RuC_{44}H_{35}ClO_3P_2$
$RuCl(O_2CPh)(CO)(PPh_3)_2$, **7**, 430

$RuC_{44}H_{38}O_2P_3$
$[RuH(CO)_2(PPh_3)_2(PH_2Ph)]^+$, **7**, 366

$RuC_{44}H_{39}F_3O_3P_2S_2$
$[Ru(SCH_2CH_2)(PPh_3)_2Cp]OTf$, **7**, 523

$RuC_{45}H_{37}ClOP_2$
$Ru(CHCHPh)Cl(CO)(PPh_3)_2$, **7**, 400, 406

$RuC_{45}H_{37}ClO_2P_2$
$Ru(C_6H_4CMeO)Cl(CO)(PPh_3)_2$, **7**, 430

$RuC_{45}H_{37}ClP_2$
$RuCl(PPh_3)_2(C_9H_7)$, **7**, 491

$RuC_{45}H_{37}Cl_2NP_2S$
$RuCl_2(CNC_6H_4Me)(CS)(PPh_3)_2$, **7**, 320

$RuC_{45}H_{37}Cl_2NP_2Se$
$RuCl_2(CSe)(CNC_6H_4Me)(PPh_3)_2$, **7**, 321

$RuC_{45}H_{37}P_2$
$[Ru(CCCCH_2)(PPh_3)_2Cp]^+$, **7**, 518

$RuC_{45}H_{38}OP_2$
$Ru(C_2COMe)(PPh_3)_2Cp$, **7**, 518

$RuC_{45}H_{45}OP_3$
$Ru(triphos)(CH_2CHCOMe)$, **7**, 457

$RuC_{46}H_{36}NO_4P_2$
$[Ru[C(O)NHCOPh)(CO)_2(PPh_3)_2]^+$, **7**, 423, 424

$RuC_{46}H_{37}ClN_2OP_2$
$RuCl[C(C_4H_3N)(C_4H_4M)](CO)(PPh_3)_2$, **7**, 347

$RuC_{46}H_{37}NOP_2SSe$
$Ru(SCSe)(CO)(CNC_6H_4Me)(PPh_3)_2$, **7**, 321

$RuC_{46}H_{37}O_2P_2SSe$
$[Ru(SeCSC_6H_4Me)(CO)_2(PPh_3)_2]^+$, **7**, 324, 325

$RuC_{46}H_{38}Cl_2N_2OP_2$
$RuCl_2[C(C_4H_4N)_2](CO)(PPh_3)_2$, **7**, 347

$RuC_{46}H_{39}ClO_3P_2$
$RuCl(CO)_2(PPh_3)_2(COC_7H_9)$, **7**, 419, 451

$RuC_{46}H_{41}P_2$
$[Ru[CCPh(C_7H_7)](dppe)Cp]^+$, **7**, 352

$RuC_{46}H_{44}O_2P_2$
$Ru(OC_6H_4PPh_2)(Ph_2PC_6H_4$-$OH)Cp^*$, **7**, 497
$RuC_{46}H_{49}ClN_2P_2$
$RuClH(CNBu^t)_2(PPh_3)_2$, **7**, 335
$RuC_{47}H_{35}F_3O_3P_2$
$Ru(CCPh)(O_2CCF_3)(CO)$-$(PPh_3)_2$, **7**, 351
$RuC_{48}H_{42}F_6NP_3$
$[Ru(NHCHPh)(PPh_3)_2Cp]PF_6$, **7**, 493
$RuC_{49}H_{36}F_6O_5P_2$
$Ru(CCHPh)(O_2CCF_3)_2(CO)$-$(PPh_3)_2$, **7**, 351
$RuC_{49}H_{38}Cl_2N_2P_2$
$Ru[CCNN(C_6H_3Cl_2)](PPh_3)_2$-Cp, **7**, 415
$RuC_{49}H_{39}Br_5P_2$
$[Ru[CCBr(C_6H_4Br)](PPh_3)_2$-$Cp]Br_3$, **7**, 351, 352
$RuC_{49}H_{41}F_6P_3$
$[Ru(CCHPh)(PPh_3)_2Cp]PF_6$, **7**, 516
$RuC_{49}H_{42}O_7P_2$
$Ru(CCCO_2Me)[C(CHCH$-$CHCO_2Me)CO_2Me]$-$(CO)(PPh_3)_2$, **7**, 413
$RuC_{49}H_{49}ClOP_2$
$Ru[C(CCBu^t)CHBu^t]Cl(CO)$-$(PPh_3)_2$, **7**, 410
$RuC_{50}H_{42}P_2$
$Ru(CH_2CCPh)(PPh_3)_2Cp$, **7**, 390, 391
$RuC_{50}H_{44}F_6O_6PS_2Se_2$
$[[Ru(PPh_3)(C_5H_4Me)]_2(Se)_2]$-$(OTf)_2$, **7**, 526
$RuC_{51}H_{42}ClNOP_2$
$Ru(CH_2PPh_3)Cl(NO)(PPh_3)$-$(C_2Ph_2)$, **7**, 338, 444
$RuC_{51}H_{42}O_5P_2$
$Ru[C(O)C_6H_4O](HOC_6H_4$-$CH_2OH)(CO)(PPh_3)_2$, **7**, 430
$RuC_{51}H_{46}O_8P_2S_2$
$Ru(O_3SC_6H_4Me)_2(OH_2)(CO)$-$(PPh_3)_2$, **7**, 315, 316
$RuC_{52}H_{40}P_2S_2$
$Ru(S_2CC_2Ph)(PPh_3)_2Cp$, **7**, 524
$RuC_{52}H_{43}ClOP_2S_2$
$Ru[C(SC_6H_4Me)SC_6H_3Me]Cl$-$(CO)(PPh_3)_2$, **7**, 430
$RuC_{52}H_{51}P_3$
$Ru(PPh_3)_3(C_4H_6)$, **7**, 455
$RuC_{52}H_{55}P_4$
$[RuH(H_2)[P(CH_2CH_2PPh_2)_3]]^+$, **7**, 372
$RuC_{52}H_{72}P_3$
$RuCl[PhP(CH_2CH_2CH_2P$-$Cy_2)_2](PhC_3CHPh)$, **7**, 463, 464
$RuC_{53}H_{43}ClO_4P_2$
$[Ru(CCHPh)(PPh_3)_2(C_9H_7)]$-$ClO_4$, **7**, 516
$RuC_{53}H_{49}Cl_2OP_3$
$RuCl_2(PPh_3)_2(Ph_2POCH_2$-$CHCMe_2)$, **7**, 448
$RuC_{53}H_{49}OP_4$
$[RuH(CO)(dppe)_2]^+$, **7**, 306
$RuC_{54}H_{44}ClP_3$
$Ru(C_6H_4PPh_2)Cl(PPh_3)_2$, **4**, 576; **7**, 428
$RuC_{54}H_{45}Cl_2P_3$
$RuCl_2(PPh_3)_3$, **12**, 1181, 1183, 1184
$RuC_{55}H_{41}F_3O_3P_2$
$Ru[C(CCPh)(CHPh)(O_2CCF_3)$-$(CO)(PPh_3)_2]$, **7**, 408
$RuC_{55}H_{45}N_2P_2$
$[Ru[CCPh(NNPh)](PPh_3)_2Cp]^+$, **7**, 352
$RuC_{55}H_{45}OP_3$
$RuH(C_6H_4PPh_2)(CO)(PPh_3)_2$, **7**, 397, 428
$RuC_{55}H_{46}BrOP_3$
$RuHBr(CO)(PPh_3)_3$, **7**, 363
$RuC_{55}H_{46}ClOP_3$
$Ru(CO)ClH(PPh_3)_3$, **1**, 320; **7**, 459, 460, 462
$RuC_{55}H_{55}NOP_2$
$[Ru(CHCHPh)(CO)(PPh_3)_2$-$(CNBu^t)_2]^+$, **7**, 401, 403
$RuC_{58}H_{52}P_4$
$Ru(CH_2CHPh)(dppm)_2$, **7**, 428
$RuC_{58}H_{54}ClNO_2P_2$
$Ru[HNCPhOC(CHPh)]Cl(CO)$-$[P(C_6H_4Me)_3]_2$, **7**, 403, 404
$RuC_{60}H_{77}P_3$
$Ru[PhP(CH_2CH_2CH_2PCy_2)_2]$-$(C_2Ph)(PhC_3CHPh)$, **7**, 463
$RuC_{60}H_{123}P_8$
$[Ru(H_2)(CCPh)(dippe)_2]^+$, **7**, 371
$RuC_{61}H_{55}OP_3$
$RuH(CCBu^t)(CO)(PPh_3)_3$, **7**, 414
$RuC_{63}H_{52}NOP_3$
$Ru(CO)(CNC_6H_4Me)(PPh_3)_3$, **7**, 330
$RuC_{66}H_{57}ClOP_4$
$RuCl(CCCPh_2OMe)(dppm)_2$, **7**, 414
$RuC_{72}H_{58}O_{12}P_4$
$Ru[C_6H_4OP(OPh)_2]_2[P(OPh)_3]_2$, **7**, 428, 429
$RuC_{98}H_{80}P_2$
$[Ru(PPh_3)_2Cp]_2[CC(Ph)C$-$(Ph)C]^{2+}$, **7**, 518
$RuCl_3$
$RuCl_3$, **12**, 1183
$RuCoC_{19}H_{10}O_7$
$Ru(CO)_4(PPh_2)Co(CO)_3$, **10**, 61
$RuCoC_{45}H_{35}O_4P_2$
$CpRu(PPh_3)_2Co(CO)_4$, **10**, 47
$RuCoC_{53}H_{40}O_5P_3$
$RuCo(PPh_2)(CO)_5(PPh_3)_2$, **7**, 366
$RuCoFeIrC_{16}H_9O_{10}P$
$FeRuCoIr(H)(PMe)(CO)_{10}Cp$, **10**, 244
$RuCoMoC_{15}H_9O_8$
$HMoRuCo(CMe)Cp(CO)_8$, **10**, 41
$RuCoNiC_{19}H_{17}O_6$
$CpRuCoNi(CO)_6(C_4Me_4)$, **10**, 166
$RuCoWC_{15}H_9O_8$
$HWRuCo(CMe)Cp(CO)_8$, **10**, 41
$RuCoWC_{19}H_{10}O_8P$
$WRuCoCp(CO)_8(PPh)$, **10**, 19, 20
$RuCo_2C_9O_9S$
$RuCo_2(S)(CO)_9$, **10**, 236, 237
$RuCo_2C_{11}O_{11}$
$RuCo_2(CO)_{11}$, **10**, 43, 52, 58, 197
$RuCo_2C_{13}H_6O_9$
$RuCo_2(CO)_9(MeC_2Me)$, **10**, 142
$RuCo_2C_{23}H_{10}O_9$
$Co_2Ru(CO)_9(PhCCPh)$, **9**, 98
$RuCo_2C_{30}H_{26}$
$[CpCo(C_{10}H_8)]_2Ru$, **7**, 610
$RuCo_2Rh_2C_{18}H_{10}O_8S$
$RuCo_2Rh_2(S)(CO)_8Cp_2$, **10**, 243
$RuCo_3C_{11}NO_{12}$
$RuCo_3(CO)_{11}(NO)$, **10**, 197
$RuCo_3C_{12}HO_{12}$
$HRuCo_3(CO)_{12}$, **10**, 43, 45, 48, 52, 58, 203
$RuCo_3C_{12}O_{12}$
$[RuCo_3(CO)_{12}]^-$, **10**, 42, 191, 192, 203
$RuCo_3C_{19}H_{19}OP$
$[RuCo_3(CO)(PPh_2)(CCBu^t)]^-$, **10**, 145
$RuCo_3C_{20}H_{20}NO_{12}$
$[Et_4N][RuCo_3(CO)_{12}]$, **10**, 51
$RuCo_3C_{30}H_{30}NO_{10}$
$[Et_4][RuCo_3(CO)_{10}(C_2Ph_2)]$, **10**, 48
$RuCo_4HgC_{16}O_{16}$
$RuCo_3(CO)_{12}(Hg)Co(CO)_4$, **10**, 341
$RuCrC_{18}H_{15}NO_2$
$Cr(CO)_2(NO)(C_5H_4CH_2CpRu$-$(C_5H_4))$, **7**, 610
$RuCuC_{15}H_{24}ClO_7P_2$
$[Cl[OP(OMe)_2]_2Ru(\eta$-C_6-$Me_6)]Cu(CO)$, **7**, 564
$RuCuC_{18}H_{36}ClP$
$RuH(PCy_3)(H)_2CuCl$, **10**, 61
$RuCuC_{50}H_{46}ClP_4$
$Ru(dppm)_2(H)_2CuCl$, **10**, 62
$RuCuC_{66}H_{48}O_{12}P_3$
$[Cu(PPh_3)_3][Ru(H_3)(CO)_{12}]$, **10**, 302
$RuFeC_8O_8$
$[Fe(CO)_4Ru(CO)_4]^{2-}$, **10**, 54
$RuFeC_{12}H_8O_6$
$Fe(CO)_3(CHCMeCMeCH)Ru$-$(CO)_3$, **10**, 54
$RuFeC_{12}H_{13}NO_5$

$Fe(CO)_2(CMeCHCHNPr^i)$-$(H)Ru(CO)_3$, **10**, 54
$RuFeC_{13}HO_{13}$
$[HFeRu(CO)_{13}]^-$, **10**, 52
$RuFeC_{13}H_2O_{13}$
$H_2FeRu(CO)_{13}$, **10**, 52
$RuFeC_{13}H_{13}NO_6$
$Fe(CO)_3(CHMeCHCHNPr^i)Ru$-$(CO)_3$, **10**, 54
$RuFeC_{14}H_{10}O_4$
$Fe(CO)Cp(CO)_2Ru(CO)Cp$, **10**, 54
$RuFeC_{14}H_{20}N_2O_5$
$Fe(CO)_3(H)((NPr^i{=}CHCH{=}NPr^i)RuMe(CO)_2$, **10**, 54
$RuFeC_{18}H_{13}NO_6$
$Fe(CO)_3(CPh{=}CHCHN{=}CMe_2)Ru(CO)_3$, **10**, 54
$RuFeC_{20}H_{18}$
$FeCp(C_5H_4)RuCp(C_5H_4)$, **7**, 616
$RuFeC_{22}H_{24}O_6$
$Fe(CO)_3(C_{16}H_{24})Ru(CO)_3$, **10**, 54, 60
$RuFeC_{24}H_{16}O_5$
$FeRu(CO)(CPh{=}CHPh)$-$(CO)_4Cp$, **10**, 54
$RuFeC_{25}H_{35}$
$[Cp^*RuCp^*FeCp]^+$, **7**, 617
$RuFeC_{26}H_{42}P_2$
$Ru[C_5H_4FeCp](PMe_3)_2Cp^*$, **7**, 399
$RuFeC_{30}H_{20}O_6$
$(CO)_3Fe(PPh_2)_2Ru(CO)_3$, **10**, 45
$RuFeC_{34}H_{20}O_6$
$Fe(CO)_3(CPhCPhCPhCPh)$-$Ru(CO)_3$, **10**, 55
$FeRu(CO)_6(C_2Ph_2)_2$, **10**, 130
$RuFeHgC_{20}H_{18}$
$[Cp(C_5H_4)FeCp(C_5H_4)Ru$-$(Hg)]^{2+}$, **7**, 618
$RuFe_2C_{12}O_{12}$
$Fe_2Ru(CO)_{12}$, **7**, 83; **10**, 37, 43, 196
$RuFe_2C_{23}H_{10}O_9$
$Fe_2Ru(CO)_9(C_2Ph_2)$, **10**, 130
$RuFe_2C_{36}H_{20}O_8$
$Fe_2Ru(CO)_8(C_2Ph_2)_2$, **10**, 130
$RuFe_3C_{11}O_{11}SeTe$
$Fe_3Ru(CO)_{11}SeTe$, **7**, 75
$RuGe_2C_7H_6Cl_6O$
$Ru(GeCl_3)_2(CO)(C_6H_6)$, **7**, 586
$RuHgC_{25}H_{30}O_{10}$
$Cp^*Ru[C_5(HgOAc)_5]$, **3**, 162
$RuHgC_{33}H_{32}Cl_2OP_2$
$Ru[C(CCPh)C(HgCl)Ph]Cl$-$(CO)(PMe_2Ph)_2$, **7**, 411
$RuHgC_{37}H_{32}Cl_3NOP_2$
$Ru(CH_2Cl)(HgCl)Cl(NO)$-$(PPh_3)_2$, **7**, 339
$RuHgC_{40}H_{30}F_3O_3P_2$
$[Ru(HgCF_3)(CO)_3(PPh_3)_2]^+$, **7**, 347
$RuHgC_{40}H_{30}F_6O_2P_2$
$Ru(CF_3)(HgCF_3)(CO)_2(PPh_3)_2$, **7**, 347, 389
$RuHgC_{44}H_{39}Cl_2NOP_2$
$Ru(CH_2HgC_6H_4Me)Cl_2(NO)$-$(PPh_3)_2$, **7**, 339
$RuHgC_{45}H_{35}F_3O_2P_2$
$RuPh(HgCF_3)(CO)_2(PPh_3)_2$, **7**, 397
$RuHg_2C_4I_2O_4$
$(HgI)_2Ru(CO)_4$, **10**, 327
$RuIrC_{33}H_{31}N_4OP$
$H(CO)(PPh_3)Ru$-(biimidazolato)Ir(cod), **10**, 39
$RuIrC_{51}H_{46}N_4OP_2$
$(H)(CO)(PPh_3)_2Ru(bim)Ir(cod)$, **8**, 374
$RuIrC_{51}H_{49}N_4OP_2$
$(H)(CO)(PPh_3)_2Ru(pz)_2Ir(cod)$, **8**, 374
$RuIrC_{51}H_{54}N_4OP_2$
$H(CO)(PPh_3)_2Ru$-(biimidazolato)Ir(cod), **8**, 305
$RuIrC_{58}H_{58}ClP_4$
(dppm)(H)Ru(H)(dppm)(Cl)Ir-(cod), **8**, 327
$RuIr_3C_{12}O_{12}$
$[RuIr_3(CO)_{12}]^-$, **10**, 191, 192
$RuIr_4C_{15}O_{15}$
$[RuIr_4(CO)_{15}]^{2-}$, **10**, 192
$RuKC_8H_5N_2O$
$K[Ru(CN)_2(CO)Cp]$, **7**, 482
$RuLiC_{14}H_{23}$
$Li[Ru(C_2H_4)_2Cp^*]$, **7**, 505
$RuLi_3C_{24}H_{66}N_6$
$[RuMe_6][Li(TMEDA)]_3$, **7**, 377
$RuLuC_{21}H_{23}O_3$
$Cp_2(THF)LuRu(CO)_2Cp$, **4**, 107; **10**, 54
$RuLuC_{27}H_{35}O_2$
$Cp^*{}_2LuRu(CO)_2Cp$, **4**, 107
$RuLuSi_4C_{33}H_{55}O_3$
$[(TMS)_2C_5H_3]_2(THF)LuRu$-$(CO)_2Cp$, **4**, 107
$RuMnC_{18}H_{14}NO_6$
$MnRu(CO)_6[PhC{=}C(H)C(H){=}NPr^i]$, **10**, 39
$RuMnC_{22}H_{11}O_5P$
$Ru(CO)Cp(H)(PPh_2)Mn(CO)_4$, **10**, 52
$RuMnC_{22}H_{16}O_5P$
$CpRu(CO)(H)(PPh_2)Mn(CO)_4$, **6**, 87, 88
$RuMnC_{25}H_{22}O_4P$
$Mn(CO)_3(CMe{=}CHMe)$-$(PPh_2)Ru(CO)Cp$, **10**, 52
$RuMnC_{54}H_{40}O_6P_3$
$Mn(PPh_3)(CO)_3(PPh_2)Ru(CO)_3$-$(PPh_3)$, **6**, 87
$RuMnC_{59}H_{44}N_2O_4P_2$
$[Ru[CCPh(N_2C_6H_4Me(CO)_4]$-$(PPh_3)_2Cp]^+$, **7**, 352
$RuMoC_{15}H_8O_5$
$Mo(CO)_3(C_5H_4C_5H_4)Ru(CO)_2$, **10**, 42
$RuMoC_{31}H_{20}O_7P_2$
$(CO)_4Mo(PPh_2)_2Ru(CO)_3$, **10**, 39
$RuMo_2C_{17}H_{10}O_7S$
$Mo_2Ru(S)Cp_2(CO)_7$, **10**, 49, 236
$RuMo_2C_{17}O_7S$
$RuMo_2Cp_2(CO)_7(S)$, **10**, 20
$RuNiC_{28}H_{23}PS_2$
$Ru(C_5H_4S)_2Ni(PPh_3)$, **10**, 61
RuO_4
RuO_4, **12**, 1157
$RuOsC_{11}H_2O_9$
$RuOs[HCCHC(O)](CO)_8$, **7**, 652
$RuOsC_{12}F_6O_8$
$RuOs(hfb)(CO)_8$, **7**, 652
$RuOsC_{14}H_{10}O_8$
$Os(CO)_4(CBu^tCH)Ru(CO)_4$, **10**, 54, 55
$RuOs_3C_{12}H_3O_{12}$
$[H_3RuOs_3(CO)_{12}]^-$, **10**, 60
$RuPdC_{16}H_{22}Cl_2N_2$
$(cymene)ClRuCl(Pz)Pd(C_3H_5)$, **9**, 350
$RuPdC_{36}H_{28}Cl_2N_2O_4P_2$
$RuCl(CO)_2(PPh_2C_5H_4N)_2PdCl$, **10**, 61
$RuReC_{14}H_9O_7$
$(CO)_5ReCH_2CH_2Ru(CO)_2Cp$, **7**, 480
$RuRe_3C_{26}H_{15}O_{11}S_3$
$Re_3Ru(S)(pyS)_2(CO)_{11}(py)$, **10**. 236
$RuRhC_{31}H_{22}O_6P_2$
$[RuRh(CO)_6(dppm)]^+$, **7**, 302
$RuRhC_{33}H_{31}N_4OP$
$H(CO)(PPh_3)Ru$-(biimidazolato)Rh(cod), **10**, 39
$RuRhC_{35}H_{29}O_3P_2$
$Ru(CO)_2(dppm)(C_7H_7)Rh(CO)$, **10**, 61
$RuRhC_{51}H_{47}N_4OP_2$
$RuRh(bim)H(CO)(PPh_3)_2(cod)$, **7**, 364
$RuRhC_{53}H_{44}ClO_3P_4$
$RuRh(CO)_3Cl(dppm)_2$, **10**, 45
$RuRhC_{58}H_{57}ClP_4$
$[Ru(dppm)_2(H)(Cl)Rh(cod)]^+$, **10**, 61
$RuRh_2C_{12}O_{12}$
$RuRh_2(CO)_{12}$, **10**, 42
$RuRh_3C_{12}HO_{12}$
$HRuRh_3(CO)_{12}$, **10**, 48
$RuRh_3C_{46}H_{31}O_{10}P_2$
$HRuRh_3(CO)_{10}(PPh_3)_2$, **10**, 48
$RuRh_5C_{52}H_{30}NO_{16}$
$[PPN][RuRh_5(CO)_{16}]$, **10**, 45
$RuRh_5C_{52}H_{30}NO_{16}P_2$
$[PPN][RuRh_5(CO)_{16}]$, **10**, 48
$RuSiC_{12}H_{21}Br$
$RuBr(CH_2TMS)(C_3H_5)Cp$, **7**, 510
$RuSiC_{16}H_{46}P_4$

$Ru[(CH_2)_2SiMe_2](PMe_3)_4$, **7**, 377, 378

$RuSiC_{21}H_{19}F_6O_7P$
$Ru(Ph_2PCH_2SiMe_2OH)(CO)_2(CO_2CF_3)_2$, **7**, 633, 634

$RuSiC_{28}H_{43}P_2$
$Ru(SiPh_2)(PMe_3)_2Cp^*$, **7**, 527

$RuSiC_{29}H_{43}O_3P_2S_3$
$Ru(SiPh_2OTf)(PMe_3)_2Cp^*$, **7**, 527

$RuSiC_{32}H_{63}ClOP_2$
$Ru(CPhCHSiEt_3)Cl(CO)(PPr^i_3)_2$, **7**, 404

$RuSiC_{46}H_{46}OP_2$
$Ru(CC\text{-}TMS)(CH_2CHCHMe)(CO)(PPh_3)_2$, **7**, 412, 413

$RuSiC_{57}H_{49}OP_3$
$Ru(SiMe_2C_6H_4PPh_2)(C_6H_4PPh_2)(CO)(PPh_3)$, **7**, 429

$RuSiC_{70}H_{63}ClOP_4$
$RuCl(C_4CPh_2O\text{-}TMS)(dppm)_2$, **7**, 354, 355

$RuSi_2C_{10}H_{26}N$
$[Ru(N)Me_2(CH_2TMS)_2]^-$, **7**, 373, 374

$RuSi_2C_{47}H_{51}ClOP_2$
$Ru[CH(CH_2TMS)CC\text{-}TMS]Cl(CO)(PPh_3)_2$, **7**, 411

$RuSi_4C_{14}H_{40}NO_2$
$[Ru(N)(O\text{-}TMS)_2(CH_2TMS)_2]^-$, **7**, 373, 374

$RuSnC_{33}H_{30}O_2$
$Ru(SnPh_3)[C(OEt)Ph](CO)Cp$, **7**, 485

$RuSnC_{46}H_{70}P_4$
$Ru[CC(C_6H_4)_2CCSnMe_3]_2(PMe_3)_4$, **7**, 415

$RuSn_3C_{20}H_{20}Cl_{12}$
$[(Cp_2Ru)_2SnCl_2][SnCl_5]_2$, **7**, 618

$RuThC_{27}H_{35}ClO_2$
$Ru(CO)_2CpThCp^*{}_2Cl$, **10**, 54

$RuThC_{27}H_{35}IO_2$
$Cp^*{}_2Th(I)RuCp(CO)_2$, **4**, 187

$RuThC_{32}H_{44}$
$RuMeCp(C_5H_4)(CH_2)ThCp^*{}_2$, **10**, 62

$RuWC_{15}H_8O_5$
$W(CO)_3(C_5H_4C_5H_4)Ru(CO)_2$, **10**, 42

$RuW_2C_{33}H_{24}O_7$
$Cp_2W_2Ru(CO)_7(C_2Tol_2)$, **10**, 99

$RuW_{11}H_2O_{40}P$
$[Ru(PW_{11}O_{39})(H_2O)]^{5-}$, **7**, 448

$RuZrC_{17}H_5IO_2$
$Cp_2ZrI[Ru(CO)_2Cp]$, **4**, 536

$RuZrC_{18}H_{14}O_3$
$Zr(CO)Cp_2(C_5H_4)Ru(CO)_2$, **10**, 27

$RuZrC_{18}H_{15}ClO_4$
$Cp(CO)_2Ru(CO_2)ZrClCp_2$, **7**, 483

$RuZrC_{18}H_{17}O_3$
$Cp(CO)_2RuCH_2OZrCp_2$, **7**, 483

$RuZrC_{18}H_{18}O_2$
$Cp_2ZrMe[Ru(CO)_2Cp]$, **4**, 536

$RuZrC_{21}H_{24}O_3$
$Cp_2Zr(OBu^t)[Ru(CO)_2Cp]$, **4**, 536; **10**, 27

$RuZrC_{23}H_{33}ClP_2$
$Cp(Me_3P)_2RuCCZrClCp_2$, **4**, 509; **7**, 415, 521

$RuZrC_{23}H_{35}ClP_2$
$Cp(Me_3P)_2RuCHCHZrClCp_2$, **7**, 415

$RuZrC_{23}H_{37}ClP$
$Cp(Me_3P)_2Ru(CH_2CH_2)ZrClCp_2$, **7**, 392

$RuZrC_{47}H_{42}Cl_2P_2$
$Cp_2Zr(Cl)(CH_2)RuCl(PPh_3)_2$, **4**, 576

$Ru_2AgSbC_{19}H_{37}F_6N_3O_{13}P_5$
$[Ru_2[Ag(MeCN)](CO)_5[(MeO)_2PNEtP(OMe)_2]_2]SbF_6$, **7**, 627

$Ru_2Al_2C_6H_{28}P_2$
$[Ru(PMe_3)]_2[(H)_2AlH(H)_2AlH(H)_2]$, **1**, 550

$Ru_2AsC_8H_6O_6$
$Ru_2(AsMe_2)(CO)_6$, **7**, 649

$Ru_2AuC_{35}H_{44}N_2O_{13}P_5$
$[Ru_2[Au(PPh_3)][(CO)_5(etmdp)_2]^+$, **7**, 627

$Ru_2BC_{15}H_{13}F_4O_3$
$[Ru_2Cp_2(CMe)(CO)_3]BF_4$, **7**, 665

$Ru_2BC_{23}H_{32}ClF_4O_2$
$[Ru_2Cp^*{}_2(Cl)(CH_2)(CO)_2]BF_4$, **7**, 668

$Ru_2BC_{27}H_{38}F_3O_3S$
$(Cp^*)_2Ru_2(CpBMe)[CF_3SO_3]$, **1**, 209

$Ru_2BC_{42}H_{44}F_4N_3$
$[Ru_2(NHPh)_3(C_6Me_6)_2]BF_4$, **7**, 559

$Ru_2BCuC_{38}H_{34}Cl_2F_4N_2O_4$
$[CuRu_2(Ph_2PpyOMe)_2(Cl_2)(CO)_2][BF_4]$, **10**, 298

$Ru_2B_2C_{50}H_{50}P$
$[[(triphos)(H)Ru]_2(BH_4)][BPh_4]$, **1**, 284, 285

$Ru_2B_9C_{20}H_{37}S$
$[(cymene)Ru]_2SB_9H_9$, **1**, 269

$Ru_2B_{10}C_{28}H_{51}Cl_2O_2P_3$
$(PMe_3)_2ClRuClRu(PPh_3)B_{10}H_8(OEt)_2$, **1**, 355

$Ru_2B_{10}C_{30}H_{56}O_4$
$[(C_6Me_6)_2RuH_2(CO)_2]RuB_{10}H_8(OEt)_2$, **7**, 796, 797

$Ru_2B_{10}C_{42}H_{48}O_6P_2$
$(PPh_3)_2Ru(MeCO_2)_3(H)_2RuB_{10}H_7$, **1**, 357

$Ru_2C_4O_{22}P_6$
$[Ru_2(P_3O_9)_2(CO)_4]^{4-}$, **7**, 633, 634

$Ru_2C_6Cl_4O_6$
$[RuCl(CO)_3(Cl)]_2$, **7**, 895
$[Ru(CO)_3(Cl)_2]_2$, **10**, 43

$Ru_2C_8O_8$
$[Ru_2(CO)_8]^{2-}$, **7**, 625, 626

$Ru_2C_9H_{10}N_4O_5$
$Ru_2(CO)_5(ada)$, **7**, 635, 636

$Ru_2C_9O_9$
$Ru_2(CO)_9$, **7**, 626, 836

$Ru_2C_{10}H_6O_8$
$Ru_2(OCMe)(OCMe)(CO)_6$, **7**, 749

$Ru_2C_{10}H_{10}NO_2$
$Ru_2Cp_2(NO)_2$, **7**, 667

$Ru_2C_{10}H_{10}S_4$
$Ru_2Cp_2(S_2)_2$, **7**, 671

$Ru_2C_{11}H_9NO_6S$
$Ru_2[SC(NMe_2)(CMe)CMe](CO)_6$, **7**, 648

$Ru_2C_{12}F_6O_8$
$Ru_2(hfb)(CO)_8$, **7**, 652

$Ru_2C_{12}H_{10}O_2S_2$
$Ru_2Cp_2S_2(CO)_2$, **7**, 671

$Ru_2C_{12}H_{10}O_8$
$Ru_2[OC(Et)][OC(Et)](CO)_6$, **7**, 646

$Ru_2C_{12}H_{12}Cl_4$
$[RuCl_2(C_6H_6)]_2$, **7**, 578

$Ru_2C_{12}H_{12}F_{12}N_2O_8P_2$
$[Ru_2[OC(Me)O]_2(CO)_4(MeCN)_2](PF_6)_2$, **7**, 633

$Ru_2C_{12}H_{12}N_2O_8$
$Ru_2(OAc)_2(CO)_4(MeCN)_2$, **7**, 643; **12**, 426, 430

$Ru_2C_{12}H_{15}O_3$
$[Ru_2(OH)_3(C_6H_6)_2]^+$, **7**, 673

$Ru_2C_{12}H_{19}N_3O_8$
$Ru_2[ON(CMe_2)]_2(CO)_5(HONCMe_2)$, **7**, 646

$Ru_2C_{13}H_{10}F_4N_3O_3P_2$
$Ru_2Cp_2[\overline{PN(F)_2NP(F)_2N}](CO)_3$, **7**, 656

$Ru_2C_{13}H_{10}O_3$
$Ru_2Cp_2(CO)_3$, **7**, 655

$Ru_2C_{13}H_{10}O_5S$
$Ru_2Cp_2(SO_2)(CO)_3$, **7**, 659

$Ru_2C_{13}H_{12}O$
$Ru_2Cp_2(C_2H_2)(CO)$, **7**, 661

$Ru_2C_{13}H_{13}NO_7$
$Ru_2[OC(NMe_2)](CMeCHMe)(CO)_6$, **7**, 646, 647

$Ru_2C_{13}H_{16}N_2O_5$
$Ru_2(CO)_5(Pr^i,Pr^i\text{-dab})$, **7**, 639, 640

$Ru_2C_{13}H_{20}N_2O_5$
$Ru_2H_2(CO)_5(Pr^iNCH_2CH_2NPr^i)$, **7**, 640, 641

$Ru_2C_{14}H_8O_4$
$Ru_2(C_{10}H_8)(CO)_4$, **7**, 655, 656

$Ru_2C_{14}H_{10}O_4$
$Ru_2Cp_2(CO)_4$, **7**, 655, 657, 660, 665

$Ru_2C_{14}H_{12}O_3$
$Ru_2Cp_2(CH_2)(CO)_3$, **7**, 665

$Ru_2C_{14}H_{13}ClO_2$
$Ru_2Cp_2Cl(CHCH_2)(CO)_2$, **7**, 663, 664

$Ru_2C_{14}H_{13}NO_7$
$Ru_2(CO)_6(MeCCHCH_2NPr^iCO)$, **7**, 638

$Ru_2C_{14}H_{13}O_3$

$[Ru_2Cp_2(Me)(CO)_3]^+$, **7**, 661, 662
$Ru_2C_{14}H_{18}N_2O_6$
$Ru_2(CO)_6(Pr^iNCH_2CH_2NPr^i)$, **7**, 640, 641
$Ru_2C_{14}H_{18}O_6S_2$
$Ru_2(SBu^t)_2(CO)_6$, **7**, 647
$Ru_2C_{15}H_{10}O_4$
$Ru_2(C_5H_4CH_2C_5H_4)(CO)_4$, **7**, 655, 664
$Ru_2C_{15}H_{12}O_3$
$Ru_2Cp_2(CCH_2)(CO)_3$, **7**, 663, 665
$Ru_2C_{15}H_{12}O_4$
$Ru_2Cp_2[CH_2C(O)](CO)_3$, **7**, 660
$Ru_2C_{15}H_{13}O_3$
$[Ru_2Cp_2(CHCH_2)(CO)_3]^+$, **7**, 663
$[Ru_2Cp_2(CMe)(CO)_3]^+$, **7**, 657, 658, 663
$Ru_2C_{15}H_{14}O_2$
$Ru_2Cp_2(CHCHCH_2)(CO)_2$, **7**, 659
$Ru_2C_{15}H_{14}O_3$
$Ru_2Cp_2(CHMe)(CO)_3$, **7**, 657, 658, 665
$Ru_2Cp_2(CO)_3(C_2H_4)$, **7**, 658, 659
$Ru_2C_{15}H_{15}O_2$
$[Ru_2Cp_2(CH_2)(CHCH_2)(CO)_2]^+$, **7**, 662
$Ru_2Cp_2(CHCHMe)(CO)_2$, **7**, 659
$Ru_2C_{15}H_{16}O_2$
$Ru_2Cp_2(CH_2CHCH_2)H(CO)_2$, **7**, 665
$Ru_2Cp_2(CH_2)(CHMe)(CO)_2$, **7**, 662
$Ru_2Cp_2(H)(CHCHMe)(CO)_2$, **7**, 664
$Ru_2Cp_2(H)(CMeCH_2)(CO)_2$, **7**, 659, 664
$Ru_2Cp_2Me(CHCH_2)(CO)_2$, **7**, 664
$Ru_2C_{15}H_{28}O_5P_4$
$Ru_2(CO)_5(Me_2PCH_2PMe_2)_2$, **7**, 626
$Ru_2C_{16}H_8N_2O_6$
$Ru_2(C_5H_4N)_2(CO)_6$, **7**, 645
$Ru_2C_{16}H_{10}N_2O_6$
$Ru_2(dan)(CO)_6$, **7**, 648
$Ru_2C_{16}H_{14}O_3$
$Ru_2Cp_2[CH(CHCH_2)](CO)_3$, **7**, 659
$RuCp[C_3H_4(RuCp(CO)_2)](CO)$, **7**, 659
$Ru_2C_{16}H_{16}NO_2$
$[Ru_2Cp_2(CHCH_2)(CO)_2-(MeCN)]^+$, **7**, 663
$Ru_2C_{16}H_{16}O_2$
$Ru_2Cp_2[CH_2C(CH_2)_2](CO)_2$, **7**, 662, 663
$Ru_2Cp_2(CH_2CHCHCH_2)(CO)_2$, **7**, 660, 663
$Ru_2Cp_2(CHCHCHMe)(CO)_2$, **7**, 660
$Ru_2C_{16}H_{16}O_3$
$Ru_2Cp_2[(CH_2)_3](CO)_3$, **7**, 665
$Ru_2Cp_2(CHEt)(CO)_3$, **7**, 659, 665
$Ru_2Cp_2(CMe_2)(CO)_3$, **7**, 659, 665
$Ru_2C_{16}H_{18}F_{12}N_6O_4P_2$
$[Ru_2(CO)_4(MeCN)_6](PF_6)_2$, **7**, 633
$Ru_2C_{16}H_{18}O_2$
$[Ru_2Cp_2(H)(CH_2CHCHCH_2)-(H)(CO)_2]^+$, **7**, 663, 664
$Ru_2C_{16}H_{19}O_6P$
$Ru_2Cp_2(CO)_3[P(OMe)_3]$, **7**, 659
$Ru_2C_{16}H_{31}O_5P_4$
$[Ru_2(COMe)(CO)_4(dmpm)_2]^+$, **7**, 629, 630
$[Ru_2Me(CO)_5(dmpm)_2]^+$, **7**, 629, 630
$Ru_2C_{16}H_{34}I_2N_2O_{12}P_4$
$Ru_2I_2(CO)_4(etmdp)_2$, **7**, 629
$Ru_2C_{16}H_{34}N_2O_{12}P_4$
$Ru_2(CO)_4(etdp)_2$, **7**, 632
$Ru_2C_{17}H_{16}O_3$
$Ru_2Cp_2(CO)_3(CH_2CHCHCH_2)$, **7**, 660
$Ru_2C_{17}H_{17}O_2$
$[Ru_2Cp_2(CMeCMeCHO)-(CO)_2]^+$, **7**, 658, 659
$Ru_2C_{17}H_{18}O_2$
$Ru_2Cp_2(CCH_2)(CMe_2)(CO)_2$, **7**, 660, 661
$Ru_2C_{17}H_{19}NO_7$
$Ru_2[OC(NMe_2)][OC[CMe-CMe(CMeCHMe)]-(CO)_5]$, **7**, 646, 647
$Ru_2C_{17}H_{19}O_2$
$[Ru_2Cp_2(CMe)(CMe_2)(CO)_2]^+$, **7**, 660, 661
$Ru_2C_{17}H_{20}O_2$
$Ru_2Cp_2(CHMe)(CMe_2)(CO)_2$, **7**, 661
$Ru_2C_{17}H_{31}O_6P_4$
$[Ru_2[C(O)Me](CO)_5(dmpm)_2]^+$, **7**, 629, 630
$Ru_2C_{17}H_{34}NO_{13}P_4$
$Ru_2(CO)_5[(MeO)_2PNEtP-(OMe)_2]_2$, **7**, 626
$Ru_2C_{17}H_{34}N_2O_{13}P_4$
$Ru_2(CO)_5(etmdp)_2$, **7**, 628, 629
$Ru_2(CO)_5[(MeO)_2PNEtP-(OMe)_2]_2$, **7**, 626, 627
$Ru_2C_{17}H_{34}N_2O_{14}P_4$
$Ru_2[OC(O)](CO)_4(etmdp)_2$, **7**, 628
$Ru_2C_{17}H_{35}N_2O_{13}P_4$
$[Ru_2H(CO)_5(etmdp)_2]^+$, **7**, 629
$Ru_2C_{17}H_{35}N_2O_{14}P_4$
$[Ru_2[OC(OH)](CO)_4(etmdp)_2]^+$, **7**, 628
$Ru_2C_{17}H_{36}N_2O_{13}P_4$
$[Ru_2H_2(CO)_5(etmdp)_2]^+$, **7**, 629
$Ru_2C_{17}H_{36}N_2O_{14}P_4$
$[Ru_2(CO)_5(H_2O)(etmdp)_2]^{2+}$, **7**, 628
$Ru_2C_{18}H_8N_6O_6$
$Ru_2(\overline{NNNC_6H_4})_2(CO)_6$, **7**, 645
$Ru_2C_{18}H_{12}O_6$
$Ru_2[C_5H_4C(O)CH_2CH_2C(O)-C_5H_4](CO)_4$, **7**, 655
$Ru_2C_{18}H_{18}$
$[Ru_2Cp_2(cot)]^{2+}$, **7**, 673
$Ru_2C_{18}H_{20}Cl_4O_4$
$[RuCl_2(MeC_6H_4CO_2Me)]_2$, **7**, 559, 561
$Ru_2C_{18}H_{24}Cl_2N_2O_4$
$Ru_2(Cl)_2[N(Pr^i)C(H)C(H)Me]_2-(CO)_4$, **7**, 895
$Ru_2C_{18}H_{36}N_2O_{12}P_4$
$Ru_2(HCCH)(CO)_4(etmdp)_2$, **7**, 632
$Ru_2C_{18}H_{54}Cl_3P_6$
$[Ru_2Cl_3(PMe_3)_6]^+$, **7**, 381
$Ru_2C_{19}H_{10}F_2O_{10}$
$Ru_2[OC(FC_6H_4)O]_2(CO)_5-(H_2O)$, **7**, 641
$Ru_2C_{19}H_{20}O_2$
$Ru_2CP_2(CHCHCHCHCMe_2)-(CO)_2$, **7**, 660
$Ru_2C_{20}H_9O_8P$
$Ru_2[PPhC(O)C_6H_4](CO)_7$, **7**, 649, 650
$Ru_2C_{20}H_{10}O_8P$
$[Ru_2(PPh_2)(CO)_8]^+$, **7**, 650
$Ru_2C_{20}H_{11}N_3O_5$
$Ru_2(C_5H_4N)(C_{10}H_7N_2)(CO)_5$, **7**, 646
$Ru_2C_{20}H_{16}O_3$
$Ru_2Cp_2(CHPh)(CO)_3$, **7**, 665
$Ru_2C_{20}H_{18}I_4$
$[CpRu(C_{10}H_8)(I)RuCp][I_3]$, **7**, 618
$Ru_2C_{20}H_{28}Cl_2N_6$
$[RuCl(N_3)(cym)]_2$, **7**, 559
$Ru_2C_{20}H_{28}Cl_4$
$[RuCl_2(C_5Me_4CH_2)]_2$, **7**, 478
$[RuCl_2(cym)]_2$, **7**, 557
$Ru_2C_{20}H_{30}Cl_2N_2O_2$
$Ru_2Cp^*_2Cl_2(NO)_2$, **7**, 667
$Ru_2C_{20}H_{30}Cl_3$
$Ru_2(Cl)_3Cp^*_2$, **7**, 477
$Ru_2C_{20}H_{30}Cl_3F_6P$
$[Ru_2(Cl)_3Cp^*_2]PF_6$, **7**, 477
$Ru_2C_{20}H_{30}Cl_4$
$Ru_2Cp^*_2(Cl)_4$, **7**, 509, 667, 668
$Ru_2C_{20}H_{30}Cl_4O$
$Ru_2Cl_4(O)Cp^*_2$, **7**, 478
$Ru_2C_{20}H_{32}Cl_4$
$[RuCl_2(C_{10}H_{16})]_2$, **7**, 461
$Ru_2C_{20}H_{34}$
$Ru_2Cp^*_2(H)_4$, **7**, 671, 672
$Ru_2C_{20}H_{34}O_4P_4$
$Ru_2(DMAD)(CO)_4(dmpm)_2$, **7**, 631
$Ru_2C_{21}H_{17}NO_3$
$Ru_2Cp_2(CO)(CNBz)(CO)_2$, **7**, 656
$Ru_2C_{21}H_{20}N_4O_7$

$Ru_2(CO)_5[py_2(MeCONH)_2\text{-}CH_2]$, **7**, 637

$Ru_2C_{21}H_{32}O$
$Ru_2Cp^*{}_2(H)_2(CO)$, **7**, 656

$Ru_2C_{21}H_{34}O_9P_4$
$Ru_2[C(O)C_2(CO_2Me)_2](CO)_4\text{-}(dmpm)_2$, **7**, 631

$Ru_2C_{22}H_{18}N_4O_6$
$Ru_2[HNC(Ph)O]_2(CO)_4\text{-}(MeCN)_2$, **7**, 644

$Ru_2C_{22}H_{20}$
$[Ru_2(C_5H_4CH_2C_5H_4)_2]^{2+}$, **7**, 673

$Ru_2C_{22}H_{20}NO_3$
$[Ru_2Cp_2[CN(Me)Bz](CO)_3]^+$, **7**, 656

$Ru_2C_{22}H_{26}O_5$
$Ru_2Cp_2(CMe_2)[C(CO_2Me)C\text{-}(CO_2Me)CHMe](CO)$, **7**, 661

$Ru_2C_{22}H_{32}Cl_2N_2S_2$
$[RuCl(SCN)(C_{10}H_{16})]_2$, **7**, 461

$Ru_2C_{22}H_{32}O_2$
$Ru_2Cp^*{}_2(H)_2(CO)_2$, **7**, 656

$Ru_2C_{22}H_{34}Cl_4$
$[RuCl_2(C_5Me_4Et)]_2$, **7**, 477

$Ru_2C_{22}H_{36}O_2$
$Ru_2Cp^*{}_2(OMe)_2$, **7**, 477, 666

$Ru_2C_{22}H_{40}N_2O_{16}P_4$
$Ru_2(DMAD)(CO)_4(etmdp)_2$, **7**, 631

$Ru_2C_{22}H_{50}$
$Ru_2Me_2(CH_2Bu^t)_4$, **7**, 377

$Ru_2C_{23}H_{12}O_8$
$Ru_2[CHCPhCHC(O)Ph](CO)_7$, **7**, 653

$Ru_2C_{23}H_{14}O_5$
$Ru_2(C_{18}H_{14})(CO)_5$, **7**, 654

$Ru_2C_{23}H_{32}ClF_3O_2$
$Ru_2Cp^*{}_2(Cl)(CH_2)[OC(CF_3)O]$, **7**, 668, 669

$Ru_2C_{24}H_{19}N_4O_4$
$[Ru_2(H)(NC_5H_4)_2(CO)_4(py)_2]^+$, **7**, 951

$Ru_2C_{24}H_{19}O_7P$
$Ru_2[OP(Ph)_2[CCBu^t](CO)_6$, **7**, 650

$Ru_2C_{24}H_{20}S_2$
$Ru_2Cp_2(S_2C_2Ph_2)$, **7**, 671

$Ru_2C_{24}H_{21}O_2P$
$Ru_2Cp_2(PPh_2)(H)(CO)_2$, **7**, 656, 657

$Ru_2C_{24}H_{27}N$
$[Ru_2CpCp^*(C_7H_9)(MeCN)]^+$, **7**, 669

$Ru_2C_{24}H_{30}Cl_2F_8$
$[RuCl(C_2F_4)Cp^*]_2$, **7**, 505

$Ru_2C_{24}H_{30}O_4$
$[Ru(CO)_2Cp^*]_2$, **7**, 478

$Ru_2C_{24}H_{30}O_7S_2$
$[Ru(CO)_2Cp^*]_2(S_2O_3)$, **7**, 488

$Ru_2C_{24}H_{32}O_3$
$Ru_2Cp^*{}_2(CH_2)(CO)_3$, **7**, 663

$Ru_2C_{24}H_{34}Cl_2$
$Ru_2Cp^*{}_2(C_4H_4)Cl_2$, **7**, 668

$Ru_2C_{24}H_{34}O_6P_2$
$Cp(Me_3P)_2Ru[C(CO_2Me)C\text{-}(CO_2Me)]Ru(CO)_2Cp$, **7**, 480

$Ru_2C_{24}H_{36}Cl_4$
$[Ru(C_6Me_6)Cl_2]_2$, **1**, 341

$Ru_2C_{24}H_{38}$
$Ru_2Cp^*{}_2(CHCH_2)(H)(C_2H_4)$, **7**, 671

$Ru_2C_{24}H_{54}$
$Ru_2Et_2(CH_2Bu^t)_4$, **7**, 377

$Ru_2C_{24}H_{72}Cl_2P_8$
$[Ru_2Cl_2(PMe_3)_8]^{2+}$, **7**, 381

$Ru_2C_{25}H_{20}O$
$Ru_2Cp_2(C_2Ph_2)(CO)$, **7**, 657, 658

$Ru_2C_{25}H_{21}O_7P$
$Ru_2[CHCPhCHC(O)Ph](CO)_6(PMe_3)$, **7**, 653

$Ru_2C_{25}H_{31}O_3$
$[Ru_2Cp^*{}_2(CCH)(CO)_3]^+$, **7**, 663

$Ru_2C_{25}H_{32}O_3$
$Ru_2Cp^*{}_2[CH_2C(O)](CO)_3$, **7**, 663

$Ru_2C_{25}H_{34}O_3$
$Ru_2Cp^*{}_2(CHMe)(CO)_3$, **7**, 660, 663
$Ru_2Cp^*{}_2(CO)_3(C_2H_4)$, **7**, 660, 663

$Ru_2C_{25}H_{35}$
$[Cp^*RuCp^*RuCp]^+$, **7**, 617, 618

$Ru_2C_{26}H_{15}O_6P$
$Ru_2(PPh_2)(CCPh)(CO)_6$, **7**, 650

$Ru_2C_{26}H_{16}O_6P_2$
$Ru_2(PPh_2)(\overline{PCHCHC_6H_4})(CO)_6$, **7**, 649, 650

$Ru_2C_{26}H_{17}O_6P$
$Ru_2(CMeC_6H_4PPh_2)(CO)_6$, **7**, 653

$Ru_2C_{26}H_{18}O_4$
$Ru_2(C_5H_4Ph)_2(CO)_4$, **7**, 655

$Ru_2C_{26}H_{20}O_3$
$Ru_2CP_2(CPh_2)(CO)_3$, **7**, 659

$Ru_2C_{26}H_{34}O_3$
$Ru_2Cp^*{}_2(CCHMe)(CO)_3$, **7**, 663

$Ru_2C_{26}H_{36}O_2$
$Ru_2Cp^*{}_2(CH_2CCHCH_2)(CO)_2$, **7**, 660
$Ru_2Cp^*{}_2(CHCHCHMe)(CO)_2$, **7**, 660

$Ru_2C_{26}H_{40}$
$Ru_2Cp^*{}_2(C_4H_4)Me_2$, **7**, 668, 669
$Ru_2Cp^*{}_2(CHCH_2)_2(C_2H_4)$, **7**, 671, 672

$Ru_2C_{26}H_{40}O_4$
$[Ru(OEt)(CO)Cp^*]_2$, **7**, 478

$Ru_2C_{26}H_{43}N$
$Ru_2Cp^*{}_2(NPr^iCMeCH_2)(H)$, **7**, 673

$Ru_2C_{26}H_{44}I_2S_2$
$Ru_2Cp^*{}_2(SPr^i)_2I_2$, **7**, 670

$Ru_2C_{26}H_{44}S_2$
$Ru_2(C_5Me_4Et)_2(SEt)_2$, **7**, 667

$Ru_2C_{26}H_{44}S_4$
$[Ru_2Cp^*{}_2(S_2)(SPr^i)_2]^+$, **7**, 670

$Ru_2C_{27}H_{15}N_7$
$Ru_2[PhCCPhC(O)NPh](CO)_6$, **7**, 655

$Ru_2C_{27}H_{17}NO_6$
$Ru_2(C_{21}H_{17}N)(CO)_6$, **7**, 655
$Ru_2(CO)_6[TolNHCCPh(C_6H_4)]$, **7**, 797

$Ru_2C_{27}H_{17}O_6P$
$Ru_2(PPh_2)(CPhCCH_2)(CO)_6$, **7**, 650, 651, 900

$Ru_2C_{27}H_{20}NO_6P$
$Ru_2(PPh_2)[PhHCC(HN)CH_2](CO)_6$, **7**, 650, 651

$Ru_2C_{27}H_{20}O_3$
$Ru_2Cp_2[CPhCPhC(O)](CO)_2$, **7**, 657, 665

$Ru_2C_{27}H_{22}N_2O_5$
$Ru_2(N_2C_8H_{11})(PhCCHPh)\text{-}(CO)_5$, **7**, 649

$Ru_2C_{27}H_{24}O$
$Ru_2Cp_2(CPhCHPh)(CHCH_2)\text{-}(CO)$, **7**, 658

$Ru_2C_{27}H_{36}O_3$
$Ru_2Cp^*{}_2(CMeCHCH_2)(CO)_3$, **7**, 660

$Ru_2C_{27}H_{43}P$
$Ru_2Cp^*{}_2(C_4H_4)(PMe_3)$, **7**, 672

$Ru_2C_{27}H_{45}P$
$Ru_2Cp^*{}_2(CHCH)_2H_2(PMe_3)$, **7**, 672

$Ru_2C_{27}H_{46}N_2O_{21}P_4$
$Ru_2(DMAD)[C(O)C(O)C_2\text{-}(CO_2Me)_2](CO)_3\text{-}(dmpm)_2$, **7**, 631

$Ru_2C_{28}H_{22}O_4$
$Ru_2(C_5H_4Bz)_2(CO)_4$, **7**, 655

$Ru_2C_{28}H_{23}O_2$
$Ru_2Cp_2(CPhCPhCHMe)(CO)_2$, **7**, 658

$Ru_2C_{28}H_{24}N_2O_2$
$Ru_2Cp_2(CNBz)_2(CO)_2$, **7**, 656

$Ru_2C_{28}H_{25}NO_7$
$Ru_2[OC(NMe_2)][CMeCHC\text{-}(Tol)CH(Tol)CH(Tol)]\text{-}(CO)_6$, **7**, 646, 647

$Ru_2C_{28}H_{38}O_2$
$Ru_2Cp^*{}_2(CMeCHCH_2CCH_2)\text{-}(CO)_2$, **7**, 662, 663

$Ru_2C_{28}H_{38}O_4P_4$
$Ru_2(PhCCPh)(CO)_4(dmpm)_2$, **7**, 631

$Ru_2C_{28}H_{38}O_6S_2$
$Ru_2(SCHC_{10}H_{18})_2(CO)_6$, **7**, 647

$Ru_2C_{29}H_{22}I_2O_4P_2$
$Ru_2I_2(CO)_4(dppm)$, **7**, 646, 647

$Ru_2C_{29}H_{26}NO_5P$
$Ru_2(C_{21}H_{17}N)(CO)_5(PMe_3)$, **7**, 655

$Ru_2C_{30}H_{20}O_6P_2$
$Ru_2(PPh_2)_2(CO)_6$, **7**, 649

$Ru_2C_{30}H_{25}O_6$
$[Ru(CO)_3(Cp)_2]_2[C(Me)\text{-}CHCH]$, **7**, 880

$Ru_2C_{30}H_{40}Cl_2F_6N_2P$

[Ru(Cl)(py)Cp*]$_2$[PF$_6$], **7**, 538
$Ru_2C_{30}H_{60}N_2O_8P_4$
Ru$_2$H[(Pr^{i}O)P(O)NEtP(OPr^i)$_2$]-(CO)$_5$(etipdp), **7**, 633
$Ru_2C_{30}H_{66}$
Ru$_2$(CH$_2$Bu^t)$_6$, **7**, 674
$Ru_2C_{31}H_{45}F_3O_3S$
[Cp*RuCp*RuCp$^+$][OTf], **7**, 618
$Ru_2C_{31}H_{67}IN_2O_{11}P_4$
Ru$_2$(I)H(CO)$_3$(etipdp)$_2$, **7**, 634
$Ru_2C_{32}H_{22}O_6P_2$
Ru$_2$(CO)$_6$(Ph$_2$PCHCHPPh$_2$), **7**, 626
$Ru_2C_{32}H_{22}O_7P_2$
Ru$_2$(CO)$_7$(dppm), **7**, 302, 626
$Ru_2C_{32}H_{24}O_6P_2$
Ru$_2$(CH$_2$)(CO)$_6$(dppm), **7**, 632
$Ru_2C_{32}H_{26}NO_6P$
Ru$_2$(PPh$_2$)[PhCC(CNBu^t)CH$_2$]-(CO)$_6$, **7**, 650, 651
$Ru_2C_{32}H_{28}F_6O_{12}$
Ru$_2$Cp$_2$[C$_8$(CF$_3$)$_2$(CO$_2$Me)$_6$], **7**, 664, 665
$Ru_2C_{32}H_{42}N_2$
[Ru(NHPh)Cp*]$_2$, **7**, 478, 667
$Ru_2C_{32}H_{50}N_2O_4P_2$
Ru$_2$(C$_{10}$H$_8$N$_2$)(CO)$_4$(PPr$^i_3$)$_2$, **10**, 339
$Ru_2C_{32}H_{52}N_2O_4P_2$
Ru$_2$(dan)(CO)$_4$(PPr$^i_3$)$_2$, **7**, 648
$Ru_2C_{32}H_{53}S_2$
Ru$_2$Cp*$_2$(SPr^i)$_2$(CCBu^t), **7**, 670
$Ru_2C_{32}H_{67}N_2O_{12}P_4$
[Ru$_2$(H)(CO)$_4$(etipdp)$_2$]$^+$, **7**, 633, 634
$Ru_2C_{32}H_{68}N_2O_{12}P_4$
Ru$_2$H$_2$(CO)$_4$(etipdp)$_2$, **7**, 635
$Ru_2C_{33}H_{24}O_6P_2$
Ru$_2$[C(O)CHCH](CO)$_5$(dppm), **7**, 632
$Ru_2C_{33}H_{25}NO_2S$
Ru$_2$Cp$_2$[SCPhCPHC(NPh)]-(CO)$_2$, **7**, 660
$Ru_2C_{33}H_{26}O_6P_2$
Ru$_2$[CH$_2$C(O)CH$_2$](CO)$_5$-(dppm), **7**, 654, 749
$Ru_2C_{33}H_{32}O_6P_2$
Ru$_2$(PPh$_2$)[PhCC[P(OEt)$_3$]-CH$_2$](CO)$_6$, **7**, 650, 651
$Ru_2C_{33}H_{34}Cl_5N_2O_{15}P_4$
[Ru$_2$Cl(CO)$_5$(etmdp)$_2$](OC$_6$-Cl$_4$O), **7**, 628
$Ru_2C_{33}H_{54}O_3P_2S_2$
Ru$_2$(SCCC$_{10}$H$_{18}$)$_2$(CO)$_3$[P-(OMe)$_3$]$_2$, **7**, 647, 648
$Ru_2C_{33}H_{67}N_2O_{12}P_4S_2$
[Ru$_2$[SC(H)S](CO)$_4$(etipdp)$_2$]$^+$, **7**, 634
$Ru_2C_{34}H_{20}O_6$
Ru$_2$(C$_4$Ph$_4$)(CO)$_6$, **7**, 655
$Ru_2C_{34}H_{26}O$
RuCp(CPhCPhCPhCPhCHO), **7**, 666
$Ru_2C_{34}H_{30}F_{12}N_6O_4P_2$
[Ru$_2$(CO)$_4$(py)$_6$](PF$_6$)$_2$, **7**, 633
$Ru_2C_{34}H_{31}NO_6PS$
Ru$_2$(PPh$_2$)[PhCC[C(EtS)N-(HBu^t)]CH$_2$](CO)$_6$, **7**, 650, 651
$Ru_2C_{34}H_{33}N_2O_6P$
Ru$_2$(PPh$_2$)[PhCC[C(EtNH)N-(HBu^t)]CH$_2$](CO)$_6$, **7**, 650, 651
$Ru_2C_{34}H_{42}$
Ru$_2$Cp*$_2$(PhCCPh)(H)$_2$, **7**, 672
$Ru_2C_{34}H_{44}Cl_2S_2$
Ru$_2$Cp*$_2$Cl$_2$(SBz)$_2$, **7**, 669
$Ru_2C_{35}H_{42}Cl_2$
Ru$_2$(CHCHCPh$_2$)Cl$_2$Cp*$_2$, **7**, 341
$Ru_2C_{36}H_{32}O_8$
Ru$_2$Cp$_2$[C$_6$Ph$_2$(CO$_2$Me)$_4$], **7**, 664, 665
$Ru_2C_{36}H_{63}N_9$
Ru$_2$(CNPr^i)$_9$, **7**, 327
$Ru_2C_{37}H_{27}IO_4P_2$
Ru$_2$(I)(CO)$_4$(dppm)(CCPh), **7**, 772
$Ru_2C_{37}H_{33}OP_2$
[Ru$_2$Cp$_2$(CH)(CO)(dppm)]$^+$, **7**, 661
$Ru_2C_{38}H_{30}Cl_5O_2P_2$
[Ru$_2$(Cl)$_3$Cl$_2$(CO)$_2$(PPh$_3$)$_2$]$^-$, **7**, 319
$Ru_2C_{38}H_{31}O_6P$
Ru$_2$[CPhCPhCH[CPhCHC-(O)Ph]](CO)$_5$(PMe$_3$), **7**, 653
$Ru_2C_{38}H_{35}OP_2$
[Ru$_2$Cp$_2$(CHCH$_2$)(CO)-(dppm)]$^+$, **7**, 661
$Ru_2C_{38}H_{46}$
[[Ru(C$_6$Me$_6$)]$_2$(C$_{14}$H$_{10}$)]$^{4+}$, **7**, 550
$Ru_2C_{38}H_{47}P$
Ru$_2$Cp*$_2$(PPh$_2$)(H)(C$_6$H$_6$), **7**, 672
$Ru_2C_{39}H_{38}N_4O_3P_2$
Ru$_2$(H)(Me$_2$pz)$_2$[Ph$_2$PC$_2$H$_4$P-(Ph)C$_6$H$_4$)](CO)$_3$, **7**, 645
$Ru_2C_{39}H_{45}F_3NO_3S$
Ru$_2$Cp*$_2$(C$_4$Ph$_2$H$_2$)(MeCN)-(CF$_3$SO$_3$), **7**, 669
$Ru_2C_{39}H_{72}N_3O_{12}P_4$
[Ru$_2$[N(CHPh)](CO)$_4$-(etipdp)$_2$]$^+$, **7**, 634
$Ru_2C_{39}H_{72}N_3O_{12}P_4S$
[Ru$_2$[PhNC(H)S](CO)$_4$-(etipdp)$_2$]$^+$, **7**, 634, 635
$Ru_2C_{40}H_{30}Cl_2F_3O_4P_2$
[Ru$_2$(O$_2$CCF$_3$)(Cl)$_2$(CO)$_2$-(PPh$_3$)$_2$]$^+$, **7**, 363
$Ru_2C_{40}H_{30}I_2O_4P_2$
Ru$_2$I$_2$(CO)$_4$(PPh$_3$)$_2$, **7**, 646, 647
$Ru_2C_{41}H_{31}O_4P_3$
Ru$_2$H(CO)$_4$(Ph$_2$PCHPPh$_2$)-(PhPC$_6$H$_4$), **7**, 649
$Ru_2C_{41}H_{73}N_2O_{13}P_4$
[Ru$_2$[N(CHPh)](CO)$_4$-(etipdp)$_2$]$^+$, **7**, 634
$Ru_2C_{42}H_{31}O_5P_3$
Ru$_2$(CO)$_5$Ph[(Ph$_2$P)$_2$C(H)PPh], **7**, 649
Ru$_2$H(CO)$_4$(Ph$_2$PCHPPh$_2$)-[PhPC$_6$H$_4$C(O)], **7**, 649
$Ru_2C_{42}H_{72}F_{12}N_8O_2$
[Ru$_2$(CO)$_2$(CNBu^t)$_8$](PF$_6$)$_2$, **7**, 633
$Ru_2C_{42}H_{72}N_8O_2$
[Ru$_2$(CO)$_2$(CNBu^t)$_8$]$^{2+}$, **7**, 328, 329
$Ru_2C_{44}H_{35}N_2O_4P_2S$
Ru$_2$[SC$_3$H$_2$N(Me)N](CO)$_4$-(PPh$_3$)$_2$, **7**, 644
$Ru_2C_{44}H_{38}N_2O_6P_2$
Ru$_2$[HNC(Me)O]$_2$(CO)$_4$-(PPh$_3$)$_2$, **7**, 644
$Ru_2C_{44}H_{58}S_2$
Ru$_2$Cp*$_2$(SPr^i)$_2$(CCTol)$_2$, **7**, 669
$Ru_2C_{44}H_{62}F_{12}O_9P_2$
[Ru(O$_2$CCF$_3$)(C$_6$H$_9$PCy$_2$)]$_2$(O)-(O$_2$CCF$_3$)$_2$, **7**, 448
$Ru_2C_{45}H_{33}O_5P$
Ru$_2$[CHCPhCH[CPhCHC-PhCHC(O)Ph]](CO)$_4$-(PMe$_3$), **7**, 653
$Ru_2C_{46}H_{36}N_2O_4P_2$
Ru$_2$(dib)(CO)$_4$(PPh$_3$)$_2$, **7**, 648
$Ru_2C_{46}H_{36}N_4O_4P_2$
Ru$_2$(pz)$_2$(CO)$_4$(PPh$_3$)$_2$, **7**, 645
$Ru_2C_{46}H_{38}N_2O_4P_2S_4$
Ru$_2$(SC$_3$H$_4$NS)$_2$(CO)$_4$(PPh$_3$)$_2$, **7**, 644
$Ru_2C_{46}H_{42}P_2$
Ru$_2$Cp$_2$(H)$_2$(PPh$_3$)$_2$, **7**, 656
$Ru_2C_{46}H_{55}S_3$
Ru$_2$Cp$_2$[CTolCHC(CTolCH -(SPr^i))CHTol](SPr^i)$_2$, **7**, 667
$Ru_2C_{47}H_{58}O_2P_2$
Ru$_2$(OMe)$_2$(dppm)Cp*$_2$, **7**, 497
$Ru_2C_{48}H_{42}F_{12}N_4O_4P_2$
[Ru$_2$(CO)$_4$(MeCN)$_4$(PPh$_3$)$_2$]-(PF$_6$)$_2$, **7**, 633, 634
$Ru_2C_{50}H_{36}O_3$
Ru$_2$CP(CPhCHPh)(CO)$_3$-(C$_4$Ph$_4$), **7**, 666
$Ru_2C_{50}H_{38}N_2O_6P_2$
Ru$_2$(NC$_5$H$_4$O)$_2$(CO)$_4$(PPh$_3$)$_2$, **7**, 644
$Ru_2C_{52}H_{42}N_2O_4P$
Ru$_2$[HNC(Ph)O][NHC(O)-C$_6$H$_4$](H)(CO)$_2$(PPh$_3$)$_3$, **7**, 644
$Ru_2C_{53}H_{45}N_2O_3P_3$
Ru$_2$(dib)(CO)$_3$(PPh$_3$)(dppe), **7**, 649
$Ru_2C_{55}H_{44}O_5P_4$
Ru$_2$(CO)$_5$(dppm)$_2$, **7**, 626, 711
$Ru_2C_{56}H_{47}O_6P_4$
[Ru$_2$[OC(Me)O](CO)$_4$ (dppm)$_2$]$^+$, **7**, 643
$Ru_2C_{58}H_{48}O_6P_4$
[Ru$_2$(CO)$_6$(dppe)$_2$]$^{2+}$, **7**, 303, 633
$Ru_2C_{58}H_{51}O_6P_4$

$[Ru_2[OC(Me)O](CO)_4$-(dppe)$_2]^+$, **7**, 643
$Ru_2C_{60}H_{50}N_4$
$Ru_2(CCPh)_2(PhNpy)_4$, **7**, 674
$Ru_2C_{60}H_{52}O_8P_4$
$Ru_2[HC(CO_2Me)CH(CO_2Me)]$-$(CO)_4(dmpm)_2$, **7**, 632
$Ru_2C_{61}H_{47}NO_4P_2$
$Ru_2(C_{21}H_{17}N)(CO)_4(PPh_3)_2$, **7**, 655
$Ru_2C_{62}H_{42}O_6$
$Ru_2(C_5Ph_4OHOC_5Ph_4)(H)$-$(CO)_4$, **7**, 657
$Ru_2C_{64}H_{52}N_4O_4P_2$
$Ru_2[OC(Ph)NH]_2[Ph_2POC$-$(Ph)N]_2Ph_2$, **7**, 674
$Ru_2C_{68}H_{68}N_4O_{12}P_2$
$Ru_2[OC[C_6H_3(OMe)_2]NH]_2$-$[Ph_2POC[C_6H_3(OMe)_2]$-$N]_2Ph_2$, **7**, 674
$Ru_2C_{68}H_{144}N_4O_{22}P_6$
$[N(Bu)_4]_4[Ru_2(P_3O_9)_2(CO)_4]$, **7**, 633
$Ru_2C_{70}H_{64}N_4O_4P_2$
$Ru_2[OC(Ph)NH]_2[C_6H_4Me)_2$-$POC(Ph)N]_2(C_6H_4Me)_2$, **7**, 674
$Ru_2C_{74}H_{54}Cl_2O_{16}$
$Ru_2Cp*_2Cl_2$ $(MeOCOCCCO_2Me)$-$[C_6(CO_2Me)_6]$, **7**, 669
$Ru_2C_{74}H_{61}Cl_2O_2P_4$
$[Ru_2Cl_2H(CO)_2(PPh_3)_4]^+$, **7**, 319, 647
$Ru_2C_{75}H_{62}F_3O_4P_4S$
$[Ru_2(H)_2[OS(CF_3)O](CO)_2$-$(PPh_3)_4]^+$, **7**, 651
$Ru_2C_{80}H_{69}O_5P_6$
$[Ru_2[OC(Me)O](CO)_3$-$(dppm)_3]^+$, **7**, 643
$Ru_2C_{86}H_{70}P_4$
$Cp(Ph_3P)_2RuCCCCRu(PPh_3)_2$-Cp, **7**, 521
$Ru_2C_{86}H_{72}P_4$
$[Cp(Ph_3P)_2Ru(CCHCHC)Ru$-$(PPh_3)_2Cp]^{2+}$, **7**, 356
$Ru_2C_{90}H_{90}N_{10}$
$[Ru_2(CNC_6H_3Me_2)_{10}]^{2+}$, **7**, 328
$Ru_2CoC_{15}H_6NO_9$
$Ru_2Co(H)(NPh)(CO)_9$, **10**, 218
$Ru_2CoRh_2C_{19}H_{14}O_8P$
$Ru_2CoRh_2(H)(PMe)(CO)_8Cp_2$, **10**, 243
$Ru_2Co_2C_{12}H_2O_{12}$
$Ru_2Co_2(H)_2(CO)_{12}$, **10**, 35, 197
$Ru_2Co_2C_{13}O_{13}$
$Ru_2Co_2(CO)_{13}$, **10**, 35, 197
$Ru_2Co_2C_{15}H_6O_{11}$
$Ru_2Co_2(C_2Me_2)(CO)_{11}$, **10**, 35
$Ru_2CuC_{21}H_{40}N_4O_{13}P_4$
$[Ru_2[Cu(MeCN)_2](CO)_5$-$(etmdp)_2]^+$, **7**, 627
$Ru_2FeC_{12}H_{10}NO_{12}P$
$FeRu_2(NH)(CO)_9[P(OMe)_3]$, **10**, 218
$Ru_2FeC_{12}O_{12}$
$FeRu_2(CO)_{12}$, **10**, 37, 43, 196
$Ru_2FeC_{46}H_{30}O_{10}P_2$
$Ru_2Fe(CO)_{10}(PPh_3)_2$, **7**, 685
$Ru_2Fe_2C_{11}O_{11}SeTe$
$Fe_2Ru_2(CO)_{11}SeTe$, **7**, 75
$Ru_2Fe_2C_{11}O_{11}Te_2$
$Fe_2Ru_2(Te)_2(CO)_{11}$, **10**, 11, 12, 232, 233
$Ru_2Fe_2C_{20}H_{10}O_8$
$Fe_2Ru(Cp)(CO)_6[CCRu$-$(CO)_2Cp]$, **7**, 22
$Ru_2Fe_4C_{22}O_{22}$
$[Fe_2Ru(CO)_{11}]_2^{2-}$, **7**, 83
$Ru_2HgC_{17}H_{34}ClN_2O_{13}P_4$
$[Ru_2(HgCl)(CO)_5(etmdp)_2]^+$, **7**, 627
$Ru_2HgC_{26}H_{78}P_8$
$Hg[RuMe(PMe_3)_4]_2$, **7**, 381
$Ru_2HgCo_6C_{24}O_{24}$
$[RuCo_3(CO)_{12}]_2(Hg)$, **10**, 341
$Ru_2InC_{14}H_{10}ClO_4$
$InCl[Ru(CO)_2Cp]_2$, **1**, 562
$Ru_2InC_{14}H_{10}Cl_2O_4$
$InCl_2[Ru(CO)_2Cp]_2$, **1**, 562
$Ru_2K_2C_{108}H_{92}P_2$
$K_2[(Ph_3P)_2Ph_2PC_6H_4RuH_2]_2$, **1**, 26, 27
$Ru_2MoC_{21}H_{10}O_8$
$CpMoRu_2(CO)_8(C{\equiv}CPh)$, **10**, 20, 21
$Ru_2MoC_{46}H_{40}P_2S_4$
$[Ru(PPh_3)Cp]_2(MoS_4)$, **7**, 524
$Ru_2NaC_4O_4Se_{16}$
$[Na[Ru(CO)_2(Se_4)_2]_2]^{3-}$, **7**, 317
$Ru_2PtC_{46}H_{34}O_6P_2$
$PtRu_2(CO)_6(PhCCPh)(dppe)$, **9**, 572
$Ru_2PtC_{62}H_{74}S_3$
$Ru_2Cp*_2(SPr^i)_2[SPt(PPh_3)_2S]$, **7**, 671
$Ru_2Rh_2C_{12}H_2O_{12}$
$H_2Ru_2Rh_2(CO)_{12}$, **10**, 39, 48
$Ru_2SbC_{17}H_{34}Cl_2N_2O_{13}P_4$
$[Ru_2(SbCl_2)(CO)_5(etmdp)_2]^+$, **7**, 627
$Ru_2Sb_2C_{36}H_{72}F_{12}N_2O_{10}P_4$
$[Ru_2(CO)_5(acetone)(etipdp)_2]$-$(SbF_6)_2$, **7**, 628
$Ru_2SiC_{24}H_{41}Cl$
$Ru_2(Cl)(CH_2)(TMS)Cp*_2$, **7**, 504, 668, 669
$Ru_2SiC_{26}H_{40}ClF_3O_2$
$Ru_2Cp*_2(Cl)(CH$-$TMS)[OC$-$(CF_3)O]$, **7**, 668, 669
$Ru_2SiC_{28}H_{44}$
$Ru_2Cp*_2Cp(TMS)$, **7**, 668, 669
$Ru_2Si_2C_{18}H_{26}O_4$
$Ru_2(C_8H_8TMS)(TMS)(CO)_4$, **7**, 654
$Ru_2Si_2C_{19}H_{24}O_5$
$Ru_2(C_8H_8SiMe_2CH_2CH_2SiMe_2)$ $(CO)_5$, **7**, 654
$Ru_2Si_2C_{19}H_{26}O_5$
$Ru_2(C_8H_8TMS)(TMS)(CO)_5$, **7**, 654
$Ru_2Si_2C_{36}H_{36}O_6P_2$
$Ru_2(Ph_2PCH_2SiMe_2)_2(CO)_6$, **7**, 633
$Ru_2Si_2C_{38}H_{66}S$
$Ru_2Cp*_2[TMS$-$CCC(CH$-$TMS)CC$-$TMS](SPr^i)$-(H), **7**, 670
$Ru_2Si_2C_{44}H_{52}$
$Ru_2Cp*_2(SiPh_2)_2(H)_2$, **7**, 672, 673
$Ru_2Si_4C_{16}H_{44}O_2$
$Ru_2(O)_2(CH_2TMS)_4$, **7**, 674
$Ru_2Sn_2C_{14}H_{18}O_8$
$[Ru(CO)_4(SnMe_3)]_2$, **7**, 838
$Ru_2TiW_5C_{78}H_{154}N_4O_{40}$
$[NBu_4]_4[Ru_2(TiW_5CpO_{18})_2$-$(CO)_4]$, **7**, 633
$Ru_2WC_{19}H_{11}NO_8$
$WRu_2(H)(NPh)(CO)_8Cp$, **10**, 218
$Ru_2WC_{20}H_{30}S_6$
$Ru_2Cp*_2(S_2)[(S)(S)_2WS]$, **7**, 670
$Ru_2ZrC_{24}H_{20}O_4$
$Cp_2Zr[Ru(CO)_2Cp]_2$, **4**, 535, 536
$ZrRu_2(CO)_4Cp_4$, **10**, 204, 210, 211
$Ru_2[P(Ph)C(O)C_6H_6](CO)_7$
$Ru_2C_{20}H_9O_8P$, **7**, 924
$Ru_3AgC_{35}H_{22}F_3O_{10}P_2$
$Ru_3(AgO_2CCF_3)(CO)_8(dppm)$, **7**, 720
$Ru_3As_3C_{47}H_{37}O_7$
$Ru_3H_2(AsPh)(CO)_7(AsPh_3)_2$, **7**, 703
$Ru_3AuC_{19}H_{15}IO_{12}P$
$Ru_4(C)(AuPEt_3)(I)(CO)_{12}$, **7**, 852
$Ru_3AuC_{58}H_{42}O_7P_3$
$Ru_3(AuPPh_3)(CO)_7(dppm)$ (CCPh), **7**, 774
$Ru_3AuOsC_{35}H_{34}F_3O_9P_4S$
$[AuRu_3[(Me_2P)_3CH](CO)_9$ $(PPh_3)][OsSCF_3]$, **10**, 299
$Ru_3Au_3C_{65}H_{48}O_{10}P_3$
$Au_3Ru_3(COMe)(CO)_9(PPh_3)_3$, **10**, 292
$Ru_3Au_3C_{68}H_{54}O_8P_3$
$Au_3Ru_3(CO)_8(C_2Bu^t)(PPh_3)_3$, **10**, 292
$Ru_3Au_3C_{74}H_{60}O_8P_3$
$Au_3Ru_3(C_{12}H_{15})(CO)_8(PPh_3)_3$, **10**, 292
$Ru_3BC_9H_4O_9$
$[Ru_3(CO)_9BH_4]^-$, **1**, 296, 297, 300, 301
$Ru_3BC_9H_5O_9$
$Ru_3BH_5(CO)_9$, **1**, 296, 297, 300, 301, 303; **7**, 708
$Ru_3BC_{31}H_{46}Cl_2F_4$
$[Ru_3Cp*_3(CH)(Cl)_2]BF_4$, **7**, 668
$Ru_3BFeC_{12}H_3O_{12}$
$H[FeRu_3(CO)_{12}BH_2]$, **1**, 297, 301

$Ru_3BFeC_{48}H_{31}NO_{12}P_2$
[PPN][$FeRu_3(CO)_{12}BH$], **1**, 297, 301
$Ru_3BWC_{16}H_7O_{11}$
H(CpW)[$Ru_3(CO)_{11}BH$], **1**, 297, 301
$WRu_3(H)(BH)(CO)_{11}Cp$, **10**, 216, 217
$Ru_3BWC_{30}H_{17}O_{11}$
H(CpW)$Ru_3(CO)_{11}$BC(Ph)C-PhH, **1**, 301, 303
WRu_3(H)[BC(Ph)=CPhH]-$(CO)_{11}$Cp, **10**, 216, 217
$Ru_3B_2C_9H_6O_9$
$HRu_3(CO)_9B_2H_5$, **1**, 308
$Ru_3B_2C_{26}H_{21}O_8P$
$HRu_3(CO)_8(PPh_3)B_2H_5$, **1**, 308, 309
$Ru_3B_2C_{45}H_{35}NP_2$
[PPN][$Ru_3(CO)_9B_2H_5$], **1**, 303
$Ru_3B_{10}C_{28}H_{58}O_2$
[$(C_6Me_6)_2Ru_2H_4$]$RuB_{10}H_8$-$(OEt)_2$, **1**, 357, 360; **7**, 796, 797
$Ru_3B_{10}C_{29}H_{58}Cl_2O_2$
$Ru_3H_2(C_6Me_6)_2(CH_2Cl_2)$-[$B_{10}H_8(OEt)_2$], **7**, 710, 723
$Ru_3B_{10}C_{30}H_{56}O_4$
$(C_6Me_6)_2Ru_2H_2(CO)_2RuB_{10}H_8$-$(OEt)_2$, **1**, 358, 360
$Ru_3BiC_9H_3O_9$
$Ru_3H_3(Bi)(CO)_9$, **7**, 701
$Ru_3C_8H_2O_8S_2$
$Ru_3H_2(S)_2(CO)_8$, **7**, 726
$Ru_3C_9ClO_9$
$[Ru_3Cl(CO)_9]^-$, **7**, 721
$Ru_3C_9HIO_9$
$Ru_3H(I)(CO)_9$, **7**, 722
$Ru_3C_9H_2O_9S$
$Ru_3H_2(S)(CO)_9$, **7**, 723
$Ru_3C_9H_2O_9S_2$
$Ru_3H_2(S)_2(CO)_9$, **7**, 726
$Ru_3C_9O_9S_2$
$Ru_3(S)_2(CO)_9$, **7**, 726
$Ru_3C_9O_9Te_2$
$Ru_3(Te)_2(CO)_9$, **7**, 726
$Ru_3C_{10}HClO_{10}$
$Ru_3H(Cl)(CO)_{10}$, **7**, 722
$Ru_3C_{10}HNO_{10}$
$Ru_3(NH)(CO)_{10}$, **7**, 855
$Ru_3C_{10}HNO_{11}$
$Ru_3(CO)_{10}(NOH)$, **10**, 93
$Ru_3C_{10}I_2O_{10}$
$Ru_3I_2(CO)_{10}$, **7**, 722
$Ru_3C_{10}NO_{12}$
$Ru_3(NO)_2(CO)_{10}$, **7**, 730
$Ru_3C_{10}O_{10}S$
$Ru_3(S)(CO)_{10}$, **7**, 726; **10**, 49
$Ru_3C_{11}ClO_{11}$
$[Ru_3Cl(CO)_{11}]^-$, **7**, 721, 722
$Ru_3C_{11}HO_{10}$
$[Ru_3H(CO)_9(CCO)]^-$, **7**, 755
$Ru_3C_{11}HO_{11}$
$[Ru_3H(CO)_{11}]^-$, **7**, 363, 688, 689, 719
$Ru_3C_{11}HO_{12}$
$[Ru_3(HCO_2)(CO)_{10}]^-$, **7**, 724
$Ru_3C_{11}H_2O_{10}$
$Ru_3H_2(CCO)(CO)_9$, **7**, 705, 749, 760
$Ru_3C_{11}H_2O_{11}$
$Ru_3H_2(CO)_{11}$, **7**, 719
$Ru_3(H)(COH)(CO)_{10}$, **7**, 719
$Ru_3C_{11}H_3NO_{11}$
$Ru_3(NOMe)(CO)_{10}$, **7**, 703
$Ru_3C_{11}H_4O_9$
$Ru_3H_2(CCH_2)(CO)_9$, **7**, 749
$Ru_3C_{11}H_5NO_{10}$
$Ru_3H(H_3NCOCH)(CO)_9$, **7**, 725, 726
$Ru_3C_{11}H_6O_{10}$
$H_3Ru_3(CO)_9(COMe)$, **7**, 748
$Ru_3C_{11}N_2O_{11}$
$Ru_3(CO)_{11}(N_2)$, **7**, 716
$Ru_3C_{11}O_{11}$
$[Ru_3(CO)_{11}]^{2-}$, **7**, 716, 719
$Ru_3C_{12}DO_{12}$
$[Ru_3(CO)_{11}(CDO)]^-$, **7**, 748
$Ru_3C_{12}HO_{13}$
$[Ru_3[C(O)OH](CO)_{11}]^-$, **7**, 748
$Ru_3C_{12}H_2O_{11}$
$Ru_3(CH_2)(CO)_{11}$, **7**, 749
$Ru_3C_{12}H_3O_{11}$
$[Ru_3(CO)_9(COCOMe)]^-$, **7**, 748
$Ru_3C_{12}H_4O_{11}$
$Ru_3H(MeCO)(CO)_{10}$, **7**, 696, 720, 748
$Ru_3C_{12}H_6O_9S_3$
$Ru_3(CO)_9(S_3C_3H_6)$, **7**, 716
$Ru_3C_{12}H_6O_{10}$
$H_2Ru_3(CO)_9$[HCC(O)Me], **7**, 748
$Ru_3C_{12}H_6O_{11}$
$Ru_3H_2(CHCO_2Me)(CO)_9$, **7**, 704
$Ru_3C_{12}H_8O_9$
$Ru_3H_3(CEt)(CO)_9$, **7**, 701
$Ru_3C_{12}H_8O_9S$
$Ru_3(H)_3(CO)_9(CSEt)$, **7**, 764
$Ru_3C_{12}H_9ClO_8S_2$
$Ru_3H_2Cl(\overline{CHSCH_2CH_2CH_2S})$-$(CO)_8$, **7**, 722
$Ru_3C_{12}H_{10}N_2O_8$
$Ru_3(CHCH_2)(Me_2NNH)(CO)_8$, **7**, 701
$Ru_3C_{12}O_{12}$
$Ru_3(CO)_{12}$, **1**, 296; **7**, 291, 301, 452, 684, 685, 711, 838; **12**, 426, 430
$Ru_3C_{13}F_3NO_{11}$
$Ru_3(CO)_{11}(CNCF_3)$, **7**, 751
$Ru_3C_{13}H_3NO_{11}$
$Ru_3(CO)_{11}(MeCN)$, **7**, 686
$Ru_3C_{13}H_3O_{11}$
$[Ru_3(CO)_{10}[CC(Me)O]]^-$, **7**, 748, 755
$Ru_3C_{13}H_3O_{13}$
$[Ru_3(CO)_{11}(CO_2Me)]^-$, **7**, 747
$Ru_3C_{13}H_6O_9$
$Ru_3H(CO)_9$[HCC(H)CMe], **7**, 791
$Ru_3H(CO)_9(MeCCCH_2)$, **7**, 791
$Ru_3C_{13}H_7NO_{10}$
$[HRu_3(CO)_{10}(CNMe_2)]^-$, **7**, 747
$Ru_3C_{13}H_7O_9S$
$HRu_3(CO)_9$(DHT), **7**, 751
$Ru_3C_{13}H_8O_9$
$Ru_3H_2(CO)_9$(MeCCMe), **7**, 791
$Ru_3C_{13}H_8O_9S_3$
$Ru_3(CO)_9(C_4H_8S_3)$, **7**, 700
$Ru_3C_{13}H_8O_{10}$
$Ru_3(H)_2(CO)_9$[CC(OMe)Me], **7**, 755
$Ru_3(H)_2(EtOCCH)(CO)_9$, **7**, 703, 751
$Ru_3C_{14}H_4N_2O_{10}$
$Ru_3(CO)_{10}(C_4H_4N_2)$, **7**, 714, 715
$Ru_3C_{14}H_5NO_9S$
$Ru_3H(pyS)(CO)_9$, **7**, 730
$Ru_3C_{14}H_6N_2O_{10}$
$Ru_3(CO)_{10}(MeCN)_2$, **7**, 686
$Ru_3C_{14}H_6O_{11}$
$Ru_3(CO)_{10}$[CCMe(OMe)], **7**, 755, 756
$Ru_3C_{14}H_8O_6S$
$Ru_3(CO)_6(S)(C_8H_8)$, **7**, 797
$Ru_3C_{14}H_9NO_9$
$Ru_3H(CO)_9[Me_2NCC(H)CH]$, **7**, 791
$Ru_3C_{14}H_{11}NO_{10}$
$H_2Ru_3(CO)_9$[=C(OEt)N-(Me)CH], **7**, 748
$Ru_3C_{14}H_{11}NO_{10}$
$Ru_3(H)_2$[EtOCN(Me)CH]-$(CO)_9$, **7**, 751
$Ru_3C_{14}H_{14}O_9P_2$
$HRu_3(CO)_9(Me_2PCHPMe_2)$, **7**, 755
$Ru_3C_{15}H_4NO_{11}$
$[Ru_3(C_5H_4NO)(CO)_{10}]^-$, **7**, 733
$Ru_3C_{15}H_4O_{10}$
$Ru_3(C_5H_4)(CO)_{10}$, **7**, 693, 699
$Ru_3C_{15}H_6O_9P$
$[Ru_3H(PPh)(CO)_9]^-$, **7**, 703, 737
$Ru_3C_{15}H_7NO_9$
$Ru_3H_2(NPh)(CO)_9$, **7**, 731
$Ru_3C_{15}H_8N_2O_9$
$Ru_3H(ampy)(CO)_9$, **7**, 733
$Ru_3C_{15}H_9IO_9$
$(C_2Bu^t)(CO)_9Ru_3(I)$, **10**, 330
$Ru_3C_{15}H_{10}O_9$
$(C_2Bu^t)(CO)_9Ru_3(H)$, **7**, 772, 773; **10**, 332
$Ru_3C_{15}H_{11}NO_9$
$Ru_3H(MeCCCHNMe_2)(CO)_9$, **7**, 705
$Ru_3C_{15}H_{12}O_{11}$
$Ru_3H(MeCOEt)(CO)_9(COMe)$, **7**, 698
$Ru_3C_{15}H_{19}O_9P_3$
$Ru_3(CO)_8[CH(PMe_2)_3]$, **7**, 717
$Ru_3C_{16}H_5NO_{10}$

$Ru_3(NPh)(CO)_{10}$, **7**, 641, 731
$Ru_3C_{16}H_6NO_{10}$
$[HRu_3(CO)_9(PhN{=}CO)]^-$, **7**, 751
$Ru_3C_{16}H_6N_2O_9$
$Ru_3H(C_7H_5N_2)(CO)_9$, **7**, 707
$Ru_3C_{16}H_9NO_{11}$
$Ru_3(CO)_{11}(CNBu^t)$, **7**, 713
$Ru_3C_{16}H_{11}O_8P$
$Ru_3(CO)_8(PMe_2Ph)$, **7**, 685
$Ru_3C_{16}H_{12}N_2O_{12}S$
$Ru_3(H)(CO)_9(S)Ru(CO)_3$-$[CH_2NMeC(NMe_2)]$, **7**, 871, 872
$Ru_3C_{16}H_{12}O_9S$
$HRu_3(CO)_9[S(C{=}CMe_2)C$-$(Me){=}CH_2]$, **7**, 750
$Ru_3C_{16}H_{13}N_2O_{11}$
$[Ru_3(CO)_{10}(NCMe)_2]$, **1**, 303
$Ru_3C_{16}H_{18}O_{16}P_2$
$Ru_3(CO)_{10}[P(OMe)_3]_2$, **7**, 713
$Ru_3C_{16}H_{19}O_9P_3$
$Ru_3(CO)_9[CH(PMe_2)_3]$, **7**, 717, 721
$Ru_3C_{17}H_6O_9S$
$Ru_3(S)(HC_2Ph)(CO)_9$, **10**, 49
$Ru_3C_{17}H_{12}N_2O_9$
$Ru_3(H)(N_2C_8H_{11})(CO)_9$, **7**, 649
$Ru_3H(NHC_6H_2Me_2NH_2)(CO)_9$, **7**, 698
$Ru_3C_{17}H_{14}O_{14}$
$Ru_3(C_9H_{14}O_6)(CO)_8$, **7**, 724
$Ru_3C_{17}H_{15}NO_9$
$Ru_3H(MeCCHCNEt_2)(CO)_9$, **7**, 705
$Ru_3C_{17}H_{16}O_6S$
$Ru_3(CO)_6(SBu^t)(C_7H_7)$, **7**, 797
$Ru_3C_{18}H_{10}O_9$
$Ru_3H_2(C{=}CMePh)(CO)_9$, **7**, 703
$Ru_3C_{18}H_{14}N_2O_8$
$Ru_3(PhCCH_2)(MeNNHMe)$-$(CO)_8$, **7**, 694
$Ru_3C_{18}H_{16}O_4S$
$Ru_3(CO)_4(S)(C_7H_9)(C_7H_7)$, **7**, 797
$Ru_3C_{18}H_{18}O_3$
$Ru_3CP_3(H)_3(CO)_3$, **7**, 665
$Ru_3C_{19}H_8O_7$
$Ru_3(C_{12}H_8)(CO)_7$, **7**, 710
$Ru_3C_{19}H_{10}O_7$
$Ru_3(CO)_7$(acenophylene), **7**, 699
$Ru_3C_{19}H_{11}NO_7$
$Ru_3(CO)_7(NPh)(C_6H_6)$, **7**, 731, 796
$Ru_3C_{19}H_{22}Cl_2O_9$
$Ru_3Cl_2(OCy)_2(CO)_7$, **7**, 722
$Ru_3C_{20}H_8N_2O_{10}$
$Ru_3(CO)_{10}$(bipy), **7**, 714, 715
$Ru_3C_{20}H_{11}O_8P$
$Ru_3H(PPh_2)(CO)_8$, **7**, 707
$Ru_3C_{20}H_{22}Cl_2O_{10}$
$Ru_3(Cl)_2(OC_6H_{11})_2(CO)_8$, **7**, 724
$Ru_3C_{20}H_{34}IN_2O_{16}P_4$
$Ru_3(I)(CO)_8[(MeO)_2PNEtP$-$(OMe)_2]_2$, **7**, 723
$Ru_3C_{20}H_{42}F_3O_{12}P_3S$
$[Ru[P(OMe)_3]_3Cp^*]OTf$, **7**, 497
$Ru_3C_{21}H_{10}NO_9$
$Ru_3(NPh)_2(CO)_9$, **7**, 731
$Ru_3C_{21}H_{11}O_9P$
$HRu_3(CO)_9(PPh_2)$, **7**, 754
$Ru_3C_{21}H_{14}O_6$
$Ru_2Cp(CO)_4[C_5H_4Ru(CO)_2Cp]$, **7**, 665
$Ru_3C_{22}H_{10}O_{12}$
$Ru_3(OC_6H_4CH{=}O)_2(CO)_8$, **7**, 698, 725
$Ru_3C_{22}H_{14}O_{12}$
$Ru_3(OC_6H_4OMe)_2(CO)_8$, **7**, 698, 725
$Ru_3C_{23}H_{24}N_2O_9$
$Ru_3(CO)_9$(Cy,Cy-db), **7**, 635
$Ru_3C_{24}H_{36}Cl_3S_4$
$[Ru_3(C_8H_{12}S)_3(Cl)_3(S)]^+$, **7**, 727
$Ru_3C_{25}H_{10}N_2O_9$
$Ru_3(NPh)_2(CO)_9$, **7**, 731
$Ru_3C_{25}H_{19}O_6P_2$
$Ru_3H_2(PhPCH_2PPh_2)(CO)_6$, **7**, 707, 708, 737
$Ru_3C_{25}H_{39}Cl_3O$
$Ru_3Cl_3(OMe)(cod)_3$, **7**, 450
$Ru_3C_{26}H_{12}N_2O_{10}$
$Ru_3(C_9H_6MO)_2(CO)_8$, **7**, 698
$Ru_3C_{26}H_{19}O_8P$
$Ru_3(Pr^iC{=}C{=}CH_2)(PPh_2)(CO)_8$, **7**, 709
$Ru_3C_{26}H_{21}O_7P$
$Ru_3(CH_2)(CH_2CCPr^i)(PPh_2)$-$(CO)_7$, **7**, 749, 792
$Ru_3H(PPh_2)(CH{=}CPr^iC{=}CH_2)$-$(CO)_7$, **7**, 709
$Ru_3C_{26}H_{21}O_8P$
$Ru_3H(CO)_8(PPh_2)(Pr^iCCMe)$, **7**, 791
$Ru_3C_{27}H_{14}NO_{10}P$
$Ru_3(PhCO)(PhPC_5H_4N)(CO)_9$, **7**, 737
$Ru_3C_{27}H_{19}NO_{10}P$
$Ru_3(PhCO)(PhPC_5H_4N)(CO)_9$, **7**, 696
$Ru_3C_{27}H_{20}N_2O_8S$
$Ru_3H(S)(CO)_8$ $(Bu^tNHCNHCMe_2CH_2)$, **7**, 730
$Ru_3C_{28}H_{17}O_8P$
$Ru_3H_2(CHCC_6H_4PPh_2)(CO)_8$, **7**, 653
$Ru_3C_{28}H_{17}O_9P$
$Ru_3(H)_2(CO)_9(CPPh_3)$, **7**, 764
$Ru_3C_{28}H_{17}O_9P_2$
$[Ru_3(PhPCH_2PPh_2)(CO)_9]^-$, **7**, 697, 707
$Ru_3C_{28}H_{18}O_9P_2$
$Ru_3H(PhPCH_2PPh_2)(CO)_9$, **7**, 707, 737, 856
$Ru_3C_{28}H_{21}NO_8$
$Ru_3(CO)_8[Me_2NCH_2CC(Me)$-$C(Ph)C(Ph)]$, **7**, 710, 752
$Ru_3C_{28}H_{25}O_8P$
$Ru_3H(CO)_8[P(Me)(Bz)Ph]$-$(CCBu^t)$, **7**, 772
$Ru_3C_{29}H_{17}O_{10}P$
$Ru_3H(Ph_2PCHCPhO)(CO)_9$, **7**, 698
$Ru_3H(Ph_3PCHCO)(CO)_9$, **7**, 705
$Ru_3C_{30}H_{17}O_{10}P$
$Ru_3(CO)_{10}[P(Ph)_2C_6H_4$-$CH{=}CH_2]$, **7**, 752, 754
$Ru_3C_{30}H_{22}O_8P_2$
$Ru_3(CO)_8(PhPCH_2PPh_2)(C_3H_5)$, **7**, 696, 787, 788
$Ru_3C_{30}H_{28}O_{12}P_2$
$Ru_3(CO)_8(PMe_2Ph)_2(MeO_2C$-$CCCO_2Me)$, **7**, 781
$Ru_3C_{30}H_{42}Cl_4$
$Ru_2Cl_4(cod)_2RuH(C_8H_{11})$-$(C_6H_6)$, **7**, 449
$Ru_3C_{30}H_{42}S_2$
$[Ru_3(S)_2(cym)_3]^{2+}$, **7**, 727
$Ru_3C_{30}H_{44}O_5S$
$Ru_3H(O)(SO_4)[C_5Me_3(CH_2)_2]$-$(Cp^*)_2$, **7**, 723
$Ru_3C_{33}H_{45}Cl_4$
$Ru_2Cl_4(cod)_2{\cdot}RuH(C_8H_{11})$-$(C_6H_6){\cdot}0.5C_6H_6$, **7**, 578
$Ru_3C_{34}H_{22}N_2O_9P_2$
$Ru_3H(C_6H_4N_2)(PhPCH_2PPh_2)$-$(CO)_9$, **7**, 697
$Ru_3C_{34}H_{24}Cl_2O_8P_2$
$Ru_3Cl_2(CO)_8$(dppe), **7**, 723
$Ru_3C_{34}H_{25}O_8P$
$Ru_3(CO)_8[CBuCC(H)C(Ph)$-$(PPh_2)]$, **7**, 752
$Ru_3C_{35}H_{22}O_{10}P_2$
$Ru_3(CO)_{10}$(dppm), **7**, 716
$Ru_3C_{35}H_{26}O_8P_2$
$Ru_3[C(O)CH_2](CH_2)(CO)_7$-(dppm), **7**, 654, 749
$Ru_3C_{36}H_{24}NO_{10}$
$Ru_3(OC_6H_4CH{=}NTol)_2(CO)_8$, **7**, 725
$Ru_3C_{36}H_{26}O_8P_2$
$HRu_3(CO)_8(Ph_2PCH{=}CH_2)$-$(Ph_2PCH{=}CH)$, **7**, 752, 753
$Ru_3C_{36}H_{50}O_6$
$Ru_3(CO)_5[Bu^tCCC(H)C$-$(H)Bu^t](C_{19}H_{30}O)$, **7**, 793
$Ru_3C_{37}H_{47}I_6O_9P_3$
$[RuI[P(OMe)_3]_3(C_4Ph_4)]I_3$, **7**, 474
$Ru_3C_{38}H_{24}O_8P_2$
$Ru_3(C_6H_4CO)(PPh_2)_2(CO)_7$, **7**, 705
$Ru_3C_{38}H_{30}O_7P_2$
$Ru_3H(CO)_7(dppm)(CCBu^t)$, **7**, 772
$Ru_3C_{39}H_{27}NO_8P_2$

$Ru_3(NPh)(CO)_8(dppm)$, **7**, 641, 731
$Ru_3C_{40}H_{25}O_9P_3$
$Ru_3(CO)_9(Ph_2PCHPPhC_6H_4P$-Ph), **7**, 684
$Ru_3C_{40}H_{44}O_{16}P_4$
$Ru_3(CO)_8[PPh(OMe)_2]_4$, **7**, 684
$Ru_3C_{41}H_{28}O_6P_2$
$Ru_3(C_4H_2Ph_2)(C_6H_4PPhCH_2P$-Ph)$(CO)_6$, **7**, 709
$Ru_3C_{41}H_{30}O_7$
$Ru_3(CO)_7[Bu^tCCC(Ph)C$-(H)Ph](PhCCPh), **7**, 793
$Ru_3C_{42}H_{30}O_6P_2$
$Ru_2(PPh_2)[OC(Ph)](CO)_5$-$(PPh_3)$, **7**, 646, 647
$Ru_3C_{43}H_{31}O_7P_3$
$Ru_3H(PPh_2)_3(CO)_7$, **7**, 736
$Ru_3C_{43}H_{34}O_{14}P_2$
$Ru_3(CO)_6(dppm)[C_2(CO_2$-$Me)_2]_2$, **7**, 710, 749, 794
$Ru_3C_{43}H_{35}O_8P_3$
$Ru_3H[PhPCH_2CMe(CH_2P$-$Ph_2)_2](CO)_8$, **7**, 700
$Ru_3C_{44}H_{36}O_7P_2$
$Ru_3(CH_2{=}CPr^iC{=}CH_2)(PPh_2)$-$(CO)_7(PPh_3)$, **7**, 708, 791
$Ru_3C_{46}H_{32}O_7P_2$
$Ru_3(CO)_7(dppm)(PhCCPh)$, **7**, 780
$Ru_3C_{46}H_{34}O_7P_2$
$Ru_3H_2(CO)_7(dppm)(PhCCPh)$, **7**, 780
$Ru_3C_{47}H_{32}O_8P_2$
$Ru_3(CO)_8(dppm)(PhCCPh)$, **7**, 780
$Ru_3C_{50}H_{35}O_6P_3$
$Ru_3(CO)_6[P(Ph)C(Ph)C(Ph)]$-$(PPh_2)_2$, **7**, 752
$Ru_3C_{50}H_{37}O_7P_3$
$Ru_3H(Ph_2PC_6H_4CO)(CO)_6$-(dppm), **7**, 710
$Ru_3C_{51}H_{38}O_7P_4$
$Ru_3(PPh)(CHPPh_2)(dppm)$-$(CO)_7$, **7**, 704
$Ru_3C_{51}H_{43}O_6P_3$
$Ru_3H(CO)_6(dppm)(PPh_3)$-(CCPh), **7**, 772
$Ru_3C_{52}H_{39}O_8P_2$
$Ru_3H(Ph_2C_6H_4CO)(CO)_7$-(dppe), **7**, 698
$Ru_3C_{55}H_{46}O_6P_4$
$Ru_3H_2(O)(CO)_5(dppm)_2$, **7**, 723
$Ru_3C_{56}H_{44}O_7P_4$
$Ru_3(O)(CO)_6(dppm)_2$, **7**, 702, 723
$Ru_3C_{58}H_{45}O_8P_4$
$[Ru_3(H)(CO)_8(dppm)_2]^+$, **7**, 720
$Ru_3C_{59}H_{57}O_5P_3$
$Ru_3(C_2Bu^t)_2(PPh_2)_2(CO)_5(Ph_2P$-$C_2Bu^t)$, **7**, 691, 692
$Ru_3C_{63}H_{40}O_7$
$Ru_3(CO)_5(C_4Ph_4CO)_2$, **7**, 458
$Ru_3C_{81}H_{66}O_6P_6$
$Ru_3(CO)_6(dppm)_3$, **7**, 711, 721
$Ru_3CoAu_3C_{66}H_{45}O_{12}P_3$
$Au_3CoRu_3(CO)_{12}(PPh_3)_3$, **10**, 292
$Ru_3CoC_{12}H_2O_{12}$
$[Ru_3Co(H)_2(CO)_{12}]^-$, **10**, 203
$Ru_3CoC_{12}H_3O_{12}$
$H_3Ru_3Co(CO)_{12}$, **10**, 43, 58
$Ru_3CoC_{12}NO_{12}$
$Ru_3Co(N)(CO)_{12}$, **10**, 197
$Ru_3CoC_{13}O_{13}$
$[Ru_3Co(CO)_{13}]^-$, **10**, 19, 20, 48
$Ru_3CoC_{14}H_7O_{12}S$
$Ru_3Co(H)(SMe_2)(CO)_{12}$, **10**, 234
$Ru_3CoC_{21}H_{20}NO_{13}$
$[Et_4N][Ru_3Co(CO)_{13}]$, **10**, 51
$Ru_3CrC_{21}H_{13}O_{11}P$
$Ru_3(CO)_8[C_6H_4Cr(CO)_3](PBu^t)$, **10**, 17
$Ru_3Cr_2C_{27}H_9O_{15}P$
$Ru_3(CO)_9[P[C_6H_5Cr(CO)_3]]$-$[C_6H_4Cr(CO)_3]$, **10**, 16
$Ru_3CuC_{83}H_{60}IN_2O_{10}P_4$
$[PPN]_2[Ru_3CuI(CO)_9(CCO)]$, **7**, 755, 757
$Ru_3FeC_{12}HNO_{12}$
$FeRu_3(H)(N)(CO)_{12}$, **10**, 225
$Ru_3FeC_{12}NO_{12}$
$[FeRu_3(N)(CO)_{12}]^-$, **7**, 32; **10**, 225
$Ru_3FeC_{13}H_2O_{13}$
$H_2FeRu_3(CO)_{13}$, **10**, 37, 41, 43, 61
$Ru_3FeC_{16}H_{18}NO_{10}P_2$
$[FeRu_3(N)(CO)_{10}[P(OMe)_3]_2]^-$, **10**, 218
$Ru_3FeC_{24}H_{24}O_8P_2$
$Ru_3H_2(PC_5H_4FeC_5H_4PPr^i{}_2)$-$(CO)_8$, **7**, 709
$Ru_3FeC_{27}H_{30}O_8P_2$
$Ru_3H_2(CH_2CHMePC_5H_4$-$FeC_5H_4PPr^i{}_2)(CO)_8$, **7**, 709
$Ru_3FeC_{37}H_{20}O_{13}P_2$
$FeRu_3(CO)_{13}(PPh_2)_2$, **7**, 55
$Ru_3Fe_2C_{14}O_{15}$
$[Fe_2Ru_3(CO)_{14}O]^{2-}$, **7**, 60, 61; **10**, 232, 233
$Ru_3Fe_2C_{17}O_{17}Te_2$
$Fe_2(CO)_6Te_2Ru_3(CO)_{11}$, **7**, 76
$Ru_3HgC_{15}H_9BrO_9$
$(C_2Bu^t)(CO)_9Ru_3(HgBr)$, **10**, 332, 339
$Ru_3HgC_{15}H_9IO_9$
$(C_2Bu^t)(CO)_9Ru_3(HgI)$, **10**, 330–2
$Ru_3HgC_{15}H_9O_9$
$C_2Bu^t(CO)_9Ru_3(HgBr)$, **10**, 330
$Ru_3HgMoC_{23}H_{14}O_{12}$
$Ru_3(CO)_9(CCBu^t)[Hg[Mo$-$(CO)_3Cp]]$, **5**, 409; **10**, 339
$Ru_3InC_{21}H_{15}O_6$
$In[Ru(CO)_2Cp]_3$, **1**, 562
$Ru_3IrC_{50}H_{36}O_8P_3$
$Ru_3Ir(H)_2(PPh_2)(PPhC_6H_4)$-$(CO)_8(PPh_3)$, **10**, 223
$Ru_3KC_{13}H_6NO_{11}$
$K[Ru_3[C(O)NMe_2](CO)_{10}]$, **7**, 748
$Ru_3Li_2C_{14}H_6O_{12}$
$Li_2[Ru_3[C(O)Me]_2(CO)_{10}]$, **7**, 748
$Ru_3MnC_{50}H_{30}NO_{13}P$
$[PPN][MnRu_3C(CO)_{13}]$, **7**, 755
$Ru_3Mo_2C_{22}H_{10}O_{12}S$
$Ru_3Mo_2(CO)_{12}Cp_2(S)$, **10**, 20
$Ru_3NiC_{14}H_8O_9$
$Ru_3NiCp(H)_3(CO)_9$, **9**, 21; **10**, 38, 40, 41
$Ru_3Ni_3C_{14}O_{13}$
$[Ni_3Ru_3(CO)_{13}C]^{2-}$, **10**, 146
$Ru_3PtC_{18}H_{14}O_{10}$
$Ru_3Pt(H)_2(CO)_{10}(cod)$, **7**, 847; **9**, 564
$Ru_3PtC_{51}H_{43}O_7P_3$
$PtRu_3(CO)_7(PPh_2)(CCBu^t)$-(dppe), **9**, 572
$Ru_3Pt_2C_{36}H_{10}O_8$
$Ru_3Pt_2(CO)_8(C_2Ph_2)_2$, **10**, 151
$Ru_3Pt_2C_{46}H_{20}O_{18}$
$Pt_2Ru_3(CO)_{18}(PhCCPh)_2$, **9**, 572
$Ru_3RhC_{14}H_9O_9$
$Ru_3Rh(H)_4(CO)_9Cp$, **10**, 211
$Ru_3RhC_{15}H_7O_{10}$
$Ru_3Rh(H)_2(CO)_{10}Cp$, **10**, 209–11
$Ru_3RhC_{67}H_{51}O_7P_4$
$Ru_3Rh(H)_2(PPh_2)(PPhC_6H_4)$-$(CO)_7(PPh_3)_2$, **10**, 223
$Ru_3Rh_2C_{25}H_{20}O_{13}P_2$
$Ru_3Rh_2(PPh)(CO)_{13}(PEt_3)$, **10**, 222
$Ru_3SiC_{16}H_{16}O_{11}$
$Ru_3H(OSiEt_3)(CO)_{10}$, **7**, 724
$Ru_3Si_2C_{20}H_{24}O_8$
$Ru_3H[TMS$-$CHCMeCC$-$(TMS)CH](CO)_8$, **7**, 710
$Ru_3WC_{21}H_{16}O_{11}$
$WRu_3(H)(CO)_{11}Cp^*$, **10**, 208
$Ru_3WC_{26}H_{21}O_{10}P$
$WRu_3(H)(PPh)(CO)_{10}Cp^*$, **10**, 222
$Ru_3ZnC_{15}H_9O_9$
$[(C_2Bu^t)(CO)_9Ru_3](Zn)$, **10**, 339
$Ru_3C_{12}O_{12}$
$Ru_3(CO)_{12}$, **10**, 48
$Ru_4AgC_{30}H_{18}O_{12}P$
$AgRu_4(H)_3(CO)_{12}(PPh_3)$, **10**, 302
$Ru_4Ag_2C_{38}H_{26}O_{12}P_2$
$Ag_2Ru_4(H)_2[Ph_2P(CH_2)_2PPh_2]$-$(CO)_{12}$, **10**, 299
$Ru_4Ag_2C_{54}H_{44}O_{12}P_2$
$Ag_2Ru_4H_2(CO)_{12}[P(Bz)_3]_2$, **10**, 289
$Ru_4AsCu_2C_{38}H_{26}O_{12}P$
$Cu_2Ru_4(H)_2[Ph_2As(CH_2)_2PPh_2]$-$(CO)_{12}$, **10**, 299

$Ru_4As_2C_{36}H_{38}O_{12}$
$[Ru_2(C_6H_4O_2)(CO)_4(AsPh_3)]_2$, **7**, 666
$Ru_4AuC_{30}H_{18}O_{12}P$
$H_3Ru_4[Au(PPh_3)](CO)_{12}$, **10**, 42
$Ru_4AuC_{31}H_{16}O_{12}P$
$(H)Ru_4(C)(AuPPh_3)(CO)_{12}$, **7**, 852
$Ru_4Au_2C_{29}H_{22}O_{12}P_2$
$Ru_4(C)(AuPMe_2Ph)_2(CO)_{12}$, **7**, 852
$Ru_4Au_2C_{48}H_{32}O_{12}P_2$
$Au_2Ru_4(H)_2(CO)_{12}(PPh_3)_2$, **10**, 289, 292
$Ru_4Au_2C_{49}H_{30}O_{13}$
$Ru_4Au_2(CO)_{13}(PPh_3)_2$, **7**, 845
$Ru_4Au_3C_{66}H_{46}O_{12}P_3$
$Au_3Ru_4(H)(CO)_{12}(PPh_3)_3$, **7**, 847; **10**, 292
$Ru_4BC_{12}HO_{12}$
$[(H)Ru_4(B)(CO)_{12}]^{2-}$, **7**, 851
$Ru_4BC_{12}H_3O_{12}$
$HRu_4(CO)_{12}BH_2$, **1**, 297, 300; **7**, 850
$Ru_4BC_{21}H_{10}O_{12}$
$Ru_4(H)[BC(Ph)C(H)(Me)]-(CO)_{12}$, **7**, 892
$Ru_4BC_{26}H_{13}O_{12}$
$(H)Ru_4(H)[BC(Ph)C(H)(Ph)]-(CO)_{12}$, **7**, 892
$Ru_4BC_{48}H_{32}NO_{12}P_2$
$[PPN][HRu_4(CO)_{12}BH]$, **1**, 304
$Ru_4C_{10}Cl_4O_{10}$
$[Ru_4(Cl)_4(CO)_{10}]^{2-}$, **7**, 869
$Ru_4C_{10}Cl_4O_{11}$
$[Ru_4(O)(Cl)_4(CO)_{10}]^{2-}$, **7**, 722
$Ru_4C_{11}Cl_2O_{11}$
$[Ru_4Cl_2(CO)_{11}]^{2-}$, **7**, 721
$Ru_4C_{11}H_3NO_{11}$
$Ru_4(H)_3(N)(CO)_{11}$, **7**, 854, 855
$Ru_4C_{11}O_{11}$
$[Ru_4(CO)_{11}]^{6-}$, **7**, 844
$Ru_4C_{11}O_{11}S_2$
$Ru_4(S)_2(CO)_{11}$, **7**, 867, 871
$Ru_4C_{12}HNO_{12}$
$Ru_4(H)(N)(CO)_{12}$, **7**, 855
$Ru_4(NH)(CO)_{12}$, **7**, 855
$Ru_4C_{12}H_3O_{12}$
$[Ru_4(H)_3(CO)_{12}]^-$, **7**, 849
$Ru_4C_{12}H_4O_{12}$
$Ru_4H_4(CO)_{12}$, **7**, 687, 689, 847; **10**, 39
$Ru_4C_{12}NO_{12}$
$[Ru_4(N)(CO)_{12}]^-$, **7**, 854
$Ru_4C_{12}O_{12}$
$[Ru_4(CO)_{12}]^{4-}$, **7**, 719, 844
$Ru_4C_{13}HO_{13}$
$[Ru_4(H)(CO)_{13}]^-$, **7**, 845
$Ru_4C_{13}H_2O_{12}$
$(H)_2Ru_4(C)(CO)_{12}$, **7**, 852
$Ru_4C_{13}H_2O_{13}$
$Ru_4(H)_2(CO)_{13}$, **7**, 845, 847; **10**, 199
$Ru_4C_{13}H_3NO_{13}$
$Ru_4(H)_3(NCO)(CO)_{12}$, **7**, 853
$Ru_4C_{13}O_{13}$
$[Ru_4(CO)_{13}]^{2-}$, **7**, 719, 844
$Ru_4C_{14}H_4O_{13}$
$Ru_4(H)(COMe)(CO)_{12}$, **7**, 899, 900
$Ru_4C_{14}H_8O_{12}S$
$Ru_4(H)_2(SMe_2)(CO)_{12}$, **7**, 865
$Ru_4C_{14}H_{10}Cl_2O_{12}$
$Ru_4(OEt)_2Cl_2(CO)_{10}$, **7**, 863
$Ru_4C_{14}O_{13}$
$Ru_4(C)(CO)_{13}$, **7**, 852
$Ru_4C_{15}H_6O_{13}S$
$Ru_4(SMe_2)(CO)_{13}$, **7**, 864
$Ru_4C_{15}H_7NO_{12}$
$Ru_4(H)(CNMe_2)(CO)_{12}$, **7**, 893
$Ru_4C_{16}H_{10}N_2O_{12}$
$Ru_4(N_2Et_2)(CO)_{12}$, **7**, 856
$Ru_4C_{16}H_{13}NO_{11}$
$Ru_4(H)_4(CO)_{11}[CN(Me)C_3H_6]$, **7**, 871
$Ru_4C_{17}H_{14}NO_{11}P$
$Ru_4(PNPr^i_2)(CO)_{11}$, **7**, 857
$Ru_4C_{18}H_6O_{12}P$
$[Ru_4(H)(PPh)(CO)_{12}]^-$, **7**, 860
$Ru_4C_{18}H_6O_{13}$
$Ru_4(C_6H_6O)(CO)_{12}$, **7**, 872
$Ru_4C_{18}H_7O_{12}P$
$Ru_4(H)_2(PPh)(CO)_{12}$, **7**, 858
$Ru_4C_{19}H_6O_{11}S$
$Ru_4(S)(C_2PhH)(CO)_{11}$, **7**, 878, 922
$Ru_4C_{20}H_8O_{12}$
$[Ru_2(CO)_4(O_2C_6H_4)]_2$, **7**, 710, 900
$Ru_4C_{20}H_{14}O_{11}S$
$Ru_4(S)[C_2CH_2C(Me)(Bu^t)]-(CO)_{11}$, **7**, 876, 902
$Ru_4C_{20}H_{18}O_{11}S_3$
$Ru_4[S(CH_2)_3]_3(CO)_{11}$, **7**, 865
$Ru_4C_{20}H_{20}S_4$
$Ru_4(S)_4Cp_4$, **7**, 522
$Ru_4C_{20}H_{40}O_{20}P_4$
$Ru_4(H)_4(CO)_8[P(OMe)_3]_4$, **7**, 846
$Ru_4C_{21}H_{14}O_{12}S$
$Ru_4[SC_2CH_2C(Me)(Bu^t)]-(CO)_{12}$, **7**, 902
$Ru_4C_{22}H_{11}O_{10}P_2$
$[Ru_4(H)(PPh)_2(CO)_{10}]^-$, **7**, 860
$Ru_4C_{22}H_{12}O_{10}P$
$Ru_4(H)_2(PPh_2)(CO)_{10}$, **7**, 857
$Ru_4C_{23}H_9O_{11}$
$Ru_4(PPh)(C_6H_4)(CO)_{11}$, **7**, 880
$Ru_4C_{23}H_{10}O_{11}P_2$
$Ru_4(PPh)_2(CO)_{11}$, **7**, 859
$Ru_4C_{23}H_{16}O_9$
$Ru_4(C_9H_7)(C_5H_9)(CO)_9$, **7**, 888
$Ru_4C_{24}H_{11}O_{12}P_2$
$Ru_4(H)(PPh)_2(CO)_{12}$, **7**, 859
$Ru_4C_{24}H_{12}O_{12}P_2$
$Ru_4(H)_2(PPh)_2(CO)_{12}$, **7**, 860
$Ru_4C_{24}H_{24}N_2O_{10}$
$Ru_4[N(Pr^i)C_3H_2Me]_2(CO)_{10}$, **7**, 894
$Ru_4C_{24}H_{28}O_4$
$[(C_6H_6)_4Ru_4(OH)_4]^{4+}$, **7**, 885
$Ru_4C_{24}H_{28}O_5$
$[(C_6H_6)_4Ru_4(O)(OH)_4]^{2+}$, **7**, 886
$Ru_4C_{24}H_{28}Se_4$
$Ru_4(Se)_4(C_5H_4Me)_4$, **7**, 526
$Ru_4C_{24}H_{36}Cl_2$
$[(C_6Me_6)RuCl_2]_2$, **1**, 320
$Ru_4C_{24}H_{36}O_8P_4$
$Ru_4(PBu^t)_4(CO)_8$, **7**, 860
$Ru_4C_{25}H_{10}O_{11}$
$[Ru_4(C_2Ph_2)(CO)_{11}]^{2-}$, **7**, 872
$Ru_4C_{25}H_{11}NO_{11}$
$Ru_4(NH)(PhCCPh)(CO)_{11}$, **7**, 856
$Ru_4C_{25}H_{12}O_{11}P_2$
$Ru_4(PPh)[P(Ph)CHCH](CO)_{11}$, **7**, 898
$Ru_4C_{25}H_{22}O_9P_2S$
$Ru_4(S)(CO)_9(PMe_2Ph)_2$, **7**, 867
$Ru_4C_{26}H_{18}O_6$
$Ru_4(C_5H_4)_2(CO)_6Cp_2$, **7**, 665, 883
$Ru_4C_{26}H_{26}N_2O_{10}$
$Ru_4[CH_2CC(H)CHNBu^t]_2-(CO)_{10}$, **7**, 892
$Ru_4C_{27}H_{12}O_{12}$
$Ru_4(H)(C_2PhCPhH)(CO)_{12}$, **7**, 880
$Ru_4C_{27}H_{19}O_{11}P$
$Ru_4(PPh_2)(OH)(C_2HPr^i)(CO)_{10}$, **7**, 875
$Ru_4C_{28}H_{15}O_{10}PTe_2$
$Ru_4(Te)_2(CO)_{10}(PPh_3)$, **7**, 869
$Ru_4C_{28}H_{27}O_{10}P$
$Ru_4(CO)_{10}(Bu^tCCCCBu^t)[P-(CCBu^t)]$, **7**, 774, 888
$Ru_4C_{29}H_{15}N_2O_{11}P$
$Ru_4(PPh)(N_2Ph_2)(CO)_{11}$, **7**, 860, 861
$Ru_4C_{29}H_{20}O_{10}P_2$
$Ru_4(H)_3(PhPCH_2PPh_2)(CO)_{10}$, **7**, 856
$Ru_4C_{30}H_{17}O_{12}P$
$Ru_4(H)_2(CO)_{12}(PPh_3)$, **7**, 847
$Ru_4C_{30}H_{25}O_6$
$[[Ru_2Cp_2(CO)_3]_2-(CMeCHCH)]^+$, **7**, 663, 664
$Ru_4C_{30}H_{32}N_2O_{10}$
$[Ru_2(CO)_4[C_5(Me)_2(NEt_2)-(O)]]_2$, **7**, 901
$Ru_4C_{31}H_{15}NO_{11}$
$Ru_4(NPh)(C_2Ph_2)(CO)_{11}$, **7**, 873
$Ru_4C_{31}H_{15}O_{11}P$
$Ru_4[P(Ph)_2(C_6H_4)C_2H](CO)_{11}$, **7**, 897, 898
$Ru_4(PPh)(C_2Ph_2)(CO)_{11}$, **7**, 874
$Ru_4C_{31}H_{19}O_{13}P$
$Ru_4(PPh_2)(C_2Bu^t)(CO)_{13}$, **7**, 874
$Ru_4C_{32}H_{15}O_{12}P$
$Ru_4[P(Ph)C_2Ph_2](CO)_{12}$, **7**, 874
$Ru_4C_{32}H_{22}O_8P_4$
$Ru_4(PPh)_2(PPhH)_2(CO)_8$, **7**, 861

$Ru_4C_{34}H_{20}O_{10}P_2$
$Ru_4(H)(PPh_2)[P(Ph)(C_6H_4)]-(CO)_{10}$, **7**, 898
$Ru_4C_{34}H_{22}O_9P_2Te_2$
$Ru_4(Te)_2(Ph_2PCH_2PPh_2)(CO)_9$, **7**, 869
$Ru_4C_{35}H_{26}O_{10}P_2$
$Ru_4(H)_4(Ph_2PCH_2PPh_2)(CO)_{10}$, **7**, 846
$Ru_4C_{36}H_{20}N_2O_{10}$
$Ru_4(NPh)_2(C_2Ph_2)(CO)_{10}$, **7**, 873
$Ru_4C_{36}H_{28}O_{10}P_2$
$Ru_4(H)_4(dppe)(CO)_{10}$, **7**, 846
$Ru_4C_{37}H_{20}P_2O_{13}$
$Ru_4(PPh_2)_2(CO)_{13}$, **7**, 838
$Ru_4C_{38}H_{20}O_{12}P_2$
$Ru_4(C_2)(PPh_2)_2(CO)_{12}$, **7**, 874, 926
$Ru_4C_{38}H_{52}N_4O_{10}$
$Ru_4[OC_5Me_2(NEt_2)_2]_2(CO)_8$, **7**, 710
$Ru_4C_{38}H_{56}O_{20}$
$[Ru_2(O_2CBu)(CO)_4[OC-(OH)Bu](O_2CBu)]_2$, **7**, 863
$Ru_4C_{39}H_{25}NO_{10}P_2$
$Ru_4(PPh_2)(NC_5H_4)[P(Ph)_2C-(H)C](CO)_{10}$, **7**, 893
$Ru_4C_{40}H_{27}O_{10}P_3$
$Ru_4(H)_2(PPh)(PPh_2)_2(CO)_{10}$, **7**, 859
$Ru_4C_{40}H_{28}O_{13}P_2$
$Ru_4(H)[C_2(CO_2Me)_2][PPh_2C-H_2PPh(C_6H_4)](CO)_9$, **7**, 896
$Ru_4C_{40}H_{60}$
$[(C_6H_4MePr^i)_4Ru_4(H)_4]^{2+}$, **7**, 885
$Ru_4C_{40}H_{60}Cl_4$
$Cp*_4Ru_4(Cl)_4$, **7**, 477, 604, 667, 883
$Ru_4C_{41}H_{28}O_{14}P_2$
$Ru_4[C_2(CO_2Me)_2](PPh_2CH_2P-Ph_2)(CO)_{10}$, **7**, 872
$Ru_4C_{44}H_{47}O_{15}P_3$
$Ru_4(H)_4(CO)_9(PMe_2Ph)[P-(OC_6H_4Me)_3][P(OCH_2)_3-CEt]$, **7**, 846
$Ru_4C_{56}H_{112}O_8P_4$
$Ru_4H_4(CO)_8(PBu_3)_4$, **7**, 644
$Ru_4C_{58}H_{40}O_{10}P_4$
$Ru_4(PPh_2)(CO)_{10}$, **7**, 856
$Ru_4C_{59}H_{42}Br_2O_9P_4$
$Ru_4[C_6H_4)PBzPPh_2]_2(Br)_2(CO)_9$, **7**, 898
$Ru_4C_{62}H_{40}O_{10}P_4$
$[Ru_2(PPh_2)(C_2PPh_2)(CO)_5]_2$, **7**, 649
$Ru_4C_{62}H_{57}O_8P_3$
$Ru_4(PPh_2)_2(C_2Bu^t)_2(CO)_8[PPh_2-(C_2Bu^t)]$, **7**, 874
$Ru_4C_{63}H_{35}O_8P$
$Ru_4(PPh)[C(Ph)CC(C_2Ph)C-(Ph)CCC(Ph)C(Ph)C(C_2-(Ph)](CO)_8$, **7**, 889
$Ru_4C_{111}H_{108}O_{22}P_6$
$[Ru_2(diop)(CO)_4]_2[O_2C(CH_2)_3-CO_2]_2(diop)$, **7**, 863
$Ru_4CuC_{30}H_{18}O_{12}P$
$CuRu_4(H)_3(CO)_{12}(PPh_3)$, **10**, 302
$H_3Ru_4[Cu(PPh_3)](CO)_{12}$, **10**, 41
$Ru_4Cu_2C_{30}H_{44}O_{12}P_2$
$Cu_2Ru_4(H)_2(CO)_{12}(PPr^i_3)_2$, **10**, 278
$Ru_4Cu_2C_{33}H_{23}O_{12}P$
$Cu_2Ru_4(H)_2(CO)_{12}[P(Bz)_2(Bz)]$, **10**, 300
$Ru_4Cu_2C_{38}H_{26}O_{12}P_2$
$Cu_2Ru_4(H)_2[Ph_2P(CH)_2)_2PPh_2]-(CO)_{12}$, **10**, 299
$Ru_4Cu_2C_{48}H_{32}O_{12}P_2$
$Cu_2Ru_4H_2(CO)_{12}(PPh_3)_2$, **10**, 289
$H_2Ru_4[Cu(PPh_3)]_2(CO)_{12}$, **10**, 41
$Ru_4Cu_2C_{49}H_{32}O_{13}P_2$
$Cu_2Ru_4(CO)_{13}(PPh_3)_2$, **10**, 289
$Ru_4Cu_2C_{54}H_{44}O_{12}P_2$
$Cu_2Ru_4(H)_2(CO)_{12}(PBz_3)_2$, **10**, 289
$Ru_4FeC_{27}H_{14}O_{11}P$
$Ru_4[PCpFeCp](C_6H_4)(CO)_{11}$, **7**, 881
$Ru_4FeC_{39}H_{22}O_{13}P_2$
$FeRu_4(C)(dppm)(CO)_{13}$, **7**, 27; **10**, 129
$Ru_4Hg_2C_{19}H_9O_{13}$
$(C_2Bu^t)(CO)_9Ru_3(Hg)_2Ru-(CO)_4$, **10**, 340
$Ru_4Ir_2C_{22}H_2O_{22}$
$[Re_4Ir_2(H)_2(CO)_{22}]^{2-}$, **10**, 199
$Ru_4Mo_2C_{24}H_{10}O_{14}S$
$Mo_2Ru_4(S)(CO)_{14}Cp_2$, **10**, 234–6
$Ru_4PtC_{21}H_{12}O_{13}$
$Ru_4Pt(CO)_{13}(cod)$, **10**, 212
$Ru_4PtC_{21}H_{14}O_{13}$
$Ru_4Pt(H)_2(CO)_{13}(cod)$, **7**, 847; **9**, 564; **10**, 212
$Ru_4Pt_2C_{18}O_{18}$
$Pt_2Ru_4(CO)_{18}$, **10**, 5, 6, 192, 193, 197
$Ru_4Pt_2C_{27}H_{26}O_{11}$
$Ru_4Pt_2(H)_2(CO)_{11}(cod)_2$, **7**, 847; **9**, 564; **10**, 212
$Ru_4Pt_2C_{42}H_{20}O_{14}$
$Pt_2Ru_4(CO)_{14}(PhCCPh)_2$, **9**, 572; **10**, 151
$Ru_4SbC_{28}H_{15}O_{10}Se_4$
$Ru_4(Se)_4(CO)_{10}(SbPh_3)$, **7**, 869
$Ru_4WC_{37}H_{33}O_{13}P_3S_2$
$(PhMe_2P)_2(CO)_9(S)Ru_4(S)W-(CO)_4(PMe_2Ph)$, **7**, 868
$Ru_4W_2C_{34}H_{25}O_{12}P_2$
$Cp*W_2Ru_4(CO)_{12}(PPh)_2$, **10**, 10
$Ru_5AsC_{25}H_9O_{13}$
$Ru_5(C_6H_4)(AsPh)(CO)_{13}$, **7**, 924
$Ru_5AuC_{35}H_{18}O_{15}P$
$Ru_5(C)(MeCO)(AuPPh_3)-(CO)_{14}$, **7**, 920
$Ru_5Au_2CrC_{29}H_{30}O_{17}P_2$
$Au_2Ru_5Cr(CO)_{17}(PEt_3)_2$, **10**, 278, 288
$Ru_5Au_2CrC_{30}H_{30}O_{17}P_2$
$Au_2Ru_5CrC(CO)_{17}(PEt_3)_2$, **10**, 278
$Ru_5C_{14}HNO_{14}$
$HRu_5(C)(CO)_{13}(NO)$, **7**, 919
$Ru_5C_{14}HO_{14}S$
$[HRu_5(S)(CO)_{14}]^-$, **7**, 922
$Ru_5C_{14}H_2O_{14}S$
$H_2Ru_5(S)(CO)_{14}$, **7**, 922
$Ru_5C_{14}NO_{14}$
$[Ru_5(C)(CO)_{13}(NO)]^-$, **7**, 919
$[Ru_5(N)(CO)_{14}]^-$, **7**, 922, 923, 939, 953
$Ru_5C_{14}O_{14}S$
$[Ru_5(S)(CO)_{14}]^{2-}$, **7**, 922
$Ru_5C_{14}O_{14}S_2$
$Ru_5(S)_2(CO)_{14}$, **7**, 909, 921, 922
$Ru_5C_{15}H_2O_{14}S$
$Ru_5(H)(C)(SH)(CO)_{14}$, **7**, 921
$Ru_5C_{15}H_2O_{14}Se$
$Ru_5(H)(C)(SeH)(CO)_{14}$, **7**, 921
$Ru_5C_{15}O_{15}S$
$Ru_5(S)(CO)_{15}$, **7**, 921, 922, 941, 946
$Ru_5C_{16}H_3O_{15}P$
$Ru_5(PMe)(CO)_{15}$, **7**, 924
$Ru_5C_{16}O_{15}$
$Ru_5(C)(CO)_{15}$, **7**, 918, 919, 934
$Ru_5C_{17}H_4O_{15}$
$H_2Ru_5(C_2H_2)(CO)_{15}$, **7**, 930
$Ru_5C_{17}H_6O_{14}S$
$Ru_5(H)(C)(SEt)(CO)_{14}$, **7**, 921
$Ru_5C_{17}H_7NO_{14}$
$Ru_4(H)(CNMe_2)(CO)_{14}$, **7**, 923
$Ru_5C_{17}H_{12}N_2O_{11}S_2$
$Ru_5(S)_2(CNMe_2)_2(CO)_{11}$, **7**, 922
$Ru_5C_{17}H_{14}O_{15}$
$HRu_5C(MeCO)(CO)_{14}$, **7**, 920
$Ru_5C_{18}H_3NO_{15}$
$Ru_5(C)(CO)_{15}(NCMe)$, **7**, 919
$Ru_5C_{19}H_5NO_{13}$
$HRu_5(C)(NC_5H_4)(CO)_{13}$, **7**, 920
$Ru_5C_{19}H_6O_{12}$
$Ru_5(C)(C_6H_6)(CO)_{12}$, **7**, 921
$Ru_5C_{19}H_{12}N_2O_{13}$
$Ru_5(CNMe_2)_2(CO)_{13}$, **7**, 923
$Ru_5C_{19}H_{12}O_{12}S$
$Ru_5(H)(S)(Me_2Cp)(CO)_{12}$, **7**, 921, 922
$Ru_5C_{20}H_8O_{13}$
$Ru_5(C)(C_6H_8)(CO)_{13}$, **7**, 921
$Ru_5C_{21}H_8N_2O_{11}S_2$
$Ru_5(S)_2(NC_5H_4)_2(CO)_{11}$, **7**, 922, 923
$Ru_5C_{21}H_{17}N_3O_{13}S_2$

$Ru_4(S)(CNMe_2)(CO)_{10}[SRu(CO)_3(CH_2NMeCNMe_2)]$, **7**, 922

$Ru_5C_{22}H_6O_{14}S$
$Ru_5(PhC_2H)(S)(CO)_{14}$, **7**, 909, 923

$Ru_5C_{22}H_{18}O_{13}S_3$
$Ru_5[(CH_2)_3S]_3(CO)_{13}$, **7**, 922

$Ru_5C_{24}H_{10}N_2O_{13}$
$HRu_5(C)(NC_5H_4)(CO)_{13}(py)$, **7**, 920

$Ru_5C_{24}H_{16}O_{11}$
$Ru_5(C)(C_6H_8)_2(CO)_{11}$, **7**, 921

$Ru_5C_{24}H_{18}N_2O_{14}$
$Ru_5(CNBu^t)_2(CO)_{14}$, **7**, 923

$Ru_5C_{24}H_{20}O_4$
$Ru_5Cp_4(CO)_4$, **7**, 665

$Ru_5C_{25}H_9O_{13}P$
$Ru_5(C_6H_4)(PPh)(CO)_{13}$, **7**, 924

$Ru_5C_{25}H_{13}O_{12}P$
$Ru_5(PPh)(CO)_{12}(PhMe)$, **7**, 924, 947

$Ru_5C_{26}H_{11}O_{13}P$
$Ru_5(H)(C)(PPh_2)(CO)_{13}$, **7**, 920

$Ru_5C_{26}H_{15}O_{12}P$
$Ru_5(PPh)(CO)_{12}(C_6H_4Me_2)$, **7**, 924

$Ru_5C_{26}H_{20}O_6$
$Ru_5(CO)_6Cp_4$, **7**, 930

$Ru_5C_{27}H_{11}O_{14}P$
$Ru_5(C)(CO)_{14}(PHPh_2)$, **7**, 920

$Ru_5C_{27}H_{20}O_7$
$Ru_5(CO)_7Cp_4$, **7**, 930

$Ru_5C_{28}H_{10}O_{16}P_2$
$Ru_5(P)(PPh_2)(CO)_{16}$, **7**, 924

$Ru_5C_{28}H_{18}O_{14}P_2$
$Ru_2(H)(PPh)[PPh(OPr)](CO)_{13}$, **7**, 924

$Ru_5C_{31}H_{16}O_{13}P_2$
$Ru_5(H)(PPh)(PPh_2)(CO)_{13}$, **7**, 924

$Ru_5C_{33}H_{15}O_{13}P$
$Ru_5(C_2Ph)(PPh_2)(CO)_{13}$, **7**, 925

$Ru_5C_{33}H_{15}O_{14}P$
$Ru_5(C)(CO)_{14}(PPh_3)$, **7**, 919

$Ru_5C_{33}H_{21}O_{12}PS$
$Ru_5(H)(C)(SEt)(CO)_{12}(PPh_3)$, **7**, 921

$Ru_5C_{34}H_{21}O_{13}PS$
$Ru_5(H)(C)(SEt)(CO)_{13}(PPh_3)$, **7**, 921

$Ru_5C_{35}H_{24}O_{12}P_2$
$Ru_5(PPh)(CCH_2Pr^i)(PPh_2)(CO)_{12}$, **7**, 925

$Ru_5C_{35}H_{28}O_6$
$Ru_3(CO)_3Cp_3(CCH_2CHC)Ru_2(CO)_3Cp_2$, **7**, 880

$Ru_5C_{36}H_{20}O_{10}P_2$
$Ru_5(H)(PPh)[C_2Ph(C_6H_4)](PPh)(CO)_{10}$, **7**, 926, 927

$Ru_5C_{37}H_{24}O_{11}P_2$
$Ru_5(PPh)(PPh_2)(CO)_{11}[C_5H_3Me(CH{=}CH_2)]$, **7**, 930

$Ru_5C_{37}H_{26}O_{11}P_2$
$Ru_5(H)_3(C)(PPh_2)(CO)_{11}(PMePh_2)$, **7**, 926, 927

$Ru_5C_{38}H_{20}O_{12}P_2$
$Ru_5(PPh)[CC(Ph)PPh_2](CO)_{12}$, **7**, 926, 927

$Ru_5C_{38}H_{21}IO_{12}P_2$
$Ru_5(H)(C_2PPh_2)I(PPh_2)(CO)_{12}$, **7**, 926

$Ru_5C_{38}H_{22}O_{12}P_2$
$Ru_5(PPh)[CCPh(C_6H_4)][PPh(OMe)](CO)_{11}$, **7**, 926, 927

$Ru_5C_{38}H_{24}O_{12}P_2$
$Ru_5(H)_2(CCH_2PPh_2)(PPh_2)(CO)_{12}$, **7**, 926, 927

$Ru_5C_{38}H_{26}O_{12}P_2$
$Ru_5(PPh)[C_2(CHCHMe)CHCH_2Me](PPh_2)(CO)_{12}$, **7**, 930, 931

$Ru_5C_{38}H_{28}O_{10}P_2S_2$
$Ru_5(CCH_2)(SMe)_2(PPh_2)_2(CO)_{10}$, **7**, 909, 928, 929

$Ru_5C_{39}H_{20}O_{13}P_2$
$Ru_5(C_2Ph)(PPh)(PPh_2)(CO)_{13}$, **7**, 926, 928

$Ru_5C_{39}H_{22}O_{13}P_2$
$Ru_5(C)(dppm)(CO)_{13}$, **7**, 919
$Ru_5(H)(C{=}CHPPh_2)(PPh_2)(CO)_{13}$, **7**, 926, 927

$Ru_5C_{39}H_{26}O_{11}P_2S_2$
$Ru_5(C_2)(SMe)_2(PPh_2)_2(CO)_{11}$, **7**, 928, 929

$Ru_5C_{39}H_{26}O_{12}P_2$
$Ru_5(H)_2(C)(CO)_{12}(dppe)$, **7**, 919, 920

$Ru_5C_{40}H_{24}O_{13}P_2$
$Ru_5(C)(CO)_{13}(dppe)$, **7**, 919

$Ru_5C_{40}H_{26}O_{10}P_4$
$Ru_5(H)(PPh)_3(PPh_2)(CO)_{10}$, **7**, 925

$Ru_5C_{40}H_{26}O_{11}P_2$
$Ru_5(C)(PPh_2)(MeC_3H_3PPh_2)(CO)_{11}$, **7**, 930, 931

$Ru_5C_{40}H_{26}O_{12}P_2S_2$
$Ru_5(C_2)(SMe)_2(PPh_2)_2(CO)_{12}$, **7**, 928, 929

$Ru_5C_{40}H_{30}O_{10}P_2S_2$
$Ru_5(CCHCH{=}CH_2)(SMe)_2(PPh_2)_2(CO)_{10}$, **7**, 928, 929

$Ru_5C_{41}H_{26}O_{11}P_2$
$Ru_5(C_6H_6)(PPh_2)_2(CO)_{11}$, **7**, 930, 931
$Ru_5[CC(PPh_2)CHCHCHMe](PPh_2)(CO)_{11}$, **7**, 930, 931

$Ru_5C_{41}H_{26}O_{13}P_2S_2$
$Ru_5(C_2PPh_2)(SMe)_2(PPh_2)(CO)_{13}$, **7**, 928, 929

$Ru_5C_{42}H_{26}O_{12}P_2$
$Ru_5[CC(PPh_2)CH_2CHCHCH_2](PPh_2)(CO)_{12}$, **7**, 909, 930, 931

$Ru_5C_{42}H_{28}O_{13}P_2$
$Ru_5(C)[Ph_2P(CH_2)_4PPh_2](CO)_{13}$, **7**, 919

$Ru_5C_{43}H_{28}O_{11}P_2$
$Ru_5(PPh_2C_2C_6H_8)(PPh_2)(CO)_{11}$, **7**, 928

$Ru_5C_{44}H_{26}O_{12}P_2S$
$Ru_5[C_2H(PPh_2)](SPh)(PPh_2)(CO)_{12}$, **7**, 926, 929
$Ru_5(C_2H)(SPh)(PPh_2)_2(CO)_{12}$, **7**, 927, 929
$Ru_5(H)(C_2PPh_2)(SPh)(PPh_2)(CO)_{12}$, **7**, 927, 929

$Ru_5C_{45}H_{26}O_{13}P_2$
$Ru_5(H)(C_2Ph)(PPh_2)_2(CO)_{13}$, **7**, 926

$Ru_5C_{45}H_{28}O_{11}P_2$
$Ru_5(C)(PPh_2)(PhC_3H_3PPh_2)(CO)_{11}$, **7**, 930, 931

$Ru_5C_{45}H_{37}O_{13}P_3$
$Ru_5(PPh_2C_2C_6H_8)(PPh_2)(CO)_{10}[P(OMe)_3]$, **7**, 909, 928, 931

$Ru_5C_{47}H_{25}N_2O_{14}P$
$Ru_5(C_2Ph)[NC(O)NCPh_2](PPh_2)(CO)_{13}$, **7**, 909, 925

$Ru_5C_{47}H_{30}N_2O_{11}P_2$
$Ru_5(C_2)(PPh_2)_2(CO)_{11}(py)_2$, **7**, 926, 928

$Ru_5C_{47}H_{38}O_9P_2S_2$
$Ru_5(CC_{11}H_{12})(SMe)_2(PPh_2)_2(CO)_9$, **7**, 928, 929

$Ru_5C_{48}H_{25}O_{12}P$
$Ru_5(C_2Ph)(PhC_2C_2Ph)(PPh_2)(CO)_{12}$, **7**, 925

$Ru_5C_{48}H_{37}O_{14}P_3$
$Ru_5(C)(PPh_2)(PhC_3H_3PPh_2)(CO)_{11}[P(OMe)_3]$, **7**, 930, 931

$Ru_5C_{54}H_{35}O_{10}P_3$
$Ru_5(PPh)(PhC_2Ph)(PPh_2)_2(CO)_{10}$, **7**, 926, 928

$Ru_5C_{56}H_{35}O_{12}P_3$
$Ru_5(C_2PPh_2)(PPh_2)(CO)_{12}(PPh_3)$, **7**, 926

$Ru_5C_{62}H_{44}O_{11}P_2$
$Ru_5(C)(CO)_{11}(dppm)_2$, **7**, 919

$Ru_5CoC_{43}H_{25}O_{12}P_2$
$CpCoRu_5(CO)_{12}(C_2Ph)(PPh)(PPh_2)$, **10**, 146

$Ru_5CrC_{18}O_{17}$
$[CrRu_5C(CO)_{17}]^{2-}$, **10**, 91

$Ru_5FeC_{39}H_{20}O_{14}$
$FeRu_5(C)(PPh_2)_2(CO)_{14}$, **10**, 129

$Ru_5FeC_{39}H_{20}O_{14}P_2$
$Ru_5Fe(CO)_{14}(PPh_2)_2C$, **7**, 27

$Ru_5FeC_{41}H_{28}O_{13}P_2S_2$
$FeRu_5(CCH_2)(SMe)_2(PPh_2)_2(CO)_{13}$, **10**, 130, 131

$Ru_5Fe_3C_{43}H_{20}O_{17}$
$Fe_3Ru_5(C)_2(PPh_2)_2(CO)_{17}$, **10**, 129

$Ru_5Mo_2C_{25}H_{10}O_{15}S$
$Mo_2Ru_5(S)(CO)_{15}Cp_2$, **10**, 234

$Ru_5PtC_{23}H_{12}O_{14}$
$PtRu_5(CO)_{14}(C)(cod)$, **10**, 147, 148

$Ru_5PtC_{42}H_{20}O_{13}$
$Ru_5Pt(C)(CO)_{13}(C_2Ph_2)_2$, **10**, 147, 148
$Ru_5Pt_4C_{28}H_{12}O_{20}$
$Ru_5Pt_4(CO)_{20}$(cod), **10**, 213
$Ru_5Pt_5C_{34}H_{26}O_{18}$
$Ru_5Pt_5(H)_2(CO)_{18}$(cod)$_2$, **7**, 847; **9**, 564; **10**, 213
$Ru_5RhC_{25}H_{15}O_{14}$
$Cp^*RhRu_5C(CO)_{14}$, **10**, 134
$Ru_5RhC_{28}H_{21}O_{11}$
$Cp^*(C_6H_6)RhRu_5C(CO)_{11}$, **10**, 134
$Ru_5RhC_{28}H_{25}O_7$
$Cp^*Cp_2RhRu_5C(CO)_7$, **10**, 134
$Ru_5SiC_{35}H_{29}O_{17}P$
$Ru_5(C)(CO)_{14}[Ph_2PCH_2CH_2Si(OEt)_3]$, **7**, 919
$Ru_6Au_2C_{29}H_{30}O_{16}P_2$
$Au_2Ru_6C(CO)_{16}(PEt_3)_2$, **10**, 278, 288
$Ru_6BC_{17}HO_{17}$
$HRu_6B(CO)_{17}$, **1**, 294, 303, 304
$Ru_6BC_{54}H_{32}NO_{18}P_2$
$[PPN][Ru_6(H_2)(C))_{18}B]$, **1**, 303, 304
$Ru_6C_{15}HO_{15}S_3$
$[Ru_6(H)(S)_3(CO)_{15}]^-$, **7**, 909, 940, 941
$Ru_6C_{15}I_2O_{14}$
$Ru_6(C)(CO)_{14}I_2$, **7**, 952
$Ru_6C_{16}NO_{16}$
$[Ru_6(C)(CO)_{15}(NO)]^-$, **7**, 936
$[Ru_6(N)(CO)_{16}]^-$, **7**, 939
$Ru_6C_{17}H_2O_{17}$
$Ru_6(H)_2(CO)_{17}$, **7**, 933, 934
$Ru_6C_{17}H_2O_{17}S$
$Ru_6(H)_2(S)(CO)_{17}$, **7**, 941
$Ru_6C_{17}O_{16}$
$[Ru_6(C)(CO)_{16}]^{2-}$, **7**, 934
$Ru_6C_{17}O_{17}S_2$
$Ru_6(S)_2(CO)_{17}$, **7**, 940, 941
$Ru_6C_{18}HO_{18}$
$[Ru_6(H)(CO)_{18}]^-$, **7**, 933, 948
$Ru_6C_{18}H_2O_{18}$
$H_2Ru_6(CO)_{18}$, **7**, 933
$Ru_6C_{18}H_2O_{20}S$
$Ru_6(S)(OH)_2(CO)_{18}$, **7**, 909, 940, 941
$Ru_6C_{18}H_3O_{16}$
$[Ru_6(C)(CO)_{15}[C(O)Me]]^-$, **7**, 936
$[Ru_6(C)(CO)_{16}Me]^-$, **7**, 936
$Ru_6C_{18}O_{17}$
$Ru_6C(CO)_{17}$, **1**, 303
$Ru_6(C)(CO)_{17}$, **7**, 930, 934, 936
$Ru_6C_{18}O_{18}$
$[Ru_6(CO)_{18}]^{2-}$, **7**, 719, 933
$Ru_6C_{18}O_{18}S$
$Ru_6(S)(CO)_{18}$, **7**, 941, 946
$Ru_6C_{19}H_5O_{15}$
$[Ru_6(C)(C_3H_5)(CO)_{15}]^-$, **7**, 936
$Ru_6C_{20}H_{12}O_{15}S_2$
$H_2Ru_6C(CO)_{15}(SEt)_2$, **7**, 937
$Ru_6C_{21}H_6O_{14}$
$Ru_6(C)(CO)_{14}(C_6H_6)$, **7**, 935
$Ru_6C_{21}H_8O_{14}$
$Ru_6(C)(CH_2CHCHCHCHCH_2)(CO)_{14}$, **7**, 935, 936
$Ru_6C_{21}H_{12}N_2O_{14}$
$Ru_6(C)(CNMe_2)_2(CO)_{14}$, **7**, 923, 939, 940
$Ru_6C_{22}H_5O_{15}Se$
$[Ru_6(C)(CO)_{15}(SePh)]^-$, **7**, 937
$Ru_6C_{22}H_6O_{17}$
$Ru_6(H)_2(C_6H_4O)(CO)_{16}$, **7**, 909, 938
$Ru_6C_{22}H_8O_{14}$
$Ru_6(C)(CO)_{14}(PhMe)$, **7**, 935
$Ru_6C_{22}H_9NO_{16}$
$Ru_6(C)(CO)_{16}(CNBu^t)$, **7**, 923, 934
$Ru_6C_{22}H_{10}O_{15}$
$Ru_6(C)(MeCH{=}CHCH{=}CHMe)(CO)_{15}$, **7**, 936
$Ru_6C_{22}H_{12}O_{14}$
$Ru_6(C)(CO)_{14}(C_6H_3Me_3)$, **7**, 935
$Ru_6C_{22}H_{12}O_{15}S$
$Ru_6(H)(S)(Me_2Cp)(CO)_{15}$, **7**, 921, 936, 941, 942
$Ru_6C_{22}H_{16}O_{15}S_3$
$Ru_6(H)(C)(SEt)_3(CO)_{15}$, **7**, 909, 937
$Ru_6C_{22}N_2O_{20}$
$[Ru_6(CN)_2(CO)_{20}]^{2-}$, **7**, 721
$Ru_6C_{23}H_{12}O_{14}$
$Ru_6(H)(CO)_{14}(C_6H_3Me_2CH_2)$, **7**, 944
$Ru_6C_{23}H_{18}N_2O_{13}S_2$
$Ru_6(S)_2(MeC_2NMe_2)_2(CO)_{13}$, **7**, 909, 941–3
$Ru_6C_{24}H_6O_{15}$
$Ru_6(H)(C)(C_2Ph)(CO)_{15}$, **7**, 936
$Ru_6C_{24}H_{10}O_{14}Se$
$Ru_6(C)(SePh)(C_3H_5)(CO)_{14}$, **7**, 937
$Ru_6C_{24}H_{12}O_{11}$
$Ru_6(C)(C_6H_6)_2(CO)_{11}$, **7**, 935
$Ru_6C_{24}H_{12}O_{14}$
$Ru_6(C)(CO)_{14}(C_6H_3Me_3)$, **7**, 935, 944
$Ru_6C_{24}H_{12}O_{15}$
$Ru_6(CO)_{15}(C_6H_3Me_3)$, **7**, 944
$Ru_6C_{24}H_{13}N_2O_{20}$
$[Ru_6H(OCNMe_2)_2(CO)_{18}]^-$, **7**, 909, 939, 940
$Ru_6C_{24}H_{15}O_{19}P$
$Ru_6(H)_2(C_6H_4O)(CO)_{15}[P(OMe)_3]$, **7**, 909, 938
$Ru_6C_{24}H_{18}N_2O_{14}S_2$
$Ru_6(S)_2(MeC_2NMe_2)_2(CO)_{14}$, **7**, 909, 941, 942
$Ru_6C_{25}H_6O_{17}S$
$Ru_6(S)(PhC_2H)(CO)_{17}$, **7**, 922, 923, 941, 942
$Ru_6C_{25}H_{16}O_6P_2$
$Ru_2(PBzPPhC_6H_4)(CO)_6$, **7**, 649, 650
$Ru_6C_{25}H_{18}O_{16}S_3$
$Ru_6[(CH_2)_3S]_3(CO)_{16}$, **7**, 922, 941, 942
$Ru_6C_{26}H_{14}O_{12}P_2$
$Ru_6(PMe)_2(C_6H_4)_2(CO)_{12}$, **7**, 909, 938, 939
$Ru_6C_{26}H_{22}N_4O_{16}S$
$Ru_6(S)(EtNCSNHEt)(EtNCNHEt)(CO)_{16}$, **7**, 940, 941
$Ru_6C_{27}H_{10}NO_{15}$
$Ru_6(NPh)_2(CO)_{15}$, **7**, 940, 947
$Ru_6C_{27}H_{10}N_2O_{15}$
$Ru_6(NPh)_2(CO)_{15}$, **7**, 731
$Ru_6C_{27}H_{10}O_{14}Se$
$Ru_6(C)(SePh)_2(CO)_{14}$, **7**, 937
$Ru_6C_{27}H_{18}O_{11}$
$Ru_6(C)(C_6H_6)(CO)_{11}(C_6H_3Me_3)$, **7**, 935
$Ru_6C_{28}H_8N_2O_{18}S_2$
$Ru_6(S)_2(NC_5H_4)_2(CO)_{18}$, **7**, 922
$Ru_6C_{28}H_8O_{20}$
$[Ru_3(CO)_8(C_6H_4O_2)]_2$, **7**, 700
$Ru_6C_{28}H_{20}O_{12}$
$Ru_6(C)(CO)_{12}(C_6H_8)(C_6H_3Me_3)$, **7**, 935
$Ru_6C_{29}H_{12}N_2O_{16}S_2$
$Ru_6(H)(S)(PhNCSNHPh)(CO)_{16}$, **7**, 940, 941
$Ru_6C_{30}H_{24}O_{11}$
$Ru_6(C)(CO)_{11}(C_6H_3Me_3)_2$, **7**, 935
$Ru_6C_{31}H_{27}N_2O_{16}$
$Ru_6(C)$(bitropyl)$(CO)_{14}$, **7**, 936
$Ru_6C_{33}H_{40}O_{13}S_4$
$Ru_6(SCH_2CMe_2CH_2)_4(CO)_{13}$, **7**, 864
$Ru_6C_{35}H_{22}N_4O_{18}S_{2M}$
$[Ru_3H(CO)_9]_2[SC(NPh)N\text{-}EtCH_2CH_2NH(S)NPh]$, **7**, 730
$Ru_6C_{36}H_{20}O_{12}P_4$
$Ru_6(PPh)_4(CO)_{12}$, **7**, 909, 939
$Ru_6C_{36}H_{22}O_{12}P_4$
$H_2Ru_6(PPh)_2(PPh)_2(CO)_{12}$, **7**, 938
$Ru_6C_{37}H_{50}O_{12}S_5$
$Ru_6(H)(SCH_2CMe_2CH_2)_4(SCH_2CMeCH_2CH_2)(CO)_{12}$, **7**, 864
$Ru_6C_{42}H_{20}O_{13}$
$Ru_6(C)(PhC_2Ph)_2(CO)_{13}$, **7**, 936
$Ru_6C_{42}H_{25}O_{12}P_5$
$Ru_6(PPh)_5(CO)_{12}$, **7**, 909, 938, 939
$Ru_6C_{44}H_{25}NO_{13}P_2$
$Ru_6(C_2H)(NC_5H_4)(PPh_2)_2(CO)_{13}$, **7**, 909, 926, 928, 938
$Ru_6C_{45}H_{78}O_{13}P_4$
$Ru_6(H)_6(P)(CO)(PBu_2)(CO)_{12}(PBu_3)_2$, **7**, 938, 939
$Ru_6C_{54}H_{101}O_{10}P_6$
$Ru_6(H)(P)(PBu_2)_2(PHBu)(CO)_{10}(PBu_3)_2$, **7**, 938, 939
$Ru_6C_{54}H_{105}O_{10}P_5$

$Ru_6(H)_6(P)(PBu_2)_2(PHBu)$-$(CO)_{10}(PBu_3)_2$, **7**, 909, 939
$Ru_6Fe_2C_{41}H_{30}O_{17}$
$Fe_2Ru_6(C_2)_2(CO)_{17}Cp^*{}_2$, **7**, 937
$Ru_6HgC_{30}H_{18}O_{18}$
$[(C_2Bu^t)(CO)_9Ru_3]_2Hg$, **10**, 327, 330, 332, 339
$Ru_6HgC_{47}H_{33}O_{17}P$
$(PPh_3)(C_2Bu^t)(CO)_8Ru_3$-$(Hg)Ru_3(CO)_9(C_2Bu^t)$, **10**, 332
$Ru_6Os_3C_{28}HO_{28}$
$[HOs_3(CO)_{10}(O_2C)Ru_6(C)$-$(CO)_{16}]^-$, **7**, 937
$Ru_6PtC_{30}H_{20}O_{18}$
$PtRu_6(CO)_{18}(H)_2(CCBu^t)_2$, **9**, 572
$Ru_6Pt_3C_{21}H_4O_{21}$
$Ru_6Pt_3(H)_4(CO)_{21}$, **10**, 199
$Ru_6Pt_3C_{22}H_2O_{22}$
$Ru_6Pt_3(H)_2(CO)_{22}$, **10**, 199, 200
$Ru_6Pt_3C_{34}H_{12}O_{20}$
$Ru_6Pt_3(H)_2(PhC{\equiv}CPh)(CO)_{20}$, **10**, 213
$Ru_6Pt_3C_{42}H_{12}O_{20}$
$Ru_6Pt_3(CO)_{20}(H)_2(C_2Ph_2)$, **10**, 38, 39
$Ru_6Pt_3C_{56}H_{30}O_{14}$
$Pt_3Ru_6(CO)_{14}(PhCCPh)_3$, **9**, 572; **10**, 151
$Ru_6Pt_3C_{60}H_{30}O_{18}$
$Pt_3Ru_6(CO)_{18}(PhCCPh)_3$, **9**, 572; **10**, 152
$Ru_7C_{17}H_7NO_{14}$
$Ru_5(H)(CNMe_2)(CO)_{14}$, **7**, 923, 947
$Ru_7C_{20}HO_{20}$
$[HRu_7(CO)_{20}]^-$, **7**, 945
$Ru_7C_{20}O_{20}S_2$
$Ru_7(S)_2(CO)_{20}$, **7**, 946
$Ru_7C_{21}O_{21}S$
$Ru_7(S)(CO)_{21}$, **7**, 941, 946
$Ru_7C_{22}H_7NO_{19}$
$Ru_7(H)(CNMe_2)(CO)_{19}$, **7**, 947, 948
$Ru_7C_{28}H_{10}N_2O_{16}$
$Ru_7(NPh)_2(CO)_{16}$, **7**, 947
$Ru_7C_{30}H_{10}O_{18}P_2$
$Ru_7(PPh)_2(CO)_{18}$, **7**, 946, 947
$Ru_7C_{34}H_{18}O_{15}P_2$
$Ru_7(PPh)_2(CO)_{15}(PhMe)$, **7**, 947
$Ru_7C_{36}H_{62}O_{12}P_4$
$Ru_7(H)_8(P)(PBu)(PBu_2)(PBu_3)$-$(CO)_{12}$, **7**, 947
$Ru_7C_{42}H_{24}O_{16}P_2$
$Ru_7(H)_3(HC_2PPh_2)(PPh_2)$-$(CO)_{16}$, **7**, 909, 947
$Ru_7Hg_2C_{34}H_{18}O_{22}$
$Ru(CO)_4[HgRu_3(CO)_9(C_2$-$Bu^t)]_2$, **10**, 334, 335
$Ru_7Pt_3C_{22}H_2O_{22}$
$Ru_7Pt_3(H)_2(CO)_{22}$, **10**, 199
$Ru_8Ag_4B_2C_{60}H_{34}Cl_2NO_{24}P_2$
$[PPN][[HRu_4(CO)_{12}BH]_2Ag_4$-$(Cl)]Cl$, **1**, 304
$Ru_8AgB_2C_{60}H_{34}NO_{24}P_2$
$[PPN][HRu_4(CO)_{12}BH]_2$, **1**, 304
$Ru_8Ag_2C_{50}H_{30}O_{24}P_2$
$[AgRu_4(H)_3(CO)_{12}]_2[Ph_2P$-$(CH_2)_2PPh_2]$, **10**, 302
$Ru_8B_2Cu_4C_{60}H_{32}Cl_2NO_{24}P_2$
$[PPN][[HRu_4(CO)_{12}B]_2Cu_4$-$(Cl)]Cl$, **1**, 304, 305
$Ru_8C_{21}H_2O_{21}$
$[Ru_8(H)_2(CO)_{21}]^{2-}$, **7**, 948
$Ru_8C_{24}H_8O_{17}S_2$
$Ru_8(S)_2(CO)_{17}(PhMe)$, **7**, 949
$Ru_8C_{26}H_7O_{19}P$
$Ru_8(P)(Bz)(CO)_{19}$, **7**, 949
$Ru_8C_{39}H_{15}O_{21}P_3$
$Ru_8(P)(PPh)(PPh_2)(CO)_{21}$, **7**, 949
$Ru_8C_{40}H_{28}O_{22}P_2$
$[Ru_4(H)_4(CO)_{11}(bdpa)$, **7**, 847
$Ru_8C_{43}H_{20}O_{17}P_2$
$Ru_8(C)_2(PPh_2)_2(CO)_{17}$, **7**, 948, 949
$Ru_8C_{48}H_{28}O_{22}P_2$
$[Ru_4(H)_4(CO)_{11}]_2(dppa)$, **7**, 847
$Ru_8PtC_{49}H_{26}O_{23}P_2$
$Ru_8Pt_2(H)_2(CO)_{23}(dppe)$, **10**, 199, 200
$Ru_8Pt_2C_{23}H_2O_{23}$
$Ru_8Pt_2(H)_2(CO)_{23}$, **7**, 848; **10**, 11, 199, 200
$Ru_9B_2C_{23}O_{23}$
$Ru_9(CO)_{23}B_2$, **1**, 294
$Ru_9C_{24}H_3O_{24}S_3$
$Ru_9H_3(S)_3(CO)_{24}$, **7**, 727
$Ru_9C_{36}H_{15}N_3O_{21}S_3$
$[Ru_3H(pyS)(CO)_7]_3$, **7**, 730
$Ru_9Hg_3C_{33}O_{33}$
$[Ru_3Hg(CO)_{11}]_3$, **7**, 720
$Ru_{10}C_{24}NO_{24}$
$[Ru_{10}(N)(CO)_{24}]^-$, **7**, 953
$Ru_{10}C_{25}HO_{24}$
$[Ru_{10}(H)(C)(CO)_{24}]^-$, **7**, 951
$Ru_{10}C_{25}H_2O_{25}$
$[H_2Ru_{10}(CO)_{25}]^{2-}$, **7**, 909, 950
$Ru_{10}C_{25}O_{24}$
$[Ru_{10}(C)(CO)_{24}]^{2-}$, **7**, 908, 950–2
$Ru_{10}C_{26}O_{24}$
$[Ru_{10}(C)_2(CO)_{24}]^{2-}$, **7**, 950, 951, 953
$Ru_{10}C_{38}H_{10}O_{22}$
$[Ru_{10}(C)_2(C_2Ph_2)(CO)_{22}]^{2-}$, **7**, 909, 953
$Ru_{10}Hg_2C_{30}Cl_4O_{28}$
$[Ru_5C(CO)_{14}(Cl)(HgCl)]_2$, **10**, 327, 328
$Ru_{11}C_{27}HO_{27}$
$[HRu_{11}(CO)_{27}]^{3-}$, **7**, 909, 953, 954
$Ru_{12}AsTlC_{58}H_{20}O_{32}$
$[AsPh_4]Tl[Ru_6C(CO)_{16}]_2$, **1**, 568
$Ru_{12}C_{49}H_{20}N_4O_{28}$
$[Ru_2(H)(NC_5H_4)_2(CO)_4(N$-$C_5H_5)_2][Ru_{10}(C)(H)$-$(CO)_{24}]$, **7**, 646
$Ru_{18}Hg_3C_{44}O_{42}$
$[Ru_9C(CO)_{21}]_2(Hg_3)^{2-}$, **10**, 340
$[Ru_{18}Hg_3(C)_2(CO)_{42}]^{2-}$, **7**, 949
$Ru_{43}H_{38}Cl_2P_2$
$Ru(Cl)_2(PPh_3)_2(nbd)$, **7**, 452

S

$SbAgRu_2C_{19}H_{37}F_6N_3O_{13}P_5$
$[Ru_2[Ag(MeCN)](CO)_5$-$[(MeO)_2PNEtP(O$-$Me)_2]_2][SbF_6]$, **7**, 627
$SbAg_{12}Au_{13}C_{180}H_{150}Br_8F_6$
$[Ag_{12}Au_{13}(PPh_3)_{10}Br_8][SbF_6]$, **10**, 256
$SbB_{10}H_{10}P$
$PSbB_{10}H_{10}$, **1**, 264
$SbB_{11}H_{11}$
$[B_{11}H_{11}Sb]^-$, **1**, 267
$SbCH_3I_2$
$MeSbI_2$, **2**, 333
$SbCH_5$
$MeSbH_2$, **2**, 328
SbC_3H_9O
Me_3SbO, **2**, 338
$SbC_3H_{11}O_2$
$Me_3Sb(OH)_2$, **2**, 338, 339
$SbC_4H_{12}I$
$[Me_4Sb]I$, **2**, 336
$SbC_4H_{12}N$
$(Me_2N)Me_2Sb$, **11**, 496
$SbC_5H_5Cl_2$
$CpSbCl_2$, **2**, 329
$SbC_6H_5Br_2$
$PhSbBr_2$, **2**, 333
$SbC_6H_5I_2$
$PhSbI_2$, **2**, 333, 334
SbC_6H_{15}
Et_3Sb, **11**, 492
SbC_8H_{19}
Bu^t_2SbH, **2**, 328
SbC_9H_{15}
$(HC{\equiv}CCH_2)_2Me_3Sb$, **11**, 497
$SbC_{10}H_{15}Cl_2$
Cp^*SbCl_2, **2**, 329
$SbC_{12}H_{10}Br_3$
Ph_2SbBr_3, **2**, 342; **11**, 497
$SbC_{12}H_{10}Cl_2$
$[Ph_2SbCl_2]^-$, **2**, 332
$SbC_{12}H_{10}Cl_3$
Ph_2SbCl_3, **2**, 342
$SbC_{12}H_{11}$
Ph_2SbH, **11**, 496
$SbC_{12}H_{23}$
$(C_6H_{11})_2SbH$, **2**, 328
$SbC_{12}H_{27}$
Bu_3Sb, **11**, 489, 490, 492, 497
$SbC_{13}H_{10}NS$
Ph_2SbSCN, **2**, 332
$SbC_{15}H_{15}$
Cp_3Sb, **2**, 329
$SbC_{15}H_{30}Br$
$(CH_2{=}C{=}CH)Bu_3SbBr$, **11**, 494
$SbC_{15}H_{32}Br$
$(CH_2{=}CHCH_2)Bu_3SbBr$, **11**, 493
$SbC_{15}H_{32}BrO_2$
$[MeOC(O)CH_2]Bu_3SbBr$, **11**, 494, 495
$SbC_{15}H_{32}I$
$(CH_2{=}CHCH_2)Bu_3SbI$, **11**, 493
$SbC_{16}H_{34}BrO_2$
$[Bu_3SbCH_2CO_2Et]^+$, **11**, 493
$SbC_{17}H_{37}O_2$
$Bu_3SbMeCH_2CO_2Et$, **11**, 493
$SbC_{18}H_{15}$
Ph_3Sb, **11**, 490, 491, 501
$SbC_{18}H_{15}Br_2$
Ph_3SbBr_2, **11**, 497
$SbC_{18}H_{15}Cl_2$
Ph_3SbCl_2, **11**, 500
$SbC_{18}H_{15}O$
Ph_3SbO, **2**, 338; **11**, 501
$SbC_{18}H_{15}S$
Ph_3SbS, **2**, 338
$SbC_{18}H_{15}S_3$
$(PhS)_3Sb$, **11**, 496
$SbC_{19}H_{34}Br$
$BzBu_3SbBr$, **11**, 493
$SbC_{19}H_{39}$
$(CH_2{=}C{=}CH)Bu_4Sb$, **11**, 494
$SbC_{19}H_{41}$
$(CH_2{=}CHCH_2)Bu_4Sb$, **11**, 493
$SbC_{20}H_{21}O_2$
$Ph_3Sb(OMe)_2$, **2**, 340
$SbC_{20}H_{43}$
$(MeCH{=}CHCH_2)Bu_4Sb$, **11**, 493, 496
$SbC_{22}H_{21}O_4$
$Ph_3Sb(OAc)_2$, **11**, 497
$SbC_{23}H_{43}$
$(Bz)Bu_4Sb$, **11**, 496
$SbC_{24}H_{20}Br$
Ph_4SbBr, **11**, 499
$SbC_{24}H_{20}I$
Ph_4SbI, **11**, 500
$SbC_{25}H_{21}O_2$
Ph_4SbO_2CH, **2**, 336
$SbC_{26}H_{23}O_2$
Ph_4SbO_2CMe, **2**, 336
$SbC_{27}H_{35}O_2$
$Mes_3Sb(OH)_2$, **2**, 338
$SbCl_3$
$SbCl_3$, **11**, 498, 501
$SbCl_5$
$SbCl_5$, **11**, 500
$SbFeC_{38}H_{31}F_6O_2P_2$
$[HFe(CO)_2(PPh_3)_2][SbF_6]$, **7**, 40
$SbFeMoC_{10}H_5F_6O_5Te_2$
$[CpMoFe(CO)_5Te_2][SbF_6]$, **7**, 77
$SbFe_2MnC_{37}H_{25}O_8P$
$MnFe_2(PPh)(CO)_8(SbPh_3)Cp$, **10**, 222
$SbFe_3C_{10}HO_9$
$Fe_3(CO)_9(Sb)(CH)$, **7**, 58
$SbFe_3C_{10}O_{10}$
$SbFe_2(CO)_{10}$, **7**, 58
$SbFe_3C_{12}ClO_{12}$
$[ClSb[Fe(CO)_4]_3]^{2-}$, **7**, 58
$SbFe_3C_{13}H_{10}O_9$
$[HFe_3(CO)_9(SbBu^t)]^-$, **7**, 58
$SbFe_3CrC_{17}O_{17}$
$[Sb[Fe(CO)_4][Fe_2(CO)_8][Cr$-$(CO)_5]]^-$, **7**, 58
$SbFe_4C_{13}HO_{13}$
$[(H)Fe_3(CO)_9[SbFe(CO)_4]]^{2-}$, **7**, 58
$SbFe_4C_{13}H_2O_{13}$
$[(H)_2Fe_3(CO)_9[SbFe(CO)_4]]^-$, **7**, 58
$SbFe_4C_{16}O_{16}$
$[Sb[Fe_2(CO)_8][Fe(CO)_4]_2]$, **7**, 58
$SbLiC_{21}H_{45}N_3$
$LiSb(C_6H_{11})_2{\cdot}PMDETA$, **1**, 11
$SbLiSi_2C_6H_{18}$
$(TMS)_2SbLi$, **2**, 30
$SbMgC_{12}H_{10}$
Ph_2SbMg, **11**, 491
$SbMnC_{14}H_{10}Cl_6O_2$
$[MnCp(CO)_2(CPh)]SbCl_6$, **6**, 63
$SbMn_2C_{13}H_7Cl_2O_7$
$Mn(CO)_5(SbCl_2)Mn(CO)_2$-(C_5H_4Me), **6**, 86
$SbMn_3C_{18}H_7ClO_{12}$
$Mn_2(CO)_{10}(SbCl)Mn(CO)_2$-$(C_5H_4Me)$, **6**, 86
$SbMoC_{39}H_{36}I_2O_3P$
$MoI_2(CO)_3(PPh_2Cy)(SbPh_3)$, **5**, 262
$SbOsC_{40}H_{33}Cl_2F_7NOP_2$
$[OsCl(CClF)(CO)(NCMe)$-$(PPh_3)_2][SbF_6]$, **7**, 347
$SbOsC_{54}H_{49}F_6O_2P_4$
$[Os(CHO)(CO)(dppe)_2][SbF_6]$, **7**, 418
$SbReC_{10}H_{10}Cl_2F_6$
$[ReCp_2Cl_2][SbF_6]$, **6**, 252
$SbReC_{32}H_{36}ClF_6N_3P_2$
$[ReCl(CNHMe)(CNMe)_2$-$(PMePh_2)_2][SbF_6]$, **6**, 203, 215
$SbRu_2C_{17}H_{34}Cl_2N_2O_{13}P_4$
$[Ru_2(SbCl_2)(CO)_5(etmdp)_2]^+$, **7**, 627
$SbRu_4C_{28}H_{15}O_{10}Se_4$
$Ru_4(Se)_4(CO)_{10}(SbPh_3)$, **7**, 869
$SbSiC_4H_{13}$

TMS-CH_2SbH_2, **2**, 328
$SbSi_2C_7H_{19}Cl_2$
$(TMS)_2CHSbCl_2$, **2**, 333
$SbSi_2C_{12}H_{23}$
$PhSb(TMS)_2$, **2**, 328
$SbSi_2C_{12}H_{24}$
$(TMS)_2SbPh$, **2**, 30
$SbSi_2C_{25}H_{48}P$
$Bu^t_3C_6H_2PSbCH(TMS)$, **2**, 326
$SbTiC_{12}H_{13}Cl_7N$
$[TiCl(MeCN)Cp_2][SbCl_6]$, **4**, 398
$SbWC_{17}H_{27}O_5S_3$
$W(CO)_5(SBu^t)(SbSBu^t)_2$, **5**, 262
$Sb_2AuC_2F_{11}O$
$[Au(CO)_2][Sb_2F_{11}]$, **3**, 48
$Sb_2B_9PdC_{16}H_{31}$
$(Me_2PhP)_2PdSb_2B_9H_9$, **1**, 267
$Sb_2B_{10}H_{10}$
$Sb_2B_{10}H_{10}$, **1**, 267
$Sb_2C_4H_{12}$
$Me_2SbSbMe_2$, **2**, 325, 326
$Sb_2C_{24}H_{20}$
$Ph_2SbSbPh_2$, **2**, 326
$Sb_2C_{32}H_{32}O_5$
$[Ph_2Sb(OAs)_2]_2O$, **2**, 342, 343
$Sb_2C_{36}H_{30}O_2$
$(Ph_3SbO)_2$, **2**, 339
$Sb_2Co_4C_{11}O_{11}$
$[Co_4(CO)_{11}Sb_2]^-$, **8**, 450
$Sb_2Fe_2C_{37}H_{25}O_7$
$Ph_2Sb[Fe(CO)_4][Fe(CO)_3$-$SbPh_3]$, **7**, 57
$Sb_2Fe_5C_{17}O_{17}$
$[Fe_3(CO)_9[SbFe(CO)_4]_2]^{2-}$, **7**, 58
$Sb_2Fe_6C_{20}O_{20}$
$[Fe_4(CO)_{12}[SbFe(CO)_4]_2]^{2-}$, **7**, 58
$Sb_2Ru_2C_{36}H_{72}F_{12}N_2O_{10}P_4$
$[Ru_2(CO)_5$(acetone)(etipdp)$_2]$-$(SbF_6)_2$, **7**, 628
$Sb_2Si_4C_{12}H_{36}$
$(TMS)_4Sb_2$, **2**, 31
$Sb_2TiC_{14}H_{16}Cl_{12}N_2$
$[Ti(MeCN)_2Cp_2][SbCl_6]_2$, **4**, 398
$Sb_3Ga_3C_{24}H_{54}Cl_6$
$[Cl_2Ga(SbBu^t_2)]_3$, **1**, 522
$Sb_3Ga_3C_{30}H_{72}$
$[Me_2Ga(SbBu^t_2)]_3$, **1**, 522
$Sb_3In_3C_{30}H_{72}$
$[Me_2In(SbBu^t_2)]_3$, **1**, 522
$Sb_3Si_6C_{21}H_{57}$
$[(TMS)_2CHSb]_3$, **2**, 328
$Sb_3Sm_3C_{64}H_{98}O$
$(Sb_3)[SmCp^*_2]_3(THF)$, **4**, 83
$Sb_4C_{24}H_{20}I_9$
$[Ph_4Sb_4I_9]^-$, **2**, 334
$Sb_4C_{40}H_{60}$
$(Cp^*Sb)_4$, **2**, 329
$Sb_4Si_8C_{28}H_{76}$
$[(TMS)_2CHSb]_4$, **2**, 328
$Sb_5Si_5C_{20}H_{55}$
(TMS-$CH_2Sb)_5$, **2**, 328
$Sb_6C_{36}H_{30}$
$(PhSb)_6$, **2**, 328
$Sb_6Pt_2C_{56}H_{59}F_6P_4$
$[Pt_2Et_3(dppm)_2]SbF_6$, **9**, 455
$ScBC_{10}H_{14}$
$Cp_2Sc(BH_4)$, **1**, 278; **4**, 113
$ScBSi_4C_{22}H_{46}$
$[(TMS)_2C_5H_3]_2Sc(BH_4)$, **1**, 278; **4**, 49
$ScC_{11}H_{13}S$
Cp_2ScMeS, **4**, 44
$ScC_{13}H_{13}$
$(C_8H_8)ScCp$, **4**, 105
$ScC_{13}H_{24}Cl_2P$
$Cp^*ScCl_2(PMe_3)$, **4**, 70
$ScC_{15}H_{15}$
$ScCp_3$, **4**, 53, 113
$ScC_{18}H_{24}$
$Sc(Bu^t_6H_3)_2$, **4**, 8
$ScC_{19}H_{22}N$
$Cp_2Sc(C_6H_4CH_2NMe_2)$, **4**, 43
$ScC_{21}H_{33}$
Cp^*_2ScMe, **4**, 84, 117
$ScC_{22}H_{32}$
$Sc(C_6Mc_5H)_2$, **4**, 8
$ScC_{22}H_{35}$
Cp^*_2ScEt, **4**, 93
$ScC_{22}H_{35}Cl_2O_3$
$Me_2CC(Ph)ScCl_2(THF)_3$, **4**, 21
$ScC_{23}H_{35}$
$Cp^*_2Sc(C_3H_5)$, **4**, 89, 93
$ScC_{23}H_{37}$
$Cp^*_2Sc(Pr)$, **4**, 93
$ScC_{25}H_{35}N$
$Cp^*_2Sc(py)$, **4**, 85, 93
$ScC_{26}H_{40}$
$Cp^*_2ScCH_2(C_5H_8)$, **4**, 85
$ScC_{34}H_{58}N_2$
$Sc(Bu^t_3C_5H_2N)_2$, **4**, 9
$ScC_{36}H_{60}$
$Sc(Bu^t_3C_6H_3)_2$, **4**, 3, 4
$Sc(Bu^t_3C_6H_3)[Bu^t_2(CMe_2CH_2)$-$C_6H_3]H$, **4**, 4
$ScSiC_{20}H_{31}$
$[Me_2Si(C_5H_3Bu^t)_2]ScH$, **4**, 61
$ScSiC_{23}H_{35}$
$[Me_2Si(C_5H_3Bu^t)_2]Sc(allyl)$, **4**, 61
$ScSiC_{23}H_{40}P$
$[Me_2Si(C_5Me_4)_2]ScH(PMe_3)$, **4**, 116
$ScSi_3C_{33}H_{62}P$
$Me_2Si(C_5Me_4)(C_5H_3CH_2CH_2P$-$Bu^t_2)ScCH(TMS)_2$, **4**, 116
$ScSi_4C_{23}H_{45}O$
$Cp_2Sc[Si(TMS)_3](THF)$, **4**, 46
$ScSi_6C_{57}HO_6$
$Sc[CH(SiMe_2C_6H_4OMe)_2]_3$, **4**, 17
$ScSi_6C_{57}H_{81}O_6$
$Sc[CH(SiMe_2C_6H_4OMe)_2]_3$, **4**, 16
$Sc_2C_{20}H_{20}Cl_2$
$[Cp_2ScCl]_2$, **4**, 36, 113
$Sc_2C_{30}H_{98}O_2$
$[Cp^*Sc(OC_6H_3Bu^t_2)Me]_2$, **4**, 70
$Sc_2C_{40}H_{58}$
$[Cp^*Sc(C_5Me_4CH_2)]_2$, **4**, 85
$Sc_2C_{40}H_{60}Te$
$(Te)[ScCp^*_2]_2$, **4**, 78
$Sc_2Si_2C_{36}H_{68}N$
$[(C_5Me_4SiMe_2NBu^t)Sc(Pr)]_2$, **4**, 118
$Sc_2Si_2C_{36}H_{74}N_2P_2$
$[(C_5Me_4SiMe_2NBu^t)Sc(PMe_3)$-$(H)]_2$, **4**, 70, 117
$SeCH_4N_2$
$NH_2C(Se)NH_2$, **11**, 528
SeC_2H_6
Me_2Se, **11**, 533
SeC_3H_3O
C_3H_3SeO, **11**, 557
SeC_3H_5O
C_3H_5SeO, **11**, 557
$SeC_3H_7O_2$
$HSeCH_2CH(NH_2)COOH$, **11**, 516
SeC_5H_6
$SeCH_2CHCHCHCH$, **11**, 546
$SeC_5H_{11}N$
$MeSeCH_2CH_2CH(NH_2)COOH$, **11**, 517
SeC_7H_5Cl
PhCSeCl, **11**, 535
$SeC_8H_5F_3O_2$
$PhSeOC(O)CF_3$, **11**, 540
$SeC_8H_8O_2$
PhSeOC(O)Me, **11**, 540
$SeC_{10}H_{10}O_4$
$(PhSe)C(Me)(COOH)_2$, **11**, 548
$SeC_{16}H_{10}F_6O_4$
$Ph_2Se(CF_3CO_2)_2$, **11**, 533
$SeC_{16}H_{16}O_6$
$Ph_2Se[(OH)CH_2CO_2]_2$, **11**, 533
$SeC_{16}H_{20}$
$C_{10}H_{15}(SePh)$, **11**, 538
$SeC_{16}H_{26}O$
$(C_{10}H_{21})PhSeO$, **11**, 557
SeF_4
SeF_4, **11**, 528, 537
SeKCN
KSeCN, **11**, 528
SeNaCN
NaSeCN, **11**, 528
SeO_2
SeO_2, **11**, 536
$Se_2C_2N_2$
$(SeCN)_2$, **11**, 528
$Se_2C_7H_6$
$PhCSe_2H$, **11**, 534
$Se_3C_{14}H_{10}$
$(PhCSe)_2Se$, **11**, 534
$Se_3Sn_3C_6H_{18}$
$(Se_8Sn_3)Me_6$, **11**, 545
$Si_5C_{20}H_{40}$
$[(CH_2)_4Si]_5$, **2**, 81
$SiAlC_{12}H_{27}$
$(Bu^t_3Si)Al$, **1**, 445
$SiAsC_{10}H_{17}$
Ph(Me)AsTMS, **2**, 29
$SiAsGaC_{26}H_{48}$

$Cp^*{}_2GaAs(TMS)_2$, **1**, 518, 519
$SiAuMnC_{39}H_{35}O_2P$
$Mn(SiMePh_2)(CO)_2Cp'Au(PPh_3)$, **10**, 48
$SiAu_3Nb_3C_8H_{19}$
$Au_3Nb_3(H)_6(C_5H_4TMS)$, **10**, 317
$SiBC_8H_{16}Cl_3$
$Bu^tCH{=}C(SiMe_2Cl)BCl_2$, **1**, 157
$SiBC_{10}H_{22}Cl_2N$
$Bu^tCH{=}C(SiMe_2Cl)B(Cl)NMe_2$, **1**, 157
$SiBC_{10}H_{22}N$
$Et_2Me_4[BNSiC_2]$, **1**, 213
$SiBC_{10}H_{22}P$
$Et_2Me_4[BPSiC_2]$, **1**, 213
$SiBC_{10}H_{22}S$
$Et_2Me_4[BSSiC_2]$, **1**, 213
$SiBC_{10}H_{22}Se$
$Et_2Me_4[BSeSiC_2]$, **1**, 213
$SiBC_{11}H_{23}S_2$
$Et_2Me_3(BS_2SiC_4H_4)$, **1**, 180
$SiBC_{18}H_{15}$
Ph_3SiB, **2**, 22
$SiBC_{19}H_{37}S$
$(BBN)CH_2CH{=}C(TMS)SPh$, **11**, 200
$SiBC_{21}H_{33}O_6$
$PhMe_2SiCH{=}CHCH_2(BO_2C_2H_2)(CO_2Pr^i)_2$, **11**, 252
$SiBC_{27}H_9F_{20}$
$(TMS)B(C_6F_5)_4$, **2**, 4
$SiBNiC_{13}H_{38}F_4P_3$
$[Ni(CH_2TMS)(PMe_3)_3][BF_4]$, **9**, 56
$SiBScC_{22}H_{46}$
$[C_5H_3(TMS)_2]_2Sc[BH_4]$, **1**, 278
$SiBUC_{19}H_{33}Cl_2N_6$
$[HB(Me_2Pz)_3]UCl_2(CH_2TMS)$, **4**, 192
$SiBZrC_{43}H_{45}F_4$
$[Zr[C(TMS)CMe_2]Cp'_2][B(C_6H_4F)_4]$, **4**, 606
$SiBZrC_{45}H_{55}O$
$[Zr(CH_2CH_2TMS)(THF)Cp'_2][BPh_4]$, **4**, 598
$SiBZrC_{52}H_{69}O$
$[Zr(CH_2TMS)(THF)Cp^*{}_2][BPh_4]$, **4**, 595
$SiB_4C_5H_{16}$
nido-$C_2B_4H_7(TMS)$, **1**, 220
$SiB_4C_6H_{18}B_2$
$(TMS)_2N_2B_4$, **1**, 261
$SiB_4SnC_6H_{16}$
$Sn(TMS)MeC_2B_4H_4$, **1**, 260
$SiB_5C_8H_{26}OP$
$(TMS\text{-}O)Bu^tC{=}PB_5H_8$, **1**, 264
$SiB_8C_5H_{22}$
arachno-$C_2B_8H_{13}(TMS)$, **1**, 228, 229
$SiB_9C_{10}H_{26}Br_5$
$(Pr^i{}_3Si)Br_5CB_9H_5$, **2**, 5
$SiB_9C_{12}H_{29}Br_5N$
$(Pr^i{}_3Si)(NCMe)Br_5CB_9H_5$, **2**, 5
$SiB_{18}C_{14}H_{32}N$
$(C_2B_9H_{11})(C_2B_9H_{11})Si(C_5H_5N)_2$, **1**, 259
$SiBaC_6H_{12}O_6$
$Ba[Si(OCH_2CH_2O)_3]$, **2**, 4
$SiCH_3Cl$
$MeClSi$, **2**, 113
$SiCH_3Cl_3$
$MeSiCl_3$, **2**, 88
$SiCH_3F_3$
$MeSiF_3$, **2**, 39
$SiCH_4Cl_2$
$MeHSiCl_2$, **2**, 113
SiC_2H_4O
$(CHO)MeSi$, **11**, 328
$SiC_2H_6Cl_2$
Me_2Cl_2Si, **2**, 88, 112
$SiC_2H_6F_2$
Me_2SiF_2, **2**, 39
SiC_2H_6S
Me_2SiS, **2**, 36
SiC_2H_7Cl
Me_2SiHCl, **2**, 113
SiC_2H_7ClO
$ClMe_2SiOH$, **2**, 113
SiC_3H_8
$Me_2Si{=}CH_2$, **2**, 49, 51
SiC_3H_9BrS
TMS-SBr, **2**, 35
SiC_3H_9Cl
TMS-Cl, **2**, 38; **11**, 338; **12**, 110
SiC_3H_9F
TMS-F, **2**, 39
SiC_3H_9I
TMS-I, **2**, 38
$SiC_4H_9F_3$
$(CF_3)TMS$, **11**, 342
$SiC_4H_9F_3O_3S$
TMS(OTf), **11**, 340
SiC_4H_9N
TMS-CN, **11**, 339, 340
$SiC_4H_{11}Cl$
$TMS\text{-}CH_2Cl$, **2**, 228
SiC_4H_{12}
Me_4Si, **12**, 152
$SiC_4H_{12}O_4$
$(MeO)_4Si$, **2**, 3
$SiC_4H_{15}NO$
$[Me_4N][SiH_3O]$, **2**, 120
SiC_4NSe
TMS-NCSe, **11**, 341
SiC_5H_{12}
$Me_2\overline{Si(CH_2)_2CH_2}$, **2**, 49
$SiC_5H_{12}O$
$CH_2{=}C(TMS)OH$, **11**, 328
$SiC_5H_{12}OTe$
$MeC(Te)O(TMS)$, **2**, 37
$SiC_5H_{14}S$
$(TMS)C_2H_4SH$, **11**, 342
$SiC_6H_5F_2$
$PhSiF_2$, **2**, 39
$SiC_6H_9F_7Se$
$(TMS)SeC_3F_7$, **2**, 36
$SiC_6H_{12}Br_2$
$BrCH_2CHCBr(TMS)$, **11**, 316
$SiC_6H_{13}Cl_3$
$Me_2CHCMe_2SiCl_3$, **2**, 90
$SiC_6H_{13}O_6$
$[Si(OCH_2CH_2O)_2OCH_2CH_2OH]^-$, **2**, 4
$SiC_6H_{14}O_3$
$(\overline{CHOCH_2CH_2O})(O\text{-}TMS)$, **11**, 330
$SiC_6H_{15}Cl$
Bu^tMe_2SiCl, **2**, 31
$SiC_6H_{15}F$
Et_3FSi, **11**, 344
SiC_6H_{16}
Et_3SiH, **11**, 344
$SiC_6H_{16}O_3$
Et_3SiO_3H, **2**, 34; **11**, 341
$SiC_7H_{15}I$
$ICH_2C(CH_2)CH_2(TMS)$, **11**, 323
$SiC_7H_{16}O$
$(CHO)Bu^tMe_2Si$, **11**, 328
$SiC_7H_{19}O_3P$
$Et_2(TMS)PO_3$, **11**, 342
SiC_8H_{18}
$Bu^t{}_2Si$, **2**, 86
$Me_2\overline{SiCMe_2CMe_2}$, **2**, 47
$SiC_8H_{18}F_2$
$Bu^t{}_2SiF_2$, **2**, 39
$SiC_9H_{11}F_4$
$[(Mes)SiF_4]^-$, **2**, 39
SiC_9H_{14}
TMS-Ph, **12**, 152, 153
$SiC_9H_{14}Te$
PhTe(TMS), **2**, 37
$SiC_9H_{18}O_2$
$AcOCH_2C(CH_2)CH_2(TMS)$, **11**, 323
SiC_9H_{21}
$[Pr^i{}_3Si]^+$, **2**, 5
SiC_9H_{22}
$Pr^i{}_3SiH$, **2**, 5
$SiC_{10}H_{14}N$
PhCN(TMS), **11**, 343
$SiC_{10}H_{16}$
$[MeC_6H_4TMS]^+$, **2**, 4
$SiC_{10}H_{22}O$
$(CHO)Pr^i{}_3Si$, **11**, 328
$SiC_{10}H_{25}N$
$(NC_4H_4)(SiPr^i{}_3)$, **11**, 343
$SiC_{11}H_{18}O_2S$
$(PhSO_2)(TMS)C_2H_4$, **11**, 342
$SiC_{11}H_{27}PTe$
$(TMS)P(Te)Bu^t{}_2$, **2**, 27
$SiC_{12}H_{10}F$
Ph_2SiF, **2**, 39
$SiC_{12}H_{20}$
Et_3SiPh, **2**, 22
$SiC_{13}H_{22}OTe$
$(TMS)O(CH_2)_4TePh$, **2**, 37
$SiC_{13}H_{27}N$
$Bu^t{}_3SiCN$, **2**, 87
$SiC_{15}H_{18}$
$Me(Bz)_2SiH$, **2**, 4
$SiC_{15}H_{19}PSe_2$
$(TMS)SeP(Se)Ph_2$, **2**, 27
$SiC_{16}H_{16}S$
Ph_3SiSH, **11**, 341

$SiC_{18}H_{15}BrS$
Ph_3SiSBr, **2**, 35
$SiC_{18}H_{16}$
Ph_3SiH, **11**, 344
$SiC_{18}H_{16}O$
Ph_3SiOH, **2**, 31
$SiC_{18}H_{22}$
$(Mes)_2Si$, **2**, 86
$SiC_{18}H_{22}F_3$
$[(Mes)_2SiF_3]^-$, **2**, 39
$SiC_{19}H_{16}O$
$(CHO)Ph_3Si$, **11**, 328
$SiC_{20}H_{15}F_3O$
$(CF_3CO)Ph_3Si$, **11**, 328
$SiC_{20}H_{30}$
$Cp^*{}_2Si$, **2**, 19
$SiC_{22}H_{18}F_2O$
$F_2C{=}C(OSiPh_3)CH{=}CH_2$, **11**, 336
$SiC_{22}H_{66}$
$Me(SiMe_2)_{10}Me$, **2**, 78
$SiC_{34}H_{102}$
$Me(SiMe_2)_{16}Me$, **2**, 78
$SiC_{38}H_{114}$
$Me(SiMe_2)_{18}Me$, **2**, 78
$SiC_{50}H_{150}$
$Me(SiMe_2)_{24}Me$, **2**, 78
$SiCaC_9H_{14}$
$TMS\text{-}C_6H_4CaH$, **1**, 118
$SiCaC_{30}H_{48}$
$Cp^*{}_2Ca(TMS\text{-}C{\equiv}CC{\equiv}C\text{-}TMS)$, **1**, 114
$SiCdFeC_{23}H_{23}BrNO_6P$
$Fe(CO)_3[Si(OMe)_3](Ph_2Ppy)\text{-}CdBr$, **10**, 338
$SiCdFeC_{24}H_{24}ClO_3P$
$Fe(CO)_3(PPh_3)(TMS)CdCl$, **10**, 338
$SiCdFeC_{26}H_{26}ClO_7P_2$
$Fe(CO)_3[Si(OMe)_3](Ph_2PCH_2\text{-}C(O)Ph)CdCl$, **10**, 338
$SiCdFeC_{31}H_{31}ClO_6P_2$
$Fe(CO)_3[Si(OMe)_3](Ph_2PCH_2P\text{-}Ph_2)CdCl$, **10**, 338
$SiCdFeC_{46}H_{43}BrO_8P_2$
$Fe(CO)_3[Si(OMe)_3](Ph_2PCH_2\text{-}C(O)Ph)CdBr(Ph_2PCH_2\text{-}C(O)Ph)$, **10**, 339
$SiCl_2$
Cl_2Si, **2**, 113
$SiCl_3I$
ICl_3Si, **11**, 342
$SiCoC_7H_9O_4$
$[(TMS)Co(CO)_4]^+$, **12**, 975
$SiCoRhC_{14}H_{11}O_6$
$Co(CO)_4Rh(SiMe_2Ph)(CO)_2$, **10**, 62
$SiCo_2C_{12}H_{11}O_6$
$TMS\text{-}CCCH_2Co_2(CO)_6$, **12**, 693
$SiCo_3C_{19}H_{24}NO$
$Co_3(CO)(N\text{-}TMS)Cp_3$, **8**, 435
$SiCo_9C_{21}O_{21}$
$[Co_9Si(CO)_{21}]^{2-}$, **8**, 455
$SiCrC_{16}H_{18}O_3$
$[(TMS)C_{10}H_9]Cr(CO)_3$, **12**, 1028
$SiCrC_{17}H_{21}NO_4$
$[[(TMS)CH_2CH_2OCH_2]\text{-}indole]Cr(CO)_3$, **12**, 1029
$SiCrC_{17}H_{22}O_4$
$[[CH_2{=}CHCH(OMe)CH\text{-}(TMS)]Ph]Cr(CO)_3$, **12**, 1034
$SiCrC_{19}H_{28}O_5$
$[(Pr^i{}_3SiO)(MeO)C_6H_4]Cr(CO)_3$, **12**, 1024
$SiCrC_{22}H_{31}NO_3$
$[(TBDMS)Me(indoline)]Cr\text{-}(CO)_3$, **12**, 1025
$SiCrC_{26}H_{25}NO_7$
$Cr(CO)_5[Ph_2(TMS)(\overline{C_3HNO})(OEt)C]$, **12**, 528, 529
$SiCrC_{11}H_{14}O_3$
$Cr(CO)_3(PhTMS)$, **5**, 488
$SiCrC_{13}H_{14}O_4$
$Cr(CO)_3(TMS\text{-}C_6H_4CHO)$, **5**, 524
$SiCrC_{22}H_{36}O_2$
$Cr[C{=}C(TMS)_2](CO)_2(C_6Me_6)$, **5**, 363
$SiCrC_{26}H_{30}O_2$
$Cr(CO)_2(HSiPh_2H)(C_6Me_6)$, **5**, 507
$SiCuLiC_9H_{11}N$
$PhMe_2SiCu(CN)Li$, **12**, 96
SiF_4
SiF_4, **2**, 39
$SiFeC_9H_{12}O_5$
$(CO)_4Fe{=}C(O\text{-}TMS)Me$, **7**, 116
$SiFeC_{12}H_{20}O_2$
$FeH(CO)_2(SiEt_3)(cbd)$, **7**, 144
$SiFeC_{13}H_{13}NO_4$
$Fe(CO)_4[Si(H)C_6H_4CH_2NMe_2]$, **7**, 28
$SiFeC_{15}H_{16}O_4$
$Fe(CO)_3[PhCH{=}CHC(TMS)\text{-}(O)]$, **7**, 147
$SiFeC_{16}H_{33}P_3$
$Fe[MeSi(CH_2PMe_2)_3](C_6H_6)$, **7**, 149, 156
$SiFeC_{24}H_{37}P_3$
$Fe(MeSi(CH_2PMe_2)_3](PhC{\equiv}CPh)$, **7**, 123, 149
$SiFeC_{28}H_{22}O_3$
$Fe(CO)_3(Ph_3SiC_7H_7)$, **7**, 129
$SiFeC_{39}H_{31}O_3P$
$HFe(CO)_3(PPh_3)(SiPh_3)$, **7**, 27
$SiFeGeC_9H_{14}O_4$
$(CO)_4Fe(Me_2SiCH_2GeMe_2)$, **7**, 30
$SiFeInC_{26}H_{26}Cl_2O_7P$
$Fe(CO)_3[Si(OMe)_3][InCl_2]\text{-}[PPh_2CH_2C(O)Ph]$, **1**, 564
$SiFePdSnC_{49}H_{46}O_6P_2$
$Fe(CO)_3[Si(OMe)_2OMe]\text{-}(dppm)Pd(SnPh_3)$, **10**, 61
$SiFeRhC_{32}H_{31}O_7P_2$
$Fe(CO)_3(dppm)[Si(OMe)_3]Rh\text{-}(CO)$, **10**, 61
$SiFe_2C_{10}H_6O_8$
$Me_2SiFe_2(CO)_8$, **7**, 29
$SiFe_2GeC_{13}H_{14}O_8$
$Fe_2(CO)_8(Me_2SiCH_2GeMe_2)$, **7**, 29
$SiFe_3C_{18}H_{22}O_9P$
$[HFe_3(CO)_9[PSiPr^i{}_3]]^-$, **7**, 50
$SiGaC_{12}H_{33}$
$Ga(CH_2TMS)_3$, **1**, 514
$SiGaC_{22}H_{24}P$
$Bu^tGaPSiPh_3$, **1**, 524
$SiGaC_{30}H_{42}N$
$Bu^t{}_2GaNBu^t(SiPh_3)$, **1**, 520
$SiGaC_{44}H_{62}P$
$Bu^t{}_2GaP(Bu^t{}_3C_6H_2)(SiPh_3)$, **1**, 519, 520
$SiGeC_{10}H_{24}N_2$
$Bu^t{}_2Me_2[\overline{SiNGeN}]$, **2**, 190
$SiGeC_{21}H_{24}$
Ph_3GeTMS, **2**, 23
$Ph_3SiGeMe_3$, **2**, 23
$SiGeC_{25}H_{24}$
$MePhHSiGePh_3$, **2**, 178
$SiGeLiC_7H_{19}$
$(TMS)Et_2GeLi$, **2**, 176
$SiGeSnC_{11}H_{30}$
$Bu^tSiMe_2GeMe_2SnMe_3$, **2**, 285
$SiGeSnC_{25}H_{36}$
$Bu^tSiMe_2GeMe_2SnPh_3$, **2**, 285
SiH_2I_2
H_2SiI_2, **11**, 339
$SiHfC_{21}H_{37}Cl$
$HfCl(SiH_2Ph)CpCp^*$, **4**, 504
$SiHgMnC_{21}H_{21}BrO_2$
$MeCp(CO)_2SiMePh_2MnHgBr$, **10**, 337
$SiInC_8H_{13}$
$In(C_5H_4TMS)$, **1**, 530
$SiInC_{12}H_{33}$
$In(CH_2TMS)_3$, **1**, 514
$SiIrC_{13}H_{27}$
$Cp^*IrH_3(TMS)$, **8**, 385
$SiIrC_{28}H_{32}NP_2$
$Ir(H_2)[N(SiMe_2)(CH_2PPh_2)_2]$, **8**, 345
$SiIrC_{35}H_{37}OP_2$
$Ir(PPh_2CH_2CH_2SiMe_2)(PPh_3)\text{-}(CO)(H)_2$, **8**, 345
$SiIrLiC_{22}H_{49}N_3$
$Cp^*IrH_2(TMS)Li(pmdeta)$, **8**, 385
$SiIr_4C_{22}H_{33}O_9P_3$
$Ir_4(CO)_9[MeSi(PEt_2)_3]$, **8**, 495
$SiKC_3H_9$
TMS-K, **2**, 22
$SiKC_{18}H_{15}O$
Ph_3SiOK, **2**, 34
$SiLiC_3H_9$
TMSLi, **12**, 96
$SiLiC_4H_9N_2$
$Li(TMS)CN_2$, **11**, 341
$SiLiC_7H_{13}$
$(TMS)C{\equiv}CCH_2CH_2Li$, **12**, 79

$SiLiC_8H_{11}$
(Me$_2$PhSi)Li, **2**, 155; **12**, 96
$SiLiC_{10}H_{16}N$
[Bz(TMS)N]Li, **11**, 5
$SiLiC_{18}H_{15}$
Ph$_3$SiLi, **2**, 22
$SiLiReC_{25}H_{20}O_2$
[Re(H)(CO)$_2$(C$_5$H$_4$SiPh$_3$)][Li], **6**, 211
$SiLiSeC_3H_9$
(TMS)SeLi, **2**, 36
$SiLiVC_{16}H_{38}O_3$
[(TMS-CH$_2$)V(OBu^t)$_3$]Li, **5**, 13
$SiLuC_{18}H_{29}O$
Cp$_2$LuCH$_2$TMS(THF), **4**, 43
$SiLuC_{24}H_{41}$
Cp*$_2$LuCH$_2$TMS, **4**, 94
$SiMgC_5H_{14}$
TMSMgEt, **1**, 87
$SiMgC_6H_{15}ClO$
(Pr^{i}O)Me$_2$SiCH$_2$MgCl, **1**, 70
$SiMg_2C_4H_{10}Br_2$
(TMS)CH(MgBr)$_2$, **1**, 71
$SiMnC_8H_8Cl_3O_2$
(C$_5$H$_4$Me)MnH(SiCl$_3$)(CO)$_2$, **6**, 143
$SiMnC_8H_9O_5$
Mn(TMS)(CO)$_5$, **6**, 33
$SiMnC_{10}H_{11}O_3$
Mn(CO)$_3$(C$_5$H$_4$SiHMe$_2$), **6**, 110
$SiMnC_{13}H_{11}O_5$
Mn(SiMe$_2$Ph)(CO)$_5$, **6**, 25
$SiMnC_{13}H_{16}Cl_5O_2$
Mn(H)(SiEt$_3$)(CO)$_2$(C$_5$Cl$_5$), **6**, 116
$SiMnC_{15}H_{15}O_6$
Mn[CH(OSiMe$_2$Ph)Me](CO)$_5$, **6**, 25
$SiMnC_{25}H_{21}O_2$
Mn(H)(SiPh$_3$)Cp(CO)$_2$, **6**, 73
$SiMnZnC_{21}H_{21}ClO_2$
MeCp(CO)$_2$SiMePh$_2$MnZnCl, **10**, 337
$SiMoC_{18}H_{24}O_3$
[(TMS-O)cod]Mo(CO)$_2$Cp, **12**, 974
$SiMo_2C_{20}H_{26}O_4P_2$
Mo$_2$(CO)$_4$(PMe$_2$)$_2$(C$_5$H$_4$Si-Me$_2$C$_5$H$_4$), **5**, 429
$SiNbC_{14}H_{21}Cl$
Cp$_2$Nb(CH$_2$TMS)Cl, **5**, 70
$SiNbC_{23}H_{53}O_2P_4$
Nb(COSiPr$^i_3$)(dmpe)$_2$(CO), **5**, 65
$SiNb_3W_9O_{40}$
SiNb$_3$W$_9$O$_{40}$$^{7-}$, **8**, 178
$SiNiC_7H_{10}Br_2O$
MeCH=CHCH$_2$O(TMS)NiBr$_2$, **12**, 962
$SiNiC_7H_{16}Br_2$
[CH$_2$=C(Me)CH$_2$(TMS)]NiBr$_2$, **12**, 962, 963
$SiNiC_9H_{20}Cl_2O$
[CH$_2$=CHCH$_2$(O-TBDMS)]-NiCl$_2$, **12**, 1028, 1029
$SiNiC_{10}H_{27}ClN_2P_2$
Ni[C(N$_2$)TMS]Cl(PMe$_3$)$_2$, **9**, 19, 94
$SiNiC_{11}H_{27}ClP_2$
NiCl(CC-TMS)(PMe$_3$)$_2$, **9**, 60
$SiNiC_{11}H_{29}ClOP_2$
Ni(COCH$_2$TMS)Cl(PMe$_3$)$_2$, **9**, 3
$SiNiC_{16}H_{26}$
Ni(cod)[C$_4$H$_2$Me$_2$(SiMe$_2$)], **9**, 124
$SiNiC_{26}H_{30}$
Ni(cod)[CPh=CHCH=CPh-(SiMe$_2$)], **9**, 124
$SiNiC_{30}H_{42}OP_2$
Ni(C$_8$H$_{13}$)[P(Mes)C(O-TMS)P(Mes)], **9**, 145
$SiNiC_{47}H_{44}P_2$
Ni[PhC$_2$(TMS)](PPh$_3$)$_2$, **9**, 133
$SiNi_2C_{80}H_{72}P_4$
[Ni(PPh$_3$)$_2$]$_2$(CH$_2$=CH)$_2$Si-(CH=CH$_2$)$_2$, **9**, 119
$SiOsC_{10}H_{29}Cl_2P_2$
Os(CH$_2$TMS)Cl$_2$(PMe$_3$)$_2$, **7**, 377, 379
$SiOsC_{16}H_{22}P_4$
Os(CH$_2$SiMe$_2$CH$_2$)(PMe$_3$)$_4$, **7**, 425
$SiOsC_{16}H_{46}P_4$
Os[(CH$_2$)$_2$SiMe$_2$](PMe$_3$)$_4$, **7**, 382, 384
$SiOsC_{58}H_{56}ClP_3$
OsHCl(CH$_2$TMS)(C$_6$H$_4$PPh$_2$)-(PPh$_3$)$_2$, **7**, 377, 379
$SiOsWC_{13}H_{38}P_3S_4$
[Os(CH$_2$TMS)(PMe$_3$)$_3$(S)$_2$W-S$_2$]$^-$, **7**, 375
$SiOs_3C_{15}H_{10}O_{11}$
Os$_3$(CH-TMS)(CO)$_{11}$, **7**, 750, 755, 757
$SiOs_3C_{15}H_{16}O_{10}$
Os$_3$H[Si(OEt)$_3$](CO)$_9$, **7**, 708
$SiOs_3C_{16}H_{16}O_{11}$
Os$_3$H[O(SiEt$_3$)](CO)$_{10}$, **7**, 708, 724
$SiOs_3C_{16}H_{16}O_{13}$
Os$_3$H[Si(OEt)$_3$](CO)$_{10}$, **7**, 696
$SiOs_3C_{18}H_{19}NO_{13}$
Os$_3$H[Si(OEt)$_3$](MeCN)(CO)$_{10}$, **7**, 708, 726
$SiOs_4C_{17}H_{10}O_{14}$
Os$_4$(H)(CO)$_{14}$(TMS), **7**, 842
$SiPbC_{27}H_{42}$
Ph$_3$PbSi(TMS)$_3$, **2**, 312
$SiPdC_{43}H_{45}P_2$
[Pd[(CH$_2$TMS)C$_3$H$_4$](PPh$_3$)$_2$]$^+$, **9**, 374
$SiReC_{19}H_{30}ClN_2O$
Re[NCH$_2$NC$_5$H$_4$]Cp*Cl(O-TMS), **6**, 251
$SiRhC_{13}H_{25}$
CpRh(C$_2$H$_4$)(SiEt$_3$)H, **2**, 115
CpRhEt(SiEt$_3$), **2**, 116
$SiRhC_{15}H_{29}$
CpRh(C$_2$H$_4$)(SiEt$_3$)Et, **2**, 116
$SiRhC_{16}H_{30}P$
Rh(C$_2$TMS)(cod)(PMe$_3$), **8**, 267
$SiRhC_{23}H_{27}F_{12}O$
[C$_5$(OSiEt$_3$)(CF$_3$)$_4$]Rh(cod), **8**, 268
$SiRuC_{12}H_{21}Br$
RuBr(CH$_2$TMS)(C$_3$H$_5$)Cp, **7**, 510
$SiRuC_{16}H_{46}P_4$
Ru[(CH$_2$)$_2$SiMe$_2$](PMe$_3$)$_4$, **7**, 377, 378
$SiRuC_{21}H_{19}F_6O_7P$
Ru(Ph$_2$PCH$_2$SiMe$_2$OH)(CO)$_2$-(CO$_2$CF$_3$)$_2$, **7**, 633, 634
$SiRuC_{28}H_{43}P_2$
Ru(SiPh$_2$)(PMe$_3$)$_2$Cp*, **7**, 527
$SiRuC_{29}H_{43}O_3P_2S_3$
Ru(SiPh$_2$OTf)(PMe$_3$)$_2$Cp*, **7**, 527
$SiRuC_{32}H_{63}ClOP_2$
Ru(CPhCHSiEt$_3$)Cl(CO)-(PPr$^i_3$)$_2$, **7**, 404
$SiRuC_{46}H_{46}OP_2$
Ru(CC-TMS)(CH$_2$CHCHMe)-(CO)(PPh$_3$)$_2$, **7**, 412, 413
$SiRuC_{70}H_{63}ClOP_4$
RuCl(C$_4$CPh$_2$O-TMS)(dppm)$_2$, **7**, 354, 355
$SiRu_2C_{24}H_{41}Cl$
Ru$_2$(Cl)(CH$_2$)(TMS)Cp*$_2$, **7**, 504, 668, 669
$SiRu_2C_{26}H_{40}ClF_3O_2$
Ru$_2$Cp*$_2$(Cl)(CH-TMS)[OC-(CF$_3$)O], **7**, 668, 669
$SiRu_2C_{28}H_{44}$
Ru$_2$Cp*$_2$Cp(TMS), **7**, 668, 669
$SiRu_3C_{16}H_{16}O_{11}$
Ru$_3$H(OSiEt$_3$)(CO)$_{10}$, **7**, 724
$SiRu_5C_{35}H_{29}O_{17}P$
Ru$_5$(C)(CO)$_{14}$[Ph$_2$PCH$_2$CH$_2$Si-(OEt)$_3$], **7**, 919
$SiSbC_4H_{13}$
TMS-CH$_2$SbH$_2$, **2**, 328
$SiScC_{20}H_{31}$
[Me$_2$Si(C$_5$H$_3$Bu^t)$_2$]ScH, **4**, 61
$SiScC_{23}H_{35}$
[Me$_2$Si(C$_5$H$_3$Bu^t)$_2$]Sc(allyl), **4**, 61
$SiScC_{23}H_{40}P$
[Me$_2$Si(C$_5$Me$_4$)$_2$]ScH(PMe$_3$), **4**, 116
$SiSnC_4H_{12}Cl_2$
(TMS)(Me)SnCl$_2$, **2**, 263
$SiSnC_8H_{18}$
(TMS)C≡CSnMe$_3$, **2**, 242
$SiSnC_{15}H_{36}$
Bu$_3$SN-TMS, **2**, 285
$SiSnC_{48}H_{40}$
Ph$_3$SnSiPh$_2$SnPh$_3$, **2**, 284
$SiTaC_{13}H_{24}Cl_3$
Cp*Ta(TMS)Cl$_3$, **5**, 118
$SiTaC_{14}H_{24}Cl_3O$
Cp*Ta[C(O)TMS]Cl$_3$, **5**, 98
$SiTaC_{14}H_{37}ClNP_2$

Ta(CHBu^t)(NTMS)Cl(PMe_3)$_2$, **5**, 88
$SiTaC_{16}H_{33}Cl_2P$
Cp*Ta(TMS)(PMe_3)Cl_2, **5**, 118
$SiTaC_{17}H_{30}Cl_3O_3$
Cp*Ta[OCH(TMS)C(O)O-Et]Cl_3, **5**, 120
$SiTaC_{19}H_{29}Cl_3NO$
Cp*Ta[OCH(TMS)C_5H_4N]Cl_3, **5**, 120
Cp*Ta[C(O)TMS(NCp)]Cl_3, **5**, 98
$SiTaC_{23}H_{53}O_2P_4$
Ta(COSiPr$^i_3$)(dmpe)$_2$(CO), **5**, 65
$SiTaC_{27}H_{34}Cl_3N_2O$
Cp*Ta[OC(TMS)NN(CPh$_2$)]-Cl_3, **5**, 120
$SiTeC_{14}H_{29}$
Bu_2Te^+C≡CCH$_2$TMS, **11**, 593
$SiTiC_4H_{11}Cl_3$
TMS-CH$_2$TiCl$_3$, **12**, 579
$SiTiC_{12}H_{14}Cl_2$
TiCl$_2$[Me$_2$Si(C_5H_4)$_2$], **4**, 345
$SiTiC_{12}H_{16}S_2$
Ti(S_2SiMe$_2$)Cp$_2$, **4**, 416
$SiTiC_{13}H_{18}F_2$
TiF$_2$Cp[C_5H_4(TMS)], **4**, 314
$SiTiC_{14}H_{33}O_4P_3$
Ti[Bu^{t}Si(CH$_2$PMe$_2$)$_3$CO$_4$], **4**, 214, 215
$SiTiC_{15}H_{21}$
[Ti[C(TMS)=CH$_2$]Cp$_2$]$^+$, **4**, 284
$SiTiC_{15}H_{25}ClP$
TiClCp$_2$[PMe$_2$(TMS)], **4**, 265
$SiTiC_{17}H_{23}N_3O$
Ti(NCO)(NCN-TMS)(C_5H_4-Me)$_2$, **4**, 400
$SiTiC_{17}H_{33}O_4P_3$
Ti(CO)$_4$(trimpsi), **4**, 216
$SiTiC_{19}H_{26}P$
Ti(SiH$_2$Ph)Cp$_2$(PMe$_3$), **4**, 265
$SiTiC_{19}H_{34}ClO$
TiCl(CH$_2$TMS)[O-(Bu$^t_2$MeC$_6$H$_2$)], **4**, 280
$SiTiC_{20}H_{28}P$
Ti(SiHMePh)Cp$_2$(PMe$_3$), **4**, 265
$SiTiC_{22}H_{32}P$
Ti(SiH$_2$Ph)Cp$_2$(PEt$_3$), **4**, 265
$SiTiC_{23}H_{41}P_3$
Ti($C_{10}H_8$)[Bu^{t}Si(CH$_2$PMe$_2$)$_3$], **4**, 214, 215
$SiTiC_{25}H_{30}P$
Ti(SiHPh$_2$)Cp$_2$(PMe$_3$), **4**, 265
$SiTiW_9V_3C_{69}H_{149}N_4O_{40}$
[NBu$_4$]$_4$[TiCp(SiW$_9$V$_3$O$_{40}$)], **4**, 308
$SiTlC_8H_{13}$
Tl(C_5H_4TMS), **1**, 531
$SiTlC_{17}H_{23}$
Tl[C_5Me_4(SiMe$_2$Ph)], **1**, 531
$SiVC_{16}H_{38}O_3$
TMS-CH$_2$V(Bu^{t}O)$_3$, **5**, 15
$SiWC_{10}H_{14}O_2$
WH(CO)$_2$(CH$_2$SiMe$_2$)Cp, **5**, 412
$SiWC_{11}H_{14}O_3$
W(CH$_2$SiMe$_2$H)(CO)$_3$Cp, **5**, 412
$SiWC_{13}H_{18}$
W(CH$_2$SiMe$_2$)Cp$_2$, **5**, 447
$SiWC_{15}H_{27}O_2P$
W(CO)$_2$(PMe$_3$)(SiH$_3$)Cp*, **5**, 409
$SiWC_{24}H_{33}F_{12}NO_2$
W[OCMe(CF$_3$)$_2$]$_2$(=CH-TMS)-(NC$_6$H$_3$Pr$^i_2$), **5**, 315
$SiWC_{28}H_{23}NO_5$
W(CO)$_5$[C(NC$_4$H$_8$)SiPh$_3$], **5**, 191
$SiWC_{29}H_{45}F_{12}NO_2$
W[CH$_2$CH(TMS)CH(TMS)]-[OCMe(CF$_3$)$_2$]$_2$(N-C$_6$H$_3$Pr2^i), **5**, 315
$SiWZrC_{28}H_{30}O_3$
Zr[C(TMS)C(H)Ph][OCW-(CO)$_2$Cp]Cp$_2$, **4**, 535
$SiYbC_8H_{21}O$
[Yb(CH$_2$TMS)(Et$_2$O)]$^+$, **4**, 25
$SiYbC_{16}H_{23}ClO$
(H)(Cl)[[Me$_2$Si(C_5H_4)$_2$]Yb-(THF)], **4**, 60
$SiZrC_{14}H_{20}N$
Cp$_2$[ZrN(TMS)CH], **12**, 341, 343
$SiZrC_{22}H_{27}N$
Cp$_2$[ZrN(TMS)C(Ph)H - CH=CH], **12**, 341, 343
$SiZrC_{26}H_{46}$
(TMS)$_4$Cp$_2$[ZrC=CC=C], **12**, 332
$SiZrC_{12}H_{14}S_5$
ZrS$_5$[(C_5H_4)$_2$SiMe$_2$], **4**, 533
$SiZrC_{13}H_{17}Cl_2$
ZrCl$_2$[CH$_2$CH$_2$[Si(Me)$_2$C$_9$H$_7$], **4**, 570
$SiZrC_{13}H_{19}Cl$
ZrCl(TMS)Cp$_2$, **4**, 504, 505
$SiZrC_{14}H_{19}ClO$
Zr(CO-TMS)(Cl)Cp$_2$, **4**, 506
$SiZrC_{14}H_{27}ClNP$
ZrCp[N(SiMe$_2$CH$_2$PPr$^i_2$)]Cl, **4**, 474
$SiZrC_{14}H_{28}NP$
ZrCp[N(SiMe$_2$CH$_2$PPr$^i_2$)]H, **4**, 474
$SiZrC_{15}H_{21}Cl$
Cp$_2$ZrCl[C(H)CH(TMS)], **4**, 579
$SiZrC_{16}H_{19}ClO$
ZrCl(O-TMS)Cp$_2$, **4**, 503
$SiZrC_{16}H_{27}N$
ZrH[NBu^t(SiHMe$_2$)]Cp$_2$, **4**, 505
$SiZrC_{16}H_{32}NP$
ZrCp[N(SiMe$_2$CH$_2$PPr$^i_2$)]Et, **4**, 474
$SiZrC_{17}H_{27}NO$
Zr(OCH$_2$SiMe$_2$NBu^t)Cp$_2$, **4**, 506
$SiZrC_{18}H_{25}ClO$
Cp'$_2$ZrCl[OCC(H)C(H)TMS], **4**, 556
$SiZrC_{18}H_{29}N$
Zr(CH$_2$CH$_2$SiMe$_2$NBu^t)Cp$_2$, **4**, 506
$SiZrC_{18}H_{29}NS_2$
Zr(S$_2$CNEt$_2$)(TMS)Cp$_2$, **4**, 504
$SiZrC_{19}H_{24}ClN$
Cp$_2$ZrCl[[CH(TMS)]C_5H_4N], **4**, 573
$SiZrC_{19}H_{30}N$
[Zr(CH$_2$CH$_2$TMS)(MeCN)-(C_5H_4Me$_2$)]$^+$, **4**, 598
$SiZrC_{19}H_{34}NP$
Zr(SiMe$_2$NBu^t)(PMe$_3$)Cp$_2$, **4**, 504, 508
$SiZrC_{20}H_{36}P$
[Zr(CH$_2$CH$_2$TMS)(PMe$_3$)-(C_5H_4Me)$_2$]$^+$, **4**, 598
$SiZrC_{21}H_{25}Cl$
Cp$_2$ZrCl[C(TMS)CHPh], **4**, 579
$SiZrC_{21}H_{28}$
Cp$_2$Zr(SMe)(CH(TMS)Ph), **4**, 574
$SiZrC_{24}H_{33}NO$
Zr[N(TMS)CHPh](THF)Cp$_2$, **4**, 508
$SiZrC_{25}H_{31}P$
ZrH(SiHPh$_2$)(PMe$_3$)Cp$_2$, **4**, 504, 505
$SiZrC_{26}H_{40}O_2$
Zr[OC(CH$_2$)SiMe$_2$C(CH$_2$)O]-Cp*$_2$, **4**, 524, 525
$SiZrC_{29}H_{32}O$
Zr[OCMePhCPhC(TMS)]Cp$_2$, **4**, 520
$SiZrC_{29}H_{36}$
ZrMe(SiHMes$_2$)Cp$_2$, **4**, 503, 505
$SiZrC_{30}H_{56}$
ZrMe(SiHMes$_2$)Cp*$_2$, **4**, 505
$SiZrC_{31}H_{32}O$
Zr(OCHMe$_2$)(SiPh$_3$)Cp$_2$, **4**, 507
$SiZrC_{31}H_{35}$
Cp$_2$Zr(H)(SiPh$_3$)(PMe$_3$), **4**, 560
$SiZrC_{31}H_{35}P$
ZrH(SiPh$_3$)(PMe$_3$)Cp$_2$, **4**, 504
$SiZrC_{31}H_{48}ClP$
ZrCl[P(TMS)(Bu$^t_3$C$_6$H$_2$)]Cp$_2$, **4**, 514
$SiZrC_{32}H_{44}Cl_2$
ZrCl$_2$Cp*$_2$[SiMe$_2$(C_5H_4)$_2$], **4**, 485
$SiZrC_{33}H_{35}N$
Zr[C(H)NBu^t](SiPh$_3$)Cp$_2$, **4**, 507
$SiZrC_{39}H_{45}ClO$
Zr(COSiPh$_3$)(Cl)Cp*$_2$, **4**, 506
$SiZrC_{39}H_{56}$
ZrMe(SiHMes$_2$)Cp*$_2$, **4**, 503, 505
$SiZrC_{40}H_{56}O$

$Zr[OC(CH_2)SiHMes_2](H)Cp*_2$, **4**, 506
$SiZr_2C_{22}H_{24}Cl_2$
$[ZrCpCl]_2[SiMe_2(C_5H_4)_2]$, **4**, 466, 469, 470
$SiZr_2C_{22}H_{24}Cl_2O$
$[ZrCpCl]_2[SiMe_2(C_5H_4)_2](O)$, **4**, 473
$SiZr_2C_{22}H_{24}S_2$
$[Zr(S)Cp]_2[(C_5H_4)_2SiMe_2]$, **4**, 531
$SiZr_2C_{23}H_{27}N$
$[CpZr(C_5H_4)]_2(N\text{-}TMS)$, **4**, 509
$SiZr_2C_{33}H_{36}$
$Cp_2ZrMe(H)_2(SiPh_2H)ZrCp_2$, **4**, 560
$Si_2Ag_2C_{12}H_{23}N_2$
$Ag_2[C(py)(TMS)_2]_2$, **3**, 77
$Si_2AsC_{12}H_{24}$
$PhAs(TMS)_2$, **2**, 29
$Si_2AsC_{25}H_{48}P$
$Bu^t{}_3C_6H_2PAsCH(TMS)_2$, **2**, 326
$Si_2AsC_{39}H_{67}$
$(Pr^i{}_3Ph)_2SiAsSiPr^i{}_3$, **2**, 30
$Si_2AsZrC_{18}H_{32}Cl$
$Zr[As(TMS)_2](Cl)(C_5H_4Me)_2$, **4**, 518
$Si_2Au_2Fe_2C_{62}H_{62}O_{12}P_4$
$[Fe(CO)_3[Si(OMe)_3]\text{-}(dppm)Au]_2$, **10**, 62
$Si_2Au_3BC_{61}H_{63}F_4P_3$
$(Ph_3PAu)_3[C(TMS)_2][BF_4]$, **3**, 15
$Si_2BMoC_{19}H_{31}NP$
$CpMo(CO)_2[P(BH_3)(Ph)[N\text{-}(TMS)_2]]$, **1**, 288, 290
$Si_2BNbC_{46}H_{40}F_4$
$[Cp_2Nb(SiPh_3)_2]BF_4$, **5**, 127
$Si_2BUC_{22}H_{41}Cl_2N_6$
$[HB(Me_2Pz)_3]UCl_2[CH\text{-}(TMS)_2]$, **4**, 192
$Si_2B_2UC_{14}H_{32}$
$[(TMS)_2C_8H_6]U(BH_4)_2$, **4**, 185
$Si_2B_3C_{14}H_{29}N_4S_2$
$Me_8Ph(B_3N_4S_2Si_2)$, **1**, 185
$Si_2B_4C_{26}H_{51}P$
$[(Bu^t)_3C_6H_2]P(TMS)_2C_2B_4H_4$, **1**, 265
$Si_2B_4GaC_{12}H_{33}$
$Bu^tGa(TMS)_2C_2B_4H_4$, **1**, 258
$Si_2B_4GaC_{20}H_{37}$
$(bipyrimidine)Bu^tGa(TMS)_2\text{-}C_2B_4H_4$, **1**, 258
$Si_2B_4GeC_8H_{22}$
$Ge(TMS)_2C_2B_4H_4$, **1**, 260
$Si_2B_4InC_{11}H_{29}$
$Me_2CHIn[(TMS)_2C_2B_4H_4]$, **1**, 532
$Si_2B_4InC_{11}H_{39}$
$(Me_2CH)In(TMS)_2C_2B_4H_{14}$, **1**, 258
$Si_2B_5C_7H_{26}P$
$(TMS)_2C{=}PB_5H_8$, **1**, 263
$Si_2B_5C_7H_{27}ClP$
$(TMS)_2CHP(Cl)B_5H_8$, **1**, 264
$Si_2B_8C_{10}H_{28}$
nido-$C_4B_8H_{10}(TMS)_2$, **1**, 229
$Si_2B_8SnC_{12}H_{32}$
$Sn[(TMS)MeC_2B_4H_4]_2$, **1**, 260
$Si_2B_9ScC_{19}H_{45}$
$[Cp*Sc(C_2B_9H_{11})CH(TMS)_2]^-$, **4**, 70
$Si_2B_{10}C_2H_{18}$
$Me_2Si_2B_{10}H_{12}$, **1**, 259
$Si_2B_{10}C_{10}H_{20}S$
$(Me_2S)[TMS)_2CH]CB_{10}H$, **1**, 227
$Si_2BaLiC_{35}H_{65}O_2$
$Li[Cp*_2BaCH(TMS)_2](THF)_2$, **1**, 115
$Si_2C_2H_6Cl_4$
$Cl_2SiMeSiMeCl_2$, **2**, 99
$Si_2C_2H_8$
$H_2SiSiH_2CH_2CH_2$, **2**, 53
$(H_2SiCH_2)_2$, **2**, 53
$Si_2C_2H_{10}$
$H_2SiMeSiMeH_2$, **2**, 99
$Si_2C_4H_{12}Cl_2$
$ClSiMe_2SiMe_2Cl$, **2**, 88
$Si_2C_4H_{12}Cl_2O$
$ClMe_2SiOSiMe_2Cl$, **2**, 113
$Si_2C_4H_{14}O_3$
$(OH)Me_2SiOSiMe_2(OH)$, **2**, 113
$Si_2C_6H_{14}Cl_4$
$Pr^iSiCl_2SiCl_2Pr^i$, **2**, 88
$Si_2C_6H_{16}$
$Me_2SiCH_2SiMe_2CH_2$, **2**, 49
$(Me_2SiCH_2)_2$, **2**, 49, 51, 53
$Si_2C_6H_{16}Cl_2$
$(ClMe_2Si)_2C_2H_4$, **11**, 338
$Si_2C_6H_{18}$
Me_6Si_2, **2**, 13, 51, 79
$Si_2C_6H_{18}O$
$(TMS)_2O$, **2**, 121
$Si_2C_6H_{18}O_2$
$(TMS)_2O_2$, **2**, 33
$Si_2C_6H_{18}S$
$(TMS)_2S$, **2**, 35; **11**, 341
$Si_2C_6H_{18}Te$
$(TMS)_2Te$, **2**, 37
$Si_2C_7H_{18}S$
$(TMS)_2CS$, **2**, 35
$Si_2C_8H_{18}$
$(TMS)_2C_2$, **2**, 5
$Si_2C_8H_{18}Cl_4$
$Bu^tSiCl_2SiCl_2Bu^t$, **2**, 89
$Si_2C_8H_{18}O$
$(CH_2{=}CH)Me_2SiOSiMe_2\text{-}(CH{=}CH_2)$, **2**, 118
$Si_2C_8H_{22}$
Et_3SiMe_2SiH, **2**, 15
$Si_2C_8H_{22}O_5$
$[Bu^t(HO)_2]_2O$, **2**, 31
$Si_2C_8H_{22}S$
$[(TMS)CH_2]_2S$, **11**, 341
$Si_2C_9H_{21}$
$[(TMS)CHCHCH(TMS)]^-$, **2**, 9
$Si_2C_{10}H_{20}$
$(TMS)CHCCCH(TMS)$, **2**, 13
$(TMS)CHCHCC(TMS)$, **2**, 13
$Si_2C_{12}H_{20}$
$(TMS)C{\equiv}CCH{=}CHC{\equiv}C(TMS)$, **11**, 324
$Si_2C_{12}H_{22}$
$(TMS)_2C_6H_4$, **11**, 342
$Si_2C_{12}H_{23}P$
$(TMS)_2PPh$, **2**, 28
$Si_2C_{12}H_{28}Cl_2$
$Pr^i{}_2SiClSiClPr^i{}_2$, **2**, 88
$Si_2C_{12}H_{30}O_3$
$(HOPr^i{}_2Si)_2O$, **2**, 31
$Si_2C_{16}H_{36}$
$Bu^t{}_2Si{=}SiBu^t{}_2$, **2**, 86
$Si_2C_{16}H_{38}O_2$
$[Bu^t{}_2(HO)Si]_2$, **2**, 31
$Si_2C_{20}H_{40}$
$(CySiBu^t)_2$, **2**, 81
$Si_2C_{24}H_{54}$
$Bu^t{}_3SiSiBu^t{}_3$, **2**, 82
$Si_2C_{24}H_{54}N_3$
$[Bu^t{}_3SiNNNSiBu^t{}_3]^-$, **2**, 25
$Si_2C_{24}H_{55}N_3$
$Bu^t{}_3SiNNNHSiBu^t{}_3$, **2**, 26
$Si_2C_{36}H_{44}$
$(Mes)_2Si{=}Si(Mes)_2$, **2**, 55, 86
$Si_2C_{36}H_{45}F_2O_2$
$[[(Mes)_2Si(F)O]_2H]^-$, **2**, 39
$Si_2C_{40}H_{38}$
$(Bu^tPh_2Si)C_8(SiPh_2Bu^t)$, **2**, 10
$Si_2CaC_7H_{19}Br$
$[(TMS)_2CH]CaBr$, **1**, 109
$Si_2CaC_{20}H_{34}O$
$Ca[C_5H_4(TMS)]_2(THF)$, **1**, 112
$Si_2CaC_{28}H_{57}NO_3$
$Cp*Ca[N(TMS)_2](THF)_3$, **1**, 118
$Si_2CaC_{29}H_{58}O_3$
$Cp*Ca[CH(TMS)_2](THF)_3$, **1**, 115
$Si_2CdC_8H_{22}$
$[(TMS)CH_2]_2Cd$, **3**, 196
Si_2CdF_6
$Cd(SiF_3)_2$, **3**, 176, 200
$Si_2CdFe_2C_{18}H_{36}O_6P_2$
$[Fe(TMS)(CO)_3(PMe_3)]_2Cd$, **10**, 54
$Si_2CdFe_2C_{48}H_{50}N_2O_{12}P_2$
$[Fe(CO)_3[Si(OMe)_3](Ph_2P\text{-}CH_2P)]_2Cd$, **10**, 339
$Si_2CdFe_2C_{64}H_{60}O_{14}P_2$
$[(CO)_3[Si(OMe)](PhFeC_6H_4\text{-}C(O)CH_2PPh_2)]_2Cd$, **10**, 339
$Si_2CdMn_2C_{42}H_{40}O_4$
$[(C_5H_4Me)Mn(SiMePh_2)\text{-}(CO)_2]_2Cd$, **6**, 144
$Si_2CdMn_2C_{42}H_{42}O_4$
$[MeCp(CO)_2SiMePh_2Mn]_2Cd$, **10**, 337
$Si_2Cd_2Fe_2C_{52}H_{52}Cl_2O_{14}P_2$
$[[Ph_2PCH_2C(O)Ph]Fe(CO)_3\text{-}(MeO)_2SiOMeCdCl]_2$, **10**, 333
$Si_2CeC_{27}H_{49}$
$Cp*_2CeCH(TMS)_2$, **4**, 84
$Si_2CoGaC_{16}H_{30}O_5$

$Ga(CH_2TMS)_2(THF)[Co(CO)_4]$, **1**, 556
$Si_2Co_3Yb_2C_{60}H_{86}O_4$
$[Cp*_2Yb]_2(CO)_4Co_3(TMS-C_5H_4)_2$, **4**, 108
$Si_2Co_4C_{13}H_6O_{11}$
$Co_4(CO)_{11}(SiMe)_2$, **8**, 448
$Si_2CrC_{14}H_{36}N_2O_2$
$Cr(O\text{-}TMS)_2(NBu^t)_2$, **5**, 291
$Si_2CrC_{30}H_{36}O_2$
$Cr(CO)_2[CC(TMS)_2](C_6Me_6)$, **5**, 508
$Si_2CuC_{21}H_{43}N_3O_2$
Cu(azasemicorrin), **12**, 407
$Si_2CuLi_2C_4H_9N$
$(TMS)_2Cu(CN)Li_2$, **12**, 99
$Si_2CuLi_2C_{17}H_{22}N$
$(PhMe_2Si)_2Cu(CN)Li_2$, **12**, 97
$Si_2Cu_2C_{12}H_{23}N_2$
$Cu_2[C(py)(TMS)_2]_2$, **3**, 77, 88
$Si_2FeC_{11}H_{18}O_4P$
$[(CO)_4FePC(TMS)_2]^-$, **7**, 37
$Si_2FeC_{11}H_{20}O$
$Fe(CO)(TMS)(=SiMe_2)Cp$, **7**, 178
$Si_2FeC_{12}H_{18}O_4$
$Fe(CO)_4(TMS\text{-}C{\equiv}C\text{-}TMS)$, **7**, 123
$Si_2FeHgMnC_{54}H_{43}O_5$
$Fe(CO)_3PPh_3SiMePh_2Hg\text{-}MnCpSiPh_3(CO)_2$, **10**, 339
$Si_2FeHgMn_2C_{60}H_{51}O_5$
$MeCpMn(CO)_2SiPh_3MnHg\text{-}Fe(CO)_3PPh_3SiMePh_2$, **10**, 337
$Si_2Fe_2C_{44}H_{30}O_8$
$[Fe(CO)_4SiPh_3]_2$, **7**, 28
$Si_2Fe_2HgC_{38}H_{44}O_6P_2$
$[Fe(SiPh_2Me)(CO)_3(PMe_3)]_2\text{-}Hg$, **10**, 54
$Si_2Fe_2Zn_2C_{68}H_{74}Cl_2O_{12}P_4$
$[Fe(CO)_3(Si(OEt)_3(PhP_2CH_2P\text{-}Ph_2)ZnCl]_2$, **10**, 338
$Si_2Fe_5C_{23}H_{10}O_{13}$
$Fe_3(CO)_9[SiFeCp(CO)_2]_2$, **7**, 29
$Si_2GeC_{56}H_{70}$
$Ge[CH_2Si(Mes)_3]_2$, **2**, 192
$Si_2HfC_{11}H_{21}Cl_3$
$HfCl_3[(TMS)_2C_5H_3]$, **4**, 546
$Si_2HfC_{16}H_{26}Cl_2$
$HfCl_2(C_5H_4TMS)_2$, **4**, 484
$Si_2Hf_2C_{36}H_{48}Cl_2$
$[Cp*Cp(Cl)Hf]_2[(SiH_2)_2C_6H_4]$, **4**, 504
$Si_2HgC_6H_{18}$
$(TMS)_2Hg$, **2**, 3; **3**, 154
$Si_2HgMn_2C_{42}H_{42}O_4$
$[MeCp(CO)_2SiMePh_2Mn]_2Hg$, **10**, 337
$Si_2IrC_{22}H_{47}$
$Cp*Ir(H)_2(SiEt_3)_2$, **8**, 385
$Si_2IrC_{31}H_{39}INP_2$
$Ir(I)(Me)[N(SiMe_2CH_2PPh_2)_2]$, **8**, 342, 345
$Si_2IrC_{34}H_{42}NP_2$
$Ir[C(CH_2)_3][N(SiMe_2CH_2P\text{-}Ph_2)_2]$, **8**, 342, 344
$Si_2KC_{26}H_{26}N$
$[(TMS)_2NK{\cdot}Toluene]_2$, **1**, 6
$Si_2LaC_{29}H_{58}O_3$
$[Cp*La[CH(TMS)_2(THF)_3]^-$, **4**, 72
$Si_2Li_2ZrC_{64}H_{65}Cl_2O_2P_2$
$[[LiCl]_2(THF)_2][ZrCp_2]\text{-}(PHSiPh_3)(PSiPh_3)\text{-}(C_5H_4C_5H_4)$, **4**, 467, 469
$Si_2LiC_6H_{18}N$
$[(TMS)_2N]Li$, **1**, 5; **11**, 5
$Si_2LiC_6H_{18}N_2$
$[SiMe_2NMe]_2Li$, **11**, 5
$Si_2LiC_6H_{18}P$
$(TMS)_2PLi$, **2**, 28
$Si_2LiRhC_{22}H_{50}N_2$
$(cod)Rh(CH_2TMS)_2Li\text{-}(TMEDA)$, **8**, 267
$Si_2LiSbC_6H_{18}$
$(TMS)_2SbLi$, **2**, 30
$Si_2LiVC_{26}H_{60}N_2$
$[(Bu^t{}_3SiN)_2VMe_2]Li$, **5**, 16
$Si_2Li_2C_6H_{18}$
$[(TMS)Li]^+{}_2$, **2**, 22
$Si_2LuC_{17}H_{34}Cl_2$
$Cp*Lu[CH(TMS)_2]Cl_2$, **4**, 73
$Si_2LuC_{32}H_{49}O$
$(C_5Me_3Ph_2)Lu(CH_2TMS)_2\text{-}(THF)$, **4**, 73
$Si_2MgC_{11}H_{29}ClO$
$MgCl[CH(TMS)_2](OEt_2)$, **1**, 72
$Si_2MgC_{20}H_{54}$
$Mg[C(TMS)_3]_2$, **1**, 85
$Si_2MgC_{28}H_{42}O_2$
$Mg[C_{14}H_8(TMS)_2](THF)_2$, **1**, 100
$Si_2Mg_2C_{23}H_{50}BrO_4$
$(TMS)_2C[MgBr(THF)_2]_2$, **1**, 73
$Si_2MnC_{14}H_{38}P_2$
$Mn(CH_2TMS)_2(dmpe)$, **6**, 131
$Si_2Mn_2SnC_{16}H_{14}Cl_8O_4$
$Cl_2Sn[Mn(SiCl_3)(CO)_2(C_5H_4\text{-}Me)]_2$, **6**, 143
$Si_2Mn_3C_{38}H_{84}N_2O_2$
$Mn_3(CH_2Bu^t)_4[OC(CH_2Bu^t)N\text{-}TMS]_2$, **6**, 130
$Si_2MoC_{14}H_{22}$
$Mo(SiMe_2SiMe_2)Cp_2$, **5**, 447
$Si_2MoC_{18}H_{29}$
$[MoH(TMS\text{-}C_2TMS)Cp_2]^+$, **5**, 368
$Si_2Mo_2C_{22}H_{28}O_4$
$Mo_2(CO)_4(TMS\text{-}C_2TMS)Cp_2$, **5**, 376
$Si_2NbC_{16}H_{26}Br$
$(C_5H_4TMS)_2NbBr$, **5**, 126
$Si_2NbC_{16}H_{26}Cl$
$(C_5H_4TMS)_2NbCl$, **5**, 127
$Si_2NbC_{16}H_{26}ClO$
$(C_5H_4TMS)_2Nb(O)Cl$, **5**, 127, 128
$Si_2NbC_{16}H_{26}ClO_2$
$(C_5H_4TMS)_2Nb(O_2)Cl$, **5**, 127
$Si_2NbC_{16}H_{26}I$
$(C_5H_4TMS)_2NbI$, **5**, 126
$Si_2NbC_{16}H_{29}$
$(C_5H_4TMS)_2NbH_3$, **5**, 76, 79
$Si_2NbC_{19}H_{35}ClF_6P_2$
$[[C_5H_4(TMS)]_2NbCl[P\text{-}(OMe)_3]]PF_6$, **5**, 127
$Si_2NbC_{19}H_{35}ClO_3$
$(C_5H_4TMS)_2NbCl[P(OMe)_3]$, **5**, 127
$Si_2NbC_{22}H_{33}O$
$Cp_2Nb(TMS\text{-}C_2\text{-}TMS)[C\text{-}(COMe)CH_2]$, **5**, 97
$Si_2NbC_{23}H_{31}ClNO$
$(C_5H_4TMS)_2NbCl(OCNPh)$, **5**, 101
$Si_2NbC_{31}H_{39}ClN$
$(C_5H_4TMS)_2Nb\text{-}(PhNCCMePh)Cl$, **5**, 101
$Si_2NbC_{46}H_{40}$
$Cp_2Nb(SiPh_3)_2$, **5**, 127
$Si_2NdC_{27}H_{49}$
$Cp*_2NdCH(TMS)_2$, **4**, 85, 130
$Si_2NiC_3Cl_6O_3$
$Ni(SiCl_3)_2(CO)_3$, **9**, 2
$Si_2NiC_6H_6F_6$
$Ni(SiF_3)_2(C_6H_6)$, **9**, 174
$Si_2NiC_7H_8Cl_6$
$Ni(SiCl_3)_2(Tol)$, **9**, 2
$Si_2NiC_7H_8F_6$
$Ni(SiF_3)_2(TolH)$, **9**, 174
$Si_2NiC_{11}H_{31}NP_2$
$NiMe[N(SiMe_2CH_2PMe_2)_2]$, **9**, 58
$Si_2NiC_{14}H_{40}P_2$
$Ni(CH_2TMS)_2(PMe_3)_2$, **9**, 56
$Si_2NiC_{18}H_{32}N_2$
$Ni(CH_2TMS)_2(py)_2$, **9**, 56
$Si_2NiC_{32}H_{58}P_2$
$Ni(CH_2SiMe_2Ph)_2(Pr^i{}_2PCH_2C\text{-}H_2PPr^i{}_2)$, **9**, 56
$Si_2NiC_{44}H_{48}P_2$
$Ni(TMS)C_2(TMS)(PPh_3)_2$, **9**, 130
$Si_2Ni_2C_{14}H_{30}Br_2$
$Ni_2Br_2[C_3H_4[CH_2(TMS)]]_2$, **9**, 143
$Si_2Ni_2C_{14}H_{30}Cl_2O_2$
$Ni_2Cl_2[C_3H_3Me(O\text{-}TMS)]_2$, **9**, 143
$Si_2Ni_2C_{30}H_{46}Cl_2O_2P_4$
$[Ni(CO)[P[PCl(Bu^t)]=\text{-}C(TMS)Ph]]_2$, **9**, 127
$Si_2Ni_2C_{44}H_{48}O_2P_2$
$[Ni(CO)[P(PPh_2)=C\text{-}(TMS)Ph]]_2$, **9**, 127
$Si_2OsC_{10}H_{22}N_3S_2$
$[Os(N)(NCS)_2(CH_2TMS)_2]^-$, **7**, 375
$Si_2OsC_{10}H_{28}O$
$Os(O)Me_2(CH_2TMS)_2$, **7**, 373
$Si_2OsC_{13}H_{27}N$
$OsN(CH_2TMS)_2Cp$, **7**, 504
$Si_2OsC_{14}H_{40}ClNP_2$
$Os(N)(CH_2TMS)_2Cl(PMe_3)_2$, **7**, 375

$Si_2OsC_{21}H_{49}N$
Os(CBu[t])$(CH_2Bu^t)_2$[N(TMS)$_2$], **7**, 361
$Si_2OsWC_9H_{25}NS_4$
Os(N)$(CH_2TMS)_2$(S)-(SMe)WS_2, **7**, 375
$Si_2OsWC_{10}H_{22}N_3S_4$
[Os(N)$(CH_2TMS)_2(S)_2$W-$(SCN)_2]^-$, **7**, 375
$Si_2Os_2C_{36}H_{36}O_6P_2$
$Os_2(Ph_2PCH_2SiMe_2)_2(CO)_6$, **7**, 633
$Si_2Os_3C_{12}Cl_6O_{12}$
$Os_3(CO)_{12}(SiCl_3)_2$, **7**, 838
$Si_2Os_3SnC_{21}H_{23}NO_{10}$
Os_3H[SnC(TMS)$_2(C_5H_4N)$]-$(CO)_{10}$, **7**, 696
$Si_2Os_3SnC_{22}H_{26}O_{13}$
Os_3[SnCH(TMS)$_2$](MeO$_2$C-CCH$_2$CO$_2$Me)$(CO)_9$, **7**, 698
$Si_2PdC_{14}H_{28}P_2$
Pd$(CH_2TMS)_2$(dmpe), **9**, 240
$Si_2PtC_{10}H_{20}O_4$
Pt[[H$_2$CCHSi(Me)$_2$]$_2$O]-$(C_2H_2O_3)$, **9**, 553
$Si_2PtC_{23}H_{40}P_2$
Pt(CH$_2$SiMe$_2$CH-TMS)-$(PMe_2Ph)_2$, **9**, 478
$Si_2PtC_{26}H_{33}OP$
Pt(PPh$_3$)[[CH$_2$CHSi(Me)$_2$]$_2$O], **9**, 553
$Si_2PtC_{32}H_{72}OP_2$
Pt(PBu$^t_3$)$_2$[[H$_2$CCHSi-(Me)$_2$]$_2$O], **9**, 553
$Si_2ReC_{13}H_{34}OP$
ReO[C(O)Me]$(CH_2TMS)_2$-(PMe_3), **6**, 233
$Si_2ReC_{16}H_{39}N_2$
Re(NBu^t)$_2$[C(H)TMS](CH$_2$-TMS), **6**, 240
$Si_2ReC_{16}H_{40}ClN_2$
Re(NBu^t)$_2$Cl$(CH_2TMS)_2$, **6**, 239
$Si_2RuC_{10}H_{26}N$
[Ru(N)Me$_2(CH_2TMS)_2]^-$, **7**, 373, 374
$Si_2RuC_{47}H_{51}ClOP_2$
Ru[CH(CH$_2$TMS)CC-TMS]Cl-(CO)$(PPh_3)_2$, **7**, 411
$Si_2Ru_2C_{18}H_{26}O_4$
$Ru_2(C_8H_8TMS)$(TMS)$(CO)_4$, **7**, 654
$Si_2Ru_2C_{19}H_{24}O_5$
$Ru_2(C_8H_8SiMe_2CH_2CH_2Si$-$Me_2)(CO)_5$, **7**, 654
$Si_2Ru_2C_{36}H_{36}O_6P_2$
$Ru_2(Ph_2PCH_2SiMe_2)_2(CO)_6$, **7**, 633
$Si_2Ru_2C_{38}H_{66}S$
Ru$_2$Cp*$_2$[TMS-CCC- (CH-TMS)CC-TMS]- (SPr^i) (H), **7**, 670
$Si_2Ru_2C_{44}H_{52}$
Ru$_2$Cp*$_2$(SiPh$_2$)$_2$(H)$_2$, **7**, 672, 673
$Si_2Ru_3C_{20}H_{24}O_8$
Ru$_3$H[TMS-CHCMeCC-(TMS)CH]$(CO)_8$, **7**, 710
$Si_2SbC_7H_{19}Cl_2$
(TMS)$_2$CHSbCl$_2$, **2**, 333
$Si_2SbC_{12}H_{23}$
PhSb(TMS)$_2$, **2**, 30, 328
$Si_2SbC_{25}H_{48}P$
Bu$^t_3$C$_6$H$_2$PSbCH(TMS), **2**, 326
$Si_2Sc_2C_{36}H_{68}N$
[(C$_5$Me$_4$SiMe$_2$NBu^t)Sc(Pr)]$_2$, **4**, 118
$Si_2Sc_2C_{36}H_{74}N_2P_2$
[(C$_5$Me$_4$SiMe$_2$NBu^t)Sc(PMe$_3$)-(H)]$_2$, **4**, 70, 117
$Si_2SmC_{26}H_{48}N$
Cp*$_2$SmN(TMS)$_2$, **4**, 80
$Si_2SmC_{27}H_{49}$
Cp*$_2$SmCH(TMS)$_2$, **4**, 121
$Si_2Sm_2C_{52}H_{88}O_5$
(OSiMe$_2$OSiMe$_2$O)[Cp*$_2$Sm-(THF)]$_2$, **4**, 93
$Si_2SnC_{10}H_{27}$
Me$_3$SnC(TMS)$_2$, **2**, 250
$Si_2TaC_{16}H_{29}$
(C$_5$H$_4$TMS)$_2$TaH$_3$, **5**, 76
$Si_2TaC_{20}H_{50}ClO_2P_4$
Ta(TMS-OCCO-TMS)-(dmpe)$_2$Cl, **5**, 65
$Si_2TaC_{22}H_{56}ClN_2P_4$
[TMS(Me)NCCN(Me)TMS]-Ta(dmpe)$_2$Cl, **5**, 66
$Si_2TaC_{26}H_{62}ClO_2P_4$
Ta(Pr$_3^i$SiOCCO-TMS)-(dmpe)$_2$Cl, **5**, 65, 89
$Si_2ThC_{25}H_{42}$
(C$_8$H$_8$)Cp*ThCH(TMS)$_2$, **4**, 186
$Si_2ThC_{27}H_{51}P$
Cp*$_2$Th(Me)P(TMS)$_2$, **4**, 168
$Si_2Th_2C_{40}H_{64}$
[[Me$_2$Si(C$_5$Me$_4$)$_2$]ThH$_2$]$_2$, **4**, 162, 190
$Si_2TiBC_{41}H_{29}F_{20}$
[TiMe[C$_5$H$_4$(TMS)]$_2$][B(C$_6$-F$_5$)$_4$], **4**, 395
$Si_2TiC_{18}H_{32}$
Cp$_2$Ti[CH$_2$(TMS)]$_2$, **12**, 594
$Si_2TiC_{11}H_{22}Cl_3$
TiCl$_3$[C$_5$H$_4$(TMS)$_2$], **4**, 291
$Si_2TiC_{11}H_{23}Cl_2N$
TiCl$_2$[N(TMS)$_2$]Cp, **4**, 300
$Si_2TiC_{15}H_{23}Cl$
TiCl(TMS)[Me$_2$Si(C$_5$H$_4$)$_2$], **4**, 397
$Si_2TiC_{16}H_{28}N_4S_2$
Ti(NSN-TMS)$_2$Cp$_2$, **4**, 399
$Si_2TiC_{18}H_{32}$
TiMe$_2$Cp[C$_5$H$_3$(TMS)$_2$], **4**, 357
$Si_2TiC_{19}H_{31}N_2$
[Ti[C(Me)=N-TMS](CN-TMS)Cp$_2$]$^+$, **4**, 366
$Si_2TiC_{19}H_{54}O_3$
Ti(OC$_6$H$_{11}$)$_3$[C$_5$H$_3$(TMS)$_2$], **4**, 307
$Si_2Ti_2C_{24}H_{44}Cl_2N_2$
[TiCl[C$_5$H$_4$(TMS)](NBu^t)]$_2$, **4**, 300
$Si_2Tl_2C_{24}H_{54}O_6S_2$
[Tl(SSi(OBu^t)$_3$]$_2$, **1**, 525
$Si_2UC_{22}H_{48}N_3OP$
(C$_8$H$_8$)U(CH$_2$TMS)$_2$(HMPA), **4**, 185
$Si_2UC_{26}H_{48}ClP$
Cp*$_2$U(Cl)P(TMS)$_2$, **4**, 168
$Si_2UC_{27}H_{49}$
Cp*$_2$UCH(TMS)$_2$, **4**, 165
$Si_2VC_8H_{24}O_2$
Me$_2$V(O-TMS)$_2$, **5**, 15
$Si_2VC_{16}H_{28}$
Cp[(TMS)$_2$Cp]V, **5**, 48
$Si_2VC_{17}H_{28}O$
Cp[(TMS)$_2$Cp]VCO, **5**, 48
$Si_2VC_{22}H_{43}P$
Cp[(TMS)$_2$Cp]VPEt$_3$, **5**, 48
$Si_2VC_{58}H_{56}BrO_2P_4$
V(TMS-OCCO-TMS)-(dmpe)$_2$Br, **5**, 6
$Si_2W_4C_{34}H_{88}O_7$
[W$_2$(OEt)$_7$(C-TMS)]$_2$, **5**, 322
$Si_2YC_{27}H_{49}$
Cp*$_2$YCH(TMS)$_2$, **4**, 84, 86, 121
$Si_2YbC_7H_{19}$
[YbCH(TMS)$_2$]$^+$, **4**, 25
$Si_2YbC_{16}H_{26}$
Yb(TMS-C$_5$H$_4$)$_2$, **4**, 30
$Si_2ZnC_8H_{23}N$
EtZnN(TMS)$_2$, **3**, 181
$Si_2ZnC_{14}H_{27}$
[Zn(CH$_2$(TMS))$_2$Ph]$^-$, **3**, 179
Si_2ZnF_6
(SiF$_3$)$_2$Zn, **3**, 176
$Si_2Zn_2C_{21}H_{33}N$
Cp[N(TMS)$_2$](ZnCp)$_2$, **3**, 180
$Si_2ZrC_{11}H_{21}Cl_3$
Zr[C$_5$H$_3$(TMS)$_2$]Cl$_3$, **1**, 96
$Si_2ZrC_{16}H_{28}ClP$
ZrCl[P(TMS)$_2$]Cp$_2$, **4**, 514
$Si_2ZrC_{16}H_{28}IP$
ZrI[P(TMS)$_2$]Cp$_2$, **4**, 515
$Si_2ZrC_{18}H_{26}O_2$
Zr(CO)$_2$(C$_5$H$_4$TMS)$_2$, **4**, 440
$Si_2ZrC_{18}H_{38}NP$
ZrCp[Ni(SiMe$_2$CH$_2$PPr$^i_2$)]-CH$_2$TMS, **4**, 474, 476
$Si_2ZrC_{22}H_{36}O$
Zr(THF)(TMS-CC-TMS)Cp$_2$, **4**, 444
$Si_2ZrC_{22}H_{46}$
Zr(TMS)[Si(TMS)$_3$]Cp$_2$, **4**, 503
$Si_2ZrC_{32}H_{38}$
Cp$_2$ZrC(TMS)CPhCPhC-(TMS), **4**, 579
$Si_2ZrC_{56}H_{50}P_2$
[ZrCp(PHSiPh$_3$)]$_2$(C$_5$H$_4$C$_5$H$_4$), **4**, 467
$Si_2Zr_2C_{24}H_{28}Cl_2O$
[ZrCl[(C$_5$H$_4$)$_2$SiMe$_2$]]$_2$O, **4**, 528
$Si_2Zr_2C_{33}H_{38}$

$Cp_2(SiH_2Ph)Zr(H)_2Zr$-
$(SiHMePh)Cp_2$, **4**, 504
$Si_2Zr_2C_{35}H_{42}$
$Cp_2Zr(SiPhMeH)(H)_2$-
$(SiPhMe_2)ZrCp_2$, **4**, 504, 560
$Si_2Zr_2C_{44}H_{44}$
$[Cp_2(Ph_2HSi)Zr(H)]_2$, **4**, 504, 560
$Si_3AgTiC_{18}H_{31}NO_2$
$Ti[C_5H_4(TMS)][C_2(TMS)]_2$-
$AgNO_2$, **10**, 27
$Si_3AlC_{54}H_{45}O_3$
$(Ph_3SiO)_3Al$, **2**, 34
$Si_3AsC_9H_{27}$
$(TMS)_3As$, **2**, 30
$Si_3AsC_{12}H_{33}Cl_2$
$(TMS\text{-}CH_2)_3AsCl_2$, **2**, 337
$Si_3BC_{26}H_{33}$
$(TMS)_2(Ph_3Si)(BC_2H_2)$, **1**, 148
$Si_3B_{18}C_4H_{22}$
$Si(SiC_2B_9H_{11})_2$, **1**, 259
$Si_3BiC_{12}H_{33}$
$(TMS\text{-}CH_2)_3Bi$, **2**, 331
$Si_3C_2H_{10}O_3$
$OSi(Me_2)OSiH_2OSiH_2$, **2**, 119
$Si_3C_3H_{15}N$
$(MeH_2Si)_3N$, **2**, 25
$Si_3C_6H_{18}O_2$
$Me_2SiOSi(Me_2)OSiMe_2$, **2**, 119
$Si_3C_9H_{24}$
$Me_2Si{=}C(TMS)_2$, **2**, 55, 67
$Si_3C_9H_{27}N$
$(TMS)_3N$, **2**, 25
$Si_3C_{11}H_{28}S$
$(TMS)_3CCHS$, **2**, 34
$Si_3C_{12}H_{30}$
$Me_2SiC(TMS)(TBDMS)$, **2**, 18
$Si_3C_{15}H_{36}$
$Me_2SiC(TMS)(SiMeBu^t{}_2)$, **2**, 18
$Si_3C_{16}H_{32}N_2$
$PhC(N\text{-}TMS)N(TMS)_2$, **2**, 25
$Si_3C_{24}H_{54}$
$(Bu^t{}_2Si)_3$, **2**, 52, 81, 82, 86
$Si_3C_{30}H_{66}$
$[(Bu^tCH_2)_2Si]_3$, **2**, 81
$Si_3C_{39}H_{60}$
$[(Mes)Bu^tSi]_3$, **2**, 81
$Si_3C_{48}H_{54}$
$[(xylyl)_2Si]_3$, **2**, 81
$Si_3C_{54}H_{66}$
$[(Mes)_2Si]_3$, **2**, 86, 95
$Si_3CdC_{10}H_{27}Br$
$CdBr[C(TMS)_3]$, **3**, 198
$Si_3CdC_{10}H_{27}Cl$
$CdCl[C(TMS)_3]$, **3**, 198
$Si_3CdC_{10}H_{28}O$
$Cd(OH)[C(TMS)_3]$, **3**, 198
$Si_3CdLiC_{33}H_{49}Br_2O_2$
$Li(THF)_2CdBr_2[C(SiMe_2Ph)_3]$, **3**, 198
$Si_3GaKC_{12}H_{34}$
$KGaH(CH_2TMS)_3$, **1**, 516
$Si_3GeC_{20}H_{42}$
$(TMS)_3CGeCp^*$, **2**, 192
Si_3H_6
$(H_2Si)_3$, **2**, 94
$Si_3InKC_{12}H_{34}$
$KInH(CH_2TMS)_3$, **1**, 516
$Si_3InPtC_{38}H_{81}P_2$
$Pt(CH_2TMS)[In(CH_2TMS)_2]$-
$[Cy_2PCH_2CH_2PCy_2]$, **1**, 566
$Si_3LiMgC_{21}H_{51}Br_2$
$Li(THF)_2(Br)_2Mg[C(TMS)_3$-
$(THF)]$, **1**, 85
$Si_3LuC_{21}H_{45}Cl$
$Cp^*Lu[CH(TMS)_2](CH_2$-
$TMS)Cl$, **4**, 73
$Si_3MgC_{10}H_{27}Br$
$(TMS)_3CMgBr$, **1**, 70
$Si_3MgC_{20}H_{46}BrN_2$
$[C_5H_3(TMS)_3]MgBr(TMEDA)$, **1**, 74
$Si_3MgC_{21}H_{48}N_2$
$MgMe[C_5H_2(TMS)_3]$-
$(TMEDA)$, **1**, 86
$Si_3NdC_{27}H_{49}$
$[Me_2Si(C_5Me_4)_2]NdCH(TMS)_2$, **4**, 61
$Si_3NiC_{35}H_{71}P_2$
$Ni[Si(TMS)(Mes){=}C{=}CPh$-
$(TMS)](PEt_3)_2$, **9**, 125
$Si_3OsC_{12}H_{33}N$
$Os(N)(CH_2TMS)_3$, **7**, 376
$Si_3OsC_{13}H_{36}ClN$
$Os(NMe)Cl(CH_2TMS)_3$, **7**, 376
$Si_3ScC_{33}H_{62}P$
$Me_2Si(C_5Me_4)(C_5H_3CH_2CH_2P$-
$Bu^t{}_2)ScCH(TMS)_2$, **4**, 116
$Si_3Sn_3C_{12}H_{36}O_3$
$[(TMS)(Me)SnO]_3$, **2**, 263
$Si_3ThC_{28}H_{52}$
$[Me_2Si(C_5Me_4)_2]Th(CH_2$-
$TMS)_2$, **4**, 162
$Si_3TiC_{19}H_{38}$
$CpTi[CH_2(TMS)]_3$, **12**, 594
$Si_3TiC_{14}H_{29}Cl_2$
$TiCl_2[C_5H_2(TMS)_3]$, **4**, 290
$Si_3TiC_{14}H_{29}Cl_3$
$TiCl_3[C_5H_2(TMS)_3]$, **4**, 290
$Si_3TiC_{17}H_{38}$
$TiMe_3[C_5H_2(TMS)_3]$, **4**, 296
$Si_3UC_{20}H_{41}$
$[(C_8H_8)U(CH_2TMS)_3]^-$, **4**, 185
$Si_3UC_{24}H_{39}Br$
$(TMS\text{-}C_5H_4)_3UBr$, **4**, 157
$Si_3UC_{24}H_{39}N_3$
$(TMS\text{-}C_5H_4)_3UN_3$, **4**, 148
$Si_3UC_{24}H_{39}O$
$O[U(TMS\text{-}C_5H_4)_3]$, **4**, 149
$Si_3UC_{25}H_{39}O$
$(TMS\text{-}C_5H_4)_3UCO$, **4**, 131
$Si_3UC_{26}H_{42}$
$(TMS\text{-}C_5H_4)_3U(CHCH_2)$, **4**, 152
$Si_3UC_{27}H_{44}N$
$(TMS\text{-}C_5H_4)_3U(CNEt)$, **4**, 131
$Si_3UC_{36}H_{45}$
$U(TMS\text{-}C_9H_6)_3$, **4**, 162
$Si_3VC_{18}H_{45}O_3$
$V[O\text{-}TBDMS]_3$, **5**, 13
$Si_3ZnC_{12}H_{33}$
$[Zn(CH_2(TMS))_3]^-$, **3**, 179
$Si_3ZnC_{16}H_{32}$
$Ph[(TMS)_3C]Zn$, **3**, 179
$Si_3ZrC_{19}H_{38}ClN_2$
$Zr[N(TMS)P(H)N(TMS)_2]$-
$(Cl)Cp_2$, **4**, 508
$Si_4AgOs_2C_{26}H_{54}N_2$
$[Ag[NOs(CH_2TMS)_2Cp]_2]^+$, **7**, 375
$Si_4AlC_9H_{27}Cl_3$
$[(TMS)_3SiAlCl_3]^-$, **2**, 23
$Si_4AlC_{14}H_{38}Cl$
$[(TMS)_2CH]_2AlCl$, **1**, 435
$Si_4AlLiC_{12}H_{36}$
$(TMS)_4AlLi$, **2**, 22
$Si_4AlNaC_{12}H_{36}$
$(TMS)_4AlNa$, **2**, 23
$Si_4Al_2C_{12}H_{32}$
$(Me_2SiCH_2)_4Al_2$, **1**, 462
$Si_4Al_4C_{16}H_{36}$
$[Bu^tSiAl]_4$, **2**, 23
$Si_4AsLiC_{14}H_{38}$
$[(TMS)_2CH]_2AsLi$, **2**, 328
$Si_4AsLiC_{47}H_{83}FO_2$
$LiAs(Pr^i{}_3C_6H_2Si)_2(F)Si(Pr^i{}_3$-
$Si)\cdot 2THF$, **1**, 11
$Si_4As_2ZrC_{22}H_{46}$
$Zr[As(TMS)_2]_2Cp_2$, **4**, 518
$Si_4Au_2C_{19}H_{40}N$
$Au_2[C_5H_4N[C(TMS)_2]_2]$, **3**, 12
$Si_4BC_{33}H_{27}F_{20}$
$(TMS)_3SiB(C_6F_5)_4$, **2**, 4
$Si_4BScC_{22}H_{46}$
$[C_5H_3(TMS)_2]_2Sc[BH_4]$, **1**, 278; **4**, 49
$Si_4B_2C_{24}H_{54}$
$(TMS)_2CB(Bu^t)C{=}C{=}C$ -
$(TMS)_2]BBu^t$, **1**, 135
$Si_4B_8C_{16}H_{44}$
$(TMS)_4Si(SiC_2B_4H_4)_2$, **1**, 259
$Si_4B_8GeC_{16}H_{44}$
$[(TMS)_2C_2B_4H_4]_2Ge$, **1**, 260
$Si_4BaC_{22}H_{42}$
$Ba[C_5H_3(TMS)_2]_2$, **1**, 112
$Si_4Ba_2C_{48}H_{108}O_4$
$Ba_2(OSiBu^t{}_3)_4$, **2**, 34
$Si_4Bi_2C_{12}H_{36}$
$(TMS)_4Bi_2$, **2**, 31
$Si_4C_8H_{24}$
$(Me_2Si)_4$, **2**, 79, 81, 93
$Si_4C_9H_{28}$
$(TMS)_3SiH$, **2**, 21
$Si_4C_{10}H_{28}O$
$(CHO)(Si)_3Si$, **11**, 328
$Si_4C_{12}H_{20}O_4$
$[CH_2{=}C(Me)]_4$
$[SiOSiOSiOSiO]$, **2**, 123
$Si_4C_{24}H_{40}$
$(Ph_2Si)_4$, **2**, 81
$Si_4C_{24}H_{56}$
$(Pr^i{}_2Si)_4$, **2**, 81, 92
$Si_4C_{28}H_{32}$
$(PhMeSi)_4$, **2**, 100

$Si_4C_{32}H_{72}$
$(Bu^s{}_2Si)_4$, **2**, 92
$Si_4C_{40}H_{60}F_4$
$(Cp*FSi)_4$, **2**, 19
$Si_4CaC_{14}H_{38}$
$Ca[CH(TMS)_2]_2$, **1**, 109
$Si_4CaC_{22}H_{42}$
$Ca[C_5H_3(TMS)_2]_2$, **1**, 112, 118
$Si_4CaC_{26}H_{62}O_3$
$Ca[CH(TMS)_2]_2(THF)_3$, **1**, 110
$Si_4CdC_{13}H_{36}Br_3O$
$(Br)_3[O\text{-}TMS]Cd[C(TMS)_3]$, **3**, 198
$Si_4CeC_{22}H_{51}N_2$
$Cp*Ce[N(TMS)_2]_2$, **4**, 72
$Si_4Cu_2Li_2C_{20}H_{56}S_2$
$Li_2Cu_2(CH_2TMS)_4(DMS)_2$, **3**, 64, 101, 120
$Si_4Cu_4C_{16}H_{44}$
$Cu_4(CH_2TMS)_4$, **3**, 58, 76, 77, 125
$Si_4Cu_4C_{36}H_{60}N_4$
$Cu_4[CH(py)(TMS)]_4$, **3**, 77
$Si_4FePdC_{37}H_{49}Cl_3O_6P_2$
$PdFe(CO)_3(dppm)[Si(O\text{-}TMS)_3Cl_3]$, **10**, 17, 18
$Si_4Fe_2C_{18}H_{36}O_6P_2$
$Fe_2(CO)_6[PC(TMS)_2]_2$, **7**, 44
$Si_4GeC_{12}H_{36}N_2$
$Ge[N(TMS)_2]_2$, **2**, 193
$Si_4GeC_{12}H_{38}N_2O_2$
$[(TMS)_2N]_2Ge(OH)_2$, **2**, 163
Si_4H_8
$(H_2Si)_4$, **2**, 93, 94
$Si_4HfC_{22}H_{46}P_2$
$Hf[P(TMS)_2]_2Cp_2$, **4**, 514
$Si_4HfC_{24}H_{47}Cl$
$HfCl[Si(TMS)_3]CpCp*$, **4**, 504
$Si_4HfC_{28}H_{48}$
$Hf[C_8H_6(TMS)_2]_2$, **4**, 627
$Si_4InKC_{16}H_{44}$
$KIn(CH_2TMS)_4$, **1**, 516
$Si_4K_4C_{12}H_{36}O_4$
$[KO\text{-}TMS]_4$, **1**, 12
$Si_4K_4C_{56}H_{68}O_4$
$[C_6H_6KOSiMe_2Ph]_4$, **2**, 34
$Si_4LaC_{24}H_{53}$
$Cp*La[CH(TMS)_2]_2$, **4**, 72
$Si_4LiC_{21}H_{51}O_3$
$(TMS)_3SiLi(THF)_3$, **2**, 22
$Si_4LiUC_{30}H_{58}Cl_2O_2$
$[(TMS)_2C_5H_3]_2U(Cl)_2Li(THF)_2$, **4**, 138
$Si_4LiVC_{24}H_{60}O_4$
$LiV(O\text{-}TBDMS)_4$, **5**, 14
$Si_4Li_4C_{42}H_{86}O_8S_2$
$[LiPhSC(TMS)_2O_2Li\cdot 2Et_2O]_2$, **1**, 17, 18
$Si_4LuRuC_{33}H_{55}O_3$
$[(TMS)_2C_5H_3]_2(THF)LuRu\text{-}(CO)_2Cp$, **4**, 107
$Si_4MgC_{14}H_{38}$
$Mg[CH(TMS)_2]_2$, **1**, 86, 87, 92
$Si_4MgC_{18}H_{48}O$
$Mg[CH(TMS)_2]_2(OEt_2)$, **1**, 88
$Si_4MgC_{22}H_{44}N_2$
$Mg[C(TMS)_2(py)]_2$, **1**, 90
$Si_4MnC_{14}H_{38}$
$Mn[CH(TMS)_2]_2$, **6**, 130
$Si_4MnC_{16}H_{20}$
$Mn(CH_2TMS)_4$, **6**, 145
$Si_4MnC_{16}H_{44}$
$[Mn(CH_2TMS)_4]^{2-}$, **6**, 128
$Mn(CH_2TMS)_4$, **6**, 138
$Si_4MnC_{18}H_{46}O$
$Mn[CH(TMS)_2]_2(THF)$, **6**, 132
$Si_4MnC_{48}H_{102}N_4O_6$
$Mn_4(CH_2Bu^t)_2(NCO)_2[OC\text{-}(CH_2Bu^t)N\text{-}TMS]_4$, **6**, 130
$Si_4Mn_2C_{22}H_{62}P_2$
$Mn_2(CH_2TMS)_4(PMe_3)_2$, **6**, 130
$Si_4Mo_2C_{20}H_{56}N_2$
$Mo_2(CH_2TMS)_4(NMe_2)_2$, **5**, 321
$Si_4NbC_{22}H_{42}ClO$
$[C_5H_3(TMS)_2]_2Nb(O)Cl$, **5**, 128
$Si_4NbC_{22}H_{42}ClO_2$
$[C_5H_3(TMS)_2]_2NbCl(O_2)$, **5**, 127
$Si_4NbC_{22}H_{45}$
$[C_5H_3(TMS)_2]_2NbH_3$, **5**, 76, 79
$Si_4NbC_{42}ClO$
$[C_5H_3(TMS)_2]_2Nb(O)Cl$, **5**, 127
$Si_4Nb_2C_{92}H_{30}O$
$[Cp_2Nb(SiPh_3)_2]_2(O)$, **5**, 127
$Si_4NdC_{22}H_{51}N_2$
$Cp*Nd[N(TMS)_2]_2$, **4**, 68
$Si_4NiC_{13}H_{36}ClOP$
$Ni(CO)[P(Cl)=C(TMS)_2]_2$, **9**, 126, 127
$Si_4NiC_{16}H_{12}N_4$
$Ni(CN\text{-}TMS)_4$, **9**, 10
$Si_4NiC_{18}H_{51}ClP_2$
$NiCl(PEt_3)[C(TMS)_2=P=C\text{-}(TMS)_2]$, **9**, 127
$Si_4NiC_{20}H_{55}P_3$
$Ni[P[CH(TMS)_2]=C(TMS)_2]\text{-}(PMe_3)_2$, **9**, 126
$Si_4NiSnC_{18}H_{46}$
$Ni(C_2H_4)_2[=Sn[CH(TMS)_2]_2]$, **9**, 112
$Si_4NiSnC_{52}H_{70}P_2$
$Ni(PPh_3)_2[HC_2SnH[CH\text{-}(TMS)_2]_2]$, **9**, 133
$Si_4NiSn_4C_{40}H_{96}N_8$
$Ni[Sn(NBu^t)_2(SiMe_2)]_4$, **9**, 14
$Si_4Ni_5C_{20}H_{38}ClO_6P_2$
$[Ni_5Cl(CO)_6[(TMS)_2CHP=PCH(Me_3Si)_2]]^{2-}$, **9**, 20
$Si_4OsC_{16}H_{44}O$
$Os(O)(CH_2TMS)_4$, **7**, 372
$Si_4Os_2C_{26}H_{52}N_4S_2$
$Os_2(SC_5H_4N)_2(N)_2(CH_2TMS)_4$, **7**, 375
$Si_4Os_3SnC_{24}H_{38}O_{10}$
$Os_3H_2[[CH(TMS)_2]SnOC[CH\text{-}(TMS)_2]](CO)_9$, **7**, 705
$Si_4PbC_{14}H_{38}$
$Pb[CH(TMS)_2]_2$, **2**, 313
$Si_4PbC_{62}H_{76}P_4$
$Pb[(Ph_2P)_2C(TMS)_2]_2$, **2**, 315
$Si_4RuC_{14}H_{40}NO_2$
$[Ru(N)(O\text{-}TMS)_2(CH_2TMS)_2]^-$, **7**, 373, 374
$Si_4Ru_2C_{16}H_{44}O_2$
$Ru_2(O)_2(CH_2TMS)_4$, **7**, 674
$Si_4Sb_2C_{12}H_{36}$
$(TMS)_4Sb_2$, **2**, 31
$Si_4ScC_{23}H_{45}O$
$Cp_2Sc[Si(TMS)_3](THF)$, **4**, 46
$Si_4SmC_{26}H_{50}O$
$[(TMS)_2C_5H_3]_2Sm(THF)$, **4**, 30
$Si_4SnC_{14}H_{38}$
$[(TMS)_2CH]_2Sn$, **2**, 285, 287
$Si_4SnC_{14}H_{38}O$
$[(TMS)_2CH]_2SnO$, **2**, 262
$Si_4Sn_2C_{16}H_{45}ClO_2$
$(HO)(TMS\text{-}CH_2)_2SnOSn\text{-}(CH_2TMS)_2Cl$, **2**, 264
$Si_4TaC_{21}H_{45}Cl_2$
$Ta[CBu^tC\text{-}TMS\text{-}CTMS](TMS\text{-}CC\text{-}TMS)Cl_2$, **5**, 106
$Si_4TaC_{22}H_{45}$
$[C_5H_3(TMS)_2]_2TaH_3$, **5**, 76
$Si_4Ta_2C_{48}H_{112}O_4$
$[TaH_2(OSiBu^t{}_3)_2]_2$, **5**, 134
$Si_4ThC_{20}H_{44}N_2$
$(C_8H_8)Th[N(TMS)_2]_2$, **4**, 186
$Si_4TiC_{14}H_{42}N_2$
$TiMe_2[N(TMS)_2]_2$, **4**, 284
$Si_4TiC_{16}H_{44}$
$Ti(CH_2TMS)_4$, **4**, 284
$Si_4TiC_{22}H_{42}Cl_2$
$TiCl_2[C_5H_3(TMS)_2]_2$, **4**, 346
$Si_4Ti_2C_{24}H_{28}Cl_2$
$[TiCl[Me_2Si(C_5H_4)_2]]_2$, **4**, 248
$Si_4UC_{20}H_{44}N_2$
$(C_8H_8)U[N(TMS)_2]_2$, **4**, 186
$Si_4YbC_{22}H_{42}Cl$
$[(TMS)_2C_5H_3]_2YbCl$, **4**, 127
$Si_4Yb_2C_{28}H_{40}Cl_2O_2$
$[[O(Me_2SiC_5H_4)_2Yb(Cl)]_2$, **4**, 61
$Si_4ZrC_{20}H_{40}$
$ZrMe[Si(TMS)_3]Cp_2$, **4**, 506
$Si_4ZrC_{24}Cl$
$ZrCl[Si(TMS)_3]CpCp*$, **4**, 506
$Si_4ZrC_{24}H_{47}Cl$
$ZrCl[Si(TMS)_3]CpCp*$, **4**, 505
$Si_4ZrC_{24}H_{50}P_2$
$Zr[P(TMS)_2]_2(C_5H_4Me)_2$, **4**, 514
$Si_4ZrC_{25}H_{47}ClO$
$Zr[COSi(TMS)_3](Cl)CpCp*$, **4**, 506
$Si_4ZrC_{25}H_{50}$
$ZrMe[Si(TMS)_3]CpCp*$, **4**, 505
$Si_4ZrC_{25}H_{51}OP$
$Zr(CO)(PMe_3)[C_5H_3(TMS)_2]_2$, **4**, 441
$Si_4ZrC_{26}H_{50}O$
$Zr[OC(CH_2)Si(TMS)_3](H)\text{-}CpCp*$, **4**, 506
$Si_4ZrC_{26}H_{51}Cl$
$Zr[CH_2CH_2Si(TMS)_3](Cl)\text{-}CpCp*$, **4**, 506

$Si_4ZrC_{28}H_{48}$
$Zr[C_8H_6(TMS)_2]_2$, **4**, 627
$Si_4ZrC_{40}H_{56}P_2$
$Zr[(PPh)_2C_6H_4][(TMS)_2C_5H_3]_2$, **4**, 514
$Si_4Zr_2C_{28}H_{48}Cl_4$
$[Zr[C_8H_6(TMS)_2]Cl]_2(Cl)_2$, **4**, 631
$Si_4Zr_2C_{29}H_{47}ClO$
$Cp_2(Cl)Zr(O)Zr[Si(TMS)_3]Cp_2$, **4**, 503
$Si_5Ba_2KSi_5C_{98}H_{95}O_9$
$KBa_2(OSiPh_3)_5(DME)_2$, **2**, 34
$Si_5C_{10}H_{30}$
$(Me_2Si)_5$, **2**, 81
$Si_5C_{25}H_{50}$
$[(CH_2)_5Si]_5$, **2**, 81
$Si_5C_{40}H_{90}$
$(Bu^i_2)Si_5$, **2**, 81
$Si_5C_{60}H_{50}$
$(Ph_2Si)_5$, **2**, 81
$Si_5GeC_{17}H_{46}$
$(TMS)_3CGeCH(TMS)_2$, **2**, 192
$Si_5Sb_5C_{20}H_{55}$
$(TMS\text{-}CH_2Sb)_5$, **2**, 328
$Si_5TiC_{70}H_{60}$
$TiSi_5Ph_{10}Cp_2$, **4**, 397
$Si_5ZnC_{17}H_{46}$
$[(TMS)_2CH][(TMS)_3C]Zn$, **3**, 179
$Si_5ZrC_{22}H_{46}$
$Zr(TMS)[Si(TMS)_3]Cp_2$, **4**, 506
$Si_5ZrC_{70}H_{60}$
$Zr[SiPh_2(SiPh_2)_3SiPh_2]Cp_2$, **4**, 503
$Si_6AgLiC_{36}H_{86}O_4$
$[Ag[C(TMS)_3]_2][Li(THF)_4]$, **3**, 106
$Si_6AlC_{21}H_{57}$
$[(TMS)_2CH]_3Al$, **1**, 446
$Si_6As_2C_{20}H_{54}$
$(TMS)_3CAsAsC(TMS)_3$, **2**, 326
$Si_6As_3Ga_3C_{24}H_{66}$
$[Br_2Ga(As(CH_2TMS)_2]_3$, **1**, 522
$Si_6As_3Li_3C_{24}H_{60}$
$[LiAs(CH\text{-}TMS)_2]_3$, **1**, 11
$Si_6As_6Ga_2C_{32}H_{88}$
$[(TMS\text{-}CH_2)_2AsGa(As(CH_2\text{-}TMS)_2)]_2$, **1**, 520, 521
$Si_6As_7Ga_5C_{60}H_{102}$
$[(PhAsH)(TMS\text{-}CH_2)_2Ga\text{-}(PhAs)_6[(TMS\text{-}CH_2)\text{-}Ga]_4]$, **1**, 522, 523
$Si_6B_2C_{20}H_{54}O$
$[(TMS)_3C]_2(B_2O)$, **1**, 171
$Si_6B_2SnC_{26}H_{47}$
$[(TMS)_2]_2Sn{=}[CB(Bu^t)C\text{-}(TMS)_2BBu^t]$, **2**, 287
$Si_6Ba_3C_{108}H_{90}O_6$
$Ba_3(OSiPh_3)_6$, **2**, 34
$Si_6BiC_{21}H_{57}$
$[(TMS)_2CH]_3Bi$, **2**, 331
$Si_6C_{11}H_{33}$
$[Si_6Me_{11}]^-$, **2**, 86
$Si_6C_{12}H_{36}$
$(Me_2Si)_6$, **2**, 80, 83, 107
$Si_6C_{22}H_{54}$
$(TMS)_3CCCC(TMS)_3$, **2**, 11, 13
$Si_6C_{24}H_{54}$
$(TMS)_6C_6$, **2**, 8; **11**, 326
$Si_6C_{27}H_{42}$
$Ph_3Si_6Me_9$, **2**, 81
$Si_6C_{32}H_{44}$
$Ph_4Si_6Me_8$, **2**, 81
$Si_6C_{42}H_{48}$
$(PhSiMe)_6$, **2**, 81
$Si_6C_{42}H_{82}O_{11}$
$(C_7H_{13}Si)_6O_7(OH)_4$, **2**, 33
$Si_6C_{72}H_{102}$
$(Pr^i_2C_6H_3Si)_6$, **2**, 24
$Si_6CdC_{20}H_{54}O_2$
$Cd[C(TMS)_2(SiMe_2OMe)]_2$, **3**, 179
$Si_6CdC_{22}H_{54}$
$[(TMS)_2(CH_2CHMe_2Si)C]_2Cd$, **3**, 196
$Si_6CeC_{33}H_{63}$
$Ce[(TMS)_2C_5H_3]_3$, **4**, 54
$Si_6CeC_{34}H_{64}$
$Ce[(TMS)_3C_8H_5]_2$, **4**, 104
$Si_6CuLiC_{28}H_{86}O_4$
$[Cu[C(TMS)_3]_2][Li(THF)_4]$, **3**, 106
$Si_6FeC_{20}H_{58}$
$[(TMS)_3CH_2]_2Fe$, **7**, 186
$Si_6FeIn_2C_{22}H_{54}Cl_2O_4$
$In_2[C(TMS)_3]_2(Cl)_2[Fe(CO)_4]$, **1**, 564
$Si_6GaTeC_{23}H_{65}$
$[(TMS)_2CH]_2GaTeSi(TMS)_3$, **1**, 525
$Si_6GeC_{42}H_{85}S$
$[[(TMS)_2CH]_3Ph][Pr^i_3C_6H_2]\text{-}Ge{=}S$, **2**, 205
Si_6H_6
$(HSi)_6$, **2**, 90
$Si_6InC_{18}H_{57}$
$[(TMS)_2HC]_3In$, **1**, 537
$Si_6In_2LiC_{28}H_{75}$
$(TMS)_3CIn(H)_2Li(THF)_2(H)In\text{-}(H)_2C(TMS)_3$, **1**, 515, 516
$Si_6KLuC_{21}H_{57}Cl$
$[(TMS)_2CH]_3Lu(Cl)K$, **4**, 19
$Si_6KLuC_{25}H_{67}ClO$
$[(TMS)_2CH]_3Lu(Cl)K(Et_2O)$, **4**, 19
$Si_6LaC_{21}H_{57}$
$La[CH(TMS)_2]_3$, **4**, 23
$Si_6Li_6C_{18}H_{54}$
$[TMS\text{-}Li]_6$, **1**, 15
$Si_6Li_{12}C_{100}H_{164}O_{23}S_6$
$[(TMS\text{-}CSO_2Ph)Li_2]_6\text{-}Li_2O{\cdot}10THF$, **1**, 17
$Si_6MgC_{20}H_{54}$
$Mg[C(TMS)_3]_2$, **1**, 87, 88, 92
$Si_6MgC_{28}H_{58}$
$Mg[(TMS)_3C_5H_2]_2$, **1**, 92
$Si_6MgC_{34}H_{74}N_2$
$Mg[C_5H_2(TMS)_3]_2(TMEDA)$, **1**, 86
$Si_6MnC_{20}H_{54}$
$Mn[C(TMS)_3]_2$, **6**, 130
$Si_6NpC_{21}H_{57}$
$Np[CH(TMS)_2]_3$, **4**, 132
$Si_6Pb_2C_{18}H_{54}$
$(TMS)_6Pb_2$, **2**, 23
$Si_6Pt_2C_{24}H_{54}O_3$
$Pt_2[[H_2CCHSi(Me)_2]_2O]_3$, **9**, 553
$Si_6PuC_{21}H_{57}$
$Pu[CH(TMS)_2]_3$, **4**, 132
$Si_6Re_2C_{24}H_{66}O_3$
$[ReO(CH_2TMS)_3]_2(O)$, **6**, 236
$Si_6Sb_3C_{21}H_{57}$
$[(TMS)_2CHSb]_3$, **2**, 328
$Si_6ScC_{57}HO_6$
$Sc[CH(SiMe_2C_6H_4OMe)_2]_3$, **4**, 17
$Si_6ScC_{57}H_{81}O_6$
$Sc[CH(SiMe_2C_6H_4OMe)_2]_3$, **4**, 16
$Si_6SmC_{33}H_{63}$
$Sm[(TMS)_2C_5H_3]_3$, **4**, 54
$Si_6SnC_{28}H_{64}$
$[(TMS)_2CH]_3SnBz$, **2**, 235
$Si_6Ta_2C_{74}H_{162}O_6$
$[Ta(silox)_3]_2(C_2)$, **5**, 134
$Si_6ThBC_{18}H_{58}N_3$
$[(TMS)_2N]_3Th(BH_4)$, **4**, 134
$Si_6ThC_{33}H_{63}$
$Th[C_5H_3(TMS)_2]_3$, **4**, 140
$Si_6TiC_{19}H_{57}N_3$
$TiMe[N(TMS)_2]_3$, **4**, 280
$Si_6UC_{18}H_{53}N_3$
$[(TMS)_2N]_2UCH_2SiMe_2N\text{-}(TMS)$, **4**, 192; **12**, 596
$Si_6UC_{19}H_{57}N_3$
$[(TMS)_2N]_3UMe$, **4**, 192
$Si_6U_2C_{48}H_{78}O$
$(O)[U(TMS\text{-}C_5H_4)_3]_2$, **4**, 148
$Si_6U_2C_{49}H_{80}O_2$
$(TMS\text{-}C_5H_4)_3UOCH_2OU\text{-}(C_5H_4TMS)_3$, **4**, 157
$Si_6YC_{21}H_{57}$
$Y[CH(TMS)_2]_3$, **4**, 17
$Si_6ZnC_{18}H_{50}O_2$
$[(HOMe_2Si)(TMS)_2C]_2Zn$, **3**, 180
$Si_6ZnC_{20}H_{54}$
$Zn[(TMS)_3C]_2$, **3**, 179
$Si_6ZnC_{22}H_{48}F_6O_4$
$Zn[C(TMS)_2(SiMe_2O_2CCF_3)]_2$, **3**, 179
$Si_6Zn_2C_{50}H_{68}O_2$
$[(Me_2PhSi)_3CZnOH]_2$, **3**, 184
$Si_7C_{14}H_{42}$
$(Me_2Si)_7$, **2**, 82
$Si_7C_{42}H_{80}O_{12}$
$Cy_7Si_7O_9(OH)_3$, **2**, 32
$Si_7VC_{42}H_{77}O_{13}$
$(C_6H_{11})_7Si_7O_{12}V(O)$, **5**, 16
$Si_7ZrC_{52}H_{92}O_{12}$
$(CySi)_7O_{12}ZrCp^*$, **2**, 32
$Si_8As_2C_{28}H_{76}$

[(TMS)$_2$CH]$_2$AsAs[CH-(TMS)$_2$]$_2$, **2**, 325
$Si_8GaC_{23}H_{65}Te$
[(TMS)$_2$CH]$_2$GaTeSi(TMS)$_3$, **1**, 526
$Si_8GaTeC_{23}H_{65}$
[(TMS)$_2$CH]$_2$GaTeSi(TMS)$_3$, **1**, 525
$Si_8Ga_2C_{28}H_{72}$
[[(TMS)$_2$HC]$_2$Ga]$_2$, **1**, 537, 538
Si_8H_8
(HSi)$_8$, **2**, 94
$Si_8In_2C_{28}H_{72}$
[[(TMS)$_2$HC]$_2$In]$_2$, **1**, 537
$Si_8K_8C_{156}H_{150}O_{14}$
K$_8$(OSiPh$_3$)$_8$(DME)$_3$, **2**, 34
$Si_8Mg_2C_{32}H_{84}O_2$
[Mg[CH(TMS)$_2$]$_2$]$_2$(dioxane), **1**, 88
$Si_8Sb_4C_{28}H_{76}$
[(TMS)$_2$CHSb]$_4$, **2**, 328
$Si_8Sn_2C_{28}H_{78}O_3$
[[(TMS)$_2$CH]$_2$Sn(OH)]$_2$O, **2**, 263
$Si_9Cd_3C_{30}H_{81}Br_4$
[[Cd[C(TMS)$_3$]]$_3$(Br)$_4$]$^-$, **3**, 198
$Si_9Mn_3C_{34}H_{89}Cl_4O$
Mn$_3$[C(TMS)$_3$]$_3$Cl$_4$(THF), **6**, 129
$Si_{12}C_{24}H_{72}$
[(TMS)$_2$Si]$_4$, **2**, 81
$Si_{12}C_{32}H_{40}$
[(TMS-CH$_2$)$_2$Si]$_4$, **2**, 81
$Si_{12}C_{32}H_{88}$
[(EtMe$_2$Si)$_2$Si]$_4$, **2**, 81
$Si_{12}Li_{24}C_{84}H_{168}N_{12}O_6$
[(TMS-CCN)Li$_2$]$_{12}$·6Et$_2$O, **1**, 17
$Si_{12}Tl_6C_{66}H_{126}$
[Tl[C$_5$H$_3$(TMS)$_2$]]$_6$, **1**, 531
$Si_{13}C_{26}H_{78}$
(Me$_2$Si)$_{13}$, **2**, 82
$Si_{16}C_{32}H_{96}$
(Me$_2$Si)$_{16}$, **2**, 82
$Si_{16}Ca_2C_{88}H_{168}O_2$
[[(TMS)$_2$C$_5$H$_3$]$_4$CaO]$_2$, **1**, 119
$Si_{20}C_{36}H_{108}$
[(TMS)$_3$SiSi]$_4$, **2**, 91
$Si_{24}Li_{24}C_{126}H_{290}N_{12}O_6$
[(TMS)$_2$C(Li$_2$)CN]$_{12}$·6Et$_2$O·C$_6$H$_{14}$, **1**, 17
$SmAlC_{12}H_{18}Cl_4$
Sm(C$_6$Me$_6$)(AlCl$_4$), **4**, 4
$SmAlC_{28}H_{50}$
Cp*$_2$Sm(Et)$_2$AlEt$_2$, **4**, 88
$SmAl_2C_{26}H_{50}O_2$
Sm(C$_5$H$_4$Bu^t)$_2$[(H)$_3$AlH-(THF)]$_2$, **1**, 551
SmC_6H_{18}
[SmMe$_6$]$^-$, **4**, 20
$SmC_{10}H_{10}$
SmCp$_2$, **4**, 32, 125
$SmC_{12}H_{13}O_2$
Cp$_2$Sm(OAc), **4**, 35
$SmC_{15}H_9$
Sm(C$_5$H$_3$)$_3$, **4**, 53
$SmC_{18}H_{44}N_2$
Sm(C$_5$Me$_4$CH$_2$CH$_2$NMe$_2$)$_2$, **4**, 31
$SmC_{19}H_{23}O$
Cp$_3$Sm(THF), **4**, 29
$SmC_{19}H_{27}O$
(C$_8$H$_8$)Sm(C$_7$H$_{11}$)(THF), **4**, 106
$SmC_{20}H_{23}O$
Cp$_2$Sm(Ph)(THF), **4**, 45
$SmC_{20}H_{30}$
SmCp*$_2$, **4**, 66, 74, 120
$SmC_{20}H_{30}N_2$
(N$_2$)(SmCp*$_2$), **4**, 81
$SmC_{20}H_{31}$
Cp*$_2$SmH, **12**, 1204
$SmC_{21}H_{37}I_2O_3$
(Bu^{t}C$_5$H$_4$)SmI$_2$(THF)$_3$, **4**, 32
$SmC_{23}H_{36}NS_2$
Cp*$_2$Sm(S$_2$CNMe$_2$), **4**, 78
$SmC_{24}H_{38}ClO$
Cp*$_2$SmCl(THF), **4**, 75
$SmC_{24}H_{38}IO$
Cp*$_2$SmI(THF), **4**, 75
$SmC_{24}H_{42}N_2O$
[Cp*$_2$Sm(NH$_2$NH$_2$)(THF)]$^-$, **4**, 82
$SmC_{25}H_{33}O$
[Me$_4$C$_2$(C$_5$H$_4$)$_2$]SmCp(THF), **4**, 60
$SmC_{27}H_{21}$
Sm(C$_9$H$_7$)$_3$, **4**, 63
$SmC_{28}H_{46}O_2$
Cp*$_2$Sm(THF)$_2$, **4**, 64, 65, 74, 120, 126
$SmC_{30}H_{38}O_3$
(C$_9$H$_7$)$_2$Sm(THF)$_3$, **4**, 63
$SmC_{30}H_{43}O$
Cp*$_2$Sm(OC$_6$HMe$_4$), **4**, 77
Cp*$_2$SmPh(THF), **4**, 89
$SmC_{30}H_{44}NO$
Cp*$_2$SmNHPh(THF), **4**, 83
$SmC_{30}H_{45}$
SmCp*$_3$, **4**, 95
$SmC_{30}H_{50}N_2$
Cp*$_2$Sm(Bu^{t}NCHCHNBu^t), **4**, 81
$SmC_{30}H_{50}O_2$
(C$_5$Me$_4$Et)$_2$Sm(THF)$_2$, **4**, 64
$SmC_{31}H_{29}O$
(C$_9$H$_7$)$_3$Sm(THF), **4**, 63
$SmC_{31}H_{45}O$
Cp*$_2$SmBz(THF), **4**, 87
$SmC_{32}H_{43}O$
Cp*$_2$SmCCPh(THF), **4**, 115
$SmC_{32}H_{51}$
Cp*$_2$Sm[Bu^{t}CHCCBu^{t}CH$_2$], **4**, 89
$SmC_{34}H_{30}N_2$
[Cp$_2$Sm(NPh$_2$)$_2$]$^-$, **4**, 41
$SmC_{34}H_{34}O_2$
(C$_{13}$H$_9$)$_2$Sm(THF)$_2$, **4**, 63
$SmC_{35}H_{41}O$
Cp*$_2$SmMe(THF), **12**, 1205
$SmC_{36}H_{27}O_2$
Cp*$_2$Sm[O(CH$_2$)$_2$Cp*](THF), **4**, 78
$SmC_{42}H_{53}O_2P$
Cp*$_2$Sm(OPPh$_3$)(THF), **4**, 67
$SmC_{44}H_{58}O_4Se_2$
[(C$_8$H$_8$)Sm(SePh)(THF)$_2$]$_2$, **4**, 101
$SmC_{48}H_{68}$
(CH$_2$CHPh)(SmCp*$_2$)$_2$, **4**, 89
$SmC_{48}H_{72}$
(CH$_2$CHCHCH$_2$CH$_2$C-HCHCH$_2$)(SmCp*$_2$)$_2$, **4**, 89
$SmC_{54}H_{70}$
(PhCCPh)(SmCp*$_2$)$_2$, **4**, 90
$SmC_{54}H_{72}$
(PhCHCHPh)(SmCp*$_2$)$_2$, **4**, 89
$SmCoC_{24}H_{40}I_2O_9$
[SmI$_2$(THF)$_5$][Co(CO)$_4$], **4**, 108
$SmLi_5C_{32}H_{72}O_8$
Li$_5$Sm(OBu^t)$_8$, **4**, 20
$SmSi_2C_{26}H_{48}N$
Cp*$_2$SmN(TMS)$_2$, **4**, 80
$SmSi_2C_{27}H_{49}$
Cp*$_2$SmCH(TMS)$_2$, **4**, 121
$SmSi_4C_{26}H_{50}O$
[(TMS)$_2$C$_5$H$_3$]$_2$Sm(THF), **4**, 30
$SmSi_6C_{33}H_{63}$
Sm[(TMS)$_2$C$_5$H$_3$]$_3$, **4**, 54
$Sm_2Al_2C_{44}H_{76}O_2$
[(Bu^{t}C$_5$H$_4$)$_2$Sm]$_2$[(H)$_3$AlH-(THF)]$_2$, **4**, 49
$Sm_2Al_2C_{48}H_{84}$
Cp*$_2$Sm[Me$_2$AlMe$_2$]$_2$SmCp*$_2$, **4**, 87
$Sm_2Bi_2C_{40}H_{60}$
(Bi$_2$)[SmCp*$_2$]$_2$, **4**, 83
$Sm_2C_{32}H_{44}Cl_2O_2$
[(MeC$_5$H$_4$)$_2$Sm(Cl)(THF)]$_2$, **4**, 36
$Sm_2C_{36}H_{62}I_2O_4$
[Cp*Sm(I)(THF)$_2$]$_2$, **4**, 64
$Sm_2C_{40}H_{60}$
Cp*$_2$Sm(CH$_2$C$_5$Me$_4$)(H)-SmCp*, **4**, 93
$Sm_2C_{40}H_{60}O$
(O)[SmCp*$_2$]$_2$, **4**, 77
$Sm_2C_{40}H_{62}$
[Cp*$_2$Sm(H)]$_2$, **4**, 90, 119
$Sm_2C_{45}H_{65}$
Cp*$_2$SmCpSmCp*$_2$, **4**, 95
$Sm_2C_{48}H_{78}N_2O_2$
(NHNH)[Cp*$_2$Sm(THF)]$_2$, **4**, 82
$Sm_2C_{50}H_{78}N_2O$
(O)[Cp*$_2$Sm(Bu^{t}CN)]$_2$, **4**, 77
$Sm_2C_{52}H_{70}N_2$
(PhNNPh)(SmCp*$_2$)$_2$, **4**, 79
$Sm_2C_{52}H_{72}N_2$
(pyCHCHpy)(SmCp*$_2$)$_2$, **4**, 80
$Sm_2C_{54}H_{70}N_2O_2$
[(PhN)OCCO(NPh)]-[SmCp*$_2$]$_2$, **4**, 79
$Sm_2C_{56}H_{70}$
(PhC$_4$Ph)(SmCp*$_2$)$_2$, **4**, 90

$Sm_2C_{78}H_{92}O_4P_2$
$(OCHCHO)[Cp^*{}_2Sm(OPPh_3)]_2$, **4**, 94
$Sm_2Si_2C_{52}H_{88}O_5$
$(OSiMe_2OSiMe_2O)[Cp^*{}_2Sm(THF)]_2$, **4**, 93
$Sm_3C_{60}H_{90}Cl_3$
$[Cp^*{}_2Sm(Cl)]_3$, **4**, 76
$Sm_3C_{84}H_{123}N_6$
$[Cp^*{}_2Sm(C_6H_{11}NC)(CN)]_3$, **4**, 76
$Sm_3Sb_3C_{64}H_{98}O$
$(Sb_3)[SmCp^*{}_2]_3(THF)$, **4**, 83
$Sm_4C_{40}H_{78}N_{14}$
$[Cp^*Sm]_4(NHNH)_6(NH_3)_2$, **4**, 68
$Sm_4C_{40}H_{82}$
$(Cp^*Sm)_4(NHNH)_2(NHNH_2)_4(NH_3)_2$, **4**, 83
$Sm_4C_{94}H_{136}O_8$
$[Cp^*{}_4Sm_2(O_2CCCO)(THF)]_2$, **4**, 80
$Sm_5C_{110}H_{172}Cl_5O_5$
$[Cp^*{}_2ClSm(Cl)SmCp^*{}_2[Me(OCH_2CH_2)_4OMe]SmCp^*{}_2]^+[Cp^*{}_2SmCl]_2Cl$, **4**, 76
$SnAs_2B_{10}C_{48}H_{52}Cl_2$
$(Ph_4As)_2B_{10}H_{12}SnCl_2$, **1**, 259, 260
$SnAs_2B_{10}C_{50}H_{58}Cl_2$
$(Ph_4As)_2B_{10}H_{12}SnMeCl_2$, **1**, 259
$SnB_2C_{17}H_{29}$
$CpSn[Et_4Me(C_3B_2)]$, **1**, 206
$SnB_2Si_6C_{26}H_{47}$
$[(TMS)_2]_2Sn{=}[CB(Bu^t)C(TMS)_2BBu^t]$, **2**, 287
$SnB_4SiC_6H_{16}$
$Sn(TMS)MeC_2B_4H_4$, **1**, 260
$SnB_8Si_2C_{12}H_{32}$
$Sn[(TMS)MeC_2B_4H_4]_2$, **1**, 260
$SnB_9C_6H_{21}$
$Me_2Sn[(MeC)_2B_9H_9]$, **1**, 260
$SnB_9C_{23}H_{33}Cl$
$Ph_3PMe[ClSn(B_9H_9C_2Me_2)]$, **1**, 260
$SnB_{11}C_{21}H_{34}Cl_2$
$[MeSnB_{11}H_{11}]PPh_3Me{\cdot}CH_2Cl_2$, **1**, 260
$SnB_{11}H_{11}$
$[B_{11}H_{11}Sn]^{2-}$, **1**, 259
$SnCH_3Br_3$
$MeSnBr_3$, **2**, 255
$SnCH_3Cl_3$
$MeSnCl_3$, **2**, 220
$SnCH_3F_3$
$(CF_3)SnH_3$, **2**, 271
$SnCH_3I_3$
$MeSnI_3$, **2**, 255
SnC_2D_6
$(CD_3)_2Sn$, **2**, 288
SnC_2HClF_6
$(CF_3)_2SnClH$, **2**, 271
$SnC_2H_2F_6$
$(CF_3)_2SnH_2$, **2**, 271
SnC_2H_6
Me_2Sn, **2**, 288
$SnC_2H_6Cl_2$
Me_2SnCl_2, **2**, 220
SnC_2H_8
Me_2SnH_2, **2**, 288
SnC_3F_9I
$(CF_3)_3SnI$, **2**, 226
SnC_3H_7Cl
$ClCH{=}CHCH_2SnH_3$, **11**, 369
SnC_3H_9
Me_3Sn, **2**, 289
SnC_3H_9Cl
Me_3SnCl, **2**, 220
$SnC_3H_9Cl_2O_2P$
$Me_3Sn[OP(O)Cl_2]$, **2**, 259, 260
SnC_3H_9F
Me_3SnF, **2**, 255
SnC_3H_{10}
Me_3SnH, **2**, 270
$SnC_3H_{10}O$
Me_3SnOH, **2**, 261
$SnC_3H_{10}O_3$
Me_3SnHO_3, **2**, 270
SnC_4F_{12}
$(CF_3)_4Sn$, **2**, 226
$SnC_4H_6Cl_2$
$(CH_2{=}CH)_2SnCl_2$, **2**, 239
SnC_4H_8
$Sn(CH_2)_4$, **2**, 252
$SnC_4H_9Cl_3$
$BuSnCl_3$, **11**, 368
Bu^tSnCl_3, **2**, 255
$SnC_4H_{10}Cl_2O$
$BuSn(OH)Cl_2$, **2**, 265
$SnC_4H_{11}I$
Me_3SnCH_2I, **11**, 369
SnC_4H_{12}
Me_4Sn, **2**, 220; **12**, 150, 368
$SnC_4H_{12}O_2$
$Me_3(MeO_2)Sn$, **2**, 290
SnC_5H_5Cl
$CpSnCl$, **2**, 289
$SnC_5H_{12}O_2$
$Me_3Sn(OCOMe)Cl$, **2**, 258
$Me_3SnOOCCH_2$, **11**, 367
$SnC_5H_{15}N$
$(Me_2N)Me_3Sn$, **2**, 290
$SnC_5H_{15}NO_4S_2$
$Me_3SnN(SO_2Me)_2$, **2**, 261
$SnC_5H_{15}O_2$
$Me_3Sn[OP(O)Me_2]$, **2**, 259
$SnC_5H_{15}O_2P$
$Me_3Sn[OP(O)Me_2]$, **2**, 259, 260
$SnC_6H_{16}O$
$(Me_3Sn)CH_2CH_2CH_2OH$, **11**, 359
$SnC_6H_{18}P_2S_4$
$Me_2Sn[SP(S)Me_2]_2$, **2**, 259, 260
$SnC_7H_8Cl_2$
$MePhSnCl_2$, **2**, 255
$SnC_7H_{16}O$
$Me_3Sn(C_4H_7O)$, **12**, 368
$SnC_7H_{19}N$
$Me_3Sn(NEt_2)$, **2**, 259
SnC_8H_{12}
$(CH_2{=}CH)_4Sn$, **2**, 239
SnC_8H_{16}
$Me_3Sn(C_5H_7)$, **11**, 374
$SnC_8H_{18}Cl_2$
Bu_2SnCl_2, **2**, 255
$SnC_8H_{19}Cl$
Bu_2SnClH, **2**, 273
$SnC_8H_{19}ClO$
$Bu^t{}_2Sn(OH)Cl$, **2**, 263
SnC_8H_{20}
Bu_2SnH_2, **2**, 273
$SnC_8H_{22}B_4Si_2$
$Sn[(TMS)_2C_2B_4H_4]$, **1**, 389
SnC_9HF_9
$(CF_3)_3SnH$, **2**, 271
$SnC_9H_{13}NO_2$
$Me_3[(NO_2)C_6H_4]Sn$, **2**, 236, 237
SnC_9H_{21}
Pr_3Sn, **2**, 289
$SnC_{10}H_{10}$
Cp_2Sn, **2**, 289
$SnC_{10}H_{13}Cl$
$PhClSn(CH_2)_4$, **2**, 251
$SnC_{10}H_{15}I_3$
$Me_5C_5SnI_3$, **2**, 249
$SnC_{10}H_{18}F_6O_6S_2$
$Bu_2Sn(OTf)_2$, **2**, 260; **11**, 367
$SnC_{10}H_{22}OS$
$Bu^t{}_2SnOCH_2CH_2S$, **2**, 222
$SnC_{10}H_{22}O_2$
$Bu^t{}_2SnOCH_2CH_2O$, **2**, 222
$SnC_{10}H_{22}S_2$
$Bu^t{}_2SnSCH_2CH_2S$, **2**, 222
$SnC_{11}H_{15}F_3O_3S$
$Cp^*Sn(OTf)$, **2**, 289
$SnC_{12}H_{12}$
Ph_2SnH_2, **2**, 273
$SnC_{12}H_{14}$
$(MeC_5H_4)_2Sn$, **2**, 288
$SnC_{12}H_{20}O_8$
$(CH_2{=}CHCH_2O_2)_4Sn$, **2**, 270
$SnC_{12}H_{22}Cl_2$
Cy_2SnCl_2, **2**, 255
$SnC_{12}H_{27}$
Bu_3Sn, **2**, 273, 289
$SnC_{12}H_{27}ClO_4$
$Bu_3Sn(ClO_4)$, **2**, 220
$SnC_{12}H_{27}D$
Bu_3SnD, **2**, 230
$SnC_{12}H_{27}O$
Bu_3SnO, **2**, 256; **11**, 367
$SnC_{12}H_{27}T$
Bu_3SnT, **2**, 271
$SnC_{12}H_{28}$
Bu_3SnH, **2**, 230, 273, 275; **11**, 358, 359
$SnC_{12}H_{28}O_3$
$Pr^iSn(OPr^i)_3$, **2**, 267
$SnC_{12}H_{31}O_2$
$Bu_3Sn(H_2O)_2$, **2**, 249
$SnC_{13}H_{27}F_3O_3S$
$Bu_3Sn(CF_3SO_3)$, **2**, 276
$SnC_{13}H_{28}Br_2$

$(Br_2CH)Bu_3Sn$, **2**, 239
$SnC_{14}H_{17}B_4N_2$
$(bipy)Sn(Me_2C_2B_4H_4)$, **1**, 389
$SnC_{14}H_{28}Cl_2O_2$
$Bu_3Sn(OCOCHCl_2)$, **2**, 259
$SnC_{14}H_{30}$
$CH_2{=}CH\text{-}SnBu_3$, **12**, 965
$SnC_{14}H_{30}O_2$
$Bu_3Sn(OCOMe)$, **2**, 259
$SnC_{15}H_{32}$
$CH_2{=}CHCH_2SnBu_3$, **2**, 224
$SnC_{16}H_{18}$
$Ph_2Sn(CH_2)_4$, **2**, 251, 270
$SnC_{16}H_{20}O_2$
$[(OH)C_4H_8]Ph_2SnOH$, **2**, 270
$SnC_{16}H_{35}O_3$
$Bu_2(CH_2OCH_2CH_2OCH_2$-$CH_2OCH_2CH_2CH_2)$-SnH, **2**, 272
$SnC_{16}H_{36}$
Bu_4Sn, **2**, 224
$SnC_{16}H_{39}N_3$
$Bu^tSn(NEt_2)_3$, **2**, 255
$SnC_{18}H_4F_{18}$
$[(CF_3)_3C_6H_2]_2Sn$, **2**, 286, 291
$SnC_{18}H_{15}$
Ph_3Sn, **2**, 289
$SnC_{18}H_{15}ClO_4$
$Ph_3Sn(ClO_4)$, **2**, 220
$SnC_{18}H_{15}F$
Ph_3SnF, **2**, 255
$SnC_{18}H_{16}$
Ph_3SnH, **2**, 230, 273; **11**, 358
$SnC_{18}H_{32}$
$PhSnBu_3$, **12**, 151, 152
$SnC_{18}H_{33}F$
Cy_3SnF, **2**, 255
$SnC_{18}H_{34}O$
Cy_3SnOH, **2**, 256
$SnC_{18}H_{37}O_6$
$(CH_2OCH_2CH_2OCH_2CH_2C$-$H_2)_3$- SnH, **2**, 272
$SnC_{20}H_{21}OPS$
$Ph_3Sn[OP(S)Me_2]$, **2**, 259, 260
$SnC_{20}H_{21}O_3P$
$Ph_3Sn[OP(O)(OMe)Me]$, **2**, 259
$SnC_{20}H_{28}O_2$
$Bu_2\overline{SnOC(Ph)C(Ph)O}$, **2**, 267
$SnC_{20}H_{30}$
$Cp^*{}_2Sn$, **2**, 249, 288, 289
$SnC_{20}H_{30}Cl_2$
$Cp^*{}_2SnCl_2$, **2**, 249
$SnC_{21}H_{19}$
$Ph_3Sn(CH{=}C{=}CH_2)$, **2**, 246
$SnC_{21}H_{20}$
$Ph_3(CH_2{=}CH)Sn$, **2**, 239
$SnC_{21}H_{32}O_6$
(PhCH)(glucopyranoside)-$SnBu_2$, **2**, 268
(PhCH)(mannopyronoside)-$SnBu_2$, **2**, 268
$SnC_{24}H_{20}$
Ph_4Sn, **2**, 224, 236
$SnC_{24}H_{27}F_3$
$[[(MeO)_2C_6H_3]C]SnF_3$, **2**, 233
$SnC_{26}H_{20}$
$Ph_3(PhC{\equiv}C)Sn$, **2**, 227
$SnC_{28}H_{60}O_{10}$
$[[[Bu^tC(O)O]Me_2Sn]_2O]_2$, **2**, 264
$SnC_{30}H_{44}O_4S_2$
$Bu_2Sn(SCH_2CO_2CHEtPh)_2$, **2**, 296
$SnC_{32}H_{20}$
$(PhC{\equiv}C)_4Sn$, **2**, 224
$SnC_{32}H_{26}N_2O_3$
$Ph_3Sn[OCOC_6H_4[N_2[(OH)$-$MeC_6H_3]]]$, **2**, 257
$SnC_{32}H_{68}$
$(C_8H_{17})_4Sn$, **2**, 224
$SnC_{40}H_{30}$
$(Ph_5C_5)SnCp$, **2**, 288
$SnC_{45}H_{69}$
$(Pr^i{}_3C_6H_2)_3Sn$, **2**, 289
$SnC_{70}H_{50}$
$(Ph_5C_5)_2Sn$, **2**, 288, 289
$SnC_{80}H_{70}$
$(Bz_5C_5)_2Sn$, **2**, 288, 289
$SnCuC_{12}H_{27}$
Bu_3SnCu, **12**, 101
$SnCuC_5H_{15}S$
$Me_3SnCu{\cdot}Me_2S$, **2**, 231
$SnCuLi_2C_{14}H_{30}N$
$Bu_3Sn(Me)CuCNLi_2$, **2**, 231
$SnFeC_{16}H_{26}$
$(Me_3Sn)_2$(ferrocene), **2**, 249
$SnFePdSiC_{49}H_{46}O_6P_2$
$Fe(CO)_3[Si(OMe)_2OMe]$-(dppm)$Pd(SnPh_3)$, **10**, 61
$SnFe_3C_{12}O_{12}$
$[Sn[Fe(CO)_4]_3]^{2-}$, **7**, 31
$SnFe_3C_{18}H_{10}O_8$
$Fe(CO)VSn[Fe(CO)_2Cp]_2$, **7**, 31
$SnFe_4C_{16}O_{16}$
$SnFe_4(CO)_{16}$, **7**, 31
$SnGeC_{21}H_{24}$
$Me_3SnGePh_3$, **2**, 284
$Ph_3SnGeMe_3$, **2**, 284
$SnGeC_{48}H_{40}$
$Ph_3SnGePh_2SnPh_3$, **2**, 284
$SnGeSiC_{11}H_{30}$
$Bu^tSiMe_2GeMe_2SnMe_3$, **2**, 285
$SnGeSiC_{25}H_{36}$
$Bu^tSiMe_2GeMe_2SnPh_3$, **2**, 285
$SnInC_{12}H_{27}O_3$
$In(OBu^t)_3Sn$, **1**, 525
$SnInMo_2C_{22}H_{27}O_{13}$
$(CO)_5MoSn(OBu^t)_3InMo(CO)_5$, **1**, 560
$SnIrC_{15}H_{29}$
$Cp^*Ir(C_2H_4)(H)(SnMe_3)$, **8**, 402
$SnIrC_{26}H_{29}Cl_3P$
$Ir(cod)(H)_2(SnCl_3)(PPh_3)$, **8**, 402
$SnKC_{15}H_{33}$
$[Bu^tCH_2]_3SnK$, **2**, 283
$SnKC_{18}H_{15}$
Ph_3SnK, **2**, 283
$SnKC_{30}H_{45}$
$(Me_2CH)_3SnK{\cdot}3Tol$, **1**, 15, 16
$SnLiC_3H_9$
Me_3SnLi, **2**, 228, 284; **11**, 363
$SnLiC_4H_{11}$
$(Me_3Sn)CH_2Li$, **11**, 365
$SnLiC_5H_{15}$
Me_5SnLi, **2**, 225
$SnLiC_6H_9$
$(CH_2{=}CH)_3SnLi$, **2**, 228
$SnLiC_{12}H_{12}O$
$(furyl)_3SnLi$, **2**, 283
$SnLiC_{12}H_{27}$
Bu_3SnLi, **11**, 363
$SnLiC_{18}H_{15}$
Ph_3SnLi, **2**, 283
$SnMnC_{23}H_{15}O_4S_2$
$Mn(S_2CSnPh_3)(CO)_4$, **6**, 89
$SnMnC_{26}H_{23}O_2$
$(C_5H_4Me)Mn(CO)_2H(SnPh_3)$, **6**, 143
$SnMnC_{27}H_{33}O_3P_2S$
$Mn(SSnPh_3)(CO)_3(PMe_3)_2$, **6**, 89
$SnMnC_{30}H_{36}NO_4$
$(NEt_4)[MnH(CO)_4(SnPh_3)]$, **6**, 68
$SnMn_2C_{24}H_{30}O_4$
$[Mn(CO)_2Cp^*]_2Sn$, **6**, 114
$SnMn_2Si_2C_{16}H_{14}Cl_8O_4$
$Cl_2Sn[Mn(SiCl_3)(CO)_2(C_5H_4$-$Me)]_2$, **6**, 143
$SnMoC_{30}H_{54}Cl_3N_6$
$Mo(SnCl_3)(CNBu^t)_6$, **5**, 310
$SnNbC_{14}H_{19}O$
$Cp_2Nb(CO)(SnMe_3)$, **5**, 128
$SnNbC_{15}H_{22}N$
$Cp_2Nb(NCMe)(SnMe_3)$, **5**, 101
$SnNb_6C_{96}H_{87}Cl_{19}$
$(Bu_3Sn)[Nb_6(C_4Ph_4)_3Cl_{19}]$, **5**, 113
$SnNiC_{29}H_{25}O$
$Ni(SnPh_3)(CO)Cp^*$, **9**, 167
$SnNiSi_4C_{18}H_{46}$
$Ni(C_2H_4)_2[{=}Sn[CH(TMS)_2]_2]$, **9**, 112
$SnNiSi_4C_{52}H_{70}P_2$
$Ni(PPh_3)_2[HC_2SnH[CH$-$(TMS)_2]_2]$, **9**, 133
$SnO_3Si_2SnC_{21}H_{23}NO_{10}$
$Os_3H[SnC(TMS)_2(C_5H_4N)]$-$(CO)_{10}$, **7**, 696
$SnOs_3Si_2C_{22}H_{26}O_{13}$
$Os_3[SnCH(TMS)_2](MeO_2C$-$CCH_2CO_2Me)(CO)_9$, **7**, 698
$SnOs_3Si_4C_{24}H_{38}O_{10}$
$Os_3H_2[[CH(TMS)_2]SnOC[CH$-$(TMS)_2]](CO)_9$, **7**, 705
$SnOs_3SnC_{12}H_2Cl_2O_{11}$
$Os_3(SnCl_2)(CH_2)(CO)_{11}$, **7**, 750
$SnPbC_{24}H_{40}O_6$
$[(Bu_3Sn)C_6H_4]Pb(O_2CMe)_3$, **11**, 467
$SnPbC_{42}H_{42}$
$(Tol)_3PbSn(Tol)_3$, **2**, 312

$SnPdC_5H_7Cl_3O$
$Pd(SnCl_3)(MeC_3H_4)(CO)$, **9**, 359
$SnPdC_{12}H_{15}Cl_3$
$Pd(SnCl_3)(MeC_3H_4)(styrene)$, **9**, 350
$SnPdC_{30}H_{31}ClP_2$
$Pd(C{\equiv}CH)(SnMe_2Cl)(dppe)$, **9**, 231
$SnPtC_{12}H_{15}Cl_3$
$Pt(SnCl_3)(C_2H_3Ph)[H_2CC(Me)CH_2]$, **9**, 547, 576
$SnPtC_{38}H_{33}Cl_3OP_2$
$PtCl(C(OSnCl_2)Me)(PPh_3)_2$, **9**, 423
$Pt(SnCl_3)(COMe)(PPh_3)_2$, **9**, 423
$SnPtC_{39}H_{39}ClP_2$
$PtMe(SnClMe_2)(PPh_3)_2$, **9**, 436
$SnPt_3C_{36}Cl_3F_{30}$
$Pt_3(Cl)_3(Sn)(C_6F_5)_6$, **9**, 475
$SnRuC_{33}H_{30}O_2$
$Ru(SnPh_3)[C(OEt)Ph](CO)Cp$, **7**, 485
$SnRuC_{46}H_{70}P_4$
$Ru[CC(C_6H_4)_2CCSnMe_3]_2(PMe_3)_4$, **7**, 415
$SnSiC_4H_{12}Cl_2$
$(TMS)(Me)SnCl_2$, **2**, 263
$SnSiC_8H_{18}$
$(TMS)C{\equiv}CSnMe_3$, **2**, 242
$SnSiC_{15}H_{36}$
$Bu_3SN\text{-}TMS$, **2**, 285
$SnSiC_{48}H_{40}$
$Ph_3SnSiPh_2SnPh_3$, **2**, 284
$SnSi_2C_{10}H_{27}$
$Me_3SnC(TMS)_2$, **2**, 250
$SnSi_4C_{14}H_{38}$
$[(TMS)_2CH]_2Sn$, **2**, 285, 287
$SnSi_4C_{14}H_{38}O$
$[(TMS)_2CH]_2SnO$, **2**, 262
$SnSi_6C_{28}H_{64}$
$[(TMS)_2CH]_3SnBz$, **2**, 235
$SnTcC_3H_9O_4$
$TcO_3(OSnMe_3)$, **6**, 162
$SnTiC_{28}H_{25}Cl$
$TiCl(SnPh_3)Cp_2$, **4**, 397
$SnUC_{33}H_{30}$
Cp_3USnPh_3, **4**, 157
$SnVC_{24}H_{15}O_6$
$(Ph_3Sn)V(CO)_6$, **5**, 6
$SnVH_8Cl_6N_2O$
$[NH_4]_2[VO(SnCl_3)Cl_3]$, **10**, 50
$SnWC_{32}H_{26}O_2$
$W(SnPh_3)(CO)_2(CHPh)Cp$, **5**, 169
$SnYbC_{33}H_{64}O_2$
$Cp^*Yb[Sn(CH_2Bu^t)_3](THF)_2$, **4**, 64
$Sn_2C_2H_6$
$SnH_3C{\equiv}CSnH_3$, **11**, 375
$Sn_2C_4H_6Br_2$
$BrMe_2SnSnMe_2Br$, **2**, 281
$Sn_2C_6H_{18}$
$Me_3SnSnMe_3$, **2**, 276, 285, 290
$Sn_2C_{10}H_{20}$
$Sn_2(CH_2)_{10}$, **2**, 252
$Sn_2C_{11}H_{22}$
$(Me_3Sn)_2C_5H_4$, **2**, 249
$Sn_2C_{12}H_{28}ClSe$
$(Pr^i{}_2SnCl)_2Se$, **2**, 294
$Sn_2C_{14}H_{16}Cl_2$
$ClMePhSnSnPhMeCl$, **2**, 281
$Sn_2C_{14}H_{24}$
$(Me_3Sn)_2C_8H_6$, **2**, 290
$Sn_2C_{16}H_{22}S_2$
$(PhS)Me_2SnSnMe_2(SPh)$, **2**, 281
$Sn_2C_{24}H_{21}ClO_2$
$(HO)Ph_2SnOSnPh_2Cl$, **2**, 264
$Sn_2C_{24}H_{54}O$
$(Bu_3Sn)_2O$, **2**, 261, 296
$(Bu^t{}_3Sn)_2O$, **2**, 261
$Sn_2C_{26}H_{56}$
$(Bu_3Sn)HC{=}CH(Bu_3Sn)$, **11**, 371
$Sn_2C_{27}H_{15}F_9N$
$(Mes)[(CF_3)_3C_6H_2]_2Sn_2N$, **2**, 282
$Sn_2C_{28}H_{20}Cl_6O_4$
$(CCl_3C(O)O)Ph_2SnSnPh_2(CCl_3C(O)O)$, **2**, 281
$Sn_2C_{30}H_{42}O$
$Bu_3SnOSnPh_3$, **2**, 261
$Sn_2C_{36}H_{30}O$
$(Ph_3Sn)_2O$, **2**, 261
$Sn_2C_{36}H_{30}S$
$[Ph_3Sn]_2S$, **2**, 293, 294
$Sn_2C_{36}H_{30}Se$
$[Ph_3Sn]_2Se$, **2**, 293
$Sn_2C_{36}H_{92}S$
$(Pr^i{}_3C_6H_2)_4[Sn_2S]$, **2**, 294
$Sn_2C_{45}H_{19}F_{36}N$
$Mes[(CF_3)_3C_6H_2]_4[Sn_2N]$, **2**, 291
$Sn_2C_{60}H_{92}Br_2$
$(Pr^i{}_3C_6H_2)_2BrSnSnBr(Pr^i{}_3C_6H_2)_2$, **2**, 281
$Sn_2C_{72}H_{102}$
$(Et_3C_6H_2)_6Sn_2$, **2**, 290
$Sn_2CoC_{39}H_{30}O_3$
$Co(SnPh_3)_2(CO)_3^-$, **8**, 5
$Sn_2CuLiC_{24}H_{54}$
$(Bu_3Sn)_2CuLi$, **2**, 231
$Sn_2FeC_{42}H_{36}O$
$FeH(CO)(SnPh_3)_2Cp$, **7**, 169
$Sn_2Fe_2W_2C_{18}Cl_2O_{18}$
$Fe_2(CO)_8[Sn[W(CO)_5]Cl]_2$, **7**, 31
$Sn_2Fe_4HgC_{16}Cl_4O_{16}$
$[(CO)_4FeSnCl_2Fe(CO)_4]_2Hg^{2-}$, **10**, 339
$Sn_2MnC_{25}H_{20}Cl_8O_2P$
$[CpMn(SnCl_3)(CO)_2(PPh_3)][SnCl_5]$, **6**, 144
$Sn_2Mn_2C_{44}H_{30}O_8S_2$
$Mn_2(SSnPh_3)_2(CO)_8$, **6**, 89
$Sn_2Nb_2C_{26}H_{38}O$
$[Cp_2Nb(SnMe_3)]_2(O)$, **5**, 128
$Sn_2NiC_{16}H_{26}$
$Ni(C_5H_4SnMe_3)_2$, **9**, 154
$Sn_2Os_3PtC_{32}H_{45}Cl_4O_{11}P$
$Os_3PtSn(H)_2Cl(SnCl_3)(CO)_{10}(OEt_2)(PCy_3)$, **10**, 217, 218
$Sn_2Pd_2C_{37}H_{30}Cl_6OP_2$
$[Pd(SnCl_3)(PPh_3)]_2CO$, **9**, 197
$Sn_2Pd_3C_{40}H_{77}O_{10}P_3$
$Pd_3Sn_2(acac)_4(CO)_2(PEt_3)_3$, **9**, 207
Sn_2PtCCl_7O
$[PtCl(SnCl_3)_2(CO)]^-$, **9**, 407
$Sn_2Pt_2C_{10}H_{14}Cl_6O_2$
$[Pt(SnCl_3)(CO)(C_3H_4Me)]_2$, **9**, 574, 576
$Sn_2Pt_2C_{20}H_{30}Cl_8F_6P_2$
$Pt_2(Cl)_2[C(CF_3){=}C(CF_3)(SnCl_3)]_2(PEt_3)_2$, **9**, 486
$Sn_2Ru_2C_{14}H_{18}O_8$
$[Ru(CO)_4(SnMe_3)]_2$, **7**, 838
$Sn_2Si_4C_{16}H_{45}ClO_2$
$(HO)(TMS\text{-}CH_2)_2SnOSn(CH_2TMS)_2Cl$, **2**, 264
$Sn_2Si_8C_{28}H_{78}O_3$
$[[(TMS)_2CH]_2Sn(OH)]_2O$, **2**, 263
$Sn_2YbC_{38}H_{82}O_2$
$Yb[Sn(CH_2Bu^t)_3]_2(THF)_2$, **4**, 19
$Sn_2Zr_2C_{26}H_{38}$
$[Zr(SnMe_3)Cp_2]_2$, **4**, 504
$Sn_3C_9H_{23}IO_2$
$[(Me_3SnOH)_2SnMe_3]I$, **2**, 261
$Sn_3C_{11}H_{10}$
$(Me_3Sn)_2C{=}CH(Me_3Sn)$, **2**, 280
$Sn_3C_{12}H_{29}ClO_2$
$(HO)Pr^i{}_2SnOSnPr^i{}_2SnCl$, **2**, 264
$Sn_3C_{23}H_{24}S_3$
$Ph_3Me(C_4Sn_3S_3)$, **2**, 294
$Sn_3C_{28}H_{64}$
$Bu_3Sn(Bu_2Sn)SnBu_2H$, **2**, 271
$Sn_3C_{32}H_{72}$
$Bu^t{}_8Sn_3$, **2**, 281
$Sn_3C_{58}H_{52}$
$(Ph_3SnCH_2)_3CH$, **2**, 235
$Sn_3C_{60}H_{78}$
$(Et_2C_6H_3)_6Sn_3$, **2**, 272, 282
$Sn_3C_{84}H_{54}$
$(C_{14}H_9)_6Sn_3$, **2**, 281
$Sn_3C_{90}H_{138}$
$(Pr^i{}_3C_6H_2)_6Sn_3$, **2**, 281
$Sn_3Pt_2C_{24}H_{40}Cl_{12}$
$[Pt(C_4Et_4)Cl_3][Pt(C_4Et_4)(SnCl_3)_3]$, **9**, 579
$Sn_3Pt_3C_{28}H_{45}Cl_9$
$[Pt(C_3H_4Me)(cod)]_2[Pt(SnCl_3)_3(C_3H_4Me)]$, **9**, 574, 576
$Sn_3RuC_{20}H_{20}Cl_{12}$
$[(Cp_2Ru)_2SnCl_2][SnCl_5]_2$, **7**, 618
$Sn_3Se_3C_6H_{18}$
$(Se_8Sn_3)Me_6$, **11**, 545
$Sn_3Si_3C_{12}H_{36}O_3$
$[(TMS)(Me)SnO]_3$, **2**, 263
$Sn_4C_8H_{24}Cl_4O_2$
$[(Me_2SnCl)_2O]_2$, **2**, 263, 264

$Sn_4C_{16}H_{36}O_{10}$
[[[MeC(O)O]Me$_2$Sn]$_2$O]$_2$, **2**, 264
$Sn_4C_{24}H_{20}S_6$
Ph$_4$[Sn$_4$S$_6$], **2**, 294
$Sn_4C_{30}H_{39}$
(Et$_2$C$_6$H$_3$)$_3$Sn$_4$, **2**, 282
$Sn_4C_{32}H_{72}$
Bu$^t_8$Sn$_4$, **2**, 282
$Sn_4C_{32}H_{72}Cl_4O_2$
[(Bu$_2$ClSu)$_2$O]$_2$, **2**, 263
$Sn_4C_{40}H_{72}Cl_{12}O_{10}$
[[[Cl$_3$CC(O)O]Bu$_2$Sn]$_2$O]$_2$, **2**, 264
$Sn_4C_{56}H_{40}Cl_{12}O_{10}$
[[[Cl$_3$CC(O)O]Ph$_2$Sn]$_2$O]$_2$, **2**, 264
$Sn_4NiSi_4C_{40}H_{96}N_8$
Ni[Sn(NBu^t)$_2$(SiMe$_2$)]$_4$, **9**, 14
$Sn_4Os_4C_{22}H_{24}O_{14}$
Os$_4$(SnMe$_2$)$_4$(CO)$_{14}$, **7**, 852
$Sn_4Os_4C_{22}H_{24}O_{16}$
Os$_4$(O)$_2$(SnMe$_2$)$_4$(CO)$_{14}$, **7**, 852
Sn_5PtCl_{15}
[Pt(SnCl$_3$)$_5$]$^{3-}$, **9**, 407
$Sn_6C_{12}D_{36}$
[(CD$_3$)$_2$Sn]$_6$, **2**, 288
$Sn_6C_{12}H_{36}$
(Me$_2$Sn)$_6$, **2**, 287
$Sn_{10}H_{22}O_2$
Bu$_2$SnOCH$_2$CH$_2$O, **2**, 268
$SrB_{10}C_8H_{21}N_3$
(MeCN)$_3$SrC$_2$B$_{10}$H$_{12}$, **1**, 116
SrC_2H_6
SrMe$_2$, **1**, 110
$SrC_{14}H_{18}O$
SrCp$_2$(THF), **1**, 112
$SrC_{28}H_{22}$
Sr^{2+}[(fluorenyl)(CH$_2$)$_2$-(fluorenyl)]$^{2-}$, **1**, 115

T

$TaAlC_{20}H_{51}ClP_4$
[Ta(CBu^t)(H)(dmpe)$_2$]-[AlMe$_3$Cl], **5**, 86
$TaAlC_{28}H_{50}$
Cp*$_2$Ta(C_2H_4·AlEt$_3$)(H), **5**, 92
$TaBC_{11}H_{13}F_4$
[TaMeCp$_2$][BF$_4$], **4**, 593
$TaBC_{12}H_{16}F_4$
[Cp$_2$TaMe$_2$][BF$_4$], **5**, 70
$TaBC_{12}H_{33}O_3P_3$
Ta(BH$_4$)(CO)$_3$(PMe$_3$)$_3$, **5**, 81
$TaBC_{12}H_{42}P_4$
Ta(BH$_4$)H$_2$(PMe$_3$)$_4$, **1**, 280; **5**, 81
$TaBC_{15}H_{24}$
Cp*CpTa(BH$_4$), **5**, 81
TaC_5H_{15}
TaMe$_5$, **5**, 69
$TaC_5H_{20}B_9$
Me$_3$Ta($C_2B_9H_{11}$), **1**, 416
$TaC_6H_{18}Cl_5P_2$
TaCl$_5$(PMe$_3$)$_2$, **5**, 102
$TaC_8H_7O_3$
CpTa(H)$_2$(CO)$_3$, **5**, 80
$TaC_8H_{14}Cl_3P$
CpTaCl$_3$(PMe$_3$), **5**, 119
$TaC_9H_{27}Cl_3P_3$
TaCl$_3$(PMe$_3$)$_3$, **5**, 102
$TaC_9H_{27}Cl_4P_3$
TaCl$_4$(PMe$_3$)$_3$, **5**, 102
$TaC_{10}H_{15}Br_4$
Cp*TaBr$_4$, **5**, 114
$TaC_{10}H_{24}B_9$
(C_5H_4Me)Me$_2$Ta($C_2B_9H_{11}$), **1**, 416
$TaC_{11}H_{15}Cl_2$
CpTa(Me$_2C_4H_4$)Cl$_2$, **5**, 113
$TaC_{12}H_{12}$
(C_6H_6)$_2$Ta, **5**, 129
$TaC_{12}H_{19}$
Ta(CH$_2$CHCHMe)(C_4H_6)$_2$, **5**, 110
$TaC_{12}H_{20}Cl_3O$
Cp*Ta(OEt)Cl$_3$, **5**, 119
$TaC_{12}H_{22}$
($C_{10}H_{21}$)[C=CHTa], **12**, 335
$TaC_{12}H_{31}P_2$
Ta(CH$_2$CH$_2$)$_2$(Et)(PMe$_3$)$_2$, **5**, 137
$TaC_{12}H_{34}Cl_2P_4$
TaH$_2$Cl$_2$(dmpe)$_2$, **5**, 73
$TaC_{12}H_{36}Cl_2P_4$
TaCl$_2$(PMe$_3$)$_4$, **5**, 102
$TaC_{12}H_{38}Cl_2P_4$
TaH$_2$Cl$_2$(PMe$_3$)$_4$, **5**, 73
$TaC_{12}H_{40}Cl_4P_4$
[TaCl$_2$(PMe$_3$)$_2$]$_2$(H)$_4$, **5**, 75
$TaC_{12}H_{41}P_4$
TaH$_5$(PMe$_3$)$_4$, **5**, 75
$TaC_{13}H_{19}Cl_2$
CpTa[CH$_2$(C_6H_{10})CH$_2$]Cl$_2$, **5**, 143
$TaC_{13}H_{21}Cl_2$
Cp*Ta(CH$_2$CHMe)Cl$_2$, **5**, 80
$TaC_{13}H_{22}$
(C_6H_{11})(C_5H_{11})[TaC=C], **12**, 344, 346
$TaC_{13}H_{25}BrN_3$
Ta(C_6H_4Me)(NMe$_2$)$_3$Br, **5**, 70
$TaC_{13}H_{26}Cl_3O_2$
Ta(CHBu^t)Cl$_3$(THF)$_2$, **5**, 138
$TaC_{14}H_{17}$
Cp$_2$Ta(CH$_2$CHCHMe), **5**, 110
$TaC_{14}H_{19}$
Cp$_2$Ta(CH$_2$CH$_2$)H, **5**, 128
$TaC_{14}H_{19}S_4$
Cp*Ta(SCHCHS)$_2$, **5**, 120
$TaC_{14}H_{21}Br_2$
Cp*Ta(MeCCMe)Br$_2$, **5**, 96
$TaC_{14}H_{21}Cl_2$
Cp*Ta(CHCHCH$_2$CH$_2$)Cl$_2$, **5**, 107
Cp*Ta(C_4H_6)Cl$_2$, **5**, 113
$TaC_{14}H_{23}S_4$
Cp*Ta(SCH$_2$CH$_2$S)$_2$, **5**, 120
$TaC_{14}H_{26}N$
Cp*Ta(MeNCH$_2$)Me$_2$, **5**, 73
$TaC_{14}H_{27}$
Cp*TaMe$_4$, **5**, 70
$TaC_{14}H_{32}Cl_2P_3$
CpTaCl$_2$(PMe$_3$)$_3$, **5**, 119
$TaC_{14}H_{35}Cl_2O_2P_4$
[Ta(H)(HOCCOH)(dmpe)$_2$-Cl]Cl, **5**, 65
$TaC_{14}H_{35}P_2$
Ta(CH$_2$CH$_2$CH$_2$Me)(CH$_2$CH$_2$)$_2$(PMe$_3$)$_2$, **5**, 138
$TaC_{14}H_{38}ClP_3$
Ta(CHBu^t)(H)Cl(PMe$_3$)$_3$, **5**, 83
$TaC_{14}H_{38}Cl_2P_3$
Ta(CHBu^t)(H)Cl$_2$(PMe$_3$)$_3$, **5**, 136
$TaC_{14}H_{38}I_2P_3$
Ta(CHBu^t)(H)I$_2$(PMe$_3$)$_3$, **5**, 136
$TaC_{15}H_{19}$
Cp$_2$Ta(MeCCH)(CH$_2$Me), **5**, 96
$TaC_{15}H_{21}S_4$
(C_5Me_4Et)Ta(SCHCHS)$_2$, **5**, 120
$TaC_{15}H_{23}Cl_2$
Cp*Ta(MeC$_4H_5$)Cl$_2$, **5**, 113
$TaC_{15}H_{30}N$
Cp*TaMe$_3$(NMe$_2$), **5**, 70, 73
$TaC_{15}H_{41}P_5$
Ta(CHPMe$_2$)(PMe$_3$)$_3$, **5**, 102
$TaC_{15}H_{43}ClP_5$
Ta(CHPMe$_2$)Cl(PMe$_3$)$_4$, **5**, 102
$TaC_{16}H_{21}O$
Cp*CpTa(CO)H, **5**, 92
$TaC_{16}H_{22}Cl_2$
(C_5H_4Pr^i)$_2$TaCl$_2$, **5**, 126
$TaC_{16}H_{23}$
CpTa(MeC$_4H_5$)(Me$_2C_4H_4$), **5**, 112
$TaC_{16}H_{23}Cl_2O_3$
Cp*TaCl$_2$(CO)$_2$(THF), **5**, 65
$TaC_{16}H_{26}Cl_3O$
Cp*Ta[C(O)CH$_2$Bu^t]Cl$_3$, **5**, 98
$TaC_{16}H_{27}$
Cp*Ta[C(CH$_2$)$_3$]Me$_2$, **5**, 73
$TaC_{16}H_{31}Cl_3P_2$
Cp*TaCl$_3$(dmpe), **5**, 119
$TaC_{16}H_{33}Cl_2P_2$
Cp*TaCl$_2$(PMe$_3$)$_2$, **5**, 119
$TaC_{16}H_{33}P_2$
[C_5Me_3(CH$_2$)$_2$]Ta(H)$_2$(PMe$_3$)$_2$, **5**, 132
$TaC_{16}H_{36}ClP_2$
Cp*TaH$_3$Cl(PMe$_3$)$_2$, **5**, 80
$TaC_{16}H_{37}$
Ta(CH$_2$Bu^t)$_2$Et$_3$, **5**, 67
$TaC_{16}H_{37}ClO_2P$
Ta(CHBu^t)(OBu^t)$_2$Cl(PMe$_3$), **5**, 138
$TaC_{16}H_{37}P_2$
Cp*TaH$_4$(PMe$_3$)$_2$, **5**, 80, 81
$TaC_{17}H_{17}$
Cp$_2$Ta(CH$_2$)Ph, **5**, 83
$TaC_{17}H_{21}Cl_2$
Cp*Ta(CHPh)Cl$_2$, **5**, 83
$TaC_{17}H_{27}Cl_2$
(C_5Me_4Bu^t)Ta(MeCCMe)Cl$_2$, **5**, 96
Cp*Ta[CH$_2$(C_5H_8)CH$_2$]Cl$_2$, **5**, 143
$TaC_{17}H_{30}ClO$
Cp*Ta(CH$_2$CHMeCHMeCH$_2$)-(OMe)Cl, **5**, 137
$TaC_{17}H_{31}$
Cp*Ta(CH$_2$CMeCH$_2$)Me$_3$, **5**, 73
$TaC_{17}H_{40}Cl_3P_2$
TaCl$_3$(=CHBu^t)(PEt$_3$)$_2$, **5**, 313
$TaC_{18}H_{23}Cl_2$
Cp*Ta(CH$_2$CHPh)Cl$_2$, **5**, 80
$TaC_{18}H_{29}$
Cp*Ta(CH$_2$CH$_2$CH$_2$CH$_2$)-(C_4H_6), **5**, 112
$TaC_{18}H_{33}O_2P_2$
Cp*Ta(CO)$_2$(PMe$_3$)$_2$, **5**, 132
$TaC_{18}H_{35}$
Cp*Ta(CH$_2$Bu^t)Me$_3$, **5**, 73

$TaC_{19}H_{31}ClP$
$(C_5H_4Pr^i)_2TaCl(PMe_3)$, **5**, 127
$TaC_{19}H_{31}Cl_2$
$Cp^*Ta[CH_2(C_7H_{12})CH_2]Cl_2$, **5**, 143
$TaC_{19}H_{32}ClO$
$Cp^*Ta(C_8H_{14})(OMe)Cl$, **5**, 137
$TaC_{20}H_{30}Cl_2$
$Cp^*_2TaCl_2$, **5**, 126
$TaC_{20}H_{31}$
$Cp^*Ta(C_4H_6)(Me_2C_4H_4)$, **5**, 112
$TaC_{20}H_{32}Cl_4O_3$
$[Cp^*TaCl_2(OH)]_2(O)$, **5**, 119
$TaC_{20}H_{32}N$
$Cp^*_2Ta(NH)H$, **5**, 80
$TaC_{20}H_{38}P$
$Cp^*Ta(CHBu^t)(CH_2CH_2)$-(PMe_3), **5**, 86
$TaC_{20}H_{43}$
$Ta(CHBu^t)(CH_2Bu^t)_3$, **5**, 143
$TaC_{21}H_{33}$
$Cp^*_2Ta(CH_2)H$, **5**, 83, 88
$TaC_{21}H_{33}O$
$Cp^*_2Ta(OCH_2)H$, **5**, 70, 102
$Cp^*_2Ta(O)Me$, **5**, 70
$TaC_{21}H_{43}P_2$
$Cp^*Ta(CBu^t)(H)(PMe_3)_2$, **5**, 88
$TaC_{22}H_{33}$
$[C_5Me_4(CH_2)_2CH]Cp^*TaH$, **5**, 86
$(C_5Me_4CH_2CHCH_2)Cp^*TaH$, **5**, 86
$Cp^*_2Ta(CCH_2)H$, **5**, 88
$TaC_{22}H_{36}Cl_4$
$[(C_5Me_4Et)TaCl_2H]_2$, **5**, 80
$TaC_{22}H_{40}ClP_4$
$(C_{10}H_8)TaCl(dmpe)_2$, **5**, 112
$TaC_{22}H_{56}ClN_2P_4$
$[Ta(CHBu^t)Cl(PMe_3)_2]_2(N_2)$, **5**, 88
$TaC_{23}H_{36}Cl_4O$
$[(C_5Me_4Et)TaCl_2]_2(H)(CHO)$, **5**, 80
$TaC_{23}H_{39}Cl_2$
$(C_5H_2Bu^t_3)Ta(Bu^tCCH)Cl_2$, **5**, 96
$TaC_{24}H_{28}Cl$
$Cp^*Ta(CHPh)(CH_2Ph)Cl$, **5**, 88
$TaC_{24}H_{29}N_2O_2$
$Cp^*Ta(OCMe_2)(CO)(bipy)$, **5**, 102
$TaC_{24}H_{32}P_2$
$Ta(C_6H_4)Ph_2(PMe_3)_2$, **5**, 97
$TaC_{25}H_{28}ClO$
$Cp^*Ta(PhCCPh)(OMe)Cl$, **5**, 96
$TaC_{26}H_{31}O_2$
$Cp^*Ta(PhCCPh)(OMe)_2$, **5**, 96
$TaC_{26}H_{35}ClN$
$Cp^*_2Ta(NPh)Cl$, **5**, 127
$TaC_{26}H_{35}ClN_3$
$Cp^*_2Ta(N_3Ph)I$, **5**, 127
$TaC_{26}H_{36}N$
$Cp^*_2Ta(NPh)H$, **5**, 80, 127
$TaC_{27}H_{27}$
$Cp'Ta(CHPh)(CH_2Ph)_2$, **5**, 83
$TaC_{28}H_{28}Cl$
$Ta(Bz)_4Cl$, **5**, 70
$TaC_{28}H_{31}F_4NP$
$Cp^*TaF_4(HNPPh_3)$, **5**, 119
$TaC_{40}H_{45}$
$Ta(CH_2C_6H_4Me)_5$, **5**, 67
$TaIrC_{14}H_{14}O_2$
$Cp_2Ta(CH_2)_2Ir(CO)_2$, **8**, 304; **10**, 28, 41, 49
$TaLi_2C_{14}H_{25}O_2S_3$
$Li_2(DME)Cp^*TaS_3$, **5**, 120
$TaLi_2C_{16}H_{31}N_2Se_3$
$Cp^*TaSe_3Li_2(TMEDA)$, **5**, 120
$TaLi_2C_{22}H_{47}N_4S_3$
$Li_2(TMEDA)_2Cp^*TaS_3$, **5**, 120
$TaLi_3C_{22}H_{39}ClO_3Se_3$
$Cp^*TaSe_3Li_3Cl(THF)_3$, **5**, 120
$TaMnC_{17}H_{16}O_2$
$TaCp_2(H)Mn(CO)_2Cp$, **10**, 28
$TaNiC_{40}H_{57}N_2O$
$Ta(TMEDA)(CH_2)(Me)$-$(OCMe_2CH_2PPh_2)_2Ni$, **10**, 28
$TaPtC_{15}H_{24}P$
$TaCp_2(CH_2)_2PtH(PMe_3)$, **10**, 28
$TaPtC_{18}H_{33}P_2$
$Cp_2TaMe(CH_2)Pt(PMe_3)_2$, **5**, 88
$TaSiC_{13}H_{24}Cl_3$
$Cp^*Ta(TMS)Cl_3$, **5**, 118
$TaSiC_{14}H_{24}Cl_3O$
$Cp^*Ta[C(O)TMS]Cl_3$, **5**, 98
$TaSiC_{14}H_{37}ClNP_2$
$Ta(CHBu^t)(NTMS)Cl(PMe_3)_2$, **5**, 88
$TaSiC_{16}H_{33}Cl_2P$
$Cp^*Ta(TMS)(PMe_3)Cl_2$, **5**, 118
$TaSiC_{17}H_{30}Cl_3O_3$
$Cp^*Ta[OCH(TMS)C(O)O$-$Et]Cl_3$, **5**, 120
$TaSiC_{19}H_{29}Cl_3NO$
$Cp^*Ta[OCH(TMS)C_5H_4N]Cl_3$, **5**, 120
$Cp^*Ta[C(O)TMS(NCp)]Cl_3$, **5**, 98
$TaSiC_{23}H_{53}O_2P_4$
$Ta(COSiPr^i_3)(dmpe)_2(CO)$, **5**, 65
$TaSiC_{27}H_{34}Cl_3N_2O$
$Cp^*Ta[OC(TMS)NN(CPh_2)]$-Cl_3, **5**, 120
$TaSi_2C_{16}H_{29}$
$(C_5H_4TMS)_2TaH_3$, **5**, 76
$TaSi_2C_{20}H_{50}ClO_2P_4$
Ta(TMS-OCCO-TMS)-$(dmpe)_2Cl$, **5**, 65
$TaSi_2C_{22}H_{56}ClN_2P_4$
[TMS(Me)NCCN(Me)TMS]Ta-$(dmpe)_2Cl$, **5**, 66
$TaSi_2C_{26}H_{62}ClO_2P_4$
$Ta(Pr^i_3SiOCCO$-TMS)-$(dmpe)_2Cl$, **5**, 65, 89
$TaSi_4C_{21}H_{45}Cl_2$
$Ta[CBu^tC$-TMS-CTMS](TMS-CC-TMS)Cl_2, **5**, 106
$TaSi_4C_{22}H_{45}$
$[C_5H_3(TMS)_2]_2TaH_3$, **5**, 76
$TaZnC_{17}H_{21}$
$(MeCp)_2TaH_2ZnCp$, **10**, 336
$TaZn_2C_{22}H_{25}$
$(MeCp)_2TaH(ZnCp)_2$, **10**, 336
$TaZn_3C_{27}H_{29}$
$Cp'_2Ta(ZnCp)_3$, **5**, 128
$Ta_2B_2C_{20}H_{36}Br_2$
$Cp^*_2Ta_2(Br)_2(B_2H_6)$, **1**, 306
$Ta_2B_4C_{10}H_{22}$
$Cp_2Ta_2(B_2H_6)_2$, **1**, 306
$Ta_2C_{12}H_{36}Cl_6P_4$
$[TaCl_2(PMe_3)_2]_2(Cl)_2$, **5**, 74
$Ta_2C_{12}H_{38}Cl_4P_4$
$[TaCl_2(PMe_3)_2]_2(H)_2$, **5**, 75
$Ta_2C_{12}H_{38}Cl_6P_4$
$[TaCl_2(PMe_3)_2]_2(H)_2(Cl)_2$, **5**, 75
$Ta_2C_{12}H_{39}Cl_5P_4$
$[TaCl_2(PMe_3)_2]_2(H)_3(Cl)$, **5**, 75
$Ta_2C_{18}H_{34}Cl_6O_2$
$Ta_2(Bu^tCCBu^t)Cl_6(THF)_2$, **5**, 93
$Ta_2C_{20}H_{30}Br_4$
$[Cp^*Ta(Br)_2]_2$, **5**, 114
$Ta_2C_{20}H_{30}Cl_6O$
$[Cp^*TaCl_3]_2(O)$, **5**, 119
$Ta_2C_{20}H_{32}Cl_4$
$(Cp^*TaCl_2H)_2$, **5**, 80
$Ta_2C_{22}H_{34}Br_6$
$[(C_5Me_4Et)Ta(Br)Br_2]_2$, **5**, 114
$Ta_2C_{22}H_{35}Cl_5$
$(C_5Me_4Et)_2Ta_2Cl_5H$, **5**, 80
$Ta_2C_{22}H_{40}Cl_6S_2$
$[Ta(Bu^tCCMe)Cl_2(Cl)(THT)]_2$, **5**, 93
$Ta_2C_{23}H_{36}Cl_4O$
$[(C_5Me_4Et)TaCl_2]_2(H)(CHO)$, **5**, 98
$Ta_2C_{24}H_{30}Cl_2O_4$
$[Cp^*TaCl(CO)_2]_2$, **5**, 65
$Ta_2Li_2C_{32}H_{54}O_4S_8$
$[Li_2(THF)_3][Cp^*Ta(S_2)_2]_2(O)$, **5**, 120
$Ta_2Li_6C_{50}H_{110}Cl_2N_{10}S_6$
$[Li_3(TMEDA)_2Cp^*TaS_3Cl]_2$-(TMEDA), **5**, 120
$Ta_2Si_4C_{48}H_{112}O_4$
$[TaH_2(OSiBu^t_3)_2]_2$, **5**, 134
$Ta_2Si_6C_{74}H_{162}O_6$
$[Ta(silox)_3]_2(C_2)$, **5**, 134
$Ta_2ZnC_{12}H_{32}$
$[(MeCp)_2TaH_2]_2Zn$, **10**, 336
$Ta_3C_{30}H_{45}Cl_4O_3$
$[Cp^*TaCl(O)]_3(O)(Cl)$, **5**, 120
$Ta_3C_{30}H_{49}Cl_2O_7$
$[[Cp^*Ta(O)]_3(O)_2Cl(H_2O)_2]Cl$, **5**, 120
$Ta_4C_{40}H_{62}O_9$
$[Cp^*Ta]_4(O)_7(OH)_2$, **5**, 120
$TbC_{21}H_{29}$
$(C_8H_8)Tb(Bu^t_2C_5H_3)$, **4**, 106
$TcBC_{12}H_{10}N_6O_3$
$Tc(HBpz_3)(CO)_3$, **6**, 156
$TcBC_{18}H_{22}N_6O_3$

$Tc[HB(Me_2C_3HN_2)_3](CO)_3$, **6**, 156
$TcBC_{29}H_{25}N_6O_2P$
$Tc(HBpz_3)(CO)_2(PPh_3)$, **6**, 156
$TcC_{26}H_{45}N_3O_2P$
$[Tc(CO)_2(N_3C_6H_{15})(PPh_3)]^+$, **6**, 156
$TcCH_3O_3$
$TcMe(O)_3$, **6**, 162
$TcC_2H_3NPS_3$
$Tc(PS_3)(CNMe)$, **6**, 159
TcC_5BrO_5
$TcBr(CO)_5$, **6**, 153, 156
TcC_5ClO_5
$TcCl(CO)_5$, **6**, 153
TcC_5IO_5
$TcI(CO)_5$, **6**, 153
$TcC_7H_3NO_5$
$[Tc(CO)_5(NCMe)]^+$, **6**, 156
$TcC_7H_3O_7$
$Tc(O_2CCF_3)(CO)_5$, **6**, 155
$TcC_7H_8F_3N_2O_5$
$Tc(O_2CCF_3)(CO)_3(en)$, **6**, 155
$TcC_7H_{11}N_2O_5$
$Tc(O_2CMe)(CO)_3(en)$, **6**, 155
$TcC_8H_5O_3$
$Tc(CO)_3Cp$, **6**, 152, 159, 160
$TcC_8H_{10}O_6$
$Tc(acac)(CO)_3(H_2O)$, **6**, 154
$TcC_9H_6ClO_3$
$[Tc(CO)_3(C_6H_6)]Cl$, **6**, 161
$TcC_9H_8O_6$
$Tc(acac)(CO)_4$, **6**, 154
$TcC_9H_9N_3O_3$
$[Tc(CO)_3(NCMe)_3]^+$, **6**, 156, 157
$TcC_9H_{10}NO_4S_2$
$Tc(S_2CNEt_2)(CO)_4$, **6**, 155
$TcC_9H_{15}N_3O_3$
$[Tc(CO)_3(N_3C_6H_{15})]^+$, **6**, 156
$TcC_{10}H_{10}Cl$
$TcClCp_2$, **6**, 159, 160
$TcC_{10}H_{11}$
$TcHCp_2$, **6**, 159
$TcC_{10}H_{15}O_3$
TcO_3Cp^*, **6**, 161
$TcC_{10}H_{18}Br_4N_2$
$TcBr_4(CNBu^t)_2$, **6**, 159
$TcC_{10}H_{31}OP_3$
$[TcH_2(H_2)(CO)(PMe_3)_3]^+$, **6**, 157
$TcC_{12}H_{10}$
$[TcPh_2]^+$, **6**, 161
$TcC_{12}H_{15}Br_2O_2$
$TcBr_2(CO)_2Cp^*$, **6**, 160
$TcC_{12}H_{15}NO_3$
$[Tc(NO)(CO)_2Cp^*]^+$, **6**, 160
$TcC_{12}H_{18}N$
$[Tc(CNMe)_6]^+$, **6**, 158
$TcC_{12}H_{19}NO_5$
$Tc(acac)(CO)_3(NHEt_2)$, **6**, 155
$TcC_{13}H_{15}O_3$
$Tc(CO)_3Cp^*$, **6**, 160
$TcC_{13}H_{18}BrN_2O_3$
$TcBr(CNBu^t)_2(CO)_3$, **6**, 159
$TcC_{13}H_{20}N$
$Tc(NPr^i{}_2C_6H_3)Me$, **6**, 162
$TcC_{15}H_{15}$
$TcCp_3$, **6**, 159, 160
$TcC_{16}H_{30}N_3OS_6$
$Tc(S_2CNEt_2)_3(CO)$, **6**, 155
$TcC_{17}H_{22}N$
$Tc(NPr^i{}_2C_6H_3)Cp$, **6**, 161
$TcC_{18}H_{20}N_6$
$[Tc(bipy)(CNMe)_4]^+$, **6**, 162
$TcC_{18}H_{26}N_2O_8P_2$
$[Tc(CO)_2(bipy)[P(OMe)_3]_2]^+$, **6**, 156
$TcC_{19}H_{22}ClO_3P_2$
$TcCl(CO)_3(PMe_2Ph)_2$, **6**, 157
$TcC_{22}H_{27}N$
$[Tc(NPr^i{}_2C_6H_3)Cp_2]^-$, **6**, 161
$TcC_{25}H_{45}N_6O$
$[Tc(NO)(CNBu^t)_5]^{2+}$, **6**, 158
$TcC_{26}H_{33}ClO_2P_3$
$TcCl(CO)_2(PMe_2Ph)_3$, **6**, 157
$TcC_{26}H_{45}BrN_6$
$[Tc(CN)Br(CNBu^t)_5]^+$, **6**, 159
$TcC_{28}H_{39}ClN_2P_3$
$[TcCl(NCMe)_2(PMe_2Ph)_3]^{2+}$, **6**, 157
$TcC_{30}H_{54}N_6$
$[Tc(CNBu^t)_6]^+$, **6**, 158
$TcC_{30}H_{66}N_2P_4$
$[Tc(CNBu^t)_2(depe)_2]^+$, **6**, 159
$TcC_{31}H_{34}NP_2$
$TcH(CNBu^t)(dppe)$, **6**, 157, 159
$TcC_{32}H_{39}O_2S_3$
$Tc[S(C_6HMe_4)]_3(CO)_2$, **6**, 155
$TcC_{34}H_{45}N_2S_3$
$Tc[S(C_6HMe_4)]_3(NCMe)_2$, **6**, 155
$TcC_{36}H_{44}NOS_3$
$Tc[S(C_6HMe_4)]_3(CO)(py)$, **6**, 155
$TcC_{37}H_{30}Cl_3OP_2$
$TcCl_3(CO)(PPh_3)_2$, **6**, 157
$TcC_{38}H_{53}N_2S_3$
$Tc[S(C_6HMe_4)]_3(CNPr^i)_2$, **6**, 155, 159
$TcC_{39}H_{30}ClO_3P_2$
$TcCl(CO)_3(PPh_3)_2$, **6**, 156
$TcC_{39}H_{31}O_3P_2$
$TcH(CO)_3(PPh_3)_2$, **6**, 157
$TcC_{43}H_{40}NO_2P_2S_2$
$Tc(S_2CNEt_2)(CO)_2(PPh_3)_2$, **6**, 155
$TcC_{43}H_{60}N_5P$
$[Tc(CNBu^t)_5(PPh_3)]^+$, **6**, 158
$TcC_{52}H_{49}N_2P_4$
$TcH(dppe)_2(N_2)$, **6**, 159
$TcC_{53}H_{49}OP_4$
$TcH(CO)(dppe)_2$, **6**, 157
$TcC_{55}H_{51}NOP_4$
$[Tc(CO)(dppe)_2(NCMe)]^+$, **6**, 157
$TcC_{55}H_{52}ClP_4$
$TcCl(=C=CHMe)(dppe)_2$, **6**, 162, 163
$TcC_{58}H_{59}ClP_4$
$[Tc(\equiv CCH_2Bu^t)Cl(dppe)_2]^+$, **6**, 162, 163
$TcC_{59}H_{60}N_2P_4$
$[Tc(CNBu^t)(dppe)_2(NCMe)]^+$, **6**, 157, 159
$TcSnC_3H_9O_4$
$TcO_3(OSnMe_3)$, **6**, 162
$Tc_2B_2C_{34}H_{44}N_{14}O_4$
$[Tc[HB(Me_2C_3HN_2)_3]$-$(CO)_2]_2N_2$, **6**, 156
$Tc_2C_4H_{12}O_4$
$Me_2(O)Tc(O)_2Tc(O)Me_2$, **6**, 162
$Tc_2C_{10}O_{10}$
$Tc_2(CO)_{10}$, **6**, 152, 156
$Tc_2C_{16}H_8O_{10}$
$[Tc(acac)(CO)_3]_2$, **6**, 154
$Tc_2C_{20}H_{20}$
$[TcCp_2]_2$, **6**, 159
$Tc_2C_{23}H_{30}O_3$
$Cp^*Tc(CO)_3TcCp^*$, **6**, 160
$Tc_2C_{25}H_{30}O_5$
$Cp^*(CO)_2Tc(CO)Tc(CO)_2Cp^*$, **6**, 160
$Tc_2C_{52}H_{80}N_4$
$Me_2(NPr^i{}_2C_6H_3)Tc(NP^i{}_2C_6H_3)_2$-$Tc(NPr^i{}_2C_6H_3)Me_2$, **6**, 162, 163
$Tc_2C_{62}H_{91}N_5$
$(NPr^i{}_2C_6H_3)_2Tc(NPr^i{}_2C_6H_3)_2Tc$-$(NPr^i{}_2C_6H_3)Me_2$, **6**, 162, 163
$Tc_3C_6Cl_6O_6$
$(CO)_3TcCl_3TcCl_3Tc(CO)_3$, **6**, 154
$Tc_3C_{15}H_{27}Br_6N_6O_3$
$[TcBr_2(NO)(CNBu^t)]_3$, **6**, 158
$Tc_3NaC_{13}H_{12}O_{13}$
$Na[Tc_3(OMe)_4(CO)_9]$, **6**, 154
$Tc_4C_{12}Cl_4O_{12}$
$[TcCl(CO)_3]_4$, **6**, 153
$Tc_4C_{12}H_4O_{16}$
$[Tc(OH)(CO)_3]_4$, **6**, 154
$TeC_2H_2Cl_4$
$ClCH=CHTeCl_3$, **11**, 588
TeC_4H_{12}
$TeMe_4$, **11**, 595
TeC_6H_6
PhTeH, **11**, 576
TeC_6H_6O
PhTeOH, **11**, 577
TeC_6H_6S
$CH_2=CH(C_4H_3S)Te$, **11**, 594
TeC_8H_{18}
Bu_2Te, **11**, 581
TeC_9H_{21}
$Bu_2Te^+Me^-$, **11**, 591
$TeC_{10}H_{21}$
$Bu_2Te^+CH_2=CH^-$, **11**, 591
$TeC_{13}H_{13}$
$Ph_2Te^+Me^-$, **11**, 591
$TeC_{14}H_{14}O_3$
$(MeOC_6H_4)_2TeO$, **11**, 583
$TeCrC_{17}H_{14}O_7$
$Cr(CO)_5[Te(C_5H_4OMe)_2]$, **5**, 262

$TeGaSi_6C_{23}H_{65}$
$[(TMS)_2CH]_2GaTeSi(TMS)_3$, **1**, 525
$TeGaSi_8C_{23}H_{65}$
$[(TMS)_2CH]_2GaTeSi(TMS)_3$, **1**, 525
TeH_2
H_2Te, **11**, 576
$TeMn_2C_{24}H_{30}O_4$
$[Mn(CO)_2Cp^*]_2Te$, **6**, 115
$TeMn_2C_{26}H_{40}O_4$
$Mn_2(TePh)_2(CO)_4Cp_2$, **6**, 116
$TeMn_3C_{21}H_{15}O_6$
$[Mn(CO)_2Cp]_3Te$, **6**, 115
$TeNaC_4H_3OS$
$(C_4H_3S)TeONa$, **11**, 583
$TeNaH$
$NaHTe$, **11**, 572
$TeSiC_{14}H_{29}$
$Bu_2Te^+C{\equiv}CCH_2Si(Me)_3$, **11**, 592
$TeSiC_{14}H_{31}$
$Bu^i_2Te^+(Me)_3SiCH{=}CHCH_2^-$, **11**, 593
$Te_8W_6C_{66}H_{40}O_{18}$
$[PPh_4]_2[W(CO)_3]_6(Te_2)_4$, **5**, 269
$ThBC_{45}H_{33}F_{20}$
$[Cp^*_2ThMe][B(C_6F_5)_4]$, **4**, 191
$ThBSi_6C_{18}H_{58}N_3$
$[(TMS)_2N]_3Th(BH_4)$, **4**, 134
$ThB_2C_2H_{12}$
$Th[BH_3Me]_2$, **1**, 290, 291
$ThB_4C_4H_{24}$
$Th(BH_3Me)_4$, **1**, 290, 291
$ThB_{18}C_{24}H_{53}$
$[Cp^*_2ThH][Co(B_9C_2H_{11})_2]$, **4**, 176
$ThC_{10}H_{10}Cl$
Cp_2ThCl, **4**, 137
$ThC_{14}H_{16}Cl_2N_2$
$(MeC_5H_3NCH_2)_2ThCl_2$, **4**, 135
$ThC_{15}H_{15}Cl$
Cp_3ThCl, **4**, 147
$ThC_{16}H_{18}$
Cp_3ThMe, **4**, 191
$ThC_{16}H_{24}Cl_2O_2$
$(C_8H_8)ThCl_2(THF)_2$, **4**, 184, 186
$ThC_{20}H_{20}$
$ThCp_4$, **4**, 160
$ThC_{20}H_{30}S_5$
$Cp^*_2ThS_5$, **4**, 167
$ThC_{21}H_3$
$[Cp^*_2ThMe]^-$, **4**, 174
$ThC_{21}H_{33}$
$[Cp^*_2ThMe]^-$, **4**, 191
$ThC_{23}H_{37}P$
$Cp^*_2Th(CH_2)_2PMe$, **4**, 171
$ThC_{24}H_{36}$
$Cp^*_2Th(C_4H_6)$, **4**, 172
$ThC_{25}H_{48}$
$Cp^*Th(CH_2Bu^t)_3$, **4**, 166
$ThC_{26}H_{40}$
$Cp^*_2Th(CH_2CMeCMeCH_2)$, **4**, 172
$ThC_{26}H_{41}ClO$
$Cp^*_2Th(OCCH_2Bu^t)Cl$, **4**, 172
$ThC_{26}H_{41}O_3$
$Cp^*_2Th(OCHBu^t)(O_2CH)$, **4**, 176
$ThC_{27}H_{21}Cl$
$(C_9H_7)_3ThCl$, **4**, 163
$ThC_{27}H_{35}ClO$
$Cp^*_2Th(COPh)Cl$, **4**, 172
$ThC_{29}H_{42}N$
$[Cp^*_2Th(C_6H_4CH_2NMe_2)]^-$, **4**, 174
$ThC_{29}H_{49}O_2$
$[Cp^*_2ThMe(THF)_2]^-$, **4**, 174
$ThC_{30}H_{48}$
$(Bu^t_2C_5H_3)_2Th(C_4H_6)$, **4**, 172
$ThC_{30}H_{52}$
$Cp^*_2Th(CH_2Bu^t)_2$, **4**, 170
$ThC_{31}H_{36}$
$Cp^*Th(Bz)_3$, **4**, 131
$ThC_{31}H_{48}$
$Cp^*Th(Bz)_3$, **4**, 166
$ThC_{33}H_{33}Cl$
$(EtC_9H_6)_3ThCl$, **4**, 164
$ThC_{34}H_{44}P_2$
$Th(Bz)_4(dmpe)$, **4**, 133
$ThC_{36}H_{28}$
$Th(C_9H_7)_4$, **4**, 162, 164
$ThC_{36}H_{44}$
$Th(CH_2C_6H_3Me_2)_4$, **4**, 132
$ThFeC_{22}H_{20}O_2$
$Cp_3ThFeCp(CO)_2$, **4**, 188; **10**, 54
$ThMgC_{27}H_{42}Cl_2O$
$(C_8H_8)Cp^*Th(Cl)_2Mg(CH_2Bu^t)(THF)$, **4**, 186
$ThNiC_{22}H_{34}O_2P_2$
$Cp^*_2Th(PH_2)_2Ni(CO)_2$, **4**, 188
$ThNiC_{36}H_{30}O_2P_2$
$Ni(CO)_2(PPh_2)_2ThCp_2$, **10**, 71
$ThPtC_{47}H_{59}P_3$
$Cp^*_2Th(PPh_2)_2Pt(PMe_3)$, **4**, 188; **10**, 71
$ThRuC_{27}H_{35}ClO_2$
$Ru(CO)_2CpThCp^*_2Cl$, **10**, 54
$ThRuC_{27}H_{35}IO_2$
$Cp^*_2Th(I)RuCp(CO)_2$, **4**, 187
$ThRuC_{32}H_{44}$
$RuMeCp(C_5H_4)(CH_2)ThCp^*_2$, **10**, 62
$ThSi_2C_{25}H_{42}$
$(C_8H_8)Cp^*ThCH(TMS)_2$, **4**, 186
$ThSi_2C_{27}H_{51}P$
$Cp^*_2Th(Me)P(TMS)_2$, **4**, 168
$ThSi_3C_{28}H_{52}$
$[Me_2Si(C_5Me_4)_2]Th(CH_2TMS)_2$, **4**, 162
$ThSi_4C_{20}H_{44}N_2$
$(C_8H_8)Th[N(TMS)_2]_2$, **4**, 186
$ThSi_6C_{33}H_{63}$
$Th[C_5H_3(TMS)_2]_3$, **4**, 140
$Th_2C_{40}H_{62}Cl_2$
$[Cp^*_2Th(H)Cl]_2$, **4**, 174
$Th_2C_{40}H_{64}$
$[Cp^*_2ThH_2]_2$, **4**, 175
$Th_2C_{52}H_{84}ClP_3$
$(C_5H_3Bu^t_2)_2Th(P_3)ThCl(C_5H_3Bu^t_2)_2$, **4**, 168
$Th_2C_{52}H_{84}P_6$
$(C_5H_3Bu^t_2)_2Th(P_6)Th(C_5H_3Bu^t_2)_2$, **4**, 168
$Th_2FeB_{18}C_{46}H_{88}$
$[Cp^*_2ThMe]_2[Fe(B_9C_2H_{11})_2]$, **4**, 174
$Th_2Si_2C_{40}H_{64}$
$[[Me_2Si(C_5Me_4)_2]ThH_2]_2$, **4**, 162, 190
$TiAgSi_3C_{18}H_{31}NO_2$
$Ti[C_5H_4(TMS)][C_2(TMS)]_2AgNO_2$, **10**, 27
$TiAlCH_2Cl_5$
$Cl_2AlMeTiCl_3$, **12**, 579
$TiAlC_{10}H_{10}Cl_4$
$TiCp_2(Cl)_2Al(Cl)_2$, **4**, 249
$TiAlC_{10}H_{14}$
$(TiCp_2)(AlH_4)$, **4**, 260, 261
$TiCp_2(H)_2AlH_2$, **1**, 546
$TiCp_2(H_2)AlH_2$, **4**, 260
$TiAlC_{13}H_{18}Cl$
$Cp_2Ti(Cl)CH_2AlMe_2$, **11**, 296, 308
$TiCp_2(CH_2)(Cl)AlMe_2$, **4**, 374
$TiAlC_{21}H_{27}$
$(TiCp_2)_2(AlH_4Me)$, **4**, 262
$TiAlC_{22}H_{36}Cl_2$
$TiCp_2(Cl)_2Al(hexyl)_2$, **4**, 248
$TiAl_2C_{10}H_{15}Cl_8$
$TiAl_2Cl_8Cp^*$, **4**, 244
$TiAl_2C_{12}H_{20}Cl_7$
$TiAl_2Cl_7EtCp^*$, **4**, 244
$TiAs_3C_{28}H_{25}$
$Ti(As_3Ph_3)Cp_2$, **4**, 410
$TiBC_{11}H_{14}$
$CH_2(C_5H_4)_2Ti(BH_4)$, **1**, 279, 280
$TiBC_{13}H_{25}$
$Ti(C_5H_3Bu_2^t)(H)_2BH_2$, **4**, 261
$TiBC_{20}H_{22}N_2O_3$
$(salen)Ti[BH_4]_2{\cdot}(THF)$, **1**, 279, 280
$TiBC_{35}H_{33}$
$[TiMeCp_2][BPh_4]$, **4**, 395
$TiBC_{44}H_{73}O_2$
$[Ti(OH)Cp^*_2(H_2O)][BPh_4]$, **4**, 402
$TiBC_{45}H_{53}$
$[TiMeCp^*_2][BPh_4]$, **4**, 367
$TiBC_{49}H_{61}O$
$[TiMeCp^*_2(THF)][BPh_4]$, **4**, 368, 394
$TiBC_{49}H_{61}S$
$[TiMeCp^*_2(THT)][BPh_4]$, **4**, 395
$TiBC_{57}H_{47}N_2$
$[Ti[N{=}C(Me)Ph](C_9H_7)_2(PhCN)][BPh_4]$, **4**, 366
$TiBC_{57}H_{62}NO$
$[Ti(OH)(HN{=}CPh_2)Cp^*_2][BPh_4]$, **4**, 399
$TiBNiC_{38}H_{38}F_4P_2S_2$
$[TiCp_2(SCH_2CH_2PPh_2)_2Ni]BF_4$, **10**, 27

$TiBRhC_{68}H_{66}P_2$
$[TiCp_2(CH_2PPh_2)_2Rh(C_8H_{12})]$-$[BPh_4]$, **4**, 371
$TiBSi_2C_{41}H_{29}F_{20}$
$[TiMe[C_5H_4(TMS)]_2][B(C_6$-$F_5)_4]$, **4**, 395
$TiB_2C_{12}H_{40}P_4$
$Ti(BH_4)_2(dmpe)_2$, **1**, 279, 280
$TiB_2C_{23}H_{38}O_2$
(bisphenoxo)$Ti(BH_4)_2$, **1**, 279, 280
$TiB_3C_4H_{22}O_2$
$Ti(BH_4)_3(DME)$, **1**, 279, 280
$TiB_3C_6H_{30}P_3$
$Ti(BH_4)_3(PMe_3)_2$, **1**, 279, 280
$TiCH_3Cl_3$
$TiMeCl_3$, **4**, 274, 280, 283
$TiCH_9Cl_3P_2$
$Ti(Me)Cl_3(PH_3)_2$, **4**, 283
$TiC_2H_5Br_3$
$TiEtBr_3$, **4**, 280
$TiC_2H_5Cl_3$
$TiEtCl_3$, **4**, 280
$TiC_2H_{15}Cl_2P_3$
$TiEtCl_2H(PH_3)_3$, **4**, 283
TiC_4H_4
TiCHCHCHCH, **12**, 760, 764
$TiC_5H_5Cl_3$
$TiCl_3Cp$, **4**, 274, 291, 292
$TiC_5H_5N_3O_9$
$TiCp(NO_3)_3$, **4**, 306, 307
$TiC_6H_7Cl_3$
$TiCl_3(C_5H_4Me)$, **4**, 291, 292
$TiC_6H_8Cl_2$
$TiMeCl_2Cp$, **4**, 293, 297
$TiC_6H_8Cl_2O$
$Ti(OMe)Cl_2Cp$, **4**, 293
TiC_6O_6
$[Ti(CO)_6]^{2-}$, **4**, 218
$TiC_7H_8Cl_2OS_2$
$TiCl_2(S_2COMe)Cp$, **4**, 310
$TiC_7H_{10}Cl_2$
$TiEtCl_2Cp$, **4**, 293, 297
$TiC_7H_{10}Cl_2S$
$TiCl_2(SEt)Cp$, **4**, 309
$TiC_7H_{11}Cl$
$TiClMe_2Cp$, **4**, 277
$TiC_7H_{11}Cl_2N$
$TiCl_2(NHEt)Cp$, **4**, 299
$TiC_7H_{11}Cl_2NO$
$TiCl_2Cp(ONMe_2)$, **4**, 302
$TiC_7H_{12}Cl_2N_2$
$TiCl_2Cp[N(H)NMe_2]$, **4**, 300
$TiC_7H_{18}Cl_3DP_2$
$Ti(CH_2D)Cl_3(dmpe)$, **4**, 283
$TiC_8H_{14}Cl_2N_2$
$TiCl_2Cp[N(Me)NMe_2]$, **4**, 300
$TiC_8H_{14}S_3$
$Ti(SMe)_3Cp$, **4**, 310
$TiC_9H_{11}ClO_2S_4$
$TiCl(S_2COMe)_2Cp$, **4**, 310
$TiC_9H_{13}Cl_2O$
$TiCl_2Cp(THF)$, **4**, 244
$TiC_9H_{13}Cl_3$
$TiCl_3(C_5HMe_4)$, **4**, 290
$TiC_9H_{14}Cl_2N_2$
$TiCl_2Cp(NNBu^t)$, **4**, 300
$TiC_9H_{15}Cl_2N$
$TiCl_2(NHBu^t)Cp$, **4**, 299
$TiC_9H_{21}ClO_3$
$ClTi(OPr^i)_3$, **12**, 83
$TiCl(OPr^i)_3$, **1**, 82
$TiC_{10}H_{10}$
Cp_2Ti, **1**, 482; **12**, 324
$TiC_{10}H_{10}Cl_2$
Cp_2TiCl_2, **1**, 69, 86; **12**, 328, 592, 1194
$TiC_{10}H_{10}F_2$
TiF_2Cp_2, **4**, 353
$TiC_{10}H_{10}F_6P_2$
$TiCp_2(PF_3)_2$, **4**, 225, 226
$TiC_{10}H_{10}I_2$
TiI_2Cp_2, **4**, 344
$TiC_{10}H_{10}N_2O_2S_2$
$Ti(NSO)_2Cp_2$, **4**, 399
$TiC_{10}H_{10}N_4$
$TiCp_2(N_2)_2$, **4**, 226
$TiC_{10}H_{10}N_4S_3$
$Ti(N_4S_3)Cp_2$, **4**, 400
$TiC_{10}H_{10}N_6$
$Ti(N_3)_2Cp_2$, **4**, 398
$TiC_{10}H_{10}O_4Se$
$Ti(SeO_4)Cp_2$, **4**, 401
$TiC_{10}H_{10}S_5$
TiS_5Cp_2, **4**, 411, 533
$TiC_{10}H_{10}Se_5$
$TiSe_5Cp_2$, **4**, 411, 412, 417
$TiC_{10}H_{11}$
Cp_2TiH, **1**, 69
$TiC_{10}H_{11}ClSe$
$TiCl(SeH)Cp_2$, **4**, 412
$TiC_{10}H_{12}S_2$
$Ti(SH)_2Cp_2$, **4**, 412
$TiC_{10}H_{12}Se$
$Ti(SeH)_2Cp_2$, **4**, 412
$TiC_{10}H_{14}Cl_2O_2$
$Ti(CH_2CH_2CO_2Et)Cl_2Cp$, **4**, 293
$TiC_{10}H_{14}O_6S$
$[TiCp_2(H_2O)_2][SO_4]$, **4**, 401
$TiC_{10}H_{15}Cl_2$
$TiCl_2Cp^*$, **4**, 244
$TiC_{10}H_{15}Cl_3$
$TiCl_3Cp^*$, **4**, 293
$TiC_{10}H_{18}Cl_4N_2$
$TiCl_4(Bu^tNC)_2$, **4**, 288
$TiC_{10}H_{20}Cl_2N_2$
$TiCl_2(Bu^tNCHCHNBu^t)$, **4**, 281
$TiC_{11}H_5Cl_2F_5S$
$TiCl_2(SC_6F_5)Cp$, **4**, 310
$TiC_{11}H_9ClO_2$
$TiCl(O_2C_6H_4)Cp$, **4**, 306
$TiC_{11}H_9Cl_3O$
$TiCl_2(ClC_6H_4O)Cp$, **4**, 308
$TiC_{11}H_{10}BrNO$
$Ti(NCO)BrCp_2$, **4**, 403
$TiC_{11}H_{10}Cl_2$
$TiCl_2[(CH_2)(C_5H_4)_2]$, **4**, 340
$TiC_{11}H_{10}Cl_2N_2$
$TiCl_2Cp(NNPh)$, **4**, 300, 302
$TiC_{11}H_{10}Cl_2O$
$TiCl_2(OPh)Cp$, **4**, 293
$TiC_{11}H_{10}F_3OP$
$TiCp_2(CO)(PF_3)$, **4**, 225
$TiC_{11}H_{10}O$
$TiCp_2(CO)$, **4**, 226
$TiC_{11}H_{11}NS_2$
$Ti(S_2CNHCp)Cp$, **4**, 310
$TiC_{11}H_{12}$
$Cp_2Ti{=}CH_2$, **4**, 274; **12**, 1211
$TiC_{11}H_{12}Cl_2N_2$
$TiCl_2Cp[N(Ph)NH_2]$, **4**, 300
$TiC_{11}H_{13}$
$[Cp_2TiMe]^+$, **12**, 1195
$TiC_{11}H_{13}Cl$
$Ti(Me)ClCp_2$, **4**, 362
$TiC_{11}H_{13}I$
$TiCp_2(Me)(I)$, **4**, 231
$TiC_{11}H_{15}$
$Ti(allyl)_2Cp$, **4**, 245
$TiC_{11}H_{15}Cl_3$
$TiCl_3(C_5Me_4CH{=}CH_2)$, **4**, 291
$TiC_{11}H_{16}O_5P_2$
$Ti(CO)_5(dmpe)$, **4**, 216
$TiC_{11}H_{17}Cl_3$
$TiCl_3(C_5Me_4Et)$, **4**, 291, 292
$TiC_{11}H_{17}NO_3$
$Ti[(OCH_2CH_2)_3N]Cp$, **4**, 307
$TiC_{11}H_{18}O_5P_2$
$Ti(CO)_5(PMe_3)_2$, **4**, 216, 217
$TiC_{12}H_{10}$
$Ti(C_{12}H_{10})$, **4**, 214
$TiC_{12}H_{10}F_6O_6S_2$
$Ti(O_3SCF_3)_2Cp_2$, **4**, 275
$TiC_{12}H_{10}N_2O_2$
$Ti(NCO)_2Cp_2$, **4**, 226, 399
$TiC_{12}H_{10}N_2S_2$
$Ti(NCS)_2Cp_2$, **4**, 399
$TiC_{12}H_{10}O_2$
$TiCp_2(CO)_2$, **4**, 224, 226, 275
$TiC_{12}H_{11}ClS_2$
$TiCl(S_2C_6H_3Me)Cp$, **4**, 310
$TiC_{12}H_{12}$
$Ti(C_6H_6)_2$, **4**, 214
$TiCp(C_7H_7)$, **4**, 242
$TiC_{12}H_{12}Cl$
$TiCl[C_2H_4(C_5H_4)_2]$, **4**, 247
$TiC_{12}H_{13}Cl$
$Ti(CHCH_2)(Cl)Cp_2$, **4**, 233
$TiC_{12}H_{13}ClO$
$TiCp_2(COMe)(Cl)$, **4**, 231
$TiC_{12}H_{14}$
$(MeC_5H_4)_2Ti$, **12**, 1196
$TiCp_2(C_2H_4)$, **4**, 231, 234
$TiC_{12}H_{14}SSe_4$
$Ti(SSe_4)(C_5H_4Me)_2$, **4**, 412
$TiC_{12}H_{14}S_5$
$TiS_5(C_5H_4Me)_2$, **4**, 411
$TiC_{12}H_{16}$
Cp_2TiMe_2, **12**, 594
$TiMe_2Cp_2$, **2**, 99; **4**, 361, 364, 365
$TiC_{12}H_{16}ClNS$
$TiCl(SCH_2CH_2NH_2)Cp_2$, **4**, 642
$TiC_{12}H_{16}S_2$
$Ti(SMe)_2Cp_2$, **4**, 378, 414

$TiC_{12}H_{17}Cl_2F_3O$
$TiCl_2(OCH_2CF_3)Cp^*$, **4**, 292
$TiC_{12}H_{17}Cl_2NS$
$TiCl(SCH_2CH_2NH_3Cl)CP_2$, **4**, 414
$TiC_{12}H_{19}Cl_3$
$TiCl_3(C_5Me_4Pr^i)$, **4**, 290
$TiC_{12}H_{28}O_4$
$Ti(OPr^i)_4$, **1**, 69; **2**, 107
$Ti(OPr^i)_4$, **12**, 83
$TiC_{13}H_{15}$
$Cp_2Ti(CH_2=CHCH_2)$, **1**, 483
$Ti(C_3H_5)Cp_2$, **4**, 255, 257, 274
$TiC_{13}H_{15}Cl$
$Cp_2TiCl(CH=CHMe)$, **12**, 593
$TiC_{13}H_{15}N$
$Cp_2TiCH_2CMe=N$, **12**, 592
$TiC_{13}H_{16}$
$TiCp_2(=CH_2)(C_2H_4)$, **4**, 377
$TiCp_2(CH_2)_3$, **4**, 377
$TiC_{13}H_{16}O$
$TiMe(OCH=CH_2)Cp_2$, **4**, 404
$TiC_{13}H_{18}O$
$Ti(C_5H_6Me)_2(CO)$, **4**, 239
$TiC_{13}H_{19}ClP$
$TiClCp_2(PMe_3)$, **4**, 265
$TiC_{13}H_{21}Cl_3$
$TiCl_3(C_5H_3Bu^t{}_2)$, **4**, 290
$TiC_{13}H_{21}N_3O_4$
$Ti(CO)_4(Me_3tacn)$, **4**, 217, 218
$TiC_{13}H_{21}O_2P_2$
$[TiCp(dmpe)(CO)_2]^-$, **4**, 215
$TiC_{13}H_{22}O_2P_2$
$TiCpH(dmpe)(CO)_2$, **4**, 215, 216, 226
$TiC_{13}H_{24}$
$TiBu^tCp^*$, **4**, 294, 296
$TiC_{13}H_{27}O_4P_3$
$Ti(CO)_4(PMe_3)_3$, **4**, 216
$TiC_{13}H_{30}O_3$
$TiMe(OBu^t)_3$, **4**, 284
$TiC_{14}H_{14}$
$Ti(C_4H_4)Cp_2$, **4**, 362
$TiC_{14}H_{16}$
$Ti(C_7H_7)(C_7H_9)$, **4**, 242
$TiCp_2(C_2Me_2)$, **4**, 237
$Ti(PhMe)_2$, **4**, 214
$TiC_{14}H_{16}N_2$
$[TiCp_2(MeCN)_2]^+$, **4**, 231, 367
$TiC_{14}H_{16}N_2O$
$Ti(NCO)(NCMe_2)Cp_2$, **4**, 227
$TiC_{14}H_{16}O_2$
$Ti(OCH=CH_2)_2Cp_2$, **4**, 404
$TiC_{14}H_{17}$
$Cp_2TiCH_2CMeCH_2$, **12**, 592
$Ti[C(Me)CHMe]Cp_2$, **4**, 253
$TiC_{14}H_{18}$
$TiMe_2[C_2H_4(C_5H_4)_2]$, **4**, 358
$TiC_{14}H_{18}N_2$
$Ti(C_5H_3NMe_2)_2$, **4**, 213
$TiC_{14}H_{19}ClN_2S$
$TiCl(NSNBu^t)Cp_2$, **4**, 399
$TiC_{14}H_{19}OP$
$TiCp_2(CO)(PMe_3)$, **4**, 226
$TiC_{14}H_{20}ClP$
$TiCl(CH=PMe_3)Cp_2$, **4**, 359
$TiC_{14}H_{21}P$
$Ti(CH_2)Cp_2(PMe_3)$, **4**, 376
$TiC_{14}H_{23}Cl_2O$
$TiCl_2Cp^*(THF)$, **4**, 245
$TiC_{14}H_{23}N_3S_6$
$Ti(S_2CNMe_2)_3Cp$, **4**, 310
$TiC_{14}H_{24}Cl_2O$
$TiCl_2(OMe)(C_5H_3Bu^t{}_2)$, **4**, 306
$TiC_{14}H_{28}P_2$
$Ti(C_4H_6)_2(dmpe)$, **4**, 283
$TiC_{14}H_{34}N_3$
$TiMe_2[N(CH_2CH_2NEt_2)_2]$, **4**, 244
$TiC_{14}H_{36}Cl_2N_6P_2$
$TiCl_2[C=P(NMe_2)_3]_2$, **4**, 281
$TiC_{15}H_{15}Cl$
$TiClCp_3$, **4**, 363
$TiC_{15}H_{17}$
$Ti(C_3H_5)[C_2H_4(C_5H_4)_2]$, **4**, 255
$TiC_{15}H_{17}NO$
$TiCp'_2(CO)(NCMe)$, **4**, 226
$TiC_{15}H_{19}$
$Ti(C_3H_3Me_2)Cp_2$, **4**, 255
$TiC_{15}H_{19}O_2$
$TiCp_2(O_2CBu^t)$, **4**, 258
$TiC_{15}H_{19}P$
$TiCp_2(Bu^tCP)$, **4**, 234
$TiC_{15}H_{20}F_2$
TiF_2CpCp^*, **4**, 314
$TiC_{15}H_{20}NO_2$
$TiCp_2(O_2CNEt_2)$, **4**, 258
$TiC_{15}H_{21}ClO_4$
$TiCl(acac)_2Cp$, **4**, 306
$TiC_{15}H_{21}P$
$TiCp_2(HCCH)(PMe_3)$, **4**, 233
$TiC_{15}H_{28}P_2$
$TiEt(C_7H_7)(dmpe)$, **4**, 242
$TiC_{15}H_{31}ClN_2$
$TiCl(CH_2Bu^t)$-$(Bu^tNCHCHNBu^t)$, **4**, 281
$TiC_{15}H_{31}O_4P_3$
$Ti(CO)_4(trmpe)$, **4**, 216
$TiC_{15}H_{32}O_3P_4$
$Ti(CO)_3(dmpe)_2$, **4**, 216
$TiC_{16}H_{14}Te_2$
$Ti(Te_2C_6H_4)Cp_2$, **4**, 416
$TiC_{16}H_{16}ClNS_2$
$TiCl(S_2CNHCp)Cp_2$, **4**, 310
$TiC_{16}H_{17}$
$Ti(C_{12}H_8Bu^t)$, **4**, 214
$TiC_{16}H_{18}N_2$
$Ti[(NH)_2C_6H_4]Cp_2$, **4**, 399
$TiC_{16}H_{19}NO$
$TiCp_2(CO)(CNBu^t)$, **4**, 226
$TiC_{16}H_{20}$
$Cp_2Ti(CH[CH_2CH_2])_2$, **12**, 594
$TiC_{16}H_{20}Cl_2$
$TiCl_2[C_2Me_4(C_5H_4)_2]^-$, **4**, 248, 340
$TiC_{16}H_{21}$
$Ti(MeCHCHCHEt)Cp_2$, **4**, 256
$TiC_{16}H_{21}NO$
$Cp_2TiOCMe_2CH_2CMe=N$, **12**, 592
$[TiCp_2(THF)MeCN]^+$, **4**, 367
$TiC_{16}H_{22}Cl_2$
$TiCl_2(C_5H_4Pr^i)_2$, **4**, 346, 347
$TiC_{16}H_{22}P_2$
$Ti(C_5H_4PMe_2)(C_7H_6PMe_2)$, **4**, 242
$TiC_{16}H_{24}$
$Ti(C_6H_6Me_2)_2$, **4**, 240
$TiC_{16}H_{24}S_2$
$Ti(SPr)_2Cp_2$, **4**, 417
$TiC_{16}H_{25}$
$Ti(allyl)_2Cp^*$, **4**, 245
$TiC_{16}H_{26}$
$Ti(C_4H_4Bu^t)_2$, **4**, 218
$TiC_{16}H_{28}P_2$
$Cp_2Ti(PMe_3)_2$, **12**, 325, 332, 333
$TiCp_2(PMe_3)_2$, **4**, 230, 248
$TiC_{16}H_{29}P$
$Ti(C_5H_7)_2(PEt_3)$, **4**, 240
$TiC_{16}H_{36}O$
$TiMe_3(OCBu^t{}_3)$, **4**, 284
$TiC_{17}H_{15}ClO_2$
$TiCl(OPh)_2Cp$, **4**, 308
$TiC_{17}H_{15}O_2$
$TiCp_2(O_2CPh)$, **4**, 258
$TiC_{17}H_{16}S_2$
$Ti(S_2C_6H_3Me)Cp_2$, **4**, 417
$TiC_{17}H_{17}ClS$
$TiCl(MeC_6H_4S)Cp_2$, **4**, 353
$TiC_{17}H_{18}Cl_2$
$TiCl_2[(CH_2)_5C_2(C_5H_4)_2]$, **4**, 350
$TiC_{17}H_{20}ClO$
$TiCl[C_2Me_4(C_5H_4)_2](CO)$, **4**, 248
$TiC_{17}H_{20}O_2$
$TiCpCp^*(CO)_2$, **4**, 224
$TiC_{17}H_{22}$
$TiCp^*(C_7H_7)$, **4**, 242
$TiC_{17}H_{25}NO_4$
$[Et_4N][TiCp(CO)_4]$, **4**, 215
$TiC_{17}H_{25}OP$
$TiCp_2(CO)(PEt_3)$, **4**, 226
$TiC_{17}H_{25}P$
$TiCp_2(C_2Me_2)(PMe_3)$, **4**, 237
$TiC_{17}H_{28}$
$TiMeCp^*[CH_2C(Me)C$-$(Me)CH_2]$, **4**, 222
$TiC_{17}H_{28}OP_2$
$TiCp_2(CO)(PMe_3)_2$, **4**, 231
$TiC_{18}H_{14}O_8$
$Ti(O_2CCH=CCO_2H)_2Cp_2$, **4**, 409
$TiC_{18}H_{16}S_2$
$Ti(C_4H_3S)_2Cp_2$, **4**, 359
$TiC_{18}H_{18}$
$Ti(CpMe_2)_2$, **4**, 239
$TiC_{18}H_{18}S_2$
$Ti(S_2C_6H_3Me)Cp(C_5H_4Me)$, **4**, 416
$TiC_{18}H_{23}$
$TiCp^*(C_8H_8)$, **4**, 266
$TiC_{18}H_{23}Cl_3$
$TiCl_3[C_5Me_4C(H)(Et)Ph]$, **4**, 291
$TiC_{18}H_{26}$

$Ti(CH_2CHBu^tCH_2)Cp(C_5H_4Me)$, **4**, 377
$TiC_{18}H_{26}Cl$
$TiCl(C_5HMe_4)_2$, **4**, 247
$TiC_{18}H_{26}O_2$
$[TiCp_2(THF)_2]^+$, **4**, 367
$TiC_{18}H_{28}$
$Ti(CH_2CHCHMe)Cp^*(C_4H_6)$, **4**, 222
$TiC_{18}H_{28}N_4S_2$
$Ti(NSNBu^t)_2Cp_2$, **4**, 399
$TiC_{18}H_{31}ClO_2P_2$
$TiClCp^*(CO)_2(dmpe)$, **4**, 222
$TiC_{19}H_{19}$
$TiBz_2Cp$, **4**, 245
$TiC_{19}H_{19}ClO$
$Cp_2TiCl(OC(=CH_2)CH_2Ph)$, **12**, 591
$TiC_{19}H_{21}OPS_3$
$TiCp'_2(S)_2P(S)(C_6H_4OMe)$, **4**, 229
$TiC_{19}H_{23}P$
$Ti(CH_2)Cp_2(PMe_2Ph)$, **4**, 387
$TiC_{19}H_{29}O_2$
$[TiMeCp_2(THF)_2]^+$, **4**, 394
$TiC_{20}H_{14}O_2$
$Ti(ind)_2(CO)_2$, **4**, 224
$TiC_{20}H_{16}$
$(C_{10}H_8)_2Ti$, **4**, 213
$TiC_{20}H_{18}N_2$
$TiCp_2(bipy)$, **4**, 239
$TiC_{20}H_{19}ClO_2$
$TiCl[Me_2C(C_6H_4O)_2]Cp$, **4**, 306
$TiC_{20}H_{20}$
$TiCp_4$, **4**, 363
$TiC_{20}H_{20}Br_2$
$[Ti(Br)Cp_2]_2$, **4**, 247
$TiC_{20}H_{20}Cl_2O$
$Ti_2Cl_2(O)Cp_4$, **4**, 401
$TiC_{20}H_{20}Cl_2Se_2$
$(TiClCp_2)_2(Se_2)$, **4**, 412
$TiC_{20}H_{20}Se_2$
$TiCp_4(Se)_2$, **4**, 227
$TiC_{20}H_{20}Se_6$
$Ti_4Cp_4(Se)_6$, **4**, 412
$TiC_{20}H_{23}$
$(TiHCp_2)_2(H)$, **4**, 260
$TiC_{20}H_{24}OP$
$[Ti(COMe)(PMe_2Ph)Cp_2]^+$, **4**, 366
$TiC_{20}H_{26}Cl_2$
$TiCl_2[C_2H_4(C_5Me_4)_2]$, **4**, 340
$TiC_{20}H_{28}ClO$
$[TiCl[C_2Me_4(C_5H_4)_2](THF)]^-$, **4**, 248
$TiC_{20}H_{30}Br$
$TiBrCp^*_2$, **4**, 247
$TiC_{20}H_{30}Cl$
$TiClCp^*_2$, **4**, 247, 249
$TiC_{20}H_{30}Cl_2$
$TiCl_2Cp^*_2$, **4**, 344
$TiC_{20}H_{30}F_2$
$TiF_2Cp^*_2$, **4**, 314
$TiC_{20}H_{30}I$
$TiCp^*_2I$, **4**, 247
$TiC_{20}H_{30}I_2$
$TiI_2Cp^*_2$, **4**, 344
$TiC_{20}H_{30}P_6$
$Ti_2Cp^*_2(P_6)$, **4**, 229
$TiC_{20}H_{30}S_3$
$TiS_3Cp^*_2$, **4**, 229, 411
$TiC_{20}H_{32}$
$TiMe_2(C_5HMe_4)_2$, **4**, 363
$TiC_{20}H_{32}S_2$
$Ti(SH)_2Cp^*_2$, **4**, 412
$TiC_{20}H_{32}Se_2$
$TiCp^*_2(SeH)_2$, **4**, 227
$TiC_{20}H_{36}O_{12}$
$Ti(dipt)_2$, **12**, 1110
$TiC_{20}H_{44}$
$Ti(CH_2Bu^t)_4$, **4**, 284
$TiC_{21}H_{20}O_3$
$Ti(allyl)(OPh)_3$, **4**, 274, 275
$TiC_{21}H_{27}$
Bz_3Ti, **4**, 244
$TiC_{21}H_{31}O_2$
$TiCp^*_2(O_2CH)$, **4**, 258
$TiC_{21}H_{32}$
$Ti(=CH_2)Cp^*_2$, **4**, 369
$TiC_{21}H_{33}$
$TiMeCp^*_2$, **4**, 251
$TiC_{21}H_{33}Cl$
$Ti(Cl)MeCp^*_2$, **4**, 357
$TiC_{21}H_{33}F_3O_5S$
$[Ti(OH)Cp^*_2(H_2O)][OTf]$, **4**, 401, 402
$TiC_{22}H_{10}F_{10}$
$Ti(C_6F_5)_2Cp_2$, **4**, 364
$TiC_{22}H_{18}F_6N_2O_6S_2$
$[TiCp_2(bipy)][OTf]_2$, **4**, 399
$TiC_{22}H_{19}Cl_2P$
$TiCl_2Cp(C_5H_4PPh_2)$, **4**, 356
$TiC_{22}H_{20}$
Cp_2TiPh_2, **12**, 1197
$TiPh_2Cp_2$, **4**, 370
$TiC_{22}H_{20}N_2$
$TiCp_2(N_2Ph_2)$, **4**, 398
$TiC_{22}H_{20}O_2$
$Ti(OPh)_2Cp_2$, **4**, 401, 402
$TiC_{22}H_{20}S_2$
$TiCp_2(SPh)_2$, **4**, 231, 417
$TiC_{22}H_{24}$
$TiMe_2[C_2H_4(ind)_2]$, **4**, 340
$TiC_{22}H_{27}Si$
$[Cp_2TiC(TMS)C(Me)(Ph)]^+$, **12**, 1195
$TiC_{22}H_{30}$
$TiMe_2[C_2H_4(C_9H_{10})_2]$, **4**, 408
$TiC_{22}H_{30}O_2$
$TiCp^*_2(CO)_2$, **4**, 224, 226
$TiC_{22}H_{33}O_2$
$TiCp^*_2(O_2CMe)$, **4**, 258
$TiC_{22}H_{34}$
$TiCp^*_2(C_2H_4)$, **4**, 234, 237
$TiC_{22}H_{34}F_6O_8S_2$
$[TiCp^*_2(H_2O)_2][OTf]_2$, **4**, 275, 401
$TiC_{22}H_{35}Cl$
$Ti(Et)ClCp^*_2$, **4**, 357
$TiC_{22}H_{35}NO_4$
$[Et_4N][TiCp^*(CO)_4]$, **4**, 215
$TiC_{22}H_{36}$
$TiMe_2Cp^*_2$, **4**, 363
$TiC_{23}H_{20}O_3$
$Ti(OPh)_3Cp$, **4**, 307
$TiC_{23}H_{20}S_3$
$Ti(SPh)_3Cp$, **4**, 310
$TiC_{23}H_{23}F_5$
$Ti(allyl)(C_6F_5)Cp(C_5H_4Bu^t)$, **4**, 274
$TiC_{23}H_{25}ClN_2O_2$
$TiClCp[(MeOC_6H_4)N=C(Me)-C(Me)=N(MeOC_6H_4)]$, **4**, 300
$TiC_{23}H_{27}ClN_4O_4$
$TiCl(C_5H_3Bu^t_2)(O_2CC_4H_3N_2)_2$, **4**, 306
$TiC_{23}H_{33}$
$Ti(CCMe)Cp^*_2$, **4**, 251, 252
$TiC_{23}H_{38}$
$Ti(Me)(Et)Cp^*_2$, **4**, 357
$TiC_{24}H_6ClD_{34}$
$TiCl(CD_2Me)_2[C_5(CD_3)_5]_2$, **4**, 364
$TiC_{24}H_{18}Cl_2O_4$
$Ti(O_2CC_6H_4Cl)_2Cp_2$, **4**, 407
$TiC_{24}H_{20}$
$TiCp_2(C_2Ph_2)$, **4**, 237, 238
$TiC_{24}H_{20}O_4$
$Ti(O_2CPh)_2Cp_2$, **4**, 406
$TiC_{24}H_{22}$
$Cp_2TiCH_2CPh=CPh$, **12**, 592
$TiC_{24}H_{24}$
$Cp_2Ti(CH_2Ph)_2$, **12**, 594
$Ti(Bz)_2Cp_2$, **4**, 362, 365
$TiPh_2(C_5H_4Me)_2$, **4**, 370
$TiC_{24}H_{24}S_2$
$Ti(CH_2SPh)_2Cp_2$, **4**, 359
$TiC_{24}H_{36}$
$Ti(CH=CH_2)_2Cp^*_2$, **4**, 359
$TiC_{24}H_{37}ClF_3NO_4S$
$[TiClCp^*_2(DMF)[OTf]$, **4**, 402
$TiC_{24}H_{38}$
$Ti(C_4H_8)Cp^*_2$, **4**, 234
$TiC_{25}H_{24}ClNS_2$
$TiCl[S_2CN(Bz)_2]Cp_2$, **4**, 416
$TiC_{25}H_{41}$
$Ti(CH_2Bu^t)Cp^*_2$, **4**, 251
$TiC_{26}H_{20}$
$[TiCp_2(CCPh)_2]^-$, **4**, 254
$TiC_{26}H_{22}N_2$
$Ti(NC_8H_6)_2Cp_2$, **4**, 398
$TiC_{26}H_{24}O_6$
$Ti(O_2CC_6H_4OMe)_2Cp_2$, **4**, 407
$TiC_{26}H_{29}Cl$
$TiClCp^*(Ph_2C_4H_4)$, **4**, 223
$TiC_{26}H_{34}Se_2$
$Ti(Se_2C_6H_4)Cp^*_2$, **4**, 416
$TiC_{26}H_{38}N_2P_2$
$Ti_2Cp_4(PMe_3)_2(N_2)$, **4**, 231
$TiC_{26}H_{42}Cl$
$TiCl(C_5H_3Bu^t_2)_2$, **4**, 247
$TiC_{26}H_{42}Cl_2$
$TiCl_2(C_5H_3Bu^t_2)_2$, **4**, 354
$TiC_{26}H_{43}N$
$TiMe[NC(H)Bu^t]Cp^*_2$, **4**, 257
$TiC_{27}H_{37}NS_4$

[NEt4][Ti(S2C6H3Me)2Cp], **4**, 310

$TiC_{27}H_{38}$
Ti(Me)(Ph)Cp*2, **4**, 364

$TiC_{27}H_{38}N$
Ti(NMePh)Cp*2, **4**, 257

$TiC_{28}H_{20}O_4$
Ti(O2CC≡CPh)2Cp2, **4**, 408

$TiC_{28}H_{28}$
Bz4Ti, **4**, 284, 287

$TiC_{28}H_{28}N_2$
TiCp2[CN(Me2C6H3)]2, **4**, 224

$TiC_{28}H_{31}P$
Ti(CHCHCPh2)(PMe3)Cp2, **7**, 341

$TiC_{28}H_{40}$
Ti(Me)(Bz)Cp*2, **4**, 364

$TiC_{28}H_{44}F_6N_2O_8S_2$
[TiCp*2(DMF)2][OTf]2, **4**, 402

$TiC_{28}H_{60}O_2$
TiMe2[O(CBu^t_3)]2, **4**, 281, 284

$TiC_{29}H_{21}Cl_2$
TiCl2(C5HPh4), **4**, 311

$TiC_{29}H_{25}OP$
TiCp2(CO)(PPh3), **4**, 226

$TiC_{29}H_{30}$
TiCpCp*(C2Ph2), **4**, 237

$TiC_{30}H_{10}S_5$
TiS5Cp2, **4**, 417

$TiC_{30}H_{30}O$
TiCpCp*(CO)(C2Ph2), **4**, 226, 237

$TiC_{30}H_{30}Se_8$
Ti6Cp6(Se)8, **4**, 227

$TiC_{30}H_{32}N_2O_2$
TiBzCp[(MeOC6H4)N=C-(Me)C(Me)=N-(MeOC6H4)], **4**, 300

$TiC_{30}H_{42}$
Ti(CpBu^t_2)2, **4**, 239

$TiC_{31}H_{28}O_3$
Ti(OCMePh2)(O2CPh)Cp2, **4**, 403

$TiC_{31}H_{30}O_6$
TiCp*(O2CPh)3, **4**, 306

$TiC_{31}H_{36}$
Ti(Bz)3Cp*, **4**, 293, 296

$TiC_{31}H_{42}O$
Ti(CpBu^t_2)2(CO), **4**, 239

$TiC_{33}H_{40}N$
Ti(N=CPh2)Cp*2, **4**, 399

$TiC_{34}H_{58}N_2$
Ti(C5H2NBu^t_3)2, **4**, 213

$TiC_{36}H_{28}$
Ti(C4Ph4)(C8H8), **4**, 241

$TiC_{36}H_{34}P_2$
Ti(CH2PPh2)2Cp2, **4**, 365

$TiC_{36}H_{60}$
Ti(C6H3Bu^t_3)2, **4**, 213

$TiC_{38}H_{30}$
Ti(C4Ph4)Cp2, **4**, 242

$TiC_{38}H_{32}O_2$
TiMe2[O(Ph2C6H3)]2, **4**, 281

$TiC_{38}H_{36}Cl_2P_2$
TiCl2[C5H4(CH2)2PPh2]2, **4**, 356

$TiC_{38}H_{48}O$
TiBz2[O(Pr^i_2C6H3)2], **4**, 285

$TiC_{40}H_{30}Cl_2$
TiCl2Cp(C5Ph5), **4**, 346

$TiC_{40}H_{31}Cl_2$
TiCl2Cp(C5Ph5H), **4**, 311

$TiC_{41}H_{44}P_2$
Ti(CH2PPh2)2CpCp*, **4**, 359

$TiC_{45}H_{39}O_4P_3$
Ti(CO)4(triphos), **4**, 217

$TiC_{50}H_{40}O_2$
Ti(Bz)2[O(Ph2C6H3)]2, **4**, 281

$TiC_{58}H_{42}Cl$
TiCl(C5HPh4)2, **4**, 247

$TiC_{58}H_{42}Cl_2$
TiCl2(C5HPh4)2, **4**, 346

$TiCl_3$
TiCl3, **12**, 1073

$TiCl_4$
TiCl4, **1**, 69; **12**, 578, 1073, 1182, 1184

$TiCoC_{14}H_{10}O_4$
TiCp2[OCCo(CO)3], **4**, 258

$TiCoC_{16}H_{27}O_7$
Ti(OBu^t)3Co(CO)4, **10**, 27

$TiCoC_{20}H_{20}Cl_2$
[CoCp2][TiCl2Cp2], **4**, 248

$TiCoC_{22}H_{26}O_6$
[TiCp2(THF)2][Co(CO)4], **4**, 259

$TiCoC_{24}H_{30}O_4$
TiCp*2[Co(CO)4], **4**, 227

$TiCo_2C_{28}H_{30}O_8$
TiCp*2[OCCo(CO)3]2, **4**, 405

$TiCo_3C_{20}H_{10}O_{10}$
TiCp2[OCCo3(CO)9], **4**, 258

$TiCo_3C_{22}H_{14}O_{10}$
TiCp'2[OCCo3(CO)9], **4**, 258

$TiCo_6C_{30}H_{10}O_{20}$
TiCp2[OCCo3(CO)9]2, **4**, 227, 405

$TiCr_2C_{10}H_{14}O_9$
[TiCp2(H2O)2][Cr2O7], **4**, 401

$TiCr_2Re_2C_{36}H_{14}O_{18}$
Re(CO)5Cr(CO)4(CTol)Ti-(CTol)Cr(CO)4Re(CO)5, **10**, 45

$TiFeC_{13}H_{23}N_3O_2$
Ti(NMe2)3Fe(CO)2Cp, **10**, 27

$TiFe_2C_{16}H_{10}O_6Te_2$
TiCp2(Te)2Fe2(CO)6, **4**, 416

$TiGeC_4H_{11}Cl_3$
Me3GeCH2TiCl3, **12**, 579

$TiGeC_{12}H_{10}Cl_2F_6$
TiCl[Ge(CF3)2Cl]Cp2, **4**, 397

$TiGeC_{29}H_{28}$
TiMe(GePh3)Cp2, **4**, 397

$TiGeC_{30}H_{28}O$
Ti(COMe)(GePh3)Cp2, **4**, 397

$TiMnC_{25}H_{22}ClO_3$
(CO)2CpMn=C(Tol)OTiCp2Cl, **12**, 567

$TiMnC_{33}H_{32}Cl_2O_2P$
TiCl2Cp[C5Me4PPh2MnCp-(CO)2], **4**, 356

$TiMn_2C_{22}H_{34}O_{10}$
Mn2(CO)9($\overline{CCH_2CH_2TiCp*_2O}$), **6**, 8

$TiMoC_{22}H_{23}O_4$
Cp2Ti(THF)(CO)Mo(CO)2Cp, **5**, 399

$TiMo_5C_5H_5O_{18}$
[TiCp(Mo5O18)]^{3−}, **4**, 308

$TiPdC_{20}H_{26}ClP$
TiCp2(CH2)(Cl)PdMe-(PMe2Ph), **4**, 387

$TiPtC_{20}H_{26}ClP$
TiCp2(CH2)(Cl)PtMe(PMe2Ph), **4**, 387

$TiPtC_{21}H_{29}P$
TiCp2(CH2)MePtMe(PMe2Ph), **4**, 387

$TiRu_2W_5C_{78}H_{154}N_4O_{40}$
[NBu4]4[Ru2(TiW5CpO18)2-(CO)4], **7**, 633

$TiSbC_{12}H_{13}Cl_7N$
[TiCl(MeCN)Cp2][SbCl6], **4**, 398

$TiSb_2C_{14}H_{16}Cl_{12}N_2$
[Ti(MeCN)2Cp2][SbCl6]2, **4**, 398

$TiSiC_4H_{11}Cl_3$
TMS-CH2TiCl3, **12**, 579

$TiSiC_{12}H_{14}Cl_2$
TiCl2[Me2Si(C5H4)2], **4**, 345

$TiSiC_{12}H_{16}S_2$
Ti(S2SiMe2)Cp2, **4**, 416

$TiSiC_{13}H_{18}F_2$
TiF2Cp[C5H4(TMS)], **4**, 314

$TiSiC_{14}H_{33}O_4P_3$
Ti[Bu^tSi(CH2PMe2)3CO4], **4**, 214, 215

$TiSiC_{15}H_{21}$
[Ti[C(TMS)=CH2]Cp2]^+, **4**, 284

$TiSiC_{15}H_{25}ClP$
TiClCp2[PMe2(TMS)], **4**, 265

$TiSiC_{17}H_{23}N_3O$
Ti(NCO)(NCN-TMS)(C5H4-Me)2, **4**, 400

$TiSiC_{17}H_{33}O_4P_3$
Ti(CO)4(trimpsi), **4**, 216

$TiSiC_{19}H_{26}P$
Ti(SiH2Ph)Cp2(PMe3), **4**, 265

$TiSiC_{19}H_{34}ClO$
TiCl(CH2TMS)[O-(Bu^t_2MeC6H2)], **4**, 280

$TiSiC_{20}H_{28}P$
Ti(SiHMePh)Cp2(PMe3), **4**, 265

$TiSiC_{22}H_{32}P$
Ti(SiH2Ph)Cp2(PEt3), **4**, 265

$TiSiC_{23}H_{41}P_3$
Ti(C10H8)[Bu^tSi(CH2PMe2)3], **4**, 214, 215

$TiSiC_{25}H_{30}P$
Ti(SiHPh2)Cp2(PMe3), **4**, 265

$TiSiV_3W_9C_{69}H_{149}N_4O_{40}$
[NBu4]4[TiCp(SiW9V3O40)], **4**, 308

$TiSi_2C_{11}H_{22}Cl_3$
TiCl3[C5H4(TMS)2], **4**, 291

$TiSi_2C_{11}H_{23}Cl_2N$
$TiCl_2[N(TMS)_2]Cp$, **4**, 300
$TiSi_2C_{15}H_{23}Cl$
$TiCl(TMS)[Me_2Si(C_5H_4)_2]$, **4**, 397
$TiSi_2C_{16}H_{28}N_4S_2$
$Ti(NSN\text{-}TMS)_2Cp_2$, **4**, 399
$TiSi_2C_{18}H_{32}$
$Cp_2Ti[CH_2(TMS)]_2$, **12**, 594
$TiMe_2Cp[C_5H_3(TMS)_2]$, **4**, 357
$TiSi_2C_{19}H_{31}N_2$
$[Ti[C(Me)=N\text{-}TMS](CN\text{-}TMS)Cp_2]^+$, **4**, 366
$TiSi_2C_{19}H_{54}O_3$
$Ti(OC_6H_{11})_3[C_5H_3(TMS)_2]$, **4**, 307
$TiSi_3C_{19}H_{38}$
$CpTi[CH_2(TMS)]_3$, **12**, 594
$TiSi_3C_{14}H_{29}Cl_2$
$TiCl_2[C_5H_2(TMS)_3]$, **4**, 290
$TiSi_3C_{14}H_{29}Cl_3$
$TiCl_3[C_5H_2(TMS)_3]$, **4**, 290
$TiSi_3C_{17}H_{38}$
$TiMe_3[C_5H_2(TMS)_3]$, **4**, 296
$TiSi_4C_{14}H_{42}N_2$
$TiMe_2[N(TMS)_2]_2$, **4**, 284
$TiSi_4C_{16}H_{44}$
$Ti(CH_2TMS)_4$, **4**, 284
$TiSi_4C_{22}H_{42}Cl_2$
$TiCl_2[C_5H_3(TMS)_2]_2$, **4**, 346
$TiSi_5C_{70}H_{60}$
$TiSi_5Ph_{10}Cp_2$, **4**, 397
$TiSi_6C_{19}H_{57}N_3$
$TiMe[N(TMS)_2]_3$, **4**, 280
$TiSnC_{28}H_{25}Cl$
$TiCl(SnPh_3)Cp_2$, **4**, 397
$TiWC_{26}H_{24}O_2$
$TiCp_2(CTol=CH_2)W(CO)_2Cp$, **10**, 27
$TiWC_{40}H_{34}O_4$
$Ti(C_5H_4Me)_2(PPh_2)_2W(CO)_4$, **4**, 410
$TiW_5C_5H_7O_{18}$
$[TiCp(W_5O_{18})H_2]^-$, **4**, 308
$TiZnC_{11}H_{12}ClI$
$TiCp_2(CH_2)ZnClI$, **4**, 376
$Ti_2AlBC_{20}H_{28}$
$(Cp_2TiH_2)_2AlH_2BH_2$, **1**, 290–2
$Ti_2AlC_{21}H_{27}$
$(TiCp_2)_2(AlH_4Me)$, **4**, 260
$Ti_2C_{10}H_{10}Cl_4O$
$(TiCl_2Cp)_2(O)$, **4**, 246
$Ti_2C_{12}H_{12}Cl_2N_4$
$[TiCl[NN(H)Ph)]]_2$, **4**, 302
$Ti_2C_{12}H_{24}Cl_8N_2O_6P_2$
$[TiCl_4[CNCH_2P(O)(OEt)_2]]_2$, **4**, 288
$Ti_2C_{18}H_{18}Cl_8N_2$
$[TiCl_4(CNC_6H_3Me_2)]_2$, **4**, 288
$Ti_2C_{20}H_{20}Cl_2$
$[TiClCp_2]_2$, **4**, 247, 275
$Ti_2C_{20}H_{20}O$
$Ti_2Cp_4(O)$, **4**, 257
$Ti_2C_{20}H_{20}Se_2$
$[Ti(Se)Cp_2]_2$, **4**, 412
$Ti_2C_{20}H_{30}Cl_4O$
$[TiCl_2Cp^*]_2(O)$, **4**, 304
$Ti_2C_{22}H_{18}Cl_4O_2$
$[TiClCp(OC_6H_4Cl)]_2$, **4**, 246
$Ti_2C_{22}H_{30}Cl_2O_2$
$[Ti(Cl)(C_7H_7)(THF)]_2$, **4**, 242
$Ti_2C_{24}H_{20}F_6O_5$
$Ti_2Cp_4(CF_3CO_2)_2(O)$, **4**, 233
$Ti_2C_{24}H_{26}N_2$
$[TiMe(NPh)Cp]_2$, **4**, 295
$Ti_2C_{24}H_{28}S_4$
$[Ti(C_5H_4Me)_2(S_2)]_2$, **4**, 413
$Ti_2C_{26}H_{32}S_4$
$(TiCp_2)[Ti(C_5H_4Pr^i)_2(S_2)_2]$, **4**, 413
$Ti_2C_{28}H_{20}O_8$
$[Ti(O_2CC\equiv CCO_2)Cp_2]_2$, **4**, 410
$Ti_2C_{28}H_{24}O_8$
$[Ti(O_2CCH=CHCO_2)Cp_2]_2$, **4**, 410
$Ti_2C_{28}H_{42}Cl_4O_4$
$[TiCl_2(C_5H_3Bu^t_2)]_2(C_2O_4)$, **4**, 306
$Ti_2C_{28}H_{60}O_{12}$
$Ti_2(OPr^i_6)(dipt)$, **12**, 1110
$Ti_2C_{40}H_{60}Se_2$
$Ti_2Cp_4^*(Se)_2$, **4**, 227
$Ti_2FeC_{38}H_{34}Cl_4N_2O$
$[[TiCp_2(PhCH=CHCN)]_2(O)]\text{-}[FeCl_4]$, **4**, 398
$Ti_2Mo_2C_{24}H_{28}O_8$
$[TiMeO_4(C_5H_4Me)_2]_2$, **4**, 405
$Ti_2Mo_2C_{36}H_{30}O_6$
$[Cp_2Ti(CO)Mo(CO)_2Cp]_2$, **5**, 399, 402
$Ti_2Mo_2C_{54}H_{70}O_4$
$Mo_2(CO)_2(COTiCp^*_2)_2Cp_2$, **5**, 399, 402
$Ti_2Si_2C_{24}H_{44}Cl_2N_2$
$[TiCl[C_5H_4(TMS)](NBu^t)]_2$, **4**, 300
$Ti_2Si_4C_{24}H_{28}Cl_2$
$[TiCl[Me_2Si(C_5H_4)_2]]_2$, **4**, 248
$Ti_2U_2W_{10}C_{20}H_{20}O_{38}$
$[Cp_2U(TiW_5O_{19})]_2^-$, **4**, 189
$Ti_2ZnC_{28}H_{32}Cl_4N_4$
$[Ti(MeCN)_2Cp_2]_2[ZnCl_4]$, **4**, 398
$Ti_3C_{25}H_{25}Cl_3O_2$
$Ti_3Cl_3(O)_2Cp_5$, **4**, 401
$Ti_3C_{51}H_{60}Cl_3N_3O_3P_3$
$[TiClCp[OP(Ph)_2=N]]_3$, **4**, 300
$Ti_4C_{20}H_{20}OS_8$
$(TiCp)_4S_8O$, **4**, 309
$Ti_4C_{20}H_{20}O_2S_8$
$(TiCp)_4S_8O_2$, **4**, 309
$Ti_4C_{20}H_{20}S_4$
$[Ti(S)Cp]_4$, **4**, 247
$Ti_4C_{24}H_{28}S_4$
$[Ti(C_5H_4Me)(S)]_4$, **4**, 309
$Ti_5C_{25}H_{25}S_6$
$Ti_5Cp_5S_6$, **4**, 309, 412
$Ti_6C_{30}H_{30}Cl_2O_6$
$Ti_6Cp_6(Cl)_2(O)_6$, **4**, 305
$Ti_6C_{30}H_{30}Cl_4O_4$
$Ti_6Cp_6(Cl)_4(O)_4$, **4**, 305
$Ti_6C_{30}H_{30}O_8$
$Ti_6Cp_6(O)_8$, **4**, 305
$Ti_7H_{19}Cl_3P_2$
$TiMeCl_3(dmpe)$, **4**, 280
$TiC_{18}H_{20}O_2$
$Ti[C_2Me_4(C_5H_4)_2](CO)_2$, **4**, 248
$TlAsRu_{12}C_{58}H_{20}O_{32}$
$[AsPh_4][Tl[Ru_6C(CO)_{16}]_2]$, **1**, 568
$TlB_2CoC_{13}H_{20}$
$TlMe_5(C_2B_2C)CoCp$, **1**, 532
$TlB_4C_{18}H_{24}F_{20}O_4Te_4$
$[(C_6H_3Me_3)_2Tl][B(OTeF_5)_4]$, **1**, 536
$TlB_9C_2H_{11}$
$[closo\text{-}TlC_2B_9H_{11}]^-$, **1**, 517
$TlC_2B_9H_{11}$, **1**, 258
$TlB_{10}C_6H_{11}F_6O_4$
$C_2B_{10}H_{11}Tl(O_2CCF_3)_2$, **1**, 227
$TlC_2F_3O_2$
CF_3COOTl, **11**, 447
$TlC_2H_3O_2$
TlOAc, **12**, 304, 311
TlC_3F_9
$(CF_3)_3Tl$, **11**, 438
TlC_3H_9
Me_3Tl, **11**, 438, 440
TlC_5H_5
CpTl, **11**, 452
TlCp, **1**, 529
TlC_7H_7S
TlSBz, **1**, 528
$TlC_9H_2F_9S$
$[(SF_3)_3C_6H_2S]Tl$, **1**, 528
$TlC_{10}H_5F_6O_4$
$PhTl(O_2CCF_3)_2$, **12**, 147
$TlC_{10}H_{11}$
$Me_2(PhCC)Tl$, **11**, 440
$TlC_{10}H_{15}$
$Tl(Cp^*)$, **1**, 531
$TlC_{11}H_9$
PhC_5H_4Tl, **11**, 452
$TlC_{12}H_{11}$
BzC_5H_4Tl, **8**, 85; **11**, 452
$TlC_{13}H_9$
(Flr)Tl, **4**, 566
$TlC_{16}H_{21}Cl_3N_3$
$[NMe_4][Tl(C_6H_4N=NPh)Cl_3]$, **1**, 513
$TlC_{18}H_{15}$
Ph_3Tl, **11**, 440
$TlC_{19}H_{24}OPS_2$
$(THF)Me_2TlS_2CPPh_2$, **1**, 525
$TlC_{40}H_{35}$
$Tl[C_5(Bz)_5]$, **1**, 531
$TlCoB_2C_{17}H_{23}$
$CpTl[Me_2C_2(BMe)_2CH]CoCp$, **1**, 206, 207
$TlCr_3C_{24}H_{15}O_9$
$Tl[Cr(CO)_3Cp]_3$, **1**, 567
TlF_6P
$Tl[PF_6]$, **10**, 2
$TlFeC_3NO_4$
$Tl[Fe(CO)_3(NO)]$, **1**, 568
$TlFe_2C_8O_8$
$[Tl[Fe(CO)_4]_2]^-$, **7**, 20
$TlGaB_{18}C_4H_{22}$

$Tl[Ga(C_2B_9H_{11})_2]$, **1**, 533

$TlGaSe_2$

$TlGaSe_2$, **1**, 524

$TlLiC_{19}H_{33}N_3$

CpTlCpLi·PMDETA, **1**, 15

$TlMoC_8H_5O_3$

$Tl[Mo(CO)_3Cp]$, **1**, 567

$TlMo_3C_{24}H_{15}O_9$

$Tl[Mo(CO)_3Cp]_3$, **1**, 567

$TlNO_3$

$TlNO_3$, **11**, 455; **12**, 309

$TlNbC_{22}H_{10}BrClF_{10}$

$Cp_2Nb(Cl)(Br)Tl(C_6F_5)_2$, **5**, 126, 127

$TlPt_3RhC_{65}H_{111}Cl_2O_3P_3$

$[TlPt_3(CO)_3(PCy_3)_3][RhCl_2(C_8H_{12})]$, **1**, 569

$TlRe_3C_{15}O_{15}$

$Tl[Re(CO)_5]_3$, **1**, 567

$TlSiC_8H_{13}$

$Tl(C_5H_4TMS)$, **1**, 531

$TlSiC_{17}H_{23}$

$Tl[C_5Me_4(SiMe_2Ph)]$, **1**, 531

$TlSiC_{18}H_{25}$

$Tl[C_5Me_4(SiMe_2Bz)$, **1**, 531

$Tl_2Au_2C_{26}H_{24}P_2S_2$

$[AuTl(Ph_2PCh_2S)]_2$, **1**, 569

$Tl_2Au_2C_{68}H_{56}P_4$

$[TlAu(C_5H_4PPh_2)_2]_2$, **1**, 531

$Tl_2B_9C_4H_{15}$

$Tl[Tl(MeC)_2B_9H_9]$, **1**, 258

Tl_2CO_3

Tl_2CO_3, **12**, 181, 309, 310

$Tl_2C_{27}H_{33}Cl_3$

$[Tl(Mes)_2][TlCl_3(Mes)]$, **1**, 514

$Tl_2C_{36}H_{48}F_{10}O_2Te_2$

$[(C_6H_2Me_3)_2TlOTeF_5]_2$, **1**, 536

$Tl_2Fe_4C_{32}H_{40}N_2O_{16}$

$[NEt_4]_2[[Fe(CO)_4]_2[TlFe(CO)_4]_2]$, **1**, 567

$Tl_2Fe_6C_{24}O_{24}$

$[Tl_2Fe_6(CO)_{24}]^{2-}$, **7**, 20

$Tl_2PtC_4N_4$

$Tl_2Pt(CN)_4$, **1**, 569

$Tl_2Re_4C_{18}O_{18}$

$Re_2(CO)_8[TlRe(CO)_5]_2$, **1**, 567

$Tl_2Si_2C_{24}H_{54}O_6S_2$

$[Tl(SSi(OBu^t)_3]_2$, **1**, 525

$Tl_3B_9C_2H_{12}O$

$Tl_3(C_2B_9H_{12})O$, **1**, 230

$Tl_4Fe_8C_{30}O_{30}$

$[Tl_4Fe_8(CO)_{30}]^{4-}$, **7**, 20

$Tl_4Fe_8C_{62}H_{80}N_4O_{30}$

$[NEt_4]_4[Tl_4Fe_8(CO)_{30}]$, **1**, 568

$Tl_6Fe_{10}C_{84}H_{120}N_6O_{36}$

$[NEt_4]_6[Tl_6Fe_{10}(CO)_{36}]$, **1**, 568

$Tl_6Si_{12}C_{66}H_{126}$

$[Tl[C_5H_3(TMS)_2]]_6$, **1**, 531

$TmC_{15}H_{15}$

$TmCp_3$, **4**, 53

U

$UAlC_{15}H_{19}$
$Cp_3U(AlH_4)$, **4**, 158
$UAl_3C_6H_6Cl_{12}$
$(C_6H_6)U(AlCl_4)_3$, **4**, 178
$UAl_3C_{12}H_{18}Cl_{12}$
$(C_6Me_6)U(AlCl_4)_3$, **4**, 178
$UAuC_2F_6O_2$
$[Au(CO)_2][UF_6]$, **3**, 48
$UBC_{15}H_{19}$
$Cp_3U(BH_4)$, **4**, 158
$UBC_{18}H_{21}N_2$
$Cp_3U(NCBH_3)(MeCN)$, **4**, 159
$UBC_{39}H_{35}$
Cp_3UBPh_4, **4**, 159
$UBSiC_{19}H_{33}Cl_2N_6$
$[HB(Me_2Pz)_3]UCl_2(CH_2TMS)$, **4**, 192
$UBSi_2C_{22}H_{41}Cl_2N_6$
$[HB(Me_2Pz)_3]UCl_2[CH(TMS)_2]$, **4**, 192
$UB_2C_{10}H_{18}$
$Cp_2U(BH_4)_2$, **4**, 146
$UB_2C_{10}H_{21}O$
$[(C_8H_8)U(BH_4)_2(OEt)]^-$, **4**, 185
$UB_2C_{18}H_{22}$
$(C_9H_7)_2U(BH_4)_2$, **4**, 163
$UB_2C_{18}H_{32}P_2$
$(C_5Me_4P)_2U(BH_4)_2$, **4**, 178
$UB_3C_5H_{17}$
$[CpU(BH_4)_3]^-$, **4**, 138, 144
$UB_3C_6H_{19}$
$(MeC_5H_4)U(BH_4)_3$, **4**, 144
$UB_3C_8H_{20}$
$[(C_8H_8)U(BH_4)_3]^-$, **4**, 185
$UB_3C_9H_{24}$
$(C_6H_3Me_3)U(BH_4)_3$, **4**, 180
$UB_3C_{10}H_{27}$
$[Cp^*U(BH_4)_3]^-$, **4**, 138
$UB_3C_{12}H_{44}P_4$
$(dmpe)_2U(BH_4)_3$, **1**, 286, 288, 289
$UB_3C_{34}H_{40}P_2$
$U(BH_4)_3(Ph_2PPy)_2$, **1**, 286, 289
$UB_4C_4H_{24}$
$U(BH_3Me)_4$, **1**, 290, 291
$UB_4C_{10}H_{40}P_2$
$(dmpe)U(BH_3Me)_4$, **1**, 290, 291
$UB_4C_{16}H_{56}P_2$
$(dmpe)_2U(MeBH_3)_4$, **1**, 290, 291
UCF_4O
$UF_4(CO)$, **4**, 131
$UC_{12}H_{20}$
$U(C_3H_5)_4$, **4**, 137
$UC_{15}H_{15}$
UCp_3, **4**, 139
$UC_{15}H_{15}Cl$
Cp_3UCl, **4**, 12, 147
$UC_{15}H_{16}$
Cp_3UH, **4**, 157
$UC_{15}H_{16}O$
Cp_3UOH, **4**, 148
$UC_{16}H_{15}NS$
Cp_3UNCS, **4**, 159
$UC_{16}H_{16}$
$[U(C_8H_8)_2]^-$, **4**, 181, 182
$UC_{16}H_{24}Cl_2O_2$
$(C_8H_8)UCl_2(THF)_2$, **4**, 184
$UC_{16}H_{28}$
$U(C_3H_4Me)_4$, **4**, 136
$UC_{16}H_{28}N_2$
$(C_8H_8)U(NEt_2)_2$, **4**, 186
$UC_{17}H_{23}Br_3O_2$
$(C_9H_7)UBr_3(THF)_2$, **4**, 163
$UC_{17}H_{35}N_3$
$CpU(NEt_2)_3$, **4**, 144
$UC_{18}H_{24}Cl_2P_2$
$(C_5Me_4P)_2UCl_2$, **4**, 178
$UC_{18}H_{24}O_4$
$(C_8H_8)U(acac)_2$, **4**, 186
$UC_{19}H_{21}N_2$
$[Cp_3U(MeCN)_2]^-$, **4**, 159
$UC_{19}H_{23}O$
$Cp_3U(THF)$, **4**, 139
$UC_{19}H_{24}$
Cp_3UBu, **4**, 141
$UC_{19}H_{25}P$
$Cp_3UCHPMe_3$, **4**, 155
$UC_{20}H_{20}$
UCp_4, **4**, 160
$UC_{21}H_{20}O$
Cp_3UOPh, **4**, 149
$UC_{22}H_{29}O$
$(MeC_5H_4)_3U(THF)$, **4**, 149
$UC_{22}H_{39}I_2O_3$
$Cp^*UI_2(THF)_3$, **4**, 164
$UC_{23}H_{20}$
Cp_3UCCPh, **4**, 155
$UC_{23}H_{29}N$
$Cp_3U[C(Me)N(C_6H_{11})]$, **4**, 153
$UC_{24}H_{27}P$
$Cp_3UCHPMe_2Ph$, **4**, 154
$UC_{25}H_{27}OP$
$Cp_3U(OCCH)PMe_2Ph$, **4**, 155
$UC_{27}H_{21}$
$U(C_9H_7)_3$, **4**, 164
$UC_{27}H_{21}Cl$
$(C_9H_7)_3UCl$, **4**, 163
$UC_{27}H_{21}I$
$(C_9H_7)_3UI$, **4**, 164
$UC_{27}H_{25}N$
Cp_3UNPh_2, **4**, 149
$UC_{27}H_{25}P$
Cp_3UPPh_2, **4**, 151
$UC_{27}H_{36}ClP_3$
$(C_5Me_4P)_3UCl$, **4**, 178
$UC_{27}H_{37}P_3$
$(C_5Me_4P)_3UH$, **4**, 178
$UC_{28}H_{24}$
$(C_9H_7)_3UMe$, **4**, 164
$UC_{28}H_{39}P_3$
$(C_5Me_4P)_3UMe$, **4**, 178
$UC_{28}H_{40}P_2$
$UMe(Bz)_3(dmpe)$, **4**, 133
$UC_{30}H_{43}OP_3$
$(C_5Me_4P)_3U(OPr^i)$, **4**, 178
$UC_{31}H_{32}NP$
$Cp_3UNC(Me)CHPMePh_2$, **4**, 156
$UC_{32}H_{40}N_2$
$Cp^*_2U(NPh)_2$, **4**, 168
$UC_{32}H_{48}$
$Cp^*_2U(methallyl)_3$, **4**, 165
$UC_{36}H_{36}OP$
$(MeC_5H_4)_3U(OPPh_3)$, **4**, 139
$UC_{36}H_{40}NP$
$Cp_3U[CN(C_6H_{11})CHPMePh_2]$, **4**, 157
$UC_{38}H_{46}Cl_2N$
$Cp^*_2UCl_2(HNPPh_3)$, **4**, 166
$UC_{40}H_{41}O_2$
$(tritox)_2U(Bz)_2$, **4**, 136
$UC_{40}H_{64}$
$[Cp^*_2UH_2]_2$, **4**, 175
$UC_{42}H_{44}N_6$
$(C_3H_5)_4U(bipy)_3$, **4**, 137
$UC_{47}H_{47}P_3$
$CpU[(CH_2)_2PPh_2]_3$, **4**, 144
$UKC_{16}H_{22}$
$(MeC_7H_8)U(MeC_7H_8)K$, **4**, 181
$ULiSi_4C_{30}H_{58}Cl_2O_2$
$[(TMS)_2C_5H_3]_2U(Cl)_2Li(THF)_2$, **4**, 138
$UMnC_{30}H_{32}OP$
$MnCp[C(OUCp_3)CHPPhMe_2)$, **6**, 56
$UMn_4C_{20}O_{20}$
$U[Mn(CO)_5]_4$, **4**, 186
$UNb_2W_{10}C_{95}H_{195}N_5O_{38}$
$[NBu_4]_5[Cp_3U(NbW_5O_{19})_2]$, **4**, 189
$USi_6C_{18}H_{53}N_3$
$[(TMS)_2N]_2\overline{UCH_2SiMe_2N}(TMS)$, **12**, 596
$USi_2B_2C_{14}H_{32}$
$[(TMS)_2C_8H_6]U(BH_4)_2$, **4**, 185
$USi_2C_{22}H_{48}N_3OP$
$(C_8H_8)U(CH_2TMS)_2(HMPA)$, **4**, 185
$USi_2C_{26}H_{48}ClP$
$Cp^*_2U(Cl)P(TMS)_2$, **4**, 168
$USi_2C_{27}H_{49}$
$Cp^*_2UCH(TMS)_2$, **4**, 165

$USi_3C_{20}H_{41}$
$[(C_8H_8)U(CH_2TMS)_3]^-$, **4**, 185
$USi_3C_{24}H_{39}Br$
$(TMS\text{-}C_5H_4)_3UBr$, **4**, 157
$USi_3C_{24}H_{39}N_3$
$(TMS\text{-}C_5H_4)_3UN_3$, **4**, 148
$USi_3C_{24}H_{39}O$
$O[U(TMS\text{-}C_5H_4)_3]$, **4**, 149
$USi_3C_{25}H_{39}O$
$(TMS\text{-}C_5H_4)_3UCO$, **4**, 131
$USi_3C_{26}H_{42}$
$(TMS\text{-}C_5H_4)_3U(CHCH_2)$, **4**, 152
$USi_3C_{27}H_{44}N$
$(TMS\text{-}C_5H_4)_3U(CNEt)$, **4**, 131
$USi_3C_{36}H_{45}$
$U(TMS\text{-}C_9H_6)_3$, **4**, 162
$USi_4C_{20}H_{44}N_2$
$(C_8H_8)U[N(TMS)_2]_2$, **4**, 186
$USi_6C_{18}H_{53}N_3$
$[(TMS)_2N]_2UCH_2SiMe_2N\text{-}TMS$, **4**, 192
$USi_6C_{19}H_{57}N_3$
$[(TMS)_2N]_3UMe$, **4**, 192
$USnC_{33}H_{30}$
Cp_3USnPh_3, **4**, 157
$UWC_{35}H_{29}O_6P$
$Cp_3UOCHCHP(Ph)_2CH_2W\text{-}(CO)_5$, **4**, 156
$U_2B_6C_7H_{31}$
$[(BH_4)_3U(C_7H_7)U(BH_4)_3]^-$, **4**, 177
$U_2C_{17}H_{21}O_8S_4$
$Cp_3U_2Me_2(SO_2)_4$, **4**, 154
$U_2C_{19}H_{21}Cl_2N_2O_2$
$[Cp_3U(MeCN)_2][UO_2Cl_2]$, **4**, 160
$U_2C_{24}H_{36}Cl_7$
$[(C_6Me_6)_2U_2Cl_7]^+$, **4**, 179
$U_2C_{28}H_{56}O_4$
$[(C_3H_5)_2U(OBu^t)_2]_2$, **4**, 137
$U_2C_{30}H_{31}$
$[Cp_3U(H)UCp_3]^-$, **4**, 157
$U_2C_{31}H_{67}N_6$
$[(NEt_2)_3U(C_7H_7)U(NEt_2)_3]^-$, **4**, 177
$U_2C_{37}H_{45}$
$[(MeC_5H_4)_3U(Me)U(Me\text{-}C_5H_4)_3]^-$, **4**, 67
$U_2C_{40}H_{64}$
$[Cp^*{}_2UH_2]_2$, **4**, 190
$U_2C_{42}H_{46}N_2$
$(N_2C_6H_4)[U(MeC_5H_4)_3]_2$, **4**, 150
$U_2C_{48}H_{62}S_2$
$[(Bu^tC_5H_4)_2U(SPh)]_2$, **4**, 138
$U_2C_{52}H_{84}Cl_2$
$[(Bu^t{}_2C_5H_3)_2U(Cl)]_2$, **4**, 138
$U_2LiC_{30}H_{36}Cl_5O_2$
$[CH_2(C_5H_4)_2]_2U_2(Cl)_5Li(THF)_2$, **4**, 161
$U_2MgC_{32}H_{28}Cl_2O_8$
$U_2(CO_2Bz)_4MgCl_2$, **4**, 133
$U_2Mg_2C_{32}H_{28}Cl_4O_{16}$
$[U(CO_2Bz)_4MgCl_2]_2$, **4**, 133
$U_2Mg_2C_{58}H_{56}Cl_4O_2$
$[U(Bz)_3(COBz)MgCl_2]_2$, **4**, 133
$U_2Mg_2C_{58}H_{56}Cl_4O_4$
$[U(Bz)_3(CO_2Bz)MgCl_2]_2$, **4**, 133
$U_2Mg_2C_{76}H_{88}Cl_2O_2P_4$
$[Cp^*U[(CH_2)PPh_2(CH_2)]]_2Mg\text{-}[CH_2PMePh_2]_2(O)_2(Cl)_2$, **4**, 166
$U_2Si_6C_{48}H_{78}O$
$(O)[U(TMS\text{-}C_5H_4)_3]_2$, **4**, 148
$U_2Si_6C_{49}H_{80}O_2$
$(TMS\text{-}C_5H_4)_3UOCH_2OU\text{-}(C_5H_4TMS)_3$, **4**, 157
$U_2Ti_2W_{10}C_{20}H_{20}O_{38}$
$[Cp_2U(TiW_5O_{19})]_2^-$, **4**, 189
$U_3B_8C_{27}H_{79}O_5$
$[U(BH_4)_2(THF)_5][(BH_4)_3U\text{-}(C_7H_7)U(BH_4)_3]$, **4**, 177
$U_3C_{34}H_{38}Br_8N_8$
$[(C_9H_7)UBr_2(MeCN)_4]_2[UBr_6]$, **4**, 163
$U_3CoC_{51}H_{45}N_6$
$[Cp_3U]_3Co(CN)_6$, **4**, 190

V

$VAlC_{18}H_{24}Cl_4$
[$(Me_3C_6H_3)_2V$]$AlCl_4$, **5**, 46
$VAuC_{24}H_{15}O_6P$
[$V(CO)_6$][$AuPPh_3$], **5**, 9
$VBC_6H_{22}P_2$
$V(BH_4)(PMe_3)_2$, **1**, 283
$VBC_{18}H_{18}F_4N$
Cp_2V[C(Ph)NMe]BF_4, **5**, 35
$VB_2C_{12}H_{40}P_4$
$V(BH_4)_2(dmpe)_2$, **1**, 280, 283
$VB_3C_6H_{30}P_2$
$V(BH_4)_3(PMe_3)_2$, **1**, 282, 283
$VB_4C_{14}H_{12}$
$(C_8H_8)V[Et_2C_2B_4H_4]$, **5**, 49
VC_4O_6
$(O_2)V(CO)_4$, **5**, 6
$VC_5H_5Cl_2O$
$CpV(O)Cl_2$, **5**, 31
$VC_5H_5I_3$
$CpVI_3$, **5**, 27
VC_5NO_6
$V(CO)_5NO$, **5**, 9
VC_6HO_6
$HV(CO)_6$, **5**, 7
$VC_6H_5NO_3$
$CpV(NO)_2(CO)$, **5**, 22
VC_6H_6
$(C_6H_6)V$, **5**, 47
$VC_6H_8Cl_3$
$MeCpVCl_3$, **5**, 30
VC_6O_6
$V(CO)_6$, **5**, 2
$VC_7H_9NO_5P$
$V(CO)_4(NO)(PMe_3)$, **5**, 9
$VC_7H_{21}N_3$
$(Me_2N)_3MeV$, **12**, 1086
$VC_8H_5N_2O_3$
$CpV(CO)_3N_2$, **5**, 21
$VC_8H_5O_3$
$CpV(CO)_3$, **5**, 20
$VC_8H_8NO_3$
$CpV(CO)_3(NH_3)$, **5**, 19
$VC_9H_5O_4$
$CpV(CO)_4$, **5**, 17, 20
$VC_9H_7O_4$
$(C_5H_7)V(CO)_4$, **5**, 6
$VC_9H_{11}Cl_2O$
$(Mes)VOCl_2$, **5**, 16
$VC_9H_{13}ClO$
CpVCl(THF), **5**, 27
$VC_9H_{14}Cl_2N$
$(Bu^tN)VCpCl_2$, **5**, 31
$VC_{10}H_{10}BrNO$
$Cp_2VBr(NO)$, **5**, 39
$VC_{10}H_{10}Br_2$
Cp_2VBr_2, **5**, 37
$VC_{10}H_{10}Cl$
Cp_2VCl, **5**, 37
$VC_{10}H_{10}Cl_2$
Cp_2VCl_2, **5**, 39
$VC_{10}H_{10}INO$
$Cp_2VI(NO)$, **5**, 38
$VC_{10}H_{10}N_6$
$Cp_2V(N_3)_2$, **5**, 39
$VC_{10}H_{10}P_2$
$(CpP)_2V$, **5**, 47
$VC_{10}H_{14}$
$(C_5H_7)_2V$, **5**, 49
$VC_{10}H_{15}OS_5$
Cp^*VOS_5, **5**, 32
$VC_{11}H_{10}N_4$
$Cp_2V(CN)(N_3)$, **5**, 39
$VC_{11}H_{10}O$
$Cp_2V(CO)$, **5**, 34
$VC_{11}H_{12}O$
$Cp_2V(CH_2O)$, **5**, 36
$Cp(C_5H_7)VCO$, **5**, 49
$VC_{11}H_{14}O$
$(C_5H_7)_2VCO$, **5**, 49
$VC_{11}H_{23}ClP_2$
$CpVCl(PMe_3)_2$, **5**, 24
$VC_{11}H_{23}Cl_2P_2$
$CpVCl_2(PMe_3)_2$, **5**, 27
$VC_{12}H_{10}F_6O_2P$
$[Cp_2V(CO)_2]PF_6$, **5**, 38
$VC_{12}H_{13}O_3S$
$CpV(CO)_3$(THT), **5**, 21
$VC_{12}H_{13}O_4$
$CpV(CO)_3$(THF), **5**, 20
$VC_{12}H_{16}$
Cp_2VMe_2, **5**, 43
$VC_{12}H_{16}S_2$
$Cp_2V(SMe)_2$, **5**, 42
$VC_{12}H_{24}Cl_3O_3$
$VCl_3(THF)_3$, **5**, 13
$VC_{12}H_{26}ClP_2$
$CpVMeCl(PMe_3)_2$, **5**, 28
$VC_{13}H_{14}$
$CpV(C_8H_9)$, **5**, 30
$VC_{13}H_{15}$
$Cp_2V(C_3H_5)$, **5**, 43
$VC_{13}H_{16}O$
$Cp(C_7H_{11})VCO$, **5**, 49
$VC_{13}H_{23}ClNO$
$(Bu^tN)VCpCl(Bu^tO)$, **5**, 31
$VC_{14}H_{14}$
$V(Tol)_2$, **5**, 46
$VC_{14}H_{20}O_3P$
$CpV(CO)_3(PEt_3)$, **5**, 27
$VC_{14}H_{22}$
$(C_7H_{11})_2V$, **5**, 48
$VC_{14}H_{23}O$
$(C_7H_{11})_2VCO$, **5**, 49
$VC_{14}H_{24}P$
$CpV(C_6H_{10})PMe_3$, **5**, 29
$VC_{15}H_{20}$
$Cp_2V[CH_2C(Me)_2CH_2]$, **5**, 43
$VC_{15}H_{31}P_2$
CpV(allyl)$Me(PMe_3)_2$, **5**, 29
$VC_{16}H_{11}O_6$
$[CpV(CO)_3]_2(H)$, **5**, 20
$VC_{16}H_{19}$
Cp_2VCCBu^t, **5**, 43
$VC_{16}H_{31}P_2$
$CpV(CHBu^t)$(dmpe), **5**, 28
$VC_{17}H_{23}$
$CpV(C_6Me_6)$, **5**, 24
$VC_{18}H_{14}$
$(C_9H_7)_2V$, **5**, 34
$VC_{18}H_{14}I$
$(C_9H_7)_2VI$, **5**, 38
$VC_{20}H_{14}O_2$
$[(C_9H_7)_2V(CO)_2]^+$, **5**, 38
$VC_{20}H_{20}Cl_2N_4$
$VCl_2(py)_4$, **12**, 1086
$VC_{20}H_{30}$
Cp^*_2V, **5**, 33
$VC_{20}H_{30}Cl$
Cp^*_2VCl, **5**, 37
$VC_{20}H_{30}INO$
$Cp^*_2VI(NO)$, **5**, 39
$VC_{20}H_{30}S_2$
$Cp^*_2V(S_2)$, **5**, 42
$VC_{21}H_{30}IO$
$Cp^*_2V(CO)I$, **5**, 38
$VC_{21}H_{30}O$
$Cp^*_2V(CO)$, **5**, 34
$VC_{21}H_{30}OS$
$Cp^*_2V(COS)$, **5**, 37
$VC_{21}H_{30}O_3$
$Cp^*_2V(CO_3)$, **5**, 35
$VC_{21}H_{33}$
Cp^*_2VMe, **5**, 43
$VC_{22}H_{20}$
$(MeC_{10}H_7)_2V$, **5**, 46
$VC_{22}H_{20}S_2$
$Cp_2V(SPh)_2$, **5**, 42
$VC_{22}H_{30}F_6O_2P$
$[Cp^*_2V(CO)_2]PF_6$, **5**, 38
$VC_{22}H_{34}Br$
$[C_5Me_4Et]_2VBr$, **5**, 37
$VC_{22}H_{38}$
$(C_5Me_4Et)_2V$, **5**, 34
$VC_{23}H_{20}P$
$CpV(Ph)PPh_2$, **5**, 24
$VC_{23}H_{33}O_2$
$Cp^*_2V(CO)[C(O)Me]$, **5**, 43
$VC_{24}H_{24}P$
$Cp_2V(CH_2)_2(PPh_2)$, **12**, 595
$VC_{25}H_{40}NO_2$
$Cp^*_2V(O_2CNEt_2)$, **5**, 40
$VC_{25}H_{42}P_3$
$Cp^*V(Bu^tCP)_3$, **5**, 26
$VC_{26}H_{30}P$

$CpV(C_2Ph_2C_4H_6)(PMe_3)$, **5**, 24
$CpV(C_4Me_2Ph_2)PMe_3$, **5**, 24
$VC_{27}H_{33}O$
$(Mes)_3V(O)$, **5**, 14
$VC_{27}H_{35}NS$
$Cp^*_2V(SCNPh)$, **5**, 36
$VC_{27}H_{37}S$
$Cp^*_2V(SC_6H_4Me)$, **5**, 42
$VC_{28}H_{10}F_{15}Ge$
$Cp_2VGe(C_6F_5)_3$, **5**, 44
$VC_{28}H_{10}F_{15}GeS$
$Cp_2V[SGe(C_6F_5)_3]$, **5**, 42
$VC_{28}H_{34}N_4$
$[(Me_2NCH_2)C_6H_4]_2V(py)_2$, **5**, 12
$VC_{30}H_{48}N_2$
$[Cp^*_2V(CNBu^t)_2]^+$, **5**, 38
$VC_{30}H_{54}N_6$
$[V(CNBu^t)_6]^{2+}$, **5**, 4
$VC_{31}H_{41}O$
$(Mes)_3V(THF)$, **5**, 13
$VC_{33}H_5F_{20}O$
$CpV(CO)(F_5C_6CCC_6F_5)_2$, **5**, 19
$VC_{42}H_{60}N_3$
$V[Bu^tNC(Mes)]_3$, **5**, 14
$VC_{44}H_{60}N_3O_4$
$[Bu^tNC(Mes)]V[Bu^tNC-(Mes)C(O)O]_2$, **5**, 14
$VC_{54}H_{48}O_2P_4$
$(CO)_2V(dmpe)_2$, **5**, 4
$VC_{54}H_{74}N_8O_2$
$(N_2)[[(Me_2NCH_2)C_6H_4]_2V-(py)]_2(THF)_2$, **5**, 12
$VC_{56}H_{51}NO_2P_4$
$[(CO)_2V(dmpe)_2(MeCN)]^+$, **5**, 4
VCl_3
VCl_3, **12**, 1086
$VCoC_{14}H_{10}O_4$
$Cp_2V[Co(CO)_4]$, **5**, 45
$VCo_3C_{20}H_{10}O_{10}$
$Cp_2V[OCCo_3(CO)_9]$, **5**, 45
$VFeC_{15}H_{15}Br_3$
$[Cp_2Fe][CpVBr_3]$, **5**, 30
$VFeC_{23}H_{26}N$
$Cp_2V[C_5H_3(CH_2NMe_2)FeCp]$, **5**, 44
$VFe_2C_{30}H_{28}$
$Cp_2V(C_{10}H_9Fe)_2$, **5**, 44
$VKC_8H_8O_3Ge$
$[CpV(CO)_3(GeH_3)]K$, **5**, 21
$VKC_{13}H_{17}$
$[CpV(C_8H_{12})]K$, **5**, 24
$VLiC_{16}H_{36}O_4$
$Li[Bu^tVO(Bu^nO)_3]$, **5**, 15
$VLiC_{91}H_{95}O_4P_6$
$[[Ph_2PCHPPh_2]_3V][Li(THF)_4]$, **5**, 12
$VLiSiVC_{16}H_{38}O_3$
$[(TMS-CH_2)V(OBu^t)_3]Li$, **5**, 13
$VLiSi_2C_{26}H_{60}N_2$
$[(Bu^t_3SiN)_2VMe_2]Li$, **5**, 16
$VLiSi_4C_{24}H_{60}O_4$
$LiV(O-TBDMS)_4$, **5**, 14
$VMnC_{14}H_{10}O_4P_3$
$CpMn(CO)P_3(CO)_3VCp$, **5**, 23
$VNaC_6O_6$
$NaV(CO)_6$, **5**, 2
$VNaC_8H_6O_3$
$[CpV(CO)_3H]Na$, **5**, 24
$VNa_2C_5HO_5$
$Na_2[HV(CO)_5]$, **5**, 3
$VNa_3C_5O_5$
$Na_3[V(CO)_5]$, **5**, 3
$VNbC_{19}H_{16}O_4$
$CpV(CO)_3(H)NbCp_2(CO)$, **5**, 23
$VPbC_{24}H_{15}O_6$
$Ph_3PbV(CO)_6$, **2**, 312; **5**, 6
$VSiC_{16}H_{38}O_3$
$TMS-CH_2V(Bu^tO)_3$, **5**, 15
$VSi_2C_8H_{24}O_2$
$Me_2V(O-TMS)_2$, **5**, 15
$VSi_2C_{16}H_{28}$
$Cp[(TMS)_2Cp]V$, **5**, 48
$VSi_2C_{17}H_{28}O$
$Cp[(TMS)_2Cp]VCO$, **5**, 48
$VSi_2C_{22}H_{43}P$
$Cp[(TMS)_2Cp]VPEt_3$, **5**, 48
$VSi_2C_{58}H_{56}BrO_2P_4$
$V(TMS-OCCO-TMS)-(dmpe)_2Br$, **5**, 6
$VSi_3C_{18}H_{45}O_3$
$V[O-TBDMS]_3$, **5**, 13
$VSi_7C_{42}H_{77}O_{13}$
$(C_6H_{11})_7Si_7O_{12}V(O)$, **5**, 16
$VSnC_{24}H_{15}O_6$
$(Ph_3Sn)V(CO)_6$, **5**, 6
$VSnH_8Cl_6N_2O$
$[NH_4]_2[VO(SnCl_3)Cl_3]$, **10**, 50
$VWC_{18}H_{11}O_7$
$(C_{10}H_8)[W(CO)_3Me][V(CO)_4]$, **5**, 24
$VZnC_{17}H_{29}Br_4O_3$
$CpVBr_2\cdot ZnBr_2\cdot 3THF$, **5**, 27
$V_2B_2C_{50}H_{52}P_4$
$[VCl(dppm)(BH_4)]_2$, **1**, 280, 283
$V_2B_2Zn_2C_{52}H_{64}P_4$
$V_2Zn_2H_4[BH_4]_2(PMePh_2)_4$, **1**, 280, 282, 283
$V_2B_4C_{12}H_{54}OP_4$
$[V(BH_4)_2(PMe_3)_2]_2O$, **1**, 282, 283
$V_2C_6H_6$
$(C_6H_6)V_2$, **5**, 47
$V_2C_{12}H_{12}$
$(C_6H_6)_2V_2$, **5**, 47
$V_2C_{12}H_{14}O_5$
$[CpV(O)]_2(O)(OCH_2CH_2O)$, **5**, 31
$V_2C_{12}H_{14}S_4$
$(MeCp)_2V_2S_4$, **5**, 32
$V_2C_{12}H_{14}S_5$
$(MeCp)_2V_2S_5$, **5**, 32
$V_2C_{15}H_{10}O_5$
$Cp_2V_2(CO)_5$, **5**, 20, 22
$V_2C_{15}H_{15}Cl_2O_2$
$Cp_3V_2Cl_2O_2$, **5**, 39, 40
$V_2C_{15}H_{15}Cl_3O$
$[Cp_2VCl](O)[CpVCl_2]$, **5**, 41
$V_2C_{16}H_{14}O_4S_2$
$Cp_2'V_2(CO)_4(S)_2$, **5**, 28
$V_2C_{16}H_{16}$
$(CpV)_2(C_6H_6)$, **5**, 26, 29
$V_2C_{16}H_{28}Cl_2P_2$
$[CpVCl(PMe_3)]_2$, **5**, 27
$V_2C_{18}H_{14}$
$(C_9H_7)_2V_2$, **5**, 47
$V_2C_{18}H_{18}$
$(CpV)_2(cot)$, **5**, 29
$V_2C_{18}H_{26}$
$[CpV[(CH_2)_4]]_2$, **5**, 29
$V_2C_{20}H_{18}$
$CpV(C_{10}H_8)VCp$, **5**, 26
$V_2C_{20}H_{22}Ge$
$(Cp_2V)_2GeH_2$, **5**, 44
$V_2C_{20}H_{30}Br_2O_3$
$[Cp^*VOBr]_2(O)$, **5**, 31
$V_2C_{20}H_{30}Br_4$
$[Cp^*VBr_2]_2$, **5**, 27
$V_2C_{20}H_{30}Cl_2S_2$
$[(Cp^*VCl)(S)]_2$, **5**, 42
$V_2C_{20}H_{30}I_2O_3$
$[Cp^*VI(O)]_2(O)$, **5**, 31
$V_2C_{20}H_{30}I_4O$
$[Cp^*VI_2]_2(O)$, **5**, 31
$V_2C_{20}H_{30}OTe_3$
$Cp^*_2V_2Te_3O$, **5**, 21
$V_2C_{20}H_{30}O_8$
$[(Cp^*O_3)V(O)]_2$, **5**, 40
$V_2C_{22}H_{38}P_6$
$[(Me_4C_5Et)V]_2(P_6)$, **5**, 26
$V_2C_{22}H_{44}P_4$
$[CpV(H)(dmpe)]_2$, **5**, 29
$V_2C_{26}H_{26}$
$[CpV(cot)]_2$, **5**, 29
$V_2C_{26}H_{28}$
$(Cp_2V)_2(H)_2(C_6H_6)$, **5**, 29
$V_2C_{26}H_{30}O_6$
$Cp^*_2V(CO)V(CO)_5$, **5**, 3, 44
$V_2C_{28}H_{60}P_6$
$[CpV(H)(dmpe)]_2(dmpe)$, **5**, 29
$V_2C_{30}H_{50}N_2O_4$
$[Cp^*V(O_2CNEt_2)]_2$, **5**, 28
$V_2C_{30}H_{50}N_4O_8$
$Cp_2V_2(O_2CNEt_2)_4$, **5**, 28, 40
$V_2C_{32}H_{20}O_8P_2$
$V_2(CO)_8(PPh_2)_2$, **5**, 7
$V_2C_{32}H_{36}O_8$
$[(MeO)_2C_6H_3]_4V_2$, **5**, 12
$V_2C_{36}H_{40}N_4$
$[[[CH_2(Me)_2C_6H_2(CH_2)]N-Bu^t]V(NBu^t)]_2$, **5**, 14
$V_2C_{54}H_{66}O$
$(Mes)_3V(O)V(Mes)_3$, **5**, 14, 16
$V_2C_{54}H_{74}N_4O_2$
$(N_2)[[(Me_2NCH_2)C_6H_4]_2V-(py)]_2(THF)_2$, **5**, 13
$V_2Co_2C_{12}H_{14}N_2O_2S_4$
$(MeCp)_2V_2S_4[Co(NO)]_2$, **5**, 32
$V_2FeC_{12}H_{17}N_2O_2S_4$
$(MeCp)_2V_2S_4Fe(NO)_2$, **5**, 32
$V_2FeC_{15}H_{14}O_3S_3$
$(C_5H_4Me)_2V_2Fe(CO)_3(S)_3$, **7**, 71, 72
$V_2FeC_{15}H_{14}O_3S_4$
$(MeCp)_2V_2S_4Fe(CO)_3$, **5**, 32

$V_2Fe(S)_2(S_2)(CO)_3Cp'_2$, **10**, 232
$V_2Fe_2C_{12}H_{14}N_2O_2S_4$
$(MeCp)_2V_2S_4[Fe(NO)]_2$, **5**, 32
$V_2IrC_{30}H_{29}ClPS_4$
$(MeCp)_2V_2S_4IrCl(PPh_3)$, **5**, 32
$V_3C_{30}H_{45}Cl_6$
$[Cp^*VCl_2]_3$, **5**, 27
$V_3C_{30}H_{45}O_6$
$[Cp^*V(O)_2]_3$, **5**, 31
$V_3C_{42}H_{54}N_6O_{12}$
$[V(CNBu^t)_6][V(CO)_6]_2$, **5**, 4
$V_3SiTiW_9C_{69}H_{149}N_4O_{40}$
$[NBu_4]_4[TiCp(SiW_9V_3O_{40})]$, **4**, 308
$V_4C_{20}H_{20}Cl_4O_4$
$[CpVCl(O)]_4$, **5**, 31
$V_4C_{20}H_{20}I_2N_2O_6$
$(CpVI)_2[CpV(NO)]_2(O)_4$, **5**, 38
$V_4C_{20}H_{20}O_4$
$Cp_4V_4(O)_4$, **5**, 40
$V_4C_{20}H_{20}S_4$
$Cp_4V_4S_4$, **5**, 32
$V_4C_{24}H_{28}S_4$
$[(MeCp)_4V_4S_4]^+$, **5**, 32
$(MeCp)_4V_4S_4$, **5**, 32
$V_4C_{40}H_{60}O_6$
$[Cp^*V]_4(O)_6$, **5**, 31
$V_4Zn_2C_{48}H_{96}Cl_{12}O_{12}$
$[V_2Cl_3(THF)_6]_2[Zn_2Cl_6]$, **12**, 1086
$V_5C_{25}H_{25}O_6$
$(CpV)_5(O)_6$, **5**, 40
$V_5C_{25}H_{25}S_6$
$Cp_5V_5(S)_6$, **5**, 32
$V_5C_{30}H_{35}S_6$
$(MeCp)_5V_5S_6$, **5**, 32
$V_6C_{25}H_{25}O_9$
$Cp_5(O)V_6(O)_8$, **5**, 40
$V_6C_{30}H_{30}O_8$
$(CpV)_6(O)_8$, **5**, 40

W

$WAlC_{12}H_{28}Cl_4O_2P_3$
$WCl(CO)(PMe_3)_3$-$(HCCOAlCl_3)$, **5**, 372
$WAlC_{13}H_{21}$
$WCp_2[(H)_2AlMe_3]$, **1**, 549
$WAlC_{18}H_{41}Cl_2NP_3$
$WH_2(PMe_3[(H)_3AlCl_2(NMe_3)]$, **1**, 549
$WAsCoFeMoC_{19}H_{16}O_7S$
$FeCoMoW(S)(AsMe_2)Cp_2$-$(CO)_7$, **10**, 37, 244
$WAs_2C_{15}H_{30}Br_2F_6O_6P_2$
$WBr_2(CO)[P(OMe)_3]_2[Me_2$-$AsC(CF_3)=C(CF_3)$-$AsMe_2]$, **5**, 262
$WAuB_9C_{32}H_{37}O_2P$
$[C_2B_9H_9Me_2]W(CO)_2$ $(\equiv CTol)Au(PPh_3)$, **10**, 8, 9
$WBC_{15}H_{15}IN_4O_4$
$WI(CO)(MeC_2Me)Tp'$, **5**, 367
$WBC_{15}H_{15}N_4O_4$
$[W(CO)(MeC_2Me)Tp']^+$, **5**, 367
$WBC_{19}H_{30}F_4N_3$
$[W(CNEt)(CNEt)_2Cp^*]BF_4$, **5**, 202, 203
$WBC_{27}H_{27}N_4O_4$
$[W(CO)(HCCPh)_2Tp']^+$, **5**, 370
$WBC_{58}H_{52}F_5P_4$
$[WF(CHCH_2Ph)(dppe)_2]BF_4$, **5**, 311
$WBRu_3C_{16}H_7O_{11}$
$H(CpW)[Ru_3(CO)_{11}BH]$, **1**, 297, 301
$WBRu_3C_{30}H_{17}O_{11}$
$H(CpW)Ru_3(CO)_{11}BC(Ph)C$-$PhH$, **1**, 301, 303
$WRu_3(H)[BC(Ph)=CPhH]$-$(CO)_{11}Cp$, **10**, 216, 217
$WB_3C_3H_{20}P$
$WH_3(PMe_3)B_3H_8$, **1**, 313, 315
$WB_7C_{20}H_{48}NO_4$
$[Bu_4N][(CO)_4WB_7H_{12}]$, **1**, 329, 331
$WB_9Co_2C_{18}H_{22}O_6$
$[[C_2B_9H_9Me_2]WCo_2(CO)_6$ $(\equiv CTol)]^-$, **10**, 8, 9
$WB_9FeC_{17}H_{22}O_5$
$[W(CO)_2(C_2B_9H_9Me_2)$-$(CTol)Fe(CO)_3]^-$, **10**, 43
$WB_9FeC_{18}H_{22}O_6$
$[W(CO)_2(C_2B_9H_8Me_2)(CHTol)$-$(CO)Fe(CO)_3]^-$, **10**, 43
$WB_9MoC_{21}H_{29}O_2$
$MoW(CO)_2(C_7H_7)(CTol)(C_2$-$B_9H_9Me_2)$, **10**, 33
$WB_9PtC_{18}H_{45}O_2P_2$
$W(CO)_2(Me_2C_2B_9H_9)Pt(PEt_3)_2$, **10**, 32
$WB_9RhC_{50}H_{52}O_2P_2$
$[C_2B_9H_9Me_2]W(CO)_2$-$(\equiv CTol)Rh(PPh_3)_2$, **10**, 8, 9
$WB_{10}CoC_{24}H_{37}O_2$
$W(C_2B_{10}H_9Me_2)(CO)_2(CTol)$-$CoCp^*$, **10**, 43
$WB_{10}FeC_{17}H_{23}O_5$
$W(CO)_2(C_2B_{10}H_{10}Me_2)$-$(CTol)Fe(CO)_3$, **10**, 43
WCH_2Cl_4
$WCl_4(=CH_2)$, **5**, 314
$WC_5H_3NO_3$
$W(CO)_3(MeCN)$, **12**, 894
$WC_5H_5ClNO_2$
$W(Cl)(NO)_2Cp$, **5**, 419
WC_6HNO_5
$W(CO)_5(CNH)$, **5**, 165
$WC_6H_3O_5$
$[WMe(CO)_5]^-$, **5**, 161
$WC_6H_6O_4$
$W(CO)_4(H_2)(C_2H_4)$, **5**, 332
$WC_6H_{10}Br_2NO$
$WBr_2(CO)(CNEt_2)$, **5**, 198
WC_6H_{18}
WMe_6, **5**, 282, 283
WC_6O_6
$W(CO)_6$, **5**, 156; **12**, 1092
$WC_7H_4O_5$
$W(CO)_5(C_2H_4)$, **5**, 332
$WC_7H_5O_2$
$W(CO)_2Cp$, **12**, 638
$WC_7H_5O_5P$
$W(CO)_5(\overline{BzCH_2})$, **5**, 244
$WC_7H_6N_2O_5$
$W(CO)_5(=NNMe_2)$, **5**, 238
WC_7H_6O
$W(CO)(HCCH)_3$, **5**, 363
$WC_7H_{10}O_3$
$W(CO)_3(H_2)(C_2H_4)_2$, **5**, 332
$WC_7H_{11}IO_2P$
$WI_2(CO)_2(PMe_3)_3(HCCH)$, **5**, 364
$WC_7H_{16}Cl_2O_2$
$W=CHBu^t(OMe)_2Cl_2$, **12**, 1211
WC_7N_7
$[W(CN)_7]^{2-}$, **5**, 307
$WC_8H_6O_3S$
$W(SH)(CO)_3Cp$, **5**, 438
$WC_8H_6O_6$
$(CO)_5W=C(OMe)Me$, **12**, 550
$WC_8H_{16}Cl_4O$
$WCl_4(MeCCMe)(OEt_2)$, **5**, 371
$WC_8H_{19}Cl_2OP$
$W=CHBu^t(O)(Me_3P)Cl_2$, **12**, 1211
$WC_8H_{19}P$
$WH_5(PMe_3)Cp$, **5**, 407
WC_8N_8
$[W(CN)_8]^{3-}$, **5**, 307
$WC_9Br_2F_{12}O$
$WBr_2(CO)(F_3CCCCF_3)_2$, **5**, 365
$WC_9H_6O_7$
$(CO)_5W=CHCO_2Et$, **12**, 395, 403
$WC_9H_8Br_2O_2$
$WBr_2(CO)_2(nbd)$, **5**, 347
$WC_9H_8O_2$
$CpW(\equiv CMe)(CO)_2$, **1**, 292
$WC_9H_8O_3$
$WMe(CO)_3Cp$, **5**, 410, 414
$WC_9H_{14}ClN$
$WClMe_3(NPh)$, **5**, 291, 296
$WC_9H_{16}O$
$WMe(CO)(C_2H_2)Cp$, **5**, 414
$WC_{10}H_8O_3$
$W(CO)_3(C_7H_8)$, **5**, 354
$WC_{10}H_{10}O_2$
$W(COMe)(CO)(C_2H_2)Cp$, **5**, 414
$WC_{10}H_{11}O_2$
$CpN(CO)_2(CH_2=CHMe)$, **12**, 969
$WC_{10}H_{11}O_3S_2$
$[W(CO)_3(S_2Me_2)Cp]^+$, **5**, 438
$WC_{10}H_{12}$
WH_2Cp_2, **5**, 443
$WC_{10}H_{12}O_2$
$WMe(CO)_2(C_2H_4)Cp$, **5**, 412
$WC_{10}H_{15}O_6P$
$W(CO)_3[P(OMe)_3](C_4H_6)$, **5**, 347
$WC_{10}H_{18}$
$WMe(C_3H_5)_3$, **1**, 84
$WC_{10}H_{26}Cl_2P_2$
$WCl_2(PMe_3)_2(C_2H_4)_2$, **5**, 334
$WC_{11}H_5IO_4$
$WI(CO)_4(CPh)$, **5**, 196
$WC_{11}H_8O_3$
$W(CHCCH_2)(CO)_3Cp$, **5**, 415
$W(CO)_3(cot)$, **5**, 347
$WC_{11}H_{10}O$
$W(CO)Cp_2$, **5**, 444
$WC_{11}H_{10}O_3$
$\overline{W(CHCHCOMe)}(CO)_2Cp$, **5**, 414
$WC_{11}H_{12}O_2$
$W(COEt)(CO)(HCCH)Cp$, **5**, 368
$WC_{11}H_{14}$
$W(H)(Me)Cp_2$, **5**, 446
$WC_{11}H_{15}I_2NO$
$WI_2(CO)(CNEt_2)Cp$, **5**, 198

$WC_{11}H_{18}OS$
WMe(O)(S)Cp*, **5**, 431
$WC_{11}H_{18}O_2$
$WMe(O)_2Cp$*, **5**, 431
$WC_{11}H_{18}S_2$
$WMe(S)_2Cp$*, **5**, 431
$WC_{11}H_{26}P_2$
$WH_3(RMe_3)_2Cp$, **5**, 407
$WC_{11}H_{27}Cl_3P_2$
$WCl_3(\equiv CBu^t)(PMe_3)_2$, **5**, 319
$WC_{11}H_{28}Cl_3P_2$
$WH(Cl)_3(=CHBu^t)(PMe_3)_2$, **5**, 319
$WC_{12}F_{12}$
$W(C_6F_6)_2$, **5**, 479
$WC_{12}H_6O_5$
$(CO)_5W$=CHPh, **12**, 391, 392, 394, 403, 415
$W(CO)_5(CHPh)$, **5**, 170
$WC_{12}H_8O_4$
$W(CO)_4$(cot), **5**, 347
$WC_{12}H_{10}O_2$
$W(CO)_2Cp_2$, **5**, 444
$WC_{12}H_{12}O_3$
$W(CO)_3(C_6H_3Me_3)$, **5**, 495
$WC_{12}H_{13}I$
$W(I)(CHCH_2)Cp_2$, **5**, 446
$WC_{12}H_{15}$
$[W(H)(C_2H_4)Cp_2]^+$, **5**, 446
$WC_{12}H_{15}NO_2$
$W(CO)_2(CNEt_2)Cp$, **5**, 200
$WC_{12}H_{15}N_3O_3$
$W(CO)_3(NCEt)_3$, **5**, 219
$WC_{12}H_{16}$
Cp_2WMe_2, **1**, 84
$WC_{12}H_{17}O_6P$
$W(CO)_3[P(OMe)_3]$(chd), **5**, 347
$WC_{12}H_{18}$
$W(C_4H_6)_3$, **5**, 346
$WC_{12}H_{20}$
$W(C_3H_5)_4$, **1**, 84
$WC_{12}H_{22}N_2OS_4$
$WO(HCCH)(S_2CNEt_2)_2$, **5**, 371
$WC_{13}H_4N_2O_5$
$(CO)_5W(CNC_6H_4NC)$, **5**, 162
$WC_{13}H_8O_6$
$(CO)_5W$=C(OMe)Ph, **12**, 389, 566
$WC_{13}H_{20}BrNO_5$
$[NEt_4][WBr(CO)_5]$, **5**, 227
$WC_{13}H_{20}ClNO_5$
$[NEt_4][WCl(CO)_5]$, **5**, 220
$WC_{13}H_{29}Cl_4N$
$NEt_4[WCl_4(\equiv CBu^t)]$, **5**, 319
$WC_{13}H_{37}ClP_4$
$WCl(PMe_3)_4(\equiv CH)$, **5**, 318
$WC_{13}H_{40}OP_4$
$WH_2(OCH_2)(PMe_3)_4$, **5**, 305
$WC_{14}H_{11}F_5N_2O_3$
$WF(CO)_3[C_6F_4CH(=NCH_2C$ - $H_2NMe_2)]$, **5**, 227
$WC_{14}H_{13}N_6O_2S$
$[W(CO)_2(CSMe)[HC(pz)_3]]^+$, **5**, 200
$WC_{14}H_{16}$
$W(PhMe)_2$, **5**, 529
$WC_{14}H_{17}O_6$
$W(CO)_3[P(OMe)_3]$(cot), **5**, 347
$WC_{14}H_{18}N_2$
$W(NCMe)_2(C_3H_5)$(Tol), **5**, 343
$WC_{14}H_{20}N_2O_6$
$[NEt_4][W(CNO)(CO)_5]$, **5**, 161, 227
$[NEt_4][W(NCO)(CO)_5]$, **5**, 227
$WC_{14}H_{32}ClNO$
$WCl(CH_2Bu^t)_2(NEt_2)(O)$, **5**, 296
$WC_{15}H_{12}O_2$
$W(CO)_2$(CTol)Cp, **5**, 198, 201, 416
$WC_{15}H_{13}O_2$
$[W(CO)_2(CHTol)Cp]^{+\cdot}$, **5**, 168
$WC_{15}H_{16}N_3O_2$
$[W(NCMe)_3(CO)_2(C_7H_7)]^+$, **5**, 360
$WC_{15}H_{18}O_3$
$W(CO)_3(C_6Me_6)$, **5**, 495
$WC_{15}H_{19}O_2P$
$W(CCC_3H_5)(CO)_2(PMe_3)Cp$, **5**, 165
$WC_{15}H_{30}Br_2O_2$
$WBr_2(OCH_2Bu^t)_2[=CH(CH_2)_3$-$CH_2]$, **5**, 314
$WC_{15}H_{30}ClO_4P_2$
$[W[OC(PMe_3)=CMeC(O)]Cl$-$(CO)_2(PMe_3)_2$, **5**, 320
$WC_{15}H_{30}P_3$
$WH_2(PMe_3)_3(Me_2C_4H_4)$, **5**, 349
$WC_{15}H_{32}Br_2O_2$
$W=CHBu^t(OCH_2Bu^t)_2Br_2$, **12**, 1212
$WC_{15}H_{33}ClO$
$W(CH_2Bu^t)_3Cl(O)$, **5**, 296
$WC_{15}H_{33}NO_3$
$W(OBu^t)_3(\equiv CNMe_2)$, **5**, 318
$WC_{15}H_{42}P_4$
$WMe(PMe_3)_4(\equiv CMe)$, **5**, 318
$WC_{15}H_{45}P_5$
$WH(CH_2PMe_2)(PMe_3)_4$, **5**, 305
$WC_{16}H_{10}Cl_2NO_4$
$[WCl_2(CO)_4(py)(CPh)]^-$, **5**, 196
$WC_{16}H_{10}O_3$
$W(CCPh)(CO)_3Cp$, **5**, 165, 166
$WC_{16}H_{23}NO_2$
$W(CO)_2(CH_2CHNEt)Cp$*, **5**, 198
$W(CO)_2(CMeNEt)Cp$*, **5**, 198
$WC_{16}H_{28}IN_2O_2$
$WI(CO)_2(CNBu^t_2)(CNEt_2)$, **5**, 197
$WC_{16}H_{38}ClF_3O_5P_4S$
$[WCl(CH_2PMe_3)(CO)_2$-$(PMe_3)_3]$OTf, **5**, 219
$WC_{17}H_{16}O_3$
$W(CO)_3(nbd)_2$, **5**, 347
$WC_{17}H_{21}ClN_2O_4$
$WCl(CO)_4$(TMEDA)(CPh), **5**, 196
$WC_{17}H_{27}Cl_2$
$WCl_2(MeCCMe)(C_5Me_4Bu^t)$, **5**, 371
$WC_{17}H_{40}Cl_2OP_2$
$WCl_2(=CHBu^t)(PEt_3)_2(O)$, **5**, 313
$WC_{17}H_{42}ClP_4$
$[WH(Cl)(\equiv CBu^t)(dmpe)_2]^+$, **5**, 320
$WC_{18}H_{10}O_5$
$W(CO)_5(CPh_2)$, **5**, 169, 170
$WC_{18}H_{21}O_2P$
$W(CO)(PMe_3)$[C(O)C(Tol)]Cp, **5**, 372
$WC_{18}H_{54}P_6$
$W(PMe_3)_6$, **5**, 304
$WC_{19}H_{15}ClN_2O_2$
$WCl(CO)_2(py)_2(CPh)$, **5**, 196
$WC_{19}H_{24}O_2P$
$[W(CO)(PMe_3)$[(Tol)C-COMe]Cp$]^+$, **5**, 372
$WC_{19}H_{25}N_2O_2S_4$
$[W(CO)(S_2CnEt_2)_2(PhCCO)]^-$, **5**, 195
$WC_{19}H_{43}NO$
$W(CH_2Bu^t)_3(NEt_2)(O)$, **5**, 296
$WC_{20}H_{22}O_4$
$W[C(OEt)Ph](CO)_3$(cod), **5**, 347
$WC_{20}H_{28}Cl_3P_2$
$WCl_3(PMe_3)_2$(PhCCPh), **5**, 371
$WC_{21}H_{44}O_3P_2$
$W(H_2)(CO)_3(PPr^i_3)_2$, **5**, 224, 225
$WC_{22}H_{31}ClP_2$
$WCl(PMe_3)_2(\equiv CMe)$(PhCCPh), **5**, 320
$WC_{23}H_{19}O$
$W(CCBu^t)(CO)(Bu^tCCH)Cp$*, **5**, 165
$WC_{23}H_{37}Cl_2P_3$
$WCl_2(PMe_3)_3$(PhCCPh), **5**, 364
$WC_{24}H_{20}$
$W(PhCCPh)Cp_2$, **5**, 368
$WC_{24}H_{23}F_{12}NO_2$
$W(CHCMe_2Ph)$(NPh)[OCMe-$(CF_3)_2]_2$, **12**, 1091
$WC_{24}H_{24}$
$W[CH_2C_6H_4CH_2]_3$, **5**, 282
$WC_{25}H_{15}O_5P$
$W[P(Ph)C(Ph)C(Ph)](CO)_5$, **9**, 540
$WC_{25}H_{39}F_6NO_2$
$W[OCMe_2(CF_3)]_2(=CHBu^t)$-$(NC_6H_3Pr^i_2)$, **5**, 314, 315
$WC_{25}H_{49}O_4$
$(Bu^tCH_2O)_4W$=Cyclopentyl, **12**, 596
$WC_{26}H_{22}IOP$
$WI(CO)(PPh_3)(C_7H_7)$, **5**, 360
$WC_{26}H_{23}OP$
$WH(CO)(PPh_3)(C_7H_7)$, **5**, 360
$WC_{26}H_{30}$
$WBz_3(\equiv CBu^t)$, **5**, 318
$WC_{27}H_{27}P$
$W(PMe_3)(HCCPh)_3$, **5**, 363
$WC_{27}H_{41}Br_2N_4$

$WBr_2(CNBu^t)_3(PhCCNBu^t)$, **5**, 372

$WC_{28}H_{20}BrO_3P$
$WBr(CO)_3(PPh_3)(CPh)$, **5**, 197

$WC_{28}H_{20}Cl_5$
$WCl_3(PhCCPh)_2Cl_2$, **5**, 371

$WC_{29}H_{35}Cl_2OP_2$
$W(OCMe_2)Cl_2(PMePh_2)_2$, **12**, 1091

$WC_{30}H_{34}Cl_2OP_2$
$W(O)(CH_2CH_2)_2Cl_2(PMePh_2)_2$, **12**, 1091

$WC_{31}H_{26}O_3P_2$
$W(CO)_3(dppe)(HCCH)$, **5**, 363

$WC_{33}H_{33}O_3$
$W(OC_6H_3Me_2)_3(C_3Et_3)$, **5**, 319

$WC_{36}H_{44}N_4$
$W_2(CPh)(C_3Ph_3)(NMe_2)_4$, **5**, 321

$WC_{37}H_{30}O_3P_2$
$W(CO)_3(CCHPh)(dppe)$, **5**, 171

$WC_{38}H_{20}O_{10}P_2$
$W_2(Co)_{10}(Ph_2PC{\equiv}CC{\equiv}CPPh_2)$, **5**, 258

$WC_{38}H_{40}N_2S_4$
$W(S_2CNEt_2)_2(Ph_2C_2)_2$, **5**, 350

$WC_{39}H_{66}N_2O_3P_2$
$W(N_2)(CO)_3(PCy_3)_2$, **5**, 235

$WC_{39}H_{66}O_3P_2$
$W(CO)_3(PCy_3)_2$, **5**, 225

$WC_{39}H_{68}O_3P_2$
$W(H_2)(CO)_3(PCy_3)_2$, **5**, 225

$WC_{41}H_{30}ClNO_5P_2$
$[PPN][WCl(CO)_5]$, **5**, 220

$WC_{41}H_{46}Cl_2OP_3$
$WCl_2({\equiv}PC_6H_2Bu^t{}_2)(CO)(PMePh_2)_2$, **5**, 246

$WC_{43}H_{40}NO_7P_3$
$[PPN][WH(CO)_4[P(OMe)_3]]$, **5**, 222

$WC_{44}H_{33}N$
$W(NCMe)(PhCCPh)_3$, **5**, 363

$WC_{45}H_{55}Cl_2OP_3$
$WCl_2({\equiv}PC_6H_2Bu^t{}_3)(CO)(PMePh_2)_2$, **5**, 244

$WC_{47}H_{43}F_6N_2O_3P_3$
$[W(CO)_3(NC_4H_4NMe)(PCy_3)_2]PF_6$, **5**, 236

$WC_{52}H_{44}Cl_2P_4$
$WCl_2(PMePh_2)_4$, **12**, 1091

$WC_{52}H_{48}N_4P_4$
$W(N_2)_2(dppe)_2$, **5**, 304

$WC_{53}H_{48}N_2OP_4$
$W(CO)(N_2)(dppe)_2$, **5**, 238

$WC_{53}H_{48}OP_2$
$W(CO)(dppe)_2$, **5**, 239

$WC_{53}H_{48}OP_4$
$WH(CO)(dppe)[C_6H_4P(Ph)-CH_2CH_2PPh_2]$, **5**, 239

$WC_{54}H_{49}FO_2P_4$
$WF(CCH_2CO_2Me)(dppe)_2$, **5**, 311

$WC_{55}H_{51}FO_2P_4$
$WF({\equiv}CCO_2Me)(dppe)_2$, **5**, 318

$WC_{55}H_{54}NOP_4$
$WH(CONMe_2)(dppe)_2$, **5**, 304

$WC_{55}H_{55}NO_2P_4$
$W(CO)(DMF)(dppe)_2$, **5**, 304

$WC_{57}H_{40}O$
$W(CO)(Ph_2C_2)_2(C_4Ph_4)$, **5**, 351

WCl_6
WCl_6, **12**, 1090

$WCoC_{13}H_8O_5$
$Co(CH_2CCH)(CO)_3W(CO)_2(Cp)$, **12**, 687

$WCoC_{37}H_{28}O_7P$
$W(CO)_4[CTolCMeCMeC(O)]Co(CO)_2(PPh_3)$, **10**, 44

$WCoC_{39}H_{31}O_4P_2$
$W(CO)_2Cp(PPh_2)(H)Co(CO)_2(PPh_3)$, **10**, 44

$WCoFeC_{12}H_{26}O_8P_3$
$WFeCo(PMe)(CO)_8(PMe_3)_2Cp$, **10**, 222

$WCoFeC_{14}H_8O_8P$
$WFeCo(PMe)(CO)_8Cp$, **10**, 37, 226, 244, 245

$WCoFeC_{19}H_{22}O_8P_2$
$W(CO)Cp(CO)(dppm)[PMeFe(CO)_4]Co(CO)_2$, **10**, 44

$WCoRuC_{15}H_9O_8$
$HWRuCo(CMe)Cp(CO)_8$, **10**, 41

$WCoRuC_{19}H_{10}O_8P$
$WRuCoCp(CO)_8(PPh)$, **10**, 19, 20

$WCoZnC_{16}H_5O_{11}$
$[Co(CO)_4]_2[WCp(CO)_3]Zn^-$, **10**, 340

$WCo_2C_{14}H_6O_8$
$WCo_2(CH)Cp(CO)_8$, **10**, 49

$WCo_2C_{21}H_{12}O_8$
$CpWCo_2(CO)_8({\equiv}CTol)$, **10**, 7, 8

$WCrC_{16}H_{13}O_3$
$CrCp(C_8H_8)W(CO)_3$, **10**, 33

$WCr_2C_{17}H_5O_{12}P$
$Cr_2W(P)(CO)_{12}Cp$, **10**, 218

$WCuPtC_{25}H_{27}O_2$
$CuPtW(CC_6H_4Me)(CO)_2CpCp^*$, **3**, 116

$WFeC_8O_8Se_2$
$[FeW(CO)_8Se_2]^{2+}$, **7**, 74

$WFeC_{14}H_{10}O_5$
$W(CO)_2Cp(MeC{=}C{=}CH_2)Fe(CO)_3$, **10**, 42

$WFeC_{16}H_{14}O_3$
$CpW(CO)_2(C_3H_4)Fe(CO)Cp$, **10**, 15, 42

$WFeC_{19}H_{10}O_5$
$Cp(CO)_2FeCCCCW(CO)_3Cp$, **5**, 165

$WFeC_{19}H_{12}O_6$
$CpWFe(CO)_6(CTol)$, **10**, 91

$WFeC_{22}H_{18}O_5$
$CpWFe(CO)_5(CTolCMeCMe)$, **10**, 96

$WFeC_{57}H_{44}O_7P_4$
$(dppm)(CO)_3W(dppm)Fe(CO)_4$, **7**, 83

$WFeRhC_{27}H_{19}O_5$
$CpWRhFe(CO)_5(C_9H_7)({\equiv}CTol)$, **10**, 8

$WFe_2C_{10}O_{10}Te_2$
$Fe_2W(CO)_{10}Te_2$, **7**, 77

$WFe_2C_{17}H_{11}O_9PS_2$
$Fe_2W(CO)_9(PMe_2Ph)S_2$, **7**, 72

$WFe_3C_{14}O_{14}$
$[WFe_3(CO)_{14}]^{2-}$, **10**, 194

$WFe_3C_{15}BrO_{15}P$
$WFe_3(P)(Br)(CO)_{15}$, **10**, 218

$WFe_3C_{19}H_{11}O_{11}PS_2$
$Fe_3W(CO)_{11}(PMe_2Ph)S_2$, **7**, 72

$WIrC_{17}H_{15}O_7$
$W(CO)_5Ir(CO)_2Cp^*$, **10**, 29

$WIrC_{47}H_{36}O_5P_3$
$WIr(CO)_5(H)(PPh_3)(PPh_2)_2$, **8**, 333

$WK_5C_7N_7$
$K_5[W(CN)_7]$, **5**, 305

$WLi_2C_5O_5$
$Li_2[W(CO)_5]$, **5**, 160

$WMoC_{14}H_9O_6$
$CpMo(CO)_2(C_3H_4C{=}OCpW(CO)_3)$, **10**, 15

$WMoC_{14}H_{10}O_4$
$Mo(CO)_2CpW(CO)_2Cp$, **10**, 29, 32

$WMoC_{17}H_{14}O_4$
$CpMo(CO)_2(C_3H_4)W(CO)_2Cp$, **10**, 15

$WMoC_{18}H_{16}O_4$
$Mo(CO)Cp(CO)(MeC_2Me)W(CO)_2Cp$, **10**, 33

$WMoC_{20}H_{13}O_7$
$Mo(CO)_5[CH(Tol)]W(CO)_2Cp$, **10**, 33

$WMoC_{24}H_{20}O_5$
$MoW(CO)_4(CTolCOMe)Cp_2$, **5**, 415

$WMoC_{28}H_{30}O_3$
$Mo(CO)_2Cp(CTolCMeCMe)(MeC_2Me)_2W(CO)$, **10**, 33

$WMoC_{35}H_{30}NO_6P_2$
$[(CO)_4Mo(PPh_2)_2W(CO)_2(CNEt_2)]^-$, **5**, 197

$WMoOs_3C_{34}H_{20}O_8$
$Cp_2MoWOs_3(CO)_8(C)(CPh)(C_2Ph)$, **10**, 105

$WMo_2Pt_4C_{47}H_{49}O_6$
$Pt_4WMo_2(CMe)(CC_6H_4Me)(CO)_6(cod)_2Cp_3$, **9**, 564

WNa_2O_4
Na_2WO_4, **11**, 413; **12**, 1178

$WNiC_{15}H_{12}O_4$
$NiCp(CO)W(C_5H_4Me)(CO)_3$, **9**, 15

$WNiC_{16}H_{20}O_2$
$W(CO)_2Cp'(CH_2CCMe_2)NiCp$, **10**, 45

$WNiC_{18}H_{19}O_2$
$W(CO)_2Cp'(CHCMeCMe)NiCp$, **10**, 45

$WNiC_{21}H_{26}O_2$

$W(CO)_2Cp(CH_2C(CH_2)_2NiCp^*$, **10**, 45
$WNiC_{22}H_{28}O_2$
NiCp*[C(Me)C(Me)CH_2]-(CO)W(CO)Cp, **9**, 19
$WNiC_{30}H_{32}O_2S_2$
NiCp*(SPh)(CO)W(SPh)-(CO)Cp', **9**, 169
$WNi_2C_{30}H_{32}O_2$
$Ni_2W(CC_6H_4Me)(CO)_2Cp_2Cp^*$, **9**, 23
$WNp_3C_5H_9$
$WNp_3(\equiv CBu^t)$, **5**, 315
$WOsSiC_{13}H_{38}P_3S_4$
$[Os(CH_2TMS)(PMe_3)_3(S)_2W$-$S_2]^-$, **7**, 375
$WOsSi_2C_9H_{25}NS_4$
$Os(N)(CH_2TMS)_2(S)$-(SMe)WS_2, **7**, 375
$WOsSi_2C_{10}H_{22}N_3S_4$
$[Os(N)(CH_2TMS)_2(S)_2W$-$(SCN)_2]^-$, **7**, 375
$WOs_2C_{20}H_{11}O_{12}PS_2$
$WOs_3(S)_2(CO)_{12}(PMe_2Ph)$, **10**, 228
$WOs_2C_{21}H_{10}O_8$
$CpWOs_2(CO)_8(C\equiv CPh)$, **10**, 21
$WOs_2C_{22}H_{13}NO_7$
$CpWOs_2(CO)_7(NCMe)(C_2Ph)$, **10**, 102
$WOs_2C_{22}H_{18}N_2O_{12}$
$W(CO)_5Os(CO)_3(Bu^tNC)Os$-$(CO)_4(Bu^tNC)$, **10**, 30
$WOs_2C_{22}H_{18}O_{18}P_2$
$W(CO)_5Os(CO)_4Os(CO)_3[P$-$(OCH_2)_3CMe]_2$, **10**, 30
$WOs_3C_{14}HO_{14}$
$[WOs_3(H)(CO)_{14}]^-$, **10**, 191, 203
$WOs_3C_{14}H_6O_{11}$
$WOs_3(H)(O)_2(CO)_9Cp$, **10**, 235
$WOs_3C_{16}H_6O_{11}$
$WOs_3(H)(CO)_{11}Cp$, **10**, 208
$WOs_3C_{16}H_6O_{11}S$
$Os_3W(CO)_{11}Cp(H)(S)$, **10**, 15
$WOs_3C_{20}H_{11}O_{12}PS_2$
$WOs_3(S)_2(CO)_{12}(PMe_2Ph)$, **10**, 228, 236
$WOs_3C_{21}H_{20}O_{10}$
$Cp^*WOs_3(CO)_9(H)(O)(CHMe)$, **10**, 96
$WOs_3C_{24}H_{10}O_{11}$
$CpWOs_3(CO)_{11}(C_2Ph)$, **10**, 21, 104
$WOs_3C_{27}H_{22}O_{11}PS_2$
$WOs_3(S)_2(CO)_{11}(PMe_2Ph)_2$, **10**, 236, 237
$WOs_3C_{28}H_{22}O_{12}P_2S_2$
$WOs_3(S)_2(CO)_{12}(PMe_2Ph)_2$, **10**, 236, 237
$WPbC_{26}H_{20}O_3$
$CpW(Co)_3PbPh_3$, **2**, 313
$WPtC_{17}H_{20}O_3$
$Cp(CO)_3WPtMe(cod)$, **9**, 563
$WPtC_{22}H_{23}O_3$
$Cp(CO)_3WPtPh(cod)$, **9**, 563
$WPtC_{57}H_{46}O_3P_4$
$W(CO)_3(C_2H)(dppm)_2Pt(C_2H)$, **10**, 47
$WReC_{17}H_7O_9$
$[W(CO)_5(CTol)Re(CO)_4]^-$, **10**, 33
$WReC_{17}H_8O_9$
$[W(CO)_5(CHTol)Re(CO)_4]^-$, **10**, 33
$WRhC_{32}H_{25}O_2$
$W(CO)_2Cp(CTolCHCPh)Rh$ (C_9H_7), **10**, 44
$WRhC_{47}H_{36}O_5P_3$
$W(CO)_4(PPh_2)_2RhH(CO)$-(PPh_3), **10**, 44
$WRuC_{15}H_8O_5$
$W(CO)_3(C_5H_4C_5H_4)Ru(CO)_2$, **10**, 42
$WRu_2C_{19}H_{11}NO_8$
$WRu_2(H)(NPh)(CO)_8Cp$, **10**, 218
$WRu_2C_{20}H_{30}S_6$
$Ru_2Cp^*_2(S_2)[(S)(S)_2WS]$, **7**, 670
$WRu_3BC_{16}H_7O_{11}$
$WRu_3(H)(BH)(CO)_{11}Cp$, **10**, 216, 217
$WRu_3C_{21}H_{16}O_{11}$
$WRu_3(H)(CO)_{11}Cp^*$, **10**, 208
$WRu_3C_{26}H_{21}O_{10}P$
$WRu_3(H)(PPh)(CO)_{10}Cp^*$, **10**, 222
$WRu_4C_{37}H_{33}O_{13}P_3S_2$
$(PhMe_2P)_2(CO)_9(S)Ru_4(S)W$-$(CO)_4(PMe_2Ph)$, **7**, 868
$WSbC_{17}H_{27}O_5S_3$
$W(CO)_5(SBu^t)(SbSBu^t)_2$, **5**, 262
$WSiC_{10}H_{14}O_2$
$WH(CO)_2(CH_2SiMe_2)Cp$, **5**, 412
$WSiC_{11}H_{14}O_3$
$W(CH_2SiMe_2H)(CO)_3Cp$, **5**, 412
$WSiC_{13}H_{18}$
$W(CH_2SiMe_2)Cp_2$, **5**, 447
$WSiC_{15}H_{27}O_2P$
$W(CO)_2(PMe_3)(SiH_3)Cp^*$, **5**, 409
$WSiC_{24}H_{33}F_{12}NO_2$
$W[OCMe(CF_3)_2]_2(=CHTMS)$-$(NC_6H_3Pr^i_2)$, **5**, 315
$WSiC_{28}H_{23}NO_5$
$W(CO)_5[C(NC_4H_8)SiPh_3]$, **5**, 191
$WSiC_{29}H_{45}F_{12}NO_2$
$W[CH_2CH(TMS)CH(TMS)]$-$[OCMe(CF_3)_2]_2(N$-$C_6H_3Pr^i_2)$, **5**, 315
$WSiZrC_{28}H_{30}O_3$
Zr[C(TMS)C(H)Ph][OCW-$(CO)_2Cp]Cp_2$, **4**, 535
$WSnC_{32}H_{26}O_2$
$W(SnPh_3)(CO)_2(CHPh)Cp$, **5**, 169
$WTiC_{26}H_{24}O_2$
$TiCp_2(CTol=CH_2)W(CO)_2Cp$, **10**, 27
$WTiC_{40}H_{34}O_4$
$Ti(C_5H_4Me)_2(PPh_2)_2W(CO)_4$, **4**, 410
$WUC_{35}H_{29}O_6P$
$Cp_3UOCHCHP(Ph)_2CH_2W$ $(CO)_5$, **4**, 156
$WVC_{18}H_{11}O_7$
$(C_{10}H_8)[W(CO)_3Me][V(CO)_4]$, **5**, 24
$WZnC_{13}H_{17}NO_3$
$Me_2N(CH_2)_3ZnWCp(CO)_3$, **3**, 187; **10**, 337
$WZrC_{18}H_{15}ClO_4$
$Cp_2(Cl)ZrOW(CO)_3Cp$, **4**, 528
$WZrC_{20}H_{25}ClO_3$
$Cp^*WO_3Zr(Cl)Cp_2$, **4**, 528
$WZrC_{20}H_{29}ClO_2PS$
$Cp(CO)_2(PMe_3)W(S)Zr$-$(Cl)Cp_2$, **4**, 532
$WZrC_{28}H_{20}O_7$
$Zr(OPh)[OC[W(CO)_5]Ph]Cp_2$, **4**, 522
$W_2AlC_{18}H_{62}P_6$
$[WH_3(PMe_3)_3]_2[(H)_2AlH(H)_2]$, **1**, 548
$W_2Al_2C_{18}H_{65}$
$[WH_3(PMe_3)_3]_2[(H)_2AlH$-$(H)_2AlH(H)_2]$, **1**, 548
$W_2Al_2C_{22}H_{74}N_2P_6$
$[WH_3(PMe_3)_3]_2[(H)_2AlH$-$(N=CHMe)_2AlH(H)_2]$, **1**, 549
$W_2As_2C_7I_2O_7$
$W(Co)_4(I)(As_2)WI(CO)_3$, **5**, 262
$W_2BC_{18}H_{19}O_4$
$W_2(MeCB(H)Et)(CO)_4Cp_2$, **1**, 292–4
$W_2C_6H_{10}O_8$
$W_2Me_2(O_2CH)_4$, **5**, 323
$W_2C_{12}H_{14}Cl_4$
$W_2Cl_4(C_5H_4Me)_2$, **5**, 440
$W_2C_{14}H_{14}S_4$
$W_2(SCHCHS)_2Cp_2$, **5**, 438
$W_2C_{16}H_{10}O_6S$
$[W(CO)_3Cp]_2(S)$, **5**, 438
$W_2C_{17}H_{12}O_6$
$[CpW(CO)_2]_2(CH_2CO)(CO)$, **5**, 165
$W_2(CO)_5(COCH_2)Cp_2$, **5**, 411
$W_2C_{18}H_4N_2O_{10}$
$(CO)_5W(CNC_6H_4NC)W(CO)_5$, **5**, 162
$W_2C_{18}H_{10}O_6$
$Cp(CO)_3WCCW(CO)_3Cp$, **5**, 165, 411, 417
$W_2C_{18}H_{16}O_4$
$Cp_2W_2(CO)_4(C_2Me_2)$, **10**, 119
$W_2C_{18}H_{17}O_4$
$[W_2(CO)_4(H)(MeC_2Me)Cp_2]^+$, **5**, 376
$W_2C_{20}H_{26}Cl_2$
$[W(Cl)(C_3H_5)(PhMe)]_2$, **5**, 529
$W_2C_{20}H_{30}Cl_4$

$W_2(Cl)_4Cp^*_2$, **5**, 440
$W_2C_{20}H_{30}O_5$
$[W(O)_2Cp^*]_2(O)$, **5**, 433
$W_2C_{20}H_{30}P_6$
$Cp^*W(P_6WCp^*)$, **5**, 426
$W_2C_{20}H_{30}S_4$
$W_2S_4Cp^*_2$, **5**, 438
$W_2C_{22}H_{30}O_2S_3$
$W_2(S)_3(CO)_2Cp^*_2$, **5**, 438
$W_2C_{23}H_{30}O_5$
$Cp^*(CO)_3WW(O)_2Cp^*$, **5**, 433
$W_2C_{26}H_{54}O_6$
$(Bu^tO)_3W{\equiv}CC{\equiv}W(OBu^t)_3$, **5**, 318
$W_2C_{27}H_{56}O_6$
$(Bu^tO)_3W{\equiv}C(CH_2)C{\equiv}W(OBu^t)_3$, **5**, 318
$W_2C_{28}H_{32}Br_4$
$W_2Br_4(PhC_2Ph)(C_5H_4Pr^i)_2$, **5**, 375
$W_2C_{28}H_{52}N_2O_6$
$W_2(OPr^i)_6(py)_2$, **12**, 1090
$W_2C_{30}H_{24}O_4$
$Cp_2W_2(CO)_4(C_2Tol_2)$, **10**, 87, 96
$W_2C_{32}H_{66}O_8$
$[(Bu^tO)_3W(OCCNMe_2)]_2$, **5**, 319
$W_2C_{36}H_{78}O_7$
$W_2(OBu^t)(OCH_2Bu^t)_6(CMe)$, **5**, 322
$W_2C_{40}H_{76}N_2O_6$
$W_2(OCH_2Bu^t)_6(py)_2$, **12**, 1090
$W_2CoMoC_{27}H_{21}O_9$
$MoW_2Co(CO)_9Cp'_3$, **10**, 244
$W_2CoNiC_{24}H_{19}O_7$
$NiCoW_2(CO)_7CpCp'_2$, **9**, 173
$W_2FeC_{17}H_{10}O_7S$
$Cp_2W_2Fe(CO)_7S$, **7**, 66
$W_2Fe_2C_{15}O_{15}S_2$
$W_2Fe_2(S)_2(CO)_{15}$, **10**, 228, 233
$W_2Fe_2C_{16}H_{11}NO_7S_2$
$Fe_2W_2(CO)_5(py)(S)_2(OPr^i)_2$, **7**, 66
$W_2Fe_2C_{22}H_{11}O_{14}PS_2$
$Fe_2W(CO)_9(PMe_2Ph)(S)[SW(CO)_5]$, **7**, 72
$W_2Fe_2Sn_2C_{18}Cl_2O_{18}$
$Fe_2(CO)_8[Sn[W(CO)_5]Cl]_2$, **7**, 31
$W_2MoC_{19}H_{15}NO_5$
$MoW_2(N)(O)(CO)_4Cp_3$, **10**, 218
$W_2MoC_{37}H_{25}O_4$
$W(C_4Ph_4)Mo(CO)_2W(CO)_2Cp$, **10**, 33
$W_2NiC_{28}H_{36}O_4$
$Cp^*_2W_2Ni(CO)_4(CMe)_2$, **10**, 119
$W_2NiC_{30}H_{24}O_4$
$W(CO)_2Cp(CTol)NiW(CO)_2Cp(CTol)$, **10**, 45
$W_2NiC_{40}H_{44}O_4$
$NiW_2(CC_6H_4Me)_2(CO)_4Cp^*_2$, **9**, 23
$W_2RuC_{33}H_{24}O_7$
$Cp_2W_2Ru(CO)_7(C_2Tol_2)$, **10**, 99
$W_2Ru_4C_{34}H_{25}O_{12}P_2$
$Cp^*W_2Ru_4(CO)_{12}(PPh)_2$, **10**, 10
$W_2ZnC_{16}H_{10}O_6$
$Zn[W(CO)_3Cp]_2$, **5**, 408
$W_3AsFeC_{25}H_{15}O_{10}$
$W_3Fe(As)(CO)_{10}Cp_3$, **10**, 218
$W_3C_{29}H_{66}Cl_2O_9$
$W_3Cl_2(OPr^i)_9(CMe)$, **5**, 322
$W_3C_{29}H_{66}O_9$
$W_3(OPr^i)_9(CMe)$, **5**, 322
$W_3CoC_{27}H_{21}O_9$
$W_3Co(CO)_9(CpMe)_3$, **10**, 7, 210
$W_3Mo_2Pt_5C_{46}H_{39}O_{10}$
$Pt_5W_3Mo_2(CMe)(CC_6H_4Me)(CC_6H_4Me)(CO)_{10}(cod)Cp_2$, **9**, 564
$W_3Pt_2C_{45}H_{36}O_6$
$W(CO)_2Cp(CTol)Pt(CTol)W(CO)_2CpPt(CTol)W(CO)_2Cp$, **10**, 46
$W_4MnNb_2C_{51}H_{108}N_3O_{22}$
$(Bu_4N)_3[Mn(CO)_3(Nb_2W_4O_{19})]$, **6**, 84
$W_4Nb_2RhC_{10}H_{15}O_{19}$
$[Cp^*Rh(Nb_2W_4O_{19})]^{2-}$, **8**, 305
$W_4PtC_{60}H_{112}O_{10}P_2$
$Pt[C_2W_2(OBu^t)_5]_2(PMe_2Ph)_2$, **9**, 494
$W_4Si_2C_{34}H_{88}O_7$
$[W_2(OEt)_7(C\text{-}TMS)]_2$, **5**, 322
$W_5C_{24}O_{24}P_4$
$[W(CO)_4][W(CO)_5]_4(P_4)$, **5**, 253, 258
$W_5Ru_2TiC_{78}H_{154}N_4O_{40}$
$[NBu_4]_4[Ru_2(TiW_5CpO_{18})_2(CO)_4]$, **7**, 633
$W_5TiC_5H_7O_{18}$
$[TiCp(W_5O_{18})H_2]^-$, **4**, 308
$W_6Te_8C_{66}H_{40}O_{18}$
$[PPh_4]_2[W(CO)_3]_6(Te_2)_4$, **5**, 269
$W_8Ir_2Nb_4C_4HO_{42}$
$[(CO)_2Ir]_2H(Nb_2W_4O_{19})_2^{5-}$, **8**, 323
$W_9Nb_2O_{19}$
$[Nb_2W_4O_{19}]^{4-}$, **8**, 249
$W_9Nb_3SiO_{40}$
$SiNb_3W_9O_{40}^{7-}$, **8**, 178
$W_9SiTiV_3C_{69}H_{149}N_4O_{40}$
$[NBu_4]_4[TiCp(SiW_9V_3O_{40})]$, **4**, 308
$W_{10}Nb_2UC_{95}H_{195}N_5O_{38}$
$[NBu_4]_5[Cp_3U(NbW_5O_{19})_2]$, **4**, 189
$W_{10}Ti_2U_2C_{20}H_{20}O_{38}$
$[Cp_2U(TiW_5O_{19})]_2^-$, **4**, 189
$W_{12}Ir_3C_{108}H_{96}O_{40}P_7$
$[(Ph_3P)_2IrH_2]_3PW_{12}O_{40}$, **8**, 404
$W_{12}Ir_3C_{132}H_{126}O_{40}P_7$
$[(PPh_3)_2Ir(cod))]_3PW_{12}O_{40}$, **8**, 323

Y

$YAlC_{14}H_{22}O$
$Cp_2Y(AlH_4)(THF)$, **4**, 49
$YBC_{14}H_{22}O$
$Cp_2Y(BH_4)(THF)$, **4**, 48
$YB_2C_{10}H_{18}$
$[Cp_2Y(BH_4)_2]^-$, **4**, 48
$YC_{14}H_{18}ClO$
$Cp_2YCl(THF)$, **4**, 119
$YC_{15}H_{15}$
YCp_3, **4**, 53
$YC_{15}H_{17}O$
$Cp_3Y(H_2O)$, **4**, 56
$YC_{18}H_{23}O$
$(C_8H_8)Y(MeC_5H_4)(THF)$, **4**, 105
$YC_{20}H_{20}Cl_2$
$[Cp_2Y(Cl)]_2$, **4**, 36
$YC_{20}H_{36}N_4$
$[Cp_2Y(NMeCH_2CH_2NMe_2)_2]^-$, **4**, 41
$YC_{25}H_{34}N$
$Cp^*_2Y(C_5H_4N)$, **4**, 86
$YC_{25}H_{41}O$
$Cp^*_2YMe(THF)$, **4**, 86, 114, 116
$YC_{29}H_{41}$
$Cp^*_2YCH_2C_6H_3Me_2$, **4**, 86
$YC_{35}H_{41}O$
$Cp^*_2YMe(THF)$, **12**, 1207
$YC_{36}H_{60}$
$Y(Bu^t{}_3C_6H_3)_2$, **4**, 7
$YC_{36}H_{66}O_4$
$[Cp^*Y(OBu^t)_2]_2$, **4**, 70
$YCaC_{40}H_{60}Cl$
$Cp^*_2Ca(Cl)YCp^*_2$, **1**, 115
$YLiC_{36}H_{56}O$
$Cp^*_2Y(CCBu^t)_2Li(THF)$, **4**, 90
$YReC_{22}H_{33}OP$
$Re(PMe_2Ph)H_4Y(THF)Cp_2$, **10**, 53
$YRe_2C_{46}H_{69}OP_4$
$Re(PMe_2Ph)_2H_7Re(PMe_2Ph)_2Y(THF)Cp_2$, **4**, 109; **10**, 53
$YSi_2C_{27}H_{49}$
$Cp^*_2YCH(TMS)_2$, **4**, 84, 86, 121
$YSi_6C_{21}H_{57}$
$Y[CH(TMS)_2]_3$, **4**, 17
$Y_2Al_2C_{32}H_{58}N_2$
$[YCp_2[(H)_2AlH_2(NEt_3)]]_2$, **1**, 551
$Y_2C_{26}H_{32}F_6O_8S_2$
$[(C_8H_8)Y(OTf)(THF)]_2$, **4**, 100
$Y_2C_{28}H_{34}O_4$
$[[O(CH_2CH_2C_5H_4)_2]Y(OH)]_2$, **4**, 61
$Y_2C_{32}H_{46}O_4$
$[(MeOCH_2CH_2C_5H_4)_2Y(H)]_2$, **4**, 47
$Y_2C_{34}H_{50}O_4$
$[(C_9H_7)Y(OBu^t)_2]_2$, **4**, 63
$Y_2C_{40}H_{60}$
$Cp^*_2Y(CH_2C_5Me_4)(H)YCp^*$, **4**, 93
$Y_2C_{40}H_{60}Cl_2$
$Cp^*_2Y(Cl)YCp^*_2Cl$, **4**, 75, 76
$Y_2C_{40}H_{62}$
$[Cp^*_2Y(H)]_2$, **4**, 94, 121
$Y_2C_{42}H_{60}N_2O$
$Cp^*Y[C_6H_4CH_2NMe(CH_2)]$-$[C_6H_4CH_2NMe(CH_2)]$-$YCp^*(THF)$, **4**, 71
$Y_5C_{33}H_{49}O_9$
$(CpY)_5(OMe)_8(O)$, **4**, 33
$YbAlC_{12}H_{16}Cl_2$
$Cp_2Yb(Cl)_2AlMe_2$, **4**, 36
$YbAlC_{20}H_{30}Cl_4$
$Cp^*_2Yb(Cl_2)AlCl_2$, **4**, 75
$YbAlC_{30}H_{53}O$
$Cp^*_2Yb(Et)AlEt_2(THF)$, **4**, 88
$YbB_3C_3H_{18}$
$Yb[BH_3Me]_3$, **1**, 288, 290
$YbB_{10}C_{16}H_{38}N_7$
$[(MeCN)_6Yb][(H_2)B_{10}H_{12}]\cdot 2MeCN$, **1**, 349, 350
$YbBeC_{31}H_{33}$
$Cp^*_2Yb(Me)BeCp^*$, **1**, 48
$YbBeC_{31}H_{48}$
$Cp^*_2Yb(Me)BeCp^*$, **4**, 67
YbC_6H_5I
PhYbI, **4**, 124
YbC_8H_5
$[YbCCPh]^+$, **4**, 27
YbC_8H_8
$Yb(C_8H_8)$, **4**, 98
$YbC_9H_7O_4$
$CpYb(O_2CCHCHCO_2)$, **4**, 34
$YbC_{10}H_{10}$
$YbCp_2$, **4**, 3, 28
$YbC_{10}H_{12}N$
Cp_2YbNH_2, **4**, 41
$YbC_{11}H_{10}N$
Cp_2YbCN, **4**, 36
$YbC_{12}F_{10}$
$Yb(C_6F_5)_2$, **4**, 14, 124
$YbC_{12}H_2F_8$
$Yb(C_6F_4H)_2$, **4**, 14
$YbC_{13}H_9O_4$
$CpYb(O_2CC_6H_4CO_2)$, **4**, 34
$YbC_{14}H_{18}NOv3$
$Cp_2Yb(NO_2)(THF)$, **4**, 36
$YbC_{14}H_{25}Cl_2O_2$
$Cp^*YbCl_2(DME)$, **4**, 68
$YbC_{15}H_{14}F_3O_2$
$Cp_2Yb[MeC(O)CHC(O)CF_3]$, **4**, 40
$YbC_{15}H_{15}$
$YbCp_3$, **4**, 29, 53
$YbC_{15}H_{18}O_2$
$Cp_2Yb(acac)$, **4**, 40
$YbC_{15}H_{21}O$
$YbMe(THF)Cp_2$, **4**, 595
$YbC_{16}H_{16}$
$[Yb(C_8H_8)_2]^-$, **4**, 103
$YbC_{16}H_{22}FO$
$(MeC_5H_4)_2YbF(THF)$, **4**, 35
$YbC_{18}H_{21}$
$Yb(MeC_5H_4)_3$, **4**, 130
$YbC_{18}H_{28}O$
$[(CH_2)_2(C_6H_8)_2]Yb(THF)$, **4**, 96
$YbC_{18}H_{31}Cl_2O_2$
$Cp^*YbCl_2(THF)_2$, **4**, 68
$YbC_{20}H_{30}$
$YbCp^*_2$, **4**, 64, 66
$YbC_{22}H_{32}O_3$
$YbC_{10}H_8(THF)_3$, **4**, 2
$YbC_{23}H_{23}N_3$
$(C_8H_8)Yb(CpN)_3$, **4**, 98
$YbC_{24}H_{32}N_2$
$[(MeC_5H_4)_2Yb(NH_2)]_2$, **4**, 41
$YbC_{24}H_{38}ClO$
$Cp^*_2YbCl(THF)$, **4**, 75
$YbC_{24}H_{38}FO$
$Cp^*_2YbF(THF)$, **4**, 74
$YbC_{24}H_{40}FO$
$Cp^*_2YbF(Et_2O)$, **4**, 74
$YbC_{24}H_{40}F_6O_2P$
$Cp^*_2YbPF_6(DME)$, **4**, 74
$YbC_{24}H_{40}O$
$Cp^*_2Yb(Et_2O)$, **4**, 113
$YbC_{24}H_{41}NO$
$Cp^*_2Yb(NH_3)(THF)$, **4**, 65
$YbC_{25}H_{40}NS_2$
$Cp^*_2Yb(S_2CNEt_2)$, **4**, 78
$YbC_{26}H_{38}NS$
$Cp^*_2Yb(SPh)(NH_3)$, **4**, 78
$YbC_{26}H_{42}O_2$
$(Bu^tC_5H_4)_2Yb(THF)_2$, **4**, 30
$YbC_{38}H_{36}OP_2$
$(C_5H_4PPh_2)_2Yb(THF)$, **4**, 31, 32
$YbC_{42}H_{44}O_2$
$(C_5H_4Ph_2)_2Yb(THF)_2$, **4**, 110
$YbFeC_{10}H_9N_3O_4$
$(MeCN)_3YbFe(CO)_4$, **7**, 85
$YbK_2C_{20}H_{36}O_2$
$[K(DME)]_2Yb(cot)_2$, **1**, 25–7
$YbMnC_{20}H_{28}I_2O_6$
$[(C_5H_4)Mn(CO)_3]YbI_2(THF)_3$, **4**, 111
$YbMnC_{25}H_{30}O_5$

$Cp^*_2YbMn(CO)_5$, **4**, 108; **10**, 48
$YbMoC_{46}H_{44}O_6$
$Yb(THF)_2[(C_5H_4Ph_2)_2Mo(CO)_4]$, **4**, 110
$YbPtC_{44}H_{50}O_2P_2$
$(THF)_2Yb(C_5H_4PPh_2)_2PtMe_2$, **9**, 444
$YbSiC_8H_{21}O$
$[Yb(CH_2TMS)(Et_2O)]^+$, **4**, 25
$YbSiC_{16}H_{23}ClO$
$(H)(Cl)[[Me_2Si(C_5H_4)_2]Yb(THF)]$, **4**, 60
$YbSi_2C_7H_{19}$
$[YbCH(TMS)_2]^+$, **4**, 25
$YbSi_2C_{16}H_{26}$
$Yb(TMS\text{-}C_5H_4)_2$, **4**, 30
$YbSi_4C_{22}H_{42}Cl$
$[(TMS)_2C_5H_3]_2YbCl$, **4**, 127
$YbSi_6C_{21}H_{57}Cl$
$[[(TMS)_2CH]_3YbCl]^-$, **4**, 18
$YbSnC_{33}H_{64}O_2$
$Cp^*Yb[Sn(CH_2Bu^t)_3](THF)_2$, **4**, 64
$YbSn_2C_{38}H_{82}O_2$
$Yb[Sn(CH_2Bu^t)_3]_2(THF)_2$, **4**, 19
$Yb_2C_{22}H_{20}F_6O_6S_2$
$[Cp_2Yb(O_3SCF_3)]_2$, **4**, 36
$Yb_2C_{22}H_{26}$
$[Cp_2YbMe]_2$, **4**, 44
$Yb_2C_{24}H_{28}Cl_2$
$[(MeC_5H_4)_2YbCl]_2$, **4**, 36
$Yb_2C_{26}H_{34}$
$[(MeC_5H_4)_2YbMe]_2$, **4**, 44
$Yb_2C_{40}H_{60}F$
$[YbCp^*_2]_2F$, **4**, 74
$Yb_2C_{40}H_{60}Te_2$
$(Te_2)[YbCp^*_2]_2$, **4**, 78
$Yb_2C_{42}H_{66}$
$[Cp^*_2YbMe]_2$, **4**, 84
$Yb_2C_{46}H_{57}O_4$
$Ph_2Yb(THF)(Ph)_3Yb(THF)_3$, **4**, 16
$Yb_2Co_3Si_2C_{60}H_{86}O_4$
$[Cp^*_2Yb]_2(CO)_4Co_3(TMS\text{-}C_5H_4)_2$, **4**, 108
$Yb_2Fe_3C_{51}H_{60}O_{11}$
$[Cp^*_2Yb]_2[Fe_3(CO)_{11}]$, **4**, 108
$Yb_2Si_4C_{28}H_{40}Cl_2O_2$
$[[O(Me_2SiC_5H_4)_2Yb(Cl)]_2$, **4**, 61
$Yb_4C_{60}H_{90}F_4$
$Cp^*_6Yb_4(F)_4$, **4**, 68
$Yb_5C_{54}H_{85}Cl_8O_3$
$Cp^*_5Yb_5(O)(Cl)_8(OEt)_2$, **4**, 68
$Yb_5C_{60}H_{90}F_9$
$Cp^*_6Yb_5(F)_9$, **4**, 74
$Yb_5C_{108}H_{140}$
$Yb[(CCPh)_2(YbCp^*_2)_2]_2$, **4**, 90

Z

$ZnBC_{25}H_{53}N_6$
$[HB(Bu^tpz)_3]ZnBu^t$, **3**, 182
$ZnB_2C_6H_{22}Cl_2P_2$
$ZnCl_2(B_2H_4{\cdot}2PMe_3)$, **1**, 311
ZnB_5CH_{11}
$(MeZn)B_5H_8$, **1**, 326
$ZnB_{10}C_8H_{36}N_2$
$[NMe_4]_2[ZnB_{10}H_{12}]$, **1**, 352
$ZnCH_2$
$ZnCH_2$, **3**, 180
$ZnCH_2I_2$
ICH_2ZnI, **11**, 180; **12**, 86
ZnC_2F_6
$(CF_3)_2Zn$, **3**, 176, 177; **11**, 162, 164
$ZnC_2H_4I_2$
$(ICH_2)_2Zn$, **12**, 71, 86
ZnC_2H_6
$ZnMe_2$, **3**, 176, 200
ZnC_2H_6O
$Et(OH)Zn$, **3**, 184
ZnC_3H_7I
$Et(CH_2I)Zn$, **11**, 162
ZnC_3H_8O
$EtZnOMe$, **3**, 180
ZnC_4H_{10}
Et_2Zn, **3**, 188; **11**, 162
ZnC_6H_{10}
$(CH_2{=}CHCH_2)_2Zn$, **11**, 164, 175
ZnC_6H_{14}
$(Me_2CH)_2Zn$, **3**, 176
ZnC_7H_7Br
$BzZnBr$, **3**, 182
ZnC_7H_{16}
$(Me_2CH)Zn(CHMeEt)$, **3**, 176
$ZnC_8H_{10}O$
$EtZnOPh$, **3**, 184
ZnC_8H_{14}
$(H_2CCHCH_2CH_2)_2Zn$, **3**, 176
$ZnC_8H_{15}N$
$Et[(Me_2N)CH_2CH_2CC]Zn$, **3**, 176
ZnC_8H_{18}
$Bu^t{}_2Zn$, **11**, 175, 176
$(MeEtCH)_2Zn$, **3**, 176
$ZnC_8H_{19}N$
$EtZnN(CHMe_2)_2$, **3**, 181
$ZnC_9H_{18}O$
$Bu^tC(ZnEt)CHMe$, **3**, 181
$ZnC_{10}H_{16}O_4$
$Zn(acac)_2$, **12**, 181
$ZnC_{10}H_{10}$
$ZnCp_2$, **3**, 180
$ZnC_{10}H_{14}$
$(CH_2CHCHCHCH_2)_2Zn$, **3**, 177
$ZnC_{10}H_{24}N_2$
$[(Me_2N)CH_2CH_2CH_2]_2Zn$, **3**, 176
$ZnC_{12}F_{10}$
$(C_6F_5)_2Zn$, **3**, 176; **11**, 162
$ZnC_{12}H_{20}N_2$
$[(Me_2N)CH_2CH_2CC]_2Zn$, **3**, 176
$ZnC_{14}H_{14}$
$(Bz)_2Zn$, **11**, 164
$ZnC_{14}H_{22}$
$[MeCC(CH_2)_4]_2Zn$, **3**, 180
$ZnC_{15}H_{33}$
$[Zn(CH_2Bu^t)_3]^-$, **3**, 179
$ZnC_{20}H_{30}$
$ZnCp^*{}_2$, **3**, 180
$ZnC_{26}C_{26}H_{30}$
$Zr(C_8H_8)(C_6H_2Me_3)_2$, **4**, 631
$ZnC_{26}H_{46}$
$Zn(Bu^t{}_2Cp)_2$, **3**, 176
$ZnCoWC_{16}H_5O_{11}$
$[Co(CO)_4]_2[WCp(CO)_3]Zn^-$, **10**, 340
$ZnCo_2C_8O_8$
$Co(CO)_4ZnCo(CO)_4$, **10**, 48, 63
$ZnCo_3C_{12}O_{12}$
$[Co(CO)_4]_3Zn^-$, **10**, 340
$ZnCo_6C_{38}H_{16}O_{22}$
$Zn[(CO)_9Co_3CC_6H_4(CH_2)_2\text{-}CO_2]_2$, **3**, 187
$ZnFe_3C_{37}H_{51}NO_6$
$(Bu_4N)[Zn[FeCp(CO)_2]_3]$, **10**, 338
$ZnLiC_3H_9$
$LiMe_3Zn$, **11**, 182; **12**, 90
$ZnLiC_{12}H_{27}$
Bu_3ZnLi, **11**, 182
$ZnMnSiC_{21}H_{21}ClO_2$
$MeCp(CO)_2SiMePh_2MnZnCl$, **10**, 337
$ZnNbC_{15}H_{17}$
Cp_2NbH_2ZnCp, **10**, 335
$ZnNb_2C_{20}H_{24}$
$[Cp_2NbH_2]_2Zn$, **5**, 81; **10**, 335
$ZnRu_3C_{15}H_9O_9$
$[(C_2Bu^t)(CO)_9Ru_3](Zn)$, **10**, 339
$ZnSi_2C_8H_{23}N$
$EtZnN(TMS)_2$, **3**, 181
$ZnSi_2C_{14}H_{27}$
$[Zn(CH_2(TMS))_2Ph]^-$, **3**, 179
$ZnSi_2F_6$
$(SiF_3)_2Zn$, **3**, 176
$ZnSi_3C_{12}H_{33}$
$[Zn(CH_2(TMS))_3]$, **3**, 179
$ZnSi_3C_{16}H_{32}$
$Ph[(TMS)_3C]Zn$, **3**, 179
$ZnSi_5C_{17}H_{46}$
$[(TMS)_2CH][(TMS)_3C]Zn$, **3**, 179
$ZnSi_6C_{18}H_{50}O_2$
$[(HOMe_2Si)(TMS)_2C]_2Zn$, **3**, 180
$ZnSi_6C_{20}H_{54}$
$Zn[(TMS)_3C]_2$, **3**, 179
$ZnSi_6C_{22}H_{48}F_6O_4$
$Zn[C(TMS)_2(SiMe_2O_2CCF_3)]_2$, **3**, 179
$ZnTaC_{17}H_{21}$
$(MeCp)_2TaH_2ZnCp$, **10**, 336
$ZnTa_2C_{12}H_{32}$
$[(MeCp)_2TaH_2]_2Zn$, **10**, 336
$ZnTiC_{11}H_{12}ClI$
$TiCp_2(CH_2)ZnClI$, **4**, 376
$ZnTi_2C_{28}H_{32}Cl_4N_4$
$[Ti(MeCN)_2Cp_2]_2[ZnCl_4]$, **4**, 398
$ZnVC_{17}H_{29}Br_4O_3$
$CpVBr_2{\cdot}ZnBr_2{\cdot}3THF$, **5**, 27
$ZnWC_{13}H_{17}NO_3$
$Me_2N(CH_2)_3ZnWCp(CO)_3$, **3**, 187; **10**, 337
$ZnW_2C_{16}H_{10}O_6$
$Zn[W(CO)_3Cp]_2$, **5**, 408
$Zn_2B_2V_2C_{52}H_{64}P_4$
$V_2Zn_2H_4[BH_4]_2(PMePh_2)_4$, **1**, 280, 282, 283
$Zn_2CH_2I_2$
$CH_2(ZnI)_2$, **11**, 182
$Zn_2C_{20}H_{42}N_2O_2$
$[MeZn(Et)(Bu^t)NC(H)C\text{-}(Me)O]_2$, **3**, 185
$Zn_2C_{24}H_{52}N_4O_2$
$[EtZnN(Bu^t)CH_2C(O)NEt_2]_2$, **3**, 181, 183
$Zn_2Co_4C_{14}O_{14}$
$[ZnCo(CO)_4]_2Co_2(CO)_6$, **10**, 333
$Zn_2Co_4C_{15}O_{15}$
$Co_2[ZnCo(CO)_4]_2(CO)_7]$, **10**, 49
$Zn_2Fe_2Si_2C_{68}H_{74}Cl_2O_{12}P_4$
$[Fe(CO)_3(Si(OEt)_3(PhP_2CH_2P\text{-}Ph_2)ZnCl]_2$, **10**, 338
$Zn_2NbC_{20}H_{21}$
$Cp_2NbH(ZnCp)_2$, **5**, 81; **10**, 335
$Zn_2Rh_2C_{38}H_{44}P_2$
$[Rh(Pr^i)_2P(Pr^i)_2]_2(H)_2(ZnBz)_2$, **10**, 341
$Zn_2Si_2C_{21}H_{33}N$
$Cp[N(TMS)_2](ZnCp)_2$, **3**, 180
$Zn_2Si_6C_{50}H_{68}O_2$
$[(Me_2PhSi)_3CZnOH]_2$, **3**, 184
$Zn_2TaC_{22}H_{25}$
$(MeCp)_2TaH(ZnCp)_2$, **10**, 336

$Zn_2V_4C_{48}H_{96}Cl_{12}O_{12}$
[$V_2Cl_3(THF)_6]_2[Zn_2Cl_6]$, **12**, 1086
$Zn_3TaC_{27}H_{29}$
Cp_2'Ta(ZnCp)$_3$, **5**, 128
$Zn_4C_{16}H_{52}N_4O_4$
$Me_4Zn_4(OCNEt_2)_4$, **3**, 185, 186
$Zn_4C_{20}H_{48}O_4$
[MeZnOBu^t]$_4$, **3**, 184
$Zn_4C_{40}H_{84}N_4O_8$
[EtZnOC(OMe)C(H)NBu^{t}Me]$_4$, **3**, 181
$Zn_4Co_{18}C_{66}O_{67}$
$Zn_4O[(CO)_9Co_3CCO_2]_6$, **3**, 187
$Zn_4Ni_2C_{30}H_{30}$
$Ni_2Zn_4Cp_6$, **9**, 173; **10**, 342
$Zn_4Ni_2C_{40}H_{50}$
$Ni_2Zn_4Cp_4Cp^*_2$, **9**, 173
$Zn_5C_{25}H_{60}S_5$
[MeZnSBu^t]$_5$, **3**, 186, 187, 202
$Zn_8Ni_4C_{180}H_{190}$
$[Zn_4Ni_2Cp_4Cp^*_5]_2$, **10**, 342
$ZrAlC_{21}H_{31}$
ZrCp$_3$(AlHEt$_3$), **1**, 548
$ZrAlC_{25}H_{38}$
[Zr[CH$_2$CH(AlEt$_2$)$_2$]Cp$_2$]Cp, **4**, 590
$ZrAl_2C_{15}H_{38}$
[Zr[CH$_2$CH(AlEt$_2$)$_2$]Cp$_2$]Cp, **4**, 599
$ZrAsRhC_{46}H_{50}O_4S_2$
[AsPh$_4$][Cp*$_2$Zr(S)$_2$Rh(CO)$_2$], **10**, 47
$ZrAsSi_2C_{18}H_{32}Cl$
Zr[As(TMS)$_2$](Cl)(C$_5$H$_4$Me)$_2$, **4**, 518
$ZrAs_2C_{34}H_{30}$
Zr(AsPh$_2$)$_2$Cp$_2$, **4**, 518
$ZrAs_2Si_4C_{22}H_{46}$
Zr[As(TMS)$_2$]$_2$Cp$_2$, **4**, 518
$ZrBC_{11}H_{17}$
Cp$_2$Zr(H)BH$_3$Me, **1**, 288, 290
$ZrBC_{16}H_{22}Cl_2N_5$
ZrCl$_2$Cp[H$_2$B(pz)$_2$](NCBu^t), **4**, 548
$ZrBC_{24}H_{33}N_2$
Zr(NBu^{t}BPhNBu^t)Cp$_2$, **4**, 508
$ZrBC_{25}H_{15}F_{15}N$
Zr[C(NBu^t)Me][MeB(C$_6$F$_5$)$_3$], **4**, 610
$ZrBC_{30}H_{16}F_{15}$
ZrMe[MeB(C$_6$F$_5$)$_3$]Cp$_2$, **4**, 593
$ZrBC_{34}H_{20}F_{15}O_2$
[Zr[C(O)Me](CO)Cp'$_2$][MeB-(C$_6$F$_3$)$_3$], **4**, 610
$ZrBC_{34}H_{24}F_{15}$
ZrMe[MeB(C$_6$F$_5$)$_3$](Me$_2$C$_5$H$_3$)$_2$, **4**, 600
$ZrBC_{35}H_{25}F_{15}N$
[Zr[C(NBu^t)(Me)]Cp$_2$][BMe-(C$_6$F$_3$)$_3$], **4**, 621
$ZrBC_{37}H_{40}N$
[Zr(Bz)(MeCN)Cp$_2$][BPh$_4$], **4**, 596
$ZrBC_{38}H_{32}F_{15}$
[Zr(H)Cp*$_2$][HB(C$_6$F$_5$)$_3$], **4**, 612
$ZrBC_{38}H_{36}NO$
[Zr[C(O)Me](MeCN)Cp$_2$]-[BPh$_4$], **4**, 610
$ZrBC_{38}H_{39}O$
[Zr(H)(THF)Cp$_2$][BPh$_4$], **4**, 594
$ZrBC_{39}H_{33}D_6N_2$
[ZrMe(CD$_3$CN)$_2$Cp$_2$][BPh$_4$], **4**, 593
$ZrBC_{39}H_{34}F_{15}$
[Zr(H)Cp*$_2$][MeB(C$_6$F$_5$)$_3$], **4**, 604, 612
$ZrBC_{39}H_{41}O$
[ZrMe(THF)Cp$_2$][BPh$_4$], **4**, 595, 601, 603
$ZrBC_{40}H_{41}O_2$
[Zr[C(O)Me](THF)Cp$_2$][BPh$_4$], **4**, 610
$ZrBC_{40}H_{43}O$
[Zr(H)(THF)Cp'$_2$][BPh$_4$], **4**, 611
$ZrBC_{40}H_{49}P_2$
[Zr(H)(PMe$_3$)$_2$Cp$_2$][BPh$_4$], **4**, 612
$ZrBC_{40}H_{51}O_2$
[Zr(OBu)(THF)Cp'$_2$][BPh$_4$], **4**, 613
$ZrBC_{41}H_{48}P$
[ZrEt(PMe$_3$)Cp$_2$'][BPh$_4$], **4**, 597, 607
$ZrBC_{42}H_{47}OS$
[Zr(SBu^t)(THF)Cp$_2$][BPh$_4$], **4**, 613
$ZrBC_{42}H_{47}O_2$
[Zr(OBu^t)(THF)Cp$_2$][BPh$_4$], **4**, 613
$ZrBC_{42}H_{53}P_2$
[Zr(H)(PMe$_3$)$_2$Cp'$_2$][BPh$_4$], **4**, 603, 612
$ZrBC_{43}H_{52}P$
[Zr(CH$_2$CHMe$_2$)(PMe$_3$)-(C$_5$H$_4$Me)$_2$][BPh$_4$], **4**, 598
$ZrBC_{44}H_{43}O$
[Zr(Ph)(THF)Cp$_2$][BPh$_4$], **4**, 606
$ZrBC_{44}H_{47}O_2$
[Zr[C(Me)C(Me)C(O)Me]-(THF)Cp$_2$][BPh$_4$], **4**, 606
[Zr[C(O)CMeCMe$_2$]-(THF)Cp$_2$][BPh$_4$], **4**, 606
$ZrBC_{44}H_{51}O$
[Zr(CH$_2$CHMe$_2$)(THF)Cp'$_2$]-[BPh$_4$], **4**, 598
$ZrBC_{44}H_{57}O_2$
[ZrMe$_2$(THF)$_2$Cp*][BPh$_4$], **4**, 615
$ZrBC_{45}H_{41}$
Zr(Bz)$_3$(Ph)BPh$_3$, **4**, 616
$ZrBC_{45}H_{45}O$
[Zr(Bz)(THF)Cp$_2$][BPh$_4$], **4**, 596
$ZrBC_{45}H_{49}O_2$
[Zr[C(O)CHCMe(Pr)]-(THF)Cp$_2$][BPh$_4$], **4**, 606
$ZrBC_{45}H_{51}N_2$
[Zr[C(NBu^t)Me](CNBu^t)Cp$_2$]-[BPh$_4$], **4**, 610
$ZrBC_{46}H_{28}F_{15}$
Zr(Bz)$_3$[BzB(C$_6$F$_5$)$_3$], **4**, 616
$ZrBC_{46}H_{65}OP_2$
[ZrMe$_2$(THF)(dmpe)Cp*]-[BPh$_4$], **4**, 615
$ZrBC_{47}H_{55}O$
[Zr[CMeC(Me)(Pr)](THF)-Cp'$_2$][BPh$_4$], **4**, 607
$ZrBC_{48}H_{31}F_{24}N_3$
[Zr(MeCN)$_3$Cp$_2$][B[(CF$_3$)$_2$-C$_6$H$_3$]$_4$], **4**, 614
$ZrBC_{48}H_{51}F_4$
[Zr(CH$_2$CCMe)Cp*$_2$][B-(C$_6$H$_4$F)$_4$], **4**, 594
$ZrBC_{48}H_{66}F_4$
[Zr(C$_5$Me$_4$CH$_2$CH$_2$-CHCHCH$_2$)Cp*][B-(C$_6$H$_4$F)$_4$], **4**, 604
$ZrBC_{49}H_{31}F_{20}P$
ZrMe(PMe$_3$)[C$_5$H$_4$CMe$_2$-(fluorenyl)][B(C$_6$F$_5$)$_4$], **4**, 593, 596
$ZrBC_{49}H_{34}F_{15}$
[Zr(Bz)$_2$[CH$_2$CH(Me)Bz]][B-(Bz)(C$_6$F$_5$)$_3$], **4**, 617
$ZrBC_{51}H_{57}F_4$
[Zr[CH$_2$C(Me)C(Me)C-(Me)CH(Me)](C$_5$H$_4$-Bu^t)$_2$][B(C$_6$H$_4$F)$_4$], **4**, 599
$ZrBC_{52}H_{53}O$
[Zr(Bz)$_2$(THF)Cp*][BPh$_4$], **4**, 615
$ZrBC_{53}H_{56}OP$
[Zr(Bz)[O(CH$_2$)$_4$PMe$_2$Ph]Cp$_2$]-[BPh$_4$], **4**, 601
$ZrBC_{58}H_{52}F_{15}O_3$
[Zr(Bz)$_3$(THF)$_3$][B(Bz)(C$_6$F$_5$)$_3$], **4**, 616
$ZrBC_{65}H_{39}F_{20}$
[(In)CH$_2$]$_2$ZrMe[CH$_2$CH$_2$-(C$_9$H$_7$)$_2$][B(C$_6$F$_5$)$_4$], **4**, 570
$ZrBSiC_{45}H_{55}O$
[Zr(CH$_2$CH$_2$TMS)(THF)Cp'$_2$]-[BPh$_4$], **4**, 598
$ZrBSiC_{52}H_{69}O$
[Zr(CH$_2$TMS)(THF)Cp*$_2$]-[BPh$_4$], **4**, 595
$ZrB_2C_{10}H_{18}$
Cp$_2$Zr(BH$_4$)$_2$, **4**, 554
$ZrB_2C_{12}H_{42}P_4$
ZrH$_2$(BH$_4$)$_2$(dmpe)$_2$, **1**, 279, 280
$ZrB_2C_{27}H_{46}O_3$
(bisphenoxo)Zr(THF)(BH$_4$)$_2$, **1**, 279, 280
$ZrB_2C_{63}H_{59}N_2$

$[ZrMe(MeCN)_2Cp_2][BPh_4]_2$, **4**, 614
$ZrB_2C_{64}H_{59}N_3$
$[Zr(MeCN)_3Cp_2][BPh_4]_2$, **4**, 614
$ZrB_2C_{72}H_{65}N_3$
$[Zr(MeCN)(Me_2C_{10}H_6N_2)Cp_2][BPh_4]_2$, **4**, 614
$ZrB_3C_5H_{17}$
$Cpzr(BH_4)_3$, **1**, 279, 280; **4**, 554
$ZrB_3C_6H_{29}P_2$
$ZrH(BH_4)_3(dmpe)$, **1**, 279, 280
$ZrB_3C_{10}H_{27}$
$Cp^*Zr(BH_4)_3$, **4**, 555
$ZrB_4C_4H_{24}$
$Zr(BH_3Me)_4$, **1**, 287, 290
$ZrB_9C_{15}H_{29}$
$ZrMeCp'_2(C_2B_9H_{12})$, **4**, 592
$ZrB_9C_{17}H_{37}O$
$Cp^*(C_2B_9H_{11})Zr(Me)(THF)$, **4**, 572
$ZrB_9C_{25}H_{49}$
$ZrMe(C_5Me_4Et)_2(C_2B_9H_{12})$, **4**, 600
$ZrB_{11}C_{12}H_{30}$
$ZrMe(CB_{11}H_{12})Cp^*$, **4**, 615
$ZrB_{11}C_{13}H_{33}$
$ZrMe_2(CB_{11}H_{12})Cp^*$, **4**, 615
$ZrB_{11}C_{14}H_{29}$
$ZrMe(CB_{11}H_{12})Cp'_2$, **4**, 591, 600
$ZrB_{11}C_{16}H_{38}O$
$[ZrMe(THF)Cp^*][CB_{11}H_{12}]$, **4**, 615
$ZrB_{11}C_{18}H_{29}$
$Zr(Bz)(CB_{11}H_{12})Cp_2$, **4**, 591, 600
$ZrB_{11}C_{18}H_{37}O$
$[ZrMe(THF)Cp'_3][CB_{11}H_{12}]$, **4**, 591
$ZrB_{11}C_{22}H_{36}N$
$[Zr(Bz)(MeCN)Cp_{2'}][CB_{11}H_{12}]$, **4**, 591
$ZrC_5H_5Cl_3$
$ZrCl_3Cp$, **4**, 544
ZrC_5H_6Cl
$Zr(H)(Cl)Cp$, **4**, 621
ZrC_6O_6
$[Zr(CO)_6]^{2-}$, **4**, 436
$ZrC_8H_{10}Cl_2$
$CpZr(allyl)Cl_2$, **4**, 565
$ZrC_9H_9Cl_2N$
$Cp(C_4H_4N)ZrCl_2$, **4**, 572
$ZrC_9H_{17}Cl_2N_4$
$ZrCpCl_2(NMeNMe_2)_2$, **4**, 547
$ZrC_{10}H_{10}$
Cp_2Zr, **12**, 324
$ZrCp_2$, **1**, 482; **4**, 466
$ZrC_{10}H_{10}BrNO_3$
$ZrBr(NO_3)Cp_2$, **4**, 537
$ZrC_{10}H_{10}ClNO_3$
$ZrCl(NO_3)Cp_2$, **4**, 537
$ZrC_{10}H_{10}Cl_2$
Cp_2ZrCl_2, **1**, 67, 71, 84; **4**, 520, 527, 582; **12**, 327, 340, 595
$ZrC_{10}H_{10}Cl_2P$
$ZrCl_2Cp(PCp)$, **4**, 570
$ZrC_{10}H_{10}F_2$
ZrF_2Cp_2, **4**, 485
$ZrC_{10}H_{10}N_6$
$Zr(N_3)_2Cp_2$, **4**, 536
$ZrC_{10}H_{11}Cl$
Cp_2ZrHCl, **4**, 556; **12**, 39, 88, 89, 325, 355
$ZrC_{10}H_{11}NO_4$
$Zr(NO_3)(OH)Cp_2$, **4**, 537
$ZrC_{10}H_{15}Cl_3$
Cp^*ZrCl_3, **12**, 330
$ZrC_{11}H_{13}Cl$
$Cp_2ZrMeCl$, **4**, 515, 576; **12**, 325, 338, 341, 343
$ZrC_{11}H_{13}ClO$
$ZrCl(OMe)Cp_2$, **4**, 520
$ZrC_{11}H_{15}Cl_2$
$ZrCl_2[C_5H_2[(CH_2)_5]Me]$, **4**, 489
$ZrC_{12}H_{11}ClS_2$
$CpZrCl[MeC_6H_3(S)_2]$, **4**, 548
$ZrC_{12}H_{14}$
$Cp_2[\overline{ZrCH_2CH_2}]$, **12**, 340, 342
$(MeC_5H_4)_2Zr$, **12**, 1196
$ZrC_{12}H_{15}Cl$
Cp'_2ZrHCl, **4**, 556
$(MeC_5H_4)_2Zr(H)Cl$, **12**, 39
$ZrC_{12}H_{15}ClO$
$ZrCl(OEt)Cp_2$, **4**, 520
$ZrC_{12}H_{16}$
Cp_2ZrMe_2, **12**, 1204, 1206, 1207
$[Zr(C_5H_4Me)_2H_2]^-$, **4**, 477
$ZrC_{12}H_{16}F_6O_9S_2$
$[Zr(H_2O)_3Cp_2][OTf]_2$, **4**, 614
$ZrC_{13}H_{16}$
$Cp_2ZrMe(CH{=}CH_2)$, **12**, 325
$ZrC_{13}H_{16}ClNOS$
$ZrCl(SOCNMe_2)Cp_2$, **4**, 525, 534
$ZrC_{13}H_{16}Cl_2O$
$ZrCl(OCH_2CHClMe)Cp_2$, **4**, 520
$ZrCl[OCH(CH_2Cl)Me]Cp_2$, **4**, 520
$ZrC_{13}H_{18}ClP$
$Cp_2ZrCl(CH_2PMe_2)$, **4**, 574
$ZrC_{13}H_{18}O$
$ZrMe(OCH_2Me)Cp_2$, **4**, 522
$ZrC_{14}H_{14}$
$Cp_2[\overline{ZrCH{=}CHCH{=}CH}]$, **12**, 340
$ZrC_{14}H_{14}F_6O_4S_2$
$ZrMe[CH(Tf)_2]Cp_2$, **4**, 538
$ZrC_{14}H_{14}O_5$
$Zr[OC(O)CH_2O\overline{CH_2C(O)C}]$-$Cp_2$, **4**, 526
$ZrC_{14}H_{16}$
$Zr(C_4H_6)Cp_2$, **4**, 446; **8**, 70
$Zr(C_7H_7)(C_7H_9)$, **4**, 434
$Zr(C_7H_8)_2$, **4**, 434, 435
$ZrC_{14}H_{17}Cl$
$Cp_2ZrCl(C_4H_7)$, **4**, 565
$ZrC_{14}H_{18}$
$Cp_2[\overline{ZrCH_2CH_2CH_2CH_2}]$, **12**, 340, 342
$ZrC_{14}H_{19}Cl$
$Cp_2Zr(Cl)CH_2CHMe_2$, **12**, 774
$ZrC_{14}H_{19}ClO$
$Cp_2ZrCl[CH(Me)OEt]$, **4**, 574
$ZrC_{14}H_{19}ClOS_2$
$[ZrClS_2CH(Me)OEt]Cp_2$, **4**, 574
$ZrC_{14}H_{20}$
Cp_2ZrEt_2, **12**, 341
$CpZr(allyl)_3$, **4**, 565
$ZrC_{14}H_{23}Cl_3O_2$
$ZrCl_3(THF)_2Cp'$, **4**, 545
$ZrC_{14}H_{44}O_5$
$Zr[OC(O)CH_2O\overline{CH_2C(O)C}]$-$Cp_2$, **4**, 526
$ZrC_{15}H_{15}Cl_7O$
$(CpZrCl)_3(Cl)_4(O)$, **4**, 550
$ZrC_{15}H_{16}$
Cp_3ZrH, **4**, 552
$ZrC_{15}H_{19}N_2$
$[Zr(Me)(MeCN)_2Cp_2]^+$, **4**, 593
$ZrC_{15}H_{20}$
$ZrCpCp^*$, **4**, 440
$ZrC_{15}H_{20}N_2O_4S_2$
$Zr(NSO)_2CpCp^*$, **4**, 537
$ZrC_{15}H_{21}Cl$
$Cp^*CpZr(H)Cl$, **12**, 39
$ZrC_{15}H_{21}O$
$[Cp_2ZrMe(THF)]^+$, **12**, 1204
$ZrC_{15}H_{22}O$
$Zr(C_8H_8)(C_3H_5)(OBu^t)$, **4**, 631
$ZrC_{15}H_{23}P$
$Zr(C_2H_4)(PMe_3)Cp_2$, **4**, 454
$ZrC_{15}H_{27}O_4P_3$
$Zr(CO)_4[MeC(CH_2PMe_2)_3]$, **4**, 436
$ZrC_{16}H_{12}F_{12}O_8S_4$
$Zr[HC(Tf)_2]_2Cp_2$, **4**, 538
$ZrC_{16}H_{14}O_2$
$Zr(O_2C_6H_4)Cp_2$, **4**, 521
$ZrC_{16}H_{16}$
$Zr(C_8H_8)_2$, **4**, 627
$ZrC_{16}H_{18}F_6O_7S_2$
$Zr(OTf)_2(THF)Cp_2$, **4**, 537, 622
$ZrC_{16}H_{19}N$
$Zr(C_8H_8)_2(NH_3)$, **4**, 627, 628
$ZrC_{16}H_{21}N_2$
$[Zr[NC(H)Me](MeCN)Cp'_2]^+$, **4**, 613
$ZrC_{16}H_{22}N$
$[Zr(CH_2Me)(MeCN)Cp']_2^+$, **4**, 597
$ZrC_{16}H_{23}O$
$[Zr(H)(THF)Cp'_2]^+$, **4**, 603
$ZrC_{16}H_{24}$
$(Me_2Cp)_2ZrMe_2$, **12**, 1203
$ZrC_{16}H_{25}P$
$Cp_2Zr(CH_2CHMe)(PMe_3)$, **4**, 576
$ZrC_{16}H_{26}Br_2$
$Cp^*(Bu^t_3H_2)ZrBr_2$, **4**, 565

$ZrC_{16}H_{28}P_2$
$Cp_2Zr(PMe_3)_2$, **12**, 334
$Zr(PMe_3)_2Cp_2$, **4**, 450, 454, 461
$ZrC_{17}H_{16}ClN$
$ZrCl(NCHPh)Cp_2$, **4**, 510
$ZrC_{17}H_{16}F_3NO_3S$
$Zr(NHPh)(OSO_2CF_3)Cp_2$, **4**, 508, 509
$ZrC_{17}H_{17}Br$
$Cp_2ZrBr(Bz)$, **4**, 575
$ZrC_{17}H_{18}O$
$Zr(OMe)PhCp_2$, **4**, 577
$ZrC_{17}H_{19}N$
$Zr(py)Cp'_2$, **4**, 619
$ZrC_{17}H_{20}N$
$Cp_2Zr(dmap)$, **12**, 333, 334
$ZrC_{17}H_{23}$
$Cp^*Zr(chpt)$, **4**, 572
$ZrC_{17}H_{24}Cl_2S_2$
$ZrCl_2S_2CH_2CH_2Cp^*Cp$, **4**, 567
$ZrC_{17}H_{25}$
$Cp_2(PMe_3)$[zirconabicyclo-[2.1.0]pentene], **12**, 340
$ZrC_{17}H_{25}N$
$Zr[C(NBu^t)Me](Me)Cp_2$, **4**, 621
$ZrC_{17}H_{25}NO_4$
$[Et_4N][Zr(Cp)(CO)_4]$, **4**, 434
$ZrC_{17}H_{25}O$
$[ZrMe(THF)Cp'_2]^+$, **4**, 602
$ZrC_{17}H_{25}P$
$Zr(PhMe)_2(PMe_3)$, **4**, 433, 434
$ZrC_{17}H_{27}O$
$[ZrMe(THF)Cp'_2]^+$, **4**, 601
$ZrC_{17}H_{27}P$
$Zr(C_4H_8)(PMe_3)Cp_2$, **4**, 451
$ZrC_{18}H_{14}Cl_2$
$Cp(Flr)ZrCl_2$, **4**, 566
$(Ind)_2ZrCl_2$, **4**, 566
$ZrC_{18}H_{16}Cl_2O_3$
$Zr[OC(O)CH_2OC_6H_4Cl]ClCp_2$, **4**, 526
$ZrC_{18}H_{19}ClO$
$ZrCl(OCH_2Bz)Cp_2$, **4**, 520
$ZrC_{18}H_{22}$
$Zr(CMeCMeCMeCMe)Cp_2$, **4**, 506
$Zr(C_8H_8)(C_5Me_4CH_2)$, **4**, 628
$ZrC_{18}H_{23}$
$Zr(C_8H_8)Cp^*$, **4**, 476, 628
$ZrC_{18}H_{24}$
$Cp_2Zr(C_4H_7)_2$, **4**, 565
$Zr(C_8H_8)(C_{10}H_{16})$, **4**, 631
$Zr(C_8H_8)Cp^*H$, **4**, 630
$ZrC_{18}H_{26}$
$Cp_2[ZrCH_2CH(Et)CH(Et)CH_2]$, **12**, 338, 339
$ZrC_{18}H_{26}Cl_2$
$ZrCl_2[C_5H_2(Me)_3]Cp^*$, **4**, 484
$ZrC_{18}H_{26}OTe$
$Zr(C_5H_4Bu^t)_2(O)(Te)$, **4**, 532
$ZrC_{18}H_{27}O$
$[Zr(CH_2Me)(THF)Cp']_2{}^+$, **4**, 597
$ZrC_{18}H_{28}$
$Cp^*[Zr(MeC_3H_4)(C_4H_60]$, **4**, 565
Cp_2ZrBu_2, **12**, 325
$ZrC_{18}H_{33}P_2$
$[Zr(H)(PMe_3)_2(C_5H_4Me)_2]^+$, **4**, 603
$ZrC_{19}H_{22}O$
$Cp_2Me_2(ZrOCH_2C_6H_4)$, **12**, 772
$ZrC_{19}H_{23}P$
$Cp_2(PMe_3)Zr(C_6H_4)$, **12**, 772
$Zr(C_6H_4)(PMe_3)Cp_2$, **4**, 442
$ZrC_{19}H_{27}O$
$[Zr(CMeCMe_2)(THF)Cp_2]^+$, **4**, 606
$ZrC_{19}H_{27}P$
$[C_6H_8Zr](PMe_3)Cp_2$, **12**, 338, 339
$ZrC_{19}H_{29}P$
$Zr(C_6H_{10})(PMe_3)Cp_2$, **4**, 451
$ZrC_{20}H_{14}O_2$
$(In)_2Zr(CO)_2$, **4**, 566
$ZrC_{20}H_{18}$
$ZrCp_2(C_{10}H_7)(H)$, **4**, 465
$ZrC_{20}H_{18}Cl_2$
$ZrCl_2[CH_2CH_2(C_9H_7)_2]$, **4**, 568
$ZrC_{20}H_{18}S_2$
$[Zr(S)Cp]_2(C_{10}H_8)$, **4**, 531
$ZrC_{20}H_{24}Cl_2$
$ZrCl_2Cp^*(C_9H_6Me)$, **4**, 484
$ZrC_{20}H_{26}Cl_2$
$ZrCl_2[C_2H_4(C_9H_{11})_2]$, **4**, 569
$ZrC_{20}H_{28}Cl_2$
$ZrCl_2[C_2H_4(C_5Me_4)_2]$, **4**, 567
$ZrC_{20}H_{29}$
$[ZrCp^*(C_5Me_4CH_2)]^+$, **4**, 604
$ZrC_{20}H_{29}O$
$[Zr[C(Me)CHMe](THF)Cp'_2]^+$, **4**, 606
$ZrC_{20}H_{29}P$
$Me_2[C_5H_4Zr](PMe_3)Cp_2$, **12**, 338
$ZrC_{20}H_{30}F_6N_2P_2$
$[Zr(PF_3)_2Cp^*_2](N_2)$, **4**, 441
$ZrC_{20}H_{30}N_4$
$ZrN_2Cp^*_2(N_2)$, **4**, 440
$ZrC_{20}H_{30}O$
$Zr(O)Cp^*_2$, **4**, 528
$ZrC_{20}H_{30}S$
$Zr(S)Cp^*_2$, **4**, 533
$ZrC_{20}H_{31}ClO$
$ZrCl(OH)Cp^*_2$, **4**, 527
$ZrC_{20}H_{32}O_2$
$Zr(OH)_2Cp^*_2$, **4**, 527
$ZrC_{20}H_{38}N_3S_6$
$CpZr(Me_2dtc)_3$, **4**, 548
$ZrC_{21}H_{22}F_3NO_2$
$Zr(CHPhNMe_2)(O_2CCF_3)Cp_2$, **4**, 526
$ZrC_{21}H_{23}N$
$Cp_2Bu^t(ZrNCC_6H_4)$, **12**, 772
$ZrC_{21}H_{25}Cl$
$ZrCl(Ph)CpCp^*$, **4**, 503
$ZrC_{21}H_{25}N$
$Zr(C_8H_8)_2(CNBu^t)$, **4**, 627, 628
$ZrC_{21}H_{25}N_2O_2S_2$
$(In)(pyrrolyl)Zr(acac)(Me_2N-CS_2)$, **4**, 566
$ZrC_{21}H_{30}N_2O$
$Zr(CO)Cp^*_2(N_2)$, **4**, 440
$ZrC_{21}H_{33}$
$[ZrMeCp^*_2]^+$, **4**, 606
$ZrC_{22}H_{18}F_6O_6S_2$
$[Zr(OTf)(bipy)Cp_2][OSO_2CF_3]$, **4**, 606
$ZrC_{22}H_{20}ClP$
$ZrCl(PPh_2)Cp_2$, **4**, 512
$ZrC_{22}H_{20}N_2$
$Zr(N_2Ph_2)Cp_2$, **4**, 511
$ZrC_{22}H_{20}N_2O_2$
$ZrCp_2Ph[ON(Ph)NO]$, **4**, 537, 578
$ZrC_{22}H_{22}$
Cp_2PhMe[zirconobicyclo-[3.1.0]hexane], **12**, 339, 340
$ZrC_{22}H_{23}OP$
$Zr(CO)(PMe_3)(C_9H_7)_2$, **4**, 441
$ZrC_{22}H_{24}$
$Cp_2Et_2(ZrC_2C_6H_4)$, **12**, 772
$ZrC_{22}H_{26}N_2$
$Zr(NC_4H_2Me_2)_2Cp_2$, **4**, 508
$ZrC_{22}H_{27}$
$Cp_2[ZrCH_2CH(Et)CH_2CH_2Ph]$, **12**, 338
$ZrC_{22}H_{30}$
$Zr(C_8H_8)Cp^*CH_2CHCHMe$, **4**, 630
$ZrC_{22}H_{30}O_2$
$Zr(CO)_2Cp^*_2$, **4**, 440, 450
$ZrC_{22}H_{33}F_3O_3S$
$ZrMe(OTf)Cp^*_2$, **4**, 591
$ZrC_{22}H_{34}$
$Cp_2[ZrCH_2CH(Bu)CH(Bu)C-H_2]$, **12**, 338, 339
$ZrC_{22}H_{34}Cl_2$
$ZrCl_2Cp(C_5H_2Bu^t{}_3)$, **4**, 484
$ZrC_{22}H_{34}P_2$
$ZrCp_2[P(C_6H_{11})H]_2$, **4**, 466
$ZrC_{22}H_{35}NO_4$
$[Et_4N][Zr(Cp^*)(CO)_4]$, **4**, 434
$ZrC_{22}H_{36}$
$Cp^*_2ZrMe_2$, **12**, 1203
$Cp^*Zr(Meallyl)_3$, **4**, 565
$ZrC_{23}H_{21}Cl$
$ZrCl(CHPh_2)Cp_2$, **4**, 520
$ZrC_{23}H_{24}$
$Cp_2[Zr(C_6H_4)(C_7H_{10})$, **12**, 772
$ZrC_{23}H_{29}P$
$Zr(PhCHC_3H_4)(PMe_3)Cp_2$, **4**, 456
$ZrC_{23}H_{41}ClP_2$
$ZrCl(CCBu^t)(CHCHBu^t)$-(dmpe)Cp, **4**, 546
$ZrC_{24}H_{23}N$
$Zr(Tol)(NCHPh)Cp_2$, **4**, 510
$ZrC_{24}H_{24}O$
$Zr(CHPh_2)(OMe)Cp_2$, **4**, 520
$ZrC_{24}H_{24}O_7S_2$
$[Zr(OSO_2C_7H_7)(H_2O)Cp_2]$-$[OSO_2C_7H_7]$, **4**, 613
$ZrC_{24}H_{30}STe$

$Zr(STeC_6H_4)(C_5H_4Bu^t)_2$, **4**, 531

$ZrC_{24}H_{30}Se_2$

$Zr(Se_2C_6H_4)(C_5H_4Bu^t)_2$, **4**, 529, 531

$ZrC_{24}H_{38}$

$Cp^*{}_2Zr(H)(C_4H_7)$, **4**, 559

$ZrC_{25}H_{31}P$

$Zr(C_4H_8)(PMe_3)(C_9H_7)_2$, **4**, 454

$ZrC_{25}H_{40}O$

$ZrH(OCHCHCHMe_2)Cp^*{}_2$, **4**, 524

$ZrC_{26}H_{26}Cl_2$

$Zr[C_5H_4[CH(Me)Ph]]_2Cl_2$, **4**, 485

$ZrC_{26}H_{28}N_2O$

$Zr(N_2Ph_2)(THF)Cp_2$, **4**, 441, 454

$ZrC_{26}H_{32}ClN$

$Cp^*[C_5H_4CH_2C(Nxyl)]ZrCl$, **4**, 583

$ZrC_{26}H_{38}Cl_2$

$Zr[C_5H_4[CHMe(C_6H_{11})]]_2Cl_2$, **4**, 485

$ZrC_{26}H_{41}IO$

$ZrI(OCHCHBu^t)Cp^*{}_2$, **4**, 524

$ZrC_{26}H_{42}Cl$

$Zr(C_5H_3Bu^t{}_2)_2Cl$, **4**, 475, 476

$ZrC_{26}H_{43}P$

$ZrH(PHCy)Cp^*{}_2$, **4**, 513

$ZrC_{27}H_{25}N_3$

$Zr(N_2Ph_2)(py)Cp_2$, **4**, 451, 511

$ZrC_{27}H_{29}P$

$Zr(PhCCPh)(PMe_3)Cp_2$, **4**, 445, 451

$ZrC_{27}H_{31}P$

$Zr(PCHCHPh)(PMe_3)Cp_2$, **4**, 451

$ZrC_{27}H_{44}Cl_2$

$ZrCl_2Cp^*(C_5H_2Bu^t{}_3)$, **4**, 484

$ZrC_{27}H_{44}O$

$Zr[OCH(CH_2CHMe_2)CH_2$-C-$H_2]Cp^*{}_2$, **4**, 524

$ZrC_{28}H_{33}P$

$Zr(Ph_2C_3H_4)(PMe_3)Cp_2$, **4**, 455

$ZrC_{29}H_{46}O$

$Zr[OCH(CH_2CHMe_2)C$ - $MeCMe]Cp^*{}_2$, **4**, 524

$ZrC_{30}H_{26}$

$Cp_2Ph_2(ZrC_2H_2C_6H_4)$, **12**, 772

$ZrC_{30}H_{26}N_2$

$Zr(CCPh)(NPhNHPh)Cp_2$, **4**, 511

$ZrC_{30}H_{34}O_2$

$(ZrEtCp_2)_2(O_2C_6H_4)$, **4**, 521

$ZrC_{30}H_{36}O_4$

$Zr(O_2CPh)_2(C_5H_4Bu^t)_2$, **4**, 526

$ZrC_{30}H_{48}O_2$

$Zr[OC(Bu^t)C(Bu^t)O]Cp^*{}_2$, **4**, 523

$ZrC_{30}H_{49}Cl_3O_5$

$[Cp^*ZrCl]_3(O)(OH)_4$, **4**, 550

$ZrC_{30}H_{50}Cl_2$

$ZrCl_2(C_5H_3Bu^t{}_2)(C_5H_2Bu^t{}_3)$, **4**, 484

$ZrC_{31}H_{28}ClO_2P$

$ZrCl[OCHBzC(O)PPh_2]Cp_2$, **4**, 525

$ZrC_{32}H_{32}ClN_2P$

$ZrCl[N(CPh_2)N(PHMes)]Cp_2$, **4**, 515

$ZrC_{32}H_{44}P_2$

$[Zr(PHCy)Cp_2]_2$, **4**, 512

$ZrC_{32}H_{48}P_2$

$[Zr(PHBu^t)(C_5H_4Me)_2]_2$, **4**, 512

$ZrC_{33}H_{33}P$

$Cp_2Zr(CH_2CHCHMe)$-$(CHPPh_3)$, **4**, 576

$ZrC_{34}H_{36}$

$[Zr(C_8H_8)]_2(C_6H_2CH_2Me_2)_2$, **4**, 631

$ZrC_{34}H_{40}$

$Zr(PhCCPh)Cp^*{}_2$, **4**, 445, 458

$ZrC_{34}H_{58}N_2$

$Zr(C_5H_2NBu^t{}_3)_2$, **4**, 434

$ZrC_{35}H_{30}P$

$CpZr(PPh_3)Ph_2$, **4**, 579

$ZrC_{35}H_{37}P_2$

$Zr(C_5H_4Bu^t)(CH_2PPh_2)_2$, **4**, 474

$ZrC_{36}H_{30}P$

$Cp_2Zr(PPh_3)Ph$, **4**, 579

$ZrC_{36}H_{31}P_2$

$ZrMeC(C_5H_4PPh_2)_2$, **4**, 574

$ZrC_{36}H_{60}$

$Zr(C_6H_3Bu^t{}_3)_2$, **4**, 433

$ZrC_{37}H_{42}O_2P_2$

$Cp^*MeZr(OBz_2P)_2$, **4**, 551

$ZrC_{38}H_{30}$

$Cp_2Zr(C_4Ph_4)$, **4**, 581

$ZrC_{38}H_{66}Cl_2$

$ZrCl_2(C_5HBu^t{}_4)(C_5H_2Bu^t{}_3)$, **4**, 484

$ZrC_{40}H_{51}ClN_2$

$Cp^*[C_5Me_4CH_2C(Nxyl)C$-$(Nxyl)]ZrCl$, **4**, 583

$Cp^*[C_5Me_4CHC(CHNxyl)N$-$xyl]ZrCl$, **4**, 583

$ZrCaC_{40}H_{60}Cl_2$

$Cp^*{}_2CaClZrClCp^*{}_2$, **1**, 120

$ZrCaC_{42}H_{60}O_2$

$Cp^*{}_2Ca(OC)_2ZrCp^*{}_2$, **1**, 115, 120

$ZrCrC_{40}H_{34}O_4P_2$

$Cp_2Zr(CH_2PPh_2)_2Cr(CO)_4$, **4**, 576

$ZrFeC_{35}H_{33}O_3P$

$Zr(OBu^t)Cp[C_5H_4(PPh_2)]Fe$-$(CO)_2Cp$, **10**, 27

$ZrFe_2C_{34}H_{40}O_4$

$Fe_2(CO)_2Cp_2(C_2O_2ZrCp^*{}_2)$, **7**, 232

$ZrInRhC_{15}H_{23}IP_2$

$[Cp_2Zr(PEt)_2Rh(In)(Me)]I$, **8**, 83

$ZrLi_2Si_2C_{64}Cl_2O_2P_2$

$[[LiCl]_2(THF)_2][ZrCp_2]$-$(PHSiPh_3)(PSiPh_3)$-$(C_5H_4C_5H_4)$, **4**, 467

$ZrLi_2Si_2C_{64}H_{65}Cl_2O_2P_2$

$[[LiCl]_2(THF)_2][ZrCp_2]$-$(PHSiPh_3)(PSiPh_3)$-$(C_5H_4C_5H_4)$, **4**, 467

$ZrMoC_{19}H_{18}O_3$

$ZrMeCp_2[(CO)Mo(CO)_2Cp]$, **4**, 593

$ZrMoC_{40}H_{30}O_4P_2$

$ZrCp_2(PPh_2)_2Mo(CO)_4$, **10**, 27

$ZrNbC_{31}H_{43}O$

$Cp_2(H)NbCHOZr(H)Cp^*{}_2$, **5**, 133

$ZrNbC_{32}H_{43}O_2$

$Cp_2(CO)NbCH_2OZr(H)Cp^*{}_2$, **5**, 133

$ZrPtC_{39}H_{48}O_2P_2$

$Cp^*ZrMe(OBz_2P)_2PtMe_2$, **4**, 551

$ZrReC_{22}H_{27}O$

$Cp_2MeZr(OCHMe)Re(Cp)_2$, **6**, 184

$ZrRhC_{37}H_{34}ClOP_2$

$Cp_2Zr(CH_2PPh_2)_2Rh(CO)Cl$, **4**, 576

$ZrRhC_{39}H_{42}ClO_4P_2$

$Cp^*Zr(OBz_2P)_2(OMe)(Cl)(Rh$-$(CO)$, **4**, 551

$ZrRhC_{53}H_{46}OP_2$

$Cp_2Zr(PPh_2)_2RhH(CO)(PPh_3)$, **10**, 44

$ZrRuC_{18}H_{14}O_3$

$Zr(CO)Cp_2(C_5H_4)Ru(CO)_2$, **10**, 27

$ZrRuC_{18}H_{15}ClO_4$

$Cp(CO)_2Ru(CO_2)ZrClCp_2$, **7**, 483

$ZrRuC_{18}H_{17}O_3$

$Cp(CO)_2RuCH_2OZrCp_2$, **7**, 483

$ZrRuC_{21}H_{24}O_3$

$Zr(OBu^t)Cp_2Ru(CO)_2Cp$, **10**, 27

$ZrRuC_{23}H_{33}ClP_2$

$Cp(Me_3P)_2RuCCZrClCp_2$, **7**, 415, 521

$ZrRuC_{23}H_{35}ClP_2$

$Cp(Me_3P)_2RuCHCHZrClCp_2$, **7**, 415

$ZrRuC_{23}H_{37}ClP$

$Cp(Me_3P)_2Ru(CH_2CH_2)$-$ZrClCp_2$, **7**, 392

$ZrRuC_{47}H_{42}Cl_2P_2$

$Cp_2Zr(Cl)(CH_2)RuCl(PPh_3)_2$, **4**, 576

$ZrRu_2C_{24}H_{20}O_4$

$ZrRu_2(CO)_4Cp_4$, **10**, 204, 210, 211

$ZrSiC_{14}H_{20}N$

$Cp_2[ZrN(TMS)CH]$, **12**, 341, 343

$ZrSiC_{22}H_{27}N$

$Cp_2[ZrN(TMS)C(Ph)H$-$CH{=}CH]$, **12**, 341, 343

$ZrSiC_{26}H_{46}$

$(TMS)_4Cp_2[ZrC{=}CC{=}C]$, **12**, 332

$ZrSiC_{13}H_{17}Cl_2$

$ZrCl_2[CH_2CH_2[Si(Me)_2C_9H_7]$, **4**, 570

$ZrSiC_{14}H_{27}ClNP$
$ZrCp[N(SiMe_2CH_2PPr^i_2)]Cl$, **4**, 474

$ZrSiC_{14}H_{28}NP$
$ZrCp[N(SiMe_2CH_2PPr^i_2)]H$, **4**, 474

$ZrSiC_{15}H_{21}Cl$
$Cp_2ZrCl[C(H)CH(TMS)]$, **4**, 579

$ZrSiC_{16}H_{32}NP$
$ZrCp[N(SiMe_2CH_2PPr^i_2)]Et$, **4**, 474

$ZrSiC_{18}H_{25}ClO$
$Cp'_2ZrCl[OCC(H)C(H)TMS]$, **4**, 556

$ZrSiC_{19}H_{24}ClN$
$Cp_2ZrCl[[CH(TMS)]C_5H_4N]$, **4**, 573

$ZrSiC_{19}H_{30}N$
$[Zr(CH_2CH_2TMS)(MeCN)-(C_5H_4Me_2)]^+$, **4**, 598

$ZrSiC_{20}H_{36}P$
$[Zr(CH_2CH_2TMS)(PMe_3)-(C_5H_4Me)_2]^+$, **4**, 598

$ZrSiC_{21}H_{25}Cl$
$Cp_2ZrCl[C(TMS)CHPh]$, **4**, 579

$ZrSiC_{21}H_{28}$
$Cp_2Zr(SMe)(CHTMSPh)$, **4**, 574

$ZrSiC_{31}H_{35}$
$Cp_2Zr(H)(SiPh_3)(PMe_3)$, **4**, 560

$ZrSiC_{32}H_{44}Cl_2$
$ZrCl_2Cp^*_2[SiMe_2(C_5H_4)_2]$, **4**, 485

$ZrSi_2C_{11}H_{21}Cl_3$
$Zr[C_5H_3(TMS)_2]Cl_3$, **1**, 96

$ZrSi_2C_{18}H_{26}O_2$
$Zr(CO)_2(C_5H_4TMS)_2$, **4**, 440

$ZrSi_2C_{18}H_{38}NP$
$ZrCp[Ni(SiMe_2CH_2PPr^i_2)]-CH_2TMS$, **4**, 474, 476

$ZrSi_2C_{22}H_{36}O$
$Zr(THF)(TMS-CC-TMS)Cp_2$, **4**, 444

$ZrSi_2C_{32}H_{38}$
$Cp_2ZrC(TMS)CPhCPhC(TMS)$, **4**, 579

$ZrSi_2C_{56}H_{50}P_2$
$[ZrCp(PHSiPh_3)]_2(C_5H_4C_5H_4)$, **4**, 467

$ZrSi_4C_{25}H_{51}OP$
$Zr(CO)(PMe_3)[C_5H_3(TMS)_2]_2$, **4**, 441

$ZrSi_4C_{28}H_{48}$
$Zr[C_8H_6(TMS)_2]_2$, **4**, 627

$ZrSi_7C_{52}H_{92}O_{12}$
$(CySi)_7O_{12}ZrCp^*$, **2**, 32

$Zr_2B_2C_{38}H_{54}O_2$
$[Cp_2\overline{ZrOCH_2B(C}_8H_{14})H]_2$, **1**, 292, 293

$Zr_2B_4C_{12}H_{54}P_4$
$Zr_2H_4(BH_4)_4(dmpe)_2$, **1**, 279, 280

$Zr_2B_5C_6H_{41}P_2$
$Zr_2H_3(BH_4)_5(PMe_3)_2$, **1**, 279–81

$Zr_2B_{18}C_{25}H_{54}$
$[Cp^*(C_2B_9H_{11})Zr]_2(CH_2)$, **4**, 572

$Zr_2C_{10}H_{10}Cl_2$
Cp_2ZrCl_2, **1**, 95

$Zr_2C_{10}H_{12}N_4O_{14}$
$[Zr(NO_3)_2(OHPCp]_2$, **4**, 537

$Zr_2C_{10}H_{14}$
$[CpZrH(H)]_2$, **4**, 580

$Zr_2C_{12}H_{14}Cl_2$
$[Zr(C_5H_4Me)_2(Cl)]_2$, **4**, 284

$Zr_2C_{20}H_{18}Cl_2$
$[ZrCp(Cl)]_2(C_5H_4C_5H_4)$, **4**, 467

$Zr_2C_{20}H_{18}Cl_2O$
$[ZrCpCl]_2(O)(C_5H_4C_5H_4)$, **4**, 471

$Zr_2C_{20}H_{18}Cl_2O_2$
$[CpZrCl(O)]_2(C_{10}H_8)$, **4**, 552

$Zr_2C_{20}H_{18}Cl_4$
$(CpZrCl_2)_2(C_{10}H_8)$, **4**, 471, 473, 552

$Zr_2C_{20}H_{18}I_2$
$(ZrCpI)_2(C_5H_4C_5H_4)$, **4**, 469

$Zr_2C_{20}H_{18}S_2$
$[Zr(S)Cp]_2(C_{10}H_8)$, **4**, 532

$Zr_2C_{20}H_{20}Cl$
$[Zr(Cl)Cp_2]_2^{2+}$, **4**, 614

$Zr_2C_{20}H_{20}Cl_2O$
$[ZrClCp_2]_2O$, **4**, 575

$Zr_2C_{20}H_{20}I_2$
$(ZrCp_2I)_2$, **4**, 467, 469

$Zr_2C_{20}H_{20}S_2$
$[Zr(S)Cp_2]_2$, **4**, 516, 532

$Zr_2C_{20}H_{38}Cl_4O_4$
$[Cp^*ZrCl_2(H_2O)_2]_2$, **4**, 551

$Zr_2C_{21}H_{21}ClO$
$(ZrClCp_2)_2(CHO)$, **4**, 523

$Zr_2C_{22}H_{18}N_2$
$[ZrCp(CN)]_2(C_5H_4C_5H_4)$, **4**, 473

$Zr_2C_{22}H_{21}P_2$
$[[ZrCp]_2(PHPh)(PPh)]^-$, **4**, 469, 470

$Zr_2C_{22}H_{24}$
$[ZrCpMe]_2(C_5H_4C_5H_4)$, **4**, 473

$Zr_2C_{22}H_{26}O$
$(ZrMeCp_2)_2O$, **4**, 528

$Zr_2C_{22}H_{32}Cl_2O_2$
$[Zr(C_7H_8)(THF)(Cl)]_2$, **4**, 434, 435

$Zr_2C_{23}H_{23}P_2$
$[[ZrCp]_2(PMePh)(PPh)]^-$, **4**, 470

$Zr_2C_{23}H_{42}Cl_3$
$Cp^*_2Zr_2(H)_3Cl_3(PMe_3)$, **4**, 562

$Zr_2C_{24}H_{22}F_6O_4$
$[Zr(O_2CCF_3)(OH)Cp_2]_2$, **4**, 525

$Zr_2C_{24}H_{22}F_6O_6$
$[Zr(O_2CCF_3)(OH)Cp_2]_2$, **4**, 526, 527

$Zr_2C_{24}H_{30}O$
$(ZrEtCp_2)_2O$, **4**, 528

$Zr_2C_{25}h_{27}Cl_2N$
$[ZrCpCl]_2(NCBu^t)(C_5H_4C_5H_4)$, **4**, 471

$Zr_2C_{26}H_{34}S_2$
$Cp_2(Et)Zr(SCH_2CH_2S)Zr-(Et)Cp_2$, **4**, 530

$Zr_2C_{27}H_{24}O$
$(Cp_2Zr)_2[(MeO)C_6H]$, **12**, 777, 780

$Zr_2C_{29}H_{31}Cl_2P$
$(ZrClCp_2)_2(PMes)$, **4**, 513

$Zr_2C_{29}H_{36}F_3NO_2$
$Cp_2Zr(CMeNBu^t)[OC(H)(CF_3)-O]Zr(Me)Cp_2$, **4**, 527

$Zr_2C_{32}H_{28}S_2$
$[Zr(SPh)Cp]_2(C_{10}H_8)$, **4**, 530

$Zr_2C_{32}H_{30}P_2$
$[ZrCp_2(PPh)]_2^-$, **4**, 469, 470
$[Zr(PHPh)Cp]_2(C_{10}H_8)$, **4**, 513

$Zr_2C_{32}H_{42}P_2$
$[ZrCp_2[P(C_6H_{11})]]_2$, **4**, 470

$Zr_2C_{32}H_{44}P_2$
$[Zr(PHCy)Cp_2]_2$, **4**, 513

$Zr_2C_{33}H_{36}O_3$
$[Zr(CH_2O)Cp_2]_3$, **4**, 447

$Zr_2C_{34}H_{44}O_2P_2$
$[Cp_2(Me_3P)Zr]_2[(MeO)_2C_6]$, **12**, 777, 780

$Zr_2C_{36}H_{30}$
$[ZrCp(CCPh)]_2(C_5H_4C_5H_4)$, **4**, 471

$Zr_2C_{36}H_{44}$
$[ZrCp^*](C_8H_6)[ZrCp^*](C_8H_8)$, **4**, 628, 629

$Zr_2C_{36}H_{52}OTe$
$[Zr(C_5H_4Bu^t)_2]_2(O)(Te)$, **4**, 532

$Zr_2C_{36}H_{52}S_2$
$[Zr(S)(C_5H_4Bu^t)_2]_2$, **4**, 533

$Zr_2C_{36}H_{52}Se_2$
$[Zr(Se)(C_5H_4Bu^t)_2]_2$, **4**, 532

$Zr_2C_{37}H_{39}$
$Cp_2Zr(CHCHC_2H_3)(CH_2PPh_2)-ZrCp_2$, **4**, 559

$Zr_2C_{38}H_{40}N_4$
$[Zr(N_2CHPh)(C_5H_4Me)_2]_2$, **4**, 511

$Zr_2C_{40}H_{66}P_2$
$[ZrH(PH_2)Cp^*_2]_2$, **4**, 513

$Zr_2C_{42}H_{61}IO$
$\overline{Zr[C[ZrICp^*_2]C(H)O}]Cp^*_2$, **4**, 522

$Zr_2C_{46}H_{70}P_2$
$Zr[PH(Bu^t_3C_6H_2)]_2Cp_2$, **4**, 514

$Zr_2C_{49}H_{48}P_4$
$[Cp_2Zr(PPh\overline{CH_2})_2]_2C$, **4**, 512

$Zr_2C_{54}H_{60}O_2$
$[Zr(C_8H_8)[OC(C_6H_2Me_3)_2]]_2$, **4**, 631, 632

$Zr_2Li_2C_{56}H_{78}O_6P_2$
$[Li(THF)_3]_2[ZrCp_2(PPh)]_2$, **4**, 467

$Zr_2SiC_{22}H_{24}Cl_2$
$[ZrCpCl]_2[SiMe_2(C_5H_4)_2]$, **4**, 466, 469, 470

$Zr_2SiC_{22}H_{24}Cl_2O$
$[ZrCpCl]_2[SiMe_2(C_5H_4)_2](O)$, **4**, 473

$Zr_2SiC_{22}H_{24}S_2$
[Zr(S)Cp]$_2$[(C$_5$H$_4$)$_2$SiMe$_2$], **4**, 531
$Zr_2SiC_{23}H_{27}N$
[CpZr(C$_5$H$_4$)]$_2$(N-TMS), **4**, 509
$Zr_2SiC_{33}H_{36}$
Cp$_2$ZrMe(H)$_2$(SiPh$_2$H)ZrCp$_2$, **4**, 560
$Zr_2Si_2C_{24}H_{28}Cl_2O$
[ZrCl[(C$_5$H$_4$)$_2$SiMe$_2$]]$_2$O, **4**, 528
$Zr_2Si_2C_{33}H_{38}$
Cp$_2$9SiH$_2$Ph)Zr(H)$_2$Zr-(SiHMePh)Cp$_2$, **4**, 504
$Zr_2Si_2C_{35}H_{42}$
Cp$_2$(SiHMePh)Zr(H)$_2$Zr-(SiMe$_2$Ph)Cp$_2$, **4**, 504, 560
$Zr_2Si_2C_{44}H_{44}$
[Cp$_2$(Ph$_2$HSi)Zr(H)]$_2$, **4**, 504, 560
$Zr_2Si_4C_{28}H_{48}Cl_4$
[Zr[C$_8$H$_6$(TMS)$_2$]Cl]$_2$(Cl)$_2$, **4**, 631
$Zr_2Si_4C_{29}H_{47}ClO$
Cp$_2$(Cl)Zr(O)Zr[Si(TMS)$_3$]Cp$_2$, **4**, 503
$Zr_2Sn_2C_{26}H_{38}$
[Zr(SnMe$_3$)Cp$_2$]$_2$, **4**, 504
$Zr_3B_6H_{66}P_4$
Zr$_3$H$_6$(BH$_4$)$_6$(PMe$_3$)$_4$, **1**, 279, 280
$Zr_3C_{30}H_{30}O_3$
[Zr(O)Cp$_2$]$_3$, **4**, 528
$Zr_3C_{43}H_{38}O_{12}$
[Zr$_3$(O)(OH)$_3$(PhCO$_2$)$_3$Cp$_3$]-[PhCO$_2$], **4**, 613
$Zr_4C_{20}H_{36}N_4O_{24}$
[Zr(NO$_3$)(OH)$_2$Cp·H$_2$O]$_4$, **4**, 537

Cumulative Subject Index

JOHN NEWTON
David John (Services), Slough, UK

This Cumulative Subject Index contains entries to the text pages of Volumes 1–12. The index covers general types of organometallic compound, specific organometallic compounds, general and specific organic compounds where their synthesis or use involves organometallic compounds, types of reaction (insertion, oxidative addition, etc.), spectroscopic techniques (NMR, IR, etc.), and topics involving organometallic compounds.

Because authors may have approached similar topics from different viewpoints, index entries to those topics may not always appear under the same headings. Both synonyms and alternatives should therefore be considered to obtain all the entries on a particular topic. Commonly used synonyms include alkyne/acetylene, compound/complex, preparation/synthesis, etc. Entries where the oxidative state of a metal has been specified occur after all the entries for the unspecified oxidation state, and the same or similar compounds may occur under both types of heading. Thus $Cr(C_6H_6)_2$ occurs under Chromium, bis(η-benzene) and again under Chromium(0), bis(η-benzene). Similar ligands may also occur in different entries. Thus a carbene–metal complex may occur under Carbene complexes, Carbene ligands, or Carbenes, as well as under the specific metal. Individual organometallic compounds may also be listed in the Cumulative Formula Index in this volume.

A

β-Abstraction reactions
  dibutylzirconocene, **12**, 325
Acenaphthalene
  Pauson–Khand reaction, **12**, 717
Acetaldehyde
  preparation
    Wacker process, **12**, 606
Acetalization
  antimony trichloride, **11**, 498
  organoantimony compounds, **11**, 498
Acetals
  conversion to ethers
    organoaluminum hydrides, **11**, 300
  α,β-unsaturated
    reaction with allylic zinc reagents, **11**, 175
Acetates, nitrodiazo-
  cyclopropanation, **12**, 399
Acetic acid
  reaction with mercury(II) nitrate, **3**, 158
Acetic acid, *cis*-(6-methyltetrahydropyran-2-yl)-
  synthesis, **12**, 889
Acetic acid, 2-vinylcyclopentyl-
  synthesis, **12**, 917
Acetomycin
  synthesis, **12**, 402
Acetone
  mercuriation, **3**, 139
Acetone, benzylidene-
  cycloaddition reactions, **12**, 927
Acetone, hexafluoro-
  reactions
    with rhodium complexes, **8**, 215
Acetone, tetrabromo-
  trapping of oxallyl cation, **12**, 350, 362
Acetonitrile, α-bromo-
  coupling reactions, **12**, 181
Acetonitrile, diazo-
  cyclopropanation, **12**, 399
Acetophenone
  dimerization
    ultrasound, **12**, 1074
Acetophenone, *o*-bromo-
  asymmetric reduction, **12**, 24
Acetophenone, bromo-
  asymmetric reduction, **12**, 24
Acetophenonebenzylimine
  asymmetric reduction, **12**, 26
Acetophenonebenzylimine, *p*-methoxy-
  asymmetric reduction, **12**, 26
Acetoxychlorination
  1,3-dienes, **12**, 821
1,4-Acetoxychlorination
  1,3-dienes, **12**, 880
Acetyl coenzyme A synthase
  nickel complexes, **9**, 91
Acetylene, *s*-alkyl-
  synthesis, **12**, 690
Acetylene, aryl-
  synthesis, **12**, 223
Acetylene, cyclopropyl-
  Pauson–Khand reaction, **12**, 716
Acetylene, diphenyl-
  hydrogenation
    catalysis, **10**, 37
  reactions with ruthenaboranes, **1**, 301
Acetylene, phenyl-
  double deprotonation
    lithiation, **11**, 46
  hydrocarboxylation, **12**, 35
  metallation, **11**, 112
Acetylene, silyl-
  hydroalumination
    stereoselectivity, **11**, 292
  insertion reactions, **12**, 156
Acetylenes, thio-
  hydroboration
    nickel catalyst, **11**, 200
Acetylenic acetals
  alkylation
    dicobalt propargylium complexes, **12**, 691
Acetylenic esters
  synthesis, **12**, 73
Acetylide complexes
  bridging
    heteronuclear metal–metal bond synthesis, **10**, 20
  tetranuclear clusters
    three unlike metals, **10**, 164, 165
Acetylide ligands
  bridging
    heteronuclear metal–metal bond synthesis, **10**, 12
Acid anhydrides
  hydrogenation, **12**, 10, 11
  methylenation, **12**, 591
Acid chlorides
  coupling with tetraalkyllead
    palladium-catalyzed, **11**, 464
  coupling with zinc–copper reagents, **11**, 168
  methylenation, **12**, 591
Acids
  α,β-unsaturated
    hydrogenation, asymmetric, **12**, 19–21
Acoragermacrone
  synthesis, **12**, 1080
Acorenone
  synthesis, **12**, 1009, 1032
Acorenone B
  synthesis, **12**, 1032
Acosamine
  synthesis, **12**, 889
Acrylamide
  cycloaddition reactions, **12**, 939
Acrylamide, dimethyl-
  hydroformylation, **12**, 31
Acrylates
  cycloaddition reactions, **12**, 926
    [3+2], **12**, 946
  β-substituted
    synthesis, **11**, 490
Acrylates, α-(hydroxyalkyl)-
  hydrogenation
    asymmetric, **12**, 22
Acrylic acid
  hydrogenation, **12**, 19
  α,β-unsaturated derivatives
    preparation, **11**, 489
Acrylic acid, α-(methoxynaphthyl)-
  hydrogenation, **12**, 19
Actinide complexes, **4**, 11–192
  alkylidene
    preparation, **12**, 596
  allyl, **4**, 136, 137
  arenes, **4**, 178–80
  bis(cyclooctatetraenyl), **4**, 182–4
    anionic, **4**, 181
  bis(pentamethylcyclopentadienyl) dihalo, **4**, 166–76
  bis(pentamethylcyclopentadienyl) halo, **4**, 164, 165
  bis(pentamethylcyclopentadienyl) hydrides, **4**, 174–6
  bis(pentamethylcyclopentadienyl) hydrocarbyls, **4**, 169–74
  carbonyls, **4**, 131
  cyclohexadienyl, **4**, 177
  cyclooctatetraenyl, **4**, 181–6
  cyclooctatetraenyl dihalo, **4**, 184–6
  cyclopentadienyl, **4**, 137–61
  cyclopentadienyl trihalo, **4**, 141–5
  dicyclopentadienyl dihalo, **4**, 145–7
  dicyclopentadienyl halo, **4**, 137, 138
  heterobimetallic, **4**, 186–90
  homogeneous catalysis, **4**, 190, 191
  hydrocarbyls, **4**, 131, 136
    heteroleptic, **4**, 133–6
    homoleptic, **4**, 131, 132
  indenyl, **4**, 162–4
  metallacarbaboranes, **1**, 373–424
  modified cyclopentadienyl, **4**, 161–76

organic synthesis, **4**, 192
pentadienyl, **4**, 177
pentamethylcyclopentadienyl, **4**, 164–76
pentamethylcyclopentadienyl dihalo, **4**, 164, 165
pentamethylcyclopentadienyl trihalo, **4**, 165, 166
ring-bridged cyclopentadienyl, **4**, 161, 162
tetrahydroborates, **1**, 286
tetrakis(cyclopentadienyl), **4**, 160, 161
tetramethylphospholyl, **4**, 178
trihydroborates, **1**, 290
tris(cyclopentadienyl), **4**, 139–41
tris(cyclopentadienyl) halo, **4**, 147–59
classical coligands, **4**, 147–51
tris(cyclopentadienyl) halo neutral ligand, **4**, 159, 160
tris(cyclopentadienyl) hydrides, **4**, 157–9
tris(cyclopentadienyl) hydrocarbyls, **4**, 151–7
Activation
C–H bonds
methylidyne triiron clusters, **7**, 275
Acyclic diene metathesis polymerization
transition-metal catalysts, **12**, 1229
Acyl anions
equivalents
lithiation, **11**, 39–41
Acyl chlorides
hydrogenation, **12**, 10, 11
Acyl complexes
triiron clusters, **7**, 280
Acyl compounds
coupling reactions, **12**, 199, 200
Acyl derivatives
coupling reactions, **12**, 218–20
cross-coupling reactions
Grignard reagents, **12**, 179
Acyl halides
decarbonylation, **12**, 376
Acylation
organomercury compounds, **11**, 408, 409
Acyloxymercuration
alkenes, **11**, 418–20
Addition polymerization
transition-metal catalyzed, **12**, 1224–6
Addition-elimination reactions
alkenation
organoantimony compounds, **11**, 491
Addition/oxidation, **12**, 1006
intramolecular, **12**, 1004
Adenosine
halogenated
coupling reactions, **12**, 197
synthesis
Stille couplings, **12**, 218
Adsorption
silica gel
cobalt complexes, Pauson–Khand reaction, **12**, 729
AF toxin IIc
intermediates
synthesis, **12**, 627
synthesis, **12**, 631
Aflatoxin
synthesis, **12**, 309
Agostic interaction
C–H bonds, **7**, 275
tetrairon butterfly methylidyne clusters, **7**, 269
Aklavinone, decarbomethoxy-
synthesis, **12**, 1030
Akuammicine
synthesis, **12**, 253, 254
Alanes, alkenyl-
allylic alkylation, **12**, 855
coupling with bromoarenes, **12**, 192
Alanes, vinyl-
coupling reactions, **12**, 190
Alanine
reaction with methylmercury cations, **3**, 156
Alanine, 4-arylphenyl-
synthesis, **11**, 233
Alanine, 3-methylpentafluorophenyl-
synthesis, **12**, 30
Alanine, phenyl-
novel derivatives
synthesis, **12**, 195
synthesis, **12**, 989
Albicanol
synthesis, **12**, 581
Alcohols
asymmetric allyl displacements
palladium complexes, **12**, 883
carbonylation, **12**, 362–9
iridium catalysts, **8**, 304
chiral
reaction titanium complexes, **4**, 274
conversion to methylene compounds
organoaluminum hydrides, **11**, 299
dehydration
catalysts, **10**, 369
oxidation
organobismuth(V) compounds, **11**, 504
polycyclic
preparation, organoboranes, **11**, 226
preparation
organoantimony compounds, **11**, 492–6
organoboron compounds, **11**, 214
trialkylstibines in, **11**, 492
reduction
organotin compounds, **11**, 359, 360
synthesis
antimony in, **11**, 495
organoboron compounds, **11**, 215, 216
organomercury compounds, **11**, 398
pentaorganylstiboranes, **11**, 493–5
zirconium mediated, **4**, 621
synthesis from aluminum alkyls, **11**, 302
synthesis from carbonyl compounds
organoaluminum hydrides, **11**, 298
Alcohols, 2-amino-
synthesis
organoniobium complexes, **5**, 142
Aldehydes
addition of diethylzinc, **11**, 171
alkenation
trialkylstibine, **11**, 489
alkynyl
aldol condensation, **12**, 691
allylation
antimony, **11**, 495
catalytic asymmetric, **11**, 321
aryl
reductive coupling, **12**, 1074
chiral
addition of zinc organometallics, **11**, 169, 170
reaction titanium complexes, **4**, 274
cycloaddition reactions, **12**, 930, 931
[3+2], **12**, 938
decarbonylation
metal carbonyls, **12**, 373–6
diastereoselective addition to Grignard reagents, **11**, 145–50
diastereoselective coupling
titanocene dichloride, **4**, 276
enantioselective addition to Grignard reagents, **11**, 150
group transfer from tetraalkyllead, **11**, 462
heteroaromatic
dimerization, **12**, 1074
homologation, **12**, 172
intramolecular coupling, **12**, 1077
α-methoxyorganolead condensation
stereoselectivity, **11**, 463
methylenation, **12**, 578, 583
Tebbe's reagent, **12**, 587
zirconium complexes, **12**, 595
nucleophilic addition
(diene)tricarbonyliron complexes, **12**, 631
preparation
metal carbonyls, **12**, 351
reaction with organozinc compounds, **3**, 190–2
reactions with π-allylmolybdenum complexes, **12**, 970, 971
reductive dimerization

molybdenum or tungsten complexes, **12**, 1090
synthesis
carbon monoxide insertion reactions, **12**, 253–7
organoboron compounds, **11**, 214
synthesis from carbonyl compounds
organoaluminum hydrides, **11**, 298
α,β-unsaturated
hydrogenation, **12**, 9, 10
Aldimines
zirconocene complexes
reductive coupling, **12**, 342
anti-Aldol products
preparation, **11**, 175
Aldol reactions
anti-selective, **11**, 257, 258
asymmetric, **11**, 256–8
boron mediated, **11**, 240–2
carbene complexes, **12**, 515–21
Mukaiyama, **11**, 263
syn-selective, **11**, 256, 257
Aldolization
boron mediation
diastereoselectivity, **11**, 241, 242
Aldoses
decarbonylation, **12**, 374
Alkadienes
conjugated
synthesis, **11**, 232
1,3-Alkadiynes, 1-trimethylsilyl-
hydroalumination, **1**, 470
Alkali metal alkoxides
addition to dialkylmagnesium compounds, **1**, 95
structures, **1**, 11
Alkali metal amides
reaction with Grignard reagents, **1**, 103
Alkali metal compounds
ethyl, **1**, 19
magnesium compounds, **1**, 105, 106
methyl, **1**, 19
Alkali metal organometallics, **1**, 1–30
amido, **1**, 2–10
antimony compounds, **1**, 10, 11
arsenic compounds, **1**, 10, 11
carbon–metal bonds, **1**, 18–29
σ-complexes, **1**, 18–25
π-complexes, **1**, 18–25
enolates
structures, **1**, 11
ethynides
structures, **1**, 20
fluorenyl, **1**, 25
group 15 complexes, **1**, 2–11
group 16 compounds, **1**, 11–14
imino, **1**, 2–10
lead compounds, **1**, 14, 15
mixed-metal compounds, **1**, 25–9
main group complexes, **1**, 29
transition metal complexes, **1**, 25–7
multimetallated, **1**, 16, 17
oxygen compounds, **1**, 11–13
phosphorus compounds, **1**, 10, 11
propynides
structures, **1**, 20
selenium compounds, **1**, 13, 14
silicon compounds, **1**, 14, 15
sulfur compounds, **1**, 13, 14
tellurium compounds, **1**, 13, 14
theory, **1**, 29, 30
thiolates, **1**, 14
tin compounds, **1**, 14, 15
Alkaline earth metal compounds
bis(pentamethylcyclopentadienyl)
complexes with Lewis bases, **1**, 114
cyclooctetraene, **1**, 117
heterobimetallic, **1**, 119
zincates, **1**, 119
Alkaline metal compounds
anionic polymerization
initiation, **1**, 109
Alkanes
dehydrogenation
iridium catalysts, **8**, 351
heterocyclic
reductive cleavage,
alkyllithium preparation, **11**, 12–15
synthesis from alkenes
organoaluminum hydrides, **11**, 297
Alkanes, 1,1-diboryl-
reactions, **11**, 208
Alkanes, α,ω-di-2-pyridyl-
synthesis, **12**, 756
Alkanes, nitro-
allylic alkylation, **12**, 867, 868
reactions with organolead compounds, **11**, 477
Alkenamides
amidomercuration
intramolecular, **11**, 424
Alkenation
addition-elimination reactions
organoantimony compounds, **11**, 491
antimony ylides, **11**, 489, 490
carbonyl compounds
molybdenum or tungsten complexes, **12**, 1090
trialkylstibine, **11**, 489
zirconium carbene complexes, **12**, 44
organoantimony reagents, **11**, 488
Alkene Growth Process
organoaluminum compounds, **1**, 478
Alkenes
acyclic
Pauson–Khand reaction, **12**, 712–15
addition compounds
magnesium, **1**, 97, 98
addition of magnesium hydride
catalysis, **1**, 86
alkylation
transmetallation–insertion reactions, **12**, 135
alkynes from
organoboron compounds in, **11**, 213
alternating copolymerization
with carbon monoxide, **12**, 1225
amino
hydroamination/cyclization, enantioselective, Ziegler–Natta catalysts, **12**, 1205
arylation
organotellurium compounds, **11**, 587
bridged bicyclic
Pauson–Khand reaction, **12**, 719–21
carbomagnesiation
zirconium catalyzed, **11**, 131
carbomercuration, **11**, 427
carbonylation, **12**, 355–62
carbozincation, **11**, 175–7
cis-disubstituted
epoxidation, **12**, 1126
cis, *trans*-epoxides from, **12**, 1129
conversion to alkanes
organoaluminum hydrides, **11**, 297
coupling
zirconium mediated, **4**, 619
coupling with pyridines
Ziegler–Natta catalysts, **12**, 1206
cyclization
allylmetal catalysis, alkene faciality, **12**, 909, 917
allylpalladium complexes, **12**, 908–10
cyclization/1-alkenyl(aryl) coupling
allylpalladium catalysts, **12**, 915, 916
cyclization/C,C-coupling
allylmetal catalysts, **12**, 913–19
cyclization/carbonylation
allylmetal catalysts, **12**, 916–19
cyclization/elimination
allylmetal complexes, **12**, 906–12
cycloaddition reactions
[3+2], **12**, 934–7, 946
deoxygenative coupling, **12**, 1091
diacetoxylation
organotellurium compounds, **11**, 583
dihydroxylation
asymmetric, **12**, 1137–73
(diene)tricarbonyliron, **12**, 626
dimerization
niobium complexes, **5**, 137, 138
1,1-disubstituted
asymmetric dihydroxylation, **12**, 1159
1,2-disubstituted

preparation by cross-coupling, **12**, 167
cis-disubstituted
asymmetric dihydroxylation, **12**, 1159
trans-disubstituted
asymmetric dihydroxylation, **12**, 1159
electron rich
Heck reactions, **12**, 276, 277
elimination/cyclization
allylmetal complexes, **12**, 906–12
epoxidation
organotellurium compounds, **11**, 583
exocyclic
preparation, organoboron compounds, **11**, 228
functionalized
hydroformylation, **12**, 28–31
fused bicyclic
Pauson–Khand reaction, **12**, 715–19
heterocyclic
asymmetric hydroboration, **11**, 248
hydroboration, **11**, 200
heterogeneous metathesis catalysts, **10**, 58
heterosubstituted
metallation, **11**, 114–24
hydroalumination, **11**, 285
hydroformylation, **12**, 27, 28
platinum complex catalysts, **9**, 443
hydrogenation
catalysis, **10**, 37
diastereoselective, **12**, 11
enantioselective, **12**, 16–22
organoaluminum reagents, **1**, 484
platinum complex catalysts, **9**, 443
hydrosilylation, **2**, 20
hydrostannation, **2**, 229–31, 276
hydrozirconation, **12**, 39
insertion reactions
carbene complexes, **12**, 533
iron complexes, **12**, 352
organotitanium complexes, **12**, 157
oxidative addition, palladium catalysts, **12**, 259–85
rhodium complexes, **8**, 87
intramolecular insertion reactions
allyl complexes, **12**, 905–21
isomerization
hydrozirconation, **12**, 40
titanium hydride catalysts, **4**, 263, 264
mercuration, **11**, 393
metal complexes
metallaboranes, **1**, 308
metallation, **11**, 97, 110–14
metallostannation, **2**, 231, 232
metathesis, **12**, 29
catalysis by rhenium(VII) alkylidene complexes, **6**, 244
rhenium(VII) alkylidene complexes, **6**, 244, 245
transition metals, **12**, 1210
monocyclic
Pauson–Khand reaction, **12**, 715–19
monosubstituted
asymmetric dihydroxylation, **12**, 1159
nitrogen-containing
metallation, **11**, 115
oligomerization
organoaluminum reagents, **1**, 484
oxygen derivatives
coupling reactions, **12**, 190, 191, 206, 207
oxygen-atom transfer
mechanisms, **12**, 1107
oxygen-containing
metallation, **11**, 116, 117
palladium complexes
bonding, **9**, 294–302
structure, **9**, 294–302
theory, **9**, 294–6
polyfunctional
coupling reactions, **12**, 191
cross-coupling with Grignard reagents, **12**, 172, 173
polymerization
lanthanide complexes, **4**, 8
metallocenes, **4**, 493
niobium complexes, **5**, 136, 137
preparation
carbonyl coupling reactions, **12**, 1074
reaction with carbene complexes
cyclobutanones from, **12**, 560–5
reactions
with organomercury compounds, **11**, 403–7
reactions with π-allyl nickel complexes, **12**, 966
reactions with metallic palladium, **9**, 293, 294
reactions with organoaluminum compounds, **1**, 495, 496
reactions with organonickel complexes, **9**, 62
reduction
(diene)tricarbonyliron, **12**, 626
silicon-containing
metallation, **11**, 117, 118
solvomercuriation, **3**, 144–8
ω-stannyl
coupling with enolates, **12**, 185
strained
addition polymerization, **12**, 1226
synthesis by ketone dimerization, **12**, 1074
substituted
insertion reactions, steric effects, **12**, 145
sulfur derivatives
coupling reactions, **12**, 191
sulfur-containing
metallation, **11**, 116, 117
synthesis
organotellurium compounds, **11**, 591, 592
synthesis from alkylidene complexes
niobium complexes, **5**, 143
synthesis from alkynes
organoaluminum hydrides, **11**, 298
terminal
asymmetric epoxidation, **12**, 1127
terminally alkyl-substituted
cyclization, allylmetal catalysts, **12**, 910–12
tetrasubstituted
asymmetric dihydroxylation, **12**, 1150, 1161
transition metal complexes
in organic synthesis, **12**, 637–81
nucleophilic attack, **12**, 6, 601–20
protection, **12**, 623–34
trisubstituted
asymmetric dihydroxylation, **12**, 1146, 1161
asymmetric epoxidation, **12**, 1127
unactivated
cycloaddition reactions, **12**, 925
unfunctionalized
epoxidation, **12**, 1108
vinylpalladation
allyl complexes, **12**, 809
Ziegler–Natta polymerization
organohafnium complexes, **4**, 618, 619
organozirconium complexes, **4**, 618, 619
Alkenes, nitro-
Michael acceptors, **11**, 174
2-Alken-1-ol, 4-amino-
synthesis
from allyl acetates, **12**, 805
Alkenols
alkoxymercuration
intramolecular, **11**, 415
Alkenyl bromides
coupling reactions, **12**, 203, 204
cross-coupling reactions
palladium-catalyzed, **12**, 182–6
transition metal-promoted, **12**, 167–9
reaction with allylic zinc reagents, **11**, 166
Alkenyl chlorides
coupling reactions, **12**, 203–5
cross-coupling reactions
palladium-catalyzed, **12**, 182
transition metal-promoted, **12**, 166

Alkenyl complexes
triiron clusters, **7**, 262
Alkenyl fluorosilane
coupling reactions, **12**, 212
Alkenyl halides
coupling reactions, **12**, 203
cross-coupling reactions
palladium-catalyzed, **12**, 182–9
transition metal-promoted, **12**, 166–9
Alkenyl iodides
coupling reactions, **12**, 204, 205
zirconium compounds, **12**, 186–9
cross-coupling reactions
transition metal-promoted, **12**, 169
Alkenyl selenides
synthesis, **12**, 230
Alkenyl siliconates
coupling reactions, **12**, 213
Alkenyl sulfides
synthesis, **12**, 230
Alkoxides
intramolecular allylic reactions, **12**, 876–8
reactions
allyl complexes, **12**, 814–17
Alkoxycarbonylation
platinum(0) catalysts, **12**, 247
transition metal complexes, **12**, 243
Alkoxymercuration
alcohols, **11**, 414–17
regioselectivity, **11**, 414
stereoselectivity, **11**, 414
Alkoxypalladation
dienes, **12**, 816
Alkyl boronates
coupling reactions, **12**, 186
Alkyl bromides
alkoxycarbonylation
transition metal complexes, **12**, 243
Alkyl halides
cross-coupling reactions, **12**, 181, 182
transition metal-promoted, **12**, 165, 166
lithiation, **11**, 7
Stille coupling, **12**, 202, 203
Alkyl iodides
alkoxycarbonylation
transition metal complexes, **12**, 243
unactivated
coupling reactions, organoboron compounds, **12**, 181
Alkyl mesylates
alkoxycarbonylation
transition metal complexes, **12**, 243
Alkylaluminum alkoxides
availability, **11**, 295
Alkylaluminum halides
availability, **11**, 295
Alkylaluminum hydrides
availability, **11**, 294
Alkylaminomercuriation
alkenes, **3**, 145
Alkylation
with alkyl halides
metal carbonyls, **12**, 351
alkylpalladium complexes, **12**, 139
carbene complexes, **12**, 515–21
diastereoselective
(diene)tricarbonyliron anions, **12**, 629, 630
intramolecular aromatic
cobalt-complexed cations, **12**, 690
malonates
nitronate anions, **12**, 354
tetraalkyllead to aldehydes, **11**, 462
stereoselectivity, **11**, 462
Alkylidenations
titanium complexes, **12**, 1083
Alkylidene complexes
group 6, **5**, 311–15
properties, **5**, 314
reactions, **5**, 314, 315
structure, **5**, 314
synthesis, **5**, 311–13
transition metals, **12**, 595, 596
triiron clusters, **7**, 282
Alkylidene transfer
carbene complexes, **12**, 584, 585, 592, 593
Alkylidenes
nucleophilic, **12**, 577–96
Alkylidyne complexes
coupling with alkyne leading to diferra–allyl clusters, **7**, 278
coupling with CO leading to C≡C functional groups on triiron clusters, **7**, 277
group 6, **5**, 315–21
properties, **5**, 318, 319
reactions, **5**, 319–21
structure, **5**, 318, 319
synthesis, **5**, 315–18
homocoupling leading to C≡C functional groups on triiron clusters, **7**, 277
iron clusters, **7**, 262, 272
μ_3-triiron clusters
preparation by alkylation of an iron carbide cluster, **7**, 268
Alkylmercuriation
alkynes, **3**, 137
Alkynals
acetals
Pauson–Khand reaction, **12**, 730
Alkyne complexes
coupling with alkylidyne leading to diferra–allyl clusters, **7**, 278
iron clusters, **7**, 259
organic synthesis, **12**, 741–66
tetranuclear clusters
three unlike metals, **10**, 165
trinuclear clusters
three unlike metals, **10**, 162, 163
Alkyne ligands
coordination modes in triiron clusters, **7**, 259
Alkyne, silyl-
carbalumination, **11**, 284
Alkynes
addition compounds
magnesium, **1**, 97, 98
alkoxymercuration, **11**, 417
alkylmercuriation, **3**, 137
aminomercuration, **11**, 423
bromoboration, **12**, 184
carbomercuration, **11**, 427
carbonylation, **12**, 355–62
carbozincation, **11**, 175–7
cocyclocarbonylation with allenes, **12**, 361, 362
conversion to alkenes
organoaluminum hydrides, **11**, 298
coupling with nitriles and ketones, **12**, 591
coupling with a μ_3-NR ligand on an iron cluster, **7**, 266
coupling with μ_3-PR groups on an iron cluster (a phosphinidene iron cluster), **7**, 266
coupling to nitriles
intramolecular, **12**, 332
cross-coupling
chemoselective, **12**, 334
cyclization/1-alkenyl(aryl) coupling
allylpalladium catalysts, **12**, 915, 916
cyclization/carbonylation
allylmetal catalysts, **12**, 916–19
cycloaddition reactions, **12**, 925, 926
[3+2], **12**, 948
arene synthesis, **12**, 747–51
cyclodimerization
nickel catalysts, **9**, 66
cyclooligomerization
mechanism, **12**, 742
cyclotrimerization
niobium complexes, **5**, 139, 140
reactions with iron atoms, **7**, 109
group 5 complexes
reductive-coupling reactions, **12**, 335
1-halo-
hydroboration, **12**, 184
hydroalumination, **11**, 285; **12**, 182
hydroboration, **11**, 196
hydrogenation
catalysis, **10**, 37
hydrometallation, **7**, 399–406
hydropalladation, **12**, 48
hydrostannation, **2**, 229–31, 276
hydroxymercuration, **11**, 413
hydrozirconation, **12**, 39, 179
insertion between μ_3-PR groups of a bis(phosphinidene)iron cluster, **7**, 264

insertion into organonickel complexes, **9**, 58
insertion into P–C bond in phosphinidene iron clusters, **7**, 264
insertion reactions
organometallic reagents, **12**, 142
organotitanium complexes, **12**, 157
palladium catalysts, **12**, 287–91
rhodium complexes, **8**, 201
intermolecular reactions
palladium catalysts, **12**, 290
intramolecular insertion reactions
allyl complexes, **12**, 905–21
palladium catalysts, **12**, 290
mercuration, **11**, 394
metallation, **11**, 110
rhodium complexes, **8**, 210
metallostannation, **2**, 231, 232
metathesis
catalysis by rhenium(VII) alkylidene complexes, **6**, 244, 245
rhenium(VII) alkylidene complexes, **6**, 244, 245
oligomerization
iron ions, **7**, 105
polymer-bound
Pauson–Khand reaction, **12**, 720
polymerization
niobium complexes, **5**, 136, 137
reactions
with π-allyl nickel complexes, **12**, 966
with $[Fe_3(\mu-H)(CO)_{11}]^-$, **7**, 262
with mercuric halides, **11**, 428
with metallic palladium, **9**, 316
with organomercury compounds, **11**, 403–7, 429
with palladium complexes, **9**, 320–3
with ruthenaboranes, **1**, 301
with tin hydrides, **11**, 361
reduction
(diene)tricarbonyliron, **12**, 626
Reppe cyclotetramerization, **9**, 122
silacyclic
synthesis, **2**, 62
silyltitanation, **4**, 277
solvomercuriation, **3**, 148–50
terminal
carboxylation, nickel complexes, **9**, 10
hydrozirconation, **12**, 182
metathesis, **12**, 222–5
transition metal complexes
coupling reactions, **12**, 333
nucleophilic attack, **12**, 619, 620
trimerization
Pauson–Khand reaction, **12**, 706
unsymmetrical
synthesis, **12**, 169
2-Alkynes
metallation, **11**, 110
Alkynes, 1-bromo-
coupling reactions, **12**, 190
hydroboration, **12**, 184
reaction with zinc–copper organometallics, **11**, 167
1-Alkynes, 1-chloro-
hydroalumination, **1**, 471
Alkynes, 1-halo-
hydroboration, **12**, 188
Alkynes, iodo-
Heck reactions, **12**, 279
Alkynes, 1-iodo-
reaction with zinc–copper organometallics, **11**, 167
Alkynes, silyl-
stability, **2**, 10
Alkynes, thio-
hydroboration
nickel catalysts, **9**, 133
Alkynic alcohols
synthesis, **11**, 494
Alkynic esters
carbocupration, **11**, 179
Alkynide complexes
group 6, **5**, 311
iron clusters, **7**, 260, 263
Alkynoates
allyl
Pauson–Khand reaction, **12**, 730
cycloadditions
Pauson–Khand reaction, **12**, 710
Alkynones
cycloadditions
Pauson–Khand reaction, **12**, 710
Alkynyl aldehydes
cobalt complexes
intramolecular coupling with allylstannanes, **12**, 698
Alkynyl halides
coupling reactions, **12**, 190, 205
cross-coupling
transition metal-promoted, **12**, 169
Alkynyl methanols, 2-furyl-
cobalt complexes
acid-catalyzed isomerization, **12**, 693
Alkynylation
organolead compounds in, **11**, 468
tetraalkyllead to aldehydes, **11**, 462, 463
Allene, 1,3-bis(silyl)-
synthesis, **2**, 12
Allene, tetramethyl-
reaction with manganese hydrides, **6**, 70
Allenes
addition reactions
Fischer carbene complexes, **12**, 535
allyl complexes, **12**, 815, 820
allyl palladium complexes from, **12**, 809, 810
allylic alkylation, **12**, 847
bromoboration, **11**, 202
captodative, **12**, 572
carboxylation
nickel catalysts, **12**, 356
chiral
synthesis, **12**, 169
cocyclocarbonylation with alkynes, **12**, 360
π-complexes
palladium(II) catalyzed nucleophilic attack, **12**, 616, 617
formation
from aldehydes, **12**, 595
hydroboration, **11**, 200
insertion into σ-alkylpalladium complexes, **12**, 272
metallation, **11**, 110
preparation by cross-coupling, **12**, 168
reaction with tin hydrides, **11**, 361
reaction wtih acyltetracarbonyl cobalt complexes, **12**, 354
reactions with iron carbonyls, **12**, 360
reactions with organothallium complexes
palladium-assisted, **12**, 147
reactions with palladium salts, **9**, 341
Allenes, 1-bromo-
coupling reactions
organozinc derivatives, **12**, 184
Allenic alcohols
synthesis, **11**, 238
Allenic esters
synthesis
transition metal complexes, **12**, 245
Allenyl bromides
coupling reactions, **12**, 223
Allenyl ligands
bridging
in heteronuclear metal–metal bond synthesis, **10**, 12
Allenylation
propargyltin compounds, **2**, 246
Allenylidene complexes
triiron clusters, **7**, 263
Allolaurinterol
synthesis, **12**, 580
Allosamizoline
synthesis, **12**, 875
Alloyohimbane
synthesis
lithiation, **11**, 19
Alloyohimbone
synthesis, **12**, 880
Allyl acetates
cyclopropanation, **12**, 396
Allyl alcohol
oxidation
palladium catalysis, **9**, 314
Pauson–Khand reaction, **12**, 712

Allyl alcohols
  synthesis
    organoselenium compounds, **11**, 536
Allyl chloride
  reactions with iron ions, **7**, 104, 106
Allyl complexes
  alkene insertion reactions
    chirality transfer, **12**, 908
  amination, **12**, 803
    mechanism, **12**, 803
  asymmetric displacements
    carbon nucleophiles, **12**, 888
    organometallics, **12**, 887, 888
  catalysts
    alkene cyclization, stereochemistry, **12**, 908–10
  dicobalt propargylium
    reactions, **12**, 692
  intramolecular insertion reactions
    alkenes and alkynes, **12**, 905–21
  metal displacement
    nucleophiles, **12**, 797–895
  in organic synthesis, **12**, 959–76
  reduction, **12**, 799
  transition metals
    nucleophilic attack, **12**, 6
Allyl compounds
  metallation, **1**, 24
Allyl diethyl phosphates
  allyl complexes, **12**, 806
Allyl enol ethers
  preparation, **12**, 584
Allyl enolium ions
  allyl complexes, **12**, 842, 843
Allyl ethers
  cyclopropanation, **12**, 396
  metallation, **11**, 116
  Pauson–Khand reaction, **12**, 712
Allyl group
  transfer from allyltriphenyllead
    radical-mediated transfer, **11**, 464
Allyl halides
  Pauson–Khand reaction, **12**, 715
Allyl homopropargyl amines
  Pauson–Khand reaction, **12**, 730
Allyl imidates
  [3.3]sigmatropic rearrangements
    palladium(II) catalyzed, **12**, 616
Allyl β-ketoesters
  decarboxylation, **12**, 863
Allyl phenyl ether
  metallation, **11**, 116
Allyl propargyl amines
  Pauson–Khand reaction, **12**, 730
Allyl propargyl ethers
  Pauson–Khand reaction, **12**, 728
Allyl selenoxides
  sigmatropic rearrangements, **11**, 557
Allyl sulfides
  metallation, **11**, 116
Allyl sulfinates
  optically active
    synthesis, **12**, 888
Allyl sulfur compounds
  rearrangements
    palladium catalyzed, **12**, 872
Allyl triflates
  reaction with organomercurials, **12**, 145
Allylamines
  cyclopropanation, **12**, 396
  geometrically pure
    synthesis, **12**, 341
  Heck reactions, **12**, 270
  hydroboration, **11**, 195
  isomerization
    palladium catalyzed, **12**, 871, 872
  nitrogen-containing, **11**, 115
  synthesis, **12**, 330, 331, 582
    organotellurium compounds, **11**, 596
Allylation
  aldehydes
    antimony, **11**, 495
  asymmetric
    nickel complexes, **12**, 964
  palladium assisted, **12**, 143
Allylboration, **11**, 236
  asymmetric, **11**, 251–6
    dicobalt propargylium complexes, **12**, 692
  double asymmetric, **11**, 253
  intramolecular, **11**, 237, 238
Allylic acetate
  rearrangement
    palladium catalyzed, **12**, 870
Allylic acetate rearrangements
  palladium(II) catalyzed, **12**, 615
Allylic acetates
  reaction with organomercurials, **12**, 143
  [3.3]sigmatropic rearrangements
    palladium(II)-induced, **12**, 615
Allylic acetoxylation
  allylpalladium chloride dimers, **12**, 818
Allylic alcohols
  enantiomerically pure chiral, **12**, 1115
  epoxidation
    metal catalyzed, **12**, 1102
    titanium tartrate catalyzed, **12**, 1119, 1120
  hydrogenation
    asymmetric, **12**, 21, 22
    enantioselective, **12**, 16
  prochiral
    Sharpless epoxidation, **12**, 1112
  reaction with aryl halides
    palladium catalysis, **12**, 267
  secondary
    hydroboration, **11**, 195
    kinetic resolution, **12**, 1112–15
    Sharpless epoxidation, **12**, 1109–20
  synthesis, **12**, 336
  tertiary
    asymmetric dihydroxylation, **12**, 1150
    synthesis, **11**, 493
Allylic alkylation
  copper catalyzed, **12**, 46
  cuprate preparation, **12**, 83
  heterogeneous catalysis, **12**, 844, 845
  intramolecular
    organometallics, **12**, 881, 882
    stabilized carbanions, **12**, 878–81
  molybdenum complexes, **12**, 890–5
    mechanism, **12**, 890
    regioselectivity, **12**, 892
  nitrogen-based leaving groups, **12**, 840, 841
  oxygen-based leaving groups, **12**, 835–40
  palladium catalyzed, **12**, 834–45
    mechanism, **12**, 834
  phase transfer catalysis, **12**, 844, 845
  regioselectivity, **12**, 834
  sulfur-based leaving groups, **12**, 841, 842
  transmetallation, **12**, 853
  tungsten complexes, **12**, 890–5
    catalysts, **12**, 894
Allylic amidation, **12**, 810
Allylic azidation, **12**, 814
Allylic carbonates
  alkylation, **12**, 839
Allylic carboxylation
  conditions, **12**, 818
  palladium catalysts, **12**, 817
Allylic compounds
  oxidative addition
    palladium(0) complexes, **9**, 343–7
Allylic dithioacetals
  coupling reactions
    with Grignard reagents, **12**, 964
Allylic esters
  reductive cleavage
    allyl complexes, **12**, 803
Allylic etherification, **12**, 814
Allylic halides
  insertion reactions
    palladium assisted, **12**, 153
  reaction with organomercurials, **12**, 143
  reactions with organothallium complexes
    palladium-assisted, **12**, 147
  reductive dimerization
    vanadium complexes, **12**, 1086
Allylic hydroxylation
  molybdenum hydrogen peroxide catalysts, **12**, 815
Allylic isomerization

asymmetric
palladium allyl complexes, **12**, 888
palladium catalyzed, **12**, 870–3
Allylic oxidation
palladium complexes, **12**, 820
Allylic reactions
asymmetric
chiral ligands, **12**, 886
Allylic sulfinates
isomerization
palladium catalyzed, **12**, 872
Allylic sulfination
features, **12**, 828
Allylic sulfones
isomerization
palladium catalyzed, **12**, 872
Allylic thionobenzoate
isomerization, **12**, 872
Allylstannanes, **12**, 698
Aluminaboranes
properties, **1**, 258
Aluminacarboranes
properties, **1**, 258, 259
Aluminadicarbadodecaborane
structure, **1**, 258
Aluminates
structures, **1**, 29
Alumination
carbon Brønsted acids, **1**, 466, 467
Aluminoxane, methyl-
Ziegler–Natta catalyst activation, **12**, 1195
Aluminoxanes
availability, **11**, 295
structures, **1**, 450–3
Aluminum
carbon bonds, **1**, 456–61
interconversions, **1**, 463–7
in organic synthesis, **1**, 485–96
reactions with nontransition metal substrates, **1**, 467–76
chemistry, **1**, 497
hydrogen bonds, **1**, 461, 462
metal bonds, **1**, 462, 463
nonmetal bonds, **1**, 462
turnings
displacement reactions, **1**, 461
Aluminum acetylides
availability, **11**, 296
Aluminum alkoxides
availability, **11**, 296
Aluminum alkyls
alkene exchange, **1**, 464
availability, **11**, 294, 295
conversion to alcohols, **11**, 302
Aluminum aryloxides
availability, **11**, 296
Aluminum aryls
availability, **11**, 295
Aluminum complexes
(μ-alkyl)-metal bonds, **1**, 552
f-block metals
(μ-H)-metal bonds, **1**, 551
group 4 metals
(μ-H)-metal bonds, **1**, 546–8
group 6 metals
(μ-alkyl)-metal bonds, **1**, 552
(μ-H)-metal bonds, **1**, 548, 549
group 7 metals
(μ-H)-metal bonds, **1**, 550
group 8 metals
(μ-H)-metal bonds, **1**, 550, 551
group 9 metals
direct bonds to aluminum, **1**, 552
group 10 metals
(μ-H)-metal bonds, **1**, 551
(μ-H)-metal bonds, **1**, 546–51
metathesis
metal alkyls, **1**, 460, 461
organomagnesium, **1**, 106, 107
transition metal bonds, **1**, 545–69
Aluminum(I) complexes
generation, **1**, 435, 436
isolation, **1**, 435, 436
Aluminum(II) complexes
generation, **1**, 435, 436
isolation, **1**, 435, 436
Aluminum(III) complexes
reactivity
steric factors, **1**, 434, 435
structures, **1**, 433
steric factors, **1**, 434, 435
Aluminum compounds
alkyl
in tetraalkylstannane preparation, **2**, 226
cross-coupling, **12**, 179
transmetallation, **12**, 179–200
Aluminum cyanides
availability, **11**, 296
Aluminum hydrides
reactions
allyl complexes, **12**, 800
Aluminum oxides
availability, **11**, 296
Aluminum reagents
availability, **11**, 294–7
reactions
mechanisms, **11**, 289–93
stereochemistry, **11**, 289–93
reactivity, **11**, 289–93
transition metal-promoted reactions, **11**, 308, 309
Aluminum, trialkyl-
in carbometallation, **12**, 157
rearrangements, **1**, 476
Aluminum, tris(trimethylsilyl)-
reaction with thioesters, **2**, 22
Aluminum, vinyl-
availability, **11**, 296
conversion to ethers, **11**, 303
Alvene
synthesis
trimethylenemethane complexes, **12**, 941
Amalgams
reduction
transition metal clusters, **10**, 327
Amaryllidacea alkaloids
synthesis
lithiation, **11**, 74
Amide alcohols
synthesis, **11**, 496
Amides
acylation
organobismuth(V) compounds, **11**, 511
addition to Grignard reagents, **11**, 151
allylic alkylation, **12**, 868, 869
amidomercuration, **11**, 424
aryl
lithiation, **11**, 42
asymmetric allyl displacements
palladium complexes, **12**, 883
formation
transition metal complexes, **12**, 249–51
methylenation, **12**, 590
reactions
allyl complexes, **12**, 810–13, 874
synthesis
acyl iron complexes, **12**, 353
metal carbonyl anions, **12**, 354
α,β-unsaturated
alkenation, trialkylstibine, **11**, 489
hydrogenation, asymmetric, **12**, 19–21
β,γ-unsaturated
synthesis, **12**, 140
Amidines
cyclic
synthesis, cross-coupling reactions, **12**, 1084
synthesis
cross-coupling reactions, **12**, 1084
Amidocarbonylation, **12**, 30
Amidomercuration, **11**, 424, 425
Amidopalladation
1,3-dienes, **12**, 826
Amine alcohols
synthesis, **11**, 496
Amine functionalized resins
mixed metal carbonyl clusters on
catalysis, **10**, 60
Amine oxides
Pauson–Khand reaction, **12**, 705
Amines
asymmetric allyl displacements
palladium complexes, **12**, 883
cyclization
catalysts, **12**, 355
homoalkylated
synthesis, **11**, 502
intramolecular displacements
allyl complexes, **12**, 873, 874
lithiation, **11**, 15
oxidation
organoselenium compounds, **11**, 535
palladium alkene complexes
nucleophilic attack, **12**, 603–5
primary

synthesis, organoboron compounds, **11**, 216, 217
propargylated tertiary
synthesis, **12**, 690
reaction of osmium tetroxide with alkenes, **12**, 1138
reactions
allyl complexes, **12**, 803–10
secondary
synthesis, organoboron compounds, **11**, 217, 218
synthesis from nitriles or enamines
organoaluminum hydrides, **11**, 300
unsaturated primary
reaction with titanium complexes, **4**, 277
Amino acids
allenic
synthesis, **12**, 852
allylic alkylation, **12**, 851, 852
cross-coupling, **12**, 169
cyclopropane-substituted
synthesis, **12**, 398
tricarbonyl chromium complexes, **5**, 485
α-Amino acids
2,4,6-triisopropylbenzenesulfonamide
oxazaborolidines from, **11**, 262
unsaturated
protected, synthesis, **12**, 70
formation
from ketene complexes, **12**, 569–71
optically active
synthesis, **12**, 271
synthesis, **11**, 249
cobalt-mediated propargylation, **12**, 696
α-Amino acids, *N*-acyldehydro-
hydrogenation
asymmetric, **12**, 16, 17
enantioselective, **12**, 16
α-Amino acids, α-alkyl-
synthesis, **12**, 569
Amino alcohols
oxidation
metal catalyzed, **12**, 1118
vicinal
synthesis, **12**, 344
Amino esters
synthesis
rhodium complexes, **8**, 201
α-Amino perfluoroacids
synthesis, **12**, 245
Amino selenium reagents
synthesis, **11**, 528
Aminomercuriation, **11**, 421–3
alkenes, **3**, 145
alkynes, **3**, 150
Aminyl radicals
production
organotin compounds, **11**, 363
Ammonium ylides
formation
reactions, **12**, 438, 439
fragmentation reactions, **12**, 439
[2,3]-sigmatropic rearrangements, **12**, 439
[1,2]-Stevens rearrangements, **12**, 439
Amphimedine
synthesis
lithiation, **11**, 79
Amphimetidine
synthesis, **12**, 215
Andrenosterone
synthesis, **12**, 73
Angucycline
synthesis, **12**, 494
Anhydrides
cyclization
catalysts, **12**, 355
decarbonylation, **12**, 377
Aniline
coupling reactions
Grignard reagents, **12**, 175
synthesis
carbene complexes, **12**, 527
transition metal complexes
lithiation, **12**, 1024
Aniline, 2-alkynyl-
intramolecular reactions
palladium(II) catalyzed, **12**, 619
Aniline, difluoro-
metallation, **11**, 123
Aniline, *o*-iodo-
coupling reactions
with alkynes, **12**, 224
Aniline, *o*-vinyl-
indoles from, **12**, 209
Anilines
lithiation, **11**, 51–3
Anisole
anti-Friedel–Crafts acylation, **12**, 773
lithiation, **11**, 48
Anisole, fluoro-
metallation, **11**, 122
Annulation
benzocycles
niobium complexes, **5**, 143, 144
Annulations
two-alkyne
carbene complexes, **12**, 500–2
Anthemol
synthesis, **11**, 105
Antheridiogens
methylenation, **12**, 578
Anthracene
magnesium compounds, **1**, 99, 100
Grignard reagent preparation, **1**, 63
Anthracene, 9,10-diphenyl-
reaction with magnesium, **1**, 99
Anthracycline
synthesis, **12**, 491
Anthramycin
synthesis, **12**, 206, 249, 274, 275
Anthranilic acids
synthesis, **12**, 252
Anthranils
reaction with enamines, **12**, 1084
Anthraquinone
coupling reactions
naphthyl triflates, **12**, 214
synthesis, **12**, 748
Anthrenus flavipes
pheromone
synthesis, **11**, 100
Antibiotic A26771B
synthesis, **12**, 881
Antimony
metal
alcohol synthesis, **11**, 495
Antimony compounds
chemistry, **2**, 321–44
Antimony dicarboxylates, triphenyl-
reactions with amines, **11**, 501
Antimony hydroxide, tetraphenyl-
synthesis, **2**, 338
Antimony oxides, triphenyl-
synthesis, **2**, 338
Antimony pentachloride
catalysts
in organic synthesis, **11**, 500
Antimony salts
catalysts
in organic synthesis, **11**, 500, 501
Antimony sulfide, triphenyl-
synthesis, **2**, 338
Antimony trichloride
acetalization, **11**, 498
alcohol synthesis, **11**, 495
catalysts
in organic synthesis, **11**, 501
Antimony ylides
alkenation, **11**, 489, 490
synthesis, **2**, 335
Antimony, pentamethyl-
synthesis, **2**, 335
Antimony, pentaphenyl-
synthesis, **2**, 335
Antimony, tris(cyclopentadienyl)-
synthesis, **2**, 329
Aphidicolin
synthesis, **12**, 245, 246, 353, 365
Aplysiatoxin
synthesis, **12**, 589
lithiation, **11**, 39
Aplysiatoxin, debromo-
synthesis
lithiation, **11**, 39
γ-Apopicropodophylline
synthesis, **12**, 285, 287
Apovincamine
synthesis, **12**, 582
Aquo complexes
organochromium
properties, **5**, 301
reactions, **5**, 302
structure, **5**, 301
organochromium complexes, **5**, 299–302
D-Arabinose
diene molybdenum complexes from, **12**, 645
Arachidonic acid
cascade derivative

synthesis, **12**, 70
Arene complexes
cine substitution, **5**, 513
diastereoselective lithiation, **5**, 510
heteronuclear clusters, **10**, 169, 170
nucleophilic addition/substitution
carbon nucleophiles, **12**, 993
nucleophilic substitution, **5**, 514
regioselectivity of lithiation, **5**, 509
transition metals
addition/acylation, **12**, 1010–12
addition/oxidation, **12**, 995–1007
cyclohexadienyl intermediates from, **12**, 995–1012
o-disubstituted, regioselectivity, **12**, 996
nucleophilic addition, regioselectivity, **12**, 1001, 1002
nucleophilic addition, regioselectivity, steric effect, **12**, 999
nucleophilic addition/ oxidation, intramolecular, **12**, 1004
nucleophilic substitution for hydrogen, **12**, 995–1007
π-Arene complexes
in organic synthesis, **12**, 1039–67
Arenes
complexation
diastereoselective, **5**, 504
electrophilic thallation, **11**, 446
hexasubstituted
synthesis, cycloaddition reactions, **12**, 747
hydrogenation
niobium complexes, **5**, 141, 142
lithiation
lithium–hydrogen exchange, **11**, 42–65
metallation, **11**, 112
ring annulation
lithiation, **12**, 1036
synthesis
cycloaddition of three alkynes, **12**, 747–51
transition metal complexes
deprotonation, diastereoselectivity, **12**, 1026
heteroatom substitution, **12**, 986, 987
ligand substitution for halides, **12**, 987
lithiation, directed by benzylic nitrogen and oxygen substituents, **12**, 1025–7
lithiation, regioselectivity, steric effects, **12**, 1022
nucleophilic addition, **12**, 979–1012
nucleophilic attack, **12**, 6
nucleophilic substitution, solvent effect, **12**, 984
nucleophilic substitution for heteroatoms on, **12**, 982–94
in organic synthesis, **12**, 1017–37
preparation, **12**, 981, 982
proton abstraction, **12**, 1017
Arenes, alkyl-
metallation, **11**, 108, 109
Arenes, allyl-
metallation, **11**, 107
Arenes, 1,2-bis(fluorosilyl)-
synthesis, **2**, 39
Argentilactone
synthesis
lithiation, **11**, 36
Aristeromycin
synthesis, **12**, 812–14, 850
Aristocularine alkaloids
synthesis
lithiation, **11**, 67
Aristolactone
synthesis, **12**, 248
Aristolone
synthesis, **12**, 411, 412
Aromatic compounds
alkylated
oxidation, organoselenium componds, **11**, 535
reactions with selenium dioxide, **11**, 536
fluoro
production, **11**, 482
Aromatic rings
microbial oxidation, **7**, 157
Aromatic substitution
phenylytterbium iodide, **4**, 124
Aromatic systems
extended
synthesis, **12**, 1077
Aroyl cyanides
formation
carbon monoxide insertion reactions, **12**, 258
Aroyl fluoride
formation
carbon monoxide insertion reactions, **12**, 258
Aroyl halides
Heck reactions, **12**, 263
Arsaarenes
synthesis, **2**, 323, 324
Arsaboranes
metal complexes, **1**, 266
properties, **1**, 265, 266
Arsaethene
synthesis, **2**, 322
Arsaethyne, aryl-
synthesis, **2**, 322
Arsenic compounds
chemistry, **2**, 321–44
Arsenic ylides
palladium(II) complexes, **9**, 229
synthesis, **2**, 335
Arsenic, pentamethyl-
synthesis, **2**, 335
Arsenic, pentaphenyl-
synthesis, **2**, 335
Arsenide complexes
triiron clusters, **7**, 275
Arsine, triphenyl-
coupling reactions
O-alkenyl derivatives, **12**, 206
Arsines, silyl-
synthesis, **2**, 29, 30
Arsoles
synthesis, **2**, 324
Arteether
synthesis, **12**, 582
Aryl bromides
coupling reactions, **12**, 192–4
Grignard reagents, **12**, 174
stannanes, **12**, 208–11
Aryl carbamates
coupling reactions
Grignard reagents, **12**, 175
Aryl chlorides
coupling reactions, **12**, 192–4
Grignard reagents, **12**, 173, 174
stannanes, **12**, 208
Aryl compounds
N,O-derivatives
coupling reactions, **12**, 213, 214
O-derivatives
coupling reactions, **12**, 196
polyfunctional
coupling reactions, **12**, 197, 214, 215
coupling reactions with Grignard reagents, **12**, 176
Aryl diazonium salts
coupling reactions
with organostannanes, **12**, 213
Aryl ethers
lithiation, **11**, 48–51
Aryl fluorides
coupling reactions with Grignard reagents, **12**, 173, 174
synthesis
organoboron compounds, **11**, 218
Aryl fluorosilanes
coupling reactions
biaryl synthesis, **12**, 212
Aryl fluorosulfonates
coupling reactions, **12**, 196
Aryl halides
coupling reactions, **12**, 192–6
Grignard reagents, **12**, 173–5
stannanes, **12**, 208–13
cyanation, **12**, 226
ligands
nucleophilic substitution with carbon nucleophiles, **12**, 988–92
lithiation, **11**, 54
Aryl iodides
coupling reactions, **12**, 194–6
Grignard reagents, **12**, 175
organostannanes, **12**, 211–13
Aryl phosphinates
preparation, **12**, 231

Aryl phosphonates
synthesis, **12**, 230
Aryl rings
metathesis, **12**, 228, 229
Aryl selenides
coupling reactions
Grignard reagents, **12**, 176
Aryl sulfides
coupling reactions
Grignard reagents, **12**, 176
Aryl sulfonates
coupling reactions, **12**, 196
Aryl sulfones
coupling reactions
Grignard reagents, **12**, 176
Aryl sulfoxides
coupling reactions
Grignard reagents, **12**, 176
Aryl thioethers
lithiation, **11**, 53
Aryl thiols
lithiation, **11**, 53
Aryl triflates
coupling reactions, **12**, 196
Grignard reagents, **12**, 175
Stille reaction, **12**, 213
Arylation
alkenes
palladium assisted, **12**, 138
organobismuth(V) compounds, **11**, 506–8
organolead compounds, **11**, 468
mechanism, **11**, 480
Arylcopper magnesium arenethiolates
structures, **3**, 104
Arylic amination
palladium-catalyzed, **12**, 229
Arylselenenyl halides
reactions, **11**, 534
Arynes
zirconocene complexes, **12**, 771–83
Ascorbic acid
allylation, **12**, 844
Aspidosperma alkaloids
synthesis, **12**, 659
Aspochalasin B
synthesis, **12**, 881
Asteriscanolide
synthesis, **12**, 793
Asymmetric amplification
organozinc reagents, **11**, 172
Asymmetric dihydroxylation, **12**, 1138–41
Asymmetric synthesis
organoboron compounds, **11**, 247–64
organomagnesium compounds, **11**, 130
organozinc compounds, **3**, 189, 190
Atactic
definition, **12**, 1196
Aucubigenone
synthesis, **12**, 735
Auracyclopentadienes
synthesis, **3**, 37
Aurates, trichloro(phenyl)-
synthesis, **3**, 32
Aurates(I), bis(alkynyl)-, **3**, 19
Aurates(I), dialkyl-
synthesis, **3**, 17
Aurates(I), dialkynyl-
synthesis, **3**, 18
Aurates(I), polyhalogenophenyl-, **3**, 18
Aurates(III), diaryl
synthesis, **3**, 33
Aurates(III), tetraalkyl-
synthesis, **3**, 42
Aurates(III), tetraaryl-
synthesis, **3**, 42
Aurones
synthesis, **12**, 256
Avermectin
fragment
synthesis, **12**, 101
Avermectin B1a
synthesis, **12**, 371
Avermectins
synthesis, **12**, 204
Aximide-1
synthesis, **12**, 62
Axisonitrile-1
synthesis, **12**, 62
1-Azaadamantane
synthesis, **1**, 147
Azabicyclo[5.3.0]decatrienones
synthesis, **12**, 413
Azabicyclo[4.3.0]nonenones
Pauson–Khand reaction, **12**, 730
Azabicyclo[3.3.0]octenones
Pauson–Khand reaction, **12**, 730
synthesis
dicobalt propargylium complexes, **12**, 698
1,2-Azaboraallene
metal complexes, **1**, 201
1,2-Azaboracyclohexanes
borane complexes
reduction, ketones, **11**, 261
Azaboranes
properties, **1**, 260, 261
transition metal derivatives, **1**, 262, 263
Azaboroles
synthesis, **1**, 161
1*H*-1,2-Azaboroles, 2,5-dihydro-
metal complexes, **1**, 212
1*H*-1,2-Azaborolides, 2,5-dihydro-
metal complexes, **1**, 212
1,2-Azaborolines
metal complexes, **1**, 212
1,2-Azaborolinides
metal complexes, **1**, 212
Azacarboranes
properties, **1**, 262
Azadecaboranes
properties, **1**, 261
Azadiboriridine
reactions, **1**, 159
Azadiboriridine, tri-*t*-butyl-
reactions, **1**, 177
Azadirachtin
synthesis, **12**, 589
Azadistanniridine
dissociation, **2**, 291
Azadodecaboranes
properties, **1**, 261
Azaferrocenes
chemistry, **7**, 195
Azagermacyclopentenes
preparation, **2**, 160
Azaindanones
synthesis
carbene complexes, **12**, 508
Azametallacyclopentene
from zirconocene aryne complexes, **12**, 772
5-Azanickelacyclopentan-4-one
synthesis, **9**, 8
Azanickelacyclopentanones
synthesis, **9**, 80
5-Aza-2,7-nonadiene, 1-acetoxy-
cyclization
palladium-catalyzed, **12**, 911
Azapenams
synthesis, **12**, 551
Azaruthenocenes
synthesis, **7**, 604
1,2,5-Azasilaborolenes
metal complexes, **1**, 213
Azasilacyclobutane
structure, **2**, 67, 68
1-Aza-3-silacyclohexadienes
synthesis, **2**, 70
1-Aza-2-sila-3-stannacyclobutane
synthesis, **2**, 70, 71
Azasilocine
synthesis, **2**, 69
Azasilole
preparation, **2**, 68, 69
Azaspiro [4.4] compounds
preparation
allylic amination, **12**, 873
1-Aza-5-stannabicyclo[3.3.3] undecane
synthesis, **12**, 42
Azazirconacyclopentenes
synthesis, **12**, 341, 343
Azetidine–vinyl-
rearrangement
palladium catalysts, **12**, 875
Azetidine-2,4-diones
formation
carbonylation, **12**, 259
Azetidines
reductive cleavage
organolithium reagents, **11**, 14
Azides
asymmetric allyl displacements
palladium complexes, **12**, 883
reactions
allyl complexes, **12**, 814
Azidomercuriation, **11**, 426
alkenes, **3**, 145
Aziridine, vinyl-
rearrangement
palladium catalysts, **12**, 875
Aziridines
cycloaddition
heterocumulenes, **11**, 500
reductive cleavage
organolithium reagents, **11**, 14
synthesis
organoboron compounds, **11**, 218
Aziridines, *N*-methoxy-
synthesis
organosilicon compounds, **11**, 340

Azirines
  photochemical reactions with chromium alkoxycarbene complexes, **12**, 572, 573
Azo compounds
  reactions with carbene complexes, **12**, 566
Azoalkanes
  triiron clusters, **7**, 266
Azoaryl sulfones, aryl-
  Heck reactions, **12**, 279
Azobenzene
  photochemical reaction with chromium alkoxycarbene complexes, **12**, 566
Azomethine ylides
  generation, **12**, 440, 447, 448
  reactions, **12**, 447, 448
Azulene, hexahydro-
  synthesis, **12**, 908
Azulene, hydro-
  lactones
    synthesis, **12**, 413
Azulenes, hexahydro-
  preparation, **12**, 617
Azulenes, hydro-
  synthesis, **12**, 401

B

B_6N_7-phenalene
  synthesis, **1**, 158
Bacimethrin
  synthesis
    lithiation, **11**, 81
Bäckvall diacetoxylation reaction, **12**, 672
Barbier reactions
  organosamarium complexes, **4**, 127
Barbier-type reactions
  organobismuth(III) compounds, **11**, 502
  organozinc compounds, **11**, 178
  trialkylstibine, **11**, 492
Barbituric acid
  reactions with organolead compounds, **11**, 475
Barium chlorides
  allyl
    preparation, **1**, 108
Barium complexes
  anthracene, **1**, 117
  arene radical anion complexes, **1**, 117
  chemistry, **1**, 57–120
Barium, allylchloro-
  reactions
    with carbonyl compounds, **1**, 109
Barium, bis(α-arylalkyl)-
  ionic structures, **1**, 111
Barium, bis(pentamethylcyclopentadienyl)-, **1**, 112
Barium, bis(triphenylmethyl)-
  complexes, **1**, 111
  formation, **1**, 111
Barium, bis(1,1,3-triphenylpropyl)-
  structure, **1**, 111
Barium, dialkynyl-
  preparation, **1**, 111
Barium, dicumyl-
  structure, **1**, 110
Barium, dicyclopentadienyl-, **1**, 112–16
  ab initio calculations, **1**, 114
Barium, difluorenyl-
  formation, **1**, 115
  ion pair structures, **1**, 115
Barium, difluorenyltetrammine-
  crystal structure, **1**, 115
Barium, diindenyl-
  formation, **1**, 115
Barium, dimethyl-
  structure, **1**, 110
Barton–McCombie reaction
  organotin compounds, **11**, 359
  organotin hydrides, **2**, 275
Basmenone, epoxy-
  synthesis, **12**, 590
9-BBN
  Suzuki reaction, **12**, 185
9-BBN-H
  synthesis, **11**, 196
9-BBN, *B*-alkyl-
  cross-coupling reaction, **11**, 233, 234
9-BBN, vinyl-
  Diels–Alder cycloadditions, **11**, 211
  synthesis, **11**, 204
Beckmann rearrangement
  organoaluminum compounds, **1**, 476
Benzannulation
  asymmetric, **12**, 490
  Fischer carbene complexes, **12**, 484
    effect of metal, **12**, 488
  heterocycles, **12**, 751
  manganese carbenes, **6**, 65
  methodology, **12**, 484–90
  photochemical
    carbene complexes, **12**, 567
  regiochemistry, **12**, 485
  synthetic applications, **12**, 491–7
Benzazaphospholine
  synthesis
    lithiation, **11**, 68
Benzazepine, tetrahydro-
  synthesis
    chromium complexes, **12**, 1049
Benzdiynes
  bis(zirconocene) complexes, **12**, 777, 780
Benzene
  carbonylation
    iridium complexes, **8**, 404
  metallation
    rhodium complexes, **8**, 210
  reactions with iron atoms, **7**, 108
  reactions with iron ions, **7**, 106
Benzene, *p*-bis(triacetoxyplumbyl)-
  preparation, **11**, 467
Benzene, 1,3-bis(trifluoromethyl)-
  metallation, **11**, 122
Benzene, *t*-butyl-
  metallation, **11**, 112
Benzene, chloro-
  reactions with iron ions, **7**, 106
Benzene, 1,3-di(*t*-butyl)-
  metallation, **11**, 112
Benzene, 1,2-diethynyl-
  cyclotrimerization, **12**, 750
Benzene, 1,2-difluoro-
  reaction with Schlosser's base, **11**, 121
Benzene, 1,4-difluoro-
  reaction with Schlosser's base, **11**, 121
Benzene, (3,3-dimethyl-1-butenyl)-
  metallation, **11**, 112
Benzene, ethyl-
  metallation, **11**, 109
Benzene, fluoro-
  metallation, **11**, 121
Benzene, fluoro(trifluoromethyl)-
  metallation, **11**, 122
Benzene, hexakis(arylsilyl)-
  synthesis, **2**, 7
Benzene, hexakis(dimethylsilyl)-
  synthesis, **2**, 7
Benzene, hexakis(trimethylsilyl)-
  synthesis, **2**, 7; **11**, 326
Benzene, (3-hydroxy-3-methyl-1-butynyl)-
  metallation, **11**, 112
Benzene, (3-hydroxy-3,4,4-trimethyl-1-pentynyl)-
  metallation, **11**, 113, 114
Benzene, (1-methylcyclopropyl)-
  metallation, **11**, 112
Benzene, nitro-
  reduction
    tellurium reagents, **11**, 577, 591
Benzene, nitroso-
  reaction with
    trialkylaluminum, **1**, 475
Benzene, pentafluoroiodo-
  reaction with Collman's reagent, **12**, 352
Benzene, perfluoroalkyl-
  preparation, **12**, 192
Benzene, trimethyl-
  trimetallation, **11**, 109
Benzeneseleninic acid
  reactions, **11**, 534
Benzeneseleninic anhydride
  reactions, **11**, 534
Benzeneseleninyl halides
  reactions, **11**, 535
Benzhydrylamine, 4,4-dimethoxy-
  oxidative addition
    allyl complexes, **12**, 805
Benzoborepine
  metal complexes, **1**, 210, 211
Benzocyclobutadiene
  synthesis
    trimerization, **12**, 750
Benzocyclobutenedione
  reaction with
    chlorotris(triphenylphosphine)cobalt, **12**, 752
Benzocyclobutenes
  strained
    synthesis, homotrimerization, **12**, 749
  synthesis
    cyclooligomerization, **12**, 748–51
    zirconocene complexes, **12**, 774, 775

1,2-Benzocyclohexanones
  asymmetric reduction, **12**, 25
1,2-Benzocyclopentanones
  asymmetric reduction, **12**, 25
Benzodiazapines
  synthesis
    intramolecular keto ester coupling reactions, **12**, 1080
1*H*-Benzo[d]-1,3-diboroles, 2,3-dihydro-
  synthesis, **1**, 206
Benzodioxaborole
  structure, **1**, 174
Benzodioxasilacyclopentene
  synthesis, **2**, 19
Benzo-1,4-dioxin
  chromium tricarbonyl complexes
    lithiation, **11**, 63
1,3-Benzodiphospholene, 2-organogermyl-
  preparation, **2**, 160
Benzofuran, 2-aryl-
  synthesis, **11**, 473
Benzofuran, dihydro-
  synthesis
    from arene ligand substitution, **12**, 986
Benzofuranones
  synthesis, **12**, 225
Benzofuran-2(3*H*)-ones
  preparation, **12**, 462
Benzofurans
  synthesis, **12**, 195, 273
    intramolecular keto ester coupling reactions, **12**, 1080
    lithiation, **11**, 54
Benzofurans, 2-vinyl-2,3-dihydro-
  synthesis, **12**, 612
Benzoic acid esters
  reductive coupling, **12**, 1074
Benzoic acid halides
  reductive coupling, **12**, 1074
Benzoic acid, *p*-bromo-
  synthesis, **12**, 363
Benzoic acids
  o-lithiation, **11**, 42
Benzonaphthyridinones
  synthesis
    lithiation, **11**, 78
Benzonitrile
  reaction with tetraalkyldialuminoxanes, **1**, 475
Benzopyrrolidinones
  preparation, **12**, 609
2,3-Benzo-1-silacyclobutene
  anionic ring-opening polymerization, **2**, 50
2*H*-1,3-Benzothiazines
  β-lactams from
    photochemical reactions, carbene complexes, **12**, 551
4*H*-1,3-Benzothiazines
  β-lactams from
    photochemical reactions, carbene complexes, **12**, 551
Benzothiazoles
  cross-coupling reactions
    Grignard reagents, **12**, 178
  synthesis
    zirconocene complexes, **12**, 774
Benzothiophene
  ring opening
    reaction with Grignard reagents, **12**, 171
  synthesis
    zirconocene complexes, **12**, 776, 777
Benzo[*b*]thiophene
  rhenium complexes, **6**, 218
  synthesis
    lithiation, **11**, 54
Benzo[*c*]thiophene 2,2-dioxides, 1,3-dihydro-
  preparation, **12**, 462
Benzothiophenes
  synthesis
    lithiation, **11**, 54
Benzotriazole
  structure, **1**, 169
Benzoxazole, 2-aryl-
  formation
    carbonylation, **12**, 258
Benzoxazoles
  synthesis, **12**, 249
Benzvalene
  ring-opening metathesis polymerization, **12**, 1223
Benzyl alcohols
  asymmetric synthesis
    chromium complexes, **12**, 1045
  lithiation, **11**, 22
Benzyl chloride
  alkoxycarbonylation
    transition metal complexes, **12**, 243
Benzyl halides
  cross-coupling reactions, **12**, 181
Benzyl ketones
  synthesis, **11**, 496
Benzylic carbanions
  tricarbonylchromium stabilized, **12**, 1054–63
Benzylic carbocations
  tricarbonylchromium stabilized, **12**, 1048–52
Benzylic halides
  alkoxycarbonylation
    transition metal complexes, **12**, 243
  reaction with zinc–copper reagents, **11**, 166
  reductive dimerization
    niobium complexes, **12**, 1087
    titanium reagents, **12**, 1073
    vanadium complexes, **12**, 1086
Benzyne
  zirconocene complexes, **12**, 771
Bervicomin
  synthesis, **12**, 446
Beryllium
  (1-*t*-butyldihydro-2-methyl-1*H*-azaborolyl)cyclopentadienyl-
    preparation, **1**, 46
  chemistry, **1**, 35–54
  dimethyl-
    band structure calculations, **1**, 40
  toxicity, **1**, 35
Beryllium borohydride, methyl-
  NMR, **1**, 38
Beryllium chloride, tetrakis-(dimethylsulfonium)-
  synthesis, **1**, 36, 37
Beryllium compounds
  alkyl
    optically active, synthesis, **1**, 52
  cyclopentadienyl-, **1**, 44–50
  NMR, **1**, 50
Beryllium hydrides, **1**, 50–2
Beryllium, acylbromo-
  synthesis, **1**, 53
Beryllium, bis(1-*t*-butyldihydro-2-methyl-1*H*-azaborolyl)-
  preparation, **1**, 46
Beryllium, bis(cycloheptadienyl)-
  NMR, **1**, 38
Beryllium, bromomethyl-
  reactions, **1**, 53
Beryllium, chloro(cyclopentadienyl)-
  reaction with potassium pentaborane, **1**, 48
  structure, **1**, 48
Beryllium, chlorodicyclopentadienyl-
  reactions, **1**, 46
Beryllium, (cyclooctadienyl)-
  NMR, **1**, 38
Beryllium, cyclopentadienylhalo-
  bonding, **1**, 50
  thermodynamic calculations, **1**, 50
Beryllium, cyclopentadienyl-(pentamethylcyclopentadienyl)-
  structure, **1**, 46
Beryllium, (di-*t*-butylphosphino)(pentamethylcyclopentadienyl)-
  synthesis, **1**, 42, 43
Beryllium, dicyclopentadienyl-
  benzene complex, **1**, 45
  NMR, **1**, 45
  ordered structure
    calculation, **1**, 44
  solution
    dipole moment, **1**, 45
    structure, **1**, 45
  structure
    rationalization, **1**, 45
  synthesis, **1**, 44–8
  vapor phase
    structure, **1**, 45
Beryllium, diethyl-
  metal organic vapor phase epitaxy, **1**, 54
Beryllium, (2-dimethylamino-*N*-methylethylamido)-hydrido-
  synthesis, **1**, 50
Beryllium, (pentamethylcyclopenta- dienyl)methyl-
  preparation, **1**, 48

Beryllium, tetra(propyn-1-yl)-(tetramethyl)-dialumino-
preparation, **1**, 39
Beryllium, tris(3-*t*-butylpyrazolyl)-hydroboratohydrido-
synthesis, **1**, 50
Betweenallene
synthesis, **2**, 65, 66
Betweenanes
synthesis, **12**, 1077
Biaryl
synthesis, **12**, 196
carbene complexes, **12**, 497
nickel-catalyzed, **12**, 174
Biaryl couplings
Grignard reagents, **11**, 133
Biaryls
synthesis, **12**, 79
organoboron compounds, **11**, 232
organomercury compounds, **11**, 401
organothallium compounds, **11**, 456
unsymmetrical biaryls, **11**, 482
Bibenzyls
synthesis
organomercury compounds, **11**, 401
Bicyclizations
multiple anion capture, **12**, 310
Bicyclo[4.4.0]decasilane
synthesis, **2**, 23
Bicyclo[6.4.0]dodecadiene
synthesis, **12**, 793
Bicyclo[2.2.1]hepta-2,5-diene
metallation, **11**, 111
Bicyclo[3.2.0]hepta-2,6-diene
metallation, **11**, 111
Bicyclo[3.2.0]hepta-3,6-diene, 1-methoxy-
reaction with propynehexakis-(carbonyl)dicobalt, **12**, 707
Bicyclo[4.1.0]heptane
metallation, **11**, 114
Bicyclo[3.2.0]hept-6-ene
metallation, **11**, 111
Bicyclo[3.2.0]hept-6-enes
Pauson–Khand reaction, **12**, 715
Bicyclo[3.3.1]nonasilane
synthesis, **2**, 23
Bicyclo[3.2.1]octa-2,6-diene
metallation, **11**, 107
Bicyclo[2.2.2]octa-1,4-diene
synthesis, **12**, 719
Bicyclo[3.3.0]octane
synthesis, **12**, 398
Bicyclo[3.3.0]octanes
synthesis
organoboranes, **11**, 227
Bicyclooctanone
synthesis, **12**, 916
Bicyclo[3.3.0]octanone
synthesis, **12**, 918
Bicyclo[3.3.0]octanone acetate
synthesis
allylpalladium catalysts, **12**, 916
Bicyclooctanone, methylene-
synthesis, **12**, 917
Bicyclo[2.2.2]octasilane
synthesis, **2**, 23
Bicyclo[2.2.2]oct-2-ene
metallation, **11**, 112
Bicyclo[3.3.0]oct-2-ene
isomerization
Pauson–Khand reaction, **12**, 717
Bicyclo[2.2.2]octenes
Pauson–Khand reaction, **12**, 721
Bicyclo[3.3.0]oct-1-enes, 8-(3-butynyl)-
Pauson–Khand reaction, **12**, 733
Bicyclo[3.3.0]octenone
synthesis, **12**, 917
Bicyclo[3.3.0]oct-1-en-3-one
synthesis
Pauson–Khand reaction, **12**, 734–6
Bicyclo[3.3.0]oct-6-en-2-one, **12**, 879
Bicyclooctenone acetate
synthesis, **12**, 918
Bicyclo[3.3.0]oct-1-en-3-one, 8-(3-butynyl)-
synthesis
Pauson–Khand reaction, **12**, 733
Bicyclooctenone, methylene-
synthesis, **12**, 918
Bicyclo[3.3.0]octenones
synthesis
dicobalt propargylium complexes, **12**, 699
Bicyclo[4.4.0]oct-1-en-3-ones
Pauson–Khand reaction, **12**, 724
Bicyclo[1.1.1]pentasilane
synthesis, **2**, 23
in-Bicyclo[4.4.4]tetradec-1-ene
synthesis, **12**, 1077
in,out-Bicyclo[4.4.4]tetradecane
synthesis, **12**, 1077
Bicyclo[7.3.1]tridecenediyne
synthesis
dicobalt propargylium complexes, **12**, 698
Biferrocenyl
structure, **1**, 200
Bifluorenylidienes
chiral
synthesis, **12**, 1092
Bilobalide
synthesis, **12**, 73, 75
Bimetallaboranes, **1**, 323
Binaphthyl
synthesis, **12**, 81
Binor-S
formation, **10**, 48
2,2'-Bi(oxazoles), 4,4',5,5'-tetrahydro-
synthesis, **12**, 887
Biphenomycin
synthesis, **12**, 193
Biphenyl, dimethyl-
dimetallation, **11**, 109
Biphenyl, tetramethyl-
tetrametallation, **11**, 109
Biphenylene
mercuriation, **3**, 142
Biphenyls
bis(tricarbonyl chromium) complexes
reduction, **5**, 495
restricted rotation, **5**, 499
Bipyridines
synthesis
from nitriles, **12**, 754
Biruthenocene, diacetyl-
preparation, **7**, 617
Biruthenocene, propionyl-
preparation, **7**, 617
Biruthenocenes
synthesis, **7**, 616, 617
Biruthenocenylene
synthesis, **7**, 616
Bisabolene
synthesis
organoboron compounds, **11**, 228
Bis(benzcyclobutene)
synthesis
zirconocene complexes, **12**, 777, 780
Biscyclizations
multiple neutral capture, **12**, 312
Bis(cyclopentadienyl) complexes, **4**, 483–9
Bismaarenes
synthesis, **2**, 323, 324
Bismaboranes
properties, **1**, 267
1-Bismaferrocene, bis(trimethyl-silyl)-3,4-dimethyl-
preparation, **7**, 197
Bismuth compounds
chemistry, **2**, 321–44
Bismuth trichloride
aldol reactional Michael reaction, **11**, 502, 503
Bismuth ylides
synthesis, **2**, 335
Bismuth, pentaphenyl-
synthesis, **2**, 335
Bismuthimines, *N*-tosyltriaryl-
preparation, **11**, 506
Bismuthines, silyl-
synthesis, **2**, 30, 31
Bismutholes
synthesis, **2**, 324
Bismuthonium ylides
in organic synthesis, **11**, 506
Bismutide complexes
triiron clusters, **7**, 275
Bisnorisocomene
synthesis, **12**, 735
Bisorganocobalt(III) complexes
cis-orientation, **8**, 67–71
Bis(oxaborolanes)
preparation, **1**, 173
Bisoxypalladation
carbomethoxylation
intramolecular, **12**, 612
Blastmycinone
synthesis, **12**, 696
Block copolymers
from ring-opening metathesis polymerization, **12**, 1214
insulating–conducting

ring-opening metathesis polymerization, **12**, 1222
polyalkenes
Ziegler–Natta catalysts, **12**, 1204, 1205
synthesis
hydrosilylation, **2**, 117
Bombykol
synthesis, **11**, 232
Bond cleavage
C=C
cumulene by $Fe_3(CO)_{12}$, **7**, 263
C=S
formation of iron carbene complexes from thioketone derivatives, **7**, 280
C≡C
during formation of trinuclear ferroles from alkynes, **7**, 279
$Et_2NC{\equiv}CMe$ on an iron cluster, **7**, 274
triiron alkyne clusters, **7**, 272
C≡O
reductive cleavage giving iron carbide clusters, **7**, 268
P–C
phosphinidene iron clusters, **7**, 264
Bond strengths
tin–tin, **2**, 290
Bonding
transition metal complexes, **12**, 4
Bonds
mercury–carbon
formation, **3**, 136–54
formation, transmetallation, **3**, 136–8, 154
1-Boraadamantane
complexes, **1**, 145
2-Boraadamantane, 2-isopropyl-
bromination, **1**, 147
Boraadamantanes
preparation, **1**, 143, 145–7
Borabenzene
adducts, **1**, 152
metal complexes, **1**, 209
preparation, **1**, 152
structure, **1**, 131, 132
3-Borabicyclo[3.3.1]nonane
preparation, **1**, 165
9-Borabicyclo[3.3.1]nonane
preparation, **1**, 138–42, 168
Borabicyclodecanes
preparation, **1**, 143
Borabicyclodecanes, *B*-alkoxy-
preparation, **1**, 143
Borabicyclononanes
preparation, **1**, 143
Borabicyclononene
synthesis, **1**, 154
Boracyclanes
preparation, **1**, 137
Boracyclanes, *b*-methoxy-
ring enlargement, **11**, 207
Boracyclopentenyl ligands
metal complexes, **1**, 203
Boraethene
metal complexes, **1**, 201
2-Boranaphthalene
preparation, **1**, 152
Borane, 2-alkene-
synthesis, **11**, 101
Borane, alkenyl-
conjugate addition, **11**, 210
oxidation, **11**, 214
Borane, alkyl-
autoxidation, **11**, 242
deprotonation, **11**, 210
Borane, alkyldibromo-
synthesis, **11**, 205
Borane, *B*-alkyldimesityl-
anions, **11**, 234
Borane, 1-alkynyldiisopropoxy-
synthesis, **11**, 202
Borane, allenyl-
chiral
synthesis, **11**, 252
rearrangement, **11**, 237, 238
Borane, allyl-
addition to aldehydes, **11**, 236–8
addition to ketones, **11**, 236–8
reactions, **11**, 235–9
reactivity, **11**, 238, 239
Borane, aryldimesityl-
synthesis, **11**, 203
Borane, β-binapthyl-
synthesis, **11**, 204
Borane, α-bromoalkyl-
synthesis, **12**, 41, 42
Borane, chloro(alkylamino)-
reactions with allenyllithium, **1**, 144
Borane, (3-chloro-1-iodo-1-propenyl)dialkyl-
preparation, **11**, 221
Borane, cyclohexenyl-
synthesis, **11**, 211
Borane, cyclotetrakis(triazolyl)-
properties, **1**, 169
Borane, dialkenylchloro-
synthesis, **11**, 197
Borane, dialkylallyl-
preparation, **11**, 101
Borane, dialkylbromo-
synthesis, **11**, 205
Borane, dialkylhalo-
mixed, **11**, 196
Borane, dialkylvinyl-
synthesis, **11**, 204
Borane, dibromoalkyl-
synthesis, **11**, 199
Borane, diiodo-
reactions, **11**, 244
Borane, diisopinocampheyl-
hydroboration, **11**, 247
Borane, dilongifolyl-
hydroboration, **11**, 247
Borane, (2-ethylapoisopinocampheyl)-
hydroboration, **11**, 247
Borane, 9-fluorenylidene-(2,2,6,6-tetramethylpiperidino)-
metal complexes, **1**, 201
Borane, 1-halo-1-alkenyl-
preparation, **11**, 221
Borane, imino-
addition reactions, **1**, 163
cycloaddition reactions, **1**, 162
reactions, **1**, 159, 160
reactions with trihaloboranes, **1**, 165
Borane, [(methoxymethoxy)-allyl]diisopinocampheyl-
allylboration, **11**, 255
Borane, methylene-
mesityl-bridged, **1**, 149, 150
Borane, monoalkylthexyl-
synthesis, **11**, 196
Borane, monoisopinocampheyl-
hydroboration, **11**, 247
Borane, oxo-
reactions, **1**, 171
Borane, pyrazolyl-
chemistry, **1**, 168
preparation, **1**, 167
structure, **1**, 132
Borane, thexyldialkyl-
synthesis, **11**, 196
Borane, thexylmonochloro-
preparation, **11**, 199
Borane, thexylmonohalogeno-
reactions, **11**, 244
Borane, trialkyl-
mixed
synthesis, **11**, 196
Borane, triallyl-
reactions with alkynes, **11**, 239
Borane, triethyl-
free radical additions, **11**, 242
Borane, triiodo-
reactions, **11**, 244
Borane, vinyl-
deprotonation, **11**, 210
dipolar cycloaddition, **11**, 212
Borane, vinyldichloro-
Diels–Alder cycloadditions, **11**, 211
Boranes
complexes
hydroboration, **11**, 193–6
disubstituted
hydroboration, **11**, 198–201
methanolysis, **11**, 213
monosubstituted
hydroboration, **11**, 196, 197
oxidation
chemoselectivity, **11**, 215
Boranes, alkenyl-
allylic alkylation, **12**, 857
reaction with vinylic epoxides
palladium-assisted, **12**, 150
ring-opening reactions
vinyloxiranes, **12**, 156
Boranes, imino-
metal complexes, **1**, 211
Boranes, methylene-
properties, **1**, 134
Boranes, pyrazolyl-
preparation, **1**, 140
Boranes, vinyl-
preparation by hydroboration of alkynes, **12**, 371
1-Bora-2-oxaadamantane
synthesis, **1**, 145
1-Bora-3-silacyclopentane
synthesis, **2**, 70
Boratabenzene
metal complexes, **1**, 209
Boratabutadiene
metal complexes, **1**, 205
Borataethene

metal complexes, **1**, 201
Borataferrocene
protonation, **1**, 204
Borataferrocenes
chemistry, **7**, 194
Borate anions, bis(brenzcatech-nicato)-
in organonickel synthesis, **9**, 33
Borates
cyclic
stabilities, **1**, 184
structures, **1**, 29
tetrahydro-
group 5 metals, metallaboranes, **1**, 280–3
Borates, alkenyl-
intramolecular transfer reactions, **11**, 227
reactivity, **11**, 229
Borates, alkynyl-
chemistry, **11**, 227–9
intramolecular transfer reactions, **11**, 227
Borates, aryl-
chemistry, **11**, 229
Borates, dihydro-
metallaboranes, **1**, 290, 291
Borates, heteroaryl-
chemistry, **11**, 229
Borates, hexahydro-
group 10 metals
metallaboranes, **1**, 328
metallaboranes, **1**, 328
Borates, monohydro-
metallaboranes, **1**, 292
Borates, tetraaryl-
coupling reactions
with enol triflates, **12**, 191
Borates, tetrahydro-
group 3 metals
metallaboranes, **1**, 278
group 4 metals
metallaboranes, **1**, 278–80
group 6 metals
metallaboranes, **1**, 283, 284
group 7 metals
metallaboranes, **1**, 283, 284
group 8 metals
metallaboranes, **1**, 284, 285
group 9 metals
metallaboranes, **1**, 284, 285
group 11 metals
metallaboranes, **1**, 286
metallaboranes, **1**, 278–86
Borates, trihydro-
actinide complexes, **1**, 290
lanthanide complexes, **1**, 288
metallaboranes, **1**, 287–90
transition metal complexes, **1**, 287, 288
Boratirene
characterization, **1**, 148
Boratricycloalkanes
preparation, **1**, 144
Borazines
chemistry, **1**, 158
phosphorus analogue, **1**, 170
reactions, **1**, 167
synthesis, **1**, 160
Borepane
preparation, **1**, 137, 138
Borepin, heptaphenyl-
synthesis, **1**, 154
Borepins
bicyclic derivatives, **1**, 155
geometry, **1**, 132
metal complexes, **1**, 210
synthesis, **11**, 204
1*H*-Borepins
metal complexes, **1**, 198
synthesis, **1**, 153, 154
Borete, 1,2-dihydro-
structure, **1**, 131, 132, 149
Boridic clusters
ruthenium, **1**, 303
Borinane, 3,5-dimethyl-
synthesis, **11**, 196
Borinanes
NMR, **1**, 171
preparation, **1**, 136
reactions, **1**, 181
Borinates
cyclic
synthesis, **11**, 205
Borinic esters
synthesis
from boronic esters, **11**, 203
from Grignard reagents, **11**, 202
Borirenes
anions, **1**, 133
chemistry, **1**, 148
metal complexes, **1**, 201
structure, **1**, 131
2-Bornene
metallation, **11**, 111
Borohydrides
reagents, **11**, 244
Borolane, *B*-allyl-2-(trimethyl-silyl)-
reactions, **11**, 251
Borolane, 2,5-dimethyl-
reduction
ketones, **11**, 261
synthesis, **11**, 247
1,2-Borolane, 3-methylene-
metal complexes, **1**, 205
Borolanes
NMR, **1**, 171
polymer-supported, **1**, 173
Borolanyl mesylate, 2,5-dimethyl-
reduction
ketones, **11**, 261
Borole
metal complexes, **1**, 202–4
Borole, 2,5-dihydro-
preparation, **1**, 150
1*H*-Borole, 2,5-dihydro-
synthesis, **11**, 203
Borole ligands
metal complexes, **1**, 198
2-Borolenes
synthesis, **1**, 202
3-Borolenes
synthesis, **1**, 202
Boroles
ammonia adducts, **1**, 203
metal complexes
slip distortion, **1**, 204
preparation, **1**, 151
1*H*-Boroles, 2,5-dihydro-
synthesis, **1**, 202
Boron carbon rings
metal complexes, **1**, 201–10
Boron compounds
fluorinated vinyl
coupling reactions, **12**, 186
transmetallation, **12**, 179–200
Boron cyclic systems
three- or four-coordinate, **1**, 129–85
Boron enolates
Aldol reactions, **11**, 240
Boron germanium rings
chemistry, **1**, 158
Boron hydrides
reactions
with allyl complexes, **12**, 799, 800
Boron lead rings
chemistry, **1**, 158
Boron nitrogen antimony rings
properties, **1**, 182
Boron nitrogen arsenic rings
properties, **1**, 182
Boron nitrogen ligands
metal complexes, **1**, 211
Boron nitrogen oxygen rings
properties, **1**, 182
Boron nitrogen phosphorus oxygen rings
properties, **1**, 184
Boron nitrogen phosphorus rings
properties, **1**, 182
Boron nitrogen rings
chemistry, **1**, 158–69
properties, **1**, 181
Boron nitrogen selenium rings
properties, **1**, 184, 185
Boron nitrogen sulfur rings
properties, **1**, 184, 185
Boron oxygen rings
four-coordinate boron, **1**, 178, 179
properties, **1**, 171–9
Boron oxygen sulfur rings
properties, **1**, 185
Boron phosphorus ligands
metal complexes, **1**, 211
Boron phosphorus oxygen rings
properties, **1**, 184
Boron phosphorus rings
four-coordinate boron, **1**, 170
properties, **1**, 169–71
Boron ring compounds
four-coordinate, **1**, 165–9
Boron rings
ligated to metals, **1**, 197–214
Boron selenium rings
properties, **1**, 179–81
Boron silicon oxygen rings
properties, **1**, 181
Boron silicon rings, **1**, 157
properties, **1**, 181
Boron sulfur rings
properties, **1**, 179–81
Boron tin rings
chemistry, **1**, 158
properties, **1**, 181
Boron trifluoride
diethyl ether adduct
Michael additions, cuprates, **12**, 73
in organocopper reactions, **12**, 104–10
Boron, fluorodimethoxy-

in oroganic synthesis, **11**, 101
Boron, vinyl-
cross-coupling reactions, **11**, 232
Boronates
alkenyl
synthesis, **11**, 205
alkyl
synthesis, **11**, 205
cyclic
synthesis, **11**, 205
Boronates, 2-alkene-
synthesis, **11**, 101
Boronates, allyl-
α-hetero-substituted, **11**, 252
Boronates, butyl-
2,3-butanediol esters
homologation, **11**, 251
Boronates, cyclopropyl-
synthesis, **11**, 212
Boronates, methoxycrotyl-
synthesis, **1**, 173
Boronates, pentenyl-
enantiomerically pure, **11**, 253
Boronic acid
transmetallation, **12**, 180
Boronic acid, 1-alkenyl-
cross-coupling reactions
thallium-accelerated
palladium-catalyzed, **11**, 456
Boronic acid, aryl-
cross-coupling reactions, **11**, 232, 233
synthesis, **11**, 203
Boronic acids, alkenyl-
in diene synthesis
palladium-assisted, **12**, 149
Boronic esters
synthesis
from trialkoxyboranes, **11**, 202
Boronic esters, allyl-
isomerization, **11**, 235
synthesis, **11**, 207
Boronic esters, allylic
synthesis, **11**, 208–11
Boronic esters, 1-chloroalkyl-
homologated, **11**, 249
Boronic esters, 1-chlorocrotyl-
synthesis, **11**, 253
Boronic esters, α-haloalkyl-
preparation, **11**, 221
Boronic esters, α-halogeno-
synthesis, **11**, 207
Boronic esters, α-iodoalkyl-
reaction with zinc dust, **11**, 235
Boronic esters, α-organyl-substituted
synthesis, **11**, 207
Boronic esters, α-phenylthio-
carbanions, **11**, 210
Boronic esters, vinyl-
Diels–Alder cycloadditions, **11**, 211
Boron(III) complexes
cyclizations with capture, **12**, 312
Boroxins
NMR, **1**, 171
Boryl compounds
dimesityl
alkylation, **11**, 210
Borylene, methyl-
preparation, **1**, 144
Bostrycin
synthesis, **12**, 582
Bovolide
synthesis, **12**, 505
Brassinolides
diol stereochemistry
biological activity, **12**, 1166
Brefeldin
synthesis
lithiation, **11**, 36
Brefeldin A
synthesis, **12**, 885, 887, 928
trimethylenemethane complexes, **12**, 942
Brevicomin
synthesis, **12**, 613
exo-Brevicomins
synthesis, **12**, 876
Bridged rings
cascade cyclization, **12**, 304
Bromine–zinc exchange
zincates, **11**, 180
Bromoform
reaction with diethylzinc and alkenes, **11**, 179
Brook rearrangement
lithiation, **11**, 24
Buckminsterfullerenes
preparation
hydrozirconation, **12**, 41
Butadiene
dimerization with hydroxy compounds, **12**, 787
oligomerization
nickel crotyl complex catalysts, **9**, 142
polymerization
mechanism, **9**, 142
1,4-*trans*-polymerization
nickel catalyst, **9**, 33
reaction with phenyl isocyanate
nickel complexes, **9**, 8
reactions with formaldehyde
palladium catalyzed, **12**, 791
reactions with iron atoms, **7**, 108
reactions with iron ions, **7**, 107
reactions with β-ketoesters
palladium catalyzed, **12**, 791
reactions with 1-nitropropane
palladium catalyzed, **12**, 791
Butadiene, hydroxy-
tricarbonyliron complex
isolation, **12**, 624, 625
1,3-Butadiene, 2-methoxy-
cyclopropanation, **12**, 404
1,3-Butadienes
symmetrical
synthesis, **12**, 170
Butadiyne, diphenyl-
hydroruthenation, **7**, 408
Butane
reactions with iron ions, **7**, 103
Butanethiolates
rhenium complexes, **6**, 220
Butanoic acid, hydroxy-
methyl ester
production, **12**, 24
Butanolides
synthesis, **12**, 398
1,2,3-Butatriene
trimerization, **12**, 766
Butatriene, 1,4-bis(trimethylsilyl)-
dimerization, **2**, 13, 22
3-Butenal, 2-methylene-
tricarbonyliron complex
isolation, **12**, 624, 625
Butenamides
synthesis
allyl complexes, **12**, 806
1-Butene
oligomerization
Ziegler–Natta catalysts, **12**, 1201
1-Butene, 3,3-dimethyl-
hydroesterification, **12**, 33
metallation, **11**, 112
1-Butene, 4-(diphenylphosphino)-
hydroformylation, **12**, 28, 29
1-Butene, 2-phenyl-
hydrogenation
asymmetric, **12**, 16
Ziegler–Natta catalysts, **12**, 1205
2-Butene
1,4-disubstituted
synthesis, allyl complexes, **12**, 806
2-Butene, 1-acetoxy-4-chloro-
synthesis, **12**, 808
3-Butene-1,2-diol
oxypalladation, **12**, 613
Butenolide
synthesis, **12**, 364, 976
Butenolides
asymmetric synthesis
Grignard reagents, **11**, 148
synthesis, **12**, 248
1,3-Butenyne
chemoselective hydrozirconation, **12**, 40
1-Buten-3-yne, 1,4-bis(trimethylsilyl)-
dimerization, **2**, 13, 22
1-Buten-3-yne, 1-methoxy-
1,3-dienes from, **12**, 43
Butorphanol
synthesis, **12**, 18
Butterfly boridic clusters, **1**, 298–301
Butterfly clusters
osmium
tetranuclear, **7**, 839
ruthenium
tetranuclear, **7**, 839
2-Butyne, hexakis(trimethylsilyl)-
synthesis, **2**, 11, 13, 22
2-Butyne-1,4-diol
cobalt complexes
reaction with thiols, **12**, 690
cyclotrimerization
nickel catalysts, **12**, 747
3-Butyne-2-one, 4-phenyl-
cycloaddition reactions, **12**, 932
γ-Butyrolactam

bicyclic
formation, carbene
complexes, **12**, 515
Butyrolactone, α-methylene-
synthesis, **12**, 248
Butyrolactones
α-methylene
synthesis, **12**, 364
synthesis, **12**, 361
Butyrolactones, α-alkylidene-
formation, **12**, 402
γ-Butyrolactones
preparation, **11**, 178
synthesis
carbene complexes, **12**, 506
γ-Butyrolactones, 2-alkylidene-
hydrogenation
asymmetric, **12**, 20

C

ε-Cadinene
  synthesis, **12**, 580
Cadmates
  structures, **1**, 29
Cadmium
  highly reactive
    preparation, **11**, 183
Cadmium clusters
  cobalt complexes, **10**, 341
  iron complexes, **10**, 338, 339
  ligand dynamics, **10**, 332
  manganese complexes, **10**, 337
  structures, **10**, 333–5
  transition metals, **10**, 323–49
Cadmium compounds
  in organic synthesis, **11**, 159–84
Cadmium, bis(trifluorosilyl)-
  stability, **3**, 176
Cadmium, dimethyl-
  photochemistry, **3**, 200–3
Cadmium, trifluoromethyl-
  organocopper compound
    preparation from, **3**, 64
Caesium, methyl-
  preparation, **1**, 20
Caesium, allyl-
  rotational barrier, **11**, 98
Calamene, 7,8-dihydroxy-
  synthesis
    lithiation, **11**, 64
Calamenene, 7,8-dihydroxy-
  synthesis, **12**, 1063
Calamenene, 8-hydroxy-
  synthesis, **12**, 1051
Calamenenes
  synthesis, **12**, 1030
Calcimycin
  synthesis
    lithiation, **11**, 39
Calcium
  co-condensation with tetrahydrofuran, **1**, 107
  ultrafine particles, **1**, 107
Calcium bromide, (1-methyl-2,2-diphenylcyclopropyl)-
  chiral, **1**, 108
Calcium carbaboranes
  preparation, **1**, 116
Calcium complexes
  anthracene, **1**, 117
  arene radical anion complexes, **1**, 117
  chemistry, **1**, 57–120
  1,3-diene
    reactions, **1**, 117
Calcium fluorides
  aryl, **1**, 108
Calcium halides
  aryl
    preparation, co-condensation reactions, **1**, 108
    solvent-free, **1**, 108
Calcium iodide, cyclopentadienyl-
  preparation, **1**, 109
Calcium, bis[1,3-bis(trimethylsilyl)cyclopentadienyl]-, **1**, 112
Calcium, bis[bis(trimethylsilyl)-methyl]-
  crystal structure, **1**, 110
Calcium, bis(*t*-butylcyclopentadienyl)-, **1**, 112
Calcium, bis(1,3-di-*t*-butylcyclopentadienyl)-, **1**, 112
Calcium, bis(methylcyclopentadienyl)-, **1**, 112
Calcium, bis(pentamethylcyclopentadienyl)-, **1**, 112
  adducts with 1,4-bis(trimethylsilyl)-1,3-butadiene, **1**, 114
  adducts with transition metal complexes, **1**, 115, 120
  solid-state structure, **1**, 113
Calcium, bis(tetraisopropylcyclopentadienyl)-
  structure, **1**, 113
Calcium, bis(trimethylsilyl)cyclopentadienyl)-, **1**, 112
Calcium, bis(triphenylmethyl)-
  formation, **1**, 111
Calcium, cyclopentadienyliodo-
  alkylated, **1**, 110
  structure, **1**, 109
Calcium, dialkynyl-
  preparation, **1**, 111
Calcium, dicyclopentadienyl-, **1**, 112–16
  adducts with ethers, **1**, 114
  unsolvated, **1**, 112
Calcium, dimethyl-
  structure, **1**, 110
Calcium, diphenyl-
  structure, **1**, 110
Calcium, (ethyltetramethylcyclopentadienyl)amide-
  structure, **1**, 119
Calcium, (pentamethylcyclopentadienyl)[bis(trimethylsilyl)methyl]-
  formation, **1**, 115
Calcium, (trimethylaluminum)-cyclopentadienyl-
  formation, **1**, 119, 120
Calicheamicin
  synthesis, **12**, 698
Calicheamicins
  DNA-cleaving
    synthesis, **12**, 77
California red scale pheromone precursor
    synthesis, **12**, 110, 111
Calixarenes
  from ruthenocenes, **7**, 614
Calomel
  reaction with alkylzinc halides, **11**, 164
Calomelanolactone
  synthesis, **12**, 748
Calyculin
  synthesis, **12**, 204
Calyculin A
  synthesis, **12**, 77
Camphene
  metallation, **11**, 112
Campholenal
  decarbonylation, **12**, 375
Camptothecin
  synthesis, **12**, 758
Canadensolide
  synthesis
    lithiation, **11**, 72
Canadine, 8-methyl-
  synthesis, **12**, 1060
Cannabinol, tetrahydro-
  synthesis
    lithiation, **11**, 52
Cannivonine, dihydro-
  synthesis, **12**, 659
Capnellanes
  synthesis, **12**, 614
Capnellene
  synthesis, **12**, 60, 61, 206, 253, 254, 369, 586, 593
ε-Caprolactone
  ethene block copolymers
    synthesis, **12**, 1205
Carbaboranes
  ab initio/IGLO/NMR, **1**, 222
  acidity, **1**, 234
  alkylation, **1**, 239
  C-alkylation, **1**, 241
  applications, **1**, 245
  boron nuclear quadrupole couplings, **1**, 226
  bridging hydrogen tautomerization, **1**, 219
  cage fluxional behavior, **1**, 232–4
  chemical properties, **1**, 232–45
  derivatives, **1**, 236–45
    B_{cage}–chalcogen bonds, **1**, 237, 238
    B_{cage}–halogen bonds, **1**, 236, 237
    B_{cage}–nitrogen bonds, **1**, 238
    B_{cage}–phosphorus bonds, **1**, 238
    C_{cage}–arsenic bonds, **1**, 240, 241
    C_{cage}–chalcogen bonds, **1**, 239–42
    C_{cage}–nitrogen bonds, **1**, 240, 241
    C_{cage}–phosphorus bonds, **1**, 240, 241
    exo-cage B–C bonds, **1**, 238, 239

diffraction, **1**, 220–2
dipole moments, **1**, 220, 226
electron diffraction, **1**, 220
electrophilic deuteration, **1**, 239
electrophilic reactions, **1**, 239
fluoride derivatives, **1**, 236
fluxionality, **1**, 233
free radical initiated reactions, **1**, 241
halogenation, **1**, 236
Hartree–Fock calculations, **1**, 220
He[I] photoelectron spectra, **1**, 226
infrared spectra, **1**, 219, 225
isomer stabilities, **1**, 220, 232–4
ligands, **1**, 374, 375
mass spectrometry, **1**, 226
metal vapor reactions, **1**, 377
microwave structure, **1**, 227
molecular orbital calculations, **1**, 218
NMR, **1**, 222–4
nucleophilic attack, **1**, 219
nucleophilic substitution, **1**, 227
partial cage opening, **1**, 238
physical properties, **1**, 220–7
polyhedral, **1**, 218–45
polymers, **1**, 244
proton affinity, **1**, 218, 220
pulse radiolysis, **1**, 227
Raman spectra, **1**, 219, 223, 225
reaction with hydrides, **1**, 235
reaction with hydrogen halides, **1**, 237
reactions with fluorides, **1**, 235
reactions with Lewis bases, **1**, 235, 236
reactions with nitrogen bases, **1**, 235, 236
rearrangement, **1**, 219
mechanisms, **1**, 233
reduction, **1**, 235
siloxane oligomers, **1**, 244
skeletal rearrangements, **1**, 232–4
spectroscopy, **1**, 222–6
spiro phosphorus derivatives, **1**, 241
structures, **1**, 218–20
synthesis, **1**, 227–31
cage fusion reactions, **1**, 229
cage growth, **1**, 228
cage growth, boron insertion reactions, **1**, 228, 229
coupled carboranes, **1**, 231
dehydrogenation, **1**, 229
from polyboranes and alkynes, **1**, 227, 228
rearrangements, **1**, 231
selective cage degradation, **1**, 229–31
tensor surface harmonic theory, **1**, 220
thermochemical characteristics, **1**, 226
thermolysis, **1**, 232
UV photoelectron spectrum, **1**, 226
UV spectroscopy, **1**, 226
vibrational spectroscopy, **1**, 224–6
Carbaboranes, glycosyl-
derivatives
B_{cage} metals, **1**, 243
C_{cage} metals, **1**, 243
deuteration, **1**, 244
levorotatory enantiomers, **1**, 244
preparation, **1**, 242
Carbaboranes, halo-
halogen exchange, **1**, 236
with boron trihalides, **1**, 236
1-Carbacephalothin
synthesis, **12**, 556
Carbacephems
synthesis, **12**, 206
Carbacyclins
analogues
synthesis, **12**, 99
coupling reactions, **12**, 191
synthesis, **12**, 846
Carbalumination, **1**, 464–6, 470
carbon–carbon bond formation, **11**, 306
discovery, **11**, 279, 280
regioselectivity, **11**, 291
scope and limitations, **11**, 283
Carbamates
reactions
allyl complexes, **12**, 812
tosylated
reaction with palladium chloride, **12**, 610
Carbanions
active methylene compounds
reactions, allyl complexes, **12**, 831–52
highly stabilized
metathesis with, **12**, 226, 227
organometallics
allyl alkylation, **12**, 852–62
α-selenium stabilized
generation, **11**, 28, 29
lithiation, **11**, 28–41
reactions, **11**, 29
selenyl
lithiation, **11**, 29, 30
selenyl allylic
lithiation, **11**, 30–2
sp^3
lithiation, **11**, 6–41
α-nitrogen, lithiation, **11**, 15–21
α-oxygen, configurational stability, **11**, 22–4
α-oxygen, enantioinduction, **11**, 22–4
α-oxygen, lithiation, **11**, 21–5
α-silicon, lithiation, **11**, 25–8
unstabilized, lithiation, **11**, 6–15
stabilized
asymmetric allyl displacements, palladium complexes, **12**, 884–7
α-sulfenyl
lithiation, **11**, 29, 30
α-sulfenyl allylic
lithiation, **11**, 30–2
α-sulfinyl
lithiation, **11**, 32–4
α-sulfinyl allylic
lithiation, **11**, 35, 36
α-sulfonyl
lithiation, **11**, 36–8
α-sulfonyl allylic
lithiation, **11**, 39
α-sulfur stabilized
generation, **11**, 28, 29
lithiation, **11**, 28–41
reactions, **11**, 29
Carbapenam
synthesis
from azetidinone, **12**, 439, 440
Carbapenem
antibiotics
synthesis, **12**, 25
Carbapenem, 1-methyl-
precursors
synthesis, **12**, 695
Carbapenems
synthesis, **12**, 206
Carbapenems, 1β-methyl-
synthesis, **12**, 22
Carbasilanes
bicyclic, **2**, 64–7
polycyclic, **2**, 64–7
spirocyclic, **2**, 64–7
Carbazole, 1a,3,4,4a-tetrahydro-
synthesis, **12**, 873
Carbazole, tetrahydro-
synthesis
lithiation, **11**, 67
Carbazomycin A
synthesis, **12**, 666
Carbazomycines
synthesis, **12**, 493
Carbene complexes
(*O*-acylimidato)
photochemical reaction with imines, **12**, 556
addition reactions, **12**, 533–7
carbene cleavage, **12**, 538–40
carbene transfer reagents, **12**, 388
cationic
formation, **12**, 393
cycloaddition reactions, **12**, 525–33
cycloheptatrienylmethyl
photolysis, **12**, 567
cyclopropyl
chemistry, **12**, 474
formation, **12**, 389, 390
group 6, **12**, 389–92
insertion reactions, **12**, 533–7
molecular orbital energy diagram, **12**, 550
in organic synthesis, **12**, 387–417, 577–96
photochemical reactions, **12**, 551–73
mechanisms, **12**, 549, 550
reactions with isocyanides, **5**, 225
sulfur stabilized
benzannulation, **12**, 487

transition metals
photochemical reactions, **12**, 549–73
triiron clusters, **7**, 280
Carbene ligands
platinum complexes, **9**, 391–426
Carbenes
alkyl(alkoxy)-
nickel complexes, **9**, 13
carbon hydrogen insertion reactions, **12**, 449–64
heteroatom stabilized, **12**, 388, 389
insertion
organomercury compounds
synthesis, **3**, 153, 154
nonheteroatom stabilized, **12**, 389
stoichiometric reagents, **12**, 389–96
Carbenes, carboranyl-
ESR spectra, **1**, 227
B-substituted, **1**, 239
Carbenes, vinyl-
cyclopropanation with, **12**, 391
Carbenium ions, siloxy-
synthesis, **2**, 5
Carbenoid
definition, **12**, 422
Carbide complexes
iron clusters, **7**, 267
Carbidic clusters
octahedral
ruthenium, **1**, 303
Carboalumination, **12**, 163
alkenyl bromides, **12**, 182
alkynes
zirconium catalyzed, **12**, 180
cuprates from, **12**, 87
Carbocupration
activated alkynes, **12**, 92
Carbocycles
five-membered ring
preparation, **12**, 472–84
four-membered ring
preparation, **12**, 471, 472
seven-membered ring
synthesis, carbene complexes, **12**, 502–4
six-membered ring
preparation, **12**, 484–502
Carbocycline
synthesis, **12**, 734
Carbocyclization
intramolecular
palladium catalysis, **12**, 284
Carbodealumination
carbon–carbon bond formation, **11**, 307
Carbodiimides
cycloaddition
aziridines or oxiranes, **11**, 500
Carbohydrates
acyl
methylenation, **12**, 588
allic alkylation, **12**, 849–51
methylenation, **12**, 587
β-Carboline, dihydro-
synthesis, **11**, 260
Carbomagnesiation
alkenes, **1**, 87
zirconium catalyzed, **11**, 131
alkynes, **1**, 67
1,3-dienes, **1**, 67
enynes, **1**, 67
unsaturated C–C bonds, **1**, 66
zirconium compounds
catalysis, **1**, 68
zirconocene catalysts, **12**, 340
Carbomercuriation
alkenes, **11**, 427
Carbometallation
alkynes, **12**, 157
Grignard reagents
titanium complexes, **12**, 157
Carbomethoxylation
bis-oxypalladation
intramolecular, **12**, 612
cyclization
vinyl triflates, **12**, 311
oxypalladation
intramolecular, **12**, 611
Carbomycin
synthesis, **12**, 643
Carbon–hydrogen bonds
activation
platinum complexes, **9**, 517, 518
Carbon dioxide
copolymerization with diynes, **9**, 5
cycloaddition reactions
[3+2], **12**, 949
in telomerization with butadiene, **12**, 790
Carbon monoxide
acylation of organozinc halides, **11**, 168
alternating copolymerization with alkenes, **12**, 1225
hydrogenation
catalysts, **10**, 42, 43
insertion into organonickel complexes, **9**, 58
insertion reactions
rhodium complexes, **8**, 203
transition metal complexes, **12**, 242–59
ligand
in heteronuclear metal–metal bond synthesis, **10**, 2
mercuriation, **11**, 392
reaction with organonickel complexes, **9**, 62
reduction
rhenium complexes, **6**, 192
reductive coupling
tantalum complexes, **5**, 65
Carbon suboxide
nickel complexes, **9**, 9
Carbon-arylation
organobismuth(V) compounds, **11**, 507, 508
Carbonates, *O*-(2-alkenyl)dithio-
reactions
allyl complexes, **12**, 827
Carbon–carbon bonds
formation
organoaluminum compounds, **11**, 306–8
Carbon–halogen bonds
formation
halodealumination, **11**, 302
Carbon–heteroatom bonds
formation
organoaluminum compounds, **11**, 303, 304
Carbon–hydrogen bonds
activation
niobium complexes, **5**, 135, 136
rhenium complexes, **6**, 186–9
formation by organoaluminum hydrides, **11**, 297–301
Carbon–metal bonds
formation
organoaluminum compounds, **11**, 306
Carbon–nitrogen bonds
formation
organoaluminum compounds, **11**, 305, 306
Carbon–oxygen bonds
formation by oxidation of organoaluminum compounds, **11**, 302, 303
Carbon–sulfur bonds
formation
organoaluminum compounds, **11**, 303, 304
Carbon–tin bonds
reaction types, **11**, 357
Carbonyl clusters
fluxionality, **10**, 193–5
heteronuclear, **10**, 188–97
fluxionality, mechanism, **10**, 191
structure, **10**, 188–93
phosphinidene bridged, **10**, 218
protonation, **10**, 197
reactivity, **10**, 195–7
Carbonyl complexes
π-acceptor ligands, **5**, 231–71
π-acceptor nitrogen ligands, **5**, 231–9
diimine ligands, **5**, 231, 232
heterocyclic ligands, **5**, 231, 232
tris(pyrazolyl)borate ligands, **5**, 232
amine ligands, **5**, 225
hydrogen bonding, **5**, 227
reactions, **5**, 227
antimony ligands, **5**, 259–62
arsenic ligands, **5**, 259–62
arsine ligands
reactions, **5**, 262
bismuth ligands, **5**, 259–62
diazene ligands, **5**, 225
diazo ligands, **5**, 235
properties, **5**, 237
reactions, **5**, 238, 239
diimine ligands
properties, **5**, 236
structures, **5**, 236
dinitrogen ligands, **5**, 235

reactions, **5**, 238, 239
dithiolene ligands
structures, **5**, 267
dithione ligands
structures, **5**, 267
group 6
alkoxide ligands, **5**, 229
alkoxides, structures, **5**, 229, 230
arsine ligands, **5**, 259
bismuthine ligands, **5**, 259
carbon dioxide ligands, **5**, 228
carbon disulfide ligands, **5**, 262
carbon disulfide ligands, reactions, **5**, 269
carbonate ligands, **5**, 228
carbonyl oxide ligands, **5**, 229
chelate ring closure, **5**, 217
chemistry, **5**, 155–202
diimine ligands, reactions, **5**, 237
dithiocarbamate ligands, **5**, 269
heterocyclic ligands, reactions, **5**, 237
monodentate phosphorus ligands, **5**, 239–47
multidentate phosphines, synthesis, **5**, 249
oxidative addition, **5**, 222
phosphaalkene ligands, **5**, 243
phosphaalkene ligands, reactions, **5**, 247
phosphaalkene ligands, structures, **5**, 246
phosphabenzene ligands, **5**, 243
phosphene ligands, **5**, 253
phosphido ligands, **5**, 240
phosphine, σ-base/π-acceptor properties, **5**, 244
phosphine, *cis–trans* isomerism, **5**, 244
phosphine, cone angle, **5**, 244
phosphine, photoelectron spectra, **5**, 245
phosphine ligands, **5**, 240, 241
phosphinidene ligands, **5**, 244, 253
phosphinidene ligands, structures, **5**, 246
phosphinidene ligands, trapping reactions, **5**, 247
phosphinine ligands, **5**, 243
phosphoranylidene ligands, **5**, 243
phosphoranylidene ligands, structures, **5**, 246
phosphorin ligands, **5**, 243
phosphorin ligands, reactions, **5**, 247
phosphorin ligands, structures, **5**, 246
photophysics, **5**, 236
reaction mechanisms, **5**, 216, 217
reaction mechanisms, flash photolysis, **5**, 217
stibine ligands, **5**, 259
substitution reactions, **5**, 216
substitution reactions, dissociative interchange, **5**, 216
sulfido ligands, **5**, 262
sulfur dioxide ligands, **5**, 262
sulfur dioxide ligands, reactions, **5**, 269
thiocarbamate ligands, **5**, 262
thiocarboxylate ligands, **5**, 262
thioketone ligands, **5**, 262
thiolate ligands, **5**, 262
thiolate ligands, reactions, **5**, 271
tris(pyrazolyl)borate ligands, reactions, **5**, 237
xanthate ligands, **5**, 262
heterocyclic ligands
properties, **5**, 236
structures, **5**, 236
imido ligands, **5**, 234
properties, **5**, 236
reactions, **5**, 238
structures, **5**, 236
imine ligands, **5**, 225
isocyanate ligands, **5**, 226
isothiocyanate ligands, **5**, 226
multidentate phosphines
reactions, **5**, 258
nitrene ligands, **5**, 234
properties, **5**, 236
reactions, **5**, 238
structures, **5**, 236
nitrile ligands, **5**, 225
nitrogen σ-donor ligands, **5**, 225–7
nitrogen ligands
properties, **5**, 237
nitrosyl ligands, **5**, 235
properties, **5**, 237
reactions, **5**, 238, 239
noncarbon σ-donor ligands, **5**, 217–30
noncarbon ligands, **5**, 215–71
oxygen ligands, **5**, 227–30
phosphene ligands, **5**, 247
reactions, **5**, 258
phosphinidene ligands, **5**, 247
reactions, **5**, 258
structures, **5**, 257
phosphorus rich ligands, **5**, 253
reactions, **5**, 259
structures, **5**, 258
polydentate phosphorus ligands, **5**, 247–59
properties, **5**, 254–8
structures, **5**, 254–8
polyphosphine ligands, **5**, 247
pyridylphosphine ligands, **5**, 247
Schiff-base ligands, **5**, 225
selenium ligands, **5**, 262–71
sulfur ligands, **5**, 262–71
tellurium ligands, **5**, 262–71
transition metals
nucleophilic attack, **12**, 6
Carbonyl compounds
alkenation, **12**, 1090, 1091
molybdenum or tungsten complexes, **12**, 1090
conversion to alcohols or aldehydes
organoaluminum hydrides, **11**, 298
conversion to methylene compounds
organoaluminum hydrides, **11**, 299
coupling reactions
imines, niobium complexes, **12**, 1088, 1089
cyclization
intramolecular, **12**, 332
hydrogermylation, **2**, 155
α-mercuriated
structures, **3**, 157–9
methylenation, **12**, 579–84
Tebbe's reagent, **12**, 586–91
reactions with π-allyl nickel complexes, **12**, 966
reactions with organoaluminum compounds, **1**, 492–4
reactions with organoaluminum reagents, **1**, 472–4
reactions with selenium dioxide, **11**, 536
reactions with stannyl anions, **11**, 364
reactions with α-stannyl carbanions, **11**, 365
reduction
diphenylstibine, **11**, 496
organotin compounds, **11**, 360, 361
tellurium reagents, **11**, 576
reductive cross-coupling, **12**, 1090
α-silyl-
in organic synthesis, **11**, 329
sterically hindered
methylenation, **12**, 579
synthesis
organoboron compounds, **11**, 216
organomercurials, **12**, 142
zirconium mediated, **4**, 620
α,β-unsaturated
hydrogenation, enantioselective, **12**, 16
reactions with organoaluminum compounds, **1**, 494
reduction, tellurium reagents, **11**, 576
Carbonyl compounds, diazo-decomposition
transition metal catalyzed, **12**, 425, 426
Carbonyl coupling reactions
molybdenum or tungsten complexes, **12**, 1090

niobium complexes, **12**, 1088
titanium complexes, **12**, 1074–80
vanadium complexes, **12**, 1086, 1087
Carbonyl fluorides
group 6
preparation, **5**, 218
Carbonyl halides
group 6
decarbonylation, **5**, 218
halide-bridged, **5**, 220
preparation, **5**, 218
properties, **5**, 220
reactions, **5**, 220
structures, **5**, 220
substituted, **5**, 218
substitution reactions, **5**, 218
oxidation state 0
preparation, **5**, 217, 218
oxidation state +1
preparation, **5**, 217, 218
oxidation state +2
preparation, **5**, 218–20
Carbonyl hydrides
group 6, **5**, 220–3
bridging hydrides, **5**, 222
dimer structures, **5**, 222
electron transfer processes, **5**, 223
hydrogenation catalysts, **5**, 223
phosphine bridged, **5**, 223
properties, **5**, 222, 223
reactions, **5**, 223
structure, **5**, 222, 223
synthesis, **5**, 221, 222
synthesis, protonation, **5**, 221
Carbonyl ligands
platinum complexes, **9**, 391–426
Carbonyl ylides
1,3-dipolar cycloaddition reactions, **12**, 447
formation, **12**, 446–9
intermolecular
formation, **12**, 446
intramolecular
formation, **12**, 447
metal stabilized, **12**, 445–9
reactions, **12**, 446–9
1,4-sigmatropic hydrogen transfer, **12**, 449
Carbonylation
alkenes, **12**, 355–62
alkyl derivatives, **12**, 367
alkylpalladium complexes, **12**, 139
alkynes, **12**, 355–62
allyl derivatives, **12**, 367
aryl derivatives, **12**, 363–6
benzyl derivatives, **12**, 367
catalysts, **10**, 47
dienes, **12**, 360
halides and alcohols, **12**, 362–9
heterocycles, **12**, 371–3
iron carbonyl catalysts, **7**, 86
ketones by, **12**, 368, 369
main group organometallic compounds, **12**, 370, 371
metal carbonyls in, **12**, 351–73
organoboranes, **11**, 225, 226
organomercury compounds, **11**, 407, 408
polymerization
transition-metal catalysts, **12**, 1227
vinyl derivatives, **12**, 363–6
Carbonylative coupling
acylstannanes, **12**, 861
organostannanes, **12**, 202
Carbopalladation
cyclization, **12**, 55
Carbopallation
enynes, **12**, 764
o-Carborane
lithiation, **11**, 6
o-Carborane, 1,2-dehydro-
preparation, **1**, 244
Carboranes
mixed-metal clusters, **10**, 167–9
polyhedral, **1**, 217–47
Carbovir
synthesis, **12**, 812, 813, 850
Carboxychlorination
allyl complexes, **12**, 825
Carboxylates
reactions
allyl complexes, **12**, 817–26
Carboxylation
metal alkoxycarbonyl complexes, **12**, 356, 357
Carboxylative cycloaddition
dimethyl benzylidenemalonate, **12**, 936
Carboxylic acid esters
allylic alkylation, **12**, 868, 869
Carboxylic acids
addition to Grignard reagents, **11**, 151
intramolecular allylic reactions, **12**, 876
synthesis
hydridocarboxylation, **12**, 355
insertion reactions, **12**, 242
organoboranes, **11**, 223
organomercury compounds, **11**, 398
transition metal complexes, **12**, 251, 252
Carboxylic compounds
reduction
organotin compounds, **11**, 360, 361
Carbozincation
alkynes and alkenes, **11**, 175–7
Carbynes
nickel complexes, **9**, 12
Cardicinic acid, 3-amino-
synthesis, **12**, 249
Cardiolite
radiopharmacy, **6**, 152
2-Carene
synthesis, **12**, 831
Carnitine
synthesis, **12**, 23
Caronaldehyde
synthesis, **12**, 399
β-Carotene
synthesis, **12**, 831
Carpenter bee pheromone
lactones
synthesis, **12**, 822
Carveol
synthesis, **12**, 83, 84
Carveyl acetate
allylic alkylation, **12**, 834
Carveyl acetone
synthesis, **12**, 863
Catalysis
heteronuclear metal–metal bonds, **10**, 33–61
homogeneous
heterometallic clusters, **10**, 35–51
synergism, **10**, 34
Catechol
six-coordinate silicon complexes
preparation, **2**, 3
Catechol monomethyl ether
reaction with aryllead compounds, **11**, 469
Catecholatoborane
isolation, **1**, 174
synthesis, **1**, 179
Catecholborane
properties, **1**, 174
reactions, **1**, 171
Catenation
organosilicon compounds, **2**, 23–5
Catharanthine
synthesis, **12**, 874, 889
CC-1065
anticancer agent
pharmacophore analogue, synthesis, **12**, 779
Cephalosporin
precursor
synthesis, **12**, 556
synthesis, **12**, 204
Cephalosporin C
synthesis, **12**, 860
Cephalosporins
synthesis, **12**, 107, 206
Cephems, 3-bromo-
coupling with organotin compounds, **12**, 204
Cerulenin, tetrahydro-
synthesis, **12**, 560
Charge-transfer compounds
from ferrocenes, **7**, 191
Chemisorption
methyl bromide
magnesium crystal surface, **1**, 62
Chemzyme, **11**, 259
Chilenine
synthesis, **12**, 1092
Chiral amplification
organozinc compounds, **3**, 192, 193
Chloramphenicol
synthesis, **12**, 1169
Chloroethylation
alkenes
palladium-assisted, **12**, 146, 153
Cholaphanes

synthesis, **12**, 115
Cholest-4-en-3-one, 6β-acetyl-
reaction with *p*-methoxy phenyllead triacetate, **11**, 471
Cholesterol, 7,8-dehydro-
metallation, **11**, 105
Cholesterol, 15β-hydroxy-
synthesis, **12**, 885
Cholesterol, 25-hydroxy-
synthesis, **12**, 856
Chorismate mutase
transition state analogue
synthesis, **12**, 590
2-Chromanols
synthesis, **12**, 143
Chromanone
formation
in insertion reactions, **12**, 460
Chromans, 2-vinyl-
synthesis, **12**, 612
Chromene, 2,2-dimethyl-
epoxidation
asymmetric, **12**, 1126
Chromenes
synthesis, **12**, 256
2-Chromenes
synthesis, **12**, 143
Chromium(0) complexes
bis(arene), **5**, 472–7
synthesis, **5**, 472–5
Chromium
dicarbon metallacarbaboranes
7-vertex, **1**, 382
12-vertex, **1**, 400
tetracarbon metallacarbaboranes
12-vertex, **1**, 423
13-vertex, **1**, 423
triple decker sandwich complexes, **5**, 475
Chromium carbonyl complexes
alkynides, **5**, 165, 166
properties, **5**, 165
reactions, **5**, 166
synthesis, **5**, 165
availability, **5**, 156
azacarbene, **5**, 185–91
calculations, **5**, 156, 157
carbenes
no heteroatom stabilization, **5**, 168–72
carbon σ-donor ligands, **5**, 161
carbyne, **5**, 192–7
charged, **5**, 159–61
reactions, **5**, 160, 161
structures, **5**, 159, 160
synthesis, **5**, 159, 160
cyanide ligands, **5**, 161
cyclopentadienyl carbyne, **5**, 198–202
isocyanide ligands, **5**, 161–5
properties, **5**, 162
reactions, **5**, 164, 165
synthesis, **5**, 162
oxacarbene, **5**, 172–84
photolysis, **5**, 157, 158
selenocarbonyl, **5**, 166, 167
spectroscopy, **5**, 156, 157
substitution reactions, **5**, 157
supported, **5**, 158, 159
thiocarbonyl, **5**, 166, 167
vinylidenes, **5**, 171
Chromium complexes, **5**, 331–80
η^3-, **5**, 335–45
η^4-, **5**, 346–52
η^5-, **5**, 352–4
η^6-, **5**, 354–8
η^7-, **5**, 358–62
acyclic pentadienyl, **5**, 352, 353
alkene chelates, **5**, 334, 335
alkenes, **5**, 332–4
preparation, **5**, 332
alkoxide, **5**, 373–6
alkoxycarbene
photolysis, **12**, 560
alkoxycarbenes
photolysis, **12**, 551
alkyl, **5**, 373–6
(alkylbenzene)tricarbonyl
deprotonation, regioselectivity, **12**, 1020
alkylidene
preparation, **12**, 595
alkyne
chromium(0), (d^6) metal centres, **5**, 363, 364
monometallic, **5**, 363–72
alkynes
intramolecular coupling, **5**, 371, 372
allyl, **5**, 342, 343
π-allyl
in organic synthesis, **12**, 976
allyl
preparation, **5**, 508
allyl dicarbonyl cyclopentadienyl, **5**, 338–41
allyl pyrazole ligands, **5**, 338
(anisole)tricarbonyl
lithiation and electrophilic quenching, **12**, 1022, 1026
(*o*-anisyl) tricarbonyl alkyl ketones
nucleophilic additions, **12**, 1043
arene tricarbonyl
addition/oxidation, **12**, 995–1004
benzylic deprotonation, substituent effects, **12**, 1056
cine/tele substitution, **12**, 990–2
deprotonation, regioselective, **12**, 1020–30
lithiated, **11**, 63, 64
lithiation, **12**, 1030, 1034
metal–metal exchange, **12**, 1019
neutral, **12**, 981
nucleophilic addition
nucleophilic addition, regioselectivity, **12**, 996–1000
preparation, **12**, 981
side-chain reactions, **12**, 1067
synthesis, **5**, 480–4
arenes, **5**, 471–533
aryl, **5**, 286
aryl tricarbonyl alkenyl ketones
enolates, **12**, 1065, 1066
aryl tricarbonyl alkyl ketones
enolates, **12**, 1065, 1066
nucleophilic additions, **12**, 1039
o-substituted, nucleophilic additions, **12**, 1040
arylaldehyde tricarbonyl
acetals, oxonium ions, **12**, 1052–4
nucleophilic additions, **12**, 1044–8
arylaldehydeimine tricarbonyl
nucleophilic additions, **12**, 1048
(aryllithium)tricarbonyl-
generation, nucleophilic catalysis, **12**, 1019
reaction with carbonyl compounds, **12**, 1019
reaction with CO_2, **12**, 1019, 1022
trapping with electrophiles, **12**, 1019, 1020
azaallyl, **5**, 345
(benzaldehyde acetals)tricarbonyl
lithiation, **12**, 1026
(benzylic alcohols)tricarbonyl
lithiation, **12**, 1026, 1027
binuclear, **5**, 373–80
bis(alkyne) cyclopentadienyl, **5**, 368–70
bis(arene)
structures, **5**, 475–7
bis(pyridine)
preparation, **5**, 531
butadienyl, **5**, 344, 345, 350–2
carbene alkynyl allylamino
Pauson–Khand reaction, **12**, 731
carbenes
Dötz annelation, **5**, 483
carbonyl diene, **5**, 346–9
carbonyl phosphine
optical polarizability, **5**, 245
reactions, **5**, 246
(chlorobenzene)tricarbonyl
o-lithiation, electrophilic trapping, **12**, 1022
cyclobutadiene, **5**, 350–2
cycloheptadienyl, **5**, 353, 354
cycloheptatriene, **5**, 354–7
cycloheptatrienyl, **5**, 358, 359
bimetallic metal–metal bonded, **5**, 362
sandwich complexes, **5**, 361
cyclohexadienyl, **5**, 353, 354
anionic, preparation, **12**, 995
cyclooctatetraene, **5**, 357, 358
cyclopentadienyl, **5**, 393–450
cyclopentadienyl arene sandwich, **5**, 479
cyclopentadienyl dichloride dimer
synthesis, **5**, 438
cyclopentadienyl diene, **5**, 349, 350

cyclopentadienyl nitrosyl, **5**, 418–20
cyclopentadienyl oxo, **5**, 431
cyclopropenyl, **5**, 344, 345
dicarbonyl cyclopentadienyl dimer
  synthesis, **5**, 396–9
dicarbonyl cyclopentadienyl dimers
  structure, **5**, 397
dicarbonyl cyclopentadienyl nitrosyl
  synthesis, **5**, 418, 419
dicarbonyl pentamethylcyclopentadienyl dimers
  reactions with sulfur, **5**, 433
dicarbonyl (thiocarbonyl) (hexaethylbenzene)
  fluxionality, **5**, 499
dienyl(alkoxy)carbenes
  photolysis, **12**, 567
dinitrosyl cyclopentadienyl halides
  synthesis, **5**, 419
17-electron
  reactions, **5**, 400
(fluorobenzene)tricarbonyl
  o-lithiation, electrophilic trapping, **12**, 1021
halides, **5**, 373–6
heteroarenes, **5**, 471–533
homoleptic allyl halides, **5**, 335, 336
homoleptic allyls, **5**, 335, 336
homoleptic dienes, **5**, 346
indenyl, **5**, 376–80
indium bonds, **1**, 557
(indoline)tricarbonyl
  lithiation, regioselectivity, **12**, 1025
mercury clusters, **10**, 336
monoalkene carbonyl phosphine, **5**, 333
monoalkyne cyclopentadienyl, **5**, 367, 368
outer sphere electron transfer reduction, **5**, 404
oxocyclobutenyl, **5**, 345
palladium carbonyls, **9**, 204
pentacarbonyl dianion
  reaction with Vilsmeir's salts, **12**, 556
pentamethylcyclopentadienyl chloride dimer
  synthesis, **5**, 439
phenyl tricarbonyl alkyl ketones
  nucleophilic addition, **12**, 1040
prop-2-ynyl, **5**, 344
(substituted arenes)tricarbonyl
  lithiation, **12**, 1027, 1028
tetracarbonyl cyclopentadienyl, **5**, 376–80
thallium bonds, **1**, 567
thiocarbonyl, **5**, 508
tricarbonyl
  carbonylation, **12**, 257
tricarbonyl acetophenone
  addition reactions, **5**, 525
  coupling reactions, **5**, 525
  reactions with 2-butenyl magnesium chloride, **5**, 525
tricarbonyl allylbenzene
  deprotonation, **5**, 521
tricarbonyl arenes
  arene exchange reactions, **5**, 482, 483
  arene exchange reactions, neighboring group effects, **5**, 483
  bond lengths, **5**, 492
  chirality, **5**, 504, 505
  conformation, **5**, 509, 515, 518
  electrochemistry, **5**, 494–7
  electrophilic attack, **5**, 517
  fluxionality, **5**, 498
  haptotropic rearrangements, **5**, 501–4
  inclusion complexes, **5**, 497
  IR spectra, **5**, 492
  ligand substitution reactions, **5**, 506–9
  lithiation, **5**, 509
  mass spectra, **5**, 494
  NMR, **5**, 493
  photochemistry, **5**, 506
  planarity, **5**, 491
  reaction with silanes, **5**, 507
  reactions, **5**, 505–29
  reduction, **5**, 495
  regioselectivity of nucleophilic attack, **5**, 513, 515
  resolution, **5**, 505
  solution conformation, **5**, 497
  spectroscopy, **5**, 492–4
  stability, **5**, 491
  stereoselective deprotonation of benzylic position, **5**, 519
  structure, **5**, 491, 492
  thermochemistry, **5**, 497
  UV–visible spectra, **5**, 493
tricarbonyl aryl, **5**, 490
tricarbonyl benzaldehyde
  addition reactions, **5**, 524
tricarbonyl (benzene)
  dipole moment, **5**, 492
  fluxionality, **5**, 497–500
  molecular orbital diagram, **5**, 492
  photoelectron spectra, **5**, 492
  reaction with $LiBHEt_3$, **5**, 511
  reduction, **5**, 495
tricarbonyl biphenylene
  conformation, **5**, 487
  NMR, **5**, 494
tricarbonyl (chlorobenzene)
  palladium-catalyzed reactions, **5**, 490
tricarbonyl *o*-cresyl alkyl ketones
  nucleophilic additions, **12**, 1042
tricarbonyl cyclopentadienyl alkyl
  photochemistry, **5**, 412
  reaction with alkynes, **5**, 414
  reactions with phosphines, **5**, 411, 413
  synthesis, **5**, 410–18
tricarbonyl cyclopentadienyl anions
  reactions with allyl halides, **5**, 415
  reactions with group 12 halides, **5**, 408, 409
  reactions with group 13 halides, **5**, 409
  reactions with group 14 halides, **5**, 409
  synthesis, **5**, 404
tricarbonyl cyclopentadienyl dimers
  decarbonylation, **5**, 396
  homolysis, **5**, 400
  photochemistry, **5**, 396
  reaction with tellurium, **5**, 434
  reactions with diphenyl disulfide, **5**, 434
  reactions with pnictogens, **5**, 423–7
  reactions with selenium, **5**, 434
  reactions with sulfur, **5**, 433
  structure, **5**, 395, 396
  synthesis, **5**, 394–6
  thermochemistry, **5**, 396
tricarbonyl cyclopentadienyl halides
  synthesis, **5**, 407
tricarbonyl cyclopentadienyl hydride
  reactions, **5**, 406
  reactions with phosphines, **5**, 406
  synthesis, **5**, 406, 407
tricarbonyl diphenylmethane
  deprotonation, **5**, 518
tricarbonyl fluorene
  deprotonation, **5**, 502
tricarbonyl (fluorobenzene)
  lithiation, **5**, 509
tricarbonyl hexaethylbenzene
  fluxionality, **5**, 499
tricarbonyl indene
  deprotonation, **5**, 501
  synthesis, **5**, 501
tricarbonyl indole
  lithiation, **5**, 511
tricarbonyl indole nucleophilic attack, **5**, 516
tricarbonyl naphthalene
  catalysis, **5**, 529
  haptotropic rearrangement, **5**, 503
  ligand substitution, **5**, 507
  lithiation, **5**, 510
  nucleophilic attack, **5**, 516
tricarbonyl (phenylanthracene)
  isomerization, **5**, 503
tricarbonyl (phenyltrimethylsilane)
  desilylation, **5**, 488
tricarbonyl polycyclic hydrocarbons, **5**, 486
tricarbonyl pyridine

synthesis, **5**, 531
tricarbonyl toluene
deprotonation lithiation, **5**, 518
tricarbonyl tris(ammine)
synthesis, **5**, 480
trimethylenemethane, **5**, 350–2
Chromium(I) complexes
bis(arene), **5**, 478
Chromium(II) complexes
alkyne, **5**, 364–6
Chromium(IV) complexes
alkynes, **5**, 371
Chromium, (acetophenone)tricarbonyl-
oxidation, **12**, 1067
Chromium, (*o*-anisaldehyde)tricarbonyl-
acetals
oxonium ions, reactions, **12**, 1053
nucleophilic additions, **12**, 1046
Chromium, (anisole)tricarbonyl-
meta acylation, **12**, 996
Chromium, (benzaldehyde)tricarbonyl-
allylboration
asymmetric, **12**, 1044, 1045
nucleophilic addition, **12**, 1044
Chromium, benzenetricarbonyl-
lithiation, **12**, 1018–20
nucleophilic addition, **12**, 995
pK_a, **12**, 1018
proton abstraction, **12**, 1018
Chromium, (benzyl alkyl ether)tricarbonyl-
reactions, **12**, 1056
Chromium, (benzyl allyl ether)tricarbonyl-
reactions, **12**, 1056
Chromium, bis(arene)
arene lability, **5**, 473
charge transfer, **5**, 478
electrochemistry, **5**, 476
NMR, **5**, 474
oxidation, **5**, 478
reactions, **5**, 477
synthesis by metal atom co-condensation, **5**, 472
Chromium, bis(benzene)
intercalations into host lattices, **5**, 478
IR spectra, **5**, 476
lithiation, **5**, 477
molecular orbital diagram, **5**, 475
photoelectron spectra, **5**, 475
reduction, **5**, 476
UV–visible spectra, **5**, 475
Chromium, bis(chlorobenzene)
halide substitution, **5**, 477
Chromium, bis(naphthalene)
structure, **5**, 473
Chromium, bis(phenanthrene)
structure, **5**, 474
Chromium, bis(toluene)
synthesis by Fischer–Hafner method, **5**, 472
Chromium(0), (chlorobenzene)tricarbonyl-
nucleophilic addition, **12**, 980
Chromium, (chlorobenzene)tricarbonyl-
reaction with oxygen nucleophiles, **12**, 983
Chromium, (cyclohexadienyl)tricarbonyl-
x-ray diffraction analysis, **12**, 995
Chromium, (1,4-dimethoxynaphthalene)tricarbonyl-
nucleophilic addition, **12**, 999
Chromium, (diphenylmercury)bis(tricarbonyl)-
reaction with butyllithium, **12**, 1019
Chromium, (fluorobenzene)tricarbonyl-
reaction with cyclopentadienyl anions, **12**, 989
substitution with carbon nucleophiles, **12**, 988
Chromium, (halobenzene)tricarbonyl-
halogen–metal exchange, **12**, 1018, 1019
Chromium, (indole)tricarbonyl-
lithiation
regioselectivity, **12**, 1029
nucleophilic addition, **12**, 998
regioselectivity, **12**, 1002
preparation, **12**, 981
Chromium, (5-methoxy-1-tetralone)tricarbonyl-
lithiation, **12**, 1030, 1031
Chromium, (7-methoxy-1-tetralone)tricarbonyl-
lithiation, **12**, 1030
Chromium, (*N*-methylindoline)tricarbonyl-
nucleophilic addition, **12**, 998
Chromium, (*N*-methyltetrahydroquinoline)tricarbonyl-
nucleophilic addition, **12**, 997
Chromium, (naphthalene)tricarbonyl-
in preparation of arene complexes, **12**, 981
lithiation, **12**, 1028
nucleophilic addition, **12**, 999
Chromium, pentacarbonyl(α-methoxybenzylidene)-
reactions with electron-deficient allkenes, **12**, 390
Chromium, pentacarbonylvinylcarbene-
cyclopropanation, **12**, 391
Chromium, phenanthrenetricarbonyl-
lithiation, **12**, 1028
thermodynamic control, **12**, 1028
Chromium, (thiophene)tricarbonyl-
lithiation, **12**, 1030
Chromium, tricarbonyl(1-tetralone)-
nucleophilic addition, **12**, 1040
2-substituted
nucleophilic addition, **12**, 1040
Chromium, tricarbonyl(acetophenone)
nucleophilic addition, **12**, 1040
Chromium, tricarbonyl(2-acetoxytetralin)-
synthesis, **12**, 1042
Chromium, tricarbonyl(*o*-anisyl propenyl ketone)-
Michael additions, **12**, 1065
Chromium, tricarbonyl(benzocyclobutanone)-
nucleophilic additions, **12**, 1042
Chromium, tricarbonyl(benzylbut-2-enyl ether)-
Wittig rearrangement, **12**, 1061
Chromium, tricarbonyl(chroman)-
deprotonation, **12**, 1059
Chromium, tricarbonyl(cyclobutabenzene)-
ring opening, **12**, 1067
Chromium, tricarbonyl(1,2-dihydronaphthalene)-
electrophilic substitution, **12**, 1064
hydroformylation, **12**, 1065
Chromium, tricarbonyl(5,6-dimethoxytetralin)-
deprotonation alkylation, **12**, 1063
Chromium, tricarbonyl(*o*-dimethylamino-1-phenethanol)-
synthesis, **12**, 1047
Chromium, tricarbonyl(*o*-dimethylaminobenzaldehyde)-
synthesis, **12**, 1047
Chromium, tricarbonyl(*o*-dimethylaminobenzophenone)-
synthesis, **12**, 1047
Chromium, tricarbonyl(5-dimethylaminoindane)-
acidity, **12**, 1057
Chromium, tricarbonyl(*N,N*-dimethylamphetamine)-
deprotonation, **12**, 1056
Chromium, tricarbonyl(ethylbenzene)-
deprotonation alkylation, **12**, 1060
reactions, **12**, 1055
Chromium, tricarbonyl(2-ethylpyridine)-
allylation, **12**, 1063
Chromium, tricarbonyl(1-ethylthiostyrene)-
reactions, **12**, 1064
Chromium, tricarbonyl(*o*-fluorobenzaldehyde)-
nucleophilic additions, **12**, 1047
Chromium, tricarbonyl(indane)-
benzylic proton exchange, **12**, 1057
Chromium, tricarbonyl(1-indanol)-
reactions, **12**, 1051
synthesis, **12**, 1042
Chromium, tricarbonyl(indanone)-
oxidation, **12**, 1067
Chromium, tricarbonyl(1-indanone)-

alkylation, **12**, 1065
kinetic resolution, **12**, 1042
reduction, **12**, 1042
Chromium, tricarbonyl(indene)-
hydroformylation, **12**, 1065
Chromium, tricarbonyl(indoxyl)-
nucleophilic additions, **12**, 1042
Chromium, tricarbonyl(isochroman)-
methylation, **12**, 1059
Chromium, tricarbonyl-(5-methoxy-1-tetralone)-
addition of crotyl magnesium chloride, **12**, 1041
Chromium, tricarbonyl(*o*-methoxyacetophenone)-
aldol reactions, **12**, 1065
Chromium, tricarbonyl(*o*-methoxybenzyl methyl ether)-
deprotonation alkylation, **12**, 1061
Chromium, tricarbonyl(4-methoxychroman)-
deprotonation alkylation, **12**, 1059
Chromium, tricarbonyl(5-methoxyindane)-
acidity, **12**, 1057
Chromium, tricarbonyl(7-methoxytetralin)-
reactions, **12**, 1057
Chromium, tricarbonyl(7-methoxy-1-tetralone)-
alkylation, **12**, 1065
Chromium, tricarbonyl(methyl*o*-methoxyphenylacetate)-
alkylation, **12**, 1060
Chromium, tricarbonyl(methyl phenylacetate)-
alkylation, **12**, 1055
Chromium, tricarbonyl(2-methyl-1-indanone)-
nucleophilic addition, **12**, 1042
Chromium, tricarbonyl-(*o*-methylaniline)-
reactions, **12**, 1067
Chromium, tricarbonyl(*o*-methyl-1-phenylpropanol)-
reactions, **12**, 1067
Chromium, tricarbonyl(*N*-methyltetrahydrobenzazepine)-
deprotonation alkylation, **12**, 1060
Chromium, tricarbonyl(*N*-methyltetrahydrobenzisoindoline)-
exo-methylation, **12**, 1057
Chromium, tricarbonyl(*N*-methyltetrahydroisoquinoline)-
alkylation, **12**, 1059
Chromium, tricarbonyl(*N*-methyltetrahydroquinoline)-
deprotonation alkylation, **12**, 1059
Chromium, tricarbonyl-(phenethanol)-
homochiral
Ritter reaction, **12**, 1049
synthesis, **12**, 1039
Chromium, tricarbonyl(phenyl propenyl ketone)-
Michael additions, **12**, 1065
Chromium, tricarbonyl(phenyl)-
phenyl tricarbonyl alkyl ketones, **12**, 1040
Chromium, tricarbonyl(*p*-phthalaldehyde)-
dimethylation, **12**, 1045
Chromium, tricarbonyl(phthalan)-
deprotonation alkylation, **12**, 1057
Chromium, tricarbonyl-(*o*-quinodimethane)-
synthesis, **12**, 1067
Chromium, tricarbonyl(styrene)-
Diels–Alder reaction, **12**, 1065
reactions, **12**, 1064, 1065
synthesis, **12**, 1044
Chromium, tricarbonyl(1-tetralol)-
preparation, **12**, 1051
Chromium, tricarbonyl-(1-tetralone)-
alkylation, **12**, 1065
oxidation, **12**, 1067
Chromium, tricarbonyl-(2-tetralone)-
nucleophilic addition, **12**, 1041
Chromium, tricarbonyl(*o*-tolualdehyde)-
acetals
oxonium ions, reactions, **12**, 1053
nucleophilic addition, **12**, 1045
Chromium, tricarbonyl(toluene)-
deprotonation, **12**, 1054, 1055
Chromium, tricarbonyl-(*p*-trifluoromethylbenz aldehyde)-
stimethylsilylcyanation, **12**, 1045
Chromium, tricarbonyl-(*o*-trimethylsilylbenz aldehyde)-
nucleophilic additions, **12**, 1048
Chromocene
oxidation, **5**, 431
reaction with butyl thiol, **5**, 434
reactions, **5**, 442
with azobenzene, **5**, 420
synthesis, **5**, 442
Chromones
synthesis, **12**, 212
Chrysanthemic acid
esters
synthesis, **12**, 408
trifluoromethylated, synthesis, **12**, 878
synthesis, **12**, 398, 406, 835, 836
Chrysanthemolactone, dihydro-
synthesis, **12**, 411
Chrysanthemonitrile
synthesis, **12**, 836, 878
Chrysomelidial
synthesis
trimethylenemethane complexes, **12**, 942
Chuangxinmycin methyl ester
synthesis
lithiation, **11**, 74
CIDNP hydrogenation
tetracarbonylhydridocobalt, **8**, 33
Cilastatin
synthesis, **12**, 406
Cinnamaldehyde
cycloaddition reactions, **12**, 930
Cinnamanilide
reductive elimination, **9**, 8
Cinnamic acid, *N*-acetamido-
hydrogenation
asymmetric, **12**, 16
Cinnamic acid, *N*-acylamino-
hydrogenation, **12**, 17
Cinnamyloxycarbonyl group
peptide protection
cleavage, **12**, 802
Citreoviral
synthesis, **12**, 889
Citronellol
hydrogenation
asymmetric, **12**, 21
Civet
synthesis, **12**, 877, 882
Claisen rearrangement
[3,3]
substrates, **12**, 589
palladium(II) catalyzed, **12**, 615
Clavicipitic acid
synthesis, **12**, 274, 275
Cluster complexes
bimetallic
coinage metals, **10**, 277
carbonyl ligands
coinage metals, **10**, 298
coinage metals, **10**, 256–98
acetylides, **10**, 313
acetylides, structures, **10**, 313
acetylides, synthesis, **10**, 313
addition metal ligand units, **10**, 303
bimetallic, **10**, 299–301
catalysis, **10**, 305
chemical properties, **10**, 302–5
cleavage of metal ligand units, **10**, 302
cleavage of metal–phosphorus bonds, **10**, 304
deprotonation, **10**, 305
dynamic behavior in solution, **10**, 298–302
edge-bridging, **10**, 260
electrochemical interconversions, **10**, 305
face-capping, **10**, 260
group 11 metal ligands, **10**, 311–17
heteronuclear, **10**, 260
intermolecular exchange of metal trialkylphosphine units, **10**, 303
ligand addition and substitution, **10**, 304
loss of carbonyl monoxide, **10**, 304
metal halides, **10**, 311
metal halides, structures, **10**, 311

metal halides, synthesis, **10**, 311
monometallic, **10**, 299
no group 11 metal ligands, **10**, 305–10
no group 11 metal ligands, structures, **10**, 310
no group 11 metal ligands, synthesis, **10**, 305
protonation, **10**, 305
stereochemical nonrigidity, metal cores, **10**, 298
structures, **10**, 260–98
synthesis, **10**, 256–60
trialkylphosphines, intermolecular exchange, **10**, 302
group 12 halides
oxidative additions, **10**, 327
transition metals
coinage metals, **10**, 255–317
trimetallic
coinage metals, **10**, 292
two different group 11 metals, **10**, 289
Cluster valence electron counts
variable, **10**, 192
Clusters
anionic
rhodium–platinum, high nuclearity, **10**, 193
bridging group 15 donor ligands, **10**, 218–28
fluxionality, **10**, 225
isomerism, **10**, 225
reactivity, **10**, 226–8
structures, **10**, 218–25
bridging group 16 donor ligands, **10**, 237
fluxionality, **10**, 235, 236
isomerism, **10**, 235, 236
reactivity, **10**, 236, 237
structures, **10**, 228–34
bridging groups, **10**, 215–18
carbon dioxide bridges, **7**, 809
chiral
properties, **10**, 244, 245
cyanide, **7**, 814
cyclic hydrocarbon ligands, **10**, 211–15
fluxionality, **10**, 214, 215
reactivity, **10**, 214, 215
structures, **10**, 212–14
cyclopentadienyl, **10**, 204–11
fluxionality, **10**, 210
reactivity, **10**, 211
structures, **10**, 204–10
diazabutadienes, **7**, 822
group 13 donor ligands, **10**, 215–18
group 14 donor ligands, **10**, 215–18
halide donors, **10**, 237–9
heteronuclear
without acyclic hydrocarbyl ligands, **10**, 187–245
imido ligands, **10**, 218
isonitriles, **7**, 814
ligands bonded through carbon and oxygen, **7**, 804–9
ligands bonded through carbon and sulfur, **7**, 809–13
metal–carbon bonds
heteroatom ligands, **7**, 803–30
monoazabutadienes, **7**, 822
naked germanium, **10**, 215
nitriles, **7**, 814
nitrogen-donor ligands bonded through carbon, **7**, 814–23
phosphaalkyne ligands, **10**, 224
phosphido ligands, **10**, 223
phosphines, **7**, 823
phosphinidene ligands, **10**, 222
phosphorus-donor ligands bonded through carbon, **7**, 823–9
thiocarbonyl ligands, **7**, 809
thioformaldehydes, **7**, 809
three or four transition metals, **10**, 239–45
isomerism, **10**, 239–44
structures, **10**, 239–44
CNDO calculations
alkyne iron clusters, **7**, 259
Cobalamin, methyl-
complexes, **8**, 49
radical reactions
cage effects, **8**, 57
Cob(III)alamin, 5'-deoxoadenosyl-
structure, **8**, 42
Cob(III)alamin, methyl-
structure, **8**, 52
Cobalamine, methyl-
methylating agent, **12**, 155
Cobalamins
axial base dissociation, **8**, 59
cation radicals, **8**, 63
distortions, **8**, 52
electrochemical reductions, **8**, 63
β-elimination, **8**, 61
flash photolysis, **8**, 57
α-halo-
radical, **8**, 64
α,β-isomerizations, **8**, 63
NMR, **8**, 66
radical reactions, **8**, 60–2
reactions
with oxidants, **8**, 62, 63
with reductants, **8**, 63, 64
S_N2 reactions
with Sn^{II}, **8**, 63
Cob(II)alamins
α side
alkylation, **8**, 48
structure, **8**, 52
Cobaloxime, methyl-
methylating agent, **12**, 155
Cobaloximes
alkali attack, **8**, 65
axial base dissociation, **8**, 52
systematics, **8**, 52
base dissociation, **8**, 51
β-carbonium ions, **8**, 65
dimethylglyoxime, **8**, 43
electrochemical reduction
equatorial variation, **8**, 64
insertion of sulfur dioxide, **8**, 65, 66
photochemistry
x-ray induced, **8**, 60
radical reactions
benzyl, **8**, 66
carbon tetrachloride involvement, **8**, 61
chromium cations, **8**, 65
decomposition, **8**, 54
insertion of oxygen, **8**, 61
N-methylsulfonamides, **8**, 66
reactions
with acids, **8**, 64
with halogens, **8**, 62
structures, **8**, 50
substituent constraints, **8**, 53
thermochemistry, **8**, 54
vibrational spectroscopy, **8**, 56
Cobalt
dicarbon metallacarbaboranes
6-vertex, **1**, 382
7-vertex, **1**, 382
8-vertex, **1**, 396
9-vertex, **1**, 396
10-vertex, **1**, 396
11-vertex, **1**, 396
12-vertex, **1**, 400, 418
monocarbon metallacarbaboranes, **1**, 378
11-vertex, **1**, 378
12-vertex, **1**, 378
tetracarbon metallacarbaboranes
6-vertex, **1**, 423
9-vertex, **1**, 423
12-vertex, **1**, 423
13-vertex, **1**, 423
tricarbon metallacarbaboranes
11-vertex, **1**, 420
Cobalt atoms
reactivity, **8**, 3
Cobalt carbonyls, **8**, 4–9
Cobalt(–I) carbonyls
chemistry, **8**, 5–7
Cobalt(0) carbonyls
chemistry, **8**, 8
Cobalt(I) carbonyls
chemistry, **8**, 8
Cobalt(II) carbonyls
chemistry, **8**, 8
vibrational spectroscopy, **8**, 8
Cobalt(III) carbonyls
chemistry, **8**, 4, 5, 8
spectroscopy, **8**, 4
vibrational spectroscopy, **8**, 8
Cobalt catalysts
alkoxycarbonylation, **12**, 243
Cobalt cluster complexes, **8**, 419–513
alkylidynetricyclopentadienyl
synthesis, **8**, 432–5
bismuth capped, **8**, 429
disulfur capped, **8**, 429
gas phase, **8**, 421
magnetic properties, **8**, 421
germanium capped, **8**, 428
heptanuclear, **8**, 437–54, 457
hexanuclear, **8**, 451–4
methylstyrene tricyclopentadienyl, **8**, 437
mixed metals, **8**, 505–13
octanuclear, **8**, 455
pentanuclear, **8**, 450
phosphorus capped, **8**, 428
tetranuclear, **8**, 438–50
arenes, **8**, 444

butterfly, **8**, 449, 450
cyclopentadienyl, **8**, 444
spectroscopy, **8**, 421
square, **8**, 445–9
tetrahedral, **8**, 438–44
triarene, **8**, 438
tricyclopentadienyl
borane ligands, **8**, 432
face-capping arene ligands, **8**, 437, 438
main group atom capped, **8**, 435–7
nitrene capped, **8**, 435
sulfur capped, **8**, 436
trinuclear, **8**, 421–38
main group capping atoms, **8**, 428, 429
Cobalt clusters
alkylidyne molybdenum, **10**, 110–12
solution fluxionality, **10**, 112
alkylidyne tungsten, **10**, 110–12
solution fluxionality, **10**, 112
alkyne molybdenum, **10**, 113–15
alkyne tungsten, **10**, 113–15
carborane tungsten, **10**, 168
iron compounds, **10**, 135–9, 144
iron dicarbide, **10**, 140, 141
ketenylidene molybdenum, **10**, 109
molybdenum compounds, **10**, 112
ruthenium compounds, **10**, 142–4
Cobalt complexes
acyl, **8**, 35–71
acyl tetracarbonyl
synthesis, **8**, 41
agostic
vibrational spectroscopy, **8**, 76
agostic hydrogen, **8**, 76, 77
alkenes, **8**, 72–4
insertion reactions, **8**, 72
nucleophilic attack, **8**, 82
rate of formation, **8**, 73
alkoxides, **8**, 21
alkyl, **8**, 35–71
alkylidene
ethene insertion, **8**, 95
isomerization, **8**, 95
synthesis, **8**, 94, 95
allyl, **8**, 78
synthesis, **8**, 78
π-allyl
in organic synthesis, **12**, 974–6
arene
synthesis, **8**, 89
arenes, **8**, 89, 90
aryl, **8**, 35–71
aryloxides, **8**, 21
benzyl, **8**, 78
synthesis, **8**, 38
π-bonded hydrocarbyl, **8**, 72–7
butadienyl, **8**, 79
isomerization, **8**, 79
cadmium clusters, **10**, 341
carbodiimides
preparation, **8**, 10
η-carbonyl, **8**, 77, 97
carbonyl semibridging, **8**, 93
catalysts
diazodecomposition, **12**, 427
chemistry, **8**, 1–101
chiral semicorrin ligands
hydrogenation, **12**, 20
cis-labilization, **8**, 4
cyclooctatetraene, **8**, 82
electrochemistry, **8**, 82
cyclooligomerization
alkynes, **12**, 744
cyclopentadienyl
cyclooligomerization, alkynes, **12**, 744
intramolecular trimerizations, **12**, 749
ring substitution, **8**, 81
synthesis, **8**, 85
cyclopentadienyl diene
PES, **8**, 79
cyclopentadienylcyclobutadiene
synthesis, **12**, 745
cyclopropenium, **8**, 78
dialkyl
cis, synthesis, **8**, 67, 68
1,4-diaryltetrazadiene, **8**, 23
(diene)cyclopentadienyl-
formation, **12**, 624, 625
dienes
metal atom vapor synthesis, **8**, 80
dienetricarbonyl
nucleophilic addition, **12**, 654
nucleophilic reactions, **12**, 651–5
theoretical analysis, nucleophilic addition, **12**, 655, 673
dienyl
bonding, **8**, 79
nucleophilic addition, **12**, 679–81
structure, **8**, 79
dihydroborates, **1**, 290
dihydrogen, **8**, 30, 31, 75, 76
T_1 measurements, **8**, 30
vibrational spectroscopy, **8**, 76
dinitrogen, **8**, 23
dinuclear
alkylidene bridged, **8**, 68
alkynes, **8**, 98
antiferromagnetic coupling, **8**, 73
mixed valence, **8**, 96
diselenaboranes, **1**, 270
dithiocarbamates, **8**, 25
epoxyacetylene
acid-promoted reactions, **12**, 688
ethyne, **8**, 74, 75
α-carbonium ions, **8**, 99
cycloaddition reactions, **8**, 99
electrochemistry, **8**, 99
electronic structure, **8**, 98
PES, **8**, 99
substituent rotations, **8**, 98, 99
synthesis, **8**, 100
vibrational spectroscopy, **8**, 74, 98, 99
ethynes
bonding, **8**, 75
NMR, **8**, 75
vibrational spectroscopy, **8**, 75
formyl, **8**, 28
gallium bonds, **1**, 555
gas phase themochemistry, **8**, 2, 3
germole, **8**, 81
halides, **8**, 22, 23
dinuclear, antiferromagnetic coupling, **8**, 22
halide lability, **8**, 22
silver assisted dissociation, **8**, 22
hexacarbonyl alkyne
dinuclear, Pauson–Khand reaction, **12**, 704
5-hexenyl
radical reactions, **8**, 61
hydride carbonyl
thermochemistry, **8**, 27
hydride formation
oxidative addition, **8**, 31, 32
hydrides, **8**, 26–71
hydrido, **8**, 26–35
thermochemistry, **8**, 26–9
hydrocarbyl, **8**, 26–71
hydroformylation cobalt catalyzed, **8**, 32
hydrogen bonds
formation, **8**, 30–2
formation, protonation, **8**, 30
indium bonds, **1**, 564
isocyanide, **8**, 96
isocyanides
cycloaddition reactions, **8**, 11
reactions, **8**, 9
synthesis, **8**, 9
ketenes
preparation, **8**, 10
η^4-ligands
cycloaddition reactions, **8**, 80–2
mercury clusters, **10**, 341
nitrosyl, **8**, 13, 14
alkane insertion, **8**, 14
alumina adsorbed, **8**, 14
bent, **8**, 13
NMR, **8**, 13
nucleophilic attack at nitrogen, **8**, 14
silica adsorbed, **8**, 14
substitution reactions, **8**, 13
synthesis, **8**, 14
vibrational spectroscopy, **8**, 13
octaethylporphyrins, **8**, 45
oxygen donors, **8**, 24
palladium carbonyls, **9**, 205
pentacarbonylpropargylium cations, preparation, **12**, 686
pentadienyl, **8**, 86

pentakis(carbonyl)propargylium(triphenylphosphine)
variable-temperature NMR, **12**, 687
phosphide
synthesis, **8**, 96
phosphines, **8**, 15–21
Arbuzov rearrangement, **8**, 20
atypical, **8**, 16–20
1,3-bidentate, **8**, 18, 19
1,4-bidentate, **8**, 19
1,5-bidentate, **8**, 19, 20
1,6-bidentate, **8**, 20
cobalt substituents, **8**, 16, 17
dinuclear, mixed valence, **8**, 18
hexadentate, **8**, 20
monodentate, **8**, 17
optically active, **8**, 17, 21
polydentate, **8**, 20
reactions, **8**, 20, 21
substitution reactions, **8**, 15, 16
substitution reactions, phosphine association, **8**, 15
tridentate, **8**, 20
phosphinidene
bonding, **8**, 97
structure, **8**, 97
polyethyne, **8**, 100
polyhydrogen, **8**, 75
polysulfides, **8**, 25
propargyliumhexakis(carbonyl)-
nucleophilic derivatives, cations, synthesis, **12**, 690
radical reactions
mutase model, **8**, 61
reactions with alkynes, **8**, 74
regioselectivity
cycloaddition reactions, **8**, 81
salen, **8**, 44
saloph, **8**, 44
alkylation, **8**, 49
selenocarbonyl groups, **8**, 77
silole, **8**, 81
solution phase thermochemistry, **8**, 4
sulfur dioxide, **8**, 25
sulfur donors, **8**, 24
tetraphenylporphyrins, **8**, 45
thermochemistry, **8**, 2–4
thiocarbonyl groups, **8**, 77, 97
thiocarbonyls, **8**, 25
thioketenes, **8**, 25, 77
thiolates, **8**, 25
transmetallation
catalyst, **12**, 158
α,β-unsaturated carbonyl
conjugate addition, **8**, 82
zinc clusters, **10**, 340
Cobalt(–I) complexes
isocyanides
vibrational spectroscopy, **8**, 9
Cobalt(I) complexes
alkyl
synthesis, **8**, 36
alkyltetracarbonyl
nucleophilic attack, **8**, 8
aryl
synthesis, **8**, 36
carbenes, **8**, 41
carbon bonds
oxidative addition, **8**, 36–9
synthesis, insertion reactions, **8**, 40, 41
synthesis, oxidative reactions, **8**, 39, 40
C–H bonds
oxidative addition, **8**, 40
insertion reactions
ketenes, **8**, 41
Cobalt(II) complexes
alkyl
synthesis, **8**, 41
Cobalt(III) complexes
isocyanides, **8**, 9–11
thiocarbonyls, **8**, 9
Cobalt compounds
allylic alkylation, **12**, 862
hydrogen bonds
bond energy, **8**, 3
methyl
bond energy, **8**, 3
Cobalt dehydrogenase
nickel complexes, **9**, 91
Cobalt ions
reactivity, **8**, 2, 3
Cobalt molybdenum complexes
propargylium cyclopentadienylpentakis(carbonyl)
dinuclear cations, preparation, **12**, 687
variable-temperature NMR, **12**, 688
Cobalt nickel
dicarbon metallacarbaboranes
7-vertex, **1**, 382
Cobalt osmium
dicarbon metallacarbaboranes
7-vertex, **1**, 382
Cobalt pentacyanide
dioxygen complexes, **8**, 12
hydrogenation, **8**, 12
Cobalt rhodium
dicarbon metallacarbaboranes
7-vertex, **1**, 382
Cobalt rhodium carbonyls
characterization, **8**, 505–10
reactivity, **8**, 505–10
synthesis, **8**, 505–10
Cobalt rhodium heptacarbonyl, **8**, 505
reaction with hydrogen, **8**, 30
Cobalt ruthenium
dicarbon metallacarbaboranes
7-vertex, **1**, 382
Cobalt tetracarbonyl
electrochemistry, **8**, 7
electron affinity, **8**, 29
hydroformylation, **8**, 33
infrared spectrum, **8**, 6
ion-pairing, **8**, 6
nucleophilic substitution, **8**, 7
oxidative addition
alkyl halides, **8**, 36
photochemistry, **8**, 27
reactions, **8**, 6, 7
synthesis, **8**, 5
Cobalt tungsten complexes
propargylium cyclopentadienylpentakis(carbonyl)
dinuclear cations, preparation, **12**, 687
Cobalt, (alkoxycarbonyl)tetracarbonyl-
synthesis, **8**, 37
Cobalt, alkylcarbonylcyclopentadienyl-
alkyl exchange, **8**, 67
Cobalt, alkylidynenonacarbonyltri-, **8**, 423–8
Cobalt, alkylidynetri-
carbonyl substitution, **8**, 425
polydentate phosphine ligands, **8**, 426
electrochemistry, **8**, 424
ferrocene substituent, **8**, 426
hydroformylation catalysts, **8**, 424
MO calculations, **8**, 423
radical anions
ligand substitution, **8**, 425
tripod complexes, **8**, 426
Cobalt, carbonylcyclopentadienyl-
anions
alkylation, **8**, 67
Cobalt, carbonylcyclopentadienyliodo-
alkylation, **8**, 67
Cobalt, carbonylcyclopentadienylmethyl-
decomposition, **8**, 67
Cobalt, carbonylcyclopentadienyl(trimethylphosphine)-
PES, **8**, 84
Cobalt, carbonyltricyclopentadienyl
synthesis, **8**, 429–31
Cobalt, chlorotris(triphenylphosphine)-
reaction with benzocyclobutenedione, **12**, 752
Cobalt, cyclopentadienyl-
electronic structure, **8**, 83
Cobalt, cyclopentadienylbis(ethene)-
cocyclization of propyne and phenylisocyanate, **12**, 757
Cobalt, cyclopentadienylbis(triethylphosphine)-
structure, **8**, 83
Cobalt, cyclopentadienylcyclobutadiene
formation, **12**, 761
Cobalt, cyclopentadienylcyclohexadiene
synthesis, **12**, 759
Cobalt, cyclopentadienyl(dimethyl- malonyl)-
cycloaddition reactions, **8**, 69
Cobalt, decacarbonyltri-, **8**, 421–3
Cobalt, (2-decynal)hexakis(carbonyl)di-
asymmetric allylboration, **12**, 692

Cobalt, (3-decynal)hexakis(carbonyl)di-
asymmetric allylboration, **12**, 693
Cobalt, dialkylcyclopentadienyl-
cyclopentadienyl exchange, **8**, 69
ligand exchange, **8**, 69
Cobalt, dicarbonylcyclopentadienyl
cyclooligomerization
alkynes, **12**, 744
PES, **8**, 83
photochemistry, **8**, 85
trimerization
alkynes, **12**, 747
Cobalt, dicarbonylpentamethylcyclopentadienyl-
structure, **8**, 83
Cobalt, dicyclopentadienyl-
charge transfer spectra, **8**, 7
ion pairing, **8**, 7
Cobalt, (dmpe)(pentamethylcyclopentadienyl)-
PES, **8**, 85
Cobalt, dodecacarbonyltetra-
adsorption, **8**, 440
NMR, **8**, 439
polydentate phosphine derivatives, **8**, 441
polydentate sulfur ligands, **8**, 441
structure, **8**, 439
synthesis, **8**, 439
thermal decomposition, **8**, 439
Cobalt, ethylidynetri-
chemisorption onto alumina, **8**, 424
Cobalt, heptacarbonyl(thioacetamide)sulfidotri-, **8**, 429
Cobalt, hexakis(carbonyl)(3-methyl-1-pentyne)-
NMR, **12**, 687
Cobalt, hexakis(carbonyl)(1,3-dibenzyloxy)propyne)di-
reaction with nucleophiles, **12**, 693
Cobalt, hexakis(carbonyl)(phenylacetylene)di-
Pauson–Khand reaction, **12**, 718
Cobalt, hexakis(carbonyl)(1-(trimethylsilyl)propyne)-
reaction with alkenes, **12**, 693
Cobalt, hydridotetracarbonyl-
thermochemistry, **8**, 26
Cobalt, ketanylidenetri-
NMR, **8**, 423
Cobalt, methyl-
cation
thermochemistry, **8**, 35
thermochemistry, **8**, 27, 35
Cobalt, nonacarbonylethylidynetri-
XPS studies, **8**, 423
Cobalt, nonacarbonylhydridotri-, **8**, 421–3
Cobalt, pentadecacarbonylhydrido-
anion, **8**, 473
Cobalt, propargylhexakis(carbonyl)di-
N-protecting group
vinblastine synthesis, **12**, 695
Cobalt, propynehexakis(carbonyl)di-
reaction with 1-methoxybicyclo[3.2.0]hepta-3,6-diene, **12**, 707
Cobalt, tetracarbonylhydrido-
Brønsted acidity, **8**, 29
Co–H bond polarity, **8**, 29
decomposition, **8**, 27
radical, **8**, 32
electronic structure, **8**, 27
synthesis, **8**, 30, 31
vibrational spectroscopy, **8**, 28
Cobalt, tetracarbonylmethyl
photochemistry, **8**, 35
Cobalt, tetracarbonyl(trialkylsilyl)-
photochemistry, **8**, 35
Cobalt, tricarbonylcyclopentadienyl-
ring slippage, **8**, 85
Cobalt, tricarbonylhydrido-
structure, **8**, 27
Cobalt, undecacarbonyl(triphenylphosphine)-
structure, **8**, 441
Cobalt, vinylidenetri-
NMR, **8**, 424
Cobalt(III), dialkyl-
cis reactions, **8**, 68–71
Cobalt(IV), alkyl-, **8**, 71
Cobalt(IV), aryl-, **8**, 71
Cobaltaazaundecaboranes
preparation, **1**, 263
Cobaltaboranes, **1**, 235
Cobaltacyclopentadienes
reaction with ethene, **12**, 759
Cobaltacyclopentane
reactions, **12**, 764
Cobaltacyclopentene
reaction with alkynes, **12**, 759
Cobaltates
tetracyano-
cyanation catalysis, **12**, 225
Cobalt(II) complexes
bis(dioximato)-
cyclopropanation, **12**, 416, 417
Cobalt(II), phthalocyaninephenyl-
methylating agent, **12**, 155
Cobalt(III) acetylacetonate
catalyst
trimethylaluminum reaction with α,β-unsaturated ketones, **12**, 156
Cobaltocene
appended to electrodes, **8**, 88
charge transfer, **8**, 89
electrochemistry, **8**, 88
electronic structure, **8**, 88, 89
synthesis, **8**, 89
Cobaltocenium salts
photochemistry, **8**, 89
Cobaltocyanides
chemistry, **8**, 11, 12
cyanides
bridging, **8**, 11
vibrational spectroscopy, **8**, 11
Cobaltocyclopentadienes
preparation, **8**, 68
from bis(ethyne) complexes, **8**, 74
reactions
with carbenes, **8**, 70
Cobaltocyclopentenes
preparation, **8**, 68
Cobaltohydrides
reactions, **8**, 32–5
Cobaltole
reaction with borane tetrahydrofuran adducts, **1**, 203
Cobinamide
β-elimination
decomposition, **8**, 57
Cob(III)yrinates
electrochemical reductions, **8**, 64
Cocaine
analogues
synthesis, **12**, 401
Codeine, 10-alkyl-
synthesis, **12**, 1057
Coenzyme B_{12}
biological function, **8**, 45, 58
chemistry, **8**, 42–66
Co–C bond
creation, **8**, 46
Co–C bond energy, **8**, 56
decomposition
acid-catalyzed, **8**, 57
cage effect, **8**, 56
radical reactions, **8**, 56
diol dehydrase model, **8**, 46
methyl transfer, **8**, 45
photochemistry, **8**, 60
structure, **8**, 42, 50
thermochemistry, **8**, 56
Coenzyme B_{12} methylmanolyl–coenzyme A mutase
models, **8**, 45
Coenzyme Q
side chain
synthesis, **12**, 800
Coenzyme Q10
synthesis, **12**, 849
Colchicine
synthesis, **12**, 832
Collinusing
synthesis, **12**, 20
Collman-type ketone formation, **7**, 116
Collman's reagent, **7**, 112
in carbonylation, **12**, 351
Colloids
bimetallic
catalysis, **10**, 60
in hydrosilylation, **2**, 115
Complex induced proximity effect
lithiation
arenes, **11**, 42
Confertin
synthesis, **12**, 120, 413
lithiation, **11**, 72
Coniferyl alcohol
synthesis, **11**, 108
Coniine hydrochloride
synthesis, **11**, 144
Coniine, *N*-methyl-
synthesis, **11**, 144
Cope rearrangements
cyclopropanes, **12**, 400
palladium(II) catalyzed, **12**, 615

Copper
  bimetallic reagents
    reactions, **11**, 182
  catalysis
    aryllead triacetate reactions, **11**, 480, 481
    dicarbon metallacarbaboranes
      12-vertex, **1**, 400
Copper acetylacetonate
  catalyst
    trimethylaluminum reaction with α,β-unsaturated ketones, **12**, 156
Copper(II) acetylacetonate
  organocopper compound
    preparation from, **3**, 63
Copper acetylides
  coupling reactions, **12**, 222
Copper(I) bromide
  organocopper compound
    preparation from, **3**, 63
Copper bronze
  catalysts
    diazodecomposition, **12**, 422
Copper(I) carboxylate
  decarboxylation, **3**, 70
Copper clusters
  metallaboranes, **1**, 304
  transition metals, **10**, 255–317
  trialkylphosphines
    edge-bridging, **10**, 261
    theoretical analysis of bonding, **10**, 261
Copper complexes
  catalysts
    cyclopropanation, **12**, 397–415
  cyclopentadienyl
    organocopper compound
      preparation from, **3**, 64
  heptahydrotriborates
    metallaboranes, **1**, 317
  zerovalent
    preparation, **12**, 63–5
Copper compounds
  alkyl
    synthesis, **3**, 71
  catalysts
    diazodecomposition, **12**, 426, 427
  perfluoroalkyl
    synthesis, **3**, 70
  vinyl
    synthesis, **3**, 70
Copper(II) halides
  organocopper compound
    preparation from, **3**, 63
Copper methyltrialkylborates
  reactions with acrylates, **11**, 227
Copper *t*-butoxide
  organocuprate preparation from, **3**, 68
Copper triflate
  catalysts
    diazodecomposition, **12**, 422
  organocopper compound
    preparation from, **3**, 62
Copper, alkynyl-
  synthesis
    metallation, **3**, 67
Copper, bis(oxazoline)-
  catalysts
    cyclopropanation, **12**, 407, 408
Copper, bromo(dimethyl sulfide)-
  organocopper compound
    preparation from, **3**, 60
Copper, mesityl-
  organocuprate preparation from, **3**, 69
Copper, perfluoro-*t*-butyl-
  synthesis, **3**, 70
Copper, phenyl-
  insolubility, **3**, 120
Copper, salicylaldimine-
  catalysts
    cyclopropanation, **12**, 406, 407
Copper, semicorrin-
  catalysts
    cyclopropanation, **12**, 407
Copper, (trifluoromethyl)-
  organocopper compound
    preparation from, **3**, 64
Copper, vinyl-
  reactions with
    cyclohexadienetricarbonyliron complexes, **12**, 660
Copper(I) chloride
  trialkyl phosphite complex
    catalyst, diazodecomposition, **12**, 426
Copper(I) complexes
  tetrahydroborates, **1**, 286
Copper(I) halide, trialkyl phosphite
  catalysts
    diazodecomposition, **12**, 422
Copper(I) salts
  co-catalyst
    transmetallation, **12**, 201
Copper(I) triflate
  catalyst
    diazodecomposition, **12**, 426
Copper(II) complexes
  bisoxazoline
    catalysts, diazodecomposition, **12**, 427
  salicyladimine
    catalysts, diazodecomposition, **12**, 427
  semicorrin
    catalysts, diazodecomposition, **12**, 427
  supported
    cyclopropanation, **12**, 399
Copper(II) sulfate
  catalysts
    diazodecomposition, **12**, 422
Copper(II), bis(*N*-benzylsalicylaldiminato)-
  cyclopropanation, **12**, 400
Corey lactone diol
  synthesis, **12**, 603
Coriolin
  synthesis, **12**, 734
    allylmetal catalysts, **12**, 919
Corticosterone, 16,17-epoxy-
  synthesis, **12**, 694
Cortisone
  synthesis, **12**, 73
Cossus cossus
  5,13-tetradecadien-1-ol
    synthesis, **11**, 106
Costa-type complexes
  alkylation, **8**, 49
  cobalt to ring isomerizations, **8**, 64
  electrochemical reductions, **8**, 64
  reactions
    with oxidants, **8**, 62
  synthesis, **8**, 44
Coumarins
  synthesis, **12**, 359
Coumarins, 3-aryl-4-hydroxy-
  synthesis, **11**, 507
Coumarins, 3-hydroxy-
  diarylation
    aryllead compounds, **11**, 473
Coumarins, 4-hydroxy-
  arylation
    aryllead compounds, **11**, 473
Coumaryl alcohol
  synthesis, **11**, 108
Coupling
  nitrene–carbene
    on triiron clusters, **7**, 281
  phosphinidene–carbene
    on triiron clusters, **7**, 281
Coupling reactions
  catalysts
    titanium complexes, **4**, 275
  highly reduced metals, **12**, 1071–92
  mixed reductive
    molybdenum or tungsten complexes, **12**, 1090, 1091
    niobium complexes, **12**, 1088, 1089
  molybdenum complexes, **12**, 1090–2
  niobium complexes, **12**, 1087–9
  organosamarium complexes, **4**, 126
  organozinc compounds, **12**, 180
  titanium catalysts, **12**, 1072–84
  tungsten complexes, **12**, 1090–2
  vanadium complexes, **12**, 1086, 1087
Crassin acetate
  synthesis, **12**, 1077
o-Cresol
  reaction with aryllead compounds, **11**, 469
Cresols
  dimetallation, **11**, 109
  lithiation, **11**, 59
Crinan
  synthesis, **12**, 285, 288
Crinipellin B
  synthesis, **12**, 63, 733
Cross-coupling reactions
  imines, **12**, 1084
  intermolecular, **12**, 1074, 1075

niobium complexes, **5**, 142, 143
palladium-catalyzed, **11**, 371, 372
Crotonates
cycloaddition reactions, **12**, 926
Crown ethers
boron-containing
synthesis, **11**, 245
organosilicon compounds, **2**, 32
Crown-thioether ligands
synthesis, **12**, 690
Cryptolestes pusillus
pheromone
synthesis, **11**, 100
Cryptone
synthesis, **12**, 62
Cryptopine, dihydro-
chromium tricarbonyl complexes
lithiation, **11**, 63
Crystal growth
organocadmium compounds, **3**, 201
Cubane
synthesis, **12**, 624
Cubanes
lithiation, **11**, 6
Cumene
metallation, **11**, 108
Cumulenes
organosilicon complexes, **2**, 12, 13
Cuparene
synthesis, **12**, 436, 437
Cuprates
acetylenic cyano
coupling with organometallic electrophiles, **12**, 76
1,4-addition reactions
stereoselectivity, **12**, 119–21
cyano
genenions, **12**, 114–18
higher order, **12**, 75–81
α-nitrogen-containing, **12**, 76
dioxenyl
synthesis, **12**, 76
electrophilic functionality
in organic synthesis, **12**, 60–73
kinetic, **12**, 81
preparation by transmetallations, **12**, 81–94
catalytic, **12**, 81–4
propargylic halozinc
addition reactions, **12**, 71
silyl
in organic synthesis, **12**, 96–100
stannyl
in organic synthesis, **12**, 100–3
thienyl cyano
synthesis, **12**, 78, 79
transmetallation
stoichiometric ligand exchanges, **12**, 85–94
Cuprates, trimethylstannyl-
reaction with acetylenic bonds, **11**, 363
Cupric chloride
in synthesis of conjugated dienes
palladium assisted, **12**, 149
Curvularin
preparation, **12**, 246
synthesis, **12**, 367
Cyanation
alkylpalladium complexes, **12**, 139
cobalt pentacyanide catalyzed, **8**, 12
Cyanide complexes
group 6, **5**, 305–8
properties, **5**, 305
reactions, **5**, 307, 308
structure, **5**, 305
synthesis, **5**, 305
Cyanides
metathesis, **12**, 225, 226
Cyanoborate process, **11**, 230
Cyanogen, seleno-
synthesis, **11**, 528
Cyclam
nickel complexes, **9**, 91
Cyclic voltammetry
thiolates, **6**, 88
Cyclidiphosphadiarsatene
synthesis, **2**, 328
Cyclization, **12**, 910–12
heteroatom substitution
arene complexes, **12**, 986, 987
palladium catalysts, **12**, 848, 849
Cyclization–anion capture cascades, **12**, 308–14
Cyclization–anion capture methodology
alkylpalladium(II) complexes, **12**, 306
Cyclization–carbonylation
thallium(I) salts, **11**, 456
Cycloaddition
aziridines
heterocumulenes, **11**, 500
Cycloaddition cascades
palladium catalyzed, **12**, 314–19
Cycloaddition copolymerization
transition-metal catalysts, **12**, 1227, 1228
Cycloaddition reactions
[3+2]
acceptors, **12**, 925–34
conditions, **12**, 924
conjugative reagents, **12**, 924–45
mechanism, **12**, 924
reagents, **12**, 924
substituted conjunctive reagents, **12**, 934–8
aromatic
metal carbenes, **12**, 413, 414
diastereoselectivity
conjunctive reagents, **12**, 938–40
intramolecular, **12**, 940, 941, 955, 956
methylenecyclopropanes, **12**, 945–57
photochemical
carbene complexes, **12**, 551–67
three alkynes
arene synthesis, **12**, 747–51
trimethylene methane complexes, **12**, 923–57
trimethylenemethanepalladium, **9**, 373–5
Cycloalka-1,2-dienes
sila substituted
synthesis, **2**, 13–15
Cycloalka-1,3-dienes
hydroboration, **11**, 247
Cycloalkanes
B-derivatives, **1**, 133–47
ring opening
palladium salts, **9**, 341–3
Cycloalkanes, 1,2-bis(methylene)-
magnesium compounds, **1**, 97
Cycloalkanes, methylene-
synthesis
Ziegler–Natta catalysts, **12**, 1206
Cycloalkenes
hydroalumination, **1**, 470
polymerization
Ziegler–Natta catalysts, **12**, 1202
2-Cycloalkenes, 1-acetoxy-4-alkenyl-
cyclization
allylpalladium catalysts, **12**, 910
ring closure
allylmetal catalysts, **12**, 908
Cycloalkenylation
palladium(II) mediated
2-trimethylsilyloxy-1,5-hexadienes, **12**, 614
Cycloalkynes
sila substituted
synthesis, **2**, 13–15
synthesis, **12**, 692
Cyclobutadiene
optically active
transition metal complexes, stability, **12**, 624
Cyclobutadienes
synthesis
cyclooligomerization, alkynes, **12**, 745
Cyclobutane, 2-alkylidene-3-methylene-
preparation, **12**, 203
Cyclobutane, methylene-
Pauson–Khand reaction, **12**, 714
polymerization
Ziegler–Natta catalysts, **12**, 1203
ring-opening polymerization
Ziegler–Natta catalysts, **12**, 1203
Cyclobutanecarboxylic acids
2-(diphenylphosphino)-
synthesis, **12**, 886
Cyclobutanes
synthesis, **12**, 94
Cyclobutanone
formation

in insertion reactions, **12**, 453
reactions with iron ions, **7**, 104
Cyclobutanones
from reaction of alkenes with carbene complexes, **12**, 560–5
preparation
carbene complexes, **12**, 471
Cyclobutene
homopolymerization
Ziegler–Natta catalysts, **12**, 1202
Cyclobut-3-ene-1,2-dione
reaction with nickel isocyanide complexes, **9**, 11
Cyclobutenediones
synthesis, **12**, 752
1,2-Cyclobutenediones
synthesis
carbene complexes, **12**, 527
Cyclobutenes
Pauson–Khand reaction, **12**, 715
Cyclobutenol, vinyl-
thermal rearrangement, **12**, 752
Cyclobutenones
insertion reactions
rhodium complexes, **8**, 218
preparation
carbene complexes, **12**, 471
Cyclocarbonylation
diene iron tricarbonyl complexes, **12**, 650
(diene)tricarbonyliron, **12**, 628
Cyclocolorenone
synthesis, **12**, 625, 696
β-Cyclodextrin
association with metallocenes, **7**, 607
inclusion complexes
metallocenes, **7**, 619
Cyclodigermazanes
synthesis, **2**, 161
Cyclodiphosphaarsine
synthesis, **2**, 327
Cyclododecadiene
synthesis, **12**, 787
Cycloheptadiene
functionalization
dienylmanganese complexes, **12**, 678
Cycloheptadienone, 4,5-*cis*-dibutyl-
synthesis, **12**, 94
Cycloheptadienones
synthesis
carbene complexes, **12**, 503
1,3,5-Cycloheptatriene
metallation, **11**, 112
Cycloheptatriene, 2-trimethylsilyl-
metallation, **11**, 118
Cycloheptatrienone
modified reactivity, **12**, 625
protection, **12**, 625
Cycloheptene
Pauson–Khand reaction, **12**, 719
2-Cycloheptene
synthesis
Heck reaction, **12**, 913
Cyclohexadiene
Pauson–Khand reaction, **12**, 719
reactions with iron atoms, **7**, 108
reactions with iron ions, **7**, 106
synthesis
cyclooligomerization, alkynes, **12**, 751, 759–64
1,3-Cyclohexadiene
cis-1,2-disubstituted
synthesis, **12**, 1007
substituted
synthesis by addition/protonation of arenetricarbonylchromium complexes, **12**, 1007–10
Cyclohexadienes
preparation
from chromium tricarbonyl arene complexes, **5**, 516
Cyclohexadienone
annulation
carbene complexes, **12**, 497, 498
1,3-Cyclohexadienone
monoallylation, **12**, 844
2,4-Cyclohexadienone
tricarbonyliron complex
isolation, **12**, 624, 625
Cyclohexadienyl complexes
transition metals
preparation, **12**, 995–1012
Cyclohexane
reactions with iron ions, **7**, 104
Cyclohexane, alkylidenemethylidene-
synthesis, **12**, 911
Cyclohexane, 1,1-dimethyl-
synthesis, **12**, 591
Cyclohexane, 3,5-dimethylidene-
synthesis, **12**, 911
Cyclohexane, methylene-
reaction with palladium dichloride, **9**, 331–5, 340
Cyclohexanes
synthesis
cyclooligomerization, **12**, 764–6
Cyclohexanone
hydrogenation
catalysts, **10**, 39
Cyclohexanone, 2-(2,7-octadienyl)-
synthesis, **12**, 791
Cyclohexene
reactions with iron ions, **7**, 104, 107
2-Cyclohexene, *cis*-1,4-dimethylamino-
synthesis, **12**, 807
Cyclohexene, 1-phenyl-
asymmetric dihydroxylation
mechanism, **12**, 1181–3
Cyclohexene, 3-(2-propynylamino)-
Pauson–Khand reaction, **12**, 734
Cyclohexene, 3-(2-propynyloxy)-
Pauson–Khand reaction, **12**, 734
1-Cyclohexene, 4-vinyl-
hydroesterification, **12**, 33, 34
Cyclohexenes
synthesis
cyclooligomerization, **12**, 764–6
2-Cyclohexenol, 3-methyl-
hydrogenation
asymmetric, **12**, 22
2-Cyclohexenone
cycloaddition reactions, **12**, 927
Cyclohex-2-en-1-one
5-substituted
synthesis by nucleophilic addition to (anisole)tricarbonylchromium, **12**, 1007, 1009
2-Cyclohexenone, 3-iodo-
reaction with zinc–copper reagents, **11**, 166
Cyclohexenone, 2-(methoxycarbonyl)-
cycloaddition reactions, **12**, 928
Cyclohexenyl acetate
allylation/carbonylation, **12**, 918
cyclization
palladium-catalyzed, **12**, 911
Cyclohexenyl phosphite
hydroformylation, **12**, 29
Cyclohexyne
transition metal complexes
coupling reactions, **12**, 338
Cycloiridadisiloxane
preparation, **8**, 350, 351
Cycloisomerization
enynes
hydropalladation, **12**, 47–52
Cyclomethylenomycin A
synthesis, **12**, 722
1,2-Cyclooctadiene, 1-*t*-butyl-
silylation, **2**, 14
1,5-Cyclooctadiene
rearrangement
palladium(II) catalyzed, **12**, 615
3,7-Cyclooctadiene, 1,2-dicarbomethoxy-
synthesis, **12**, 793
Cyclooctatetraene
bridging ligand
chromium complexes, **5**, 450
magnesium compounds, **1**, 98
reaction with potassium and ytterbium, **1**, 25
Cyclooctatetraenes
ring-opening metathesis polymerization, **12**, 1223
synthesis
cyclooligomerization, alkynes, **12**, 747
Cyclooctene
Pauson–Khand reaction, **12**, 719
reaction with carbon dioxide

nickel complexes, **9**, 7
Cyclooctenone, 2-(methoxycarbonyl)-
cycloaddition reactions, **12**, 928
Cyclooctenylcarboxylic acid
β,γ-unsaturated
synthesis, **9**, 7
Cyclooligomerization
alkynes
mechanism, **12**, 742
metal mediated, **12**, 741–66
Cyclopalladation
palladium(II) complexes, **9**, 242–8
Cyclopentadiene
reactions with iron anions, **7**, 107
reactions with iron atoms, **7**, 108
Cyclopentadiene, 5-fluoro-
synthesis
cyclopentadienylthallium, **11**, 455
Cyclopentadieneone imines
synthesis, **12**, 756
Cyclopentadienes
Pauson–Khand reaction, **12**, 716
preparation
carbene complexes, **12**, 475–81
Cyclopentadienone complexes
by carbonylation of metallacyclopentadienes, **12**, 744
Cyclopentadienone imines
synthesis
cyclooligomerization, alkynes, **12**, 745
Cyclopentadienones
preparation
carbene complexes, **12**, 475
tricarbonyliron complex
isolation, **12**, 624, 625
Cyclopentadienyl complexes
alkenyl, **4**, 485
alkyl, **4**, 484
heteronuclear clusters, **10**, 93
phosphino, **4**, 485
silyl, **4**, 484
substituted, **4**, 484–9
Cyclopentadienylmercury compounds
structures, **3**, 160–2
Cyclopentane, 1-(3-acetoxy-2-propenyl)-
synthesis, **12**, 912
Cyclopentane, 3-alkenyl-4-vinyl-
synthesis, **12**, 915
Cyclopentane, 2,5-dimethyl-1-methylene-
synthesis, **12**, 936
Cyclopentane, divinyl-
synthesis, **12**, 910
Cyclopentane, methylene-
synthesis
Ziegler–Natta catalysts, **12**, 1206
Cyclopentane, methylidenestyryl-
synthesis, **12**, 906
Cyclopentane, methylidenevinyl-
synthesis, **12**, 906
Cyclopentane, vinyl-
synthesis, **12**, 917
Cyclopentanes
preparation
carbene complexes, **12**, 473, 474
synthesis
Heck reaction, **12**, 272
Cyclopentanone
formation, **12**, 350
from diazo ketones, **12**, 452–4
from enamines, **12**, 350
Cyclopentanone, amino-
addition reactions, **12**, 68, 69
Cyclopentanone, methylene-
addition reactions, **12**, 68
synthesis, **12**, 918
Cyclopentanone, 2-pentylidene-
hydrogenation
asymmetric, **12**, 20
Cyclopentanones
preparation
metal carbonyls in, **12**, 351
synthesis
cross-coupling reactions, **12**, 1084
Cyclopentapyrans, **12**, 537
synthesis
carbene complexes, **12**, 513
Cyclopentasilane, permethyl-
synthesis, **2**, 80, 81
Cyclopentene
homopolymerization
Ziegler–Natta catalysts, **12**, 1202
hydrooligomerization
Ziegler–Natta catalysts, **12**, 1202
Pauson–Khand reaction, **12**, 716
1,3-polymerization
mechanism, **12**, 1202
reactions with iron ions, **7**, 107
Cyclopentene, 1-aceteoxy-
Pauson–Khand reaction, **12**, 718
Cyclopentene, 2-allyl-1-(*N*-tosylamino)-
cyclization, **12**, 609
Cyclopentene, 3-(3-butynyl)-
Pauson–Khand reaction, **12**, 731
Cyclopentene, 3,4-diacetoxy-
allylic alkylation
palladium-catalyzed, **12**, 836
Cyclopentene, 1-(4-pentynyl)-
Pauson–Khand reaction, **12**, 731, 732
Cyclopentenediones
preparation
carbene complexes, **12**, 472, 473
Cyclopentenes
imino-
synthesis, **12**, 328
Cyclopent-4-ene-1,2,3-trione, 4,5-diphenyl-
reaction with nickel isocyanide complexes, **9**, 11
Cyclopentenone
cycloaddition reactions
[3+2], **12**, 934
synthesis, **12**, 359, 966
Cyclopentenone, 3-alkyl-4-(hydroxyalkyl)-
synthesis, **12**, 730
Cyclopentenone, 5-(hydroxymethyl)-
Pauson–Khand reaction, **12**, 712
2-Cyclopentenone
cycloaddition reactions, **12**, 927
2-Cyclopentenone, 4-hydroxy-
hydrogenation
asymmetric, **12**, 22
Cyclopentenones
alkoxy
preparation, carbene complexes, **12**, 474, 475
disubstituted
synthesis, **12**, 648
from Pauson–Khand reaction, **12**, 704
synthesis, **12**, 73, 360
stereospecific, **12**, 939
2-Cyclopentenones
4,5-disubstituted
Pauson–Khand reaction, **12**, 722
2-Cyclopenten-1-ones, 3-methyl-
preparation, **12**, 614
Cyclopentynes
transition metal complexes
coupling reactions, **12**, 338
Cyclophane, octafluoro-
preparation
reductive dimerization, **12**, 1073
Cyclophanes
chromium tricarbonyl complexes, **5**, 487, 488
synthesis, **12**, 1077
meta-Cyclophanes
synthesis, **11**, 109
Cyclophosphaarsine
synthesis, **2**, 327
Cyclophosphazene, fluoro-
reaction with lithiated ruthenocenes, **7**, 610
Cyclopolyarsines
synthesis, **2**, 328
Cyclopolymerization
dialkenes
Ziegler–Natta catalysts, **12**, 1201
Cyclopolysilanes
electronic structure, **2**, 96
Cyclopoly(silapropynylenes)
preparation, **2**, 64
Cyclopolystibines
synthesis, **2**, 328
Cyclopropanation
carbene complexes, **12**, 387–417
catalysts, **12**, 415–17
cationic iron carbene complexes, **12**, 394
copper and rhodium catalysts, **12**, 397–415
diazocarbonyl compounds, **12**, 398–402
diazomethane, **12**, 396, 397

dienes
regiocontrol, **12**, 391
(diene)tricarbonyliron, **12**, 627
dirhodium tetraacetate, **12**, 400
enantioselective, **12**, 394
intermolecular
carbene complexes, **12**, 390, 391
diastereoselectivity, **12**, 405
enantioselectivity, **12**, 405–9
regioselectivity, **12**, 404, 405
stereoselectivity, **12**, 403, 404
intramolecular, **12**, 399, 400, 457
carbene complexes, **12**, 391, 392
chemoselectivity, **12**, 411
enantioselective, **12**, 409–11
iron carbenoid complexes, **7**, 119
iron complexes, **7**, 178
mechanism, **12**, 395, 414, 415
organoantimony compounds, **11**, 498, 499
palladium-catalyzed, **12**, 316
stereoselectivity, **12**, 394
zinc carbenoids, **11**, 178, 179
Cyclopropane
donor–acceptor-substituted
preparation, **12**, 390
reactions with iron ions, **7**, 103
synthesis
cascade cyclization, **12**, 305–7
Cyclopropane, *gem*-dibromo-
ring opening carbonylation
nickel carbonyl induced, **9**, 4
Cyclopropane, (2,2-dimethyl-methylene)-
cycloaddition reactions, **12**, 950, 955
Cyclopropane, (1,1-diphenyl-methylene)-
cycloaddition reactions, **12**, 953
nickel phosphine catalysts, **12**, 953
Cyclopropane, (2,2-diphenyl-methylene)-
cycloaddition reactions, **12**, 953
Cyclopropane, isopropylidene-
cycloaddition reactions, **12**, 951
Cyclopropane, methylene-
allyl complexes, **12**, 846, 847
cycloaddition reactions
diastereoselectivity, **12**, 955
Pauson–Khand reaction, **12**, 714
substituted
cycloaddition reactions, **12**, 949–54
Cyclopropane, 1-methylene-2-vinyl-
cycloaddition reactions, **12**, 953
Cyclopropane, 2-methyl-methylene-
cycloaddition reactions, **12**, 950
palladium phosphine complex catalysts, **12**, 950
Cyclopropane, phenyl-
metallation, **11**, 113
Cyclopropane, 2-phenyl-methylene-
cycloaddition reactions, **12**, 953
Cyclopropane, siloxy-
transmetallation onto palladium, **12**, 228
Cyclopropane, silyloxy-
ring opening, **11**, 160
Cyclopropane, 2,2,3,3-tetramethyl-methylene-
cycloaddition reactions, **12**, 950
Cyclopropane, 2-trimethylsilyl-methylene-
cycloaddition reactions, **12**, 954
Cyclopropane, vinyl-
allyl complexes, **12**, 846, 847
reactions with organothallium complexes
palladium-assisted, **12**, 147
synthesis, **12**, 836
Cyclopropaneamino acids
synthesis, **12**, 399
Cyclopropanecarboxylates, 1-nitro-
synthesis, **12**, 399
Cyclopropanecarboxylic acid, 2,3-bis(diphenylphosphino)-1-methyl-
synthesis, **12**, 886
Cyclopropane-1-carboxylic acids, 1-amino-
synthesis, **12**, 399
Cyclopropaneethylidene-
cycloaddition reactions
palladium phosphine complex catalysts, **12**, 950
Cyclopropanes
activated
homoconjugate addition to, **12**, 352, 378
donor–acceptor-substituted
synthesis, **12**, 398
lithiation, **11**, 6
mercuration, **11**, 393
metallation, **11**, 113, 114
solvomercuriation, **3**, 151, 152
synthesis
carbomercuration, **11**, 427
organomercury compounds, **11**, 409
vinylic
ring opening, **12**, 140
Cyclopropanes, methylene-
cycloaddition reactions, **12**, 945–57
[3+2], **12**, 946
acceptors, **12**, 946–9
conditions, **12**, 945, 946
mechanisms, **12**, 945, 946
reagents, **12**, 945, 946
Cyclopropanone mixed ketals
synthesis, **12**, 72, 73
Cyclopropenation
carbene complexes, **12**, 388
metal carbenes, **12**, 412
Cyclopropene
zirconocene complexes
reaction with alkynes, **12**, 339, 340
Cyclopropene, 3,3-dimethyl-
cotrimerization with methyl propenoate, **12**, 766
insertion reactions
rhodium complexes, **8**, 218
1-Cyclopropene-3-carboxylates
ring opening, **12**, 412
Cyclopropenes
Pauson–Khand reaction, **12**, 715
Cyclopropenone
insertion reactions
rhodium complexes, **8**, 218
spirocyclic ketal
reaction with Gilman cuprates, **12**, 93
Cyclopropenone, diphenyl-
addition to isothiocyanates
nickel carbonyl promoted, **9**, 4
Cyclopropenones
synthesis
organomercury compounds, **11**, 410
Cyclosarkomycin
synthesis, **12**, 722
Cyclosilanes
anion radicals
ESR, **2**, 92–4
bonding, **2**, 94, 95
NMR, **2**, 94, 95
polymerization, **2**, 100
ring-opening, **2**, 100
separation, **2**, 79
strain energy, **2**, 94
structures, **2**, 81, 82
Cyclosiloxanes
copolymerization, **2**, 123, 124
synthesis, **2**, 33
Cyclosiloxanes, dimethyl-
anionic polymerization, **2**, 119–21
cationic polymerization, **2**, 121–3
acid activated clays, **2**, 121
condensation polymerization
catalysts, **2**, 125
emulsion polymerization, **2**, 123
Cyclotelluroetherification, **11**, 586
Cyclotetramers
from alkynes
reactions with iron atoms, **7**, 108
Cyclotetrasilane, tetramethyl-
synthesis, **2**, 79
Cyclotetrasilanes
synthesis, **2**, 23
Cyclotetrastannane, octa-*t*-butyl-
reactions, **2**, 282
Cyclotetrastibine
synthesis, **2**, 328
Cyclotriarsine
synthesis, **2**, 327

Cyclotriboroxane
  properties, **1**, 171
Cyclotrigermanes
  photodecomposition, **2**, 184
Cyclotrigermazanes
  synthesis, **2**, 161
Cyclotrimerization
  alkynes
    mechanism, **12**, 742
Cyclotrimers
  from alkynes
    reactions with iron atoms, **7**, 108
Cyclotrisilanes
  production, **2**, 79
  synthesis, **2**, 23
Cyclotristibine
  synthesis, **2**, 327
Cytochrome P-450
  mimics, **12**, 1099
Cytosines
  synthesis
    Stille couplings, **12**, 218

D

Damascenones
  synthesis, **12**, 100
Damascone
  synthesis, **11**, 102
Damascones
  synthesis, **12**, 100
Damsin
  precursor synthesis, **12**, 696
  synthesis, **12**, 413
Danishefsky's diene
  α-oxidation, **11**, 336
  reactions, **11**, 336
  sigmatropic rearrangement reactions, **11**, 336
Daunomycinone
  A-ring
    synthesis, **12**, 584
Daunomycinone, 4-demethoxy-
  synthesis, **12**, 491
Decaborates, dodecahydro-
  metallaboranes, **1**, 352–8
Decaborates, hexadecahydro-
  metallaboranes, **1**, 349
Decaborates, tetradecahydro-
  group 8 metals
    metallaboranes, **1**, 350
  group 9 metals
    metallaboranes, **1**, 351
  group 10 metals
    metallaboranes, **1**, 351, 352
  group 11 metals
    metallaboranes, **1**, 352
  group 12 metals
    metallaboranes, **1**, 352
  metallaboranes, **1**, 350–2
Decaphosphaferrocenes
  preparation, **7**, 197
Decarbonylation
  acyl halides, **12**, 376
    rhodium complexes, **8**, 199
  acylplatinum(II) complexes, **9**, 408
  aldehydes
    metal carbonyls, **12**, 373–6
  anhydrides, **12**, 377
  metal carbonyls, **12**, 373–7
Decarboxylation
  tin carboxylates, **2**, 233
2,7,9-Decatrienes, 1-acetoxy-
  cyclization
    allylpalladium catalysts, **12**, 913
1-Decene
  asymmetric dihydroxylation, **12**, 1151, 1154
Deemulsifiers
  silicones
    petroleum industry, **2**, 126
Defucogilvocarcin V, 1-*O*-methyl-
  synthesis, **12**, 223, 496
Dehalogenation
  organolanthanide(II) complexes, **4**, 126
Dehydrogenative silylation, **11**, 343
  anilines, **11**, 342
Demercuration
  solvomercuration, **11**, 411–30
Dendrobine
  synthesis, **12**, 331, 332, 734, 890
Dendrolasin
  synthesis, **12**, 368, 861
Deoxygenation
  organolanthanide(II) complexes, **4**, 126
Deoxygenative coupling
  carbonyl compounds, **12**, 1091
Deoxygrenolicin
  synthesis, **12**, 610
ipso-Desilylation
  arylsilanes, **11**, 326
Desmosterol
  synthesis, **11**, 101
Detergent industry
  silicones, **2**, 125, 126
Deuterium
  hydrogen equilibration
    catalysis, **10**, 37
Deuterium labeling
  hydrozirconation, **12**, 41
Dewar furan
  synthesis, **12**, 624
Dextromethorphan
  synthesis, **12**, 18
Diacetoxylation
  1,3-dienes
    allyl complexes, **12**, 823
Dialkenes
  cyclization
    Ziegler–Natta catalysts, **12**, 1206
  hydrocyclization
    Ziegler–Natta catalysts, **12**, 1207
Diamines
  chiral osmium complexes
    asymmetric dihydroxylation, **12**, 1139
  symmetric
    synthesis, **12**, 344
  synthesis
    organomercury compounds, **11**, 422
  vicinal
    synthesis, organoniobium complexes, **5**, 143
Diamond–square–diamond rearrangements
  metallaboranes, **1**, 277, 347
Diarsaboranes
  metal complexes, **1**, 266
  properties, **1**, 266
Diarsadodecaboranes
  properties, **1**, 266
Diarsaferrocenes
  preparation, **7**, 197
1,3,2-Diazaboracyclohexanes
  synthesis, **1**, 164
Diazaboranes
  properties, **1**, 261, 262
1*H*-1,3,2-Diazaborole, 2,3-dihydro-
  preparation, **1**, 161
1,3,2-Diazaborolidines
  synthesis, **1**, 161
Diazabutadiene
  complexes
    conformation, **5**, 217
Diazadecaboranes
  properties, **1**, 262
Diazadiboranaphthalenes
  synthesis, **1**, 164, 165
1,3,2,4-Diazadiboretidines
  metal complexes, **1**, 211
Diazadisilacyclohexadiene
  synthesis, **2**, 68
1,3-Diazadistannetane
  preparation, **2**, 282
1,1'-Diazaferrocenes
  preparation, **7**, 195
2,3-Diaza-5-norbornene
  Pauson–Khand reaction, **12**, 720
Diazapentaboranes
  synthesis, **1**, 160
1,3-Diaza-2-silacyclopentane
  reactions, **2**, 69
Diazepam
  synthesis, **9**, 245; **12**, 249, 250
Diazepinones
  synthesis, **12**, 551
Diazetidinones
  synthesis
    reaction of azo compounds with carbene complexes, **12**, 566
Diazines
  DMG containing
    lithiation, **11**, 79–81
Diaziridine
  synthesis
    tungsten alkoxycarbene complexes, **12**, 566
Diazo amides
  intramolecular insertion reactions
    catalytic, **12**, 456
  stabilities, **12**, 423
Diazo compounds
  decomposition
    acid promoted, **12**, 423, 424
    catalysts, **12**, 422
    transition metal catalyzed, **12**, 424–6
  heteroatom ylides from, **11**, 490
  stabilities, **12**, 423–6
  synthesis
    organoboranes, **11**, 222

Diazo esters
insertion reactions
catalytic intramolecular, **12**, 455
stabilities, **12**, 423
Diazo ketones
intramolecular insertion reactions
catalytic, **12**, 452–4
stabilities, **12**, 423
Diazocarbonyl compounds
cyclopropanation, **12**, 398–402
Diazodecomposition
cyclopropanation, **12**, 388
metal carbene complexes, **12**, 421–64
transition metal catalysts, **12**, 426–30
Diazomethane
cyclopropanation, **12**, 396, 397
reactions with iron atoms, **7**, 108
1,4-Diazonia-2,5-diboratacyclohexanes
synthesis, **1**, 166
Diazonium salts
Heck reactions, **12**, 263
Dibenzoboroles
synthesis, **1**, 156
Dibenzo[a,c]ferrocene
preparation, **7**, 200
5*H*-Dibenzo[b,f]silepinium ions, 5-methyl-
preparation, **2**, 58
Dibenzothiophene
rhenium complexes, **6**, 218
Dibismadodecaboranes
properties, **1**, 267
1,1'-Dibismaferrocene, 2,2',5,5'-tetramethyl-, **7**, 197
9,10-Diboraanthracene, 9,10-dihydro-, **1**, 210
1,2-Diborabenzene
chemistry, **1**, 153
1,3-Diborabenzene
metal complexes, **1**, 209
1,4-Diborabenzene
metal complexes, **1**, 209
2,6-Diborabicyclo[3.1.0]hex-3-ene
synthesis, **1**, 154
1,2-Dibora-3,5-cyclohexadienes
chemistry, **1**, 153
metal complexes, **1**, 209
Diboracyclohexene
dimerization, **1**, 145
1,4-Dibora-2-cyclohexenes
dehydrogenative complex formation, **1**, 209
Diboracyclohexenes
preparation, **1**, 153
1,3-Dibora-2,5-disilacyclohexane
synthesis, **2**, 70
Diborafulvenes
metal complexes, **1**, 199, 208
1,4-Diboranaphthalene
metal complexes, **1**, 210
Diborane
hydroboration, **11**, 193–6
Diborane(2)
metallaboranes, **1**, 312
Diboranes(4)
metallaboranes, **1**, 311
1,2-Diboratabenzene
metal complexes, **1**, 209
Diborates, heptahydro-
metallaboranes, **1**, 304
Diborates, hexahydro-
group 5 metals
metallaboranes, **1**, 306
group 8 metals
metallaboranes, **1**, 306, 307
metallaboranes, **1**, 306, 307
Diborates, pentahydro-
group 6 metals
metallaboranes, **1**, 308
group 8 metals
metallaboranes, **1**, 308
group 9 metals
metallaboranes, **1**, 309
group 10 metals
metallaboranes, **1**, 310
metallaboranes, **1**, 307–10
1,3-Diboretane
preparation, **1**, 135
Diborete
stabilities, **1**, 150
1,2-Diborete, 1,2-dihydro-
preparation, **1**, 150
1,3-Diborete, 1,3-dihydro-
preparation, **1**, 150
Diboretene
structure, **1**, 131
Diboriranide
preparation, **1**, 150
1,2-Diborolane, 4-methylene-
metal complexes, **1**, 204, 205
1*H*-1,3-Diborole, 2,3-dihydro-
metal complexes, **1**, 205
1,3-Diborolenes
metal complexes, **1**, 205
1,3-Diborolenyl ligands
metal complexes, **1**, 198, 205
1,3-Diboroles
preparation, **1**, 152
1*H*-1,3-Diborolyl, 2,3-dihydro-
metal complexes, **1**, 205
Dibromides
debromination
tellurium reagents, **11**, 579, 592
vicinal
elimination reactions, alkenes from, **11**, 491
Dicarbide complexes
hexanuclear clusters, **7**, 272
Dicarbonyl compounds
alkylation, **12**, 844
1,2-Dicarbonyl compounds
aryl
coupling reactions with alkynes, **12**, 1089
1,3-Dicarbonyl compounds
carbon-arylation, **11**, 507
α,ω-Dicarbonyl compounds
cyclization, **12**, 1077
Dicarbonylation
metal carbonyls, **12**, 368
1,4-Dicarbonyls
synthesis, **12**, 228
Dicobalt
dicarbon metallacarbaboranes
7-vertex, **1**, 382
9-vertex, **1**, 396
Dicobalt complexes
bridged, **8**, 93–8
hydride bridged, **8**, 93
Dicobalt diiridium carbonyls, **8**, 510
Dicobalt heptacarbonyl
hydrogenation, **8**, 30
Dicobalt octacarbonyl, **8**, 90–2
carbonyl exchange, **8**, 91
chemisorbed, **8**, 92
cobalt tetracarbonyl synthesis from, **8**, 6
Co–Co bond energy, **8**, 30
derivatives
charge transfer synthesis, **8**, 91
disproportionation, **8**, 31
phosphine derivatives
structure, **8**, 92
reactions
base assisted, **8**, 91
structure, **8**, 90
substitution reactions, **8**, 91
synthesis, **8**, 91
trimerization
alkynes, **12**, 747
Dicobalt tetrarhodium carbonyls, **8**, 510
Dicranenones
synthesis, **12**, 453
Dictymene, epoxy-
synthesis, **12**, 590
Dicytopterene A
synthesis, **12**, 889
Dicytopterene C
synthesis, **12**, 889
Diels–Alder cycloaddition
(diene)tricarbonyliron, **12**, 627
Diels–Alder reaction
asymmetric aza-, **11**, 262
hetero-, **11**, 262
titanium complexes
catalysts, **4**, 275
zirconium catalyzed, **4**, 622
Diene aldehydes
allylboration, **11**, 255
Dienes
acetoxy-
nickel catalyzed cyclization, **12**, 910
acyl
synthesis, **12**, 354
addition compounds
magnesium, **1**, 97, 98
alkoxymercuration, **11**, 416
amidomercuration, **11**, 425
aminomercuration, **11**, 422
carbonylation, **12**, 360
complexation
protection, **12**, 623–34
conjugated acyclic
Pauson–Khand reaction, **12**, 713
cyclic
cyclopropanation, **12**, 400
cyclization
tin radical-initiated, **11**, 362
transition metal catalyzed, **12**, 791, 792
zirconocene, **12**, 329
cycloaddition
catalytic metal catalyzed, **12**, 793
cyclopropane synthesis
cascade cyclization, **12**, 305, 306

diacetoxylation
  organotellurium
    compounds, **11**, 584
diiodo-
  synthesis, **12**, 335
dimerization
  C=Y trapping reagents, **12**, 790
  transition metal allyl complexes, **12**, 785
dimerization with hydridic trapping reagents, **12**, 790
dimetallation, **11**, 105
electron-rich
  synthesis, **12**, 186
fused ring synthesis
  cascade cyclization, **12**, 303
hydroboration, **11**, 196
hydrogermylation, **2**, 154
hydroxymercuration, **11**, 412
insertion reactions
  carbene complexes, **12**, 533
isomerization
  catalysis, **10**, 38
ligand liberation in iron complexes
  nucleophilic addition, **12**, 623
  oxidation, **12**, 623
  photochemical reduction, **12**, 623
metallation, **11**, 103–5, 110–14
oligomerization
  linear, **12**, 742, 743
  palladium catalysts, **9**, 371–3
polyfluoro-
  preparation, **12**, 188
polymerization
  transition metal catalyzed, **12**, 1225
protection
  complexation to iron, **12**, 626–9
reaction with acyltetracarbonyl cobalt complexes, **12**, 354
reaction with titanium complexes, **4**, 274
reaction with transition metal complexes, **12**, 138
reactions with organothallium complexes
  palladium-assisted, **12**, 147
reactions with palladium salts, **9**, 331–5, 340
reductive dimerization, **12**, 788
silyl-
  preparation, **11**, 314
solvomercuration, **11**, 427
spirocyclizations
  multiple insertion cascades, **12**, 300, 301
synthesis, **12**, 335
  alkenylboronic acids in, **12**, 149
  vinylsilanes, **12**, 153
telomerization
  enantioselective, **12**, 791
  palladium catalysts, **9**, 371–3
  palladium-catalyzed,
    mechanism, **12**, 786
transition metal complexes
  nucleophilic attack, **12**, 6, 637–81
  organic synthesis, **12**, 637–81
  preparation of optically active, **12**, 629
  protection, **12**, 623–34
  stability of optically active, **12**, 629
trapping of oxallyl cation, **12**, 350
unstable
  protection, **12**, 624
1,2-Dienes
  reactions with palladium salts, **9**, 341
1,3-Dienes
  acetoxychlorination, **12**, 821
  1,4-acetoxychlorination, **12**, 880
  allyl complexes, **12**, 820
  allyl complexes from, **12**, 815
  allylic alkylation, **12**, 847, 848
  amidopalladation, **12**, 826
  arylation
    palladium-assisted, **12**, 151
  cyclizations
    allylpalladium catalysts, **12**, 912
  cycloaddition reactions, **12**, 928
    [3+2], **12**, 948
  cyclopropanation, **12**, 390
  diacetoxylation
    allyl complexes, **12**, 823
  1,4-difunctionalization
    allylic alkylation, **12**, 835
  exocyclic
    synthesis, **12**, 327
  Heck reactions, **12**, 264
  hydrosulfinylation
    palladium catalysts, **12**, 829
  monoepoxides
    amidation with potassium phthalimide, **12**, 810
  oxidation
    allyl complexes, **12**, 807, 808
  palladium halide addition
    allyl complexes, **12**, 808, 809
  stereoselective synthesis
    from propargylsilane, **12**, 43
  sulfonylpalladation, **9**, 336
  synthesis
    organomercury compounds, **11**, 401
    zirconium mediated, **4**, 621
1,4-Dienes
  cyclic
    synthesis, **12**, 906
  vinylpalladation, **12**, 809
α,ω-Dienes
  cyclization
    organoniobium complexes, **5**, 143
Dienol
  transition metal complexes
    substitution, **12**, 632, 633
Dienyl acetate
  cyclization
    palladium-catalyzed, **12**, 909
  cyclization/elimination
    allylmetal catalysts, **12**, 906
  transition metal complexes
    substitution, **12**, 633
Dienyl complexes
  transition metals
    nucleophilic attack, **12**, 6, 637–81
    organic synthesis, **12**, 637–81
    protection, **12**, 623–34
Diferra–allyl clusters, **7**, 278, 280
Diferrocenes
  mixed valence properties, **7**, 192–4
2,5-Digermaazolidines
  preparation, **2**, 160
1,3-Digermacyclobutanes
  preparation, **2**, 140
Digermacyclobutenes
  substituted
    formation, **2**, 141
Digermadiphosphorinane
  preparation, **2**, 160
Digermaphospholanes
  preparation, **2**, 160
Digermenes
  chemistry, **2**, 201, 202
Digermirane
  synthesis, **2**, 140
Digermiridine
  preparation, **2**, 159
5,6-DiHETE
  synthesis, **12**, 627
11,12-DiHETE
  synthesis, **12**, 627
Dihydrogen complexes
  group 6, **5**, 223–5
    agostic interactions, **5**, 225
    neutron structures, **5**, 225
    rotational tunneling, **5**, 224
Dihydroxylation
  alkenes
    asymmetric, **12**, 1099
  asymmetric
    catalytic, **12**, 1141–5
    double diastereoselection, **12**, 1164
    enantioselectivity, **12**, 1152–8
    kinetic resolution, **12**, 1167
    mechanism, **12**, 1152–6
    organic synthesis, **12**, 1168–73
    stoichiometric, **12**, 1138–40
    substrates, **12**, 1158–67
  osmium tetroxide, **12**, 1141
Diimines
  reaction with dialkylzincs, **11**, 171
Diiridacyclobutene complexes
  formation, **8**, 395, 396
Diiridium complexes
  bridging ligands, **8**, 329
Diiron nonacarbonyl
  chemistry, **7**, 4
  cyclobutadiene synthesis, **12**, 745
  oxallyl cations, **12**, 350

Diiron octacarbonyl
  anion
    chemistry, **7**, 5
Diketene
  coupling reactions with
    Grignard reagents, **12**, 171
Diketones
  mercuriation, **3**, 140
1,3-Diketones
  asymmetric reduction, **12**, 24
β-Diketones
  reactions with organolead
    compounds, **11**, 471–4
Diltiazem
  synthesis, **12**, 1169
Dimagnesium organometallics
  equilibrium, **1**, 76
Dimamp
  dopa synthesis, **12**, 14, 16
Dimanganese decacarbonyl
  aldehyde derivatives, **6**, 8
  bond energies, **6**, 4
  bonding model, **6**, 3
  chemical properties, **6**, 5, 6
  chemistry, **6**, 2–10
  derivatives, **6**, 7–9
  mass spectrometry, **6**, 4
  photolysis, **6**, 3
  physical properties, **6**, 2–4
  reaction with depe, **6**, 8
  reaction with dppm, **6**, 9
  sulfur ligands, **6**, 8
  thiocarbonyl derivatives, **6**, 7
  trifluorophosphine derivatives,
    6, 8
Dimedone
  diarylation
    organolead compounds, **11**, 471
Dimetallamacrocycles
  polydentate ligands, **10**, 18
Dimethyl benzylidenemalonate
  cycloaddition reactions
    [3+2], **12**, 934
Dimethyl ether
  reactions with iron anions, **7**, 107
Dimethyl selenide *N*-chloro-
    succinimide
  reactions, **11**, 533
α,ω-Dinitriles
  cyclooligomerization, **12**, 756
Dinitrogen
  fixation
    vanadium(III) complexes,
      5, 13
Diols
  activation
    asymmetric dihydroxy-
      lation, **12**, 1168
  synthesis
    asymmetric dihydroxy-
      lation, **12**, 1168
Diorganoaluminum(II) complexes
  synthesis, **1**, 446
Diorganoantimony(V) compounds
  synthesis, **2**, 342, 343
Diorganoantimony halides
  synthesis, **2**, 332
Diorganoarsenic(V) compounds
  synthesis, **2**, 341, 342
Diorganoarsenic halides
  synthesis, **2**, 332
Diorganobismuth halides
  synthesis, **2**, 332
Diorganocadmium compounds, **3**,
    196, 197
Diorganogold(III) halides
  synthesis, **3**, 33, 34
Diorganogold(III) pseudohalides
  synthesis, **3**, 33, 34
Diorganomercurials
  preparation of organozinc
    compounds, **11**, 164
Diorganothallium compounds
  in organic synthesis, **11**, 444–6
Diorganozinc compounds, **3**, 176,
    177
Diorganylgold(III) complexes
  group 15 donor ligands, **3**,
    36–9
  group 16 donor ligands, **3**,
    34–6
Diosmacyclopropane
  synthesis, **7**, 300
Diosmium decacarbonyl
  synthesis, **7**, 633–5
Diosmium nonacarbonyl
  ligand stabilized, **7**, 626–33
Diosmium octacarbonyl
  synthesis, **7**, 625
2,9-Dioxabicyclo[3.3.1]nonane,
    1,3-dimethyl-
  synthesis, **12**, 453
1,3,2,5-Dioxaboraphosphorinanes
  properties, **1**, 184
1,3,2-Dioxaborinanes
  properties, **1**, 176
  synthesis, **1**, 175
Dioxaborolanes
  preparation, **1**, 173
1,3,2-Dioxaboroles
  preparation, **1**, 172
1,3-Dioxa-2-silacyclohexane, 4,4-
    dimethyl-
  conformation, **2**, 62
Dioxasilacyclopentanes
  synthesis, **2**, 59
Dioxastannolanes
  structures, **2**, 295
1,3,2-Dioxastannolanes
  derived from carbohydrates, **2**,
    268
1,3,2-Dioxastannolenes
  in organic synthesis, **2**, 267
  preparation, **2**, 267
  reactions, **2**, 268, 269
  structures, **2**, 268
Dioxolanes
  preparation, **12**, 447
1,3-Dioxolanes, 2-trimethyl-
    silyloxy)-, **11**, 330
1,3-Dioxoles
  preparation, **12**, 447
Dipalladation
  tetraethylxylenediamines, **9**,
    245
Dipeptides
  synthesis, **12**, 569
Diphenylselenium
    bis(trifluoroacetate)
  reactions, **11**, 533
Diphosphaboranes
  properties, **1**, 264
Diphosphaborirane
  properties, **1**, 169
1,3,2,4-Diphosphadiboretane
  metal complexes, **1**, 211, 212
Diphosphadistannetanes
  isolation, **2**, 291
1,1'-Diphosphaferrocene, 3,3',4,4'-
    tetramethyl-
  crystal structure, **7**, 196
Diphosphahexaboranes
  properties, **1**, 264
Diphosphaundecaboranes
  properties, **1**, 264
Diphosphenes
  chromium tricarbonyl
    complexes, **5**, 490
Diphosphine, stannyl-
  preparation, **2**, 291
Diplatinum
  dicarbon metallacarbaboranes
    9-vertex, **1**, 396
1,3-Dipolar cycloaddition
  (diene)tricarbonyliron, **12**, 627
Dipropargyl
  cyclopolymerization
    Ziegler–Natta catalysts, **12**,
      1226
Diquinane
  synthesis, **12**, 790
Diquinanes
  synthesis, **12**, 723
Directed metallation groups
  heteroatom-based
    lithiation, **11**, 48–55
Directed *ortho*-metallation
    reactions, **11**, 72–5
  heteroaromatics
    lithiation, **11**, 70–81
Dirhenium decacarbonyl, **6**,
    168–73
  oxidation, **6**, 171
  reactions with alkenes, **6**, 171
  reactions with alkynes, **6**,
    171–3
  substitution reactions, **6**,
    169–71
Dirhodium
  dicarbon metallacarbaboranes
    7-vertex, **1**, 382
Dirhodium carboxamides
  catalysts
    diazodecomposition, **12**,
      428
Dirhodium diiridium carbonyls, **8**,
    510
Dirhodium(II) carboxamides
  catalysts
    cyclopropanation, **12**, 408,
      409
Dirhodium(II) carboxylates
  catalysts
    cyclopropanation, **12**, 408,
      409
    diazodecomposition, **12**,
      428
Dirhodium(II) complexes
  catalysts
    cyclopropanation, **12**,
      397–415
Dirhodium(II) tetraacetate
  catalysts
    diazodecomposition, **12**,
      422
    cyclopropanation, **12**, 400

Diruthenium
  dicarbon metallacarbaboranes
    7-vertex, **1**, 382
Diruthenium decacarbonyl
  synthesis, **7**, 633–5
Diruthenium nonacarbonyl
  ligand stabilized, **7**, 626–33
Diruthenium octacarbonyl
  synthesis, **7**, 625
Diselenadiboretane
  preparation, **1**, 171
Diselenaundecaboranes
  properties, **1**, 270
Diselenides
  reactions, **11**, 533
Disila-1-azetine
  synthesis, **2**, 68
1,4-Disilabarrelene
  synthesis, **2**, 64
Disilabenzene
  matrix isolation, **2**, 61
1,2-Disilacyclobutane
  synthesis, **2**, 17, 47
Disilacyclobutanes
  properties, **2**, 51–3
Disilacyclobutenes
  properties, **2**, 53, 54
Disilacyclodeca-3,8-dienes
  preparation, **2**, 64
Disilacycloheptenes
  synthesis, **2**, 63
Disilacyclohexadiene
  synthesis, **2**, 48
Disilacyclohexenes
  synthesis, **2**, 61
Disilacyclooctadiene
  synthesis, **2**, 53, 54
Disilacyclopentanes
  synthesis, **2**, 59
1,3-Disilacyclopentene
  synthesis, **2**, 18
Disilacyclopentenes
  synthesis, **2**, 59
Disilacyclopropanes, **2**, 48
Disilacyclopropanimine
  synthesis, **2**, 48
1,4-Disila(Dewar benzene)
  synthesis, **2**, 64, 65
1,2-Disila-*closo*-dodecaborane, 1,2-dimethyl-
  preparation, **1**, 259
Disilahexadienes
  synthesis, **2**, 61
Disilane, hexamethyl-
  rotation barriers, **2**, 79
Disilazanes
  synthesis, **2**, 25
Disilenes
  polymerization, **2**, 98, 99
  synthesis, **2**, 25
Disiliranes — *see* Disilacyclopropanes
Disiloxanols
  condensation polymerization, **2**, 124
Disparlure
  synthesis, **12**, 1168, 1173, 1174
    lithiation, **11**, 33
1,3-Distannacyclobutane
  structure, **2**, 250
Distannacyclodecane
  preparation, **2**, 252
Distannane, 1,2-dicarboxylato-
  preparation, **2**, 281
Distannane, hexaaryl-
  preparation, **2**, 278
  properties, **2**, 290
Distannane, hexabenzyl-
  preparation, **2**, 278
Distannanes
  from organotin formates, **2**, 277
  palladium catalyzed reactions, **2**, 280
  reactions with alkyllithium, **2**, 280
Distannoxane, tetraalkyl-
  in organic synthesis, **2**, 264
  NMR, **2**, 263
  structure, **2**, 263
Distannoxane, tetrabutyldichloro-
  oligomerization with dibutyltin dichloride, **2**, 265
Distannoxane, tetramethyldiacetato-
  structure, **2**, 264
Distibaferrocenes
  preparation, **7**, 197
Ditechnetium decacarbonyl, **6**, 152
Diterpenoid quinones
  synthesis, **12**, 494
Dithiaboracyclohexanes
  spectra, **1**, 181
Dithiaboranes
  properties, **1**, 269
Dithiaborinanes
  isomerization, **1**, 181
1,3-Dithia-2-borolane, 2-alkyl-
  transmetallation, **11**, 220
Dithiacyclohexane
  reaction with triosmium clusters, **7**, 810, 812
1,3-Dithiane
  rhenium pentacarbonyl complexes
    preparation, **6**, 218
1,4-Dithiane
  rhenium pentacarbonyl complexes
    preparation, **6**, 218
1,3-Dithiane, 2-lithio-2-methyl-
  reaction with (chlorobenzene)tricarbonylchromium, **12**, 988
1,3-Dithiane, 2-methyl-2-phenyl-
  ylides, **12**, 436
1,3-Dithiane, 2-phenyl-
  ylides, **12**, 436
Dithianes
  benzylic cross-coupling
    transition metal-promoted, **12**, 166
Dithiastannolanes
  structures, **2**, 295
Dithioacetals
  cross-coupling with monothiophthalimide, **12**, 1092
  reductive coupling, **12**, 1092
2,2'-Dithiophene
  formation, **12**, 79
Ditin, hexamethyl-
  radical cation
    properties, **2**, 290–3
Diynes
  cyclization
    zirconocene, **12**, 327
  oligomerization
    linear, **12**, 742, 743
  reductive coupling
    titanocene, **12**, 327
  reductive cyclization
    palladium catalyzed, **12**, 54
1,4-Diynes
  synthesis
    dicobalt propargylium complexes, **12**, 693
1,5-Diynes
  synthesis
    dicobalt propargylium complexes, **12**, 693
1,6-Diynes
  cyclooligomerization, **12**, 748
Dodecahedrane
  synthesis, **12**, 736
Dodecatriene
  synthesis, **12**, 787
3-Dodecen-12-olide
  synthesis, **11**, 100
β-Dolabrin
  synthesis, **12**, 210
Dolichodiol
  synthesis, **12**, 879
Dolichols
  synthesis, **12**, 21, 22
L-Dopa
  asymmetric synthesis, **12**, 16
Dötz annelation
  chromium tricarbonyl phosphabenzene complexes, **5**, 531
Dynemicin
  structure, **12**, 697
Dynemicin A
  synthesis
    lithiation, **11**, 79

E

Early transition metal effect, **11**, 281
Eburnamonine
  synthesis, **12**, 398
Ecgonine methyl ester, anhydro-
  synthesis, **12**, 401
Echinosporin
  synthesis, **12**, 245, 246
β-Effect
  tin chemistry, **11**, 357
Effective atomic number rule
  ruthenium clusters, **7**, 909
Egomaketone
  synthesis, **12**, 368, 861
Egostane
  synthesis, **12**, 871
EHMO calculations
  alkyne iron clusters, **7**, 259
Eicosatetraenoic acid
  synthesis
    lithiation, **11**, 33
Eisch–Piotrowski reagent
  stoichiometric reagent, **11**, 296
Elaeokanine
  synthesis, **11**, 144
Elaeokanine B
  synthesis
    lithiation, **11**, 33
Elastomers
  liquid silicones, **2**, 127, 128
Eldanolide
  synthesis, **12**, 506
18-Electron rule
  iron complexes, **7**, 102
  oxidation state, **12**, 2
19-Electron complexes
  iron cyclopentadienyl, **7**, 199
Electrooptics
  liquid crystalline polysiloxanes, **2**, 131
Electrophilic attack
  transition metal complexes, **12**, 6
Ellagic acid
  synthesis, **12**, 199
Ellipticine
  synthesis, **12**, 193
Ellipticine alkaloids
  synthesis
    lithiation, **11**, 74
Enamides
  hydrogenation
    asymmetric, **12**, 17, 18
    enantioselective, **12**, 16
Enamines
  allylic alkylation, **12**, 867
  conversion to amines or imines
    organoaluminum hydrides, **11**, 300
  in cyclopentenone formation, **12**, 350
  dicobalt cationic complexes
    iminium ions, **12**, 692
  Heck reactions, **12**, 277
  synthesis, **12**, 585, 590
Enantiomer recognition
  organozinc compounds, **3**, 192, 193
1,4-Endoxines
  deoxygenation
    titanium complexes, **4**, 276
Ene ethers
  metallation, **11**, 116
Ene reactions
  magnesium, **11**, 131
  silyl ethers, **11**, 330
Ene thioethers
  metallation, **11**, 116
Enediynes
  antibiotics
    synthesis, **12**, 222
  13-keto-synthesis
    dicobalt propargylium complexes, **12**, 698
  polycyclization
    palladium catalyzed, **12**, 54, 55
1,5-Eneynes, 3,4-dioxy-
  synthesis, **12**, 693
Enisoprost
  synthesis, **12**, 45
Enol acetates
  reactions with organomercurials, **12**, 142
  reactions with organothallium complexes
    palladium-assisted, **12**, 147
Enol ester lactones
  synthesis, **12**, 248
Enol ethers
  allylic alkylation, **12**, 865–7
  asymmetric dihydroxylation, **12**, 1150
  cyclic
    arylation, **12**, 133, 150
    synthesis, **12**, 1080
  displacement reactions, **12**, 142
  Heck reactions, **12**, 276
  silylated
    synthesis, **11**, 499
Enol fluorosulfonates
  coupling reactions, **12**, 207
Enol mesylates
  coupling reactions, **12**, 207
Enol phosphates
  coupling reactions, **12**, 190
Enol stannane
  allylic alkylation, **12**, 866
Enol triflates
  carbonylative coupling with stannanes, **12**, 206
  coupling reactions, **12**, 190
  cyanation, **12**, 225
  Sonogashira coupling, **12**, 223
  Stille reaction, **12**, 206
Enolates
  coupling
    organopalladium compounds, **12**, 163
  coupling with ω-stannyl alkenes, **12**, 185
  metathesis, **12**, 227
  tin
    in organic synthesis, **11**, 368
Enols
  hydrogenation
    asymmetric, **12**, 18
    enantioselective, **12**, 16
Enone, β-chloro-
  coupling reactions
    with vinyl boronates, **12**, 182
Enones
  bicyclic
    synthesis, **12**, 327, 328
  cycloaddition reactions, **12**, 927
α-Enones
  synthesis
    trialkylstibine, **11**, 489
α,β-Enones
  (tricarbonyl)iron complexes
    preparation, **12**, 633
Enones, β-chloro-
  coupling reactions, **12**, 185
Enones, α-iodo-
  coupling reactions
    organozinc compounds, **12**, 187
Enoxysilacyclobutanes
  reactions with aldehydes, **11**, 333
Enynes
  acyclic all-carbon
    Pauson–Khand reaction, **12**, 723–8
  acyclic heteroatom-containing
    Pauson–Khand reaction, **12**, 728–31
  cyclic
    Pauson–Khand reaction, **12**, 731–4
  cyclization
    tin radical-initiated, **11**, 362
    zirconocene, **12**, 326
  cycloisomerization
    hydropalladation, **12**, 47–52
    palladium catalyzed, **12**, 49
  cyclopropane synthesis
    cascade cyclization, **12**, 306, 307
  fused ring synthesis
    cascade cyclization, **12**, 303
  hydroxyacylation, **12**, 690
  methoxyacylation, **12**, 690
  reductive coupling
    zirconocene, **12**, 326
  reductive cyclization

palladium catalyzed, **12**, 52, 53
spirocyclizations
multiple insertion cascades, **12**, 302
1,3-Enynes
dicobalt cationic complexes
reaction with trimethylsilyl enol ethers, **12**, 692
monohydroboration, **11**, 201
oligomerization
linear, **12**, 742, 743
reaction with dicarbonyl manganese complexes, **12**, 694
1,5-Enynes
synthesis
propargyl dicobalt complexes, **12**, 692
1,6-Enynes
synthesis
dicobalt propargylium complexes, **12**, 699
α,ω-Enynes
cycloaddition, **12**, 760
synthesis
dicobalt propargylium complexes, **12**, 698
Enynoates
conjugated
synthesis, **12**, 69
Enzymes
in resolution of chiral chromium tricarbonyl arene complexes, **5**, 505
Ephedrine
synthesis, **12**, 1045, 1048
chromium complexes, **12**, 1049
Ephrine, phenyl-
synthesis, **12**, 23
13-Epiforskolin, 6,7-dideoxy-
synthesis, **12**, 108
Epinephrine
synthesis, **12**, 23
Episantalene
synthesis, **12**, 579
Epiverbenalol
synthesis, **12**, 632
Epoxidation
alkenes
catalysts, **10**, 50
allylic alcohols
(diene)tricarbonyliron, **12**, 627
asymmetric
metal catalyzed, organic synthesis, **12**, 1097–132
Mukaiyama condition, **12**, 1124
oxo-based asymmetric catalysts, **12**, 1120–31
peroxo-based, **12**, 1108–20
salen manganese complexes
mechanism, **12**, 1124, 1125
methods, **12**, 1123, 1124
unfunctionalized alkenes, **12**, 1108
Epoxides
carbonylation, **12**, 371
conversion to alcohols or aldehydes
organoaluminum hydrides, **11**, 298
formation
from diols, **12**, 1173
hydrogenation, **12**, 10, 11
opening
metal catalyzed, **12**, 1098
reaction with metal carbonyl anions, **12**, 354
reaction with silyl cobaltates, **12**, 354
reactions with
organoaluminum complexes, **1**, 469, 487–90
reductive cleavage
organolithium reagents, **11**, 12
synthesis, **12**, 202
trans
from *cis*-alkenes, **12**, 1129
unsaturated
reactions with titanium complexes, **4**, 275
Equilibrium step growth polymerization
transition-metal catalysts, **12**, 1229
Ergoline
C-ring
synthesis, **12**, 890
synthesis, **12**, 880
Erythronolide A, dihydro-
synthesis, **11**, 255
ESCA
iron carbide clusters, **7**, 268
Esperamicin
synthesis, **12**, 698
Esters
α-azido-β-hydroxy
synthesis, **12**, 1169
enol ethers from
titanocyclobutane complexes, **4**, 277
α-iodo-
coupling reactions, organozinc compounds, **12**, 187
α-keto
asymmetric reductions, **12**, 23
β-keto
α-acylamino, dynamic kinetic resolution, **12**, 26
asymmetric reduction, **12**, 23
α-substituted, dynamic kinetic resolution, **12**, 25, 26
methylenation, **12**, 584
Tebbe's reagent, **12**, 588–90
reaction with α,β-unsaturated imines
niobium complexes, **12**, 1089
reactions with α-stannyl carbanions, **11**, 366
γ-substituted α,β-unsaturated
asymmetric dihydroxylation, **12**, 1166
synthesis
acyl iron complexes, **12**, 353
hydridocarboxylation, **12**, 355
insertion reactions, **12**, 242–8
metal carbonyl anions, **12**, 354
nitrosotricarbonylferrates, **12**, 354
unsaturated
synthesis, **12**, 788
α,β-unsaturated
alkenation, trialkylstibine, **11**, 489
asymmetric hydrogenation, **12**, 19–21
cycloaddition reactions, **12**, 926
[3+2] cycloaddition reactions, **12**, 946, 947
cyclopropanation, **12**, 390
methylenation, **12**, 585
β,γ-unsaturated hydrogenation
diastereoselective, **12**, 13
Esters, α-halo-
anions
preparation, organoboranes, **11**, 222
Estradiol, 17-ethynyl-
synthesis, **12**, 694
17-Estradiol
cobalt carbonyl-labeled, **12**, 700
17-Estradiols, 16-substituted
synthesis, **12**, 691
Estrone
synthesis, **12**, 750
Estrone methyl ether
synthesis, **12**, 459
Ethane
reactions with iron ions, **7**, 103, 107
Ethane, bis(diphenylphosphino)-
nickel complexes
catalysts, cross-coupling reactions, **12**, 165
Ethane, 1,2-diamino-1,2-diphenyl-
synthesis
titanium complexes, **12**, 1083
Ethane, 1,2-dibromo-
for zinc activation, **11**, 160
Ethanol, 2-bromo-2-phenyl-
preparation, **12**, 248
Ethanol, 1-(*o*-bromophenyl)-
production, **12**, 24
Ethanol, 2-methoxy-
O-phenylation, **11**, 509
Ethanol, 2-phenoxy-
O-phenylation, **11**, 509
Ethene
adsorption onto palladium, **9**, 293
dimerization
nickel catalysts, **9**, 40
Pauson–Khand reaction, **12**, 712
polymerization
Ziegler–Natta catalysis, **12**, 1194
propene block copolymer
synthesis, **12**, 1205

reactions with iron atoms, **7**, 108
reactions with iron ions, **7**, 104, 107

Ethene oxide
reactions with iron ions, **7**, 106

Ethene, 1,1-difluoro-
reaction with phenyl isocyanate, **9**, 8

Ethene, tetrakis(trimethylsilylethynyl)-
synthesis, **2**, 10

Ethene, tetrakis(trimethylsilylmethyl)-
structure, **2**, 9

Ethene, 1-trimethylsilyl-1-bromo-
coupling reactions, **12**, 184

Ethene, tris(trimethylstannyl)-
preparation, **2**, 280

Ethers
aryl
asymmetric dihydroxylation with osmium tetroxide, **12**, 1146
synthesis, **12**, 985
lithiation, **11**, 21
reactions with organoaluminum compounds, **1**, 491
silyl enol
in organic synthesis, **11**, 333–8
preparation, **11**, 333
synthesis from acetals, ketals or orthoformates
organoaluminum hydrides, **11**, 300
synthesis from aluminum vinyls, **11**, 303
thioenol
vinylic substitution, **12**, 142

Ethers, alkyl silyl, **11**, 329, 330

Ethyl acrylate
ethene block copolymers
synthesis, **12**, 1205
hydroformylation, **12**, 30

Ethyl bromoacetate
cross-coupling reactions, **12**, 181

Ethyl cyanoacetate
arylation, **12**, 226

Ethyl diazoacetate
decomposition
catalysts, **12**, 422
stability, **12**, 423

Ethyl oleate
metathesis, **12**, 1213

Ethyl propiolate
carbozincation, **11**, 176

Ethylene glycol
oxidation
iridium catalysts, **8**, 304

Ethylene, ethylsulfonyl-1-nitro-
reaction with zinc–copper reagents, **11**, 166

Ethylenediamine, benzyldimethyl-
dilithiation, **1**, 17

Ethylenediamine, *N*-lithio-
in organic synthesis, **11**, 5

Ethylenediamine, *N*-lithio-*N,N',N'*-trimethylene-
in organic synthesis, **11**, 5

Ethyne
carbalumination, **11**, 283
coupling reactions
aryl-1,2-dicarbonyl compounds, **12**, 1089
hydrogenation
organoaluminum reagents, **1**, 484
oligomerization
organoaluminum reagents, **1**, 484
polymerization
nickel ylide catalysts, **9**, 36
transition-metal catalyzed, **12**, 1226
reactions with iron atoms, **7**, 108
reactions with iron ions, **7**, 104
reactions with organoaluminum compounds, **1**, 495, 496
trimerization
cobalt assisted, **8**, 89
cobalt complexes, **8**, 89

Ethyne, bis(trimethylsilyl)-
cotrimerization with diynes, **12**, 749

Ethyne, bis(trimethylstannyl)-
trimerization
arenes from, **12**, 749

Ethyne, phenyl-
oligomerization
nickel catalyst, **9**, 30
1,2,4-triphenylbenzene from nickel catalysts, **12**, 747

Eupolauramine
synthesis, **12**, 198

Europium
dicarbon metallacarbaboranes
12-vertex, **1**, 418

Exaltan
synthesis, **12**, 838

Exchange reactions
lithium–tellurium, **11**, 15
lithium–tin, **11**, 14

F

α-Farnesene
synthesis, **12**, 855
Fascaplysin
synthesis
organoboron compounds, **11**, 232
Fenestrane
synthesis
dicobalt propargylium complexes, **12**, 699
Fenestranes
synthesis
allylmetal catalysts, **12**, 918
Fenske–Hall methods
iridium complexes, **8**, 345
Ferracarbaboranes, indenyl-
chemistry, **1**, 387
reactions, **1**, 375
Ferrates, hydridotetracarbonyl-
reducing agents, **12**, 355
Ferrates, tetracarbonyl-
reactions, **12**, 351–3
Ferrilactams
synthesis, **12**, 371
Ferrocene, bis(diphenylphosphino)-
crystal structure, **7**, 189
Ferrocenes
bridged, **7**, 189–91
chemistry, **7**, 185–98
crystal structures, **7**, 186
dissociation energy, **7**, 186
formation, **7**, 185
half-open, **7**, 187, 188
lithiated, **11**, 64, 65
open, **7**, 187, 188
reactions, **7**, 185
redox properties, **7**, 191, 192
solid-state compounds, **7**, 191, 192
stability, **7**, 185
substituted
synthesis, **12**, 134
[2.2]Ferrocenophane-1,13-diyne
averaged valence species, **7**, 193
Ferrocenophanes
multiply bridged, **7**, 190
[1]Ferrocenophanes
phosphorus-bridged, **7**, 189
[2]Ferrocenophanes, **7**, 189
[3]Ferrocenophanes, **7**, 190
[4]Ferrocenophanes, **7**, 190
Ferroles, **7**, 249
dinuclear complexes, **7**, 259
fluxional process of trinuclear complexes
relevance to dinuclear complexes, **7**, 267
formation by photolysis of diferra–allyl clusters, **7**, 280
preparation, **7**, 136
transformation to five-membered heterocycles, **7**, 267
trinuclear
formation by coupling between bis(μ_3-alkylidyne) clusters and an alkyne, **7**, 278
trinuclear clusters, **7**, 259
Ferromagnets
from ferrocenes, **7**, 191
Ferruginine
synthesis, **12**, 401
Fire retardants
organotin compounds, **2**, 296
Fischer carbene complexes
N-acylamino
preparation, **12**, 470
benzannulation, **12**, 484
electronic absorption spectra, **12**, 549
preparation, **12**, 470
Fischer complexes
thermal chemistry, **12**, 469–540
Fischer synthesis
iron carbenoid complexes, **7**, 116
Fischer-type carbene complexes
cyclopropanation, **12**, 389
nickel, **9**, 12
Fischer–Hafner preparations
niobium complexes, **5**, 129
tantalum complexes, **5**, 129
Fischer–Tropsch chemistry, **5**, 133, 134
Fischer–Tropsch mechanism
tetrairon cluster model system for methanation, **7**, 268
triiron methylidyne cluster model system for methanation, **7**, 276
Five-membered rings
synthesis
intramolecular allylic alkylation, **12**, 879
FK-506
analogue
synthesis, **12**, 78
synthesis
lithiation, **11**, 36
FKANAL
synthesis, **12**, 88
Flavanones
dehydrogenation
thallium(I) salts, **11**, 456
Flexibiliene
synthesis, **12**, 857
Fluorenone
synthesis
lithiation, **11**, 42
Fluorenones
synthesis, **12**, 192
Fluorodemercuriation
aromatic mercury compounds, **3**, 160
Fluoxetine
synthesis, **12**, 23
Foam control
detergent industry
silicones, **2**, 126
petroleum industry
silicones, **2**, 126
Formamidines
chiral
synthesis, lithiation, **11**, 18
lithiation, **11**, 17
Formic acid
reactions
allyl complexes, **12**, 801, 802
Francium, 2,5-dihydro-
Pauson–Khand reaction, **12**, 718
Fredericamycin
synthesis
lithiation, **11**, 60
Fredericamycin A
synthesis, **12**, 496
Frenolicin
synthesis, **12**, 357, 1034
Frenolicin, deoxy-
synthesis, **12**, 492, 1002
zirconocene complexes, **12**, 777, 779
Fridamycin E
synthesis
lithiation, **11**, 49
Friedel–Crafts acylation
arene complexes, **5**, 517
Frontalin
synthesis, **12**, 186, 613, 1168
Frullanolide
synthesis, **12**, 365
Fukinone, dehydro-
synthesis, **12**, 580
Fulvalene
metallocenes, **7**, 617
Fulvene, 6,6-dimethyl-
reaction with rhodium complexes, **8**, 215
Fulvenes
Pauson–Khand reaction, **12**, 716
Fumaraldehyde
tricarbonyliron complex, **12**, 633
Functionalities, **11**, 581
Functionalized polymers
Ziegler–Natta catalysts, **12**, 1203, 1204
Furan
coupling with Grignard reagents, **12**, 170
coupling reactions, **12**, 229
Furan, 3-alkylidene-
synthesis, **12**, 878
Furan, 3-bromo-

synthesis, **12**, 963
Furan, dihydro-
coupling with Grignard reagents, **12**, 170
Pauson–Khand reaction, **12**, 716
Furan, 2,5-dihydro-
formation
Heck reaction, **12**, 280
hydroformylation, **12**, 31
Furan, 3-formyltetrahydro-
synthesis, **12**, 31
Furan, halo-
cross-coupling reactions
Grignard reagents, **12**, 178
Furan, methylenetetrahydro-
cycloaddition reactions, **12**, 930
Furan, 2-methyltetrahydro-
cuprates, **12**, 81
Furan, 2-phenyl-
synthesis
intramolecular keto ester coupling reactions, **12**, 1080
Furan, 2-stannyl-5-alkoxy-
coupling reactions, **12**, 209
Furan, tetrahydro-
formation
cyclopropanation, **12**, 391
preparation, **12**, 611
reductive cleavage
organolithium reagents, **11**, 14
synthesis
alkoxypalladation, **12**, 825
Furanether B
synthesis, **12**, 722
Furanoid terpenes
synthesis, **12**, 825
Furanones
propargylated
synthesis, **12**, 691
3(2*H*)-Furanones
preparation, **12**, 446, 453
Furanones, aryl-
synthesis, **12**, 211
Furans
alkylation
propargyl dicobalt complexes in, **12**, 690
2,3-*cis*-disubstituted-2,3-dihydro-
synthesis of zirconocene complexes, **12**, 775, 776
DMG-containing
lithiation, **11**, 72
metallation, **11**, 119
non-DMG-containing
lithiation, **11**, 71
synthesis, **12**, 336
cobalt carbene complexes, **12**, 505
intramolecular keto ester coupling reactions, **12**, 1080
Furans, dihydro-
synthesis, **12**, 402
Furans, 2-vinyltetrahydro-
preparation, **12**, 613
Furo[3,2-b]byridines
synthesis, **12**, 224
Furochromone
synthesis
lithiation, **11**, 72
Furukawa reaction
organozinc compounds, **11**, 178
Fuscol
synthesis, **12**, 580
Fused rings
synthesis
cascade cyclization, **12**, 303
Fusicoccin
synthesis
dicobalt propargylium complexes, **12**, 698

G

Gabaculine
  synthesis, **12**, 657, 805
Gadolinium
  dicarbon metallacarbaboranes
    7-vertex, **1**, 382
    12-vertex, **1**, 400
Gallaboranes
  properties, **1**, 258
Gallacarboranes
  properties, **1**, 258, 259
Gallane, dimethyl-
  structure, **1**, 515
Gallane, phosphino-
  preparation, **1**, 519
Gallates
  structures, **1**, 29
Gallium carbide
  structure, **1**, 506
Gallium complexes
  anionic, **1**, 514
  arenes, **1**, 534–6
  arsenic ligands, **1**, 518–24
  cage compounds, **1**, 522
  carboboranes, **1**, 532–4
  cationic, **1**, 514
  chemistry, **1**, 503–40
  cyclopentadienyl, **1**, 529–31
  group 6 metals
    direct bonds to gallium, **1**, 553
  group 7 metals
    direct bonds to gallium, **1**, 553–5
  group 9 metals
    direct bonds to gallium, **1**, 555, 556
  group 10 metals
    direct bonds to gallium, **1**, 557
  halogen complexes, **1**, 511–13
  hydrides, **1**, 514–17
  low oxidation state, **1**, 528–36
  metal–metal bonds, **1**, 537–40
  metal bonds, **1**, 553–7
  nitrogen ligands, **1**, 518–24
  oxygen ligands, **1**, 524–8
  phosphorus ligands, **1**, 518–24
  pseudohalogen complexes, **1**, 511–13
  selenium ligands, **1**, 524–8
  sulfur ligands, **1**, 524–8
  tellurium ligands, **1**, 524–8
  transition metal bonds, **1**, 545–69
  trialkyl, **1**, 504–11
  triaryl, **1**, 504–11
Gallium hydride
  amine adducts, **1**, 515
Gallium, trialkyl
  complexes, **1**, 509
Gallium, trimesityl-
  structure, **1**, 507
Gallium, trimethyl-
  crown ether complexes, **1**, 510
Gelsemine
  synthesis, **12**, 285, 289
Gephyrotoxin, perhydro-
  synthesis, **12**, 104
Geraniol
  asymmetric dihydroxylation, **12**, 1149, 1150
  hydrogenation
    asymmetric, **12**, 21
Geraniol, 4,5-didehydro-
  synthesis, **11**, 105
Germacyclobutene
  isolation, **2**, 141
2-Germa-1,3-dioxolans
  stabilities, **2**, 170
Germadiphospholane
  preparation, **2**, 160
Germaimines
  chemistry, **2**, 202–4
  precursors, **2**, 158
Germanaboranes
  properties, **1**, 259, 260
Germanacarboranes
  properties, **1**, 260
Germanazene
  synthesis, **2**, 161
Germane, trimethyl-
  vinylic
    cuprates from, **12**, 62, 63
Germanes, arylhydro-
  dehydrocoupling, **2**, 156
Germanes, halo-
  hypervalent, **2**, 148
  properties, **2**, 148–50
  synthesis, **2**, 145
Germanes, hydro-
  reactivity, **2**, 152
  thermolysis, **2**, 157
Germanes, polyhalo-
  alkylation, **2**, 147
  arylation, **2**, 147
Germanes, silyl-
  synthesis, **2**, 23
Germanium centered radicals, **2**, 186–90
  formation, **2**, 187
Germanium compounds
  doubly bonded, **2**, 199–205
  highly coordinated, **2**, 206
  transmetallation-insertion reactions, **12**, 154
Germanium, bis(bis(trimethylsilyl)dicarbahexaborane)-
  preparation, **1**, 260
Germanones
  chemistry, **2**, 204
Germaoxaziridine
  synthesis, **2**, 159
Germaphosphenes
  chemistry, **2**, 204
  precursors, **2**, 158
  synthesis, **2**, 158
Germathiaphosphiranes
  preparation, **2**, 159
Germathiaphospholanes
  preparation, **2**, 160
Germathiazolidines
  preparation, **2**, 160
Germathiones
  chemistry, **2**, 204
Germatranes
  synthesis, **2**, 171
Germenes
  chemistry, **2**, 200
Germocines
  synthesis, **2**, 171
Germylenes
  addition to unsaturated compounds, **2**, 196
  chemical reactivity, **2**, 193–9
  chemistry, **2**, 190–9
  complexes with Lewis bases, **2**, 198
  condensation
    halogermanes from, **2**, 147
  insertion reactions
    σ-single bonds, **2**, 193
  oxidation, **2**, 198
  preparation, **2**, 190–2
  properties, **2**, 193
  structure, **2**, 193
Gibberellic acid
  synthesis, **12**, 580
Gibberellins
  synthesis, **12**, 578
Giese reaction
  organotin compounds, **11**, 361
Gilvocarcin V
  synthesis, **12**, 496
Gingerol
  synthesis, **12**, 627
D-Glucofuranose, 1,2-*O*-isopropylidene-
  dehydrogenation, **7**, 700
Glycals
  arylation, **12**, 133
    arylstannanes, **12**, 150
Glycerol
  asymmetrically deuterated, **11**, 250
Glycerol, 2,3-*O*-isopropylidene)-1-(diphenylphosphine)-
  reaction with hexakis(carbonyl)(phenylacetylene)-dicobalt, **12**, 718
Glycine
  Schiff base derivatives
    allylic alkylation, **12**, 851
Glycine, methylenecyclopropyl-
  synthesis, **12**, 839
Glycinoeclepin A
  synthesis
    Stille coupling, **12**, 207
Glycols
  monophenylation
    organobismuth(V) compounds, **11**, 509
  vicinal

oxidative cleavage, **11**, 504
Glycosidation
allyl complexes, **12**, 817
hafnium promoted, **4**, 621
zirconium promoted, **4**, 621
Glycosides
benzonaphthopyranones
synthesis, **12**, 150
synthesis
glycal arylation, **12**, 133
C-Glycosides
synthesis, **12**, 589
Glyoxylates
cross-coupling reactions, **12**, 1075
Gnididione
synthesis
lithiation, **11**, 72
Gold
applications, **3**, 2
dicarbon metallacarbaboranes
12-vertex, **1**, 400, 418
reclamation, **3**, 1
Gold clusters
bimetallic, **10**, 256
heterometallic, **3**, 43
homometallic, **3**, 43
transition metals, **10**, 255–317
trialkylphosphine
theoretical analysis of bonding, **10**, 261
trimetallic, **10**, 295
trimetallic
trialkylphosphines, **10**, 302
Gold complexes, **3**, 1–49
alkenes
synthesis, **3**, 44
alkyne
synthesis, **3**, 44
carbene
reactions, **3**, 44–7
synthesis, **3**, 44–7
carbonyl
properties, **3**, 47–9
synthesis, **3**, 47–9
isocyanide
properties, **3**, 49
synthesis, **3**, 49
mercury clusters, **10**, 343–9
polynuclear, **3**, 6
trifluoromethyl-
chemical vapor deposition, **3**, 33
Gold(I) complexes
acetylides
polymeric, **3**, 6
alkyl phosphine
synthesis, **3**, 3
alkynyl
applications, **3**, 6
dinuclear, **3**, 13
aryl, **3**, 8
synthesis, **3**, 19
cyclopentadienyl phosphine, **3**, 9
fluorenyltriphenyl phosphine, **3**, 11
halogenoalkyl
synthesis, **3**, 4
imidazolyl, **3**, 11
iminoalkyl, **3**, 11
phenylacetylides, **3**, 6
two gold–carbon bonds, **3**, 17–19
ylides, **3**, 19–31
dinuclear, **3**, 23–31
mononuclear, **3**, 20–3
polynuclear, **3**, 23–31
Gold(II) complexes
ylides, **3**, 19–31
dinuclear, **3**, 23–31
mononuclear, **3**, 20–3
polynuclear, **3**, 23–31
Gold(III) complexes
alkyl, **3**, 31
carbenes, **3**, 43
Gold(I) complexes
clusters
metallaboranes, **1**, 299
Gold(III) complexes
diaryl
synthesis, **3**, 37
monoaryl
synthesis, **3**, 32
one gold–carbon bond, **3**, 31–3
two gold–carbon bonds, **3**, 33–9
three gold–carbon bonds, **3**, 39–42
four gold–carbon bonds, **3**, 42, 43
ylides, **3**, 19–31, 43
dinuclear, **3**, 23–31
mononuclear, **3**, 20–3
polynuclear, **3**, 23–31
Gold compounds
ferrocenyl, **3**, 11
Gold(III) nitrate, dimethyl-
synthesis, **3**, 36
Goniofufurone
synthesis, **12**, 357
Graft copolymers
from ring-opening metathesis polymerization, **12**, 1215
Grandisol, 2-fluoro-
synthesis, **12**, 582
Grifolin
synthesis, **12**, 859
Grignard reactions
1,4-addition, **11**, 132
Grignard reagents
addition to alkenes, **1**, 66
addition to alkynes
copper catalysis, **1**, 68
addition to cyclopropenes, **1**, 66
alkynes
protection, **1**, 70
allylic alkylation, **12**, 854
analysis, **1**, 78
benzylic
addition to α,β-unsaturated carbonyl compounds, **1**, 82
enantioselective, carbonyl compounds, **1**, 82
reactions, **1**, 82
benzylic di-
preparation, **1**, 71
carbonation
rate constants, **1**, 78
carbonyl compounds, **1**, 80–4
chiral
reactions, **1**, 82
chiral sparteine ligands
structure, **1**, 73
five-coordinate
solid-state structure, **1**, 74
six-coordinate
solid-state structure, **1**, 74
coupling reactions with nickel complexes, **12**, 964, 965
regioselectivity, **12**, 964
crown ether, **1**, 77
cubane di-
preparation, **1**, 72
di-
cyclization, **1**, 76
preparation, **1**, 71, 72
1,1-di-
preparation, **1**, 71
solid-state structure, **1**, 73
1,2-di-
preparation, **1**, 71
1,3-di-
preparation, **1**, 71
electron transfer, **1**, 78
2-enyl
interconversions, **1**, 77
fluorination, **1**, 78
o-functionalized aromatic
preparation, **1**, 70
halogenated
preparation, **1**, 70
in organic cynshtesis, **11**, 129
in organometallic compound synthesis, **1**, 83, 84
in synthesis of dimetallic organometallics, **1**, 83
in synthesis of metallocycles, **1**, 84
in synthesis of metallocyclic organometallics, **1**, 84
in synthesis of organotin compounds, **2**, 224
inversion at α-C atoms, **1**, 77
LAXS studies, **1**, 75
matrix isolation studies, **1**, 62
methyl functionalized
preparation, **1**, 70
^{25}Mg NMR, **1**, 76
phenols
protection, **1**, 70
poly-
preparation, **1**, 71
polymeric poly-
preparation, **1**, 72
preparation
added ligands, **1**, 60
bromine–magnesium exchange, **1**, 65
carbanion intermediates, **1**, 63
commercial, **1**, 58
direct reaction, **1**, 58–63
exchange reactions, **1**, 63–6
from bromoanisoles, **1**, 65
from iodopyridines, **1**, 65
functionalized, **1**, 70
kinetics, **1**, 60
magnesium activation, **1**, 59
mass transfer, **1**, 60
mechanism, **1**, 60, 61
mercury compounds, **1**, 65

metal–halogen exchange, **1**, 64
metal–hydrogen exchange, **1**, 63
organotellurium compounds, **1**, 66
oscillating rates of reaction, **1**, 60
pyridyl sulfoxides, **1**, 65
radical anion intermediates, **1**, 61
radical intermediates, **1**, 60
reaction with unsaturated bonds, **1**, 66–9
retention of stereochemistry, **1**, 61
solvents, **1**, 60
transmetallation, **1**, 65
Walborsky adsorption model, **1**, 61
zero-order kinetics, **1**, 60
reaction with π-allylnickel complexes, **12**, 960
reactions, **1**, 78–84
aromatic ketones, **1**, 81
coupling, **1**, 83
enolization of ketones, **1**, 81
heterolytic, **1**, 80
SET mechanisms, **1**, 80
transition metal compounds, **1**, 83
α,β-unsaturated carbonyl compounds, **1**, 82
skeleton rearrangements, **1**, 79, 80
solid-state
physical properties, **1**, 72–4
structures, **1**, 72–8
structures, four-coordinate, **1**, 73
structures, intramolecularly coordinated oxygen donors, **1**, 74
solutions
structures, **1**, 75–7
spectroscopy, **1**, 77
stereochemical changes, **1**, 77
theoretical studies, **1**, 78
thermal rearrangement, **1**, 79
thermochemistry, **1**, 78
titanium tetrachloride
alkylation, **1**, 83
tri-
preparation, **1**, 71
triphenylmethyl
solid-state structure, **1**, 73
unsolvated, **1**, 62
unsolvated neopentyl
solid-state structure, **1**, 73
zirconium tetrachloride
alkylation, **1**, 84
Griseoviridin
synthesis, **12**, 808
Grosshemin
synthesis, **12**, 581
Group 4 complexes
heterobimetallic, **10**, 25–7
bridged, **10**, 27
unbridged, **10**, 27
Group 5 complexes
heterobimetallic, **10**, 28
Group 6 clusters
group 7 compounds, **10**, 88–90
group 8 compounds, **10**, 90–109
carbides, **10**, 90, 91
group 9 compounds, **10**, 109–19
group 10 compounds, **10**, 119
mixed-metal, **10**, 87, 88
Group 6 complexes
group 6 bonds
bridged, **10**, 33
group 7 bridged, **10**, 33–42
group 8 bridged, **10**, 42, 43
group 9 bridged, **10**, 43–5
group 10 bridged, **10**, 45–7
group 11 bridged, **10**, 47
group 12 bridged, **10**, 47
heterobimetallic, **10**, 29–47
bridged, **10**, 33–47
unbridged, **10**, 30–2
hexaorgano
without carbonyl ligands, **5**, 282
without carbonyl ligands, properties, **5**, 283
without carbonyl ligands, reactions, **5**, 283
without carbonyl ligands, structure, **5**, 283
homonuclear metal–metal bonds, **5**, 321–3
properties, **5**, 323
reactions, **5**, 323
structure, **5**, 323
synthesis, **5**, 321–3
oxo
synthesis, **5**, 289, 290
tetraorgano
without carbonyl ligands, **5**, 282
without carbonyl ligands, properties, **5**, 283
without carbonyl ligands, reactions, **5**, 283
without carbonyl ligands, structure, **5**, 283
Group 7 clusters
group 6 compounds, **10**, 88–90
group 8 compounds, **10**, 124–6
group 9 compounds, **10**, 127, 128
group 10 compounds, **10**, 128
Group 7 complexes
group 6 bridged, **10**, 33–42
group 7 bridged, **10**, 48
group 8 bridged, **10**, 48–52
group 9 bridged, **10**, 52, 53
group 10 bridged, **10**, 53
group 11 bridged, **10**, 53
heterobimetallic, **10**, 47–53
bridged, **10**, 48–53
unbridged, **10**, 48
Group 8 clusters
alkylidyne group 9 compounds, **10**, 132–44
alkyne group 9 compounds, **10**, 135–44
carbide group 9 compounds, **10**, 132–4
group 6 compounds, **10**, 90–109
group 7 compounds, **10**, 124–6
group 9 compounds, **10**, 132–46
group 10 anionic carbides, **10**, 146
group 10 compounds, **10**, 146–54
group 10 neutral carbides, **10**, 146, 147
ketenylidene group 9 compounds, **10**, 132–4
methoxymethylidyne group 9 compounds, **10**, 135
methylene group 10 compounds, **10**, 147–9
methylidyne group 10 compounds, **10**, 147–9
mixed-metal, **10**, 129–32
vinylidene group 9 compounds, **10**, 135–44
Group 8 complexes
group 6 bridged, **10**, 42, 43
group 7 bridged, **10**, 48–52
group 8 bridged, **10**, 54, 55
group 9 bridged, **10**, 60, 61
group 10 bridged, **10**, 61
group 11 bridged, **10**, 61, 62
group 12 bridged, **10**, 61, 62
heterobimetallic, **10**, 53–62
bridged, **10**, 54–62
unbridged, **10**, 54
Group 9 clusters
alkylidyne
trinuclear, **10**, 154, 155
alkylidyne group 8 compounds, **10**, 132–44
alkyne
tetranuclear, **10**, 155
trinuclear, **10**, 154, 155
alkyne group 8 compounds, **10**, 135–44
carbide group 8 compounds, **10**, 132–4
carbide mixed-metal, **10**, 154, 155
group 6 compounds, **10**, 109–19
group 7 compounds, **10**, 127, 128
group 8 compounds, **10**, 132–46
group 10 compounds, **10**, 156, 157
ketenylidene group 8 compounds, **10**, 132–4
methoxymethylidyne group 8 compounds, **10**, 135
mixed-metal, **10**, 154, 155
vinylidene group 8 compounds, **10**, 135–44
Group 9 complexes
group 6 bridged, **10**, 43–5
group 7 bridged, **10**, 52, 53
group 8 bridged, **10**, 60, 61
group 9 bridged, **10**, 63, 64
group 10 bridged, **10**, 64
group 11 bridged, **10**, 68
group 12 bridged, **10**, 68
heterobimetallic, **10**, 62–8
bridged, **10**, 63–8
unbridged, **10**, 62, 63
Group 10 clusters
group 6 compounds, **10**, 119
group 7 compounds, **10**, 128

group 8 anionic carbides, **10**, 146
group 8 compounds, **10**, 146–54
group 8 neutral carbides, **10**, 146, 147
group 9 compounds, **10**, 156, 157
methylene group 8 compounds, **10**, 147–9
methylidyne group 8 compounds, **10**, 147–9
Group 10 complexes
group 6 bridged, **10**, 45–7
group 7 bridged, **10**, 53
group 8 bridged, **10**, 61
group 9 bridged, **10**, 64
group 10 bridged, **10**, 69–71
group 11 bridged, **10**, 71
group 12 bridged, **10**, 71
heterobimetallic, **10**, 69–71
bridged, **10**, 69–71
unbridged, **10**, 69
Group 11 complexes
group 6 bridged, **10**, 47
group 7 bridged, **10**, 53
group 8 bridged, **10**, 61, 62
group 9 bridged, **10**, 68
group 10 bridged, **10**, 71
group 11 bridged, **10**, 71
group 12 bridged, **10**, 71
Group 12 complexes
group 6 bridged, **10**, 47
group 8 bridged, **10**, 61, 62
group 9 bridged, **10**, 68
group 10 bridged, **10**, 71
group 11 bridged, **10**, 71
Guaianolide
synthesis, **12**, 722
Guanidine, tetramethyl-
methylenation, **12**, 591
Gymnopusin
synthesis, **12**, 192
lithiation, **11**, 60

H

Hafnium
dicarbon metallacarbaboranes
7-vertex, **1**, 382
12-vertex, **1**, 400
Hafnium complexes, **4**, 439–61
allyl, **4**, 564, 565
bonding, **4**, 564, 565
reactivity, **4**, 565
synthesis, **4**, 564, 565
bidentate ligands, **4**, 547–50
nitrogen-donor ligands, **4**, 547, 548
oxygen-donor ligands, **4**, 550
sulfur-donor ligands, **4**, 548, 549
bis(arene), **4**, 433, 434
bis(cycloheptatriene), **4**, 434
bis(cyclooctatetraene), **4**, 627, 628
bis(cyclopentadienyl), **4**, 501–38
bis(tetrahydroborates), **4**, 554
bridging cyclopentadienyl, **4**, 552
bridging fulvalenes, **4**, 552
carbon-donor ligands, **4**, 546
carbonyl
heteroleptic, **4**, 434–6
homoleptic, **4**, 436
oxidation states 0 and below, **4**, 434–6
cycloheptatrienyl, **4**, 572
cyclooctatetraene, **4**, 627–31
cyclooctatetraene dialkyl, **4**, 631
cyclooctatetraene dihalide, **4**, 631
cyclopentadienyl
applications, **4**, 546, 547
fluorenyl, **4**, 566
indenyl, **4**, 440, 566, 567
methylated cyclopentadienyl, **4**, 545
mixed sandwich compounds, **4**, 628–30
monocyclooctatetraene, **4**, 628–31
monocyclopentadienyl, **4**, 544–52
bonding, **4**, 544, 545
structure, **4**, 544, 545
synthesis, **4**, 544, 545
mono(tetrahydroborates), **4**, 554
oxidation state 0 and below, **4**, 433–6
oxidation state +3, **4**, 465–79
oxidation state +4, **4**, 543–85
oxoclusters, **4**, 550, 551
permethylcyclopentadienyl, **4**, 440
polymetallic systems, **4**, 550–2
pyridine
structure, **4**, 441
silylated cyclopentadienyl, **4**, 546
tetrahydroborates, **4**, 554–6
bonding, **4**, 554, 555
reactivity, **4**, 555, 556
structure, **4**, 554, 555
synthesis, **4**, 554, 555
tris(cyclopentadienyl), **4**, 552
tris(tetrahydroborates), **4**, 554, 555
zirconium bonds, **10**, 25
Hafnium(III) complexes
anions, **4**, 476–9
reduction, **4**, 476
spectroscopy, **4**, 476
structure, **4**, 479
synthesis, **4**, 476–9
neutral mononuclear, **4**, 474–6
chemical properties, **4**, 474
spectroscopy, **4**, 474
structure, **4**, 475, 476
synthesis, **4**, 474
Hafnium(IV) complexes
alkoxides
chemical properties, **4**, 522
spectroscopy, **4**, 521
structure, **4**, 522
synthesis, **4**, 519–21
alkylideneamido
chemical properties, **4**, 510
spectroscopy, **4**, 510
structure, **4**, 510
synthesis, **4**, 510
amides
chemical properties, **4**, 509
spectroscopy, **4**, 507, 508
structure, **4**, 508, 509
synthesis, **4**, 507, 508
antimonido
spectroscopy, **4**, 518
synthesis, **4**, 518
arsenido
spectroscopy, **4**, 518
synthesis, **4**, 518
aryloxides
chemical properties, **4**, 522
spectroscopy, **4**, 521
structure, **4**, 522
synthesis, **4**, 519–21
bis(cyclopentadienyl)
germyl, **4**, 503, 504
germyl, hydrogenolysis, **4**, 505
germyl, insertion reactions, **4**, 506, 507
germyl, salt elimination, **4**, 505
germyl, spectroscopy, **4**, 504, 505
germyl, structure, **4**, 504
germyl, thermolysis, **4**, 505
silyl, **4**, 503, 504
silyl, hydrogenolysis, **4**, 505
silyl, insertion reactions, **4**, 506, 507
silyl, salt elimination, **4**, 505
silyl, spectroscopy, **4**, 504, 505
silyl, structure, **4**, 504
silyl, thermolysis, **4**, 505
stannyl, **4**, 503, 504
stannyl, hydrogenolysis, **4**, 505
stannyl, insertion reactions, **4**, 506, 507
stannyl, salt elimination, **4**, 505
stannyl, spectroscopy, **4**, 504, 505
stannyl, structure, **4**, 504
stannyl, thermolysis, **4**, 505
carboxylates
chemical properties, **4**, 526
spectroscopy, **4**, 526
structure, **4**, 526
synthesis, **4**, 525, 526
chalcogenido-bridged dimers, **4**, 531, 532
dinuclear
amido-bridged, **4**, 509
imido-bridged, **4**, 509
dithiocarbamato, **4**, 534, 535
enediolates
chemical reactivity, **4**, 525
spectroscopy, **4**, 525
structure, **4**, 525
synthesis, **4**, 523, 524
enolates
chemical reactivity, **4**, 525
spectroscopy, **4**, 525
structure, **4**, 525
synthesis, **4**, 523, 524
heterobimetallic phosphido-bridged
chemical reactivity, **4**, 518
spectroscopy, **4**, 517
structure, **4**, 518
synthesis, **4**, 516, 517
heterotrimetallic phosphido-bridged
chemical reactivity, **4**, 518
spectroscopy, **4**, 517
structure, **4**, 518
synthesis, **4**, 516, 517
hydrazido
chemical properties, **4**, 511
spectroscopy, **4**, 511
structure, **4**, 511
synthesis, **4**, 510, 511
hydrazonato
chemical properties, **4**, 511
spectroscopy, **4**, 511
structure, **4**, 511
synthesis, **4**, 510, 511

hydroxides, **4**, 527
oxoanions, **4**, 536–8
oxo-bridged dimers, **4**, 527, 528
chemical properties, **4**, 528
spectroscopy, **4**, 528
structure, **4**, 528
phosphido
chemical properties, **4**, 515, 516
spectroscopy, **4**, 514
structure, **4**, 514, 515
synthesis, **4**, 512, 513
polychalcogenido ligands, **4**, 533, 534
pseudohalides, **4**, 536–8
transition metals, **4**, 535, 536
Hafnocene
living polymerization catalyst
two alkene monomers, **12**, 1205
Hafnocenes
alkene complexes, **4**, 442–6
alkenyl, **4**, 579–82
alkoxy hydride, **4**, 559
alkyl hydride, **4**, 559
alkyne complexes, **4**, 442–6
ansa-derivatives, **4**, 567–70
characterization, **4**, 567–9
reactivity, **4**, 569, 570
synthesis, **4**, 567–9
butadiene complexes, **4**, 446
oxidative coupling, **4**, 460
structures, **4**, 451
carbonyl derivatives
synthesis, **4**, 440
dialkyl, **4**, 575–7
addition reactions, **4**, 576
catalysis, **4**, 577
insertion reactions, **4**, 576
reactivity, **4**, 575, 576
synthesis, **4**, 575, 576
diaryl, **4**, 577–9
diazo complexes, **4**, 441
dihalides, **4**, 483
dihydrides, **4**, 562
reactions with aldehydes, **4**, 564
reactions with alkynes, **4**, 563
reactions with ammonia, **4**, 564
reactions with azides, **4**, 564
reactions with carbon monoxide, **4**, 563
reactions with ketones, **4**, 564
reactions with main-group elements, **4**, 562
reactions with water, **4**, 564
reactivity, **4**, 562–4
dinitrogen complexes, **4**, 440
hydride chloride, **4**, 556
reactions with alkenes, **4**, 556
reactions with alkynes, **4**, 556
reactions with carbon dioxide, **4**, 557
reactions with carbon monoxide, **4**, 557
reactions with nitriles, **4**, 557
reactions with organophosphorus compounds, **4**, 558
reactions with oxygen-containing substrates, **4**, 558, 559
reactivity, **4**, 556–9
hydrides, **4**, 556–64
iminoacyl, **4**, 582–5
reactivity, **4**, 584, 585
monoalkyl, **4**, 573–5
NMR, **4**, 453
oxidative addition, **4**, 455–61
intermolecular, **4**, 457–61
intramolecular, **4**, 455, 456
phosphine complexes, **4**, 441
polymetallic hydrides, **4**, 561, 562
reactions, **4**, 454–61
spectroscopy, **4**, 452–4
structures, **4**, 450–4
trialkylsilyl hydrides, **4**, 560, 561
x-ray crystallography, **4**, 450, 451
ylides
reactions, **4**, 458
Hagemann's ester
reaction with aryllead triacetates, **11**, 471
Halides
carbonylation, **12**, 362–9
conversion to hydrocarbons
organoaluminum hydrides, **11**, 301
reactions
allyl complexes, **12**, 817–26
reactions with
organoaluminum compounds, **1**, 491, 492
Haloacetoxylation
diols, **12**, 1173
Haloboration, **11**, 201, 202
Haloetherification
π-allylmolybdenum complexes, **12**, 973
Haloform
preparation
organoboranes, **11**, 223
Halogeno complexes
without carbonyl ligands, **5**, 283–6
Halolactonization
π-allylmolybdenum complexes, **12**, 973
Halomercuriation, **11**, 428
Halostachine
synthesis, **12**, 1045
Halostachine, *o*,*N*-dimethyl-
synthesis, **12**, 1045
Halostachine, *o*-methyl-
synthesis, **12**, 1045
Hapalindole
synthesis, **12**, 462
Hastanecine
synthesis, **12**, 411, 412
Heck arylations
intramolecular, **12**, 280
Heck reaction
allylmetal–alkene cyclization, **12**, 913, 914
intermolecular, **12**, 260–74
intramolecular
optically active phosphine ligands, **12**, 285
regioselectivity, **12**, 300
organomercury compounds, **11**, 403
palladium catalysis, **12**, 259
Heck vinylation
intramolecular, **12**, 280
Helepuberinic acid ester, dehydro-
synthesis, **12**, 800
Heliannuol A
synthesis
lithiation, **11**, 38
Heliotridane
synthesis, **12**, 372
Hemicarcerand
binding with ruthenocene, **7**, 607
Hemiisotactic
definition, **12**, 1196
Hemiisotactic polymers
Ziegler–Natta catalysts, **12**, 1199
1,6-Heptadiene-4-(trimethylsiloxy)-
homopolymerization, **12**, 1203
Hepta-1,6-diyne, 1,7-bis(trimethylsilyl)-
reaction with carbon monoxide, **2**, 11
Heptafulvene
synthesis
carbene complexes, **12**, 516
Heptafulvenes
synthesis, **12**, 629, 630
3-Heptanol, 4-methyl-
synthesis, **12**, 802
4-Heptanone
reactions with iron ions, **7**, 106
Heptasilaparacyclophane
preparation, **2**, 64
1-Heptene
Pauson–Khand reaction, **12**, 712
6-Heptene, 1-acetoxy-2-methylene-
cyclization
allylpalladium catalysts, **12**, 912
palladium catalysts, **12**, 911
1,6-Heptenyne
Pauson–Khand reaction, **12**, 724
1,6-Heptenyne, 1-[2-phenylcyclohexoxy]-
Pauson–Khand reaction, **12**, 727
1,6-Heptenyne, 7-[2-phenylcyclohexoxy]-
Pauson–Khand reaction, **12**, 727
Heptenynes
3,4-disubstituted
Pauson–Khand reaction, **12**, 725
1-Heptyne
reactions with iron ions, **7**, 105

5-HETE
methyl ester
synthesis, **12**, 631
Hetero-Diels–Alder cyclocondensation
(diene)tricarbonyliron, **12**, 627
Heteroannulation
allylpalladium complexes, **12**, 141
Heteroarenes
synthesis
cyclooligomerization, **12**, 753–9
Heteroaromatic compounds
alkylated
oxidation, organoselenium componds, **11**, 535
reactions with selenium dioxide, **11**, 536
lithiation, **11**, 70–81
Heteroaryl compounds
coupling reactions, **12**, 197–9
Stille reaction, **12**, 215–18
Heteroaryl derivatives
coupling reactions
Grignard reagents, **12**, 177, 178
Heteroaryl rings
metathesis with, **12**, 228, 229
Heteroatoms
coordination to, **12**, 131
Heteroboranes
properties, **1**, 257–71
Heterocumulenes
cycloaddition
aziridines or oxiranes, **11**, 500
Heterocycles
carbonylation, **12**, 371–3
nitrogen-containing
synthesis, **12**, 147
Heterocyclic compounds
metallation, **11**, 119–21
Heterodienes
transition metal complexes
reactions, **12**, 633, 634
Heterodinuclear complexes
metal–metal bonds, **10**, 23–71
Heteroferrocenes
chemistry, **7**, 194–8
Heteronuclear clusters
acyclic hydrocarbyl ligands, **10**, 85–171
synthesis, **10**, 5
three unlike metals, **10**, 157–66
trinuclear alkylidyne, **10**, 157–9
Heteronuclear metal–metal bonds
synthesis, **10**, 1–21
addition reactions, **10**, 6–9
bridge assisted reactions, **10**, 11–19
condensation reactions, **10**, 9–11
displacement of neutral ligands, **10**, 2–6
ligand substitution, **10**, 2–6
metal–metal exchange reactions, **10**, 19–21
reagent assisted displacement, **10**, 6
Heteronuclear multiple metal–metal bonds
synthesis, **10**, 2
Hexaborates, hexahydro-
group 8 metals
metallaboranes, **1**, 328
Hexaborates, nonahydro-
group 8 metals
metallaboranes, **1**, 327
group 10 metals
bimetallaboranes, **1**, 327
metallaboranes, **1**, 327
5,9-Hexacosadienoic acid
synthesis, **12**, 821
1,4-Hexadiene
reactions
arylpalladium halides, **12**, 809
1,5-Hexadiene
cyclization
Ziegler–Natta catalysts, **12**, 1206
cyclopolymerization
Ziegler–Natta catalysts, **12**, 1201, 1204
reactions with iron ions, **7**, 104
1,5-Hexadienes
[3.3]sigmatropic rearrangement, **12**, 615
1,5-Hexadienes, 2-trimethylsilyloxy-
cycloalkenylation, **12**, 614
1,5-Hexadiyne
homotrimerization, **12**, 749
Hexane-3,4-diol
synthesis, **12**, 1140
Hexanoic acid lactone, 5-hydroxy-2-methyl-
synthesis, **12**, 822
Hexarhodium hexadecacarbonyl
catalysts
diazodecomposition, **12**, 428, 429
Hexasilacyclooctyne
synthesis, **2**, 13
Hexasilaprismane
synthesis, **2**, 24
1,3,5-Hexatriyne
derivatives
cyclization, **12**, 750
Hexatriyne, 1,6-bis(trimethylsilyl)-
synthesis, **2**, 10
5-Hexen-1-amine, 2,2-dimethyl-
hydroamination/cyclization
enantioselective, **12**, 1205
1-Hexene
amination by diethylamine
palladium catalyst, **9**, 311
2-Hexenoate, 5-hydroxy-4-phthalimido-
synthesis
allyl complexes, **12**, 810
5-Hexen-1-ols
cyclization, **12**, 611
Hex-1-en-5-yne
Pauson–Khand reaction, **12**, 723
Hexoses, 2-deoxy-
synthesis, **11**, 253
Hexyne
hydrogenation
catalysis, **10**, 38
High nuclearity clusters
alkylidyne, **10**, 159–61
Highly reduced metals
coupling reactions, **12**, 1071–92
Hippadine
synthesis, **12**, 198
Hirsutene
synthesis, **12**, 411, 412, 579
allylmetal catalysts, **12**, 919
chiral auxiliary mediated Pauson–Khand process, **12**, 734
trimethylenemethane complexes, **12**, 943
Hirsutic acid
synthesis, **12**, 734
Histrionicotoxin
analogue
synthesis, **12**, 823
Histrionicotoxin, depentylperhydro-
synthesis, **12**, 874
Histrionicotoxin, dipentylperhydro-
synthesis, **12**, 823
Histrionicotoxin, perhydro-
synthesis, **12**, 104, 874
HIV protease inhibitors
synthesis
pinacol dimerization, **12**, 1086
HMPA
lithiation
in organic synthesis, **11**, 4
Hoberg reagent
stoichiometric reagent, **11**, 296
Homoallylic alcohols
asymmetric epoxidation, **12**, 1117
hydrogenation
asymmetric, **12**, 21, 22
enantioselective, **12**, 16
synthesis, **11**, 235, 237, 251, 493, 502; **12**, 966, 970
Homoallylic ethers
synthesis, **12**, 581
Homobenzylic alcohols
synthesis, **11**, 493
Homocitronellol
synthesis, **12**, 22
Homocoupling
in cross-coupling
nickel catalysis, **12**, 164
Homoenolates
coupling
organopalladium compounds, **12**, 163
metathesis with, **12**, 228
Homoferrocenes, **7**, 188
Homogeraniol
hydrogenation
asymmetric, **12**, 22
Homologation
organoboron compounds, **11**, 249–51
Homopropargylic alcohols
preparation, **11**, 181
synthesis, **11**, 238, 494
Hosomi–Sakurai reactions, **11**, 321
Humulene
synthesis, **12**, 857, 882, 890

Hydraindanone
  synthesis, **12**, 329
Hydrangenol
  synthesis
    lithiation, **11**, 60
Hydrazines
  oxidation
    organoselenium
      compounds, **11**, 535
Hydrazones
  coupling reactions
    intramolecular, **12**, 332
  oxidation
    organoselenium
      compounds, **11**, 535
Hydrazones, *N*-aroyl-
  asymmetric reduction, **12**, 26
Hydrazonese
  asymmetric reduction, **12**, 26,
    27
Hydride elimination
  allyl complexes, **12**, 802, 803
α-Hydride elimination
  transition metal complexes, **12**,
    4, 5
β-Hydride elimination
  alkene formation, **12**, 133
  organotransition metal
    complexes, **12**, 131
Hydrides
  addition to arene ligands, **12**,
    991
  asymmetric allyl displacements
    palladium complexes, **12**,
      882
  reactions
    with allyl complexes, **12**,
      799–803
Hydrido-carbonyl clusters, **10**,
    197–203
  fluxionality, **10**, 200–2
  reactivity, **10**, 202, 203
  structures, **10**, 197–200
Hydridocarboxylation
  metal hydrides, **12**, 355, 356
Hydridodemercuriation, **3**, 165
Hydroalkylation
  α,β-unsaturated carbonyl
    compounds, **11**, 596
Hydroalumination, **1**, 456,
    458–60, 470
  alkynes, **12**, 180, 182
  cuprates from, **12**, 87
  discovery, **11**, 279, 280
  regioselectivity, **11**, 291
  scope and limitations, **11**, 284
  stereoselectivity, **11**, 292
Hydroamination
  cyclization of aminoalkenes
    enantioselective using
      Ziegler–Natta catalysts,
      12, 1205
  organolanthanide catalysts, **4**,
    120
Hydroboration
  alkenes
    (diene)tricarbonyliron, **12**,
      627
    organolanthanide catalysts,
      4, 121, 122, 129
  alkoxy-directed
    esters, **11**, 195
  allyl compounds, **11**, 195
  asymmetric synthesis, **11**,
    247–9
  disubstituted boranes, **11**,
    198–201
  enantioselective
    rhodium catalysts, **11**, 248
  functional alkenes, **11**, 196
  mechanism, **11**, 201
  monosubstituted boranes, **11**,
    196, 197
  organoboron compounds
    synthesis, **11**, 193–201
  reduction of enol ethers, **12**,
    589
  thioalkynes
    nickel catalysts, **9**, 133
  transition metal promoted, **11**,
    199
Hydrocarbons
  μ-hydrido-bridged
    synthesis, **12**, 1077
  metal clusters, **7**, 747–97
  metallation, **11**, 96–114
    enthalpies, **1**, 63
    magnesium amides, **1**, 103
  organomercury compounds
    preparation, **11**, 392, 393
  polylithiated, **1**, 21
  synthesis from halides
    organoaluminum hydrides,
      11, 301
  unsaturated
    electrocatalytic carboxyla-
      tion, **9**, 9
Hydrocarboxylation, **12**, 33–5
  asymmetric, **12**, 35
  regioselective, **12**, 33–5
Hydrocarbyl ligands
  acyclic
    heteronuclear clusters, **10**,
      85–171
Hydrocyanation
  nickel catalysts, **9**, 122
Hydrocyclization
  dienes
    yttrium complexes,
      catalysts, **4**, 116
Hydrodemercuriation
  hydroperoxides, **3**, 146
Hydrodenitrogenation
  niobium complexes, **5**, 140,
    141
  tantalum complexes, **5**, 103
Hydrodesulfurization
  clusters, **7**, 809
  coal
    rhenium complexes, **6**, 218
  iridium complexes, **8**, 381
  molybdenum cyclopentadienyl
    complexes as model for,
    5, 435
  rhodium complexes
    catalysts, **8**, 200
  thiophene
    catalysts, **10**, 51
Hydroesterification, **12**, 33–5
  asymmetric, **12**, 35
  regioselective, **12**, 33–5
Hydroformylation
  alkenes
    rhodium complexes, **8**, 201
  asymmetric
    cobalt phosphine
      complexes, **8**, 17
  catalysts, **10**, 44–7
  cobalt catalyzed
    Heck–Breslaw mechanism,
      8, 32–4
  cobalt phosphine complexes
    catalysts, **8**, 20
  diastereoselective, **12**, 27–31
  enantioselective, **12**, 31–3
  ethene
    clusters, **7**, 804
  Heck–Breslaw mechanism
    cobalt catalyzed, **8**, 32–4
  isomerization
    cobalt complexes, **8**, 34
  phosphine degradation
    cobalt complexes, **8**, 34
  phosphine-directed
    intramolecular, **12**, 29
  platinum complex catalysts, **9**,
    461
  regioselective, **12**, 27–31
  styrene
    catalysts, **10**, 44
  transition metal hydrides
    catalysts, **12**, 27–33
Hydroformylation-amidocar-
    bonylation
  pentafluorostyrene, **12**, 30
Hydrogenation
  acylcobalt carbonyls, **8**, 34
  alkylpalladium complexes in,
    12, 139
  asymmetric
    cobalt phosphine
      complexes, **8**, 17
  carbon monoxide
    catalysts, **10**, 42, 43
  catalysis, **10**, 37–43
    chromium tricarbonyl arene
      complexes, **5**, 528
  chemoselective, **12**, 9–11
  cobalt hydrogen pentacyanide,
    8, 12
  cobalt phosphine complexes,
    8, 19
  diastereoselective, **12**, 11–13
  enantioselective, **12**, 14–27
    cyclopentadienyl tricar-
      bonyl manganese
      complexes, **6**, 113
    Ziegler–Natta catalysts, **12**,
      1205
  ethene
    samarium complexes,
      catalysis, **4**, 114
  hydridotetrakis(trialkyl
      phosphate)cobalt, **8**, 35
  iridium catalysts, **8**, 304
  isotope effect
    tetracarbonylhydridocobalt,
      8, 33
  ketenes
    cobalt complexes, **8**, 34
  oxygenated compounds
    clusters, **7**, 804
  palladium complexes
    alkene alkylation, **12**, 135
  phosphines
    bidentate, **8**, 18
  photocatalyzed
    hexene, **10**, 39

radical
cobalt tetracarbonyl hydride, **8**, 32
regioselective, **12**, 9–11
regioselectivity
tetracarbonylhydridocobalt, **8**, 33
selective
catalysts, **10**, 357
titanium hydride catalysts, **4**, 262, 263
transition metal catalysts, **12**, 9–27

Hydrogermylation
unsaturated compounds, **2**, 154

Hydromagnesiation
titanium catalysis
addition to alkenes, **1**, 69

Hydromercuriation
alkynes, **3**, 137

Hydrometallation
alkynes, **7**, 399–406
transition metal alkyl complexes, **12**, 39–46

Hydrooligomerization
asymmetric alkenes
Ziegler–Natta catalysts, **12**, 1200

Hydropalladation
enynes
cycloisomerization, **12**, 47–52
multiple insertion cascades
enynes, **12**, 302
organic synthesis, **12**, 46–55

Hydroperoxides
alkyl
preparation, organoboron compounds, **11**, 214
stoichiometric reagents in epoxidations, **12**, 1101
t-butyl
in epoxidations, **12**, 1101
cumyl
in epoxidations, **12**, 1101
trityl
in epoxidations, **12**, 1101

Hydroperoxymercuriation
alkenes, **3**, 145

Hydroquinones
oxidation
organoselenium compounds, **11**, 534
synthesis, **12**, 751, 752

Hydroruthenation
diphenylbutadiyne, **7**, 408

Hydrosilation
catalysts, **10**, 49, 50
iridium catalysts, **8**, 304

Hydrosilylation
alkenes, **2**, 20, 21
organolanthanide catalysts, **4**, 121
cross-linking, **2**, 118
homogeneous catalysis, **2**, 115
inhibitors
cross-linking, **2**, 118
mechanism, **2**, 114–16
organofunctional
polysiloxanes, **2**, 129
organofunctional silane synthesis, **2**, 116, 117
oxygen co-catalysis, **2**, 115
rhodium catalysis
mechanism, **2**, 115
silanes, **11**, 343
siloxane synthesis, **2**, 114–18
substituent effects, **2**, 116

Hydrostannation, **2**, 271
alkenes and alkynes, **2**, 229–31
azoarenes, **2**, 276
carbonyl compounds, **2**, 276
double bonds, **2**, 275
effect of pressure, **2**, 230
initiation by organoboranes, **2**, 230
initiation by oxygen, **2**, 230
mechanism, **2**, 229

Hydrostannolysis, **2**, 271

Hydroxyalkylation
alkenes, **12**, 136

Hydroxyalkylselenides
reduction, **11**, 551

Hydroxylamines
oxidation
organoselenium compounds, **11**, 535

Hydroxylation
asymmetric
alkenes, **12**, 1137–73

Hydroxymercuriation
alkenes, **11**, 411–13
regiochemistry, **11**, 411
stereochemistry, **11**, 411

Hydroxysemperoside deglucoside
synthesis, **12**, 606

Hydrozirconation, **12**, 163
alkenes and alkynes, **12**, 39
alkynes, **12**, 179, 325
1,3-butenyne
chemoselective, **12**, 40
cuprate preparation by, **12**, 83
organozinc halides, **12**, 44
polyunsaturated compounds, **12**, 39
terminal alkynes, **12**, 182

Hypoglycine A
synthesis, **12**, 839

Hypoxanthine
synthesis
Stille couplings, **12**, 218

I

Ibogamine
  synthesis, **12**, 614, 889
Ibuprofen
  synthesis, **11**, 109; **12**, 31, 33, 35
    lithiation, **11**, 55
Illudol
  synthesis, **12**, 762
Imeluteine
  synthesis, **12**, 192
Imidazole
  cyanocuprates, **12**, 79
  reaction with triosmium clusters, **7**, 820
  reactions
    annulation, **12**, 761
Imidazole, iodo-
  synthesis, **12**, 963, 964
Imidazoles
  synthesis
    carbene complexes, **12**, 514
Imidazolidinone, stannyl-
  lithiation, **11**, 21
Imidazoline
  lithiation, **11**, 44
Imidazolines
  β-lactams from photochemical reactions
    carbene complexes, **12**, 551
  optically active
    reactions with alkoxycarbene complexes, **12**, 554
Imides
  cyclic
    addition to Grignard reagents, **11**, 152
  reactions
    allyl complexes, **12**, 810
Imido complexes
  group 6
    bonding, **5**, 296, 297
    oxo, synthesis, **5**, 291–6
    reactions, **5**, 298, 299
  triiron clusters, **7**, 263
  without carbonyl ligands, **5**, 289–99
Imido selenium reagents
  synthesis, **11**, 528
Imidoyl chlorides
  coupling reactions
    with stannanes, **12**, 220
Imines
  aryl
    dimerization, **12**, 1083
  asymmetric reduction, **12**, 26, 27
  coupling reactions with carbonyls
    niobium complexes, **12**, 1088, 1089
  cycloaddition reactions, **12**, 933, 934
    [3+2], **12**, 938
  dimerization
    niobium complexes, **12**, 1088
  nucleophilic addition
    (diene)tricarbonyliron complexes, **12**, 632
  reaction with carbene complexes
    photochemical, **12**, 551–6
  reactions with Grignard reagents, **11**, 134, 135
  reduction
    organoboron compounds, **11**, 260
    tellurium reagents, **11**, 577
  reductive coupling, **12**, 1083, 1084
  Schiff bases
    chromium tricarbonyl complexes, **5**, 528
  synthesis from nitriles or enamines
    organoaluminum hydrides, **11**, 300
  trifluoroiodo
    Heck reactions, **12**, 279
Imines, alkoxy-
  addition of allylic zinc reagents, **11**, 171
Imines, *N*-diphenylphosphinoyl-
  reaction with dialkylzincs, **11**, 171
Imines, vinyl-
  (tricarbonyl)iron complexes
    preparation, **12**, 633
Iminidicarbonates
  reactions
    allyl complexes, **12**, 812
Iminium ligands
  clusters, **7**, 815
Iminium ylides
  formation, **12**, 440
  reactions, **12**, 440
Immunoassay
  dicobalt propargylium complexes, **12**, 700
Inandenin-12-one
  synthesis, **12**, 845, 874
Indacarboranes
  properties, **1**, 258, 259
Indacene
  synthesis
    Heck reaction, **12**, 913
Indanes
  synthesis, **12**, 273
Indanone
  formation
    aromatic substitution transformation, **12**, 463
Indates
  structures, **1**, 29
Indene
  Pauson–Khand reaction, **12**, 717
Indene, allyl-
  preparation, **12**, 869
Indene, hexahydro-
  synthesis, **12**, 908
Indenes
  preparation
    carbene complexes, **12**, 475–81
Indirubin
  synthesis, **11**, 491
Indirubin, dimethyl-
  synthesis, **11**, 491
Indium complexes
  anionic, **1**, 514
  arenes, **1**, 534–6
  arsenic ligands, **1**, 518–24
  cage compounds, **1**, 522
  carboboranes, **1**, 532–4
  cationic, **1**, 514
  chemistry, **1**, 503–40
  cyclopentadienyl, **1**, 529–31
  direct bonds to metals, **1**, 557–67
  group 6 metals
    direct bonds to indium, **1**, 557–60
  group 7 metals
    direct bonds to indium, **1**, 560, 561
  group 8 metals
    direct bonds to indium, **1**, 562–4
  group 9 metals
    direct bonds to indium, **1**, 564, 565
  group 10 metals
    direct bonds to indium, **1**, 566
  halogen complexes, **1**, 511–13
  hydrides, **1**, 514–17
  low oxidation state, **1**, 528–36
  metal–metal bonds, **1**, 537–40
  nitrogen ligands, **1**, 518–24
  oxygen ligands, **1**, 524–8
  phosphorus ligands, **1**, 518–24
  pseudohalogen complexes, **1**, 511–13
  selenium ligands, **1**, 524–8
  sulfur ligands, **1**, 524–8
  tellurium ligands, **1**, 524–8
  transition metal bonds, **1**, 545–69
  trialkyl, **1**, 504–11
  triaryl, **1**, 504–11
Indium, trimesityl-
  structure, **1**, 507
Indium, tris(2,4,6-tris(trifluoromethyl)-phenyl)-
  structure, **1**, 507
Indole
  coupling reactions, **12**, 229
    allenylzinc reagents, **12**, 198

propargylation
dicobalt complexes, **12**, 690
reactions
annulation, **12**, 761
Stille reaction, **12**, 215
synthesis
carbene complexes, **12**, 538
Indole alkaloids
synthesis, **12**, 146, 1002, 1003
Indole, 2-acylethyl-
synthesis
intramolecular keto ester coupling reactions, **12**, 1080
Indole, 4-bromo-
synthesis, **12**, 963, 964
Indole, 2-carboxyethyl-
synthesis
intramolecular keto ester coupling reactions, **12**, 1080
Indole, 3-(1,1-dimethylpropargyl)-
synthesis, **12**, 690
Indole, β-formyl-
reaction with triosmium clusters, **7**, 819
Indole, hexahydro-
synthesis, **12**, 911
Indole, 2-isopropenyl-2,3-dihydro-
synthesis, **12**, 873
Indole, octahydro-
synthesis
organoboron compounds, **11**, 218
Indole, perhydro-
synthesis
organoboron compounds, **11**, 218
Indole, 2-phenyl-
synthesis
intramolecular keto ester coupling reactions, **12**, 1080
Indole, tetrafluoro-
synthesis, **12**, 30
Indolecarbamates
lithiation, **11**, 74
Indoles
DMG containing
lithiation, **11**, 73–5
phenylation
organobismuth(V) compounds, **11**, 510
preparation, **12**, 619
3-propargylation, **12**, 690
reactions with chromium alkoxycarbene complexes
photochemical, **12**, 555
substituted
synthesis, zirconocene complexes, **12**, 778
synthesis, **12**, 195, 210, 273
intramolecular keto ester coupling reactions, **12**, 1080
lithiation, **11**, 54
palladium promoted, **12**, 210
transmetallation, **12**, 135
Indoline, 2,2-dimethyl-
preparation, **12**, 609
Indolines
substituted synthesis
zirconocene complexes, **12**, 778
2(3*H*)-Indolinones
preparation, **12**, 462
Indolyl acetate, hexahydro-
synthesis, **12**, 918
Indoxyls
synthesis, **12**, 225
Insertion reactions
alkenes
palladium-assisted, **12**, 139
alkyltransition metal complexes
organic synthesis, **12**, 131–58
carbon–hydrogen
aromatic chemoselectivity, **12**, 462, 463
chemoselectivity, **12**, 457
diastereoselectivity, **12**, 459
enantioselectivity, **12**, 459–61
regioselectivity, **12**, 457–9
Grignard reagents
titanium complexes, **12**, 157
intermolecular
catalytic, **12**, 450, 451
intramolecular
catalytic, **12**, 451–4
intramolecular aryl groups into thio-enol ethers, **12**, 150
metal carbene complexes, **12**, 421–64
rhodium complexes, **8**, 201–4
silicon–hydrogen, **12**, 464
steric and electronic effects
organotransition metal complexes, **12**, 133
Invictolide
synthesis, **12**, 328, 329
Iodonium salts
Heck reactions, **12**, 279
Iodonium salts, vinyl-
coupling reactions
palladium catalysts, **12**, 220
Ion exchange resins
mixed metal carbonyl clusters on
catalysis, **10**, 60
β-Ionol
synthesis, **11**, 104
Ionomycin
synthesis, **12**, 588
Ips paraconfusus
pheromone
synthesis, **11**, 108
Ipsdienol
synthesis, **11**, 107
Ipsenol
synthesis, **11**, 107
Ireland–Claisen rearrangement
organosilicon compounds, **11**, 321
Iridabenzene
preparation, **8**, 400
Iridaboranes
isocloso
orthoboronated, **1**, 345
preparation, **8**, 322
Iridacyclic complexes
chemistry, **8**, 394–400
Iridacyclobutanes
preparation, **8**, 395
Iridacyclobutene complexes
formation, **8**, 394
Iridacyclohexadiene
formation, **8**, 400
Iridacyclopentadiene
chemistry, **8**, 395
Iridacyclopentenone complexes
preparation, **8**, 400, 401
Iridaindanones
preparation, **8**, 312, 398
Iridalactone
preparation, **8**, 398
Iridapyran complexes
structure, **8**, 400
Iridatetraboranes, **1**, 313
Iridathiabenzene complexes
preparation, **8**, 381
Iridium
dicarbon metallacarbaboranes
11-vertex, **1**, 396
12-vertex, **1**, 400, 418
films
preparation, **8**, 304
monocarbon metallacarbaboranes, **1**, 378
10-vertex, **1**, 378
Iridium cluster complexes, **8**, 419–513
gas phase, **8**, 421
heptanuclear, **8**, 504, 505
hexanuclear, **8**, 500–4
mixed metals, **8**, 505–13
pentanuclear, **8**, 500
sulfur dioxide, **8**, 498
tetranuclear, **8**, 490–9
phosphines, **8**, 494
spectroscopy, **8**, 421
trithiane, **8**, 497
trinuclear, **8**, 487–90
phosphido, preparation, **8**, 487
pyridine, structure, **8**, 487
Iridium clusters
iron compounds, **10**, 142
tungsten compounds, **10**, 118, 119
Iridium complexes
adducts, **8**, 305–23
aldehydes
square-planar, **8**, 315
alkoxy
square-planar, **8**, 315
alkyl
square-planar, **8**, 314
allyl, **8**, 368
aromatic, **8**, 393, 394
benzimidazolato
square-planar, **8**, 317
bicyclocarbenes, **8**, 381
biimidazole, **8**, 374
bimetallic, **8**, 324–36
bridging methylene, **8**, 324
bridging oxo, **8**, 325
diphosphine-bridged, **8**, 324–9
diyne-bridged, **8**, 332
hydroxide bridging, **8**, 325
isocyanide, **8**, 326
methylene-bridged, **8**, 330

pyrazole-bridged, **8**, 330
pyrrolido-bridged, **8**, 330
silylene bridging, **8**, 324
tetraphenylporphyrin-bridged, **8**, 330
thiolato-bridged, **8**, 330, 331
trichlorostannates, **8**, 331
binuclear
cyclopentadienyl, **8**, 387–90
dicyclohexylphosphido-bridged, **8**, 372
boropyrazole, **8**, 351
bridging allyl, **8**, 372
buckminsterfullerene, **8**, 391
butatrienyl, **8**, 309
carbamoyl, **8**, 351
carbenes, **8**, 309
square-planar, **8**, 315
carboalkoxo
square-planar, **8**, 316
carborane
square-planar, **8**, 320
catalysis, **8**, 304, 305
C–H bond activation, **8**, 377–9
chemistry, **8**, 303–406
18-crown-6
square-planar, **8**, 320
cyclohexadienyl, **8**, 393
cyclooctadiene
binuclear, **8**, 372–5
4-coordinate, **8**, 366–9
5-coordinate, **8**, 370
6-coordinate, **8**, 370
cyclooctenyl
six-coordinate, **8**, 370
cyclopentadienyl, **8**, 377–93
substituted, **8**, 392, 393
cyclopentadienyl alkene, **8**, 379, 380
cyclopentadienyl allyl, **8**, 380
cyclopentadienyl dichloro
dimeric complexes, **8**, 385–7
cyclopentadienyl diene, **8**, 380, 381
cyclopentadienyl phosphine, **8**, 383, 384
cyclopentadienyl polyhydrides, **8**, 384, 385
cyclopentadienyl thiophene, **8**, 380, 381
diazo compounds
reactions, **8**, 311
dienes, **8**, 365–76
dihydrides, **8**, 351–4
β-diketonates, **8**, 311
dioxygen adducts
square-planar, **8**, 318
ditolylformamidinato, **8**, 374
fullerenes, **8**, 306
fulvalene-bridged, **8**, 388
heterobimetallic, **8**, 332–6
hydrido allyl, **8**, 378
hydridohydroxyacetyl, **8**, 349, 350
hydropyridinato bridged, **8**, 373, 374
hydroxy
square-planar, **8**, 315
hydroxymethyl, **8**, 349
imidazole
square-planar, **8**, 318
iminooxosulfuranes
square-planar, **8**, 320
indenyl, **8**, 391
indium bonds, **1**, 565
indole, **8**, 394
2-mercaptopyridinate, **8**, 374
mercury clusters, **10**, 341, 342
metallaboranes, **1**, 342
methoxycarbonyl, **8**, 348, 349
methylene-bridged, **8**, 379
nitrogen donors
square-planar, **8**, 317, 318
norbornadiene, **8**, 376
oxidative addition
hydrogen, **8**, 306–9
organic halides, **8**, 309
silanes, **8**, 306–9
oxygen donors
square-planar, **8**, 315, 316
phenoxy
square-planar, **8**, 316
phosphido-bridged, **8**, 330, 388, 389
binuclear, **8**, 372
phosphines
dissociation mechanisms, **8**, 306
intermolecular scrambling, **8**, 306
water-soluble, **8**, 306
phosphites, **8**, 305
phosphorus ylides, **8**, 311
platinum-bridged, **8**, 327
polyhydride, **8**, 400–6
poly(1-pyrazolyl)borato-tridentate, **8**, 351
pyrazolato, **8**, 374
square-planar, **8**, 317
selenocarbonyl
square-planar, **8**, 320
silver, **8**, 305
six-coordinate, **8**, 347–65
monohydrido, **8**, 347–51
square-planar, **8**, 305–23
adducts, carbon disulfide, **8**, 306
alkyl, **8**, 311–14
aryl, decarboxylation, **8**, 312
dinitrogen, oxidative addition, **8**, 310
diphosphetenes, **8**, 310
halides, **8**, 305–11
phosphines, **8**, 305–11
sulfido-bridged, **8**, 389, 390
sulfur dioxide
square-planar, **8**, 320
tetrafluorobenzobarrelene, **8**, 376
thiaboranes, **1**, 268
triflates
square-planar, **8**, 315
vinylidene, **8**, 376
Iridium(I) complexes
bis(((dialkylphosphino)methyl)-dimethylsilyl)amine, **8**, 342
five-coordinate, **8**, 338–47
methene, **8**, 342
methoxymethyl, **8**, 339
nitrosyl, **8**, 339
phosphide, **8**, 342
phosphonate, **8**, 338
square-planar
alkene adducts, **8**, 319, 320
alkyne adducts, **8**, 319, 320
sulfur ligands
square-planar, **8**, 320
tetradentate ligands, **8**, 341
tridentate ligands, **8**, 340
trifluoromethyl, **8**, 340
triphos, **8**, 340
vinylidene, **8**, 342, 344
Iridium(III) complexes
acyl, **8**, 354–8
alkyl, **8**, 354–8
arylhydrazido, **8**, 347
arylhydrazine, **8**, 347
benzodiazole, **8**, 363
carbonylchloro
oxidative addition, **8**, 359, 360
chalcogenides, **8**, 360–3
dicarbon tetrasulfide, **8**, 362
dimethyl
preparation, **8**, 355
diselenium, **8**, 361
disulfur monoxide, **8**, 362
five-coordinate, **8**, 345, 346
hydrogen sulfate, **8**, 364
methylidene, **8**, 345
octaethylporphyrin hydrides, **8**, 345
peroxo, **8**, 364
porphyrin, **8**, 365
six-coordinate, **8**, 364, 365
sulfates, **8**, 364
sulfur trifluoride, **8**, 360
tetrathio, **8**, 361
tetrazene, **8**, 347
Iridium dihydrides
chemistry, **8**, 402–4
Iridium hexahydrides, **8**, 406
Iridium pentahydrides, **8**, 406
Iridium tetrahydrides
chemistry, **8**, 405
Iridium trihydrides, **8**, 404, 405
Iridium, (bromo)undecacarbonyl-tetra-
synthesis, **8**, 492
Iridium, dodecacarbonyltetra-, **8**, 490
adsorption, **8**, 491
catalyst, **8**, 491
water–gas shift reaction, **8**, 492
hydrogenation of carbon monoxide
catalyst, **8**, 491
NMR, **8**, 495
reactions, **8**, 492
structure, **8**, 490
Iridium, pentadecacarbonylhexa-, **8**, 500
Iridodiol
synthesis, **12**, 879
Iridomyrmecin
synthesis, **12**, 329, 330
Irisquinone
synthesis, **12**, 274, 276
Ironsulfur clusters, **7**, 284
Iron
dicarbon metallacarbaboranes
6-vertex, **1**, 382
7-vertex, **1**, 382
8-vertex, **1**, 396

9-vertex, **1**, 396
10-vertex, **1**, 396
12-vertex, **1**, 400
monocarbon metallacarbaboranes
11-vertex, **1**, 378
tetracarbon metallacarbaboranes
8-vertex, **1**, 423
10-vertex, **1**, 423
11-vertex, **1**, 423
13-vertex, **1**, 423
tricarbon metallacarbaboranes
11-vertex, **1**, 420
Iron carbonyl clusters
indium bonds, **1**, 563
Iron carbonyl complexes
dinuclear
phosphido, **7**, 41–7
phosphine, **7**, 41–7
tetranuclear
phosphido, **7**, 54, 55
phosphine, **7**, 54, 55
trinuclear
phosphido, **7**, 49–53
phosphine, **7**, 49–53
Iron carbonyl hydrides
chemistry, **7**, 6–8
mononuclear cationic, **7**, 6
polynuclear, **7**, 7, 8
Iron carbonyl phosphine complexes
anionic, **7**, 39, 40
hydrido, **7**, 39, 40
Iron carbonyls
reactive fragments, **7**, 8–11
condensed phase radical chemistry, **7**, 8, 9
gas phase anion radicals, **7**, 9, 10
gas phase cation radicals, **7**, 10, 11
neutral, **7**, 8
supported, **7**, 11, 12
Iron clusters
acyl tungsten, **10**, 91
alkylidyne molybdenum, **10**, 93
alkylidyne ruthenium, **10**, 129, 130
alkylidyne tungsten, **10**, 91–3
alkyne molybdenum, **10**, 97
alkyne platinum, **10**, 151, 152
alkyne ruthenium, **10**, 130, 131
carborane tungsten, **10**, 168
cobalt compounds, **10**, 135–44
cobalt dicarbide, **10**, 140, 141
iridium compounds, **10**, 142
nickel compounds, **10**, 149, 150
phosphido tungsten, **10**, 91
rhodium compounds, **10**, 134, 141, 142
ruthenium carbide, **10**, 129
ruthenium compounds
C_3 ligands, **10**, 132
ruthenium dicarbide, **10**, 131
sulfido tungsten, **10**, 92
tungsten
C_2 ligands, **10**, 96, 97
vinylidene ruthenium, **10**, 130, 131
Iron–cobalt
dicarbon metallacarbaboranes
7-vertex, **1**, 382
Iron complexes
η^2–η^3 interconversions, **7**, 126
η^3–η^4 interconversions, **7**, 133
η^4–η^5 interconversions, **7**, 156–61
η^6-ligands, **7**, 198–204
acrylyl, **7**, 128
acyclic pentadienyl
nucleophilic additions, **12**, 666, 667
acyl
preparation, **12**, 351
acyl cyclopentadienyl, **7**, 167–76
acyls, **7**, 124
alkene cyclopentadienyl, **7**, 181–5
alkenes, **7**, 120–3
chelated, **7**, 125
alkenylidene, **7**, 127
alkyl cyclopentadienyl, **7**, 167–76
alkynes, **7**, 123
alkynyl cyclopentadienyl, **7**, 169
allyl, **7**, 127–9
chelated, **7**, 130–3
π-allyl
ambiphilic, **12**, 968
electrophilic, **12**, 968
nucleophilic, **12**, 968
organic synthesis, **12**, 966–8
synthesis, **12**, 967
synthesis from allylic substrates, **12**, 967
synthesis from dienes, **12**, 967
allylic alkylation, **12**, 862
arene cyclopentadienyl
addition/oxidation, **12**, 1004–6
cations, **7**, 199–203
preparation, **12**, 982
arenes, **7**, 198, 199
aryl, **7**, 110
aryl cyclopentadienyl, **7**, 168
azaborolyl, **7**, 165
benzo[*b*]thiepin, **7**, 138
bis(arene) cations, **7**, 203, 204
bis(boratabenzene), **7**, 204
bis(diene), **7**, 155, 156
bis(pyrrolyl)
reaction with triosmium clusters, **7**, 820
boroles, **7**, 138
cadmium clusters, **10**, 338, 339
carbenes
benzannulation, **12**, 489
cyclopropanation, **12**, 392–5
procatalysts and cyclopropanation, **12**, 395–409
carbenoid, **7**, 116–20
carbonyl anions, **7**, 3–6
carbonyls, **7**, 3–6
cationic carbene
generation, **12**, 482
cationic pentadienyl
reaction with zinc–copper reagents, **11**, 166
cyclobutadiene, **7**, 136
cyclobutadienophane, **7**, 137
cycloheptadiene
hydride abstraction, **12**, 672
ligand exchange, **12**, 671
cycloheptadienyl
nucleophilic additions, **12**, 670–3
cycloheptadienyls
crystal structures, **7**, 162
cyclohexadiene
skeletal rearrangements, **7**, 139
cyclohexadienyl
nucleophilic addition, **12**, 655–68
rotation barriers, **7**, 163
cyclooctadienyl
reactions with nucleophilies, **12**, 673
cyclooctatriene
Friedel–Crafts reaction, **12**, 673
cyclopentadienone, **7**, 136
cyclopentadienones, **7**, 138
cyclopentadienyl
η^2, η^4 ligands, **7**, 181
alkyne, **7**, 183
allene, **7**, 183
carbenoids, **7**, 176–81
dienes, **7**, 185
phosphallyl, **7**, 185
cyclopropenyl, **7**, 127
1,4-diazadienes, **7**, 125
dicarbonyl cyclopentadienyl dimer
catalyst in ligand substitution, **5**, 506
diene-carbenoid
reactions, **7**, 153–5
dienes, **7**, 139–46
dienetricarbonyl
aldol condensation, **12**, 629
nucleophilic addition, **12**, 649, 650
nucleophilic addition, **12**, 649–51
adjacent unsaturated functionalities, **12**, 630–2
optically pure, **12**, 655
dienetricarbonyl anions
diastereoselective alkylation, **12**, 629, 630
dienyldicarbonyl
nucleophilic addition, **12**, 655–73
dienyltricarbonyl
nucleophilic addition with stereocontrol, **12**, 657
resolution, **12**, 657
dihydropyridines, **7**, 138
dinuclear
μ-η^2, η^3-ligands, **7**, 250
acyl, **7**, 234–6
alkenes, **7**, 236
alkyls, **7**, 234–6
alkynes, **7**, 236
allyl, **7**, 236

allylcyclopentadienyl, **7**, 250
bridging η^1, η^1, η^2, η^2-ligands, **7**, 249, 250
bridging η^1, η^5-ligands, **7**, 249
bridging acetylide ligands, **7**, 248
bridging acyls, **7**, 234, 235
bridging alkenyl ligands, **7**, 247, 248
bridging alkenylidenes, **7**, 245, 246
bridging alkoxycarbonyls, **7**, 234, 235
bridging alkyls, **7**, 236–40
bridging allyl ligands, **7**, 249
bridging carbenes, **7**, 236–40
bridging carbenes, coupling of unsaturated hydrocarbons, **7**, 239
bridging carbenes, reactivity, **7**, 238
bridging carbynes, **7**, 241–5
bridging diazadienes, **7**, 235
bridging esters, **7**, 234, 235
bridging heteroatom analogues, **7**, 234, 235
bridging ketenes, **7**, 234, 235
bridging ketones, **7**, 234, 235
bridging methylene, **7**, 236
bridging methylene, reactivity, **7**, 238
butatrienes, **7**, 250
carbenes, **7**, 234–6
catalysis, alkene hydroformylation, **7**, 251
catalysis, radical chain additions to alkenes, **7**, 251
diallyl, **7**, 250
diboranaphthalenes, **7**, 250
dienes, **7**, 236
flyover complexes, **7**, 250
hydrocarbon ligands, **7**, 231–51
iron–iron bonds, **7**, 62–9
vinyl, **7**, 247
vinylidene, reactions, **7**, 245
vinylidene, synthesis, **7**, 245
dodecahydrooctaborates, **1**, 334
enonetricarbonyl
optically pure, **12**, 657
fluoroarene cyclopentadienyl cations, nucleophilic substitution, **12**, 993
gallium bonds, **1**, 555
halobenzene cyclopentadienyl nucleophilic substitution for halide, **12**, 992
heptafulvenes, **7**, 138
heteroalkenes, **7**, 123, 124
heteroallyl, **7**, 130
heterocyclobutadienes, **7**, 137
heterodienes, **7**, 146–8
iminoacyls, **7**, 124
indium bonds, **1**, 562
lactones, **7**, 130
mercury clusters, **10**, 338, 339
methylenecyclopropene, **7**, 133
oxadienes, **7**, 125
palladium
catalysts, hydrogenation, **10**, 38
palladium carbonyls, **9**, 205
pentadienyls, **7**, 161–3
pentafulvalenes, **7**, 138
phosphaallyl, **7**, 130
reactive dienes, **7**, 136–9
silyldienetricarbonyl
acylation, **12**, 628
tetraenes, **7**, 148–53
tetranuclear
with iron–iron bonds, **7**, 70–2
tetranuclear clusters
metallaboranes, **1**, 298–301
thallium bonds, **1**, 567, 568
thiepin, **7**, 138
thiophene dioxide, **7**, 138
transmetallation
catalyst, **12**, 158
tricarbonyl
diene protection, **12**, 660
trienes, **7**, 148–53
trimethylenemethyl, **7**, 134–6
trinuclear
with iron–iron bonds, **7**, 70–2
trinuclear clusters
metallaboranes, **1**, 296
unstable dienes, **7**, 136–9
vinyl
protonation, **12**, 393
vinylketene
synthesis, **7**, 138
xylylene, **7**, 138
zinc clusters, **10**, 338
Iron compounds
anions
gas phase reactions, **7**, 107, 108
antimony ligands, **7**, 56–8
arsenic ligands, **7**, 55, 56
atoms
co-condensation reactions, **7**, 108, 109
gas phase reactions, **7**, 107, 108
reactions, **7**, 103–9
benzathiadiazole ligands, **7**, 34
bismuth ligands, **7**, 58–60
boron ligands, **7**, 16–18
cadmium ligands, **7**, 82
carbonyl phosphine, **7**, 39
catalysis, **7**, 85–7
copper ligands, **7**, 79–81
dinuclear
nitrogen ligands, **7**, 33
diorganyl, **7**, 114–16
ethylenediamine ligands, **7**, 33
germanium ligands, **7**, 29, 30
gold ligands, **7**, 79–81
group 11 metals, **7**, 79–81
group 12 metals, **7**, 82
group 13 ligands, **7**, 16–20
group 14 ligands, **7**, 21–31
group 15 ligands, **7**, 31–60
group 16 ligands, **7**, 60–78
halide ligands, **7**, 78, 79
homoleptic, **7**, 110
indium-containing, **7**, 19
ions
gas phase reactions, **7**, 103–7
reactions, **7**, 103–9
isothiocyanates, **7**, 61
ketenylidene ligands, **7**, 21–7
lead ligands, **7**, 31
mercury ligands, **7**, 82
mononuclear
hydrocarbon ligands, **7**, 101–205
monoorganyl, **7**, 110–14
naked carbon atoms, **7**, 21–7
nitrido
pentanuclear, **7**, 35
tetranuclear, **7**, 35
nitrido ligands, **7**, 34
nitrogen bridges, **7**, 31
nitrogen ligands, **7**, 31–5
nitrosyl ligands, **7**, 32
oxygen ligands, **7**, 60, 61
phosphine ligands, **7**, 36
phosphorus ligands, **7**, 35–55
polynuclear
hydrocarbon ligands, **7**, 259–86
polythioethers, **7**, 62
post transition series elements, **7**, 79–82
pyridine ligands, **7**, 33
selenium ligands, **7**, 74–6
silicon ligands, **7**, 27–9
silver ligands, **7**, 79–81
sulfur ligands, **7**, 61
without iron–iron bonds, **7**, 61, 62
tellurium ligands, **7**, 76–8
thallium-containing, **7**, 20
thiolato carbonyl, **7**, 62
thiols, **7**, 61
tin ligands, **7**, 30, 31
transition metals, **7**, 82–5
without hydrocarbon ligands, **7**, 1–87
zinc ligands, **7**, 82
Iron isocyanides
chemistry, **7**, 12–15
gas phase, **7**, 15
halide ligands, **7**, 15
homoleptic, **7**, 12
nitrogen-donor ligands, **7**, 13
phosphine ligands, **7**, 13–15
sulfur ligands, **7**, 15
synthesis from iron carbonyls, **7**, 12, 13
Iron–molybdenum sulfur clusters, **7**, 73
Iron–nickel
dicarbon metallacarbaboranes
8-vertex, **1**, 396
Iron pentacarbonyl
disubstitution, **7**, 38, 39
monosubstitution, **7**, 35
Iron porphyrins
alkyl
preparation, **7**, 113
aryl
preparation, **7**, 113
Iron–sulfur protein

Fe_4S_4 clusters, 7, 284
Iron tetracarbonyl
anion
chemistry, **7**, 5
Iron thiocarbonyl compounds
chemistry, **7**, 16
Iron tricarbonyl
chemistry, **7**, 3
complexes
demetallation, **12**, 745
cyclobutadiene synthesis, **12**, 745
cyclooligomerization
alkynes, **12**, 744
directing substituent
stereocontrolled bond formation, **12**, 629–32
reactivity, **7**, 3
spectroscopy, **7**, 3
structure, **7**, 3
Iron, (7-acylcycloheptatriene)tricarbonyl-
alkylation, **12**, 629, 630
Iron, (benzene)dicarbaborane)-
synthesis, **1**, 377
Iron, bis(cyclopentadienyl-dinitrosyl-
properties, **7**, 232
Iron, bis[dicarbonylbis(pentamethylcyclopentadienyl)]-
properties, **7**, 232
thermolysis, **7**, 232
Iron, bis(dicarbonylcyclopentadienyl)-
analogues, **7**, 233
catalysis, **7**, 250
electrochemistry, **7**, 232
photochemistry, **7**, 234
properties, **7**, 232
pyrolysis, **7**, 232
thermolysis, **7**, 232
Iron, bis[dicarbonyl(pentamethylcyclopentadienyl)]-
catalysis, **7**, 250
Iron, (cyclobutadiene)tricarbonyl-
photolysis, **12**, 624
Iron, (1,3-cycloheptadien-6-one)-
methylation, **12**, 629, 630
Iron, (1,3-cyclohexadiene)tricarbonyl-
alkylation, **12**, 629
Iron, cyclohexadienyl-dicarbonyl(triphenylphosphine)-
reaction with Grignard reagents, **12**, 658
Iron, cyclohexadienyltricarbonyl tetrafluoroborate
reaction with organocadmium reagents, **11**, 183
Iron, dicarbonyl(cyclopentadienyl)(alkenes)-
cations
chemistry, **12**, 617
Iron, dicarbonyl(cyclopentadienyl)hydrido-, **7**, 163–7
preparation, **7**, 163
Iron, (*p*-dichlorobenzene)cyclopentadienyl-
cation
nucleophilic substitution, **12**, 993
Iron, (*o*-dichlorobenzene)cyclopentadienyl-
nucleophilic substitution, **12**, 992
Iron, dienetricarbonyl-
diastereoselective alkylation, **12**, 629
Iron, dihydridotetrakis(carbonyl)-
dissociation constant, **12**, 351
Iron, (2-formyl-1,3-butadiene)tricarbonyl-
nucleophilic addition, **12**, 625, 631, 632
Iron, (2-hydroxymethyl-1,3-butadiene)tricarbonyl-
ionization, **12**, 633
Iron, (methyl 3,5-hexadienoate)tricarbonyl
alkylation, **12**, 629, 630
Iron, (nitrobenzene)cyclopentadienyl-
cation
nucleophilic substitution, **12**, 993
Iron, octacarbonylmethylenedi-
carbonylation, **7**, 239
hydrogenation, **7**, 239
reaction with phosphinimines, **7**, 239
reaction with water/sodium iodide, **7**, 239
Iron, tetracarbonyldihydrido-
chemistry, **7**, 6, 7
Iron, tetracarbonylhydrido-
anion
chemistry, **7**, 7
Iron, tricarbonylbutadiene-
reactions with nucleophiles, **12**, 649, 650
Iron, tricarbonyl(cyclobutadiene)-
preparation, **12**, 624
Iron, tricarbonylcyclohexadiene-
reactions with nucleophiles, **12**, 650, 653
Iron, tricarbonyldicyclopentadienylethylidenedi-
photochemistry, **7**, 240
Iron, tricarbonyl(5,6,7,8-η-ergosterol)-
reactions, **12**, 626
Iron, tricarbonyl(tropone)-
cycloaddition, **12**, 625, 626
Michael addition, **12**, 625, 626
reaction with diazoalkanes, **12**, 625
Iron, tris(dibenzoylmethide)-
catalyst
cross-coupling of alkenyl bromides, **12**, 168
Iron(II) acetylacetonate
catalyst
trimethylaluminum reaction with α,β-unsaturated ketones, **12**, 156
Iron(II) complexes
π-alkenes
nucleophilic attack, **12**, 617, 618
π-alkynes
nucleophilic attack, **12**, 620
arene cyclopentadienyl
cationic, **12**, 981
Iron(III) acetylacetonate
catalyst
trimethylaluminum reaction with α,β-unsaturated ketones, **12**, 156
γ-Irons
synthesis, **12**, 580
Isatins
lithiation, **11**, 51
Isobenzofuran, alkyl-
synthesis, **12**, 585
Isobenzofurans, 1,3-dihydro-1,3-disubstituted
stereoselective synthesis, **12**, 1057
Isobutene
reactions with iron ions, **7**, 104
Isobutyronitrile, lithium
reaction with (chlorobenzene)tricarbonylchromium, **12**, 988
Isocarbacyclin
synthesis, **12**, 454
Isocarbacyclins
preparation, **11**, 165
synthesis, **11**, 472
lithiation, **11**, 40
Isocarbonyl ligands
iron stannylene complexes, **7**, 283
Isocomene
synthesis, **12**, 735
Isocoumarins
synthesis, **12**, 147
Heck reaction, **12**, 261
Isocyanates
alkene functionalization, **9**, 7
coupling with alkenes
nickel complexes, **9**, 8
cycloaddition
aziridines or oxiranes, **11**, 500
polymerization, **12**, 1224, 1225
Isocyanide complexes
group 6, **5**, 308–11
properties, **5**, 309
reactions, **5**, 309–11
structure, **5**, 309
synthesis, **5**, 308, 309
Isocyanide ligands
platinum complexes, **9**, 391–426
Isocyanides
insertion into organonickel complexes, **9**, 61
polymerization, **12**, 1224, 1225
Isocyclocolorenone
synthesis, **12**, 696
Isoflavanones
synthesis, **11**, 507; **12**, 134
Isolobophytolide
synthesis, **12**, 881, 1077
Isomerization, **12**, 693
catalysis, **10**, 37
iron carbonyl catalysts, **7**, 86, 87
Isomünchnones
meosionic oxazolium ylides, **12**, 432
reactions, **12**, 448, 449
Isonitrile complexes

triiron clusters, **7**, 263, 264
Isonitriles
insertion reactions
carbene complexes, **12**, 534
reaction with manganese carbenes, **6**, 66
Isooxazolidines
synthesis
chromium complexes, **12**, 1048, 1049
Isopimaridiene
insertion reactions, **12**, 452
Isoprene
telomerization with alcohols, **12**, 787
telomerization with amines, **12**, 788
Isoquinoline
cross-coupling reactions
Grignard reagents, **12**, 177
Isoquinoline alkaloids
synthesis
hydrogenation, **12**, 17
Isoquinoline, tetrahydro-
synthesis
chromium complexes, **12**, 1049
Isoquinolinecyclobutenediones, tetrahydro-
synthesis, **12**, 211
Isoquinolines
β-lactams from photochemical reactions
carbene complexes, **12**, 551
synthesis, **12**, 273
transmetallation, **12**, 135
Isoquinolines, tetrahydro-
synthesis
lithiation, **11**, 19
Isoquinolones
synthesis
Heck reaction, **12**, 261
Isoquinuclidines
synthesis, **12**, 889
3-Isorauniticine
synthesis
allylmetal catalysts, **12**, 919
Isoretronecanol
synthesis, **12**, 372, 436
Isorobustin
synthesis, **11**, 473
Isospecific polymerization
Ziegler–Natta catalysts, **12**, 1197
Isotactic
definition, **12**, 1196
Isothiazoles
synthesis, **12**, 332
Isothiocyanates
reactions with Grignard reagents, **11**, 139
Isourea, 2-allyl-
allylic amidation, **12**, 810
Isovelleral
synthesis, **12**, 245
Isoxazoles
reaction with enamines, **12**, 1084
Isoxazolidine, iodomethyl-
synthesis, **12**, 115, 116
Isoxazolines
synthesis, **11**, 212
Δ^4-Isoxazolines
synthesis
carbene complexes, **12**, 528
Ispdienol
synthesis, **12**, 630
Itaconic acid
hydrogenation
asymmetric, **12**, 19

J

Jaborosalactone
  synthesis, **12**, 871
Japanese hop ether
  synthesis, **12**, 722
Jasmonate, dihydro-
  synthesis, **12**, 879
Jasmone
  synthesis, **12**, 722
Jasmone, dihydro-
  synthesis, **12**, 453
    cobalt-mediated propargylation, **12**, 696
cis-Jasmone
  synthesis, **12**, 352, 378
Jatraphone
  synthesis, **12**, 369
epi-Jatraphone
  synthesis, **12**, 369
Jatrophone
  synthesis, **12**, 108, 109, 206, 253, 254
Joubertiamine, *O*-methyl-
  synthesis, **11**, 473; **12**, 663
Juliá epoxidation
  asymmetric
    chalcones, **12**, 1098
Juvabione
  synthesis, **12**, 1007

K

Kainic acid
synthesis, **12**, 735
Kalloide
synthesis
lithiation, **11**, 72
Kaurane
synthesis, **12**, 750
Ketals
conversion to ethers
organoaluminum hydrides, **11**, 300
Ketene complexes
from carbene complexes, **12**, 550
transition metals
nucleophilic attack, **12**, 569–73
Ketene silyl acetals
alkylation
allylic acetates, **12**, 868
chiral
asymmetric α-amination, **11**, 337
Ketenes, vinyl
nucleophilic addition
transition metal complexes, **12**, 633
transition metal complexes
reactions, **12**, 633, 634
Ketenimines
cycloaddition reactions
[3+2], **12**, 949
reactions with metal carbenes, **12**, 392
Ketenimines, allenyl-
synthesis, **12**, 593
Ketenylidene complexes
cationic triiron cluster intermediates, **7**, 275
iron carbide clusters, **7**, 272
tetrairon intermediates
nucleophilic addition
leading to acyl clusters, **7**, 268
triiron clusters, **7**, 270
Ketenylidene ligands
bridging
in heteronuclear metal–metal bond synthesis, **10**, 13
Ketimines
zirconocene complexes
reductive coupling, **12**, 342, 345
Keto amides
intramolecular coupling, **12**, 1078–80
Keto esters
intramolecular coupling, **12**, 1078–80
β-Keto esters
reactions with organolead compounds, **11**, 471–4
Ketocarbenes
1,3-dipolar addition, **12**, 402
Ketone enamines
reactions with organolead compounds, **11**, 477, 478
Ketone enolates
allylic alkylation, **12**, 862
reactions with organolead compounds, **11**, 477, 478
Ketones
alkenation
trialkylstibine, **11**, 489
alkylation
dicobalt propargylium complexes, **12**, 691
allylic alkylation, **12**, 862–5
aluminum alkylenation, **1**, 474
α-amino
asymmetric reduction, **12**, 23
aryl
reductive coupling, **12**, 1074
synthesis using Collman's reagent, **12**, 352
t-butyl phenyl
asymmetric reduction, **12**, 25
α-chloro-
reactions with allylstannanes, **12**, 202
cycloaddition reactions, **12**, 931–3
dehydrogenation
organoselenium compounds, **11**, 534
deoxygenative coupling, **12**, 1091
diaryl
cross-coupling reactions, **12**, 1075
diastereoselective addition to Grignard reagents, **11**, 145–50
dimerization
strained alkene synthesis, **12**, 1074
enantioselective addition to Grignard reagents, **11**, 150
α-fluoro-
synthesis, **12**, 863
hydrogenation
asymmetric, **12**, 23–6
α-hydroxy
optically active, asymmetric dihydroxylation, **12**, 1150
β-hydroxy-
synthesis, **11**, 492
intramolecular coupling, **12**, 1077
α-iodo-
synthesis, **11**, 337
mercuriation, **3**, 140
α-methoxylation
organosilicon compounds, **11**, 337
α-methyl allyl
synthesis, **11**, 496
methylenation, **12**, 578–83
Tebbe's reagent, **12**, 586, 587
zirconium complexes, **12**, 595
α-phenyl
synthesis, **11**, 337
preparation
metal carbonyls, **12**, 351
organoboranes, **11**, 222
prochiral
enantioselective reduction, organoboron compounds, **11**, 258
propargylation, **12**, 691
propynyl
synthesis, **11**, 497
reactions with π-allylmolybdenum complexes, **12**, 970, 971
reduction
asymmetric, **12**, 23–5
dienetricarbonyliron complexes, **12**, 631
reductive dimerization
molybdenum or tungsten complexes, **12**, 1090
ring expansion
organomercury compounds, **11**, 410
selective dioxolanation, **11**, 340
α-sulfonyloxylation, **11**, 337
synthesis
acyl iron complexes, **12**, 353
by carbonylation, **12**, 368, 369
carbon monoxide insertion reactions, **12**, 253–7
organoantimony compounds, **11**, 496, 497
organoboron compounds, **11**, 214, 216
organomercury compounds, **11**, 398
α,β-unsaturated
asymmetric hydrogenation, **12**, 9, 10, 19–21
[3+2] cycloaddition reactions, **12**, 927, 928, 947
hydroboration, **11**, 198
methylenation, **12**, 582, 586
synthesis, organoantimony compounds, **11**, 491

Ketones, amido-
  synthesis
    Grignard reagents, **11**, 152
Ketones, α-hydroxymethylene
  reactions with organolead compounds, **11**, 474
Ketones, methyl vinyl
  cycloaddition reactions, **12**, 927
Khellin
  synthesis, **12**, 492
    lithiation, **11**, 72
Kihizimycin
  synthesis, **12**, 587
Kinetic resolution
  chromium tricarbonyl arene complexes, **5**, 505
Kläui's ligand, **5**, 230
Kocheshkov reaction
  stannic chloride, **2**, 245
Koumidine
  synthesis, **12**, 582
Koumine
  synthesis, **12**, 582, 587
Kubas's compound
  preparation, **5**, 223
Kynurenine
  synthesis
    Stille coupling, **12**, 219

L

Lacrimin
  synthesis, **12**, 212
β-Lactam antibiotics
  synthesis, **12**, 696
Lactams
  dehydrogenation
    organoselenium compounds, **11**, 534
  synthesis
    rhodium complexes, **8**, 201
β-Lactams
  bicyclic
    synthesis, **12**, 441
  formation
    in insertion reactions, **12**, 456
    transition metal complexes, **12**, 249–51
  from imines and carbene complexes
    photochemical, **12**, 551–6
  α-methylene
    synthesis, **12**, 96, 365
  ring opening
    palladium salts, **12**, 140
  synthesis, **12**, 363
    chromium complexes, **12**, 1048, 1049
    cobalt-mediated propargylation, **12**, 696
    sulfur ylides, **12**, 436, 437
Lactones
  bicyclic
    synthesis, **12**, 357, 361
  dehydrogenation
    organoselenium compounds, **11**, 534
  formation
    in insertion reactions, **12**, 455
  methylenation
    Tebbe's reagent, **12**, 588
  α-methylene
    preparation of substituted, **12**, 617, 618
    synthesis, **12**, 582
  synthesis, **12**, 147, 363
    carbene complexes, **12**, 504–6
    hydridocarboxylation, **12**, 355
    insertion reactions, **12**, 242–8
    intramolecular alkoxycarbonylation, **12**, 247
  β,γ-unsaturated
    synthesis, **12**, 140
β-Lactones
  synthesis
    palladium catalysts, **12**, 357
γ-Lactones, 2-propylidene-
  hydrogenation
    asymmetric, **12**, 21
Lanosterol
  inactivators
    synthesis, **12**, 583
Lanthanide atoms
  reactions with arenes, **4**, 8, 9
Lanthanide complexes, **4**, 11–192
  alkenyl, **4**, 26, 27
  alkynyl, **4**, 26, 27
  allyl, **4**, 27, 28
  aluminum tetrahydride derivatives, **1**, 551
  anionic tetrakis(cyclopentadienyl), **4**, 58
  arene, **4**, 97
  arsolyl, **4**, 96
  bis(cyclooctatetraenyl), **4**, 104
    anionic, **4**, 102–4
  bis(pentamethylcyclopentadienyl), **4**, 64–8
    hydrazido ligands, **4**, 82
    hydrides, **4**, 92–4
    hydrocarbyls, **4**, 83–92
    sulfur ligands, **4**, 78
    tellurium ligands, **4**, 78
  bis(pentamethylcyclopentadienyl) alkoxides, **4**, 77
  bis(pentamethylcyclopentadienyl) aryloxides, **4**, 77
  bis(pentamethylcyclopentadienyl) fluorides, **4**, 74
  bis(pentamethylcyclopentadienyl) halo, **4**, 73–94
  carbonyls, **4**, 13
  cyclooctatetraenyl, **4**, 98–106
  cyclooctatetraenyl halo, **4**, 99–102
  cyclopentadienyl, **4**, 28–58
  cyclopentadienyl dihalo, **4**, 32–5
  cyclopentadienyl halo, **4**, 28
  dicyclopentadienyl, **4**, 28–32
  dicyclopentadienyl allyl, **4**, 45
  dicyclopentadienyl aluminum hydrides, **4**, 49
  dicyclopentadienyl borohydride, **4**, 48
  dicyclopentadienyl carboxylates, **4**, 39
  dicyclopentadienyl halo, **4**, 35–50
  dicyclopentadienyl hydrides, **4**, 47–50
  dicyclopentadienyl hydrocarbyls, **4**, 43–6
  fluorenyl, **4**, 63
  fullerenes, **4**, 106
  heptamethylindenyl, **4**, 63
  heterobimetallic, **4**, 107–12
  heteronuclear NMR, **4**, 113
  homogeneous catalysis, **4**, 113–22
  hydrocarbyls, **4**, 13–26
    anionic homoleptic, **4**, 20, 21
    heteroleptic, **4**, 21–6
    neutral homoleptic, **4**, 13–19
  indenyl, **4**, 63
  metallaboranes, **1**, 349
  metallacarbaboranes, **1**, 373–424
  octahydrotriborates
    metallaboranes, **1**, 316
  pentadienyl, **4**, 95, 96
  pentamethylcyclopentadienyl, **4**, 64–95
  pentamethylcyclopentadienyl dihalo, **4**, 68–73
  pentamethylcyclopentadienyl halo, **4**, 64
  phospholyl, **4**, 96
  ring-bridged cyclopentadienyls, **4**, 59–61
  tetrahydroborates, **1**, 286
  tricyclopentadienyl, **4**, 51–5
  tricyclopentadienyl aquo, **4**, 56
  tricyclopentadienyl halo, **4**, 57, 58
  trihydroborates, **1**, 288
  tris(pentamethylcyclopentadienyl), **4**, 95
  zero oxidation state, **4**, 1–9
    alkenes, **4**, 1, 2
    alkynes, **4**, 1, 2
    arenes, **4**, 2–8
    arenes, matrix isolation studies, **4**, 2
    bis(arene), **4**, 3–8
    bis(arene), bonding, **4**, 5, 6
    bis(arene), magnetism, **4**, 6, 7
    bis(arene), reactions, **4**, 7, 8
    bis(arene), redox reactions, **4**, 7
    bis(arene), ring displacement reactions, **4**, 7
    bis(arene), structure, **4**, 3
    bis(arene), synthesis, **4**, 3
    naphthalene, **4**, 2
Larendamycin
  analogues
    synthesis, **11**, 232
Lasiodiplodin
  synthesis
    lithiation, **11**, 33
Lasiol
  synthesis, **12**, 623
Laurenan
  synthesis, **12**, 589
Lauthisan
  synthesis, **12**, 590
Lead compounds
  alk-1-ynyl tricarboxylates
    stability, **11**, 465
  aryl tricarboxylates
    chemistry, **11**, 465
  transmetallation–insertion reactions, **12**, 154
  tributyl(α-methoxyalkyl)

preparation, **11**, 463
Lead, alk-1-ynyl tricarboxylates
preparation, **11**, 468
Lead, allyltriphenyl-
allyl group transfer
radical-mediated transfer, **11**, 464
Lead, aryl triacetates
reactions
copper catalysis, **11**, 480, 481
replacement of lead
by azides, **11**, 479
by iodide, **11**, 479
Lead, aryl tricarboxylates
reactions
acid catalysis, **11**, 481–3
Lead, (2,4-dimethoxyphenyl) triacetate
preparation, **11**, 466
Lead, furyl triacetate
preparation, **11**, 467
Lead, (*p*-methoxyphenyl) triacetate
preparation, **11**, 466
structure, **11**, 465
Lead, phenyl triacetate
preparation, **11**, 467
Lead, tetraalkyl-
coupling with acid chlorides
palladium-catalyzed, **11**, 464
coupling reactions
with acyl chlorides, **12**, 221
group transfer to aldehydes, **11**, 462, 463
reaction with silver nitrate, **3**, 71
Lead, tetraethyl-
group transfer to aldehydes, **11**, 462
Lead, thienyl triacetate
preparation, **11**, 467
Lead, *p*-tributylstannylphenyl triacetate
preparation, **11**, 467
Lead, (2,4,6-trimethoxyphenyl) triacetate
preparation, **11**, 466
Lead, vinyl tricarboxylates
preparation, **11**, 468
stability, **11**, 465
Lepicidin
synthesis, **12**, 79, 80, 204
Leucine-enkephalin
analogues
synthesis, **12**, 17
Leukotriene B_4
preparation, **12**, 189
Leukotriene E_4
acetylenic analogues
synthesis, **12**, 694
Levallorphan
synthesis, **12**, 18
Levamisole
synthesis, **12**, 23
LIDAKOR
applications, **11**, 96
Ligands
anionic, **12**, 2
cationic, **12**, 3
classes, **12**, 2, 3
neutral, **12**, 2
Limaspermine
synthesis, **12**, 660
Lipstatin, tetrahydro-
synthesis, **12**, 97
Liquid crystalline polysiloxanes
properties, **2**, 131
structural types, **2**, 131
Liquid silicone elastomers, **2**, 127
Lithiation
benzenetricarbonylchromium, **12**, 1018–20
rings
organic synthesis, **12**, 1017–37
tricarbonyl cyclopentadienyl manganese complexes, **6**, 110
Lithium aluminum hydride
titanium trichloride system
coupling reactions, **12**, 1073
Lithium amides
homochiral
in organic synthesis, **11**, 6
in organic synthesis, **11**, 4
structure, **1**, 5
Lithium aminoborohydrides
synthesis, **11**, 244
Lithium *N*-benzyltrimethyl-silylamide
in organic synthesis, **11**, 5
Lithium bis(trimethylsilyl)amide
in organic synthesis, **11**, 5
Lithium *t*-butyl(diphenylmethyl-silyl)amide
in organic synthesis, **11**, 5
Lithium chloride
Stille reaction, **12**, 201
Lithium compounds
magnesium
preparation, **1**, 105
Lithium cyclohexylisopropylamide
in organic synthesis, **11**, 5
Lithium dialkylborohydrides
synthesis, **11**, 205
Lithium diethylamide
in organic synthesis, **11**, 5
Lithium diisopropylamide
in organic synthesis, **11**, 4
Lithium 1-(dimethylamino)-naphthalenide
in organic synthesis, **11**, 6
Lithium dimethylbenzylamide
in organic synthesis, **11**, 5
Lithium di-(2,4,6-triisopropyl-phenyl)ethyl-
reactions, **11**, 244
Lithium hexamethyldisilazane
in organic synthesis, **11**, 5
Lithium magnesium compounds
alkynyl
preparation, **1**, 105
benzyl
preparation, **1**, 105
Lithium monoorganylborohydrides
synthesis, **11**, 205
Lithium morpholide
in organic synthesis, **11**, 5
Lithium naphthalenide
in zinc activation, **12**, 183
zerovalent copper compounds from, **12**, 63
Lithium naphthalide
in organic synthesis, **11**, 4
Lithium *t*-octyl-*t*-butylamide
in organic synthesis, **11**, 5
Lithium organobarates
formation, **1**, 119
Lithium piperidide
in organic synthesis, **11**, 5
Lithium pyrrolidide
in organic synthesis, **11**, 5
Lithium reagents
in organic synthesis, **11**, 3–6
transmetallation, **12**, 164–79
Lithium 2,2,6,6-tetramethyl-piperidide
in organic synthesis, **11**, 5
Lithium 2-thienylcyanocuprate
preparation, **12**, 64
Lithium trialkylalkynylborate
reactions
with allyl bromide, **12**, 858
Lithium tri-*t*-butylberyllate
synthesis, **1**, 42
Lithium triorganozincates
preparation, **11**, 182
Lithium zincates
reaction with
dibromocyclopropanes, **11**, 180
Lithium, alkenyl-
structure, **11**, 97
Lithium, alkyl-
unstabilized sp^3
preparation using
organolithium reagents, **11**, 7
Lithium, allyl-
isomerization, **11**, 98
structures, **1**, 24
Lithium, aryl-
anions
synthesis, **11**, 42–81
synthesis
lithium–halogen exchange, **11**, 66–70
Lithium, benzyl-
synthesis, **11**, 55–61
Lithium, butyl-
in organic synthesis, **11**, 3
structure, **1**, 19, 20
superbase reagent, **11**, 94
Lithium, *s*-butyl-
in organic synthesis, **11**, 3
Lithium, *t*-butyl-
in organic synthesis, **11**, 3
structure, **1**, 19, 20
Lithium, diarylimido-
structure, **1**, 3
Lithium, diarylimino-
structure, **1**, 3
Lithium, 2,4-dimethyl-2,4,-pentadienyl-
structure, **11**, 103
Lithium, heteroaryl-
anions
synthesis, **11**, 42–81
Lithium, indenyl-
structure, **1**, 24
Lithium, methyl-
in organic synthesis, **11**, 3
structure, **1**, 20
tetramer
structure, **1**, 30

Lithium, 2-methyl-2,4-pentadienyl-
structure, **11**, 103
Lithium, pentadienyl-
structure, **11**, 103–5
Lithium, phenyl-
in organic synthesis, **11**, 3
Lithium, phenylallyl-
conformation, **11**, 108
Lithium, tributyltin-
preparation, **2**, 283
Lithium, tri(2-furyl)tin-
structure, **2**, 283
Lithium, trimethylsilyl-
synthesis, **2**, 22
Lithocene
anion
structure, **1**, 22
Living polymerization
alkenes
Ziegler–Natta catalysts, **12**, 1205
definition, **12**, 1210
two alkene monomers
Ziegler–Natta catalysts, **12**, 1205
Loganin
synthesis
trimethylenemethane complexes, **12**, 942
Loganine
synthesis, **12**, 735
LTA_4
synthesis, **12**, 631
asymmetric epoxidation, **12**, 1126
LTB_4
synthesis, **12**, 631
Luminescent probes
polymers end-capped
ring-opening metathesis polymerization, **12**, 1215
Lutidinium ylides, **7**, 590
Lycoramine
synthesis, **11**, 473
γ-Lycorane
synthesis, **12**, 761
Lycoranes
synthesis, **12**, 826
Lycoridine
synthesis, **12**, 285, 289
Lysergic acid
synthesis, **12**, 756

M

Macbecin I
  synthesis, **12**, 112
Macrocycles
  preparation
    chromium tricarbonyl complexes, **5**, 528
  synthesis
    intramolecular allylic alkylation, **12**, 881
    Stille reaction, **12**, 206
Macrocyclic lactones
  synthesis, **12**, 845
Madelung indole synthesis
  lithiation, **11**, 59
Magnesacycles
  dimerization, **1**, 92
Magnesacyclohexane
  dimerization, **1**, 92
Magnesacyclopentanes
  formation, **1**, 87
Magnesates
  orthomagnesiation
    cubane, **1**, 64
  structures, **1**, 29
Magnesiation
  ethene, **1**, 95
Magnesium
  activated
    preparation, **11**, 130
  activation
    Grignard reagent preparation, **1**, 59
  active
    formation from magnesium hydride, **1**, 59
  addition compounds
    1,4-diphenyl-1,3-butadiene, **1**, 97, 98
    unsaturated organic molecules, **1**, 96–100
  co-condensation
    with fluorobenzene, **1**, 59
    with organic halides, **1**, 62
  highly active, **1**, 100
  single-crystal surface, **1**, 59
  slurries
    Grignard reagent preparation, **1**, 59
  transmetallation
    allylic alkylation, **12**, 853
  turnings
    Grignard reagent preparation, **1**, 59
  ultrasonic irradiation
    Grignard reagent preparation, **1**, 59
Magnesium aluminum compounds
  alkynyl, **1**, 107
Magnesium carbide
  structure, **1**, 98
Magnesium compounds
  anthracene
    complexes with oxygen and nitrogen ligands, **1**, 100
    crystal structures, **1**, 99
    ethene insertion, **1**, 100
    Grignard reagent preparation, **1**, 59
    in organometallic synthesis, **1**, 100
    polymer supported, **1**, 99
    radical anion complexes, **1**, 99
    reactivity, **1**, 100
    solid-state NMR, **1**, 100
  chemistry, **1**, 57–120
  diaryl
    preparation, **1**, 86
  dimetallic
    preparation, **11**, 175
  diorganyl
    addition to alkenes, **1**, 87
    addition to cyclopentenes, **1**, 87
    alkyl bridges, **1**, 88
    3-coordinate complexes, **1**, 88
    4-coordinate complexes, **1**, 88
    crown ether adducts, **1**, 93
    crown ether complexes, **1**, 89
    cryptand adducts, **1**, 93
    cryptand complexes, **1**, 89
    electron transfer reactions, **1**, 95
    ether cleavage reactions, **1**, 96
    fluxionality, **1**, 93
    gas-phase structures, **1**, 94
    inner complexes, **1**, 90
    intramolecular coordination, **1**, 90
    intramolecularly coordinated crown ethers, **1**, 90
    ionic disproportionation, **1**, 89
    ionization, **1**, 89, 90
    molecular association, solutions, **1**, 92
    NMR, **1**, 92
    physical properties, **1**, 87–94
    polyether complexes, **1**, 89
    preparation, **1**, 85–7
    properties, **1**, 85–96
    reactions, **1**, 94–6
    reactions, organic substrates, **1**, 95
    reactions with carbon monoxide, **1**, 95
    reactions with crown ethers, **1**, 95
    redistribution of organic ligands, **1**, 86
    rotaxane threaded complexes with crown ethers, **1**, 93
    rotaxane threaded structures, **1**, 89
    solid-state, structures, **1**, 87
    solutions, properties, **1**, 92
    solvation by ethers, **1**, 88
    structures, **1**, 87–94
    thermal decomposition, **1**, 94
  organo-
    ion pair formation, **1**, 93
  organolithium
    reactivity, **1**, 105
Magnesium hydrides
  organyl
    hydromagnesiation, octene, **1**, 101
Magnesium lithium bromide
  tris(trimethylsilyl)methyl, **1**, 105, 106
Magnesium reagents
  transmetallation, **12**, 164–79
Magnesium, alkylhalo-, **1**, 70
Magnesium, allenylchloro-
  preparation, **1**, 70
Magnesium, allyl-
  isomerization, **11**, 98
Magnesium, allylchloro
  structure, **1**, 74
Magnesium, bis(fluorenyl)-
  preparation, **1**, 86
Magnesium, bis(indenyl)-
  preparation, **1**, 86
Magnesium, bis(trimethylsilyl)-
  synthesis, **2**, 22
Magnesium, bis[tris(trimethylsilyl)methyl]-
  solid-state
    structure, **1**, 87
  structure, **1**, 87, 88
  sublimation, **1**, 85
Magnesium, butyl(*s*-butyl)-
  preparation, **1**, 85, 86
Magnesium, chloro(pentamethylcyclopentadienyl)-
  in organometallic synthesis, **1**, 84
Magnesium, cyclopentadienylhalo-
  structure, **1**, 74
Magnesium, cyclopentadienylhydrido-
  formation, **1**, 101
Magnesium, cyclopentadienylneopentyl-
  gas-phase electron diffraction, **1**, 94
Magnesium, dialkynylbis(TMEDA)-
  structure, **1**, 91
Magnesium, dicyclopentadienyl-
  ab initio LCAO–MO calculations, **1**, 94

adducts, **1**, 93
preparation, **1**, 86
structures, **1**, 91
Magnesium, diethylamidohydride
hydromagnesiation of alkenes, **1**, 103
Magnesium, dimethyl-
addition to alkoxyketones, **1**, 95
Magnesium, dineopentyl-
mean bond dissociation energies, **1**, 94
Magnesium, diphenyl-
activation, **1**, 86
halogen exchange of crown ether complexes, **1**, 96
intramolecular metallation of crown ether complex, **1**, 96
Magnesium, diphenylvinylene-
synthesis, **1**, 97
Magnesium, tetrakis[(diphenylvinylene)tetrahydrofuran]-
structure, **1**, 91
Magnesium, tetrakis[(naphthalenediyl)tetra- hydrofuran]-
structure, **1**, 91
Magnesium, tetrakis-[(phenylene)tetrahydrofuran]-
structure, **1**, 91
Magydardienediol
synthesis, **12**, 581
Malonates
reaction with acyltetracarbonyl cobalt complexes, **12**, 354
Malonates, alkylidene-
reaction with organozinc compounds, **11**, 175
Malonates, diazo-
cyclopropanation, **12**, 399
Malonic acid
reactions with organolead compounds, **11**, 475–7
Malononitrile
anions
arylation, **12**, 226
Malyngolide
synthesis, **12**, 371
Manganacycles
synthesis, **6**, 39
Manganates
hexacarbonyl-
physical properties, **6**, 13, 14
properties, **6**, 13, 14
reactions, **6**, 14
synthesis, **6**, 13, 14
pentacarbonyl-
physical properties, **6**, 10–12
properties, **6**, 10–13
reactions, **6**, 12, 13
synthesis, **6**, 10–12
Manganese
dicarbon metallacarbaboranes
7-vertex, **1**, 382
12-vertex, **1**, 400
tricarbon metallacarbaboranes
11-vertex, **1**, 420
Manganese anions
alkylation, **6**, 27, 28
Manganese clusters
cyclopentadienyl tricarbonyl, **6**, 112
Manganese complexes
(6-substituted-cyclohexadienyl) tricarbonyl
addition/oxidation, **12**, 1006
acyclic dienyltricarbonyl
preparation, **12**, 675
acyl, **6**, 22–7
addition, **6**, 24
decarbonylation, **6**, 28
hydrosilation, **6**, 25
photochemistry, **6**, 25
reduction, **6**, 24
synthesis, **6**, 22–4
synthesis of manganese anion acylation, **6**, 22
alkene–alkylidene, **6**, 67
alkenes
nucleophilic attack, **6**, 99
preparation, **6**, 103
alkenyl
reactions, **6**, 43
alkyls, **6**, 27–43
carbomethoxylation, **6**, 39
cleavage by halogens, **6**, 41
conjugate addition, **6**, 39
electrophilic cleavage, **6**, 41
insertion reactions, **6**, 36–41
insertion reactions, alkenes, **6**, 39
insertion reactions, carbon dioxide, **6**, 36–9
insertion reactions, carbon monoxide, **6**, 36–9
insertion reactions, sulfur dioxide, **6**, 41
insertion reactions, tellurium, **6**, 41
IR spectroscopy, **6**, 44
NMR, **6**, 44
nucleophilic transfer, **6**, 43
photochemical reactions, **6**, 41, 42
physical properties, **6**, 44–8
reaction with butadiene, **6**, 40
reaction with metal hydrides, **6**, 36
reaction with pentadiene, **6**, 40
reactivity, **6**, 35–43
reductive elimination, **6**, 43
spectroscopy, **6**, 44
synthesis, **6**, 27–35
theoretical calculations, **6**, 45, 46
thermal decomposition, **6**, 28
thermochemistry, **6**, 44, 45
thermolysis, **6**, 35
alkynes, **6**, 56–9
formation, **6**, 103
preparation, **6**, 103
alkynyl
reactions, **6**, 43
allyl
agostic hydrogen interactions, **6**, 100
formation, **6**, 104, 105
nucleophilic attack, **6**, 99
reactions, **6**, 42
antimony ligands, **6**, 85–8
arene, **6**, 94–6
cations, **6**, 95
nucleophilic attack, **6**, 94
substituted, nucleophilic attack, **6**, 95
arene cyclopentadienyl
addition/oxidation, **12**, 1006, 1007
arene tricarbonyl
cationic, **12**, 981
preparation, **12**, 982
arsenic ligands, **6**, 85–8
aryl, **6**, 48–55
reactions, **6**, 53–5
synthesis, **6**, 48–53
bis(cyclooctadiene)
formation, **6**, 103
bismuth ligands, **6**, 85–8
bridging sulfuryl ligands, **6**, 89
bridging thionyl ligands, **6**, 89
cadmium clusters, **10**, 337
carbenes, **6**, 59–68
benzannulation, **12**, 489
photochemical benzannulation, **12**, 567
reactions, **6**, 65–8
synthesis, **6**, 59–64
carbon bonds, **6**, 21–78
carbonyl, **6**, 1–16
17-electron, **6**, 9, 10
carbonyl halides, **6**, 1–16
properties, **6**, 14–16
reactions, **6**, 16
synthesis, **6**, 15
carbonyl pseudohalides
properties, **6**, 16
carbynes, **6**, 59–68
reactions, **6**, 65–8
synthesis, **6**, 59–64
cycloheptadienyl
nucleophilic attack, **6**, 96, 97
phenyllithium additions, **12**, 676
cyclohexadienyl
cationic, reaction with hydrides, **6**, 97
nucleophilic attack, **6**, 97
reaction with nucleophiles, **6**, 97
cyclopentadienyl, **6**, 109–16
cyclopentadienyl polycyclic hydrocarbons, **6**, 110
cyclopentadienyl tricarbonyl, **6**, 110, 111
acid–base chemistry, **6**, 115
anionic, **6**, 113
arsenic ligands, **6**, 112
bridging cyanides, **6**, 112
bridging hydride, **6**, 112
bridging vinyl, **6**, 112
chemistry, **6**, 115, 116
derivatives, **6**, 111–15
dinuclear, **6**, 116
haptotropic rearrangement, **6**, 110

monosubstituted, **6**, 112, 113
nitrogen ligands, **6**, 112
phosphine, **6**, 113
phosphorus ligands, **6**, 112
photochemical reactions, **6**, 111
photoinduced oxidative addition, **6**, 116
photolysis, **6**, 115
radical species, **6**, 112
redox chemistry, **6**, 115
selenium ligands, **6**, 112
sulfur ligands, **6**, 112
dienes
anions reaction with electrophiles, **6**, 100
[2+2] cycloaddition, **6**, 101
dienyl
nucleophilic additions, **12**, 673–9
(dienyl)
nucleophilic attack, **6**, 97–9
dienyldicarbonylnitroso
preparation of cations, **12**, 677
reactions with nucleophiles, **12**, 677
dienyltricarbonyl
reactions with nucleophilies, **12**, 675
dihydrogen, **6**, 73
electrophilic attack, **6**, 99–101
fluorenyl tricarbonyl, **6**, 111
fluorophosphines, **6**, 86
formyl, **6**, 25–7
decarbonylation, **6**, 26
thermal stability, **6**, 26
α-haloalkyl, **6**, 42
(halobenzene) tricarbonyl cation
nucleophilic substitution for halide, **12**, 994
(hexamethylbenzene)-
nucleophilic attack, **6**, 94
hydrides, **6**, 68–78
acidity, **6**, 71, 72
alkene reduction, **6**, 70
basicity, **6**, 71, 72
dinuclear, **6**, 74–8
hydride migration, **6**, 70
insertion reactions, carbon dioxide, **6**, 74
insertion reactions, ethyne, **6**, 74
photolysis, **6**, 73
properties, **6**, 71–3
Raman spectroscopy, **6**, 78
reactions, **6**, 70, 71
synthesis, **6**, 68, 69
theoretical studies, **6**, 73
thermochemistry, **6**, 72, 73
trinuclear, **6**, 74–8
hydrocarbon, **6**, 93–105
α-hydroxyalkyl-
synthesis, **6**, 33
indium bonds, **1**, 560
isonitrile, **6**, 119–23
π acidity, **6**, 120
alkylation, **6**, 120
allyl transfer, **6**, 120
electron transfer, **6**, 120
insertion reactions, **6**, 120
ligand substitution catalysis by PdO, **6**, 119
NMR coupling constant, **6**, 120
substitution, photochemical, **6**, 119
ketene imines, **6**, 66
ketoacyl
reaction, **6**, 25
ligand bond energies, **6**, 111
mercury clusters, **10**, 337, 338
(methylenecyclohexadienyl)
reaction with electrophiles, **6**, 101
nitrogen ligands, **6**, 84, 85
nitrosyl, **6**, 119–23
17-electron, **6**, 123
insertion reactions, **6**, 123
insertion reactions, carbon monoxide, **6**, 123
reduction, **6**, 123
nonmetallic elements, **6**, 83–90
oxygen ligands, **6**, 84, 85
palladium carbonyls, **9**, 205
pentacarbonyl halides, **6**, 14
pentadienyl
formation, **6**, 104, 105
reaction with amines, **6**, 98
reactions, **6**, 42
perfluoroalkyl
synthesis, **6**, 28
phosphido-bridged, **6**, 86
phosphido-bridged heterobimetallic, **6**, 87
phosphine sulfides, **6**, 87
phosphinedithiocarbonates, **6**, 89
phosphines
deprotonation, **6**, 87
phosphorus ligands, **6**, 85–8
three-electron donor, **6**, 85
phosphorus(V) ligands, **6**, 85
photochemical transformations, **6**, 102
preparations, **6**, 103–5
reactions, **6**, 103–5
selenium ligands, **6**, 88–90
silane, **6**, 73, 74
photoacoustic calorimetry, **6**, 74
sulfur ligands, **6**, 88–90
cyclic voltammetry, **6**, 88
tellurium ligands, **6**, 88–90
tricarbonyl(cyclohexadienyl)-
synthesis, **6**, 94
tricarbonyl(diene)
anionic, chemistry, **6**, 99
tricarbonyl(pentadienyl)
cycloaddition reactions, **6**, 102
valence photoelectron spectra, **6**, 74
vinyl, **6**, 56–9
zinc clusters, **10**, 337
Manganese(I) complexes
aluminum tetrahydride derivatives, **1**, 550
Manganese(II) complexes
alkyl, **6**, 128–32, 138–41
reaction with isocyanates, **6**, 130
alkyl ylide, **6**, 130
allyl
synthesis, **6**, 142
anionic alkyl, **6**, 128, 129, 139, 140
anionic aryl, **6**, 128, 129, 139, 140
aryl, **6**, 128–32, 138–41
bis(cyclopentadienyl), **6**, 134–8
cationic alkyl, **6**, 141
cationic aryl, **6**, 141
cyclopentadienyl carbonyl, **6**, 141–4
photolysis, **6**, 143
cyclopentadienyl carbonyl hydride, **6**, 142
dialkyl
ether adducts, **6**, 132
internal Lewis bases, **6**, 132
Lewis base adducts, **6**, 130
reaction with phosphines, **6**, 130
reaction with tetramethylethylenediamine, **6**, 131
six-coordinate adducts, **6**, 132
isonitriles, **6**, 133, 144
mono(cyclopentadienyl), **6**, 133, 134
neutral alkyl, **6**, 129–32, 140, 141
neutral aryl, **6**, 129–32, 140, 141
Schiff bases, **6**, 144
Manganese(III) complexes
salen
epoxidation, **12**, 1120–9
epoxidation mechanism, **12**, 1124, 1125
synthesis, **12**, 1120, 1121
Manganese compounds
higher oxidation states, **6**, 127–46
Manganese hydrides, **6**, 21–78
addition to organic unsaturation, **6**, 28–33
aldehyde complexes, **6**, 33
Manganese rhenium decacarbonyl
chemistry, **6**, 4, 5
Manganese vapor
reaction with benzene, **6**, 68
Manganese, cycloheptadienyltricarbonyl-
activation to nucleophile addition, **12**, 679
preparation, **12**, 675
Manganese, cyclohexadienyltricarbonyl-
nitrosyl activation, **12**, 1006
preparation
from arenemanganesetricarbonyl complexes, **12**, 96
Manganese, tricarbonyl(cycloheptatrienyl)-
nucleophilic attack, **6**, 96
Manganocenes
amine adducts, **6**, 136
chemistry, **6**, 134
electron diffraction, **6**, 135
electronic spectra, **6**, 135
hydrogen ligands, **6**, 138
Lewis base adducts, **6**, 136–8
nitrogen adducts, **6**, 137

NMR, **6**, 135
open, **6**, 135
oxygen ligands, **6**, 138
phosphene adducts, **6**, 137
photoelectron spectra, **6**, 135
reactions, **6**, 138
reactions with isonitriles, **6**, 136
spin-state, **6**, 135
Manganocenophane
preparation, **6**, 136
Manicone
synthesis
dicobalt propargylium complexes, **12**, 694
Manoalide
synthesis
lithiation, **11**, 72
Marmelo oxides
synthesis, **12**, 825
Material science
organolanthanides, **4**, 129, 130
Maytansine
synthesis, **12**, 85
Menogaril
synthesis, **12**, 491, 497
Menthol
synthesis, **12**, 21
Mercurial compounds, phenyl-
synthesis, **3**, 138, 140
Mercurials, alkyl-
reaction with alkenes, **12**, 133
Mercurials, aryl-
radical reactions, **3**, 167
Mercuriation
aromatic, **11**, 394
organic halides, **11**, 390
Mercuric acetate
in synthesis of conjugated dienes
palladium-assisted, **12**, 149
Mercuric chloride
thiophilic agent, **12**, 150
Mercury
dicarbon metallacarbaboranes
12-vertex, **1**, 400
Mercury(II) carboxylates
decarboxylation, **3**, 142
Mercury chloride, methyl-
reduction, **3**, 166
Mercury clusters
chromium complexes, **10**, 336
cobalt complexes, **10**, 341
gold complexes, **10**, 343–9
iridium complexes, **10**, 341, 342
iron complexes, **10**, 339
ligand dynamics, **10**, 332
manganese complexes, **10**, 337, 338
molybdenum complexes, **10**, 336, 337
nickel complexes, **10**, 342
niobium complexes, **10**, 335, 336
osmium complexes, **10**, 340
photochemistry, **10**, 330, 331
platinum complexes, **10**, 342, 343
rhenium complexes, **10**, 338
ruthenium complexes, **10**, 339, 340
structures, **10**, 333–5
transition metals, **10**, 323–49
extrusion of mercury, **10**, 327
tungsten complexes, **10**, 337
Mercury complexes
palladium carbonyls, **9**, 207
Mercury compounds
aryl and vinyl
carbonylation, **12**, 370
Mercury(II) compounds
aryl
structures, **3**, 159, 160
methyl-
interaction with biologically important molecules, **3**, 156
Lewis acids, **3**, 155
structures, **3**, 155–7
Mercury(II) fluorosulfonate
reaction with carbon monoxide, **3**, 154
Mercury–hydrogen exchange
aliphatic, **3**, 139, 140
aromatic, **3**, 141, 142
Merrifield's resin
Pauson–Khand reaction, **12**, 720
Merulidial
synthesis, **12**, 50
Mesembrine
synthesis, **11**, 473; **12**, 587
Mesitol
reaction with *p*-methoxyphenyl lead triacetate, **11**, 469
reaction with phenylethynyl-lead triacetate, **11**, 470
Mesitylene
trimetallation, **11**, 109
Metal acyl enolates
organic synthesis, **12**, 378–81
Metal alkoxycarbonyl complexes
carboxylation, **12**, 356, 357
Metal carbene complexes
organic synthesis, **12**, 421–64
Metal carbonyls
anions
reactions, **12**, 351–5
in organic synthesis, **12**, 349–81
Metal clusters
acetylide, **7**, 772–87
η^1-acyl, **7**, 748, 749
η^2-alkene, **7**, 765–8
η^1-alkyl, **7**, 747, 748
alkylidene, **7**, 749–56
alkylidyne, **7**, 757–65
alkyne insertion, **7**, 764
deprotonation, **7**, 763
protonation, **7**, 763
reactions, **7**, 757
reductive elimination, **7**, 764
spectroscopy, **7**, 765
structure, **7**, 765
synthesis, **7**, 757
alkyne, **7**, 772–87
allenyl
dimetallic, **7**, 788
allyl
dimetallic, **7**, 788
aminocarbene, **7**, 752
bimetallic
in zeolites, catalysis, **10**, 60
carbene, **7**, 751
catalysis, **10**, 33
condensation with clusters, **10**, 10, 11
(dialkylamino)carbene, **7**, 752
formyl, **7**, 748
heterometallic
homogeneous catalysis, **10**, 35–51
hydrocarbon ligands, **7**, 747–97
η^2-hydrocarbon ligands, **7**, 765–87
η^3-hydrocarbon ligands, **7**, 787–92
η^4-hydrocarbon ligands, **7**, 793, 794
η^5-hydrocarbon ligands, **7**, 794
η^6-hydrocarbon ligands, **7**, 795, 796
η^7-hydrocarbon ligands, **7**, 797
η^8-hydrocarbon ligands, **7**, 797
iminomethylidene, **7**, 752
immobilized mixed
catalysis on functionalized supports, **10**, 58–60
catalysts, **10**, 51
ketenylidene, **7**, 749–56
methoxycarbyne, **7**, 751
methylene, **7**, 749
preparation, **12**, 1223
trifluoromethyliminomethylene, **7**, 754
vinyl, **7**, 768–70
vinylidene, **7**, 749–56, 768–70
Metal organic chemical vapor deposition
gallium, indium, or thallium compounds, **1**, 504
Metal organic vapor phase epitaxy
organoberyllium compounds, **1**, 53
Metal powders
reactive
magnesium reactions, **1**, 100
Metal vapor synthesis
niobium complexes, **5**, 129–33
tantalum complexes, **5**, 129–33
Metallaalkenes
in alkene metathesis, **12**, 1210
molybdenum-based
ring-opening metathesis polymerization initiators, **12**, 1213
Metallaboranes, **1**, 275–371
arene-ligated systems, **1**, 357, 358
3 boron atoms, **1**, 313
4 boron atoms, **1**, 319–22
5 boron atoms, **1**, 323–7
6 boron atoms, **1**, 327, 328
7 boron atoms, **1**, 329, 330
8 boron atoms, **1**, 331–7
9 boron atoms, **1**, 337–49
10 boron atoms, **1**, 349–62
group 8 metals, **1**, 358
group 9 metals, **1**, 361
group 10 metals, **1**, 362
group 11 metals, **1**, 362
11 boron atoms, **1**, 363
12 boron atoms, **1**, 363

group 8 metals, **1**, 363
group 10 metals, **1**, 363
13 boron atoms, **1**, 363
14 boron atoms, **1**, 364
16 boron atoms, **1**, 364
17 boron atoms, **1**, 364, 365
18 boron atoms, **1**, 365
20 boron atoms, **1**, 365
chemistry, **1**, 275–365
metal-rich, **1**, 293–304
phosphine-ligated systems, **1**, 355–7
Metallacarbaboranes, **1**, 373–424
bonding, **1**, 374–6
bridged complexes, **1**, 391
chromium cyclopentadienyl, **5**, 450
clusters, **1**, 375, 376
cobaltocenium salts, **1**, 392
diborole hybrid complexes, **1**, 381, 419
dicarbon, **1**, 381–419
8–11 vertices, **1**, 394–9
8–11 vertices, bonding, **1**, 398
8–11 vertices, structure, **1**, 398
8–11 vertices, synthesis, **1**, 394–7
hydrocarbon complexes, **1**, 388
icosahedral
12-vertex, **1**, 399–417
lanthanidees, **1**, 419
in medicine, **1**, 416
metal–metal bonds, **1**, 411
metal extraction, **1**, 416
monocarbon, **1**, 377–80
multisandwich complexes, **1**, 387
oxidative fusion, **1**, 388, 398, 422
properties, **1**, 376
reactions, **1**, 377
reactivity, **1**, 374–6
reagents for organic synthesis, **1**, 393
sandwich complexes, **1**, 375, 376
stabilizing organic groups, **1**, 416
structure, **1**, 374–6
synthesis, **1**, 376, 377
metal-promoted oxidative fusion, **1**, 377
tetracarbon, **1**, 421–4
tetradecker sandwich complexes, **1**, 381
tricarbon, **1**, 419–21
triple-decker sandwich complexes, **1**, 381
12-vertex
alkylidyne-bridged, **1**, 411
applications, **1**, 415–17
cage rearrangement, **1**, 414
carbene-bridged, **1**, 411
carbyne-bridged, **1**, 411
catalysis, **1**, 415
physical properties, **1**, 414
spectroscopy, **1**, 414
synthesis, **1**, 411–14
13-vertex cages, **1**, 417–19
wedged, **1**, 398, 399
closo-Metallacarbaboranes
synthesis, **1**, 207, 208
nido-Metallacarbaboranes
synthesis, **1**, 207, 208
Metallacycles, tricyclic
synthesis
zirconocene complexes, **12**, 774, 775
Metallacyclic complexes
niobium and tantalum, **5**, 103–9
Metallacyclobutanes
in alkene metathesis, **12**, 1210
formation, **12**, 390, 594
Metallacycloheptatriene
cyclotrimerization
alkynes, **12**, 742
Metallacyclopentadienes
cyclooligomerization
alkynes, **12**, 743–5
Metallaoxetanes
alkene epoxidation, **12**, 1107
Metallasilacyclopropanes
synthesis, **2**, 71
Metallation
C–H bonds
rhodium complexes, **8**, 205–15
crown ethers
organomagnesium compounds, **1**, 64
Hauser bases, **1**, 64
Metallocene(IV) dihalides, **4**, 483–93
α-Metallocene-carbinols, **7**, 611–13
α-Metallocene-carbonium ions, **7**, 611–13
Metallocenes
cryptands, **7**, 613
electron donor/acceptor complexes, **7**, 608
enantiomerically pure synthesis, **12**, 1200
radiochemistry, **7**, 619
reactions with dichloromercury, **7**, 618
reactions with iodine, **7**, 618
reactions with Lewis acids, **7**, 618
reactions with tetrachlorostannane, **7**, 618
reorientation barriers, **7**, 607
ring rotation, **7**, 607
single-site cationic catalysts
functionalized polymers, **12**, 1203
ansa-Metallocenes, **4**, 489–93
from bridged bis(fluorenes), **4**, 492, 493
from bridged bis(indenes), **4**, 492, 493
from substituted bis(cyclopentadienes), **4**, 490–2
Ziegler–Natta catalysts, **12**, 1197
Metallocenophanes
synthesis, **7**, 615, 616
Metallocycles
bicyclic
reductive coupling of enynes, **12**, 326
Metallocyclopropenes
formation, **12**, 333
Metalloporphyrins
metal–indium bonds, **1**, 566, 567
metal–thallium bonds, **1**, 569
Metallostannation
alkenes and alkynes, **2**, 231, 232
Metathesis
Al–C and Al–H bonds, **11**, 285
Metathesis polymerization
niobium complexes, **5**, 138, 139
tantalum complexes, **5**, 138, 139
titanium catalyzed, **4**, 388–90
Metathesis reactions
cyclopropanation, **12**, 392
organometallic reagents, **12**, 221–31
organopalladium or organonickel compounds, **12**, 163
Metazocine
synthesis, **12**, 18
Methane
reactions with iron anions, **7**, 107
reactions with iron ions, **7**, 103, 106
trimethylene complexes
natural product synthesis, **12**, 941–5
Methane, borylstannyl-
reactions, **11**, 208
Methane, boryltrimethylsilyl-
reactions, **11**, 208
Methane, diazo-
decomposition
transition metal catalyzed, **12**, 425
Methane, dihalo-
synthesis
organoboranes, **11**, 223
Methane, diiodo-
reaction with diethylzinc, **11**, 162
Methane, dilithio-, **1**, 21
Methane, diphenyldiazo-
cyclopropanation, **12**, 399
Methane, nitrodiazo-
cyclopropanation, **12**, 399
Methane, phenyldiazo-
cyclopropanation, **12**, 399
Methane, polyaurio-
synthesis, **3**, 16
Methane, tetraaurio-
hypercoordinate, **3**, 15
Methane, tetrakis[(tricyclohexylphosphine)gold(I)]-
synthesis, **3**, 16
Methane, tetrakis(trimethylstannyl)-
NMR, **2**, 222
Methane, tetrasilyl-
synthesis, **2**, 21
Methane, (*p*-tolylsulfonyl)diazo-
cyclopropanation, **12**, 399
Methane, triaurio-
synthesis, **3**, 15
Methane, π-trimethylene-
complexes, **12**, 971

Methanol
  carbonylation
    rhodium complexes, **8**, 201
  homologation
    catalysts, **10**, 48
Methyl acetate
  homologation
    catalysts, **10**, 51
Methyl acrylate
  ethene block copolymers
    synthesis, **12**, 1205
  hydroformylation, **12**, 31
Methyl chrysanthemates
  synthesis, **12**, 878
Methyl coenzyme M reductase
  nickel complexes, **9**, 91
Methyl dihydrojasmonate
  synthesis, **12**, 453
Methyl farnesoate
  asymmetric dihydroxylation, **12**, 1148, 1149
Methyl itaconate
  hydrogenation
    asymmetric, **12**, 20
Methyl jasmonate
  synthesis, **12**, 865, 885, 887
Methyl methacrylate
  syndiospecific polymerization
    Ziegler–Natta catalysts, **12**, 1204
Methyl vinhaticoate
  synthesis, **12**, 446
Methylamine, bis(trimethylsilyl)-
  synthesis, **2**, 17
Methylamine, trimethylsilyl-
  synthesis, **2**, 17
Methylenation
  carbonyl compounds, **12**, 579–84
    Tebbe's reagent, **12**, 586–91
Methylene compounds
  synthesis from alcohols or carbonyls
    organoaluminum hydrides, **11**, 299
Methylene compounds, active
  reactions
    allyl complexes, **12**, 831–52
Methylene homologation
  zinc carbenoids, **11**, 179–81
α-Methylene lactones
  synthesis, **12**, 87
Methylenomycin B
  synthesis, **9**, 4; **12**, 359, 722
Methylseleno compounds
  synthesis, **11**, 530
Mevinolin
  synthesis
    lithiation, **11**, 33
Michael addition
  acyliron tetracarbonyl anions, **12**, 352
  dienetricarbonyliron complexes, **12**, 632
  organozinc compounds, **11**, 173–5
Michael addition reactions
  carbene complexes, **12**, 521–5
Michael reaction
  aldol reations
    bismuth trichloride, **11**, 502, 503
Micrococcinic acid
  synthesis, **12**, 216
Migratory deinsertion
  transition metal complexes, **12**, 4, 5
Migratory insertion
  transition metal complexes, **12**, 4, 5
Minaprine
  synthesis
    lithiation, **11**, 79
Misoprostol
  synthesis, **12**, 85
Mitomycin A
  synthesis, **12**, 493
Mitsunobu reaction
  intramolecular
    benzodihydropyran from, **12**, 490
Mixovirescine M_2
  preparation, **12**, 189
Modhephene
  synthesis, **12**, 579
Molecular orbital calculations
  iron carbide clusters, **7**, 269
Molybdenocene
  reactions, **5**, 443
Molybdenum
  dicarbon metallacarbaboranes
    12-vertex, **1**, 400, 418
  monocarbon metallacarbaboranes
    12-vertex, **1**, 378
Molybdenum carbonyl complexes
  alkynides, **5**, 165, 166
    properties, **5**, 165
    reactions, **5**, 166
    synthesis, **5**, 165
  availability, **5**, 156
  azacarbene, **5**, 185–91
  calculations, **5**, 156, 157
  carbenes
    no heteroatom stabilization, **5**, 168–72
  carbon σ-donor ligands, **5**, 161
  carbyne, **5**, 192–7
  charged, **5**, 159–61
    reactions, **5**, 160, 161
    structures, **5**, 159, 160
    synthesis, **5**, 159, 160
  cyanide ligands, **5**, 161
  cyclopentadienyl carbyne, **5**, 198–202
  isocyanide ligands, **5**, 161–5
    properties, **5**, 162
    reactions, **5**, 164, 165
    synthesis, **5**, 162
  oxacarbene, **5**, 172–84
  photolysis, **5**, 157, 158
  selenocarbonyl, **5**, 166, 167
  spectroscopy, **5**, 156, 157
  substitution reactions, **5**, 157
  supported, **5**, 158, 159
  thiocarbonyl, **5**, 166, 167
  vinylidenes, **5**, 171
Molybdenum clusters
  alkylidyne cobalt, **10**, 110–12
    solution fluxionality, **10**, 112
  alkylidyne iron, **10**, 93
  alkylidyne nickel, **10**, 119–24
  alkylidyne osmium, **10**, 93–6
  alkylidyne rhodium, **10**, 116–18
  alkyne cobalt, **10**, 113–15
  alkyne iron, **10**, 97
  carborane tungsten, **10**, 168
  cobalt compounds, **10**, 112
  ketenylidene cobalt, **10**, 109
  phosphido rhodium, **10**, 116–18
  rhenium alkyne, **10**, 89
  ruthenium compounds, **10**, 97, 98
  sulfido rhodium, **10**, 116–18
  tungsten, **10**, 87
Molybdenum complexes, **5**, 331–80
  η^3-, **5**, 335–45
  η^4-, **5**, 346–52
  η^5-, **5**, 352–4
  η^6-, **5**, 354–8
  η^7-, **5**, 358–62
  acyclic pentadienyl, **5**, 352, 353
  alkene chelates, **5**, 334, 335
  alkenes, **5**, 332–4
  alkoxide, **5**, 373–6
  alkyl, **5**, 373–6
  alkylidene
    preparation, **12**, 595
  alkyne
    Mo(0),*(d^6)*, **5**, 363, 364
    monometallic, **5**, 363–72
  alkynes
    intramolecular coupling, **5**, 371, 372
  allyl, **5**, 342, 343
  π-allyl
    decomplexation, **12**, 644, 647
    directed reactions, **12**, 973, 974
    in organic synthesis, **12**, 968–74
    reactions with Brønsted acids, **12**, 972, 973
    reactions with electron deficient alkenes, **12**, 972
    reactions with electrophiles, **12**, 970–3
    reactions with oxidizing electrophiles, **12**, 973
    synthesis, **12**, 969, 970
    synthesis from allylic substrates, **12**, 969, 970
    synthesis from molybdenum diene complexes, **12**, 970
  allyl carbonyl cyclopentadienyl nitrosyl
    cations, **5**, 342
  allyl dicarbonyl cyclopentadienyl, **5**, 338–41
  allyl pyrazole ligands, **5**, 338
  allylic alkylation, **12**, 890–5
  arenes, **5**, 471–533
  azaallyl, **5**, 345
  binuclear, **5**, 373–80
  bis(alkyne) cyclopentadienyl, **5**, 368–70
  bis(arene), **5**, 478, 479

bis(cyclopentadienyl)
(acetonitrile)
synthesis, **5**, 446
bis(cyclopentadienyl) carbon disulfide
reactions with alkynes, **5**, 450
bis(cyclopentadienyl) dichloro
reactions with nucleotides, **5**, 448
bis(cyclopentadienyl) dihalides
electrochemistry, **5**, 445
synthesis, **5**, 445
bis(cyclopentadienyl) dihydrido
reaction with tin halides, **5**, 448
bis(cyclopentadienyl) disulfide and tetrasulfide, **5**, 449
bis(cyclopentadienyl) dithiolate
synthesis, **5**, 450
bis(cyclopentadienyl) oxide
synthesis, **5**, 448
bis(cyclopentadienyl) phosphiridene
synthesis, **5**, 448
bis(cyclopentadienyl) triphenylphosphine
synthesis, **5**, 448
bis(cyclopentadienyl)dialkyl
synthesis, **5**, 446
bis(cyclopentadienyl)dihydride
reactions, **5**, 443
bis(cyclopentadienyl)tetrakis-(carbonyl)(2-butyne)
dinuclear cation preparation, **12**, 686
bis(cyclopentadienyl)tetrakis-(carbonyl)
reactions, **12**, 689
bis(cyclopentadienyl)tetrakis-(carbonyl)propargylium
dinuclear cation preparation, **12**, 686
dinuclear NMR studies, **12**, 687
butadiene, **5**, 346
butadienyl, **5**, 344, 345, 350–2
carbonyl diene, **5**, 346–9
carbonyl phosphine
NMR, **5**, 245
cationic pentadienyl
reaction with zinc–copper reagents, **11**, 166
18 + δ complexes, **5**, 403
coupling reactions, **12**, 1090–2
cyclic carbene, **5**, 411
cyclobutadiene, **5**, 350–2
cycloheptadienyl, **5**, 353, 354
cycloheptatriene, **5**, 354–7
cycloheptatrienyl, **5**, 358, 359
bimetallic metal–metal bonded, **5**, 362
sandwich complexes, **5**, 361
cyclohexadiene
nucleophilic additions, **12**, 646
reactions with nucleophiles, **12**, 642
cyclohexadienyl, **5**, 353, 354
cyclooctatetraene, **5**, 357, 358
cyclopentadienone, **5**, 352
preparation, **12**, 647
cyclopentadienyl, **5**, 393–450
cyclopentadienyl diene, **5**, 349, 350
cyclopentadienyl nitrosyl, **5**, 418–20
cyclopentadienyl oxo, **5**, 431
cyclopropenyl, **5**, 344, 345
dicarbonyl cyclopentadienyl diethyldithiocarbamate
synthesis, **5**, 438
dicarbonyl cyclopentadienyl dimer
synthesis, **5**, 396–9
dicarbonyl cyclopentadienyl dimers
bond strength, **5**, 396
reaction with alkynes, **5**, 399
reaction with CO, **5**, 397
reaction with diazoalkanes, **5**, 399
reaction with isonitriles, **5**, 399
reaction with lithium acetylides, **5**, 399
reaction with phosphaalkynes, **5**, 399
reaction with phosphines, **5**, 397
structure, **5**, 397
dicarbonyl cyclopentadienyl nitrosyl
synthesis, **5**, 418, 419
dicarbonyl cyclopentadienyl trihalides
decarbonylation, **5**, 441
dicarbonyl (diphenylphosphine) cyclopentadienyl acyl
acyl migration, **5**, 413
dicarbonyl propene cyclopentadienyl tetrafluoroborate
synthesis, **5**, 406
dicarbonylcyclopentadienyl
stereodirector, **12**, 644
dicarbonyldienecyclopentadienyl
addition of cyanides, **12**, 641
alkylation, **12**, 642
hydride addition, **12**, 640
nucleophilic attack, **12**, 638–49
reaction with $NOPF_6$, **12**, 643
reaction with vinylcuprates, **12**, 641
reactions with nucleophiles, **12**, 639–49, 673
dienyl
hydroboration, **12**, 649
dinitrosyl cyclopentadienyl halides
synthesis, **5**, 419
(17-ethylnylestradiol)tetrakis-(carbonyl)bis(cyclopentadienyl)
dinuclear structure, **12**, 687
formyl
synthesis, **5**, 405
fulvalene(-2-butynyl)tetrakis-(carbonyl)
dinuclear structure, **12**, 687
halides, **5**, 373–6
heteroarenes, **5**, 471–533
homoleptic allyl halides, **5**, 335, 336
homoleptic allyls, **5**, 335, 336
homoleptic dienes, **5**, 346
indenyl, **5**, 376–80
indium bonds, **1**, 557
isocarbonyl
synthesis, **5**, 399
manganese compounds, **10**, 88
mercury clusters, **10**, 336, 337
monoalkene carbonyl phosphine, **5**, 333
monoalkyne cyclopentadienyl, **5**, 367, 368
nitrosyl cyclopentadienyl dihalides
synthesis, **5**, 419, 420
oxocyclobutenyl, **5**, 345
palladium carbonyls, **9**, 204
pentamethylcyclopentadienyl tetrachloride
synthesis, **5**, 440
photochemical reactions with oxazines, **12**, 554
preparation, **12**, 638, 639
prop-2-ynyl, **5**, 344
tetracarbonyl cyclopentadienyl, **5**, 376–80
thallium bonds, **1**, 567
tricarbonyl benzene
reaction with iodine, **5**, 495
tricarbonyl cyclopentadienyl alkyl
photochemistry, **5**, 412
reactions with phosphines, **5**, 411, 413
synthesis, **5**, 410–18
tricarbonyl cyclopentadienyl anions
reaction with acyl chlorides, **5**, 413
reaction with group 12 halides, **5**, 408, 409
reaction with group 14 halides, **5**, 409
reactions with arene diazonium salts, **5**, 423
reactions with group 13 halides, **5**, 409
reactions with propargyl halides, **5**, 415
reactions with sulfur trichloride, **5**, 426
reactions with vinyl halides, **5**, 416
synthesis, **5**, 404
tricarbonyl cyclopentadienyl benzyl
carbonyl insertion reaction, **5**, 413
photolysis, **5**, 412
tricarbonyl cyclopentadienyl chloride
reactions with thallium pentafluorophenyl thiolate, **5**, 437
tricarbonyl cyclopentadienyl dimers

bond strength, **5**, 396
decarbonylation, **5**, 396
disproportionation, **5**, 406
photochemical disproportionation, **5**, 403
photochemistry, **5**, 396
reactions with diethyl zinc, **5**, 408
reactions with diphenylphosphine, **5**, 430
reactions with disulfides, **5**, 436
reactions with ethylene sulfide, **5**, 435
reactions with nitrobenzene, **5**, 421
reactions with pnictogens, **5**, 423–7
reactions with quinones, **5**, 403
reactions with sulfur, **5**, 435
reactions with tetraphenyldiphosphine, **5**, 430
structure, **5**, 395, 396
synthesis, **5**, 394–6
thermochemistry, **5**, 396
tricarbonyl cyclopentadienyl halides
synthesis, **5**, 407
tricarbonyl cyclopentadienyl hydride
photochemistry, **5**, 406
reactions, **5**, 406
reactions with phosphines, **5**, 406
synthesis, **5**, 406, 407
tricarbonyl cyclopentadienyl iodide
reaction with isonitriles, **5**, 407
tricarbonyl cyclopentadienyl methyl
reaction with lithium triethylborohydride, **5**, 414
reactions with dicarbonyl cyclopentadienyl tolylcarbyne tungsten complexes, **5**, 415
tricarbonyl cyclopentadienyl (phenylacetylide)
synthesis, **5**, 417
tricarbonyl cyclopentadienyl tetrafluoroborate
reactions, **5**, 404
tricarbonyl pentamethylcyclopentadienyl dimers
reactions with sulfur, **5**, 435
trimethylenemethane, **5**, 350–2
Molybdenum(II) complexes
alkyne, **5**, 364–6
Molybdenum(IV) complexes
alkynes, **5**, 371
Molybdenum hexacarbonyl
catalysts
diazodecomposition, **12**, 430
Molybdenum indium complexes
dienedicarbonyl
nucleophilic additions, **12**, 648
Molybdenum, π-allyldicarbonylcyclopentadienyl-
functional group interconversions, **12**, 642
preparation, **12**, 638
Molybdenum, bis(arene)
reactions with azulene, **5**, 479
reactions with fulvenes, **5**, 479
Molybdenum, bis(naphthalene)
synthesis, **5**, 479
Molybdenum, pentacarbonyl(α-methoxybenzylidene)-
reaction with dienes, **12**, 390
Molybdenum, trichlorooxobis-(tetrahydrofuran)-
carbonyl alkenation, **12**, 1090
Molybdenum, tris(methyldiphenylphosphine)(methyldiphenylphosphine)-
synthesis, **5**, 529
Monarch butterflies
sex pheromone
synthesis, **12**, 822, 841, 842, 845
Monocyclizations
multiple anion capture, **12**, 309, 310
multiple neutral capture, **12**, 311
Monomorine
synthesis, **11**, 153
Monomorine I
synthesis, **12**, 115, 116
Monoorganothallium compounds
in organic synthesis, **11**, 446–51
preparation, **11**, 446, 447
properties, **11**, 446, 447
reactions, **11**, 447–51
Monooxygenases
oxidative transformations, **12**, 1098
Morphine
synthesis, **12**, 1057
Morpholine *N*-oxide, *N*-methyl-
asymmetric dihydroxylation
osmium tetroxide, **12**, 1141
Mössbauer spectroscopy
iron carbide clusters, **7**, 268, 270
triiron ketenylidene clusters, **7**, 271
Mukaiyama cross-Aldol reactions
zirconium catalyzed, **4**, 622
Multilayer sandwich compounds
synthesis, **1**, 199, 200
Multiple insertion cascades, **12**, 299–319
Mundaserone
synthesis, **12**, 285, 289
Muricatacin
synthesis, **12**, 1168
Muscalure
synthesis, **11**, 101
Muscarine
synthesis, **12**, 453
Muscone
synthesis, **11**, 172; **12**, 120, 838

N

Nagilactone F
synthesis, **12**, 109
Nanaomycin A
synthesis, **12**, 610, 752
zirconocene complexes, **12**, 777, 779
Nanocrystallites
semiconductors
synthesis, **3**, 203
Naphthalene
tetrasubstituted derivative
synthesis
zirconocene complexes, **12**, 776, 778
Naphthalene, 1-amino-
dilithiation, **1**, 17
Naphthalene, dihydro-
Pauson–Khand reaction, **12**, 719
Naphthalene, dimethanooctahydro-
homopolymerization
Ziegler–Natta catalysts, **12**, 1202
Naphthalene, 2-ethenyl-6-methoxy-
hydroformylation, **12**, 31
Naphthalene, hexahydro-
synthesis
cyclooligomerization, **12**, 764
Naphthalene, octahydro-
synthesis, **12**, 908
Naphthalene, 1,2,3,4-tetrahydro-
synthesis, **12**, 273, 274
Naphthalene, 2-vinyl-6-methoxy-
hydrocarboxylation
asymmetric, **12**, 35
Naphthalynes
zirconocene complexes, **12**, 776, 778
1-Naphthols, 2,3-disubstituted
synthesis
organoniobium complexes, **5**, 142
Naphthoquinone, iodoacyl-
synthesis
zirconocene complexes, **12**, 777, 779
Naphthyl triflates
coupling reactions, **12**, 214
2-Naphthylethene, 6-methoxy-
hydroesterification
regioselective, **12**, 33
Naproxen
synthesis, **12**, 19, 31, 33, 35
Natta–Breslow reagent
stoichiometric reagent, **11**, 296
Nazarov cyclization
silicon-directed, **11**, 319
β-Necrodol
synthesis, **12**, 580, 895
Negamycin
synthesis, **12**, 605
optically active, **12**, 358
Neocarzinostatin
synthesis, **12**, 203, 313
Neocarzinostatin chromophore A
synthesis
dicobalt propargylium complexes, **12**, 697
Neoisostegane
synthesis, **12**, 20
Neomanoalide
synthesis, **12**, 860
Neopentane
reactions with iron anions, **7**, 107
Neopentyl iodides
cross-coupling
transition metal promoted, **12**, 165
Neophylidene
ring-opening metathesis
polymerization initiator, **12**, 1214
Nerol
asymmetric dihydroxylation, **12**, 1149, 1150
hydrogenation
asymmetric, **12**, 21
Nicholas reaction
dinuclear-propargylium complexes, **12**, 686
Nickel(0) catalysts
pyridone synthesis, **12**, 757
Nickel
dicarbon metallacarbaboranes
7-vertex, **1**, 382
11-vertex, **1**, 396
12-vertex, **1**, 400
monocarbon metallacarbaboranes, **1**, 378
10-vertex, **1**, 378
12-vertex, **1**, 378
tricarbon metallacarbaboranes
7-vertex, **1**, 420
11-vertex, **1**, 420
Nickel acetylacetonate
catalyst
trimethylaluminum reaction with α,β-unsaturated ketones, **12**, 156
Nickel(II) acetylacetonate
catalysts
diazodecomposition, **12**, 430
Nickel alkoxides
carbonylation, **9**, 3
Nickel alkyls
carbonylation, **9**, 3
Nickel carbonyl
reaction with diboroles, **9**, 177
Nickel catalysts
cross-coupling, **12**, 165
cyclooligomerization
alkynes, **12**, 747
cyclotrimerization
2-butyne-1,4-diol synthesis, **12**, 747
ligandless
coupling reactions, **12**, 170
Nickel cluster complexes, **9**, 19–24
carbenes, **9**, 23, 24
carbonyl, **9**, 15, 19–22
carbynes, **9**, 23, 24
cyclopentadienyl, **9**, 170–4
homonuclear, **9**, 171, 172
heterometallic, **9**, 95–8
heteronuclear
cyclopentadienyl, **9**, 172–4
homometallic, **9**, 93–5
homonuclear
synthesis, metal vapor techniques, **9**, 171
hydrido, **9**, 19, 22
iodo-bridged, **9**, 19, 22
isocyanide, **9**, 22, 23
methylene-bridged, **9**, 15
Ni–C σ-bonds, **9**, 93–8
trinuclear
paramagnetic, **9**, 171
sulfide-bridged, **9**, 171
Nickel clusters
alkylidyne molybdenum, **10**, 119–24
alkylidyne tungsten, **10**, 119
iron compounds, **10**, 149, 150
ruthenium
hydrogenation catalysts, **10**, 38
ruthenium compounds, **10**, 150, 151
Nickel complexes, **9**, 107–82
acyl, **9**, 124
aldehydes, **9**, 127
alkaneimidoyl, **9**, 125
alkenes, **9**, 113–25
thermal decomposition, **9**, 114
alkoxyvinyl, **9**, 47
alkyl, **9**, 37–40
alkyl aryl, **9**, 47
alkyl-methylallyl, **9**, 143
alkyl phenyl ketene
preparation, **9**, 9
alkynes, **9**, 129–36
physical studies, **9**, 129–31
theory, **9**, 129–31
allenes, **9**, 121
allyl, **9**, 136
alkene cyclization/β-elimination, **12**, 906
alkene cyclizations, **12**, 910
mass spectroscopy, **9**, 137
molecular orbital calculations, **9**, 137
physical studies, **9**, 137
pyrazolyl gallates, **9**, 140
reactions, **9**, 137–44
theory, **9**, 137

thermolysis, **9**, 140
allyl trifluoroacetate catalyst
isocyanide living
polymerization, **12**, 1225
π-allyl
organic synthesis, **12**, 959–76
allylic, **9**, 132
allylic alkylation, **12**, 862
anionic alkynes, **9**, 133
arene, **9**, 33
π-arenes, **9**, 174–6
reactions, **9**, 174–6
synthesis, **9**, 174–6
aryl, **9**, 37–40
aryl azides, **9**, 125
benzyl, **9**, 145, 146
benzyne, **9**, 133, 134
binuclear, **9**, 15–19
carbonyl, **9**, 15, 16
methylenediphosphine bridged, **9**, 16
tetramethyleneethane bridged, **9**, 15
biological significance, **9**, 91, 92
bis(borole)
NMR, **1**, 204
π-bonded ligands, **9**, 30–6
boracyclopentadienyl, **9**, 156
boranes, **9**, 180–2
bridging alkynes, **9**, 15
butadiene, **9**, 114
carbenes, **9**, 1–24, 47
carbon σ-bonded, **9**, 29–98
carbon dioxide, **9**, 5–10, 128
carbon disulfide, **9**, 129
carbon suboxide, **9**, 122
carbonyl, **9**, 1–24
carboranes, **9**, 180–2
carboxymethyl butadienes, **9**, 121
catalysis
allylmetal insertion reactions in alkenes, **12**, 906–8
carbomagnesiation, **1**, 68
cross-coupling, mechanism, **12**, 164
crotyl, **9**, 142
cyclam, **9**, 9
cycloalkenyl, **9**, 142
cyclobutadiene
cationic, reduction, **9**, 148
cyclooctenyl, **9**, 124
cyclopentadienyl, **9**, 149
cyclic voltammetry, **9**, 152
electrochemistry, **9**, 152
photolysis, **9**, 163
cyclopropylium, **9**, 146, 147
DAD-Ar_2, **9**, 118
decamethylbimetallic cyclopentadienyl, **9**, 154
dialkynes, **9**, 134, 135
diarsaboranes, **1**, 266
diazabutadiene, **9**, 113, 118
diboranes, **1**, 311
diethyl-1,4-dimethyl-1,4-diboranaphthalene, **9**, 124
dimetallic
alkynes, **9**, 135, 136
1,1-dimethylsilole, **9**, 124
dinuclear
aminocarbyne, **9**, 17, 18
carbenes, **9**, 18, 19
diazoalkanes, **9**, 19
isocyanide, **9**, 17, 18
nitrosyl, **9**, 17, 18
diphosphaallyl, **9**, 144, 145
diphosphines, **9**, 113
diphosphorus, **9**, 136
19-electron, **9**, 162
electron pair donor ligands, **9**, 14
ethene, **9**, 110–13
ethoxycarbonyl
preparation, **9**, 4
Fischer-type carbenes, **9**, 111
fullerene, **9**, 125
fulvalene, **9**, 151
heteroalkenes, **9**, 125–9
heteroallyl, **9**, 144, 145
heterocyclobutadiene, **9**, 149
indenyl, **9**, 156
indium bonds, **1**, 566
isocyanides, **9**, 1–24
isocyanoethanol, **9**, 11
isoquinolone, **9**, 12
ketenes, **9**, 120, 128
ketones, **9**, 127
ketophosphonates, **9**, 128
ligand-free, **9**, 30–6
η^2-ligands, **9**, 108–36
physical studies, **9**, 109, 110
theory, **9**, 109, 110
XPES studies, **9**, 109
η^3-ligands, **9**, 136–47
η^4-ligands, **9**, 147–9
reactions, **9**, 148, 149
synthesis, **9**, 148, 149
theory, **9**, 147
η^5-ligands, **9**, 149–74
dinuclear, **9**, 167–70
gas-phase studies, **9**, 150
heterodinuclear, **9**, 169, 170
ion cyclotron resonance spectroscopy, **9**, 150
magnetic studies, **9**, 150–2
mass spectroscopy, **9**, 150
physical studies, **9**, 150–4
spectroscopy, **9**, 150–2
theory, **9**, 150–4
thiolato bridged, **9**, 170
η^2-ligands
ab initio calculations, **9**, 109
mercury, **9**, 11
mercury clusters, **10**, 342
methylenecyclopropane, **9**, 119
monoalkynes, **9**, 132, 133
monoazadienes, **9**, 118
naphthalene, **9**, 123
norbornadiene, **9**, 119
pentafluorophenyl, **9**, 122
pentafulvalenes, **9**, 168
phosphaalkene, **9**, 126
phosphaalkynes, **9**, 136
phosphine cyclopentadienyl, **9**, 160
phosphinoenolates, **9**, 8
polyalkynes, **9**, 134, 135
polydentate arsenic ligands, **9**, 122
polymetallic
alkynes, **9**, 135, 136
reactions with carbon dioxide and alkynes, **9**, 121
silylene, **9**, 125
step-growth polymerization, **12**, 1226, 1227
tethered alkenes, **9**, 117
trimethylene methane
cycloaddition reactions, **12**, 923–57
trimethylsiloxyallyl, **9**, 143
vinyl halides, **9**, 119
xylyl, **9**, 146
zinc clusters, **10**, 342
Nickel(0) complexes
dinuclear
dmpm-bridged, **9**, 16
dppm-bridged, **9**, 16
ketenes, **9**, 9
polyalkene, **9**, 119
reactions with alkenes, **9**, 80
reactions with isocyanates, **9**, 80
trimethoxysilylalkene, **9**, 119
trimethylsilylalkene, **9**, 119
Nickel(II) complcxcs
π-allyl, **12**, 960–6
coupling reactions with main group organometallic reagents, **12**, 964, 965
homoenolate synthons, **12**, 961
synthesis, **12**, 960
trimethylsilyl-substituted, **12**, 962
allyl halides
carbocyclic, **12**, 961
as diene synthons, **12**, 961
as isoprene synthons, **12**, 961
π-allyl halides
coupling reactions with organohalide electrophiles, **12**, 961–4
carbonyl
structure, **9**, 2
methyl(aryloxide)
reactions, **9**, 4
1-siloxy substituted π-allyl chloride
synthesis, **12**, 960
transmetallation
organozinc reagents, **12**, 156
Nickel effect, **11**, 281
ethene dimerization, **9**, 34
Nickel osmium clusters
hydrogenation cataysts, **10**, 38
Nickel platinum clusters
synthesis, **10**, 2, 3
Nickel polydecker complexes
π-bonded, **9**, 176, 177, 180
physical properties, **9**, 177
synthesis, **9**, 177–80
theory, **9**, 177
Nickel powder
preparation, **9**, 30
Nickel sandwich complexes
diborolene, **9**, 176, 177, 180
fulvalene, **9**, 155, 156

polysubstituted cyclopentadienyl, **9**, 155, 156
Nickel tetracarbonyl
photolysis, **9**, 2
preparation, **9**, 1
reactions with alkyl isocyanides, **9**, 10
substitution reactions, **9**, 2
Nickel triple decker complexes
slipped, **9**, 177
Nickel ylides, **9**, 30–6
catalyst
ethene polymerization, **9**, 36
Nickel, bipyridyl(cyclooctadienyl)-
synthesis, **9**, 119
Nickel, dicarbonyldicyclopentadienyldi-
chemistry, **9**, 167, 168
Nickel, dichloro[bis(diphenylphosphino)propane]-
catalyst
cross-coupling reactions, **12**, 165
Nickel, dichloro(diphenylphosphinomethane)-
catalyst
dienylsilane synthesis, **12**, 173
Nickel, tris(norbornene)-
structure, **9**, 119
Nickelacarbocyclic complexes
chemistry, **9**, 73–80
properties, **9**, 78–80
reactions, **9**, 78–80
synthesis, **9**, 73–8
Nickelacycles
formation from carbon dioxide, **9**, 6
Nickelacyclic complexes
diazabutadiene, **9**, 80
Nickelacyclobutane
synthesis, **9**, 30
Nickelacyclobutenones
formation, **9**, 15
Nickelaheterocyclic complexes
chemistry, **9**, 80–91
Nickela-2-silacyclobutenes
preparation, **2**, 71
Nickel–carbon bonds
enthalpies, **9**, 30
Nickel–cobalt polydecker complexes, **9**, 177
Nickelocene
chemistry, **9**, 149, 154–6
electrochemistry, **9**, 152–4
pentamethyl, **9**, 156
peralkylation, **9**, 151
reactions with dialkylphosphine sulfides, **9**, 163
reactions with ylides, **9**, 165
Nickel–platinum complexes
synthesis, **9**, 98
Nickel–zinc clusters, **9**, 173
Nigericin
synthesis, **12**, 611
Niobiacyclobutadienes, **5**, 106
Niobiacyclobutanes, **5**, 103–5
Niobiacyclobutenes, **5**, 105, 106
Niobiacyclopentadienes, **5**, 108
Niobiacyclopentanes, **5**, 106, 107
Niobiacyclopentenes, **5**, 107
Niobium complexes
η^3-, **5**, 110
η^4-, **5**, 110–13
η^5-, **5**, 114–28
η^6-, **5**, 129–32
η^7-, **5**, 132
acyl, **5**, 97–102
aldehydes, **5**, 102
alkenes, **5**, 89–92
alkoxide hydrides, **5**, 75
alkyl, **5**, 67–73
alkylidene
NMR, **5**, 87
preparation, **12**, 595
alkylidenes, **5**, 83–9
IR spectra, **5**, 87
reactivity, **5**, 88, 89
spectroscopy, **5**, 87
structure, **5**, 87
synthesis, **5**, 83–6
alkylidynes, **5**, 83–9
reactivity, **5**, 88, 89
spectroscopy, **5**, 87
structure, **5**, 87
synthesis, **5**, 83–6
alkynes, **5**, 92–7
coupling, **12**, 336
naphthols from, **12**, 337
arenes, **5**, 102, 103
aryl, **5**, 67–73
benzyne, **5**, 92–7
binary carbonyls, **5**, 60
bis(cyclopentadienyl), **5**, 121
bis(cyclopentadienyl) alkene, **5**, 91, 92
bis(cyclopentadienyl) alkyne, **5**, 96, 97
bis(cyclopentadienyl) halides, **5**, 126
bis(cyclopentadienyl) nitrogen donors, **5**, 127
bis(cyclopentadienyl) oxygen donors, **5**, 127
bis(cyclopentadienyl) phosphorus donors, **5**, 127
bis(cyclopentadienyl) silyl-, **5**, 127
bis(cyclopentadienyl) sulfur donors, **5**, 128
boron hydrides, **5**, 121
butadiene, **5**, 112, 113
carbocyclic compounds, **5**, 110–33
carbon dioxide, **5**, 101
carbon disulfide, **5**, 101
carbonyls, **5**, 59–65
chemistry, **5**, 57–144
cluster compounds, **5**, 120
coupling reactions, **12**, 1087–9
cyclobutadiene, **5**, 113
cycloheptatrienyl, **5**, 133–44
cyclometallated trimethylphosphine, **5**, 102
cyclopentadienyl alkyl, **5**, 70
reactivity, **5**, 70–3
cyclopentadienyl alkyne, **5**, 93–7
cyclopentadienyl aryl, **5**, 70
reactivity, **5**, 70–3
cyclopentadienyl carbonyl, **5**, 62–5
cyclopentadienyl carbonyl hydrides, **5**, 80
cyclopentadienyl halide hydrides, **5**, 80
cyclopentadienyl hydrides, **5**, 76
ene compounds, **5**, 110–33
enyl ligands, **5**, 110–33
formyl, **5**, 98
heterobimetallic, **5**, 120; **10**, 28
heterobimetallic hydrides, **5**, 81
hydrides, **5**, 73–81
imine, **5**, 100, 101
iminoacyl, **5**, 97–102
in catalysis, **5**, 133–44
in organic synthesis, **5**, 133–44
isocyanates, **5**, 101
isocyanides, **5**, 65, 66
ketenes, **5**, 101
ketenimines, **5**, 101
ketones, **5**, 102
mercury clusters, **10**, 335, 336
metal–metal bonds, **5**, 128
monokis(cyclopentadienyl), **5**, 114–21
monokis(cyclopentadienyl) alkenes, **5**, 89
monokis(cyclopentadienyl) alkyne, **5**, 96
monokis(cyclopentadienyl) halide, **5**, 114
monokis(cyclopentadienyl) imido, **5**, 119
monokis(cyclopentadienyl) nitrogen donors, **5**, 119
monokis(cyclopentadienyl) oxygen donors, **5**, 119
monokis(cyclopentadienyl) phosphorus donors, **5**, 119
monokis(cyclopentadienyl) selenium donors, **5**, 120
dinuclear, **5**, 120
trinuclear, **5**, 120
monokis(cyclopentadienyl) silyl, **5**, 118
monokis(cyclopentadienyl) sulfide donors
dinuclear, **5**, 120
trinuclear, **5**, 120
monokis(cyclopentadienyl) sulfur donors, **5**, 120
nitriles, **5**, 100, 101
phosphine hydrides, **5**, 73
pyridine, **5**, 102, 103
silaacyl, **5**, 98
substituted carbonyls, **5**, 59–65
zinc clusters, **10**, 335
Niobium pentachloride
methyl lithium system
reductive dimerization of benzylic halides, **12**, 1087
Niobium, tetrachlorobis(tetrahydrofuran)-
reaction with benzaldehyde *N*-trimethylsilyl imine, **12**, 1088
Niobium, trichlorocyclopentadienyl-
cocyclization of α,ω-diynes

with terminal alkynes, **12**, 750
Nitratomercuriation
alkenes, **3**, 145
Nitrene
iron clusters
formation of aniline derivatives, **7**, 281
Nitrene clusters
nucleophilic addition to triiron clusters, **7**, 280
Nitrene complexes
triiron clusters, **7**, 263
Nitrene ligands
clusters, **7**, 814
Nitrenes
nucleophilic addition to iron clusters giving nitrene–acyl clusters, **7**, 281
Nitrido complexes
group 6
synthesis, **5**, 291
Nitrile complexes
triiron clusters, **7**, 263, 264
Nitriles
allylic alkylation, **12**, 868, 869
aminomercuriation, **11**, 424
conversion to amines or imines
organoaluminum hydrides, **11**, 300
cyclization
intramolecular, **12**, 332
α,α-dichloro-β-hydroxy-
preparation, **11**, 492
dimerization, **12**, 1084
niobium complexes, **12**, 1088
insertion reactions
carbene complexes, **12**, 534
methylenation, **12**, 591
preparation
organoboranes, **11**, 222
reactions with Grignard reagents, **11**, 139
reductive coupling, **12**, 1083, 1084
α,β-unsaturated
alkenation, trialkylstibine, **11**, 489
cycloaddition reactions, **12**, 927
Nitrilium ylides
formation, **12**, 440
nitrogen-hydrogen insertion, **12**, 441
reactions, **12**, 440
Nitro arenes
reduction
tellurium reagents, **11**, 577
Nitro compounds
allylic
alkylation, **12**, 841
reductive coupling, **12**, 1083, 1084
Nitroaromatics
reductive transformations
catalyzed by cluster species, **7**, 280
Nitrogen compounds
reduction
organotin compounds, **11**, 360
synthesis
organomercury compounds, **11**, 401
Nitrogen ylides
metal stabilized, **12**, 438–41
palladium(II) complexes, **9**, 229
Nitrogen-arylation
organobismuth(V) compounds, **11**, 510
Nitromercuriation, **11**, 426
Nitronates
alkylation malonates, **12**, 354
Nitrones
reactions with Grignard reagents, **11**, 136
synthesis
halomercuration, **11**, 429
Nitrophenylseleno compounds
synthesis, **11**, 530
Nitroso compounds
reactions
with carbene complexes, **12**, 539
Nitrosoamines
lithiation, **11**, 16
Nitrous oxide
nickel complexes, **9**, 9
NMR
^{13}C
alkylidyne complexes, iron clusters, **7**, 272, 273
alkyne complexes, iron clusters, **7**, 260, 261
carbide complexes, iron clusters, **7**, 269, 270
Nonaborates, pentadecahydro-
group 8 metals
metallaboranes, **1**, 338
group 9 metals
metallaboranes, **1**, 338, 339
group 11 metals
metallaboranes, **1**, 339
metallaboranes, **1**, 338, 339
Nonaborates, tridecahydro-
metallaboranes, **1**, 340–3
Nonaborates, undecahydro-
group 8 metals
metallaboranes, **1**, 344–6
group 9 metals
metallaboranes, **1**, 347, 348
group 10 metals
metallaboranes, **1**, 349
metallaboranes, **1**, 344–9
3,8-Nonadienoic acid
synthesis, **12**, 787
3-Nonen-2-ol
oxypalladation, **12**, 613
Nonlinear optical materials
from ferrocenes, **7**, 191
Nor-alcohols
synthesis, **11**, 496
Norbornadiene
cyclization with alkenes, **12**, 766
cycloaddition reactions, **12**, 925
dimerization
catalysts, **10**, 48, 49
insertion/carbonylation, **12**, 359
Pauson–Khand reaction, **12**, 720
reactions with iron ions, **7**, 106
Norbornadiene, 2-trimethylsilyl-
metallation, **11**, 118
Norbornene
homopolymerization
Ziegler–Natta catalysts, **12**, 1202
metallation, **11**, 111
Pauson–Khand reaction, **12**, 719
polymerization
by ring-opening metathesis polymerization, **12**, 1212
reaction with dimethylaminopyridine
titanium catalyzed, **4**, 389
Norbornene, 1-methyl-
reaction with titanacycles, **4**, 389
Norbornenes
functionalized
ring-opening metathesis polymerization, **12**, 1217
Norbornen-2-ones
cycloadditions
Pauson–Khand reaction, **12**, 710
Norcarane
metallation, **11**, 114
11-Nor-9-carboxy-Δ^9-tetrahydrocannabinol
synthesis, **12**, 251
11-Nor-9-carboxy-Δ^9-tetrahydrocannabinol methyl ester
synthesis, **12**, 245
Noriceane, 3,5-dehydro-
reaction with rhodium complexes, **8**, 218
Normanicone
synthesis
dicobalt propargylium complexes, **12**, 694
Nortricyclane
metallation, **11**, 113
Nucleophilic addition
arene metal complexes, **12**, 979–1012
Nucleophilic attack
coordinated ligands, **12**, 6
Nucleosides
preparation, **12**, 145
Stille reaction, **12**, 216
synthesis
allylic amidation, **12**, 812
glycal arylation, **12**, 133
transmetallation, **12**, 135
Nudaurelia cytherea cytherea
sex attractant
synthesis, **11**, 100
Nystatin A_1
synthesis, **12**, 107

O

Obtusenyne
  synthesis, **12**, 589
Ochratoxin B
  synthesis
    lithiation, **11**, 50
β-Ocimene
  metallation, **11**, 105
Octaborates
  conjoined
    metallaboranes, **1**, 337
Octaborates, dodecahydro-
  group 7 metals
    metallaboranes, **1**, 333
  group 8 metals
    metallaboranes, **1**, 334
  group 9 metals
    metallaboranes, **1**, 335
  group 10 metals
    metallaboranes, **1**, 336, 337
  group 11 metals
    metallaboranes, **1**, 337
  metallaboranes, **1**, 333–7
Octaborates, octahydro-
  metallaboranes, **1**, 337
Octaborates, tetradecahydro-
  group 8 metals
    metallaboranes, **1**, 332
  group 9 metals
    metallaboranes, **1**, 332
  metallaboranes, **1**, 331–3
1,7-Octadiene
  reactions with iron ions, **7**, 104
2,7-Octadiene, 1-acetoxy-
  allylation/carbonylation
    allylmetal catalysts, **12**, 917
  cyclization
    allylmetal catalysts, **12**, 910
2,7-Octadiene, 1-amino-
  synthesis, **12**, 788
2,6-Octadiene, 1,8-bis(silyl)-
  synthesis, **12**, 792
2,7-Octadiene, 1-(*N,N*-dimethylamino)-
  synthesis, **12**, 790
2,6-Octadiene, 1-silyl-
  synthesis, **12**, 792
2,7-Octadienyl acetate
  cyclization
    allylpalladium catalysts, **12**, 915
  synthesis, **12**, 917
2,7-Octadienyl acetate, 4-alkyl-4-hydroxy-
  cyclization
    allylmetal catalysts, **12**, 907
Octadienyl sulfones
  synthesis, **12**, 789
Octastannacubane
  preparation, **2**, 282
Octatetrayne, bis(*t*-butyldiphenylsilyl)-
  synthesis, **2**, 10
1,3,7-Octatriene
  synthesis, **12**, 789
2-Octen-1,8-diol, 3,7-dimethyl-
  synthesis, **12**, 849
Oct-2-en-7-yne, 1-acetoxy-
  allylation/carbonylation
    allylmetal catalysts, **12**, 917
Oct-2-en-7-ynyl acetate
  cyclization
    allylpalladium catalysts, **12**, 915
Okadaic acid
  synthesis
    lithiation, **11**, 36
Oligodecker structures
  electron count, **1**, 201
Oligomerization
  alkenes
    Ziegler–Natta catalysts, **12**, 1200
  butadiene
    catalysts, **10**, 49
  linear
    alkynes, **12**, 742, 743
  organolanthanide catalysts, **4**, 114
  rhodium isocyanide catalysts, **8**, 163
Oligomers
  iron compounds, **7**, 284
Oligosilanes
  electron delocalization, **2**, 91–5
  electronic spectra, **2**, 91, 92
  oxidation potentials, **2**, 91, 92
Oligostannanes
  cyclic
    preparation, **2**, 281
Olivin
  synthesis, **12**, 491
Oncinotine
  synthesis, **12**, 845
Onnamide
  synthesis, **12**, 204
Onnamide A
  synthesis, **12**, 79, 80
Oplopanone
  synthesis, **12**, 62
8-*epi*-Oplopanone
  synthesis, **12**, 62
Oplopanone, anhydro-
  synthesis, **12**, 62
Oppolzer's magnesium
  ene reactions, **1**, 79
Optically active macromolecules
  Ziegler–Natta catalysts, **12**, 1200, 1201
Organic synthesis
  organolanthanide complexes, **4**, 122–9
  organoscandium complexes, **4**, 121, 122, 129
  organoyttrium complexes, **4**, 122–9
Organoactinide compounds
  Ziegler–Natta catalysts, **12**, 1195
Organoalkali compounds
  organoboron compounds from, **11**, 202, 203
Organoaluminum complexes
  ab initio calculations, **1**, 441–3
  bonding, **1**, 441–55
  chemical trapping, **1**, 443–5
  coordination, **1**, 454, 455
  monomeric tricoordinate
    bonding, **1**, 447–9
  oligomers, **1**, 449, 450
  preparation, **1**, 455–67
  spectroscopy, **1**, 441–3
  structure, **1**, 441–55
  subvalent
    structure, **1**, 441–7
Organoaluminum(I) complexes
  structures, **1**, 445–7
Organoaluminum compounds
  in organonickel complex synthesis, **9**, 41
  oxidation
    carbon–oxygen bonds, **11**, 302–3
  reactions with metal salts, **11**, 286
  transmetallation, **12**, 131
    to palladium, **12**, 155
Organoaluminum hydrides
  carbon–hydrogen bond formation, **11**, 297–301
Organoaluminum magnesium compounds
  anthracene, **1**, 107
Organoaluminum reagents
  in organic synthesis, **1**, 438, 439; **11**, 278–309
  reactions with transition metal compounds, **1**, 478–84
Organoantimony compounds
  in organic synthesis, **11**, 487–511
  pentavalent
    oxidizing agents, **11**, 497
Organoantimony dihalides
  synthesis, **2**, 333–5
Organoantimony hydrides
  synthesis, **2**, 328
Organoantimony(V) halides
  catalysts
    in organic synthesis, **11**, 499, 500
Organoargentates
  anionic mononuclear heteroleptic
    solid-state structures, **3**, 105–8
  anionic mononuclear homoleptic
    solid-state structures, **3**, 105–8
  aryl

bonding, **3**, 129
stereochemistry, **3**, 129
solid-state
structures, **3**, 100–9
synthesis, **3**, 73
Organoarsenic(V) compounds
synthesis, **2**, 343
Organoarsenic dihalides
synthesis, **2**, 333–5
Organoarsenic hydrides
synthesis, **2**, 328
Organobarium compounds
dialkyl, **1**, 110, 111
diaryl, **1**, 110, 111
ion pairs, **1**, 111
Organobarium halides
preparation, **1**, 107–9
Organoberyllium complexes
alkenyl, **1**, 36–9
alkoxy, **1**, 42, 43
alkyl, **1**, 36–9
alkynyl, **1**, 36–9
amino, **1**, 42, 43
anionic, **1**, 42
anthracene, **1**, 36
hydrolysis, **1**, 36, 37
aryl, **1**, 36–9
bis(dienyl), **1**, 38
coordination, **1**, 39–41
cyclopentadienyl, **1**, 37
structure, **1**, 48–50
phosphino, **1**, 42, 43
radical, **1**, 40
spirobicyclic, **1**, 36
thio, **1**, 42, 43
x-ray crystal structure, **1**, 39
Organobismuth compounds
in organic synthesis, **11**, 487–511
Organobismuth(III) compounds
in organic synthesis, **11**, 502, 503
Organobismuth(V) compounds
arylation, **11**, 506–8
in organic synthesis, **11**, 503–11
oxidation, **11**, 504, 505
Organobismuth dihalides
synthesis, **2**, 333–5
Organobismuth hydrides
synthesis, **2**, 328
Organoboranes
chiral
α-pinene, **11**, 260
cross-coupling reactions, **11**, 231–4
mechanism, **11**, 231
palladium-catalyzed, **11**, 231
organozinc compound preparation, **11**, 164
synthesis
transmetallation, **11**, 202–4
Organoborates
reactions with electrophiles, **11**, 227–30
Organoboron compounds
abstraction of α-hydrogen, **11**, 241
autoxidation, **11**, 214, 215
bimolecular homolytic substitution, **11**, 241
carbanions
in organic synthesis, **11**, 234, 235
ractions, **11**, 208–11
chemistry, **11**, 192–264
Diels–Alder cycloaddition, **11**, 211
four coordinate boron
carbon–carbon formation, **11**, 220–42
α-halo-
rearrangements, **11**, 221–5
halogenolysis, **11**, 218, 219
oxidation, **11**, 214–16
protonolysis, **11**, 213, 214
radical reactions, **11**, 241–3
safety, **11**, 192
synthesis, **11**, 193–213
synthetic applications, **11**, 213–46
transmetallation, **11**, 220; **12**, 131
to palladium, **12**, 146–50
α,β-unsaturated
addition reactions, **11**, 211–13
Organocadmium compounds, **3**, 175–203
adducts, **3**, 197
alkyl
adducts, **3**, 177, 178
in electronics industry, **3**, 200–3
fluoroalkyls, **3**, 196
in organonickel complex synthesis, **9**, 41
physical studies, **3**, 200
theoretical studies, **3**, 200
zinc–heteroatom bonds, **3**, 198, 199
Organocalcium clusters, **1**, 58, 108
Organocalcium compounds
amides, **1**, 118
Schlenk-type equilibrium, **1**, 118
biradicals, **1**, 118
calcium–tellurium exchange, **1**, 109
cuprates, **1**, 120
dialkyl, **1**, 110, 111
diaryl, **1**, 110–12
Schlenk-type equilibrium, **1**, 109
Organocalcium(I) compounds
chemistry, **1**, 58
Organocalcium halides
polymeric, **1**, 108
preparation, **1**, 107–9
mechanism, **1**, 108
reactions
with organic compounds, **1**, 109
Organocalcium hydrides
aryl
formation, **1**, 118
properties, **1**, 118
Organocalcium oxides
preparation, **1**, 118
Organocerium compounds
Grignard reagent reactivity, **11**, 130
Organochromium complexes
aquo, **5**, 299–302
reactions, **5**, 302
synthesis, **5**, 300
pentaaquo
UV–visible spectra, **5**, 301
Organocobaloximes
photolysis, **8**, 61
Organocobalt complexes
Co–C bond energy, **8**, 54
dinuclear, **8**, 90–101
ligands, **8**, 4–26
η^3-ligands, **8**, 78
η^4-ligands, **8**, 79–82
η^5-ligands, **8**, 83–9
Organocobalt(III) complexes
bonds
resonance effects, **8**, 53
chemistry, **8**, 42–71
coenzyme B_{12} mimics, **8**, 42–66
Organocobalt compounds
transmetallation
to palladium, **12**, 155
Organocopper compounds, **3**, 57–129
alkenyl
solid-state structures, **3**, 95
alkyl
solid-state structures, **3**, 77–80
alkynyl
bonding, **3**, 124
heteroleptic, solid-state structures, **3**, 96–100
homoleptic, solid-state structures, **3**, 96–100
amine stabilized cationic
solid-state structures, **3**, 110–15
amine stabilized neutral
solid-state structures, **3**, 110–15
aryl
functionalized heteroatom-containing substituents, solid-state structures, **3**, 86–8
noncoordinating substituents, solid-state structures, **3**, 82–6
solid-state structures, **3**, 82–95
bonding, **3**, 75–129
cyclopentadienyl
structures, **3**, 115
ferrocenyl
structures, **3**, 117
heteroatom
in organic synthesis, **12**, 94–103
heteroleptic
solid-state structures, **3**, 89–95
metal salt complexes, **3**, 66
organic synthesis, **12**, 60–124
solution composition studies, **12**, 121–4
solvent effects, **12**, 118, 119
phosphine stabilized cationic
solid-state structures, **3**, 110–15
phosphine stabilized neutral

solid-state structures, **3**, 110–15
phosphorylides
solid-state structures, **3**, 115
polymeric
structures, **3**, 120, 121
solid-state
structures, x-ray analysis, **3**, 77–117
solution
structures, **3**, 125
stereochemistry, **3**, 121–9
structure, **3**, 75–129
synthesis, **3**, 60–75
electrochemical, **3**, 71
transmetallation, **3**, 60–4
thermal stability, **3**, 75–7
transmetallation from
to palladium, **12**, 154
Organocopper(I) compounds
organic synthesis, **12**, 103–13
Organocuprates
anionic mononuclear heteroleptic
solid-state structures, **3**, 105–8
anionic mononuclear homoleptic
solid-state structures, **3**, 105–8
anionic polynuclear heteroleptic
solid-state structures, **3**, 108, 109
anionic polynuclear homoleptic
solid-state structures, **3**, 108, 109
aryl
bonding, **3**, 129
stereochemistry, **3**, 129
conjugate addition to unsaturated carbonyl componds, **3**, 127
neutral heteroleptic
solid-state structures, **3**, 101–5
neutral homoleptic
solid-state structures, **3**, 101–5
organocopper compound preparation, **3**, 64
solid-state
structures, **3**, 100–9
synthesis
interaggregate exchange, **3**, 69
metallation, **3**, 67–9
self-assembly, **3**, 69
Organoditin compounds
in organic synthesis, **11**, 358–63
Organogermanium compounds
alkali metal bonds
synthesis, **2**, 174
applications, **2**, 206–9
in organic synthesis, **2**, 206
biological activity, **2**, 207, 208
carbon bonds, **2**, 139–44
chemical properties, **2**, 143, 144
intracyclic, **2**, 140
synthesis, **2**, 139
chemistry, **2**, 137–209
Grignard reagents, **2**, 176
group 12 element bonds, **2**, 176
group 13 element bonds, **2**, 177
group 14 element bonds, **2**, 177
group 15 element bonds, **2**, 158–66
complexes, **2**, 162
functionalized, **2**, 161
heterocyclic, **2**, 159
insertion reactions, **2**, 163
group 16 element bonds, **2**, 166–74
halogen bonds
chemistry, **2**, 145–50
hydrogen bonds, **2**, 150–7
alkenylation, **2**, 153
alkylation, **2**, 153
halogenation, **2**, 152
preparation, **2**, 151, 152
metal bonds, **2**, 174–86
organofunctional, **2**, 142
organometallic precursors
high-performance materials, **2**, 208, 209
physical properties, **2**, 209
spectroscopy, **2**, 209
transition metals, **2**, 179
transmetallation, **12**, 131
Organogermylcuprates
in organometallic chemistry, **2**, 176
Organogold complexes
σ-bonded
reactions, **3**, 3–17
structure, **3**, 3–17
synthesis, **3**, 3–17
Organohafnium complexes
bis(cyclopentadienyl)
dicationic, **4**, 614
bis(cyclopentadienyl) acyl, **4**, 610
bis(cyclopentadienyl) alkenyl, **4**, 605–7
bis(cyclopentadienyl) alkoxides
cationic, **4**, 613, 614
bis(cyclopentadienyl) alkyl
cationic, **4**, 590–605
cationic, reactivity, **4**, 600–5
cationic, structures, Lewis base-free, **4**, 599, 600
cationic, structures, Lewis base-stabilized, **4**, 595–8
cationic, synthesis, **4**, 590–4
bis(cyclopentadienyl) allyl
cationic, **4**, 590–605
cationic, reactivity, **4**, 600–5
cationic, structures, Lewis base-free, **4**, 599, 600
cationic, structures, Lewis base-stabilized, **4**, 595–8
cationic, synthesis, **4**, 590–4
bis(cyclopentadienyl) aryl, **4**, 605–7
bis(cyclopentadienyl) benzyl
cationic, **4**, 590–605
cationic, reactivity, **4**, 600–5
cationic, structures, Lewis base-free, **4**, 599, 600
cationic, structures, Lewis base-stabilized, **4**, 595–8
cationic, synthesis, **4**, 590–4
bis(cyclopentadienyl) carbonyl, **4**, 610
bis(cyclopentadienyl) hydrides, **4**, 611–13
bis(cyclopentadienyl) iminoacyl, **4**, 610
bis(cyclopentadienyl) pyridyl, **4**, 607, 608
cationic, **4**, 589–622
applications, **4**, 618–22
monocyclopentadienyl hydrocarbyl
cationic, **4**, 615
in organic synthesis, **4**, 619–22
tetraaza-macrocycle hydrocarbyl
cationic, **4**, 617
tris(hydrocarbyl)
cationic, **4**, 616, 617
Organohalohydrogermanes
synthesis, **2**, 152
Organohydrogermanes
reactions with transition metal complexes, **2**, 157
synthesis, **2**, 151
Organoiron compounds
transmetallation
to palladium, **12**, 155
Organolanthanide compounds
Ziegler–Natta catalysts, **12**, 1195
Organolanthanide(III) compounds
catalysts
living polymerization of alkenes, **12**, 1205
Organolanthanides
material science, **4**, 129, 130
Organolead compounds
carbon bonds
formation, **2**, 306–8
chemistry, **2**, 305–17
cyclopentadienyl
photolysis, **2**, 313
reactions, **2**, 315
formation, **2**, 306–17
main group metals, **2**, 311, 312
α-methoxy
aldehyde condensation, stereoselectivity, **11**, 463
molecular orbital calculations, **2**, 305
nitrogen, **2**, 311
NMR, **2**, 306
in organic synthesis, **2**, 317; **11**, 461–83
oxygen, **2**, 309
selenium, **2**, 309–11
sulfur, **2**, 309–11
transition metal complexes, **2**, 312, 313
formation, **2**, 307
tricarboxylates
synthesis by metal–metal exchange, **11**, 467, 468

Organolead(II) compounds
preparation, **2**, 313–16
Organolead(IV) compounds
tricarboxylates
chemistry, **11**, 465–83
Organolead halides
chemistry, **2**, 308, 309
Lewis acids, **2**, 308
Organolead pseudohalides
chemistry, **2**, 308, 309
Organolead radicals
reactions, **2**, 313
Organolithium complexes
amido
calculations, **1**, 30
inorganic salt complexes, **1**, 10
Lewis base donor complexes, **1**, 6
monomeric, structure, **1**, 9
structure, **1**, 5
arsenides, **1**, 10
imino, **1**, 2
calculations, **1**, 30
nitrogen compounds, **1**, 5
phosphides, **1**, 10
solvation, **1**, 19
structures
theory, **1**, 29
Organolithium compounds
in cyclopentadienyl alkyl nickel complex synthesis, **9**, 66
in organic synthesis, **11**, 1–81
in organonickel complex synthesis, **9**, 41
in organotin compound synthesis, **2**, 225
transmetallation, **12**, 131
to palladium, **12**, 154
Organolithium reagents
organocopper compound preparation from, **3**, 63
Organomagnesium alkoxides
properties, **1**, 101, 102
Organomagnesium alumoxanes, **1**, 106
Organomagnesium amides
crystal structures, **1**, 103
properties, **1**, 101–4
Organomagnesium clusters, **1**, 58
formation, **1**, 62
Organomagnesium complexes
in cyclopentadienyl alkyl nickel complex synthesis, **9**, 66
in organonickel complex synthesis, **9**, 41
Organomagnesium compounds
heterobimetallic compounds, **1**, 104–7
nitrogen heterocycles
paramagnetic, **1**, 103
in organic synthesis, **11**, 129–53
organoboron compounds from, **11**, 202, 203
Schlenk equilibrium, **1**, 66, 75
selenocyanates, **1**, 104
thiocyanates, **1**, 104
transmetallation
to palladium, **12**, 154
Organomagnesium(I) compounds
chemistry, **1**, 58
Organomagnesium diazenides
properties, **1**, 104
Organomagnesium halides
chemistry, **1**, 58–84
organocopper compound
preparation from, **3**, 63
preparation, **1**, 58–72
Organomagnesium hydrides
properties, **1**, 101
Organomagnesium reagents
clusters, **1**, 59
organocuprate preparation from, **3**, 65
Organomagnesium tris(pyrazolyl)hydroborates
preparation, **1**, 104
Organomanganese compounds
cuprates from, **12**, 82
ketone synthesis from, **12**, 221
Organomanganese(II) compounds
in organic synthesis, **6**, 128–38
Organomanganese(III) compounds
in organic synthesis, **6**, 138–44
Organomanganese(IV) compounds
in organic synthesis, **6**, 145, 146
Organomanganese(V) compounds
in organic synthesis, **6**, 146
Organomanganese(VII) compounds
in organic synthesis, **6**, 146
Organomanganese transients, **6**, 46–8
gas phase studies, **6**, 47, 48
spectroscopy, **6**, 46, 47
Organomercury compounds, **3**, 135–69
alkyl
reaction with alkenes, **12**, 133
alkylation, **11**, 402–7
allenic
preparation, **11**, 390
dialkyl
preparation, **11**, 390
diaryl
preparation, **11**, 390
dimerization, **11**, 401, 402
halogenation, **11**, 396, 397
metathesis, **12**, 221
nitrogen-containing
preparation, **11**, 394
in organic synthesis, **11**, 389–430
in organonickel complex synthesis, **9**, 41
polydentate Lewis acids, **3**, 162–5
preparation, **11**, 390–5
transmetallation, **11**, 391
propargylic
preparation, **11**, 390
properties, **3**, 155–69
protonolysis, **11**, 395
metal hydride reduction, **11**, 396
radical reactions, **3**, 166
structures, **3**, 155–69
sulfonyl, **11**, 429
transmetallation, **12**, 131
to palladium, **12**, 132
Organomercury salts
radical anions, **3**, 167
Organometallate complexes
without carbonyl ligands, **5**, 288, 289
Organometallic alkali metals
magnesium alkoxides, **1**, 105
Organometallic compounds
allyl alkylation, **12**, 852–62
main group
carbonylation, **12**, 370, 371
neutral homoleptic
without carbonyl ligands, **5**, 282, 283
without carbonyl ligands, **5**, 281–323
Organometallic halides
properties, **1**, 109, 110
reactivity, **1**, 109, 110
structures, **1**, 109, 110
Organometallic polymers
synthesis, **5**, 395
Organonickel complexes
allyl alkyl, **9**, 65, 66
allyl aryl, **9**, 65, 66
cyclopentadienyl alkyl, **9**, 66–73
properties, **9**, 70–3
reactions, **9**, 70–3
synthesis, **9**, 66–70
cyclopentadienyl aryl, **9**, 66–73
properties, **9**, 70–3
reactions, **9**, 70–3
synthesis, **9**, 66–70
properties, **9**, 64, 65
reactions, **9**, 56–63
alkyl group displacement, **9**, 62, 63
aryl group displacement, **9**, 62, 63
Ni–C bond generated, **9**, 58–62
Ni–C bond not involved, **9**, 56, 57
structures, **9**, 64, 65
synthesis, **9**, 41–56
Organonickel compounds
transmetallation, **12**, 131
to palladium, **12**, 156–8
Organoosmium chemistry
Gmelin, **7**, 293, 294
starting materials, **7**, 294
Organoosmium complexes
binucler
synthesis, **7**, 299
clusters
synthesis, **7**, 299
crystal structures, **7**, 291
cyclopentadienyl
synthesis, **7**, 299
mixed-metal clusters, **7**, 300
mononuclear
synthesis, **7**, 299
synthesis, **7**, 299
trinuclear
synthesis, **7**, 299
Organoosmium compounds
annual surveys, **7**, 293
chemistry, **7**, 291–4
Organopalladium carbonyls
properties, **9**, 196
Organopalladium(II) complexes
decomposition, **9**, 265–7

isomerization, **9**, 265–7
metathesis, **9**, 261–5
reactions, **9**, 261–71
Organopalladium compounds
insertion reactions
alkenes, **12**, 149
transmetallation, **12**, 131, 143
Organoplatinum(IV) polymers
preparation, **9**, 464
Organopolyboranes
preparation, **1**, 245, 246
reactions, **1**, 246, 247
Organopolyboron hydrides
chemistry, **1**, 245–7
physical properties, **1**, 245
Organopolysilanes
chemistry, **2**, 77–107
Organopotassium compounds
in organic synthesis, **11**, 93–124
Organorhenium compounds
high valent, **6**, 231–52
Organorhodium compounds
transmetallation
to palladium, **12**, 155
Organoruthenium complexes
arenes
synthesis, **7**, 296
binuclear
synthesis, **7**, 297
carbonyl
synthesis, **7**, 296
clusters
synthesis, **7**, 298
cyclopentadienyl
synthesis, **7**, 296
diene
synthesis, **7**, 296
mixed-metal clusters
synthesis, **7**, 298
trinuclear clusters
synthesis, **7**, 297
Organoruthenium compounds
annual surveys, **7**, 293
chemistry, **7**, 291–4
crystal structures, **7**, 291
starting materials, **7**, 294
Organoselenium chemistry, **11**, 515–58
Organoselenium complexes
ate, **11**, 518
Organoselenium compounds
in living systems, **11**, 516
structures, **11**, 516–22
synthesis, **11**, 538
Organoselenium monitored reactions
inorganic reagents, **11**, 527
organic reagents, **11**, 528–33
Organoselenium reagents
electrophilic
cyclization, **11**, 542
synthesis, **11**, 529
Organoselenocyanates
synthesis, **11**, 530
Organosilicon compounds, **2**, 1–39
alkali metals
synthesis, **2**, 22
alkaline earth metals
synthesis, **2**, 22
halogen
synthesis, **2**, 38, 39
in organic synthesis, **11**, 313–45
oxygen, **2**, 31–4
peroxides
synthesis, **2**, 33, 34
polyazo
synthesis, **2**, 25, 26
selenium
synthesis, **2**, 34–7
spectroscopy, **2**, 3
sulfur
synthesis, **2**, 34–7
tellurium
synthesis, **2**, 34–7
transmetallation, **12**, 131
trioxides
synthesis, **2**, 33, 34
Organosilver compounds, **3**, 57–129
alkenyl
synthesis, **3**, 71, 72
alkyl
solid-state structures, **3**, 77–80
synthesis, **3**, 71
alkynyl
heteroleptic, solid-state structures, **3**, 96–100
homoleptic, solid-state structures, **3**, 96–100
synthesis, **3**, 71, 75
amine stabilized cationic
solid-state structures, **3**, 110–15
amine stabilized neutral
solid-state structures, **3**, 110–15
aryl
functionalized heteroatom-containing substituents, solid-state structures, **3**, 86–8
noncoordinating substituents, solid-state structures, **3**, 82–6
solid-state structures, **3**, 82–95
synthesis, **3**, 71, 73
bonding, **3**, 75–129
imidazoyl
synthesis, **3**, 73
phosphine stabilized cationic
solid-state structures, **3**, 110–15
phosphine stabilized neutral
solid-state structures, **3**, 110–15
phosphorylides
solid-state structures, **3**, 115
polymeric
structures, **3**, 120, 121
solid-state
structures, x-ray analysis, **3**, 77–117
solution
structures, **3**, 125
stereochemistry, **3**, 121–9
structure, **3**, 75–129
synthesis, **3**, 60–75
thermal stability, **3**, 75–7
trifluoromethyl
synthesis, **3**, 72
Organosilver lithium compounds
diaryl
synthesis, **3**, 74
Organosodium compounds
in organic synthesis, **11**, 93–124
Organostannanes
coupling reactions
nickel complexes, **12**, 965
Organostannyl radicals
properties, **2**, 285–90
Organostannylenes
properties, **2**, 285–90
types, **2**, 285–9
Organostrontium compounds
dialkyl, **1**, 110–12
diaryl, **1**, 110, 111
Organostrontium(I) compounds
chemistry, **1**, 58
Organostrontium halides
preparation, **1**, 107–9
Organotellurium compounds
in deprotection of organic functionalities, **11**, 582, 583
free radical chemistry, **11**, 594, 595
in organic synthesis, **11**, 571–96
preparation, **11**, 572–6
tellurium-free compounds from, **11**, 587–90
Organothallium compounds
insertion reactions
alkenes, **12**, 146
in organic synthesis, **11**, 437–58
organocopper compound preparation from, **3**, 64
in organonickel complex synthesis, **9**, 41
transmetallation, **12**, 131, 146
to palladium, **12**, 146–50
Organotin alkoxides
acidolysis, **2**, 229
chemistry, **2**, 261–70
in organic synthesis, **11**, 367
Organotin amides
acidolysis, **2**, 229
Organotin carbonates
properties, **2**, 259
Organotin carboxylates
chemistry, **2**, 256–9
decarboxylation, **2**, 233
dialkyl
structures, **2**, 258
Lewis acid properties, **2**, 258
nitrogen containing, **2**, 257
NMR, **2**, 258
structures, **2**, 257
Organotin clusters, **2**, 265
Organotin compounds
acyl
Wittig reactions, **2**, 238
alkyl
chemistry, **2**, 234, 235
alkynyl
Diels–Alder reactions, **2**, 238
preparation, **2**, 242
reaction with trialkyl-boranes, **2**, 253
allyl
preparation, **2**, 243–9

reaction with electrophiles, **2**, 244
reactions with enophilies, **2**, 247–9
reactions with radicals, **2**, 245, 246
reactions with singlet oxygen, **2**, 270
transmetallation, **2**, 244
applications, **2**, 296, 297
aryl
reaction with electrophiles, **2**, 236
structure, **2**, 236–8
biological activity, **2**, 297
bond energies, **2**, 219
catalysis, **2**, 297
copper bonds, **2**, 284
cyclopentadienyl-
structure, **2**, 249
derivatives of phosphoric acids, **2**, 259
derivatives of sulfur acids, **2**, 260
direct synthesis, **2**, 232, 233
halogen bonds
in organic synthesis, **11**, 368
halogens
in organic synthesis, **11**, 358, 359
handling, **11**, 358
magnesium bonds, **2**, 284
metal bonds, **2**, 277–85
molecular orbital calculations, **2**, 219
molecular weight measurements, **2**, 222
Mössbauer spectroscopy, **2**, 219–23
nitrogen, **2**, 290–3
nitrogen bonds
in organic synthesis, **11**, 368
NMR, **2**, 219–23
Karplus relationships, **2**, 220
solid state, **2**, 221
optically active, **2**, 234
in organic synthesis, **11**, 355–81
palladium-catalyzed reactions, **2**, 235
phosphorus, **2**, 290–3
photoreactions, **11**, 381
physical data, **2**, 219
selenium, **2**, 293–6
separation, **2**, 271
sterically hindered, **2**, 235
sulfur, **2**, 293–6
tellurium, **2**, 293–6
tin bonds, **2**, 277–85
toxicity, **11**, 358
transmetallation, **12**, 131
trifluoromethyl halides
preparation, **2**, 226
vinyl
formation, **2**, 238–41
NMR, **2**, 239
reaction wtih electrophiles, **2**, 241
structure, **2**, 239
Organotin esters
oxyacids, **2**, 256–60
Organotin formates
decomposition, **2**, 277
Organotin halides
chemistry, **2**, 254–6
complexes, **2**, 255
structures, **2**, 256
Lewis acid properties, **2**, 255
structures, **2**, 255
synthesis, **2**, 254
Organotin hydrides
bonded to polymers, **2**, 272
decarboxylation of acids, **2**, 275
in organic synthesis, **2**, 273; **11**, 358–63
properties, **2**, 270–6
reactions with isocyanides, **2**, 275
reactions with nitriles, **2**, 275
reactions with organic halides, **2**, 274
reactions with selenides, **2**, 274
reactions with sulfides, **2**, 274
reactions with tellurides, **2**, 274
reactions with thiono esters, **2**, 275
relative reactivities, **2**, 273
water soluble, **2**, 272
Organotin hydroxides
chemistry, **2**, 261–70
Organotin oxides
chemistry, **2**, 261–70
in organic synthesis, **11**, 367
reactions with carboxylic acids, **2**, 256
Organotin oxycarbonates
preparation, **2**, 257
properties, **2**, 259
Organotin peroxides
chemistry, **2**, 261–70
properties, **2**, 270
Organotin radical ions
properties, **2**, 290
Organotin radicals
reactions, **2**, 278
Organotitanium ate compounds
cuprates from, **12**, 83
Organotitanium compounds
coupling reactions
palladium-catalyzed, **12**, 222
transmetallation, **12**, 131
to palladium, **12**, 156–8
Organoytterbium compounds
in organonickel complex synthesis, **9**, 41
Organozinc compounds, **3**, 175–203
addition reactions, **11**, 169–77
to carbonyl derivatives, **11**, 169–72
adducts
in electronics industry, **3**, 200, 201
alkyl
adducts, **3**, 177, 178
alkyl–zinc–oxygen bonds, **3**, 184, 185
alkyl–zinc–selenium groups, **3**, 186
alkyl–zinc–sulfur groups, **3**, 186
allylic
addition to alkenes, **11**, 175
amides, **3**, 181, 182
bulky silyl ligands, **3**, 179, 180
carbamates
synthesis, **3**, 185
carbenes, **3**, 180
carbynes, **3**, 180
cyclopentyl-
preparation, **11**, 175
diselenocarbamates, **3**, 186
dithiocarbamates, **3**, 186
in electronics industry, **3**, 200–3
heterometallic, **3**, 187, 188
in organic synthesis, **3**, 188–95
photodissociation, **3**, 188
physical studies, **3**, 188
preparation, **11**, 160–4
transmetallation, **11**, 163, 164
reactivity, **11**, 165–78
substitution reactions, **11**, 165–8
theoretical studies, **3**, 188
transmetallation, **12**, 131
to palladium, **12**, 155
zinc reagents, **12**, 156
unsaturated ligands, **3**, 180
zinc–halide bonds, **3**, 182
Organozinc reagents
addition to chromium tricarbonyl benzaldehyde complexes, **5**, 524
coupling reactions, **12**, 180
cuprates from, **12**, 65–73
transmetallation to copper, **12**, 92
Organozirconium complexes
bis(cyclopentadienyl)
dicationic, **4**, 614
bis(cyclopentadienyl) acyl, **4**, 610
bis(cyclopentadienyl) alkenyl, **4**, 605–7
bis(cyclopentadienyl) alkoxides
cationic, **4**, 613, 614
bis(cyclopentadienyl) alkyl
cationic, **4**, 590–605
cationic, reactivity, **4**, 600–5
cationic, structures, Lewis base-free, **4**, 599, 600
cationic, structures, Lewis base-stabilized, **4**, 595–8
cationic, synthesis, **4**, 590–4
bis(cyclopentadienyl) allyl
cationic, **4**, 590–605
cationic, reactivity, **4**, 600–5
cationic, structures, Lewis base-free, **4**, 599, 600
cationic, structures, Lewis base-stabilized, **4**, 595–8
cationic, synthesis, **4**, 590–4
bis(cyclopentadienyl) aryl, **4**, 605–7
bis(cyclopentadienyl) benzyl

cationic, **4**, 590–605
cationic, reactivity, **4**, 600–5
cationic, structures, Lewis base-free, **4**, 599, 600
cationic, structures, Lewis base-stabilized, **4**, 595–8
cationic, synthesis, **4**, 590–4
bis(cyclopentadienyl) carbonyl, **4**, 610
bis(cyclopentadienyl) hydrides, **4**, 611–13
bis(cyclopentadienyl) iminoacyl, **4**, 610
bis(cyclopentadienyl) pyridyl, **4**, 607, 608
cationic, **4**, 589–622
applications, **4**, 618–22
monocyclopentadienyl hydrocarbyl
cationic, **4**, 615
in organic synthesis, **4**, 619–22
tetraaza-macrocycle hydrocarbyl
cationic, **4**, 617
tris(hydrocarbyl)
cationic, **4**, 616, 617
Organozirconium compounds
cuprates from, **12**, 88
transmetallation, **12**, 131
alkenyl group, **12**, 157
to palladium, **12**, 155
Orgonozinc reagents
transmetallation
to copper, **12**, 92
Oroganomercury hydrides
structures, **3**, 165, 166
Orthoboronation, **1**, 343
Orthoformates
conversion to ethers
organoaluminum hydrides, **11**, 300
Orthomagnesiation
cubane, **1**, 64
Oshima–Lombardo reagent
alkylidenation, **12**, 578–85
Osmacarbocycles, **7**, 424–6
Osmacycles, **7**, 424–30
Osmacyclohexadienones, **7**, 808
Osmium
dicarbon metallacarbaboranes
7-vertex, **1**, 382
11-vertex, **1**, 396
12-vertex, **1**, 400
monocarbon metallacarbaboranes, **1**, 378
Osmium clusters
acetylide platinum, **10**, 153, 154
acetylide tungsten
tetranuclear, **10**, 104–6
trinuclear, **10**, 102
trinuclear, reactivity, **10**, 102–4
alkylidyne molybdenum, **10**, 93–6
alkylidyne ruthenium, **10**, 129, 130
alkylidyne tungsten, **10**, 93–6
trinuclear, **10**, 100
alkyne ruthenium, **10**, 130, 131
alkyne tungsten
tetranuclear, **10**, 101
trinuclear, **10**, 100
bisalkylidyne osmium
tetranuclear, **10**, 101
carbomethoxymethylidynes, **7**, 805
decanuclear, **7**, 950
anionic carbonyls, **7**, 950
carbido, **7**, 950–3
hydrido carbonyls, **7**, 950
formyl-methylidenes, **7**, 805
heptanuclear, **7**, 945–7
carbido, **7**, 946
carbonyl, **7**, 945
carbonyl hydrides, **7**, 945
hydrocarbon ligands, **7**, 947
nitrogen ligands, **7**, 946, 947
phosphine ligands, **7**, 946, 947
sulfur ligands, **7**, 946, 947
hexanuclear, **7**, 930–45
arsine ligands, **7**, 938, 939
carbido, **7**, 934–7
carbonyl hydrides, **7**, 930–3
carbonyls, **7**, 930–3
heteroatom ligands, **7**, 937–44
hydrocarbon ligands, **7**, 944, 945
nitrogen ligands, **7**, 939, 940
oxygen ligands, **7**, 937, 938
phosphine ligands, **7**, 938, 939
selenium ligands, **7**, 940–4
sulfur ligands, **7**, 940–4
high nuclearity, **7**, 907–53
synthesis, **7**, 908
medium nuclearity, **7**, 907–53
nonanuclear, **7**, 949, 950
octanuclear, **7**, 947–9
carbonyl hydrides, **7**, 947, 948
carbonyls, **7**, 947, 948
pentanuclear, **7**, 909–30
arsenic ligands, **7**, 921–5
benzene-coordinated, **7**, 921
carbido, **7**, 918–21
carbonyl, **7**, 909–18
carbonyl hydrides, **7**, 909–18, 920
heteroatom ligands, **7**, 921–30
hydrocarbon ligands, **7**, 930
nitrogen ligands, **7**, 921–5
oxidative addition, **7**, 920
phosphinoalkyne, **7**, 925–30
phosphorus ligands, **7**, 921–5
reduction, **7**, 920
selenium ligands, **7**, 921–5
sulfur ligands, **7**, 921–5
polynuclear, **7**, 300
tetranuclear, **7**, 835–902
acetylide dianion, **7**, 874
allenyl, **7**, 900
aminoalkyne, **7**, 901
aminocarbyne, **7**, 893
benzyne, **7**, 880
bond lengths, **7**, 838
bridging carbon–group 13 element donor ligands, **7**, 892
bridging carbon–group 15 element donor ligands, **7**, 893–8
bridging carbon–group 16 element donor ligand, **7**, 899–902
bridging carbon–heteroatom donor ligands, **7**, 892–902
bridging carbon–nitrogen donor ligand, **7**, 893–6
bridging carbon–oxygen donor ligand, **7**, 899–901
bridging carbon–phosphorus donor ligand, **7**, 896–8
bridging carbon–sulfur donor ligands, **7**, 901, 902
carbene, **7**, 871, 879
carbon ligands, **7**, 871–9
η^3-carbon ligands, **7**, 879, 880
η^4-carbon ligands, **7**, 880–2
η^5-carbon ligands, **7**, 882–4
η^6-carbon ligands, **7**, 885–7
carbon ligands, **7**, 888
carbon monosulfide, **7**, 902
carbonyl, **7**, 840–70
carbonyl, bridging group 13 ligands, **7**, 850, 851
carbonyl, bridging group 14 ligands, **7**, 852
carbonyl, bridging group 15 ligands, **7**, 853–62
carbonyl, bridging group 16 ligands, **7**, 862–70
carbonyl, bridging group 17 ligands, **7**, 869, 870
carbonyl, bridging heavier group 15 ligands, **7**, 862
carbonyl, bridging hydride ligands, **7**, 845–8
carbonyl, bridging ligands, **7**, 850–70
carbonyl, bridging nitrogen ligands, **7**, 853–6
carbonyl, bridging oxygen ligands, **7**, 862–4
carbonyl, bridging phosphorus ligands, **7**, 856–61
carbonyl, bridging selenium ligands, **7**, 868, 869
carbonyl, bridging sulfur ligands, **7**, 864–8
carbonyl, bridging tellurium ligands, **7**, 868, 869
carbonyl anions, **7**, 844, 845
carboxylate, **7**, 878
cluster shapes, **7**, 838–40
cyclobutyne, **7**, 876
cyclopentadienylidene, **7**, 884
diphenylacetylene, **7**, 873
ethyne, **7**, 872
hydrides, **7**, 871

hydrido carbonyl anions, **7**, 849, 850
hydrido carbonyl cations, **7**, 850
indenyl, **7**, 872, 888
methoxyalkylidyne, **7**, 899
N-methylpyrrolidine, **7**, 871
phenylphosphine, **7**, 850, 880
structures, **7**, 838–40
styrenyl, **7**, 873
synthesis, **7**, 836, 837
synthesis, allene, **7**, 889
synthesis, allylvinyl, **7**, 889
synthesis, carbene, **7**, 891
synthesis, carbon ligand coupling, **7**, 888–92
synthesis, carboxylate, **7**, 890
synthesis, cyclobutadienyl, **7**, 889
synthesis, diyne, **7**, 890
vinylidene, **7**, 875, 876
trinuclear, **7**, 683–737
alkynyl ligands, **7**, 694
allyl, **7**, 696
amido, **7**, 731
aminoalkyne, **7**, 696
anionic, **7**, 719–21
antimony ligands, **7**, 735
arsenic ligands, **7**, 735
bridging transoid butadiene, **7**, 697
bridging vinyl ligands, **7**, 694
catalyzing substitution, **7**, 710
comparison with ruthenium clusters, **7**, 689
cyclic thioethers, **7**, 707
cyclohexa-1,3-diene, **7**, 705
diazomethane, **7**, 694, 696
dichloromethane, **7**, 710
dimethylthietane, ring opening, **7**, 697
diyne, **7**, 694
doubly-bridging 2-electron donating ligands, **7**, 692–4
doubly-bridging 3-electron donating ligands, **7**, 694–7
doubly-bridging 4-electron donating ligands, **7**, 697, 698
doubly-bridging 5-electron donating ligands, **7**, 698
doubly-bridging 6-electron donating ligands, **7**, 699, 700
doubly-bridging ligands, **7**, 691–700
doubly-bridging ligands donating more than 6 electrons, **7**, 700
electron counting, **7**, 689, 690
fluxionality, **7**, 717, 718
furyl, **7**, 692
hydrido, **7**, 719–21
ligands coordinated through heteroatoms, **7**, 711–37
methylindolyl, **7**, 693
methylpyrrolyl ligands, **7**, 692, 693
methylthienyl, **7**, 692
nitrogen ligands, **7**, 730
nitrosyl ligands, **7**, 730
oximes, **7**, 696
phenyl azide, **7**, 696
phosphabenzene, **7**, 708
phosphorus ligands, **7**, 735
pyridine-2-thione, **7**, 730
reactions with thioamides, **7**, 728
reactions with thioureas, **7**, 728
selenium ligands, **7**, 726
silane, **7**, 696
stannene, **7**, 696
stannyne, **7**, 696
sulfur dioxide, **7**, 694
sulfur ligands, **7**, 726
tellurium ligands, **7**, 726
tetrahydrothiophene, **7**, 696
thermal substitution, **7**, 685
thioformyl, **7**, 706
triply-bridging 4-electron donating ligands, **7**, 702–5
triply-bridging 5-electron donating ligands, **7**, 705–8
triply-bridging 6-electron donating ligands, **7**, 708, 709
triply-bridging ligands, **7**, 701–10
triply-bridging ligands donating more than 6 electrons, **7**, 709, 710
unsaturated 46-electron clusters, **7**, 689
UV irradiation, **7**, 685
vinylidene ruthenium, **10**, 130, 131
Osmium complexes
β-acetyl ethyl, **7**, 392
acetylides, **7**, 411–16
reactivity, **7**, 415, 416
synthesis, **7**, 412–15
acyclic pentadienyl, **7**, 465–8
acyl, **7**, 417–24
alkoxyalkylidenes, **7**, 343
alkoxycarbonyls, **7**, 422, 423
alkoxymethyl, **7**, 387
alkyl, **7**, 372–93
high-valent, **7**, 372–8
nitrido, **7**, 372
oxo, **7**, 372
alkylidene, **7**, 336–48
alkylidyne, **7**, 356–62
reactivity, **7**, 360–2
synthesis, **7**, 356–60
α-alkynyl, **7**, 390
alkynyl vinyl, **7**, 407
γ-aminoalkyl, **7**, 393
ammine aniline, **7**, 588
ammine arene
electrochemical oxidation, **7**, 589
ammine arenes, **7**, 588
ammine benzene
catalytic hydrogenation, **7**, 589
ammine 2,4-cyclohexadien-1-one, **7**, 588
ammine 2,5-cyclohexadien-1-one, **7**, 588
ammine furan, **7**, 589
ammine naphthalene, **7**, 589
ammine pyrene, **7**, 589
ammine pyrrole, **7**, 589
ammine tetramethylbenzene, **7**, 588
ammine thiophene, **7**, 589
ammine *p*-xylene, **7**, 588
arenes
group 14 element bonds, **7**, 586
aroyl, **7**, 417–24
aryls, **7**, 393–9
homoleptic, **7**, 393
benzene cyclohexadienyl protonation, **7**, 563
benzene iodo
reaction with organolithium reagents, **7**, 562
binuclear
alkyl, **7**, 674
arene, **7**, 674
aryl, **7**, 674
carbonyl, **7**, 625–51
carbonyl, acetylide-bridged, **7**, 650
carbonyl, alkenediyl, **7**, 652
carbonyl, alkenes, **7**, 652
carbonyl, dienediyl ligands, **7**, 655
carbonyl, dienyl ligands, **7**, 655–66
carbonyl, η^3-enyl ligands, **7**, 653, 654
carbonyl, hydrocarbon ligands, **7**, 652–66
carbonyl, sulfur-bridged, **7**, 647
carbonyl, σ-yl-type ligands, **7**, 652, 653
cyclopentadienyl, **7**, 666–73
cyclopentadienyl, d^5–d^5 systems, **7**, 667–73
cyclopentadienyl, d^5–d^6 systems, **7**, 667–73
cyclopentadienyl, d^6–d^6 systems, **7**, 666, 667
metal–metal bonds, **7**, 625–74
bis(arene) dihalo
half sandwich, **7**, 556–8
carbomoyls, **7**, 423, 424
carbonyl hydrides, **7**, 362–72
dihydrogen, **7**, 370–2
nitrogen donors, **7**, 368
phosphines, **7**, 363
ternary, **7**, 362, 363
carbonyls, **7**, 300, 301
carboxamides, **7**, 423, 424
chalcocarbonyls, **7**, 319–27
cycloaddition, **7**, 326
electrophilic attack, **7**, 325
migratory insertion, **7**, 326
nucleophilic attack, **7**, 324
reactivity, **7**, 324–7
reduction, **7**, 327
synthesis, **7**, 320–4
β-chloroalkyl, **7**, 392
cumulenylidene, **7**, 353–6

cyclopentadienyl alkyl, **7**, 480–2
cyclopentadienyl halides, **7**, 479
cyclopentadienyl hydroxymethyl, **7**, 483
α-diazoalkyls, **7**, 389
dithiomethoxycarbonyls, **7**, 422, 423
enoyl, **7**, 420, 421
enynyl, **7**, 407–11, 462–4
formyls, **7**, 417, 418
α-functionalized alkyls, **7**, 387–9
β-functionalized alkyls, **7**, 392, 393
γ-functionalized alkyls, **7**, 392, 393
haloalkylidenes, **7**, 346–8
heteroatom substituted alkylidenes, **7**, 343–6
hydrocarbon alkylidenes, **7**, 336–42
hydrocarbon ligands, **7**, 441–68
hydrocarbonyls, **7**, 422
hydrocarbyl, **7**, 372–430
hydroxymethyl, **7**, 387
imido-alkyl, **7**, 375
iminoacyl, **7**, 376, 421, 422
isonitriles, **7**, 327–36
ketenes, **7**, 804
keto alkyls, **7**, 389
mercury clusters, **10**, 340
monohalomethyl, **7**, 387
mononuclear
carbon ligands, **7**, 299–336
perfluoroalkyl, **7**, 389–91
phosphinomethyl, **7**, 391
β-phosphonio ethyl, **7**, 392
phosphonioalkyls, **7**, 389
porphyrins
catalysts, diazodecomposition, **12**, 430
β-silyl alkyl, **7**, 392
β-stannyl alkyl, **7**, 392
sulfinyl alkyls, **7**, 389
thioacyls, **7**, 422
thioalkyl vinyl, **7**, 406
thioformyls, **7**, 422
trimethylenemethane, **7**, 464, 465
trinuclear clusters
metallaboranes, **1**, 296
vinyl, **7**, 399–411
vinylidene, **7**, 348–56
ynyl, **7**, 462–4
Osmium(0) complexes
alkenes, **7**, 441–5
alkynes, **7**, 441–5
arene carbonyl dichloro, **7**, 568
carbonyls
binuclear, **7**, 301
group 15, 16, 17 ligands, **7**, 302–6
mononuclear, **7**, 301
dienes, **7**, 452–5
ethene, **7**, 442
Osmium(I) complexes
carbonyls
group 15, 16, 17 ligands, **7**, 306, 307
Osmium(II) complexes
acyclic allyl, **7**, 459, 460
alkenes, **7**, 445–9
alkyl
polyphosphine, **7**, 380
alkyl carbonyl, **7**, 378–87
alkynes, **7**, 445–9
aluminum tetrahydride derivatives, **1**, 551
arene carbonyl dichloro, **7**, 568
aryl carbonyl, **7**, 396
aryls, **7**, 395–9
carbonyl
bidentate phosphines, **7**, 308
carboxylates, **7**, 316
functionalized phosphines, **7**, 311
macrocycles, **7**, 317
multidentates, **7**, 317
nitrogen donors, **7**, 312
oxygen donors, **7**, 314
phosphido, **7**, 311
poly(azolyl)borates, **7**, 314
pyridinethiolates, **7**, 314
Schiff bases, **7**, 316
sulfonates, **7**, 316
tridentate phosphines, **7**, 310
carbonyl halides, **7**, 307
carbonyl phosphines, **7**, 308
carbonyls
group 15, 16, 17 ligands, **7**, 307–19
dienes, **7**, 449–52
hydrido alkenes, **7**, 448
mercury aryls, **7**, 397
silaethenes, **7**, 449
trivalent carbonyls, **7**, 319
Osmium(IV) complexes
carbonyls, **7**, 319
Osmium(VI) complexes
perfluorophenyl, **7**, 394
Osmium(VI) glycolate amine complexes
reaction of osmium tetroxide with alkenes, **12**, 1138
Osmium(VI) glycolates
asymmetric dihydroxylation, **12**, 1138
Osmium nickel clusters
hydrogenation cataysts, **10**, 38
Osmium tetroxide
asymmetric dihydroxylation, **12**, 1138
Osmium, decacarbonyldihydrido(triphenylphosphine)tris-
fluxionality
volume of activation, **7**, 688
Osmocene
chemistry, **7**, 603–20
conformation, **7**, 605
decamethyl-
electronic absorption spectra, **7**, 606
oxidation, **7**, 609
protonation, **7**, 609
dimer, **7**, 609
electronic absorption spectra, **7**, 606
free energy of ionization, **7**, 605
IR spectra, **7**, 606
macrocyclic derivatives, **7**, 613, 614
methyloxy-
structural phase transition, **7**, 607
NMR, **7**, 606
oxidation, **7**, 608, 672
PES, **7**, 606
Raman spectra, **7**, 606
substitution reactions, **7**, 610, 611
synthesis, **7**, 604
thermodynamic data, **7**, 606
[3]Osmocenophane
synthesis, **7**, 616
β-Oxabenzocyclohexanone
asymmetric reduction, **12**, 25
8-Oxabicyclo[3.2.1]oct-6-enes
Pauson–Khand reaction, **12**, 721
Oxaboranes
properties, **1**, 267
Oxaborolane
reactions, **1**, 173
Oxadiboririane
preparation, **1**, 171
Oxadisilacyclobutanes
properties, **2**, 55
Oxamagnesacycles
structures, **1**, 90
Oxametallaboranes
properties, **1**, 267
Oxametallacycles
in epoxidation, **12**, 1100
from zirconocene aryne complexes, **12**, 772
Oxanickelacyclopentanones
synthesis, **9**, 90
7-Oxanorbornadiene
Pauson–Khand reaction, **12**, 720
Oxaosmacycles, **7**, 426, 427
1,3-Oxaphospholes
synthesis
carbene complexes, **12**, 513
1-Oxa-3-platinacyclobutane
reactions, **9**, 481
Oxaruthenacycles, **7**, 426, 427
Oxasilacyclobutanes
properties, **2**, 55
Oxasilacyclobutenes
properties, **2**, 55
Oxasilacyclopentenes
synthesis, **2**, 59
Oxasiletanides
Peterson olefination, **11**, 327, 328
Oxathiastannolanes
structures, **2**, 295
Oxazaborolidines
ketones
reduction, **11**, 259, 260
Oxazaphospholidines
reduction, **11**, 261
Oxazines
reactions with molybdenum complexes
photochemical, **12**, 554
Oxaziridines

asymmetric epoxidation, **12**, 1098
Oxazoles
cross-coupling reactions
Grignard reagents, **12**, 178
formation
from diazocarbonyl compounds, **12**, 440
Oxazoles, 2-aryl-
synthesis, **12**, 209
N-Oxazolidinoyl activators
lithiation, **11**, 20
2-Oxazolidones, 4-vinyl-
synthesis, **12**, 874
Oxazolines
cross-coupling reactions
Grignard reagents, **12**, 178
formation
carbonylation, **12**, 258
lithiation, **11**, 42
reactions with molybdenum complexes
photochemical, **12**, 555
Oxazolines, (*o*-diphenylphosphinophenyl)-
synthesis, **12**, 887
Oxazolines, thienyl-
synthesis, **12**, 887
N-Oxazolinoyl activators
lithiation, **11**, 20
Oxetanes
reductive cleavage
organolithium reagents, **11**, 12
ring opening
palladium salts, **12**, 139
Oxidation
alkenes
catalysts, **10**, 50
iron carbonyl catalysts, **7**, 87
organoaluminum compounds
carbon–oxygen bonds, **11**, 302, 303
organobismuth(V) compounds, **11**, 504, 505
organotellurium compounds, **11**, 583, 584
Oxidation state
18-electron rule, **12**, 2
Oxidations
metal catalyzed
mechanisms, **12**, 1100–7
organic synthesis, **12**, 1097–132
Oxidative addition
alkene insertion
palladium catalysts, **12**, 259–85
allylic compounds
to palladium(0) complexes, **9**, 343–7
methyl iodide
rhodium complexes, **8**, 190
nickel complex synthesis, **9**, 47–53
regioselectivity, **12**, 172
transition metal complexes, **12**, 4
transmetallation, **12**, 161–231
Oxidative addition-insertion reactions
transition metal alkyl complexes, **12**, 241–91
Oxidative catalysis
substitution reactions
cyclopentadienyl tricarbonyl manganese complexes, **6**, 111
Oxides
immobilized mixed metal clusters on
catalysis, **10**, 52–8
2,3-Oxidosqualene-lanosterol cyclase
inhibitor, **12**, 78
Oxilorphan
synthesis, **12**, 18
Oximes
oxidation
organoselenium compounds, **11**, 535
Oxindoles
synthesis, **12**, 256, 312
Oxiranes
cycloaddition
heterocumulenes, **11**, 500
deoxygenation
tellurium reagents, **11**, 578, 591
isomerization
superbase, **11**, 120
reactions with iron atoms, **7**, 108
Oxiranes, alkenyl-
oxidative addition
allyl complexes, **12**, 807
Oxiranes, silyloxy-
in oxidation of silyl enol ethers, **11**, 336
Oxiranes, vinyl-
alkenylboranes
nickel bis(acetylacetonate), **12**, 156
Oxo complexes
group 6
properties, **5**, 296
reactions, **5**, 298
structure, **5**, 296
without carbonyl ligands, **5**, 289–99
Oxo transfer
mechanisms, **12**, 1104–7
Oxonium ylides
formation, **12**, 442, 443
metal stabilized, **12**, 441–5
oxygen–hydrogen insertion, **12**, 445
reactions, **12**, 442, 443
[2,3]-sigmatropic rearrangements, **12**, 442, 443
[1,2]-Stevens rearrangement, **12**, 444
sulfur–hydrogen insertion, **12**, 445
Oxy–Cope reactions
palladium(II) catalyzed, **12**, 615
Oxygen-arylation
organobismuth(V) compounds, **11**, 508–10
α-Oxylithio rearrangements, **11**, 24, 25
Oxymercuriation
alkenes, **3**, 145
Oxypalladation
alkenes
intramolecular, **12**, 610
carbomethoxylation
intramolecular, **12**, 611
Oxythallation
cyclononadiene, **11**, 447

P

Pacifigorgiol
  synthesis, **12**, 587
Palitoxin
  synthesis, **12**, 189
Palladacycles
  formation, **9**, 250
Palladacycloalkane complexes
  formation, **9**, 240
Palladacyclobutane, **12**, 868
  formation, **9**, 237
Palladacyclobutarene
  structure, **9**, 237
Palladacyclopentadienes
  reaction of dimethyl butynedioate with cyclic alkenes, **12**, 760
Palladacyclopentane
  decomposition, **9**, 274
Pallada(IV)cyclopentane complexes
  decomposition, **9**, 273
Palladium
  carbon σ-bonds
    formation, **9**, 233–61
  catalysis
    acid chloride coupling with tetraalkyllead, **11**, 464
  catalysts
    in benzylic zinc halide reactions, **11**, 167
    cross coupling reactions, **11**, 371, 372
  dicarbon metallacarbaboranes
    12-vertex, **1**, 400, 418
  metallic
    reactions with alkenes, **9**, 293, 294
    reactions with alkynes, **9**, 316
  reactions with arenes, **9**, 381
  transmetallation to, **12**, 132–55
Palladium acetate
  transmetallation to, **12**, 154
    lithium salt additives, **12**, 133
Palladium(II) acetate
  catalysts
    cyclopropanation, **12**, 396, 397
Palladium catalysis
  cyclohexadiene synthesis, **12**, 764
Palladium catalysts
  alkoxycarbonylation, **12**, 243
  cross-coupling reactions, **12**, 165
Palladium chloride
  transmetallation to
    lithium salt additives, **12**, 133
Palladium chloride, allyl-
  dimers
    allylic carboxylation, **12**, 818
Palladium chloride, (3-chloro-2-methylenecycloalkyl)-
  dimers
    hydroxide–methanol cleavage, **12**, 803
Palladium clusters
  alkyne
    synthesis, **9**, 320
  asymmetric carbonyl caps, **9**, 202
Palladium complexes
  alkenes, **9**, 293–316
    electrochemistry, **9**, 300
    electronic spectroscopy, **9**, 296
    intramolecular attack by allylanilines, **12**, 609–14
    NMR, **9**, 297–300
    nucleophilic attack, **12**, 603–5
    nucleophilic attack on alkene, **9**, 310
    nucleophilic attack by oxygen nucleophiles, **12**, 606
    reactions, **9**, 310–16
  π-alkenes
    [3.3]sigmatropic rearrangements, **12**, 615, 616
  alkenes
    substitution reactions, **9**, 310
    substitution reactions, alkene, **9**, 310
    synthesis, **9**, 302–10
    thermodynamics, **9**, 300–2
    vibrational spectroscopy, **9**, 296, 297
  σ-alkyl
    preparation, **12**, 139
  alkynes, **9**, 316–23
    bonding, **9**, 316, 317
    structure, **9**, 316, 317
    synthesis, **9**, 317–20
    theory, **9**, 316, 317
  alkynyl, **9**, 230–2
    dinuclear, **9**, 231
    reactions, **9**, 231, 232
    structures, **9**, 231, 232
    synthesis, **9**, 230, 231
  allyl
    alkenes, cyclization/1-alkenyl(aryl) coupling, **12**, 915, 916
    allylation, **12**, 831–4
    asymmetric displacements, **12**, 882, 890
    catalysts, alkene cyclization, stereochemistry, **12**, 908
    1,3-dienes, cyclization, **12**, 912
    intramolecular displacements, **12**, 873–82
    polynuclear, **9**, 337
  π-allyl-
    preparation, **12**, 135, 138
  allylic, **9**, 320–3
    allyl transfer, **9**, 349
    bonding, **9**, 323–8
    charge reversal, **9**, 357, 358
    conjugated dienes, addition reactions, **9**, 370, 371
    halide exchange, **9**, 349, 353, 354
    halogenation, **9**, 349, 356, 357
    NMR, **9**, 325–8
    nucleophilic attack, **9**, 349
    nucleophilic attack at carbon, **9**, 349, 360–71
    nucleophilic attack at carbon, regiochemistry, **9**, 361–7
    nucleophilic attack at carbon, stereochemistry, **9**, 367, 368
    nucleophilic attack at palladium, **9**, 350–2
    oxidation, **9**, 349, 355, 356
    photochemical decomposition, **9**, 354, 355
    photoelectron spectroscopy, **9**, 324, 325
    reactions, **9**, 349–71
    reactions with acids, **9**, 349
    reactions with carbon monoxide, **9**, 349, 358, 359
    reactions with dienes, **9**, 349
    reduction, **9**, 349, 357, 358
    structure, **9**, 323–75
    synthesis, **9**, 328–48
    theory, **9**, 323, 324
    thermal decomposition, **9**, 349, 354, 355
  arene, **9**, 381–3
  butadiene
    NMR, **9**, 299
  carbene ligands, **9**, 193–219
  carbenes, **9**, 216–19
    heterocyclic, synthesis, **9**, 217
    novel, **9**, 219
    structures, **9**, 218, 219
    synthesis, **9**, 216–18
  carbon σ-bonded, **9**, 225–76
  carbon π-bonded, **9**, 291–383
  carbonyl halides
    carbonyl-bridged, **9**, 196
    chloride-bridged, **9**, 196
    synthesis, **9**, 195, 196
  carbonyl ligands, **9**, 193–219
  carbonyls
    carbonylation, allylic compounds, **9**, 208
    catalysis, **9**, 207, 208

catalysis, reactions of the carbonyl ligands, **9**, 207, 208
metallic derivatives, **9**, 204–7
catalysis
allylmetal insertion reactions, alkenes, **12**, 906–8
catalysts
cyclopropanation, **12**, 396, 397
diazodecomposition, **12**, 428
coordination to heteroatoms, **12**, 139
cyclobutadiene
synthesis, **9**, 308
1,5-cyclooctadiene
vibrational spectroscopy, **9**, 296
cyclopentadienyl, **9**, 375–80
bridged, synthesis, **9**, 377, 378
reactions, **9**, 378–80
synthesis, **9**, 375–7
decanuclear carbonyls, **9**, 204
diarsaboranes, **1**, 266
dibenzylideneacetone
NMR, **9**, 299
dicyclopentadiene
nucleophilic attack, **9**, 310
dinuclear carbonyls, **9**, 196–8
A-frame, **9**, 197
carbonyl-bridged, **9**, 197
water soluble, **9**, 198
4,7-diphenyl-1,10-phenanthroline
electrochemistry, **9**, 300
diselenaboranes, **1**, 270
distibaundecaboranes, **1**, 267
ditelluraboranes, **1**, 270
divinylcyclohexane
structure, **9**, 311
dodecanuclear carbonyls, **9**, 204
geranyl
amination, **12**, 804
heptahydrotriborates, **1**, 316, 317
hexadecanuclear carbonyls, **9**, 204
hexanuclear carbonyls, **9**, 202, 203
high nuclearity carbonyls, **9**, 203, 204
homoleptic carbonyls, **9**, 194
catalysis, theoretical studies, **9**, 194, 195
stability, theoretical studies, **9**, 195
theoretical studies, **9**, 194, 195
intramolecular chelation
stabilization, **12**, 602, 603
iron
catalysts, hydrogenation, **10**, 38
isocyanide ligands, **9**, 193–219
isocyanides, **9**, 208–15
trinuclear, **9**, 209
multinuclear carbonyls, **9**, 198–204
multiple insertion cascades, **12**, 300
neryl
amination, **12**, 804
norbornadiene
vibrational spectroscopy, **9**, 297
octanuclear carbonyls, **9**, 203
palladium migration, **12**, 138
pentanuclear carbonyls, **9**, 202
radicals, **9**, 272
reaction with alkynes, **9**, 251–4, 320–3
reaction with benzene, **9**, 259
reaction with carbon monoxide, **9**, 255–7
reaction with diazoacetonitrile, **9**, 259
reaction with diisocyanoarenes
oligomers, **9**, 258
reaction with hexafluorodiazopropane, **9**, 259
reaction with isocyanides, **9**, 257, 258
step-growth polymerization, **12**, 1226, 1227
tetranuclear carbonyls, **9**, 201, 202
degradation, **9**, 201
synthesis, mechanism, **9**, 201
trieicosanuclear carbonyls, **9**, 204
trimethylene methane
cycloaddition reactions, **12**, 923–57
trinuclear carbonyls, **9**, 198–201
coordination chemistry, **9**, 199
oxidative addition reactions, **9**, 200
reactions with alkynes, **9**, 200
structure, **9**, 198, 199
synthesis, **9**, 198, 199
vinyl
preparation, **12**, 135
vinyl ether
structure, **9**, 305
Palladium(0) complexes
alkyne
synthesis, **9**, 317, 318
diene
synthesis, **9**, 307
dispersed on polystyrene
catalysts, **12**, 806
isocyanides, **9**, 208, 209
monoene
synthesis, **9**, 302, 303
Palladium(I) complexes
alkyne
synthesis, **9**, 319
carbon σ-bonded, **9**, 226, 227
diene
synthesis, **9**, 307
hydrido-bridged, **9**, 226
isocyanides, **9**, 210
bridging isocyanides, **9**, 210
Palladium(II) complexes
alkyl, **9**, 233–72
alkyne
synthesis, **9**, 320
π-alkynes
nucleophilic attack, **12**, 619
π-allyl
anion capture, **12**, 310
aryl, **9**, 233–72
cyclobutadiene
synthesis, **9**, 309, 310
cyclooligomerization
alkynes, **12**, 751
diene
synthesis, **9**, 308, 309
diorgano
reactions, **9**, 266, 267
formation
from palladium(0) complexes, **9**, 233–7
from palladium metal, **9**, 233–7
organoaluminum reagents, **9**, 238
organolithium reagents, **9**, 239
organomagnesium reagents, **9**, 240
organomercury reagents, **9**, 238
organometal reagents, **9**, 238–41
organosilicon reagents, **9**, 238
organotin reagents, **9**, 238
isocyanides, **9**, 211–15
multiple insertions, **9**, 213, 214
structures, **9**, 215
monoalkyl
stability, **9**, 265
monoene
cationic, 5-coordinate, synthesis, **9**, 306
neutral, 4-coordinate, synthesis, **9**, 304, 305
neutral, 5-coordinate, synthesis, **9**, 306
synthesis, **9**, 304–6
nucleophilic attack on monoalkenes, **12**, 602–17
reaction with acids, **9**, 267, 268
reaction with alcohols, **9**, 267, 268
reaction with alkenes, **9**, 248–51
reaction with amines, **9**, 267, 268
reaction with carbon dioxide, **9**, 271
reaction with carbon disulfide, **9**, 271
reaction with halogens, **9**, 268–70
reaction with molybdenum peroxide, **9**, 268
reaction with nitric oxide, **9**, 271
reaction with organohalides, **9**, 268–70
reaction with oxygen, **9**, 271
reaction with peracids, **9**, 267, 268

reaction with peroxides, **9**, 267, 268
reaction with sulfur dioxide, **9**, 271
reaction with thiols, **9**, 267, 268
reduction, **9**, 270
structures, **9**, 271, 272
vinyl alcohol
synthesis, **9**, 305
Palladium(IV) complexes
chemistry, **9**, 272–6
hydrocarbyl
isolation, **9**, 272
intramolecular coordination, **9**, 273–5
phenyl
formation, **9**, 274
reactions, **9**, 276
structures, **9**, 276
sulfur donor ligands, **9**, 275, 276
tripodal nitrogen ligands, **9**, 275, 276
Palladium compounds
π-allyl
silyl-substituted, **12**, 153
Palladium dichloride
complexes
cyclotrimerization, alkynes, **12**, 747
Palladium salts
reactions with dienes, **9**, 329
reactions with monoenes, **9**, 328, 331–5
Palladium surfaces
carbon monoxide binding, **9**, 194
Palladium trifluoroacetate
reactions with alkenes, **9**, 331, 336
Palladium(II) ylides
carbon σ-bonded, **9**, 227–30
structures, **9**, 230
Palladium-catalyzed dimerization, **12**, 787
Palladium, bis(dibenzylacetone)-
catalyst
alkene cyclization, **12**, 906
Palladium, dichloro(bis(diphenylphosphino)ferrocene)-
catalyst
cross-coupling reactions, **12**, 165
Palladium(IV), pentafluorophenyl-
synthesis, **9**, 272
Palladium, [phenyl(trimethylsilyl)polystyrylphenyl)methyl]-
catalyst, **12**, 806
Palladium, tetrakis(triphenylphosphine)-
catalyst
alkene cyclization, **12**, 906
cycloaddition reactions, **12**, 950
cycloaddition reactions
carbon dioxide, **12**, 949
Palladium, trimethylenemethane-
cycloaddition reactions, **9**, 373–5
Palladium, (trimethylenemethane)diphosphine-
Fenske–Hall calculations, **12**, 934
Palytoxin
synthesis, **11**, 232
Pancratistatin
synthesis, **11**, 50
Paniculide B
synthesis, **12**, 606
Pantolactone
synthesis, **12**, 23
Pantolactone, keto-
reduction
asymmetric, **12**, 23
Paralobesia viteana
pheromone
synthesis, **11**, 100
Parham cycliacylation
lithiation, **11**, 67
Pauson–Khand cyclization–carbonylation
alkynylsilanes, **11**, 325
Pauson–Khand reaction, **8**, 68; **12**, 703–36
additives, **12**, 705
alkene regiochemistry, **12**, 707
electronic effects, **12**, 710
carbene complexes, **12**, 484
cobalt alkyne complexes, **8**, 99
dicobalt propargylium complexes, **12**, 698, 699
functional groups, **12**, 706
intermolecular, **12**, 712–23
synthetic applications, **12**, 722, 723
intramolecular, **12**, 704, 723–36
adsorption on chromatographic supports, **12**, 705
mechanism, **12**, 706–11
nonracemic, **12**, 718
pressure, **12**, 704
promoters, **12**, 705
regiochemistry, **12**, 707
scope, **12**, 705, 706
synthetic applications, **12**, 734–6
Pedersen's complex
synthesis, **5**, 127
Pellitorine
synthesis, **12**, 265, 830
Penams, 6-vinylidene-
synthesis, **12**, 77
Penicillin
synthesis
from cephalosporins, **12**, 436
Penicillin G
potassium salt
synthesis, **12**, 819
Pentaarsaferrocene
preparation, **7**, 197
nido-Pentaboranes, **1**, 327
structure, **1**, 319
Pentaborates, decahydro-
group 7 metals
metallaboranes, **1**, 323
group 9 metals
metallaboranes, **1**, 323
metallaboranes, **1**, 323
Pentaborates, heptahydro-
group 8 metals
metallaboranes, **1**, 327
group 9 metals
metallaboranes, **1**, 327
metallaboranes, **1**, 327
Pentaborates, nonahydro-
group 8 metals
metallaboranes, **1**, 323
group 9 metals
metallaboranes, **1**, 324
metallaboranes, **1**, 323, 324
Pentaborates, octahydro-
group 8 metals
metallaboranes, **1**, 326
group 9 metals
metallaboranes, **1**, 326
group 11 metals
metallaboranes, **1**, 326
group 12 metals
metallaboranes, **1**, 326
metallaboranes, **1**, 326
Pentaborates, pentahydro-
metallaboranes, **1**, 327
Pentad content
polymers
definition, **12**, 1196
Pentadecker complexes
metal complexes, **1**, 199, 200
Pentadiene
hydrogenation
catalysis, **10**, 37
1,4-Pentadiene
reactions
arylpalladium halides, **12**, 809
1,4-Pentadiene, 2-methyl-
reactions
arylpalladium halides, **12**, 809
Pentalene
synthesis
trimethylenemethane complexes, **12**, 943
Pentalene, hexahydro-
enantiospecific preparation, **12**, 909
Pentalenene, **12**, 60
synthesis, **12**, 473, 735
Pentalenic acid
synthesis, **12**, 329, 330, 735
Pentalenolactone
synthesis
trimethylenemethane complexes, **12**, 943
Pentalenolactone E
synthesis, **12**, 456
Pentalenolactone E methyl ester
synthesis, **12**, 735
allylmetal catalysts
Pentalenolactone P methyl ester
synthesis, **12**, 245, 246
4-Pentanal, 2,2-dimethyl-
hydroesterification, **12**, 33
Pentaorganylstiboranes
alcohol synthesis, **11**, 493–5
ketones from, **11**, 496, 497
Pentaphosphaferrocenes
preparation, **7**, 197
Pentasilacycloheptyne
synthesis, **2**, 13, 63
Pentastannanes
structure, **2**, 284
Pentastanna[1.1.1]propellane
preparation, **2**, 282
Pentazocine

synthesis, **12**, 18
4-Pentenamide
hydroformylation, **12**, 29
4-Penten-1-amine
hydroamination/cyclization
enantioselective, **12**, 1205
4-Penten-1-amine, *N,N*-diisopropyl-
homopolymerization, **12**, 1203
4-Penten-1-amine, 2,2-dimethyl-
hydroamination/cyclization
enantioselective, **12**, 1205
4-Penten-1,3-diols
intramolecular oxypalladation, **12**, 612
1-Pentene
asymmetric oligomerization
Ziegler–Natta catalysts, **12**, 1200
1-Pentene, 5-benzylamino-
hydroformylation, **12**, 29
2-Pentene, 2-bromo-4-methyl-
synthesis
hydrozirconation, **12**, 41
4-Pentene, 1-(*t*-butyldimethylsiloxy)-
homopolymerization, **12**, 1203
1-Pentene, 4-methyl-
asymmetric oligomerization
Ziegler–Natta catalysts, **12**, 1200
2-Pentenes, 1-amino-5-hydroxy-
synthesis
allylic amination, **12**, 807
Pent-2-enoic acid, 5-aryl-5-hydroxy-
synthesis, **11**, 493
4-Penten-1-ol
reaction palladium(II), **12**, 613
Pentenolactone
synthesis, **12**, 582
4-Pentenylamine, 3-hydroxy-
reaction with palladium chloride, **12**, 610
Pentyne
hydrogenation
catalysis, **10**, 37
1-Pentyne
dimerization
rhodium complexes, **8**, 211
reactions with iron ions, **7**, 104
2-Pentyne
reactions with iron ions, **7**, 104
Peptides
formation
from ketene complexes, **12**, 569–71
synthesis, **12**, 366
allylic alkylation, **12**, 851, 852
Peptides, dehydro-
hydrogenation
asymmetric, **12**, 17
Periplanone A
synthesis, **12**, 586
Permethrenic acid
synthesis, **12**, 398
Permethrinic acid
synthesis, **12**, 399
Peroxo complexes
alkyl
structural characteristics, **12**, 1101
Peroxo transfer
mechanisms, **12**, 1101–7
Peroxymercuriation, **11**, 417, 418
intramolecular
hydroperoxides, **3**, 147
Perseleninic acid
reactions, **11**, 535
Personal care industry
organofunctional siloxanes, **2**, 130
Peterson olefination
organosilicon compounds, **11**, 327, 328
Peterson reaction
lithiation, **11**, 26–8
Petroleum industry
silicones, **2**, 126
Phenacyl halides
homocoupling, **12**, 181
Phenanthrenes
preparation
reductive coupling carbonyl compounds, **12**, 1074
Phenanthridines
synthesis, **12**, 192
Phenanthrols
synthesis, **12**, 192
1-Phenethylamine, *N*-acetyl-
synthesis, **12**, 1053
Phenol
coupling reactions
Grignard reagents, **12**, 175
reactions with iron ions, **7**, 105
Phenol, 3,5-di-*t*-butyl-
reaction with 2,4,6-trimethoxyphenyllead triacetate, **11**, 469
Phenol, 2,4-dimethyl-
reaction with aryllead compounds, **11**, 469
Phenol, 2,6-dimethyl-
reaction with aryllead compounds, **11**, 469
Phenol, dimethyl-
dimetallation, **11**, 109
Phenol, 4-methoxy-2,6-dimethyl-
reaction with aryllead compounds, **11**, 469
Phenol, 4-nitro-
reaction with pentaphenylbismuth, **11**, 508
Phenol, thio-
dimetallation, **11**, 124
Phenol, 3,4,5-trimethyl-
reaction aryllead compounds, **11**, 469
Phenolic compounds
carbon-arylation
organobismuth(V) compounds, **11**, 507
oxidation
organobismuth(V) compounds, **11**, 505
Phenols
C-arylation
aryllead triacetates, **11**, 469, 470
dimetallation, **11**, 124
ortho-alkoxy
synthesis of carbene complexes, **12**, 500
ortho-amino
synthesis of carbene complexes, **12**, 500
oxidation
organoselenium compounds, **11**, 534
synthesis
aryllead tricarboxylates, **11**, 481
cyclooligomerization of nickel catalysts, **12**, 752
organoboron compounds, **11**, 214
Phenols, *o*-allyl-
intramolecular oxypalladation
Wacker oxidation, **12**, 612
Phenoxides
reactions
allyl complexes, **12**, 814–17
Phenylation
O-
organobismuth(V) compounds, **11**, 509
N-
organobismuth(V) compounds, **11**, 510
[2]Phenylene, bis(trimethylsilyl)-
synthesis, **12**, 750
[3]Phenylene, bis(trimethylsilyl)-
synthesis, **12**, 750
Phenylseleno compounds
synthesis, **11**, 530
Pheromones
synthesis, **12**, 173, 184
Phodophyllotoxin, deoxy-
synthesis, **12**, 20
Phosphaalkenes
reactions
with carbene complexes, **12**, 540
Phosphaalkyne complexes
triiron clusters, **7**, 263, 264
Phosphaalkynes
coupling with triiron bis(alkylidyne) clusters, **7**, 280
Phosphaboranes
properties, **1**, 263, 264
transition metal complexes, **1**, 265
Phosphacarboranes
properties, **1**, 265
Phosphadecaboranes
properties, **1**, 264
1*H*-1,2,5-Phosphadiborole, 2,5-dihydro-1-phenyl-
metal complexes, **1**, 214
1,2,5-Phosphadiborolenes
metal complexes, **1**, 214
Phosphadicarbadodecaboranes
properties, **1**, 265
Phosphadodecaboranes
properties, **1**, 264
Phosphaferrocenes
preparation, **7**, 196
Phosphagermirenes
preparation, **2**, 159
synthesis, **2**, 140
Phosphanorbornadienes
synthesis, **12**, 30, 31

1,2,5-Phosphasilaborolenes
  metal complexes, **1**, 213
Phosphasilaboroles
  formation, **1**, 169
Phosphasilirene
  synthesis, **2**, 68
Phosphete, dihydro-
  synthesis, **12**, 591
Phosphine complexes
  group 6, **5**, 302–5
    properties, **5**, 304
    reactions, **5**, 305
    structure, **5**, 304
    synthesis, **5**, 302–4
Phosphine oxides
  Pauson–Khand reaction, **12**, 705
Phosphine, cyclohexyl(*o*-anisyl)-methyl-
  dopa synthesis, **12**, 16
Phosphine, stannyl-
  addition to alkenes, **2**, 291
  addition to alkynes, **2**, 293
Phosphine, tri-*t*-butylstannyl-
  lithium derivative
    reaction with tri-*t*-butylphenyl-dichlorophosphine, **2**, 291
Phosphine, tris(2-furyl)-
  coupling reactions
    O-alkenyl derivatives, **12**, 206
  palladium complexes
    coupling reactions, **12**, 214
Phosphines
  metallation, **7**, 427–9
  reactions
    allyl complexes, **12**, 830, 831
  reductive elimination
    organonickel complexes, **9**, 62
Phosphines, silyl-
  synthesis, **2**, 27
Phosphines, triaryl
  preparation, **12**, 231
Phosphinidene complexes
  iron clusters, **7**, 263
    coupling with alkynes, **7**, 266
  nucleophilic addition to iron clusters giving phosphinidene–acyl clusters, **7**, 281
Phosphinidene ligands
  bridging
    in heteronuclear metal–metal bond synthesis, **10**, 12, 15
Phosphole, phenyl-
  reaction with triosmium clusters, **7**, 811
Phospholes
  synthesis, **12**, 30, 31
    cyclooligomerization, **12**, 743
Phosphonates, α-diazo-β-keto-
  synthesis, **12**, 453
Phosphonic acid, γ-amino-
  microbial metabolites
    synthesis, **12**, 805
Phosphonium salts, alkenyl-
  preparation from enol triflates, **12**, 231
Phosphorus
  reaction with cobalt phosphine complexes, **8**, 17
Phosphorus compounds
  synthesis
    organomercury compounds, **11**, 401
Phosphorus ylides
  homologation of carbonyl compounds, **12**, 578
  metal stabilized, **12**, 432
  palladium(II) complexes, **9**, 227, 228
N-Phosphorylamides
  lithiation, **11**, 19
Photochemical reactions
  metal carbene complexes, **12**, 549–73
    mechanisms, **12**, 549, 550
Phthalazine alkaloids
  asymmetric dihydroxylation
    osmium tetroxide, **12**, 1147, 1148
Phthalides
  synthesis
    lithiation, **11**, 67
Phthalimide, *N*-aryl-
  synthesis, **12**, 249
Phthalimide, monothio-
  cross-coupling with dithioacetal, **12**, 1092
Phycocyanin
  synthesis
    cobalt-mediated propargylation, **12**, 696
Phyllanthocin
  synthesis, **12**, 29
Phyllocladane
  synthesis, **12**, 750
    trimethylenemethane complexes, **12**, 944
Phyllocladene
  synthesis, **12**, 614
Physostigmine
  synthesis, **12**, 583
Phytoalexins
  synthesis, **12**, 212
Phytochrome
  synthesis
    cobalt-mediated propargylation, **12**, 696
Picrotaxane
  synthesis, **12**, 848
Picrotoxinin
  synthesis, **12**, 51
Pinacol
  cross-coupling reactions, **12**, 1075, 1086, 1087
  dimerization
    niobium complexes, **12**, 1088
    vanadium complexes, **12**, 1086
Pinacols
  preparation
    carbonyl coupling reactions, **12**, 1074
Pine bark beetle
  sex pheromone
    synthesis, **12**, 859
α-Pinene
  conversion to β-pinene, **11**, 103
  organoboranes, **11**, 260
Pinocarvenol
  synthesis, **12**, 815
Pinocarvone
  synthesis, **12**, 815
Piperidine, methyl-
  synthesis, **12**, 29
Piperidines
  synthesis
    organoboron compounds, **11**, 218
Piperidines, α,α'-dialkyl-
  synthesis by lithiation, **11**, 17
2-Piperidone
  acylation
    organobismuth(V) compounds, **11**, 511
Plastics
  modification
    organofunctional siloxane polymers, **2**, 130
3-Platinaazetidine
  preparation, **9**, 478
Platinacycloalkanes, **9**, 477–83
  preparation, **9**, 478–80
  properties, **9**, 481
  reactions, **9**, 481–3
  structures, **9**, 481, 500
Platinacyclobutane sulfones
  synthesis, **9**, 480
Platinacyclobutane sulfoxides
  synthesis, **9**, 480
Platinacyclobutanes
  photolysis, **9**, 505
  preparation, **9**, 497, 498
  reactions, **9**, 504
  structure, **9**, 481
  thermal decomposition, **9**, 505
  thermolysis, **9**, 505
2-Platinacyclobutanone
  preparation, **9**, 479
Platinacyclobutanones
  structure, **9**, 481
Platinacyclohexane
  preparation, **9**, 478
Platinacyclopentadiene
  preparation, **9**, 480, 493
Platinacyclopentane
  preparation, **9**, 478, 505
Platinacyclopentanediones
  preparation, **9**, 480
Platinatetraboranes, **1**, 317
Platinum
  dicarbon metallacarbaboranes
    8-vertex, **1**, 396
    10-vertex, **1**, 396
    11-vertex, **1**, 396
    12-vertex, **1**, 400
  monocarbon metallacarbaboranes, **1**, 378
    10-vertex, **1**, 378
    11-vertex, **1**, 378
Platinum clusters
  acetylide osmium, **10**, 153, 154
  acetylide ruthenium, **10**, 153, 154
  alkyne iron, **10**, 151, 152
  alkyne ruthenium, **10**, 151, 152
  tungsten chains, **10**, 120–2

tungsten stars, **10**, 122, 123
Platinum complexes
acyl, **9**, 560
alkenes, **9**, 534–64
alkynes, **9**, 564–72
allenes, **9**, 558–60
allenyl, **9**, 577
preparation, **9**, 577
structure, **9**, 577
allyls, **9**, 573–9
preparation, **9**, 573–5
reactions, **9**, 576
structure, **9**, 575, 576
theory, **9**, 576
arenes, **9**, 582
preparation, **9**, 582
reactivity, **9**, 582
structure, **9**, 582
arynes, **9**, 569
preparation, **9**, 569
reactivity, **9**, 569
structure, **9**, 569
benzyl
preparation, **9**, 577
reactivity, **9**, 577
structure, **9**, 577
bimetallic alkenes, **9**, 561–3
bimetallic alkynes, **9**, 570–2
butadienyl
preparation, **9**, 577
reactivity, **9**, 577
structure, **9**, 577
C,N-carbenes, **9**, 402–21
C,O-carbenes, **9**, 402, 421–3
carbenes, **9**, 419–26
preparation from electrophilic attack on coordinated DF_3, **9**, 420
preparation from nitrogen ylides, **9**, 420
preparation from thiolate amines, **9**, 402, 421
carbon σ-bonded, **9**, 431–518
carbon π-bonded, **9**, 533–82
carbonyl anions
polynuclear, **9**, 397–9
polynuclear, magnetic moments, **9**, 398
polynuclear, mass spectrometry, **9**, 398
polynuclear, NMR, **9**, 398
polynuclear, reactions, **9**, 399
polynuclear, spectroscopy, **9**, 398
polynuclear, structures, **9**, 398
carbonyl ligands, **9**, 392–411
carbonyl, isocyanide and carbene ligands, **9**, 391–426
catalysts
cyclopropanation, **12**, 396, 397
diazodecomposition, **12**, 428
cumulenes
preparation, **9**, 558, 559
reactivity, **9**, 558, 559
structure, **9**, 558, 559
cyclic 5-membered N,O-carbenes, **9**, 425
cyclobutadiene, **9**, 579
preparation, **9**, 579
reactivity, **9**, 579
structure, **9**, 579
cyclopentadienyl, **9**, 579–82
diarsaboranes, **1**, 266
diselenes, **9**, 560
fullerenes, **9**, 561
preparation, **9**, 561
reactivity, **9**, 561
structure, **9**, 561
gallium bonds, **1**, 557
heterobimetallic alkenes
preparation, **9**, 563
reactivity, **9**, 563
structure, **9**, 563
heterobimetallic alkynes
preparation, **9**, 571, 572
reactivity, **9**, 571, 572
structure, **9**, 571, 572
heterocyclopentadienyl, **9**, 581, 582
preparation, **9**, 581, 582
structure, **9**, 581, 582
heterometallic alkenes
preparation, **9**, 563, 564
reactivity, **9**, 563, 564
structure, **9**, 563, 564
heterometallic alkynes
preparation, **9**, 572
reactivity, **9**, 572
structure, **9**, 572
homobimetallic alkenes
preparation, **9**, 561, 562
reactivity, **9**, 561, 562
structure, **9**, 561, 562
homobimetallic alkynes
preparation, **9**, 570, 571
reactivity, **9**, 570, 571
structure, **9**, 570, 571
homometallic alkenes
preparation, **9**, 563
reactivity, **9**, 563
structure, **9**, 563
homometallic alkynes
preparation, **9**, 572
reactivity, **9**, 572
structure, **9**, 572
hydrocarbyls, **9**, 577
indium bonds, **1**, 566
isocyanides, **9**, 411–19
ketenes
preparation, **9**, 559, 560
reactivity, **9**, 559, 560
structure, **9**, 559, 560
mercury clusters, **10**, 342, 343
metallated arsenic ligands
preparation, **9**, 507–9
reactions, **9**, 514, 515
metallated ligands, **9**, 507–17
preparation, **9**, 507–12
properties, **9**, 512–14
reactions, **9**, 514–17
structures, **9**, 512–14
metallated nitrogen ligands
preparation, **9**, 510, 511
reactions, **9**, 515–17
metallated phosphorus ligands
preparation, **9**, 507–9
reactions, **9**, 514, 515
monoalkenes
monodentate, **9**, 534–51
polydentate, **9**, 551–3
monoalkynes, **9**, 564–9
N,N-carbenes, **9**, 423, 424
N,O-carbenes, **9**, 425, 426
O,O-carbenes, **9**, 426
palladium carbonyls, **9**, 205
phosphaalkenes, **9**, 560
phosphaalkynes, **9**, 569, 570
preparation, **9**, 569, 570
reactivity, **9**, 569, 570
structure, **9**, 569, 570
phosphaboranes, **1**, 260, 265
polyalkenes
conjugated, **9**, 560, 561
conjugated, preparation, **9**, 560, 561
conjugated, reactivity, **9**, 560, 561
conjugated, structure, **9**, 560, 561
polyenes
unconjugated, **9**, 553–8
polymetallic alkenes, **9**, 563, 564
polymetallic alkynes, **9**, 572
propargyl
preparation, **9**, 577
reactivity, **9**, 577
structure, **9**, 577
S,S-carbenes, **9**, 426
trimethylenemethane, **9**, 578, 579
preparation, **9**, 578, 579
reactivity, **9**, 578, 579
structure, **9**, 578, 579
Platinum(0) complexes
carbonyl clusters
NMR, **9**, 395
reactions, **9**, 395–7
replacement of carbonyl by sulfur dioxide, **9**, 396
spectroscopy, **9**, 394, 395
tetranuclear, NMR, **9**, 395
carbonyl ligands
monomeric, **9**, 392
polynuclear, **9**, 392–7
polynuclear, structures, **9**, 393, 394
polynuclear, synthesis, **9**, 392, 393
isocyanides
mononuclear, **9**, 411
polynuclear, **9**, 411–13
monoalkenes
monodentate, **9**, 535–40
monodentate, displacement reactions, **9**, 538
monodentate, dissociation, **9**, 538
monodentate, oxidative addition, **9**, 540
monodentate, preparation, **9**, 535–7
monodentate, protonation, **9**, 538
monodentate, reactions, **9**, 538–40
monodentate, structure, **9**, 537, 538
polydentate, **9**, 551
monoalkynes, **9**, 565–8
preparation, **9**, 565, 566
reactions, **9**, 568
structure, **9**, 566–8
polyenes

unconjugated, **9**, 553, 554
unconjugated, preparation, **9**, 553
unconjugated, reactions, **9**, 554
unconjugated, structure, **9**, 553
Platinum(I) complexes
carbon σ-bonded, **9**, 432, 433
carbonyls, **9**, 399–401
ligand substitution, **9**, 401
triangular dppm-bridged clusters, **9**, 401–4
cyclopentadienyl
preparation, **9**, 580, 581
reactivity, **9**, 580, 581
structure, **9**, 580, 581
dppm-bridged dimers, **9**, 432
isocyanides, **9**, 413–15
IR spectroscopy, **9**, 414
NMR, **9**, 414
metal–metal bonds, **9**, 432
Platinum(II) complexes
alkenyls, **9**, 483–9
preparation, **9**, 483–6
properties, **9**, 487
reactions, **9**, 487–9
structure, **9**, 487
alkyls, **9**, 433–66
carbon bonds, stability, **9**, 445
coordinated ligands, reactions, **9**, 466
dppm, **9**, 441, 442
electrophilic cleavage, **9**, 458
homolytic cleavage, **9**, 457
insertion reactions, **9**, 460–2
isomerization, **9**, 458–60
luminescence spectra, **9**, 453
NMR, **9**, 445
oxidative addition, **9**, 454, 463–6
photoelectron spectroscopy, **9**, 452
photoisomerization, **9**, 460
preparation, **9**, 433–44
preparation, bimetallic compounds, **9**, 444
preparation, displacement of weakly bound ligands, **9**, 439–41
preparation, insertion reactions, **9**, 442, 443
preparation, main group organometallics, **9**, 436–9
preparation, oxidative addition, **9**, 434–6
properties, **9**, 445–53
reactions, **9**, 453–66
reductive elimination, **9**, 453
spectroscopy, **9**, 445–53
structure, **9**, 445–53
substitution, **9**, 458–60
thermal decomposition, **9**, 453–8
x-ray diffraction, **9**, 445
alkynyls, **9**, 490–4
preparation, **9**, 490, 491
properties, **9**, 491–3
reactions, **9**, 493, 494
structure, **9**, 491–3
aryl, **9**, 433–66
carbon bonds, stability, **9**, 445
coordinated ligands, reactions, **9**, 466
dppm, **9**, 441, 442
electrophilic cleavage, **9**, 458
homolytic cleavage, **9**, 457
insertion reactions, **9**, 460–2
isomerization, **9**, 458–60
luminescence spectra, **9**, 453
NMR, **9**, 445
oxidative addition, **9**, 454, 463–6
photoelectron spectroscopy, **9**, 452
preparation, **9**, 433–44
preparation, bimetallic compounds, **9**, 444
preparation, displacement of weakly bound ligands, **9**, 439–41
preparation, insertion reactions, **9**, 442, 443
preparation, main group organometallics, **9**, 436–9
preparation, oxidative addition, **9**, 434–6
properties, **9**, 445–53
reactions, **9**, 453–66
reductive elimination, **9**, 453
spectroscopy, **9**, 445–53
structure, **9**, 445–53
substitution, **9**, 458–60
thermal decomposition, **9**, 453–8
x-ray diffraction, **9**, 445
bis(alkynyl)
preparation, **9**, 490
bis(isocyanide), **9**, 417, 418
electrochemistry, **9**, 418
carbon σ-bonded, **9**, 433–96
carbonylation, **9**, 460
carbonyls, **9**, 404–11
18-electron, **9**, 410
catalysts
epoxidation, **12**, 1108
chloro
hydroformylation catalysts, **9**, 460
hydrogenation catalysts, **9**, 460
cyclopentadienyl
preparation, **9**, 580, 581
reactivity, **9**, 580, 581
structure, **9**, 580, 581
dicarbonyls, **9**, 409
binuclear, **9**, 410, 411
dithiocarbamate
alkyl group isomerization, **9**, 460
double carbonylation, **9**, 462
haloalkyls, **9**, 467–77
preparation, **9**, 467–71
preparation, displacement of weakly bound ligands, **9**, 468–70
preparation, insertion reactions, **9**, 471
preparation, main group organometallics, **9**, 467, 468
preparation, oxidative addition, **9**, 467
properties, **9**, 472, 473, 490
Pt–C bond stability, **9**, 472
reactions, **9**, 473–7
reactions, other metal complexes, **9**, 473–5
spectroscopy, **9**, 473, 494
structure, **9**, 472, 473
x-ray diffraction, **9**, 472
haloaryls, **9**, 467–77
preparation, **9**, 467–71, 481
preparation, displacement of weakly bound ligands, **9**, 468–70
preparation, insertion reactions, **9**, 471
preparation, main group organometallics, **9**, 467, 468
preparation, oxidative addition, **9**, 467
properties, **9**, 472, 473
Pt–C bond stability, **9**, 472–94
reactions, **9**, 473–7
reactions, other metal complexes, **9**, 473–5
spectroscopy, **9**, 473
structure, **9**, 472, 473
x-ray diffraction, **9**, 472
Hg bonds, **9**, 444
hydride-bridged
preparation, **9**, 444
hydroxy
reactions, **9**, 461
isocyanides, **9**, 415–19
α-ketoacyl, **9**, 434
methyl
chemical vapor deposition, **9**, 457
monoalkenes
monodentate, **9**, 541–9
monodentate, displacement reactions, **9**, 547
monodentate, dissociation, **9**, 547
monodentate, migratory alkene insertion, **9**, 548
monodentate, NMR, **9**, 544
monodentate, nucleophilic addition, **9**, 548
monodentate, pentacoordinate, **9**, 542–4
monodentate, preparation, **9**, 541–4
monodentate, reactions, **9**, 547–9
monodentate, structure, **9**, 544, 545
monodentate, theory, **9**, 550, 551
polydentate, **9**, 551–3
monoalkynes, **9**, 568, 569
preparation, **9**, 568

reactions, **9**, 569
structure, **9**, 568
theory, **9**, 569
monocarbonyl, **9**, 405–8
cis–trans isomerization, **9**, 405
electronic spectroscopy, **9**, 406
ligand replacement reactions, **9**, 407
migratory insertion reactions, **9**, 407, 408
spectroscopy, **9**, 406
structures, **9**, 406
synthesis, **9**, 405, 406
monoisocyanides, **9**, 416, 417
pentafluorophenyl
preparation, **9**, 468
polyenes
unconjugated, **9**, 554–8
unconjugated, diene displacement, **9**, 557
unconjugated, insertion reactions, **9**, 558
unconjugated, migration reactions, **9**, 558
unconjugated, nucleophilic addition, **9**, 557
unconjugated, pentacoordination, **9**, 554
unconjugated, preparation, **9**, 554
unconjugated, reactions, **9**, 556–8
unconjugated, redox reactions, **9**, 558
unconjugated, structure, **9**, 554–6
preparation
organotin compounds, **9**, 438
tetrakis(isocyanide), **9**, 419
thienyl
preparation, **9**, 484
trifluoromethyl
catalysts, **9**, 475, 476
preparation, **9**, 467
tris(isocyanide), **9**, 419
ylides, **9**, 495, 496
preparation, **9**, 495
properties, **9**, 495
reactions, **9**, 496
structure, **9**, 495
ytterbium bonds, **9**, 444
zirconium bonds, **9**, 444
Platinum(III) complexes, **9**, 496, 497
Platinum(IV) complexes, **9**, 497–507
carbonyls, **9**, 411
cyclopentadienyl
preparation, **9**, 581
reactivity, **9**, 581
structure, **9**, 581
glycinato, **9**, 506
isocyanides, **9**, 419
monoalkenes
polydentate, **9**, 553
preparation, **9**, 497–500
oxidative addition to platinum(II) complexes, **9**, 497
properties, **9**, 500–4
structures, **9**, 500–4
Platinum ruthenium clusters
selective hydrogenation catalysts, **10**, 38, 39
Platinum, halotrimethyl-
reactions, **9**, 498–500
Platinum(0), dichlorocarbonyl(trialkylphosphine)-
reduction, **9**, 392
Platinum(0), tricarbonyltetrakis(triphenylphosphine)-
reactions, **9**, 397
Platinum(0), tricarbonyltris(di-*t*-butylphenylphosphine)-
reactions, **9**, 397
Platinum(II), dichlorodicarbonyl-
reactions with alkynes, **9**, 409
Platinum(II) compounds
catalysts
diazodecomposition, **12**, 428
Plinol C
synthesis, **12**, 580
Plumbaboranes
properties, **1**, 259, 260
Plumbacarboranes
properties, **1**, 260
Plumbanes, silyl-
synthesis, **2**, 23
Plumbation
electrophilic, **11**, 466
Plumbyl radicals
reactions, **2**, 313
Poly(2-pyridones)
preparation, **12**, 1227
Poly(2-pyrones)
preparation, **12**, 1227
Polyacetylene
from ring-opening metathesis polymerization, **12**, 1222
synthesis, **12**, 1226
Polyalkenes
functionalized end groups, **12**, 1204
isotactic
Ziegler–Natta catalysts, **12**, 1197, 1198
Poly(α-alkenes)
microstructures, **12**, 1196
Polybenzvalene
soluble
isomerized, **12**, 1223
Polyboranes
rearrangements
tensor surface harmonic theoretical studies, **1**, 232
Poly(butadiene)
hydroxytelechelic
preparation, **12**, 1217
Polycyclization
dienynes
palladium catalyzed, **12**, 54, 55
Polycyclizations
multiple insertion cascades, **12**, 300–4
Polycyclobutene
melting point, **12**, 1202
Polycyclopentene
melting point, **12**, 1202
Polydentate bridging ligands
in heteronuclear metal–metal bond synthesis, **10**, 17–19
Poly(dimethyl-co-methylvinyl)-siloxanes
equilibrium
NMR, **2**, 123
Polydimethylsiloxanes
industrial applications, **2**, 125–8
ring chain equilibrium, **2**, 119
Polyenes
alkoxymercuration, **11**, 416
asymmetric dihydroxylation, **12**, 1166
cyclic skipped
synthesis, **12**, 1077
hydroxymercuriation, **11**, 412
metallation, **11**, 103
preparation
ring-opening metathesis polymerization, **12**, 1222
solvomercuriation, **11**, 427
synthesis
organomercury compounds, **11**, 401
vinylsilanes, **12**, 153
zirconium mediated, **4**, 621
Poly(ethene-*block*-methyl methacrylate)
synthesis, **12**, 1205
Polyferrocenes
mixed valence properties, **7**, 192–4
Polygermanes
branched, **2**, 177
cyclic, **2**, 178
linear, **2**, 177
Polyhedral skeletal electron pair theory, **1**, 331; **10**, 261, 289
alkyne iron clusters, **7**, 259
metallaboranes, **1**, 277
ruthenium clusters, **7**, 909
tetrairon carbide clusters
butterfly vs. trigonal bipyramidal structures, **7**, 268
Polymerization
alkenes
lanthanide complexes, **4**, 8
metallocenes, **4**, 493
organolanthanide catalysts, **4**, 117
insertion-isomerization type
Ziegler–Natta catalysts, **12**, 1202
Polymers
metal-terminated, **12**, 1204
synthesis
transition metals in, **12**, 1209–29
Polymethylene
formation
diazomethane decomposition, **12**, 425
Polymethylhydrosiloxane
reactions
allyl complexes, **12**, 801
Polynorbornene

melting point, **12**, 1202
Poly(*p*-phenylene)
preparation, **12**, 1225
Poly(1,4-phenylenevinylene)
preparation
ring-opening metathesis polymerization, **12**, 1223
Polypropene
alkene-terminated atactic
synthesis, **12**, 1204
alkene-terminated isotactic
synthesis, **12**, 1204
isotactic
production, **12**, 1197
Polypryazole
ligands
cyclopropanation, **12**, 415
Polypyrroles
linear
synthesis, **12**, 224
Polysilacyclobutasilazanes
preparation, **2**, 50
Polysilanes
cage, **2**, 88–91
synthesis, **2**, 23, 24
chemical modification, **2**, 101
crystallinity, **2**, 102, 103
cyclic, **2**, 79–87
photolysis, **2**, 86, 87
reactions, **2**, 83–6
synthesis, **2**, 79–81
electrical conductivity, **2**, 106, 107
high polymers, **2**, 96–107
molecular weight, **2**, 97
oligomers
linear, **2**, 78, 79
optical properties, **2**, 106
photoconductivity, **2**, 106, 107
photolysis, **2**, 105
physical properties, **2**, 102, 103
polycyclic, **2**, 88–91
polymers
synthesis, **2**, 96–100
as silicon carbide precursors, **2**, 107
spectroscopy, **2**, 3
structure, **2**, 102, 103
surface energies, **2**, 107
thermochromism, **2**, 103–5
thermolysis, **2**, 107
ultraviolet
absorption, **2**, 103–5
irradiation, **2**, 105
Polysilanes, aryl
dearylation, **2**, 101
Polysilanes, permethyl-
oligomers
linear, **2**, 78
rotational barriers, **2**, 79
Polysilastyrene
forming, **2**, 107
Polysiloles
synthesis, **2**, 59
Polysiloxanes
functionalization
hydrosilylation, **2**, 117
organofunctional, **2**, 128–32
equilibrium polymerization, **2**, 128
heterocondensation, **2**, 129
synthesis by hydrosilylation, **2**, 129
Polystannane hydrides
preparation, **2**, 271
Poly(vinyl chloride)
stabilization
organotin compounds, **2**, 296
Porphyrin
benzannulation, **12**, 492
Porphyrin complexes
chiral
synthesis, asymmetric epoxidation, **12**, 1130, 1131
cyclopropanation, **12**, 416
Porphyrin, octaethyl-
metal–indium bonds, **1**, 566
metal–thallium bonds, **1**, 569
Porphyrin, tetraphenyl-
metal–indium bonds, **1**, 566
metal–thallium bonds, **1**, 569
Porphyrins
cobalt complexes
reactions, with oxidants, **8**, 63
germanium derivatives, **2**, 161
synthesis, **12**, 1077
transmetallation, **12**, 135
Potassium bisoctatetraene calcate
preparation, **1**, 117
Potassium *t*-butoxide
superbase reagent, **11**, 94
Potassium enoxyborates
preparation, **12**, 865
Potassium phthalimide
amidation
1,3-diene monoepoxides, **12**, 810
Potassium selenocyanates, **11**, 528
Potassium-graphite
preparation of highly activated zinc, **11**, 160
Potassium, alkenyl-
structure, **11**, 97
Potassium, allyl-
reactions, **11**, 97–103
rotational barrier, **11**, 98
Potassium, butyl-
structure, **11**, 95
Potassium, 2,4-dimethyl-2,4-pentadienyl-
structure, **11**, 103
Potassium, heptatrienyl-
conformation, **11**, 104
Potassium, methyl-
preparation, **1**, 20
Potassium, 2-methyl-2,4-pentadienyl-
conformation, **11**, 104
Potassium, pentadienyl-
structure, **11**, 103
Potassium, pentamethylcyclopentadienyl-
pyridine solvate
structure, **1**, 22
Potassium, phenylallyl-
structure, **11**, 108
Potassium, prenyl-
reaction with steroidal iodides, **11**, 101
Potassium, 2-tetrahydrofuryl-
structure, **11**, 96
Potassium, trineopentyltin-
structure, **2**, 283
Potassium, triphenyltin-
structure, **2**, 283
Prazepam
synthesis, **9**, 245
Precapnelladiene
synthesis, **12**, 590
Preclamol
synthesis, **12**, 271
Pregna-5,16-diene, 3β,20-dihydroxy-
hydroformylation, **12**, 29
Prelog–Djerassi lactone
synthesis, **12**, 672, 825
Previtamin D_3
synthesis, **12**, 223
Prezizaene
synthesis, **12**, 187
Promonensin B
synthesis, **12**, 170
Propanal, 2-aryl-
oxidation, **12**, 33
Propanal, 2-trifluoromethyl-
synthesis, **12**, 30
Propanal, 3-trifluoromethyl-
synthesis, **12**, 30
Propane
reactions with iron ions, **7**, 103
Propane, 1-amino-
reactions with iron ions, **7**, 105
Propanoic acid, 2-formyl-
ethyl ester
synthesis, **12**, 30, 31
Propargyl acetate
complexes
reaction with organoaluminum reagents, **12**, 693
Propargyl alcohols
chiral
dicobalt complexes, reactions, **12**, 688
ether derivatives
dicobalt complexes, reactions, **12**, 688
Propargyl ligands
bridging
in heteronuclear metal–metal bond synthesis, **10**, 12
Propargyl nitriles
synthesis
dicobalt propargylium complexes, **12**, 693
Propargyl selenoxides
sigmatropic rearrangements, **11**, 557
Propargylamines
activation
triruthenium clusters, **7**, 816
Propargylic cations
stabilization
dicobalt hexacarbonyl group, **12**, 685
Propargylic esters
synthesis
transition metal complexes, **12**, 245
Propargylic halides

  reaction with zinc–copper organometallic compounds, **11**, 166
Propargylic tosylates
  reaction with zinc–copper organometallic compounds, **11**, 166
Propargylium complexes
  cyclopropyl cations
    reactivity, **12**, 688
  dicobalt
    applications, **12**, 694–700
    coupling with aromatic compounds, **12**, 690
    coupling with carbon nucleophiles, **12**, 690–3
    coupling with noncarbon nucleophiles, **12**, 690
    elimination reactions, **12**, 689
    enol derivatives, reactions, **12**, 691, 692
    highly electrophilic, reactions, **12**, 689
    intermolecular reactions, **12**, 694–6
    intramolecular reactions, **12**, 697, 698
    nucleophilic coupling
      organic synthesis, **12**, 694–8
    Pauson–Khand reactions, **12**, 698, 699
    proton loss, **12**, 689
    reactions, **12**, 689–94
    reduction, **12**, 693, 694
  dicobalt cationic
    reaction with alkyls, **12**, 693
  dinuclear, **1**, 685–700
    preparation, **12**, 686–9
    reactivity, **12**, 686–9
    structure, **12**, 686–9
  radical reactions, **12**, 693, 694
  vinyl-substituted
    reactivity, **12**, 688
[1.1.1]Propellane
  lithiation, **11**, 6
Propellane
  oxidative demetallation, **12**, 761
[3.3.3]-Propellanes
  synthesis, **12**, 956
Propene
  asymmetric oligomerization
    Ziegler–Natta catalysts, **12**, 1200
  ethene block copolymer
    synthesis, **12**, 1205
  insertion into ethyl–metal bonds
    enantiofacial selectivity, **12**, 1198
  isospecific polymerization
    Ziegler–Natta catalysts, **12**, 1197
  polymeriztion
    vanadium complexes, **5**, 45
    zirconocenes, **4**, 493
  reactions with iron anions, **7**, 107
  reactions with iron atoms, **7**, 108
  reactions with iron ions, **7**, 103, 104, 107
  syndiospecific polymerization, **12**, 1199
Propene, bromotrifluoro-
  zinc reagents, **11**, 161
Propene, 1,3-dichloro-
  reaction with organozinc compounds, **11**, 165
1-Propene, 3-methoxy-1-phenyl-
  hydrozirconation, **12**, 40
Propene, 2-methyl-
  reactions with iron ions, **7**, 107
Propene, 2-phenyl-
  hydroesterification
    asymmetric, **12**, 35
Propene, 3,3,3-trifluoro-
  hydroformylation, **12**, 30
Propiolamide
  carbozincation, **11**, 177
Propionate, 3,7-dimethyl-2,7-octadienyl-
  synthesis, **12**, 849
Propionitriles, β-aryl-β-hydroxy-
  synthesis, **11**, 494
Propranolol
  synthesis, **12**, 23
Propylidenation
  carbonyl compounds, **12**, 1090
Propynals
  cobalt complexes
    aldol reactions, **12**, 696
Prop-2-yn-1-ol
  trimerization, **12**, 747
Prostacyclin
  analogues
    synthesis, **12**, 581
Prostacycline
  synthesis
    allyl complexes, **12**, 817
Prostaglandin
  analogues
    methylenation, **12**, 581
    synthesis, **12**, 690
  endoperoxide analogues
    synthesis, **12**, 139
  synthesis
    carbene complexes, **12**, 481
    organosilicon compounds, **11**, 317
Prostaglandin B_1
  synthesis, **12**, 185
Prostaglandin B_1 methyl ester
  synthesis, **11**, 232
Prostaglandin B_1 methyl ester, 15-dehydro-
  synthesis, **12**, 265, 267
Prostaglandin E_1
  15-methyl analogue
    synthesis, **12**, 89
Prostaglandin E_1, 11-deoxy-, **12**, 880
Prostaglandin E_2
  synthesis, **12**, 141
Prostaglandin E_2 methyl ester
  synthesis, **12**, 391
Prostaglandin E_2 methyl ester, 11-deoxy-
  synthesis, **12**, 865
Prostaglandin F_1, 13-dehydro-esters
    synthesis, **12**, 68, 69
Prostaglandin $F_{2\alpha}$, 12-methyl-
  synthesis, **12**, 616
Prostaglandins
  analogues
    synthesis, **12**, 68
  preparation, **12**, 606
  synthesis, **12**, 89, 311, 864
Prostaglandins, 13-dehydro-
  synthesis, **12**, 68
Prostanoids
  synthesis, **12**, 722
Protease
  inhibitors
    synthesis, **12**, 104
Prothracacin
  synthesis, **12**, 249
Protiodesilylation
  oxidative demetallation
    cobalt complexes, **12**, 761
Protoberberines
  synthesis, **12**, 750
Protonolysis
  organonickel complexes, **9**, 62
Pseudoephedrine
  synthesis, **12**, 1045
    chromium complexes, **12**, 1049
Pseudoephedrine, *N*-methyl-
  synthesis, **12**, 1056
Pseudoguaianolide
  synthesis, **12**, 722
Pseudoguaianolides
  lactones, **12**, 413
Pseudohalogen complexes
  without carbonyl ligands, **5**, 283–6
Pseudomeconine
  synthesis, **12**, 247
Pseudopeptides
  synthesis
    organocopper reagents, **12**, 113
Pseudoscopine
  synthesis, **12**, 811
Pseudotropine
  synthesis, **12**, 811
Pseudouridine, 2'-deoxy-
  synthesis, **12**, 277, 278
Pterosin Z
  synthesis, **12**, 748
allo-Pumilliotoxin 339B
  synthesis, **12**, 874
Punctaporonin B
  synthesis, **12**, 881
Purine nucleosides
  synthesis, **12**, 72
2*H*-Pyran, 4-methyl-3,6-dihydro-
  metallation, **11**, 119
4*H*-Pyran
  metallation, **11**, 119
4*H*-Pyran, 4-methyl-
  metallation, **11**, 119
Pyran, dihydro-
  coupling with Grignard reagents, **12**, 170
Pyran, 3,4-dihydro-
  formation
    Heck reaction, **12**, 280
Pyran, divinyl-
  synthesis, **12**, 790
Pyran, divinyltetrahydro-
  synthesis, **12**, 791
Pyran, tetrahydro-

cis-annelated, **12**, 877
formation
cyclopropanation, **12**, 391
synthesis
alkoxypalladation, **12**, 825
Pyranones
synthesis, **12**, 758, 759
Pyrans
preparation, **12**, 611
synthesis, **12**, 758, 759
cyclooligomerization, **12**, 764
Pyrans, 2-vinyltetrahydro-
preparation, **12**, 613
Pyrazdiyne
zirconocene complexes, **12**, 783
Pyrazine, substituted
synthesis
zirconocene complexes, **12**, 783
Pyrazine, tetralkyl-
synthesis
titanium complexes, **12**, 1084
Pyrazines
synthesis, **12**, 756
cyclooligomerization, **12**, 753–6
Δ^2-Pyrazolines
synthesis, **12**, 627
Pyrazolo[1,5-a]pyridines
synthesis
carbene complexes, **12**, 513, 528
Pyrazolylborato ligands
clusters, **10**, 166, 167
Pyrazyne
zirconocene complexes, **12**, 783
Pyrenophorin
synthesis, **12**, 842
Pyrethroid insecticides
synthesis, **12**, 398
Pyrethroids
synthesis, **12**, 406, 624, 627
Pyridazine
lithiation, **11**, 79
Pyridazines
halogenated
coupling reactions, **12**, 224
Pyridazines, halo-
cross-coupling reactions
Grignard reagents, **12**, 178
Pyridine
cross-coupling reactions
Grignard reagents, **12**, 177
orthometallation
triosmium clusters, **7**, 820
Pyridine, 2-allyloxy-
isomerization
palladium catalyzed, **12**, 872
Pyridine, 3-aryl-
synthesis, **12**, 192
Pyridine, 2-(ω-cyanoalkyl)-
synthesis, **12**, 756
Pyridine, 2-(dialkylamino)-
synthesis, **12**, 754
Pyridine, dihydro-
synthesis
carbene complexes, **12**, 509
Pyridine, 1,4-dimethyldihydro-
metallation, **11**, 119
Pyridine, 1,4-dimethyl-1,2,3,6-tetrahydro-
metallation, **11**, 119
Pyridine, iminodihydro-
synthesis, **12**, 756–8
Pyridine, 2-imino-1,2-dihydro-
synthesis from carbodiimides, **12**, 758
Pyridine, 1-methyl-1,4-dihydro-
metallation, **11**, 119
Pyridine, 1,2,5,6-tetrahydro-
synthesis, **11**, 144
Pyridine, 2-(thiomethyl)-
synthesis, **12**, 754
Pyridines
coupling
zirconium mediated, **4**, 619
DMG containing
lithiation, **11**, 76–9
halogenated
coupling reactions, **12**, 224
metallation, **11**, 119
ortho-substituted
coupling with alkenes, **12**, 1206
polyhalogenated
coupling with Grignard reagents, **12**, 178
reactions with Grignard reagents, **11**, 141–4
synthesis
carbene complexes, **12**, 507–9
cross-coupling reactions, **12**, 1084
cyclooligomerization, **12**, 753–6
Pyridinium salts
reaction with zinc organometallics, **11**, 171
3-Pyridinols
synthesis
lithiation, **11**, 78
Pyridinophanes
preparation, **12**, 178
Pyridone, methyldihydro-
synthesis, **12**, 29
Pyridones
synthesis, **12**, 224, 225, 756–8
4-Pyridones, 2,3-dihydro-
regiocontrolled synthesis
Grignard reagents, **11**, 143
Pyrido[2,1-*b*]quinazoline
synthesis, **12**, 249, 250
Pyridyne
zirconocene complexes, **12**, 780, 782
3,4-Pyridynes
synthesis
lithiation, **11**, 78
Pyrimidine
lithiation, **11**, 81
Pyrimidines, halo-
cross-coupling reactions
Grignard reagents, **12**, 178
α-Pyrone
synthesis
nickel complexes, **9**, 5
2-Pyrone, tetraethyl-
synthesis, **9**, 5
Pyrones
bicyclic
synthesis, **12**, 758
cycloaddition reactions, **12**, 928
synthesis
carbene complexes, **12**, 515
Pyrrole
activation
triruthenium clusters, **7**, 819
coupling reactions, **12**, 229
cyanocuprates, **12**, 79
reactions
annulation, **12**, 761
Pyrrole, 2-alkenyl-
synthesis, **12**, 167
Pyrrole, 2-cyano-5-amino-
synthesis
cyclooligomerization, alkynes, **12**, 745
Pyrrole, ethylenebis[3,4-diphenyl]-tetrahydro-
osmium oxide
asymmetric dihydroxylation, **12**, 1140
Pyrrole, *N*-methyl-
metallation, **11**, 119
Pyrrole, *N*-phenyl-
metallation, **11**, 119
Pyrroles
DMG-containing
lithiation, **11**, 72, 73
synthesis, **12**, 341, 756
carbene complexes, **12**, 507–9, 534
coupling reactions, niobium complexes, **12**, 1089
cross-coupling reactions, **12**, 1084
cyclooligomerization, **12**, 743
trapping of oxallyl cation with, **12**, 350
Pyrrolidine
synthesis, **12**, 358
cycloaddition reactions, **12**, 933
Pyrrolidine, benzylidene-
synthesis, **12**, 909
Pyrrolidine, 2,5-dialkyl-
synthesis, **12**, 822
Pyrrolidine, 2-(diphenylhydroxymethyl)-
preparation, **11**, 259
Pyrrolidine, (*N*-diphenylphosphino)-(2-diphenylphosphinoxymethyl)-
synthesis, **12**, 887
Pyrrolidine, divinyl-
synthesis, **12**, 910
Pyrrolidine, methylidenevinyl-
synthesis, **12**, 906
Pyrrolidine, propenyl-
synthesis, **12**, 909
Pyrrolidine, vinyl-
synthesis, **12**, 916
Pyrrolidines
lithiation, **11**, 17
synthesis
aminomercuration, **11**, 423
organoboron compounds, **11**, 218
Pyrrolidinyl acetate
synthesis

allylpalladium catalysts, **12**, 916
1-Pyrrolines
synthesis
carbene complexes, **12**, 514
Δ'-Pyrrolines
synthesis
cross-coupling reactions, **12**, 1084
Pyrrolinone
synthesis
carbene complexes, **12**, 510–13
Pyrrolizidine alkaloids
synthesis, **12**, 362, 436
Pyrrolizidines
synthesis
organoboron compounds, **11**, 218
Pyrrolizidines, dioxo-
synthesis
cross-coupling reactions, **12**, 1084
Pyrromethanenones
synthesis
cobalt-mediated propargylation, **12**, 696

Q

Q-particles
  semiconductors
    synthesis, **3**, 203
Quadracycline
  reaction with rhodium complexes, **8**, 216
Quadricyclane
  reactions with iron ions, **7**, 106
Quadrilure
  synthesis, **12**, 101
Quadrone
  synthesis, **12**, 614, 734
Quadrupledecker complexes
  transition metals, **1**, 199–201
Quantum dots
  semiconductors
    synthesis, **3**, 203
Quaterruthenocenylene
  synthesis, **7**, 616
Queen bee substance
  synthesis, **12**, 269
Quercus lactone
  synthesis, **11**, 101
Quinidine acetate, dihydro-
  osmium complexes
    asymmetric, dihydroxylation, **12**, 1138, 1139
Quinidine, *p*-chlorobenzoyldihydro-
  osmium complexes
    asymmetric dihydroxylation, **12**, 1138, 1139, 1141
Quinine acetate, dihydro-
  osmium complexes
    asymmetric dihydroxylation, **12**, 1138, 1139
o-Quinodimethane, α-hydroxy-
  Diels–Alder reaction, **11**, 245
Quinol vinyl ethers
  rearrangement, **12**, 594
Quinoline
  cross-coupling reactions
    Grignard reagents, **12**, 177
Quinoline, decahydro-
  synthesis
    organoboron compounds, **11**, 218
Quinoline, 2-fluoro-
  lithiation, **11**, 78
Quinoline-3-carboxylates, 2-substituted-1,4-dihydro-4-oxo-
  synthesis, **12**, 318
Quinolines
  halogenated
    coupling reactions, **12**, 224
  β-lactams from
    photochemical reactions, carbene complexes, **12**, 551
  lithiation, **11**, 51
  synthesis
    cross-coupling reactions, **12**, 1084
    Heck reaction, **12**, 270
Quinolone
  antibacterials
    synthesis, **12**, 215
Quinolones
  synthesis, **12**, 212, 256
    Heck reaction, **12**, 261
Quinones
  acylation
    acyl iron complexes, **12**, 353
  antibiotics
    synthesis, **12**, 211
  polycyclic
    synthesis, **12**, 1030
  synthesis, **12**, 751, 752
    cyclooligomerization, alkynes, **12**, 744
Quinophenylselenoimines
  synthesis from phenols, **11**, 534
Quinquearyls
  synthesis, **12**, 195

R

Rabelomycin, deoxy-
  synthesis, **12**, 1030
Raft clusters
  synthesis, **10**, 5
Reagents, **11**, 576
Realger structures
  P_8 complexes of iron, **7**, 284
Receptor binding
  propargyl dicobalt complexes, **12**, 690
Recifeiolide
  synthesis, **12**, 800
Redox condensation
  carbonylmetallate reaction with metal carbonyl complexes, **10**, 4
Redox-active polymers
  synthesis, **12**, 1223
Reduction
  alkenes
    organolanthanide complexes, **4**, 129
  alkylpalladium complexes, **12**, 139
  enantioselective
    organoboron compounds, **11**, 258–61
  iron carbonyl catalysts, **7**, 85, 86
  tellurium reagents, **11**, 576–81
Reductive carbonylation
  organosamarium complexes, **4**, 126
Reductive coupling
  aryl ketones and aldehydes, **12**, 1074
  unsaturated compounds
    transition metal complexes, **12**, 323–46
Reductive cyclization
  by dichlorozirconocene, **12**, 325
Reductive dechalcogenation
  molybdenum complexes, **12**, 893
Reductive dimerization
  benzylic halides
    titanium reagents, **12**, 1073
Reductive elimination
  organonickel complexes
    phosphines, **9**, 62
  organotransition metal complexes, **12**, 131
  transition metal complexes, **12**, 4
Reformatsky reaction
  organozinc compounds, **11**, 159, 170
Reformatsky-type reactions
  organolanthanide catalysts, **4**, 124
γ-Resorcylic acid
  propargyl dicobalt complexes, **12**, 690
γ-Resorcylic acid, 6-methyl-
  propargyl dicobalt complexes, **12**, 690
Reverse-Brook rearrangement
  lithiation, **11**, 24
Rhenacyclobutane
  synthesis, **6**, 243
Rhenacyclopentene
  synthesis
    mechanism, **6**, 243
Rhenium
  dicarbon metallacarbaboranes
    12-vertex, **1**, 400
Rhenium carbonyls, **6**, 168–76
Rhenium clusters
  carbido, **6**, 175
  carbonyl, **6**, 173–6
  high nuclearity, **6**, 175, 176
Rhenium complexes
  acetylides, **6**, 193, 194
  acyl, **6**, 196–200
    coupling reactions, **6**, 196–9
    decarbonylation, **6**, 196
    deprotonation, **6**, 199–201
  alcohols, **6**, 217, 218
  aldehyde, **6**, 213, 214
  alkenes, **6**, 204–6
  alkoxides, **6**, 217, 218
  alkyl monocyclopentadienyl oxo
    synthesis, **6**, 247
  alkylidene, **6**, 240–5
  alkylidyne, **6**, 240–5
  alkyls, **6**, 189–95
    deprotonation, **6**, 191
    hydride abstraction, **6**, 191, 192
  alkyne, **6**, 207, 208
  allyl, **6**, 205, 208, 209
  amidino, **6**, 215
  amido, **6**, 215
  arene, **6**, 210
  arsenic ligands, **6**, 214–16
  aryldiazenido, **6**, 215
  azobenzene, **6**, 215
  benzene, **6**, 210
  butyne
    synthesis, **6**, 207
  carbamoyl, **6**, 215
  carbenes, **6**, 200–3
  carbides, **6**, 193, 194
  carboxylic acids, **6**, 200
  carbyne, **6**, 203, 204
  cyclohexadiene, **6**, 210
  cyclopentadienyl, **6**, 210–13
    reactivity, **6**, 210, 211
    ring slippage, **6**, 212, 213
    synthesis, **6**, 210, 211
  diarylamidines, **6**, 215
  diimine tricarbonyl, **6**, 215
  dinitrogen, **6**, 216
  diphosphene, **6**, 216
  disulfides, **6**, 220
  enolates, **6**, 190
  ethene, **6**, 201
  ethyne
    bromination, **6**, 208
  fluoro, **6**, 176
  group 7 metals
    direct bonds to gallium, **1**, 554
  group 8 metals
    direct bonds to gallium, **1**, 555
  halides, **6**, 176, 177
  high oxidation states
    alkoxides, **6**, 248
    alkyl cyclopentadienyl, **6**, 248, 249
    amides, **6**, 248
    biscyclopentadienyl, **6**, 252
    cyclohexadienyl, **6**, 252
    hydrido cyclopentadienyl, **6**, 250
    monocyclopentadienyl, **6**, 245–52
    pseudohalides, **6**, 248
    thiolates, **6**, 248
  hydrazido, **6**, 215
  hydrides, **6**, 180–6
    deprotonation, **6**, 183
    hydride abstraction, **6**, 183
    nucleophilicity, **6**, 184
    oxidation, **6**, 184
    properties, **6**, 181–3
    reactivity, **6**, 183–6
    synthesis, **6**, 180, 181
  π-hydrocarbons, **6**, 204–13
  hydroxide, **6**, 223
  indium bonds, **1**, 560
  ketenyl, **6**, 202
  ketones, **6**, 213, 214
  Lewis acids, **6**, 177–9
  low valent, **6**, 167–223
  mercury clusters, **10**, 338
  nitride, **6**, 176, 177
  nitrites, **6**, 177
  nitrogen ligands, **6**, 214–16
  nitrosyl, **6**, 216
  oxo, **6**, 221–3
  oxygen ligands, **6**, 217–23
  pentacarbonyl
    tetrafluoroborate, **6**, 177
  phosphide, **6**, 216
  phosphido ligands, **6**, 215
  phosphinimines, **6**, 177
  phosphorus ligands, **6**, 214–16
  polysulfides
    synthesis, **6**, 251, 252
  propargyl, **6**, 207
  Schiff bases, **6**, 215
  selenium ligands, **6**, 217
  sulfides
    synthesis, **6**, 251, 252
  sulfur ligands, **6**, 217–23
  tetrasulfido, **6**, 220
  thallium bonds, **1**, 567

1,4,7-triazacyclononane, **6**, 215
tridentate cyano, **6**, 215
trishomocubane, **6**, 216, 223
vinyl
deprotonation, **6**, 206
vinyl ligands, **6**, 194, 195
vinylidene, **6**, 200–3
Rhenium(III) complexes
aluminum tetrahydride derivatives, **1**, 550
Rhenium(IV) complexes, **6**, 232
alkylidene, **6**, 242–4
alkylidyne, **6**, 242–4
alkynes
synthesis, **6**, 232
dimers, **6**, 250, 251
imido cyclopentadienyl, **6**, 251
phosphines, **6**, 232
Rhenium(V) complexes, **6**, 232–5
alkylidene, **6**, 242–4
alkylidyne, **6**, 242–4
dihydrogen, **6**, 235
imido
synthesis, **6**, 234, 235
nitrido
synthesis, **6**, 235
oxo, **6**, 232–4
Rhenium(VI) complexes, **6**, 235–7
alkylidene, **6**, 242–4
alkylidyne, **6**, 242–4
imido
synthesis, **6**, 236
oxo
synthesis, **6**, 236
sulfido
synthesis, **6**, 236
Rhenium(VII) complexes, **6**, 237–40
alkylidenes, **6**, 240–2
alkene metathesis, **6**, 244, 245
alkylidyne, **6**, 240–2
alkylidynes
alkene metathesis, **6**, 244, 245
imido
synthesis, **6**, 239, 240
oxo
synthesis, **6**, 237–9
oxo imido
synthesis, **6**, 240
Rhenium metallaalkenes
initiators
ring-opening metathesis polymerization, **12**, 1214
Rhenium pentacarbonyl
alkylation, **6**, 189
Rhenium radicals
stabilization, **6**, 169
Rhenium, (cyclopentadienyl)-(nitroso)(triphenylphosphine)-
alkylation, **6**, 190, 191
Rhenium, (dichloromethane)-cyclopentadienyl(nitroso)-(triphenyhlphosphine)-
tetrafluoroborate, **6**, 177–9
Rhenium(VI), hexamethyl-
reactions, **6**, 235
Rhenium–carbon bonds
multiple, **6**, 200–4
Rhenium–rhenium bonds
17-electron, **6**, 168, 169
homolysis, **6**, 168, 169
Rhodacyclopentadienes
synthesis, **12**, 748
Rhodium
dicarbon metallacarbaboranes
6-vertex, **1**, 382
7-vertex, **1**, 382
10-vertex, **1**, 396
11-vertex, **1**, 396
12-vertex, **1**, 400
monocarbon metallacarbaboranes
11-vertex, **1**, 378
12-vertex, **1**, 378
Rhodium cluster complexes, **8**, 419–513
bisalkylidyne
synthesis, **8**, 463
bismethylidyne
synthesis, **8**, 463, 464
columnar, **8**, 479, 480
cyclopentadienyl, **8**, 462
face capping arene ligand, **8**, 466
face capping cyclopentadienyl ligand, **8**, 466
dinitrosyl, **8**, 464
disulfur capped, **8**, 461, 465
face-sharing octahedra, **8**, 479
gas phase, **8**, 421
magnetic properties, **8**, 421
globular
high-pressure synthesis, **8**, 480
interstitial rhodium atoms, **8**, 480–2
heptanuclear, **8**, 477
carbonyl iodo-, **8**, 477
interstitial nitride, **8**, 477
hexafluorobutyne
synthesis, **8**, 465
hexanuclear, **8**, 472–7
arsines, **8**, 477
dimethylsulfide, **8**, 474
interstitial carbon, **8**, 474
interstitial carbon, synthesis, **8**, 475
interstitial nitrogen, **8**, 476
NMR, **8**, 476
phosphido, **8**, 476
phosphines, **8**, 474
phosphines, synthesis, **8**, 476
polyhapto ligands, **8**, 474
protonation, **8**, 476
interstitial antimony atoms, **8**, 485
interstitial arsenic atoms, **8**, 485
interstitial disulfur atoms, **8**, 486
interstitial main group atoms, **8**, 483–6
interstitial phosphorus atoms, **8**, 485
interstitial sulfur atoms, **8**, 485
isocyanides, **8**, 465
large, **8**, 479–86
medium, **8**, 479–86
mixed metals, **8**, 505–13
NMR, **8**, 486
octanuclear, **8**, 478
oxo capped, **8**, 464
pentanuclear, **8**, 471
alkyl, **8**, 471
phosphanidene, **8**, 478
phosphido
synthesis, **8**, 459
sulfur dioxide
synthesis, **8**, 469
tetrahydride, **8**, 469
tetrakis(pentamethylcyclopentadienyl)tetrahydride, **8**, 470
tetranuclear
phosphido ligands, **8**, 471
Si-tripod, **8**, 469
square, **8**, 470
tricarbonyltricyclopentadienyl, **8**, 462
triindenyl, **8**, 465
trimethylphosphite
neutron diffraction, **8**, 458
reactivity, **8**, 458
trinuclear, **8**, 457–66
unsaturated metal aggregate species, **8**, 486, 487
Rhodium clusters
alkylidyne molybdenum, **10**, 116–18
alkylidyne tungsten, **10**, 116–18
iron compounds, **10**, 134, 141, 142
phosphido molybdenum, **10**, 116–18
phosphido tungsten, **10**, 116–18
ruthenium compounds, **10**, 134
sulfido molybdenum, **10**, 116–18
sulfido tungsten, **10**, 116–18
Rhodium complexes
acac, **8**, 118
solid state, **8**, 118
acyl, **8**, 185–217
alkenes, **8**, 226–34
group 15 donor ligands, **8**, 228–31
group 16 donor ligands, **8**, 226, 227
group 17 donor ligands, **8**, 226, 227
alkenyl, **8**, 185–217
alkyl, **8**, 185–217
alkynes, **8**, 235–8
alkynyl, **8**, 185–217
allyl, **8**, 273–6
alkene insertion reactions, **12**, 906
terminally alkyl-substituted alkenes, cyclization, **12**, 910–12
arenes, **8**, 271, 272
aryl, **8**, 185–217
azulene, **8**, 174
bis(diphenylarsino)methane, **8**, 161
carbenes, **8**, 220–5
carbon bonds only, **8**, 266–8
carbon dioxide, **8**, 161, 162
carbon disulfide, **8**, 161, 162
carbonyl, **8**, 116–22
binary, **8**, 116

group 15 donor ligands, **8**, 123–61
group 15 donor ligands, liquid crystals, **8**, 123
oxygen donors, **8**, 116–19
paramagnetic, **8**, 117
photolysis, **8**, 118
carbonyl arsenic donor ligands, **8**, 158–61
carbonyl azines, **8**, 132, 135
carbonyl benzotriazole, **8**, 134
carbonyl bibenzimidazole, **8**, 136
carbonyl bicyclic aminophosphorane, **8**, 154
carbonyl bidentate nitrogen donor ligands, **8**, 124–34
carbonyl biimidazole, **8**, 136
carbonyl bipyridine, **8**, 127
carbonyl bipyridyl, **8**, 128
carbonyl bipyrimidine, **8**, 129, 130
carbonyl bis(dimethylphosphino)methane, **8**, 143
carbonyl bis(diphenylphosphino)bipyridine, **8**, 149
carbonyl bis(diphenylphosphino)cyclopropane, **8**, 149
carbonyl bis[(diphenylphosphino)ethyl]-benzylamine, **8**, 157
carbonyl bis(diphenylphosphino)furan, **8**, 151
carbonyl bis(diphenylphosphino)methane, **8**, 143
carbonyl [bis(diphenylphosphino)methyl]pyridine, **8**, 146, 155, 157
carbonyl bis(diphenylphosphino)pyridine, **8**, 157
carbonyl bis(pyrazolyl)-methanes, **8**, 127
carbonyl cyclamphosphine oxide, **8**, 133
carbonyl cytidine, **8**, 123
carbonyl diaminonaphthalene, **8**, 130
carbonyl difurfulylpyrole, **8**, 131
carbonyl diimines, **8**, 128
carbonyl dimethyladenine, **8**, 126
carbonyl diphenylamino-quinoline, **8**, 130
carbonyl (diphenylphosphinophenyl)amine, **8**, 156
carbonyl dipyridyl sulfide, **8**, 130
carbonyl formamidinate, **8**, 128
carbonyl guanine, **8**, 128
carbonyl guanosine, **8**, 123
carbonyl hydrospirophosphoranes, **8**, 154
carbonyl imidazole, **8**, 126
carbonyl indolate–pyridine, **8**, 130
carbonyl inosine, **8**, 123
carbonyl mercaptothiazolinate, **8**, 133
carbonyl metallamacrocycles, **8**, 151
carbonyl methyladenine, **8**, 126
carbonyl methyltrispyrazolyl-gallate, **8**, 134
carbonyl naphthyridinone, **8**, 126
carbonyl nitrogen donor ligands
tetradentate, **8**, 134–6
tridentate, **8**, 134–6
carbonyl nitrogen donors, **8**, 123
carbonyl oxalate, **8**, 117
carbonyl phenazine, **8**, 128
carbonyl phenylacetamide, **8**, 133
carbonyl phenyl(4-butyl-2-pyridyl)amine, **8**, 126
carbonyl phenyl(2-pyridyl)amine, **8**, 126
carbonyl phenylthioacetamide, **8**, 133
carbonyl phosphide bridged, **8**, 138
carbonyl phosphorus donor ligands
monodentate, **8**, 138–43
tetradentate, **8**, 156–8
tridentate, **8**, 156–60
carbonyl phosphorus nitrogen donors
bidentate, **8**, 152–6
carbonyl phosphorus oxygen donors
bidentate, **8**, 152–6
carbonyl phosphorus sulfur donors
bidentate, **8**, 152–6
carbonyl phosphorus–phosphorus ligands
bidentate, **8**, 143–51
carbonyl picolinate, **8**, 132
carbonyl porphyrin, **8**, 136
carbonyl pyrazine, **8**, 127
carbonyl pyrazole, **8**, 124
carbonyl pyridine-2-thiol, **8**, 133
carbonyl sapphyrin, **8**, 136
carbonyl selenium, **8**, 120–2
carbonyl sulfur, **8**, 120–2
carbonyl sulfur dioxide, **8**, 151
carbonyl superoxide, **8**, 128
carbonyl tellurium, **8**, 120–2
carbonyl tetrathiosquarate, **8**, 121
carbonyl tetrazole, **8**, 134
carbonyl thiocaprolactamate, **8**, 133
carbonyl thiolate, **8**, 120, 138
carbonyl triazene, **8**, 128
carbonyl 1,2,4-triazole, **8**, 135
carbonyl trispyrazolylmethane, **8**, 134
carbonyl tropolone, **8**, 117
carbynes, **8**, 220–5
catalysis
allylmetal insertion reactions, alkenes, **12**, 906–8
catalysts
diazodecomposition, **12**, 428–30
hydrogenation, **12**, 17
chemistry, **8**, 115–281
conjugated dienes, **8**, 240–4
group 16 donor ligands, **8**, 245–9
group 17 donor ligands, **8**, 245–9
cumulenes, **8**, 235
cyclodextrins, **8**, 245
cyclooctadiene, **8**, 245–68
cyclooctadienone, **8**, 270
cyclopentadienyl, **8**, 167–85
dinuclear, bridging, **8**, 171
photolysis, **8**, 168
substituted, **8**, 170
cyclopentadienyl amino acids, **8**, 180
cyclopentadienyl hydride, **8**, 175
dicyclopentadienyl
microwave synthesis, **8**, 170
dienyl, **8**, 272–81
dienylcyclopentadienyl
nucleophilic addition, **12**, 679
dihapto ligands, **8**, 226–40
diketone, **8**, 117
dithiocarbamate, **8**, 121
dithiocarbonate, **8**, 121
duroquinone, **8**, 269
enyl, **8**, 272–81
fluorocarbon, **8**, 199
fulvene
heterometallic, **8**, 172, 173
group 15 donor ligands, **8**, 249–65
halocarbonyl, **8**, 116
haloisocyanide, **8**, 163, 164
heptadienes, **8**, 268
heteroalkenes, **8**, 239, 240
heteroalkynes, **8**, 239, 240
heterocumulene, **8**, 239, 240
heterodienes, **8**, 270
hexadiene
synthesis, **8**, 269
hfacac, **8**, 118
homotropilidenes, **8**, 268
hydride, **8**, 169
hydrocarbon ligands, **8**, 167–281
hydroperoxo, **8**, 170
indenyl, **8**, 170
indium hexadiene, **8**, 270
isocyanates, **8**, 239
isocyanide, **8**, 163–7
anions, association, **8**, 164
binary, **8**, 163, 164
group 15 donor ligands, **8**, 164–7
group 16 donor ligands, **8**, 164–7
vibrational spectra, **8**, 164
isothiocyanate, **8**, 239, 240
metallaboranes, **1**, 340–2
metallacyclic, **8**, 217–20
metallation
carbon–halogen bonds, **8**, 190–200
methyl
preparation, **12**, 157

monohapto ligands
metathesis, **8**, 185–90
transmetallation, **8**, 185–90
norbornadiene, **8**, 245–68
nucleophilic attack
carbon monoxide and alkenes, **8**, 204, 205
pentadienyl, **8**, 277–81
pentamethylcyclopentadienyl, **8**, 175–85
pentamethylcyclopentadienyl dicarbonyl
preparation, **8**, 184
polyoxo, **8**, 119
porphyrins
metallation, **8**, 212
tetrafluorobutadiene, **8**, 245–68
tetrahapto ligands, **8**, 240–70
thiaboranes, **1**, 268
thioacac, **8**, 121
thiocarbonyl, **8**, 161, 162
vinylsilanes, **8**, 268
without hydrocarbon ligands, **8**, 116–67
zinc clusters, **10**, 341
Rhodium(III) complexes
porphyrins
catalysts, diazodecomposition, **12**, 430
Rhodium tetracarbonyl, **8**, 116
Rhodium, bis(triphenylphosphine)hydrido(azaundecaborane)-
preparation, **1**, 262
Rhodium, dicyclopentadienyl-(hexafluorobutyne)-, **8**, 465
Rhodium, dodecacarbonyltetra-, **8**, 466
adsorption, **8**, 467
carbonyl ligand substitution, **8**, 468
catalyst, **8**, 467
electrochemistry, **8**, 469
hydroformylation catalyst, **8**, 467
ligand substitution, **8**, 469
NMR, **8**, 466
phosphite derivatives, **8**, 468
phosphine derivatives
structure, **8**, 468
phosphite derivatives
structure, **8**, 468
reaction with carbon monoxide and hydrogen, **8**, 468
reaction with phosphine ligands, **8**, 470
triphos derivative, **8**, 469
tripod derivative, **8**, 469
adsorption, **8**, 469
Rhodium, hexacarbonyltriphosphine complexes, **8**, 457
Rhodium, hexadecacarbonylhepta-
structure, **8**, 477
Rhodium, hexadecacarbonylhexa-
adsorption, **8**, 472
isoelectronic series, **8**, 473
preparation, **8**, 467
reaction with surface water, **8**, 472
synthesis, **8**, 472
Rhodium, nonacarbonyl-(trithiacyclohexane)-tetra-
structure, **8**, 469
Rhodium, nonadecacarbonylcarbidoocta-
synthesis, **8**, 478
Rhodium, octacarbonyldi-
reaction with carbon monoxide, **8**, 466
Rhodium, octacarbonyl(triethyl phosphite) (tripod)
structure, **8**, 468
Rhodium, octacosacarbonyldicarbido-, **8**, 483
Rhodium, pentacosacarbonyldicarbido-, **8**, 483
Rhodium, pentacosacarbonyl-dinitrogentetradeca-
structure, **8**, 484
Rhodium, pentacosacarbonyl-hydrido-
anion, **8**, 479
Rhodium, pentacosacarbonyl-hydridotetradeca-
NMR, **8**, 482
Rhodium, pentadecacarbonyl-hydrido-
anion
structure, **8**, 473
Rhodium, pentadecacarbonylpenta-
structure, **8**, 471
Rhodium, phosphite hydride
synthesis, **8**, 457
Rhodium, tetracosacarbonyl-trihydridotrideca-
acidity, **8**, 482
Rhodium, tricarbonyltris(pentamethylcyclopentadienyl)tri-
structure, **8**, 463
Rhodium, tricontacarbonyldicarbido-, **8**, 483
Rhodium, tricontacarbonyl-dodeca-, **8**, 479
anion
reaction with hydrogen, **8**, 473
fragmentation, **8**, 471
Rhodium, tricosacarbonyl-hydridodinitrogen-dodeca-
synthesis, **8**, 484
Ribonolactone
synthesis, **12**, 734
Ribonolactone, 2-deoxy-
synthesis
organosilicon compounds, **11**, 317
Rieke's magnesium
Grignard reagent preparation, **1**, 61
preparation, **1**, 59
Ring closure reactions
organotellurium compounds, **11**, 585–7
Ring-laddering
alkali metal organometallics, **1**, 2
Ring-opening metathesis polymerization
alkene geometry, **12**, 1217
chain tacticity, **12**, 1217
chain transfer, **12**, 1215–17
cycloheptene
alkene carbene tungsten complexes, **5**, 334
initiators, **12**, 1210–15
monomers, **12**, 1217
novel polymers by, **12**, 1222, 1223
reaction scope, **12**, 1215–22
solvents, **12**, 1217
stereoregulating initiators, **12**, 1220
stereochemistry, **12**, 1217–22
termination, **12**, 1215–17
transition metals in, **12**, 1209–23
Ring-opening polymerization
methylenecyclobutane
Ziegler–Natta catalysts, **12**, 1203
Ring-stacking
alkali metal organometallics, **1**, 2
Ritter reaction
N-propargylation of dicobalt complexes, **12**, 690
Routiennocin
synthesis, **12**, 371
lithiation, **11**, 72
Royleanone, 12-*O*-methyl-
synthesis, **12**, 494
Rubidium, methyl-
preparation, **1**, 20
Ruppuran
properties, **7**, 619
Ruthenaalkenes
initiators
ring-opening metathesis polymerization, **12**, 1214
Ruthenacarbocycles, **7**, 424–6
Ruthenacycles, **7**, 424–30
Ruthenacyclopentadiene
dinuclear, **7**, 443
Ruthenacyclopentadienes
synthesis, **7**, 668
Ruthenacyclopentadienones, **7**, 808
Ruthenium
dicarbon metallacarbaboranes
6-vertex, **1**, 382
7-vertex, **1**, 382
10-vertex, **1**, 396
11-vertex, **1**, 396
12-vertex, **1**, 400
monocarbon metallacarbaboranes, **1**, 378
12-vertex, **1**, 378
Ruthenium catalysts
cross-coupling
alkenyl bromides, **12**, 168
Ruthenium clusters
acetylide platinum, **10**, 153, 154
alkylidyne iron, **10**, 129, 130
alkylidyne osmium, **10**, 129, 130
alkylidyne tungsten, **10**, 93–6
alkyne iron, **10**, 130, 131
alkyne osmium, **10**, 130, 131
alkyne platinum, **10**, 151, 152
azide, **10**, 93

doubly-bridging 5-electron donating ligands, **7**, 698
doubly-bridging 6-electron donating ligands, **7**, 699, 700
doubly-bridging ligands, **7**, 691–700
doubly-bridging ligands donating more than 6 electrons, **7**, 700
electron counting, **7**, 689, 690
flash photolysis, **7**, 685
fluxionality, **7**, 717, 718
hydrido, **7**, 719–21
ketenylidene, **7**, 705
ligands coordinated through heteroatoms, **7**, 711
methylpyrrolidene, **7**, 706
nitrogen ligands, **7**, 730
nitrosyl ligands, **7**, 730
organoisocyanate, **7**, 704
oxygen ligands, **7**, 723
phenylene, **7**, 703
phosphabenzene, **7**, 704
phosphorus ligands, **7**, 735
pyridine-2-thione, **7**, 730
pyridonato, **7**, 707
reactions with thioamides, **7**, 728
reactions with thioureas, **7**, 728
selenium ligands, **7**, 726
sulfur ligands, **7**, 726
tellurium ligands, **7**, 726
thermal substitution, **7**, 685
triply-bridging 4-electron donating ligands, **7**, 702–5
triply-bridging 5-electron donating ligands, **7**, 705–8
triply-bridging 6-electron donating ligands, **7**, 708, 709
triply-bridging ligands, **7**, 701–10
triply-bridging ligands donating more than 6 electrons, **7**, 709, 710
unsaturated 46-electron clusters, **7**, 689
UV irradiation, **7**, 685
vinylidene ligands, **7**, 703
tungsten compounds, **10**, 98–100
vinylidene iron, **10**, 130, 131
vinylidene osmium, **10**, 130, 131
weak metal–metal bonding, **7**, 909
Ruthenium complexes
β-acetyl ethyl, **7**, 392
acetylacetone cyclopentadienyl, **7**, 496
acetylide cyclopentadienyl, **7**, 485, 486, 513–17
addition to electrophilic alkenes, **7**, 518–20
acetylides, **7**, 411–16
reactivity, **7**, 415, 416
synthesis, **7**, 412–15
acyclic pentadienyl, **7**, 465–8
acyls, **7**, 417–24
alkanediyl cyclopentadienyl
thermal and photochemical decomposition, **7**, 481
alkene cyclopentadienyl, **7**, 504–12
alkenyl cyclopentadienyl
synthesis, **7**, 502
alkoxide cyclopentadienyl, **7**, 476–8
alkoxo cyclopentadienyl trihydrido, **7**, 500
alkoxyalkylidenes, **7**, 343
alkoxycarbene cyclopentadienyl, **7**, 513–17
alkoxycarbonyls, **7**, 422, 423
alkoxymethyl, **7**, 387
alkyl, **7**, 372–93
high-valent, **7**, 372–8
nitrido, **7**, 372
oxo, **7**, 372
alkyl cyclopentadienyl
synthesis, **7**, 502
alkyl cyclopentadienyl phosphines
preparation, **7**, 482
alkylidene, **7**, 336–48
alkylidyne, **7**, 356–62
reactivity, **7**, 360–2
synthesis, **7**, 356–60
alkyne cyclopentadienyl, **7**, 485, 486, 504–17
α-alkynyl, **7**, 390
alkynyl vinyl, **7**, 407
allenylidene cyclopentadienyl, **7**, 517, 518
allyl cyclopentadienyl, **7**, 483–5, 504–12
amino acids, **7**, 560
γ-aminoalkyl, **7**, 393
ammine isonicotinamide, **7**, 590
anthracene, **7**, 591
arene chloro
Michaelis–Arbuzov reactions, **7**, 563
arenes
η^2-, **7**, 587–90
η^4-, **7**, 591
acetylide, **7**, 579–86
addition of alkynes, **7**, 531
alkylidene, **7**, 579–86
allenylidene, **7**, 579–86
allyl, **7**, 575–9
azavinylidene, **7**, 579–86
bis(alkene), **7**, 575–9
carbenes, **7**, 579–86
cumulene, **7**, 579–86
dienes, **7**, 575–9
dienyl, **7**, 575–9
group 14 element bonds, **7**, 586
low hapticity, **7**, 587–92
trimethylenemethane, **7**, 575–9
tripodal oxygen-donor ligands, **7**, 560
vinyl, **7**, 579–86
vinylidene, **7**, 579–86
aroyl, **7**, 417–24
aryls, **7**, 393–9
homoleptic, **7**, 393
benzene
reaction with α-pyridone, **7**, 560
benzene chloro
phosphines, reaction with sodium borohydride, **7**, 562
benzene *o*-xylylene, **7**, 562
benzylidene cyclopentadienyl, **7**, 480, 485
benzyne, **7**, 592
binuclear
alkyl, **7**, 674
arene, **7**, 673
aryl, **7**, 674
bisphenylacetylide, **7**, 674
carbonyl, alkyne-bridged, **7**, 664
carbonyl, allyl, **7**, 658
carbonyl, allylimine-bridged, **7**, 636
carbonyl, arsenic-bridged, **7**, 649
carbonyl, benzotriazole-bridged, **7**, 645
carbonyl, benzylisonitrile, **7**, 656
carbonyl, bisoximato-bridged, **7**, 646
carbonyl, bridging carboxylato, **7**, 641–5
carbonyl, bridging carbyne group, **7**, 662
carbonyl, butadiene, **7**, 660
carbonyl, carbene-bridged, **7**, 658, 660
carbonyl, chloro-bridged, **7**, 646
carbonyl, cyclooctatetraenes, **7**, 654
carbonyl, cyclopentadienyl, **7**, 655
carbonyl, diazomethane, **7**, 654
carbonyl, dienediyl ligands, **7**, 655
carbonyl, dienyl ligands, **7**, 655–66
carbonyl, 1,2-diiminobenzene-bridged, **7**, 648
carbonyl, diphenylphosphido-bridged, **7**, 656
carbonyl, diphosphazane-bridged, **7**, 626
carbonyl, dppm-bridged, **7**, 632
carbonyl, η^3-enyl ligands, **7**, 653, 654
carbonyl, ethylidyne, **7**, 660
carbonyl, ethylidyne-bridged cation, **7**, 663
carbonyl, fulvene, **7**, 654
carbonyl, hydrocarbon free, **7**, 625–51
carbonyl, hydrocarbon ligands, **7**, 652–66
carbonyl, methyl-bridged carionic, **7**, 661
carbonyl, methyne-bridged carion, **7**, 661
carbonyl, nitrogen-bridged, **7**, 648
carbonyl, phenylacetylide-bridged, **7**, 650

carbonyl, phosphinato-bridged acetylide-bridged, **7**, 650
carbonyl, phosphorus-bridged, **7**, 649
carbonyl, poly(pyrazolyl)-borato ligands, **7**, 656
carbonyl, pyrazolato-bridged, **7**, 645, 646
carbonyl, pyridine-2-carbaldehydeimine-bridged, **7**, 637
carbonyl, pyridyl-bridged, **7**, 645
carbonyl, selenium-bridged, **7**, 647
carbonyl, sulfur-bridged, **7**, 647
carbonyl, trimethylenemethane-bridged, **7**, 662
carbonyl, vinyl-bridged carion, **7**, 662
carbonyl, σ-yl-type ligands, **7**, 652, 653
cyclopentadienyl, **7**, 666–73
cyclopentadienyl, aza-allyl, **7**, 673
cyclopentadienyl, benzene-bridged, **7**, 672
cyclopentadienyl, d^5–d^5 systems, **7**, 667–73
cyclopentadienyl, d^5–d^6 systems, **7**, 667–73
cyclopentadienyl, d^6–d^6 systems, **7**, 666, 667
cyclopentadienyl, nitrosyl-bridged, **7**, 667
cyclopentadienyl, sulfido-bridged, **7**, 667
cyclopentadienyl, sulfur-bridged, **7**, 669
cyclopentadienyl, tetrahydride-bridged, **7**, 671
cyclopentadienyl, thiolato-bridged, **7**, 670
diazadiene, **7**, 635–41
metal–metal bonds, **7**, 625–74
monoazadiene, **7**, 635–41
triazenido-bridged, **7**, 644

bis(arene), **7**, 550–96
alkenes, **7**, 562
alkyl, **7**, 566, 567
bifunctional ligands, **7**, 565, 566
bridging ligands, **7**, 565, 566
carbon monoxide derivatives, **7**, 561–4
ethene, **7**, 562
hydrides, **7**, 566, 567
isocyanide derivatives, **7**, 561–4
monocyclic, **7**, 550–2
monomeric ligand-stabilized alkyls, **7**, 568–75
monomeric ligand-stabilized hydrides, **7**, 568–75
nitrogen-donor ligands, **7**, 558–60
oxygen-donor ligands, **7**, 558–60
phosphine derivatives, **7**, 561–4
phosphines, **7**, 562
polycyclic, **7**, 550–2
selenium-donor ligands, **7**, 564, 565
sulfur-donor ligands, **7**, 564, 565

bis(arene) dichloro
photolability, **7**, 557

bis(arene) dihalo
half sandwich, **7**, 556–8
heterobinuclear pyrazolato, **7**, 565
heterodinuclear, **7**, 566
reaction with pyrazoles, **7**, 565

carbene cyclopentadienyl, **7**, 485, 486
carbene(alkylidene) cyclopentadienyl, **7**, 520, 521
carbenes
benzannulation, **12**, 489
carbide cyclopentadienyl, **7**, 485, 486
carbomoyls, **7**, 423, 424
carbonyl cyclopentadienyl cations, **7**, 482, 483
carbonyl hydrides, **7**, 362–72
dihydrogen, **7**, 370–2
nitrogen donors, **7**, 368
phosphines, **7**, 363
ternary, **7**, 362, 363
carbonyls, **7**, 300, 301
carboxamides, **7**, 423, 424
catalysts
diazodecomposition, **12**, 430
chalcocarbonyls, **7**, 319–27
cycloaddition, **7**, 326
electrophilic attack, **7**, 325
migratory insertion, **7**, 326
nucleophilic attack, **7**, 324
reactivity, **7**, 324–7
reduction, **7**, 327
synthesis, **7**, 320–4
β-chloroalkyl, **7**, 392
chloromethyl cyclopentadienyl, **7**, 489
cumulenylidene, **7**, 353–6
cyclobutadiene, **7**, 474–6
cyclobutadiene carbonyl
formation and CO displacement from, **7**, 474
cyclobutadienyl
hydride addition, **7**, 475
cyclobutenyl cation
ring-opening, **7**, 476
cyclobutenyl cyclopentadienyl, **7**, 520
cyclodiene tricarbonyl, **7**, 596
cyclodienes, **7**, 593–6
cyclodienyl, **7**, 593–6
cyclodienyl tricarbonyl, **7**, 596
cyclohexadiene, **7**, 556
cyclopentadienyl
antimony-donor ligands, **7**, 490–7
arsenic-donor ligands, **7**, 490–7
σ-bonded carbon, reactions, **7**, 501
cyclometallated, **7**, 501–4
dihydrogen, **7**, 497–501
diphenylacetylene, **7**, 531
halide-bridged dimers, deformation, **7**, 477
nitrogen-donor ligands, **7**, 490–7
oxidation, **7**, 478
oxygen-donor ligands, **7**, 490–7
phosphorus-donor ligands, **7**, 490–7
reaction with androsterone, **7**, 540
reaction with benzoic acid, **7**, 532
reaction with chlorocyclohexane, **7**, 540
reaction with cholesterol, **7**, 540
reaction with cyclohexene, **7**, 539
reaction with 7-dehydrocholesterol, **7**, 540
reaction with dithiahex-3-yne, **7**, 517
reaction with ergosterol, **7**, 540
reaction with hydroquinone, **7**, 532
reaction with methylindoles, **7**, 534
reaction with oestradiol, **7**, 540
reaction with oestrone 3-methyl ether, **7**, 534
reaction with phosphides, **7**, 487
reaction with polystyrene, **7**, 532
reaction with pyrazolylborates, **7**, 496
selenium-donor ligands, **7**, 522–7
silicon bonds, **7**, 527–30
stilbene, **7**, 531
structure, **7**, 477
styrene, **7**, 531
sulfur-donor ligands, **7**, 522–7
cyclopentadienyl anthracenophane, **7**, 536
cyclopentadienyl arenes, **7**, 530–40
cyclopentadienyl benzothiophene, **7**, 538
cyclopentadienyl bisvinyl, **7**, 489
cyclopentadienyl σ-bonded carbon, **7**, 497–504
cyclopentadienyl chrysene, **7**, 534
cyclopentadienyl cyclopropylcarbene, **7**, 480, 485
cyclopentadienyl diaminocarbene, **7**, 485

cyclopentadienyl diazabutadiene, **7**, 495
cyclopentadienyl dibenzo-*p*-quinodimethane, **7**, 536
cyclopentadienyl dibenzothiophene, **7**, 538
cyclopentadienyl dicarbonyl, **7**, 478–88
cyclopentadienyl diene, **7**, 483–5, 504–12
cyclopentadienyl dienyl, **7**, 504–12
cyclopentadienyl diphosphene, **7**, 487
cyclopentadienyl dithiocarbamates, **7**, 523, 524
cyclopentadienyl enyl, **7**, 521
cyclopentadienyl fluorene, **7**, 534
cyclopentadienyl formyls, **7**, 483
cyclopentadienyl halide, **7**, 476–80
  protonation, **7**, 498
cyclopentadienyl halide nitrosyl, **7**, 488
cyclopentadienyl halo
  reduction, **7**, 500
cyclopentadienyl hydride, **7**, 479, 480, 497–504
cyclopentadienyl hydrido
  protonation, **7**, 498, 499
cyclopentadienyl hydroxycarbonyls, **7**, 483
cyclopentadienyl hydroxymethyls, **7**, 483
cyclopentadienyl isocyanides, **7**, 495
cyclopentadienyl isoquinoline, **7**, 537
cyclopentadienyl monoenes, **7**, 483–5
cyclopentadienyl neopentyl, **7**, 502
cyclopentadienyl nitrosyl phosphines, **7**, 489
cyclopentadienyl pentaphenylcyclopentadienyl, **7**, 490
cyclopentadienyl phosphaacyl, **7**, 487
cyclopentadienyl polycyclic arenes, **7**, 534
cyclopentadienyl polyselenium, **7**, 526, 527
cyclopentadienyl polysulfur, **7**, 526, 527
cyclopentadienyl pyridine, **7**, 537
cyclopentadienyl rubrene, **7**, 536
cyclopentadienyl selenolato, **7**, 522
cyclopentadienyl silaethene, **7**, 529
cyclopentadienyl sulfoxides, **7**, 524–6
cyclopentadienyl sulfur dioxide, **7**, 524
cyclopentadienyl thioethers, **7**, 522, 523
cyclopentadienyl thiolates, **7**, 522, 523
cyclopentadienyl thiols, **7**, 522
cyclopentadienyl thiones, **7**, 523
cyclopentadienyl thiophene, **7**, 522, 523, 538
cyclopentadienyl thiosulfinato, **7**, 524–6
cyclopentadienyl vinylidene, **7**, 513–17
cyclopentadienyl vinylvinylidene, **7**, 517, 518
[2^2](1,4)cyclophane, **7**, 557
cyclophanes, **7**, 552–5
cylopentadienyl σ-bonded carbon, **7**, 501–4
p-cymene, **7**, 556, 560
α-diazoalkyls, **7**, 389
dichloro bis(cymene)
  electrochemical reduction, **7**, 557
dichloro bis(methyl *o*-toluate)
  reaction with chiral amines, **7**, 559
12,15-dihydro-4,5,7,8-tetramethyl[2^2](1,4)cyclophane, **7**, 557
dinuclear acetylide cyclopentadienyl, **7**, 521
dithiomethoxycarbonyls, **7**, 422, 423
enoyl, **7**, 420, 421
enynyl, **7**, 407–11, 462–4
formyls, **7**, 417, 418
α-functionalized alkyls, **7**, 387–9
γ-functionalized alkyls, **7**, 392, 393
β-functionalized alkyls, **7**, 392, 393
half-sandwich arenes, **7**, 556–86
half-sandwich cyclopentadienyl, **7**, 476–530
haloalkylidenes, **7**, 346–8
heptahydrodiborates
  metallaboranes, **1**, 304
heteroatom substituted alkylidenes, **7**, 343–6
hexaphenylcyclohexadienyl
  synthesis, **7**, 474
hydrocarbon alkylidenes, **7**, 336–42
hydrocarbon ligands, **7**, 441–68
hydrocarbonyls, **7**, 422
hydrocarbyl, **7**, 372–430
hydroxymethyl, **7**, 387
imido-alkyl, **7**, 375
iminoacyl, **7**, 376
iminoacyls, **7**, 421, 422
indenyl, **7**, 491
  reaction with oxygen, **7**, 516
indium bonds, **1**, 562
isonitriles, **7**, 327–36
keto alkyls, **7**, 389
8-membered rings, **7**, 593–6
7-membered rings, **7**, 593–6
mercury clusters, **10**, 339, 340
methyl 1,4-dihydro-*o*-toluate, **7**, 557
monohalomethyl, **7**, 387
mononuclear
  carbon ligands, **7**, 299–336
mononuclear half-sandwich compounds
  synthesis, **7**, 668
naphthalene, **7**, 591
nickel clusters
  hydrogenation catalysts, **10**, 38
oxapentadienyl, **7**, 465
pentadienyl, **7**, 467
perfluoroalkyl, **7**, 389–91
α-phellandrene, **7**, 556
phosphinomethyl, **7**, 391
β-phosphonio ethyl, **7**, 392
phosphonioalkyls, **7**, 389
β-silyl alkyl, **7**, 392
β-stannyl alkyl, **7**, 392
sulfinyl alkyls, **7**, 389
tetranuclear clusters
  metallaboranes, **1**, 300
tetraphenyloxapentadienyl
  ring-opening, **7**, 474
thallium bonds, **1**, 568
thiaboranes, **1**, 268
thioacyls, **7**, 422
thioalkyl vinyl, **7**, 406
thioformyls, **7**, 422
thiophenes, **7**, 555, 556
triaqua bis(arene)
  deprotonation, **7**, 558
  irradiation, **7**, 558
triaqua bis(benzene)
  cation, **7**, 558
  water exchange, **7**, 558
trimethylenemethane, **7**, 464, 465
trinuclear
  nitrene-capped, **7**, 641
tris(methyl cyanide) benzene
  cations, preparation, **7**, 559
vinyl, **7**, 399–411
vinylidene, **7**, 348–56
ynyl, **7**, 462–4
zinc clusters, **10**, 339
zirconium bonds, **10**, 25
Ruthenium(0) complexes
  alkenes, **7**, 441–5
    without carbonyl ligands, synthesis, **7**, 444
  alkynes, **7**, 441–5
  anionic hydrido dienes, **7**, 455
  arene dihalo phosphines, **7**, 570–5
  arene halo
    nitrogen-donor ligands, **7**, 575
  benzyne, **7**, 445
  carbonyls
    binuclear, **7**, 301
    group 15, 16, 17 ligands, **7**, 302–6
    mononuclear, **7**, 301
  cyclopentadienones, **7**, 457, 458
  dienes, **7**, 452–5
  enones, **7**, 457, 458
  quinodimethanes, **7**, 455, 456
  quinones, **7**, 457, 458
  xylylenes, **7**, 455, 456
Ruthenium(I) complexes
  carbonyls

group 15, 16, 17 ligands, **7**, 306, 307
Ruthenium(II) complexes
acyclic allyl, **7**, 459, 460
alkenes, **7**, 445–9
alkenes hydrazine, **7**, 450
alkenes pyrazole, **7**, 451
alkyl
polyphosphine, **7**, 380
alkyl carbonyl, **7**, 378–87
alkynes, **7**, 445–9
aluminum tetrahydride derivatives, **1**, 550
amino acid alkenes, **7**, 450
ammine alkenes, **7**, 445
aqua alkenes
ring-opening metathesis polymerization, **12**, 1214
aqua dienes, **7**, 452
arene dihalo phosphines, **7**, 570–5
aryl carbonyl, **7**, 396
aryls, **7**, 395–9
carbonyl
bidentate phosphines, **7**, 308
carboxylates, **7**, 316
functionalized phosphines, **7**, 311
group 15, 16, 17 ligands, **7**, 307–19
macrocycles, **7**, 317
multidentates, **7**, 317
nitrogen donors, **7**, 312
oxygen donors, **7**, 314
phosphido, **7**, 311
poly(azolyl)borates, **7**, 314
pyridinethiolates, **7**, 314
Schiff bases, **7**, 316
selenium donors, **7**, 317
sulfonates, **7**, 316
sulfur donors, **7**, 317
tridentate phosphines, **7**, 310
carbonyl halides, **7**, 307
carbonyl phosphines, **7**, 308
carboxylato dienes, **7**, 451
dienes, **7**, 449–52
hydrido dienes, **7**, 451
mercury aryls, **7**, 397
metallaboranes, **1**, 340–2
pyrrolyl, **7**, 397
tetrafluorobarrelene, **7**, 450
trivalent carbonyls, **7**, 319
Ruthenium(IV) complexes
allyl, **7**, 461, 462
carbonyls, **7**, 319
octaethylporhyrinate, **7**, 395
Ruthenium films
from ruthenocene, **7**, 620
Ruthenium osmium
dicarbon metallacarbaboranes
7-vertex, **1**, 382
Ruthenium platinum clusters
selective hydrogenation
catalysts, **10**, 38, 39
Ruthenium rhodium
dicarbon metallacarbaboranes
7-vertex, **1**, 382
Ruthenium tetroxide
reaction with alkenes, **12**, 1157
Ruthenium, (cyclopentadienyl)-(cyclopentadiene)-
electronic absorption spectra, **7**, 606
Ruthenocene
applications, **7**, 620
chemistry, **7**, 605
coordination to palladium, **7**, 614
coordination to platinum, **7**, 614
decamethyl-
electronic absorption spectra, **7**, 606
oxidation, **7**, 609
protonation, **7**, 609
electronic absorption spectra, **7**, 606
free energy of ionization, **7**, 605
gas-phase electron transfer barrier, **7**, 605
IR spectra, **7**, 606
macrocycle bridged, **7**, 613
macrocyclic, **7**, 614
macrocyclic derivatives, **7**, 613, 614
mass spectra, **7**, 606
methyloxy-
structural phase transition, **7**, 607
nitro-
nonlinear optical properties, **7**, 607
NMR, **7**, 606
oligomers
preparation, **7**, 617
oxidation, **7**, 608
pentamethyl-
mercuration, **7**, 611
PES, **7**, 606
polycyclic
synthesis, **7**, 604
properties, **7**, 605
Raman spectra, **7**, 606
reaction with arenes, **7**, 530
recoil reactions, **7**, 619
rotation
barrier, **7**, 605
^{103}Ru-labeled, **7**, 619
structure, **7**, 605
substituents
electronic effects, **7**, 606
substituted
structure, **7**, 605
synthesis, **7**, 604
substitution reactions, **7**, 610, 611
synthesis, **7**, 604
thermodynamic data, **7**, 606
Ruthenocene phosphazenes
ring-opening polymerization, **7**, 611
Ruthenocene, acetyl-
biological properties, **7**, 619
Ruthenocene, alkynyl-
synthesis, **7**, 610
Ruthenocene, vinyl-
polymers, **7**, 617
Ruthenocenes
catalytic activity, **7**, 620
estrogenic activities, **7**, 619
mercuriation, **3**, 141
permercuriation, **3**, 162
[1.1]Ruthenocenophane
oxidation, **7**, 609
[3]Ruthenocenophane
synthesis, **7**, 615
[4]Ruthenocenophane
synthesis, **7**, 615
Ruthenocenophanes
NMR, **7**, 615
[1.1]Ruthenocenophene
dication, **7**, 615

S

Samarium
  dicarbon metallacarbaboranes
    12-vertex, **1**, 400, 418
Samarium complexes
  alkyl dihalides
    cuprates from, **12**, 85
Samarium, bis(pentamethylcyclopentadienyl)hydrido-
  dimer
    polymerization catalyst, **12**, 1204
Samarium, bis(pentamethylcyclopentadienyl)-methyl(tetrahydrofuran)-
  catalyst
    living polymerization, alkenes, **12**, 1205
Samarium(III) complexes
  alkyl
    transmetallation to copper, **12**, 84
  aluminum tetrahydride derivatives, **1**, 551
Samin
  synthesis, **12**, 265
Sanninoidea exitiosa
  sex attractant
    synthesis, **11**, 106
α-Santalene
  synthesis, **11**, 101
α-Santalol
  synthesis, **11**, 101
Sarain A
  synthesis, **12**, 99
Sarkomycin
  synthesis, **12**, 453, 454, 879
Sativene
  synthesis, **12**, 790
Sawada reaction
  organozinc compounds, **11**, 178
Scandium
  dicarbon metallacarbaboranes
    12-vertex, **1**, 400
Scandium atoms
  reactions with arenes, **4**, 8, 9
Scandium complexes, **4**, 11–192
  alkenyl, **4**, 26, 27
  alkynyl, **4**, 26, 27
  allyl, **4**, 27, 28
  arene, **4**, 97
  arsolyl, **4**, 96
  bis(cyclooctatetraenyl), **4**, 104
    anionic, **4**, 102–4
  bis(pentamethylcyclopentadienyl), **4**, 64–8
    hydrides, **4**, 92–4
  bis(pentamethylcyclopentadienyl) halo, **4**, 73–94
  carbonyls, **4**, 13
  cyclooctatetraenyl, **4**, 98–106
  cyclooctatetraenyl halo, **4**, 99–102
  cyclopentadienyl, **4**, 28–58
  cyclopentadienyl dihalo, **4**, 32–5
  cyclopentadienyl halo, **4**, 28
  dicyclopentadienyl, **4**, 28–32
  dicyclopentadienyl allyl, **4**, 45
  dicyclopentadienyl aluminum hydrides, **4**, 49
  dicyclopentadienyl borohydride, **4**, 48
  dicyclopentadienyl carboxylates, **4**, 39
  dicyclopentadienyl halo, **4**, 35–50
  dicyclopentadienyl hydrides, **4**, 47–50
  dicyclopentadienyl hydrocarbyls, **4**, 43–6
  fullerenes, **4**, 106
  heterobimetallic, **4**, 107–12
  heteronuclear NMR, **4**, 113
  homogeneous catalysis, **4**, 113–22
  hydrocarbyls, **4**, 13–26
    anionic homoleptic, **4**, 20, 21
    heteroleptic, **4**, 21–6
    neutral homoleptic, **4**, 13–19
  indenyl, **4**, 63
  pentadienyl, **4**, 95, 96
  pentamethylcyclopentadienyl, **4**, 64–95
  pentamethylcyclopentadienyl dihalo, **4**, 68–73
  pentamethylcyclopentadienyl halo, **4**, 64
  phospholyl, **4**, 96
  ring-bridged cyclopentadienyls, **4**, 59–61
  tricyclopentadienyl, **4**, 51–5
  tricyclopentadienyl halo, **4**, 57, 58
  tris(pentamethylcyclopentadienyl), **4**, 95
  zero oxidation state, **4**, 1–9
    alkenes, **4**, 1, 2
    alkynes, **4**, 1, 2
    arenes, **4**, 2–8
    arenes, matrix isolation studies, **4**, 2
    bis(arene), **4**, 3–8
    bis(arene), bonding, **4**, 5, 6
    bis(arene), magnetism, **4**, 6, 7
    bis(arene), reactions, **4**, 7, 8
    bis(arene), redox reactions, **4**, 7
    bis(arene), ring displacement reactions, **4**, 7
    bis(arene), structure, **4**, 3
    bis(arene), synthesis, **4**, 3
    naphthalene, **4**, 2
Schlenk equilibrium
  displacement
    diorganylmagnesium compounds, **1**, 85
  Grignard reagents
    thermodynamic parameters, **1**, 76
Schlosser's base
  structure, **11**, 96
Scolytus multistriatus
  ephromone
    synthesis, **11**, 101
Scopadulic acid B
  synthesis, **12**, 304, 305
Scopine
  synthesis, **12**, 811
Selenaboranes
  properties, **1**, 270, 271
Selenadistanniranes
  preparation, **2**, 281
1,2,5-Selenasilaborolenes
  metal complexes, **1**, 213
Selenides
  alkylation, **11**, 552
  complexation, **11**, 552
  halogenation, **11**, 552
  heterosubstituted
    synthesis, **11**, 541
  oxidation, **11**, 553, 554
    to selenoxides, **11**, 554
  reactions
    allyl complexes, **12**, 826, 827
  synthesis, **11**, 527, 538
Selenides, alkenyl
  coupling with Grignard reagents, **12**, 171
Selenium
  electronegativity, **11**, 518
  introduction into organic compounds, **11**, 522
  metallic
    reactions, **11**, 532
  in Periodic Table, **11**, 518
  physical properties, **11**, 516–22
  radionucleotides, **11**, 517, 518
  removal from organic compounds, **11**, 522
Selenium compounds
  silyl, **2**, 36, 37
  structure, **11**, 519, 520
  synthesis
    organomercury compounds, **11**, 400
Selenium dioxide
  hydrogen peroxide mixtures
    reactions, **11**, 537
  reactions, **11**, 536
Selenium halides
  synthesis, **11**, 528
Selenium tetrafluoride
  reactions, **11**, 537
Selenoacetals
  oxidation

organoselenium compounds, **11**, 535
synthesis, **11**, 540
Selenoalkyl compounds
structure, **11**, 520–2
Selenoalkyl lithium
synthesis, **11**, 543
Selenoalkyl metals
structure, **11**, 521
Selenoalkyl radicals, **11**, 521
Selenocarbenium ions, **11**, 521
Selenocarbonyl compounds
synthesis, **11**, 539, 541, 545
Selenocarboxylic acids
synthesis, **11**, 539
Selenocysteine
occurrence, **11**, 516
Selenoketones
α-phenyl-
synthesis, **11**, 337
Selenol esters
synthesis, **11**, 530, 545
Selenolcarbonates
synthesis, **11**, 545
Selenones
reactivity, **11**, 557
synthesis from selenides, **11**, 554
Selenonium salts
reactivity, **11**, 552
synthesis, **11**, 552
Selenophene
reaction with triosmium clusters, **7**, 811
ring opening
reaction with Grignard reagents, **12**, 171
Selenophenes
rhenium complexes, **6**, 219
Selenopseudo ureido compounds
synthesis, **11**, 530
Selenoxides
elimination reactions, **11**, 555
reactions, **11**, 533
reactivity, **11**, 554
synthesis from selenides, **11**, 554
Semiconducting amorphous thin layers
organogermanium compounds, **2**, 209
Semiconductors
nanocrystallites
synthesis, **3**, 203
organocadmium compounds, **3**, 200
SEN 215
synthesis, **12**, 249
Sendaverine
synthesis, **12**, 249, 250
Septicine
synthesis, **12**, 191
Serricornin
synthesis, **12**, 802, 883
Serrulactic acid, dihydroxy-
synthesis, **12**, 1041, 1051
Sesamin
synthesis, **11**, 473
Sesbanine
synthesis
lithiation, **11**, 79
Sharpless asymmetric dihydroxylation
alkenes, **12**, 1138
Sharpless dihydroxylation
asymmetric
allylsilanes, **11**, 320
Sharpless epoxidation
allylic alcohols, **12**, 1109–20
molecular sieves, **12**, 1111
reaction variables, **12**, 1110, 1111
substrates, **12**, 1112–18
Sharpless kinetic resolution, **11**, 326
Shell Higher Olefin Process
nickel catalysts, **9**, 52
nickel ylide catalysts, **9**, 36
Shikimic acid
methyl ester
synthesis, **12**, 657
precursor
synthesis, **12**, 824
Side chain liquid crystal polymers
properties, **2**, 131
Sila-drugs
synthesis, **2**, 70
1-Silaallene
synthesis, **2**, 18
Silaaziridine
synthesis, **2**, 67
Silabenzenes
Dewar form, **2**, 51
production, **2**, 56
synthesis, **2**, 61
Silabicyclobutanes
isomerism, **2**, 64
Silaboranes
properties, **1**, 259
Silacarboranes
properties, **1**, 259
Silacarborundanes
synthesis, **2**, 67
Silacycloallenes
preparation, **2**, 64
Silacyclobutadienes
synthesis, **2**, 51
Silacyclobutanes
insertion reactions, **2**, 49
physical properties, **2**, 49
properties, **2**, 49, 50
pyrolysis, **2**, 49
ring-opening, **2**, 49
synthesis, **2**, 49
Silacyclobut-2-ene
platinum complex, **2**, 50
Silacyclobutenes
synthesis, **2**, 48, 50
Silacycloheptatrienes
synthesis, **2**, 63
1-Sila-2,5-cyclohexadiene
pyrolysis, **2**, 61
Silacyclohexadienes
synthesis, **2**, 60
Silacyclohexanes
synthesis, **2**, 60
Silacyclohexenes
synthesis, **2**, 60
Silacyclooctadienes
synthesis, **2**, 70
1-Silacyclopenta-2,4-diene
synthesis, **2**, 57
Silacyclopentadienes, **11**, 345
properties, **2**, 57–9
reactions, **2**, 58, 59
synthesis, **2**, 48
transition metal complexes, **2**, 59
Silacyclopentanes
properties, **2**, 55, 56
1-Silacyclopent-2-enes
formation, **2**, 56
1-Silacyclopent-3-enes
formation, **2**, 56
Silacyclopent-3-enes
metathesis, **2**, 56
Silacyclopentenes
reactions, **2**, 56, 57
synthesis, **2**, 56
Silacyclopropane, vinyl-
preparation, **2**, 47
Silacyclopropanes, **2**, 46, 47
Silacyclopropanimine
rearrangement, **2**, 67
Silacyclopropenes, **2**, 47, 48
Siladicarbadodecaboranes
preparation, **1**, 259
Silafulvenes
zirconocene complexes, **2**, 71
Silafuran, methylene-
synthesis, **2**, 62
2-Silaindan, 2,2-dichloro-
synthesis, **2**, 56
Silaindene
synthesis, **2**, 18
Silane
polymerization
titanium hydride catalysts, **4**, 264, 265
Silane, acyl-
reaction with silyl cyanides, **2**, 15
Silane, allyl-
addition to tricarbonyl chromium carbonium ions, **5**, 522
synthesis, **2**, 8, 9; **12**, 962
Silane, amino-
synthesis, **2**, 25
Silane, aryl-
synthesis, **2**, 7, 8
Silane, chloro-
hydrolysis, **2**, 113, 114
synthesis, **2**, 38
Silane, chlorotrimethyl-
Michael additions
cuprates, **12**, 72
Silane, dichloro-
condensation with alkali metals, **2**, 96–8
Silane, dimethyldichloro-
synthesis, **2**, 112
Silane, ethynyltrimethyl-
trimerization, **2**, 10
Silane, fluoro-
synthesis, **2**, 38
Silane, 1-hydroxyalkylvinyl-
hydroformylation, **12**, 30
Silane, 3-hydroxy-2-formylalkyl-
synthesis, **12**, 30
Silane, iodo-
synthesis, **2**, 38
Silane, 1-naphthylphenylmethyl-
diazodecomposition, **12**, 464
Silane, phenyl-
dehydrocoupling
vanadocene, **5**, 45
Silane, propargyl-
1,3-dienes from, **12**, 43

Silane, tetra-1-alkynyl-
organoboration, **2**, 65, 66
Silane, tetrahalo-
spectroscopy, **2**, 3
Silane, tetramethoxy-
synthesis, **2**, 3
Silane, thio-
synthesis, **2**, 34, 35
Silane, trimethylbromo-
synthesis, **2**, 38
Silane, trimethylhalo-
in oroganocopper reactions, **12**, 110–14
Silane, trimethylvinyl-
hydroesterification, **12**, 34
Silane, tris(trimethylsilyl)-
synthesis, **2**, 21
Silane, vinyl-
insertion reactions
dienes, **12**, 149
transmetallation, **12**, 142
synthesis, **2**, 5, 6
Silanecarbonitrile oxide, trimethyl-
synthesis, **2**, 17
Silanes
allyl
in organic synthesis, cuprates from, **12**, 94
carbacyclic, **2**, 45–71
carbacyclic acetylenic, **2**, 63
dehydrogenative coupling, **2**, 99, 100
dienyl
synthesis, **12**, 167
organofunctional
synthesis by hydrosilylation, **2**, 116, 117
reducing agents, **11**, 343–5
vinyl
in organic synthesis, cuprates from, **12**, 94
Silanes allylic
synthesis, **12**, 167
Silanes, acyl-
in organic synthesis, **11**, 328, 329
synthesis, **12**, 219
Silanes, 2-alkenyltrimethyl-
metallation, **11**, 117
Silanes, alkyl-
transmetallation
palladium assisted, **12**, 152
Silanes, alkynyl-, **11**, 324, 325
Silanes, (allenylmethyl)-
reactions, **11**, 318
Silanes, allyl-
coupling reactions, **12**, 190
electrophilic substitution reactions, **11**, 317
osmylation, **11**, 320
oxidation, **11**, 320
preparation, **11**, 317–23; **12**, 170
reaction with propargyl dicobalt complexes, **12**, 692, 693
reactions, **11**, 317–23
reactions with oxocarbenium ions, **11**, 319
synthesis, **11**, 317; **12**, 169
Silanes, allyltrimethyl-
metallation, **11**, 117
Silanes, amino-
in organic synthesis, **11**, 342, 343
Silanes, aryl-
in organic synthesis, **11**, 326
transmetallation
palladium assisted, **12**, 152
Silanes, chlorotrimethyl-
for zinc activation, **11**, 160
Silanes, crotyl
chiral
reactions, **11**, 321
Silanes, dienyl-
synthesis, **12**, 173
Silanes, difluoro-
coupling reactions
with vinyl triflates, **12**, 207
Silanes, α,β-epoxy-
in organic synthesis, **11**, 326, 327
Silanes, heteroaryl-
in organic synthesis, **11**, 326
Silanes, keto-
in organic synthesis, **11**, 328, 329
Silanes, β-keto-, **11**, 329
Silanes, phenyltrimethyl-
metallation, **11**, 118
Silanes, propargyl-, **11**, 324, 325
allene precursors, **11**, 325
Silanes, trifluoro-
coupling reactions
with vinyl triflates, **12**, 207
Silanes, vinyl-
carbonyl–ene reactions, **11**, 316
electrophilic substitution reactions, **11**, 315
formation, **12**, 595
hydroalumination, **1**, 471
oxidative cleavage, **11**, 317
preparation, **11**, 314, 315; **12**, 594
reactions, **11**, 315–17
Silanesalkynyl-
optically active
synthesis, **12**, 169
Silanols
synthesis, **2**, 31, 32
Silanols, vinyl-
asymmetric epoxidation, **12**, 1117
Silanones
synthesis, **2**, 33
Silaoxetene
synthesis, **2**, 55
Silapropellanes
synthesis, **2**, 65
Silapyran
synthesis, **2**, 62
Silascaphanes
synthesis, **2**, 67
Silaspiroalkanes
pyrolysis, **2**, 65
3-Sila-6-stannacyclohexa-1,4-diene
synthesis, **2**, 70, 71
3-Silathietane, 3,3-dimethyl-
synthesis, **2**, 70
Silatriazoline
synthesis, **2**, 67, 68
Silatrimethylenemethane complexes
synthesis, **2**, 18
Silazanes
transmetallation
palladium assisted, **12**, 152
Silaziridines
synthesis, **2**, 17
Silene, diethynyl-
synthesis, **2**, 50
Silene, dimethyl (di-*t*-butylmethylsilyl)(trimethylsilyl)-
bond length, **2**, 18
Silenes
synthesis, **2**, 17
Silepins — *see* Silacycloheptatrienes
Silicenium cations
stability, **2**, 58
Silicon carbide
precursors
polysilanes, **2**, 107
Silicon compounds
allylic alkylation, **12**, 862
transmetallation, **12**, 200–21
Silicon hydrides
reactions
allyl complexes, **12**, 801
Silicon tethering
in organic synthesis, **11**, 330–2
Silicon–aluminum compounds
synthesis, **2**, 22, 23
Siliconates, tetrafluoro-
synthesis, **2**, 38
Silicon–boron compounds
synthesis, **2**, 22, 23
Silicon–carbon bonds, **2**, 3–20
double, **2**, 17, 18
Silicone polymers
cost, **2**, 3
Silicones
chemistry, **2**, 111–32
in detergent industry, **2**, 125, 126
elastomers
liquid, **2**, 127, 128
in petroleum industry, **2**, 126
Silicon–hydrogen componds
synthesis, **2**, 20, 21
Silicon–metal compounds
synthesis, **2**, 22, 23
Silicon–nitrogen bonds
multiple, **2**, 26, 27
Silicon–nitrogen compounds
synthesis, **2**, 25–7
Silicon–phosphorus bonds
double
synthesis, **2**, 29
reaction with organic acid chlorides, **2**, 28
Silicon–pnictide compounds
synthesis, **2**, 27–31
Silicon–sulfur bonds
double, **2**, 36
Silipramine
synthesis, **2**, 70
Siliranes — *see* Silacyclopropanes
Siliranes, methylene-
reactions
palladium catalyzed, **12**, 957
Silirenes — *see* Silacyclopropenes

Siloles — *see* Silacyclopentadienes
Siloxanes
  linear
    spectroscopy, **2**, 3
  monomer feedstocks
    polymerization, **2**, 119–25
    synthesis, **2**, 112–18
    synthesis, direct process, **2**, 112, 113
    synthesis, direct process, promoters, **2**, 113
    synthesis, mechanism, **2**, 112
  polymers
    organofunctional, industrial applications, **2**, 129–32
    organofunctional, synthesis, **2**, 128, 129
  silanol terminated linear
    condensation polymerization, **2**, 124, 125
  transmetallation
    palladium assisted, **12**, 152
Siloxides
  synthesis, **2**, 34
Siloxyl ligands
  bridging
    in heteronuclear metal–metal bond synthesis, **10**, 17
Silthianes
  linear
    synthesis, **2**, 35
Silver borofluoride
  transmetallation, **3**, 72
Silver clusters
  transition metals, **10**, 255–317
  trialkylphosphines
    edge-bridging, **10**, 261
    theoretical analysis of bonding, **10**, 261
Silver complexes
  ruthenium clusters, **1**, 304
Silver nitrate
  reaction with tetraalkyllead, **3**, 71
  transmetallation, **3**, 74
Silver(I) reagents
  reactions
    with platinum(II) haloaryl complexes, **9**, 473
Silver salts
  cocatalyst
    transmetallation, **12**, 201
Silver tetrafluoroborate
  heteronuclear metal–metal bond syntheis, **10**, 2
Silver, perfluoroisopropyl-
  synthesis, **3**, 72
Silver, phenyl-
  structure, **3**, 120
Silver, styrenyl-
  synthesis, **3**, 72
Silyl cobaltates
  homologation reactions, **12**, 354
Silyl cyanides
  synthesis, **2**, 15–17
Silyl enol ethers
  conjugate addition, **11**, 335
  cycloaddition reactions, **11**, 335, 336
  heterosubstituted
    synthesis, **11**, 337
  Lewis acid induced reactions, **11**, 333
  nucleophilicty, **11**, 337
  oxidative cyclization, **11**, 338
Silyl ethers
  hydrogenolysis
    Pauson–Khand reaction, **12**, 717
Silylation
  catalysts, **10**, 49, 50
Silylcarbocyclization
  catalysts, **10**, 50
Silylene, bis(pentamethylcyclopentadienyl)-
  synthesis, **2**, 19
Silylenes
  in direct synthesis of methylchlorosilanes, **2**, 113
  polymerization, **2**, 98, 99
  synthesis, **2**, 19, 20
Silylenol ethers
  alkylation
    dicobalt propargylium complexes, **12**, 691
  allylic alkylation, **12**, 866
  synthesis, **12**, 585
Silylformylation
  catalysts, **10**, 50
Silylium ions
  preparation, **2**, 4, 5
Silylketene acetals
  aldol condensation
    dicobalt propargylium complexes, **12**, 692
  reactions with dienyliron tricarbonyl complexes, **12**, 663
Simmons–Smith reaction
  organozinc compounds, **3**, 194, 195; **11**, 159, 178
Sinapyl alcohol
  synthesis, **11**, 108
Sirenin
  synthesis, **12**, 871
Six-membered rings
  synthesis
    intramolecular allylic alkylation, **12**, 880
SM 9064
  synthesis, **12**, 631
Sodiocuprates
  in organic synthesis, **12**, 114
Sodium bis(methoxycarbonyl)-methanide
  reaction with dienetricarbonylcobalt complexes, **12**, 652
Sodium *t*-butoxide
  structure, **1**, 11, 12
  superbase reagent, **11**, 94
Sodium carbonylplatinates
  $Na_2[Pt_{30}(CO)_{60}]$
    synthesis, **9**, 397
Sodium cobalt tetracarbonyl
  preparation, **12**, 354
Sodium magnesium compounds
  alkynyl
    preparation, **1**, 105
  phenyl
    preparation, **1**, 105
Sodium oxychloride
  epoxidation
    salen manganese complexes, **12**, 1123, 1124
Sodium selenocyanates
  synthesis, **11**, 528
Sodium vanadium hexacarbonyl
  preparation, **5**, 2
Sodium, allyl-
  reactions, **11**, 97–103
  rotational barrier, **11**, 98
Sodium, butyl-
  structure, **11**, 95
Sodium, indenyl-
  structure, **1**, 24
Sodium, methyl-
  structure, **1**, 20
Sodium, pentamethylcyclopentadienyl-
  pyridine solvate
    structure, **1**, 22
Solenopsin-a
  synthesis, **11**, 144
Solvato complexes
  without carbonyl ligands, **5**, 283–6
Solvent effect, **12**, 984
Solvomercuriation, **3**, 144–52
  demercuration, **11**, 411–30
Sonication
  Pauson–Khand reaction, **12**, 729
Sonogashira couplings
  carbonylative, **12**, 224
Sonogashira protocol, **12**, 222
Sorbate, γ,δ-epoxy-
  amidation, **12**, 811
Sparteine
  synthesis, **12**, 886
Spatol
  synthesis, **12**, 722
Spermine alkaloids
  synthesis
    organoboron compounds, **11**, 245
Spherosiloxanes
  synthesis, **2**, 32, 33
Sphondin
  synthesis, **12**, 492
Spiroacetals
  synthesis, **12**, 584
Spirocyclizations
  multiple insertion cascades, **12**, 300–2
Spiroenone
  addition reactions, **12**, 88
Spiroketals
  synthesis, **12**, 357
Spiro[4.4]nona-1,3-diene
  metallation, **11**, 112
Spiropentane
  metallation, **11**, 114
Spiro[2.2]pentane
  metallation, **11**, 114
Squalene
  asymmetric dihydroxylation, **12**, 1148, 1149
Stannaboranes
  properties, **1**, 259, 260
Stannacarboranes
  properties, **1**, 260
Stannacycloalkanes

structures, **2**, 250–4
synthesis, **2**, 225
Stannacyclobutane
preparation, **2**, 251
reaction with oxygen, **2**, 251
Stannacyclohexanes
preparation, **2**, 251
Stannacyclopentanes
preparation, **2**, 251
Stannacyclopropene
preparation, **2**, 250
Stannaethenes
structure, **2**, 250
Stannaindacenes
preparation, **2**, 254
Stannane, [2-(acetoxymethyl)-3-allyl]-tri-*n*-butyl-
cycloaddition reactions, **12**, 930
Stannane, alkyl-
properties, **2**, 234–54
Stannane, alkynyl-
alkynyl-, **2**, 242
Stannane, allyl-
allyl-, **2**, 243–9
Stannane, aryl-
aryl-, **2**, 236–8
Stannane, methylphenyl-
preparation, **2**, 225
Stannane, pentamethyl-
preparation, **2**, 225
Stannane, vinyl-
formation, **2**, 238–41
preparation, **2**, 225
Stannanes
allylic
transmetallations, cuprates, **12**, 86
carbonylative coupling with enol triflates, **12**, 206
coupling reactions, **12**, 200
functionalized
in organic synthesis, **11**, 375–81
Stannanes, acyl-
allylic alkylation, **12**, 861
in organic synthesis, **11**, 377, 378
Stannanes, alkenyl-
allylic alkylation, **12**, 858
in organic synthesis, **11**, 369–72
preparation, **11**, 369
reactions, **11**, 370–2
transmetallation, **11**, 370, 371
Stannanes, (1-alkenyl)tributyl-
alkene cyclization, **12**, 915
Stannanes, α-alkoxy-
preparation, **12**, 540
Stannanes, alkyl-
functionalized at remote positions, **11**, 380
Stannanes, alkynyl-
cross-coupling
palladium-catalyzed, **11**, 375
Stannanes, allenyl-
synthesis, **11**, 374
Stannanes, allyl
allylic alkylation, **12**, 860
Stannanes, allyl-
coupling with halides, **11**, 373
in organic synthesis, **11**, 372–4
preparation, **11**, 372, 373
reactions, **11**, 373, 374
reactions with electrophiles, **11**, 373, 374
transmetallation, **11**, 373
Stannanes, α-aminoalkyl-
preparation, **11**, 379
Stannanes, aryl-
allylic alkylation, **12**, 858
in organic synthesis, **11**, 369–72
reaction with alkenes, **12**, 151
reactions, **11**, 370–2
transmetallation, **11**, 370, 371
Stannanes, 2-azaallyl-
preparation, **11**, 379
Stannanes, α-diazo-
in organic synthesis, **11**, 379
Stannanes, trialkylchloro-
reaction with organozinc compounds, **11**, 165
Stannanes, tributyl-
in organic synthesis, **11**, 358
Stannanes, triphenyl-
in organic synthesis, **11**, 358
Stannanes, vinyl-
conjugate addition to α,β-unsaturated enals, **12**, 965
lithium–tin exchange, **12**, 60
synthesis
hydrozirconation, **12**, 41
organoboron compounds, **11**, 243
Stannanimine
preparation, **2**, 291
Stannanone, dialkyl-
preparation, **2**, 264
Stannethene
formation, **2**, 291
Stannocene, decaphenyl-
structure, **2**, 289
Stannocenes
structure, **2**, 288
α-Stannyl alcohols
in organic synthesis, **11**, 375, 376
β-Stannyl alcohols
preparation, **11**, 376
γ-Stannyl alcohols
preparation, **11**, 377
Stannyl anion, trimethyl-
nucleophilicity, **2**, 228
Stannyl anions
addition to multiple bonds, **11**, 363, 364
in organic synthesis, **11**, 363, 364
reactions, **11**, 363, 364
substitution, **11**, 363
α-Stannyl carbanions
in organic synthesis, **11**, 365, 366
Stannyl cuprates
in organic synthesis, **2**, 228
reaction with alkenes and alkynes, **2**, 231
structures, **2**, 284
Stannyl enol ethers
alkylation, **12**, 866
α-Stannyl ketones
in organic synthesis, **11**, 378
β-Stannyl ketones
preparation, **11**, 378
γ-Stannyl ketones
in organic synthesis, **11**, 379
Stannyl lithium, tributyl-
in organic synthesis, **11**, 363
Stannyl lithium, trimethyl-
in organic synthesis, **11**, 363
Stannyl radicals
EPR, **2**, 278, 289
properties, **2**, 289, 290
Stannylamines
metathesis
onto arylpalladium compounds, **12**, 229
Stannylene
reactions with boranediyl-borirane, **2**, 287
Stannylene complexes
iron, **7**, 283
Stannylene, bis[2,4,6-tris(trifluoromethyl-phenyl)]-
reactions, **2**, 291
Stannylene, dimethyl-
structure, **2**, 288
transfer agent for, **2**, 281
Stannylmetallic compounds
reaction with organic halides, **2**, 227
Star clusters
synthesis, **10**, 8
Star-of-David disorder
substituted triruthenium dodecacarbonyls, **7**, 711
Stemodane
synthesis, **12**, 762
Stemodin
synthesis
cobalt-mediation propargylation, **12**, 696
Step-growth polymerization
transition-metal promoted, **12**, 1226–9
Stereoblock
definition, **12**, 1196
Stereocontrolled bond formation
iron tricarbonyl, **12**, 629–32
Sterepolide
synthesis, **12**, 50
Steroid receptor binding
dicobalt propargylium complexes, **12**, 700
Steroids
labeling
propargyl dicobalt complexes, **12**, 690
side chain-modified
synthesis, **12**, 842
synthesis, **12**, 186
tricarbonyl chromium complexes, **5**, 486
Sterpurene
synthesis, **12**, 473
Stibaarenes
synthesis, **2**, 323, 324
Stibaboranes
properties, **1**, 267
Stibide complexes
triiron clusters, **7**, 275
Stibine, alkyldiphenyl-
lithio derivatives
alkenation, **11**, 491
Stibine, diphenyl-

carbonyl compound reduction, **11**, 496
Stibine, trialkyl-
alcohol preparation, **11**, 492
alkenation
carbonyl compounds, **11**, 489
Stibine, tributyl-
alkyl aryl sulfone synthesis, **11**, 497
Stibine, triphenyl-
dechlorination, **11**, 501
Stibines, silyl-
synthesis, **2**, 30, 31
Stiboles
synthesis, **2**, 324
Stibonic acids, arene
synthesis, **2**, 344
Stibonium bromide, allenyltributyl-
synthesis, **11**, 494
Stibonium bromide, benzyltrialkyl-
homobenzylic alcohol synthesis, **11**, 493
Stigmasterol
osmylation, **12**, 1166
Stilbene
hydrogenation
catalysis, **10**, 37
Stilbenes
reaction with osmium tetroxide complexes, **12**, 1152
synthesis
carbonyl coupling reactions, niobium complexes, **12**, 1088
Heck reaction, **12**, 261
reductive coupling of carbonyl compounds, **12**, 1074
Stille coupling
allyltin compounds, **2**, 245
chromium arene complexes, **12**, 490
palladium catalyzed, **5**, 410
Stille reaction
organotin compounds, **2**, 235
zinc chloride, **12**, 201
Streptonigrin
synthesis, **12**, 215
lithiation, **11**, 79
organoboron compounds, **11**, 232
Strontium complexes
anthracene, **1**, 117
arene radical anion complexes, **1**, 117
chemistry, **1**, 57–120
Strontium, bis(pentamethylcyclopentadienyl)-, **1**, 112
Strontium, bis(triphenylmethyl)-
formation, **1**, 111
Strontium, dialkynyl-
preparation, **1**, 111
Strontium, dicyclopentadienyl-, **1**, 112–16
Strontium, dimethyl-
structure, **1**, 110
Structural parameters
alkyne complexes
iron clusters, **7**, 260, 261
iron alkylidyne clusters, **7**, 272, 273
iron carbide clusters, **7**, 269, 270
Styrene
asymmetric epoxidation, **12**, 1127
asymmetric hydroformylation, **12**, 31
cyclopropanation
metal carbenes, **12**, 415
deuteration
Ziegler–Natta catalysts, **12**, 1205
hydroboration, **11**, 248
hydroformylation
catalysts, **10**, 44
hydrogenation
palladium catalysts, **9**, 294
Pauson–Khand reaction, **12**, 714
synthesis
Heck reaction, **12**, 261
reductive dimerization of carbonyl compounds, **12**, 1090
transition metal complexes
nucleophilic addition, **12**, 1012
Styrene oxide
synthesis
via asymmetric catalysis, **12**, 1099
Styrene, bromo-
synthesis, **12**, 210
Styrene, 4-isobutyl-
hydrocarboxylation
asymmetric, **12**, 34, 35
Styrene, *p*-isobutyl-
hydroformylation, **12**, 31
Styrene, *cis*-β-methyl-
asymmetric epoxidation, **12**, 1121
Styrene, nitro-
reaction with zincates, **11**, 182
Styrene, pentafluoro-
hydroesterification, **12**, 35
hydroformylation/amidocarbonylation, **12**, 30
Styrene, 2,4,6-trimethyl-
hydroesterification, **12**, 33
Subergorgic acid
synthesis, **12**, 111, 112
Succinates, arylmethyl-
synthesis, **12**, 20
Succinates, methyl-
synthesis, **12**, 19
Sulfates
cyclic
synthesis, **12**, 1169
Sulfides
asymmetric allyl displacements
palladium complexes, **12**, 883
oxidation
metal catalyzed, **12**, 1118
reactions
allyl complexes, **12**, 826, 827
Sulfides, alkenyl phenyl
coupling reactions with Grignard reagents, **12**, 171
Sulfilimines
photochemical reactions with chromium alkoxycarbene complexes, **12**, 572
reactions
with carbene complexes, **12**, 539
Sulfinates
asymmetric allyl displacements
palladium complexes, **12**, 884
cleavage, **12**, 800
reactions
allyl complexes, **12**, 827–9
Sulfinatomercuriation
alkenes, **3**, 145
Sulfites
cyclic
synthesis, **12**, 1171
vinyl cyclic
synthesis, **12**, 371
Sulfolane
silylium ions in, **2**, 4
Sulfonamides
allylic amidation, **12**, 810
N-propargylation, **12**, 690
Sulfonates, polyfluoroalkyl
coupling reactions, **12**, 214
Sulfones
alkyl aryl
synthesis, **11**, 497
allylic
alkylation, stereochemistry, **12**, 842
dilithiation, **1**, 17
β-hydroxy-
reductive cleavage, **12**, 800
β-keto
desulfonylation, tellurium reagents, **11**, 581
preparation
organoboranes, **11**, 222
synthesis
organomercury compounds, **11**, 400
α,β-unsaturated
cycloaddition reactions, **12**, 928
Sulfones, alkenyl-
cross-coupling with Grignard reagents, **12**, 171
Sulfones, *t*-butyl
lithiation, **11**, 53
Sulfones, diaryl
lithiation, **11**, 53
Sulfones, α-diazo-β-keto-
synthesis, **12**, 453, 454
Sulfonium ylides
reactions, **12**, 434–6
[2,3]-sigmatropic rearrangements, **12**, 435, 436
stable
formation, **12**, 433, 434
[1,2]-Stevens rearrangements, **12**, 436
Sulfonium ylides, benzyl-
formation, **12**, 434
Sulfonium ylides, dimethyl-
photochemical reactions with chromium alkoxycarbene complexes, **12**, 572

Sulfonium ylides, phenyl-
formation, **12**, 433
Sulfonylation
diols
synthesis, **12**, 1168
Sulfonylpalladation
1,3-dienes, **9**, 331, 336
Sulfoxides, *t*-butyl
lithiation, **11**, 53
Sulfoximines
lithiation, **11**, 41
Sulfoxonium ylides
stable
formation, **12**, 433, 434
Sulfur compounds
synthesis
organomercury compounds, **11**, 400
Sulfur dioxide
insertion reactions
rhodium complexes, **8**, 203
Sulfur ylides
metal stabilized, **12**, 433–7
palladium(II) complexes, **9**, 229
photochemical reactions with chromium alkoxycarbene complexes, **12**, 572
Sultams, *N*-enoyl-
cyclopropanation
palladium acetate catalysis, **12**, 397
Sunscreen preparations
organofunctional siloxanes, **2**, 130
Superbase reagents
organosodium and organopotassium compounds, **11**, 94
Superbases
alkali metal organometallics, **1**, 13
structure, **11**, 95, 96
unimetal, **11**, 96
Supinidine
synthesis, **12**, 436
Suprofen
synthesis, **12**, 33
Suzuki coupling
binaphthyl synthesis, **12**, 81
Suzuki reaction
cross-coupling reactions, **12**, 179
mechanism, **12**, 180
Sydnones
synthesis
transmetallation, **12**, 135
Syndiotactic
definition, **12**, 1196
Syndiotactic polymers
Ziegler–Natta catalysts, **12**, 1199
Synergism
catalysis
homogeneous, **10**, 34

T

Taberpsyshine
  synthesis, **12**, 582
Tacticity
  polyalkenes, **12**, 1196
Talaromycin B
  synthesis, **12**, 583
Tamoxifen
  analogues
    synthesis, **12**, 660
  synthesis, **12**, 204
Tantacyclobutadienes, **5**, 106
Tantacyclobutanes, **5**, 103–5
Tantacyclobutenes, **5**, 105, 106
Tantacyclopentadienes, **5**, 108
Tantacyclopentanes, **5**, 106, 107
Tantacyclopentatrienes, **5**, 109
Tantacyclopentenes, **5**, 107
Tantalacycles
  in ring-opening metathesis polymerization, **12**, 1213
Tantalacyclobutanes
  in ring-opening metathesis polymerization, **12**, 1213
Tantalum
  dicarbon metallacarbaboranes
    12-vertex, **1**, 400
Tantalum complexes
  η^3-, **5**, 110
  η^4-, **5**, 110–13
  η^5-, **5**, 114–28
  η^6-, **5**, 129–32
  η^7-, **5**, 132
  acyl, **5**, 97–102
  aldehydes, **5**, 102
  alkenes, **5**, 89–92
  alkoxide hydrides, **5**, 75
  alkyl, **5**, 67–73
  alkylidene
    preparation, **12**, 595
  alkylidenes, **5**, 83–9
    IR spectra, **5**, 87
    NMR, **5**, 87
    reactivity, **5**, 88, 89
    spectroscopy, **5**, 87
    structure, **5**, 87
    synthesis, **5**, 83–6
  alkylidynes, **5**, 83–9
    reactivity, **5**, 88, 89
    spectroscopy, **5**, 87
    structure, **5**, 87
    synthesis, **5**, 83–6
  alkynes, **5**, 92–7
    reaction with hydrazones, **12**, 344, 346
    reactions with unsaturated compounds, **12**, 335
  arenes, **5**, 102, 103
  aryl, **5**, 67–73
  benzyne, **5**, 92–7
  binary carbonyls, **5**, 60
  bis(cyclopentadienyl), **5**, 121
  bis(cyclopentadienyl) alkene, **5**, 91, 92
  bis(cyclopentadienyl) alkyne, **5**, 96, 97
  bis(cyclopentadienyl) halides, **5**, 126
  boron hydrides, **5**, 121
  butadiene, **5**, 112, 113
  carbocyclic compounds, **5**, 110–33
  carbon dioxide, **5**, 101
  carbon disulfide, **5**, 101
  carbonyls, **5**, 59–65
  in catalysis, **5**, 133–44
  chemistry, **5**, 57–144
  cluster compounds, **5**, 120
  cyclobutadiene, **5**, 113
  cycloheptatrienyl, **5**, 133–44
  cyclometallated trimethylphosphine, **5**, 102
  cyclopentadienyl alkyl, **5**, 70
    reactivity, **5**, 70–3
  cyclopentadienyl alkyne, **5**, 93–7
  cyclopentadienyl aryl, **5**, 70
    reactivity, **5**, 70–3
  cyclopentadienyl carbonyl, **5**, 62–5
  cyclopentadienyl carbonyl hydrides, **5**, 80
  cyclopentadienyl halide hydrides, **5**, 80
  cyclopentadienyl hydrides, **5**, 76
  ene compounds, **5**, 110–33
  enyl ligands, **5**, 110–33
  formyl, **5**, 98
  heterobimetallic, **5**, 120; **10**, 28
  heterobimetallic hydrides, **5**, 81
  hexahydrodiborates, **1**, 306
  hydrides, **5**, 73–81
  imine, **5**, 100, 101
  iminoacyl, **5**, 97–102
  isocyanates, **5**, 101
  isocyanides, **5**, 65, 66
  ketenes, **5**, 101
  ketenimines, **5**, 101
  ketones, **5**, 102
  metal–metal bonds, **5**, 128
  monokis(cyclopentadienyl), **5**, 114–21
  monokis(cyclopentadienyl) alkenes, **5**, 89
  monokis(cyclopentadienyl) alkyne, **5**, 96
  monokis(cyclopentadienyl) halide, **5**, 114
  monokis(cyclopentadienyl) imido, **5**, 119
  monokis(cyclopentadienyl) nitrogen donors, **5**, 119
  monokis(cyclopentadienyl) oxygen donors, **5**, 119
  monokis(cyclopentadienyl) phosphorus donors, **5**, 119
  monokis(cyclopentadienyl) selenium donors, **5**, 120
  monokis(cyclopentadienyl) silyl, **5**, 118
  monokis(cyclopentadienyl) sulfur donors, **5**, 120
  nitriles, **5**, 100, 101
  in organic synthesis, **5**, 133–44
  phosphine hydrides, **5**, 73
  pyridine, **5**, 102, 103
  substituted carbonyls, **5**, 59–65
  zinc clusters, **10**, 336
Taxodione
  synthesis, **12**, 497
Taxol
  side chain
    synthesis, **12**, 1169, 1173
  synthesis, **12**, 1077
Tazettine
  synthesis, **12**, 285, 288
Tebbe's reagent
  alkylidenation, **12**, 577–96
  stoichiometric reagent, **11**, 296
Technetium
  chemistry, **6**, 151–64
  handling, **6**, 151
  isotopes, **6**, 151
  in nuclear medicine, **6**, 152
  occurrence, **6**, 151
  production, **6**, 152
Technetium carbonyls
  chemistry, **6**, 152–4
Technetium complexes
  alkyl, **6**, 162
  arenes, **6**, 161
    heart-imaging agents, **6**, 161
  carbon multiple bonds, **6**, 162
  formal oxidation states, **6**, 164
  halocarbonyl, **6**, 153
  hydrido carbonyl, **6**, 153
  imido, **6**, 162
  nitrogen compounds, **6**, 155–7
  NMR, **6**, 164
  phosphorus compounds, **6**, 155–7
  tris(arylimido) alkyl
    preparation, **6**, 162
  x-ray crystal structures, **6**, 163
Technetium(I) complexes
  carbonyl, **6**, 156
  carbonyl β-diketonates, **6**, 154
  dithiocarbamate, **6**, 155
  halocarbonyl, **6**, 156
  hexakis(isocyanide), **6**, 158
Technetium(III) complexes
  14-electron, **6**, 155
Technetium cyclopentadienyl complexes, **6**, 159–61

brain-imaging agents, **6**, 160
Technetium isocyanides, **6**, 158, 159
binary, **6**, 158
crystal structure, **6**, 158
dealkylation, **6**, 159
derivatives, **6**, 158, 159
heart-imaging agents, **6**, 159
substitution
associative mechanisms, **6**, 159
Technetium(III) isocyanides
14-electron, **6**, 159
Teleocidin A1
synthesis, **12**, 1003
Teleocidin A_2
synthesis, **12**, 107
Tellimagradin
synthesis
lithiation, **11**, 67
Telluraboranes
properties, **1**, 270, 271
Telluradigermiranes
synthesis, **2**, 169
Telluradistanniranes
preparation, **2**, 281
Tellurides
vinylic
cyanocuprates reaction with, **12**, 91
Tellurides, vinyl 2-thienyl
in organic synthesis, **11**, 594
Tellurium compounds
silyl, **2**, 36, 37
synthesis
organomercury compounds, **11**, 400
Tellurium reagents
inorganic
in organic synthesis, **11**, 571, 572
in organic synthesis, **11**, 576–96
Tellurium, tetraalkyl-
in organic synthesis, **11**, 595
Tellurolactonization
unsaturated carboxylic acids, **11**, 585
Telluronium methylides, dibutyl
stabilization, **11**, 591
Tellurophene
ring opening
reaction with Grignard reagents, **12**, 171
Ter-thiophenes
preparation, **12**, 178
Terpenoids
synthesis
allylic alkylation, **12**, 849
allylmetal catalysts, **12**, 918
α-Terpinene
metallation, **11**, 105
Terruthenocenylene
synthesis, **7**, 616
Tetraazaborolines
preparation, **1**, 162
1,4,8,11-Tetraazacyclotetradecane
nickel complexes, **9**, 91
Tetraazadiborinanes
properties, **1**, 185
Tetraazamacrocycles
synthesis
organoboron compounds, **11**, 245
Tetraborane(8) carbonyl
electron diffraction, **1**, 245
Tetraboranes
metallaboranes, **1**, 313
Tetraborates, dihydro-
metallaboranes, **1**, 322
Tetraborates, heptahydro-
metallaboranes, **1**, 321
Tetraborates, hexahydro-
group 8 metals
metallaboranes, **1**, 321
group 9 metals
metallaboranes, **1**, 321
metallaboranes, **1**, 321
Tetraborates, nonahydro-
group 5 metals
metallaboranes, **1**, 319
group 8 metals
metallaboranes, **1**, 319
metallaboranes, **1**, 319
Tetraborates, octahydro-
group 8 metals
metallaboranes, **1**, 319
group 9 metals
metallaboranes, **1**, 320, 321
metallaboranes, **1**, 319–21
Tetraborates, tetrahydro-
metallaboranes, **1**, 322
Tetraborocine
preparation, **1**, 154
Tetradecenyl acetate
synthesis
dicobalt propargylium complexes, **12**, 693
Tetraenes
acyclic
cyclisomerization, **12**, 792
carbacyclization
palladium catalysts, **12**, 816
Tetragallium tetraiodide
structure, **1**, 527
Tetragermadimercurocane
preparation, **2**, 176
Tetrairon tridecacarbonyl
anion
chemistry, **7**, 6
Tetralladia cations
formation, **9**, 272
Tetramethyltin radical cation
structure, **2**, 290
Tetraorganobismuthines
synthesis, **2**, 325, 326
Tetraorganodiarsines
synthesis, **2**, 325, 326
Tetraorganostibines
synthesis, **2**, 325, 326
Tetraosmium hexadecacarbonyl
synthesis, **7**, 842, 843
Tetraosmium pentadecacarbonyl
synthesis, **7**, 841, 842
Tetraosmium tetradecacarbonyl
synthesis, **7**, 840, 841
Tetrapyrroles
precursors
synthesis, **12**, 696
Tetraquinanedione
synthesis, **12**, 733
Tetrarhenium clusters, **6**, 173–5
Tetrasilacyclohexyne
synthesis, **2**, 13; **11**, 325
Tetrastannahexathiaadamantane
preparation, **2**, 294
Tetrastannanes
structure, **2**, 284
catena-Tetrastibine
synthesis, **2**, 328
Tetrathiafulvalene
alkynic analogue synthesis
alkynedicarboxaldehyde cobalt complexes, **12**, 693
Tetrazamacrocycles
synthesis, **12**, 551
Tetrazole
lithiation, **11**, 44
Tetrazolyl ethers
coupling reactions
Grignard reagents, **12**, 175
Tetrin-A
synthesis, **12**, 881
Tetronomycin
synthesis, **12**, 613
Tetroxide, **12**, 1147, 1148
Textile industry
organofunctional siloxanes, **2**, 130
Thallacarboranes
properties, **1**, 258, 259
Thallation
arenes, **11**, 456
Thallation–carbonylation
arenes, **11**, 447
Thallation–palladation
indole synthesis, **11**, 451
Thallium complexes
arenes, **1**, 534–6
arsenic ligands, **1**, 518–24
carboboranes, **1**, 532–4
cyclopentadienyl, **1**, 529–31
group 6 metals
direct bonds to thallium, **1**, 567
group 7 metals
direct bonds to thallium, **1**, 567
group 8 metals
direct bonds to thallium, **1**, 567, 568
group 9 metals
direct bonds to thallium, **1**, 569
group 10 metals
direct bonds to thallium, **1**, 569
group 11 metals
direct bonds to thallium, **1**, 569
low oxidation state, **1**, 528–36
metal bonds, **1**, 567–9
metal–metal bonds, **1**, 537–40
nitrogen ligands, **1**, 518–24
oxygen ligands, **1**, 524–8
phosphorus ligands, **1**, 518–24
selenium ligands, **1**, 524–8
sulfur ligands, **1**, 524–8
tellurium ligands, **1**, 524–8
transition metal bonds, **1**, 545–69
Thallium compounds, **1**, 503–40
anionic, **1**, 514
aryl
carbonylation, **12**, 370
cationic, **1**, 514
halogen complexes, **1**, 511–13

hydrides, **1**, 514–17
pseudohalogen complexes, **1**, 511–13
trialkyl, **1**, 504–11
triaryl, **1**, 504–11
Thallium(I) compounds
in organic synthesis, **11**, 452–5
Thallium(II) compounds
in organic synthesis, **11**, 437–51
Thallium(III) compounds
aryl
preparation, **11**, 445
Thallium hexafluorophosphate
heteronuclear metal–metal bond syntheis, **10**, 2
Thallium hydroxide
cross-coupling reactions
vinylboron compounds, **11**, 232
Thallium(III) nitrate
in organic synthesis, **11**, 455
Thallium polymers
electroactive
preparation, **11**, 444
Thallium salts
coupling reactions, **12**, 181
Thallium(III) salts
reoxidation
transmetallation to palladium, **12**, 146
Thallium, cyclopentadienyl-
in organic synthesis, **11**, 452–5
Thermolysis
organonickel complexes, **9**, 62
β-Thiabenzocyclohexanones
asymmetric reduction, **12**, 25
Thiaboranes
properties, **1**, 268, 269
Thiacarboranes
properties, **1**, 269
β-Thiacycloalkanones
asymmetric reduction, **12**, 25
Thiacyclohexane
reaction with triosmium clusters, **7**, 810, 812
1,2,5-Thiadiborole, 2,5-dihydro-
metal complexes, **1**, 214
1,2,5-Thiadiborolene ligands
metal complexes, **1**, 199
1,2,5-Thiadiborolenes
metal complexes, **1**, 214
Thiadisilacyclohexane
synthesis, **2**, 70
Thiadistanniranes
preparation, **2**, 281
Thiaferrocene
preparation, **7**, 197
Thiagermirane, 3-alkylidene-
formation, **2**, 169
Thiametallacarbaboranes
preparation, **1**, 379
4-Thiaprostoglandin E_1
synthesis, **12**, 865
1,2,5-Thiasilaborolenes
metal complexes, **1**, 213
1-Thia-2-silacyclobut-3-ene
synthesis, **2**, 55
Thiazines
β-lactams from
photochemical reactions of carbene complexes, **12**, 551
Thiazole
Stille reaction, **12**, 215
Thiazolines
β-lactams from
photochemical reactions of carbene complexes, **12**, 551
optically active
synthesis, **12**, 554
Thienamycin
synthesis, **12**, 25, 97, 98, 372, 441
cobalt-mediated propargylation, **12**, 696
optically active, **12**, 358
Thioacetals
oxidation
organoselenium compounds, **11**, 535
Thioamides
acylation
organobismuth(V) compounds, **11**, 511
Thiocarbonyl ylides
formation
reactions, **12**, 449
metal stabilized, **12**, 445–9
Thiocyanates
aryl
synthesis, organomercury compounds, **11**, 400
Thioesters
S-alkyl
alkylidenation, **12**, 585
Thioethane
ring opening, **7**, 811
Thioethers
aryl
preparation, **12**, 984
Thioformamido ligands
clusters, **7**, 813
Thioglycollates
dialkyltin isooctyl
stabilizers, **2**, 295
Thioketones
oxidation
organoselenium compounds, **11**, 535
α-phenyl-
synthesis, **11**, 337
reduction
tellurium reagents, **11**, 577
Thiols
esters
methylenation, **12**, 590
oxidation
organobismuth(V) compounds, **11**, 505
organoselenium compounds, **11**, 535
Thiophene
coupling reactions, **12**, 229
cross-coupling reactions
Grignard reagents, **12**, 177
reaction with triosmium clusters, **7**, 811
ring opening
reaction with Grignard reagents, **12**, 171
Stille reaction, **12**, 215
Thiophene, dihydro-
reaction with triosmium clusters, **7**, 811
Thiophene, 2,3-dihydro-
rhenium complexes, **6**, 220
Thiophene, 2-lithio-
synthesis, **12**, 78
Thiophene, tetrahydro-
rhenium complexes, **6**, 219
rhenium pentacarbonyl complexes
preparation, **6**, 218
Thiophenes
alkylation
propargyl dicobalt complexes in, **12**, 690
DMG containing
lithiation, **11**, 72
metallation, **11**, 119
non-DMG containing
lithiation, **11**, 71
rhenium complexes
hydrodesulfurization of coal, **6**, 218
synthesis, **12**, 335
Thiourea ligands
clusters, **7**, 813
Three-membered rings
synthesis
intramolecular allylic alkylation, **12**, 878
Threonine
synthesis, **12**, 26
Thromboxane B_2
synthesis, **12**, 851, 876
Thujaplicin
synthesis, **12**, 210
Tiglane
synthesis, **12**, 329, 330
Tin acetate, ethyl α-tributyl-
coupling reactions, **12**, 209
Tin acetate, tri-*n*-butyl-
catalyst
cycloaddition reactions, **12**, 930
Tin alkoxides
reactions
allyl complexes, **12**, 817
Tin complexes
palladium carbonyls, **9**, 207
Tin(IV) complexes
cyclizations with capture by, **12**, 312–14
Tin compounds
carbon bonds
formation, **2**, 224–33
chemistry, **2**, 217–97
transmetallation, **12**, 166–9
Tin(II) compounds
cyclopentadienyl, **2**, 288
Tin hydrides
polymer supported
in organic synthesis, **11**, 358
reactions
allyl complexes, **12**, 800, 801
Tin reagents
anionic
in organic synthesis, **11**, 363–6
carbocationic
in organic synthesis, **11**, 369
cationic

in organic synthesis, **11**, 366–9
classification, **11**, 357
Tin thiolates
coupling reactions, **12**, 230
Tin, acetonyltributyl-
coupling reactions, **12**, 209
Tin, acyl-
synthesis, **12**, 219
Tin, α-alkoxymethyl-
coupling reactions, **12**, 209
Tin, alkyl trialkoxides
reactions, **2**, 267
Tin, alkyl tricarboxylates
hydrolysis, **2**, 265
Tin, allyltributyl-
synthesis, **2**, 224
Tin, bis(trialkyl) oxide
exchange reactions, **2**, 262
use in organic synthesis, **2**, 262
Tin, bis(triallyl) oxide
NMR, **2**, 261
Tin, bis(tri-*t*-butyl) oxide
reactions, **2**, 261
Tin, bis(tributyl) oxide
timber protection, **2**, 296
Tin, bis[(trimethylsilyl)methyldi-carbahexaborane]-
preparation, **1**, 260
Tin, bis(triphenyl) sulfide
structure, **2**, 293
Tin, butyl trichloride
hydrolysis, **2**, 265
Tin, cyanomethyltributyl-
coupling reactions, **12**, 209
Tin, dialkyl hydrides
properties, **2**, 271
Tin, dibutyl oxide
reaction with dicarboxylic acids, **2**, 258
Tin, dibutyldichloride
structure, **2**, 255
Tin, dibutylditriflate
preparation, **2**, 260
Tin, dibutyl(4,7,10-trioxaundecyl) hydride
separation, **2**, 272
Tin, hexaalkyldi-
in organic synthesis, **11**, 358
Tin, hydroxymethyl-
coupling reactions, **12**, 209
Tin, methyltribromide
structure, **2**, 255
Tin, methyltriiodide
structure, **2**, 255
Tin, tetra(allylperoxy)-
preparation, **2**, 270
Tin, tetrabutyl-
synthesis, **2**, 224
Tin, tetramethyl-
structure, **2**, 234
Tin, tetraoctyl-
synthesis, **2**, 224
Tin, tetraphenyl-
structure, **2**, 236
synthesis, **2**, 224
Tin, tributyl hydride
reactions, **2**, 273
Tin, tributyl tritide
preparation, **2**, 271
Tin, tri(4,7-dioxaoctyl) hydride
solubility, **2**, 272
Tin, trifluoromethyl hydride
preparation, **2**, 271
Tin, trimesitylfluoride
structure, **2**, 255
Tin, trimethyl hydrotrioxide
preparation, **2**, 270
Tin, trimethylfluoride
structure, **2**, 255
Tin, trimethylhydroxide
complexes, **2**, 261
Tin, trimethyl(methylperoxy)-
preparation, **2**, 290
Tin, triphenylchloride
structure, **2**, 255
Tin, triphenylfluoride
structure, **2**, 255
Tin(II), cyclopentadienyl chloride
preparation, **2**, 289
Tin(II), cyclopentadienyl-
compounds, **2**, 288
structures, **2**, 288
Tin(II), di(bistrimethylsilyl-methyl)-
reactions, **2**, 286
structure, **2**, 285
Tin(II), di(2,4,6-tris-(trifluoromethyl)phenyl)-
preparation, **2**, 286
Tin(II), pentamethylcyclopenta-dienyl triflate
structure, **2**, 289
Tin(IV), dialkyl hydroxide
chemistry, **2**, 262–5
Tin(IV), dialkyl oxide
chemistry, **2**, 262–5
Tin(IV), monoalkyl hydroxide
reactions, **2**, 265, 266
Tin(IV), monoalkyl oxides
reactions, **2**, 265, 266
Tin(IV), trialkylhydroxides
chemistry, **2**, 261, 262
Tin(IV), trialkyloxides
chemistry, **2**, 261, 262
Tirandamycic acid
synthesis, **12**, 588
Tischenko products
preparation, **8**, 316
Titanacycles
in ring-opening metathesis polymerization, **12**, 1213
Titanacyclobutanes
reactions, **4**, 387
in ring-opening metathesis polymerization, **12**, 1211, 1213
Titanacyclobutenes
reactions, **4**, 384
Titanacyclopentadiene
reactions of alkynes and alkenes, **12**, 760
1-Titana-3-stannacyclobutane
preparation, **2**, 250
Titanium
dicarbon metallacarbaboranes
7-vertex, **1**, 382
Titanium(IV) alkoxide
catalysts
Sharpless epoxidation, **12**, 1110
Titanium(III) aluminohydride, biscyclopentadienyl
synthesis, **1**, 546
Titanium complexes
bis(arene), **4**, 213, 214
bis(cyclopentadienyl)
properties, **4**, 342–51
reactions, **4**, 346–57
structures, **4**, 342–51
bis(naphthalene), **4**, 213
carbenes
alkylidenations, **12**, 1083
carbonyl, **4**, 215–18
catalysts
coupling reactions, **12**, 1072–84
chemistry, **4**, 419
chlorohydrido
coupling reactions, **12**, 1073
heteroleptic carbonyl, **4**, 215–17
homoleptic carbonyl, **4**, 218
homoleptic diene, **4**, 218
mono(arene), **4**, 214
oxidation state 0, **4**, 213–18
oxidation state +2, **4**, 221–67
oxidation state +3, **4**, 221–67
pentadienyl, **4**, 239, 240
silyl, **4**, 265
Ti–C σ-bonds, **4**, 357–97
trichloro(dimethoxyethane)
coupling reactions, **12**, 1073
zirconium bonds, **10**, 27
Titanium(II) complexes
alkene, **4**, 234–7
alkyl, **4**, 222
alkyne, **4**, 237–9
reactions, **4**, 237–9
structure, **4**, 237
synthesis, **4**, 237
arene, **4**, 240–2
carbonyl, **4**, 224–30
cycloheptatrienyl, **4**, 242
monocyclopentadienyl, **4**, 222–4
oxidation reactions, **4**, 239
phosphines, **4**, 230–4
reactions, **4**, 231–4
synthesis, **4**, 230
Titanium(III) complexes, **4**, 244–67
alkyl
without anionic ligands, **4**, 244
arsenic bonds, **4**, 265, 266
bis(cyclopentadienyl)
alkyl, **4**, 251, 252
allyl, **4**, 255–7
σ-bonded alkenyl, **4**, 253–5
σ-bonded alkynyl, **4**, 253–5
halides, **4**, 247–50
synthesis, **4**, 247–66
bis(cyclopentadienyl) hydrides, **4**, 259–65
heteronuclear, **4**, 260, 261
homonuclear, **4**, 259, 260
cyclooctatetraenyl, **4**, 266, 267
hydrides
hydrogenation catalysts, **4**, 262, 263
mono(cyclopentadienyl)
synthesis, **4**, 244–7
phosphorus bonds, **4**, 265, 266
Ti–N bonds, **4**, 257

Ti–O bonds, **4**, 257–9
Titanium(IV) complexes
alkoxo, **4**, 401–6
alkyl
properties, **4**, 283, 284
reactions, **4**, 284–7
structures, **4**, 283, 284
sulfur-containing, **4**, 359
synthesis, **4**, 357–61
without anionic π-ligands, **4**, 280–7
without anionic π-ligands, synthesis, **4**, 280–3
alkylidene, **4**, 374–90
stoichiometric reactions, **4**, 374–88
arenethiolato, **4**, 414
benzoato, **4**, 407
bis(cyclopentadienyl), **4**, 311–418
bis(cyclopentadienyl) halides, **4**, 311–57
synthesis, **4**, 311–42
carboxylates, **4**, 406–10
cationic acetonitrile, **4**, 398
chalcogenides, **4**, 411–13
cyclooctatetraene, **4**, 419
dithiocarbamato, **4**, 416, 419
heterocyclic compounds, **4**, 400, 401
hydrochalcogenides, **4**, 411–13
hydroselenido, **4**, 412
hydrosulfido, **4**, 412
hydroxo, **4**, 401–6
isocyanides, **4**, 288
isothiocyanates, **4**, 399
mono(cyclopentadienyl), **4**, 290–310
mono(cyclopentadienyl) halides, **4**, 290–3
reactions, **4**, 292, 293
structures, **4**, 291, 292
synthesis, **4**, 290, 291
monocyclopentadienyl Schiff base, **4**, 303, 304
noncyclopentadienyl π-ligands, **4**, 419
oxa, **4**, 401–6
polyselenides, **4**, 412
polysulfides, **4**, 411
salicylato, **4**, 409
Schiff bases, **4**, 400, 401
thiolato, **4**, 414
Ti–As bonds, **4**, 410
Ti–C σ-bonds, **4**, 293–8
cationic, **4**, 367
cationic nitriles, **4**, 366
photolysis, **4**, 364
properties, **4**, 361–3
protolysis, **4**, 365, 369
reactions, **4**, 296–8, 363–74
reactions with iminoboranes, **4**, 372
reduction, **4**, 365
structures, **4**, 361–3
synthesis, **4**, 293–6
thermolysis, **4**, 363, 370
Ti–Ge bonds, **4**, 397
Ti–N bonds, **4**, 397–400
Ti–N σ-bonds, **4**, 299–302
Ti–O bonds, **4**, 304–9, 401–10
Ti–P bonds, **4**, 410
Ti–S bonds, **4**, 309, 310, 411–18
Ti–Se bonds, **4**, 411–18
Ti–Si bonds, **4**, 397
Ti–Sn bonds, **4**, 397
Ti–Te bonds, **4**, 411–18
Titanium compounds
allylic alkylation, **12**, 862
catalysts
Grignard reagents, reaction with alkynes, **1**, 69
Grignard reagents, reaction with dienes, **1**, 69
Titanium metallacycles, **4**, 374–90
stoichiometric reactions, **4**, 374–88
Titanium tartrates
Sharpless epoxidation, **12**, 1119
Titanium tetrachloride
alkylidenation, **12**, 578
zinc system
coupling reactions, **12**, 1073
Titanium trichloride
lithium aluminum hydride system
coupling reactions, **12**, 1073
magnesium system
coupling reactions, **12**, 1073
Titanium, bis(cyclopentadienyl)-bis(isocyanide)-
synthesis, **4**, 224
Titanium, bis(cyclopentadienyl)-diphenyl-
propene polymerization, **12**, 1197
Titanium, dialkyldicyclopentadienyl-
alkylidenation, **12**, 594
Titanium, dicarbonylbis(cyclopentadienyl)-
ligand exchange reactions, **4**, 225, 226
oxidative reactions
formation of Ti–N bonds, **4**, 229, 230
formation of Ti–O bonds, **4**, 226, 227
formation of Ti–P bonds, **4**, 229, 230
formation of Ti–S bonds, **4**, 227–9
formation of Ti–Se bonds, **4**, 227–9
properties, **4**, 224, 225
synthesis, **4**, 224
Titanium, tribenzyl-
synthesis, **4**, 244
Titanocene
alkylated
aluminum tetrahydride derivatives, **1**, 548
catalyst
asymmetric reduction, **12**, 27
reductive coupling
unsaturated compounds, **12**, 324
Titanocene, bis(trimethylphosphine)-
synthesis, **12**, 325
Titanocene, dichloro-
transmetallation
with zincates, **11**, 182
Titanocene, dimethyl-
protonolysis
Ziegler–Natta catalysts, **12**, 1196
Titanocene-1,1'-binaphth-2,2'-diolate
catalyst
asymmetric reduction, **12**, 27
Titanocenes
antitumor agents, **4**, 353
bimetallic
synthesis, **4**, 311
biphenyl-bridged, **4**, 342
dichloride
electrochemical oxidation, **4**, 355
reduction, **4**, 354
synthesis, **4**, 311
dichlorides
reaction with amino acids, **4**, 408
difluoride
reactions, **4**, 353
dihalides
crystal structures, **4**, 346
dimethyl
synthesis, **4**, 357
enolates, **4**, 404
heterocyclic
synthesis, **4**, 314
isodicyclopentadienyl ligands, **4**, 333
phenoxides
preparation, **4**, 402
stereodifferentiating, **4**, 332
trimethylene-bridged, **4**, 342
ansa-Titanocenes
enantiomers
resolution, **4**, 407
methylation, **4**, 358
NMR, **4**, 346
structures, **4**, 346–57
synthesis, **4**, 340
Titanocyclobutane
cleavage, **4**, 377
Titanocyclobutanes
reactivity, **4**, 381, 419
Titanocyclopentadiene
hexahydronaphthalene synthesis, **12**, 764
Titanocyclopentanes
synthesis, **4**, 380
Titanoimines
arylation/vinylation, **12**, 230
TMEDA
lithiation
in organic synthesis, **11**, 4
Toluene
lithiation, **11**, 55
reactions with iron atoms, **7**, 109
Toluene, fluoro-
metallation, **11**, 121
Toluidines
lithiation, **11**, 59
Tomamycin
synthesis, **12**, 249
Tosyl cyanide

reaction with alkylzinc halides, **11**, 165
Trachelanthamidine
synthesis, **12**, 436
Transcyanation
organolanthanide complexes, **4**, 129
Transfer reactions
divalent carbon
organomercury compounds, **11**, 409–11
Transition metal clusters
amalgam reductions, **10**, 327
hydrides
group 12 metals, **10**, 326
redox chemistry
group 12 metals, **10**, 327–30
zinc, cadmium, and mercury, **10**, 323–49
Transition metal complexes
alkenes
protection, **12**, 623–34
alkyl
in organic synthesis, **12**, 39–46, 59–124, 131–58, 241–91
π-alkynes
nucleophilic attack, **12**, 619, 620
allyl
alkene cyclization/C,C-coupling, **12**, 913–19
alkene cyclization/carbonylation, **12**, 916–19
alkene cyclization/elimination, **12**, 906–12
intramolecular insertion reactions, alkenes and alkynes, **12**, 905–21
in organic synthesis, **12**, 785–93
boryl, **1**, 290–3
dienes
protection, **12**, 623–34
dienyl
protection, **12**, 623–34
metal bonds, **1**, 545–69
metallacarbaboranes, **1**, 373–424
unsaturated compounds
reductive coupling, **12**, 323–46
Transition metal hydrides, **12**, 9–35
Transition metals
in organic synthesis, **12**, 1–6
in polymer synthesis, **12**, 1209–29
Transmetallation
alkyltransition metal complexes
in organic synthesis, **12**, 131–58
allylic alkylation, **12**, 852
boron
to palladium, **12**, 148
cuprate-induced, **12**, 85
definition, **12**, 5
germanium compounds
to palladium, **12**, 152–4
Grignard reagents
titanium complexes, **12**, 157
insertion reactions
phase transfer conditions, **12**, 150
lead compounds
to palladium, **12**, 152–4
lithium or magnesium reagents, **12**, 164–79
mercury to palladium, **12**, 132–46
catalysis, **12**, 142–6
organocopper compound preparation, **3**, 64–7
organocopper compounds
to palladium, **12**, 154
organolithium compounds
to palladium, **12**, 154
organomagnesium compounds
to palladium, **12**, 154
organotellurium compounds, **11**, 592–4
oxidative addition, **12**, 161–231
silicon compounds
to palladium, **12**, 152–4
tin compounds
to palladium, **12**, 150, 151
tin or silicon reagents, **12**, 200–21
to palladium, **12**, 132–55
zirconium to aluminum, **12**, 46
zirconium to copper, **12**, 45
zirconium, zinc, aluminum, or boron reagents, **12**, 179–200
Trialkylsilyl azides
chemistry, **11**, 341
Trialkylsilyl cyanides
chemistry, **11**, 339, 340
Trialkylsilyl halides
in organic synthesis, **11**, 338, 339
Trialkylsilyl peroxides
chemistry, **11**, 340, 341
Trialkylsilyl selenols
chemistry, **11**, 341, 342
Trialkylsilyl stannanes
chemistry, **11**, 342
Trialkylsilyl thiols
chemistry, **11**, 341, 342
Trialkylsilyl trifluoromethanesulfonates
chemistry, **11**, 340
Triarylstannyl radicals
preparation, **2**, 278
1,2,4-Triazene
synthesis, **12**, 754
Triazenes
synthesis
cyclooligomerization, **12**, 753–6
1,3,5-Triazine, 1,3,5-trimethylhexahydro-
adduct with dimethylzinc, **3**, 177
1,2,3-Triboracyclopentenyl ligands
metal complexes, **1**, 198
Triboranes
metallaboranes, **1**, 313–18
Triborates, heptahydro-
group 8 metals
metallaboranes, **1**, 316
group 9 metals
metallaboranes, **1**, 316
group 10 metals
metallaboranes, **1**, 316
group 11 metals
metallaboranes, **1**, 317
metallaboranes, **1**, 316, 317
Triborates, hexahydro-
group 8 metals
metallaboranes, **1**, 317
metallaboranes, **1**, 317
Triborates, octahydro-
group 5 metals
metallaboranes, **1**, 315
group 6 metals
metallaboranes, **1**, 315
group 8 metals
metallaboranes, **1**, 315
group 9 metals
metallaboranes, **1**, 316
group 10 metals
metallaboranes, **1**, 316
metallaboranes, **1**, 315, 316
Triborates, trihydro-
metallaboranes, **1**, 318
Tributylstannyl radicals
reactivity, **2**, 273
Trichodermol
synthesis, **12**, 97, 98, 663
Trichodiene
synthesis, **12**, 662
Trichothecenes
synthesis, **12**, 660
Tricobalt rhodium carbonyls, **8**, 510
9-Tricosene
synthesis, **11**, 101
Tricyclo[5.3.0.0^{2,6}]deca-4,8-dien-3,10-diones
synthesis, **12**, 624
Tricyclo[4.1.0.0^{2,4}]
synthesis
Pauson–Khand reaction, **12**, 715
Tricyclo[4.1.0.0^{2,7}]heptanes
lithiation, **11**, 6
Tricyclo[5.2.1.0^{4,10}]decane
synthesis, **12**, 733
Tricyclo[5.3.0.0^{2,6}]dec-4-en-3-ones
synthesis, **12**, 715
Tricyclo[6.2.2.0^{2,7}]dodeca-3,5,9,11-tetraene
tricarbonyliron complex
isolation, **12**, 624, 625
Tricyclo[2.2.1.0^{2,6}]heptanes
metallation, **11**, 113
Tridecanolide, 13-methyl-
synthesis, **12**, 803
Trienes
cyclopropane synthesis
cascade cyclization, **12**, 305, 306
hydrogermylation, **2**, 154
Triflates
chemistry, **11**, 340
Triiron dodecacarbonyl
chemistry, **7**, 4
Triiron dodecarbonyl
catalysts
diazodecomposition, **12**, 430
Triiron undecacarbonyl
anion

chemistry, **7**, 6
Trimethoprim
synthesis
lithiation, **11**, 81
Trimethylamine oxide
in cluster substitution reactions, **7**, 902
in heteronuclear metal–metal bond synthesis, **10**, 2, 6
Trimethylene methane complexes
nickel
cycloaddition reactions, **12**, 923–57
palladium
cycloaddition reactions, **12**, 923–57
Trimethylsilyl cyanide
synthesis, **2**, 15
Trimethylsilyl diazomethane
chemistry, **11**, 341
Trimethylstannyl anions
formation, **2**, 285
Triophamine
synthesis, **12**, 103
Triorganoantimony bis(sulfonates)
synthesis, **2**, 341
Triorganoantimony(III) compounds
synthesis, **2**, 329
Triorganoantimony(V) compounds
synthesis, **2**, 338–41
Triorganoantimony dicarboxylates
synthesis, **2**, 341
Triorganoarsenic(III) compounds
synthesis, **2**, 329
Triorganoarsenic(V) compounds
synthesis, **2**, 337
Triorganobismuth(III) compounds
synthesis, **2**, 329
Triorganobismuth(V) compounds
synthesis, **2**, 341
Triorganostannates
structures, **1**, 14
Triorganothallium compounds
in organic synthesis, **11**, 437–43
preparation, **11**, 437, 438
properties, **11**, 437, 438
reactions, **11**, 438–43
Triorganozincates
halogen–zinc exchange, **11**, 182
Triosmium clusters, **7**, 804–30
ethene sulfide, **7**, 810
phosphaalkynes, **7**, 826, 830
phosphabenzene, **7**, 826, 830
reactions with alkynes, **7**, 808
silyl-acetylide, **7**, 826, 830
Triosmium dodecacarbonyl
disubstituted, **7**, 714
EXAFS, **7**, 685
halogenation, **7**, 723
hydrogenation, **7**, 687
IR, **7**, 685
isocyanides, **7**, 717
NMR, **7**, 685
Raman studies, **7**, 685
reactions with halides, **7**, 721
structure, **7**, 685
substituted
structures, **7**, 711
synthesis, **7**, 711
trisubstituted, **7**, 715
Triphenylene
synthesis, **11**, 121
1,3,5,2,4,6-Triphosphatriborinane
metal complexes, **1**, 211, 212
Triple-decker sandwich molecules
ruthenium, **7**, 617
Tripledecker complexes
34 electron, **1**, 199
boron rings
metal complexes, **1**, 198, 199
doubly opened, **1**, 199
slipped, **1**, 199
Triptoquinone
synthesis, **12**, 251, 252
Triquinacene
synthesis, **12**, 733, 735
Triquinane
angularly fused
synthesis, **12**, 731
synthesis, **12**, 399
Triquinane sesquiterpenoids
synthesis
allylmetal catalysts, **12**, 918, 919
Triquinanes
angularly fused
synthesis, **12**, 723
Trirhenium clusters, **6**, 173–5
Trirhodium iridium carbonyls, **8**, 510
Triruthenium clusters, **7**, 804–30
reactions with alkynes, **7**, 808
silyl-allenyl, **7**, 826, 830
tetramethylallene episulfide, **7**, 810
Triruthenium dodecacarbonyl
disubstituted, **7**, 714
EXAFS, **7**, 685
hydrogenation, **7**, 689
IR, **7**, 685
isocyanides, **7**, 717
NMR, **7**, 685
photolysis, **7**, 684
Raman studies, **7**, 685
reactions with 1,4-diazabutadiene, **7**, 635
reactions with halides, **7**, 721
structure, **7**, 685
substituted
structures, **7**, 711
synthesis, **7**, 711
trisubstituted, **7**, 715
Tris(diethylamino)sulfonium difluoromethylsilicate
coupling reactions
alkenyl iodides, **12**, 205
Triselena[3]ferrocenophane
preparation, **7**, 189
1,3,5-Trisilabenzene
metal complexes, **2**, 62
Trisilacyclobutanes
properties, **2**, 54
1,3,5-Trisilacyclohexane
reactions with iron ions, **7**, 106
synthesis, **2**, 62
Trisilacyclopentanes
synthesis, **2**, 59
Trisilacyclopentenes
synthesis, **2**, 59
Trisilylamines
synthesis, **2**, 25
Trisodium pentacarbonylvanadate
preparation, **5**, 3
Trisporol B
preparation, **12**, 189
Tristannacyclodecane
preparation, **2**, 252
1,5,9-Tristannacyclododecane
structure, **2**, 250
1,2,4-Tristannacyclopentanes
preparation, **2**, 295
Tristannanes
structure, **2**, 284
Tristannoxanes
reduction to distannanes, **2**, 278
Tristannoxanes, tetraalkyl-
in organic synthesis, **2**, 264
preparation, **2**, 263
catena-Tristibine
synthesis, **2**, 328
Tris(trimethylsilyl)amine
synthesis, **2**, 5
Trithia[3]ferrocenophane
preparation, **7**, 189
1,3,5-Trithiane
rhenium pentacarbonyl complexes
preparation, **6**, 218
Trithiatristannaadamantanes
structure, **2**, 294
Tropine
synthesis, **12**, 811
Tropine, 6,7-dehydro-
preparation, **12**, 350
Tropolones
synthesis, **12**, 210
Tropone, 2-phenyl-
cycloaddition reactions, **12**, 936
Tropones
cycloaddition reactions, **12**, 930
[3+2], **12**, 934–7
synthesis, **12**, 405
Tropones, aryl-
synthesis, **12**, 194
Tropylidene, siloxy-2-cyano-
synthesis, **2**, 15
Tryptamine
propargylation
dicobalt complexes, **12**, 690
Tryptamine, *N*-carbomethoxy-
synthesis, **12**, 690
Tryptamine, *N*-methyl-*N*'-carbomethoxy-
synthesis, **12**, 690
Tryptophols
synthesis
lithiation, **11**, 59
Tungsten
dicarbon metallacarbaboranes
12-vertex, **1**, 400, 418
Tungsten carbene complexes
alkynyl allylamino
Pauson–Khand reaction, **12**, 731
Tungsten carbonyl complexes
alkynides, **5**, 165, 166
properties, **5**, 165
reactions, **5**, 166
synthesis, **5**, 165
availability, **5**, 156

azacarbene, **5**, 185–91
calculations, **5**, 156, 157
carbenes
no heteroatom stabilization, **5**, 168–72
carbon σ-donor ligands, **5**, 161
carbyne, **5**, 192–7
charged, **5**, 159–61
reactions, **5**, 160, 161
structures, **5**, 159, 160
synthesis, **5**, 159, 160
cyanide ligands, **5**, 161
cyclopentadienyl carbyne, **5**, 198–202
isocyanide ligands, **5**, 161–5
properties, **5**, 162
reactions, **5**, 164, 165
synthesis, **5**, 162
oxacarbene, **5**, 172–84
photolysis, **5**, 157, 158
selenocarbonyl, **5**, 166, 167
spectroscopy, **5**, 156, 157
substitution reactions, **5**, 157
supported, **5**, 158, 159
thiocarbonyl, **5**, 166, 167
vinylidenes, **5**, 171
Tungsten clusters
acetylide osmium
tetranuclear, **10**, 104–6
trinuclear, **10**, 102
trinuclear, reactivity, **10**, 102–4
acyl iron, **10**, 91
alkylidyne cobalt, **10**, 110–12
solution fluxionality, **10**, 112
alkylidyne iron, **10**, 91–3
alkylidyne nickel, **10**, 119
alkylidyne osmium, **10**, 93–6
trinuclear, **10**, 100
alkylidyne rhodium, **10**, 116–18
alkylidyne ruthenium, **10**, 93–6
alkyne cobalt, **10**, 113–15
alkyne osmium
tetranuclear, **10**, 101
trinuclear, **10**, 100
bisalkylidyne osmium
tetranuclear, **10**, 101
carborane cobalt, **10**, 168
carborane iron, **10**, 168
carborane molybdenum, **10**, 168
hydrocarbyl imido, **10**, 106–8
imido ruthenium
synthesis, **10**, 106
iridium componds, **10**, 118, 119
iron
C_2 ligands, **10**, 96, 97
phosphido iron, **10**, 91
phosphido rhodium, **10**, 116–18
platinum chains, **10**, 120–2
platinum stars, **10**, 122, 123
propargyl, **10**, 108, 109
ruthenium compounds, **10**, 98–100
sulfido iron, **10**, 92
sulfido rhodium, **10**, 116–18
Tungsten complexes, **5**, 331–80
η^3-, **5**, 335–45
η^4-, **5**, 346–52
η^5-, **5**, 352–4
η^6-, **5**, 354–8
η^7-, **5**, 358–62
acyclic pentadienyl, **5**, 352, 353
alkene chelates, **5**, 334, 335
alkenes, **5**, 332–4
alkoxide, **5**, 373–6
alkyl, **5**, 373–6
alkylidene
preparation, **12**, 595
alkyne
monometallic, **5**, 363–72
tungsten(0),(d^6), **5**, 363, 364
alkynes
intramolecular coupling, **5**, 371, 372
allyl, **5**, 342, 343
π-allyl
in organic synthesis, **12**, 968–74
allyl dicarbonyl cyclopentadienyl, **5**, 338–41
allyl pyrazole ligands, **5**, 338
allylic alkylation, **12**, 890–5
arenes, **5**, 471–533
azaallyl, **5**, 345
binuclear, **5**, 373–80
bis(alkyne) cyclopentadienyl, **5**, 368–70
bis(arene), **5**, 478, 479
bis(cyclopentadienyl)
dihydrido
reactions with carboxylic acids, **5**, 449
bis(cyclopentadienyl) dimethyl
elimination reactions, **5**, 446
reactions with trityl cations, **5**, 446
bis(cyclopentadienyl) methyl
reaction with phosphines, **5**, 446
bis(toluene)
reaction with trimethylphosphine, **5**, 529
butadienyl, **5**, 344, 345, 350–2
carbonyl diene, **5**, 346–9
coupling reactions, **12**, 1090–2
cyclic carbene, **5**, 411
cyclobutadiene, **5**, 350–2
cycloheptadienyl, **5**, 353, 354
cycloheptatriene, **5**, 354–7
cycloheptatrienyl, **5**, 358, 359
bimetallic metal–metal bonded, **5**, 362
sandwich complexes, **5**, 361
cyclohexadienyl, **5**, 353, 354
cyclooctatetraene, **5**, 357, 358
cyclopentadienyl, **5**, 393–450
cyclopentadienyl diene, **5**, 349, 350
cyclopentadienyl nitrosyl, **5**, 418–20
cyclopentadienyl oxo, **5**, 431
cyclopropenyl, **5**, 344, 345
dicarbonyl cyclopentadienyl dimers
structure, **5**, 397
synthesis, **5**, 396–9
dicarbonyl cyclopentadienyl (ethyl isonitrile) anion
alkylation, **5**, 412, 430
dicarbonyl cyclopentadienyl nitrosyl
synthesis, **5**, 418, 419
dicarbonyl pentamethylcyclopentadienyl dimers
reactions with sulfur, **5**, 438
dinitrosyl cyclopentadienyl halides
synthesis, **5**, 419
formyl
synthesis, **5**, 405
halides, **5**, 373–6
heteroarenes, **5**, 471–533
homoleptic allyl halides, **5**, 335, 336
homoleptic allyls, **5**, 335, 336
homoleptic dienes, **5**, 346
indenyl, **5**, 376–80
isopropylcyclopentadienyl dichloride dimers
reactions, **5**, 440
mercury clusters, **10**, 337
monoalkene carbonyl phosphine, **5**, 333
monoalkyne cyclopentadienyl, **5**, 367, 368
monohydroboranes, **1**, 292
nitrosyl cyclopentadienyl dihalides
synthesis, **5**, 419, 420
oxocyclobutenyl, **5**, 345
palladium carbonyls, **9**, 204
prop-2-ynyl, **5**, 344
rhenium allenylidene, **10**, 88
tetracarbonyl cyclopentadienyl, **5**, 376–80
tricarbonyl arenes
oxidation, **5**, 495
tricarbonyl benzene
reaction with $LiBHEt_3$, **5**, 511
tricarbonyl chlorobenzene
nucleophilie substitution of halide, **5**, 512
tricarbonyl cyclopentadienyl alkyl
photochemistry, **5**, 412
reactions with phosphines, **5**, 411, 413, 430
synthesis, **5**, 410–18
tricarbonyl cyclopentadienyl anion
reaction with group 12 halides, **5**, 408, 409
synthesis, **5**, 404
tricarbonyl cyclopentadienyl anions
reaction with group 14 halides, **5**, 409
reactions with group 13 halides, **5**, 409
reactions with sulfur trichloride, **5**, 426
tricarbonyl cyclopentadienyl benzyl
photolysis, **5**, 412
tricarbonyl cyclopentadienyl dimers
decarbonylation, **5**, 396

photochemistry, **5**, 396
reactions with pnictogens, **5**, 423–7
structure, **5**, 395, 396
synthesis, **5**, 394–6
thermochemistry, **5**, 396
tricarbonyl cyclopentadienyl halides
synthesis, **5**, 407
tricarbonyl cyclopentadienyl hydride
photochemistry, **5**, 406
reaction with dicyclopentadienyl tin, **5**, 409
reactions, **5**, 406
reactions with alkynes, **5**, 415
reactions with phosphines, **5**, 406
synthesis, **5**, 406, 407
tricarbonyl cyclopentadienyl iodide
reaction with isonitriles, **5**, 407
tricarbonyl cyclopentadienyl methyl
oxidation, **5**, 410
reaction with ethyne, **5**, 414
tricarbonyl cyclopentadienyl tetrafluoroborate
reactions, **5**, 404
tricarbonyl cyclopetnadienyl dimers
reactions with disulfides, **5**, 438
trimethylenemethane, **5**, 350–2
zinc clusters, **10**, 337
Tungsten(II) complexes
alkyne, **5**, 364–6
Tungsten(IV) complexes
alkynes, **5**, 371
aluminum tetrahydride derivatives, **1**, 548
Tungsten hexacarbonyl
reaction with dithioacetals, **12**, 1092
Tungsten hexachloride
butyl lithium systen
reductive dimerization of carbonyl compounds, **12**, 1090
Tungsten, dichlorotetrakis(methyldiphenylphosphine)-
reaction with ketones and alkenes, **12**, 1091
Tungsten, hexakis(isopropyloxy)bis(pyridine)di-
reaction wtih acetone, **12**, 1090
Tungsten, hexaphenoxy-
in organometallic synthesis, **1**, 84
Tungsten, pentacarbonyl(α-methoxybenzylidene)-
reaction with alkenes, **12**, 389
Tungstenaalkenes
preparation
alkylidene transfer, **12**, 1213
in ring-opening metathesis polymerization, **12**, 1211
Tungstenocene
reactions, **5**, 443
Tylosin
synthesis, **12**, 643, 672
Tyloxin
synthesis, **12**, 825

U

Ubiquinone 10
  synthesis, **12**, 849
Ultrasonic radiation
  magnesium
    Grignard reagent preparation, **1**, 60
Ultrasonics
  zinc activation, **11**, 161
Ultrasound
  reductive coupling
    acetophenone, **12**, 1074
Unsaturated compounds
  reductive coupling
    transition metal complexes, **12**, 323–46
Uracil
  reactions
    annulation, **12**, 761
Uracil, 5-bromo-
  cross-coupling, **12**, 218
Uracil, 5-iodo-
  coupling reactions
    with alkynes, **12**, 224
  cross-coupling, **12**, 218
Uranium complexes
  alkylidene
    preparation, **12**, 596
Urea, *N*-4-pentenyl-
  reaction with palladium chloride, **12**, 610
Urea, seleno-
  synthesis, **11**, 528
Ureidomercuration
  intramolecular, **11**, 425
Urethanes
  lithiation, **11**, 17
Uridine, 5-iodo-
  coupling reactions, **12**, 198
  Stille coupling, **12**, 217
UV-PES
  alkyne iron clusters, **7**, 259
δ-Valerolactone
  ethene block copolymers
    synthesis, **12**, 1205

V

Valilatone
  synthesis, **12**, 371
Vanadium
  dicarbon metallacarbaboranes
    7-vertex, **1**, 382
Vanadium complexes
  adducts, **5**, 48, 49
  alkyl, **5**, 11–17
  alkylidene
    preparation, **12**, 595
  arene-bridged binuclear, **5**, 26
  arenes, **5**, 45–50
  aryl, **5**, 11–17
  binary nitrosyl carbonyl, **5**, 9, 10
  bis(arene), **5**, 45–7
    physical properties, **5**, 46, 47
    properties, **5**, 45, 46
    reactivity, **5**, 46
    synthesis, **5**, 45, 46
  bis(cyclopentadienyl), **5**, 33–45
  carbonyl, **5**, 2–10
  carbonyl alkyls, **5**, 8
    NMR, **5**, 8
    spectroscopy, **5**, 8, 9
  carbonyl hydrides, **5**, 7
  catalysts
    epoxidation, **12**, 1103
  coupling reactions, **12**, 1086, 1087
  cycloheptatrienyl, **5**, 49, 50
  cyclohexadienyl, **5**, 48
  cyclooctatetraenyl, **5**, 49, 50
  dienes, **5**, 45–50
  dienyl, **5**, 48, 49
  divalent, **5**, 12
  half-sandwich, **5**, 47
  heteroatom substituted arenes, **5**, 47, 48
  heterobimetallic, **10**, 28
  imido-amido alkyl, **5**, 16
  mixed ligand systems, **5**, 48, 49
  mono(arene), **5**, 47
  monocyclopentadienyl, **5**, 27–32
    ligands other than CO, **5**, 24, 25
    multidecker, **5**, 26
    polynuclear carbonyls, **5**, 22–4
  monocyclopentadienyl alkenes, **5**, 24
  monocyclopentadienyl alkynes, **5**, 24
  monocyclopentadienyl hexadienyl, **5**, 24
  nitrosyl carbonyl monocyclopentadienyl, **5**, 22
  open-metallocene, **5**, 48, 49
  palladium carbonyls, **9**, 205
  pentadienyl, **5**, 48
  polyene, **5**, 45–50
  tetrahydroborates, **1**, 283
Vanadium(I) complexes
  alkyl, **5**, 11
  monocyclopentadienyl, **5**, 17–26
Vanadium(II) complexes
  alkyl
    bonding, **5**, 11, 12
    structure, **5**, 11, 12
    synthesis, **5**, 11, 12
  bis(cyclopentadienyl)
    properties, **5**, 33, 34
    synthesis, **5**, 33, 34
  monocyclopentadienyl, **5**, 27–30
  monocyclopentadienyl alkyl, **5**, 28, 29
  monocyclopentadienyl alkylidene, **5**, 28, 29
  monocyclopentadienyl carboxylato, **5**, 28
  monocyclopentadienyl cyclooctatetraenyl, **5**, 29, 30
  monocyclopentadienyl halide, **5**, 27
  monocyclopentadienyl hydrides, **5**, 28, 29
  monocyclopentadienyl sulfido, **5**, 28
  monocyclopentadienyl thiolato, **5**, 28
Vanadium(III) complexes
  alkyl
    carbon dioxide insertion, **5**, 14
    heteroleptic, **5**, 14, 15
    homoleptic, **5**, 13, 14
    homoleptic, structure, **5**, 13
    homoleptic, synthesis, **5**, 13
    isocyanide insertion, **5**, 14
    proton transfer reactions, **5**, 13
    reactivity, **5**, 13, 14
  bis(cyclopentadienyl), **5**, 37–44
  monocyclopentadienyl, **5**, 27–30
  monocyclopentadienyl alkyl, **5**, 28, 29
  monocyclopentadienyl alkylidene, **5**, 28, 29
  mono(cyclopentadienyl carboxylato, **5**, 28
  monocyclopentadienyl cyclooctatetraenyl, **5**, 29, 30
  monocyclopentadienyl halide, **5**, 27
  monocyclopentadienyl hydrides, **5**, 28, 29
  monocyclopentadienyl sulfido, **5**, 28
  monocyclopentadienyl thiolato, **5**, 28
Vanadium(IV) complexes
  alkyl, **5**, 15–17
  aryl, **5**, 15–17
  bis(cyclopentadienyl), **5**, 37–44
  monocyclopentadienyl, **5**, 30–2
  monocyclopentadienyl halide imido, **5**, 31
  monocyclopentadienyl halides, **5**, 30, 31
  monocyclopentadienyl sulfido, **5**, 32
  monocyclopentadienyl thiolato, **5**, 32
Vanadium(V) complexes
  alkyl, **5**, 15–17
  aryl, **5**, 15–17
  bis(cyclopentadienyl), **5**, 37–44
  imidoalkyl, **5**, 16
  monocyclopentadienyl, **5**, 30–2
  monocyclopentadienyl halides, **5**, 30, 31
  monocyclopentadienyl sulfido, **5**, 32
  monocyclopentadienyl thiolato, **5**, 32
Vanadium compounds
  chemistry, **5**, 1–50
Vanadium hexacarbonyl, **5**, 2–7
  bonding, **5**, 2, 3
  CO displacement reactions, **5**, 4
    with phosphines, **5**, 4
  disproportionation, **5**, 3, 4
  redox reactions, **5**, 5–7
  structure, **5**, 2, 3
  synthesis, **5**, 2, 3
Vanadium trichloride
  lithium aluminum hydride system
    reductive dimerization, **12**, 1073
    reductive dimerization, benzylic halides, **12**, 1086
Vanadium(II), bis(indenyl)-
  oxidation, **5**, 38
Vanadium, bis(indenyl)-
  synthesis, **5**, 34
Vanadium, dichlorotetrakis-(pyridine)-
  reductive dimerization
    benzylic halides, **12**, 1086
Vanadium, tetracarbonylcyclopentadienyl-, **5**, 17–22
  bonding, **5**, 17, 18
  CO replacement, **5**, 18
  decarbonylation
    photoinduced, **5**, 22
  photolysis, **5**, 19
  reaction with chiral bidentate amines, **5**, 20

reaction with isocyanides, **5**, 21
reactivity at metal center, **5**, 18–22
spectroscopy, **5**, 21
structure, **5**, 17, 18
synthesis, **5**, 17, 18
Vanadium, trichlohexakis-(tetrahydrofuran)di-dizinc hexachloride
pinacol dimerization, **12**, 1086
Vanadium(IV), tris(dialkylamino)-methyl-
pinacol dimerization, **12**, 1086
Vanadocene
acetylides, **5**, 42–4
alkoxo, **5**, 40, 41
alkyl, **5**, 42–4
alkylidenes, **5**, 42–4
alkylidynes, **5**, 42–4
allyl, **5**, 42–4
amido, **5**, 42
aryloxo, **5**, 40, 41
carbon monoxide adducts, **5**, 34
carbonyl complexes, **5**, 34, 35
carboxylato, **5**, 40
coordination chemistry, **5**, 38, 39
coordination compounds, **5**, 34–7
dialkyl, **5**, 42–4
N,N-dialkyl carbamato, **5**, 40
diazo-
synthesis, **5**, 36
dichloride, **5**, 45
disulfido, **5**, 42
dithioarsenate, **5**, 42
dithiocarbamate, **5**, 42
dithiophosphonate, **5**, 42
germylene-bridged, **5**, 44
halides, **5**, 37
antitumor activity, **5**, 39, 40
heterobimetallic, **5**, 44, 45
hydrides, **5**, 42–4
imido, **5**, 42
oxo, **5**, 40, 41
polynuclear, **5**, 44, 45
polysulfide, **5**, 42
pseudohalide derivatives
antitumor activity, **5**, 39, 40
properties, **5**, 39
synthesis, **5**, 39
reactions with alkenes, **5**, 35
reactions with carbonyl sulfide, **5**, 37
reactions with heteroallenes, **5**, 36, 37
reactions with isocyanates, **5**, 39
reactions with isocyanides, **5**, 35
reactions with ketene imines, **5**, 36
reactions with nitrilium salts, **5**, 35
reactions with nitrosyl, **5**, 35
reactions with paraformal-dehyde, **5**, 36
reactions with phenyl isothiocyanate, **5**, 36
reactions with thiobenzo-phenone, **5**, 36
reactions with thioisocyanates, **5**, 39
reactions with thioketenes, **5**, 36
reactions with unsaturated molecules, **5**, 35, 36
ring-substituted derivatives, **5**, 37
sulfido, **5**, 42
synthesis, **5**, 33
thiolato, **5**, 42
Vanadocene, decamethyl-
synthesis, **5**, 33
Vancomycin
synthesis, **12**, 193
Verbenalol
synthesis, **12**, 632
Vernolepin
synthesis, **12**, 361
α-Vetispirene
synthesis, **12**, 104, 105
Vilsmeir's salts
reaction with chromium pentacarbonyl dianion, **12**, 556
Vinblastine
models
synthesis, propargylcobalt complexes, **12**, 695
synthesis, **12**, 874
Vincamine
synthesis, **12**, 582
Vincristine
models
synthesis using propargyl-cobalt complexes, **12**, 695
Vindoline
synthesis, **12**, 498, 579
Vineomycinone B2 methyl ester
synthesis, **12**, 212
Vinyl boronates
coupling reactions
with β-chloroenones, **12**, 182
Heck insertion reaction, **12**, 181
Vinyl epoxides
allyl complexes, **12**, 816, 819
allylic alkylation, **12**, 845, 846
Vinyl esters
Pauson–Khand reaction, **12**, 715
Vinyl ethers
cyclopropanation, **12**, 400
Pauson–Khand reaction, **12**, 715
Vinyl halides
acyclic
alkoxycarbonylation, **12**, 246
carbonylation, **12**, 364
displacement reactions, **12**, 142
Pauson–Khand reaction, **12**, 715
Vinyl oxetanes
allylic alkylation, **12**, 846
Vinyl triflates
acyclic
alkoxycarbonylation, **12**, 246
cyclization/carbomethoxyla-tion, **12**, 311
Vinylation
organolead compounds in, **11**, 468
tetraalkyllead to aldehydes, **11**, 462
Vinylcarbenes
cyclopropenation, **12**, 412
Vinyldiazo compounds
[3+4] annulations, **12**, 400, 401
[3+2] annulations, **12**, 401, 402
Vinylic epoxides
reaction with alkenylboranes
palladium-assisted, **12**, 150
ring opening
palladium salts, **12**, 139
Vinylic halides
reactions with organothallium complexes
palladium-assisted, **12**, 147
Vinylic hydrogen substitution
metal assisted, **12**, 133
Vinylic organomercurials
palladium-assisted reaction with alkenes, **12**, 135
Vinylic substitution
metal assisted, **12**, 134
Vinylidene bridged clusters
synthesis, **10**, 19, 20
Vinylidene complexes
iron clusters, **7**, 260, 262, 263
tetranuclear clusters
three unlike metals, **10**, 165
trinuclear clusters
three unlike metals, **10**, 163, 164
Vinylpalladation
alkenes
allyl complexes, **12**, 809
Viologens
ruthenocene-containing, **7**, 611
Visnagan
synthesis, **12**, 492
Vitamin A
polyene analogues
synthesis, **12**, 842
Vitamin B_6
synthesis, **12**, 756
Vitamin B_{12}
nomenclature, **8**, 42
Vitamin B_{12r}
nomenclature, **8**, 42
Vitamin B_{12s}
nomenclature, **8**, 42
Vitamin D_3
1α-,25-dihydroxy-
synthesis of side-chain analogs, **12**, 66
synthesis, **12**, 889
Vitamin D
synthesis, **12**, 204
Vitamin D skeleton
synthesis, **12**, 313
Vitamin E
side-chain
synthesis, **12**, 22, 885
Vitamin K_1
synthesis, **12**, 860

W

Wacker oxidation
  in organic synthesis, **12**, 606, 607
  palladium alkene complexes
    nucleophilic attack, **12**, 606
  regiochemistry, **12**, 607
Wade's rules
  tripledecker complexes, **1**, 198
Wasps
  pheromones of common
    synthesis, **12**, 594
Water–gas shift catalyst
  platinum complexes, **9**, 393
Water–gas shift reaction
  catalysts, **10**, 43
  iron carbonyls, **7**, 85
Wilkinson's catalyst
  decarbonylation
    aldehydes, **12**, 374
Withaperuvin C
  synthesis, **12**, 871
Wittig methylenation, **12**, 586
[2,3]-Wittig rearrangements
  lithiation, **11**, 24
Wittig–Horner reaction
  iron complexes, **7**, 126
Wittig–Horner–Emmons reactions
  iron complexes, **7**, 148
Wurtz reaction
  organotin compound synthesis, **2**, 226
Wurtz-coupling
  zinc organometallic
    compounds, **11**, 162

X

Xanthine
reaction with methylmercury cations, **3**, 156
Xylenes
dimetallation, **11**, 109
o-Xylylene
tricarbonyliron complex
isolation, **12**, 624, 625

Y

Z

Zealeranone
  preparation, **12**, 246, 247
Zearalenone
  synthesis, **12**, 211, 366
Zeise's dimer
  catalysts
    diazodecomposition, **12**, 428
Zeise's salt
  preparation, **12**, 602
Zeolites
  bimetallic clusters in catalysis, **10**, 60
Zeranol
  propargyl dicobalt complexes, **12**, 690
Ziegler–Natta catalysis, **12**, 1193–207
Ziegler–Natta catalysts
  generation, **12**, 1195, 1196
  homogeneous, **12**, 1194
  mechanism, **12**, 1194, 1195
    agostic interaction, **12**, 1194
    -alkyl elmination, **12**, 1195
    -hydrogen elimination, **12**, 1195
    transmetallation, **12**, 1195
  metallocene-based, **12**, 1194
  organic synthesis with, **12**, 1205–7
  stereoselective polymerization, **12**, 1196–9
Ziegler–Natta reaction
  alkyl group transfer
    from aluminum to titanium, **12**, 157
Ziegler nickel effect
  mechanisms, **1**, 483, 484
  organoaluminum compounds, **1**, 478
Ziegler–Natta catalysis
  mechanisms, **1**, 482, 483
Ziegler–Natta catalysts
  zirconocenes, **4**, 493
Ziegler–Natta polymerization
  titanium catalyzed, **4**, 390–7
Ziegler–Rienäcker cyclization
  alkadienes, **1**, 470
Zimelidine
  synthesis, **12**, 805
Zinc
  bimetallic reagents
    preparation, **11**, 182
    reactions, **11**, 182
    reactivity, **11**, 182
  transmetallation from
    allylic alkylation, **12**, 854
Zinc carbenoids
  reactivity, **11**, 178–81
Zinc clusters
  cobalt complexes, **10**, 340
  iron complexes, **10**, 338
  ligand dynamics, **10**, 332
  manganese complexes, **10**, 337
  nickel complexes, **10**, 342
  niobium complexes, **10**, 335
  rhodium complexes, **10**, 341
  ruthenium complexes, **10**, 339
  structures, **10**, 333–5
  tantalum complexes, **10**, 336
  transition metals, **10**, 323–49
  tungsten complexes, **10**, 337
Zinc(II) complexes
  cyclizations with capture by, **12**, 314
  porphyrin catalysts
    diazodecomposition, **12**, 430
Zinc compounds
  dimetallic
    preparation, **11**, 175
  in organic synthesis, **11**, 159–84
  organomagnesium, **1**, 106
  transmetallation, **12**, 179–200
Zinc dust
  alkylidenation, **12**, 578
Zinc electrodes
  sacrificial, **11**, 174
Zinc enolates
  preparation, **3**, 180, 181
Zinc ester homoenolates
  synthesis, **12**, 72–4
Zinc homoenolates
  preparation, **11**, 160
Zinc malonates
  reaction with alkynes, **11**, 176
Zinc reagents
  allenic
    preparation, **11**, 169
  allylic
    reactivity, **11**, 165
  propargylic
    preparation, **11**, 169
Zinc, allenyl-
  coupling reactions, **12**, 188
Zinc, bis(alkenyl)-
  synthesis
    organoboron compounds, **11**, 220
Zinc, bis[3-dimethyl-(amino)propyl]-
  structure, **3**, 176
Zinc, bis(2,4-pentadienyl)-
  tetramethylethylenediamine adduct, **3**, 177
Zinc, bis(trifluoromethyl)-bis(pyridine)-
  structure, **3**, 177
Zinc, bis[tris(trimethylsilyl)-methyl]-
  preparation, **3**, 179
Zinc, cyclopropenyl-
  coupling reactions, **12**, 188
Zinc, cyclopropyl-
  coupling reactions
    alkenyl iodides, **12**, 187
Zinc, α-(dialkoxyboryl)alkyl-
  preparation, **11**, 210
Zinc, dialkyl-
  reaction with alkynes, **11**, 176
Zinc, dialkynyl-
  preparation, **3**, 180
Zinc, diallyl-
  addition to strained alkenes, **11**, 175
  preparation, **11**, 164
Zinc, diaryl-
  organocopper compound
    preparation from, **3**, 64
Zinc, dibenzyl-
  preparation, **11**, 164
Zinc, diethyl-
  organosilver compound
    synthesis from, **3**, 71
  reaction with diiodomethane, **11**, 162
Zinc, dimethyl-
  adduct with 18-crown-6, **3**, 177
  adduct with sparteine, **3**, 177
  preparation, **3**, 176
Zinc, diphenyl-
  adduct with crown ethers, **3**, 178
  reaction with Lewis bases, **3**, 178
Zinc, 3-furylbromo-
  coupling reactions, **12**, 184
Zinc, (iodomethyl) iodide
  cuprates from, **12**, 86
Zinc, (iodomethyl)iodo-
  reaction with alkynylcopper, **11**, 180
Zinc, perfluoroalkyl-
  in coupling reactions, **12**, 183
Zincates
  preparation, **3**, 179; **11**, 182
  reactions, **11**, 182
  structures, **1**, 29
Zincocene
  structure, **3**, 180
Zirconacycles
  cationic
    alkene insertion reactions, **12**, 1206
Zirconacyclopropane
  carbomagnesiation
    catalysis, **1**, 67
Zirconaindene
  from zirconocene aryne complexes, **12**, 772
Zirconaindenes
  synthesis
    zirconocene complexes, **12**, 776
Zirconium
  dicarbon metallacarbaboranes
    7-vertex, **1**, 382
    12-vertex, **1**, 400
Zirconium complexes, **4**, 439–61

chemical properties, **4**, 528
spectroscopy, **4**, 528
structure, **4**, 528
phosphido
chemical properties, **4**, 515, 516
spectroscopy, **4**, 514
structure, **4**, 514, 515
synthesis, **4**, 512, 513
polychalcogenido ligands, **4**, 533, 534
pseudohalides, **4**, 536–8
selenates, **4**, 529–31
tellurates, **4**, 529–31
thiolates, **4**, 529–31
transition metals, **4**, 535, 536
triscyclopentadienyl
aluminum tetrahydride derivatives, **1**, 548
Zirconium compounds
transmetallation, **12**, 179–200
Zirconium, bis(cyclopentadienyl)-bis(*t*-butyl methacrylate)-
methyl methacrylate
polymerization, **12**, 1204
Zirconium, bis(cyclopentadienyl)dimethyl-
syndiospecific polymerization
methyl methacrylate, **12**, 1204
Zirconium, bis(cyclopentadienyl)methyl- (tetrahydrofuran)-
cation
syndiospecific polymerization of methyl methacrylate, **12**, 1204
Zirconium, bis(1,2-dimethycyclopentadienyl)dimethyl-
catalyst
ring-opening polymerization, methylenecyclobutane, **12**, 1203
Zirconium, chlorocyclopentadienylhydrido-(pentamethylcyclopentadienyl)-
hydrozirconation, **12**, 39
Zirconium, chlorodicyclopentadienylhydrido-
hydrozirconation, **12**, 39
Zirconium, chlorohydridobis-(methylcyclopentadienyl)-
hydrozirconation, **12**, 39
Zirconobicycles
imine complexes, **12**, 330, 331
Zirconocene
alkene complexes
synthesis, **12**, 338
alkyne complexes
stability, **12**, 334
aryne complexes, **12**, 771–83
benzyne complexes, **12**, 771
carbomagnesiation
catalysis, **1**, 67
cyclic alkyne complexes
coupling, **12**, 338
cyclobutene complexes
reactions with unsaturated compounds, **12**, 340
ethene complexes
synthesis, **12**, 340, 342
imine complexes
formation, **12**, 342, 344
generation, **12**, 341, 343
phosphine complexes
diene cyclization, **12**, 329
reductive coupling
unsaturated compounds, **12**, 324
synthesis, **12**, 325
thioalkyne complexes
reactions, **12**, 333
unsaturated molecule complexes, **12**, 773
Zirconocene(IV) cationic complexes, **12**, 780
Zirconocene hydrochloride
in hydridocarboxylation, **12**, 355
Zirconocene isobutyl chloride
ketone synthesis, **12**, 774
Zirconocene oxide
generation, **4**, 528
Zirconocene sulfide
generation, **4**, 533
Zirconocene, chlorohydrido-
hydrozirconation, **12**, 325
Zirconocene, diaryl-
thermolysis, **12**, 771
Zirconocene, dibutyl-
synthesis, **12**, 325
Zirconocene, dimethyl-
activation
Ziegler–Natta catalysts, **12**, 1196
Zirconocene, vinyl-
preparation, **12**, 88
Zirconocene, vinylmethyl-
β-hydrogen abstraction, **12**, 325
Zirconocenes
aldimine complexes, **4**, 449
alkene complexes, **4**, 442–6
alkenyl, **4**, 579–82
alkoxy hydride, **4**, 559
alkyl hydride, **4**, 559
alkyne complexes, **4**, 442–6
oxidative addition, **4**, 460
butadiene complexes, **4**, 446
oxidative coupling, **4**, 460
structures, **4**, 451
carbene complexes
synthesis, **4**, 461
carbonyl derivatives
synthesis, **4**, 440
dialkyl, **4**, 575–7
addition reactions, **4**, 576
catalysis, **4**, 577
insertion reactions, **4**, 576
reactivity, **4**, 575, 576
synthesis, **4**, 575, 576
diaryl, **4**, 577–9
diazo complexes, **4**, 441
dihalides, **4**, 483
dihydrides, **4**, 562
reactions with aldehydes, **4**, 564
reactions with alkynes, **4**, 563
reactions with ammonia, **4**, 564
reactions with azides, **4**, 564
reactions with carbon monoxide, **4**, 563
reactions with ketones, **4**, 564
reactions with main-group elements, **4**, 562
reactions with water, **4**, 564
reactivity, **4**, 562–4
dinitrogen complexes, **4**, 440
halogenation, **4**, 454, 455
heteroalkene complexes, **4**, 447–50
oxidative coupling, **4**, 457
heteroalkyne complexes, **4**, 447–50
oxidative coupling, **4**, 457
hydride chloride, **4**, 556
reactions with alkenes, **4**, 556
reactions with alkynes, **4**, 556
reactions with carbon dioxide, **4**, 557
reactions with carbon monoxide, **4**, 557
reactions with nitriles, **4**, 557
reactions with organophosphorus compounds, **4**, 558
reactions with oxygen-containing substrates, **4**, 558, 559
reactivity, **4**, 556–9
hydrides, **4**, 556–64
imido complexes, **4**, 509
iminoacyl, **4**, 582–5
reactivity, **4**, 584, 585
ketene complexes, **4**, 447
monoalkyl, **4**, 573–5
NMR, **4**, 453
oxidative addition, **4**, 455–61
intermolecular, **4**, 457–61
intramolecular, **4**, 455, 456
phosphaalkyne complexes, **4**, 449
phosphine complexes, **4**, 441
polymetallic hydrides, **4**, 561, 562
protonolysis, **4**, 454, 455
reactions, **4**, 454–61
spectroscopy, **4**, 452–4
structures, **4**, 450–4
thioaldehyde complexes, **4**, 448
trialkylsilyl hydrides, **4**, 560, 561
x-ray crystallography, **4**, 450, 451
ylides
reactions, **4**, 458
Ziegler–Natta catalysts, **4**, 493
ansa-Zirconocenes
cyclobutene-bridged, **4**, 486
derivatives, **4**, 567–70
characterization, **4**, 567–9
reactivity, **4**, 569, 570
synthesis, **4**, 567–9
oxidation, **4**, 471

Zirconocycles
diene cyclization, **12**, 329, 330
Zirconocyclopentane, 2,3-diethyl-
synthesis, **12**, 338, 339
Zirconocyclopentanes
synthesis, **12**, 338, 339
Zirconocyclopropene
phosphine-stabilized
reaction with alkynes, **12**, 334